Xin-She Yang · Simon Sherratt · Nilanjan Dey ·
Amit Joshi
Editors

# Proceedings of Sixth International Congress on Information and Communication Technology

ICICT 2021, London, Volume 3



Springer

*Editors*
Xin-She Yang
Middlesex University
London, UK

Simon Sherratt
The University of Reading
Reading, UK

Nilanjan Dey
JIS University
Kolkata, India

Amit Joshi
Global Knowledge Research Foundation
Ahmedabad, India

ISSN 2367-3370 ISSN 2367-3389 (electronic)
Lecture Notes in Networks and Systems
ISBN 978-981-16-1780-5 ISBN 978-981-16-1781-2 (eBook)
https://doi.org/10.1007/978-981-16-1781-2



This Springer imprint is published by the registered company Springer Nature Singapore Pte Ltd.
The registered company address is: 152 Beach Road, #21-01/04 Gateway East, Singapore 189721, Singapore

# Preface

The Sixth International Congress on Information and Communication Technology was held during 25–26 February 2021, digitally, on ZOOM and organised by Global Knowledge Research Foundation. The associated partners were Springer and InterYIT IFIP. The conference will provide a useful and wide platform both for display of the latest research and for exchange of research results and thoughts. The participants of the conference will be from almost every part of the world, with background of either academia or industry, allowing a real multinational multicultural exchange of experiences and ideas.

A great pool of more than 1150 papers were received papers for this conference from across 89 countries among which around 350 papers were accepted and will be presented through digital platform during the two days. Due to overwhelming response, we had to drop many papers in hierarchy of the quality. Total 42 technical sessions will be organised in parallel in 2 days along with few Keynotes and Panel Discussions. The conference will be involved in deep discussion and issues which will be intended to solve at global levels. New technologies will be proposed, experiences will be shared and future solutions for design infrastructure for ICT will also be discussed. The final papers will be published in 4 volumes of proceedings by Springer AISC Series.

Over the years this congress has been organised and conceptualised with collective efforts of a large number of individuals. I would like to thank each of committee members and the reviewers for their excellent work in reviewing the papers. Grateful acknowledgements are extended to the team of Global Knowledge Research Foundation for their valuable efforts and support.

I look forward to welcome you on the 6th Edition of this ICICT Congress 2021.

London, UK — Xin-She Yang
Reading, UK — Simon Sherratt
Kolkata, India — Nilanjan Dey
Ahmedabad, India — Amit Joshi

# Contents

# Editors and Contributors

## About the Editors

**Xin-She Yang** obtained his D.Phil. in Applied Mathematics from the University of Oxford, and subsequently worked at the Cambridge University and the National Physical Laboratory (UK) as Senior Research Scientist. He is currently Reader in Modelling and Optimization at Middlesex University London and Adjunct Professor at Reykjavik University (Iceland). He is also elected as Bye-Fellow at the Cambridge University and the IEEE CIS Chair for the Task Force on Business Intelligence and Knowledge Management. He was included in the "2016 Thomson Reuters Highly Cited Researchers" list.

**Simon Sherratt** was born near Liverpool, England, in 1969. He is currently Professor of Biosensors at the Department of Biomedical Engineering, University of Reading, UK. His main research area is signal processing and personal communications in consumer devices, focusing on wearable devices and health care. Professor Sherratt received the 1st place IEEE Chester Sall Memorial Award in 2006, the 2nd place in 2016 and the 3rd place in 2017.

**Nilanjan Dey** is Assistant Professor at the Department of Information Technology, Techno India College of Technology, India. He has authored/edited more than 75 books with Springer, Elsevier, Wiley, CRC Press and published more than 300 peer-reviewed research papers. Dr. Dey is Editor-in-Chief of the International Journal of Ambient Computing and Intelligence; Series Co-Editor of Springer Tracts in Nature-Inspired Computing (STNIC); and Series Co-Editor of Advances in Ubiquitous Sensing Applications for Healthcare, Elsevier.

**Amit Joshi** is Director of the Global Knowledge Research Foundation and the International Chair of InterYIT at the International Federation of Information Processing (IFIP, Austria). He has edited more than 40 books for Springer, ACM and other reputed publishers. He has also organized more than 50 national and international conferences and workshops in association with the ACM, Springer and IEEE in India, Thailand and the UK.

## Contributors

**Mohamed Artan Abdi** Computer Engineering Department, Istanbul Commerce University, Istanbul, Turkey

**Sayed Abdul Kadir Al-Nahiun** American International University-Bangladesh (AIUB), Dhaka, Bangladesh

**Aisha Abdulahi** College of Computer Engineering and Sciences, Prince Sattam bin Abdulaziz University, Alkharj, Kingdom of Saudi Arabia

**Zaid Ameen Abduljabbar** Hubei Engineering Research Center on Big Data Security, School of Cyber Science and Engineering, Huazhong University of Science and Technology, Wuhan, China

**Zaid Ameen Abduljabbar** University of Basrah, Basrah, Iraq

**Zaid Ameen Abduljabbar** Neusoft Institute Guangdong, Guangdong, China

**Md. Abdur Rahman** American International University-Bangladesh (AIUB), Dhaka, Bangladesh

**Enas Wahab Abood** University of Basrah, Basrah, Iraq

**David Ebo Adjepon-Yamoah** Ashesi University, Berekuso, Ghana

**David Ebo Adjepon-Yamoah** PMB 3 CT Cantonment, Accra, Ghana

**Ahmed Abdulelah Jasim Al-Ashoor** University of Basrah, Basrah, Iraq

**Khamis A. Al-karawi** University of Diyala, Diyala, Baqubah, Iraq

**Hasan Abbas Al-Mohammed** Computer Science and Engineering Department, Qatar University, Doha, Qatar

**Adnan Yehia Al-mutawkkil** Sana'a University, Sana'a, Yemen

**Tibra Al-smadi** Department of Computer Information System, University of Science and Technology, Irbid, Jordan

**Haya Alaskar** College of Computer Engineering and Sciences, Prince Sattam bin Abdulaziz University, Alkharj, Kingdom of Saudi Arabia

**Najla Albarrak** Central Bank of Kuwait, Safat, Kuwait

**Najla Albarrak** Cranfield University, Cranfield, Bedfordshire, UK

**Abdulrahman S. Alenizi** Business School, University of Salford, Manchester, UK

**Shaikah Alhewaidi** College of Computer Engineering and Sciences, Prince Sattam bin Abdulaziz University, Alkharj, Kingdom of Saudi Arabia

**Fattah Alizadeh** University of Kurdistan Hewler, Erbil, Iraq

**Keerthi Reddy Alla** Koneru Lakshmaiah Education Foundation, Guntur, Andhra Pradesh, India

**Giner Alor-Hernández** Tecnológico Nacional de México, Instituto Tecnológico de Orizaba, Veracruz, México

**Nour Alqudah** Department of Computer Information System, University of Science and Technology, Irbid, Jordan

**Nour Alqudah** Department of Computer Information System, Jordan University of Science and Technology, Irbid, Jordan

**Hessa Alsanousi** Central Bank of Kuwait, Safat, Kuwait

**Hessa Alsanousi** Cranfield University, Cranfield, Bedfordshire, UK

**Sarah Abdalrahman Alshqaqi** Sana'a University, Sana'a, Yemen

**Mustafa A. Al Sibahee** Shenzhen Institute of Huazhong University of Science and Technology, Shenzhen, China

**Mustafa A. Al Sibahee** Iraq University College, Basrah, Iraq

**Pilita A. Amahan** Occidental Mindoro State College, San Jose Occidental Mindoro, Philippines

**Fabio Andreassi** Guglielmo Marconi University, Rome, Italy

**Daria A. Andrianova** Saint Petersburg, Yandex N.V., Russia

**Maki Arame** Polytechnic University of Japan, Kodaira, Tokyo, Japan

**Maki Arame** Kumamoto University, Kumamoto-city, Kumamoto, Japan

**David Alejandro Vergara Arango** Politécnico Colombiano Jaime Isaza Cadavid, Medellín, Colombia

**Dimitar Arnaudov** Technical University - Sofia, Sofia, Bulgaria

**Karl Anthony S. Arnejo** Premier Research Institute of Science and Mathematics, Mindanao State University-Iligan Institute of Technology, Iligan City, Philippines

**Alberto Arteta Albert** Department of Computer Science, Troy University, Alabama, USA

**Konstantinos G. Arvanitis** Agricultural University of Athens, Athens, Greece

**Yalmar Ponce Atencio** Escuela Profesional de Ingeniería de Sistemas, Universidad Nacional José María Arguedas, Andahuaylas, Apurímac, Peru

**Rasha Awawdeh** Department of Computer Information System, University of Science and Technology, Irbid, Jordan

**Karlo Babić** University of Rijeka, Department of Informatics, Rijeka, Croatia

**Karlo Babić** Center for Artificial Intelligence and Cybersecurity, University of Rijeka, Rijeka, Croatia

**K. Babulu** JNTUV, Kakinada, Dwarapudi, Vizianagaram, Andhra Pradesh, India

**Wafa Baccouch** Research Laboratory of Biophysics and Medical Technologies, Higher Institute of Medical Technologies of Tunis, University of Tunis El Manar, Tunis, Tunisia

**Shouherdho Banerjee Akash** American International University-Bangladesh (AIUB), Dhaka, Bangladesh

**David P. Barros-Piedra** Universidad Politécnica Salesiana, Cuenca, Ecuador

**Gee Jay C. Bartolome** Department of Engineering, Cavite State University-CCAT Campus, Cavite, Philippines

**Slobodan Beliga** University of Rijeka, Department of Informatics, Rijeka, Croatia

**Slobodan Beliga** Center for Artificial Intelligence and Cybersecurity, University of Rijeka, Rijeka, Croatia

**Cinzia Bellone** Faculty of Applied Sciences and Technology, Guglielmo Marconi University, Rome, Italy

**Ayyoub El Berbri** Moulay Ismail University, Meknes, Morocco

**German V. Arevalo Bermeo** Universidad Politécnica Salesiana, Quito, Ecuador

**Matteo Bertocco** Department of Information Engineering, Università degli Studi di Padova, Padova, Italy

**Alberto Bianchi** Carel Industries Spa, Padova, Italy

**Milon Biswas** Department of Computer Science and Engineering, Bangladesh University of Business and Technology, Dhaka, Bangladesh

**Nuria Gómez Blas** Departamento de Sistemas Informáticos, Universidad Politécnica de Madrid, Madrid, Spain

**Roberto Bodo** Department of Information Engineering, Università degli Studi di Padova, Padova, Italy

**Adonis Bogris** Department of Informatics and Computer Engineering, University of West Attica, Aghiou Spiridonos, Egaleo, Greece

**Nikolay V. Boldyrikhin** Don State Technical University, Rostov-on-Don, Russia

**Trevor Bostic** The MITRE Corporation, McLean, VA, USA

**Joel Botai** Department of Technology, Central University of Technology, Free State, Bloemfontein, South Africa

**Joel Botai** South Africa Weather Service, Pretoria, South Africa

**Dalila Boughaci** USTHB, Computer Science Department, LRIA Laboratory, University of Science and Technology Houari Boumediene, Houari, Algeria

**Ahmed Boumehdi** National School of Computer Science and Systems Analysis (ENSIAS), Information, Communication and Embedded Systems (ICES), Rabat, Morocco

**Myriam Bounhas** Emirates College of Technology, Abu Dhabi, United Arab Emirates

**Myriam Bounhas** LARODEC Research Laboratory, ISGT, Tunis University, Tunis, Tunisia

**John G. Breslin** National University of Ireland Galway, Galway, Ireland

**Seddik Bri** Moulay Ismail University, Meknes, Morocco

**Ronna ten Brink** The MITRE Corporation, McLean, VA, USA

**John Patrick D. Buco** Research and Extension, Cavite State University-CCAT Campus, Cavite, Philippines

**P. Bujjibabu** Aditya Engineering College (A), Surampalem, Andhra Pradesh, India

**Katherine C. Bustamante-Cacao** IEEE, Madrid, Spain

**Michael Cabanillas-Carbonell** Universidad Privada del Norte, Lima, Peru

**Mary Huaman Carrión** Escuela Profesional de Ingeniería de Sistemas, Universidad Nacional José María Arguedas, Andahuaylas, Apurímac, Peru

**Felipe Castro-Medina** Tecnológico Nacional de México, Instituto Tecnológico de Orizaba, Veracruz, México

**Juan José Oré Cerrón** Escuela Profesional de Ingeniería de Sistemas, Universidad Nacional José María Arguedas, Andahuaylas, Apurímac, Peru

**Yumi Cha** Department of Mechanical and System Design Engineering, Hongik University, Seoul, Korea

**Charidimos Chaintoutis** Department of Informatics and Telecommunications, National and Kapodistrian University of Athens, Panepistimiopolis, Ilissia, Athens, Greece

**Alan Chalker** Ohio Supercomputer Center, Columbus, OH, USA

**Eang Teng Chan** Tunku Abdul Rahman University College, Kuala Lumpur, Malaysia

**Paul A. Chasi-Pesántez** Universidad Politécnica Salesiana, Cuenca, Ecuador

**Dewitt Scott Chavez-Ponce** Universidad Nacional de San Agustín de Arequipa, Arequipa, Perú

**Larissa V. Cherckesova** Don State Technical University, Rostov-on-Don, Russia

**Suhyun Choi** Department of Mechanical and System Design Engineering, Hongik University, Seoul, Korea

**Alae Chouiekh** Laboratory of Multimedia Signal and Communications Systems, National Institute of Posts and Telecommunications, Rabat, Morocco

**Franco Cicirelli** CNR - National Research Council of Italy - Institute for High Performance Computing and Networking (ICAR), Rende, CS, Italy

**António Leça Coelho** Laboratório Nacional de Engenharia Civil, Lisboa, Portugal

**Marijke Coetzee** University of Johannesburg, Auckland Park, South Africa

**Richard Flores Condori** Escuela Profesional de Ingeniería de Sistemas, Universidad Nacional José María Arguedas, Andahuaylas, Apurímac, Peru

**Edwin J. Coronel-González** Universidad Politécnica Salesiana, Cuenca, Ecuador

**Walter Daems** FTI-CoSys Lab, University of Antwerp, Antwerp, Belgium

**Walter Daems** Flanders Make Strategic Research Centre, Lommel, Belgium

**Mohammed Ait Daoud** Faculty of Sciences Ben M'sick, LTIM, Hassan II University of Casablanca, Casablanca, Morocco

**Biswajit Kumar Dash** Department of Electrical and Computer Engineering, University of Texas Rio Grande Valley, Edinburg, TX, USA

**Erunika Dayaratna** Department of Statistics and Computer Science, Faculty of Science, University of Peradeniya, Peradeniya, Sri Lanka

**Edgar I. De la Cruz** Universidad de Las Fuerzas Armadas ESPE, Sangolquí, Ecuador

**Eduardo O. dela Cruz Jr.** Arellano University, Manila, Philippines

**Nikos Dimitriou** National Centre for Scientific Research "Demokritos", Athens, Greece

**Habiba Drias** USTHB, Computer Science Department, LRIA Laboratory, University of Science and Technology Houari Boumediene, Houari, Algeria

**Rui P. Duarte** Polytechnic Institute of Viseu, Viseu, Portugal

**Rui P. Duarte** CISeD - Research Centre in Digital Services, Viseu, Portugal

**Hans Ehm** Infineon Technologies AG, Neubiberg, Germany

**Bilel Elayeb** Emirates College of Technology, Abu Dhabi, United Arab Emirates

**Bilel Elayeb** RIADI Research Laboratory, ENSI, Manouba University, Manouba, Tunisia

**Mostafa Elgendy** University of Pannonia, Veszprem, Hungary

**Ramón Santana Fernández** Autonomous University of Baja California, Ensenada, BC, Mexico

**Annamaria Ficara** University of Palermo, Palermo, Italy

**Annamaria Ficara** University of Messina, Messina, Italy

**Salvatore Filippone** Cranfield University, Cranfield, Bedfordshire, UK

**Salvatore Filippone** University of Rome Tor Vergata, Roma, RM, Italy

**Giacomo Fiumara** University of Messina, Messina, Italy

**Eric Franz** Ohio Supercomputer Center, Columbus, OH, USA

**Soni Asmaul Fuadi** PT PLN (Persero) Corporate University, UPDL Banjarbaru, Banjarbaru, South Borneo, Indonesia

**Teona Gelashvili** Tallinn University of Technology, Tallinn, Estonia

**Malahat Ghoreishi** LUT University, Lappeenranta, Finland

**Mandeep Kaur Ghumman** Department of ECE, Chandigarh University, Mohali, Punjab, India

**Yoshiko Goda** Kumamoto University, Kumamoto-city, Kumamoto, Japan

**Ramiro Gonçalves** UTAD (Universidade de Trás-Os-Montes E Alto Douro), Vila Real, Portugal

**Brian Greaves** University of Johannesburg, Auckland Park, South Africa

**Hanane Grissette** LISAC Laboratory, Faculty of Sciences, Sidi Mohamed Ben Abdellah University, Fez, Morocco

**Luis F. Guerrero-Vásquez** Universidad Politécnica Salesiana, Cuenca, Ecuador

**Mustapha Hakdaoui** Faculty of Sciences Ben M'sick, LGAGE, Hassan II University of Casablanca, Casablanca, Morocco

**Yanling Han** College of Information Technology, Shanghai Ocean University, Shanghai, China

**Junko Handa** Polytechnic University of Japan, Kodaira, Tokyo, Japan

**Junko Handa** Kumamoto University, Kumamoto-city, Kumamoto, Japan

**Ari Happonen** LUT University, Lappeenranta, Finland

**Ali N. Hasan** Department of Electrical Engineering, Higher Colleges of Technology, Abu Dhabi, United Arab Emirates

**Zhonghua Hong** College of Information Technology, Shanghai Ocean University, Shanghai, China

**David Hudak** Ohio Supercomputer Center, Columbus, OH, USA

**Manuel J. Ibarra** Escuela Profesional de Ingeniería Informática y Sistemas, Universidad Nacional Micaela Bastidas de Apurímac, Abancay, Apurímac, Peru

**El Hassane Ibn El Haj** Laboratory of Multimedia Signal and Communications Systems, National Institute of Posts and Telecommunications, Rabat, Morocco

**Youssef Ibnatta** Department of Electrical Engineering, NEST Research Group ENSEM, Hassan II University, Casablanca, Morocco

**Mary Ann E. Ignaco** Department of Computer Studies, Cavite State University-CCAT Campus, Cavite, Philippines

**Rashedul Islam** American International University-Bangladesh (AIUB), Dhaka, Bangladesh

**Mahmoud Ismail** Infineon Technologies AG, Neubiberg, Germany

**Makoto Itoh** University of Tsukuba, Tsukuba-city, Ibaraki, Japan

**Dongsik Jang** Department of Industrial Management Engineering, Korea University, Seoul, Republic of Korea

**Andrzej Jarynowski** Interdisciplinary Research Institute, Wroclaw, Poland

**Leydi M. Jiménez** Universidad de Las Fuerzas Armadas ESPE, Sangolquí, Ecuador

**Balkrishan Jindal** Department of CE, Yadavindra College of Engineering, Talwandi Sabo, Punjab, India

**Du Jing** Naval Aviation University Aviation Foundation, Yantai, China

**Doug Johnson** Ohio Supercomputer Center, Columbus, OH, USA

**Jorge Joo-Nagata** Universidad Metropolitana de Ciencias de La Educación, Santiago, Chile

**A. Jorge Morais** Universidade Aberta, Lisboa, Portugal

**A. Jorge Morais** LIAAD - INESC TEC, Lisboa, Portugal

**Rhuzel Leinster Dean J. Julian** Research and Extension, Cavite State University-CCAT Campus, Cavite, Philippines

**Ulises Juárez-Martínez** Tecnológico Nacional de México, Instituto Tecnológico de Orizaba, Veracruz, México

**M. Shamim Kaiser** Institute of Information Technology, Jahangirnagar University, Savar Union, Bangladesh

**Nurul Kharismawati Kamaludin** Masters of Government Science, Jusuf Kalla School of Government, Muhammadiyah University Yogyakarta, Special Region of Yogyakarta, Kasihan, Bantul, Indonesia

**M. Kamaraju** Gudlavalleru Engineering College, Seshadri Rao Knowledge Village, Gudlavalleru, Krishna (dt), Andhra Pradesh, India

**Ivan G. Kamenev** Federal Research Center of Computer Science and Control Dorodnitsyn, Computing Centre of the Russian Academy of Sciences, Moscow, Russia

**Ivan G. Kamenev** National research university, Higher school of economics, Moscow, Russia

**Nikhil Kandibanda** Koneru Lakshmaiah Education Foundation, Guntur, Andhra Pradesh, India

**Jiho Kang** Machine Learning Big Data Institute, Korea University, Seoul, Republic of Korea

**Priyanka Katta** Koneru Lakshmaiah Education Foundation, Guntur, Andhra Pradesh, India

**Mohammed Khaldoun** Department of Electrical Engineering, NEST Research Group ENSEM, Hassan II University, Casablanca, Morocco

**Ohnmar Khin** University of Mandalay, Mandalay, Myanmar

**Sarthak Khurana** Department of Computer Science and Engineering, National Institute of Technology Raipur, Raipur, India

**Youngho Kim** Department of Industrial Management Engineering, Korea University, Seoul, Republic of Korea

**Satoshi Kitazaki** National Institute of Advanced Industrial Science and Technology, Tsukuba-city, Ibaraki, Japan

**Silma Koekemoer** School of IT, Nelson Mandela University, Gqeberha, South Africa

**Dexuan Kong** University of Electronic Science and Technology of China, Chengdu, SC, China

**Marcel Konstantinov** Infineon Technologies AG, Neubiberg, Germany

**Kimon Kontovasilis** National Centre for Scientific Research "Demokritos", Athens, Greece

**Dennis A. Korochentsev** Don State Technical University, Rostov-on-Don, Russia

**K. G. Korotkov** Saint-Petersburg Scientific-Research Institute for Physical Culture, St. Petersburg, Russia

**Veronica O. Kravchenko** Don State Technical University, Rostov-on-Don, Russia

**Swarna Kuchibhotla** Koneru Lakshmaiah Education Foundation, Guntur, Andhra Pradesh, India

**Salam Labidi** Research Laboratory of Biophysics and Medical Technologies, Higher Institute of Medical Technologies of Tunis, University of Tunis El Manar, Tunis, Tunisia

**Ferdinand C. Lacuata** Arellano University, Manila, Philippines

**Cecilia Sik Lanyi** University of Pannonia, Veszprem, Hungary

**Dongryun Lee** Department of Mechanical and System Design Engineering, Hongik University, Seoul, Korea

**Juhyun Lee** Department of Industrial Management Engineering, Korea University, Seoul, Republic of Korea

**Junseok Lee** MICube Solution, Seoul, Republic of Korea

**Wai Sze Leung** University of Johannesburg, Auckland Park, South Africa

**De Jian Li** Beijing Smart-Chip Microelectronics Technology Co., Ltd, Haidian District, Beijing, P.R. China

**Mauro Lima** Polytechnic Institute of Viseu, Viseu, Portugal

**Antonio Liotta** University of Bozen-Bolzano, Bozen-Bolzano, Italy

**Erasmus(Xin) Liu** Neusoft Institute Guangdong, Guangdong, China

**Jianyun Liu** College of Information Technology, Shanghai Ocean University, Shanghai, China

**Dimitrios Loukatos** Agricultural University of Athens, Athens, Greece

**Dimitrios Loukatos** National Centre for Scientific Research "Demokritos", Athens, Greece

**Songfeng Lu** Shenzhen Institute of Huazhong University of Science and Technology, Shenzhen, China

**Songfeng Lu** Hubei Engineering Research Center on Big Data Security, School of Cyber Science and Engineering, Huazhong University of Science and Technology, Wuhan, China

**Arif Ridho Lubis** School of Industrial Engineering, Telkom University, Bandung, Indonesia

**Arif Ridho Lubis** Department of Computer Engineering and Informatics, Medan City, Indonesia

**Muharman Lubis** School of Industrial Engineering, Telkom University, Bandung, Indonesia

**Muharman Lubis** Department of Computer Engineering and Informatics, Medan City, Indonesia

**Lyvia Winiyanti Lumingkewas** School of Industrial Engineering, Telkom University, Bandung, Indonesia

**Lyvia Winiyanti Lumingkewas** Department of Computer Engineering and Informatics, Medan City, Indonesia

**Nikita G. Lyashenko** Don State Technical University, Rostov-on-Don, Russia

**Asdrúbal López-Chau** Universidad Autónoma del Estado de México, Centro Universitario UAEM Zumpango, Zumpango, Estado de México, México

**Everardo Gutiérrez López** Autonomous University of Baja California, Ensenada, BC, Mexico

**Luis Fernando de Mingo López** Departamento de Sistemas Informáticos, Universidad Politécnica de Madrid, Madrid, Spain

**Mohamed Maaroufi** Research in Electric Energy and Command, Mohammadia Engineering School, Mohamed V University, Rabat, Morocco

**Justine Shane S. Macarat** Premier Research Institute of Science and Mathematics, Mindanao State University-Iligan Institute of Technology, Iligan City, Philippines

**Fardeen Mahbub** American International University-Bangladesh (AIUB), Dhaka, Bangladesh

**Rajitha Manellanga** Department of Statistics and Computer Science, Faculty of Science, University of Peradeniya, Peradeniya, Sri Lanka

**Ioannis Manolopoulos** National Centre for Scientific Research "Demokritos", Athens, Greece

**Javier E. Martinez-Ledesma** TelecomAustro, Cuenca, Ecuador

**Domingos Martinho** ISLA Santarem, Largo Candido Dos Reis, Santarem, Portugal

**Sanda Martinčić-Ipšić** University of Rijeka, Department of Informatics, Rijeka, Croatia

**Sanda Martinčić-Ipšić** Center for Artificial Intelligence and Cybersecurity, University of Rijeka, Rijeka, Croatia

**Julio Huanca Marín** Escuela Profesional de Ingeniería de Sistemas, Universidad Nacional José María Arguedas, Andahuaylas, Apurímac, Peru

**Muthoni Masinde** Department of Technology, Central University of Technology, Free State, Bloemfontein, South Africa

**Katleho L. Masita** Department of Electrical and Electronic Engineering Technology, University of Johannesburg, Gauteng, South Africa

**Ryuichi Matsuba** Kumamoto University, Kumamoto-city, Kumamoto, Japan

**Tajim Md. Niamat Ullah Akhund** Department of Computer Science and Engineering, Daffodil International University, Dhaka, Bangladesh

**Youssef El Mdari** Faculty of Sciences Ben M'sick, LTIM, Hassan II University of Casablanca, Casablanca, Morocco

**Shatha Melhem** Department of Computer Information System, University of Science and Technology, Irbid, Jordan

**Ana Meštrović** University of Rijeka, Department of Informatics, Rijeka, Croatia

**Ana Meštrović** Center for Artificial Intelligence and Cybersecurity, University of Rijeka, Rijeka, Croatia

**Nguyen Binh Minh** Vietnamese-German University, Binh Duong, Vietnam

**Patrick Moder** Infineon Technologies AG, Neubiberg, Germany

**José Manuel Valencia Moreno** Autonomous University of Baja California, Ensenada, BC, Mexico

**José F. Morgado** Polytechnic Institute of Viseu, Viseu, Portugal

**Irene Moulitsas** Cranfield University, Cranfield, Bedfordshire, UK

**Ahmad Mustafa** Department of Computer Information System, University of Science and Technology, Irbid, Jordan

**Abhinav Muthavarapu** Koneru Lakshmaiah Education Foundation, Guntur, Andhra Pradesh, India

**Jií Mírovský** Faculty of Mathematics and Physics, Institute of Formal and Applied Linguistics, Charles University, Prague, Czech Republic

**Abdelwahed Namir** Faculty of Sciences Ben M'sick, LTIM, Hassan II University of Casablanca, Casablanca, Morocco

**Fabio Naselli** Architecture Department, Epoka University, Tirana, Albania

**Ivan Nedyalkov** Blagoevgrad, Bulgaria

**Joaquim Neto** Universidade Aberta, Lisboa, Portugal

**Joaquim Neto** Laboratório Nacional de Engenharia Civil, Lisboa, Portugal

**El Habib Nfaoui** LISAC Laboratory, Faculty of Sciences, Sidi Mohamed Ben Abdellah University, Fez, Morocco

**Huu Phuoc Dai Nguyen** Doctoral School on Safety and Security Science, Óbuda University, Budapest, Hungary

**Thuyet That Vo Nguyen** Vietnamese-German University, Binh Duong, Vietnam

**Tran Quang Nhu** Vietnamese-German University, Binh Duong, Vietnam

**Christian Nigro** Independent computing professional, Rende, Italy

**Libero Nigro** DIMES, University of Calabria, Rende, CS, Italy

**Firdaus Adi Nugroho** School of Industrial Engineering, Telkom University, Bandung, Indonesia

**Firdaus Adi Nugroho** Department of Computer Engineering and Informatics, Medan City, Indonesia

**Achmad Nurmandi** Department of Government Affairs and Administration, Jusuf Kalla School of Government, Universitas Muhammadiyah Yogyakarta, Special Region of Yogyakarta, Kasihan, Bantul, Indonesia

**Achmad Nurmandi** Master of Government Science, Universitas Muhammadiyah Yogyakarta, Special Region of Yogyakarta, Bantul, Indonesia

**Achmad Nurmandi** Masters of Government Science, Jusuf Kalla School of Government, Muhammadiyah University Yogyakarta, Special Region of Yogyakarta, Kasihan, Bantul, Indonesia

**Bayan Obaid** College of Computer Engineering and Sciences, Prince Sattam bin Abdulaziz University, Alkharj, Kingdom of Saudi Arabia

**Jeff Ohrstrom** Ohio Supercomputer Center, Columbus, OH, USA

**Jorge O. Ordoñez-Ordoñez** Universidad Politécnica Salesiana, Cuenca, Ecuador

**Sameh Oueslati** Research Laboratory of Biophysics and Medical Technologies, Higher Institute of Medical Technologies of Tunis, University of Tunis El Manar, Tunis, Tunisia

**Acheampong Owusu** University of Ghana Business School, Legon, Accra, Ghana

**Ingmar Pappel** Amphora Information Management LLC, Tallinn, Estonia

**Ingrid Pappel** Tallinn University of Technology, Tallinn, Estonia

**Sangsung Park** Department of Big Data Statistics, CheongJu University, Cheong Ju, Republic of Korea

**Shinyoung Park** Department of Mechanical and System Design Engineering, Hongik University, Seoul, Korea

**Jun Peng** Department of Electrical and Computer Engineering, University of Texas Rio Grande Valley, Edinburg, TX, USA

**Obed Kwame Adzaku Penu** University of Ghana Business School, Legon, Accra, Ghana

**Ismael S. Pesántez-Romero** Universidad Politécnica Salesiana, Cuenca, Ecuador

**Milan Petrović** University of Rijeka, Department of Informatics, Rijeka, Croatia

**Esteban González Peña** Politécnico Colombiano Jaime Isaza Cadavid, Medellín, Colombia

**Paulina Phoobane** Department of Technology, Central University of Technology, Free State, Bloemfontein, South Africa

**Irina A. Pilipenko** Don State Technical University, Rostov-on-Don, Russia

**Jorge E. Giraldo Plaza** Politécnico Colombiano Jaime Isaza Cadavid, Medellín, Colombia

**Lucie Poláková** Faculty of Mathematics and Physics, Institute of Formal and Applied Linguistics, Charles University, Prague, Czech Republic

**Vitaly M. Porksheyan** Don State Technical University, Rostov-on-Don, Russia

**Ario Dwi Prabowo** PT PLN (Persero) Corporate University, UPDL Banjarbaru, Banjarbaru, South Borneo, Indonesia

**Danang Eko Prastya** Department of Government Affairs and Administration, Jusuf Kalla School of Government, Universitas Muhammadiyah Yogyakarta, Special Region of Yogyakarta, Kasihan, Bantul, Indonesia

**Gabriel E. Pulla-Lojano** Universidad Politécnica Salesiana, Cuenca, Ecuador

**Francesco Pupo** DIMES, University of Calabria, Rende, CS, Italy

**Roberto Quispe Quispe** Escuela Profesional de Ingeniería de Sistemas, Universidad Nacional José María Arguedas, Andahuaylas, Apurímac, Peru

**Milton Saulo Raimundo** Universidade de São Paulo, São Paulo, SP, Brazil

**Pietro Ranucci** Architecture Department, Roma TRE University, Rome, Italy

**Raja Rashidul Hasan** American International University-Bangladesh (AIUB), Dhaka, Bangladesh

**Pavel V. Razumov** Don State Technical University, Rostov-on-Don, Russia

**Josep Reyes-Rojas** Universidad Privada del Norte, Lima, Peru

**Maria Teresa N. Rivera** Arellano University, Manila, Philippines

**Lisbeth Rodríguez-Mazahua** Tecnológico Nacional de México, Instituto Tecnológico de Orizaba, Veracruz, México

**Jacson J. Rodríguez** Universidad de Las Fuerzas Armadas ESPE, Sangolquí, Ecuador

**Jonathan A. Romero** Universidad de Las Fuerzas Armadas ESPE, Sangolquí, Ecuador

**Peter G. Romerosa** Arellano University, Manila, Philippines

**Adil Saadi** Moulay Ismail University, Meknes, Morocco

**Mohammed Sadik** Department of Electrical Engineering, NEST Research Group ENSEM, Hassan II University, Casablanca, Morocco

**Olga A. Safaryan** Don State Technical University, Rostov-on-Don, Russia

**Thalendra Sahu** Department of Computer Science and Engineering, National Institute of Technology Raipur, Raipur, India

**Salahudin** Department of Government Science, Universitas Muhammadiyah Malang, Tegalgondo, Malang, Indonesia

**Salahudin** Government Science, Universitas Muhammadiyah Malang, Tegalgondo, Malang, Indonesia

**Salahudin** Government Science, Universitas Muhammadiyah Malang, Tlogomas Street, Tegalgondo, Malang, Indonesia

**Edgar R. Salazar** Universidad de Las Fuerzas Armadas ESPE, Sangolquí, Ecuador

**Zohra Sbai** College of Computer Engineering and Sciences, Prince Sattam bin Abdulaziz University, Alkharj, Kingdom of Saudi Arabia

**Zohra Sbai** National Engineering School of Tunis, Tunis El Manar University, Tunis, Tunisia

**Benedikt Schnell** Technical University of Munich, Munich, Germany

**Benedikt Schnell** Infineon Technologies AG, Neubiberg, Germany

**Becca Scollan** The MITRE Corporation, McLean, VA, USA

**K. V. Semenov** University ITMO, St. Petersburg, Russia

**Robert E. Settlage** Virginia Tech, Blacksburg, VA, USA

**Bayar Shahab** University of Kurdistan Hewler, Erbil, Iraq

**Aakanksha Sharaff** Department of Computer Science and Engineering, National Institute of Technology Raipur, Raipur, India

**Mohammed Q. Shatnawi** Department of Computer Information System, Jordan University of Science and Technology, Irbid, Jordan

**Chong Fei Shen** Beijing Smart-Chip Microelectronics Technology Co., Ltd, Haidian District, Beijing, P.R. China

**Seungwon Shin** Department of Mechanical and System Design Engineering, Hongik University, Seoul, Korea

**Thokozani Shongwe** Department of Electrical and Electronic Engineering Technology, University of Johannesburg, Gauteng, South Africa

**Saurabh Shukla** National University of Ireland Galway, Galway, Ireland

**Thandar Shwe Sin** University of Mandalay, Mandalay, Myanmar

**Navtej Singh** Department of CSE, Punjabi University, Patiala, Punjab, India

**Basel Solaiman** ITI Department: Image and Information Processing, IMT-Atlantique, Technopole Brest Iroise, Brest Cedex, CS, France

**Salma Sraidi** Research in Electric Energy and Command, Mohammadia Engineering School, Mohamed V University, Rabat, Morocco

**Jeff Stanley** The MITRE Corporation, McLean, VA, USA

**Jan Steckel** FTI-CoSys Lab, University of Antwerp, Antwerp, Belgium

**Jan Steckel** Flanders Make Strategic Research Centre, Lommel, Belgium

**Faez Syahroni** Master of Government Science, Universitas Muhammadiyah Yogyakarta, Special Region of Yogyakarta, Bantul, Indonesia

**Dimitris Syvridis** Department of Informatics and Telecommunications, National and Kapodistrian University of Athens, Panepistimiopolis, Ilissia, Athens, Greece

**Cuauhtémoc Sánchez-Ramírez** Tecnológico Nacional de México, Instituto Tecnológico de Orizaba, Veracruz, México

**Zahia Tamen** USTHB, Computer Science Department, LRIA Laboratory, University of Science and Technology Houari Boumediene, Houari, Algeria

**Mui Joo Tang** Tunku Abdul Rahman University College, Kuala Lumpur, Malaysia

**Xiao Ke Tang** Beijing Smart-Chip Microelectronics Technology Co., Ltd, Haidian District, Beijing, P.R. China

**Subhasis Thakur** National University of Ireland Galway, Galway, Ireland

**Nguyen Duc Thinh** Vietnamese-German University, Binh Duong, Vietnam

**Mia Amor C. Tinam-isan** Premier Research Institute of Science and Mathematics, Mindanao State University-Iligan Institute of Technology, Iligan City, Philippines

**Masashi Toda** Kumamoto University, Kumamoto-city, Kumamoto, Japan

**Metin Turan** Computer Engineering Department, Istanbul Commerce University, Istanbul, Turkey

**Thavavel Vaiyapuri** College of Computer Engineering and Sciences, Prince Sattam bin Abdulaziz University, Alkharj, Kingdom of Saudi Arabia

**Alexandra Valiton** The MITRE Corporation, McLean, VA, USA

**Ricardo Vardasca** ISLA Santarem, Largo Candido Dos Reis, Santarem, Portugal

**Lemuel Clark P. Velasco** Premier Research Institute of Science and Mathematics, Mindanao State University-Iligan Institute of Technology, Iligan City, Philippines

**Erik Verreycken** FTI-CoSys Lab, University of Antwerp, Antwerp, Belgium

**Erik Verreycken** Flanders Make Strategic Research Centre, Lommel, Belgium

**Luis Daniel Villanueva-Montoya** Universidad Nacional de San Agustín de Arequipa, Arequipa, Perú

**Hien Vo** Vietnamese-German University, Binh Duong, Vietnam

**Rossouw Von Solms** School of IT, Nelson Mandela University, Gqeberha, South Africa

**Hui Wang** Beijing Smart-Chip Microelectronics Technology Co., Ltd, Haidian District, Beijing, P.R. China

**Jing Wang** College of Information Technology, Shanghai Ocean University, Shanghai, China

**Jing Wang** Key Laboratory of Fisheries Information, Ministry of Agriculture, Shanghai, China

**Lina Wang** Naval Aviation University Aviation Foundation, Yantai, China

**Qiang Wang** University of Electronic Science and Technology of China, Chengdu, SC, China

**Zhong Wu** University of Electronic Science and Technology of China, Chengdu, SC, China

**Kun Xiao** The University of Electronic Science and Technology of China, Chengdu, P. R. China

**Ke Xu** Shenzhen Institute of Huazhong University of Science and Technology, Shenzhen, China

**Elias Yaacoub** Computer Science and Engineering Department, Qatar University, Doha, Qatar

**Li Xin Yang** Beijing Smart-Chip Microelectronics Technology Co., Ltd, Haidian District, Beijing, P.R. China

**Shuhu Yang** College of Information Technology, Shanghai Ocean University, Shanghai, China

**Samet Yilmaz** Department of Computer Science, Hunan University, Changsha, China

**Ren Ying** Naval Aviation University Aviation Foundation, Yantai, China

**Abdellah Yousfi** Faculty of Law, Economics and Social Sciences, Souissi-Rabat Mohammed V University, Rabat, Morocco

**Vasyl Yurchyshyn** Department of Computer Systems Software, National Technical University of Ukraine "Igor Sikorsky Kyiv Polytechnic Institute", Kyiv, Ukraine

**Yaroslav Yurchyshyn** Department of Computer Systems Software, National Technical University of Ukraine "Igor Sikorsky Kyiv Polytechnic Institute", Kyiv, Ukraine

**Mohammed Mohammed Zayed** Sana'a University, Sana'a, Yemen

**Alexander A. Zelensky** Don State Technical University, Rostov-on-Don, Russia

**Yanhong Zhang** Naval Aviation University Aviation Foundation, Yantai, China

**Yun Zhang** College of Information Technology, Shanghai Ocean University, Shanghai, China

**Yun Zhang** Key Laboratory of Fisheries Information, Ministry of Agriculture, Shanghai, China

**Yuting Zhang** China Mobiles, Shanghai, China

**Huiping Zhou** University of Tsukuba, Tsukuba-city, Ibaraki, Japan

**Ruyan Zhou** College of Information Technology, Shanghai Ocean University, Shanghai, China

**Haiyi Zhu** College of Information Technology, Shanghai Ocean University, Shanghai, China

**Rajnai Zoltán** Doctoral School on Safety and Security Science, Óbuda University, Budapest, Hungary

# Refining the Efficiency of R-CNN in Pedestrian Detection

**Katleho L. Masita, Ali N. Hasan, and Thokozani Shongwe**

**Abstract** The use of pre-trained deep convolutional neural networks for the purpose of enhancing the performance of detectors such as the region-based convolutional neural networks has demonstrated an exceptional role in the field of pedestrian detection. There have been various methods that have been investigated with the intent of improving the detection results of pedestrian detectors. In particular, most deep convolutional neural network techniques have shown advanced results in recent experimentation and studies. The primary reason for this improvement in performance based on deep neural networks is the ability of deep networks to learn substantial mid-level and high-level image features and also generalize well in complex images. In this study, an advanced deep learning technique based on R-CNN is refined and investigated for better performance in pedestrian detection on four specific pedestrian detection databases. The experiments involve the application of a deep learning feature extraction model in conjunction with the R-CNN detector. The deep learning feature extraction models employed are the AlexNet, VGG16 and the VGG19. The architecture of the R-CNN is re-modelled for enhanced performance by incorporating deep stacking networks and altering the activation function from sigmoid to Rectified Liner Unit (ReLU). The results of the experiments across the four investigated datasets provide significant findings about the performance of R-CNN detector when trained on different pre-trained deep convolutional neural networks with the application of deep stacking networks and ReLU activation function.

K. L. Masita (✉)
Department of Electrical and Electronic Engineering Technology, University of Johannesburg, Gauteng 2006, South Africa
e-mail: 201301860@student.uj.ac.za

A. N. Hasan
Department of Electrical Engineering, Higher Colleges of Technology, Abu Dhabi, United Arab Emirates
e-mail: ahasan1@hct.ac.ae

T. Shongwe
Department of Electrical and Electronic Engineering Technology, University of Johannesburg, Gauteng 2006, South Africa
e-mail: tshongwe@uj.ac.za

 
X.-S. Yang et al. (eds.), *Proceedings of Sixth International Congress on Information and Communication Technology*, Lecture Notes in Networks and Systems 216,
https://doi.org/10.1007/978-981-16-1781-2_1

**Keywords** Pedestrian detection · Deep stacking networks · Rectified Linear Unit · Convolutional neural networks · Region-based neural networks (R-CNN)

## 1 Introduction

Deep learning features have positively enhanced performance in pedestrian detection in recent years, and a number of CNN-based techniques have subsequently been developed [1–4]. For most of the methods developed, however, there are still some shortcomings. (1) They usually use convolutional characteristics without distinction, resulting in sub-optimal outcomes, such as occlusion and hard negative samples because of the distraction from the complex context. Visual activity from the background continues to confuse the detectors, contributing to a significant number of fail detections of occluded pedestrians and false alarms due to cluttered surroundings. (2) Deepest convolutional layer with a rough resolution and broad receptive area is mostly used for prediction, which is inefficient to find small objects, given its high-level semantic characteristics. (3) Most approaches use a large deep network and require a level of refining to enhance detection accuracy. The time of estimation is sacrificed to guarantee precision, rendering such approaches inadequate for real-time use.

In general, researchers use deep models to explain higher-level representations of object instances, which are useful in distinguishing pedestrians in traffic scenarios. Sadly, this method leaves out a number of small-scale pedestrians as well as details on the position of large-scale pedestrians. Owing to the intrinsic features of CNN, the main outlets cannot be outlined, and the central spatial location cannot be illuminated. Convolution is a spatial process that obtains spatial image knowledge by adding a convolution kernel to the local image. The local activity of CNNs results in its failure to collect images from a regional point of view. It is still always a struggle to build an effective framework for the identification of pedestrians.

Among the numerous tasks of computer vision that have profited by Deep Learning models in various papers in literature is pedestrian detection. This article proposes an effective technique of using pre-trained deep convolutional neural network models for improving the performance of an R-CNN detector in pedestrian detection. The pre-trained deep CNN models used for enhancing the R-CNN detector performance are applied as classification models in the overall architecture of the R-CNN detector. These pre-trained deep CNN models include the original AlexNet, VGG16 and VGG19 models. The various R-CNN detectors based on the different classification models are trained and tested on four different pedestrian datasets. The training databases contrast by a measure of images they contain, just as the quantity of pedestrians depicted in every frame. The results of the experiments clearly show a few principles regarding the behaviour of the R-CNN detector when trained on different classification models.

This paper is structured in the following manner: Sect. 2 presents the background in pedestrian detection and the strategies and techniques that have been connected to it. Section 3 defines the principal details of the pre-trained CNN models used for classification. Section 4 explains the R-CNN detector, the region proposals extraction process, the proposed R-CNN training workflow as well as how the R-CNN detector is improved to produce good results during detection. Section 5 discusses the experiments, results and the process of training and testing the R-CNN detector. Section 6 outlines and examines the results of the experiments. Section 6 is the summary and conclusion of the paper.

## 2 Pedestrian Detection Methods

In 2019, a number of software engineering researchers from the Beijing Union University developed a pedestrian detection model based on HOG (Histogram of Oriented Gradients) features and SVM (Support Vector Machine) to realize Vehicle–Human–Environment Interaction. The pedestrian detection framework adopted in this article extracts features by means of the HOG and then employs the SVM as the classifier to train the classification model and ultimately fulfil the pedestrian detection task. Experimental results suggest that the HOG extraction model is efficient for optical flow adjustments and can correctly classify pedestrians with specific illumination [5].

In the article "Pedestrian Detection Using R-CNN Object Detector" [6], AlextNet is employed as the benchmark classification model to investigate and improve the R-CNN object detector's performance in the scenario of two pedestrian detection datasets. In this research work, VGG16 and VGG19 are used as the primary classification models to investigate and improve the performance of the R-CNN object detector. This paper is an extension of the AlexNet-based R-CNN object detector's technical contributions from the paper "Pedestrian Detection Using R-CNN Object Detector" and included the results from the article for comparison with new experiments and methods with additional data. There are two additional pedestrian detection scenarios explored, which are based on two globally recognized pedestrian detection datasets. A comparison between the three detectors is then made, namely, VGG16-based R-CNN detector, VGG19-based R-CNN detector and AlexNet-based R-CNN object detector.

In 2019, Zahid Ahmed et al. [1] developed a deep learning-based pedestrian detection model utilizing depthwise separable convolution and Single Shot Detector structure that consists of distinctive activation maps utilizing OpenCV. The goal of the study was to develop a stable, powerful and efficient deep learning-based pedestrian detection system for synchronous applications. The experimental findings demonstrated that the designed model performs especially well in overwhelming backgrounds and furthermore considers the way that real-life situations have progressively altering sceneries and backgrounds and with all things considered, this model gives both a precise and quick detection of pedestrians adaptable for real-time operations.

# 3 Improving R-CNN Detector

## 3.1 *Proposed Architectural Innovation of the R-CNN Detector Base Network*

Training a DNN is a challenging task in terms of computing. This is because the traditional teaching approaches used at the training level are the stochastic gradient descent learning algorithmic norm, which is extremely difficult to parallelize through computers. This suggests that learning on a significant scale is non-trivial. For example, through applying a solitary but extraordinary GPU computer to prepare DNN-based object detectors is conceivable, with a large amount of long stretches of image training databases yielding unusual results. It turns out to be increasingly complex to ramp up this implementation with a large amount of training data.

By integrating the Deep Stacking Neural Networks (DSNs), which were developed with the need to solve the problem of learning scalability and which are one of the strategic and revolutionary solutions to the problem of computational errors during DNN training, we expand the classification capabilities of the classification network [7]. The Deep Stacking Networks architecture that we model and integrate into the base network is as seen in Fig. 1 below.

In addition to this, we modify the performance element of the R-CNN architecture by employing the ReLU function rather than the sigmoid function as the basis for the activation function [7] (Fig. 2).

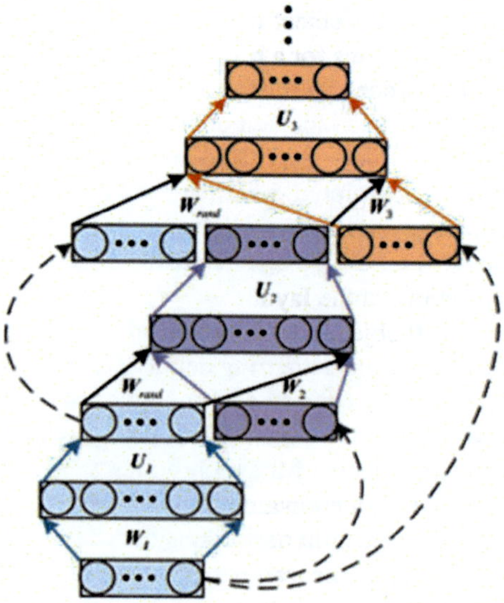


**Fig. 1** Deep stacking networks

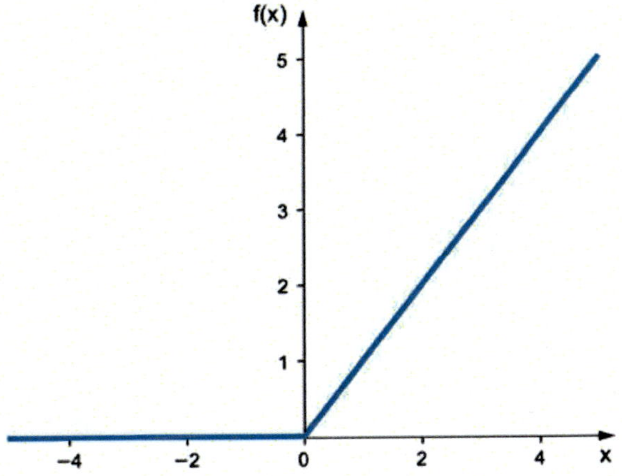


**Fig. 2** ReLU function

### 3.2 *Proposed R-CNN Training Workflow*

As discussed above, the proposed system involves an integration of a pre-trained deep CNN classification model with the R-CNN model through a concept called Transfer learning. This is achieved in the manner indicated in the flow chart in Fig. 3

The proposed system as illustrated in Fig. 3 specifies how to calibrate a pre-trained deep convolutional neural network for performing classification on a different dataset of images. The specified pre-trained deep CNN model has already been trained on a dataset that involves over a million images and can perform classification into one thousand object categories. The network is equipped with learned rich feature representations for a broad scope of images.

The process of Transfer learning is executed with the pre-trained CNN classifier as indicated in the flow chart above. The pre-trained deep CNN model is simply utilized as the starting point for learning a new task. The process of fine-tuning or calibrating a CNN model through transfer learning is more convenient, faster and simpler than training a network with randomly initialized weights from the ground up.

Owing to the layout of the last three layers of the pre-trained deep CNN model for 1000 object categories, the layers must be tweaked for the current classification task. The layers are then moved to the current classification task by substituting the last three layers of the pre-trained network with a fully connected layer, a softmax layer and an output layer for classification. The new fully connected layer options are then defined by the latest dataset. This is done by setting the fully connected layer to possess an equivalent size as the number of classes found in the new data. The new classification network, which contains the transferred layers and new layers, is then trained. The trained classifier is utilized in the R-CNN framework for training a category-specific object detector. The process improves both the training speed and R-CNN detector's performance.

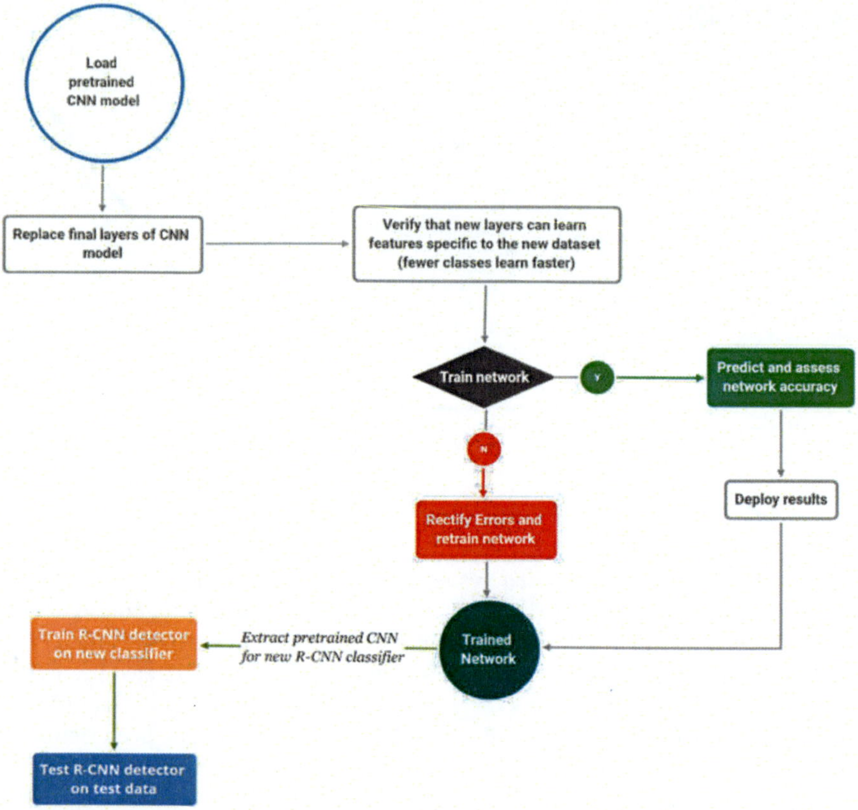


**Fig. 3** Proposed R-CNN training workflow

# 4 Training and Testing the R-CNN Detector

## *4.1 Classification Models Applied*

The proposed pedestrian detection model is built on the Alexnet, VGG16 and VGG19 convolutional neural network (CNN) models integrated with the R-CNN detector. The AlexNet, VGG16 and VGG19 models are individually used as the base classification models [8, 9].

## 4.2 Datasets

The datasets used for training and testing are the Penn-Fudan dataset (170 pictures), the KTH Multiview Football dataset (771 pictures), Chinese University of Hong Kong Campus 01 (CUHK01) with 3884 pictures and Chinese University of Hong Kong Campus 02 (CUHK02) with 7264 pictures [10–13].

## 4.3 Training Data Pre-Processing

In preparing the datasets with a region of interest, part of the image labelling is done by an automated custom algorithm, and the other part is done manually. This is because the automated custom algorithm is not effective enough to label the entire datasets, thus, the balance of image labelling is performed manually.

The automated custom algorithm for training data prep-processing is the Aggregated Channel Features (ACF) Detector automated algorithm. The fundamental architecture of the aggregate channel features is the channel. The most customary sort of channel is the colour channels of the image, with Gray-scale and RGB being the regular ones. Apart from colour channels, numerous dissimilar channel types have been created to encode various kinds of data for more complex challenges. Normally, channels may be described as a recorded map of the initial picture, whose pixels are processed from equal patches of initial pixels [6]. Distinctive channels may be processed with a linear or non-linear transformation of the preliminary photo. To permit sliding window detection, the ameliorations are forced to be translationally invariant.

The duration of time it takes to label all the images of the CUHK01 dataset is 19 h, and the CUHK02 dataset takes 36 h for labelling all the target objects in the images. The KTH Football dataset takes 14 h for image labelling, and the Penn-Fudan Pedestrian dataset takes 9 h for image labelling.

Training data is created from the ground-truth objects to be used for training the R-CNN object detector. A sampling factor of 2 is used for sampling the ground-truth data. The essence of sampling is that it alleviates overtraining the object detector on corresponding samples.

As part of the training phase, 60% of the ground-truth dataset is utilized as the training sample for all of the four selected datasets. The 40% samples which remain are employed to test the R-CNN detectors.

### 4.4 Performance Measures

**Average Precision**

Average precision is a measure that joins recall and precision for positioned recovery results. For one data requirement, the average precision is the mean of the exactness scores after each significant document is recovered.

$$Average\ Precision\ (AP) = \frac{\sum_r P@r}{R} \tag{1}$$

where r is the rank of each significant document, R is the aggregate number of applicable documents and P@r is the exactness of the top-r recovered documents. Mean average precision is a decent framework measure for a few reasons. Right off the bat, it is a single value measure, which is advantageous for utilizing especially for a number of diverse Information Retrieval frameworks. Also, it is delicate since its count utilizes the total data of applicable documents: the aggregate number of significant documents in the total record accumulation and the ranked position of the documents in a resultant list. Finally, it is solid since it considers the complete information about the important documents. A mean average precision often requires a huge amount of effort to calculate and its value is conceptual and cannot be clarified expressly.

**Recall**

The recall is the capacity to remember all applicable instances in a classification model. By marking it as Positive (True Positive) relative to the forecasts made, Recall measures how many of the real positives the model catches.

$$Recall = \frac{TruePositive}{TruePositive + FalseNegative} \tag{2}$$

**F-Measure**

The F-measure combines precision and recall, and it is the harmonic mean of precision and recall; the conventional F-measure or balanced F-score is measured in the following way:

$$F - measure = 2 \times \frac{precision \times recall}{precision + recall} \tag{3}$$

# 5 Experimental Results

There are three cutting-edge models of the R-CNN object detection technique, which are selected for comparison and have been previously published for performance comparison on different datasets. One method is the AlexNet-based R-CNN object detection method defined as AlexNet-R-CNN, the other is a VGG16-based R-CNN technique defined as VGG16-R-CNN and the final method is a VGG19-based R-CNN method called the VGG19-R-CNN.

The experimental outcomes are illustrated in Figs. 4, 5, 6 and 7 as shown in the section below. When measuring the AP, the y-axis shows the precision and the x-axis depicts the recall.

As observed in Fig. 4 above, the AlexNet-based R-CNN detector achieves a higher average precision percentage than VGG16 based and VGG19-based R-CNN detectors. The AP achieved by AlexNet-R-CNN is 43.53%, the AP results achieved by VGG16-R-CNN is 14.99% and the AP achieved by the VGG19-R-CNN detector is 22.48%. This makes a 28.53 percentage difference between the AlexNet-R-CNN detector and the VGG16-R-CNN detector. It also makes a 21.04 percentage difference between the AlexNet-R-CNN detector and the VGG19-R-CNN detector. Thus, the AlexNet-R-CNN detector by far outcompetes the VGG16 and VGG19-based detectors on the Penn-Fudan Pedestrian dataset. It is important to note that within the first fraction of instances of recall, the VGG16-R-CNN and VGG19-R-CNN line graphs rise sharply, more like an exponential graph then it settles and begins to gradually decline from the second fraction of recall and throughout the remaining instances of the experiment. On the other hand, the Alexnet-R-CNN line graph begins with a very high performance of approximately 99% AP and then declines rapidly on the first two fractions of recall; between 0.2 and 0.4 instances of recall, the line graph slightly picks up after which it assumes a declining trend for the remaining fractions of recall. Two observations can be made from these AP results. The first one is that

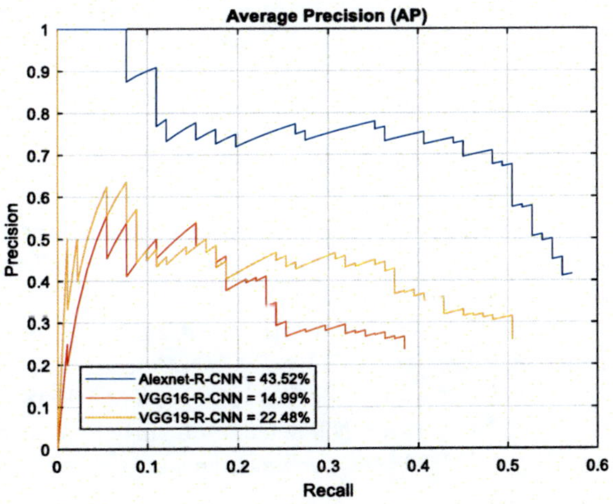


**Fig. 4** Average precision results on the Penn-Fudan Pedestrian dataset

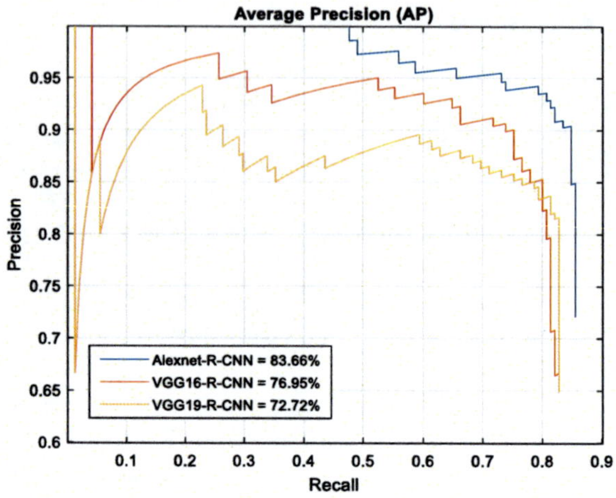


**Fig. 5** Average precision results on the KTH Football dataset

**Fig. 6** Average precision results on the CUHK01 Pedestrian dataset

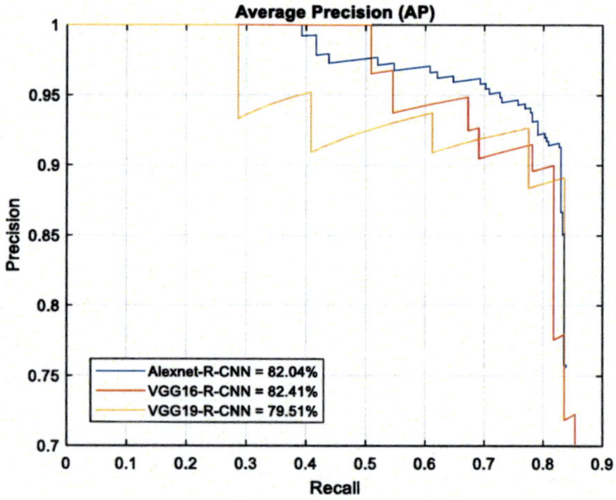


the Alexnet-R-CNN detector, being the least deep neural network, performs better on this arguably small Penn-Fudan Pedestrian dataset of 177 images. The second is that there is a significant difference in performance between the VGG16-R-CNN detector and the VGG19-R-CNN detector on this dataset despite a minor difference in the design architecture of the two detectors.

On the KTH Football dataset, the AlexNet-R-CNN detector once again performs better than both the VGG16-R-CNN detector and VGG19-R-CNN detector. The performance of AlexNet-R-CNN reaches 83.66% in average precision on this dataset, while the VGG16-R-CNN detector obtains an average precision of 76.95% and the VGG19-R-CNN detector achieves an average precision of 72.72%. The difference in

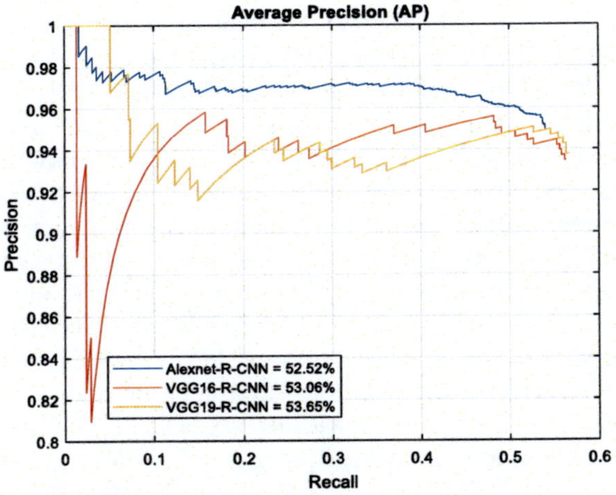


**Fig. 7** Average precision results on the CUHK02 Pedestrian dataset

performance between the AlexNet-R-CNN detector and the VGG16-R-CNN detector is 6.71% and the difference in performance between the AlexNet-R-CNN and the VGG19-R-CNN detector is 10.94%. Between the VGG16-R-CNN detector and the VGG19-R-CNN detector, the difference in performance is only 4.23%. As can be observed in Fig. 5, the two line graphs for VGG16-R-CNN and VGG19-R-CNN show a similar graphical pattern of ascent and descent and they both stop at approximately the same instance of recall. On the other hand, the AlexNet-R-CNN line graph clearly begins with a very high performance of approximately 99% AP and then gradually declines until it stops at approximately 0.85 instance of recall. The AlexNet-R-CNN line graph clearly lies above both the VGG16-R-CNN and VGG19-R-CNN line graphs, and thus, the AlexNet-R-CNN detector performs better than the other two detectors. The observations made from the results in this experiment are significant. The KTH Football dataset is more than four times bigger than the Penn-Fudan dataset, yet, the AlexNet-R-CNN detector still outcompetes the other two detectors, while it is the least deep neural network. It is noteworthy, however, that the difference in performance between the three detectors is not very wide compared to the results on the Penn-Fudan Pedestrian dataset. The KTH Football database is made up of 771 images, while this is much bigger than the Penn-Fudan Pedestrian dataset, and the KTH Football dataset can still be classified and regarded as a considerably small dataset.

As can be observed in Fig. 6, the VGG16-R-CNN detector accomplished the highest average precision percentage compared to the Alexnet-R-CNN and VGG19-R-CNN detectors. VGG16-R-CNN attains an average precision of 82.41%, the average precision attained by the Alexnet-R-CNN detector is 82.04% and the AP achieved by VGG19-R-CNN is 79.51%. The performance percentage difference between the VGG16-R-CNN detector and the Alexnet-R-CNN detector is only 0.37%,;also, the performance difference between VGG16-R-CNN detector and the

VGG19-R-CNN detector is 2.90%. On the other hand, the percentage difference between the VGG19-R-CNN detector and the Alexnet-R-CNN detector is 2.53%. It is clear that the VGG16-R-CNN detector and the AlexNet-R-CNN are converging in performance on the CUHK01 dataset. It is also important that all the three detectors are performing with significantly high detection precision on this dataset, which can be attested to the size and quality of the dataset rather than just the CNN depths of the individual detectors. It can be clearly seen from the graphical trends that the VGG16-R-CNN and the VGG19-R-CNN have a similar trend, which is significant of their CNN architectural similarity, and their difference in precision levels is significant of their different CNN depths. Lastly, the difference in performance between the three detectors is less wide compared to the results of the previous datasets.

As illustrated in Fig. 7, the VGG19-R-CNN detector attains a higher average precision percentage compared to the Alexnet-R-CNN and VGG16-R-CNN detectors. VGG19-R-CNN reaches an average precision of 53.65%, the average precision reached by the Alexnet-R-CNN detector is 53.06% and the AP achieved by Alexnet-R-CNN is 52.52%. The performance percentage difference between the VGG19-R-CNN detector and the Alexnet-R-CNN detector is 1.13%, also, the performance difference between VGG19-R-CNN detector and the VGG16-R-CNN detector is only 0.59%; hence, there happens to be no important performance difference between these two detectors on this dataset. There is a noticeable but gentle decrease in the precision for AlexNet-R-CNN line graph from 0.1 recall all the way to 0.5 recall, while for the VGG19-R-CNN and the VGG16-R-CNN line graphs, there is a trend that indicates a gentle but constant rise in precision between 0.1 instance of recall and 0.5 instance of recall. The main observation from these results is the fact that at this point, the performance results of the VGG19-R-CNN detector and the VGG16-R-CNN detector are beginning to converge. There is no critical or further improvement in the performance between the VGG19-R-CNN detector and the VGG16-R-CNN detector. Additionally, the VGG19-R-CNN detector and the VGG16-R-CNN detector are both starting to outperform the Alexnet-R-CNN detector in performance results.

Table 1 below displays the efficiency of pedestrian detectors across four pedestrian detection datasets with respect to Average Precision (AP), Recall and F-Measurement.

## 6 Conclusion

In this study, a deep learning training workflow is investigated on three different pre-trained deep learning classification models and a common object detection technique called the R-CNN for investigating three different pedestrian detectors. The three classification models are the VGG16, VGG19 and the AlexNet model. The investigated pedestrian detectors are each trained across four different pedestrian detection datasets. The three pedestrian detectors use the edge boxes algorithm to extract region proposals from images. The main objectives of the experiments are to

**Table 1** Comparison of the R-CNN detectors on the four different datasets

| Dataset | Performance measures | Penn-Fudan | KTH Football | CUHK01 | CUHK02 |
|---|---|---|---|---|---|
| AlexNet-R-CNN | AP | 43.52 | 83.66 | 82.04 | 52.52 |
| | Recall | 0.58 | 0.86 | 0.84 | 0.54 |
| | F-measure | 0.48 | 0.78 | 0.79 | 0.68 |
| VGG16-R-CNN | AP | 14.99 | 76.95 | 82.41 | 53.06 |
| | Recall | 0.38 | 0.83 | 0.85 | 0.57 |
| | F-measure | 0.29 | 0.73 | 0.76 | 0.70 |
| VGG19-R-CNN | AP | 22.48 | 72.72 | 79.51 | 53.65 |
| | Recall | 0.51 | 0.83 | 0.84 | 0.57 |
| | F-measure | 0.35 | 0.72 | 0.77 | 0.70 |

primarily investigate the influence of the depth of the three pedestrian detectors on their performance and the influence of the size of the datasets on the performance of the detectors trained on the individual datasets. The outcomes as per the conducted experiments are that deeper neural network pedestrian detectors tend to perform better on bigger pedestrian databases than on smaller pedestrian databases. At the same time, deep learning pedestrian detectors generally tend to possess a more stable performance on bigger datasets than on smaller datasets. It is noteworthy, however, that the CUHK02 pedestrian dataset is much bigger than the CUHK01 pedestrian dataset, yet, all three R-CNN detectors achieve higher performance on the CUHK01 pedestrian dataset than on the CUHK02 dataset. This observation obviously opens a question about the influence of the quality of the dataset images on the performance of the detector trained on it. This is another aspect of the study that can be open for further investigation in future research. The quality of the dataset has to be classified according to specific standards and benchmarks based on the aspects of shading, occlusion, capture angle, etc. This will help to appropriately verify and qualify the performance of any given detector trained on any given dataset. Furthermore, this will help in ensuring that the certainty and reliability of the performance of pedestrian detection systems are heightened and enhanced for a safer and better object detection technology.

## References


1. Ahmed ZZ, Iniyavan R (2019) Enhanced vulnerable Pedestrian detection using deep learning. In: 2019 International Conference on communication and signal processing (ICCSP), Chennai, India, pp 0971–0974. https://doi.org/10.1109/ICCSP.2019.8697978
2. Jin Y, Wen Y, Liang J (2020) Embedded real-time Pedestrian detection system using YOLO optimized by LNN. In: 2020 International Conference on electrical, communication, and computer engineering (ICECCE), Istanbul, Turkey, pp 1–5. https://doi.org/10.1109/ICECCE49384.2020.9179384

3. Harishankar V, Karthika R (2020) Real time Pedestrian detection using modified YOLO V2. In: 2020 5th International Conference on communication and electronics systems (ICCES), COIMBATORE, India, pp 855–859
4. Zhu K, Li L, Hu D, Chen D, Liu L (2019) An improved detection method for multi-scale and dense pedestrians based on Faster R-CNN. In: 2019 IEEE International Conference on signal, information and data orocessing (ICSIDP), Chongqing, China, pp 1–5. https://doi.org/10.1109/ICSIDP47821.2019.9173020.
5. Nan M, Li C, JianCheng H, QiuNa S, JiaHong L, GuoPing Z (2019) Pedestrian detection based on HOG features and SVM realizes vehicle-human-environment interaction. In: 2019 15th International Conference on computational intelligence and security (CIS), Macao, Macao, pp 287–291
6. Masita KL, Hasan AN, Paul S (2018) Pedestrian detection using R-CNN object detector. In: Proceedings of 2018 IEEE Latin American Conference on computational intelligence, Guadalajara, Mexico
7. Masita KL, Hasan AN, Shongwe T (2020) Deep learning in object detection: a Review. In: 2020 International Conference on artificial intelligence, big data, computing and data communication systems (icABCD), Durban, South Africa, pp 1–11. https://doi.org/10.1109/icABCD49160.2020.9183866
8. Krizhevsky A, Sutskever I, Hinton G (2012) ImageNet classification with deep convolutional neural networks. Adv Neural Inf Proc Syst 1106–1114
9. Simonyan K, Zisserman A (2014) Very deep convolutional networks for large-scale image recognition. Department of Engineering Science, University of Oxford, Visual Geometry Group
10. Kazemi V, Burenius M, Azizpour H, Sullivan J (2013) Multi-view body part recognition with random forests. In: Proceedings of BMVC 2013
11. Liming Wang, Jianbo Shi, Gang Song, and I-fan Shen, "Object detection combining recognition and segmentation". Lecture Notes in Computer Science, vol. 4843, pp. 189–199, Springer Link, 2007.
12. Wei L, Rui Z, Xiaogang W (2012) Human reidentification with transferred metric learning, ACCV
13. Li W, Wang X (2013) Locally aligned feature transforms across views. In: 2013 IEEE Conference on computer vision and pattern recognition, Portland, OR, pp 3594–3601 https://doi.org/10.1109/CVPR.2013.461
14. Chen Y, Jhong S, Li G, Chen P (2019) Thermal-based Pedestrian detection using faster R-CNN and region decomposition Branch. In: (2019) International Symposium on intelligent signal processing and communication systems (ISPACS). Taipei, Taiwan 2019:1–2. https://doi.org/10.1109/ISPACS48206.2019.8986298

# Towards Spark-Based Deep Learning Approach for Fraud Detection Analysis

**Alae Chouiekh and El Hassane Ibn El Haj**

**Abstract** Fraud activity is a major concern in the telecommunication business domain. Thus, it's very important to analyze the huge amount of data available in the network in order to detect early potential fraud behaviors and take countermeasures accordingly. In this paper, we used a novel approach using deep learning techniques to analyze datasets from a real mobile communication career and to extract and analyze features from labeled/unlabeled data. Our dataset was extracted from the business support system (BSS) containing 2.5 million call details records (CDR) of active subscribers. The proposed method integrates deep learning techniques with Apache Spark framework for parallel training execution. We found that our approach performed better than traditional machine learning with an F1 score of 91% and enhance the training speed of our deep learning model significantly when using the spark platform. Thus, the use of this proposed model can remarkably decrease the cost related to the unauthorized use of telecommunication services.

**Keywords** Fraud · Telecommunication · Deep learning · Machine learning

## 1 Introduction

Fraud activity is a worldwide problem affecting many industries such as banking [1], credit card transactions [2], and mobile telecommunication [3], leading to substantial annual revenue losses [4]. Telecommunication is a fast-growing industry worldwide and voice-related fraud activities represent a serious threat to mobile communications, especially with the increasing number of mobile devices. A survey from the Communications Fraud Control Agency shows that the estimated annual fraud losses incurred by the telecom industry were $38.1 billion in 2015. Fraud can be defined as the use of telecommunication networks with the intention of avoiding the payment of services such as voice calls, SMS, MMS, data usage, or any other subscription provided by the telecom provider. Fraud detection problems involve always huge

A. Chouiekh (✉) · E. H. Ibn El Haj
Laboratory of Multimedia Signal and Communications Systems, National Institute of Posts and Telecommunications, Rabat, Morocco



X.-S. Yang et al. (eds.), *Proceedings of Sixth International Congress on Information and Communication Technology*, Lecture Notes in Networks and Systems 216,
https://doi.org/10.1007/978-981-16-1781-2_2

datasets, constantly growing and requiring online analysis [5], including calls events, SMS events, recharge events, or any other subscription activity related to different services. The processing and manipulation of the dataset constitute a big challenge for telecom providers to extract fraudulent events in the near real time and generate alerts as soon as fraudulent traffic is detected. This gives the operator the opportunity to prevent additional revenue loss and stop any suspicious users to acquire services illegally. Therefore, there is an urgent need to design more accurate models to improve the quality of service and enhance network security in order to address the fraud activity time. Several fraud detection methods have been suggested in the telecommunication business domain, and they are related either to some machine learning techniques [6–8] or data mining algorithms [9]. In this work, we investigated the use of a new approach based on a distributed deep learning implementation relying on spark's memory computation for our training. Spark is the state-of-the-art open-source framework designed for data-parallel computations and maintaining data in memory rather than in disk in order to speed up the training of deep models [10]. Arguably, when it comes to fraud activities, most telecom providers are strived to make informed and swift decisions to attain high user satisfaction and avoid revenue loss. Deep learning has proved successful at different classification tasks in different research areas: such as computer vision [11], natural language processing or NLP [12], different object detection tasks [13], and some dimensionality reduction algorithms [14]. More specifically, the deep learning techniques use deep architectures of many layers to extract unsupervised features avoiding the labor work of the engineering process. This method becomes more challenging and requires much more tuning, especially when dealing with a huge amount of datasets. In the present work, we investigated a real-time fraud framework based on deep learning models trained over the open-source Apache Spark platform using its standalone cluster mode [15]. The main goal of this research is to detect and report fraudulent events by analyzing customer's pattern. This could be performed by analyzing the CDRs of any customer. More precisely, using the features present in the log of events, each profile could be classified as a fraudster or not during a specific timestamp. In the next section, we present the events used by our machine learning techniques:

- Number of Call Events which include: the number of local or international calls with corresponding charging units, max call duration, and the number of calls per day.
- Number of used Voucher Events which describe the amount of recharges per day during different timeslots.
- Number of subscribed services/offers per day for each customer.
- Number of Data Events including the downloading/uploading volume with corresponding charging units.
- Number of sent SMS Events per day for each customer.

The CDRs are primarily used for charging purpose. Indeed, there is a charging rule applied to the user according to a given specific service. Additionally, the collected CDRs information will be used in helping to analyze suspicious behaviors in the network. The analysis of CDRs is the best way for the telecom provider to detect skeptical cases in the network such as not charged calls (especially with long duration), the excessive number of recharges with a big amount of vouchers, and different type of patterns related to the fraud detection use cases. This will prevent a fraudster from getting any useful information in order to escape from countermeasure mechanisms. In this work, we demonstrated the effectiveness of the deep learning model and showed how the accuracy could be improved significantly when compared to the traditional machine learning techniques such as GBC(Gradient Boosting Classifier), SVM, or Random forest. Moreover, the training duration of the deep learning model was decreased when compared to our previous work [16] as a result of using the Apache Spark implementation compared to a single machine computation mode using the serial algorithm implementation. The rest of this paper is following this structure: Sect. 2 describes our proposed deep learning architecture based on Apache Spark implementation. Section 3 evaluates the results of the proposed model and compares it with the state-of-the-art machine learning algorithms using a real telecommunication dataset. Section 4 is summarizing all of our conclusions.

## 2 Proposed Spark-Based Deep Learning Architecture

Building a deep learning model is usually computationally costly during the training or prediction phase, especially when using a serial algorithm implementation to process a huge amount of data [17]. To tackle these challenges, we have suggested a novel approach based on Apache Spark technology to boost the training speed. Figure 1 shows the proposed framework including dataset collection and deep learning model training over Spark standalone cluster.

After the CDRs are collected, they were loaded to the operator billing system and the corresponding charging of users is applied. The same CDRs were also prepared to be transferred to the spark ecosystem in .CSV format. Of note, the Apache Spark contains two main components: spark driver or master and different spark workers. The spark master initiates the network configuration and distributes the partial model training to different workers. The model training is described as follows:

(a) The Spark driver initiates the network configuration and hyper parameter settings.
(b) The Data is split into different subsets following the Training Master Configuration and mini-batch sizes.
(c) Iterate over the data split. For each split of the training data:
    - Distribute the configuration and hyper parameters from the master to all workers: momentum, stochastic gradient descent (SGD) optimization algorithm, and Learning rate.
    - Different sub-models training jobs are running on different workers: Average the parameters and return the averaged results with a complete model training to the master node.

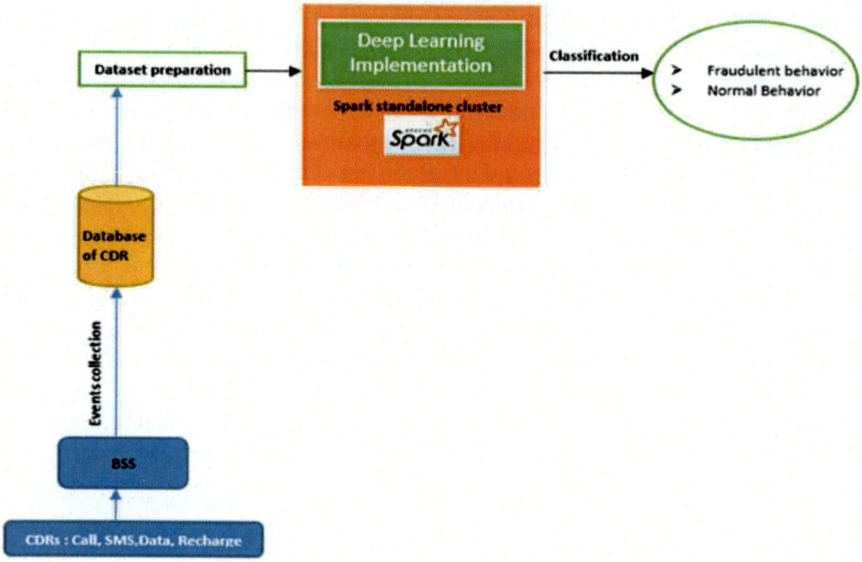


**Fig. 1** Proposed fraud detection framework using deep learning over Apache Spark ecosystem

(d) Training is finished, and the master is having the trained network after aggregating/collecting the results from the workers.

We used multiple sets of partitions: training, validation, and testing sets, and taking the averaged results over these sets (we excluded the result from a single, which can be biased and misleading). Figure 2 shows an example of the training workflow using different workers.

The training of a partial model was performed as shown in Fig. 3 (see below).

The hyper parameters tuning techniques are mandatory during the process of parameters optimization; in this work, we decided to use the optimized random search process [18]. The best performance of hyper parameters selection is reported following different tests executed over different periods of time. The final setup is given in Table 1.

The rectifier linear unit (RELU) has many benefits, the most important one is the faster training time [19], it outperformed many other methods such as Tanh-function or the sigmoid function. It is important to emphasize that the used dataset was divided initially into different parts: the training dataset represents 50%, the testing dataset represents 25%, and finally, the validation dataset represents 25%.

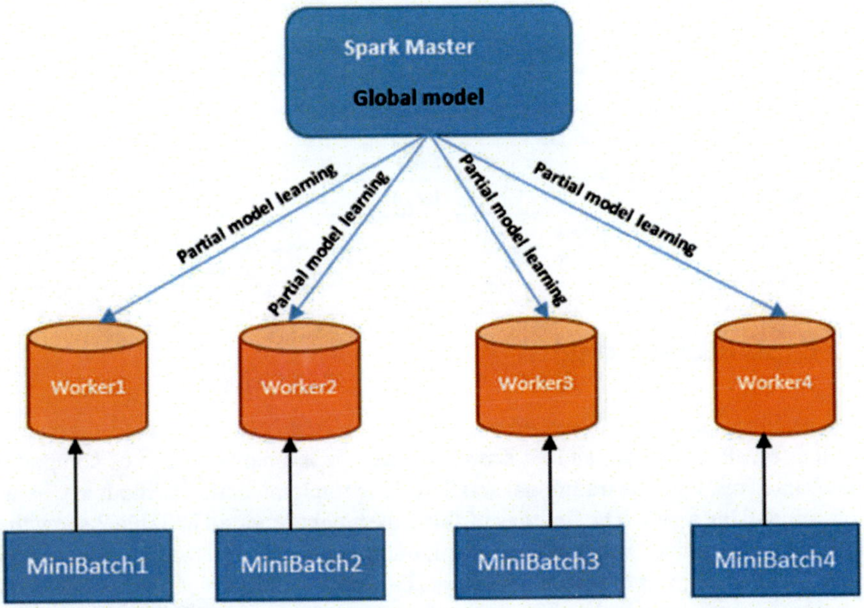


**Fig. 2** Distributed deep learning training on spark

**Fig. 3** Partial model learning

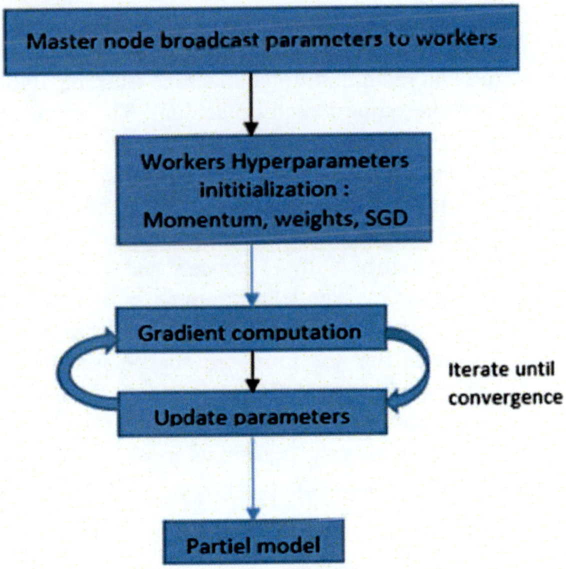

**Table 1** Parameter setup

| Parameter | Value |
|---|---|
| Mini-batch size | 200,000 |
| Learning rate | 0.1 |
| Optimization algorithm | SGD |
| Activation function | RELU |
| Momentum | 0.9 |
| Architecture | 4 layers of 200 neurons each |

# 3 Experimental Results

The performance of our proposed model has been evaluated using different metrics such as recall, precision, and F1 score. The first experiments consist of comparing our framework with the traditional serial training implementation without any parallelization using spark. The findings of this experiment reported in Table 2 show that the best performance was obtained using our proposed model with an F1 score of 91%. In addition, this experiment captures the speedup that we can expect from our system. In fact, the speedup efficiency of our model's learning increased six times when compared to a serial training implementation without using the spark platform. This confirms the benefit of using data-parallel computations using different workers over the spark ecosystem.

In order to examine further the overall performance of our proposed model, we explored two main traditional machine learning algorithms as follows: the Random forest and the support vector machines. The same datasets already presented at the beginning of this research have been used. During this work, an SVM model was implemented, and we have adopted the radial basis function kernel which achieved the best performance. Moreover, given that fraudulent events are minor in our datasets, we adopted K-fold cross validation approach to avoid the overfitting problem. The dataset was trained with 11 subsets and 1 set was used for evaluating the fraud detection accuracy. The learning duration was calculated for the two conventional machine learning algorithms. Our findings are reported in Table 3. The presented results showed that the proposed spark-based deep learning algorithm and serial deep learning implementation outperformed our traditional machine learning algorithm. And that our spark-based deep learning was still achieving a good learning speed outperforming the traditional machine learning algorithm. It is noteworthy to mention that our serial deep learning model without using spark yielded also better

**Table 2** Performance results

| Model type | Precision | Recall | F1-Score | Learning duration (min) |
|---|---|---|---|---|
| Proposed model | 0.86 | 0.88 | 0.91 | 35 |
| Serial training implementation | 0.81 | 0.82 | 0.88 | 221 |

**Table 3** Performance benchmarking of our proposed solution with some traditional machine learning algorithms

| Model type | Precision | Recall | F1-Score | Learning duration (min) |
|---|---|---|---|---|
| SVM | 0.83 | 0.88 | 0.82 | 30 |
| RF | 0.86 | 0.91 | 0.85 | 32 |
| Proposed model | 0.86 | 0.88 | 0.91 | 35 |
| Serial training implementation | 0.81 | 0.82 | 0.88 | 221 |

results in terms of F1 score when compared to the conventional machine learning algorithms, but with high computational cost. Our results showed that the learning duration is very short by using our spark-based implementation with SVM, RF, and our proposed model (30 min, 32 min, and 35 min, respectively) when compared to the non-spark-based implementation (221 min), which means that the learning process is ~ 6 times faster using the spark platform. This result indicates that our methods are efficient in detecting early fraudulent events (in the near real time) and thus avoiding additional revenue losses. The normal training method without spark costs several hours. Therefore, this method could be used also for offline training processes, and not only for short decision-making tasks such as the fraud detection use case.

## 4 Conclusion and Discussion

In this work, we have demonstrated how the performance of deep learning algorithm outperformed the conventional machine techniques to predict fraudulent behaviors in the mobile telecommunication ecosystem. Particularly, we found that our approach can be used as an excellent tool to detect early fraudulent events. Our results indicate that Apache Spark performs extremely well within the implementation of the deep learning algorithm compared to the serial implementation without any parallel computing as evidenced by the short learning speed over spark-based implementation shown in Figs. 1 and 2. Thus, it is possible that our novel methodology based on Apache Spark platform can be used by mobile telecommunication providers to tackle fraud problems. The work presented here suggests that our approach is not only useful in detecting early fraud problems in mobile communication, but it can also be useful for other entities where fraud is a real threat. For instance, our approach could be applied to many research areas such as financial transactions and insurance fraud. It is important to mention that the parameter tuning of our approach can be improved further in order to improve the performance of our deep learning model.

## References

1. Wei W, Li J, Cao L, Ou Y, Chen J (2013) Effective detection of sophisticated online banking fraud on extremely imbalanced data. World Wide Web 16:449–475
2. Sahin Y, Bulkan S, Duman E (2013) A cost-sensitive decision tree approach for fraud detection. Expert Syst Appl 40(15):5916–5923
3. Volinsky C (2014) Telecommunications fraud detection, using social networks for. Encyclopedia of Social Network Analysis and Mining. Springer, New York, pp 2107–2111
4. Chang YC et al (2017) Mining the networks of telecommunication fraud groups using social network analysis
5. Hilas CS, Mastorocostas PA, Rekanos IT (2015) Clustering of telecommunications user profiles for fraud detection and security enhancement in large corporate networks: a case study. Appl Math Inf Sci 9(4):1709
6. Murynets I et al (2014) Analysis and detection of SIMbox fraud in mobility networks. INFOCOM, 2014 Proceedings IEEE. IEEE
7. Olszewski D (2014) Fraud detection using self-organizing map visualizing the user profiles. Knowl-Based Syst 70:324–334
8. Akoglu L, Christos F (2013) Anomaly, event, and fraud detection in large network datasets. In: Proceedings of the sixth ACM international conference on Web search and data mining. ACM
9. Sharma A, Panigrahi PK (2013) A review of financial accounting fraud detection based on data mining techniques. arXiv preprint arXiv:1309.3944
10. Zaharia M, Chowdhury M, Franklin MJ, Shenker S, Stoica I (2010) Spark: cluster computing with working sets. In: USENIX conference on hot topics in cloud computing, pp 10–10
11. Kaiming H et al (2016) Deep residual learning for image recognition. In: Proceedings of the IEEE conference on computer vision and pattern recognition
12. Mikolov T et al (2013) Distributed representations of words and phrases and their compositionality. Advances in neural information processing systems
13. Shaoqing R et al (2015) Faster R-CNN: towards real-time object detection with region proposal networks. Advances in neural information processing systems
14. Hinton GE, Ruslan RS (2006) Reducing the dimensionality of data with neural networks. Science 313(5786):504–507
15. Spark A (2016) Apache spark: lightning-fast cluster computing. http://spark.apache.org
16. Chouiekh A, Haj EHIE (2018) ConvNets for fraud detection analysis. Procedia Computer Science 127:133–138
17. Wu X, Zhu X, Wu GQ, Ding W (2014) Data mining with big data. IEEE Trans Knowl Data Eng 26(1):97–107
18. Bergstra J, Bengio Y (2012) Random search for hyperparameter optimization. J Mach Learn Res 281–305
19. Kaiming H et al (2015) Delving deep into rectifiers: surpassing humanlevel performance on imagenet classification. In: Proceedings of the IEEE international conference on computer vision
20. Chouiekh A (2017) Machine learning techniques applied to prepaid subscribers: case study on the telecom industry of Morocco. In: 2017 Intelligent Systems and Computer Vision (ISCV). IEEE

# Enhanced Dual-Band Bandstop Filter with Stepped-Impedance Resonators Using a Computer-Aided Optimization Technique

**Ayyoub El Berbri, Adil Saadi, and Seddik Bri**

**Abstract** This work presents a designed and optimized dual-band bandstop filter (DBBSF) with stepped-impedance resonators (SIR). This enhanced filter functions at the 5G NR mobile network bands centered at 1.5 GHz and 3.5 GHz, respectively. An adjusted optimization method is employed to meet the design specifications. The birding of optimization is switched from high-accuracy models to low-accuracy models by using this method. The numerical results of this enhanced filter indicate a great performance at the bands n50 and n78.

**Keywords** Bandstop filter · Dual-band filter · Optimization · 5G

## 1 Introduction

5G is the next generation of mobile communication systems industrialized by 3GPP [1]. It is a logical continuation of previous mobile radio systems and is largely based on 4G (LTE). 5G technology essentially consists of a new radio interface "5G New Radio" (5G NR), which is very different from 4G in the following characteristics:

- Up to 100 times higher data rates to handle the rapidly increasing mobile data volume and consequently higher system capacity (1000 times higher data transmission capacity).
- Shorter latency when transmitting data in the 5G system and hence shorter reaction time of telecommunication services to customers.
- A greater number of terminals connected simultaneously (up to 1 million per km$^2$) as part of the Internet of Things (IoT).
- New frequency bands below and above 6 GHz promising fully available frequency ranges.

Unlike the last systems, 5G allows more reliable connections with guaranteed transmission speeds and shorter latency times, longer terminal autonomy and network

A. E. Berbri (✉) · A. Saadi · S. Bri
Moulay Ismail University, Meknes, Morocco



X.-S. Yang et al. (eds.), *Proceedings of Sixth International Congress on Information and Communication Technology*, Lecture Notes in Networks and Systems 216,
https://doi.org/10.1007/978-981-16-1781-2_3

segmentation according to the application (network splitting, network networks). virtual parallel mobile radio communications). In this work, a filter is designed and optimized to function in the two 5G NR bands n50 and n78 [1].

The bandstop filter is widely utilized in microwave applications typically adopted in receivers and transmitters. It is a filter preventing the passage of part of the frequencies. It can be composed of a high-pass filter and a low-pass filter whose cut-off frequencies are often close but different, the cut-off frequency of the low-pass filter is systematically lower than the cut-off frequency of the high-pass filter. Or with another simpler way, although it has a less selective frequency response, would be to place what is known as a "trap circuit" [2, 3]. We use in this paper DBBSF with SIR for dual-band utilization [4].

In this work, we optimize a DBBSF with SIR planned for 5G NR bands n50 = (1432–1517 GHz) and (n78 = (3300–3800 GHz), which are centered at 1.5 GHz and 3.5 GHz respectively [1]. We employ an adjusted Tuning Space Mapping (TSM) for the optimization. The optimal numerical results reveal great performances at the two bands n50 and n78.

## 2 DBBSF with SIRs

Figure 1a displays DBBSF with SIRs, these SIRs help to make the filter perform at the two frequencies. Figure 1b displays the equivalent circuit of this filter with lumped elements [4].

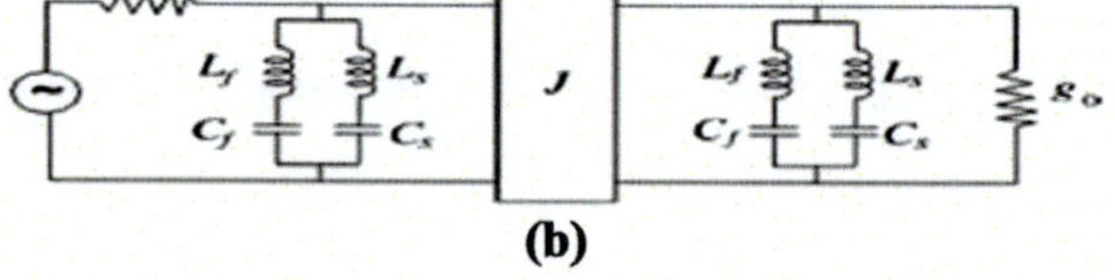


**Fig. 1** **a** The DBBSF with SIR **b** Equivalent circuit with lumped elements [4]

## 3 TSM

TSM is an optimization method that moves the optimization burden from a fine model (accurate) to a sororate model (Tuning model). The tuning model is based on the fine model data.

The main optimization problem [5–9] is formulated as follows:

$$\overset{*}{x} f = \arg\min_{x} U(R_f(x)) \tag{1}$$

The tunning model $R_t^{(i)}$ is be optimized with respect to $x_t$:

$$x_{t,1}^{(i)} = \arg\min_{x_t} U(R_t^{(i)}(x_t)) \tag{2}$$

## 4 Results and Discussion

The design parameters are $x = [L_0\ L_1\ L_2\ L_3]^T$ mm. Figure 2 displays the fine model of this DBBSF with SIR as shown in Fig. 2, which is simulated using a substrate of $\varepsilon r = 4$, height $H = 1.4$ mm and loss tangent = 0.0004.

From Fig. 3, we can observe that the two stopbands are centered at 2.06 / 5 GHz with the bandwidths ([1.8 GHz–2.52 GHz]/ [4.48 GHz–5.14 GHz]). These results do not meet the design specification of this filter, for a major performance, optimization is required.

The data file of the divided fine model Fig. 4 is inserted in the tuning model using the corresponding elements. The tuning parameters in the cheap circuit, that we optimize in this step, are $x_t = [L_0\ L_{t1}\ L_{t2}\ L_{t3}]^T$mm. The parameters obtained after the optimization represent the dimensions added to meet the design specification, it's transformed directly to the fine model dimensions.

Figure 5 displays the results of the fine model after one iteration of the optimization method, and the results are consistent with design specifications. The two

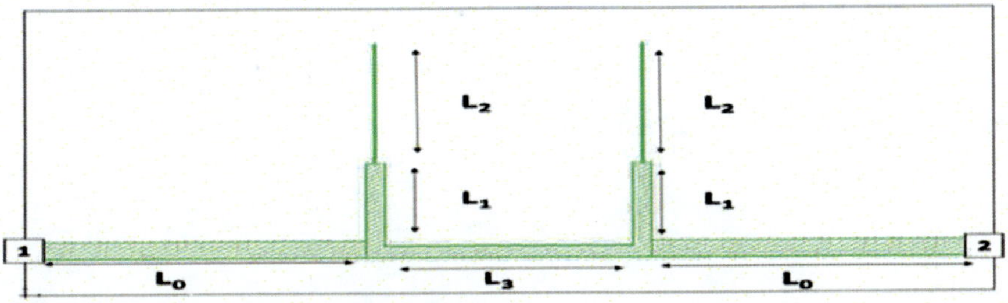


**Fig. 2** The fine model of the DBBSF with SIR

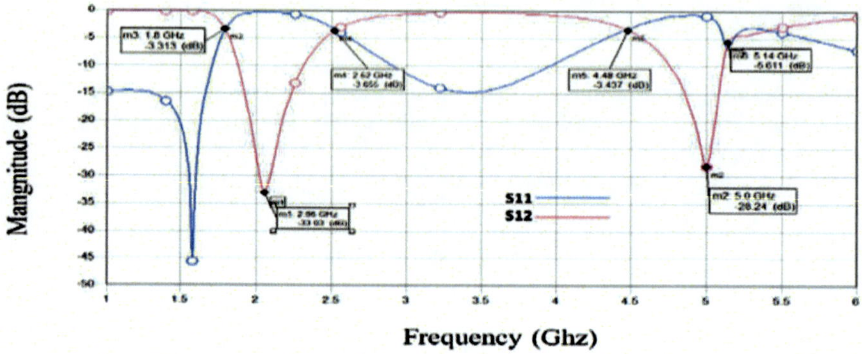


Fig. 3 The Initial fine model response (Parameter $S_{11}$)

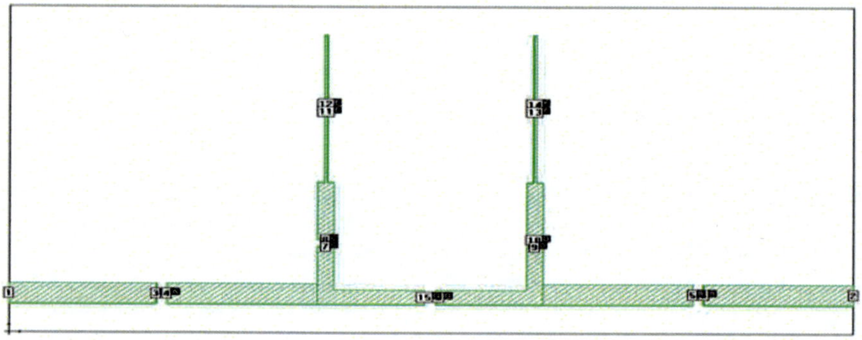

Fig. 4 The fine model divided

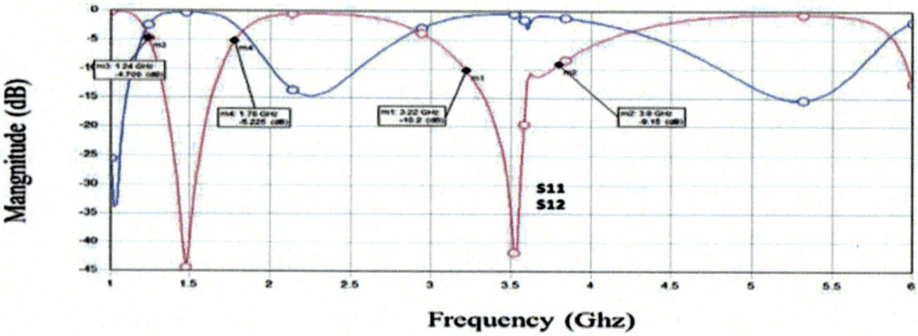


Fig. 5 The optimized fine model response

mid-stopband frequencies were 1.5 GHz and 3.5 GHz with respective 3 dB-down bandwidths [1.24 GHz–1.78 GHz] and [3.22 GHz–3.8 GHz]. And the rejection level was more than 43 dB at the center frequencies.

## 5 Conclusion

In this work, we designed a DBBSF with SIR for 5G NR. The filter performance is further enhanced using an adjusted Tuning Space Mapping to perform for more complex microwave circuits such as our design. The numerical results show that the 3 dB-down bandwidths are [1.24 GHz–1.78 GHz] and [3.22 GHz–3.8 GHz]. The method used gives fast optimization with rigorous results delivered after only one iteration.

## References

1. 3GPP specification series: 38series.url: https://www.3gpp.org/DynaReport/38-series.htm Accessed 02 Oct 2020
2. Uchida H, Kamino H, Totani K, Yoneda N, Miyazaki M, Konishi Y, Makino S, Hirokawa J, Ando M (2004) Dual-band-rejection filter for distortion reduction in RF transmitters. IEEE Trans Microw Theory Tech 52(11):2550–2556
3. Cameron RJ, Yu M, Wang Y (2005) Direct-coupled microwave filters with single and dual stopbands. IEEE Trans Microw Theory Tech 53(11):3288–3297
4. Chin K, Yeh J, Chao S (2007) Compact dual-band bandstop filters using stepped-impedance resonators. IEEE Microwave Wirel Compon Lett
5. Meng J, Koziel S, Bandler JW, Bakr MH, Cheng QS (2008) Tuning space mapping: a novel technique for engineering optimization. In: 2008 IEEE MTT-S International Microwave Symposium Digest, Atlanta, GA, USA
6. Echeverria D, Hemker PW (2008) Space mapping and defect correction. Int Math J Comput Methods Appl Math 5(2):107–136
7. Ismail MA, Smith D, Panariello A, Wang Y, Yu M (2004) EM-based design of large-scale dielectric-resonator filters and multiplexers by space mapping. IEEE Trans Microwave Theory Tech 52(1):386–392
8. Simpson TW, Peplinski J, Koch PN, Allen JK (2001) Metamodels for computer-based engineering design: survey and recommendations. ENG COMPUT 17(2):129–150
9. Bandler JW, Biernacki RM, Chen SH, Grobelny PA, Hemmers RH (1994) Space mapping technique for electromagnetic optimization. IEEE T Microw Theory 42(12):2536–2544

# Self-Presentation and Social Anxiety on Facebook

Eang Teng Chan and Mui Joo Tang

**Abstract** Social networking sites, like Facebook, have given everyone the opportunity to tell "their stories" online, share their happiness, sorrow and thoughts almost in real time through the social networking sites, and it is all about online identity, about how the people present themselves on the online world, or social networking sites. The online identity of certain people might not be the same as their real-world identity, because what they represent on the Web is different from what they represent in the real world; it might be a "true identity" or "ideal identity". This study is to explore how the users manage their online identities in virtual interaction. Social Comparison theory is used in the theoretical framework of this research. This research is carried out by using survey questionnaire on social media users aged 22–35 year who live in Klang Valley, Kuala Lumpur, Malaysia. The result indicates that most of the participants present their ideal identities than the real identities on Facebook, and they are somehow heavily affected by social anxiety due to social comparison on Facebook.

**Keywords** Social media · Self-presentation · Social anxiety

## 1 Introduction

The rise of social media and social networking sites has made it faster and easier for people to access information around the world, making it easier to digest and communicate with each other even by having small chunks of information. It is taking over people's lives, turning the people's lives from the real world to the online world; it is almost always possible to find people who are busy playing with their mobile or smart device, whether to update their lifestyle on social networking sites or read through other people updates. However, in social media, users might be more focused, and obsessed with creating the perfect digital version of themselves; while it's not really representing their self in the real world, they are just trying to

E. T. Chan (✉) · M. J. Tang
Tunku Abdul Rahman University College, Kuala Lumpur, Malaysia



X.-S. Yang et al. (eds.), *Proceedings of Sixth International Congress on Information and Communication Technology*, Lecture Notes in Networks and Systems 216,
https://doi.org/10.1007/978-981-16-1781-2_4

compare themselves with the others, with social anxiety, that being feared to judge, and rejected by other users. This research is to find out the relationship between self-presentation and social anxiety of Facebook users.

## 2 Literature Review

### 2.1 *Belonging, Personality, and Self-Presentation on Facebook*

Social networking sites offer features that allow users to "make new friends" and "stay in touch with (past and present) friends". Self-presentation refers to the users trying to control or shape others through their presentation on social media. It involves expressing themselves and their behaviors in order to create an image of expectation, and in other word, self-presentation can be known as a sort of impression management [1]. The motivation of self-presentation may be important in online social media and online behavior as it is a form of motivation that can be described as a person's level of concern about what other people think of him, and how necessary it is for him/her to project some sort of self-image from the consequences [2]. Individuals with a strong motivation of self-presentation believe that it is extremely important to show themselves to others in a particular way. This behavior can be observed in social networking sites, such as Facebook, where users can share their personal information through Facebook's Feeds or/and, Facebook profiles. In addition, individuals (the users) often claim that they present real images, (the true identity) on the Internet, but the objective measurements and the judge's scores indicate that these images are slightly idealized, at the expense of accuracy [3]. The researches on "network identity" and "virtual identity" on social networking sites are mostly "digital profiles" that allow individuals to express their interests in the external environment and can be established when they log in [4].

One of the important features of the emerging model of this kind of self-presentation and identity is that people tend to be different people, or have different online roles that are different from "real life". The absence of an entity's online self-presentation enables people to hide their unwanted physical characteristics; anonymity enables individuals to re-create their biography and personality. As a result, the re-modeling and anonymity of the social networking sites allow people to reinvent themselves by creating new identities [5]. Carmon [6] found out that the females who call themselves "smart" or "nice" in the offline society are more likely to post "fun" or "interesting" or "social" on social networking sites as theirs remark, and girls with low self-esteem are more likely to describe themselves as "sexy" and "crazy" than high self-esteem girls. In addition, Pew survey indicates that 56% of American social media users with online profiles have posted false information on social media sites. The male users posted fabricated information more often than the female users [7].

In a nutshell, self-presentation is important to be identified in this research, as it might affect the clarity of a person's self-esteem and self-concept. Moreover, self-reflection also occurs in this stage because when the users present themselves in the way that are not acceptable to the others, they might receive negative feedbacks, and it may not satisfy the user's self-esteem and self-concept, and causes social anxiety.

## 2.2 *Self-Expression, Self Esteem, and Self-Concept on Facebook*

Self-expression is defined as "the expression of one's own personality—the assertion of one's individual traits [8]", which refer to how the users express their individuality through social media. Danah Boyd, a senior researcher at Microsoft, stated: "Selfies are essentially just a form of self-expression, albeit a new one" during her study on social media habit. She also said it's just the same for selfies and getting dressed before going out, where clothes and selfies are useful tools for identity negotiation. Selfies are ubiquitous in social media, with some of the world's most popular celebrities joining in, and selfie is the way to capture every single moment in their life that means something to them, and is also a way to discover and understand us [9].

Kathryn Wilson, Stephanie Fornasier, and White Katherin M [10] opined self-esteem is defined as "the subjective evaluation a person makes and maintains about himself or herself and the extent of belief in their capability, worth, and significance, which is conveyed through their attitudes and verbal behaviour". In addition, self-esteem can be understood as one of the most stable traits that develops over time and a fluid state that responds to everyday events and circumstances. Long-term or occasional exposure to upward comparisons on social media sites may have a negative impact on people's self-evaluation and self-esteem. Daily and long-term use of social media may have a negative impact on trait self-esteem, while occasional use of social media may also affect stately self-esteem. In fact, studies have shown that people who use Facebook regularly are more likely to be depressed and less happy, which can be known as social anxiety.

Social media have a great impact not only on the interaction between people but also on how people view and perceive them [11]. A recent survey found that one-third of Facebook users are unhappy with their lives after visiting a particular network [12]. Instead, most of the users will post or write certain things to express their emotion and what they had been through daily on their Facebook feed, in others to give them a self-esteem boost, and at the same time, they also care for the "likes" they received. If the "likes" they received was much lower than others, they might delete the post due to their self-esteem. It makes the users feel good when they see people react and like what they have posted. Even when they see a new notification pop up, their brain could release a neurotransmitter called dopamine, which controls the brain's pleasure and reward centers [13]. Adam Alter [13] explains, "The more dopamine is released, the more the brain realizes that it's triggered by a reward". The addictive could boost up the users' self-esteem as well, as the user might feel that they are being accepted by the society [13] (Alter 2017). Other than that, according to Andrew T. Stephen and Keith Wilcox [14], their research has shown that social networking sites boost the self-esteem of users who focus on close friends (strong tie relationships). They explain that people are more concerned about how their close friends receive information when they publish their information online. In addition, they found that increased self-esteem from using social networking sites also affected self-control, which are referring to the self-presentation of the users [14].

However, Chou and Edge [15] describe, when the users scrolling through Facebook's news feed and reading other people's life updates, Facebook users tend to pay attention to other people's happy life events, which may give people the impression that others are leading a better life than themselves. Therefore, Facebook users will compare the ideal life of others to events in real life, and real-life events are not always fun and exciting. Furthermore, in Philip Verduyn's research, he divided Facebook use into two categories: active and passive. The findings suggest that passive use of Facebook can increase jealousy and negatively affect people's happiness. The findings are consistent with previous research because jealousy also stems from social comparisons [16].

## 2.3 Social Comparison of Users on Facebook

When the users see information about other users, they tend to associate and link it to themselves, and this social comparison provides them with self-evaluative information that can be used to make positive or negative self-judgments [17]. When they perceived the information as a negative judgment, they might involve in social anxiety or social phobia.

Facebook is a platform that provides all kinds of tools to search, observe, and review other people's data. Individuals have certain social goals, and that's, for example, you can do a simple search on Facebook, rather than looking for an offline interaction [18]. In summary, Facebook not only creates an environment for self-optimization but also provides fast access to other self-optimization opportunities. As a result, these Facebook activities may stimulate individuals to make frequent, undirected, upward and/or horizontal social comparisons automatically, especially for those who spend more time/frequently browsing Facebook. It's due to the Facebook platform that promotes self-disclosure that users may reveal highly personal information that they don't usually disclose [19]. According to Steers and his team's research, they had predicted that the longer time the users spend on Facebook, the more chances they have to make social comparisons with their friends [19].

Referring to Theodore Roosevelt's quotes, "Comparing is the thief of joy", however, comparing is also the enemy of joy, which will bring the users to negative effects, such as social anxiety, depression, and others [20]. Normally, their interactions with other people have quickly turned into their own version of an idealized version of where they think they should be living or how to live their lives. Once they have given these standards an inappropriate meaning or an exaggerated importance, they may begin to experience the symptoms of anxiety, depression, or frustration [20].

In Social Comparison Theory, people are motivated to evaluate their own opinions, to know more about their abilities, and when they can't assess their opinions and abilities, they tend to think of themselves as other people [21]. In addition, Festinger [21] also hypothesizes, "to the extent that objective and non-social means are not available, people evaluate their opinions and abilities by comparing respectively, with opinions and abilities of others", where if a person cannot compare his or her ideas and abilities in a realistic or physical environment, then the person will tend to deduce a subjective view [21].

### 2.4 Social Anxiety on Facebook

Marissa Maldonado in her studies shows the anxiety and concern of how others perceive them can result in long-term stress that may lead to health problems, including mental health issues [22] (Maldonado 2016). Furthermore, Kross and Verduyn [23] predict that there was a direct link between heavy use of social media, particularly Facebook, and anxiety, depression, and jealousy [23]. In addition, Woods and Scott [24] found out that those users who are emotionally invested in social media, when they don't use social media, would become isolated and frustrated, which in turn increases the risk of anxiety and depression. Moreover, in Shaw, Timpano, Tran, and Joormann [25], they observe that users with high social anxiety symptoms are more passive in the form of Facebook, which is to observe other users' profiles without interacting with other users. They also hypothesize that passively using Facebook can increase social anxiety symptoms, because it can lead to personal relationships with personal concerns, such as being rejected, and having to do it with other people. Ruoyun Lin and Sonja Utz [26] had found out that when it rained in one place, people felt more depressed, which lead to posting negative status on Facebook, followed by negative comments that could lead to those who were friends with the users on Facebook, but who lived farther away, where it didn't rain, feeling worse. In the same way, when the friends are happy, people have more positive emotions, at least in their state of mind.

## 3 Methodology

### 3.1 Subjects/Participants

250 social media users aged 22–35 years are included as samples in this research. Purposive sampling is used to only allow participants who have access to the Internet and those who are social media users in Klang Valley area, Kuala Lumpur.

## 3.2 Research Design

The research method that is used in this research is the quantitative research method, which is the survey questionnaire. The survey questionnaire will cover **FOUR** main areas, which are the demographics, general usage of Facebook, the evaluation of respondents on their self-presentation on Facebook, and the impact of social anxiety on Facebook.

**Table 1** Time socialized on Facebook

| Time used | Percentage (%) |
|---|---|
| An hour or less | 18 |
| 5 h or less | 42 |
| 10 h or less | 27 |
| All day long | 13 |

**Table 2** Frequency to update Facebook feeds

| Frequency to update Facebook | Percentage (%) |
|---|---|
| Daily | 20 |
| Every few days | 53 |
| Every few weeks | 17 |
| Evert few months | 8 |
| Every few times in a year | 2 |

**Table 3** Activities posted on Facebook

| Activities on Facebooks | Percentage (%) |
|---|---|
| Sharing funny videos | 94.8 |
| Check in | 69 |
| Posting photos | 94 |
| Expressing feelings | 42.8 |
| Tagging friends | 59/2 |
| Just scroll through the feeds | 0.4 |

**Table 4** Self-Presentation and User-perception on Facebook

| | | |
|---|---|---|
| Are you presenting your true identity or ideal identity on Facebook? | True identity | Ideal identity |
| | 15% | 85% |
| Do you think self-presentation is important on Facebook to satisfy your self-esteem? | Yes | No |
| | 92% | 8% |
| My Facebook profile has accurately reflected who I am in real life | 18% | 82% |

# 4 Result

## 4.1 *Social Media Usage*

Tables 1, 2, and 3 indicate that Facebook users are active on Facebook. This can be shown through the frequency they update their Facebook feed with the highest percentage going to every few days, daily, every few weeks, every few months, and every few times in a year. Majority of users spend about 5–10 h per day on Facebook on sharing funny videos, posting photos, checking in, tagging friends, expressing feeling, and just scroll through the feeds.

## 4.2 *Self Presentation and User-Perception on Facebook*

Table 4 shows that 85% of the respondents present their ideal identity whereas only 15% present their true identity on Facebook. At the same time, majority of them view that self-presentation is important on Facebook to satisfy their self-esteem with the aggregate of 92% agreement. Besides that, most of the respondents agree that their Facebook profile has not accurately reflected who they are in real life. This concludes online activities are no longer separated from our real life but are an integral part of it [27].

## 4.3 *Social Anxiety on Facebook*

Table 5 shows that respondents are inclined to the agreement toward all the statements used to test the level of social anxiety on Facebook. This has indirectly shown some positive relationship between the Facebook users and the level of social anxiety they have when using Facebook. The users might overthink whether or not they should share something. Sometimes they can be looking at something for quite some time, debating if it's worth putting up. Sometimes when they do decide to share something, they would change their mind a few minutes later and delete it hoping no one saw that embarrassing thing they posted [28] (Alison 2016). With social media,

**Table 5** Social anxiety on Facebook

| Statement | Strongly disagree (%) | Disagree (%) | Neutral (%) | Agree (%) | Strongly agree (%) |
|---|---|---|---|---|---|
| I felt moment of sudden fear, terror or fright when I'm browsing FB | 8 | 22 | 18 | 26 | 26 |
| I felt anxious, worried or nervous when my friends are not socialising with me on FB | 11 | 18 | 17 | 27 | 27 |
| I have thoughts of being rejected, humiliated or offending others on FB | 11 | 17 | 18 | 26 | 28 |
| I'm self-conscious about my appearance even when I'm already presenting the ideal sides of myself on FB | 9 | 19 | 17 | 27 | 28 |
| I'm worried if I make a fool of myself on FB | 11 | 17 | 15 | 27 | 30 |
| I'm easily hurt or mind by what others have said or done to judge me on FB | 11 | 16 | 18 | 24 | 31 |
| I often feel that other FB users are better than me | 8 | 19 | 17 | 23 | 33 |
| I have a sense of isolation whether I am alone or I am with many FB friends | 11 | 17 | 17 | 26 | 29 |
| I spend a lot of time preparing what to say or post, or how to act on FB | 11 | 16 | 18 | 27 | 28 |

we are actually handing ourselves an object that will decrease our mental well-being, and cause social anxiety and social phobia on social media, Facebook. Even sitting in a room by the user themselves, they might be able to get anxious just looking at Facebook on their computer screen. It's like the modern-day version of social anxiety.

## 5 Conclusion and Discussion

Based on the collected data, those users who are presenting ideal identity are mainly females, because females are more emotional, while most of the male users are presenting their true identity. For future research, the study can focus on the variable of gender to see the differences in their self-presentation and social anxiety on Facebook. Besides that, the age group can be open for all age ranges as long as they are Instagram users. This is because different age groups might have differences in self-presentation and level of social anxiety too. At the same time, there might be some potential influence of social values on the use of self-presentational behaviors and social anxiety, hence, it would be interesting to explore whether culture has an influence on the self-presentation behaviors. The collected data have shown that there is a certain relationship between self-presentation and social anxiety that is caused by social comparison among the Facebook user. The likes, comments, and posts they share on social media can often seem inconsequential, but they matter. They tap into some of the very elements that make us human, our addictions, desires, anxieties, and joys. If they don't receive what they're expecting, it would cause social anxiety among them as they are afraid of having done something wrong. At the same time, they are afraid to be humiliated, judged, and rejected as well. As a result, some of the users would make changes of their presentation on Facebook, to cater and fit in to others' interests. All of these could be due to social comparisons, which has led to the feelings of insecurity, particularly on Facebook.

**The authors acknowledged the raw materials provided by May Yap.**

## References

1. Terry M, Shepperd JA (2013) Self-Presentation. https://doi.org/10.4135/9781412956253.n494 Accessed 02 Dec 2018
2. Schneider FW, Gruman JA, Courts LM (2012) Applied social psychology: understanding and addressing social and practical problems, 2nd edn. Sage Publications, Thousand Oaks
3. Toma C, Hancock JT (2011) A new twist on love's labor: self-presentation in online dating profile. In: Wright KB & Webb LM (eds) Computer-mediated communication in personal relationship. Peter Lang Publishing, New York, pp 41–55
4. Jordan K, Hauser J, Foster S (2003) The augmented social network: building identity and trust into the next generation internet. https://doi.org/10.5210/FM.V818.1068

5. Zhao S, Grasmuck S, Martin J (2008) Identity construction on Facebook: digital empowerment in anchored relationship. Comput Hum Behav 24(5)https://doi.org/10.1016/j.chb.2008.02.012
6. Carmon I (2010) What Facebook does to teenage girls. https://jezebel.com/5682488/what-does-facebook-do-to-teenage-girls Accessed 02 Dec 2018
7. Lenhart A, Madden M (2007) Teens, privacy and online social networks. http://www.pewinternet.org/2007/04/18/teens-privacy-and-online-social-networks/ Accessed 02 Dec 2018
8. Orehek E, Human LJ (2017) Self-expression on social media: do tweets present accurate and positive portraits of impulsivity, self-esteem, and attachment style? Pers Soc Psychol Bull 43(1):60–70. https://doi.org/10.1177/0146167216675332
9. Dahl M (2016) Selfie-esteem: teens say selfies give a confidence boost. https://www.today.com/parents/selfie-esteem-teens-say-selfies-give-confidence-boost-2D12164198 Accessed 01 Dec 2018
10. Wilson K, Fornasier S, White Katherin M (2010) Psychological predictors of young adults' use of social networking sites. Cyberpsychology Behav Soc Networking. https://doi.org/10.1089/cyber.2009.0094
11. Phung A (2016) Social media and self-esteem. https://www.academia.edu/19628824/Social_Media_and_Self-esteem Accessed 01 Dec 2018
12. Mierzwa J, Jurjewicz H (2016) Does Facebook, Twitter, Instagram influence well-being and self-esteem among early adolescents?.Studia Socialia Cracoviensia 8, 1(14):137–152
13. Alter A (2017) Irresistible: the rise of addictive technology and the business of keeping us hooked, 1st edn. Penguin Random House, New York, United States
14. Wilcox K, Stephen AT (2012) Are close friends the enemy? online social networks, self-esteem, and self-control. J Consum Res https://doi.org/10.1086/668794
15. Chou HTG, Edge N (2012) "They are happier and having better lives than I am": the impact of using Facebook on perceptions on others' Cyberpsychology Behavior, and Social Networking. 15(2). https://doi.org/10.1089/cyber.2011.0324
16. Verduyn P, Lee D (2015) Passive Facebook usage undermines affective well-being: experimental and longitudinal evidence. J Exp Psychol Gen 144(2):480–488 https://doi.org/10.1037/xge0000057
17. Feinstein BA, Hershenberg R (2013) Negative social comparison on Facebook and depressive symptoms: rumination as a mechanism. Psychol Pop Media Cult 2(3):161–170
18. Morrison CM, Gore H (2010) The relationship between excessive internet use and depression: a questionnaire-based study of 1,319 young people and adults. Psychopathology 43(2):121–126. https://doi.org/10.1159/000277001
19. Steers MN, Wickham RE, Acitteli LK (2014) Seeing everyone else's highlight reels: how Facebook usage is linked to depressive sympton. J Soc Clin Psychol 33(8). https://doi.org/10.1521/jscp.2014.33.8.701
20. Ross DJ (2016) Comparison is the enemy of joy. http://mindbodyhealth.us/2016/10/04/comparison-enemy-joy/ Accessed 03 Dec 2018
21. Festinger L (1954) A theory of social comparison processes. J Hum Relat 7(2):117-140 https://doi.org/10.1177/001872675400700202
22. Maldonado M (2018) The anxiety of Facebook. https://psychcentre.com/lib/the-anxiety-of-facebook/ Accessed 02 Dec 2019
23. Kross E, Verduyn P (2013) Facebook use predicts declines in subjective well-being in young adults. PLoS ONE 8(8):1–6. https://doi.org/10.1371/journal.pone.0069841
24. Woods HC, Scott H (2016) Sleepyteens: social media use in adolescence is associated with poor sleep quality, anxiety, depression and low self-esteem. J Adolesc https://doi.org/10.1016/j.adolescence.2016.05.008
25. Shaw SM, Timpano KR, Tran TB, Joormann J (2015) Correlates of Facebook usage patterns: the relationship between passive Facebook use, social anxiety symptons, and brooding. Comput Hum Behav https://doi.org/10.1016/j.chb.2015.02.003
26. Ruoyun L, Utz S (2015) The emotional responses of browsing Facebook: happiness, envy, the role of tie strength. Comput Hum Behav https://doi.org/10.1016/j.chb.2015.04.064

27. Chamorro-Premuzic T (2015) How different are your online and offline personalities? https://www.theguardian.com/media-network/2015/sep/24/online-offline-personality-digital-identity Accessed 02 Apr 2019
28. Alison (2016) Using social media when you have social anxiety. https://themighty.com/2016/10/using-social-media-when-you-have-social-anxiety/ Accessed 02 Apr 2019

# Automatic Regional Estimation of Myocardial Strain Using Deep Learning

**Wafa Baccouch, Sameh Oueslati, Basel Solaiman, and Salam Labidi**

**Abstract** Although the ejection fraction remains essential in terms of diagnosis and prognosticity, it still leads to many errors and low reproducibility. Thus, quantifying local deformities of the myocardium is a crucial step of major interest for the diagnosis and follow-up of such a heart condition. The study of this quantitative parameter, in magnetic resonance imaging (MRI), has resulted in several methods. The objective of this paper is to review some relevant methods of calculating regional deformations based on deep learning such as tagged MRI, coded displacement imaging (DENSE) and strain-encoded imaging (SENC). Discussion of the results of some research has shown that monitoring segment deformation automatically makes the task more accurate and allows for better treatment planning by offering the ability to distinguish akinetic myocardial segments that have permanently lost their contractility from those likely to recover their contractile function after revascularization.

**Keywords** Myocardial strain · Cardiac MRI · Automatic quantification · Deep learning

## 1 Introduction

The heart is a vital organ whose main function is to propel oxygenated blood throughout the body [1]. Thus, possible malfunctions can harm all other systems of the body and therefore have fatal consequences. In clinical practice, imaging studies that provide the physician with information on the overall and regional functioning of the myocardium [2] are most often used to assess cardiac function and in particular left ventricular function.

W. Baccouch (✉) · S. Oueslati · S. Labidi
Research Laboratory of Biophysics and Medical Technologies, Higher Institute of Medical Technologies of Tunis, University of Tunis El Manar, 1006 Tunis, Tunisia

B. Solaiman
ITI Department: Image and Information Processing, IMT-Atlantique, Technopole Brest Iroise, 83818, 29238 Brest Cedex, CS, France
e-mail: basel.solaiman@imt-atlantique.fr

 
X.-S. Yang et al. (eds.), *Proceedings of Sixth International Congress on Information and Communication Technology*, Lecture Notes in Networks and Systems 216,
https://doi.org/10.1007/978-981-16-1781-2_5

Although conventional magnetic resonance imaging (MRI) is considered the reference modality of cardiac exploration in terms of analysis of ventricular wall movement and evaluation of myocardial function, visual estimation of the movement of myocardial segments, based on acquired cine-MRI sequences, remains highly subjective, approximate and therefore highly dependent on radiologist experience [3]. It also leads to high inter- and intra-observer variability. In this context, the development of regional assessment tools for the myocardial function is a major challenge and could be important in identifying the anomaly at an early stage.

In the literature and in order to bring more objectivity to the diagnosis, a range of approaches offering objective, regional and above all automatic evaluation has been proposed. In the first section, we will begin with the characterization of myocardial deformation, describing the different possible directions by which measurements of deformation values and deformation velocity are calculated. In the following section, we will cite the most relevant methods of assessing myocardial deformation in MRI. We dedicate the third section to the study of the various automatic methods based on deep learning dedicated to estimating this quantitative parameter. In the last section, we will discuss the results of some recent research and the contribution of automating the process of calculating the deformation of myocardial segments.

## 2 State of the Art of Automatic Methods for Cardiac Strain and Strain Rate Evaluation in MRI

### *2.1 Characterization of Myocardial Deformation*

The myocardial deformation technique measures, as the name suggests, allow to measure the deformity of the heart muscle. This deformation is expressed as a percentage and takes as a reference point over time the cardiac geometry at the t-0 time of the heart cycle. The deformation according to a dimension (1-D) is calculated as follows:

$$S(t) = \frac{L(t) - L_0}{L_0} \tag{1}$$

where L0 represents the initial size of the myocardial segment considered at time t-0 and L (t) the lengthening of this segment at time (t). Myocardial deformation is not uniform throughout the myocardium and depending on the properties of the tissue, it can vary considerably [4]. Measurements of deformation and shear values are generally made in three directions, radial, longitudinal and circumferential direction, as shown in Fig. 1.

Being mainly related to myocardial thickening, radial deformation is considered the most important of myocardial deformations. Longitudinal deformation describes the upward movement of the base of the left ventricle (LV) and circumferential

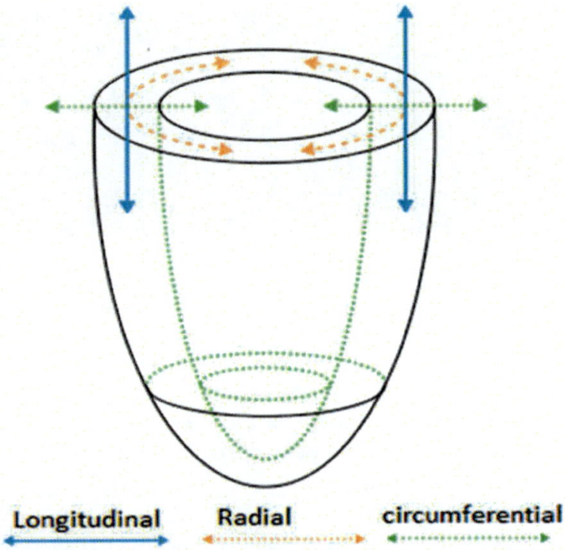


**Fig. 1** Representation of left ventricular contraction in the three orthogonal axes: longitudinal, circumferential and radial [5]

deformation indicates the shortening of the "hoop" [4]. In clinical practice, before any qualitative or quantitative analysis, the LV is segmented into 17 segments according to American Heart Association (AHA) standards. By calculating local deformation in relation to each segment, this allows us to accurately identify which of the myocardial segments is pathological.

## 2.2 *The Different Methods of Assessing Myocardial Strain in MRI*

MRI offers a range of approaches that allow accurate identification of such a contraction abnormality [6]. These approaches include tagged MRI [7]. Before integrating this method into the clinical routine, the doctor surgically implants the physical markers in the myocardium and then tracks their movements by external imaging. This technique is impractical and non-reproducible as implanted markers influence the movement of cardiac segments, which impairs measurement accuracy. In tagged MRI, the quantification of myocardial deformation is non-invasive. Physical markers are replaced by others created at the level of myocardial tissues.

This creation is done by means of locally induced disturbances of magnetization with selective radiofrequency saturation of several thin label planes in a plane perpendicular to the imaging plane before the acquisition of the image. During myocardial contraction, the labels follow the movement of the tissues. It is therefore possible to reconstruct the intra-myocardial movement thanks to their location. This technique allows the assessment of myocardial viability. However, disturbances produce

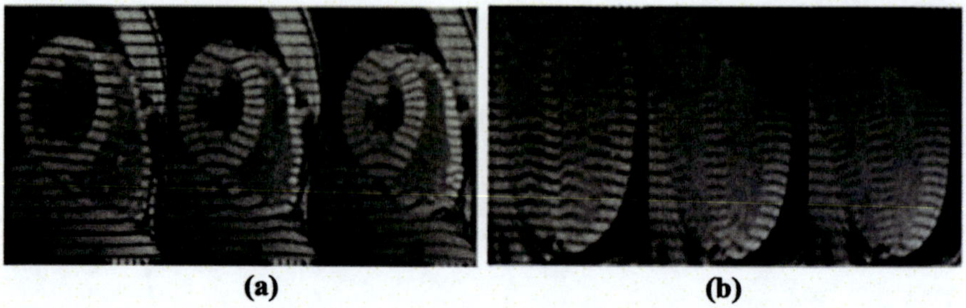

**Fig. 2** Two sets of small axis cuts **a** and large axis **b** in cine-MRI tissue marking

reduced signal intensity regions that are manifested by dark lines in acquired images, which impairs the quality of the image. Moreover, this approach is subjective and remains limited by the poor quality of images. Figure 2 shows two sets of marked cardiac MRI images.

Other methods include coded displacement imaging with stimulated echoes (DENSE) [8]. It is an advanced tagged-MRI technique that allows, as the name suggests, encoding tissue movement directly in the heart signal phase [9]. The process is based on three radiofrequency (RF) pulses responsible for generating a stimulated echo [10]. To achieve a two-dimensional displacement for each pixel that is part of the study region, simply repeat the sequence with a coding in the orthogonal direction [10]. Currently, there is also the so-called DENSE pulse sequence by applying a balanced free precession in a permanent regime. The results of the experiment showed that this sequence offers a good quality myocardial deformation characterization. However, it is characterized by a low signal-to-noise ratio (SNR).

This is explained by the fact that coding gradually disappears during the heart cycle, the same problem with tags. As a final example, we cite the technique called strain-encoded MRI (SENC). It is a technique also based on the principle of tagging. It uses parallel and non-orthogonal pre-saturation planes at the plane of the image. In conventional follow-up imaging, the orthogonal pre-saturation shots used are the origin of the black lines ("tags") on the image itself. The deformation of these marks during the heart cycle provides access to detailed quantification of segmental ventricular function. At the end of the acquisition, the resulting images represent an outset of tissue deformation mapping (myocardial shortening in the longitudinal direction) during the systole.

### *2.3 Automatic Methods of Assessing Myocardial Deformation Using Deep Learning*

Deep learning is a new field of machine learning (ML) research, which was introduced with the aim of bringing the ML closer to one of its original objectives: artificial intelligence (AI). It relies on multi-level learning algorithms to model complex relationships between data. Higher-level features and concepts are thus defined in terms of lower levels, and such a hierarchy of functionality is called deep architecture [11]. In cardiac MRI, deep learning has been used in several research studies to perform tasks performed manually in clinical routines such as segmentation of heart cavities for calculating the left ventricular ejection fraction (LVEF) [12] or other tasks such as measuring myocardial thickening, measuring myocardial tension and cardiac strain. In our study, we will look at work that automatically studies regional myocardial deformation using deep learning as a tool. In the search for "Edward Ferdian" [13] et al. deep learning was combined with the short-axis marked MRI technique. Two neural networks were used: one to estimate radial deformation and the other to estimate the circumferential deformation of the left ventricle (LV).

In order to estimate the location and movement of myocardial markers, already placed at anatomically consistent locations regardless of the locations of the overlying tags, the team used spatial and temporal characteristics. Space features were extracted and learned using the convolutional neural network (CNN) architecture, while temporal behavior was learned using a recurrent neural network (RNN). The in-depth learning algorithm is responsible for first delineating the region of interest (ROI) containing the LG, then reframing and resampling the return on investment for each image to detect and track myocardial cues on all images. This allows the deformations to be calculated according to the movement of the reference points. In other research such as that of "Esther Puyol-Anton" [14] et al. a fully automated pipeline for estimating LV deformation includes segmentation, motion tracking and estimation of longitudinal, radial and circumferential deformations from cardiac magnetic resonance imaging (CMR). The principle is well explained in Fig. 2.

The left ventricular segmentation task was performed by a fully convoluted network (FCN) followed by a motion correction step to correct breathing-induced movements using a proposed iterative recording algorithm used in the work. The next step is to follow the movement and then generate the layers and segments to measure the local deformations studied. In [15], deep learning was a strategy for performing automatic myocardial segmentation based on explicit active B-spline surfaces, which are initiated using an annular model. To assess myocardial deformation, the researchers used a non-rigid image recording technique. In other research [16], a deep learning approach for automated quantification of myocardial segmental deformation has been developed from cine-MRI images. The principle is to locate the central point of the LV using an architecture (FCN). Then, the ROI containing LV is extracted from all cardiac sections. These areas of interest are then used to segment the LV.

The evaluation of myocardial deformation was carried out by a Laplace-based approach to track points on the left ventricular wall by looking for solutions from the Laplace equation of each of the two successive images during the cardiac cycle. After follow-up, the deformation estimate is carried out using the Lagrangian technique-based approach.

## 3 Results and Discussions

In this section, we present some results obtained using deep learning-based approaches that allow an automatic and regional estimation of segmental myocardial deformation. In [13], the deep learning algorithm locates the region containing the LV. The result is given in Fig. 3.

The Fig. 4 shows that deep learning can provide a result very close to that of reference. A comparison was made between errors resulting from the deformable rope method without using manual correction to calculate inter-observer variability and average observer deformation taken as a reference. These errors were more important than the automated RNN-CNN prediction. The performance of the proposed method has been proven by comparison with the method of encoding displacement with stimulated echoes or tracking characteristics. In [13], the deformation values obtained by the proposed automatic method were compared to the deformity analysis obtained by an expert cardiologist. The level of agreement was defined as excellent being above

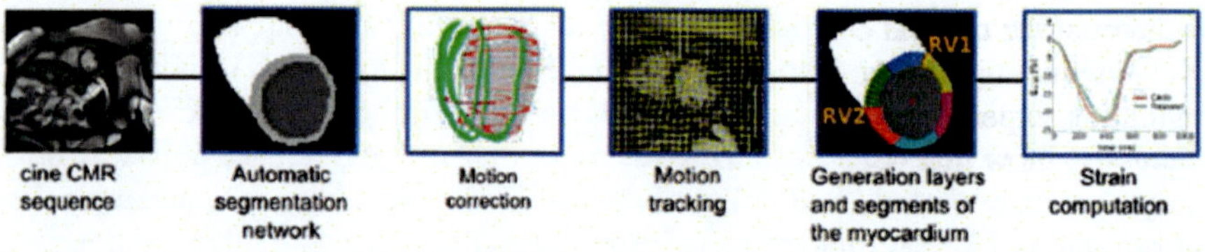


**Fig. 3** The principle of automatic estimation of myocardial deformation from cine-MRI images using CNN [14]

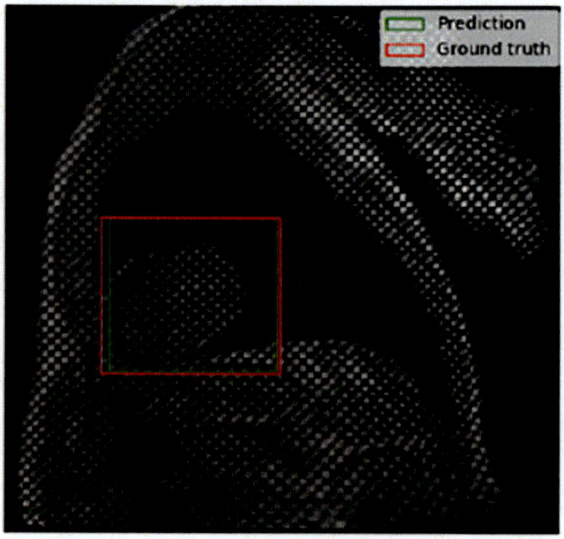


**Fig. 4** The principle of automatic estimation of myocardial deformation from cine-MRI images using CNN [13]

74%. The results show an excellent agreement between the automatic method and the reference method. In contrast, significant dispersion was observed for the peak radial deformation.

The latter is consistent with the greater intra- and inter-observer variability observed in the assessment of radial deformation in the literature. This approach allows the analysis of myocardial wall movement automatically and more accurately by eliminating manual post-processing steps and reducing inter- and intra-observer variation. In [15], the results show that observers analyzed 77% of the apical segments, 95% of the middle section segments and 100% of the basal section segments, while the automatic strategy was able to identify all ventricular segments in all cases. In the evaluation of deformations and from manual or automatic segmentation, myocardial monitoring was carried out and the resulting deformation curves were compared. The results showed that the two components of radial and circumferential deformation were able to distinguish between healthy regions of the myocardium, the akinetic segments having permanently lost their contractility and those likely to recover their contractile function after revascularization. This provides information on viability. The proposed method was considered accurate and robust for clinical practice because it overcomes several limitations of manual analysis. In [16], validation of the deformation estimate was achieved through two validation experiments using synthetic ghost images. All the results of the work discussed show that deep learning is a very promising tool and allows accurate and early detection of abnormal heart contractions from the estimation of myocardial deformities.

## 4 Conclusion

In this paper, we began by characterizing myocardial deformation by determining the different possible directions that segmental deformities can be calculated to better understand the tissue deformation mechanism of the heart muscle. In a second step, we cited the different approaches proposed in the literature to estimate myocardial deformation in MRI. Finally, we detailed some automatic methods based on deep learning by discussing the contribution of this artificial intelligence tool to the precise identification of the contractile anomaly and the determination of its severity. This provides an opportunity to assess the myocardial viability of the affected segments. However, previous research was based on methods based on the midline or radial line, which are not suitable for noisy images and produce large errors due to the lack of solid edges. These methods also involve circular symmetry for the heart, which can increase the error. The results obtained and statistical analyses obtained in this work have shown that the cine-MRI can eliminate the need for labeled MRI for estimating myocardial deformation.

## References

1. Lalande A (2012) IRM cardio-vasculaire, des séquences d'acquisition aux paramètres physiologiques. HAL 18–22
2. Tran TT et al (2017) Empirical mode decomposition and monogenic signal based approach for quantification of myocardial infarction from MR images. 2–12
3. Wafa B, Sameh O, Basel S, Salam L (2020) Quantification of left ventricular function in MRI: a review of current approaches. ATSIP Conference, IEEE xoplore, 1–6
4. Chitiboi E (2017) Magnetic resonance imaging of myocardial strain: a review of current approaches. J Magn Reson Imaging 1–18
5. Binda C (2014) Evaluation d'un nouvel indice de contractilité myocardique indépendant des conditions de charge en échographie 2D Strain speckle tracking: application clinique à la correction chirurgicale des valvulopathies, thesis, 1–51
6. Waterton J et al (1985) Magnetic resonance (MR) cine imaging of the human heart. Br J Radiol 711–716
7. Shehata ML (2009) Myocardial tissue tagging with cardiovascular magnetic resonance. J Cardiovasc Magn Reson 1–12
8. Aletras AH (1999) DENSE: displacement encoding with stimulated echoes in cardiac functional MRI. J Cardiovasc Magn Reson 247–252
9. Jane Cao J, Ngai N, Duncanson L, Cheng J, Gliganic K, Chen Q (2018) A comparison of both DENSE and feature tracking techniques with tagging for the cardiovascular magnetic resonance assessment of myocardial strain. J Cardiovasc Magn Reson 1–9
10. Scatteia A, Baritussio A, Bucciarelli-Ducci C (2017) Strain imaging using cardiac magnetic resonance, Cross Mark 1–12
11. Litjens G (2017) A survey on deep learning in medical image analysis. Med Image Anal 1–29
12. Huaifei H (2014) Automatic segmentation pf left ventricule in cardiac MRI using local binary Fitting model and dynamic programming techniques, PloS ONE 1–17
13. Ferdian E (2020) Fully automated myocardial strain estimation from cardiovascular MRI–tagged images using a deep learning framework in the UK biobank, Radiol Card Imaging 1–10
14. Puyol-Ant´on E (2018) Fully automated myocardial strain estimation from cine MRI using convolutional neural networks. Int Symp Biomed Imaging 1–5
15. Morais P (2017) Fully automatic left ventricular myocardial strain estimation in 2D short-axis tagged magnetic resonance imaging. Phys Med Biol 1–22
16. Hammouda K (2020) A new framework for performing cardiac strain analysis from cine, 1–15. www.nature.com/scientificreports/

# Adversarial LSTM-Based Sequence-to-Sequence Model for Drug-Related Reactions Understanding

**Hanane Grissette and El Habib Nfaoui**

**Abstract** Social networks become an extensive source used to automatically report treatment and drug issues. Few approaches have been proposed in this matter, namely multi-word medical concept identification and drug-related descriptions that denoted vast connections for understanding patients' shared experiences. Therefore, the paper explores the problem of mutual forms of drug interactions including beneficial drug effects. We developed an LSTM-based encoder–decoder model that leverages and memorizes concepts' details and relationships to form a drug description that compactly with highly informed embeddings seeks multi-word drug-related descriptions from online generated narratives. Meanwhile, we apply adversarial training to regularize sequential embeddings of distributed biomedical n-grams representation from controlled medical vocabulary such as PubMed. It consists of a semi-supervised featuring medical concepts from online generated narratives at sentence-level, especially to create a feature descriptor to select possible drug-related descriptions. The model achieves an accuracy of 86% in assessing false positives and negatives of patients' statements. Whereby the model shows a well-understanding of real-world medication-related descriptions with significant improvements over baseline neural network approaches.

**Keywords** Drug reaction · Encoder–decoder · LSTM · Sentiment analysis

## 1 Introduction

Social network platforms play a crucial role in spreading information across the world and, in this sense, play a large role in global pandemic combat. Massive online everyday discussions are sharing and spreading information, misinformation, and harmful experiences with massive data analysis on Twitter, Instagram, YouTube, Reddit, and Gab. An estimated 1,342,1037 original discussions opened in this period

H. Grissette (✉) · E. H. Nfaoui
LISAC Laboratory, Faculty of Sciences, Sidi Mohamed Ben Abdellah University, Fez, Morocco
e-mail: hanane.grissette@usmba.ac.ma

 
X.-S. Yang et al. (eds.), *Proceedings of Sixth International Congress on Information and Communication Technology*, Lecture Notes in Networks and Systems 216,
https://doi.org/10.1007/978-981-16-1781-2_6

with 465,721 comments from 3,734,815 different users have been reported in one month by authors in [1]. A 57% of respondents are concerned about the amount of misinformation about COVID-19 others may be getting where 81% of social media users said that they were seeing banners, pop-ups, and upfront notices about the coronavirus.[1] This pandemic period is widely known by tremendous patients' narratives on social networks, characterized by vigorous activity and fresh dynamism into global pandemic combat. Refer to authors in [2], a good example can be seen with the peak of searches for information on the Internet and social media platforms in China preceding the peak of incidence in COVID-19 cases by 10–14 days, with which Internet and social media networks searches have a demonstrated correlation with the incidence of disease.

This study models the underlying drug reaction detection and considers another aspect of drug-related events including beneficial drug reactions. Many existing methods fail to define positive reactions that deal with huge errors in other related computation steps such as extracting the speaker's subjectivity of a given drug reaction [3], which results in false positives and negatives. The failure of defining drug reactions from patients' narratives on social network factors is studied and an optimal level of drug reaction abstraction is proposed. Thus, a discussion opened to define the relation between the two models is discussed, leading to a new cost model. Thereafter, the experiments proved the efficiency of joint LSTM-based components parameters and combined the collection of medical information on adverse incidences related to a particular target (drug, disease, etc.). Finally, the proposed model shows the great ability of comprehending multiple-related medical concept meanings, which memorizes dependencies through multiple LSTM components to seek medication-related indicators for detecting drug interactions on social networks.

The remainder of this paper is structured as follows. Section 2 provides a summary of the literature review of various Deep Neural Networks-based sentiment analysis approaches and further highlights medical concept normalization methods. Section 3 presents the proposed LSTM-based Encoder–Decoder approach. Section 4 offers large social network experimentation outputs convolving large biomedical embeddings within LSTM-based Encoder–Decoder discriminator. Section 5 summarizes this paper and provides some perspectives.

## 2 Literature Background

In this section, we present a summary of contemporary studies performed in the context of online medication-related entity detection. We seek to demonstrate its role to enhance the performance of sentiment analysis and the applicability of SA in the context of drug reaction identification.

[1] https://www.ofcom.org.uk/research-and-data/tv-radio-and-on-demand/news-media/coronavirus-news-consumption-attitudes-behaviour/interactive-data.

## 2.1 Deep Neural Network Methods

Several works demonstrated how the real-time monitoring of mood conditions led to an improvement in designing information and emotional sphere of the patient. Integrating the sentiment analysis in performing efficient real-time monitoring of patients' conditions and experiences is of critical importance. Authors in [4] explored the value of patient intelligence on pharmacovigilance in social media, especially defined as adverse outcomes caused by medications. They considered temporal changes in the number of tweets to decide the emerging topic. Potential systematic pathway to ADR monitoring from social media used the Medline, Embase, Scopus, and Web of Science databases and the Google Scholar search engine to define medical corresponding names and attributes. But, few works of medical entity extraction were done for recognizing patient discussions of drugs and events. Limited labeled training data fails to cover the entire input space. Indeed, the semantically richer nature of multi-word medical concepts, nontechnical, and descriptive language of patients generated narratives [5]. As an example, "what happens during a seizure ?" the term seizure may refer to a medical condition, disease, or descriptive situation. Where it may appear in many positive patients' statements as much as appears in negative patients' statements. In our work, we leveraged semi-supervised learning to unlock new paths towards reducing limited sufficient annotated sentiment resource problems and the scarcity of labeled data for training a robust model [6]. But they may be very vulnerable to adversarial attacks with noises and diversity in medication-related text. Adversarial training is a method for learning robust deep networks. DNN models can be improved in training with the adversarial information using a manifold-based influence measure effective detection as well as allowing for targeted attacks.

In our study, we mainly focus on investigating the surge of learning from heterogeneous unlabeled examples spread every second via social network data available in combination with typically smaller sets of labeled data biomedically controlled.

## 2.2 Medical Concepts Normalization

As discussed in Sect. 1, these platforms are becoming an increasingly preferable place for sharing personal health-related narratives, and discussing various drug problems, e.g., adverse drug reactions. Adverse drug reaction (ADRs) becomes a significant public health concern by various communities and researchers [7]. Drug reaction normalization consists of annotating the presence or absence of ADR mentions, with a mutation to formal ADR description in the Unified Medical Language System (UMLS). Conditional random fields (CRFs) approach is widely used to extract drug-related mentions and is mainly based on deterministic features. Other researchers solved that by involving machine learning-based approaches to leverage a variety of features, including words' semantic similarities. But, due to limited training data, it weakly converges for detecting informal drugs and preventing drug reactions from highly informal language. Indeed, it usually collected for specific targets, e.g., drugs or diseases, and in a given period. In this sense, many benchmark datasets have been collected from online patient-generated posts, we will discussed in details in the experiemnts section. Deep learning has potential advantages of leveraging contextual and semantic information into unique dense representation, deducing possible relationships in terms of similarities, reasoning correlate concepts, and multi-word expressions in long-term dependencies, which further reached acceptable performance with fewer training examples than the other baselines. For example, [8] proposed to use the recurrent neural network (RNN) based model to label words in an input sequence with ADR membership tags. Embeddings have shown great success in representing the chemical, biological, and biomedical information and we have exploited to automatically detect drug reaction membership and denote connections for candidates items. Because of the sensitivity of detecting adverse drug reactions to public health effects, it mainly required more credible settings and parameters. In this study, we demonstrate the ability of LSTM components to leverage and memorize token details and relationships to form drug description that compactly with highly informed embeddings seeks drug reactions multi-word descriptions from online generated narratives.

## 3 Proposed Approach

The goal is to standardize common terminology through individual processes formed for more personalization and awareness. In the future, we aim to design a dynamic system for reporting and Medication Errors learning. It consists of an extended transition-based Medical concept recognition. Furthermore, we validate the effectiveness of distracting drug reaction components on retrieving the relevant positiveness and negativeness. The next neural network layer for the sentiment generation process is shared between medical distributed-based normalization in terms of embedding and drug-related terms identification. Indeed, Medical distributed-based normalization incorporates primarily MEDLINE database references and abstracts of life sciences and biomedical topics and PubMed. As depicted in Fig. 2, the approach has been involved in many steps: (1) Collecting and normalizing data using keywords-based crawling system is created for the collection of Twitter posts. In this study, we focused on identifying in the COVID-19 contexts. We used a list of COVID-19-related Keywords, e.g., *Corona* and *Chloroquine*. Thus, we are interested in getting information attributed by : ['id', 'created_at', 'source', 'original_text', 'retweet_count', 'ADR', 'original_author', 'hashtags', 'user_mentions', 'place', 'place_coord_boundaries']. Firstly, we used drug-grouped datasets such as COVID-19, Chloroquine, Remdesivir, and distinct groups. Each group contains 20938, 3299, and 2918 original tweets, respectively. (2) Sequence-to-sequence model used n-grams biomedical embedding Matrix to construct corresponding medical vectors by involving distributed biomedical at sentence-level generation [9]. A distributed representation of 700 dimensions that are trained on PubMed and

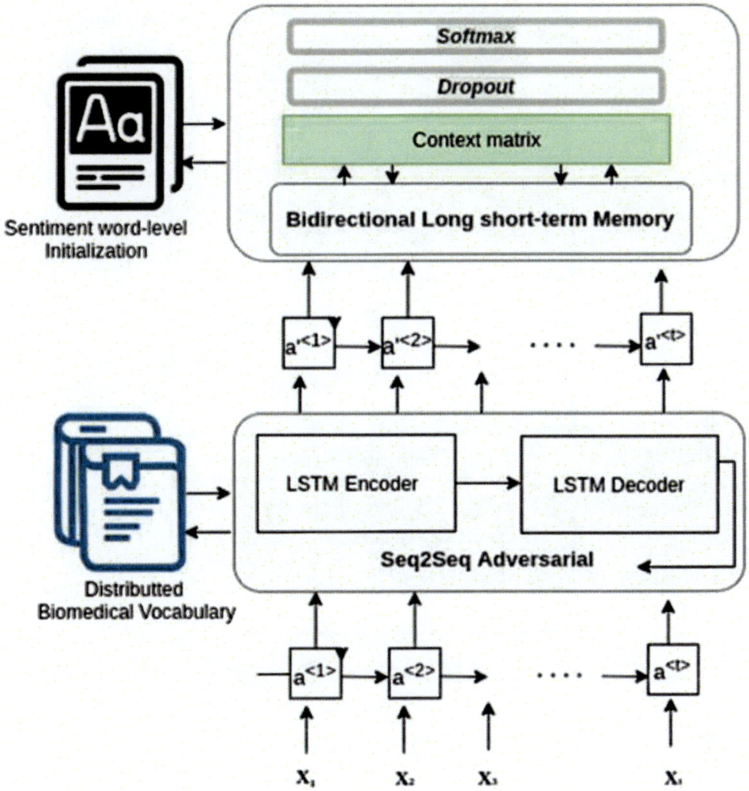


**Fig. 1** The whole model description

Clinical notes MIMIC-III whereby achieved an accuracy of 0.848. (3) Adversarial Long Short-Term Memory (LSTM)-based Encoder–decoder model to map the input sequence to a vector of a fixed dimensionality, and then another LSTM to decode the target sequence from the vector of possible drug reaction target within the text. (4) Bidirectional Sentiment model is created for an extrinsic evaluation to measure the model's ability to correct false positives and negatives through efficient drug reaction discrimination from real-world patients' descriptions. Also, Twitter sentiment analysis probes using additional salient Twitter features to summarize online criteria and particularities.

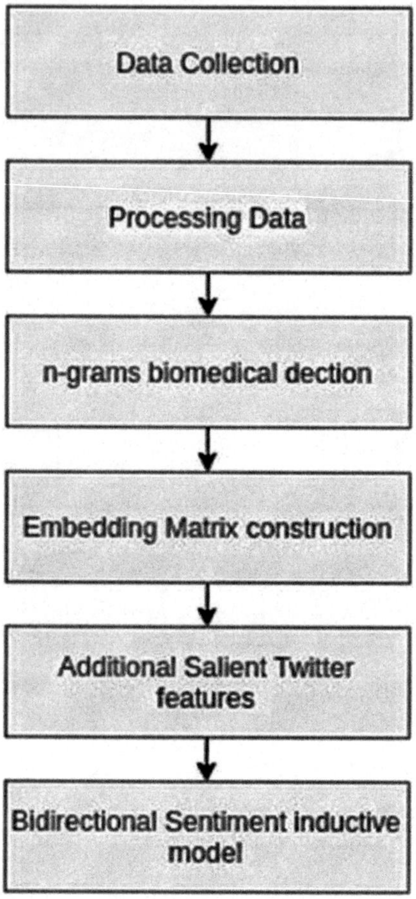


**Fig. 2** The whole model description

# 4 Implementation and Experiments

Medical sentiment concerns the patient's health status, medical conditions, and treatments. They frequently necessitate an analysis at the features extraction level, which can be difficult without natural understanding and sufficient background knowledge of the formal medical contexts. One of the most potent ideas in deep learning is that transfer knowledge has been learned from different data and for a given task. This is a form of transfer learning. Transfer learning makes sense when you have a lot of data for the problem you are transferring from and usually relatively less data for the problem you are transferring to. In this study, we aim at imitating and retrieving the corresponding real-word drug-related entities in the formal medical ontologies by adopting a distributed representation of two databases PubMed and

**Table 1** COVID-19-related tweet and their drug reaction spans from real-world descriptive entities

| Tweet | Medical concepts | Drug reaction span |
|---|---|---|
| @covid19crusher poor clinical outcome morocco lead switch hydroxychloroquine March twentythree average time covid19a remdesivir people died | Covid-19<br>hydroxychloroquine<br>remdesivir | Died |
| Laura ingraham continues pimp hydroxychroloquine magic bullet covid19 fda cautioned use treatment doctors irresponsibly went fox said sideeffects lie fog | Hydroxychroloquine<br>bullet | fog |
| Make tik tok video misguide people spit food item deliver hide information affect people bea patient lupus rationing rx severe patient gt50 improve amp discharge hospital | gt50<br>lupus | Spit food<br>amp<br>rx<br>gt50 |
| Fda approves widespread use drug covid19 limit effectiveness grave side effect base stu tired tolerant | Base stu | Tired<br>tolerant |
| Addict take #hydroxychroloquine follow 500ml glass dettol per presidential instruction lie comfortable position cover clean white sheetcall undertaker | Hydroxychroloquine<br>dettol | Addict |

MIMIC-III Clinical notes. PubMed distributed vocabulary involved a 28,714,373 documents, an overall of 181,634,210 sentences, and 4,354,171,148 tokens. whereas MIMIC-III Clinical notes trained on 2,083,180 documents, 41,674,775 sentences, and 539,006,967 tokens, respectively. The second case is to inform our model about drug reaction structural grams and their sentimental assessments, we enrich the embedding framework by using labeled pre-trained lexicons such (i) [10], they annotate a set of more than 3500 posts to online health forums of breast cancer, Crohn's disease, and different allergies, respectively. They selected a set of diseases that, according to medical experts, shows high heterogeneity concerning both the degree of medical understanding of the diseases and the profile of the users. (ii) Social Media Mining for Health (SMM4H)-2017 shared task [11]. It consists of mining drugs-related information such as automatic adverse drug reactions (ADRs) detection from medication-mentioning tweets and normalization of ADR expressions. It was characterized by massive training data that consisted of 15 717 annotated tweets for classification tasks and 10,260 and 6650 ADR phrases and identifiers for ADRs detection and normalization, respectively.

The overall framework of this study is summarized in Fig. 1; as shown in this figure, this framework involves three stages: (1) build an enhanced sentence-level distributed biomedical representation that involved the earlier described distributed vocabulary from two comprehensive corpora: PubMed and the clinical notes from MIMIC-III Clinical Database, (2) exploit contextual embedding through LSTMs Encoder–Decoder mechanism to infer Drug reactions from texts, and (3) sentiment induction based on Bidirectional-LSTM. The main task is to extract drug multi-word concern from the whole context of the sentence, and the model generates drug reaction membership tags in form of phrases using multiple LSTM components with self-attention as a discriminator among these artificial neural networks. Specifically, the LSTMs Encoder–Decoder structure uses an encoder to hold the biomedical embeddings information that represents drug reaction as n-gram phrase in tweets reporting ADRs. The embedding matrix leverages contextual and medical information meaning from the sentence context in order to define the multi-word expression of drug reaction individual meaning in the field of distributional semantics configuration, which clarifies sentence's similarity in the same contextual target space, which further share semantically common drug description meanings. The decoder is the second network that convolves the output of the encoder and gives the fittest drug reaction membership tag to the actual input or intended drug reactions. Also, the encoder considers the input of the generator as an input sequence. This approach contributes to the formulation of drug-related phrases containing tags that have related meanings to given multi-word expressions.

We conduct many training developments to achieve good recognition accuracy and we used large patient narratives datasets from Twitter. First, an intrinsic evaluation has been done on the biomedical distributed configuration space using clinical sentence pair similarity tasks. Moreover, the proposed approach has been evaluated on pre-trained BioSentVec of clinical sentence pair similarity tasks. We conducted a comparison using two benchmarks datasets dedicated for biomedical evaluation: (1) the 100 sentence pairs from BIOSSES[2] and the 1068 sentence pairs from MedSTS[3], respectively. As result, our framework embedding achieved an accuracy of 0.85 and 0.83, respectively, on both datasets. We also evaluated the accuracy of the proposed model with only PubMed vocabulary or MIMIC-III Clinical notes or by combining them jointly with BioSentVec. We both achieved an accuracy of 0.817, 0.350, and 0.795 comparing with BIOSSES, whereas MedSTS got an accuracy of 0.750, 0.759, and 0.767, respectively.

We have conducted an extrinsic evaluation by exploiting obtained representations in a fine-tuning task that demonstrated higher performance in terms of discriminating false positives and negatives. Since authors in [5, 12] proved how drug reaction presence in text fails to retrieve the relevant sentimental status. They frequently need examination at the features extraction level, which can be challenging with limited

[2] https://tabilab.cmpe.boun.edu.tr/BIOSSES/DataSet.html.

[3] https://arxiv.org/ftp/arxiv/papers/1808/1808.09397.pdf.

sufficient background knowledge and formal medical contexts. These explain why sentiment analysis systems are able to perform sentiment analysis towards a given entity and services reasonably well but fare poorly on clarifying sentiment towards a set of words that may refer to an adverse drug reaction existence in texts. To this end, we chose to validate this study by designing an additional task, and we developed a BiLSTM-based classifier that took the output of Encoder–decoder representations to induce sentiment value from large patient self-reports about the COVID-19 pandemic. It is designed mainly to deploy the dynamic embeddings of medical concepts dependencies on huge shared data. We chose to use the popular sentimental lexicons to amalgamate the full range of emotional state related to medication-related experiences in a given configuration space in terms of positivity and negativity, we used three sentiment lexicons: (i) SenticNet [13]: A knowledge base based on Sentic computing framework dedicated to sentiment analysis. (ii) SentiWordNet [14]: SentiWordNet is a public lexical vocabulary that is based on extending WordNet opinion terms by using semi-supervised learning. (iii) AFINN [15]: AFINN vocabulary consists of a list of English words that have been manually labeled.

In such cases, recent contributions have taken the advantage of these manually labeled dictionaries and performed an increased SA approaches by incorporating different sentimental annotation aspects, e.g., [6] used a simple word matching, they showed that the new word list may perform better than ANEW, though not as good as the more elaborate approach found in SentiStrength. Other contributions improve these dictionaries by combining a set of lexicons, e.g., [15] provides a set of normative emotional ratings for a large number of words in the English language rated in terms of pleasure, arousal, and dominance. Pleasure, or affective valence, ranges from pleasant to unpleasant, arousal ranges from calm to excited, and dominance/control ranges from controlled to in control.

The model proved a great understanding of medical items within text. From the previous study, we illustrated some examples of the COVID-19-related tweet and their drug reaction spans from real-world descriptive entities in Table 1. This contribution adjoins and attaches meanings with conceptual medical targets to imitate such a drug description and select from the text. Thereafter, the results raised the need to coordinate efforts to include the collection of information on adverse incidences related to the drug-related issue, which which consider the ambiguity level of medication-related components within text that decrease the comprehension and preventability of adverse drug reactions (ADRs). From our experiments, the model shows on high chance to define drug-related issue or misinformation through analogical study. For each dataset group, we tracked 209 107 76 possible drug issue descriptions.

The evaluation models have slightly different results on different Twitter data groups. The BiLSTM-based model outperforms other neural network baselines and achieved an accuracy of 0.86, where got in low performance with other bio-medical knowledge combinations (only with sentence-based embeddings from Pubmed or clinical notes MIMIC III). With higher number of epochs, they respectively got an accuracy of 0.81 and 0.79. However, stacked-LSTM-based and BiLSTM-based model consistently improved the sentiment classification performance, especially they have resulted in great performance when we exploit proposed configuration on an online datasets and have shown a well-understanding to the bio-medical components.

## 5 Conclusion

Everyday patients discussions exhibit a huge amount of biological concepts citations, and medication-related reports are affectively and semantically associated in different contexts. Neural Networks addresses such issues in the context of deep learning for enhancing Natural Language Processing (NLP) capabilities through exploring the richer nature of multi-word concepts. In this study, we propose to design a LSTM-based Encoder–decoder to mine drug-related events from online patient narratives. It relies on distributed medical configuration policy in order to effectively mimic corresponding nontechnical and descriptive real-world expressions to formal medical formats from controlled medical vocabularies.

## References

1. Cinelli M, Quattrociocchi W, Galeazzi A, Michele Valensise C, Brugnoli E, Lucia Schmidt A, Zola P, Zollo F, Scala A (2020) The COVID-19 social media infodemic. Sci Rep
2. González-Padilla DA, Tortolero-Blanco L (2020) Social media influence in the COVID-19 pandemic
3. Hanane G, Habib Nfaoui EL, Bahir A (2017) Sentiment analysis tool for pharmaceutical industry and healthcare. Trans Mach Learn Artif Intell
4. Sarker A et al (2015) Utilizing social media data for pharmacovigilance: a review
5. Grisstte H, Nfaoui E (2019) Daily life patients sentiment analysis model based on well-encoded embedding vocabulary for related-medication text. In: Proceedings of the 2019 IEEE/ACM international conference on advances in social networks analysis and mining, ASONAM '19. New York, NY, USA. Association for Computing Machinery, pp 921–928
6. Grissette H, Nfaoui EH (2020) Enhancing convolution-based sentiment extractor via dubbed N-gram embedding-related drug vocabulary. Netw Model Anal Health Inf Bioinf 9(1):42
7. O'Connor K, Pimpalkhute P, Nikfarjam A, Ginn R, Smith KL, Gonzalez G (2014) Pharmacovigilance on twitter? Mining tweets for adverse drug reactions. In: AMIA ... Annual Symposium proceedings/AMIA Symposium. AMIA Symposium
8. Cocos A, Fiks AG, Masino AJ (2017) Deep learning for pharmacovigilance: Recurrent neural network architectures for labeling adverse drug reactions in Twitter posts. J Am Med Inf Assoc
9. Chen Q, Peng Y, Lu Z (2018) BioSentVec: creating sentence embeddings for biomedical texts. CoRR. ArXiv:abs/1810.09302
10. Carrillo-de Albornoz J, Rodrguez-Vidal J, Plaza L (2018) ediseases dataset
11. Sarker A et al (2018) Data and systems for medication-related text classification and concept normalization from Twitter. J Am Med Inf Assoc
12. Grissette H, Nfaoui EH (2019) A conditional sentiment analysis model for the embedding patient self-report experiences on social media. Adv Intell Syst Comput
13. Cambria E, Olsher D, Rajagopal D (2014) SenticNet 3: A common and common-sense knowledge base for cognition-driven sentiment analysis. In: Proceedings of the national conference on artificial intelligence

14. Baccianella S, Esuli A, Sebastiani F (2010) SENTIWORDNET 3.0: an enhanced lexical resource for sentiment analysis and opinion mining. In: *Proceedings of the 7th international conference on language resources and evaluation, LREC 2010*
15. Nielsen F (2011) AFINN

# Structural Organization of the Server System of Multiuser Device for Home Health Control

**K. G. Korotkov and K. V. Semenov**

**Abstract** The organizational principles of the server system of multiuser device for home health control based on the Electrophotonic Imaging technology are presented. We discuss the prospects of implementing a new Internet of Things (IoT) scenario for personal monitoring of health status through a cell phone. The system is based on the Electrophotonic Imaging (EPI) method which is based on recording patterns of light created by emission stimulated by an electromagnetic field and computer processing of recorded images. In 2021, the Bio-Well device's mobile version is released, working with the phone through Bluetooth. As practice has shown, many users want to modify the system, using their logo and design, a set of specific programs in the package, often with the formulation of individual recommendations, for example, on the applied aromatic oils, supplements, or diets. This demand required creating a server-based multiuser system that allows for flexible system configuration for specific user groups with automatic tracking of access and individual parameters. This article describes the algorithmic structure of this system.

**Keywords** Home health monitoring · e-health · Server structure

## 1 Introduction

Nowadays, home health monitors are becoming more and more common. Millions of people regularly measure blood pressure and blood sugar; handheld devices and cell phones allow them to monitor the dynamics of pulse changes, calorie intake, distance traveled, and other parameters necessary for wellness and sports activities. In 2019, the FDA issued a series of guidance that presented an explicit criterion for classifying a particular device as a wellness device that does not require medical certification [1]. The guidance explains that a general wellness product should be

K. G. Korotkov (✉)
Saint-Petersburg Scientific-Research Institute for Physical Culture, St. Petersburg, Russia

K. V. Semenov
University ITMO, St. Petersburg, Russia

 
X.-S. Yang et al. (eds.), *Proceedings of Sixth International Congress on Information and Communication Technology*, Lecture Notes in Networks and Systems 216,
https://doi.org/10.1007/978-981-16-1781-2_7

used for maintaining or encouraging a general state of health or a healthy activity. The document outlines two categories for wellness devices that fall outside of FDA regulation.

The first group of devices is those that make general claims and do not make any reference to diseases or conditions. In particular, they may contain claims related to relaxation or stress management, mental acuity, or self-esteem management.

The second group of general wellness devices may have claims to help users reduce the risk of certain diseases or help users better live with specific diseases or conditions.

One of the methods widely used for evaluating general health is the method of Electrophotonic Imaging (EPI) based on analyzing the glow of fingers in an electromagnetic field [2]. The Electrophotonic Imaging (EPI) method is developed for recording electron and photon emission stimulated by an electromagnetic field and computer processing of recorded images. 10 microseconds impulse stimulates a response from the subject in the form of the gas plasma (that is why in physical terms this approach is known as Gas Discharge Visualization—GDV [3, 4]). Emitted photons are measured by a CCD device, and digital images are transmitted to a computer for data processing. We calculate the emitted photons' energy, emission intensity, and several parameters from the two-dimensional images. The EPI technique is useful in evaluating health levels [4–8] and monitoring the reactions of a person to different influences [9–12]. Psychologists working with Russian Olympic and Paralympic teams use this method to analyze athletes' training and competitive activities [13]. Specialized algorithms formed by numerous clinical trials calculate parameters and generate some conclusions. In the last modification of the device (commercial name Bio-Well, www.bio-well.com), a robust software located on the server process data from the end-user, allowing not to depend on the user's computer performance and to upgrade programs regularly. More than 3000 users in 60 countries since 2014 use this system.

In 2021, the Bio-Well device's mobile version is released, working with the phone through Bluetooth (Fig. 1). As practice has shown, many users want to modify the system, using their logo and design, a set of specific programs in the package, often with the formulation of individual recommendations, for example, on the applied aromatic oils, supplements, or diets. This demand required creating a server-based multiuser system that allows for flexible system configuration for specific user groups with automatic tracking of access and individual parameters. This article describes the algorithmic structure of this system.

## 2 System Description

The system uses the principle of three-level architecture, synonymous with three-tier or multitier architecture, consisting of a user application, an application server, and the database server. The system uses the principle of three-level architecture, synonymous with three-tier or multitier architecture. The first level is the actual

Fig. 1 Bio-Well mobile device

end-user application, it has no direct connections to the database (according to security requirements), is not loaded with basic business logic (according to scalability requirements), and does not store the state (according to reliability requirements).

The first level has a standard logic: authorization interface, algorithms of encryption, permissibility and format compliance verification, basic operations (sorting, grouping, calculation of values) with data already loaded to the terminal, as well as printing and communication with the Bio-Well device and sending data to the server.

The second level holds the application server and most of the business logic. Fragments exported to terminals, triggers, and stored procedures submerged in the third level remain outside it. The second level connects user connections with the database.

The third level is a relational database MySQL version 5.1 with stored procedures, triggers, a table of registered users, users' cards, and data on experiments.

The server on CentOS with the Apache and PHP 5.X web server hold the application server and the database server. This software provides freedom, combined with great flexibility and performance. If needed, we may migrate the server part to other platforms at the lowest cost.

The application server has the following different functions:

- Parameters calculation;
- Business logic modules;
- Network exchange with the users.

Modules calculating parameters are written in C++ language and presented as executable files. Each module takes parameters as command-line arguments and places the results in the specified folder. We built modules for all platforms and installed them using the GCC compiler (GNU Compiler Collection). The sources describe the CMake utility [14] that creates the project files for the user platform and compiler. The server stores EPI images as PNG files, so the sources also include the LIBPNG and ZLIB libraries [15].

Modules of the business logic serve all the applications and accountable for sharing with users and the database server. For ease of correction and additions, for these modules, we used PHP 5.X. The entire set of modules forms a web service, which provides user applications with the ability to work with the system using different exported functions. SOAP protocol (http://en.wikipedia.org/wiki/SOAP) provides interaction with user applications using the WSDL language for the description [16]. Requirements for a user application:

- Support for platforms: Windows, Mac OS, Linux, and major mobile platforms.
- Work with the device through the exchange of a USB port or Bluetooth.
- Support multiple languages.

Based on these basic requirements for developing user applications, we have chosen the Qt library [17] cross-platform software development tools in the programming language C++.

The need for flexible system modification to meet individual user groups' requirements demanded developing a specialized software environment that allows for tracking all available configurations and providing an access hierarchy. In the code of the Bio-Well system, we are using conditional operators that determine the functionality depending on the configuration of the system. Accordingly, building the project considers which option is selected, and the compiler makes exactly the required executable file. Next, the installers assembling program also selects the required option and builds the necessary executable file. However, this system has several drawbacks:

- As the number of these configurations increases, the readability of the code becomes more complicated.
- When developing new program options, like implementing new Apple or Windows upgrades or adding new languages, it is always necessary to consider all different options and test all the configurations. It often happens that the basic Bio-Well program is operating without any problems, but other modifications of the program require careful manual configuration.
- It is necessary to have separate software update branches for each variant and upload them to the server.

A standard solution to the problem is to use plug-ins. Essentially, the main application file implements basic functionality, and the application of separate modules (usually DLL or others) may add functionality. There are millions of such applications, such as Google Chrome. The basic version is quite limited, but it can solve a vast number of tasks with extensions. The advantages are obvious:

- As the individual modules are loaded at the stage of packing the main program, it is easy to change these modules' composition.

- There is no strict dependency on the core program. It is easy to debug and develop different parts of the program.
- Everything is implemented within the same programming language (usually) and works fast enough.
- Adding new functionality does not lead to rework of the application kernel itself.
- The Qt library has extensive functionality to develop such modules.

The minuses are not entirely evident:

- All modules are platform dependent. The modules are built in the same way as the program itself, separately for each platform, respectively, complicating their support.
- The modules may have some bugs that can lead to critical failures in the main application.

Nowadays, more and more developers are on the way to creating their original platforms. A system initially develops some platform, with the implemented application kernel having its DBMS with basic functionality plus built-in programming language. Accordingly, the so-called "configurations" are implemented for each separately solved task in the built-in language providing the functionality required for this configuration. Such applications include Microsoft Office and various CAD (AutoCAD, ArchiCAD).

The main advantages of such a system are as follows:

It is possible to implement any functionality and any interface (that is, the appearance of the application) for absolutely any task.

- 100% code portability executed in the built-in language.
- Individual modules cannot interfere with the kernel because they run in the built-in programming language.

Figure 2 demonstrates the simplified functional scheme of the implemented system. For each group of users, a different configuration is created, which differs in a set of functions, both for the program and for the device (logo, color, box-type selection, etc.). Each user has an individual login and password; a personal cabinet keeps all the data. Its chip number identifies each device. The software automatically determines the user's configuration and the device, checks the subscription status, and in case all the attributes match, open the account login. In violation of any conditions, the program automatically generates corresponding notifications: "Check the subscription status" or "There are ten days left till the end of your subscription." After the end of the subscription, the user can view his/her database but cannot do readings. A user of a particular configuration cannot enter another configuration's software environment, but there are demo accounts available for all users. This system allows easily add and change configurations, does not depend on the user's type of device, and works on mobile devices of any kind and desktop computers.

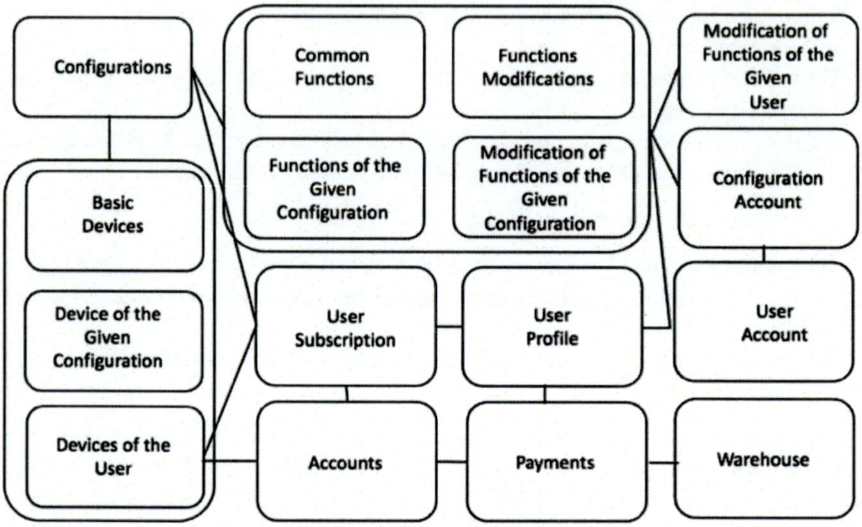


**Fig. 2** The simplified functional scheme of the Bio-Well server system

## 3 Conclusion

Bio-Well complex has been operating since 2014. During this time, it has demonstrated the stability of operation, the flexibility of settings, ease of use and maintenance, and the possibility of unlimited expansion of system modifications. More than 3000 users in 60 countries use the system, regular seminars and conferences are held, and scientific articles are published based on the results of scientific research in various countries. The introduction of Bio-Well, working through a cell phone, will advance this technology to the level of the Internet of Things (IoT) scenario for personal monitoring of wellness and health status through a cell phone.

## References

1. https://www.regulations.gov/docket?D=FDA-2014-N-1039. Accessed 01 Sep 2021
2. www.Bio-Well.com. Aaccessed 01 Sep 2021
3. Muehsam D, Chevalier G, Barsotti T (2015) An overview of biofield devices. Glob Adv Health Med 4:42–51
4. Korotkov KG (2017) The energy of health. Amazon.com publishing
5. Korotkov KG (2018) Review of EPI on medicine and psychophysiology published in 2008–2018. Int J Complement Altern Med 11(5):311–315
6. Korobka IE, Yakovleva TG, Korotkov KG, Belonosov SS (2018) Electrophotonic imaging technology in the diagnosis of autonomic nervous system in patients with arterial hypertension. J Appl Biotechnol Bioeng 5:112–118

7. Buck KH, Novelli C, Costa FT et al (2018) O uso da bioeletrografia na comparação entre mulheres com câncer de mama, mulheres saudáveis sedentárias e mulheres praticantes de corrida. Centro de Pesquisas Avançadas em Qualidade de Vida. 8(2):9–11
8. Yakovleva TG, Buntseva OA, Belonosov SS et al (2015) Identifying patients with colon neoplasias with gas discharge visualization technique. J Altern Complement Med 21:720–724
9. Kushwah KK, Nagendra HR, Srinivasan TM (2015) Effect of integrated Yoga program on energy outcomes as a measure of preventive health care in healthy people. Cent Eur J Sport Sci Med 12(4):61–71
10. Shiva KK, Nagendra HR, Srinivasan TM, Marimuthu P (2016) Electrophotonic imaging-based analysis of diabetes. Int J Altern Complement Med 4(5):134–137
11. Korotkov KG, Churganov OA, Gavrilova EA (2019) Influence of drinking structured water on human psychophysiology. J Appl Biotechnol Bioeng 6(4):171–177
12. Pokrovskii VM, Nechepurenko AA, Tarasov DG et al (2019) Sinoatrial Node Pacemaker Cell Pool Dynamics Upon Synchronization With Vagus Nerve Rhythm. J Appl Biotechnol Bioeng 6(3):114–116
13. Drozdowski A, Gromova I, Korotkov K, Shelkov O, Akinnagbe F (2012) Express-evaluation of the psycho-physiological condition of Paralympic athletes. J Sports Med 3:215–222
14. http://cmake.org. Accessed 01 Sep 2021
15. http://libpng.sourceforge.net/index.html and http://zlib.net Accessed 01 Sep 2021
16. http://en.wikipedia.org/wiki/Web_Services_Description_Language. Accessed 01 Sep 2021
17. http://qt.nokia.com. Accessed 01 Sep 2021

# Analyzing the Most Influencing Factors in Limiting the Spread of COVID-19 Disease

**Rasha Awawdeh, Shatha Melhem, Nour Alqudah, Tibra Al-smadi, and Ahmad Mustafa**

**Abstract** COVID-19 disease is an infectious disease caused by Coronavirus. Studying COVID-19 influencing factors has become essential by many researchers in various scientific disciplines. This study analyzes factors that may impact the total COVID-19 infection and death cases in selected countries. The factors we study are population density, age, healthcare factors, temperature, humidity, wind speed, and government responses. We utilize exploratory data analysis techniques and statistical measures to understand the relationships between variables. The results show that the population density factor, age, and government measures are strongly correlated with the total infection and death cases, whereas, temperature, humidity, and wind speed factors are not correlated with cases. This study may help in containing the disease and limiting its spread.

**Keywords** COVID-19 · Factor analysis · Exploratory data analysis

R. Awawdeh (✉) · S. Melhem · N. Alqudah · T. Al-smadi · A. Mustafa
Department of Computer Information System, University of Science and Technology, Irbid, Jordan
e-mail: riawawdeh18@cit.just.edu.jo

S. Melhem
e-mail: szmelhem18@cit.just.edu.jo

N. Alqudah
e-mail: nsalqudah18@cit.just.edu.jo

T. Al-smadi
e-mail: tralsmadi18@cit.just.edu.jo

A. Mustafa
e-mail: ammustafa@just.edu.jo



X.-S. Yang et al. (eds.), *Proceedings of Sixth International Congress on Information and Communication Technology*, Lecture Notes in Networks and Systems 216,
https://doi.org/10.1007/978-981-16-1781-2_8

## 1 Introduction

In late 2019, COVID-19 appeared in Wuhan, China. Since then, the epidemic has become a severe public health problem worldwide [6, 1]. As conditions in China stabilized, the spread has become out of control in many other places such as in parts of Europe and the United States. Many health sectors have lost the ability to treat a large number of corona cases despite their strength. Our study aims to provide information that may help in reducing the spread of COVID-19.

This study focuses on finding the most influencing factors in the COVID-19 epidemic. These factors could play a significant role in increasing its prevalence, and some of these factors may have a direct relationship in reducing or controlling the spread of the virus. So, we analyze five factors. The first two are population density and age. Both have shown a significant impact on the total number of deaths among the studied countries. The third factor is healthcare, which we divide into sub-factors, including the local public health spending factor, the number of medical personnel (doctors and nurses), hospital bed availability, cardiovascular factors, and diabetes. Temperature, humidity, and wind speed is the fourth factor that we analyzed, judging from the rumors about that weather is affecting the spread of coronavirus. The last factor is the government measures.

## 2 Related Work

Huang et al. [8] presented the mechanisms the Chinese government has taken after discovering the first case in Wuhan, which limited the spread of COVID-19, at least for a short time. These mechanisms include home quarantine, the closure of affected regions, hang of public grouping, cluster segregation of infected people, protraction of holidays, and prevention of travel measures. Regarding hospital facilities and human resources (HR) support, Huang et al. mentioned that the government started building temporary hospitals to accommodate a large number of patients, and by sending medical teams to Wuhan to tackle the shortage of medical teams. All the measures paid off in reducing the virus spread. However, if China took the needed measures five days earlier from January 23, the virus impact would have reduced by two-thirds. In [4], Dowd et al. showed that the patient's health state is related to the possibility of death regardless of age. Patients who have tuberculosis, diabetes, hypertension, COPD, or who smoke have a higher likelihood of death. Also, men are at a higher risk of death than women.

Several approaches have used Statistical and Machine learning methods to study the spread of the virus [10, 9]. Qiu et al. [10] concluded that each new infection case in Wuhan produces more cases outside Wuhan within one week in cities with population flows from Wuhan. Liu et al. [9] studied the risk factors for COVID-19 development and progression using t-test, U-test Mann–Whitney, and regression analysis.

Remuzzi et al. [11] discussed the spread of coronavirus in Italy. Authors in [5] suggested that suppression would minimally require social distancing of the whole population, household quarantine of their family members, and home isolation of cases in the UK and US.

# 3 Dataset

## *3.1 Government Responses*

We used the Oxford COVID-19 Government Response Tracker (OxCGRT) dataset [7]. The data is collected by over one hundred students and academics of Oxford University from public sources of data, like news and government media, from all over the world. The indicators are two types: Ordinal, which measures the policies based on a scale of severity, and Numeric, which is measures values in USD. OxCGRT dataset has 18 indicators, which included containment and closure measures (C1-C8), Economic response measures (E1-E4), and Health systems measures (H1-H5).

## *3.2 Population Density and Age*

We used data from "Our world in data website" [2] to analyze population density and age factors, it contains a number of confirmed cases, deaths, and testing worldwide and relies on data from the European CDC. It has 32 features about countries, total cases, total deaths, population density, and the median age of cases. So, CDC data provides a global perspective on the evolving pandemic.

## *3.3 Healthcare Factors*

We used "World Bank data—World development indicators: Health Systems" [12]. This data explains health spending per capita by country, as well as midwives, doctors, nurses, and specialist surgical staff per capita to explore whether the health factors affect the spreading of the virus.

## *3.4 Temperature, Humidity, and Wind Speed*

Finally, we used "COVID-19 global weather data" on Kaggle [3]. The data contains average temperature, percent humidity, wind speed, and hours of sunlight in the

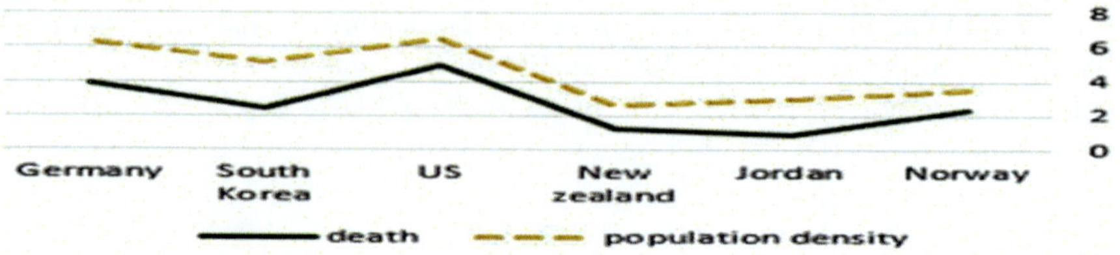


**Fig. 1** Population density versus death cases (in log-scaled)

capital city of 156 countries. The ranges of date from 22/January/2020 to 21/March /2020 were initially collected for usage in the COVID-19 Week 1 challenge.

## 4 Analysis

COVID-19 pandemic spreads worldwide, hitting cities of the world in waves: starting in Wuhan, then Milan, and New York. No city seems immune to infection cases. However, some cities are more vulnerable to the spreading of pandemic and its destructive impacts than others. This section studies the effect of different factors that exacerbate this pandemic on the number of infection and death cases.

### *4.1 Population Density*

Population density is a major factor that researchers have investigated. Figure 1 illustrates the impact of population density on death cases for six countries; Switzerland, Norway, Germany, the US, South Korea, and Jordan. It shows that the higher the population density rate, the higher the mortality rate. It implicitly means that the spreading of this virus in regions with high density is more than in others. However, population density is just one factor that shows how vulnerable these places are to this epidemic.

### *4.2 Age*

Figure 2 illustrates that the median age for infection cases is relatively close to 46 in Germany, 38 in the US and New Zealand, 40 in Norway, 43 in Switzerland, and 24 in Jordan. Studying the relationship between older median ages and death cases, we conclude that the United States, New Zealand, and Germany have a higher number of death cases, as shown in Fig. 2. The age of older cases implies a higher mortality rate.

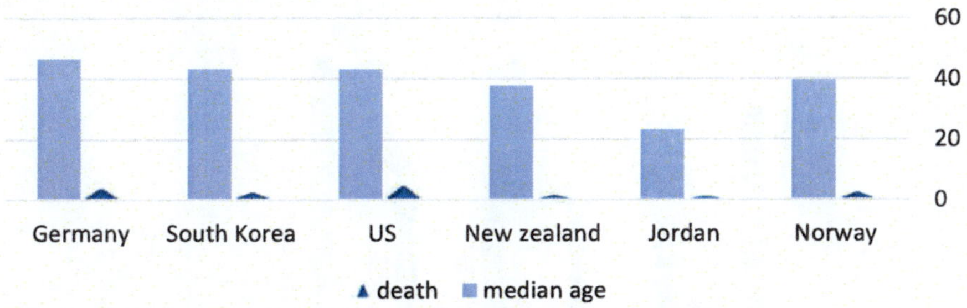


**Fig. 2** Number of death cases according to median age

## *4.3 HealthCare*

**The local public health spending factor**: Figure 3 shows that despite the high rate of government health spending in both the USA and Germany, both have the highest rates of the increasing number of infected and dead. So, we do not see the relationship between the public health spending factor and the increasing number of infections and deaths.

**The number of medical personnel**: Figures 4 and 5 show that the USA has a smaller percentage of doctors and nurses while having a high number of infected

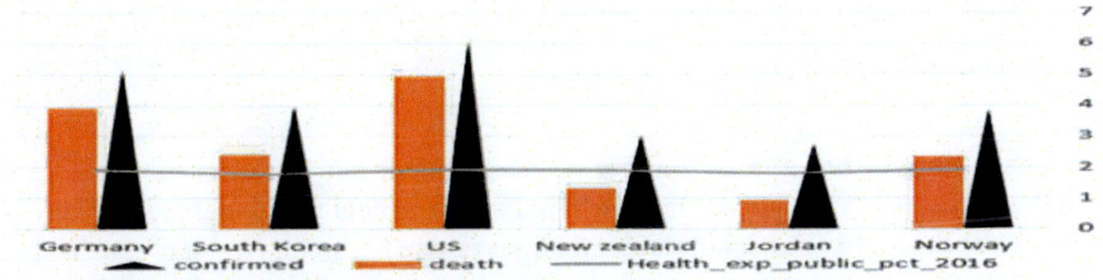


**Fig. 3** Local public health spending factor versus the number of confirmed cases and deaths

**Fig. 4** The number of physicians

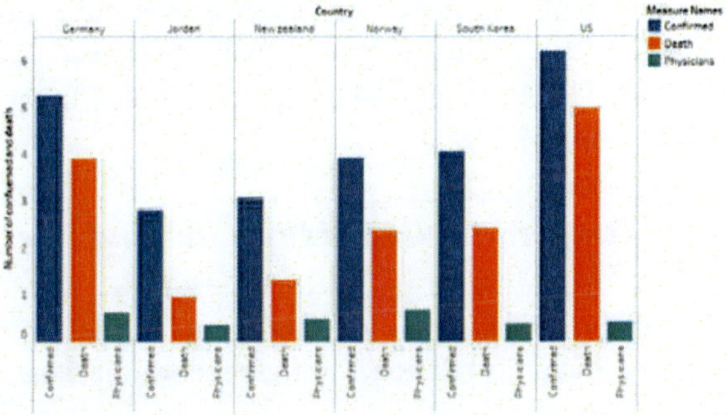

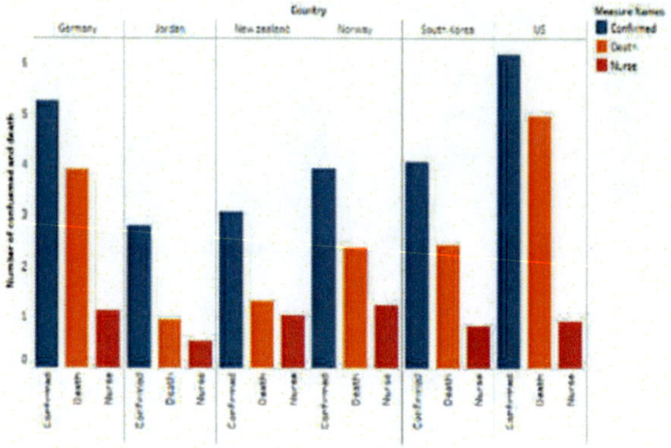

Fig. 5 The number of nurses

Fig. 6 Hospital bed availability

and dead. Norway has a large number of medical staff but less number of infected and dead. Nevertheless, we see that the number of medical personnel in Germany is high, while the number of deaths and infections is also high. So, we conclude that the number of medical personnel shortages might not be a strong reason for the increase in numbers.

**The hospital bed availability**: From the illustration below in Fig. 6, we find that the availability of beds in hospitals may be an essential factor in influencing the number of infected and dead. As we see in both the US and Germany, the availability of beds in hospitals is small and the number increases significantly.

## *4.4 Temperature/Humidity and Wind Speed*

**The effect of Humidity on total cases**: From Fig. 7, it seems there is no clear location–humidity relation. The figure shows that humidity is relatively low in Jordan,

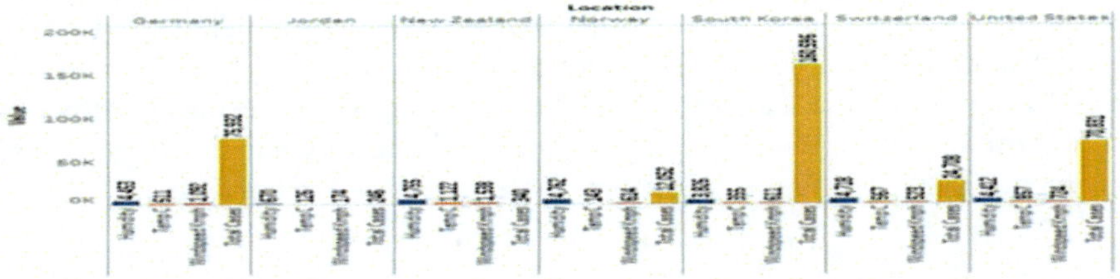

**Fig. 7** Temperature, humidity, and wind speed versus total cases per country

while humidity is always high in Norway. However, the number of cases in Norway represents 25% of the total cases in the remaining countries, which means there is no clear association between the humidity degree and the total number of cases.

**The effect of temperature and wind speed on total cases**: It seems that there is a substantial spread in Korea when the temperature is cold, but its spread does not quickly stop even when the temperature increases in Korea. Corona spread in Europe, as Germany started with relatively high, medium temperature (around 20 °C). After analysis, it appears that temperatures are not correlated with the spreading of the virus (but its confirmed cases may become fewer). Also, the wind speed is relatively low in Jordan, while wind speed is always high in New Zealand, and wind speed does not matter in coronavirus spread as shown in Fig. 7.

**The effect of Humidity, Temperature, and Wind speed on total death**: Figure 8 shows the total number of deaths and how the weather affects the exacerbation of the Coronavirus. The death rate is high in the first stage of the spread of COVID-19 in each country, but many countries are approaching about 2–3% of the death rate. It also shows that the number of deaths is high in Korea and the United States. However, the temperature in Korea and the United States decreased at the end of February 2020, whereas it was high at the end of March 2020, and the number of deaths is high. This shows that temperature is not correlated with mortality rate.

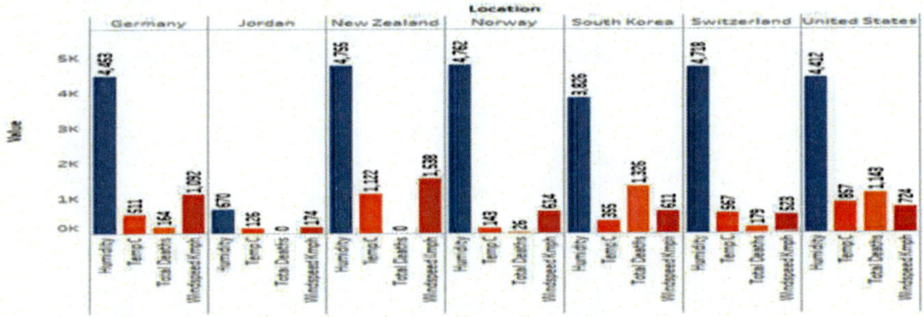


**Fig. 8** The effect of temperature, humidity, and wind speed on total death per country

On the other hand, humidity is also not related to the high mortality rate; it is high in all countries. Also, it seems that the wind speed is relatively high in Germany and New Zealand, but there is no relationship between wind speed and the prevalence of coronaviruses. So, we concluded that the population rate in countries and age control the mortality and prevalence rates more than temperature and humidity.

## *4.5 Government Responses*

Governments are taking a series of procedures and measures in their response to fighting the spread of coronavirus; these measures (indicators) are described in the OxCGRT dataset.

OxCGRT describes governments responses by using an index that takes the ordinal Containment and closure indicators (C1–C7), Income support (E1), and Public information campaign (H1) and gave each indicator a specific value and subtracted half-point if the indicator is general rather than targeted, then rescaled the final cost of the system by the maximum value to create a score between 0 and 100. This index is called Stringency Index. We choose six countries based on the number of cases: the United States, Taiwan, Jordan, Norway, New Zealand, and Switzerland. As a preliminary stage, we started trying to find a relationship between the number of confirmed cases and the stringency index. Figure 9 shows that the highest score in the stringency index is scored in Jordan in March 2020 closing measures when the first case appeared on the 3rd day of March. The stringency index was 11.1 on 15/March/2020, the second case recorded, so the stringency index increased to 33.3. On 18/March/2020, the stringency index reached the maximum with total cases of 35, and kept at its maximum until 20/April/2020. On 11/May/2020, the index turned into 83.6, and the total number of cases was 576.

In New Zealand, the first case was on 28/February/2020 with an index of 15.5, increased to 19.4 the next day after the second case was confirmed, and jumped to 37.7 on 18/March/2020 when the number of cases increased to 20. On 20/March/2020, the total cases increased to 39, and the stringency index slightly increased. It kept rising until it arrived at 97.4 on 27/April/2020 with confirmed cases equal to 1,122. On 11/May/2020, it decreased to 88.1 with total cases of 1,142. Although the number

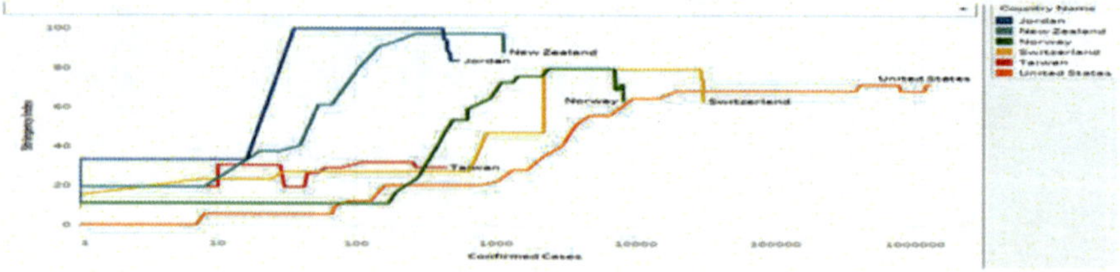


**Fig. 9** Government responses based on the stringency index

of populations in New Zealand is much less than Jordan, it has the highest number of cases. The bottleneck is about the high jump of the stringency index of Jordan after the second case. The straight spike from 33 to 100 one time contributed to curbing patients increase for months, unlike the New Zealand government, which fluctuated in reaching the highest level of stringency index. Norway was similar to Jordan in the early stages with an index of 11.1, but it continued at this level from 27/February/2020 to 9/March/2020 although the number of cases increased and became 169. It started to rise gradually to reach its maximum stringency index of 79.6 on 24/March/2020 with total cases of 2.371, but it did not last very long. On 20/April/2020, the stringency index continued decreasing until it reached 63.2 which is very low, although the number of cases was rising until 8.099 on 11/May/2020. In Switzerland, the stringency index curve went to its maximum at a value up to 79.5 with total cases of 2.200 on 17/March/2020 and started to decrease on 27/April/2020. However, the number of cases was 28.978. This increase of cases is due to the repeated low increasing of stringency index at the early stages of the spread of COVID-19. The United States did not respond within January in 2020 despite that the number of cases was six. In February 2020 with total cases of 8, it started to respond with an index of 5.5, and it maintained this value throughout February month in 2020. Although the number of cases was increasing daily, the stringency index remained slowly rising until it reached 71.6 on 8/April/2020 with total cases of 398,809. In Taiwan, the situation is different, although it is one of the countries where Corona spread early. Its stringency index reached only 32.2 as the maximum, but the number of cases on 11/May/2020 is 438. From previous observations, we noticed that the early response to the spread of COVID-19 is vital.

We observe that the most critical factor affecting both cases and deaths is the population density with a correlation equal to 0.85 and 0.81 for cases and death. The age also takes a robust positive relationship equal to 0.71 and 0.76. So, the age factor is considered an effective one because older cases take a higher mortality rate. On the other hand, some healthcare factors have a weak correlation with total cases, e.g., diabetes prevalence in each country, which equals to 0.06, the CVD death cases equal to 0.16, and bed availability equal to 0.08. However, diabetes prevalence affected the death cases more than confirmed cases. People with diabetes are more likely to die in a pandemic. As the previous figures showed, there is no correlation between total cases/death and humidity and wind speed. However, the temperature weakly correlates with both cases and death (i.e., equal to 0.15 and 0.23). It means that there is a relationship but only in some situations. Finally, the government response factor has a negative correlation, which means that the total cases decrease as the government responses increase.

## 5 Conclusion

In this paper, we observed factors that may affect the total confirmed and death cases. The population density and age factor showed a significant robust positive relationship. Furthermore, the early government response, regardless of the type of measures taken by the government, affected positively in reducing the daily number of cases. Several factors related to public health were also studied, including the local public health spending factor, the number of medical personnel (doctors and nurses), and hospital bed availability. It was observed that there is no relationship between the public health spending factor and the increasing number of cases and deaths. On the other hand, temperature, wind speed, and humidity did not affect the spread of the disease.

## References


1. Coronavirus Update (Live) (2021) 84,087,380 Cases and 1,830,453 Deaths from COVID-19 Virus Pandemic-Worldometer.Worldometers.info. https://www.worldometers.info/coronavirus/?mc_cid=5fc7bf99f2&mc_eid=39f59cd803. Accessed 1 Jan (2021)
2. Coronavirus Source Data (2021) Our world in data. https://ourworldindata.org/coronavirus-source-data. Accessed 1 Jan 2021
3. COVID19_Global_Weather_Data.Kaggle.com. https://www.kaggle.com/winterpierre91/covid19-global-weather-data. Accessed 1 Jan 2021
4. Dowd J, Andriano L, Brazel D et al (2020) Demographic science aids in understanding the spread and fatality rates of COVID-19. Proc Natl Acad Sci 117(18):9696–9698
5. Gandhi KRR, Murthy KVR, Prasada Rao SSP, Casella F (2020) Non-pharmaceutical interventions (NPIs) to reduce COVID-19 mortality. SSRN Electron J
6. Gong X, Wang Z, Chen R, Li T, Zeng D, Li M (2020) Clinical features of familial clustering in patients infected with 2019 novel coronavirus in Wuhan, China. Virus Res 286:198043
7. Hale T, Webster S, Petherick A, Phillips T, Kira B (2020) Coronavirus government response tracker. University of Oxford, Blavatnik School of Government
8. Huang Y, Lin C, Wang P, Xu Z (2020) Saving China from the Coronavirus and economic meltdown: experiences and lessons. SSRN Electron J
9. Liu W, Tao Z, Wang L et al (2020) Analysis of factors associated with disease outcomes in hospitalized patients with 2019 novel coronavirus disease. Chin Med J (Engl) 133(9):1032–1038
10. Qiu Y, Chen X, Shi W (2020) Impacts of social and economic factors on the transmission of coronavirus disease 2019 (COVID-19) in China. J Popul Econ 33(4):1127–1172
11. Remuzzi A, Remuzzi G (2020) COVID-19 and Italy: what next? The Lancet 395(10231):1225–1228
12. World Bank WDI 2.12 - Health Systems. Kaggle.com. https://www.kaggle.com/danevans/world-bank-wdi-212-health-systems. Accessed 1 Jan 2021

# Performance Comparison of Two Java-Based Actor Systems

**Franco Cicirelli, Christian Nigro, Libero Nigro, and Francesco Pupo**

**Abstract** This paper belongs to a research work that develops a deterministic actor framework in Java named Theatre, which addresses the modeling, analysis, and implementation of time-critical, cyber-physical systems. Theatre distinguishes from other actor systems by being control-based, time-sensitive, and centered on not thread-based actors. Although its main orientation to real-time embedded systems, Theatre can also be exploited for building high-performance applications on nowadays multi-core machines with shared memory. This paper reports an experience of performance comparison of Theatre with the Akka actor toolkit, through a parallel and scalable version of the Iterated Prisoner's Dilemma multi-agent system. Akka is a well-known library implemented in Scala and Java, which efficiently supports classical actors and it is best suited to untimed distributed applications. The experimental work confirms the lightweight nature of Theatre and its concurrent/parallel lock-free computational model, are also a key for achieving high execution performance, even better than that provided by Akka and similar toolkits, when large-scale systems are considered.

**Keywords** Actors · Asynchronous message passing · Java · Theatre · Akka · Multi-core machines · High-performance computing · Scalable parallel iterated prisoner's dilemma

---

F. Cicirelli (✉)
CNR - National Research Council of Italy - Institute for High Performance Computing and Networking (ICAR), 87036 Rende, CS, Italy
e-mail: f.cicirelli@icar.cnr.it

C. Nigro
Independent computing professional, Rende, Italy
e-mail: christian.nigro21@gmail.com

L. Nigro · F. Pupo
DIMES, University of Calabria, 87036 Rende, CS, Italy
e-mail: l.nigro@unical.it

F. Pupo
e-mail: f.pupo@unical.it

 
X.-S. Yang et al. (eds.), *Proceedings of Sixth International Congress on Information and Communication Technology*, Lecture Notes in Networks and Systems 216,
https://doi.org/10.1007/978-981-16-1781-2_9

# 1 Introduction

In the last years, a paradigm shift emerged in the development of concurrent/parallel applications designed to exploit the computing potential of today's multi-core machines with shared memory. In fact, the traditional concurrency model based on threads and locks [1, 2] (in the forms of semaphores or monitor locks) used to prevent interference among threads competing in the use of shared data variables, is more and more criticized and felt unsuited to ensure correctness and scalability to concurrent software systems. Not only do locks contribute to a degradation in the program execution performance, but pairing them with general complex and difficult to reason upon thread interleavings, can be error-prone to the point of causing well-known problems of deadlock, starvation, etc.

On the contrary, the actor model [3] is growing in popularity as an effective alternative for programming concurrent systems without locks. Actors are modular concurrent units with a hidden data status that can only be modified by receiving and processing asynchronously exchanged messages. Many actor implementations currently exist in such languages like Java, C++, Scala, etc. Java, in particular, proves to be highly adequate to host actors because of its object-oriented support and the availability of features such as multi-threaded programming, object serialization, computational reflection, distributed systems programming, powerful library of high-level concurrency mechanisms.

The original contribution of this paper consists in reporting an experimental work which was carried out for comparing the execution performance of two Java-based actor systems on a given benchmark: Theatre [2, 4, 5] and Akka [6]. Although the two systems address different application domains, with Theatre which is more devoted to predictable and time-sensitive systems (including real-time and simulated-time systems), and Akka which is more directed to general-purpose mostly untimed distributed systems, both can be exploited to build complex parallel systems. As a significant benchmark, a scalable version of the Iterated Prisoner's Dilemma problem (IPD) [2, 7] was used, which helped highlighting some general problems arising when converting a Theatre-based actor program into an equivalent Akka one, like how to share large data structures (e.g., for statistical data collection) among concurrent actors, and how to exploit a same (pseudo) random generator in a group of actors. Experimental results confirm the lightweight character of Theatre actors and its concurrent/parallel architecture, are a key for achieving good execution performance, improving that achievable by Akka.

The remainder of this paper is as follows. First, some background information about the Actors model and some related works are presented. Then the parallel IPD case study is described, along with its scalable message load and synchronization burden. After that the achieved experimental results are discussed. Finally, conclusions are drawn with an indication of future work.

## 2 Background and Related Work

Actors were invented by Carl Hewitt [8] in the seventies of the last century, as autonomous concurrent software entities (agents) interacting each other by asynchronous message passing, for Artificial Intelligence applications. Actors were turned into an object-oriented framework by Gul Agha in his seminal PhD work [3]. The actor model is characterized by its minimal semantics: each actor has a mailbox holding the incoming messages, and a thread which extracts, if there are any, one message at a time from the mailbox and processes (reacts to) it. Actor names reduces to mailbox addresses. Actor names can be transmitted by messages and should ensure location transparency of communications. Message processing is atomic (*macro-step semantics*) and can modify the local data status, can send messages to known actors (acquaintances, including itself for proactivity), can create new actors. Actor behaviour is asynchronous and message delivery is intrinsically non-deterministic (messages can be consigned out of order with respect to the sending order). Absence of shared data among actors, implies message arguments should be passed *by-value* only.

Despite the conceptual semantic model of actors, many implementations exist which introduce variations that also depend on the adopted programming language and library ecosystem. Often an *actor system* has first to be established which ensures actors' creation receives configuration information and ultimately concrete concurrency support. Message arguments are often exchanged also *by-reference* for efficiency reasons, thus avoiding copying operations during interactions, but ensuring data sharing problems between actors do not arise.

In this paper, the focus is on Java-based implementations of actors [9]. Some notable frameworks are the following. Kilim [10] is characterized by the adoption of ultra-light-weight actors with memory isolation through a type system. A bytecode post-processor (*weaver*) replaces normal Java threads by more efficient ones, handled by a (possibly) customizable scheduler. Messages are a special category with controlled ownership. They can be mutable and be passed by-reference (zero-copy messaging) without causing data races. ActorFoundry [11] started as a genuine Java implementation of the Actors model, then moved [12] to the use of a weaver as in Kilim to provide similar efficient mechanisms. Jetlang [13] is a library of high-performance in-memory message passing, which can be used with JVM languages like Scala, Clojure or Groovy. It uses a notion of *fibers*, typed channels with publish/subscribe pattern. It ensures message delivery in a best-effort manner. Actor4J [14] aims specifically at improving the performance of message passing by a thread pool. A group of actors is mapped to a same thread, and the actors' queues are moved to the hosting thread. Two queue levels are used: a conflict-free not thread-safe used for intra-thread communications; a protected queue for inter-thread communications.

## 2.1 Akka Concepts

Akka [6] is a widespread, popular and efficient actor implementation in both Scala and Java, whose users include large companies like Intel, Twitter, LinkedIn, etc. Akka is felt as one of the most powerful performing actor frameworks over the JVM. Its design was influenced by Erlang [14], e.g., in the view of actors as lightweight processes which must be scheduled onto a thread to execute, and for the hierarchical structure (*supervision*) used for error handling. Actor naming is based on the ActorRef proxy, which ensures actor location transparency. A fundamental concept is concerned with the *Dispatcher* component, which dynamically exploits the computing services of a reusable thread pool to activate actors, thus allowing message delivery and processing, provided actors have messages in their mailbox. An actor system is first established in an application, which allows to create and configure user-defined actors. The Dispatcher can possibly be replaced. The default is based on the ForkJoinPool executor of Java (JDK 8 or subsequent versions).

Akka is currently experimented in the realization of *reactive systems* through the concept of *reactive streams*, addressing, e.g., Internet-of-Things applications by means of a *microservices* architecture.

## 2.2 Theatre Concepts

Theatre [2, 4, 5] is an original, control-based, actor implementation in Java. A reflective, possibly customizable, control layer is adopted which (transparently) regulates message scheduling and dispatching. Theatre favors the development of time-critical cyber-physical systems, by ensuring determinism [15] in the message delivery order. Theatre, though, can also be exploited for high-performance parallel computing [2], which is the focus of this paper. A parallel system is a *federation of theatres* (Java threads). Each theatre hosts a collection of actors plus an instance of the control layer. A transport layer component allows the exchange of inter-theatre messages. A time-server component is used to keep updated (through hidden control messages) a global notion of time (real or simulated time) or to discover the termination condition of an untimed concurrent system. Actor naming rests on Java object references, which enable location transparent communications. Actors are highly lightweight: they share the single thread associated with the control form. Dynamically, actors can migrate from a theatre to another. Execution performance is enhanced by the fact that a Theatre system is completely lock-free.

Two concurrency levels exist: cooperative concurrency within each theatre, where the control form dispatches one message at a time. Only when the current message processing is finished, the event-loop of the control form is re-entered, new sent messages are scheduled, and a next message is selected and dispatched. Actors belonging to different theatres, e.g. allocated to distinct cores, can instead be executed in parallel. Due to the cooperative concurrency schema, resources among the actors of a same theatre, can be safely shared without interference (similarly to actor4J [14]).

## 2.3 *Actor Programming Style*

To give a flavor of the actual actor programming style in Java, Fig. 1a, b show an excerpt of the Player class used in the Iterated Prisoner's Dilemma case study, respectively, in Theatre and in Akka.

As one can see from Fig. 1a, Theatre actors rely on Java reflection. Methods devoted to message processing (*message servers*) have the annotation @Msgsrv. Sending a message to an acquaintance actor act, is accomplished as: act.send("msgname"[, args]). Each actor act can be moved to a given theatre by an act.move(tID) operation.

In Akka, first public static classes for the expected messages are introduced, then the createReceive() method returns a receive buffer where for each matched message, a corresponding "msgsrv" method is invoked through a lambda expression.

A theatre system is configured by instantiating the Theatre class. For each theatre, its unique ID, the transport layer, time-server and the control form instances to use are specified. The run() method of the master theatre creates the initial actors and sends them their init() messages (Java constructors are not used). Finally, all the control forms are activated.

In an Akka program main method, first an actor system is established, e.g., thus:

```
ActorSystem system = ActorSystem.create("IPD_actor_system");
```

then actors are created as follows:

```
public class Player extends Actor{
   private local data and acquaintances
   @Msgsrv public void init( params ){
         initialize local data from params
   }//init
   @Msgsrv public void play(){
         do player moves in this step
   }//play
   @Msgsrv public void adapt(){
         adapt player behavior for next step
   }//adapt
}//Player
```

```
public class Player extends AbstractActor{
   private local data and acquaintances
   public static class Init{ local vars
      public Init( params ){ init local vars }
   }//Init
   public static class Play{}//Play
   public static class Adapt{}//Adapt
   public Receive createReceive() {
      return receiveBuilder()
         .match( Init.class, msg->init( params ))
         .match( Play.class, msg->play() )
         .match( Adapt.class, msg->adapt() )
         .build();
   }//createReceive
   //message servers
   private void init( params ){...}
   private void play(){...}
   private void adap(){...}
}//Player
```

**Fig. 1** **a** Theatre player actor class. **b** Akka player actor class

```
player[i][j] = system.actorOf( Props.create(Player.class));
```

where the Props class furnishes to actors some configuration data, including the Dispatcher to use. A message can be sent to a player ActorRef p (with the *at-most-once* message delivery) by the tell (*fire-and-forget*) operation: p.tell(new Player.Play(), Actor.noSender()) in the case the message sender identity is not necessary to transmit.

## 3 The Iterated Prisoner's Dilemma Case Study

The Prisoner's Dilemma (PD) is a competitive game [16] where two players have independently to decide, without any interaction, between C(ooperate or D(efect, knowing in advance that each decision implies a given gain/punishment on the basis of what was the partner decision. In standard PD the payoff of each possible pair of decisions, is as follows: (D,D)→(1,1), (D,C)→(5,0), (C,D)→(0,5), (C,C)→(3,3). It is evident that players acting "rationally" would conclude the best decision is (D,D). In fact, it would be very risky to choose C with the partner which decides D. In these cases, the cooperating player would achieve 0 and the defecting player would get the maximum benefit 5. The outcome (D,D) proves to be a Nash equilibrium for the PD game [16]. But socially, and from this derives the "dilemma" term, the choice (C,C) would be the best because it would distribute more benefits to players. Although the basic PD admits only one result, things become more uncertain when PD is iterated a number of times not known to players. It has been experimentally demonstrated by Axelrod et al. in [7], that in a population of 256 players randomly assigned to a squared matrix of $16 \times 16$ cells, the Iterated PD (IPD) can in some case tend definitely to cooperation, when each player takes its decisions against a persistent neighborhood of partners, thus reproducing some typical behavior found in biology or in today social networks. In [2] the IPD was used as a modeling example using Theatre, with the goal of showing that cooperation can also emerge in large and scalable populations, with a possible speedup with respect to a mono-core realization. In this paper, the parallel IPD (PIPD) is also used as a case study but differently from the solution in [2] which served to test specifically the parallel simulation control form of Theatre using a time-stepped simulation, also considering that Akka does not have a simulation support, the PIPD is solved using a new untimed algorithm, so as to allow a performance comparison of actors and message passing in Theatre and Akka.

First PIPD was instrumented in Theatre using an untimed organization. Then the solution was manually ported, with some necessary adaptation, to Akka.

### *3.1 A PIPD Theatre Model*

Players are stored into an $NxN$ matrix (population). Each player initially receives a strategy $(y, p, q)$, where $p$ is the probability this player will choose C knowing that the opponent had chosen C in the last game; similarly, $q$ is the probability this player chooses C knowing its partner had chosen D in the last game; $y$ is the probability of choosing C without memory of the opponent previous choice. PIPD proceeds by steps (also called *periods*). At each step, every player plays 4 times with each of its partners defined by a neighborhood relation. By convention, the partners of a player are retrieved at its north, east, south, and west. However, in a case the neighborhood is defined by Manhattan rules (the player matrix acts as a Persistent Toroidal Grid-PTG), in a different scenario the neighbors can be established randomly at the beginning of an execution (Persistent Random Network-PRN). In a third case, neighbors are defined randomly at the start of each period (Temporary Random Network-TRN). The payoff earned by each player at each period, averaged over the 16 carried games, is stored in a period_payoff data structure. PD naturally depends on a *reciprocity principle*: when A plays against B along a certain direction, B also is playing with A in the dual direction. Therefore, player A can determine the payoff of both A and B along a direction, and store it accordingly. Since actors are encapsulated units, no data sharing is possible. A data parallel schema was then devised, where some basic data structures (period_payoff, player strategies < y,p,q >, last choice of each partner in previous game of the same period, and so forth) are placed in a class G of global data, and ensuring that actors, without the use of locks, always accesses common data in a safe and consistent way.

The PIPD model is composed of the Manager and Player actor classes plus the classes G and Main. The model is split into $P$ theatres, where $P$ is the number of available cores ($P = 4$ in the experimental framework). Each theatre is then assigned a squared subblock of $(N/p)$x$(N/p)$ players, where $p = \sqrt{P}$. $P$ Manager actors are responsible of managing the corresponding portions of the player territory. One manager is elected as the master one. When a slave manager finishes initializing all its dependent players, or terminates the computation of the current step, it informs the master manager by a *done* message. When all the managers have terminated their actions at current step, the master manager commands moving to the next step. Steps are continued until a max value STEPS is reached. Basically, every manager handles two phases at each step. In the first phase, all its assigned players play standard PD with its partners. In the second phase, players adapt their behavior (strategy $(y, p, q)$) on the basis of the best performing player in the neighborhood. When a player finishes its playing or adapting phase, it sends a *done* message to its manager. To avoid conflicts on global shared data, at each individual game, the player who belongs to the highest priority theatre (or manager), i.e., having the lesser ID, actually conducts the game and modify global data, with the partner which skips this game. Another design issue concerned the update of statistical data at the end of each period. This task was assigned to the master manager only.

For statistical coherence with a mono-core IPD realization [2], all the players of a same theatre/manager, share a same (pseudo) random generator (RG). Toward this, a not thread-safe RG class was implemented, one instance of which is passed to all the players governed by a given manager. Sharing this object is completely safe due to the cooperative concurrency existing within a same theatre.

### 3.2 *Porting the PIPD Model to Akka*

The PIPD model was adapted to work with Akka (see also Fig. 1). Since Akka actors run on distinct threads, independently from the belonging manager region, the precedence constraint which establishes the player, in a game, which actively plays was first based on the manager ID and for the same manager it is based on the player unique identifier. A second adaptation was concerned with the random generator object shared by all the players assigned to a manager. Of course, the not thread-safe RG class developed for Theatre cannot be used by multiple actors/threads. On the other hand, the use of a thread-safe class as Random of Java, is unthinkable because a Random object would introduce so much contention to the point of determining an intolerable performance degradation. The solution was the use of a SplittableRandom object which, as stated in the Java APIs, with "high probability", would deliver almost the same behavior of a shared generator within a manager/region.

Whereas the PIPD model effectively uses $P$ threads (one per core), the corresponding Akka model is open to exploit to the larger extent the parallelism available on a multi-core machine, including that arising from the hardware hyperthreading. The Akka model is managed (compiled and executed) through the mediation of the maven Java project builder tool.

## 4 Experimental Results

The two PIPD models, based on Theatre and on Akka, were executed with $P = 4$ (number of cores) and $N$ (dimension of the player matrix) which varies from 16 to 512, by power of 2. The number of $STEPS$ of each experiment was set to 3000 as in [2, 7]. The expected output as reported in [2, 7] was checked preliminarily on all the models, thus retrieving that under PTG/PRN the PIPD evolves toward cooperation, whereas under TRN it evolves toward defection. Then, for each scenario, the elapsed time (ET) required by the two versions of the model (actually the average of five runs) was measured and the relative speedup $Akka_{ET}/Theatre_{ET}$ computed. All the experiments were carried out on a Win 10 machine, Intel Core i7-7700 CPU@3.60 GHz, 32 GB Ram, using Java 8, Akka 2.6.10 with maven 3.5.1. The experiments were also repeated using an optimized version of the Theatre model, where the number of internal messages within theatres is minimized. In fact, there is no need for each player to send back a done() message at the end of a play() or

**Table 1** Performance of Theatre PIPD versus Akka PIPD (4 cores)

| $N$ | $Akka_{ET}(msec)$ | $Theatre_{ET}(msec)$ | $Speedup$ | $Theatre_{ET}^{Opt}(msec)$ | $Speedup^{Opt}$ |
|---|---|---|---|---|---|
| 16 | 576 | 434 | 1.33 | 334 | 1.72 |
| 32 | 1524 | 1109 | 1.37 | 809 | 1.88 |
| 64 | 5500 | 3765 | 1.46 | 2479 | 2.22 |
| 128 | 23,450 | 13,474 | 1.74 | 9324 | 2.52 |
| 256 | 94,129 | 56,129 | 1.68 | 40,453 | 2.33 |
| 512 | 368,186 | 222,572 | 1.65 | 162,720 | 2.26 |

adapt() message server. A shared and safe counter per theatre/manager was used which is reset before a play or adapt phase, then is incremented at the end of a play()/adapt(). Only the last player in a theatre actually sends a done() message to its relevant manager. Such a solution cannot be adopted in Akka without the use of some form of locking. Experimental results are collected in Table 1.

As one can see from Table 1, the basic Theatre model improves the performance achieved by the Akka model. The optimized Theatre model, though, performs even better by delivering a relative speedup of 2.52 when the matrix size $N$ is 128, as a consequence of the reduced message passing load. It is to be noted that the PIPD model suffers from a sequential part, at each period, due to statistical data updating in the master manager.

## 5 Conclusions

Theatre [2, 4, 5] is an actor framework in Java designed for modeling, analysis, and implementation of cyber-physical systems with timing constraints. Theatre can also be used on a multi-core machine for high-performance computing. This paper reports an experience comparing the execution performance of Theatre and Akka [6] using a parallel, scalable and untimed version of the Iterated Prisoner's Dilemma [2, 7]. The experience confirms the actor and message passing architecture of Theatre are capable of delivering a good execution performance, better than Akka.

On-going work is directed to:

- Improving the support of Theatre for parallel computing;
- Implementing in Java and Rust a deterministic version of Theatre [15] for cyber-physical systems;
- Porting to Theatre a message receive mechanism like that of Akka, which is based on streams and lambda expressions, and checking the execution benefits which can be gained by replacing the current use of Java reflection during message sending and dispatching, which contributes to clean code and programming elegance;
- Experimenting with the use of Theatre actors in the development of a microservice reactive architecture to be exploited, e.g., in Internet-of-Things applications.

## References


1. Lee EA (2006) The problem with threads. Computer 39:33–42
2. Nigro L (2021) Parallel theatre: an actor framework in Java for high-performance computing. Simul Model Pract Theory 106 https://doi.org/10.1016/j.simpat.2020.102189
3. Agha G (1986) Actors: a model of concurrent computation in distributed systems. PhD Thesis, MIT Artificial Intelligence Laboratory
4. Nigro L, Sciammarella PF (2018) Qualitative and quantitative model checking of distributed probabilistic timed actors. Simul Model Pract Theory 87:343–368
5. Cicirelli F, Nigro L, Sciammarella PF (2020) Seamless development in Java of distributed real-time systems using actors. Int J Sim Process Modelling 15(1/2):13–29
6. Java Akka https://doc.akka.io/docs/akka/current/actors.html?language=java
7. Cohen MD, Riolo RL, Axelrod R (1999) The emergence of social organization in the Prisoner's Dilemma: how context-preservation and other factors promote cooperation, Santa Fe Institute Working Paper 1999–01–002, also Psycology
8. Hewitt C, Bishop P, Steiger R (1973) A universal modular Actor formalism for artificial intelligence. In: 3rd International Joint Conference on artificial intelligence (IJCAI), pp 235–245
9. Karmani RK, Shali A, Agha G (2009) Actor frameworks for the JVM platform: a comparative analysis. In: 7th International Conference on principles and practice of programming in Java (PPPJ '09), pp 11–20
10. Srinivasan S, Mycroft A (2008) Kilim: isolation typed actors for Java. InL Proceedings of the European Conference on object oriented programming (ECOOP)
11. Astley M (1998) The ActorFoundry: a Java-based actor programming environment. University of Illinois at Urbana-Champaign, Open Systems Laboratory
12. Frantz C, Nowostawski M, Purvis M (2010) Multi-agent platforms and asynchronous message passing: frameworks overview. Tech Rep
13. Rettig M (2020) Jetlang https://github.com/jetlang. Accessed Nov 2020
14. Bauer DA, Mäkiö J (2018) Actor4j: a software framework for the actor model focusing on the optimization of message passing. In: AICT 2018, 14th Advanced International Conference on Telecommunications. IARIA
15. Cicirelli F, Nigro L (2020) Using deterministic theatre for energy management in smart environments. Chapter of forthcoming Springer book "Intelligent Sustainable Systems", Springer ASST Series on Advances in Sustainability Science and Technology, ISSN:2662–6829
16. Wooldridge M (2009) An introduction to multi-agent systems. 2nd edn. John Wiley & Sons

# E-government Research in Africa: A Systematic Review

**Acheampong Owusu and Obed Kwame Adzaku Penu**

**Abstract** E-government systems have been implemented by many governments the world over as a move for citizens to access public services via single window digital platforms. In this study, a systematic review of research that has been done on e-government within the context of Africa was conducted by synthesizing scholarly papers published between 2010 and 2020. A total of 77 papers were retrieved from the targeted electronic databases including ScienceDirect, Google Scholar, Taylor and Francis, Emerald, Elsevier, Online Wiley, Scopus, and Web of Science based on a search criterion. Further inclusion and exclusion criteria downsized the papers to 39 which were used for the actual review. Major findings were classified into research themes, theories/frameworks, research methods, and areas that could be explored in future research. Three dominant research themes identified were on adoption, implementation challenges and challenges with access to e-government services, and the General theme on e-government. In terms of theories/frameworks used, most papers did not use any model; also, dominant research methods are the quantitative method, followed by the qualitative method. This study serves as a roadmap for understanding the research literature on e-government research in Africa.

**Keywords** E-government · ICT · Systematic review · Africa

## 1 Introduction

Information and communication technologies (ICT) have changed the operational and structural processes of public sector institutions in both the Global North and South [5]. ICTs have become a part of everyday tasks [12]. One of such changes brought about by ICTs is its institutionalization in government processes in what

A. Owusu (✉) · O. K. A. Penu
University of Ghana Business School, Legon, Accra, Ghana
e-mail: aowusu@ug.edu.gh

O. K. A. Penu
e-mail: okapenu@st.ug.edu.gh



X.-S. Yang et al. (eds.), *Proceedings of Sixth International Congress on Information and Communication Technology*, Lecture Notes in Networks and Systems 216,
https://doi.org/10.1007/978-981-16-1781-2_10

is termed "Electronic Government". Electronic government (E-Government or e-government) refers to the use of information technology (IT) to render services to citizens and businesses as well as civil society groups at the local, national, and international levels [10, 15]. For many governments, the essence of deploying e-government systems is to make public services more accessible to citizens [18]. Aside from efficiency in accessing public services, e-government systems create benefits by reducing manual and paperwork-based processes, ensuring continuous service delivery availability to citizens, improved response time, and avoidance of mistakes and errors in rendering such public services [17]. In other jurisdictions, especially in developing economies, e-governance bridges the digital divide and increases the adoption rates of government and public services and policies among citizens [21]. It further serves as a means to check acts of corruption and poor public management, and the cost-effective way of interaction between government agencies and their elements by delivering access to services and information via digital platforms [20].

According to Gupta et al. [6], the offer of e-government service makes citizens and businesses access services such as e-transportation, e-procurement, e-tax, e-participation, e-health among others. This suggests that the concept of e-government can be implemented for different facets of a country's economy (transportation, health, procurement, taxes, energy, agriculture, etc.). However, in recent years, there has been limited amounts of research and knowledge regarding e-government [13].

Arguably, no systematic review has been done on papers that discuss e-government in Africa holistically. Authors like Adjei-Bamfo et al. [1] for example, systematically reviewed the literature on "the role of e-government in sustainable public procurement in developing countries" but limit their review to sustainable public procurement (SPP). This necessitates the need to undertake a systematic review that holistically looks at papers done on e-government in Africa without a focus on a specific area. This study, therefore, attempts to fill this gap in research by presenting level research done on e-government in the African context.

The rest of the paper is structured as follows: a methodology section that shows the method applied in carrying out this systematic review follows. In this same section, the criteria for inclusion and exclusion of papers that are used in the final analyses for the study are presented. Next is the section that shows the findings from the analysis of the papers. This section touches on the central themes regarding e-government research in Africa, and the methods employed in research regarding e-government in Africa. There is also a presentation on the types of e-government technologies, models, and frameworks used in undertaking the studies. The final section concludes the study and presents the limitations and recommendations for future research regarding this particular study.

## 2 Methodology

In this research, papers are synthesized to identify emerging themes, methods, and gaps for future research regarding e-government research.

### 2.1 Data Sources

The focus of the search done was on papers published over the past decade (January 2010–May 2020). The papers were sourced from electronic databases including ScienceDirect, Google Scholar, Taylor and Francis, Emerald, Elsevier, Online Wiley, Scopus, and Web of Science. The focus of the collected papers was on only peer-reviewed journal articles and conference papers, as such, book chapters, blogs, thesis, and magazines were eliminated from the study.

### 2.2 Inclusion and Exclusion Criteria

A total of 77 papers were downloaded from the targeted databases. These 77 papers were then scrutinized to ensure they were usable in the study. For a paper to be usable in the study, it must discuss the concept of e-government, have a focus on Africa, be a peer-reviewed journal article or a conference paper (doctorate, masters, and undergraduate thesis, magazines, textbooks, and reports were not added to the list of papers for the study) and publications must be between 2010 and 2020. After the papers were scrutinized for inclusion and exclusion, there were 39 papers remaining which formed the final number of papers that were analyzed for the current study.

### 2.3 Search and Selection Process

The search process for the papers was done under the guidance of the inclusion and exclusion criteria enumerated above. The processes for searching and selecting the papers began with the identification of keywords and terms related to e-government. The keywords used were identified to be based on the terms adopted by Rana et al. (2015). The words were: "'electronic government', 'e-government', 'e-gov', 'egov', 'digital democracy', 'online government'", and perming them with the words "'adoption', 'acceptance', 'usage', 'implementation', 'impact', and 'diffusion'" by using the logical AND, and OR functions available in the databases that were searched. The search included the restriction between the years 2010 and 2020.

## 3 Findings and Discussion

The subsections presented under this section are the year of publication, themes derived from the research, theories, research methods, and themes for future research.

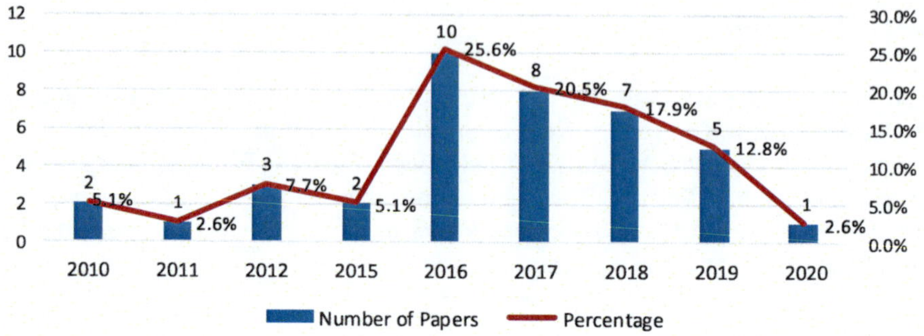


**Fig. 1** Year of publication

## 3.1 *Year of Publication and Outlets*

This section was used to classify the papers according to the year in which they were published. The classification indicates a stable rise in the number of publications from 3 in the year 2012 with a slight decline to 2 papers in the year 2015. Before 2012, there were 2 papers in 2010 and a decline to 1 paper in 2011. The number, however, shot up to 10 papers in 2016, but again went down steadily to 8 papers in the year 2017 and 7 papers in 2018. There was a further reduction to 5 papers in 2019 and 1 paper in 2020. It is, however, worth noting that papers used in the analysis were collected up until May 2020; hence, the year 2020 could witness an increase in the number of papers into the later part of that year. In Fig. 1, the number of papers according to the years in which they were published is presented.

## 3.2 *Research Themes*

Under this sub-section, the papers were analyzed and grouped to identify the central issue that they sought to discuss, or each study intends to investigate. Analysis of the papers identified 13 themes. These are papers that discussed the design, adoption, benefits, impact or effect, maturity, implication, neglect, investigations (forensics), sustainability, and management of e-government. Other themes that were identified include implementation and access challenges regarding regards to access to e-government platforms or services. Papers that talked about e-government, in general, were classified as general. For example, regarding challenges with regards to access, Agangiba et al. [3] conducted a qualitative study with a focus on Ghana to examine the experiences of persons who have visual impartments' ability to access e-government service to inform stakeholders toward improving inclusivity and accessibility of E-government services for persons who have some form of disability. Another example of Investigations (Forensics) of e-government is the study by Kigwana et al. [8] in proposing a digital forensic investigation framework for e-government structure.

**Table 1** Research themes

| Theme | Number of papers | Percentage (%) |
|---|---|---|
| Design | 1 | 3 |
| Adoption | 10 | 26 |
| Benefits | 1 | 3 |
| Impact/Effect | 4 | 10 |
| Challenges (Implementation) | 6 | 15 |
| Maturity | 1 | 3 |
| Implications | 1 | 3 |
| Challenges (Access) | 5 | 13 |
| Neglect | 1 | 3 |
| Investigations (Forensics) | 1 | 3 |
| Sustainability | 1 | 3 |
| Management | 1 | 3 |
| General | 6 | 15 |
| Total | 39 | 100 |

Table 1 shows that the three dominant themes discussed in the research area pertain to adoption, implementation challenges and challenges with access to e-government services, and the General theme on e-government. Adoption papers were 10 (26%), papers on implementation challenges were 6 (15%), there were also 6 (15%) papers that discussed e-government in general while papers that discussed access challenges were 5 (13%).

### 3.3 Nature of the Platform

Papers were also categorized to determine the nature of the e-government platform they sought to discuss. The findings of the categorization show that the 2 leading areas were papers that discussed e-government in general hence were categorized as "General platform" (41.0%). This was followed by papers that were done on ICTs (12.8%), and then papers that discussed Single window platforms for government services (5.1%). The breakdown of the nature of the platform is presented in Table 2.

### 3.4 Geographical Context of the Study

The papers were also analyzed to identify the countries within Africa that they focused on. This was collectively termed as the "Geographical Context of the Study".

**Table 2** Nature of the platform

| Type of platform | Number of papers | Percentage (%) |
|---|---|---|
| General platform | 16 | 41.0 |
| Physical infrastructure | 1 | 2.6 |
| National registration platform | 1 | 2.6 |
| Payment platform | 1 | 2.6 |
| ICTs | 5 | 12.8 |
| Records management systems | 1 | 2.6 |
| Single window platform for government services | 2 | 5.1 |
| e-Citizen portal | 1 | 2.6 |
| Independent E-government portals | 1 | 2.6 |
| e-filing and e-payment | 1 | 2.6 |
| Mobile phone aided e-Government service | 1 | 2.6 |
| e-Government Project Management platform | 1 | 2.6 |
| e-Government websites | 1 | 2.6 |
| e-Public Procurement | 1 | 2.6 |
| National examination information platform | 1 | 2.6 |
| e-Centers | 1 | 2.6 |
| DF investigators (DFi) Tools on e-Government platforms | 1 | 2.6 |
| e-Filling (Tax filling) | 1 | 2.6 |
| e-Learning | 1 | 2.6 |
| Total | 39 | 100.0 |

The findings show that the papers that were reviewed were done mainly within the context of "More than One African Country" and "South Africa". Following this were papers done with the focus on individual African countries such as Ghana, Kenya, Zambia, Nigeria, Uganda, Gambia, Ethiopia, Rwanda, Malawi, and Cameroon. Papers that were placed in the category of Africa and Other Continents are those that focused on at least one African country and any other country outside of Africa. Papers that were done with a focus on "More than One African Country" were 7 (17.9%), with 7 (17.9%) papers on South Africa. There were 4 (10.3%) papers on Zambia, 4 (10.3%) on Kenya, 3 (7.7%) papers on Ghana, 3 (7.7%) papers on Tanzania, and 2 (5.1%) papers on Nigeria. Papers that were placed in the "Africa and Other Continents" category were 2 (5.1%). The remaining countries (Gambia, Ethiopia, Rwanda, Uganda, Malawi, and Cameroon) also had 1 (2.6%) paper each. Figure 2 is a graph of the publications according to the geographical context of the study.

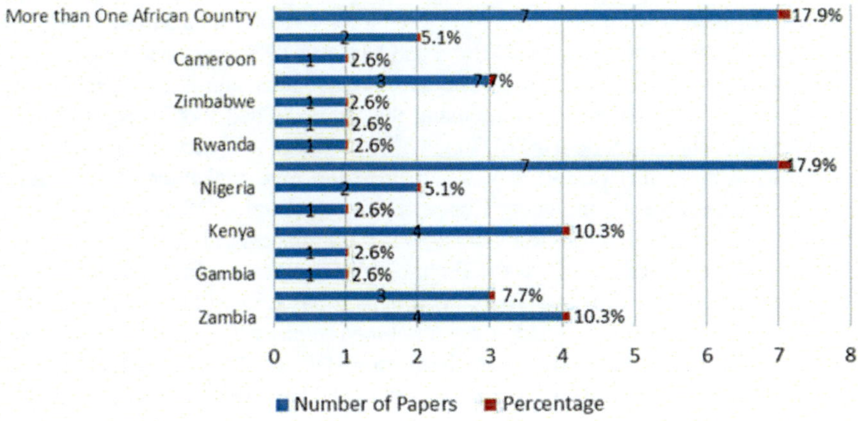


**Fig. 2** Geographical context of the study

## 3.5 *Underlying Theories/Frameworks/Models*

Papers were also classified according to the theories, models, or frameworks that were used in undertaking the studies as shown in Fig. 3. Analysis of the papers brought out nine underlying lenses as a blueprint for undertaking the studies. These are "The Technology Acceptance Model (TAM)", "Unified Theory of Acceptance and Use of Technology (UTAUT)", "Delone and McLean IS Success Model", "Unified Theory of Acceptance and Use of Technology 2 model (UTAUT2), Sen's Capability Approach, Contextual Interaction Theory", "Conceptual models and papers that combined models (Combination of Models)". Also, papers that did not use any model were classified in this study as "Not applicable".

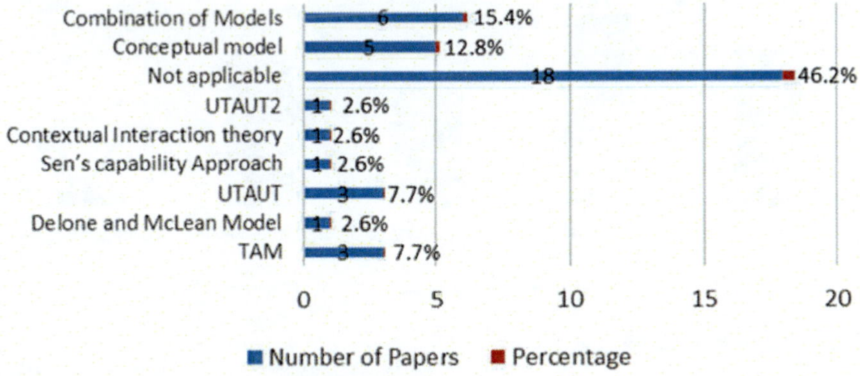


**Fig. 3** Underlying Theories/Frameworks/Model

3 (7.7%) papers were done using the Technology Acceptance Model (TAM), 1 (2.6%) paper was done using the Delone and McLean IS Success Model, 3 (7.7%) papers were done using the Unified Theory of Acceptance and Use of Technology (UTAUT), 1 (2.6%) paper was done using the Unified Theory of Acceptance and Use of Technology 2 model (UTAUT2), 1 (2.6%) paper was done using the Sen's Capability Approach, 1 (2.6%) was done using the Contextual Interaction Theory, 5 (12.8%) papers were done using Conceptual models and 6 (15.4%) papers were done by combining two or more models. 18 (46.2%) papers did not use any model. Some examples of models such as the Technology-Organisation Environment (TOE) framework with a decomposed Theory of Planned Behaviour (TPB) and the Technology Acceptance Model (TAM) were combined in one study (see [23]). Another example is the combination of the Modernization Theory (MT) to complement the Political, Economic, Social, and Technological (PEST) analysis framework (see [7]).

### 3.6 Underlying Methods

Papers were also analyzed to identify the methodologies used in undertaking the studies. Five classifications were identified under this section. These are the qualitative, quantitative, mixed-method, literature review, and conceptual studies. Qualitative papers were 10 (25.6%), quantitative papers were 12 (30.8%), mixed-method papers were 7 (17.9%), literature review had 9 (23.1%) papers and the conceptual paper was 1 (2.6%). These findings are an indication that most of the papers were done using the quantitative method, followed by the qualitative method, literature review, mixed-method, and conceptual papers. Figure 4 shows the papers based on the research methods that were used.

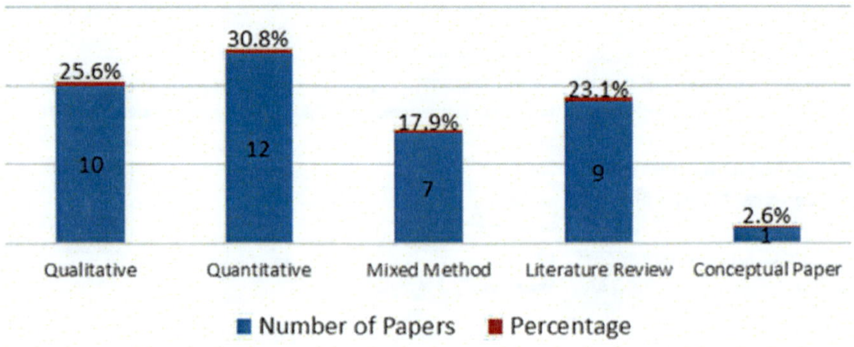


**Fig. 4** Papers by methods

# 4 Identified Gaps for Future Research

Under this section, the researchers review these gaps and put them under themes. This is valuable in guiding future research on e-government, especially within the context of Africa. In all, three future research themes were identified in this study. The first was on the need to use other methodologies and sample sizes. Second is the need for the theorization of e-government studies and, the third is the need for more studies on challenges in implementing and accessing e-government platforms.

## *4.1 Using Other Methodologies and Larger Sample Sizes*

Regarding the adoption of other methodologies and the use of larger sample sizes, researchers like Bwalya and Mutula [4] using the mixed-methods research approach, explored the contextual factors that influence the design of the e-government systems and proposes a conceptual framework with due reference to local contextual characteristics. They, however, suggest that a lot more research that takes into consideration larger sample sizes can be carried out to provide a better understanding of the influencing factors for e-government adoption at the individual and institutional levels. Similarly, Kim et al. [9], also suggest that studies intended to be done in the future can provide more generalized explanations through larger samples sizes on why certain sequences generate certain outcomes, and why certain interventions that are provided during the maturity stages of e-government implementation may either succeed or fail. Further, a study by Mishra and Geleta [11] investigate how perceived service delivery mediates the relationship between citizens' satisfaction and e-government. Even though their study points to a non-significant direct effect of citizen's satisfaction but identifies a complete mediation of perceived service delivery on the effect of e-government and call for future studies to focus on extended sample size. Additionally, researchers like Otieno and Omwenga [14] and Yavwa and Twinomurinzi [22] aver that e-government studies have been done using small sample sizes which limits the findings of the study in terms of generalization and recommend further study in other regions and larger sample sizes.

## *4.2 Theorization of E-government Studies*

Regarding the theorization of e-government studies, a researcher such as Ifinedo [7] for example explores the drivers of e-government maturity in two Latin America and Sub-Saharan Africa using the Modernization theory (MT) and the political, economic, social, and technological (PEST) analysis framework. Even though the study clearly enunciates the relevance of the PEST analysis framework for investigating innovation in technology across different geographical contexts, the study suggests studies in the future to draw and expand on other contemporary theoretical frameworks and concepts in discussing e-government adoption. Researchers like Agangiba and Kabanda [2] also recommend that future research be done using theoretical models that can provide a better understanding of the relationship between persons with disabilities (PWDs) society and, technology in making e-government more accessible. There are also further calls regarding the theorization of e-government research by Bwalya and Mutula [4] toward developing adoption models at individual and institutional levels to explicitly explore the influencing factors for the adoption, testing, and validation of the models as reference points for implementations by governments.

### *4.3 More Studies on Challenges in Implementing and Accessing E-government Platforms*

Twizeyimana et al. [19] investigate the implementation challenges of e-government in Rwanda to digitalize all government to consumer and government to business platforms. Even though the study significantly identifies six main categories of implementation challenges, they recommend future studies to comparatively explore the development, management, and governance of e-government in other African countries. Another avenue for future research is the challenges related to setting up such huge public–private-partnership (PPP) initiatives regarding e-government studies as there is the need to understand the project from the view of citizens with a focus on the use and impact. Similarly, Agangiba and Kabanda [2] after performing a systematic review on E-government accessibility research trends in developing countries, call for similar researchers to investigate the perceptions of PWDs regarding accessibility to e-government services which will provide a more holistic solution that befits the context of developing countries. Also, Bwalya and Mutula [4] call for research future research that explores the difficulties of using the e-government systems.

## 5 Conclusion, Implications, and Limitations

This study undertakes a systematic review of e-government publications in Africa by performing a categorization and analysis of 39 papers to determine the current level of the study done over the past ten years. This study focuses on three major objectives: (1) to identify the themes discussed in e-government research in Africa (2) to identify the research methods that have been used in e-government research in Africa, and (3) to identify the areas for future research regarding e-government research in Africa. Additionally, we present the year, in which the papers were published, levels of analysis, and the models and frameworks used in undertaking the studies.

This study, having explored the objectives it set at the beginning of the paper, has some implications. First, the paper contributes to the literature on e-government within the context of Africa. Second, the study through the identified research gaps provides areas for the researchers to consider in future research. The implications of this study to policymakers and practitioners are that it can better inform them about the adoption and impact of e-government. Also, further research can be explored in terms of research such as the use of frameworks, and methodologies in e-government research. In terms of lapses, what tends to be lacking is research on e-government backed by frameworks/theory/models since most of the studies that have been done have not used any framework/theory or model.

One of the limitations has to do with the number of papers that were analyzed for the study as the focus was only on peer-reviewed journals and conference papers. A lot more document types such as reports can be reviewed to provide a wider perspective of the study. Also, papers used in the study were exclusively extracted from ScienceDirect, Google Scholar, Taylor and Francis Emerald, Elsevier, Online Wiley, Scopus, and Web of Science. Future studies can consider databases like IEEE Xplore and ACM. However, with these limitations mentioned, we believe that the journal databases and the literature used are pertinent to meeting the objectives of the study.

## References


1. Adjei-Bamfo P, Maloreh-Nyamekye T, Ahenkan A (2019) The role of e-government in sustainable public procurement in developing countries: A systematic literature review. Resour Conserv Recycl 142:189–203
2. Agangiba M, Kabanda S (2016) E-Government accessibility research trends in developing countries. In: MCIS, p 8
3. Agangiba M, Kabanda S, Brown I (2018) Experiences of persons with disabilities in accessing e-government services in Ghana. Services in Ghana (2018). Research Papers. 69
4. Bwalya KJ, Mutula S (2016) A conceptual framework for e-government development in resource-constrained countries: the case of Zambia. Inf Dev 32(4):1183–1198
5. Furtado V, Lima L, Chagas DA, Pinheiro V, Caminha C, Furtado E, Mafra M (2019) E-Totem, digital locative media to support E-participation in cities. Int J Electron Gov Res (IJEGR) 15(3):1–20
6. Gupta KP, Singh S, Bhaskar P (2016) Citizen adoption of e-government: a literature review and conceptual framework. Electron Gov Int J 12(2):160–185
7. Ifinedo P (2012) Drivers of e-government maturity in two developing regions: focus on Latin America and Sub-Saharan Africa. JISTEM-J Inf Syst Technol Manage 9(1):05–22
8. Kigwana I, Kebande VR, Venter HS (2017) A proposed digital forensic investigation framework for an e-government structure for Uganda. In: 2017 IST-Africa Week Conference (IST-Africa) (pp 1–8). IEEE
9. Kim HJ, Pan G, Pan SL (2007) Managing IT-enabled transformation in the public sector: a case study on e-government in South Korea. Gov Inf Q 24(2):338–352
10. Kumar S, Baishya K, Sadarangani PH, Samalia HV (2020) Cultural influence on e-government development. Electron J Inf Syst Eval 23(1):17–33
11. Mishra SS, Geleta AT (2020) Can an e-government system ensure citizens' satisfaction without service delivery? Int J Public Adm 43(3):242–252

12. Moreno L, Martínez P (2019) Accessibility compliance for e-government websites: laws, standards, and evaluation technology. Int J Electron Gov Res (IJEGR) 15(2):1–18
13. Norris DF, Lloyd BA (2006) The scholarly literature on e-government: characterising a nascent field. Int J Electron Gov Res (IJEGR) 2(4):40–56
14. Otieno I, Omwenga E (2015) Citizen-centric critical success factors for the implementation of e-government: a case study of Kenya Huduma centres. In: 2015 IST-Africa Conference (pp 1–9). IEEE
15. Purwanto A, Zuiderwijk A, Janssen M (2020) Citizen engagement with open government data. Transforming government: people, process and policy 14(1):1–30
16. Rana NP, Dwivedi YK (2015) Citizen's adoption of an e-government system: validating extended social cognitive theory (SCT). Gov Inf Q 32(2):172–181
17. Singh H, Grover P, Kar AK, Ilavarasan PV (2020) Review of performance assessment frameworks of e-government projects. Transforming Gov People Process Policy 14(1):31–64
18. Thompson N, Mullins A, Chongsutakawewong T (2020) Does high e-government adoption assure stronger security? Results from a cross-country analysis of Australia and Thailand. Gov Inf Q 37(1):1–9
19. Twizeyimana JD, Larsson H, Grönlund Å (2018) E-Government in Rwanda: implementation, challenges and reflections. Electron J E-Government 16(1):19–31
20. Vragov R, Kumar N (2013) The impact of information and communication technologies on the costs of democracy. Electron Commer Res Appl 12(6):440–448
21. Weerakkody V, El-Haddadeh R, Al-Sobhi F, Shareef MA, Dwivedi YK (2013) Examining the influence of intermediaries in facilitating e-government adoption: an empirical investigation. Int J Inf Manage 33(5):716–725
22. Yavwa Y, Twinomurinzi H (2018) Impact of culture on e-government adoption using UTAUT: a case Of Zambia. In: 2018 International Conference on eDemocracy and e-government (ICEDEG) (pp 356–360). IEEE
23. Ziba PW, Kang J (2019) Factors affecting the intention to adopt e-government services in Malawi and the role played by donors. Inf Dev 1–21

# Control to Manipulate Robotic Arms Using Augmented Reality

**Edgar I. De la Cruz, Edgar R. Salazar, Jonathan A. Romero, Leydi M. Jiménez, and Jacson J. Rodríguez**

**Abstract** This work presents an application for mobile devices in augmented reality, as a tool for teaching technological learning in the handling of industrial robotic arms, advanced control algorithm techniques are implemented, the simulation is performed from the application allowing the user, select the different desired trajectories, visualize the behavior of the robotic arm in the monitoring of the proposed path, in addition to animations to understand and become familiar with the operation of the same, as well as the visualization of control errors in each simulation performed. The application developed is focused on training and assistance of industrial robotic arms allowing users to interact and manage the structure of the robot in the educational sector, within an environment of friendly and intuitive augmented reality, providing new technologies with digital information immersion simulation programs. The tests conducted on the mobile application demonstrate the intuition and ease of use with the user, in understanding the operation and functioning of programmable robotic manipulators.

**Keywords** Augmented reality · Robotic manipulator · Advanced controls · 3D animation · Control by numerical methods · Unity 3D

E. I. De la Cruz (✉) · E. R. Salazar · J. A. Romero · L. M. Jiménez · J. J. Rodríguez
Universidad de Las Fuerzas Armadas ESPE, Sangolquí, Ecuador
e-mail: eide@espe.edu.ec

E. R. Salazar
e-mail: ersalazar@espe.edu.ec

J. A. Romero
e-mail: jaromero12@espe.edu.ec

L. M. Jiménez
e-mail: lmjimenez23@espe.edu.ec

J. J. Rodríguez
e-mail: jjrodriguez9@espe.edu.ec



X.-S. Yang et al. (eds.), *Proceedings of Sixth International Congress on Information and Communication Technology*, Lecture Notes in Networks and Systems 216,
https://doi.org/10.1007/978-981-16-1781-2_11

# 1 Introduction

In recent years education has been acquiring new milestones, forms and strategies in teaching, the new immersive technology tools increase the interest in students, improving their learning experience [1], these types of tools allow interaction, immersion and feedback within the virtual environment, naturally encouraging the interest in discovering things [2]. The applications in AR, and VR in the last decade increasing enormously, the main axis of these technological tools, is the interaction with the user within a virtual environment, the main difference in this type of technological tools, is that virtual reality allows the immersion of the user through the use of virtual reality glasses or helmets, creating digital contexts that imitate the real world within a virtual world, unlike augmented reality that allows to superimpose digital information in the real world, interacting with virtual content through a technological device without leaving the physical world [3].

The applications in virtual environments are numerous, allowing the construction, reconstruction of any type of object in 3D without the limitation of dimensions. Applications in AR can be grouped into two main groups depending on the specific tasks: the first group, applications that are focused as training assistants [4, 5], being applications that offer assistance and step-by-step guidance in processes for manual assemblies, comparing the facilities of AR assistants with audio and video-based assistants, allowing to reduce the number of errors that occur in manufacturing production, as well as operating failures due to erroneous maneuvers, reducing the effort spent in memorizing the steps of execution, likewise optimizing the time spent in training or courses on site in the production area [6, 7]. AR assistants have a wide and varied range of uses, allowing users to interact with different processes, in the educational, industrial, and economic fields, among others [8].

The second group, the applications that are focused on entertainment, in the work developed in [9] is an application that is oriented to the teaching of the metabolism of the human body, especially to the metabolism of glucose, this application consists of different and varied animations that provide an immersive environment close to reality, allowing a better understanding of the metabolism process, since the food enters by mouth, and makes its respective journey through the digestive system, at work [10, 11] is an AR game that is based on vision, for interaction through non-contact movements, in which users make dynamic gestures with hands and feet to interact with virtual elements that are on the scene in front of a physical camera, activating or selecting the predefined interaction events at work [12–14], in the work [15] an analysis of the technological, pedagogical characteristics of games for mobile devices in AR focused on education, in which they describe the characteristics of the game Calory Battle RA.

This work developed is intended to facilitate the learning of industrial manipulators through an application in AR. The application consists of animations and visualization of the components of the manipulators, as well as multimedia files using the recognition of 2D QR markers, in addition to the implementation of advanced control algorithms the same that through the simulation allows to observe the behavior of the structure of the robot, in which the operating end follows the trajectories desired by the user.

This work consists of IV sections, the first section includes the Introduction, the second section is the development of the application, specifying the marker recognition technique used, the characteristics of the AR environment for the mobile application, the law of advanced control by numerical methods, the third section shows the results obtained by the user when using the application on a mobile device, with the respective analysis of the results obtained and, finally, in the fourth section the conclusions of the work done.

## 2 Development

In Fig. 1 presents the flow chart for the development of the proposed work and the creation of the AR application, it considers four main stages, which develop a specific task, plus one or two processes that allow to execute the workflow of the application to be used in the smartphone:

(i) ***Layer 1.1*** allows 3D development using Blender software, which is a program for developing 3D objects, animation, rendering, etc. [16]. The 3D model of the robot arm with their respective axes of rotation are made to match each of the links, as shown in Fig. 2.

(ii) ***Layer 1.2*** describes the recognition technique used, 2D recognition allows text and image detection. This image is loaded into the Target Manager of the Vuforia portal, with the aim of creating a database, later to be downloaded as a file directly compatible with Unity.

(iii) ***Layer 2*** this layer describes the action of the buttons within the environment of the mobile application as shown in Table 1, which will allow different actions and navigation in the environment, which is simple but very practical and intuitive.

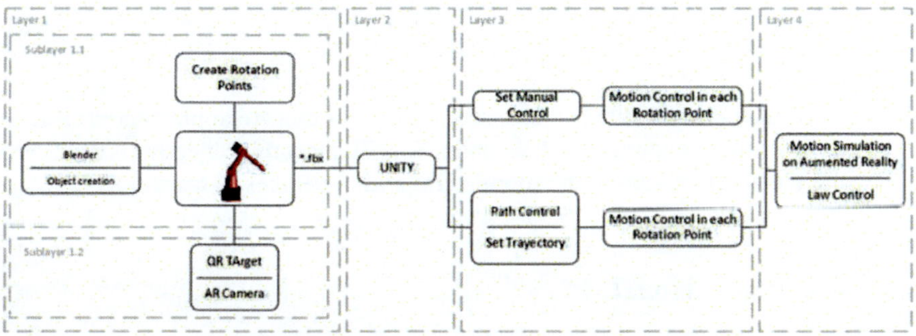


**Fig. 1** Mobile application development workflow diagram

**Fig. 2** 3D model of the robotic arm in Blender with its rotation point for the second link

**Table 1** Icons

| Name | Icon | Action |
|---|---|---|
| Introduction |  | Navigate to the introduction scene |
| Control |  | Navigate to the control scene |
| Video |  | Play video |
| Animation |  | Play animation |
| Play |  | Play the path control |
| Graphics |  | View the trend graphs of the controller |
| Return |  | Return to the previous scene |
| Exit |  | Close application |

(iv) ***Layer 3***: this layer details the control algorithms that are implemented through programming codes in the AR application. To meet this objective is described the kinematic model of the robotic arm with 3 degrees of freedom.

## *2.1 Kinematic Model*

In Fig. 3 shows a robotic arm with generalized coordinates and whose point of interest is at its operating end h, l is the length of each link and q is its angle of rotation of

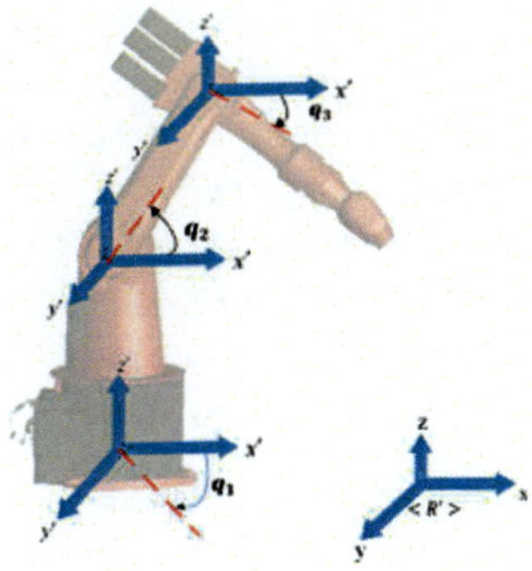

**Fig. 3** Robotic arm and coordinate space

each link, its kinematic model is given by:

$$\begin{cases} h_x = l_2\, C_{q2}\, C_{q1} + l_3\, C_{q2,q3}\, C_{q1} \\ h_y = l_2\, C_{q2}\, S_{q1} + l_3\, C_{q2,q3}\, S_{q1} \\ h_z = l_2\, S_{q2} + l_3\, S_{q2,q3} \end{cases}$$

The first-order differential equation is considered (1).

$$\dot{h} = f(h, v, t) \text{ with } h(0) = h_0 \tag{1}$$

## 2.2 *Control*

With reference to the publication [17], the controller is implemented based on numerical methods; in Eq. (1) the output of the system to be controlled is represented by $\mathbf{h}$, $\dot{\mathbf{h}}$ is the first derivative, $t$ the time and $\mathbf{u}$ the control action. are the values of $\mathbf{h}(t)$ in the discrete-time $h(k)$, where $T_0$ represents the sampling time and $k \in \{0, 1, 2, 3, 4, 5...\}$.

The use of numerical methods to calculate the evolution of the system is based mainly on the approximation of the system of states at the instant $k+1$, if the state of the system and the action of control are known over time at the instant $k$ are known. This approach is called Euler's method.

$$h(k+1) = h(k) + T_0\, f(h, u, t) \tag{2}$$

Therefore, the discrete model can be expressed through expression:

$$h(k+1) = h(k) + T_0\, J(q(k))v(k) \tag{3}$$

To make the tracking error tend to zero, the following expression is used:

$$h(k+1) = h_d(k) - W(h_d(k) - h(k)) \tag{4}$$

In which, $\mathbf{h}_d$ is the desired path, $\mathbf{W}$ is the diagonal matrix and its values $0 < \text{diag}(w_{hx}, w_{hy}, w_{hz}) < 1$ are parameters for the proposed controller.

Considering (1) and (2), the system can be rewritten as (1) and (2), the system can be rewritten as $Au = b$.

$$\underbrace{J(q(k))}_{A}\underbrace{v(k)}_{u} = \underbrace{\frac{h_d(k+1) - W(h_d(k) - h(k)) - h(k)}{T_o}}_{b} \tag{5}$$

Finally, the viable solution method is to formulate it as a constrained linear optimization problem.

$$\frac{1}{2}\parallel v\parallel_2^2 = \min$$

(v) ***Layer 4*** this layer allows the values of the rotation angles for each link obtained by means of the control algorithms, to be incorporated into the 3D model to finally generate the movement, which complies with the trajectory entered by the user.

## 3 Results Obtained

This section shows the interface of the AR application and its usability as a technological tool for handling industrial robots, as well as the simulation of a control algorithm that allows the user to follow the desired path using different images.

Figure 4 shows the main scene when the Smartphone's camera is focused on the QR code. To exit the AR application press "exit". Pressing on "introduction" presents the introduction environment. Figure 5 shows the event that takes place when you click on "animation", which consists of an animation of the parts that make up the robotic arm.

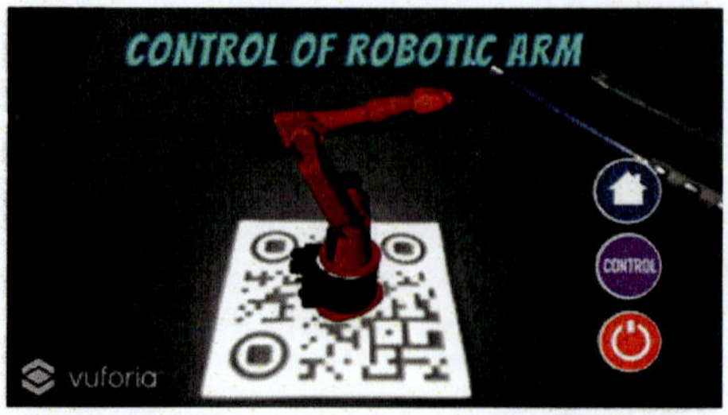


**Fig. 4** Main scene of the AR application when the QR code is previously focused

**Fig. 5** Animation of the parts that make up the robotic arm

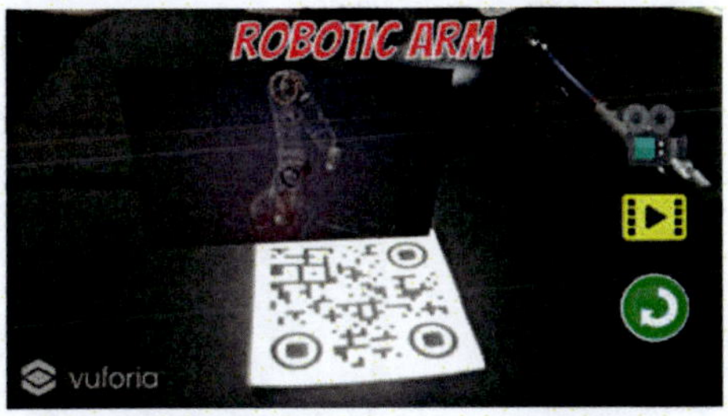


**Fig. 6** Reproduction of the video in AR

Figure 6 shows an introductory video about robotic arms when clicking on "video", this video contains: general characteristics, constitution, and mathematical modeling.

To control the robotic arm press the "control" icon, then a scene containing 3 sliders is displayed. To be able to make manual movement on the robotic arm, move the 3 sliders that are presented in the environment, as shown in Fig. 7.

For the robot to perform the trajectory control, first it is necessary to focus the image of the desired trajectory next to the QR image of the robotic arm, when the image is recognized, its trajectory appears in the increased environment, as seen in Fig. 8 the trajectory was recognized in the form of a heart. Clicking on "play" starts the trajectory control, and the operating end of the robot arm follows the heart-shaped trajectory as shown in Fig. 9.

Figure 10 shows the x, y, and z position errors in the AR application and Fig. 11 shows the errors in the Matlab software.

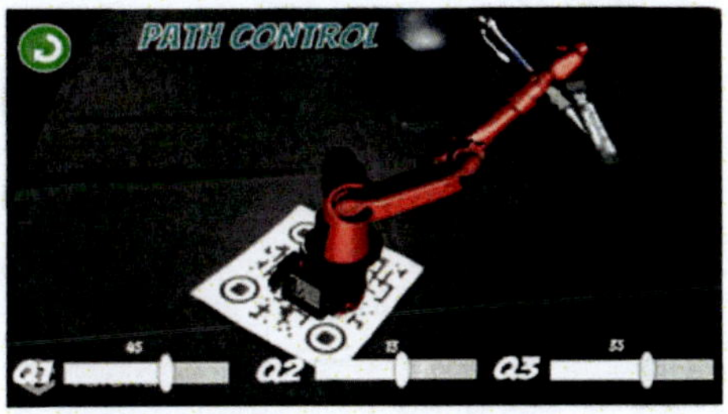


**Fig. 7** Manual movement of the robot arm when changing the parameters q1, q2 and q3

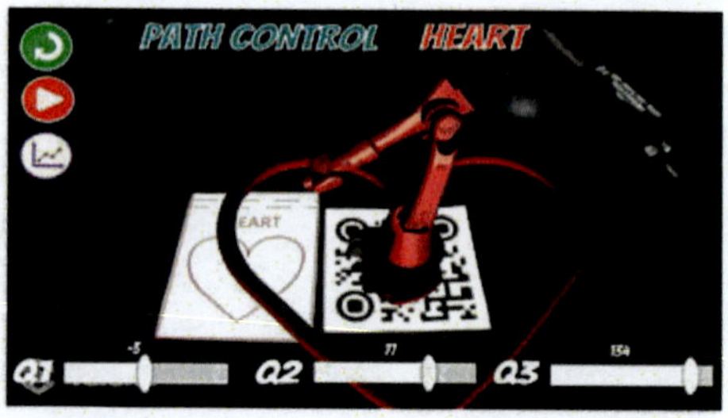


Fig. 8 Visualization of the robotic arm's heart-shaped trajectory

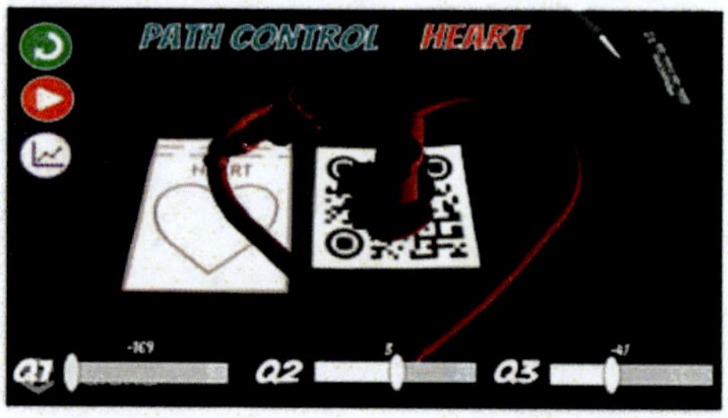


Fig. 9 Robot arm movement in the heart-shaped path when applying control

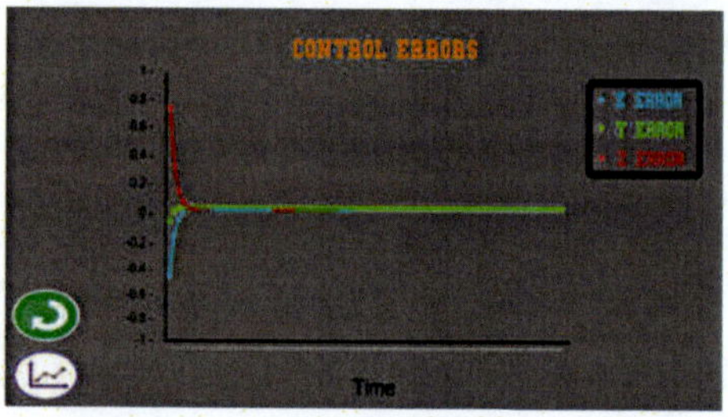


Fig. 10 Position errors x, y, and z in the heart-shaped path within the AR

Figure 12 shows the infinite path in the AR application. Figures 13 and 14 show the graph of x, y, and z position errors in the AR application and Matlab software, respectively.

The validity of the usability of the application of augmented reality, for the management of industrial robots, in a timely manner, in the management and training of industrial robots of three degrees of freedom, are indicated by the results obtained, for which it was used the method of summary evaluation of the SUS, whose weighting characteristics are described in the paper [18].

The sample of the survey was 15 students from the Universidad de las Fuerzas Armadas ESPE of the Electronic Engineering career, who were asked 10 questions

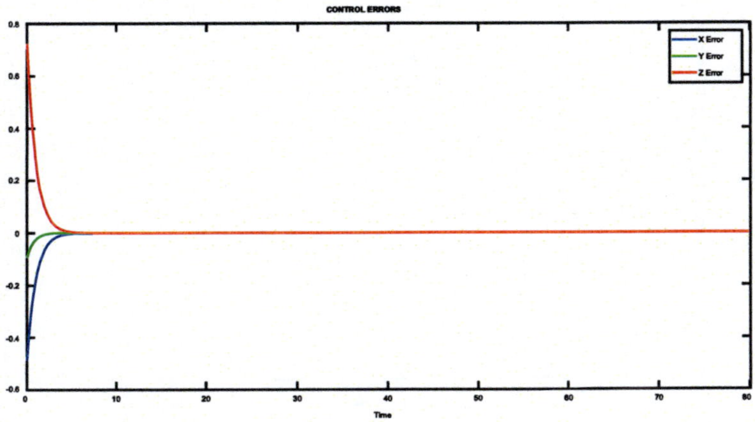


**Fig. 11** Position errors x, y, and z in the Matlab heart-shaped path software

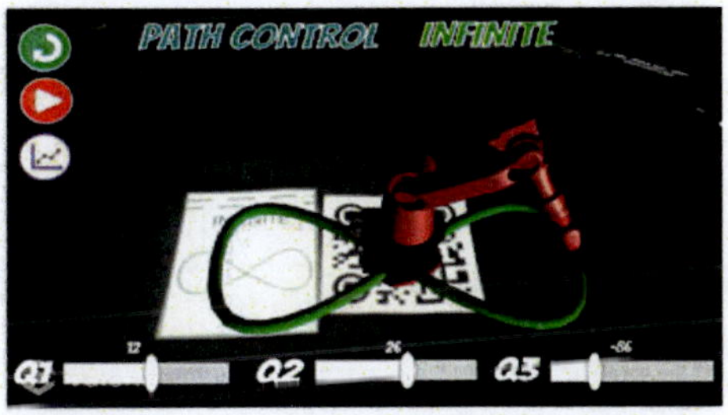


**Fig. 12** Robot arm movement in the path in the form of infinite when applying control

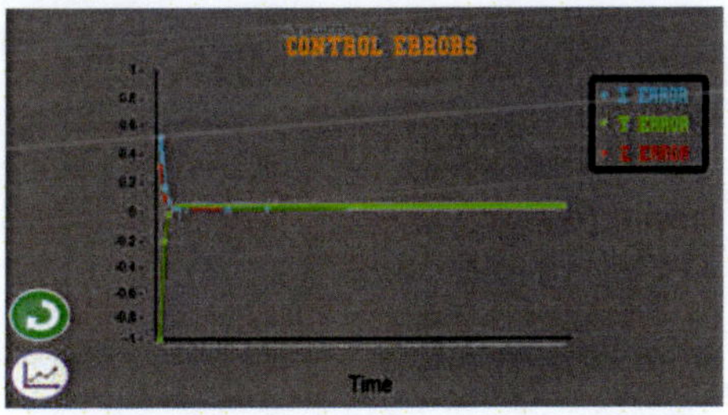


**Fig. 13** Position errors x, y, and z in the infinite path within the AR

according to their experience after using the AR application. Table 2 shows the scale of styles that generates only one number, which represents a composite average of the usability of the AR application for Smartphones.

For the calculations made, the even questions have a weight from 4 to 0, for the odd questions the weight is from 0 to 4, where 0 is the worst and 4 is the best, each

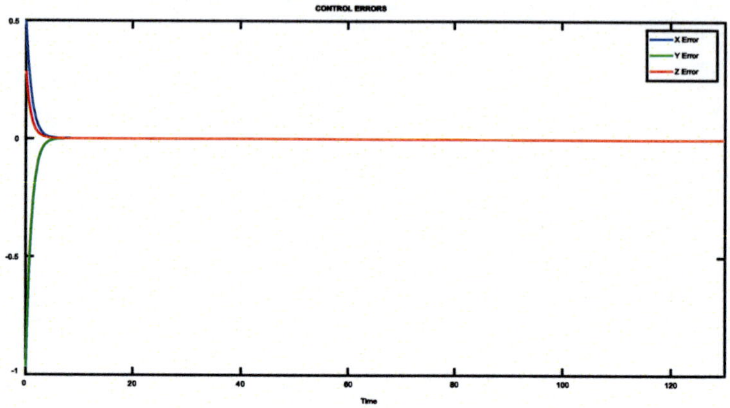


**Fig. 14** Position errors x, y, and z in Matlab infinity path software

**Table 2** Results obtained from the SUS survey

| Questions | Score | | | | | Operation |
|---|---|---|---|---|---|---|
| You would like to use the application frequently | | | | 3 | 12 | 3.87 |
| Considers that the application is complex | 13 | 2 | | | | 3.87 |
| It was easy to use the AR application | | | | 4 | 11 | 3.73 |
| Need help or assistance using the application | 15 | | | | | 4 |
| The application's functionalities are correctly integrated | | | 3 | | 12 | 3.6 |
| There is flexibility in the use of the application | | | 2 | 2 | 11 | 0.4 |
| You think most people would quickly learn to use the AR application | | | | 1 | 14 | 3.93 |
| The use of the application is difficult | 12 | 2 | 1 | | | 3.73 |
| The user interaction was friendly when using the AR | | | | 2 | 13 | 3.87 |
| You consider it necessary to have previous knowledge to use the AR application | 13 | 2 | | | | 3.87 |
| Total | | | | | | 34.87 |

of these values are multiplied by their respective number of answers and finally the arithmetic average of each question correspondingly. The result obtained of the 10 questions has a total sum of 34.87, the score of usability SUS is calculated by means of multiplication of 2.5 to the total sum obtained, obtaining a percentage of 87.18%, result that represents a high usability of the developed technological tool, so that the application in AR is feasible for the handling in the training and assistance of industrial manipulators.

## 4 Conclusions

The work done shows through the SUS survey, which is an immersive and easy-to-use application. Thanks to its great usability, this application contributes to the training of users on the subject of robotic arms. The document presents an augmented reality application for Smartphones that allows scanning images and then interacting with the environment, in the specific case of manipulating the links of the robotic arm, in addition to knowing its general characteristics through 3D animations and videos, as well as 3d movement of the robotic arm, the results obtained in the simulation of the mobile application allow to validate the correct operation of both the controller and the movement of the robotic arm, since the robotic arm complies with the trajectories designated by the user and that the errors of position lead to zero.

**Acknowledgements** The authors would like to thank the Universidad de las Fuerzas Armadas ESPE, for the support in developing this work.

## References

1. Pantelidis V (2007) Reasons to use virtual reality in education. Themes Sci Technol Edu 2:59–70
2. Billinghurst M (2002) Augmented reality in education. In: New horizons for learning, pp 3–4
3. Kaufmann H (2003) Collaborative augmented reality in education. In: Institute of software technology and interactive systems. Vienna University of Technology, pp 2–4
4. Loch F, Quint F, Brishtel L (2016) Comparing video and augmented reality assistance in manual assembly. In: 12th international conference on intelligent environments, pp 147–150
5. Boonbrahm P, Kaewrat C (2014) Assembly of the virtual model with real hands using augmented reality technology. Lect Notes Comput Sci 8525:329–338
6. Schwald B, De Laval B (2003) An augmented reality system for training and assistance to maintenance in the industrial context. J WSCG 11(1–3):1–8
7. Haritos T, Macchiarella N (2005) A mobile application of augmented reality for aerospace maintenance training. In: 24th digital avionics systems conference, pp 1–9
8. Bottecchia S, Cieutat J, Jessel J (2010) Augmented reality system for collaborative tele assistance in the field of maintenance through internet. In: Proceedings of the 1st augmented human international conference, pp 1–7
9. Barrow J, Forker C, Sands A, O'Hare D, Hurst W (2019) Augmented reality for enhancing life science education. In: Visual

10. Lv Z, Halawani A, Feng S, Réhman S, Li H (2015) Touch-less interactive augmented reality game on vision-based wearable device. Pers Ubiquit Comput 19(3–4):551–567
11. Laine T (2018) Mobile educational augmented reality games: a systematic literature review and two case studies. Computers 7(19):2–28
12. Lee K (2012) Augmented reality in education and training. Tech Trends 56:13–21
13. Bower M, Howe C, McCredie N, Robinson A, Grover D (2014) Augmented reality in education—cases, places and potentials. Edu Media Int 1:1–15
14. Chicaiza E, De la Cruz I, Andaluz V (2018) Augmented reality system for training and assistance in the management of industrial equipment and instruments. Int Sympos Vis Comput 11241:675–686
15. Boletsis C (2013) The table mystery: an augmented reality collaborative game for chemistry education. In: Lecture notes in computer science, vol 8101, pp 86–95
16. Blender (2019) https://www.blender.org/ Last accessed 1 Nov 2019
17. Andaluz V, Molina M, Erazo Y, Ortiz J (2017) Numerical methods for cooperative control. Intell Robot Appl 10463:879–898
18. Bangor A, Kortum P, Miller J (2008) An empirical evaluation of the system usability scale. J. Hum. Comput. Interact. 574–594

# Analysis of Website Quality and City Government's Twitter Accounts in East Java Province, Indonesia

## (Case Study: Surabaya, Malang, Mojokerto, Pasuruan and Probolinggo)

**Danang Eko Prastya, Achmad Nurmandi, and Salahudin**

**Abstract** This study aims to determine how local government websites' quality in implementing sound governance principles, transparency, accountability, effectiveness, efficiency, and website features. Also, it explains how the local's government official Twitter account conveys information to the public. This research is descriptive-interpretive. The research objects were the official website and city governments' Twitter accounts from Surabaya, Malang, Mojokerto, Pasuruan, and Probolinggo. The researchers collected data through surveys and observations. Website quality assessment was carried out by giving a score on the indicators of features, transparency, accountability, effectiveness, and efficiency of the website with 40 analysis units. Meanwhile, Twitter contents were analyzed using the Nvivo 12 Plus with the NCapture feature to systematically document and analyze data. The findings showed that the Surabaya city government website obtained the highest score of 85% (High Quality). The Pasuruan city government's website had the lowest score of 42.5% (Low Quality), while the average quality website went to the municipal government of Mojokerto and Probolinggo. Only the Pasuruan city government had a website with a low-quality category. The Surabaya, Malang, and Pasuruan municipal governments' Twitter content had similarities in the public's issues. The problems surrounding COVID-19 were discussed in the three city government's Twitter accounts. The issues discussed in the Twitter accounts of the Probolinggo and Mojokerto municipal governments were more about the city government's programs

D. E. Prastya (✉) · A. Nurmandi
Department of Government Affairs and Administration, Jusuf Kalla School of Government, Universitas Muhammadiyah Yogyakarta, Special Region of Yogyakarta, Brawijaya Street, Kasihan, Bantul, Indonesia
e-mail: danang.eko.psc20@mail.umy.ac.id

A. Nurmandi
e-mail: nurmandi_achmad@umy.ac.id

Salahudin
Department of Government Science, Universitas Muhammadiyah Malang, Tlogomas StreetKarang Ploso, Tegalgondo, Malang, Indonesia
e-mail: salahudinmsi@umm.ac.id



X.-S. Yang et al. (eds.), *Proceedings of Sixth International Congress on Information and Communication Technology*, Lecture Notes in Networks and Systems 216,
https://doi.org/10.1007/978-981-16-1781-2_12

for the community. Growing issues also increased activity on the city government's Twitter accounts.

**Keywords** E-Government · Website quality · Social Media · Information · Public Service

## 1 Introduction

Providing accurate and transparent information is the government's job [1]. As a form of information technology in supporting government tasks is the stipulation of presidential instruction no. 3 of 2003 concerning national policies and strategies for the use of e-government. A Blueprint for e-Government Application Systems for Local Governments has been published by the Department of Communication and Information Technology. The emergence of e-government begins at four levels, creating a visible website to the public [2]. Websites' availability is considered clear evidence of the government's seriousness in organizing e-government [3]. Therefore, creating information websites for government agencies is an essential part of implementation [4].

A website's existence may provide an opportunity for the public to find out about all activities or government policies and increase the community's interest and chances in accessing public services. In supporting the development and implementation of e-government, the website can be used as an estuary of all the government's wealth of information [5]. Knowing the website's performance is one way to monitor and identify deficiencies and opportunities to improve quality and performance. Masyur's research on 30 Indonesian local government websites showed that the average page speed was 59% or grade E. The quality of local government websites based on the Google-generated index was still deficient. Low website quality could affect internet users' interest in visiting government websites [4]. A popular website could be seen from statistics on the number of visitors to the site [6].

Meanwhile, website transparency and accountability also play an essential role in providing government performance information. Financial reports and government performance, such as APBD data, performance accountability, are still minimally submitted on local government websites. Forty-seven percent of local governments do not provide information on their websites' economic and performance data [7]. Developing a website may be done by standardizing the content of the website's minimum requirements [8]. Therefore, in implementing e-government, it is necessary to assess local government websites [9].

## 2 Literature Review

### 2.1 *The Use of Website on E-government*

Technology use in government is known as the E-Government, where technology plays a role in facilitating the management of information and services [10]. E-government emerged due to Information and Communication Technology (ICT) developments to provide government services to stakeholders [11]. Through today's fast-growing world, ICT enables governments to deliver services more efficiently by facilitating information sharing across the globe. ICTs provide access to information such as the internet, digital or wireless networks, and other systems [12]. There are three components to e-government—first, ICT, which includes the internet, web-based applications, devices, and cellular; second, stakeholders and consumers of electronic services. The third is the result component. These three components can increase interaction between stakeholders and empower the community through transparent access to information to increase government operations' efficiency and effectiveness [13].

With the internet's growth, the government's public service model has gradually changed from traditional manual methods to online e-government websites [14]. E-Government services are usually provided through a Web portal developed and managed by the government [15]. Government use of websites can increase transparency, foster trust in government, reduce corruption in the public sector, and strengthen accountability and transparency [16]. Government web portals serve a different kind of function. Some of them focus on access to information. Access to integrated, open government and citizen electronic participation transactions is also provided through government-wide access to data, e-petitions, electronic consultations, etc.[17]. Through the website, the government can provide services according to community needs. Sometimes people use it to obtain information and exchange information. Some use it for transactions. Therefore, complete service information on the website needs to be considered by the government [15].

### 2.2 *Measurement of the Quality of Government Website*

A quality government website plays an essential role in building trust and ensuring successful e-government and delivery services [18]. Some factors affect e-government websites' quality, including accessibility, usability, content, citizen participation, services, features, security, and privacy [19]. The existence of concrete information can also build trust, loyalty, and security for its users [20, 21]. Throughout addition, images, fonts, the right colors, download speed, and link speed when accessed, attractive page designs reflect government entities' identity [22]. Therefore the evaluation approach is essential for improving the quality of the website [23].

A method to evaluate and choose the evaluation criteria for websites is different and still requires more theoretical justification. Most previous approaches have focused on the importance of primary content or a specific set of website results. They either use subjective factors, such as ease of access, clarity of text, quality of presentation, etc., as performance indices or use manual survey questionnaires to determine a specific rating index for websites, such as accessibility, speed, browsability, and website content [24]. Various criteria have been used to assess the quality of the website. Henriksson [25] evaluates e-government websites using a list of criteria: security, privacy, usability, content, and citizen participation. Barnes and Vidgen had also developed an eQual approach to assess e-government websites [26]. Meanwhile, features, transparency, accountability, effectiveness, and efficiency can also be used to evaluate government websites [27, 28]. Its web-government features must also display an overview, organizational structure, geography, maps, regional regulations and policies, news, and guest books [29, 30].

### 2.3 *Social Media on Government: Official Twitter Accounts*

Social media is seen as a means of interaction and provides more opportunities for the public to interact with government functions. Today's social media offer significant benefits in shaping society and relationships with the government. Users can easily participate, share, interact, discuss, collaborate, and create content with various platforms such as YouTube, Facebook, WhatsApp, Instagram, Twitter, and others [31]. It is limited to the relationship between policymakers and policymakers and new interactions that can be mutually beneficial through social media [32]. Therefore, the current use of ICT has theoretically given birth to new construction in government and society [33].

Using social media in government tends to have a significant impact on improving public services. This increase can encourage public participation Information on government programs can be conveyed using social media [34]. As one of the social media, Twitter has been used by the government to convey information quickly, such as disaster alertness, and to get a public response [35, 36]. Through microblogging, the social network, users of Twitter can quickly read, respond, and send messages called tweets [37]. Twitter's existence has been widely accepted as it provides direct and broad access to many of the most recent news operations while still helping to disseminate incident reporting [38].

## 3 Method

This research is descriptive-interpretive, namely research with the stages of describing, describing, interpreting the problems, and then drawing conclusions from the issues. The research objects were the official website and the city governments'

**Table 1** Score and Value

| Score (%) | Features | Transparency | Accountability | Effective | Efficient | Overall value |
|---|---|---|---|---|---|---|
| 0–25 | Incomplete | Not transparent | Not accountable | Not effective | Not efficient | No quality |
| 26–50 | Less complete | Less transparent | Less accountable | Less effective | Less efficient | Low quality |
| 51–75 | Quite complete | Quite transparent | Quite accountable | Quite effective | Quite efficient | Average quality |
| 76–100 | Complete | Transparent | Accountable | Effective | Efficient | High quality |

Twitter accounts of Surabaya, Malang, Mojokerto, Pasuruan, and Probolinggo. Data collection was done by survey and observation of the website and Twitter accounts. Website quality assessment was carried out by giving a score on the indicators of features, transparency, accountability, effectiveness, and efficiency of the website with 40 analysis units. In calculating the assessment, a particular score was given by referring to the assessment scale analysis. Score analysis showed the process of assessing policy alternatives by creating and using indicators to assess policy alternatives developed following the objectives formulated and, if necessary, giving weighting to indicators that are considered more important than other indicators [39]. In calculating the comparative assessment, a specific score would be given by referring to the assessment scale analysis [8]. If there are data available, it will be given a score of 1; if the data is not available, it will be given 0. The formula for calculating the weight of the score is as follows (Table 1).

$$\text{Score weighted} = \frac{\text{Total website scoring results } \times 100\%}{\text{Total score}}$$

After the weighted score is obtained then an assessment is carried out according to the following Table 1.

The city government website's access speed was also calculated using Google Tools, namely, Page Speed Insights. Website loading speed was assessed on a percentage scale, slow (0–49%), average (50–89%), and fast (90–100%). Meanwhile, Twitter content was analyzed using the Nvivo 12 Plus with the NCapture feature to systematically document and analyze data [40].

# 4 Results

## 4.1 Website Quality on City Government

Forty analysis units evaluated government websites in Surabaya, Malang, Mojokerto, Pasuruan, and Probolinggo to assess the website's quality. The assessment provided a score on the website's features, transparency, accountability, effectiveness, and efficiency. The results of the scores on five-city government websites in the province of East Java are as follows (Fig. 1).

The figure above explains that the Surabaya city government website gets the highest score with a percentage value of 85%. Simultaneously, the Pasuruan website has the lowest score, with a percentage value of 42.5%. The websites of the City of Surabaya and Malang are included in high-quality websites. Meanwhile, Mojokerto and Probolinggo have average quality websites. Only Pasuruan has a website with a low-quality category. The Mojokerto city government owns the website feature with a high score with a percentage value of 35%, and Pasuruan achieves the lowest score of 22.5%. The Surabaya and Malang websites' transparency has the highest score with a value of 22.5%, while Pasuruan has the lowest score of 5%. Surabaya's city has the highest accountability score, with 7.5%, and the lowest score of 2.5% obtained by Mojokerto and Pasuruan. Surabaya and Malang city governments in website effectiveness and efficiency get high scores with a percentage value of 22.5%. Meanwhile, the effectiveness and efficiency of Pasuruan get the lowest score with 12.5%.

The ease with which people can access the website needs to be considered by the local government. Google data shows that 53% of Indonesian Internet users leave a website that lasts more than 3 s [41]. When the loading speed is long, the public or visitors will feel annoyed and immediately close the website. The following figure shows the speed of access of five-city governments in the East Java Province (Fig. 2):

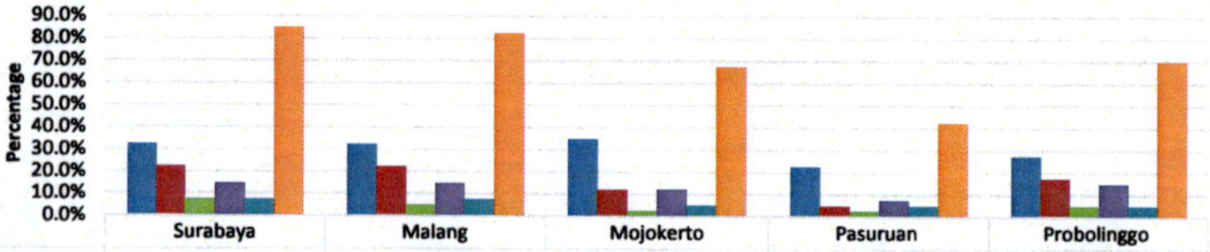


| | Surabaya | Malang | Mojokerto | Pasuruan | Probolinggo |
|---|---|---|---|---|---|
| Features | 32.5% | 32.5% | 35% | 22.5% | 27.5% |
| Transparency | 22.5% | 22.5% | 12.5% | 5% | 17.5% |
| Accountability | 7.5% | 5% | 2.5% | 2.5% | 5% |
| Effectiveness | 15% | 15% | 12.5% | 7.5% | 15% |
| Efficiency | 7.5% | 7.5% | 5% | 5% | 5% |
| Score | 85% | 82.5% | 67.5% | 42.5% | 70% |

**Fig. 1** City government website quality score

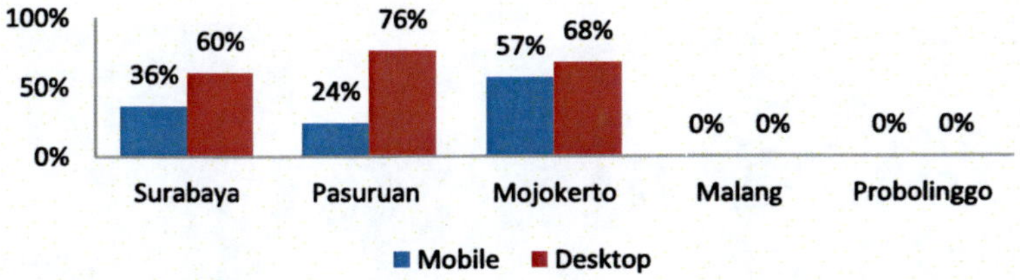


**Fig. 2** Website access speed

The picture above shows that the Pasuruan city government website has a high score for access speed via desktop compared to Surabaya and Mojokerto. Meanwhile, Mojokerto obtains a high score for mobile speed compared to Surabaya and Pasuruan. Despite getting a high score via mobile, the access speed of the Mojokerto website is relatively average. Meanwhile, access via the desktop website of Mojokerto, Surabaya, and Pasuruan is also the average speed. Therefore, the city government is expected to increase the speed of access to visit the website easily.

## *4.2 Social Media on Government: Official Twitter*

Local governments have used Twitter to deliver information and media for interaction with the community [34]. Local governments can socialize work programs and policies to the public in real time, and the public can respond directly. The municipal governments of Surabaya *(@SapawargaSby)*, Malang *(@PemkotMalang)*, Mojokerto *(@humas_kota)*, Pasuruan *(@pemkot_pasuruan)*, and Probolinggo *(@humaskotaprob)* have used Twitter as a means of communication and interaction with the community. In the five official Twitter accounts of the city government in East Java Province, only the Surabaya and Malang cities have a blue tick or are verified by Twitter. Each city government's Twitter provides information that is diverse and related to the community. The Nvivo 12 Plus analysis results using the Wordcloud feature describe the issues that are informed to the public. Issues that develop on Twitter can be seen in the following Fig. 3.

The picture above explains that the city government's Twitter accounts of Surabaya, Malang, and Pasuruan has similarities in the issues that are communicated to the public. Issues surrounding the handling of COVID-19 were discussed on the three Twitter accounts. The municipal governments of Surabaya, Malang, and Pasuruan urge local communities to obey and obey health protocols. The public is invited to remain vigilant and remain optimistic in the face of a pandemic. Meanwhile, the issues discussed in the Twitter of the Probolinggo and Mojokerto municipal governments are more about the city government's programs for the community. Work programs such as infrastructure development, protecting the environment, and the municipal government agenda are informed through the municipal government Twitter accounts. Issues are developing in the city government and also increasing

**Fig. 3** Wordcloud Twitter for the City Governments of Surabaya, Malang, Mojokerto, Pasuruan, and Probolinggo

Twitter activity. Surabaya's city government in 2020 tweeted 833 tweets, with 146 tweets occurring in March. This increase occurred due to the outbreak of COVID-19 cases in Indonesia. The Surabaya municipal government's Twitter activity can be seen in the following Fig. 4.

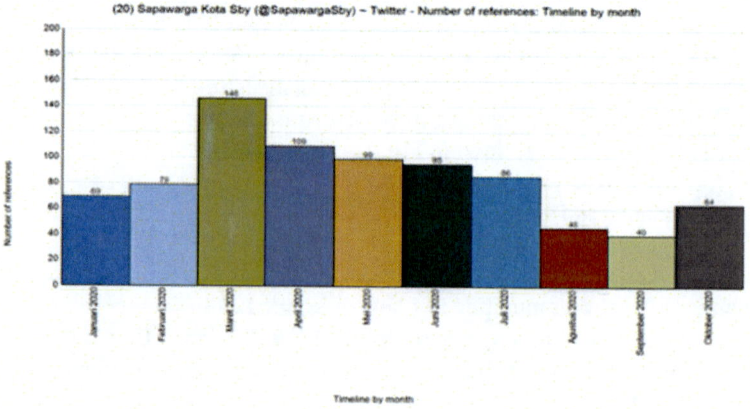


**Fig. 4** Tweet activity@sapawargaSby

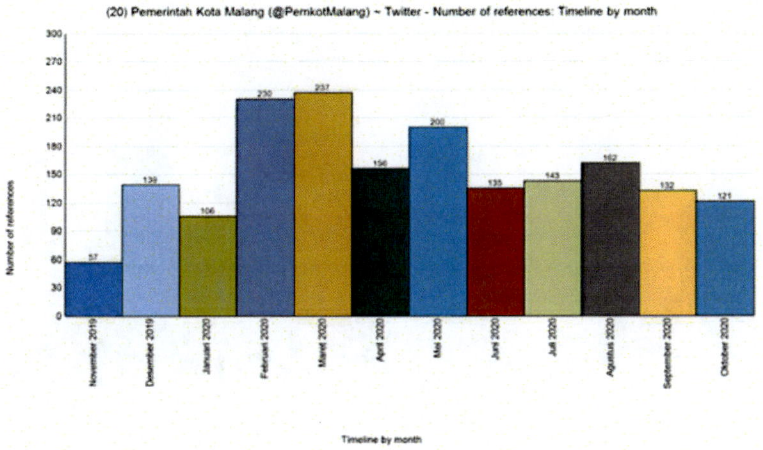


**Fig. 5** Tweet activity@PemkotMalang

The increase in Twitter activity also occurred in the Malang city government regarding the COVID-19 issue in Indonesia. It was active within one year from November 2019 to October 2020, having 1818 tweets. Most tweets occurred in March 2020, with 237, as shown in the following Fig. 5.

Meanwhile, the Pasuruan municipal government's Twitter account has not been active in the last two years since no tweet activity starting from May 2018–March 2019. It was starting to be active and consistent tweets from January to September 2020. The highest number of tweets occurred in February 2020, with 81 tweets. In comparison, the least was only one tweet in November 2018 and 2019. There were 518 tweets, and there was a fluctuation in the Pasuruan city government's Twitter activity, as seen in the picture below (Fig. 6).

The Mojokerto's city government through @humas_kota from 2015 to 2020 also looks inactive from July 2017 to September 2019. Of 354 tweets, there were the highest tweets in April–June 2016 and the lowest tweet in July–September 2016. This condition shows that the Mojokerto municipal government's Twitter activity is not active and consistent from 2015–2020. There is fluctuation inactivity, as shown in the following Fig. 7.

The Probolinggo City Government's Twitter activity through (@humaskotaprob) from 2017 to 2020 was seen to be inactive from June 2017 to February 2019 and then continued from April 2019 to February 2020. Even though there was an increase in activity from January to March, there was a decrease from April to May 2017. Of the total 90 tweets in 2017–2020, the highest tweet was in March 2020, and the lowest was in March 2020. This condition shows that the Probolinggo city government's Twitter activity was not active and consistent from 2015 to 2020. There is a fluctuation in action, as shown in the following Fig. 8.

Apart from being a medium of information for the public, Twitter is also used by the city government to communicate and coordinate with local institutions or

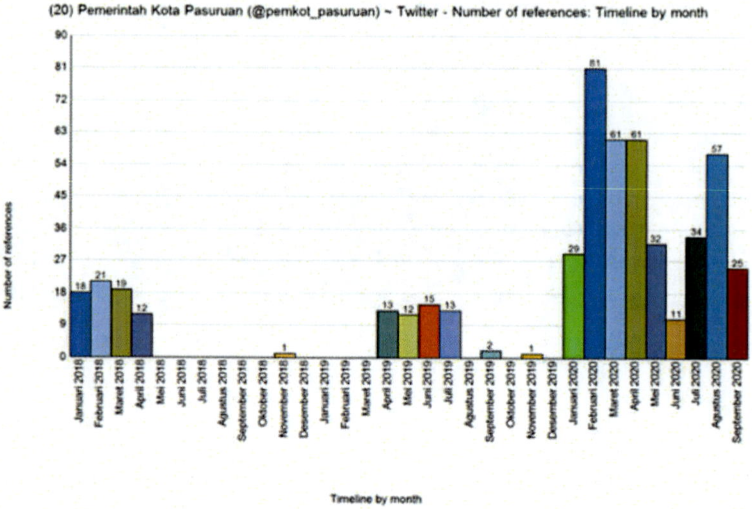


**Fig. 6** Tweet activity(@pemkot_pasuruan)

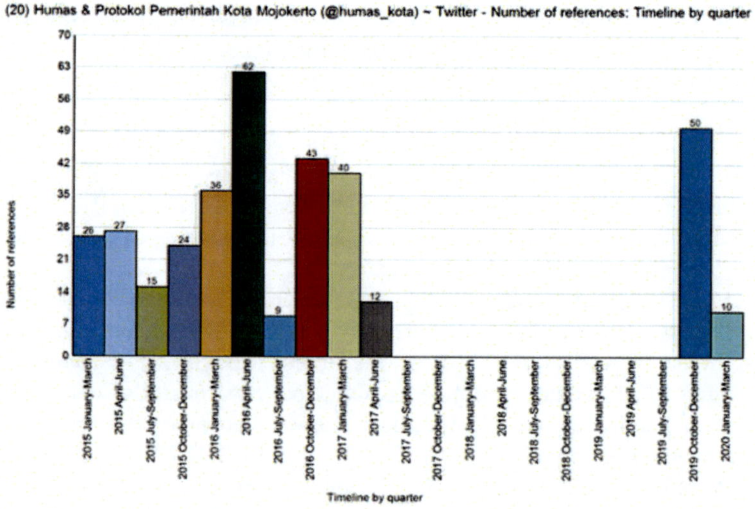


**Fig. 7** Tweet activity@humas_kota

agencies in supporting the performance of the city government. The interaction and communication of city governments in providing services to the community can be seen from the following cluster analysis results (Fig. 9).

The Surabaya's city government is active in interacting with the social service, the population and civil registration office, the education office, and the food and agriculture security service. Meanwhile, Malang's city government interacts and

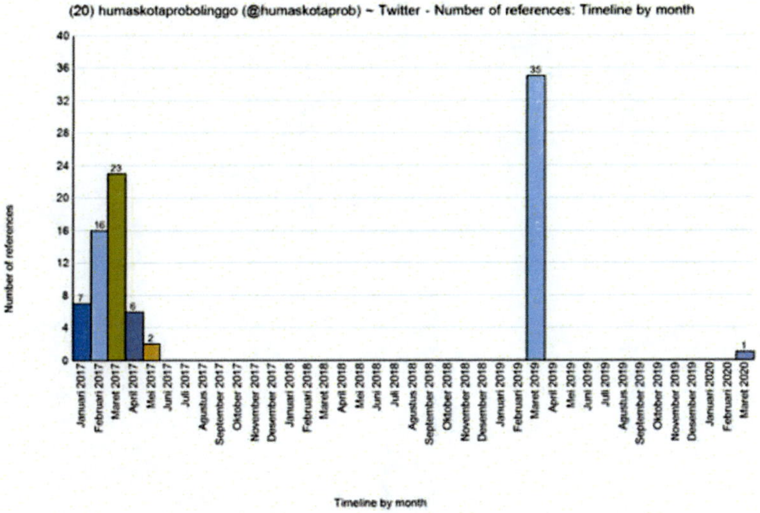


**Fig. 8** Tweet activity@humaskotaprob

communicates more with agencies related to transportation services, regional disaster management, local public works, and the East Java provincial government. Unlike the Surabaya and Malang municipal governments, the Pasuruan's city government is more active in interacting with health services. The municipal government is more intense in interacting with the city red cross and @ your heart.

## 5 Conclusion

The use of websites in local government is an implementation of e-government that is implemented in Indonesia. As an information and communication medium, having a quality website is an essential point in achieving the success of e-government by local governments. The city government's website in the province of East Java shows that the Municipal Governments of Surabaya and Malang have websites with high-quality categories. However, only the Pasuruan City Government has a website that is low quality. Using the website, the city government has also used Twitter to interact and information with the public. The city government is active in providing information on work programs and issues developing in the region. Therefore, Twitter can build communication patterns and coordination among government agencies in providing services to the community.

This research's limitation is that it only examines the official city government website's quality based only on five indicators, namely features, transparency, accountability, effectiveness, and efficiency, even though many other indicators can be used to measure city quality government websites. Besides, the research was only

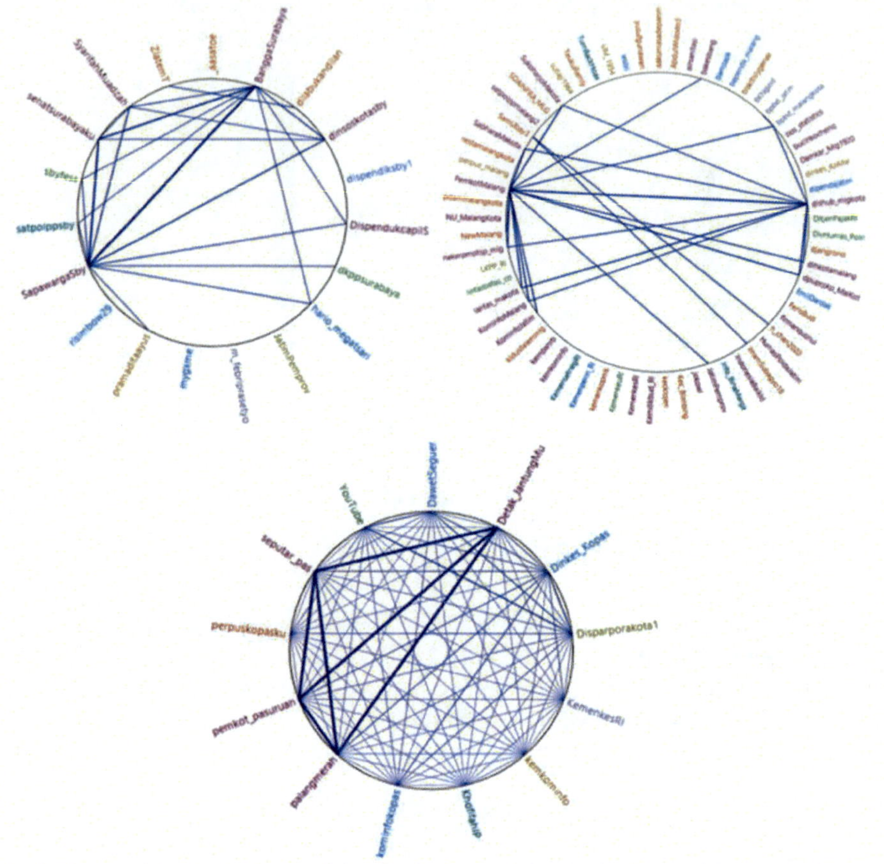


**Fig. 9** Cluster Analysis @sapawargaSby, @PemkotMalang, @pemkot_pasuruan

conducted on five-city governments in East Java province, and in the future, it can be carried out in all cities in East Java Province and Indonesia. On the other hand, the use of Twitter by the city government is only used as a medium of information and communication. Tweets from interactions with the community can be used to make a policy.

## References


1. Cahyono TA, Susanto TD (2019) Acceptance factors and user design of mobile e-government website (Study case e-government website in Indonesia). Procedia Comput Sci 161:90–98
2. dan Informatika DK (2004) Cetak Biru (Blueprint) Sistem Aplikasi E-Government: Bagi Lembaga Pemerintah Daerah. Jakarta

3. Kanter CH, Purnama AP (2015) Peran Website Pemerintah Kota Manado dalam Mendukung Terwujudnya Good Governance. J Penelit dan Pengemb Komun dan Inform 19(3)
4. Masyhur F (2014) Kinerja Website Resmi Pemerintah Provinsi di Indonesia official website performance local government in Indonesia. J Pekommas 17(1):9–14
5. Yunita NP, Aprianto RD (2018) KONDISI TERKINI PERKEMBANGAN PELAKSANAAN E-GOVERNMENT DI INDONESIA : ANALISIS WEBSITE. In: Seminar Nasional Teknologi Informasi dan Komunikasi, pp 329–336
6. Hernikawati D (2016) Analisis Popularitas Website Tingkat Kementerian. J Masy Telemat dan Inf 7(2):79–88
7. Martani D, Fitriasari D, Annisa (2013) Transparansi Keuangan dan Kinerja pada Website Pemerintah Daerah Kabupaten/Kota di Indonesia. In: Proceeding PESAT (Psikologi, Ekonomi, Sastra, Arsitektur & Teknik Sipil), vol 5, pp 70–80
8. Sulistiyo D, Negara HP, A.W YF (2008) Analisis kajian standarisasi isi situs web pemerintah daerah kabupaten/kota. Seminar Nasional Informatika (SEMNASIF) 1(5)
9. Sitokdana MNN (2015) Evaluasi Implementasi eGovernment Pada Situs Web Pemerintah Kota Surabaya, Medan, Banjarmasin, Makassar dan Jayapura. J Buana Inform 6(4):289–300
10. Muttaqin MH, Susanto TD (2019) The effect of website components on user trust in increasing the interest to use public administration service on E-government website. Proc 2019 Int Conf Comput Sci Inf Technol Electr Eng ICOMITEE 1:30–36
11. Kaur A, Kaur H (2017) Exploring E-governance service delivery dimensions: a study of registrar of companies. Indian J Public Adm 63(4):557–566
12. Dutta A, Devi MS, Arora M (2017) Census web service architecture for e-governance applications. ACM Int Conf Proc Ser Part F1280:1–4
13. Cumbie BA, Kar B (2016) A study of local government website inclusiveness: the gap between E-government concept and practice. Inf Technol Dev 22(1):15–35
14. Huang J, Guo W, Fu L (2019)Research on E-government Website satisfaction evaluation based on public experience. In: 4th international conference on education and social development 206–213
15. Jiang, Ji (2014) E-government web portal adoption: the effects of service quality.e-Service J 9(3):43
16. Porumbescu GA (2016) Linking public sector social media and e-government website use to trust in government. Gov Inf Q 33(2):291–304
17. Henman P, Graham T (2018) Webportal vs. google for finding government information on the web: from a website-centric approach to a web ecology perspective. Inf Polity 23(4):361–378
18. Tan CW, Benbasat I, Cenfetelli RT (2008) Building citizen trust towards e-government services: do high-quality websites matter? In: Proceedings of the 41st Annual Hawaii International Conference on System Sciences (HICSS 2008), 217
19. Paul S, Das S (2020) Accessibility and usability analysis of Indian e-government websites. Univers Access Inf Soc 19(4):949–957
20. Sugandini D et al (2013) Web quality, satisfaction, trust and its effects on government. Int J Qual Res 12(4):885–904
21. Syaifullah S, Soemantri DO (2016) Pengukuran Kualitas Website Menggunakan Metode Webqual 4.0 (Studi Kasus: CV. Zamrud Multimedia Network). J Ilm Rekayasa dan Manaj Sist Inf 2(1):19–25
22. Warjiyono W, Hellyana CM (2018) Pengukuran Kualitas Website Pemerintah Desa Jagalempeni Menggunakan Metode Webqual 4.0. J Teknol Inf dan Ilmu Komput 5(2):139
23. Abdel-Basset M, Zhou Y, Mohamed M, Chang V (2018) A group decision-making framework based on neutrosophic VIKOR approach for e-government website evaluation. J Intell Fuzzy Syst 34(6):4213–4224
24. Wu J, Guo D (2015) Measuring E-government performance of provincial government website in China with slacks-based efficiency measurement. Technol Forecast Soc Chang
25. Henriksson A, Yi Y, Frost B, Middleton M (2007) Evaluation instrument for e-government websites. Electron Gov Int J 4(2):204–226

26. Barnes SJ, Vidgen RT (2006) Data triangulation and web quality metrics: A case study in e-government. Inf Manag 43(6):767–777
27. Isni AR (2018) ANALISIS KUALITAS WEBSITE PEMERINTAH DAERAH PADA KABUPATEN SESUMATERA BARAT. J Akunt 6(3)
28. Cheisviyanny C, Helmy H, Dwita S (2018) Analisis Kualitas Website Pemerintah Daerah Kabupaten/Kota di Provinsi Sumatera Barat. Simp Nas Keuang Negara 6(3):1087–1104
29. Effendi PM, Susanto TD (2019) Test of citizens' physical and cognitive on Indonesian e-government website design. Procedia Comput Sci 161:333–340
30. Kurniawan F, Rakhmawati NA, Abadi AN, Zuhri M, Sugiyanto WT (2017) Indonesia local government information completeness on the web. Procedia Comput Sci 124:21–28
31. Cahyono AS (2016) PENGARUH MEDIA SOSIAL TERHADAP PERUBAHAN SOSIAL MASYARAKAT DI INDONESIA. Publiciana 9(1)
32. Linders D (2012) From e-government to we-government: Defining a typology for citizen coproduction in the age of social media. Gov Inf Q 29(4):446–454
33. Gagliardi D, Misuraca G, Niglia F, Pasi G (2019) How ICTs shape the relationship between the state and the citizens: exploring new paradigms between civic engagement and social innovation. In: Proceedings of the 52nd Hawaii international conference on system sciences
34. Novianti E, Nugraha AR, Komalasari L, Komariah K (2020) "PEMANFAATAN MEDIA SOSIAL DALAM PENYEBARAN INFORMASI PROGRAM PEMERINTAH ( Studi Kasus Sekretariat Daerah Kabupaten Pangandaran )", AL MUNIR. J Komun dan Penyiaran Islam 11(1):48–59
35. Chatfield AT, Scholl HJJ, Brajawidagda U (2013) Tsunami early warnings via Twitter in government: Net-savvy citizens' coproduction of time-critical public information services. Gov Inf Q 30(4):377–386
36. Carley KM, Malik M, Landwehr PM, Pfeffer J, Kowalchuck M (2016) Crowd sourcing disaster management: the complex nature of Twitter usage in Padang Indonesia. Saf Sci 90:48–61
37. Kosasih I (2016) Peran Media Sosial Facebook dan Twitter dalam Membangun Komunikasi (Persepsi dan Motifasi Masyarakat jejaring Sosial Dalam Pergaulan). Lembaran Masy J Pengemb Masy Islam 2(1):29–42
38. Heravi BR, Harrower N (2016) Twitter journalism in Ireland: sourcing and trust in the age of social media*. Inf Commun Soc 19(9):1194–1213
39. Dwiyanto A (2006) Transparansi Pelayanan Publik. Graha Ilmu, Yogyakarta
40. Edwards-Jones A (2014) Qualitative data analysis with NVIVO. J Educ Teach 40(2):193–195
41. Gevelber L (2016) Mobile search intent: New consumer behavior data. https://www.thinkwithgoogle.com/, 2016. [Online]. Available: https://www.thinkwithgoogle.com/marketing-strategies/app-and-mobile/mobile-search-consumer-behavior-data/. Accessed 29 Oct 2020

# How Does Local Government Use the Website and Twitters as Public Services Tools? A Case Study of Aceh Province Government, Indonesia

**Faez Syahroni, Achmad Nurmandi, and Salahudin**

**Abstract** Currently, many local government websites have a unique appearance, but the quality of the information content presented in the process is not always updated. Thus, the use of the website is seen only as a complement to government services. This study aims to analyze the implementation of the use of the website and Twitter account of the City Government in Aceh Province. This study uses a qualitative research method that examines a case with data analysis techniques using NVivo12 Plus software. The findings show that in general the five websites already have website content that refers to regulations, only the Banda Aceh City Government is in the high quality website category while the other four City Government websites are of low quality. The assessment of the website is not only on its quality, but also on the level of access speed. The government websites of Sabang, Langsa, Lhokseumawe and Subulussalam are included in the fast category, only the Banda Aceh City Government has an average website that is not fast enough. In addition, when viewed from the use of Twitter quality, the Banda Aceh City Government's twitter account has a very good quality value, because it has a blue tick and is always active in presenting the latest news. Meanwhile, the twitter accounts of four other city governments have not maximized their twitter activity in presenting news.

**Keywords** Analysis · Website usability · Twitter · Municipal government · Indonesia

F. Syahroni (✉) · A. Nurmandi
Master of Government Science, Universitas Muhammadiyah Yogyakarta, Special Region of Yogyakarta, Brawijaya, Kasihan, Bantul, Indonesia
e-mail: faez.syahroni.psc20@mail.umy.ac.id

A. Nurmandi
e-mail: nurmandi_achmad@umy.ac.id

Salahudin
Government Science, Universitas Muhammadiyah Malang, Karang Ploso, Tlogomas, Tegalgondo Malang, Indonesia
e-mail: salahudinmsi@umm.ac.id

 
X.-S. Yang et al. (eds.), *Proceedings of Sixth International Congress on Information and Communication Technology*, Lecture Notes in Networks and Systems 216,
https://doi.org/10.1007/978-981-16-1781-2_13

# 1 Introduction

By providing information about government operations and facilitating electronic communications with community members, information and communication technology (ICT) promises to improve government transparency, accountability, and public participation [1]. Constraints on public services can be abolished by E-Government. In particular, to increase the quality and efficacy of public services related to the collection, management and provision of policy information and diverse data from or to all members of the public. Improving the quality of public services in the government bureaucracy, which is currently known as e-Government, by using information technology [2].

The website is one of the products of e-government as a way of growing interest and opportunity for the population in the provision of public services and community input. To achieve an open, accountable and efficient administration. The government can provide a variety of information to the public business, and/or fellow governments through the website. The website is the source of all the local government's wealth of information that is used to promote the creation and implementation of e-government [3].

A website with good content, slow access or a poor system of navigation. In this case, website creators must be aware that the technology owned by the audience is very diverse, from the simplest to the most advanced, so that all of them ensure that they can easily access the existing website [2]. Social media platforms across all government accounts, along with Flickr and YouTube, over an average period of time. The first government Twitter account was one from the city of Stockholm, the capital of Sweden, which was registered in August 2008. A study conducted in 2014 found that when it comes to posting government content, Twitter is the most popular social media application and the one with the highest amount of activity in terms of users who like content or follow accounts [4].

On this basis, the researcher wants to do more comparisons to find out how local government website standards are applied to the principles of good governance, namely openness, accountability, efficiency and effectiveness, as well as to analyze the features of the website and twitter accounts of five municipalities in Aceh province. easy details.

# 2 Literature Review

## *2.1 Government Behavior*

Services and platforms for communication to make life easier and more convenient for citizens and businesses; and (b) knowledge sharing with citizens and businesses to allow the advancement of new online tools and services. The government of the United Kingdom hopes to increase competitiveness and efficiency and minimize government duplication by reusing and exchanging ICT assets and creating an atmosphere for transparent and safe ICT solutions [5].

### 2.2 *How ICT Effecting Behavior*

In the context of e-government adoption, a variety of research models have been used to forecast and describe user actions. Many of these models concentrate on the implementation of information technology (IT) and information systems (IS). From technology adoption theories, the theoretical concept of e-government is derived. The Theory of Rational Behavior, one of the earliest models, emerged in the field of psychology and was later expanded to establish the Expected Behavior Theory [6].

### 2.3 *ICT Performance on Services*

A broad range of technology implementation models are used in electronic government (hereinafter e-government), leaving researchers and policy makers confused about which one to use. Nine well-known theoretical models of the adoption of information technology were tested in this study and 29 separate constructs were described. Using data collected from 377 respondents from seven selected cities in India, a unified model of e-government adoption (UMEGA) was developed and validated [7].

## 3 Research Methods

This study uses qualitative research methods, in [8], states that this approach helps researchers to investigate a phenomenon so that it is better understood, used in a limited sense to analyze an event. The focus of this research is the analysis of the municipal government website and twitter account. Municipal official Twitter accounts were analyzed using NVivo 12 Plus with the NCapture function to record and analyze information systematically [9]. To be carefully and thoroughly studied and examined [10]. By visiting the City Government website, data is obtained. The calculation of website output using a Google provided instrument, Page Speed Insights. On a percentage scale, website loading speed is rated.

Through evaluation using scoring, the efficiency of five city government websites in Aceh Province is assessed. Score analysis uses indicators to evaluate policy changes developed in line with the objectives developed and, where appropriate, by weighting indicators deemed more relevant than other indicators [11]. A certain score will be given with reference to the evaluation scale analysis in calculating a

**Table 1** Score and value

| Score | Indicators |
|---|---|
| 0–25 | Not quality |
| 26–50 | Less quality |
| 51–75 | Quality |
| 76–100 | Very quality |

comparative rating [12]. If the information is available/available a score of 1 will be given if the data is not available/not available, then a score of 0 will be given if the data is not available/not available. The score weight estimation formula is as follows:

$$\text{The weighted Score} = \frac{\text{Total website scoring results}}{\text{Total score}} \quad (1)$$

After the weighted score is received, an evaluation according to the following Table 1 is carried out.

## 4 Results and Discussion

### *4.1 Analysis of the Quality and Speed of Access to City Government Websites in Aceh Province*

Based on the assessment and research on the website of the City Government of Banda Aceh, Langsa, Lhokseumawe, Sabang and Subulussalam were assessed by 40 review units. Evaluation is done by giving a score on website features, transparency, accountability, performance and effectiveness. On five city government websites in Aceh Province, the following scores were obtained in Fig. 1.

The picture above shows that the Banda Aceh City Government website received the highest score with a percentage value of 85%. While the website of the Subulussalam City Government has the lowest score with a percentage value of 20%. Only the website of the Banda Aceh City Government is in the high quality category, while the websites of the other four municipal governments are still in the low category.

The website feature with a high score is owned by the City Government of Banda Aceh with a percentage value of 30% and the lowest score of 7.5% is achieved by the City Government of Subulussalam. The transparency of the websites of Banda Aceh and Langsa City had the highest score with a score of 22.5%, while Sabang City had the lowest score of 5% and the Municipal Government of Subulussalam with a score of 0%. The city of Banda Aceh has the highest accountability score with a score of 10% and the lowest score of 0% is obtained by the municipal government of Lhokseumawe, Sabang and Subulussalam. Banda Aceh and Langsa city governments in terms of website effectiveness and efficiency get high scores with a percentage

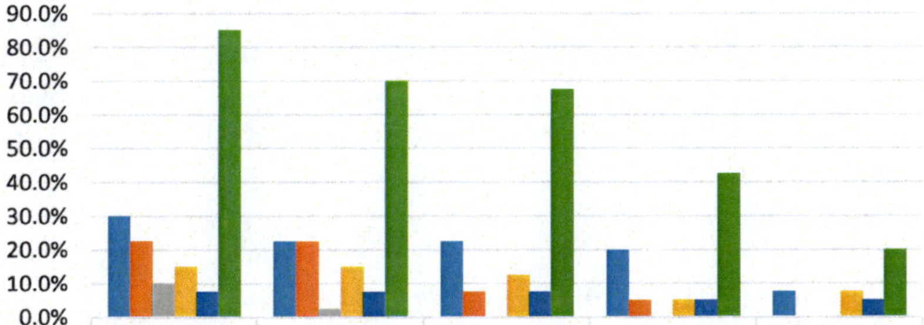


| | Banda Aceh | Langsa | Lhokseumawe | Sabang | Subulussalam |
|---|---|---|---|---|---|
| Features | 30.0% | 22.5% | 23% | 20.0% | 7.5% |
| Transparency | 22.5% | 22.5% | 7.5% | 5% | 0.0% |
| Accountability | 10.0% | 3% | 0.0% | 0.0% | 0% |
| Effectiveness | 15% | 15% | 12.5% | 5.0% | 8% |
| Efficiency | 7.5% | 7.5% | 8% | 5% | 5% |
| Score | 85% | 70.0% | 67.5% | 42.5% | 20% |

**Fig. 1** City government website quality in Aceh province, Indonesia

value of 22.5%. While the effectiveness and efficiency of the city of Sabang obtained the lowest score with 10%.

Google data reveals that 53% of internet users in Indonesia will leave a website that is longer than 3 s. When the loading speed is long, people or visitors will feel annoyed and automatically close the website. The website access speed of five city governments in Aceh Province can be seen in Fig. 2 below.

The image above shows that the Sabang City Government website has a high score for desktop access speed. Meanwhile, compared to the Municipal Governments of Banda Aceh, Langsa, Sabang, and Subulussalam, the Lhokseumawe Municipal Government scores high for the speed of cellular connectivity. The website access speed of the Lhokseumawe City Government is relatively slow despite getting high

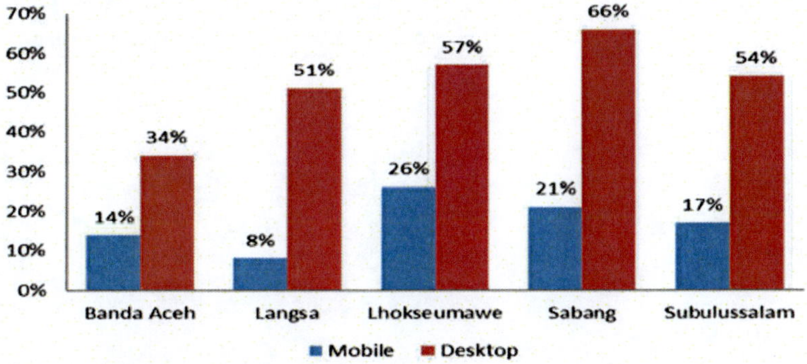


**Fig. 2** City government access speed in Aceh province, Indonesia

scores via cellphones. Meanwhile, access by the Municipal Government of Lhokseumawe, Langsa and Subulussalam via the desktop website is an average speed. However, the city government of Banda Aceh is sluggish. The city government is also expected to increase the speed of access so that individuals can visit the website quickly.

## 5 Use of City Government Twitter Accounts

The local government must socialize the project and work policies to the community in a real way and the community can respond directly. Twitter has been used as a medium for contact and interaction with groups by the municipalities of Banda Aceh, Langsa, Lhokseumawe, Sabang and Subulussalam. The official twitter accounts of five city governments in Aceh Province in Table 2 are as follows.

Only Banda Aceh City has a blue tick or is checked by Twitter out of the five official Twitter city governments in Aceh Province. Every Tweet from the city government provides a variety of community-related information. The results of the study of NVivo 12 Plus using the Wordcloud function define the problems that the public is aware of. Issues that arise in the Twitter municipal government can be seen in the following Fig. 3.

The picture above illustrates that the twitter of the City Government of Banda Aceh, Langsa, Lhokseumawe and Sabang have similar issues that are informed to the public, namely the issue of activities, work programs and policies in running the wheels of government. Issues that play a role include government-community

**Table 2** City government twitter account table

| No | City government | Twitter account |
|---|---|---|
| 1 | Banda aceh | @humasbna |
| 2 | Langsa | @humaslangsakota |
| 3 | Lhokseumawe | @pemko_Lsw |
| 4 | Sabang | @humassabang |
| 5 | Subulussalam | @KSubulussalam |

**Fig. 3** Word cloud analysis of city government twitter account in Aceh province, Indonesia

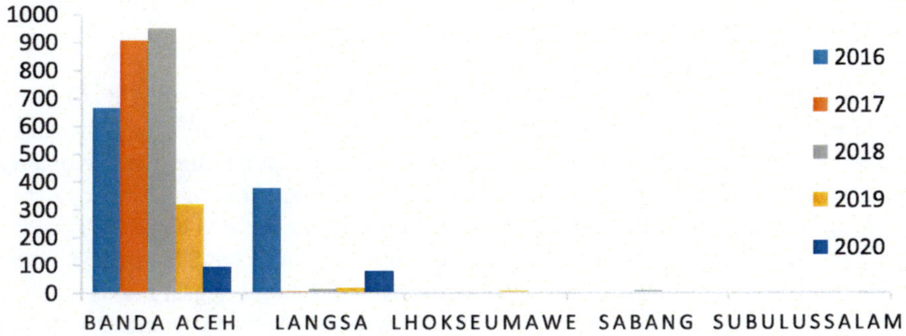


**Fig. 4** City government twitter account activity in Aceh province from year to year

cooperation, development, education and religion. Meanwhile, the issues discussed in the twitter of the City of Subulussalam are more of events or events that occur such as floods and strong winds that cause power outages.

The issue that is developing in the city government also increases the twitter activity of the city government. This increase proves that the government is actively working on various kinds of activities and other activities. City Government twitter activity in Aceh Province can be seen in the following Fig. 4.

From the picture above it can be narrated that, the activity of the Banda Aceh city government and the Langsa city government has always been active since the last five years until 2020. The twitter activity of the Banda Aceh municipal government from 2016 has continued to increase until 2018, the peak of the highest increase with a value of 951 tweets. However, it is very unfortunate that the twitter activity of the Banda Aceh municipal government has decreased drastically in the following years 2019 and 2020. Then, the highest level of activity of the Langsa City government twitter account only occurred in 2016 with a value of 378 Tweets. From 2017 to 2019, the Langsa municipal government twitter was not very active. However, in 2020 the local government of Langsa again presented the information on its Twitter account, even though it was only limited to news with a value of 79 tweets.

In contrast to the twitter of the municipal government of Banda Aceh and Langsa, the activities of the municipal government of Lhokseumawe and Sabang appear less active. The last Twitter activity of the Sabang municipal government was active in 2018 with a value of 9 tweets, while the Lhokseumawe municipal government twitter account was last active in 2019 with a value of 7 tweets.

If the twitter account activity of the municipal government of Banda Aceh, Langsa, Lhokseumawe and Sabang has been active in the last five years. Only the twitter of the Subulussalam Municipality Government via @KSubulussalam had a bad or inactive activity level from 2016 to 2020 with a total of 6 tweets. This has a very bad impact on the government in increasing public trust in performance and information disclosure in the embodiment of Good Governance.

## 6 Conclusion

Based on the findings of discussion and debate research, the Banda Aceh City Government website is included in the high-quality website category, while the Langsa and Lhokseumawe Municipal Government websites are included in the average quality score category. The website of the Sabang City Government is in the low category with less quality. Only the Municipal Government of Subulussalam has an Unqualified website. The assessment of the website is not only on its quality, but also on the level of access speed. The government websites of Sabang, Langsa, Lhokseumawe and Subulussalam are already included in the average speed website category. Only the city government of Banda Aceh has an average website that is not fast enough. In addition, Twitter has been used as a medium of information and communication between city governments and the public. From the activity of the City Government Twitter account in Aceh Province, it can be concluded that the Twitter accounts of the Banda Aceh City Government and Langsa are always active in presenting information. The Banda Aceh City Government received the highest Twitter activity score in 2018 with 951 tweets, while the Langsa City Government Twitter account also received the highest activity score in 2016 with 378 tweets. Unlike the twitter accounts of the Banda Aceh City Government and the Langsa City Government, the activity of the Lhokseumawe Municipal Government's twitter account was last active in 2019 and the last Twitter account of the Sabang City Government was active in 2018. Only the Subulus-Pemkot's Twitter account has been inactive for the last five years.

## References


1. Feeney MK, Brown A (2017) Are small cities online? Content, ranking, and variation of U.S. municipal websites. Gov Inf Quart 34(1):62–74. https://doi.org/10.1016/j.giq.2016.10.005
2. Oktariyanda TA, Rahaju T (2018) E-government strategy of Surabaya city government through e-rt/rw to improve the quality of public service. J Phys Conf Ser 953(1). https://doi.org/10.1088/1742-6596/953/1/012161
3. Yunita NP, Aprianto RD (2018) Kondisi Terkini Perkembangan Pelaksanaan E-Government Di Indonesia: Analisis Website. Seminar Nasional Teknologi Informasi Dan Komunikasi, 2018(Sentika), 329–336. https://fti.uajy.ac.id/sentika/publikasi/makalah/2018/40.pdf
4. Flores CC, Rezende DA (2018) Twitter information for contributing to the strategic digital city: towards citizens as co-managers. Telematics Inform 35(5):1082–1096. https://doi.org/10.1016/j.tele.2018.01.005
5. Carter L, Weerakkody V, Phillips B, Dwivedi YK (2016) Citizen adoption of e-government services: exploring citizen perceptions of online services in the United States and United Kingdom. Inf Syst Manag 33(2):124–140. https://doi.org/10.1080/10580530.2016.1155948
6. Salahudin NA, Jubba H, Qodir Z, Jainuri P (2020) Islamic political polarisation on social media during the 2019 presidential election in Indonesia. Asian Aff 51(3):656–671. https://doi.org/10.1080/03068374.2020.1812929
7. Dwivedi YK, Rana NP, Janssen M, Lal B, Williams MD, Clement M (2017) An empirical validation of a unified model of electronic government adoption (UMEGA). Gov Inf Q 34(2):211–230. https://doi.org/10.1016/j.giq.2017.03.001

8. Mohajan HK (2018) Qualitative research methodology in social sciences and related subjects. J Econ Dev Environ People 7(1):23–48
9. Edwards-Jones A (2014) Qualitative data analysis with NVIVO. J Educ Teach 40(2):193–195. https://doi.org/10.1080/02607476.2013.866724
10. Aspers P, Corte U (2019) What is qualitative in qualitative research. Qual Sociol 42(2):139–160. https://doi.org/10.1007/s11133-019-9413-7
11. Dwiyanto A (2006) Transparansi Pelayanan Publik. dalam Agus Dwiyanto, ed. 2006. Mewujudkan Good Governance Melalui Pelayanan Publik
12. Negara HP, Aw YF (2008) Analisis Kajian Standarisasi Isi Situs Web Pemerintah Daerah Kabupaten. Kota (semnasIF):55–62

# How Does the Local Government Provide Official Website and Twitter Quality? Case Study of Three Local Governments of Indonesia

**Nurul Kharismawati Kamaludin, Achmad Nurmandi, and Salahudin**

**Abstract** This study aims to determine the quality of city government websites in applying the principles of good governance. In this case, it can be measured through four aspects of assessment which include, assessment of website features, in which it is divided into several elements, namely, at a glance, local government, geography, maps of areas and resources, regional regulations or policies, as well as guest books and news. Then transparency, in this case seen from government activity reports, government performance reports, and local government financial reports. then accountability, which is assessed from several indicators, namely, in the form of regional government budget reports, grants, and social assistance reports, then timely presentation of financial reports and audit opinion from the BPK. Furthermore, the assessment of effective and efficient, effectiveness can be assessed from Interactivity, Navigation System. While efficiency is assessed from three elements, namely; The website does not have blank pages, important information contained on thc main page and the website does not contain intermediate pages. This research is a qualitative descriptive study, then the object of this research is based on three city government websites in Central Java Province, namely, the cities of Salatiga, Tegal and Surakarta. The final score obtained from the three city governments was different, but the cities of Salatiga and Surakarta got the same score, namely 50%, while the score obtained by the city of Tegal was 52.2%. Even so, the three city governments fall into the category of fairly high quality assessment.

N. K. Kamaludin (✉) · A. Nurmandi
Masters of Government Science, Jusuf Kalla School of Government, Muhammadiyah University Yogyakarta, Special Region of Yogyakarta, Brawijaya Street, Kasihan, Bantul, Indonesia
e-mail: nurul.kharisma.psc20@mail.umy.ac.id

A. Nurmandi
e-mail: nurmandi_achmad@umy.ac.id

Salahudin
Government Science, Universitas Muhammadiyah Malang, Tlogomas Street, Karang Ploso, Tegalgondo, Malang, Indonesia
e-mail: salahudinmsi@umm.ac.id

 
X.-S. Yang et al. (eds.), *Proceedings of Sixth International Congress on Information and Communication Technology*, Lecture Notes in Networks and Systems 216,
https://doi.org/10.1007/978-981-16-1781-2_14

**Keywords** E-Government · Official website · Official twitter · Transparency · Accountability · Effective · Efficient

## 1 Introduction

The government is an organization whose job is to run the government wheels. the task of the government is to provide good services to the community. To make this happen, the government needs to make improvements to service quality so that it gets a response from the public. Because society is a source of information that can improve the performance of the government. Currently, the flow of technology and information developments is so fast that it is a necessity for every group in business, education, and government.

Since the issuance of Presidential Instruction No. 3 of 2003 concerning the policy for developing E-Government. the government is required to be able to take advantage of advances in Information and Communication Technology (ICT) through the development of E-Government to improve the ability to process, distribute, distribute quality information and services to the public [1]. The Republic of Indonesia Presidential Instruction (Inpres) was formed with the consideration organizing good governance and improving effective and efficient public services, so there is a need for strategic policies in E-Government development [2]. ICTs offer governments in developing countries an effective resource for serving citizens and stakeholders through an electronic government strategy or E-Government, the terms E-Government and e-Governance are used widely to represent the use of ICTs in public sector organizations [3]. E-Government is defined as the delivery of information from the government-based online via the Internet or other means, in order to carry out tasks and functions effectively and efficiently [2]. Meanwhile, E-Governance is a tool or application that functions as government information technology [4]. So the government's efforts to improve quality performance services are by providing online-based services. so that the website is a form of developing information technology that is effective in publications. In addition, the website is a means for the public to access information [5]. Apart from the website, the government also needs public feedback for its performance assessment. One of the ways to get feedback from the public is by using social media, from which the social media used is Twitter [6]. This study aims to determine the quality of the city government website. The object of this research is the three official websites and twitter of three city governments in Central Java, namely Salatiga, Tegal, and Surakarta. In addition, what makes the difference in this study is that the previous study examined only the quality measurement based on the website, whereas this study measures the quality of the quality city government website from the assessment of the official city government website and Twitter. This research is aimed at three city governments in Central Java, namely, the city of Salatiga, the city of Tegal, and Surakarta. The arrangement of the manuscripts in this study includes the title, author's name, address, abstract, introduction, literature review, method, results, and conclusions.

## 2 Literature Review

### *2.1 The Use of Website on E-Government*

The emergence of E-Government is due to the development of Information and Communication Technology (ICT) with the aim of providing government services to stakeholders [7]. In addition, the use of technology in the government sector is also known as E-Government, in which technology plays a role in facilitating the management of information and services [8]. The current popularity of the internet can change the model of government services to the public or the public gradually, namely those that used to do manual services but now services through websites that have been provided or online services [9].

In E-Government services are usually delivered through a web portal that is built and then managed by the government [10]. With the use of the website by the government so as to increase transparency, so as to grow public trust in the government, in addition to reducing corruption in the public sector, thereby strengthening the accountability of the government [11]. The provision of access to electronic transactions that exist throughout the government that is integrated and aimed at the government in terms of openness and electronic participation of citizens through government-wide access to data, e-petitions, electronic consultation, and so on [12].

### *2.2 E-governance*

Electronic Governance (E-governance) is the implementation of functions and achievement of governance outcomes using Information and Communication Technology (ICT). UNESCO defines that "E-governance is the use of ICT by various community actors to improve access to information and build their capacity" [13]. E-governance, or electronic government (E-Government), is defined as the government's use of web-based technologies and applications, or electronic services, which improve access and delivery of government services and information to citizens [14]. People, business, government, and other related entities of E-governance as "the use of information communication technology by governments to increase transparency" [15]. E-governance is an important factor in the development process and service delivery in public and private institutions. However, government units delegating to Kenya must demonstrate ICT readiness for e-governance to be successful in their efforts to achieve this goal [16]. To view e-governance as the use of government information communication technology in increasing openness, raising awareness, and ensuring transparency in the management of government businesses. This can be seen on the Internet or other digital means through information in the form of online services [17]. The implementation of e-governance is an important mechanism in increasing citizen participation, monitoring and evaluating government projects, ensuring government accountability and transparency, and transfer of information from one sector to another [18].

## 2.3 Public Services

Public services are defined as products of the public bureaucracy that are accepted by the public or users of public services as a whole [19]. In Law Number 25 of 2009 which discusses Public Services in Indonesia, public services are service providers made by the government relating to budgeting, administration, resources, and implementation of public services [20]. Governments of several countries have recognized the need to use E-Government to facilitate public sector reform and good governance by applying creative and sustainable ICT technologies both in government policies and in relation to their relations with citizens and the private sector [21]. Public services as contact with and distribution of services to the general public [22]. To improve public services it is important for political bodies to do so. In practice, the public expects the government to continue to increase the standard of public services. This is in line with the fulfillment of community rights and obligations and the realization of state duties through optimal public services. This is in line with the fulfillment of community rights and obligations and the realization of State responsibility through optimal public services [23].

## 2.4 Measuring Quality of Government Website

A quality government website has an important role in building trust and ensuring the successful implementation of E-Government and other delivery [17]. Then there are factors that affect the quality of the E-Government website including accessibility, usability, content, community participation, services, features, security and privacy [24]. The availability of clear information can build loyalty so as to provide security to its users [19]. Various criteria are used in evaluating the quality of the website [21]. In evaluating websites, E-Government there are several criteria used which include: security, privacy, usability, content, and participation. Apart from that the features, transparency, accountability, effectiveness, and efficiency can be used in measuring the websites of the government [23].

## 2.5 Quality of E-Government Service

The current development of social media, especially users, is no longer seen as mere reactions, but can provide opportunities for people to interact more in a function of government [25]. In addition, the use of social media from government plans is able to significantly increase public services. The improvement of these services is considered to be able to encourage the growth of participation in the community [26]. With the effort to realize good governance, therefore the government should provide quality services to the community. Service quality can be measured by several methods, one of which is quality (SERVQUAL) which is then divided into five parts, namely, tangible, reliability, responsiveness, assurance, and empathy [27]. Furthermore, the website quality assessment consists of units of analysis, assessment categories, and criteria from assessment on the website which include: website features, divided into several elements, namely, at a glance, local government, geography, map of the area, resources, regional policy regulations, guest books, and news. Then transparency, seen from activity reports, performance reports, and local government financial reports. Furthermore, accountability is assessed from several indicators, namely, regional government budget reports, grants and social assistance reports, timely presentation of financial reports, and BPK audit opinion. Then an effective and efficient website. Effectiveness in the value of Interactivity, Navigation System. While efficiency is rated on 3 elements, namely, the website does not have a blank page, important information is on the main page, and the website does not go through an intermediary page [28].

## 3 Method

This research is a descriptive-interpretive study by describing, interpreting, and making conclusions. Descriptive research means telling or expressing intentions, meanwhile interpretant has the character of an impression, opinion or view, therefore interpretive descriptive research is a type of assessment by describing or telling opinions or views on the object. Besides that, the objects that are the aim of this research are the official websites and Twitter of three city governments in Central Java, namely the city government of Salatiga, Tegal city, and also the city of Surakarta. Therefore, data collection was carried out by means of surveys and observations on the Web site and Twitter. The stages of assessing the quality of the website are carried out by giving a score on 5 indicators which include features on the website, transparency, accountability, effectiveness, and efficiency, then seen from the 40 existing analysis units (Table 1).

This research was conducted by surveying and observing each of the 3 (three) websites of city governments in Central Java province. So that data collection is done

**Table 1** City government website and twitter

| No | Local government | Website | Twitter account |
|---|---|---|---|
| 1 | Salatiga | http://salatiga.go.id/ | @HumasSalatiga |
| 2 | Tegal | https://tegalkota.go.id/v2/index.php?lang=id | @Portaltegalkota |
| 3 | Surakarta | https://surakarta.go.id/ | @Pemkot_solo |

**Table 2** Page speed insights scoring

| Score | Rating |
|---|---|
| 0–49 | Slow |
| 50–89 | Average |
| 90–100 | Fast |

*Source* https://web.dev/performance-scoring/

by looking for websites from the city government. To measure website performance, it is measured using tools provided by Google, namely Page *Speed Insights*. Then seen from the website loading speed on a percent scale (Table 2).

Assessment of three local government websites in Central Java in measuring the quality of municipal government websites by using a suspension assessment. The analysis of the scores shows the assessment process for policy alternatives by creating and using assessment indicators to assess policy alternatives that have been developed and are in accordance with the objectives that have been formulated, if necessary, give weight to the indicator value which is considered important from other indicators. To calculate the comparative assessment, a certain score is given, namely, referring to the assessment scale analysis [28]. If there is data available (available) then a score of 1 can be given, but if there is no data available or (absent) then the value is 0. The following is the formula for calculating the weight of the score, which is as follows.

$$\text{Weight score : number of scoring the results of local government website} \times 100\% \text{ Total score}$$

After the weighted score is obtained, then an assessment is carried out according to the Table 3.

In addition to the official city government website account, one of the objects in this study is the official Twitter of the three city governments in Central Java province. Then the Twitter account is analyzed using the Nvivo 12 plus application and using the NCapture feature so that it can be systematically documented and analyzed [29].

**Table 3** Score rating levels

| Score (%) | Overall value |
|---|---|
| 0–25 | Not qualified |
| 26–50 | Less quality |
| 51–75 | Quite qualified |
| 76–100 | Quaified |

# 4 Result

## *4.1 Website Quality on City Government*

Stages in measuring the quality of the quality websites of the city of Salatiga, Tegal City, and Surakarta city are by using the assessment on 40 units of analysis. Then the assessment is seen from giving the lowest score, namely, (1) features, (2) transparency, (3) accountability, (4) (effectiveness), and (5) is efficiency. The results of the scores from the three East Java city government websites are as follows (Fig. 1).

The picture above shows that the website of the city government of Salatiga and Surakarta got a score of 55%, while the website of the city government of Tegal got 52.5%. Although they obtained different scores, the three of them were said to be of sufficient quality in the assessment standards. In the feature assessment, the highest score was owned by the City Government of Tegal with a percentage of 17.5%, then Surakarta was 12.5% and the lowest was Salatiga. For website transparency, the highest score was Surakarta and then Tegal, then Salatiga, 7.5%. Furthermore, for accountability, the highest score was Surakarta, namely 12.5%, Tegal city 10%, while Salatinga 0. Then the effectiveness assessment, Salatiga and Surakarta cities got the same score, namely 17.5%, while Tegal city 15%. And in the efficiency assessment, the city of Salatiga scored 5%, while the cities of Tegal and Surakarta both had a score of 2.5% (Figs. 2 and 3).

The ease with which people can access the website needs to be considered by the city and local governments, because the official website has a very important role in communicating or in conveying information to the community. Data from Google shows that 53% of people or internet users in Indonesia will leave a website that

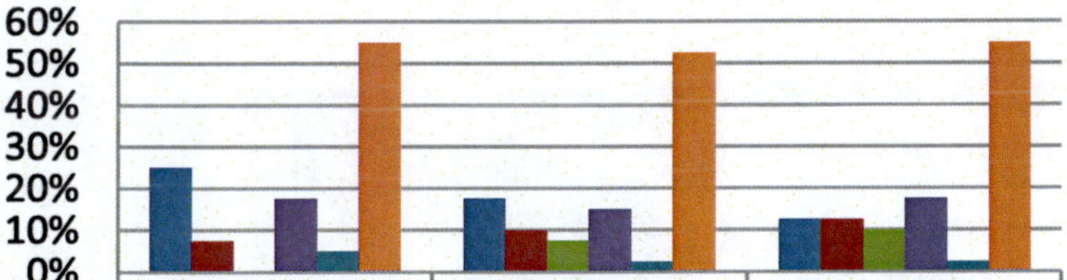


| | Kota Salatiga | Kota Tegal | Kota Surakarta |
|---|---|---|---|
| Fiture | 25% | 17.5% | 12.5% |
| Transparency | 7.5% | 10% | 12.5% |
| Accountability | 0 | 7.5% | 10% |
| Effective | 17.5% | 15% | 17.5% |
| Efficient | 5% | 2.5% | 2.5% |
| Weight Score | 55% | 52.5% | 55% |

**Fig. 1** City Government Website Quality Score

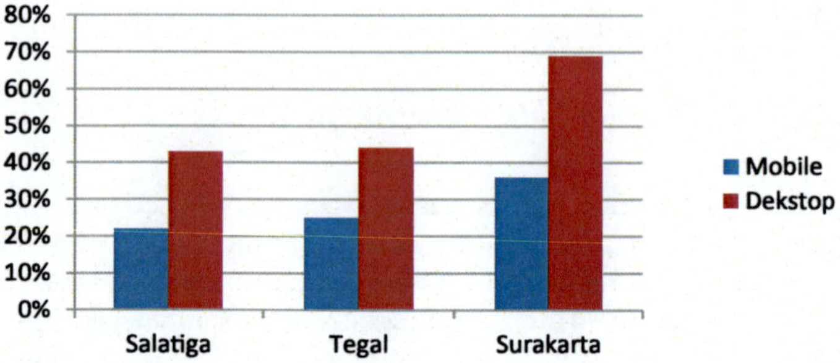


**Fig. 2** Website access speed

**Fig. 3** Twitter city government of Salatiga, Tegal city and Surakarta city

loads more than 3 s [30]. For the website speed of the three city governments in Central Java, it can be seen in the following Fig. 4.

From the picture above shows the website of the Surakarta city government gets the highest value of access speed via desktop, namely 69%, while the city government of Tegal is 44% and the city of Salatiga as much as 43%. Then the speed of accessing

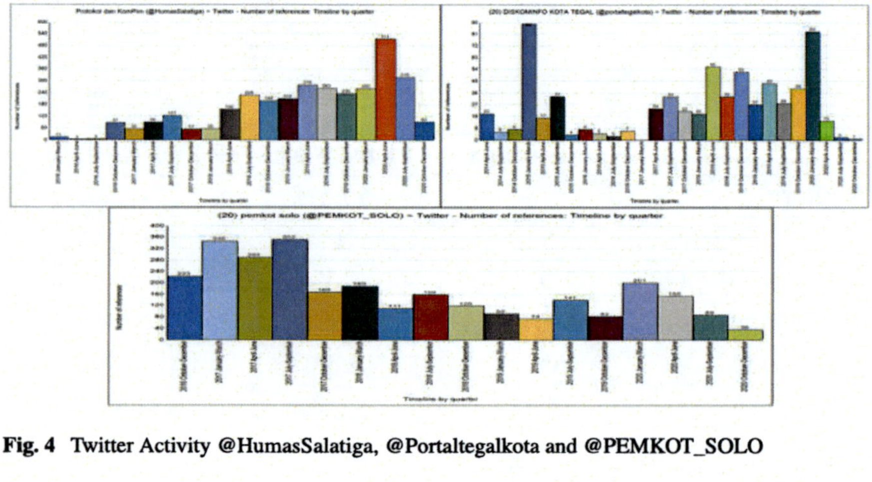

**Fig. 4** Twitter Activity @HumasSalatiga, @Portaltegalkota and @PEMKOT_SOLO

using a mobile, the city government of Surakarta also received the highest score at 36%, when compared to the Tegal city government which received 25%, and the Surakarta city government which was 22%. The city government expected to be able to increase the speed of the website in accessing so that people can easily visit the local municipal government website.

## 4.2 City Government Official Twitter Account

To make improvements in services, the government needs *feedback* from the community, therefore one way to get *feedbak* from community is by using social medians, one of which is Twitter [6]. By using Twitter, the public and the government will find it easier to find out about information and performance from the government. Therefore the city government of Salatiga, Tegal city, and Surakarta city have used Twitter accounts as a means of communication and interaction with the community.

The picture above explains that the Twitter of the Salatiga, Tegal, and Surakarta has different information. The Salatiga City Government is bigger in providing information about Covid-19. While the city of Tegal is more informed regarding the announcement of the results of the CPNS and Covid 19 selection. Then the city of Solo (Surakarta) is more on information related to PNFest 2020 by informing and inviting to enliven the PNFest event. Therefore, the issue that develops in the city government can increase the Twitter activity of the city government.

The picture above shows the Twitter account activity of three city governments, the first being the Salatiga city government in 2016–2020, the Salatiga city government tweeting 3273 tweets. The most-tweeted in April–June 2020 was 514, this increase occurred during President Joko Widodo's working visit to Central Java in the context of handling and overcoming the coronavirus by visiting the Covid-19 handling post. Furthermore, the Tegal city government through the @portaltegalkota account if viewed from 2014 to 2020 looks inactive from April to June 2017. Then the most tweets were in January-March 2015 and January–March 2020. In 2020 the city Tegal has 101 tweets, most of which are in January and March, while the fewest are from October to December. Therefore, it can be assessed that the Twitter activity of the Tegal municipal government is not very active and consistent from April 2014 to December 2020. Then the Surakarta city government twitter activity through (@PEMKOT_SOLO), in the picture above is said to be inconsistent, when viewed from December 2016 to December 2020. The number of tweets at least is 36 tweets in October and December 2020, while July–September 2017 as many as 352 tweets. In 2020 the municipality of Surakarta received 481 tweets, viewed from January to December.

## 5 Conclusion

Based on the results of the assessment regarding the quality of the websites of the Salatiga city government, Tegal city and Surakarta city described previously, it can be concluded that the websites of the three local governments have different scores, namely the city of Salatiga getting 55%, while the city of Tegal obtained 52.5% and the city of Surakarta as much as 55%. Although it has a different final score, the website of the city government is considered of sufficient quality. However, the official website and Twitter of the local government have little feedback or feedback from the community. The benefits of using the website by the government are that it makes easier to provide online-based public services and can support the quality of government performance, then the benefits obtained by the community are easier to get information and services by the government. The inclusion of research from the three city government websites in Central Java, in obtaining the highest percentage value, is seen from the criteria in the website assessment method. Furthermore, after obtaining the required data, the data is added up as a whole consisting of features, transparency, accountability, effectiveness, and efficiency.

## References

1. Napitupulu DB (2016) Kualitas layanan E-Government (Sebuah Analisa Di Pemkot X Dengan Pendekatan E-Govqual Dan Ipa). J Penelit Pos dan Inform 6(2):153. https://doi.org/10.17933/jppi.2016.060203
2. Warjiyono W et al (2018) Pengukuran Kualitas Website Pemerintah Desa Jagalempeni Menggunakan Metode Webqual 4.0. J Efisiensi 5(1):139. https://doi.org/10.25126/jtiik.201852666
3. Pratiwi ADE, Wardani NH (2019) Analisis Kualitas Situs Web Pemerintah Daerah Pelayanan Perijinan Terpadu Terhadap Kepuasan Pengguna Menggunakan Model E-Govqual (Studi Kasus: Dinas. Teknol Inf dan 3(9):9177–9185
4. Kagalwala T (2018) Status of nurses in India. Indian Pediatr 55(10):919–920. https://doi.org/10.1007/s13312-018-1410-5
5. Kuswantoro A (2015) Efisiensi: Kajian Ilmu Administrasi. J Efisiensi XIII(1):76
6. Rosdiana R, Eddy T, Zawiyah S, Muhammad NYU (2019) Analisis Sentimen pada Twitter terhadap Pelayanan Pemerintah Kota Makassar, pp 87–93
7. Kaur A, Kaur H (2017) Exploring E-governance service delivery dimensions: a study of registrar of companies. Indian J Public Adm 63(4):557–566. https://doi.org/10.1177/0019556117726711
8. Muttaqin MH, Susanto TD (2019) The effect of website components on user trust in increasing the interest to use public administration service on E-Government website. In: Proceedings 2019 International Conference on Computer Science Information Technology Electronic Engineering ICOMITEE 2019, vol 1, pp 30–36. https://doi.org/10.1109/ICOMITEE.2019.8921171
9. Huang J, Guo W, Fu L (2019) Research on E-Government website satisfaction evaluation based on public experience. In: 4th international conference on education and social development, pp 206–213. https://doi.org/10.12783/dtssehs/icesd2019/29844
10. Jiang, Ji (2014) E-Government web portal adoption: the effects of service quality. e-Service J 9(3):43. https://doi.org/10.2979/eservicej.9.3.43

11. Porumbescu GA (2016) Linking public sector social media and E-Government website use to trust in government. Gov Inf Q 33(2):291–304. https://doi.org/10.1016/j.giq.2016.04.006
12. Henman P, Graham T (2018) Webportal vs google for finding government information on the web: from a website-centric approach to a web ecology perspective. Inf Polity 23(4):361–378. https://doi.org/10.3233/IP-180071
13. Kumarwad LL, Kumbhar RD (2016) E-governance initiatives in Maharashtra (India): problems and challenges. Int J Inf Eng Electron Bus 8(5):18–25. https://doi.org/10.5815/ijieeb.2016.05.03
14. Solvak M, Unt T, Rozgonjuk D, Võrk A, Veskimäe M, Vassil K (2018) E-governance diffusion: population level e-service adoption rates and usage patterns. Telemat Inform 36:39–54. https://doi.org/10.1016/j.tele.2018.11.005.
15. Ali H, Ali T, Matar Z, Jawad F (2015) Citizens' Acceptance and Readiness towards Adopting E-Participation Tools in Kingdom of Bahrain. Int J Infonomics 8(2):1029–1036. https://doi.org/10.20533/iji.1742.4712.2015.0121
16. Gichuki J, Wanyembi PG, Kibaara D (2017) Level of E-governance readiness in Kenya: a case of Laikipia County 1(I):24–34
17. Đurek V, Ređep NB (2016) Review on e-readiness assessment tools. Cent Eur Conf Inf Intell Syst pp 161–169
18. Harcourt P (2020) E-governance governance application; improving the administrative efficiency of public sector in rivers state, Nigeria Abstract, pp 6–18
19. Hidayati N (2016) E-Goverment Dalam Pelayanan Publik (Studi Kasus tentang Faktor-faktor Penghambat Inovasi Layanan E-Samsat Jatim di Kabupaten Gresik). Kebijak. dan Manaj. Publik 4(3):1–8
20. Eko Priyo Purnomo Zulfiani Ayu Astutik GO (2019) International conference on public organization asia pacific society for public affairs (apspa) khon kaen province, thailand, 28–30 august 2019. Int Conf Public Organ Asia Pac Soc Public Aff 28–30
21. Ashaye OR, Irani Z (2019) The role of stakeholders in the effective use of E-Government resources in public services. Int J Inf Manage 49(January):253–270. https://doi.org/10.1016/j.ijinfomgt.2019.05.016
22. Engin Z, Treleaven P (2019) Algorithmic government: automating public services and supporting civil servants in using data science technologies. Comput J 62(3):448–460. https://doi.org/10.1093/comjnl/bxy082
23. Budiana HR, Bajari A, Mulyawan R (2020) Transformasi nilai Sabilulungan dalam aktivitas komunikasi pelayanan publik di Pemerintahan Kabupaten Bandung. J Kaji Komun 8(1):42. https://doi.org/10.24198/jkk.v8i1.23768
24. Yaghi A, Al-Jenaibi B (2018) Happiness, morality, rationality, and challenges in implementing smart government policy. Public Integr 20(3):284–299. https://doi.org/10.1080/10999922.2017.1364947
25. Linders D (2012) From E-Government to we-Government: Defining a typology for citizen coproduction in the age of social media. Gov Inf Q 29(4):446–454. https://doi.org/10.1016/j.giq.2012.06.003
26. Lestari, IS, Moon JY (2014) Association for Information Systems AIS Electronic Library (AISeL) Understanding citizen participation in governmental social media: a case study of the Indonesian regional police
27. Widyarto WO, Djamal N, Adhim F (2018) Analisis Kualitas Pelayanan Publik dengan Metode Fuzzy-Service Quality (F-Servqual) dan Index Potential Gain Customer Value (IPGCV). J Sist dan Manaj Ind 2(2):101. https://doi.org/10.30656/jsmi.v2i2.769
28. Cheisviyanny C, Helmy H, Dwita S (2018) Analisis Kualitas Website Pemerintah Daerah Kabupaten/Kota di Provinsi Sumatera Barat. Simp Nas Keuang Negara 6(3):1087–1104
29. Edwards-Jones A (2014) Qualitative data analysis with NVIVO. J Edu Teach 40(2):193–195. https://doi.org/10.1080/02607476.2013.866724
30. Gevelber L (2016) Mobile search intent: New consumer behavior data. https://www.thinkwithgoogle.com/

# Cloud Computing Adoption-Based Digital Open Government Services: Challenges and Barriers

**Abdulrahman S. Alenizi and Khamis A. Al-karawi**

**Abstract** The governments of the Arab Gulf have given strong guidelines on growing the use of the cloud in the public sector as part of their digital strategy. The adoption of cloud is identified as the most significant investment pillar within digital transformation initiatives. Cloud is a model or architecture and a modern framework for computation. And the Service-Oriented Design is the central architecture. Cloud Computing has emerged as a core resource-sharing computing network that incorporates storage, apps, frameworks, and business processes. Digital open Governance (DOG) plays a critical position across every enterprise, and digital open governance systems are assisted by clouds of multiple levels. Cloud has a range of resources that are combined and reused. Since digital open governance utilizes distributed services, this needs a lot of infrastructure. Cloud platforms are beneficial in reducing infrastructure and software expenses. This paper explains how cloud services can be used in digital open governance applications to reduce hardware and platform costs, improve network protection, etc. Digital open government cloud-based project, in Kuwait, is considered to be of great importance for free government agencies. In spite of the fact that it has evolved phenomenally, there are still many adoption challenges faced by various open governmental data cloud-based digital services; problems require careful treatment and consideration by any government agency which intends to implement it. The current paper discovers the different features of the degree to which these barriers obstruct the adoption and utilize of cloud computing in the Kuwaiti public area, and also examines the impact of the barriers on the participants' perception of security. The practical proof that was seen by participant's offered strong responses to the set online questionnaire from Kuwaiti users, which comprised IT-staff from several government sectors.

**Keywords** Digital open government · Cloud · Security · Users' perceptions · Information and Communication Technologies (ICTs) · Public Sector

A. S. Alenizi (✉)
Business School, University of Salford, Manchester, UK
e-mail: a.alenizi@edu.salford.ac.uk

K. A. Al-karawi
University of Diyala, Diyala, Baqubah, Iraq

 
X.-S. Yang et al. (eds.), *Proceedings of Sixth International Congress on Information and Communication Technology*, Lecture Notes in Networks and Systems 216,
https://doi.org/10.1007/978-981-16-1781-2_15

## 1 Introduction

In our present time, improvement of Information and Communication Technology (ICT) is one of the recognizable features, and essentially similar each other new turn of events, it has a major impact concerning change and modernization of people's life by and large. To me more unequivocal, the move of ICT has an affected change in the customers' masterminded exertion with their more open governments, developing their longings, totally, in this aspect. Right when the digital open government and open data systems were begun, certain affiliation relationships around the world moved from the standard occurrence of public relationship to open data structures, owing to the authentication of the enormity of giving high-type and able relationship for its customers. Regardless, the execution of digital open government made a way for more prominent undertaking and demand difficulties like creative and money-related ones [1, 2]. Creative ICT like scattered figuring progress could help with beating burdens; appropriated planning recommends a central change in the mechanical atmosphere improvement of the public region and the manner in which governments do their digital affiliations [3, 4]. In made countries like USA, Japan, and EU, cloud movement has been driven by driving exercises, and it has made an astounding advancement toward changing into the bleeding edge period of digital open occupant looked after affiliations [4]. The strategy is to use basically versatile, inescapable, locale self-managing Information Technology resources to make total cycles and copy the open government data which plan with the customers' cravings to refresh the relationship between open government with more direct, adaptable, consolidated, and straightforwardness enrolling. Notwithstanding, the cloud-based digital open government saw as a vigilant change in the councils likewise shows a customer-driven digital affiliations condition focusing to grow customers' speculation. Kuwait, the country composed in the point of convergence of the Middle East region, is proceeding to a complete progression to cloud-digital open government [5]. Kuwait grasped the central piece of cloud computing concerning the digital open government and introduced the fundamental "Public Cloud Platform", made to prompt solid new developments and improvement in the digital open government applications. Nowadays, certain indisputable government units and associations complete cloud-based responses to offer better open digital relationship to their customers; still, an unfathomable number of Kuwaiti nationals don't yet utilized cloud-hosted digital open government and still, rely on paper printouts related to their official trades [6]. Fitting improvement is taken in this paper for reacting to the sales: Do the blocks and burdens of digital open government task impact the use of cloud-based digital open government relationship from the Kuwaiti customers' and government's viewpoint? The inevitable postponed outcome of this assessment passed on that some Computerized open government barriers truly affect the sort toward cloud-based public advanced affiliations similar to the nonattendance of security and care. In spite of detailing, a piece of the security concerns identified with this assessment setting, appropriately these can be thought of while determining another speculative model.

## 2 Cloud Building Blocks for Digital Open Government

The term cloud storage is comparatively recent and is definite by the IEEE Computing Society as a model in which knowledge is continuously kept on servers on the internet and momentarily stored on clients, like desktops, entertainment centers, laptops, notebooks, handhelds, etc. Cloud computing building blocks are embedded in hardware and software architecture and networking devices that allow creative infrastructure scaling and virtualization [7, 8]. However, the next technology developments are more complex and continue dynamic provisioning to be handled in a broad cluster inside the infrastructure. There are also consequences for the next generation technology architecture to allow efficient use of capital and fault tolerances in the enterprise. Cloud infrastructures have the ability to incorporate positive performance behavior. While sharing a broad infrastructure will average the complexity of individual workloads, it is impossible to determine the exact output characteristics of the program at any given time. As any shared infrastructure, a number of individual workloads can have an effect on the usable CPU, Networks, and I/O resources resulting in inconsistent output behavior of the integrated applications. The definition of the cloud is a wide-scale dispersed computing model. Where a pool of virtualized, flexible, and manageable storage, computing power, networks, and services can be offered to customers on demand through the Internet [9]. There are many benefits from using the cloud such as Ease of implementation, cost savings, access to IT capabilities, and renewable computing [10].

### 2.1 *Cloud Challenges*

Cloud infrastructure, like most Internet technologies, has challenges, though. Many of the challenges of cloud infrastructure for digital open government deployment derive from its novelty and the relative growth of the cloud services industry.

**Free standards and interoperability**

There is a shortage of standards in the usage and application of cloud infrastructure. Users should be able to migrate among cloud service suppliers with a minimum of threat and expense [11].

**Security and privacy**

Another field of great interest to policymakers is the security and privacy of knowledge stored in cloud storage environments. Significant attention must be paid when implementing cloud computing to process knowledge that is crucial to national security [12].

**Business continuity**

The growing possibility of data loss due to incorrect backups or device errors is beyond consumer reach, rendering it another obstacle to maintain business continuity. Governments need to consider business continuity threats to ensure that appropriate steps such as good agreements, disaster retrieval, and business continuity strategies are in place [13].

**Internet dependency**

As cloud infrastructure systems dependent entirely on cloud computing. Accessibility and pace of the Internet as a transporter between customer and service supplier, speed, and affordability would be a matter of concern[14].

# 3 Digital Open Government Challenges

Despite the way that the affiliations have advanced various endeavors in such a way, yet, the achievement of digital open government programs is dependent on the customers' receptiveness to utilize and get the digital open government using new Information and Communication Technology (ICT)[15]. In spite of advanced still, there are many challenges as follows:

## *3.1 Technological Challenges*

The part, here, explains the major tangles that face modernized open government from the particular perspective as shown in the following [16].

**IT-infrastructure**: There are many examples about this challenge such as inadequate networking capacity, insufficient integration across systems, unwell updated hardware and software, incompatibility, and complexity of the current systems.

**Security**: An example of this challenge is security by lack of transactions, Failure to trust Internet and government electronic services, and deficiency of protective facilities in the public sector [2].

**Availability**: An instance of these challenges is lack of ability to include resources and information on request and poor reaction to citizens' demands and unsuccessful delivery of e-services.

**Accessibility**: An example of this challenge is difficulty in using the Program for people with disabilities and the reach of the Internet is minimal.

**Website design**: An example of this challenge is restricted languages for the display of material on the website and perception of ease of usage.

### 3.2 Users Aspects

The following tends to the basic difficulties identified with the users' aspects [17].

**Lack of awareness**: An instance of these challenges is lack of awareness and advantages of digital open government systems and lack of orientation campaigns to endorse digital open government.

**ICT skills**: an example of these challenges is lack of IT skills between digital open government users (i.e., users, employees, IT-staff) and the absence of practical information on safe online activities.

### 3.3 Social Challenges

Social problems associated with the gathering of digital open government cannot be genuinely seen as a mechanical hindrance; nevertheless they focus on the social consequences of the overhauled IT. An example of these challenges is ethnic and tribal faith, language issues, and resistance shift [18].

### 3.4 Financial Challenges

The main challenge is lack of budget or high cost [18]. An instance of these challenges is high repair and running costs and high budget security solutions.

## 4 Methodology

In this assessment, considering the coronavirus pandemic, a quantitative assessment framework using an online examination structure was used to get the fundamental data and to pick the evaluation targets. An audit review was picked since it eats up less degree of time, is more moderate, respondents can remain secretive and it can likewise offer a data inflow from a basic model with a minimum level of researcher bias [19]. The investigation was relied on to set up the level of how Kuwait inhabitants see troubles concerning enduring cloud-based digital open occupant kept up affiliations moreover the impact of these challenges on the certain security. Comparative challenges have in like manner been concentrated from the viewpoint of the affiliations providers. IT-staff or employees in this assessment are specialists of a particular section of the public zone, by virtue of which it is imperative to assessments the viewpoints on IT-staff about the issue. Authorities have portrayed the innovative information technology of dissipated figuring on the focal page of the online graph since it is a fundamental original thought in non-present day countries. The online survey interface was progressed to the proposed study test through different correspondence prompts (for instance E-mail and social media). The examination test has referenced Kuwaiti customers who are web customers having a focal perception of digital open inhabitant kept up affiliations and passed on figuring. The second occurrence of IT-laborers, for instance, site experts, programming/structure fashioners, and designers, respondents have moved closer for responding through.

## 5 Result and Finding

### 5.1 *Summary of the Questionnaire*

As explained over, the survey includes four sections put down in the Arabic language for better arrangement and accessibility, since Arabic is consistently given in the language of the occupants of Kuwait residences. Regardless totally determined, and confirmed the consent of the focal solicitation on the shield material page by telling that their consent is unequivocal by the "yes" answer to a sales. For around two months, it remained open "on the web", to all Kuwaiti customers around the world, close to the realization of which the responses were 100. This model appraisal fuses its two class' staff and Kuwaiti customers. While 80 of the individuals were ordinary Kuwaiti customers, 10 individuals from them were delegates from different government relationship in Kuwait. Nevertheless, 10 residents' comments were refused due to inaccurate or missing replies. A whole of 90 responses were free for the data evaluation, for seeing the degree to where the digital open government challenges impact the confirmation of cloud relationship in the public space from the perspective of the Kuwaiti customers what's more its effect on clear security. The use of the Internet is just expected of those of the internet generation. In addition to the experience of information technology staff in emerging and delivering such services. Furthermore, as the expansion of cloud-based facilities in Kuwait is at an early phase, respondents will be the main potential stakeholders. Knowing their perceptions and attitudes is therefore crucial to improvement in cloud-based digital open government facilities. The outcomes of the analysis are detailed below. Along these lines, it is essential to comprehend the acknowledgments and points of view, since it is fundamental for refreshing the cloud-based DOG Services.

### 5.2 *Participants Details*

The population description of the two different surveys (i.e., Kuwaiti residents and information technology staff) is outlined in Tables 1 and 2. In this layout, creators have recorded as of late certain profiles like age, coaching level, open information level, and security care level. Concerning the age factor, the appraisal demonstrated

**Table 1** General data for information technology staff in public area

| | Description | Percent (%) | Participants |
|---|---|---|---|
| Participate age | 25–35 | 55 | 49 |
| | 36–45 | 36 | 32 |
| | 46–55 | 6 | 5 |
| Qualification<br>Security awareness stage | Diploma | 28 | 26 |
| | Bachelor | 61 | 55 |
| | Postgraduate | 10 | 9 |
| | Medium | 66 | 55 |
| | Advanced | 27 | 25 |

**Table 2** Information for Kuwait users

| | Description | Percent (%) | Participants |
|---|---|---|---|
| Participate age | 18–25 | 36 | 32 |
| | 26–35 | 51 | 45 |
| | 36–45 | 9 | 8 |
| | 46–55 | 3.2 | 3 |
| | 56 + | 2 | 2 |
| Qualification | Diploma | 32 | 29 |
| | Bachelor | 57 | 51 |
| | Postgraduate | 10 | 10 |
| Security awareness stage | Medium | 77 | 69 |
| | Advanced | 23 | 21 |

that the best level of everybody test were those of the ages fundamentally between (18–45) years of age, which was unsurprising with the best common bit of Kuwaiti digital clients. The greater part of the respondents that took an interest had a drawn-out certification as getting ready capacity level which counted with the overall distribution of digital clients in Kuwait. On an overall explanation, the clients' respondents were self-seen as people with medium digital information and practically youngster to medium-security care. On separating, it was seen that the essential capability in the IT-staff test was self-obvious proof with a huge load of digital information and security care.

# 6 Explanation of the Research Questions

Thinking about the outcomes, there are various difficulties comparatively as obstructions to digital open government confirmation and get together, explicitly: mechanical, social, human focuses, and monetary; these still ruin the insistence and usage of scattered enlisting associations in Kuwait's public area as per the point of view of the people. Regardless, the public position district meets with faces hindrances from those clients who anticipate more fundamental guidelines of the public digital associations (for example administrative) when showed up distinctively corresponding to the digital associations of the private region. Considering certain appraisal reviews, specialists have discovered nine restrictions that come in the strategy for attestation of digital open government. As such, respondents were mentioned to review their own recognition from the degree of centrality for each square by analyzing the difficulties on the off chance that it doesn't address one (0), basic one (1), or gigantic one (2).

## *6.1 Users' Views of Cloud-Based DOG Hurdles*

Tables 3 and 4 indicate that all nine obstacles were chosen as either a really significant or significant obstacle, although none of the nine were chosen as not a problem for all the applicants.

**Table 3** Really significant challenges to accept Cloud-Based application from users' viewpoint

| Challenge | Really significant challenge | |
|---|---|---|
| | Number | Percent (%) |
| IT-infrastructure | 11 | 10 |
| Security | 20 | 18 |
| Availability | 10 | 9 |
| Accessibility | 8 | 7.2 |
| Website design | 5 | 4.5 |
| Lack of awareness | 14 | 12.6 |
| IT skills | 7 | 6.3 |
| Culture | 11 | 9.9 |
| Lack of budget/high cost | 4 | 3.6 |

**Table 4** Significant challenge for embracing cloud applications from customer viewpoints

| Challenge | Significant challenge | |
|---|---|---|
| | Number | Percent (%) |
| IT-infrastructure | 10 | 9 |
| Security | 19 | 17.1 |
| Availability | 8 | 7.2 |
| Accessibility | 7 | 6.3 |
| Website design | 10 | 9 |
| Lack of awareness | 14 | 12.6 |
| IT skills | 7 | 6.3 |
| Culture | 12 | 10.8 |
| Lack of budget/high cost | 3 | 2.7 |

**Table 5** Really significant challenge to consider cloud from the IT-staff viewpoint

| Challenge | Significant challenge | |
|---|---|---|
| | Number | Percent (%) |
| IT-infrastructure | 10 | 9.9 |
| Security | 18 | 16.2 |
| Availability | 9 | 8.1 |
| Accessibility | 7 | 6.3 |
| Website design | 9 | 8.1 |
| Lack of awareness | 15 | 13.5 |
| IT skills | 8 | 7.2 |
| Culture | 12 | 10.8 |
| Lack of budget/high cost | 2 | 1.8 |

## 6.2 *IT-Employees' Views of Cloud-Based DOG Obstacles*

The study of the IT-employee experience of the obstacles at the two stages of awareness is outlined in Tables 5 and 6. Again, lack of knowledge and protection are the two most common problems, as highlighted in the next segment.

## 6.3 *Respondents' Views of the Really Significant Challenges*

Analyzing the two main challenges viewed by people and IT-staff as "really significant", it can be found that the security of cloud-based applications and their advantages in the digital open government sense was popular (18%) and (16.2%), respectively. This signifies that advertising efforts can play a major role in the improvement

**Table 6** Significant challenge to accept cloud services from IT-Employees viewpoint

| Challenge | Significant challenge | |
|---|---|---|
| | Number | Percent (%) |
| IT-infrastructure | 10 | 9 |
| Security | 17 | 15.3 |
| Availability | 7 | 6.3 |
| Accessibility | 9 | 8.1 |
| Website design | 11 | 9.9 |
| Lack of awareness | 14 | 12.6 |
| IT skills | 6 | 5.4 |
| Culture | 13 | 11.7 |
| Lack of budget/high cost | 3 | 2.7 |

of digital open government programs. The Lack of awareness issues of knowledge and services offered by digital open government platforms come as a second "really significant", challenge to popularity (12.5.6%) and (12.5%). There is an agreement among scholars that this is a serious technological challenge and a well-documented viewpoint for the introduction of digital open government systems around the world.

## 6.4 Assessment of Challenges Perceptions

This segment seeks to compare the key roles of Kuwait residents and IT-staff on challenges to the adoption and usage of cloud-based digital open government applications. It is clear from the preceding parts that there are a variety of problems that are similar to each. Second, for each of the survey groups, the Security of, and profit from, cloud-based public resources was deemed a "really significant" challenge and rated it as the most relevant on the list. This agreement between the two study groups suggests that it is a primary challenge to be tackled, taking into account a high degree of importance. Next, the two survey communities concluded that lack of awareness was listed as the second main challenge in the "really significant" ranking. Finally, there was a gap in the next most relevant "really significant" task of cloud-based digital open government adoption, representing the individual's viewpoint on sample categories. In the case of information technology staff within government departments, the vulnerability of the IT facilities was perceived as the next graded obstacle to enabling them to deliver successful public services. In regard to the experiences of everyday people, the atmosphere of community has an influence on the effective usage of creative government programs.

# 7 Conclusion

The introduction of new technology to an enterprise could present some of the risks associated with deployment and usage. There are however, several reasons that inspire decision-makers to embrace and utilize this technology. This paper explored the implementation of cloud infrastructure in the sense of digital open government. The advancement of digital structures for open government that is based on the advancement of the cloud is the continuous voice of online digital organizations and free information around the world. Especially in non-modern countries that seek to improve digital open government. The fundamental objective of this work is to eliminate the challenges of digital open government and how it impacts the assertion of cloud-based digital organizations with respect to the public territory and also how much these problems affect clear security. The findings of the survey were focused on the feedback and opinions of senior executives and users in Kuwait. This was done to help the complicity dynamics model of digital open government, and also to improve the bottom line of cloud-based digital open government offered to Kuwaiti society. Problems were found based on a proper composition review and were examined by examination. The results of the scheme showed that security, lack of awareness, and culture issues play a fundamental role in the establishment and use of cloud-digital organizations in an digital open government. It also showed that non-specific points are of great importance among other research-related safety issues.

# References


1. Lallmahomed MZ, Lallmahomed N, Lallmahomed GM (2017) Factors influencing the adoption of e-Government services in Mauritius. Telematics Inform 34:57–72
2. Alkhwaldi A, Kamala MA, Qahwaji RS (2018) Analysis of cloud-based e-government services acceptance in Jordan: challenges and barriers
3. Abu-Shanab E (2014) Antecedents of trust in e-government services: an empirical test in Jordan. Transforming Government: People, Process and Policy
4. Almarabeh T, Kh. Majdalawi Y, Mohammad H (2016) Cloud computing of E-government. Scientific Research Publishing, Communications and Network
5. ALMutairi NN, Thuwaini SF (2015) Cloud computing uses for E-government in the middle east region opportunities and challenges. Int J Bus Manag10:60
6. Weerakkody V, El-Haddadeh R, Al-Sobhi F, Shareef MA, Dwivedi YK (2013) Examining the influence of intermediaries in facilitating e-government adoption: An empirical investigation. Int J Inf Manage 33:716–725
7. Velte AT, Velte TJ, Elsenpeter R (2019) Cloud computing: a practical approach. TMH Pub.
8. Ali O, Soar J, Yong J (2014) Impact of cloud computing technology on E-government. In: International conference on information and software technologies, pp 272–290
9. Foster IZ, Raicu Y, Lu I (2008) Shiyong, Cloud computing and grid computing 360-degree compared. In: grid computing environments workshop, GCE'08, pp 1–10
10. Bhardwaj S, Jain L, Jain S (2010) Cloud computing: A study of infrastructure as a service (IAAS). Int J Eng Inform Technol 2:60–63
11. ATSE X (2010) Cloud Computing: opportunities and challenges for Australia. In: Report of a study by the Australian academy of technological sciences and engineering (ATSE). Melbourne

12. Mukherjee K, Sahoo G (2012) A novel methodology for security and privacy of cloud computing and its use in E-governance. In: 2012 world congress on information and communication technologies, pp 561–566
13. Sallehudin H, Razak R, Ismail M, Fadzil A, Baker R (2019) Cloud computing implementation in the public sector: factors and impact. Asia-Pac J Inform Technol Multimed 7:27–42
14. Sahu BL, Tiwari R (2012) A comprehensive study on cloud computing. Int J Adv Res Comput Sci Softw Eng 2:33–37
15. Al-Rashidi H (2013) The role of internal stakeholders and influencing factors during the phases of e-government initiative implementation. Brunel University, School of Information Systems, Computing and Mathematics
16. Savoldelli A, Codagnone C, Misuraca G (2014) Understanding the E-government paradox: Learning from literature and practice on barriers to adoption. Gov Inf Q 31:S63–S71
17. Ndou VD (2004) E–Government for developing countries: opportunities and challenges. Electron J Inform Syst Dev Countries 18:1–24
18. Ali M, Weerakkody V, El-Haddadeh R (2009) The impact of national culture on E-government implementation: a comparison case study
19. Gray DE (2013) Doing research in the real world. Sage

# Teacher Agency: Navigating the Virtual Learning Environment in the New Normal

**Peter G. Romerosa, Ferdinand C. Lacuata, Maria Teresa N. Rivera, and Eduardo O. dela Cruz Jr.**

**Abstract** The COVID-19 pandemic has brought about unprecedented change in our social lives, particularly in schooling. This led to the development of diverse types of online learning management systems reflecting teachers' efforts in redefining the roles of education amidst the global crisis. Using a case study method, this research examined the initial implementation of the virtual learning environment of a particular school in the Philippines on two folds: first, the conditions that created, facilitated, and enhanced the prospects for learning through the digital mediated instruction; and second, how teachers navigated challenges and opportunities in teaching in the new normal. This examination led to the researchers' theorizing them as active social actors who are deemed to be co-constructors in the new education landscape. The results revealed that the exercise of teacher agency in multi-sited settings resulted in the development of relevant teaching pedagogies and formation of support groups among teachers and parents, and consequently, has given shape to the diverse learning management models. It sought to suggest that by critically assessing teacher agency during the crisis education and the context where agency was situated as well as exercised, we can draw on-the-ground insights that can help us formulate informed education policies for appropriate instructional management systems.

**Keywords** COVID-19 pandemic · Teacher agency · New normal · Virtual learning environment · Learning management system

## 1 Introduction

When the Philippine government officially announced that the Philippines was already under the pandemic during the first quarter of 2020, schools were compelled to implement policies to end the Academic Year 2019–2020 earlier, allowing students

P. G. Romerosa (✉) · F. C. Lacuata · M. T. N. Rivera · E. O. dela Cruz Jr.
Arellano University, Manila, Philippines

E. O. dela Cruz Jr.
e-mail: eduardo.delacruz@arellano.edu.ph

 
X.-S. Yang et al. (eds.), *Proceedings of Sixth International Congress on Information and Communication Technology*, Lecture Notes in Networks and Systems 216,
https://doi.org/10.1007/978-981-16-1781-2_16

to comply with unsettled requirements on conditional terms. The subsequent months were fraught with uncertainties and issues as to whether education should continue, the Philippine education system was prepared in terms of teachers' capacity and infrastructural readiness for digital mediated instruction, and the ways to reduce social divide in the new learning environment given the onslaught of the pandemic in all social spheres.

However, the growing number of COVID-19 cases, shutdown of businesses, travel restrictions, limited physical mobility in public places, job losses, and extension of lockdowns had caused serious impacts on school participation, generating polarized debates among education stakeholders. A petition for academic freeze or cancelation of the current academic activities was advanced by teachers particularly those in the public sector due to the rising cases, digital gap, lack of preparations and materials for learning modalities, and lack of mass testing as well as contact tracing. On the other hand, other sectors opposed such a call. Private schools cautioned that cancelation of classes may lead to unemployment, especially that private schools' income is largely dependent on enrollment.

Meanwhile, President Rodrigo Duterte signed on July 17 Republic Act No. 11480 which amends Republic Act 7797, rescheduling the opening of classes between June and August. This move put an end on the issue of the cancelation of current academic year.

Consequently, most private schools adopted the virtual learning modality and commenced their classes in August, while public schools moved the opening of classes from August 24 to October 5 to provide more time for the preparation of learning modules as their primary learning modality. Other schools opted to implement blended learning or the combination of the modular approach and online instruction depending on their current situations.

The choice to adopt alternative delivery modes of learning is contingent on complex and interdependent factors such as institutional infrastructures and resources, geographical and socio-economic background of stakeholders (i.e., students and teachers), existing funds of knowledge on types of delivery modalities (virtual learning environment, modular approach, blended learning etc.). Regardless of the modes of delivery that the school adopts, teachers are at the forefront of the changing education paradigm from the traditional face-to-face teaching to virtual instruction. With the heightened disparity of students' access to education in the new normal, teachers are called to exercise agency reflecting strategies that contextualize imposed learning platforms.

Hence, this research prompted the authors to explore and recognize the contributions of teachers as frontliners in providing basic education in the new normal. While Information Technology (ICT) Experts and computer engineers focus on the development of effective learning management systems, educationists are the ones to implement them; and thus, it is worthy to hear their voices and learn from their experiences to make education in the pandemic era possible and meaningful.

### 1.1 The PLMS as the Primary Virtual Instructional Delivery System

The impositions of community quarantines and other health protocols as a response to the pandemic have prompted schools, colleges, and universities to adopt the teaching–learning management systems to continue their operations, shaping their own education in the new normal. For the basic education department (pre-school, elementary, junior high school, and senior high school), Arellano University has implemented the Pegasus Learning Management System (PLMS), which serves as the primary digital teaching–learning modality where the teachers and students interact through the virtual environment. In practice, the PLMS uses both synchronous and asynchronous learning modalities. Synchronous learning means online or distance learning that occurs in real time, usually requires a scheduled class meeting. On the other hand, asynchronous learning does not require real-time meeting; rather, learning materials (video lessons, modules, lectures prepared through PowerPoint presentations, rubrics, and assessment activities) as well as announcements are accessed by the students through online ways.

### 1.2 The PMLS as a Negotiated Learning Platform

From the outset of the pandemic, basic education teachers were instructed to make video lessons as an initial step in shifting instruction from the traditional face-to-face classroom to the digital borderless platform. The video lesson contains salient elements: intended learning outcomes, discussion of the content and activities, and assessment tools. Thus, it reflects a concise instructional delivery that is usually seen in a traditional face-to-face teaching. As part of the protocol, the video lessons were reviewed by the evaluation committee, then returned to the teachers to integrate comments and suggestions for revisions. Eventually, the video lessons were uploaded to the PLMS. However, even within this context, video lessons have become the main source of learning. Moreover, the use of video lessons has set a common digital teaching pedagogy and other supplemental online and offline strategies; and thus, creating diverse teaching–learning models.

## 2 Main Objective and Research Problems

The study aimed to describe the experiences of teachers in navigating the mandated virtual learning platform by the university (PLMS). Specifically, it answered the following questions.

1. Under what conditions create, facilitate, and enhance the prospects for learning through the teaching virtual teaching models?

2. How is teacher agency exercised, constrained, and facilitated in the virtual learning environment?
3. What virtual teaching–learning models are constructed on-the-ground as a response to the mandated virtual learning platform by the university?
4. What alternatives are drawn to further improve the institutional virtual learning delivery system?

## 3 Research Methodology

This study utilized a qualitative case study approach, a methodology that provides tools for researchers to explore complex phenomena within the teachers' contexts, using a variety of data sources. It also involves the use of a variety of lenses to uncover multiple facets of the phenomenon to be understood [1]. For Yin [2], case study design is utilized to answer "hows" and "whys" questions to be able to unpack the socio-cultural conditions that are relevant in understanding the investigated phenomenon. Moreover, this design does not manipulate behaviors; and thus, objectively provides the researcher to draw insights and meanings primarily from the experiences and perspectives of the participants [3].

To generate data from the participants, the researchers conducted online focus group discussions among teachers from the basic education department (preschool, elementary, junior high school, and senior high school). The participants were selected according to years of experience (novice and seasoned) and subjects handled, ensuring that all disciplines are equally represented. Online follow-up interviews were undertaken to clarify key issues that needed further elaborations.

## 4 Conditions Facilitating the Virtual Learning Environment

While the PLMS has served a point of convergence, shaping a common online pedagogical practice using video lessons and a system of monitoring and assessing students' outputs, it has also become a space for stakeholders, particularly teachers to make creative improvisations or strategies (*diskarte*), allowing them to leverage the imposed pedagogical structure to meet specific needs of their students. This implies that complex conditions facilitate or constrain diverse responses, elucidating the context of negotiations and implementations of the PLMS.

## 4.1 Student Context and Parental Involvement

Prior to the regular virtual meetings using the PLMS, teachers used to meet students and their parents particularly in preschool and elementary levels in other social media platforms such as Google Meet, Zoom, FB messenger, Google Classroom, and the like. This method has established the grounds for teacher-student/parent connection. For example, primary students who had been accustomed to this practice before the regular use of the PLMS aspired to see their teachers frequently during synchronous and asynchronous sessions. This has resulted in the creation of models 1 and 2, which represent a peculiar practice in the early grades. As teacher A has put in,

> My regular presence in online meetings is important so that I get to establish my image as a teacher to my students. This sets me apart from other characters they see in social media.

On the other hand, learners in the higher levels are more comfortable with teaching models that encourage a certain degree of flexibility for autonomy. In another case, teachers were able to circumnavigate their class schedule by assessing what works in practice and heeding to the request of their parents. This has facilitated in formulating model 2, enabling students to learn two to three subjects only within the day so that students' computer exposures are minimized, and other time is spent for offline activities.

## 4.2 The Nature of PLMS and Online Learning

### 4.2.1 The Affordances of PLMS and Online Learning

Embedded in the PLMS are features that structure online pedagogy and teacher-student interaction, and distribute supplementary learning materials, the use of assessment tools, documenting learning processes, assigning as well as reporting of grades. These imperatives can enhance or limit teachers' creativity. For example, the PLMS contains mostly objective types of tests in the form of multiple choice, matching type, true or false, identification, and essay, which are presented in standardized formats across all levels (pre-school to college). For example, by default, the multiple choice type of test in the PLMS has a minimum of five choices. For the elementary grades, four choices are considered standard. Likewise, in a progressive preschool curriculum, examinations are usually done through observations and other performance-based activities.

Moreover, skill-oriented subjects in the higher levels especially in SHS such as machining, programming, cookery, and so forth require assessment tools beyond the given types of tests available in the PLMS. Further, since the PLMS is accessed through internet connectivity, access to resources, participation in online class discussion as well as submission of tests and other requirements are strongly mediated by the amount of, and speed of connections as well as other network issues. Moreover, socio-economic backgrounds and geographical locations of students play a crucial role in determining what type of remote learning is most applicable in addressing these situations. This builds on the argument that digital pedagogies are not neutral [4, 5]. Consequently, schooling during the pandemic may widen the disparity among students and families. Thus, this situation calls for the need for more discussions that will seek for equitable alternatives for learners. At the micro level, this implies that teachers should be trained to be more agentic and resourceful in terms of developing alternative modalities and strategies to integrate students into the new educational landscape.

## 4.3 Teacher Agency

The pandemic has heightened the limitations of formal schooling concerning crisis intervention. This spurred a major paradigm shift in teaching, allowing teachers to create pedagogical possibilities that are sensitive and applicable to the needs of students and their families during the pandemic period. The following narratives highlight how teachers deployed strategies in navigating opportunities and constraints in teaching in the new normal.

Preschool Teacher A

> At first, I was anxious about getting the attention of my pupils because online teaching is much different from face-to-face teaching, which requires physical interactions, movements, and direct observations for preschoolers. The shift to online learning has affected the strategies and activities we use in class. For example, I tend to be more expressive with my actions, facial expressions, and manners of speaking to get my pupils' attention and encourage them to express themselves.
>
> The feeling of anxiety is heightened by the presence of parents or guardians around while I teach. At a certain point, I feel that I expose my vulnerabilities to the families of my pupils. So, what I did was to involve them. I realized that when parents are involved, they get to know their child better, they cooperate well with teachers, and suggest ways on how to improve teacher-parent communication for the improvement of pupil attention.

Preschool Teacher B

> In terms of strategies, I discerned that online teaching validates that the most effective way to evaluate pupils' performance is done through one-on-one assessment. Although this is quite tedious and requires longer time, it nevertheless allows me to discover their specific needs, capacities, and behaviors that are not usually observed when they are in a group or class.

Preschool teachers were able to apply established routines of progressive education from the traditional face-to-face teaching to the new learning environment. However, these routines need to be redesigned according to their pupils' home context.

Elementary Teacher A

> Since teaching using the video lesson is less interactive than the traditional Math teaching, during the facetime, I provide more problem exercises. We use the Zoom whiteboard or sometimes I share my Google meet screen so that my students get to see how I manually solve the problem. This period is important because students can ask questions which allows me to further elaborate some key points.

## Elementary Teacher B

> After watching the video lesson, we ensure that there is an online meeting to enhance the discussion of the topic. However, when the video lesson is insufficient or not available, I make my own PowerPoint presentations. I prefer that teachers are given a chance to make our own lessons so that we can exercise strategies (*diskarte*) in preparing videos or PPTs for our lessons. PowerPoint presentations are crucial for online teaching.

## Junior High School Teacher

> We formed a network through group messengers and Facebook pages among ourselves to help one another so that we can share resources (teaching materials, PowerPoint presentations, and video lessons) and update each other about important information that we need to know. Our network also allows us to learn from each other about how to navigate different online platforms like Zoom, Google meet, and Facebook messenger room, and make video lessons that we can use for online class discussions. Usually, digital natives or mostly younger teachers are the ones who facilitate webinars on updating technical skills. While some seasoned teachers struggle to acquire technical skills, we feel how they try their best to learn and adapt. In this sense, the new working and learning environment enables us to reinforce our empathy and sensitivity for each other.

## Senior High School Teacher

> Using what I have learned from my Competency-Based Training (CBT), I use the stationing strategy for my cookery class. This strategy allows me to group my students according to their needs and abilities based on the results of the pre-assessment. In the first station, students are provided with lectures on the focus competencies in the pre-assessment. After developing the required competencies, learners are instructed to move to the second station, where verbal tests and recitations are done. In the third station, students are intensively guided for them to acquire the needed competencies for the laboratory activities. The fourth station, which is also the last stage is the virtual laboratory where students apply the targeted competencies, which we tackled from the previous stations.
>
> Since the meeting is per station, it is usually held once a week; and thus, the issues of limited loads and poor internet connectivity are reduced. Also, during asynchronous sessions, students are guided with checklists for their offline activities.

Drawing on the notion of agency by Biesta and Tedder [6], teachers are viewed as informed actors who "critically shape their responses to problematic situations". This concept of agency presupposes that as social actors, teachers navigate their environment in ways that allow them to leverage the situation towards their own agenda.

From the abovementioned narratives, agency was exercised to:

1. Develop teaching strategies: Pedagogical practices were constructed, allowing teachers to discover relevant ways to increase student online participation,

improve discussions, make practical student assessment, maximize time for online meeting, and look for reliable digital resources. It is important to note that the development of these pedagogical practices was meant to integrate students with limited internet connectivity into the mainstream instructional delivery system. One example of this was a senior high school teacher who grouped his students according to their readiness and provided them with a set of tasks for asynchronous sessions while he was giving an online lecture to a particular group.

2. Form support networks: Support groups among teachers and parents were formed to facilitate mentoring and sharing of resources. Because participants in the virtual working and learning environment share the same emotional experiences and life issues, knowledge sharing has been democratized where the source of information and expertise is no longer determined by age or seniority. Moreover, sensitivity and empathy were heightened because of the need to understand and reach out each other's plight and needs.

In this sense, agency is achieved because of the interplay between individual efforts, available resources, and contextual factors as these come together in particular situations [6], 137. This means that the teacher's capacity to navigate constraints (limited physical movements, poor internet connectivity, less advanced gadgets, limited teaching resources) and opportunities (free trainings and access to online resources, family support, and help from professional networks) is a mediated by his or her own capital (cultural, material, social, and symbolic), disposition, institutional resources, community values, and structural factors.

While teacher agency is central in creating spaces for negotiations for autonomy, it is therefore important to examine the motivations and consequences of exercising agency and calibrate its impacts. From this vantage point, we can draw insights from their experiences and improvisations that can inform future decision making.

## 5 Derived Virtual Teaching Models

Teachers' on-the-ground experiences illustrate that their responses to institutional mandates, and structural changes are not linear or predictable. The diversity of responses illuminate that teachers make sense of the social conditions that influence the way they exercise their capacities. This led to the development of five online teaching models with processes were derived, as follows:

**Model 1 Processes**: First, concept notes and initial activities are provided through the PLMS (asynchronous). Second, a video lesson is viewed where the topic is discussed thereafter. As mentioned earlier, the video lesson illustrates how the teacher delivers the lesson from the introduction of the topic (i.e., motivation) to the application of knowledge or skills (asynchronous). Third, an online meeting is set up via Google Meet, Zoom, and other online meeting platforms where the teacher further elaborates key points in the discussion, and addresses misconceptions from the synchronous sessions. Fourth, additional activities are provided to reinforce the application of the targeted skills. The processes are spread out through the week from Monday to Friday. This model with a teaching process from asynchronous to synchronous direction is applicable for elementary, JH, SHS teaching and learning.

**Model 2**: This model is a variation of model 1. Video watching, follow-up discussion about the video, and answering of the DLP activities happen within the day, covering two to three subjects or limited number of subjects only. This means that synchronous and asynchronous sessions are conducted on the same day. However, the submission of the exercises is extended. This synchronous-asynchronous model that happens in one day is likewise applicable for elementary, JH, SHS teachers and learners.

**Model 3**: This model also follows the same processes as model 1. In addition to the given procedures, the teacher holds an online meeting before the asynchronous session to explain instructions for parents and/or students. This model variant is usually applied for pre-school and elementary teachers.

**Model 4**: In case the video lesson is not available, the teacher conducts a synchronous class. The face-to-face teaching model is transported to online teaching, using a PowerPoint presentation as the primary tool in delivering the lesson. This variant is applicable for elementary, JHS, but mostly observed among SHS teachers.

**Model 5**: When the video lesson is insufficient, the teacher holds online classes like that of model 4. In this model, the video lesson is used as a reference, which is applicable to elementary and JHS, but usually practiced by SHS teachers.

## 6 Structuring Free Rein: Resolving Constraints

See Table 1.

## 7 Conclusion

Teachers' on-the-ground experiences about the virtual instructional delivery system reflect an appropriation of the policy that seeks to illuminate, direct, and manage the diverse ways of negotiations and deployment of strategies (*diskarte*), leveraging teaching–learning outcomes toward teachers' agenda. Within these insights, implementing policies may be (re)calibrated and (re)contextualized to integrate teachers' identified creative strategies. However, it is necessary to note that outcomes of teacher agency may either be enabling or disabling; thus, it requires a deeper examination of how positioned actors navigate contextual factors that likewise either constrain or facilitate teacher agency in terms of creating informed policies. The exploration of teacher agency allows us further to reexamine the impacts of giving teachers' free rein for pedagogical structural changes vis-a-vis institutional implementing policies,

**Table 1** Constraints and alternatives

| | |
|---|---|
| **Constraint 1**: The centralized and standardized administration of the periodic examinations posed problems in terms of<br>(1) limiting the flexibility of the tests to integrate needs and concerns of students and the context of other campuses; and<br>(2) delays in test production and errors in the encoding of tests. On the other hand, the idea to localize periodical examinations per campus is predicted to cause PLMS overload and login problems | **Alternative**: An item banking system across learning areas is recommended. This system contains test items for every subject generated from all teachers handling the same subject. A test committee is then in-charge to develop the test items based on the submitted questions in the item bank. Once edited and approved by the teachers concerned, the test will be ready for administration. Additionally, training shall be conducted on<br>a. how to prepare tests through LMS, and<br>b. creative performance-based assessments for the online learning modality |
| **Constraint 2**: On the one hand, the development of the centralized video lessons facilitates constraints as well as opportunities but provides a repository of resources that are available for online teaching. It also allows for sharing of expertise especially for science teachers and other teachers whose orientation is exclusive in each discipline. For example, a biology teacher may not be qualified to teach physics; and thus, the physics teacher may share his or her video lesson with the biology teacher, and vice versa<br>On the other hand, the adoption of the centralized video lessons may limit teacher's creativity and flexibility *(diskarte)*, in terms of contextualizing the lesson according to intended learning outcomes and corresponding teaching strategies, and assessment practices that are appropriate to a particular group of students. During the facetime, the video lesson is enhanced, (re)contextualized, and nuanced, which draws additional time resources from the teacher | **Alternative**: While the development of the centralized video lessons is deemed to be imperative as starting directional point especially for neophyte teachers during the first quarter of the academic year, diverse ways of preparing the lessons may also be integrated as supplemental instructional devices such as actual teaching using a PPT or creating one's own on video lesson. In choosing other alternatives, a local committee composed of the principal, assistant principal, coordinators, and teachers in each department shall be organized and mobilized |
| **Constraint 3**: In consonance with teachers' *diskarte*, free rein is explicitly and implicitly practiced as far as teachers' flexibility and autonomy in preparing their lessons (i.e., video lesson). This involves deciding on what competencies or intended learning outcomes, teaching–learning strategies, learning assessment tools from learning activities, and references to be included and utilized in the lesson. However, it is also crucial that these lessons have to be checked in line with the Most Essential Learning Competencies (MELC) of the Department of Education (DepEd), particularly lessons of teachers who opt to make their own videos and do the actual online teaching | **Alternative**: A local committee may be created and mobilized to ensure that localized video lessons and lesson plans of those who will do actual online teaching are checked in line with the MELC and institutional standards. Webinars may be considered to capacitate teaching on areas that need further improvement |

as well as DepEd's MELC mandates. To regulate instructional management protocol for ensuring desired and expected learning outcomes, the extent of free rein by each teacher has to be still exercised within these established policies and mandates. This **free rein exercise** involves the creation protocols and processes based on existing practices as facilitated by the suggested committees in every local setting.

## References


1. Baxter PE, Jack S (2010) Qualitative case study methodology: study design and implementation for novice researchers. https://www.researchgate.net/publication/228621600_Qualitative_Case_Study_Methodology_Study_Design_and_Implementation_for_Novice_Researchers
2. Yin R (2003) Case study research: design and methods. Applied social science research methods, vol 5. Sage
3. Harrison H, Birks M, Franklin R, Mills J (2017) Case study research: foundations and methodological orientations. Forum Qualitative Sozialforschung/Forum: Qualitative Social Research, 18(1), Art. 19, http://nbn-resolving.de/urn:nbn:de:0114-fqs1701195
4. Peters MA (2020) Reimagining the new pedagogical possibilities for universities post-Covid-19. Accessed from https://www.researchgate.net/publication/342467820_Reimagining_the_new_pedagogical_possibilities_for_universities_post-Covid-19. https://doi.org/10.1080/00131857.2020.1777655
5. Selwyn N (2010) The new connectivities of digital education. In Apple M, Ball S, Grandin LA (eds) The Routledge international handbook of the sociology of education. Routledge, London, pp 90–98
6. Biesta G, Tedder M (2007) Agency and learning in the lifecourse: towards an ecological perspective. Stud Edu Adults 39:132–149

# Secure Trusted Operating System Based on Microkernel Architecture

**De Jian Li, Hui Wang, Xiao Ke Tang, Li Xin Yang, Chong Fei Shen, and Kun Xiao**

**Abstract** Currently, trusted execution environment technologies are widely used to enhance the security of connected devices such as mobile phones and tablets. The trusted operating system is the core component of a trusted execution environment solution, but the trusted operating system itself may face security challenges. The main reason is that the trusted operating system needs to provide different services for different application scenarios, which makes its function more complicated and further increases the code size. In most trusted execution environment solutions, the trusted operating system is based on Monolithic kernel architecture. This will inevitably face security issues such as software defects and lack of isolation between components. In this paper, we propose a scheme for design secure trusted operating system based on a microkernel architecture. In our scheme, we take the modified take-grant model as the system security model and design a discretionary access control mechanism based on the capability system. We implemented our secure trusted operating system on the i.MX6Q platform. The test results show that it works properly and meets the requirements of the GP TEE specifications.

D. J. Li (✉) · H. Wang · X. K. Tang · L. X. Yang · C. F. Shen
Beijing Smart-Chip Microelectronics Technology Co., Ltd, Haidian District, 66 Xixiaokou Road, Beijing 100192, P.R. China
e-mail: lidejian@sgitg.sgcc.com.cn

H. Wang
e-mail: wanghui9@sgitg.sgcc.com.cn

X. K. Tang
e-mail: tangxiaoke@sgitg.sgcc.com.cn

L. X. Yang
e-mail: yanglixin2@sgitg.sgcc.com.cn

C. F. Shen
e-mail: shenchongfei@sgitg.sgcc.com.cn

K. Xiao
The University of Electronic Science and Technology of China, Xiyuan Ave, West Hi-Tech Zone, Chengdu 611731, P. R. China
e-mail: xiaokun@uestc.edu.cn

 
X.-S. Yang et al. (eds.), *Proceedings of Sixth International Congress on Information and Communication Technology*, Lecture Notes in Networks and Systems 216,
https://doi.org/10.1007/978-981-16-1781-2_17

**Keywords** Trusted Operating System · System Service · Discretionary Access Control · Capability · Microkernel

## 1 Background

GlobalPlatform (abbreviated as GP) is the world's leading organization for the development of smart card multi-application management specifications. Its members include IC manufacturers, connected device vendors, software vendors, operators, and third-party testing organizations. In support of companies such as ARM, GP began in 2011 drafting Trusted Operating Environment (Trusted Execution Environment, abbreviated as TEE) specification. Its core idea is to ensure that all sensitive data is stored and handled in the isolated trustworthy environment [1]. TEE provides a safe operating environment for the authorized security software, also known as "trusted applications". TEE ensures end-to-end security through the implementation of data protection and access control, its architecture is shown in Fig. 1(a).

Because of ARM's leadership in embedded processors, TEE is widely used in connected devices. Apple Inc. implemented its own TEE solution named Secure Enclave based on ARM TrustZone technology. Secure Enclave is pre-installed on products such as iPhone and iPad, which can provide secure storage and secure computing service for fingerprint recognition application or human face recognition application. At present, the fingerprint recognition application of mainstream Android mobile phones is also implemented based on TEE solution.

According to GP TEE specification, two independent operating environments are implemented based on the hardware platform: Rich Execution Environment (REE) and Trusted Execution Environment (TEE). The general-purpose operating system (Linux, Android, or other commercial operating systems) usually runs in REE. Such general-purpose operating system is called Rich OS. While the trusted operating system (abbreviated as Trusted OS) runs in TEE [2]. The security mechanism of TEE includes two aspects.

(1) Security zone isolation: TEE and REE are independent of each other. The hardware mechanism can ensure the isolation of code execution in the two environments, and can ensure the confidentiality, integrity or authenticity of the code and data stored in TEE.
(2) The chain of trust: TEE measures the execution code step by step from the ROT (root of trust, such as root key, ROM code, etc.) to build a chain of trust and ensure the credibility of the code.

TEE is composed of a series of hardware and software security mechanisms. Unlike other trusted computing solutions such as Hardware Security Module (HSM) or Trusted Platform Module (TPM), GP TEE specification does not require an external security chip or security coprocessor. Instead, it designs a trusted hardware device based on ARM TrustZone Technology [3]. TEE can monitor and control the operation of the system from the hardware platform to achieve enhanced security [4].

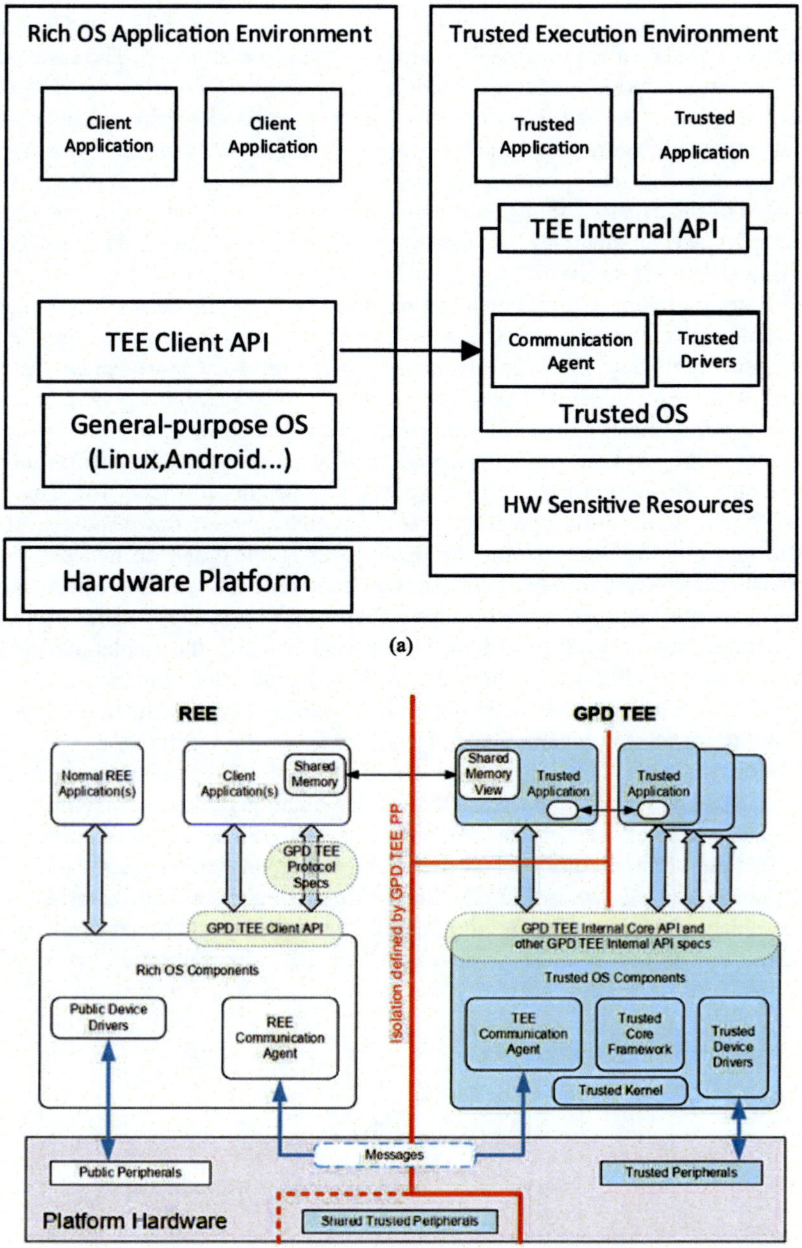


(a)

(b)

**Fig. 1** **(a)** Overview of TEE architecture **(b)** Software system of TEE and REE

GP TEE specification defines how to build software system of TEE and REE, its' architecture is shown in Fig. 1(b). It is a typical client/server model. The application of the general-purpose operating system is client application (abbreviated as CA). The application of the Trusted OS is called trusted application (abbreviated as TA). TA is acted as a server to provide security services to CA, for example, cryptographic operation, secure storage, trusted communication, etc. In order to define the basic functions of the Trusted OS, the communication interface between CA and TA, the API for TA development, GP has announced TEE Internal Core API specification and TEE client API specification.

TEE internal core API specification specifies the basic functions that a Trusted OS needs to implement and the programming interfaces provided to the TA developers, including TA lifecycle management, memory management, trusted storage management, trusted peripheral access, cryptographic operations, clock management, arithmetic operations, and many other aspects [5].

The operating system generally consists of two parts: kernel space running in privileged mode and user space running in a non-privileged model [6]. Since the kernel is the core of the operating system, it manages all the resources in the system and provides services for the upper-layer applications, so it needs to run in kernel space. According to the difference of the functions provide by the kernel, different architectures can be adopted to implement the operating system. In different architectures, the way of providing system services and the implementation of device drivers are different, which affects the overall performance of the operating system. Currently, the most popular operating system architectures include Monolithic kernel, Microkernel, and Hybrid kernel (as shown in Fig. 2).

Monolithic kernel is the earliest proposed architecture. Operating systems such as Unix, MS-DOS, and Linux are typical implementation of this architecture. In this architecture, all the basic services of operating system run in the kernel space as part of the kernel, such as memory management, interrupt management, hardware drivers, file systems… (as shown in Fig. 2). Since all system services run in kernel space, they can access other system services directly without any additional overhead. So, the kernel can provide a set of system call interface for the application, which offers

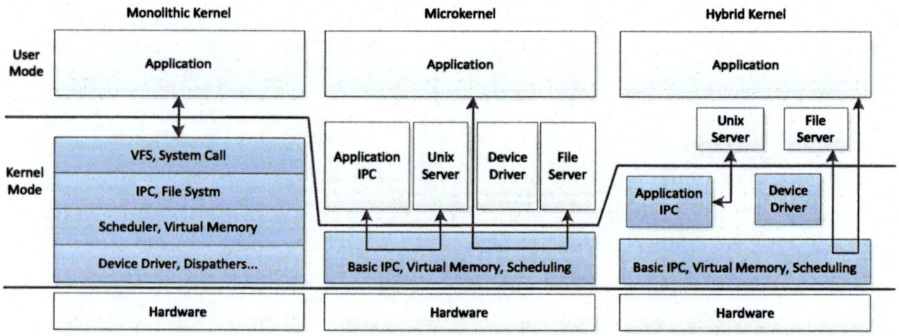


**Fig. 2** The architecture of monolithic kernel, microkernel and hybrid kernel [8]

many functions provided by system services. However, this architecture also has its own shortcomings. Firstly, all system services are parts of the kernel, the amount of the kernel code naturally becomes very large. It will bring great challenges to the compilation, debugging, and maintenance of the kernel, and it is difficult to verify the correctness of the kernel. In addition, since all services are worked in kernel space, the dynamic extension of system services is difficult. Although Linux can provide dynamic loading mechanism for kernel modules [7], this mechanism also increases the complexity.

Microkernel was proposed in the late 1980s in order to solve the expansion problem and the code maintenance problem of Monolithic kernel. QNX and L4 microkernel series are typical implementation of this architecture. The principle of the architecture is to reduce the service provided by the kernel, so that the kernel only provides the most basic system services, such as scheduling, IPC, and IO control, while other system services are all worked in the user space (as shown in Fig. 2). As the function of the kernel decreases, the amount of the kernel code drops linearly, which solves the problem of code maintenance. In addition, because system services are worked in the user space, extending system services is as simple as adding applications, so its scalability is also improved. Moreover, since the kernel and system services run in different address spaces, the hardware protection mechanism can effectively isolate the kernel, thereby improving the reliability and stability of the kernel.

However, because system services are putted into the user space. When the application needs to access the system service, the IPC mechanism must be used, thereby the performance of the system call must be reduced. As shown in Fig. 2, if the application needs to access the service provided by File Server, the system call will transfer system control to the kernel, then the kernel schedules in order to let File Server can execute. After that File Server completes the processing and then transfer system control to the kernel again. Finally, the kernel resumes the execution of the application. The whole process needs to perform task context switching twice and address space switching twice. While the same procedure in Monolithic kernel only needs to perform context switching once. Thus, in Microkernel, the overhead of system calls will increase greatly.

Hybrid kernel architecture in fact belongs to the first generation of Microkernel, which lets certain system services to work in the kernel space in order to reduce the cost of system call. But Hybrid kernel can't achieve the expected goal. Only few operating systems implement this architecture such as Mach and Windows NT.

## 2 Related Work

At present, academia and industry have carried out research on the implementation of TEE based on TrustZone technology. Some of the implementation schemes have been applied in connected device, including: Trust Kernel of Shanghai Jiaotong

University [9], Qualcomm's QSEE [10], G&D's MobiCore [11], Trustonic's Kinibi [12], Sierraware's SierraTEE [13], and Linaro's OP-TEE [14].

These implementations require a general-purpose operating system to work in the normal world of ARM processors, which keep the existing application ecosystem. The Trusted OS worked in the secure world of ARM processors and the trusted application forms the software system of TEE. In summary, the implementation of TEE requires many considerations: trusted boot mechanism, hardware and software isolation mechanism, Trusted OS, trusted applications, and communication agent component in general-purpose operating system. Take the OP-TEE project as an example, and its architecture is shown in Fig. 3(a).

As the application areas of TEE continue to expand, the complexity of their software systems continues to increase. Today the Trusted OS is already a full-function operating system, which also needs scalability in order to support different trusted applications. The interaction with general-purpose operating system is also needed. The rapid expansion of functions let the code size of the Trusted OS increase sharply. The Trusted OS in OP-TEE project is optee_os. The 1.0.0 version of op_tee includes nearly 65,000 lines of code(abbreviated as LOC), while LOC of the latest 4.0 version has increased to nearly 16 million. According to statistics, good quality engineering code may have 2 defects per kLOC [15].

Different application areas also require more demands on the functions of the Trusted OS, such as network communication, biometrics identification, remote management, and so on. For this reason, GP is constantly improving its TEE specification. In recent years, several specifications for different services have been added based on TEE internal core API specification and TEE client API specification, as shown in the following table.

The constantly modification and update of current Trusted OS increases the probability of vulnerabilities [16]. Some instances of attacks using Trusted OS vulnerabilities have been reported in recent years. The fingerprint recognition system of the Samsung Galaxy S5 mobile phone uses TEE to store fingerprint data and encrypt the data. However, in April 2015, it discovered that there are system vulnerabilities in its Trusted OS. When the attacker gains root privileges of the Linux kernel, they can access the fingerprint device to obtain the fingerprint directly, and then use the fingerprint data to impersonate the user login [17].

Current trusted operating systems are usually based on Monolithic kernel architecture. Taking optee os as an example, its kernel, the TEE framework component, the device driver, the cryptographic component, the secure storage component, and the network communication component all work in the kernel space without any isolation measures between them. This architecture can bring some advantages such as the development is convenient and the performance is optimized overall, but it will bring certain security risks. For example, when some error occurs in network communication it may damage other components or the kernel, and even make system collapse.

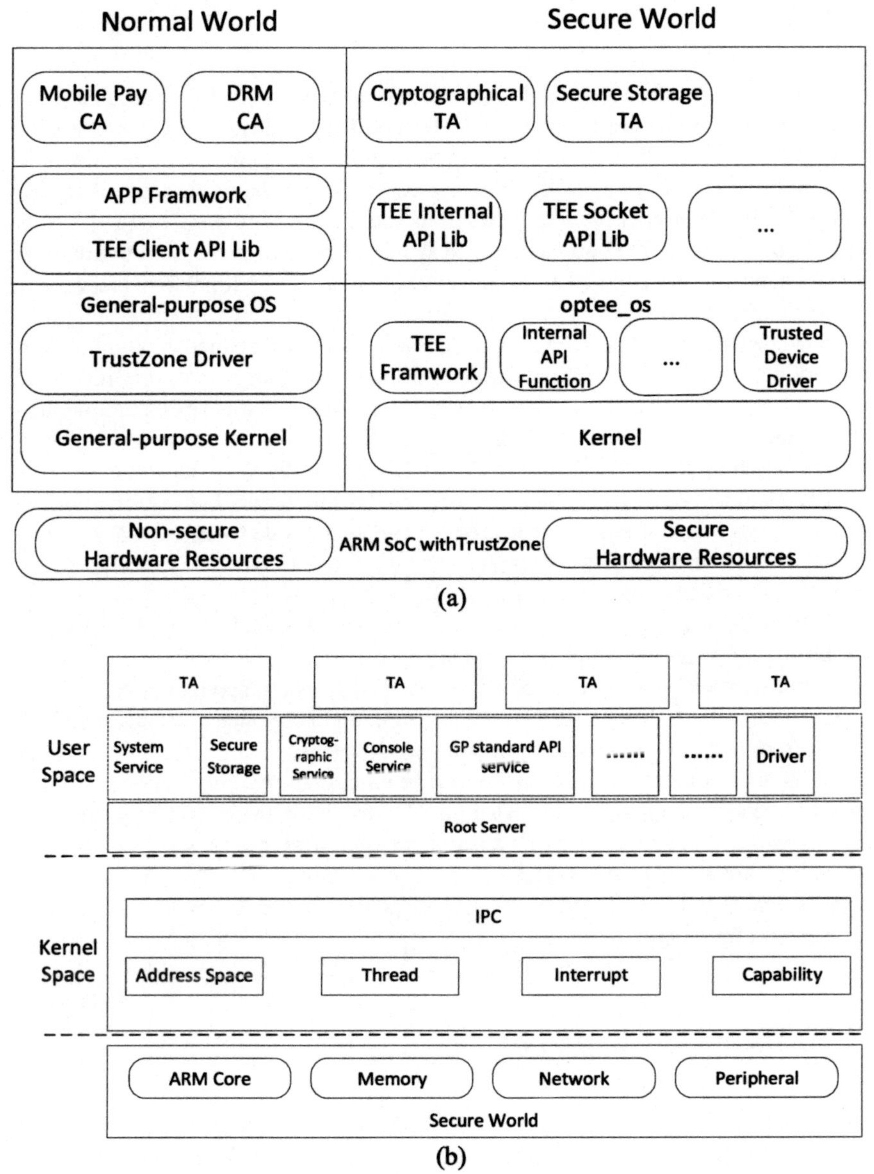


**Fig. 3** **(a)** The architecture of OP-TEE **(b)** The architecture of secure trusted OS based on microkernel architecture

# 3 Our Work

In the operating system, components that can bypass the security policy (and therefore must be fully trusted) are considered as Trusted Computing Base (abbreviated as TCB) [18]. If the Trusted OS is based on Monolithic kernel architecture, the entire operating system is considered as TCB. Researches have shown that the greater the scope of TCB, the higher the likelihood of a systemic security problem [19]. On the other hand, if you want to build a secure system, you need to make TCB trustworthy, which means that the code of TCB cannot contain defects. Modularity is a reliable way to handle the complexity of a software system because it allows to a complex system to be divided into more manageable parts. This will reduce the likelihood of software defects. However, modularizing the kernel does not help us, because in the same address space without isolation, even if the component is split small enough, there is no mechanism to implement boundary protection. It can be said that the kernel is atomic. of. So, the key to make TCB trustworthy is to make the kernel as small as it can be verified by exhaustive test or formal methods. The minimum kernel only contains code that must be executed in privileged mode to implement a generic mechanism, and any functionality that can be implemented by non-privileged mode code should remain outside the kernel [20].

Microkernel can also be used as a software isolation technology based on the thin privileged software layer. The privileged mode code of the operating system based on Microkernel architecture is very small. All system services such as device drivers, protocol stacks, and file systems, are running in user space. It can be said that Microkernel architecture can support complex system software work in user space. In user space, system services can be restricted that it can only access resources given by the system designer. It is also possible to divide the operating system into multiple isolated components by means of a modular design idea, so any component failure will not damage the kernel or other components. For example, the device driver is unable to access memory outside its address space or beyond its authority, so it cannot affect other components.

## *3.1 Trusted OS Architecture*

We have designed our secure Trusted OS based on Microkernel architecture. In our Trusted OS, the kernel only provides the most basic general mechanism without implementing any specific services. The general mechanism mainly includes thread management and scheduling, isolation protection between components (in the form of address space), and inter-process communication (abbreviated as IPC) mechanism. The system service of the Trusted OS, such as secure storage, cryptographic service, and GP standard API service all work in user space. While adopting to the idea of modular design, each service is designed as an individual component. Through hardware and software mechanisms, components can be isolated from each other.

The miniaturization of the kernel and the user mode system services ensure the isolation and reliability of the system. On the other hand, we take the modified take-grant model as the system security model, implement the discretionary access control mechanism based on the capability system. All resources and kernel services are controlled by access control mechanism, it will strengthen the isolation between the system components and applications. The architecture of our Trusted OS is shown in Fig. 3(b).

In our Trusted OS, only the kernel work in kernel space and only has basic functions which are closely related to hardware operation. LOC of the kernel is restricted to about 6000, it will greatly reduce the likelihood of software defects. Trusted OS consists of the kernel running in kernel mode (privileged mode), the root server, and system services running in user mode (non-privileged mode), which together provide services for the trusted application (TA) running in the user mode.

The root service is the only user mode thread created directly by the kernel. Its main work consists of two parts. The first is to abstract the hardware resources in the system into forms that other threads in the system can access, such as converting memory resources into memory management services, converting external storage resources into secure storage services, and converting CPU resources into process management services. So, root server is assumed as system resource manager. The second is to configure components of our Trusted OS, such as which services should be provided and then what resources should be provided to the system service.

Because of the limitation of cost and power consumption, connected devices often use flash memory devices without protection mechanism as the external storage device. So, both TEE and REE have to share the same external storage device. In our trusted storage service implementation, encryption technology is still adopted to ensure its security, but the security risk of the encryption key management is optimized.

System services for trusted operating systems are all implemented as components of our Trusted OS. We use Interface Description Language (abbreviated as IDL) based on XML description to define the component's service interface and shield the TA from the communication flows between the components. We also refer to GP TEE internal core API specification, TEE internal API specification, TEE Sockets API specification, and other service specifications to offer GP standard API service. It will help our Trusted OS to be compatible with existing trusted applications.

## 3.2 Trusted OS Kernel

As mentioned above, the main function of the kernel is to manage hardware resources. Hardware resources for connected devices mainly include computing resources, interrupt resources, and memory resources. All the resources and functionality of the kernel are abstract as kernel object. Each kind of kernel object implement abstract for particular resources or functionality. Some particular kernel functions are used to handle particular kind of kernel object. For example, the kernel implements abstraction of processor resources through thread control block object, thereby supporting code run in user mode and schedule as a thread. The authorized thread can switch to the kernel space only through system call interface, and then handle the kernel object through some kernel functions in order to access system resources or kernel functions. The kernel object types supported by our kernel include address space objects (frame, untyped Memory, page table, page directory), thread control block object (tcb_t), endpoint object (endpoint_t), event object(notification_t), interrupt management object(irq control, irq handler), capability object (capability, cnode, cspace). The architecture of our kernel is shown Fig. 4(a).

The kernel object is implemented as a structure, taking the endpoint object as an example, which is defined as follows:

```
wqwq typedef struct _endpoint
{
  list_t waiters;
  uint32_t state;
  object_t obj;
}DEFAULT_ALINMENT endpoint_t;
```

The kernel only implements the most basic general mechanism and does not include memory management service, so the kernel object is created by user mode threads which can request memory management service offered by root server. For security reason, although kernel objects are created by user mode threads, it is not allowed for user mode threads to directly access (read or modify) kernel objects. Threads can only access kernel objects through kernel functions. User mode thread needs to switch to kernel mode through system call interface, and then corresponding kernel functions can be executed. So the kernel functions which can be triggered by user mode thread, the operations which can be used to kernel object will be strictly restricted.

The security requirement of the kernel is to protect the confidentiality and integrity of kernel objects, so we use a hybrid/neutral security policy to ensure that no unauthorized access (read or modify) to kernel objects occurs. The security model adopts the take-grant model which belongs to discretionary security model, and we make some modifications to take-grant model in order to control implicit authority distribution. The discretionary access control mechanism is implemented based on the capability system. The security design of our kernel is done using a formal approach.

Our kernel provides IPC mechanism to let the thread communicate with other thread or system services, which is also used to access the kernel objects and trigger events between the threads. It is the only way that the user mode thread can trigger kernel functions or access system resources. Therefore, the discretionary access control mechanism is realized by controlling the communication authority between threads and between the thread and the kernel object. A user mode thread that has the capability to a kernel object can initiate IPC process only if its capability also

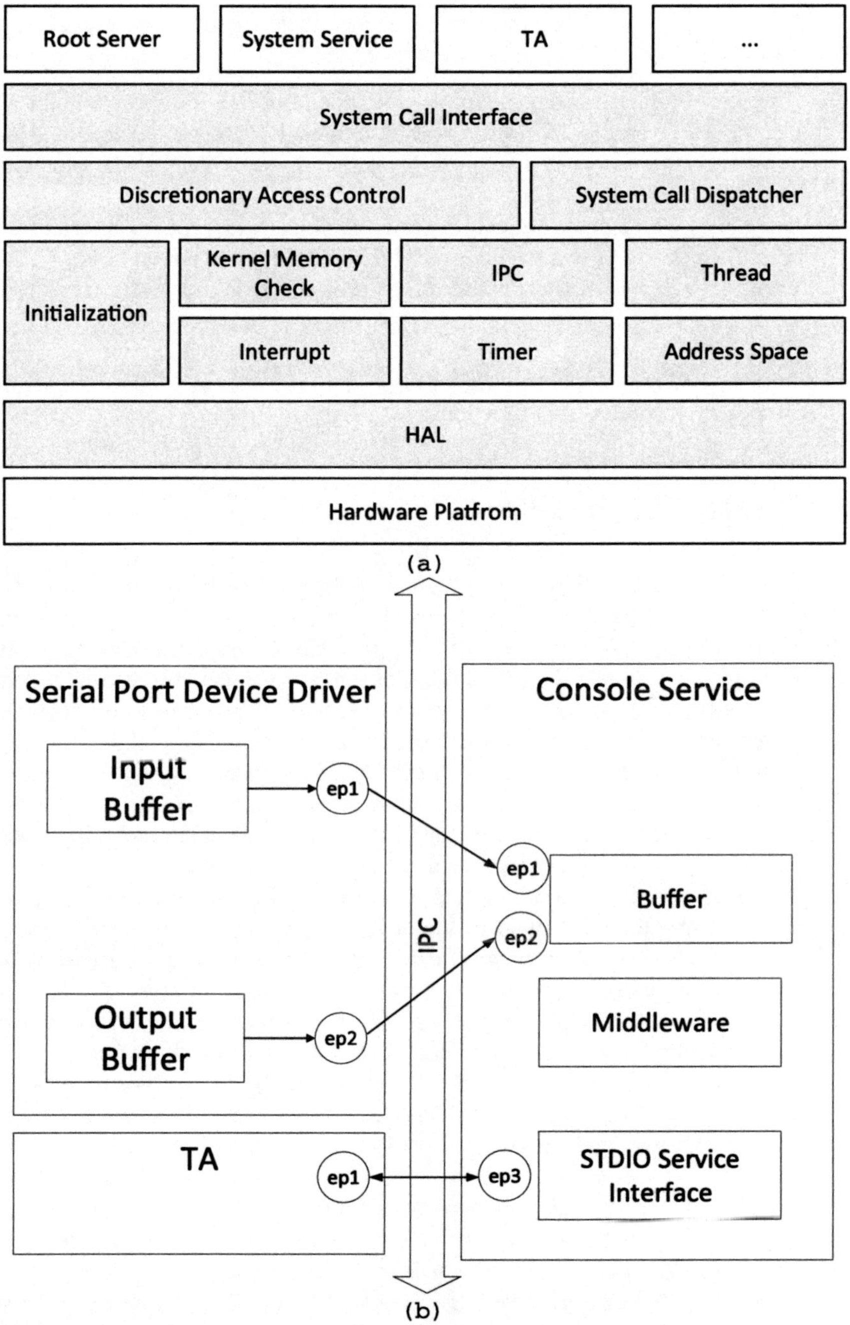


**Fig. 4** **(a)** The architecture of Trusted OS kernel **(b)** The Framework of Trusted OS System Services

includes the corresponding authority. In our kernel, there are three different types of IPC process.

(1) The thread which has the capability to endpoint object with proper authority initiates IPC process. It will allow two threads to communicate and the sent/received message is passed/ from another thread which has the capability to the same endpoint object.
(2) The thread which has the capability to event object with proper authority initiates IPC process. It will trigger event or wait for another thread to trigger this event. Obviously, another thread must have the capability to the same event object.
(3) The thread which has the capability to other objects with proper authority initiates IPC process. It will trigger the system call flow and finally trigger the execution of some kernel functions.

## *3.3 Trusted OS System Service*

System services of trusted operating systems can be divided into device drivers and trusted services. Device drivers are responsible for converting hardware resources into device-independent interfaces. Trusted services include trusted storage, cryptographic service, GP standard API service, console service, etc. The trusted service requests service of the kernel and the device in order to access hardware resources, manage resources through internal mechanisms and corresponding algorithms, and provide system service interface to facilitate TA development.

The device driver is responsible for control hardware through device interface. Device interfaces are generally defined by the device manufacturer, including device state control, device interrupts, and data transfer with the device. Usually, device sharing is a security issue in the system. If multiple tasks try to access a physical device at the same time, the state of the device is prone to errors. Therefore, we have two solutions. For safety–critical devices, the device driver is strictly restricted to allow only one trusted service access for a certain period. For non-safety–critical devices, if there is a need for device reuse, the device driver will use buffer mechanism to virtualizes multiple devices, and in fact multiple tasks access device orderly.

Most device drivers need to handle device interrupts. In our Trusted OS, the processing of device interrupts is divided into two phases, one is the interrupt response phase, and the other is the event response phase. The interrupt response is done by the kernel and the event response is done by the device driver. After the interrupt occurs and the scene is saved, the interrupt response module of the kernel reads the interrupt register to obtain the interrupt number, and reads the device register to obtain the interrupt-related parameters, and then the kernel triggers the interrupt event through IPC mechanism. The interrupt event is registered by the thread where the device driver is located, and the corresponding operation is performed after waiting for the event to be triggered in the thread.

We will use the console service as an example to introduce the general framework of system services. The console service can capture the text entered by the user through the serial port and output the printed text. The service will request the service of serial port device driver. Figure 4(b) depicts the relationship between the trusted application, the console service, and the serial port device driver.

In order to realize the formatted output of the character stream on the serial port, the console service implements the interpretation of the escape character, maintains the state for the virtual terminal, and sends each character to the serial port device driver through the endpoint object (ep1). After that, the serial port driver controls the hardware to send the character. In order to obtain the text entered by the user, the console service requests data from the serial port device driver through the endpoint object (ep2). After that, the console service converts the data into a user's control command. The trusted application requests the standard I/O interface provided by the console service through the endpoint object (ep3).

## 4 Evaluation

### *4.1 Trusted OS Kernel*

We implemented our secure Trusted OS on the i.MX6Q platform (as shown in Fig. 5(a)). LOC of the kernel is restricted to about 6000, and the release version of our kernel occupies approximately 67.1 KB of static memory space and requires about 168.9 KB of runtime dynamic memory space. In order to test the function of the kernel, we add test function in the code of root:

(1) read the startup information that kernel transfer to the root server
(2) map a data section through RAM capability
(3) traverse code object to find out all RAM capability and display its information
(4) create a thread monitoring error information and process it
(5) enable serial port receiving function and data receiving interrupt
(6) create a thread in order to listen to serial port interrupt.

The test code has covered most of the functions provided by the kernel, the test result is shown in Fig. 5(b).

Performance test is also conducted for scheduling, thread switching, interrupt handling, and access control. In our test, a timer is provided by hardware platform to calculate the duration of execution of each function. Since the functional test code used before have already covers most of the functionality of the kernel, it will be used in performance test too.

i.MX6Q is a processor of ARMv7-A architecture, which provides hardware modules related to performance testing, called PMU (Performance Monitoring Unit). The PMU contains a processor cycle counter that directly reflects the processor cycle

(a)

(b)

**Fig. 5** **(a)** Test environment **(b)** Functional test result of trusted OS kernel

consumption, so we use this counter directly when performing performance measurements. We set the processor's frequency to 792 MHz, which is 792 cycles of about 1 microsecond. The test result is shown in Fig. 6(a).

## *4.2 Trusted OS System Service*

For system services, we did not specifically write test cases, but adopted the TEE standard API compliance test program named xtest, which is provided by OP-TEE project. Figure 6(b). shows that our Trusted OS can pass the TEE standard API compliance test.

| Kernel Function | No.of tests | Total Cycles | Average Cycles | Max Cycles |
|---|---|---|---|---|
| Scheduling | 30 | 15831 | 527 | 880 |
| | 40 | 19270 | 481 | 880 |
| | 50 | 22219 | 444 | 880 |
| Thread Switching | 20 | 7784 | 389 | 457 |
| | 25 | 9714 | 388 | 457 |
| | 30 | 11628 | 387 | 457 |
| Interrupt handling | 30 | 78569 | 2618 | 3444 |
| | 40 | 104702 | 2617 | 3444 |
| | 50 | 130309 | 2606 | 3444 |
| Access Control | 30 | 38209 | 1273 | 3590 |
| | 40 | 66734 | 1668 | 4038 |
| | 50 | 90989 | 1819 | 5071 |

(a)

(b)

**Fig. 6** **(a)** Performance test result of trusted OS kernel **(b)** TEE standard API compliance test

## 5 Conclusions

We implemented secure Trusted OS based on the Microkernel architecture and the modified take-grant security model. The test results show that it works properly and meets the requirements of the GP TEE specifications. It can also be seen that the access control mechanism and the microkernel architecture bring a certain degree of performance loss to our Trusted OS, but the performance loss is generally within the controllable range. How to improve the performance of our Trusted OS and verify its security is the direction we need to continue to work.

## References


1. Ekberg JE, Kostiainen K, Asokan N (2013) Trusted execution environments on mobile devices. In: Proceedings of the 2013 ACM SIGSAC conference on computer and communications security. ACM, pp 1497–1498
2. GlobalPlatform (2018) TEE System Architecture, version 1.0 [EB/OL]. https://globalplatform.org/specs-library/?filter-committee=tee
3. Arm A (2009) Security technology-building a secure system using TrustZone technology. ARM Technical White Paper
4. Sabt M, Achemlal M, Bouabdallah A (2015) Trusted execution environment: what it is, and what it is not. In: 14th IEEE international conference on trust, security and privacy in computing and communications
5. GlobalPlatform (2018) TEE Internal Core API Specification, version 1.2 [EB/OL]. https://globalplatform.org/specs-library/?filter-committee=tee
6. Roch B (2004) Monolithic kernel versus microkernel: TU Wien 10(4):1–6
7. De Goyeneche JM, De Sousa EAF (1999) Loadable Kernel Modules. IEEE Softw 16(1):65–71
8. Golftheman (2008) Monolith-, Micro- and a Hybrid Kernel [EB/OL]. https://upload.wikimedia.org/wikipedia/commons/d/d0/OS-structure2.svg:Wikipedia
9. Li W, Ma M, Han J et al (2014) Building trusted path on untrusted device drivers for mobile devices. In: Proceedings of 5th Asia-Pacific workshop on systems. ACM, p 8
10. Qualcomm (2018) Snapdragon security [EB/OL]. https://www.qualcomm.com/solutions/mobile-computing/features/security
11. Lee Y, Kim J, Lim KS et al (2018) Secure mobile device structure for trust IoT. J Supercomput 74(12):6646–6664
12. Trustonic. Trustonic Kinibi technology [EB/OL]. https://developer.trustonic.com/discover/technology.
13. Sierraware (2014) Open virtualization build and boot guide for arm V7 AND arm V8, Version 2.5.0.1 [EB/OL]. http://www.openvirtualization.org/SierraTEE_build_and_boot_guide.pdf
14. Mendoza S (2018) Open portable trusted execution environment [EB/OL]. https://www.op-tee.org/
15. Malaiya YK, Denton J (1997) Estimating defect density using test coverage. Rapport Technique CS-98–104 6, Colorado State University
16. Atamli-Reineh A, Borgaonkar R, Balisane RA et al (2016) Analysis of trusted execution environment usage in samsung KNOX. In: Proceedings of the 1st workshop on system software for trusted execution. ACM, p 7.
17. Kim D, Jung Y, Toh KA et al (2016) An empirical study on iris recognition in a mobile phone. Expert Syst Appl 54:328–339
18. Heiser G (2005) Secure embedded systems need microkernels. USENIX; Login 30(6):9–13
19. Heiser G, Elphinstone K, Kuz I et al (2007) Towards trustworthy computing systems: taking microkernels to the next level. ACM SIGOPS Oper Syst Rev 41(4):3–11
20. Heiser G, Elphinstone K (2016) L4 microkernels: The lessons from 20 years of research and deployment. ACM Trans Comput Syst (TOCS) 34(1):1

# The Case of Fabric and Textile Industry: The Emerging Role of Digitalization, Internet-of-Things and Industry 4.0 for Circularity

**Malahat Ghoreishi and Ari Happonen**

**Abstract** In the circular economy (CE) world, design is a vital tool to transform waste to wealth to enhance material recycling for new products (e.g., design for modularity, design for disassembly, etc.). Industry 4.0 technologies and digital innovations have recently provided positive impact towards CE transition, for instance through data-analysis of products lifecycle. In this regard, a close collaboration is required between industries, consumers and policy makers for a more detailed and precise data on product lifecycle and waste collection as well as consumer behavior. Accordingly, digitalization can help in tracking and tracing products through transferring the real-time data on products' location, condition, availability, etc., which can boost product-life extension through sensors and digital platforms. In order to have a more efficient and effective CE solution, close collaboration is required among various actors of textile circular ecosystem, where the consumers are the core of such ecosystem. In this sense, IoT plays an essential role by transferring information and data to all the actors and improving their awareness of new circular opportunities. This paper focuses on how utilizing IoT will increase the opportunities of textile circulation towards a higher effective ecosystem and maximizing products value.

**Keywords** Industry 4.0 · Circular Economy · Internet-of-Thing · Textiles · Circularity · Sustainability · Sensors

## 1 Introduction

Fast fashion has been recently one of the significant concerns of environmental experts which makes huge pollution worldwide. Sustainability solutions have a very slow implementation to prevent social and environmental negative impacts

M. Ghoreishi (✉) · A. Happonen
LUT University, Lappeenranta, Finland
e-mail: Malahat.ghoreishi@lut.fi

A. Happonen
e-mail: Ari.happonen@lut.fi

 
X.-S. Yang et al. (eds.), *Proceedings of Sixth International Congress on Information and Communication Technology*, Lecture Notes in Networks and Systems 216,
https://doi.org/10.1007/978-981-16-1781-2_18

of textile industry with the growing of "middle-class" population. Therefore, there is a vital need of collaboration between all segments of the fashion industry ecosystem, in particular the role of policymakers and governments to create sustainable supportive frameworks. The Global Fashion Agendas chief sustainability officer Morten Lehman stated in 2019 that *"While it's encouraging to see 12.5% of the global fashion market taking concrete action toward circular business models, we must urgently address major roadblocks collaboratively to pave the way for a systemic shift towards circularity"*. Ellen MacArthur Foundation [1] reports that less than 1% of the used materials in clothes production are recycled. In addition, the amount of used clothes collection for recycling is under 15%, whereas 73% end up to the landfills or being burned. Currently, fashion industries have realized the opportunities that circular business models can bring to the textile companies, for example increasing the value chain by recycling the used materials instead of wasting finite resources. Implementing circular solutions of collecting and recycling fibers would bring EURO 80 billion value to the world economy by 2030 [1].

According to European Parliament [2], 2–10% of environmental impact in Europe is due to clothes industry. Additionally, European Commission reported in 2013 that textile industries alone generate approximately 16 million tons of waste in EU annually. According to the report, this high amount of textile-based waste is generated due to the lack of transparency and poor level of data sharing between all the actors of the clothes industry ecosystem, such as manufacturers, retailers, suppliers, and specially consumers. From waste point of view, huge amounts of water and chemicals are required in all stages of producing raw materials, spinning, weaving, and dyeing fabrics, as can be seen in Fig. 1.

Due to the growth in new styles and fashion as well as increasing population in middle-class, clothes production has been increased enormously over the past 15 years [1]. On the other hand, studies on customers' behavior show that huge

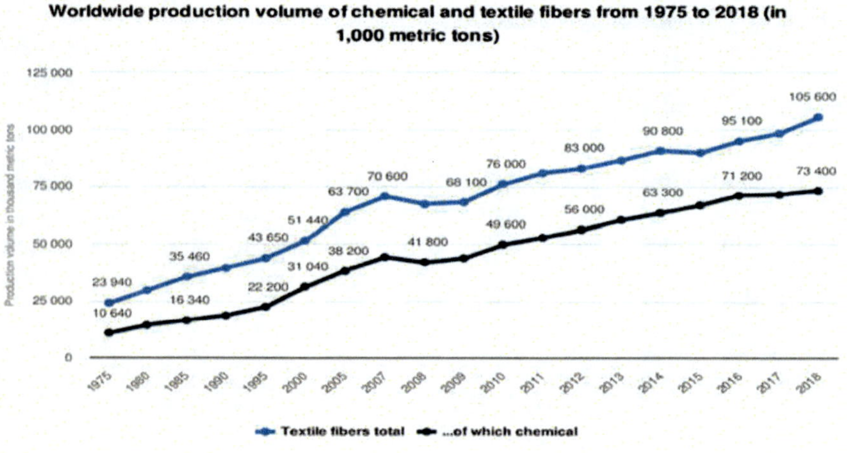


**Fig. 1** Periodical production of textile fibers and chemical (*Source* Garside 2019)

amount of clothes are thrown away before they are damaged or unfunctional [3]. In other words, willingness to pay and purchase the clothes are higher than consumption of the products. Thus, customers play the key role in contributing to high environmental footprints. Moreover, apparel industry is highly based on linear model in all phases of producing, distributing and disposing of their products. Current system is highly extracting large volumes of materials to produce fabrics and garments from non-renewable resources which will be used in a short period and will end up huge landfill waste. It is discussed if the textiles and clothing industries implement eco-friendly designs in their processes, fashion industry will turn hugely to a more sustainable economy model [4].

## 2 Circular Economy in Textile Industry

The circular economy (CE) has recently gained attentions of policy makers and businesses as a solution to enhance sustainability goals. In a CE world when products, materials, and components face their end-of-life, they will be used as raw material in processes of new material production to reduce waste landfills which is called closed-loop and is in contrast with linear economy model, "take-make-dispose" [5]. The CE has contributed significantly on industrial development by minimizing waste and optimizing finite resources so that materials maintain their economic value for possible longer time which leads to a more sustainable process [6]. Such outstanding benefits of CE can help companies with their quest to be constantly competitive and to minimize their environmental footprints. Circular strategies such as reuse, restore, refurbish, repair, etc., enhance productivity and resource efficiency [7]. Circular solutions offer positive effect on value creation as well as *"customer value proposition, supply chain, value networks of the companies and capturing the value of new offerings"* [8].

As mentioned before, consumer has an enormous role in achieving sustainability goals in the fashion industry. Consumer attitudes on use, wash, repair, rental, and loan as well as quality consideration in purchasing clothes rather than several purchases are highly affective factors in transition toward circularity in fashion industries. In addition, customer collaboration with textile industries regarding recycling and returning the old and used clothes to the companies are considered as high impact factor in moving toward CE. Therefore, having the right mindset, knowledge, and consumption guidelines is highly important in value creation based on CE principles. Accordingly, it is essentially required to have close collaboration among all the segments in the value network to help pave the way for circular innovations for companies.

Circular business models such as rent (clothes rent to slow the loop), design products for re-use or recycle (circular fashion), design for better quality that can convince customers to buy less but higher quality (slow fashion) [2] have been recognized as the solutions for the negative footprints made by textile industries. According to Smits et al. [9], textile industries can transform their linear business strategies to circular strategy based on three approaches as follows:

**Table 1** Principles of circular fashion

| | The production perspective | | Customer perspective |
|---|---|---|---|
| Design principles | Source and production principles | Lifetime extension | Behavioral and usage model |
| Purpose | Local source and production | Reuse, recycle or compost all remains | |
| Longevity | Zero toxicity | Supportive service for long life | Use, wash and repair with care |
| Resource efficiency | Efficiency consideration | Wide and high-quality collaboration | Rent, loan, swap, secondhand or redesign instead of new purchase |
| Recyclability | Good ethics work mood | | |

*Source* Brismar [10]

- Material flows circular strategies, e.g., reuse wastes as resources and generate value.
- Servitization strategies with focuses on non-ownership and more as service.
- Efficient use of resources which leads to sufficiency.

In this regard and considering the circular feasibility of mentioned business for fabric industries, Brismar [10] identified sixteen key principles for circular fashion framework which aims to support sustainable textile and fashion industry as illustrated in Table 1.

Understanding and identifying the business ecosystem approach for textile circulation is essential since circular ecosystem consists of various networks including different companies, individuals, organizations, and digital technologies to enable the circularity in the ecosystem and life cycle. Fontell and Heikkila [11] introduce the circular business ecosystems as business ecosystem approach as follow: *"Circular business ecosystems are business ecosystems, which together create products, solutions and services based on the principles of a circular economy and apply circular business models in their way of operating and doing business"*. Circular business ecosystem of textiles is illustrated in Fig. 2.

In this system, the goal is to reduce additional consumption, efficient, and on-time maintenance services for the products as well as preventing waste generation in order to save resources and enhance businesses. As can be seen from Fig. 2, the major dominant business models in textile circular ecosystem are repair, reuse, and recycle. Furthermore, users consisting of induvial consumers, professionals or organizations are the core in textile circulation ecosystem and make it a significant consumer-oriented ecosystem. The main goal in all the loops is to maintain material flows and create value where the loops strongly depend on users' attitude and behavior. According to Fontell and Heikkila [11] the role of retailers has been increasing in re-using textiles. It is discussed that their role will be more involved in near future not only in offering new clothes based on virgin materials, but also clothes which

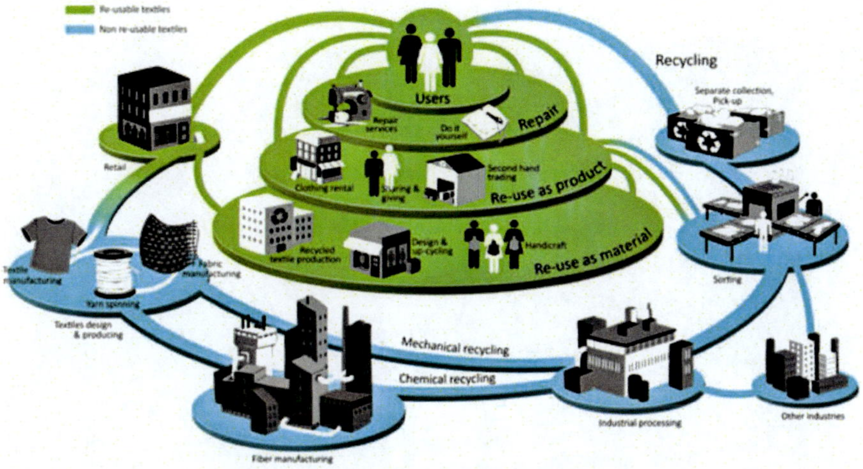


**Fig. 2** Circular ecosystem of textiles (*Source* Fontell and Heikkila [11])

are made by recycled fabrics, upcycled or second-hand. Retailers are capable to help increase sustainable knowledge of consumers and help them in making more sustainable choices due to their close interactions with end-users and customers. Van den Berg and Bakker [12] mention that product design plays key role in CE, therefore the first step in moving to CE is to design products that are first of all "be loved, liked, or trusted longer". Consumers will keep their favorite clothes for longer time and in some occasions till the products are broken down. Table 2 shows the circular product characteristics introduced by Van den Berg and Bakker [12].

Based on Table 2, in textile industries, clothes should be designed with better quality fabrics and fibers which enables reuse and repair of textiles as products or as materials. Repurposing and redesigning pre and post-consumer textiles aim to eliminate textile wastes at design stages. If the fabrics and garments are in acceptable, good condition they can be re-purposed or up-cycled by transforming materials into new products with similar or higher quality or can be transferred to the next users. Design for modularity and standardization in fashion industry is related to clothes design which matches various combinations. Clothes designs should take into account the easiness for the users to care and repair the products by providing guidelines and instructions for users. Moreover, providing spare parts or related information on maintenance services while selling products support design for ease of maintenance in textile industries. In design for upgradability, clothes are designed for adaptable usage, e.g., use of the same clothes for several occasions. Buttons and zippers are made of different materials than garments in textiles, therefore designing for reassembly or disassembly helps in easy removals of such materials from the products and makes it easier to re-use fabrics. Finally regarding the recycling of textiles and garments, it is important that consumers have the determination to choose if the clothes can be reused or should be recycled. All the principles above help to

**Table 2** Product design strategies in circular economy

| Circular Economy Design systems and products to recover resources and value | | | | | |
|---|---|---|---|---|---|
| Only non-destructive | Future proof Last long Use long | Last long | Performance | Reliability | Durability |
| | | Use long | Roadmap fit; Upgradability | Adaptability; Timeless design | Anticipation legislation (e.g. toxicity, recyclability ) |
| | Disassembly allow to service, remake and recycle | Connections | Quick and easy disconnect | Limit use and diversity of fasteners | Limit use and diversity tools |
| | | Product architecture | Simplify product architecture | Allows ease of access to components | Clarity of disassembly sequences |
| | Maintenance Reuse of products | Maintenance | Ease of cleaning | Ease of repair/upgrade | Allow onsite repair or upgrade |
| | | Lifetime prognostics | Online monitoring for quality, testing, maintenance and billing | | |
| | Remake Reuse of parts | Modularity | Use modular components | Standardize interfaces | Back- & Forwards compatibility |
| | | Reliability assessment | Allow for easy read out of components | | |
| | | (Recover) Logistics | Product can easily be returned | Spare part harvesting | Local production |
| Destructive and non-destructive | Recycle Reuse of materials | Materials | Avoid the use of (non-compliant) coatings | Only use materials that can be recycled | Use preferred/pure materials |
| | | Electronics | Easy/fast detection of materials | Use SMD components | |
| | | Connections | Avoid fixed connections | Break down by (shredding/disassembly) to: • Pieces of uniform composition • Pieces of relatively large size (>1cm) | |

Adapted from Van den Berg and Bakker [12]

keep the material flow, however in current infrastructure it is difficult to handle the whole flow, especially regarding collection and sort of discarded clothes.

## 3 Digital Solutions for Textile Circularity

Despite of all the advantages that circular economy models can bring to the economy, society, and environment, transition toward circularity might be of high challenge and costly for various businesses [13]. Antikainen et al. [14] discuss that Industry 4.0 (the fourth industrial revolution) such as IoT (Internet-of-Things), AI (artificial intelligence), Big Data, and 3D printers (three-dimensional printers) can pave the way of moving toward circular models for businesses. Pillsbury and Lübben [15] mention that *"digital networks and interaction can create an integrated system of actors, assets and stakeholders where not only supply chains can be real-time tuned with the factory but also retail channels and even products and final customers can communicate and exchange data within the system"*. Such technologies can support products' visibility by providing precise information and real-time data through intelligent sensors on products' condition, location, availability, and assets. Industry 4.0 technologies should make it possible to produce eco-system of "smart factories", "smart networks", and "smart products". In such eco-systems, the improved data transformation on information about markets and end-users' demand helps in the complexity on decision-making for the companies. Zhong et al. [16] define the roles of Industry 4.0 in CE as follows:

- Design: Smart design, smart prototyping, smart controller, smart sensors
- Maintenance: Real-time control and monitoring, Collaborative decision-making
- Control: Big Data analytics, data-enabled prediction, and data-based modeling
- Scheduling: Marketing, warehouse management, transports
- Monitoring: Technology makes the real-time sharing of information possible.

According to Ghoreishi and Happonen [17], Industry 4.0 technologies are capable of transferring tremendous amount of data. Such technologies connect real-time data on product design (e.g., resource use to reduce the consumption) which helps businesses in decision-making in the production process. In addition, monitoring products through digital technologies help companies in predicting maintenance and future requirements for maintenance services so that organizations can prevent extra costs related to the maintenance and services will be done just at the time they are needed. Moreover, digitalization connects data on supply and demand through shared platforms and apps which transfer information between organizations and customers [18].

## 4 IoT and Circular Fashion

Industry 4.0 paradigm plays a significant role in circulation in apparel industries due to the mass data generation in all the activities within the ecosystem. Ingemarsdotter et al. [19] define the opportunities that IoT can bring to support circular strategies as shown in Table 3.

Different sensors, platforms, and applications of machine learning can connect more efficient interaction among the actors of the ecosystem to accelerate circulation.

**Table 3** Integration of IoT opportunities and CE strategies

| CE strategies | IoT application |
|---|---|
| Servitization business models | Helps in enabling performance-based service contracts by the monitoring system performance<br>Enables more attractive service models |
| Maintenance business model | Facilitates maintenance and adaptation through precise data records on installed products and parts<br>Enables predictive and condition-based maintenance services |
| Reuse and/or remanufacturing business model | Enables tracing and tracking used products and materials<br>Enables more accurate estimation on products' lifetime |
| Design for durability business model | Reducing faults in product design through data collection and analysis on material condition |

Utilizing data to track and trace the product brings the opportunity to manage life cycle of the product and to enhance predictive maintenance, re-use, and recycling processes. Hence, the openness of data and sharing precise data and information among different actors of circular textile ecosystem is essential to achieve circular goals. Huge data on fiber composition and chemicals used for dyeing and manufacturing the yarn can be analyzed faster with the reduction in flaws of prototyping and manufacturing processes. This way helps resource optimization and efficiency as well as accelerating the recycling of materials and other circulation processes. Moving toward "smart factory" via implementing innovative technologies creates significant sustainable benefits and higher efficiency worldwide [20]. Figure 3 illustrates different implications of Industry 4.0 technologies in fashion industry.

Utilizing RFID tags that can hold all the relative information regarding the materials and chemicals in the products can help the circulation in the ecosystem. In addition, implementing these tags will help in tracking product availability in shops and analysis of customers' behavior and demands. Hence, these tags can help transferring precise and real-time data without retailers involvement to the manufacturers so that they can make better decisions in manufacturing products and provide services for the end-users. Furthermore, since there is enormous data to be collected on life-cycle assessment of the products, digital technologies can enhance transparency, availability, and reliability of data [21]. However, the circulation of such data within an ecosystem has been challenging due to the lack of data availability.

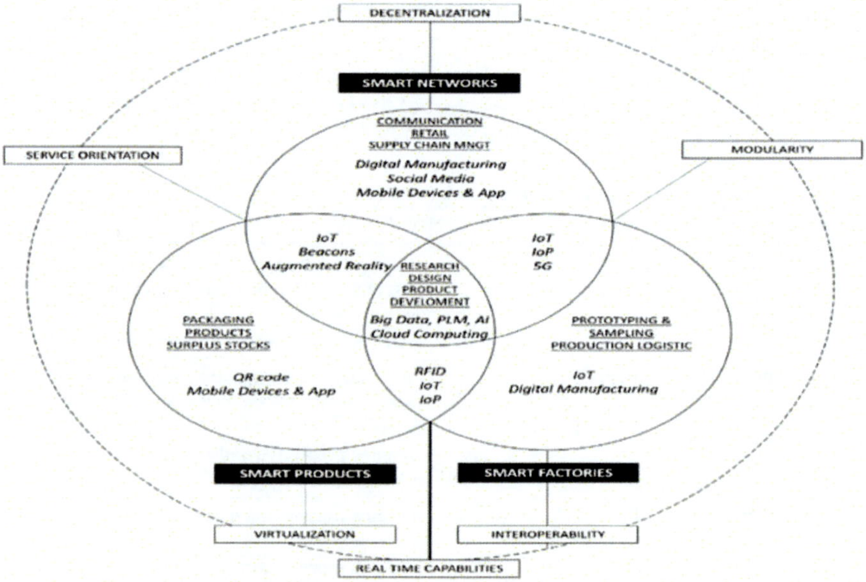


**Fig. 3** Fashion principles and Industry 4.0 technologies (*Source* Bertola and Teunissen [21])

Traditional fashion business models were based on design, for which industry 4.0 technologies can create smart networks. In this regard, digital platforms are the fundamentals in circulation of textiles. Through these platforms which are programmed by machine learning algorithms, consumers can rent, borrow and resell clothes which enhances the re-use business model in circular economy of textiles. Moreover, AI-based platform and IoT sensors can help in optimization and management of warehouses. Data on availability of the products in the storages and warehouses will transfer to the managements and helps in predicting consumers' need and therefore, prevents extra production. In a smart fashion system, which is built upon IoT, Blockchain, and AI, products are produced based on the quantity of markets' demand. Traceability of such products makes it possible to track the entire lifecycle and to customize, optimize and upgrade products according to users' demand.

The modern embedded technology age buzzword IoT is gaining an increase of market value with 2020 $1.7 trillion reaches [22]. The IoT has gained focus of smart manufacturing industries recently in the processes of high-speed monitoring and controlling data transformation in industrial systems. Implementing IoT in such systems accelerates interaction between devices and human-to-device interactivity in a more trustable and faster way. The IoT is described as an ecosystem in which devices are connected to each other or to the Internet to monitor the status of objects, gathering data, and transferring useful information through applications via Internet networks. Smart objects, Radio Frequency (RF) technologies as well as machine-to-machine communication are the main building blocks of the IoT [23–25]. To enable interactions and communications between physical objects and devices, the IoT needs some essential components such as Radio Frequency Identification (RFID). RFID tags are known as Auto-ID technology to help identify the objects, e.g., following high value items in supply chains [26]. By use of RFID tags, objects can communicate even in wireless situations to transfer information that helps in monitoring data on smart objects. Implementing sensors in the smart objects in addition to RFID tags makes it possible to measure data on quantity changes, temperature, and such types of information [21]. As can be seen in Fig. 3, in the factory level IoT is used in production sphere through fast prototyping and sampling through which the production quantity will be more precise and closer to the market demand and prevents overproduction. The IoT applications which are used in retail help in communication with customers and by use of sensors they can analyze and track the lifecycle of products. Moreover, the IoT can help in smart product design and development as well as smart packaging. For example, customers can track their products easily via mobile application, where the application utilizes sensors related augmented data, collected from the smart packages. However, to enable this function, (1) correct smart solutions in supply chains management, Big Data, and assets fleet-level analysis [27] (2) warehousing decarbonization [28], (3) possibilities and challenges to implement digital solutions, and 4) artificial intelligence-based approaches as enablers for highest level circularity solutions [29] are required to be considered in ecosystem. The integrated architectural model in Fig. 3 presents the 6 design principle (Modularity, Interoperability, Decentralization, Service orientation, Virtualization and Real-time capabilities) for the implementation of Industry 4.0 technologies into the context of fashion industry

[30]. Combination of such applications will lead to a more circular system that is more cost-efficient and have less footprints.

Furthermore, integration of IoT and CE accelerates logistics, supply chain management, and asset management with sufficient knowledge of infrastructure and product reusability, etc. For example, IoT can be used as sensors in devices that are reusable, and the materials can be recycled. Products that are utilized by sensors such as QR, can enable the process of recycling the components of products when they face end-of-life [31]. In this case, reverse logistics will help the used products to retain their value and return in the cycle of life. Since one of the key roles of CE is to make durable and long-lasting materials, IoT can help in transforming precise data on assets' condition to improve the quality of products. Finally, IoT can speed up data collection in product as a service CBMs which has a huge contribution in CE and reduces negative environmental impacts.

## 5 Conclusions and Discussions

The emerging role of Industry 4.0 has recently revolutionized almost all business and consumer life activities. In near future, smart connected clothes with digital passports will support the precise data on materials, designs, lifecycles as well as consumers' demands and recycling methods. In product level, tracking the garments and clothes helps companies in waste management, e.g., by identification and fabric content recognition, digital technologies can boost the process of recycling and waste management, hence help the textile industries in their circularity processes. IoT sensors together with AI-based platforms and Blockchain applications can help all the actors of the textile industry ecosystem to transfer real-time data and knowledge on the product's lifecycle and circular solutions in each process. In addition, such technologies can help in upcycling and recycling of the second-hand clothes through digital platforms which can help products to retain their value and simultaneously reduce the consumption of virgin materials. IoT and AI can help in fast prototyping, prediction, and reducing the number of faults in apparel industry, therefore boost CE by reducing the wastes from prototyping and improving and optimizing the energy efficiency.

In this paper, we focused on the opportunities and advantages of integrating digital technologies for textile industries, specially IoT. The study proved that there is lack of literature and scientific studies conceptually and empirically on this specific area. There is huge interest and opportunities for qualitative and quantitative studies in this field to analyze how apparel businesses identify their circular ecosystem and how digital solutions can help them to reach their circular and sustainable goals. Therefore, further research in this field is highly recommended. Since the availability of the data plays an important role in implementation of digital technologies, further research on challenges and barriers that textile industries face in integration of Industry 4.0 and CE is suggested. Finally, policies and regulations are required to be followed by all the actors of the ecosystem in order to reach sustainability goals in a secure manner.

## References


1. Ellen MacArthur Foundation (2017) A new textiles economy: redesigning fashion's future
2. Perliament E (2019) Environmental impact of the textile and clothing industry What consumers need to know
3. Laitala K (2014) Consumers' clothing disposal behaviour - a synthesis of research results. Int J Consum Stud 38:444–457. https://doi.org/10.1111/ijcs.12088
4. Moorhouse D, Moorhouse D (2017) Sustainable design : circular economy in fashion and textiles sustainable design : circular economy in fashion and textiles. Des J 6925:S1948–S1959. https://doi.org/10.1080/14606925.2017.1352713
5. MacArthur E (2013) Towards the circular economy volume 1. Ellen MacArthur Found. https://doi.org/10.1162/108819806775545321
6. Tukker A (2015) Product services for a resource-efficient and circular economy—a review. J Clean Prod 97:76–91. https://doi.org/10.1016/j.jclepro.2013.11.049
7. Blomsma F, Kjaer L, Pigosso D, McAloone T, Lloyd S (2018) Exploring Circular Strategy Combinations - Towards Understanding the Role of PSS. Procedia CIRP. 69:752–757. https://doi.org/10.1016/j.procir.2017.11.129
8. Urbinati A, Latilla V, Symposium DC-II (2018) undefined: the role of product design in circular economy business model. Researchgate Net
9. Smits H, Cunningham G, Wilting J, De Roos Feinberg C, van 't Hof K (2015) Business models and circular strategies for textiles
10. Brismar D (2019) Green Strategy | Sustainable and Circular Fashion Consulting—Towards a more sustainable and circular fashion industry, https://www.greenstrategy.se/en/
11. Fontell P, Heikkila P (2017) Model of circular business ecosystem for textiles
12. Van den Berg MR, Bakker CA (2015) A product design framework for a circular economy. PLATE (Product Lifetimes Environ Conf Proc) 365–379
13. Nascimento DLM, Alencastro V, Quelhas OLG, Caiado RGG, Garza-Reyes JA, Lona LR, Tortorella G (2019) Exploring Industry 4.0 technologies to enable circular economy practices in a manufacturing context: A business model proposal. J Manuf Technol Manag 30:607–627. https://doi.org/10.1108/JMTM-03-2018-0071
14. Antikainen M, Uusitalo T, Kivikytö-Reponen P (2018) Digitalisation as an Enabler of Circular Economy. Procedia CIRP. 73:45–49. https://doi.org/10.1016/j.procir.2018.04.027
15. Dutzler H, Germany PS, Schrauf S, Geissbauer R, Strategy P, Müller SR, East SM, Bohsali S, Australia P, Burns P, Spooner N, Vermeire P, Brazil P, Valiño R, Canada P, Wetmore M, Denmark P, Vedsø J, Nieminen K, France P, Romac B, Le Bellac V, India P, Mukhtyar K, Italy P, Caragnano G, Japan P, Kawai T, Okamoto K, Scharff C, Mexico P, Urioste CZ, Khurana A, Mulders M, Poland P, Dziurdzia M, Singapore P, Spain P, Isasi CK, Theron P, Sweden P, Lindblad F, Baldinger G, Atkin N, Jukes D, Shaw D, Pillsbury S, Jaruzelski B (2018) How industry leaders build integrated operations ecosystems to deliver end-to-end customer solutions
16. Zhong RY, Xu X, Klotz E, Newman ST (2017) Intelligent Manufacturing in the Context of Industry 4.0: a review. Engineering 3:616–630. https://doi.org/10.1016/J.ENG.2017.05.015
17. Ghoreishi M, Happonen A (2020) Key enablers for deploying artificial intelligence for circular economy embracing sustainable product design: Three case studies. In: 13TH INTERNATIONAL ENGINEERING RESEARCH CONFERENCE (13TH EURECA 2019). AIP Publishing, p 050008
18. Jabbour AB, Jabbour CJC, Filho M, Roubaud D (2018) Industry 4.0 and the circular economy: a proposed research agenda and original roadmap for sustainable operations. Ann Oper Res 270:273–286 https://doi.org/10.1007/s10479-018-2772-8

19. Ingemarsdotter E, Jamsin E, Balkenende R (2020) Opportunities and challenges in IoT-enabled circular business model implementation – A case study. Resour Conserv Recycl 162:105047. https://doi.org/10.1016/j.resconrec.2020.105047
20. Kortelainen H, Happonen A, Kinnunen SK (2016) Fleet service generation—challenges in corporate asset management. In: Lecture Notes in Mechanical Engineering. Pleiades Publishing, pp 373–380
21. Bertola P, Teunissen J (2018) Fashion 4.0. Innovating fashion industry through digital transformation. Res J Text Appar 22:352–369. https://doi.org/10.1108/RJTA-03-2018-0023
22. Wen Z, Hu S, Clercq DD, Beck MB, Zhang H, Zhang H, Fei F, Liu J (2018) Design, implementation, and evaluation of an Internet of Things ( IoT ) network system for restaurant food waste management. Waste Manag 73:26–38. https://doi.org/10.1016/j.wasman.2017.11.054
23. Gubbi J, Buyya R, Marusic S, Palaniswami M (2013) Internet of Things ( IoT ): A vision, architectural elements, and future directions. Futur Gener Comput Syst 29:1645–1660. https://doi.org/10.1016/j.future.2013.01.010
24. Borgia E (2014) The internet of things vision: Key features, applications and open issues. Comput Commun 54:1–31. https://doi.org/10.1016/j.comcom.2014.09.008
25. Bai TDP, Rabara SA (2015) Design and Development of Integrated, Secured and Intelligent Architecture for Internet of Things and Cloud. Computing. https://doi.org/10.1109/FiCloud.2015.23
26. Hämäläinen H, Salmela E, Happonen A (2007). RFID on Item Level Tagging in Supply Chain with High-Valued Products. https://doi.org/10.5281/ZENODO.3376601
27. Kärri T, Marttonen-Arola S, Kinnunen S-K, Kortelainen H, Kunttu S, Hanski J, Ahonen T, Valkokari P, Uusitalo T, Horn S (2017) Fleet-based industrial data symbiosis: Data to Business Knowledge (D2BK) model. https://cris.vtt.fi/en/publications/fleet-based-industrial-data-symbiosis-data-to-business-knowledge
28. Minashkina D, Happonen A (2020) Decarbonizing warehousing activities through digitalization and automatization with WMS integration for sustainability supporting operations. E3S Web Conf 158, 03002 https://doi.org/10.1051/e3sconf/202015803002
29. Ghoreishi M, Happonen A (2020) New promises AI brings into circular economy accelerated product design: A review on supporting literature. In: E3S Web of Conferences
30. Hermann M, Pentek T, Otto B, Pentek T (2015) Design Principles for Industrie 4.0 Scenarios: A Literature Review Competence Center Corporate Data Quality (CC CDQ) View project Industrial Data Space View project Design Principles for Industrie 4.0 Scenarios: A Literature Review
31. Moktadir MA, Ali SM, Kusi-Sarpong S, Shaikh MAA (2018) Assessing challenges for implementing Industry 4.0: Implications for process safety and environmental protection. Process Saf Environ Prot 117:730–741 https://doi.org/10.1016/j.psep.2018.04.020

# Quality Assessment of Text Data Using C-RNN

**Aakanksha Sharaff, Sarthak Khurana, and Thalendra Sahu**

**Abstract** As the Internet came into existence, the new medium of communications was established, these include Emails, SMS, calls over the Internet, Social media, etc. In short duration, these mediums turned out to be quite popular among the masses. The world soon began to flood with textual data. This research is dedicated to performing an assessment on the widely available textual data. The necessity of assessing texts is since not all the data present in the world is beneficial for the common man. Textual data is unpredictable, unstructured, and noisy in nature. Thus, assessing texts is not an easy task. To further increase this complexity, this research uses SMS data for its purpose. SMS is a very popular medium of communication in the early 2000s and even today in the form of text-based chat applications. This generation is an age of smartphones, with the rapid increase in the number of smartphone users, the daily traffic of SMS data keeps increasing at an exponential rate. Quality assessment of SMS data has become extremely challenging due to the complexitics and changing behavior of messages imposed by spammers. This research proposed a novel approach of quality assessment of textual data using a deep learning Convolutional Recurrent Neural Network (C-RNN) model, for quality assessment of textual data, it is generally used in image processing. It is a hybrid neural network; consists of Convolutional Neural Network (CNN) and Recurrent Neural Network (RNN) both. Thus, making it more powerful than the other two.

**Keywords** Short Message Service (SMS) · Convolutional Recurrent Neural Network (C-RNN) · Convolutional Neural Network (CNN) · Recurrent Neural Network (RNN)

A. Sharaff (✉) · S. Khurana · T. Sahu
Department of Computer Science and Engineering, National Institute of Technology Raipur, Raipur, India
e-mail: asharaff.cs@nitrr.ac.in

 
X.-S. Yang et al. (eds.), *Proceedings of Sixth International Congress on Information and Communication Technology*, Lecture Notes in Networks and Systems 216,
https://doi.org/10.1007/978-981-16-1781-2_19

## 1 Introduction

With the revolution of technology in the past 50 years due to computers, most of the work, which was performed handwritten and manually, then started to shift to the virtual world of computers. This caused huge collections of textual data in the systems, but memory is an expensive resource of computation, moreover, most of this data is not useful for the common user. This unwanted data wastes memory and time of the user. This research uses SMS data to perform the "Quality Assessment" operation. Messages can be classified into two categories: spam and ham. Spam messages are unsolicited, usually political, or commercial messages sent to a large group of recipients. On the other hand, ham messages are useful messages that contain useful information. A common user tendency is to read ham messages while discarding or ignoring spam messages. These spam messages affect social stability and unity. According to a survey performed on 12,000 Indian users by the online community platform LocalCircles, around 96% users received unwanted text messages every day. Nearly half of the respondents get between four and seven spam messages in a day. Therefore, spam messages have become a serious concern in this technological world. Effective filtration of these short messages is vital for social balance and people's ordinary life [1].

Quality Assessment of SMS data is becoming more challenging day by day due to complexity imposed by spammers and changing behavior of spam messages. Different methods used for SMS spam filtration, such as Naive Bayes (NB), Support Vector Machines (SVM), k-Nearest Neighbor (KNN), Artificial Neural Networks (ANN), Decision Tree, Random Forest with different datasets showed that the models that use NB and SVM [2] give the highest accuracy among others [3]. However, methods like Logistic Regression, Decision Tree, and Bayesian classification still have some complex calculation problems, hence these are time-consuming models. To solve this problem of complex calculations, Convolution Neural Networks (CNN) are introduced [4]. C-RNN is a hybrid of both RNN and CNN, which combines the properties of both CNN and RNN. This research combines the process of word vectorization and feature extraction method with C-RNN to analyze the quality assessment of SMS.

The research paper is organized as follows: Sect. 2 is discussing the related work behind this study. Section 3 introduces a C-RNN model designed to train the word embedding. Section 4 presents the whole research and analysis of the obtained result. Section 5 is used to summarize the research result of this work.

## 2 Related Work

### 2.1 *Text Feature Extraction and Representation Techniques*

Text feature extraction consists of popularly three methods: Bag of Words, term frequency and Inverse term frequency (TF-IDF), and word2vec. These three methods are proposed by Waykole et al. [5]. The bag of words is the simplest, flexible, and most used method, among others. It forms a feature set from all the words for a particular instance [6]. A sequential bag of words model to classify human action is proposed by Liu et al. [7]. The major problem with the bag of words model is that the words which have higher frequency become dominant and these words do not need to provide much information for the model. Due to which domain-related which might be important can be ignored or discarded. This problem is solved by **TF-IDF** by rescaling the frequency of words by considering the frequency of words in all the documents [8]. **Word2Vec** is used to construct vectors from words.

### 2.2 *Text Quality Assessment Techniques*

Text analyzing methods are vastly used in email classification, information filtering, text message classification, and on search engines. Nowadays mostly supervised learning models such as KNN, SVM, decision tree, random forest, and neural networks are used for text classification [9, 10]. Convolutional neural networks and recurrent neural networks for text classification are proposed by Radhika et al. [11]. A method to combine SVM and RF for text classification is proposed by Kumar et al. [12]. It gives improved results over individuals. A method using Naive Bayesian, decision tree, and KNN to classify emails based on feature selection technique is proposed by Sharaff et al [13]. An overview of machine learning models for email classification is proposed by Dada et al. [14].

### 2.3 *Convolutional Neural Network*

Deep learning has developed rapidly as a new field over the past decade. It has attracted many researchers' attention. The convolutional neural network can reduce the problem of complexity of networks effectively and the no. of training parameters [15]. CNN has been used along with Bi-LSTM model for text classification to improve the performance of various classifier in text [16].

### 2.4 Recurrent Neural Networks

Recurrent Neural Networks (RNNs) are used to bring an interesting twist to the basic neural network. In traditional models of neural networks, the initial number of inputs must be determined before giving inputs [17]. The Recurrent Neural Network remembers the past and its decisions are influenced by what it has learned from its past decisions. The main feature of RNN is its hidden layer which remembers some information from its experience.

### 2.5 Convolutional Recurrent Neural Network

The Convolutional Recurrent Neural Networks is the combination of two of the best neural networks. Convolutional Recurrent Neural Network (C-RNN) involves a CNN followed by an RNN [18]. A C-RNN method to make a music recommender system based on genre is proposed by Gunawan et al. [19].

## 3 Proposed Method

The research tries to bring forth a novel approach to classify textual data using deep learning techniques which are generally used in the field of data science. It uses C-RNN architecture for the fulfillment of this purpose. The overall methodology of the research work is shown in Fig. 1.

### 3.1 Dataset

The dataset is taken from "UCI SMS Spam Dataset [20]". This dataset is a tagged collection of 5574 messages out of which 4827 are legitimate messages and 747 messages are illegitimate messages. The messages are labeled as ham and spam [21]. Table 1 presents the summary of dataset.

### 3.2 Data Pre-Processing

This dataset gathered must be preprocessed to input to the C-RNN architecture. The steps involved in this process is presented in Fig. 2. Pre-processing starts from Message Segmentation, Word Vectorization, Word Embeddings to Tokenized Vectors, and Tokenized Vectors to Embedding Matrix.

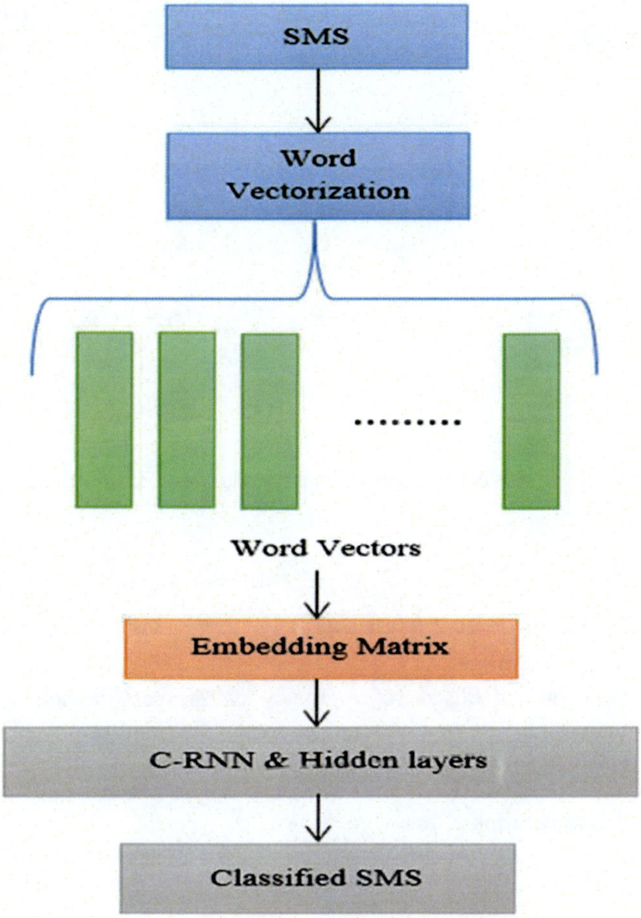


**Fig. 1** Proposed model

**Table 1** Summary of dataset

| | |
|---|---|
| Total SMS | 5574 |
| Categories | Spam and Ham |
| Ham | 4827 (87%) |
| Spam | 747 (13%) |

## *3.3 Proposed Deep Learning Model*

This research tries to utilize multiple networks in conjunction with each other to achieve maximum accuracy as its output. The main highlight of these networks is

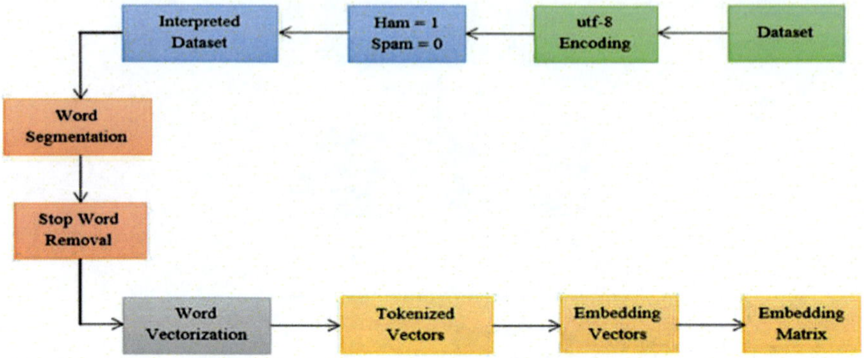


**Fig. 2** Summary of data pre-processing

the C-RNN model. C-RNN architecture is a hybrid architecture, that contains both CNN and RNN models.

### *3.4 Other Classification Methods*

Before, the deduction of this novel approach for the classification of texts, this research went through various phases of understanding and testing many famous and mostly used algorithms for this purpose. Support Vector Machines, Naïve Bayes Classifier, Neural Networks (RNN and CNN), and Logistic Regression has been used to analyze the classification of message.

## 4 Evaluation and Results

After the completion of model development, it is necessary to perform the evaluation of the model and to determine adequate results from them. Before, fitting the embedding matrix to the embedding layer of the deep learning model, the matrix is divided into two parts: one part for training the data and the other part for testing the trained model. 10-fold cross-validation technique has been used in this research.

The accuracy for each epoch is generated. The accuracy of the model is in the range of 98.80–99.88%. The accuracy of the final epoch is 99.68% and is shown in Fig. 3 and Table 2.

$$\text{Average Accuracy} = \text{Sum of all accuracies divided by no. of epochs}$$
$$\text{Average Accuracy is } 99.532\%$$

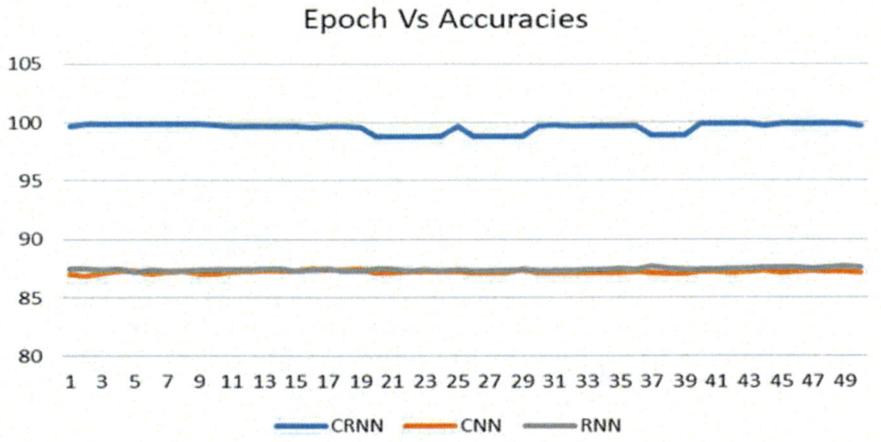


**Fig. 3** Epoch versus Accuracy

**Table 2** Summary of Evaluation Results

| Techniques | Type | Precision (%) | Recall (%) | F1-score (%) | Accuracy (%) |
|---|---|---|---|---|---|
| SVM | ML | 99.58 | 97.88 | 98.72 | 97.77 |
| Naïve Bayes | ML | 100 | 96.40 | 98.16 | 96.76 |
| Logistic Regression | ML | 99.79 | 94.54 | 97.09 | 94.29 |
| CNN | DL | 87.61 | 99.27 | 93.08 | 87.16 |
| RNN | DL | 87.57 | 99 | 93.35 | 87.61 |
| Proposed Method (C-RNN) | DL | 99.95 | 99.67 | 99.81 | 99.68 |

This research also computes other parameters to completely justify the correctness of its novel approach and is shown in the following Table 2.

## 5 Conclusion and Future Work

The proposed model is a textual data classifier, which is developed from the C-RNN architecture of neural networks. The proposed model has good efficiency and noise invulnerability. The efficiency of the proposed model is subjected to various parameters and hyper-parameters. Some of the parameters that can affect efficiency are: No. of Epochs, Verbose, Learning Rate, Optimizers, Loss, Embedding Dimension, Window Size, Test and Train Split, etc. Thus, in the upcoming future one can try various combinations of these different parameters in order to achieve high accuracy. One can also try to utilize the advantages of Bayesian deep learning concepts also. This model can be effectively used in the field of Inbox Filtration for SMS or Emails.

## References


1. Tekkanat E, Topaloglu M (2016) The study of customer attitudes towards advertisements. Procedia Econ Finance 39:32–38
2. Du Y (2019) Bayesian deep convolutional neural network for SMS quality analysis. J Vis Commun Image Represent 102677
3. Sharaff A (2019) Spam detection in SMS based on feature selection Techniques. In: Abraham A, Dutta P, Mandal J, Bhattacharya A, Dutta S (eds) Emerging technologies in data mining and information security. advances in intelligent systems and computing, vol 813. Springer, Singapore. https://doi.org/10.1007/978-981-13-1498-8_49
4. Mohammad RMA (2020) A lifelong spam emails classification model. Appl Comput Inf
5. Waykole RN, Thakare A (2018) A review of feature extraction methods for text classification. IJAERD 5(04)
6. Elshourbagy M, Hemayed E, Fayek M (2016) Enhanced bag of words using multilevel k-means for human activity recognition. Egyptian Inf J 17(2):227–237
7. Liu H, Tang H, Xiao W, Guo Z, Tian L, Gao Y (2016) Sequential Bag-of-Words model for human action classification. CAAI Trans Intell Technol 1(2):125–136
8. Bhagvati C (2018) Word representations for gender classification using deep learning. Procedia Comput Sci 132:614–622
9. Mirończuk MM, Protasiewicz J (2018) A recent overview of the state-of-the-art elements of text classification. Exp Syst Appl 106:36–54
10. Alsmadi I, Alhami I (2015) Clustering and classification of email contents. J King Saud Univ Comput Inf Sci 27(1):46–57
11. Radhika K, Bindu KR (2018) A text classification model using convolution neural network and recurrent neural network. Int J Pure Appl Math 15:1549–1554
12. Kumar HK, Harish BS (2018) Sarcasm classification: a novel approach by using content based feature selection method. Procedia Comput Sci 143:378–386
13. Sharaff A, Nagwani NK, Swami K (2015) Impact of feature selection technique on email classification. Int J Knowl Eng 1(1):59–63
14. Dada EG, Bassi JS, Chiroma H, Adetunmbi AO, Ajibuwa OE (2019) Machine learning for email spam filtering: review, approaches and open research problems. Heliyon 5(6):e01802
15. Samatin Njikam AN, Zhao H (2020) Chartec-net: an efficient and lightweight character-based convolutional network for text classification. J Electr Comput Eng
16. Jang B, Kim M, Harerimana G, Kang SU, Kim JW (2020) Bi-LSTM model to increase accuracy in text classification: combining word2vec cnn and attention mechanism. Appl Sci 10(17):5841
17. Du J, Vong CM, Chen CP (2020) Novel Efficient RNN and LSTM-like architectures: recurrent and gated broad learning systems and their applications for text classification. IEEE Trans Cybern
18. Tsironi E, Barros P, Weber C, Wermter S (2017) An analysis of convolutional long short-term memory recurrent neural networks for gesture recognition. Neurocomputing 268:76–86
19. Gunawan AA, Suhartono D (2019) Music recommender system based on genre using convolutional recurrent neural networks. Procedia Comput Sci 157:99–109
20. Almeida TA, Hidalgo JMG, Yamakami A (2011) Contributions to the study of SMS spam filtering: new collection and results. In: Proceedings of the 11th ACM symposium on document engineering, pp 259–262
21. Sharma D, Sharaff A (2019) Identifying spam patterns in sms using genetic programming approach. In: 2019 international conference on intelligent computing and control systems (ICCS). IEEE, pp 396–400

# Casablanca Smart City Project: Urbanization, Urban Growth, and Sprawl Challenges Using Remote Sensing and Spatial Analysis

**Youssef El Mdari, Mohammed Ait Daoud, Abdelwahed Namir, and Mustapha Hakdaoui**

**Abstract** From the years 2000, the idea of Smart Cities became the new tendency around the world; cities like Barcelona, Nice, and Singapore based their management on digitized data. Attracted by this new global tendency Moroccan emerging country launched the Casablanca Smart City project. Considered as the economic capital with 30% of the national GDP and the most populous city in the country, Casablanca has the potential to become a Smart City and lead the other African cities in the same path. However, the project is facing tremendous challenges; the most concerning ones are Rapid Urbanization and Population growth. With 3.4 million people in the urban area and more than 4.3 million in its agglomeration, Casablanca has experienced an unprecedented population growth in the last decades [1]. The urban growth and sprawl, added to the failing strategies to control them have resulted in an environmental and urban crisis. GIS and Remote Sensing are key technologies for the decision makers to help achieve sustainable development and raise Casablanca to the level of a Smart City. This paper addresses the smart city challenges of Casablanca with regards to Urban Growth and Sprawl, Built-up Density, and Urbanization. The following analysis is one of the few to raise those aspects in Morocco.

**Keywords** Smart City · Rapid Urbanization · Urban Growth · Urban Sprawl · Population Growth · GIS · Remote Sensing

## 1 Introduction

Faced with the challenges of fast urbanization, urban growth, urban sprawl, and demographic growth specific to emerging countries, the concept of smart cities (or intelligent cities) is particularly significant in Africa, in the current case the Kingdom

Y. E. Mdari (✉) · M. A. Daoud · A. Namir
Faculty of Sciences Ben M'sick, LTIM, Hassan II University of Casablanca, Casablanca, Morocco

M. Hakdaoui
Faculty of Sciences Ben M'sick, LGAGE, Hassan II University of Casablanca, Casablanca, Morocco



X.-S. Yang et al. (eds.), *Proceedings of Sixth International Congress on Information and Communication Technology*, Lecture Notes in Networks and Systems 216,
https://doi.org/10.1007/978-981-16-1781-2_20

of Morocco. Through its economic capital, Casablanca, Morocco aims to become a reference in smart cities in the continent.

Urbanization is an expansion in the number of individuals living in metropolitan areas as an eventual outcome of financial improvement under explicit conditions [2]. The uncontrolled spread of metropolitan advancement into neighboring locales, or Urban sprawl, is a significant attribute of provincial urbanization [3]. Natural climate and metropolitan socio-economic advancement unequivocally influence metropolitan land use development [4].

As reported by [5], in Morocco, serious natural and metropolitan planning issues are produced by demographic growth and fast urbanization rate, especially in Casablanca. According to [6], Geographic Information Systems (GIS) and Remote Sensing (RS) are widely used methods for assessing urban sprawl and exploiting spatial data.

This study analyzes the transformation overgone by Casablanca over 33 years (1986–2019) considering urban growth, urban sprawl, built-up density. Are also studied urban expansion and its impact on the city forms [7] and the Smart City Casablanca Project. Data from populous census and satellite imageries are exploited in the following work thanks to classification tools and the analysis of spatial data.

## 2 Related Works

Here, the available studies about urban growth, urban sprawl, and smart cities are quickly examined. Available research studied urban growth and sprawls in several fields [8], along with the city's form [9], expansion, and trail of sprawl [10]. The expanding metropolitan populace squeezes the peri-urban land, which had prompted broad examination on the spatial advantage of multi-story building [11].

As expressed by the European Smart cities commission, the degree of smartness of European metropolitan areas is determined by governance, mobility, utilities, facilities, building, and environment [12].To be illustrative, the city of London is defined a smart city as reported by the 2018 cities motion index generated by the IESE [13],

### 2.1 *Smart Cities Definition and Characteristics*

Two models of conceptualization of the Smart City can be recognized. The Techno-centric approach (advanced by the ICT area) and the Holistic way that deals with social, economical, ecological, human, cultural, and technological factors. Today, many appellations such as the green city, the innovative city, and the efficient city are related to the Smart city concept. A popular model of six axes was established to help identifying a smart city [14] (Fig. 1).

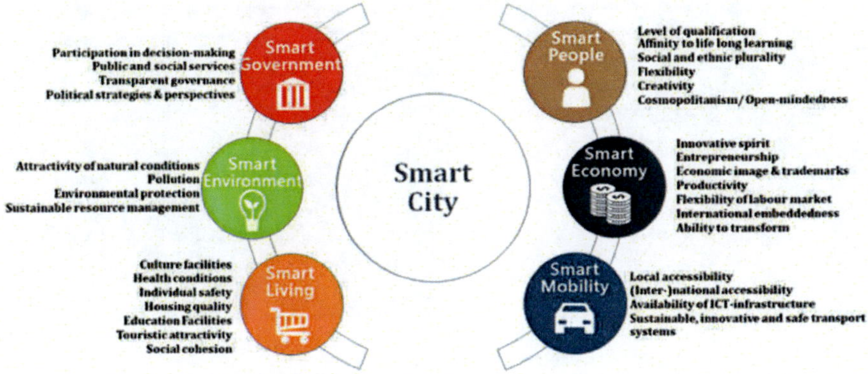


**Fig. 1** Smart city characteristics and factors

## 2.2 *Casablanca Smart City Model*

The city of Casablanca presented its smart city project during the event "Casablanca Smart City Expo". This project was awarded by the American Institute of Electrical and Electronic Engineering "Smart City Initiative". It is based on a concept of smart, frugal, social, and sustainable city that places the inhabitants at the heart of the project (Fig. 2).

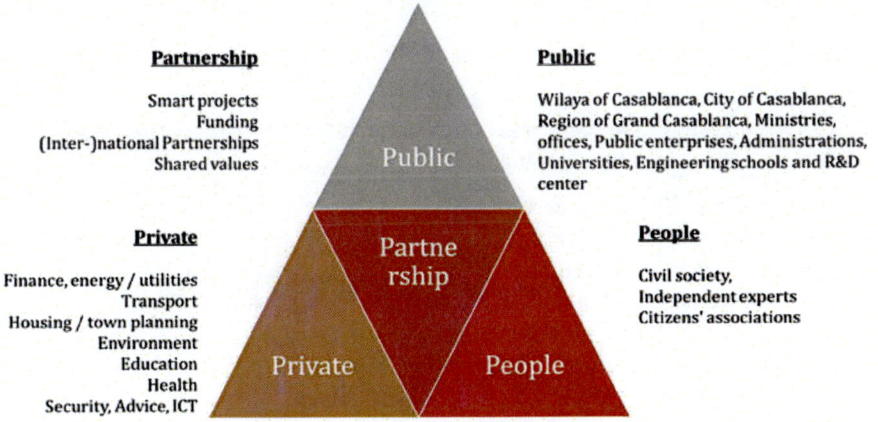


**Fig. 2** E-Madina 4P model (Public, Private, People and Partnership) [15]

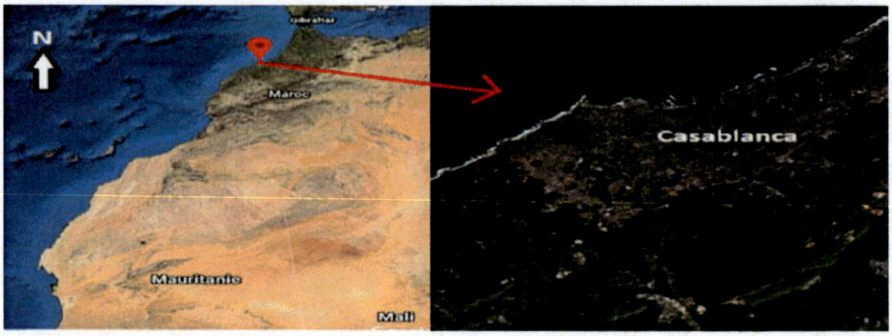


**Fig. 3** Study area: the city of Casablanca

# 3 Study Area

Casablanca is the economic and financial heart of Morocco holding 60% of the foreign trade. It accounts for 30% of the GDP. The city of Casablanca has an estimated 3,359,818 inhabitants and 4,270,750 individuals in its agglomeration. By concentrating 48% of investments, the city sees a door opening in terms of redevelopment. Casablanca has become the first city of Morocco and one of the most important agglomerations in Africa, while consolidating its position as a regional and continental hub (Fig. 3).

# 4 Spatial Data

To achieve this study, the work was based on a mutli-time series dataset, accessible satellite imageries, and topographic maps:

- Landsat TM (1986) / Sentinel-2A (2019).
- Aerial photographs—Topographic maps.
- Demographic and census data (1982 and 2014).
- Urban Development plan of Casablanca—Field observations (Fig. 4).

# 5 Methodology

## *5.1 Data Processing Workflow*

The current work is founded on a combination of non-spatial and spatial data from remote sensing over several periods of time. The methodology utilized to acquire the land cover change maps, urban growth, and urban sprawl is illustrated in Fig. 5.

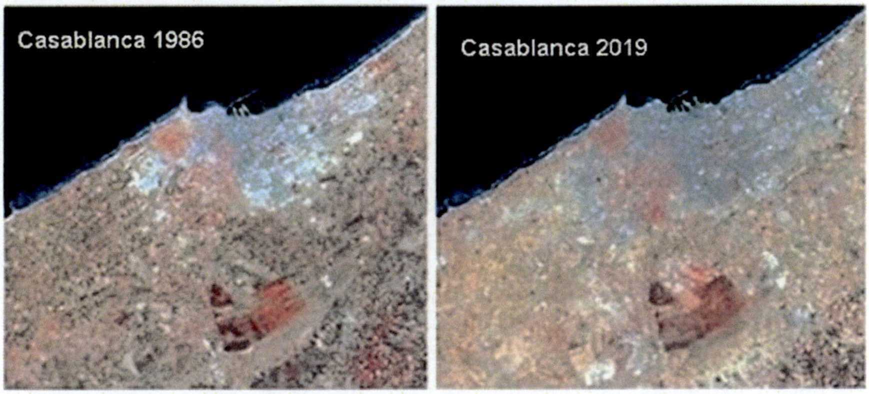


**Fig. 4** Satellite imageries of Casablanca Landsat 1986 and Sentinel-2A 2019

### *5.2 Satellite Imagery Classification*

Satellite imagery classification is a widely accepted method for change detection. Land cover maps are generated from the satellite image classification. In the current case, the Minimum Distance Classification (MDC) method is used. The components classified with the MDC method are sets of vectors (not individual vectors like other classifying methods). The units of land use classification defined in the study area are: built-up (very dense built-up, dense built-up, moderately dense built-up), forests, green spaces, and bare soil.

The formula of Kappa coefficient is:

$$k = \frac{N\sum_{i-1}^{r} X_{ii} \; - \; \sum_{i-1}^{r}(X_i * X_i)}{N^2 - \sum_{i-1}^{r}(X_{i+} * X_{+i})} \tag{1}$$

## 6 Results Analysis

### *6.1 Land Use Cover Maps and Change Detection*

The city of Casablanca experienced a tremendous expansion in the metropolitan area and changes in built-up density. The urban growth and urban sprawl have been increasing all over the study period and can be seen in the crown shape of the new built-up. The generated results of the classification overall accuracies for 1986 is 86% and 2019 is 87% and the Kappa coefficient for 1986 is 81% and 2019 is 83%. The results of overall accuracy and coefficient Kappa combined with field observation

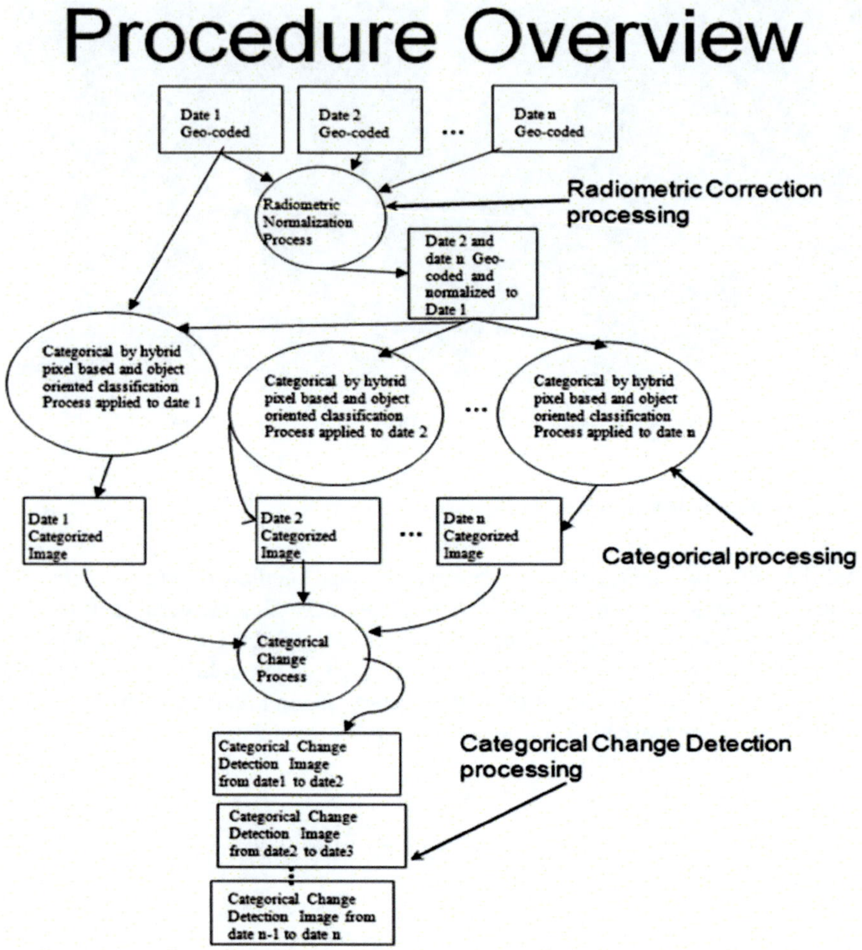


**Fig. 5** Spatial and temporal changes flowchart [7]

validate the results generated by the imagery classification method, as shown by the land use maps, during the study period (1986–2019) (Fig. 6).

According to the results, urban areas have experienced a tremendous change over the study period, the building zones increased by 144% from 1986 to 2019. The bare soil has decreased by 27% while green spaces and forest have decreased by 4.55% and 31% (Table 1).

The principal reason for the wide expansion in the metropolitan area is the increasing population in the city of Casablanca. The interest for new housing is also increasing. The Urban Sprawl Index (USI) is determined by calculating the ratio of annual urban growth (4.52%) and annual population growth (1.75): USI = 2.58. In the case of Casablanca, urbanization grows faster than population.

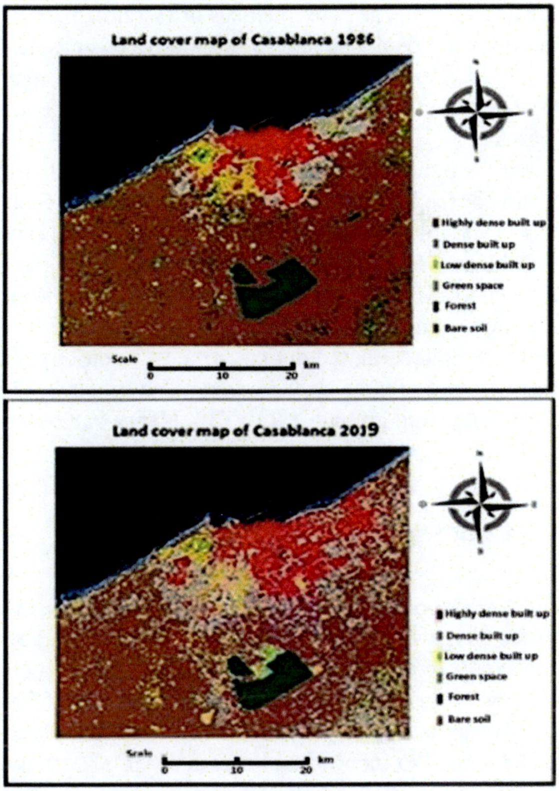


**Fig. 6** Land cover maps of Casablanca between 1986 and 2019

**Table 1** Class area from 1986 to 2019 [7]

| | 1986 | | 2019 | |
|---|---|---|---|---|
| | ha | % | ha | % |
| Very dense built-up | 4227,12 | 6 | 8480.59 | 12 |
| Dense built-up | 3384,27 | 5 | 14,811.3 | 21 |
| Low dense built-up | 3482,1 | 5 | 3862.72 | 6 |
| Green spaces | 653.58 | 1 | 623.78 | 1 |
| Forest | 4392.09 | 6 | 3028.78 | 4 |
| Bare soil | 53,205.80 | 77 | 38,537.79 | 56 |
| Studied area surface | 69,344.96 | 100 | 69,344.96 | 100 |

This study is the reality check for the urban planners and deciders; it shows how the city of Casablanca struggles to maintain the global Sustainability Standards; as shown before, the fast urbanization and the population growth cripples the Smart city aspirations. The city of Casablanca is still in the phase of ensuring a sustainable life for its citizens. The consequences of fast urbanization on the Casablanca smart city initiative are demonstrated in this study: this city lacks several criteria that belong to smart cities. It needs to make a shift toward a smart city and its citizens too must take that shift and get rid of the old habits and behaviors, and start acting like true "urbans".

At the political and legislative level, laws, and strategies need to ensure the right frame for the Smart City Casablanca Initiative to flourish. It should be noted that the city of Casablanca has all the factors it needs to become a smart city: the population of Casablanca is open-minded and connected, access to smartphones and the Internet is granted to all, and the financial means to acquire the technological tools are available, the implementation of the tramway and new electric buses, the intelligent method of waste collection, the optimization of electricity and water consumption, etc. All these factors with the right planning, and strategies that combine the conceptual side and the existing ground reality, Casablanca can become a real smart city.

## 7 Conclusion

The foundation of the Smart City project is the ability to converge all the data related to the city in one large digitalized platform. It is however hard to collect this data, especially when the urban areas continue to grow and spread in uncontrollable ways. During the last three decades, the urban space of Casablanca has changed tremendously as the city has doubled in size and population. Urban planners and policy makers need to monitor and assess urban growth and sprawl as well as the transformation experienced by the city in order to collect existing and new city data. This paper reveals the alarming environmental crisis and mobility problems faced by suburban citizens. Remote sensing and GIS provide key information to study urbanization phenomena and how the urban growth and sprawl affect the city forms. To achieve smart city aspirations, a lot of work needs to be done at both levels, top to bottom and bottom to top. Politicians, decision makers, and citizens need to collaborate and more inclusive and participatory approaches are needed. Smart cities should not be driven only by the economic aspect but by the idea of a better and sustainable living space for the inhabitants and future generations.

## References


1. HCP : Le Grand Casablanca compte 4.270.750 habitants. https://aujourdhui.ma/societe/hcp-le-grand-casablanca-compte-4-270-750-habitants-118242. Last accessed 28 Nov 2020
2. Pawan: Urbanization and its causes and effects: a review. Int J Res Sci Innov IJRSI. 3
3. Belanche D, Casaló LV, Orús C (2016) City attachment and use of urban services: benefits for smart cities. Cities 50:75–81
4. Angel S, Parent J, Civco DL, Blei A, Potere D (2011) The dimensions of global urban expansion: estimates and projections for all countries, 20–2050. Plan 75:53–107

5. El Garouani A, Mulla DJ, El Garouani S, Knight J (2017) Analysis of urban growth and sprawl from remote sensing data: case of Fez. Morocco Int J Sustain Built Environ 6:160–169
6. Calkins HW (1972) An information system and monitoring framework for plan implementation and the continuing planning process
7. Youssef E, Abdelouahed N, Mustapha H, Monitoring and analysis of urban growth and sprawl and built up density using remote sensing data: case of Casablanca. Morocco. Editors. 42
8. Alnsour JA (2016) Managing urban growth in thecityofAmman. Jordan Cities 50:93–99
9. Altieri L, Cocchi D, Pezzi G, Scott EM, Ventrucci M (2014) Urban sprawl scatterplots for Urban Morphological Zones data. Ecol Indic 36:315–323
10. Disperati L, Virdis SGP (2015) Assessment of land-use and land-cover changes from 1965 to 2014 in Tam Giang-Cau Hai Lagoon, central Vietnam. Appl Geogr 58:48–64
11. Beckett H (1942) Population density and the heights of buildings. Trans Illum Eng Soc 7:75–80
12. European smart cities 4.0 (2015) http://www.smart-cities.eu/?cid=01&ver=4. Last accessed 29 Nov 2020
13. Berrone P, Ricart JE, Duch A, Carrasco C (2019) IESE cities in motion index 2019. Servicio de Publicaciones de la Universidad de Navarra
14. Giffinger R, Pichler-Milanović N (2007) Smart cities: ranking of European medium-sized cities. Centre of Regional Science, Vienna University of Technology
15. e-MadinalThe Smart City Cluster. http://www.e-madina.org/. Last accessed 29 Nov 2020

# Facial Expressions Classification on Android Smartphone for a User

**Thandar Shwe Sin and Ohnmar Khin**

**Abstract** In this century, smartphone has become one of the important things that most of the people brings these powerful machines with extensive battery life, qualified built-in cameras, some kinds of sensors and other multifunctional tools that may be useful in everyday routine. Also, the worldwide situation about this time, the issue of facial expression evaluation by a computer is considered as one of the most interesting topics because facial expression identification systems have been applied for many fields, for instance, assessing the customer, as an assistance in health care systems especially for those patients who suffer autism. In this research, an automatic facial expressions identification system for a user is developed as an application to the Android platform. In order to facilitate the application, some existing tools are used to develop with required computations. To work a system fast and stable, the existing SDKs from *Affectiva* have been decided to apply. The proposed system points out the way to classify the basic emotional conditions, namely anger, joy, sadness, fear, disgust, and surprise for a user.

**Keywords** *Affectiva* SDK · Emotion classification · Facial expressions · Image processing

## 1 Introduction

One of the distinctive manners that have provided immensely to the development and growth of humankind is the capability of communication with each other with strong languages. Nevertheless, the impressive percentage of what is communicated has been spread through nonverbal communications not through those languages.

There are many kinds of application for automatic facial expression classification go along with all the research about this system development. In these centuries, during the development of AI, the growth of robotic systems has been becoming popular. Considering that nonverbal communications through facial expression have

T. Shwe Sin (✉) · O. Khin
University of Mandalay, Mandalay, Myanmar
e-mail: thandarshwesin@mu.edu.mm



X.-S. Yang et al. (eds.), *Proceedings of Sixth International Congress on Information and Communication Technology*, Lecture Notes in Networks and Systems 216,
https://doi.org/10.1007/978-981-16-1781-2_21

included one of the essential parts of human interactions. Obviously saying, psychiatrists and psychologists can get benefit from these systems in order that facial expression recognition is crucial for their jobs particularly in emotion recognition cases. For instance, for those kinds of patients such as autism or schizophrenia suffering emotion perception deficits. Other well-known applications related with facial expression could be for visual aiding such as in educational field, in gaming, in remote teaching (LMS—learning management system).

This proposed system is intended to create an Android application that can be able to detect a use'r emotion and his or her appearance based on image, by using the existing *Affectiva* SDK and emotion detection library, which offers the permission for students to use in academic researches. This system also opens to improve in the future—to use this system for classifying emotions on multiple faces.

## 2 Related Work

There are many researchers who are trying to build automatic system for this interesting topic about facial expressions in different approaches. Some automatic system has been built to evaluate facial expressions with pictures and others are trying with video and some even in real time.

In 1973, Kanade implemented the first automatic facial parameters evaluation system. Extracted humans' facial features were compared to apply in the Kanade's algorithm, which has the ability to evaluate several facial parameters automatically. However, the identification success rate for his program was only 45–75%.

The most recent related work using the solution namely *Affectiva*, was implemented by Martin Magdin [1], under Faculty of Natural Sciences, Department of Computer Science, at Constantine the Philosopher University, Nitra. His program was developing for real-time emotion recognition using a webcam. They started to research in this field at the beginning of 2015 in the Faculty of Computer Science, FNS CPU in Nitra and they made publication on January 2017. Their evaluation success rate was 84.27%. The main difference with our proposed system was that their main goal is developing the FER system on the windows platform with the help of webcam, whereas, our proposed system is intended to implement on the android platform using the smartphone.

## 3 Automatic System for Facial Expression Evaluation—Theoretical Background and State of the Art

There are several applications of automatic facial expression recognition for HCI, which has usually emphasized on developing computers that have interactions with human in a more natural manner [2], much closer enough to the manner of interaction between human beings. Mehrabian [2] mentioned that the nonverbal communication, mostly the facial expressions of the person who is speaking supports 55%, and at the same time, the verbal communication and vocal part (i.e., spoken words) support only 38% as the impact of the spoken message as a whole.

According to Darwin, who already assumed that all human beings, and even animals show emotion through remarkably similar behaviors [3], and consequently, he exhibited more than 40 emotional states manifestation, which are his own detailed representations [4]. Ekman and Friesen also concluded that certain emotions are universal recognition through different cultures [5], and later on came up with strong evidence [6] to refute the critics.

Due to the exhibition of facial expression has become popular, also the classification of face expressions or emotions becomes difficult task. Therefore, some of the researchers have decided to limit the imaginable changes of face expressions based on two kinds of approach:

1. Coding System of Facial Action (FACS) [7]
2. Parameters of Facial Animation (FAPs) [8]

There are six basic universal emotions: angry, disgust, fear, joy, surprise, and sadness and one more plus—neutral; which were evaluated by Ekman [9] in the case study description of cross-cultural: "universal categories of emotional expressions" in 1975. In 1978, Ekman and Friesen developed the FACS to describe the AUs which represents an observable facial muscle movement's activity. Another approach, FAPs was designated with MPEG4 [8] standardized parameters, which is developed a standard for audio and video coding format group. FAPs are also concerned with the FACS. In the case of FAPs, motion of head and movement of facial expression are illustrated and a related value decides the amount of deformation. After that all the face models are represented by defining the set of normalizations.

Base on the historical point of view, there are several algorithms for facial expressions recognition category and that system is really complex and costly. This type of facial expression evaluation has to concern with those three basic problems: Face detection in a chaotic environment, Extraction of facial landmarks to be easier in evaluation stage, and Classification and face expression evaluation.

Beyond that time, researchers observe for further improvement by combining various approaches such as integrating features, model pre-training, multiple scales training and so on. Later on, they can develop complete solutions namely SDK (Software Development Kit), lots of libraries and supported API (Application Programming Interface) for the individual steps.

In this proposed system, the *Affectiva* SDK is decided to use as a solution for facial expression evaluation task. *Affectiva* is an automotive AI company focusing on the category of emotion AI, which has been founded since 2009 at Media Lab of MIT University. The *Affectiva*'s humanize technology is particularly used in the advertising companies, including Coca-Cola. As well, their solution is being used in educational field to know whether students are engaging or not. To get the training dataset, the researchers of the *Affectiva* exposed advertising videos to viewers and captured their reaction during watching time, with the help of embedded camera (webcam).

### 3.1 State of the Art

To reduce the complexity and estimate the flow of movement for future upgrade, researchers observe various popular approaches namely APIs, SDKs, and libraries.

The first proposed solution for FER system was the Microsoft Azure emotion detection technology. As the brand, Microsoft, has been popular for its powerful tools, they were expected to be a suitable one to start, however, it didn't fit for this project. The first reason is that it is an online API and not an offline library. Therefore, the users need to have a strong internet connection over Wi-Fi, otherwise, the data package will be consumed. Another problem is that the tool has limited server requests and afterwards a payment is required to use that API.

Because of these restrictions, another alternative was necessary to solve the proposed system, which could be applied as library in developing of Android applications. Therefore, the alternative called *Affectiva* which has both an offline library and an online API. It can classify six basic emotions: angry, joy, fear, surprise, disgust, and sadness; and also allows to apply additional tools. Furthermore, other human biosignals features can also be recognized from the benefits of the *Affectiva*. The significant reason of using the *Affectiva* SDK and library is that it has been supported very well to Android, and it allows free to use for academic researchers. Moreover, we can facilitate our application for future improvement.

## 4 Methodology for Facial Expression Evaluation System

This proposed system is intended to develop by *Affectiva* which has a reliable assessment and possible improvement on Android platform.

The earlier structure of the system was capturing a subject's face that may be in a picture or video frame or in real-life time within an acceptable size. After that, it accepts the face and exports it to the next step.

The solution works to detect a subject's face is performed by using the detection algorithm, namely Viola-Jones [10], which is implemented in OpenCV library. Afterwards, the facial landmark to each face is detected by 34 identified landmarks. After that the extracted facial points, the rotation of head and the critical units between the eyes for each subject are expressed in the SDK. The main point of the facial landmark extraction from the accepted subject's face is to collect the most critical information in an effective manner and computation time. HOG (histogram of Oriented Gradients) is drawn out facial features from the identified picture area by the points of face orientation. In case of emotional expressions such as anger, joy, sadness, surprise and fear are based on emotional action units' combinations and this encoding was supported by EMFACS (Emotion Facial Action Coding System). EMFACS is an abbreviated version of FACS, in which EMFACS scores only those facial actions which are related to emotions obtained from FACS final test. Therefore, EMFACS saves the coding times. The purpose of the classifiers is to identify some observed information and get the desired result which tried to divide this information into a related class or category. The popular algorithms are used in classification process by the *Affectiva* company, called Support Vector Machine and Random Forests, and later on, the deep neural network; which are trained on more than 10,000 face images collected from around the world which are encoded manually.

As in common, the Android applications are using such programming languages mainly in Java and XML for UI. According to Android project framework, the Main Activity java class contains the set-up of the UI views, considering the permissions and some functions that simulate the buttons to do actions.

The smartphone for running the proposed system is a Xiaomi Mi 5X, which has been used for over 3 years during for the time of this work so that its operational power is partly limited.

The *Affectiva* SDK namely *affdex*, supports to classify emotions and detect facial expressions. The recommended hardware requirements for SDK are:

- Processor: 2 GHz,
- RAM: 1 GB,
- Free disk space: 950 Mb,
- Supported windows: 7 and above.

The components included in the SDK were integrated with the C++ written applications (library) and.Net framework. Furthermore, in addition to Android (Java), the *Affectiva* SDK also supports several different platforms such as Linux (C++), MacOS (Objective-C), Unity, Windows, iOS (Objective-C), and Web.

## 5 Experiment and Evaluation of Success Rate

To implement a new system to detect emotions, database is one of the most critical components. If the research used only one database for system testing, it would not have been enough robust and reliable to use in the real world. The benefit of *Affectiva* library is that it can operate on different remarked databases. Several databases, that are commonly remarked by professionals [11], have been used for this proposed system are

- Bao Face Database,
- Yale Face Database,
- NIST Mug-shot Images Database,
- Vision Group of Essex University Face Database,
- JAFFE Database.

## 5.1 System Architecture

The user interface of the proposed system is easy to understand and it has enough ability to determine the subject's emotional state. As in Fig. 1 (main layout), all controls of the system are placed in the main window. The main part of the system is classifying emotions of the subject on a static image. In the main layout, there are 4 action buttons and the result panel on left side of phone screen which shows the result of classified emotions and appearance (Gender, Age, and Ethnicity).

Firstly, the picture is selected from the "SELECT NEW IMAGE" button and it is automatically uploaded for identification. The "ROT LEFT" and "ROT RIGHT" buttons are implemented for rotating the selected image to +90 or –90°. The "REAL-TIME CAMERA" button is implemented for beta testing to connect with real-time camera, facilitated for future development. The program can exit by pressing the default backward button on phone screen.

The intention of the proposed system was evaluation and identification have to be accomplished on individual databases. Each of these databases has various conditions, in which means that different light conditions, chaotic background, rotated head positions, and so on. However, the evaluation results must be a comprehensive one and it was conducted moderately through the mentioned databases.

The success percentage rate for each database has been calculated by the following formula,

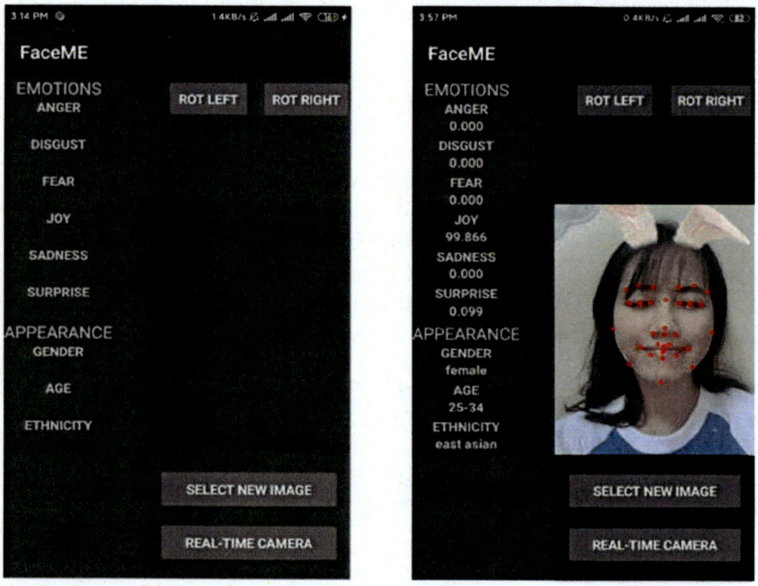


**Fig. 1** Application User Interface on Smartphone

$$100 \times \frac{N}{T},$$

where $N$ is the numbers of successful tested images and $T$ is the total number of images included in each database. All images trained to android app from each database are downloaded and collected in the embedded application in Xiaomi Mi 5X namely gallery.

## *5.2 Result and Discussion*

The details of implementation on each database are presented in Table 1. Throughout the testing implementation, assessing the application designed using the *Affectiva* library was perfectly operated with 5 reliable databases [11], which has been mentioned earlier, within a short time. The reason for only few percentage failures occurred because (as in Fig. 2) some subjects did not express enough emotional expressions and the head rotation of some subjects are more than 15°.

Currently, the only problem to advance in our solution is dependent on a subject's head rotation in which means that if the subject's face is either rotate or tilt more than

**Table 1** Description of tested databases

| Database | Type of database | | Total images | Success rate (%) |
|---|---|---|---|---|
| Bao face database | Spontaneous database<br>Most Subjects are from Asia<br>Images have bad illumination and low resolution | | 149 | 85 |
| Yale face database | Posed database<br>All images are in grey scale with different light conditions<br>Subjects are divided into 4 sections: winking, wearing glasses, no glasses and different emotional expressions | | 165 | 100 |
| NIST mug-shot images database | Posed database<br>All images are in grey scale | Frontal Face Images | 1418 | 100 |
| | | Left/Right rotated Images | 1230 | 0 |
| Vision group of essex university face database | Spontaneous database<br>Subjects include both male and female<br>Images are in low resolution | | 7900 | 98 |
| JAFFE database | Posed database<br>All Subjects are Japanese female<br>Images are in grey scale | | 213 | 95 |
| Total images | | | 11,075 | |
| Average success rate (%) | | | 85 | |

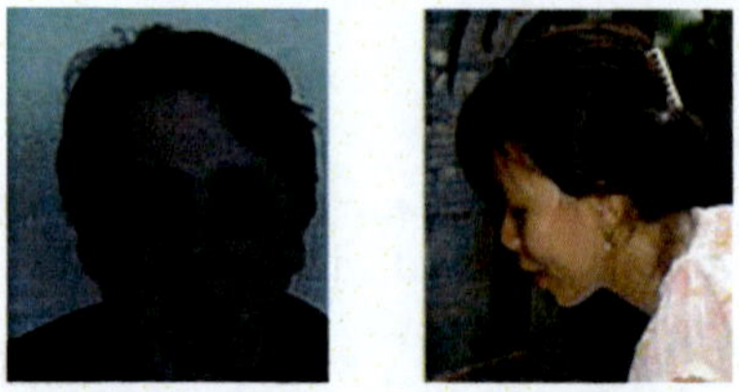

**Fig. 2** Selection of images on which emotion could not be classified

15°, our solution could not detect face, and so that the subsequent conditions cannot perform. To solve this problem the *Affeciva* library needs to train their algorithms with face images from various angel points of face view.

## 6 Conclusion and Future Development

During recent years, the capabilities and the tools are combined perfectly to facilitate developing applications to serve people as solutions for their difficulties. Among them, facial expressions classification using a smartphone has a great impact in several areas of our daily life. In particular, as an assistance in health care system.

During making this experiment, we didn't target only on settling the success rate of the applied solution. At the time of implementing and testing, the application was also evaluated whether it can solve the problem—to classify emotional state—under different light conditions, different chaotic s, various head rotation, and so on.

Due to the time limitation available to develop this project, a prototype with testing only on a single user face has been built for the case of emotions classification. Therefore, classifying emotions on multiple faces within computational speed in operation capability is a further improvement that has been develop in future.

Although the actual operation results did not need to be verified statistically, the average success rate should be known to check whether the application could perform well or not. This solution has been thoroughly tested and we have got the total average success rate with 85%.

## References


1. Magdin M (2017) Real time facial expression recognition using webcam and SDK affectiva. Int J Interact Multimedia Artif Intell
2. Mehrabian A (2008) Communication without words. In Communication theory, pp 193–200
3. https://blogs.scientificamerican.com/observations/the-evolution-of-emotion-charles-darwins-little-known-psychology-experiment/
4. Keltner D (2009) Born to be good: the science of a meaningful life. WW Norton & Company, New York
5. Ekman P, Friesen WV (1971) Constants across cultures in the face and emotions. J Personality Soc Psychol 17(2):124–129

6. Ekman P (1994) Strong evidence for universals in facial expressions: a reply to Russell's mistaken critique. Psychol Bull 115(2):268–287
7. Ekman P, Friesen WV (1977) Manual for the facial action coding system. Consulting Psychologists Press
8. MPEG Video and SNHC (1998) Text of ISO/IEC FDIS 14 496–3: audio. In: Atlantic City MPEG Mtg, Oct 1998, Doc ISO/MPEG N2503
9. Ekman P et al (1975) Unmasking the face: a guide to recognizing emotions from facial cues
10. https://en.wikipedia.org/wiki/Viola%E2%80%93Jones_object_detection_framework
11. McDuff D, Kaliouby ER, Senechal T, Amr M, Cohn JF, Picard R (2013) *Affectiva*-mit facial expression dataset (AM-FED): naturalistic and spontaneous facial expressions collected 'in-the-wild'. In: Paper presented at the IEEE computer society conference on computer vision and pattern recognition workshops, pp 881–888. https://doi.org/10.1109/CVPRW.2013.130

# A Survey in IoT for Healthcare Applications

**Sarah Abdalrahman Alshqaqi, Abdualmajed Ahmed Al-Khulaidi, Adnan Yehia Al-mutawkkil, and Mohammed Mohammed Zayed**

**Abstract** IoT connects smart objects also it is considered a strong concept in modern technology, it is used in various fields, including industry, education, entertainment, agriculture, as well as health and others. IoT acts a significant role in health, whether for patients, doctors, hospitals, or health insurance companies to decrease the costs and advance the performance. There are many scientific researches that dealt with IoT and its role in health. However, comprehensive content for the Internet-of-Things from several directions is not included. For this reason, this research aims to discuss all the aspects and fields that IoT is used in healthcare.

**Keywords** Internet-of-Things · Sensors · RFID · NFC · WSN · IoT framework · Healthcare

## 1 Introduction

Internet-of-Things is a communication of electronic objects, actuators, and sensors through a connection of the Internet to permit data exchanging between objects, applications, servers, users, and service providers [1]. Internet-of-Things is a revolution in industry that effects various manufacturing and distributes changes in business models and at any area in life such as healthcare. In the area of healthcare manufacturing, Health Information Technology (HIT), is the recent most important technology in information management of health, helps exchanging information of health securely between service providers, customers, caregivers, and patients in computerized areas.

S. A. Alshqaqi (✉) · A. A. Al-Khulaidi · A. Y. Al-mutawkkil · M. M. Zayed
Sana'a University, Sana'a, Yemen

A. A. Al-Khulaidi
e-mail: khulaidi@su.edu.ye

M. M. Zayed
e-mail: mzayed@su.edu.ye

 
X.-S. Yang et al. (eds.), *Proceedings of Sixth International Congress on Information and Communication Technology*, Lecture Notes in Networks and Systems 216,
https://doi.org/10.1007/978-981-16-1781-2_22

In [2] In fact, Internet-of-Things explains problem more efficiently than old solutions as well progresses using IoT in diverse areas such as healthcare. Newly, in smart hospitals, the smart healthcare is considered one of the greatest significant areas. Embedded sensors in user bodies play a significant part in saving the life of patients at an accurate time. And also, support doctors to track patients in another place in the world.

In this paper, we focused on the most important studies interested in Internet-of-Things (IoT), First, we will define Internet-of-Things (IoT), then, essential component of IoT, next, applications in IoT, after that, diverse purposes of Internet-of Things in health care systems, then, key advantages of Internet-of-Things in health care systems, next, some parameters are used in the healthcare system, then, sensors in healthcare system, after that, some opportunities of Internet-of-Ihings in healthcare. Finally, IoT framework.

## 2 Internet-of-Things (IoT)

This section concentrates on explicating IoT definitions.

### 2.1 IoT

Internet-of-Ihings is a big bump after Internet [3], next generation of the Internet in development technology [4]. IoT, first invented in 1999, by Kevin Ashton [5] used in management of supply chain [6]. There is no standard meaning for IoT [7]. IoT has converted all parts of our lives [8], allows people and smart devices [9, 4, 10], devices are real objects in this physical world [11, 6] identifying [12], sensing, actuating [13] that are distributed randomly [14] to connect anytime, anywhere [15] providing the ability to share information [13] to accomplish some objective [12], based on standard communication protocols [5, 4]. IoT is a group of sensors, actuators, as hardware [6], processors, and computers by a variety of communication networks [16]. In [9, 13] the IoT depending on initial estimates, more than 40 billion devices in addition the market of Internet-of-Things may reach around $1.7 trillion in 2020 [15].

### 2.2 Essential Component of IoT

In this technology, the sensors [11] used in health care are called Wireless medical sensors [14] then are embedded in Internet-of-Things, like wireless sensor networks (WSNs), middleware [13], barcodes [17], smart detection [4], radio frequency identification (RFID) [18, 19] near field communication (NFC), low energy wireless communications, cloud computing [4, 13].

During the first days, Radio Frequency Identification (RFID) was the least requirement for the devices to be considered as part of Internet-of-Things environment, from now on RFID is considered as the first enabler in Internet-of-Things, and the core of networking to construct IoT [9, 11]. RFID has used for identification which communicates the identity of device or user in wireless by radio waves as a serial number [11], recording, and localization purposes [9], automatic identification for any device that connects to the IoT environment [6]. The important components of Radio Frequency Identification are reader, tag, access controller, antenna, software, and server [11]. Radiofrequency identification, short-range communication technology that tag of Radio Frequency Identification connects through Radio Frequency Identification reader through radiofrequency electromagnetic. Tags might hold diverse forms of data, but then the form of data usually used for Internet-of-Things applications is electronic product code (EPC). Electronic Product Code, generally unique identifier to the device. The unique identifiers confirm which device traced through tags of Radio Frequency Identification have individual identities in the Internet-of-Things [12]. But, these days, IoT systems communicate devices by various communication protocols and wireless technology, such as Wi-Fi, Near Field Communication (NFC), and ZigBee.

For example, In Near Field Communication (NFC) is a new technology which is built on the Radio Frequency Identification standard [12] in 13.56 MHz, needing distance 4 cm. Near Field Communication gives easier and suitable way to the users to perform communications, transmit digital content, also join automated objects. Each Near Field Communication tag holds Unique Identification (UID) which connected with the tag. Near Field Communication technology is often joined into smart phones that can transmit data with each another when transported together [12]. The Middleware, Internet-of-Things is consisting of huge diverse objects producing enormous quantities of data. Middleware is between Internet-of-Things data, hardware, and applications which developers generate to exploit IoT. Therefore, IoT middleware supports taking together different data and devices which allows creators to produce new services for Internet-of-Things [12].

In advanced technology, Wireless Sensor Technology (WSN) is the most used [6, 20], data of sensors which is shared with nodes of sensors [12] then sent to central system. The components of WSN are (1) WSN hardware such as nodes, (2) Wireless Sensor Technology communication stack interact with outside world over the Internet for performing as gateway for Wireless Sensor Technology, (3) WSN middleware combines infrastructure, service-oriented architecture in addition network of sensors [6], comprise sensors and also hold gateways which gather data then send to the server [12].

Integration of Wireless Sensor Technology and Radio Frequency Identification.

In [4] several kinds of smart sensors that developed depending on physical principles of electromagnetic and biosensor. Sensor's data in Internet-of-Things may be taught then joined for specific purpose such as in making of the decision. Such as Radio Frequency Identification joined sensors, tags of sensors, independent tag, sensor devices, and Radio Frequency Identification reading systems, integration of Radio Frequency Identification and sensors permits IoT in industrial services or other services. Internet-of-Things integrating with RFID also WSN consequently support to progress Internet-of-Things applications in healthcare, and smart transportation [6], smart city, smart infrastructure.

### 2.3 Applications in IoT

Applications of healthcare for the Internet-of-Things can possibly bring complete patient care in numerous situations, containing in hospital, nursing homes, and at home [16]. IoT permits small objects for sensing, processing and gives the capabilities of communication to design, and to develop the sensors, embedded objects, and other things to know environment [21]. The objective is to surrounded technology into the background of everyday life [6]. The IoT has significant technical, social in addition to economic significance [7]. Though, in past periods, it has covered wide range of applications such as healthcare, utilities, transport [6], supply chains, social applications [12], smart homes such as monitoring system and alarm system of safe and reliable [22], and smart cities [9], in personal life and business level [23], self-driving cars [24]. The amount of new IoT products and applications has grown in current years [25]. Health care or medical care signifies one of the greatest application fields in Internet-of-Things [8, 26], in medical care area carry great suitable to both patients and doctors for current disease monitoring and analysis increase healthcare quality [10, 17]. In monitoring system, the technology of wearable sensor technology is more used [8], Additionally initiate smart systems, for example, mobile healthcare in addition monitoring systems in smart healthcare [8], such as heart rate monitors and progressive devices can be able to monitor specialized implants, like advanced hearing aids [3, 27] insulin pump is the most used. Numerous medical systems such as fitness programs and care for elderly [15], like artificial provider for heart rate [28, 16] wearable heart monitors [27, 29]. Numerous nation state and system of government are concerned in developing the standards of Internet-of-Things because it can take great economic benefits in the future [30, 31].

### 2.4 Diverse Purposes of Internet-of-Things in Health Care Systems

1. Gathering life parameters in Continuous way: it covers blood pressure, there are tools which are collect and store information in all day.
2. Collection of life parameters in periodical way: it includes temperature of body, then collect also store reports but at specific times.
3. Collection of specific parameters in continuous way related with shared illnesses: contain percentage of blood sugar and body water, use tools are collect and store information in all day.

4. Collection of specific parameters in periodical way related with shared illnesses: include percentage of blood sugar, body water, use smart devices through joining to system gather and store reports in specific times.
5. Monitoring: each thing (user, device of health) then interacts with WSN devices which observed, measured and controlled 24/7.
6. Remote assistance: services of healthcare provide in remote way over devices of the internet like healthy houses, management f medication and telemedicine.
7. Management of data: through involving internet of things, each healthcare data (support, analysis, treatment, even events of daily life) use through whole collection and management of data.
8. Transfer smart contents to the person: system has ability for modifying the profile of the person then send to screen of the user.
9. Enterprises combination: combined integrating information can be obtained. Consequently, permits authorized persons to exchange all information of medical anywhere [6, 29].

## 2.5 *Key Advantages of Internet-of-Things in Health Care Systems*

1. Reduce costs: through the capability to observe patients remotely, the cost of visiting can be decreased. In addition, the first equipment was of home care, various users can be hospitalized and check at their homes.
2. Results of treatment: monitoring is reliable is automatic, then data are stored in cloud also sent to doctors.
3. Management of disease: when health signs of person are described, then illness can be identified and treated before it advances develop.
4. Decrease of errors: complete, exact data which are integrated automatically can decrease the cost of financial and errors of medical.
5. Satisfaction of patient: ensuring the requirements of patient, accuracy of data, treatment is specific time, decrease of costs, reduction of repetitive visiting, then there is a positive effect to the user.
6. Managing of medication: internet of things supports patients to exact usage of medicines in addition to the drugstores and centers of healthcare to avoid drug waste [29, 32, 33].

## 2.6 *Some Parameters Are Used in Healthcare System*

In healthcare systems, the system design can be designed to track specific parameters or set of parameters. Some systems which are designed to monitor user or group parameters are:

1. Monitoring systems for blood, to monitor blood sugar level.
2. Monitoring systems for temperature, monitoring is important and reported to the doctors.
3. Healthcare systems for elderly, some members of family need to monitor remotely so internet of things devices can use for remote monitoring [33].

## 2.7 *Sensors in Healthcare System*

The sensor, an important element for Internet-of-Things, measures parameters of health. There are popular sensors such as:

1. Accelerometers: in [8] to sense blood level especially for elderly. Absolutely, (MMA7260QT) accelerometer to detect movement of body.
2. Gyroscope: detects angular velocity and support in warning healthcare professionals when a human fall.
3. Magnetometer: in detects and supports in detecting the relative orientation, use in combination with gyroscope and accelerometer that notice human fall in addition use in elderly care devices.
4. ECG Sensors: in [14] Electrocardiogram sensors use to detect and measure activities of electrical within the heart.
5. Temperature Sensors: in [17] that use for temperature of user body [23]. Such as, LM35, TMP36 and TMP75 [15, 24, 29, 31].
6. Heart Rate Sensors: that use to measure the heart rate [33].

## 2.8 *Some Opportunities of Internet-of-Things in Healthcare*

1. To acquire, analyze and share health data of patient.
2. For treatment, which can be given in more detailed level.
3. For communication, between individual devices with the whole healthcare system [15].

## 2.9 *IoT Framework*

Ubiquitous healthcare which emergent technology which rises effectiveness, correctness, and accessibility of treatment for medical [27]. Ubiquitous healthcare system is offering suitable service for both patients and doctors, in addition to make it easy to detect health condition of patient. So, users can monitor their health without the need to visit a clinic or a hospital.

Figure 1 shows Traditional U-healthcare system architecture and Fig. 2 shows Layers of Ubiquitous health care system IoT. Architecture is: (1) body area network

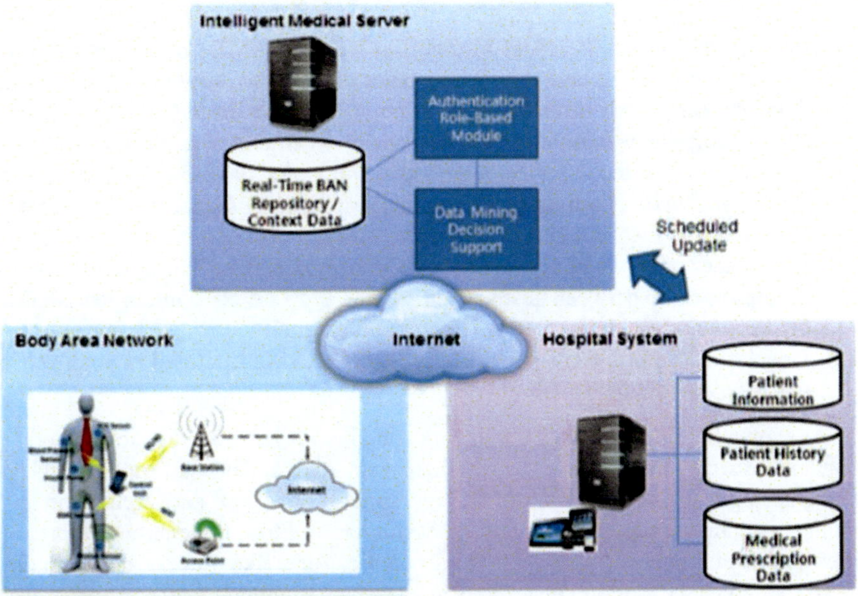


**Fig. 1** Traditional U-healthcare system architecture

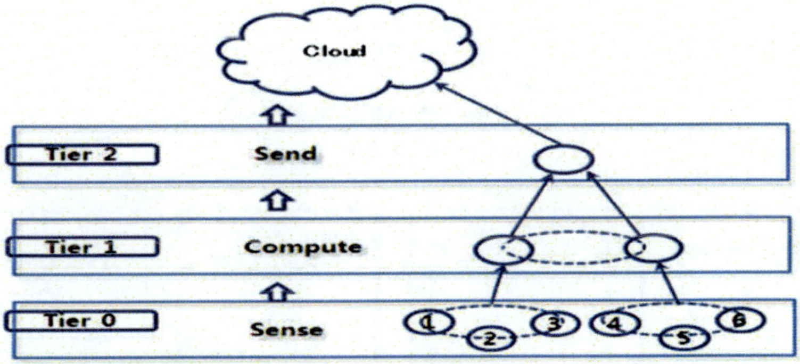


**Fig. 2** Tiers of U-healthcare system IoT

(BAN)—wearable body sensor network—personal monitoring devices (2) intelligent medical server (3) hospital System.

1. Body Area Network, the sensors which are connected to user's body with the aim of capture signals, contain blood pressure and temperature of body. It contains wireless body area network and personal monitoring devices. Wireless Body Area Network (WBAN) is designed as wear or bio-sensors in body of user.

Sensors gather data from body of patient then send to central node as electromagnetic waves. Personal Monitoring Devices (PMD) can be installed on PC or mobile or Personal digital assistance device. PMD contracts information from Wireless Body Area Network through Zigbee or Bluetooth thus Personal Monitoring Devices holds logics to control whether for sending information to Intelligent Medical Server or not, send from personal monitoring devices to Intelligent Medical Server by any technology as SMS. Thus, PMD is a service requester and IMS is a service provider.

2. IMS is the backbone of system, collects information from Body Area Network. It helps as a hub between hospital and patient. Data stored in Intelligent Medical Server and available to staff and doctors in hospital.
3. Hospital System is third part, data is recorded, saved, altered or removed by patients, doctors.

As shown in Fig. 3 [34]. IoT framework contains a layer of application, support of service and an application layer, layer of network, layer of devices. Application layer contains applications of Internet-of-Things. support of service and application layer include general support abilities and exact support abilities. In general support abilities that use in diverse Internet-of-Things applications, like process and store data. Even though exact support capabilities meet the requirement of several applications then network layer and device layer.

- Application layer

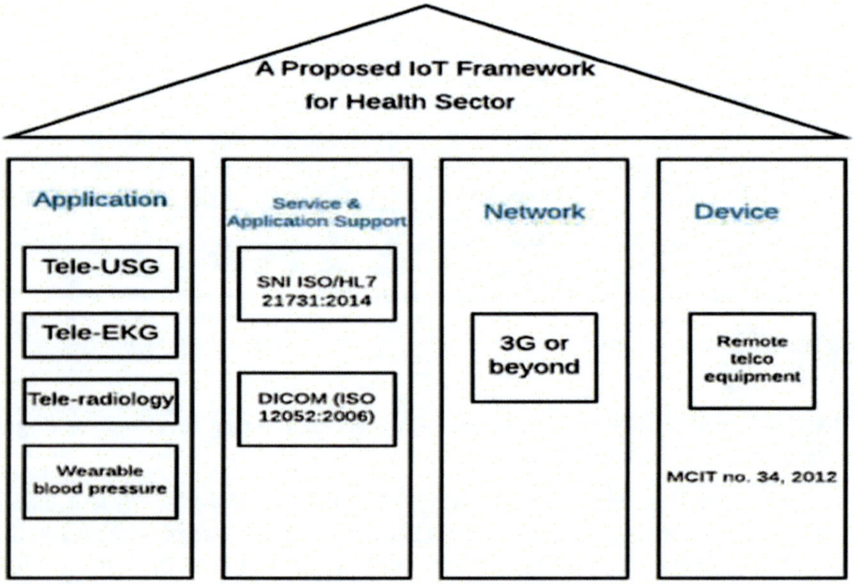


**Fig. 3** Proposed IoT framework in the health sector in Indonesia

Equipment of IoT in health area is connected with diseases. This equipment holds teleECG to process patients with illnesses of heart, teleradiology, that process patients with breathing for providing teleconsultation between doctors and specialist doctors.

- In service and application support

  a. Health Level Seven is international standard-development organization, combination of 31countries.Health Level Seven generates wide-ranging standard which used for interoperability in health area.Health Level Seven came from 7th level in OSI model.
  b. Digital imaging and communications in medicine describes format of medical image that exchange requirement of quality and data intended for medical usage. It is used as a protocol of interface from device to PC.

- Network Layer detect network and data connectivity. It communicates with the sensor equipment. such as: RFID, and Bluetooth. Teleradiology used as telehealth system. For example, in wearable blood pressure monitoring device for connection is used to exchange data between device and computer.
- Device Layer such as a remote telco equipment.

Architecture that helps in construction and maintaining solutions of industrial to consumers [19]. Thus, it is four layers. Sensor layer contains huge number of sensors, controllers, and actuators which represent the lowest layer. Applications is the top layer, that offers user interface for customers or users to run the system, and for external agencies. The middle layers are technological implementations to transport, process data to generate services. As shown in Fig. 4. the layers are:

1. sensor layer the lowest, contains intelligent devices and sensors, that is a core to IoT. The sensors can fit into or onto the devices. The arrangement of sensors differs by goal of the usage.
2. gateways and networks layer: play an important part in enabling IoT and growth in future.

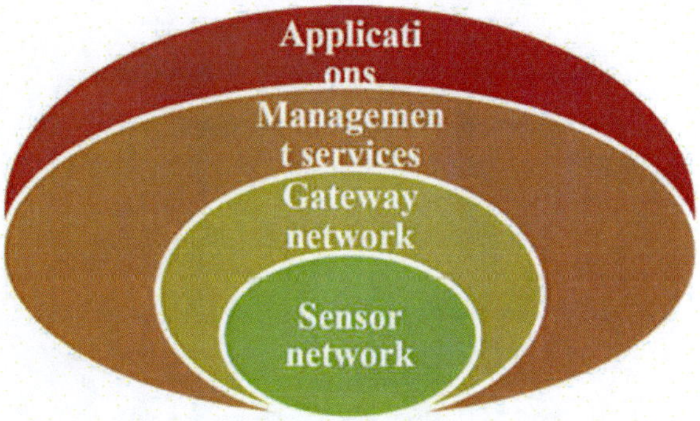


**Fig. 4** Proposed architecture of IoT in industrial

3. management service layer, it interacts as residence in addition work for all processes of companies, and information assets of organization are kept here, also, warehouses, decision support systems for managers and customers.
4. Application Layer, permits persons collects information from a variety of devices, like PCs and phones, and modified interfaces like control console provide for patient in smart bed in hospital. Thus, diversity of institutions, organizations, individuals or industries can contact internet of things with appropriate applications for different purposes.

In Fig. 5, contains of these layers are [35]:

1. Data access layer: like time, place through all levels of agricultural products, containing sensors, RFID data, then capture and process by video recording. The levels usually consist of time in (farming, storage, sale, manufacturing, logistics).
2. Data transmission layer: exchange data taken from the previous layer to servers in next layer, using technologies containing wireless sensor network, Wi-Fi, Ethernet.
3. Application service support platform: contains of engines and services.

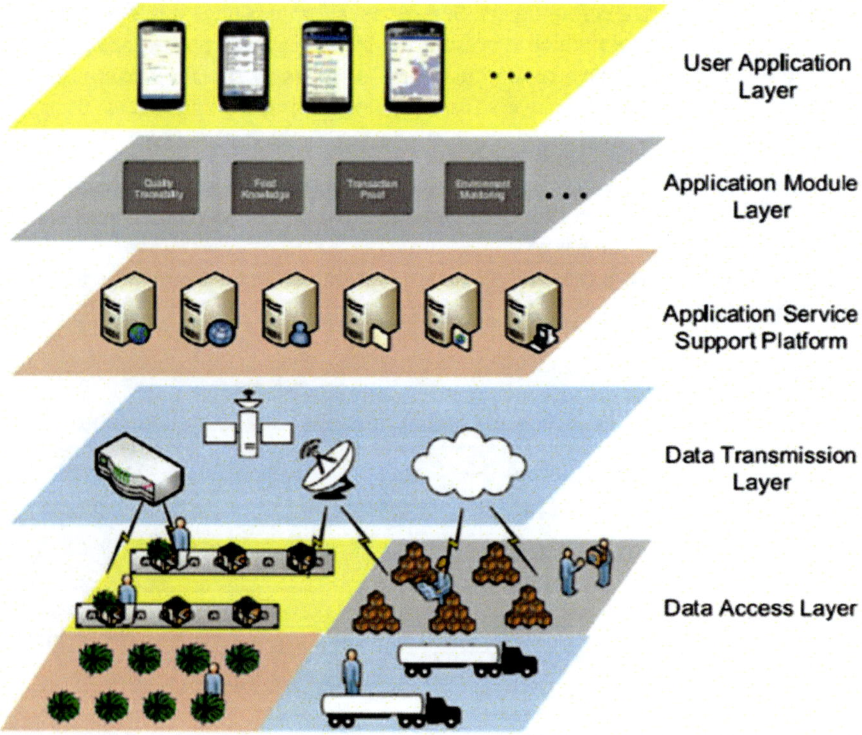


**Fig. 5** Proposed architecture of the IoT

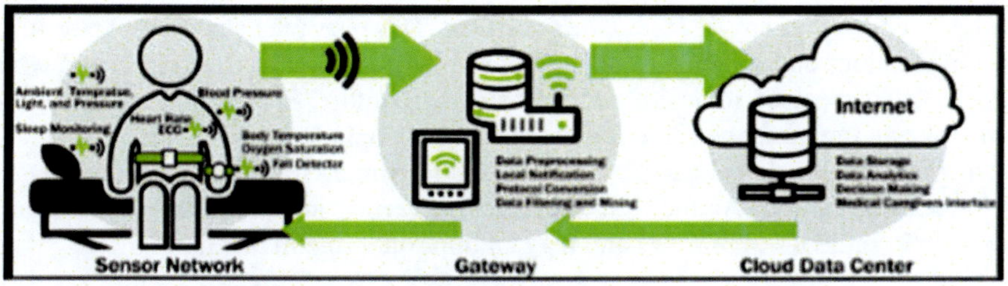


**Fig. 6** General architecture of healthcare using Internet-of-Things

4. Application module layer: contains multiple modules so user choices one or numerous modules with the purpose of achieve the requirement of businesses in agriculture. Such as, persons influence to trace the quality of products through quality module. Persons influence to obtain food 's information through knowledge food module. Persons influence to ensure the truth of operation through transaction proof module.
5. User application layer: executed by modules of application for aim of users. The users of target separated into these categories: supervisors, creators, providers then customers.

According to [36] various layers are included in IoT differently according to the purpose to use IoT like: layer of edge technology, layer of access gateway, layer of middleware, and layer of application, several studies that propose architecture for healthcare, for example in Fig. 6.

1. Sensor network layer: gathers biomedical data such as temperature of body. The sensors in user's body to trace and manage person's health.
2. Gateway layer: sensors get data in addition at the same time send to a node such as phones then processed for extract meaningful information.
3. layer of cloud data center: data that acquired from gateway layer is used here to analysis or other goals.

In general, in [37] the same architecture is used in IoT for business that including:

1. Things/device layer: contain sensors, actuators, processors, antenna and software that runs on physical thing to manage and operate the functionality.
2. Connectivity layer: contain protocols that permit communication between devices.
3. Application layer: contains process management, application platform, analytics and data management.

## 3 Conclusion

Thanks to the concept of IoT, which has played a major role in several areas, especially in hospitals and health, whether it contributed to helping patients or the doctor to know or discover any disease at the same time. This research helped us to know IoT, its applications and advantages, types of sensors used, and the tires associated with the IoT that are used in health, industry also agriculture. Surrounded technologies will take an important part to bring healthcare to people in remote places and monitoring system which make available a continuous stream of accurate data for better healthcare decisions. As the technology for gathering, analyzing and transmitting data, IoT continues to improve and play an important role in this time.

## References

1. Yuan YS (2020) A study of Internet Of Things enabled healthcare acceptance in Malaysia
2. Rahouma KH (2020) Challenges and solutions of using the social Internet of Things in healthcare and medical solutions—a survey
3. Xican C (2015) Low power sensor design for IoT and mobile healthcare applications
4. Shancang Li (2015) The internet of things: a survey.
5. Bandyopadhyay D (2011) Internet of Things—applications and challenges in technology and standardization
6. Jayavardhana (2013) Internet of Things (IoT): a vision, architectural elements, and future direction
7. Laplante PA (2015) A structured approach for describing healthcare applications for the Internet of Things
8. Subasi A (2018) IoT based mobile healthcare system for human activity recognition
9. Salim F (2015) Urban computing in the wild: a survey on large scale participation and citizen engagement with ubiquitous computing, cyber physical systems, and Internet of Things
10. Yang Y (2016) Lightweight distributed secure data management system for health internet of things
11. Madakam S (2015) Internet of Things (IoT): a literature review
12. Whitmore A (2015) The Internet of Things—a survey of topics and trends
13. Biljana L (2015) A review of Internet of Things for smart home: challenges and solutions
14. Khader M (2017) Simplified AES algorithm for healthcare applications on Internet of Thing
15. Strielkina A (2017) Modelling of healthcare IoT using the queuing theory
16. Laplante PA (2016) The Internet of Things in healthcare potential applications and challenges
17. Strielkina A (2018) Availability models for healthcare IoT systems: classification and research considering attacks on vulnerabilities
18. Takano Y (2016) Unconnected component inclusion technique for patent network analysis: case study of Internet of Things-related technologies
19. Gupta N (2017) Implementation of an IOT framework for smart healthcare
20. Ugrenovic D (2015) CoAP protocol for web-based monitoring in IoT healthcare application
21. Mora H (2017) An IoT-based computational framework for healthcare monitoring in mobile environments
22. WANG Dong-lai (2016) The Internet of Things the design and implementation of smart home control system
23. Aimad K (2016) Data quality in internet of things: a state-of-the-art survey
24. Molanes RF (2018) Deep learning and reconfigurable platforms in the Internet of Things
25. Ahlgren B (2016) Internet of Things for smart cities: interoperability and open data
26. Alharam AK (2017) Complexity of cyber security architecture for IoT healthcare industry: a comparative study
27. Gelogo YE (2015) Internet of Things (IoT) framework for u-healthcare system
28. Alansari Z (2017) The Internet of things adoption in healthcare applications
29. Alansari Z (2016) The Internet of Things adoption in healthcare applications

30. Manuel D (2016) State-of-the-art, challenges, and open issues in the integration of Internet of things and cloud computing
31. Xu LD (2014) Internet of Things in industries: a survey
32. Bi Z (2014) Internet of Things for enterprise systems of modern manufacturing
33. Kumar N (2017) IoT architecture and system design for healthcare systems
34. Ariyanti S (2018) A proposed the Internet of Things (IoT) framework for health sector in Indonesia
35. Liu Y (2016) An Internet-of-Things solution for food safety and quality control: a pilot project in China
36. Jangra P (2018) A design of real-time multilayered smart healthcare monitoring framework using IoT
37. Schoder D (2018) Introduction to the Internet of Things

# The Role of Big Data in Social Science: A Case Study Using Hadoop

**Nour Alqudah and Mohammed Q. Shatnawi**

**Abstract** The research on big data in the social sciences and its impact has received much interest from practitioners and policy-makers. Data science can help find the answer to research questions in the social sciences because data is the lifeblood of the decision-making process; it is also the raw material for the accountability process. New data sources, new technologies, and new analytical approaches can make evidence-based decision-making more efficient and flexible. The analysis of this data plays a large role in discussing the challenges facing our societies today. This research provides an analysis of the six factors that influence happiness—GDP per capita, social support, life expectancy, freedom, corruption, and generosity. In this research, the World Happiness Report was studied in 2019, where its survey of the state of global happiness ranks 156 countries through their citizens' happiness. Depending on six factors. This study focuses on analyzing factors that affect happiness and satisfaction with life.

**Keywords** Big data · Big data analytics · Social science · Hadoop · CDH · Hue · World Happiness Report · Social support · Life expectancy

## 1 Introduction

These days, human society has begun to record its activities and perform its social life using different digital media, where every aspect of our life is gadget oriented, there is a massive amount of data that has sprung from different digital sources. This is called big data; it is the power of the next generation in the IT industry. The big data revolution depends on the availability of different types of data that were not

N. Alqudah (✉) · M. Q. Shatnawi
Department of Computer Information System, Jordan University of Science and Technology, Irbid, Jordan
e-mail: nsalqudah18@cit.just.edu.jo

M. Q. Shatnawi
e-mail: mshatnawi@just.edu.jo

 
X.-S. Yang et al. (eds.), *Proceedings of Sixth International Congress on Information and Communication Technology*, Lecture Notes in Networks and Systems 216,
https://doi.org/10.1007/978-981-16-1781-2_23

available before and in which the exploitation of this data has become very important by researchers, companies, and different sectors [1].

Many previous research works indicated that big data has the potential to generate value for business and also to meet social challenges [2]. At the business level, organizations have realized that they are getting many benefits through big data analytics. These analytical results are based on the most influential organizations in marketing, new revenue opportunities, and better customer service [3]. In the social sciences sector, big data and its use have raised several questions, including questions regarding how data is collected and used in ethical ways and how big data affects data on individuals and societies [4]. There is an expectation that big data will be a social good that must be secured and used for the transparency of services and the development of a new culture, wherein recent research regarding big data there is an increasing section of big data capacity to support social science research [5].

Big data will continue to reach all corners of society as the tremendous growth of digital information is available for many social sciences and social impact applications such as health care, education, and environmental protection. On the other hand, data analysis is widespread and has multiple forms and trends, such as predictive and descriptive analyzes and diagnostic analyzes [6].

In social sciences, understanding user responses to social networks in terms of user behaviors and disseminating information is valuable because opinions can influence purchase decisions or news. Understanding user opinions understand how societies interact with one another. Big data is a powerful source of research in this field that may provide a better understanding of human behavior and social phenomenon in a way that was never possible [7]. Data analysis is not only about addressing and understanding critical social problems, but it is also essential for the social good in general, which is defined as a procedure or application that benefits society [8, 9].

This study focuses on analyzing factors that affect happiness and satisfaction the life, as it analyzes the World Happiness Report was studied in 2019. The degrees of happiness and their rankings use data from a Gallup survey. The results are based on answers to the main question, known as the Cantril Ladder, for life assessment. This was asked for a question in the questionnaire, where respondents are asked to think about a ladder and evaluate life from 10 to 0, as the best life 10 and the worst possible life is 0. A random sample was taken to answer this question from different continents and countries [10, 11].

This research was built on these questions. How do factors affect the happiness of the regions? How do per capita GDP and social support influence the Healthy Life Expectancy factor? Is corruption perception of government which makes a person sad? Is GDP which makes a person happy?

The remaining parts of the paper are organized as follows: Sect. 2 presents related works. Section 3 presents the experiment, result, and discussion. Sect. 4 presents conclusion.

## 2 Related Work

Many sectors use big data in their applications, in the social sector, Wojcik et al. [12] focus on how political orientation affects subjective well-being by studying large samples of online survey takers, US politicians, Twitter users, and LinkedIn users; their findings illustrate the nuanced relationship between political ideology, self-enhancement, and happiness,

Park et al.[13] study they used big data to study how well-being is related to time trends by measuring language expressed on social media and tested in identifying individual attitudes, whether past, present, or future, and linking them with gender, age, five adult personality, life satisfaction, and depressive symptoms.

Kishore and Camarushy [14] discussed Hadoop's concept and found that big data analyzes with pigs and hives help identify problems and help solve them. They also analyzed crime data in Boston, which could improve the crime situation.

Bailón [15] focuses on examining how data research can transform policy-making, focusing on how it can help improve communication and government in policy-related areas, and how sociologists can reconsider old questions at the intersection of new techniques and disciplinary methods. Sociologists need to embrace big data, gain literacy in research that makes it possible, understand the methods needed to analyze data, and build reliable interpretations based on a common language compatible with other disciplines. This study's implications for understanding how a social world is vast and how to build reliable stories of what big data reveals and its impact on society.

McFarland et al. [16] suggest that there may be a possibility of constructive synthesis across models in the term "forensic social science approach," where they find that big data represents the emergence of a new class of data derived from digital records from an amazing set of social phenomena. Everyone is focused on collecting digital social information on vast areas of human beings. All of this is used to understand a particular social theory and understand its results.

Archenaa and Mary Anita [17] talked about how to make use of the big data produced by health care and the government because these data have a significant role in predicting the occurrence of emergencies before they occur, can provide adequate care to citizens by predicting the study and analysis of different categories of citizens. They found that using Hadoop plays an influential role in conducting real-time analysis of big data. Golder and Macy [18] using millions of tweets across 84 countries to examine the change of psychological processes revealed individual-level diurnal and seasonal mood rhythms using. On the other hand, Liu et al. [19] predicted life satisfaction. Where they showed that negative emotional expressions on Facebook within the past 9–10 month.

Nelson et al. [20] The Nelson formula relies on the use of qualitative analysis to understand big data. It recommended using wisdom and social experience to interpret data and summaries, develop new hypotheses, and test these hypotheses instead of using standard computational methods to manipulate big data, and Gerber [21] uses 25 types of crime in the US-based on the geographic areas occupied by individuals

**Table 1** Columns description

| Column | Describe |
|---|---|
| Country (region) | Name of the country |
| Ladder | Cantril Ladder is a measure of life satisfaction |
| SD of Ladder | Standard deviation of the ladder |
| Positive affect | Measure of positive emotion |
| Negative affect | Measure of negative emotion |
| Social support | Social support contributed to the calculation of the Happiness Score |
| Freedom | freedom contributed to the calculation of the Happiness Score |
| Corruption | Perception of Corruption contributes to Happiness Score |
| Generosity | Perception of Corruption contributes to Happiness Score |
| Log of GDP per capita | GDP contributes to the Happiness Score calculation |
| Healthy life expectancy | Life expectancy contributed to the calculation of the Happiness Score |

who use Twitter and identify discussion topics about a city in the United States to predict crime.

## 3 Experiment, Result, and Discussion

### *3.1 Data Set Description*

The World Happiness Report 2019 data used in this study, where its survey of the state of global happiness ranks 156 countries through their citizens' happiness. Depending on six factors, each individual's gross domestic product, average life expectancy, Freedom, Generosity, and trust describe the government as corrupt. These factors help assess happiness in each country. The Dystopia Happiness Scale is measured using the Dystopia Scale (1.85)+ the remaining or unexplained value for each country [11]. The data have 11 columns that are described in Table1.

### *3.2 Preprocessing*

The data was quickly processed using Hadoop operations that handle big data analysis. The lost data was handled by filling all the empty values with a zero value. Besides, unused columns were deleted. All of these operations ensure that results are obtained quickly and efficiently.

### 3.3 Experiment

In this experiment, a Cloudera Distribution environment, including Apache Hadoop (CDH), was used. Hue is a CDH component known as a web editor that allows us to interact with data stores to explore, browse, and import quickly and efficiently within practical seconds to get a result. Data is queried with different editing programs such as impala, hive, pig, java, and spark through it. The data has been downloaded to Hadoop, and the file has been uploaded inside the Hadoop distributed file system (Hdfs). Then upload into Hue and run some SQL commands.

### 3.4 Result and Discussion Analyzing Based on the Country

In Fig. 1, 25 countries were selected; we find that Nigeria, a country on the continent of Africa, has achieved a higher rate of healthy life expectancy than the rest of the countries in existence

Figure 2, 25 countries were selected; we find that Chile is a South American country and has achieved a higher life satisfaction rate than the rest of the countries in existence.

**Relation of Factors with Healthy/Life Expectancy Factor**

*Ladder*

The above scatter plot shows the relation of Ladder with Healthy/Life Expectancy. The positive correlation can be clearly observed in this scatter plot.

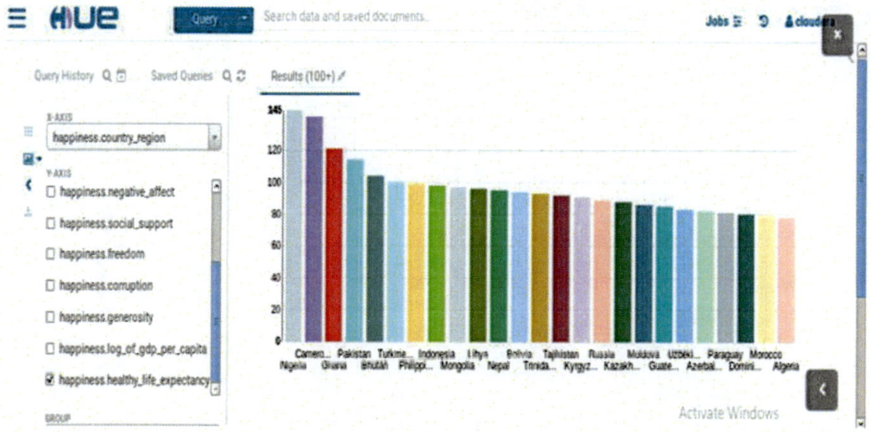


**Fig. 1** A country analysis based on healthy life expectancy

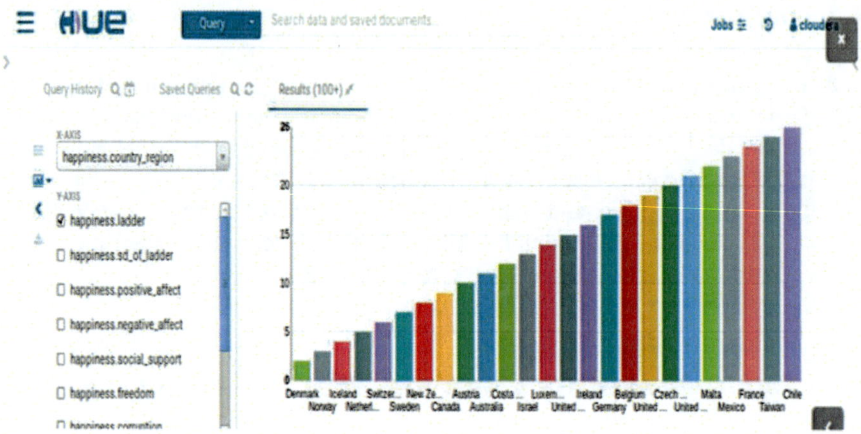


**Fig. 2** A country analysis based on ladder

*Log of GDP per capita*

A scatter plot above illustrates the relationship of the GDP to a healthy/average life expectancy. A high positive correlation can be observed.

*Social Support*

A scatter plot above illustrates the relationship of Social Support to a healthy/average life expectancy. Again, a positive correlation appears.

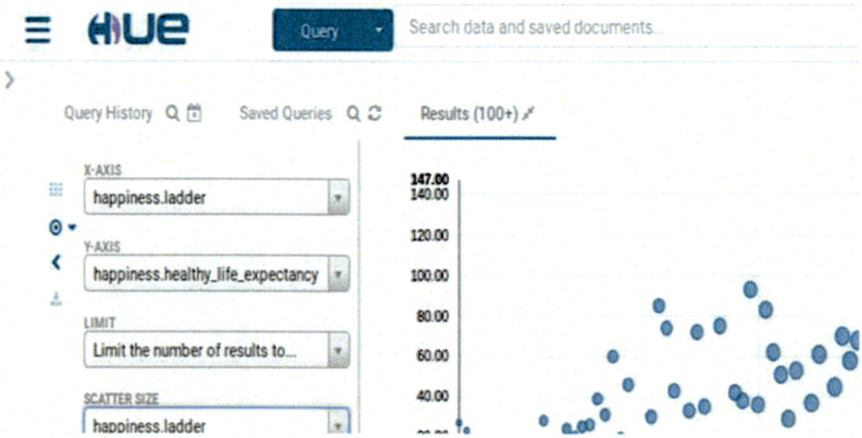


**Fig. 3** Ladder impact on Healthy/Life Expectancy

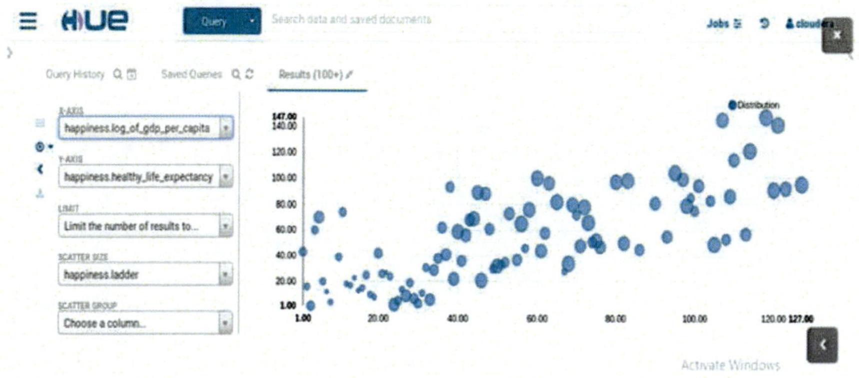


**Fig. 4** GDP impact on health/life expectancy

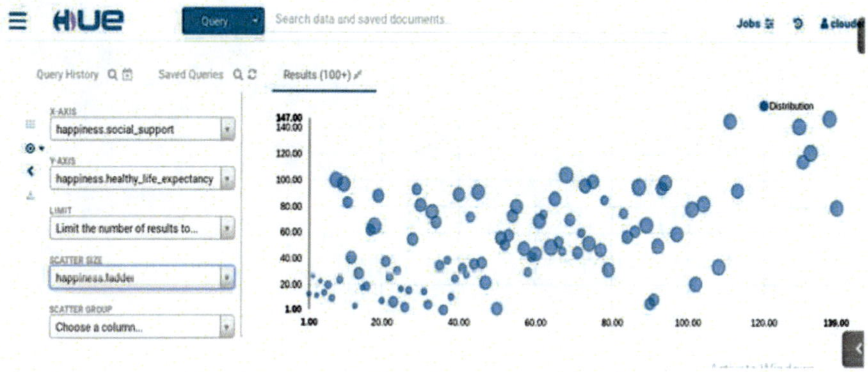


**Fig. 5** Social Support effect on health/life expectancy

A result from Figs. 3, 4, and 5 shows that factors that tend to support humanity and try to boost society's development tend to increase the Healthy Life Expectancy. These factors include GDP, Social Support, and Ladder.

**Happiness Score**

*Corruption impact*

The plot scattered above illustrates the relationship of corruption to the outcome of happiness. From Fig. 6 it can be seen that corruption does not significantly affect the ladder of happiness.

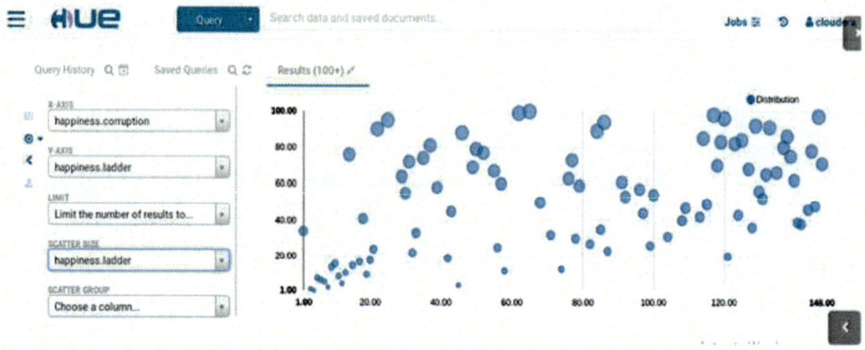


Fig. 6 Corruption impact

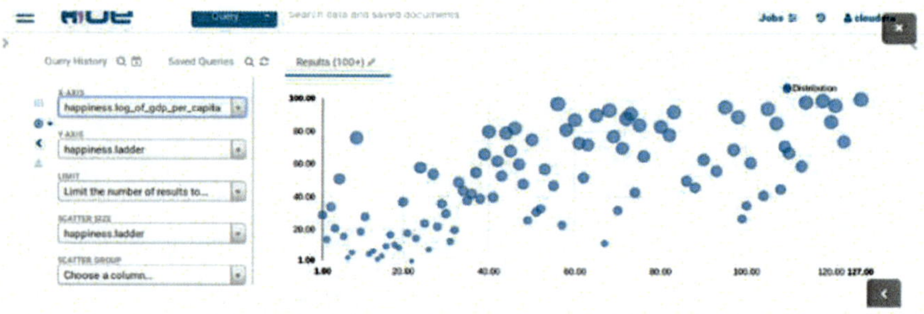


Fig. 7 GDP impact

*GDP impact*

The plot scattered above illustrates the relationship of GDP to the outcome of happiness. From Fig. 7 it can be seen that GDP significantly affects the ladder of happiness.

## 4 Conclusion

This research aims to study the factors that affect happiness, and the global happiness data for 2019 used, which is a study of the state of global happiness that occupied 156 countries through its citizens' happiness based on six factors. New analytical approaches can make evidence-based decision-making more efficient and flexible. By analyzing this data, a set of facts was found that could help practitioners and policy-makers as the analysis of this data plays a significant role in discussing the factors facing our societies today and their impact. The widespread criticism of the "World Happiness Report" appears to be correct. Great focus on GDP and strongly linked features such as family and life expectancy. As a result, we found that the average life expectancy of health, GDP per capita, and social support are the essential variables when anticipating the happiness ladder. Also, we find that corruption does not significantly affect the ladder of happiness.

It goes well with the common wisdom that money makes humans happy up to a certain threshold. Having a good social network is essential, and a family tends to provide for this. Higher life expectancy and health make the human-less anxious about how they survive.

## References


1. Jin X et al (2015) Significance and challenges of big data research. Big Data Res 2(2):59–64
2. Kim K, Lee S (2018) How can big data complement expert analysis? A value chain case study. Sustainability 10(3):709
3. Choi T-M, Chan HK, Yue X (2016) Recent development in big data analytics for business operations and risk management. IEEE Trans Cybern 47(1):81–92
4. Wamba SF et al (2015) How 'big data' can make big impact: Findings from a systematic review and a longitudinal case study. Int J Prod Econ 165:234–246
5. Lytras MD et al (2015) An emerging–social and emerging computing enabled philosophical paradigm for collaborative learning systems: toward high effective next generation learning systems for the knowledge society. Elsevier
6. Lytras MD, Raghavan V, Damiani E (2017) Big data and data analytics research: from metaphors to value space for collective wisdom in human decision making and smart machines. International J Semant Web Inf Syst (IJSWIS) 13(1):1–10
7. Connelly R et al (2016) The role of administrative data in the big data revolution in social science research. Soc Sci Res 59:1–12
8. Lytras MD, Visvizi A (2020) Big data research for social science and social impact. Multidisciplinary Digital Publishing Institute
9. Card NA (2015) Applied meta-analysis for social science research. Guilford Publications
10. Twenge JM The sad state of happiness in the United States and the role of digital media. World Happiness Report
11. Helliwell JF, Huang H, Wang S (2019) Changing world happiness. World Happiness Report
12. Wojcik SP et al (2015) Conservatives report, but liberals display, greater happiness. Science 347(6227):1243–1246
13. Park G et al (2017) Living in the past, present, and future: Measuring temporal orientation with language. J Pers 85(2):270–280
14. Kishore VC, Camarushy AS (2019) Analysis of Boston's crime data using apache pig
15. González-Bailón S (2013) Social science in the era of big data. Policy Internet 5(2):147–160
16. McFarland DA, Lewis K, Goldberg A (2016) Sociology in the era of big data: the ascent of forensic social science. Am Sociol 47(1):12–35
17. Archenaa J, Anita EM (2015) A survey of big data analytics in healthcare and government. Proc Comput Sci 50:408–413
18. Golder SA, Macy MW (2011) Diurnal and seasonal mood vary with work, sleep, and daylength across diverse cultures. Science 333(6051):1878–1881
19. Liu P et al (2015) Do Facebook status updates reflect subjective well-being? Cyberpsychol Behav Soc Netw 18(7):373–379
20. Nelson LK et al (2018) The future of coding: a comparison of hand-coding and three types of computer-assisted text analysis methods. Sociol Meth Res 0049124118769114
21. Gerber MS (2014) Predicting crime using Twitter and kernel density estimation. Dec Support Syst 61:115–125

# Territorial Planning, Smart Cities, and New Governance

**Cinzia Bellone, Fabio Andreassi, Fabio Naselli, and Pietro Ranucci**

**Abstract** Everyday life, as well as retraceable in the current scientific literature, tells us that "we are" nowadays experiencing an epochal moment in the great transformation of communication and interrelationships systems between people, places, and communities through a new network of "flows" compared that, non-existent, of "only" 50 years ago. The networks of "functional relationship nodes", nowadays, structure the territory and open the system to shaping—and planning/programming—of smart cities and smart lands. Indeed, together with the physical networks, those "invisible networks" shape a new urbanity of the territory, making it as an "open system" and "smart system" concerning the reference context by improving and enhancing social and cultural interaction and circulation of information and innovations. This research aims to define that medium of connection between the new urban/spatial models (smart-city and smart-land) and the governance process, in compliance with the notions of sustainability. Even more, we want to shape a way of governing that allows the more and more numerous actors to manage the transformations within the urbanized area. We want to achieve the definition of a new process of urban "construction" which could, and should, be subordinated to an increasingly better-defined relationship between decision-makers and citizens, by bringing into play topics such as "democracy, technology, technocracy". A complementary aim is to explore the relationship between the aforementioned terms: each of them represents the development of processes capable of generating renewed urban and social paradigms able of increasing, or not, the quality of urban life.

C. Bellone (✉)
Faculty of Applied Sciences and Technology, Guglielmo Marconi University, Rome, Italy
e-mail: c.bellone@unimarconi.it

F. Andreassi
Guglielmo Marconi University, Rome, Italy
e-mail: fandreassi@uniss.it

F. Naselli
Architecture Department, Epoka University, Tirana, Albania
e-mail: fnaselli@epoka.edu.al

P. Ranucci
Architecture Department, Roma TRE University, Rome, Italy



X.-S. Yang et al. (eds.), *Proceedings of Sixth International Congress on Information and Communication Technology*, Lecture Notes in Networks and Systems 216,
https://doi.org/10.1007/978-981-16-1781-2_24

**Keywords** New governance · Smart city · Smart land · Territorial planning · Governance and technocracy

# 1 Introduction

## *1.1 Scientific Background*

The territory is nowadays marked around the world by the blast of many cities (not only the largest ones) follow out low-density settlement models that generated episodes of sub-urbanization, then creating fractures which are not easily recoverable through the planning along with conventional urban planning tools.[1]

> The main causes of extended sub-urbanization are evident: planning deregulation, new location preferences for both economic activities and families, growing processes of territorial specialization and of spatial segregation. On the other hand, it is more difficult to identify the solutions to put a limit to settlement dynamics which, if projected in the medium-long term, draw up an undesirable scenario, in total contradiction to the principles of sustainability assumed to guide settlement development by all the countries of the European Union. ([2]: translation by the authors).

The framed new urban layout also depends on that transformation process which sees the urban functions (especially the higher ones) place themselves within unconsolidated urban systems, not yet defined nor bordered but, despite this, strongly interconnected by networks and linked to the overcoming of the traditional hierarchical organization between the different parts of the physical and social realities in the extended urban territory.

The new way of "using" resources, and the new configuration of various human activities, allows us to interpret the city no longer as a merely "space of places" nevertheless as a "space of flows", starting in the establishment of the "networked city" [3]. City into which the movements to be made to reach the house, the job, and the urban services constitute the matrix of the new city [4].

Thus, the Christaller-Losch-Beckmann reticular scheme of, on which we have based our choices per years, turns out to be inadequate to return the dense sequence model of multi-diversities present in the reality; since it is hierarchical and considers only the economic aspects (scale economies and transportation costs) [5]. The networks of "functional relationship nodes", therefore, structure the territory and open the system to shaping—and planning/programming—of smart cities and smart lands. Indeed, together with the physical networks, those "invisible networks" shape a new urbanity of the territory, making it as an "open system" and "smart system" concerning the reference context (both the local-city and the wide-land); by improving and enhancing social and cultural interaction and the circulation of information and innovations.

[1] This paper presents the most recent advancement of research already published in Bellone et al. [1]. ISBN: 978–3-030–00,978-6, ISSN: 2194–5357 with peer review.

The new (invisible) "networks" make it possible to create complementary and synergic relationships, reaching high levels in the competitiveness of cities. The role of planning, consequently, is questioned by the new way of "functioning" of the productive city, with very little possibility of predictability, and profound changes in the social and spatial configuration of both the material and immaterial internal relationships of the city community.

### 1.2 Aims and Scope

Scholars try to interpret the up-to-dated complexity of the current urban-form by trying to arrive at the definition of suitable new governance capable of restoring spatial and social welfare while respecting the concepts of local and global sustainability.

Thanks to the role played by the concept of "civitas", the European city has always maintained a clear and defined "urban-form" aimed at an efficient social-cultural organization. The dislocation of the "parts of the city" and the urban functions, usually hierarchically organized, thus also mirrored the definition of urban mobility: "the layout of urban settlements influences the decisions concerning whether to move or not, related to the destinations to achieve and the choice of the modes of conveyance" [6].

The inter-connections between functions, in the contemporary city, derive from the characteristics of integration and attractiveness, they are not anymore necessarily based on proximity and proximity, "*the physical agglomeration nature of the city is lost and the once-clear boundary between urban and rural is dissolved*" [6]. To manage the complexity of the new territorial reality, it is necessary to determine scientific theoretical models that are also pragmatic, political, and cultural.[2]

"In the age of networks, new ways of orientation for urban governance actions must therefore be conceived and developed. The planning itself will no longer take place according to a top-down process within a single public entity, but in the context of changing networks, sometimes fixed-term, in any case in continuous restructuring, forming, and evolution" [7].

This research aims to define that medium of connection between the new urban/spatial models (smart-city and smart-land) and the governance process, in compliance with the notions of sustainability. Even more, we want to shape a way of governing that allows the more and more numerous actors to manage the transformations within the urbanized area. We want to achieve the definition of a new process of urban "construction" ("formation over the time" of the spaces and functions of the city) which could, and should, be subordinated to an increasingly better-defined

[2] Defined by many as "urban policies", they have had attention, both nationally and internationally, for many years, starting from documents such as the *Green Book on the Urban Environment* of 1990, up to the two documents *Europa 2000* and *Europa 2000 +* (1992–1994)—to name but a few—the *European Space Development Scheme* has been finalised.

relationship between decision-makers and citizens, by bringing into play topics such as "democracy, technology, technocracy".

A complementary aim is to explore the relationship between the aforementioned terms: each of them represents the development of processes capable of generating renewed urban and social paradigms able of increasing, or not, the quality of urban life.

## 2 Two Themes

### *2.1 Smart City*

What the Smart city is, seems to be now clear, even if in its multifaced understandings. The definition we can find more in the scientific literature today indicates that it is a new arrangement of the urban structure which, in order to ensure the quality of life and sustainability of the environment, hinges on the development of both physical infrastructures (transport) and intangible assets (IoT). The city becomes an "efficient machine" where networks and flows (invisible and of data) allow decision-makers to organize and shape the territory.

In Italy, the most current references, as well as the numerous initiatives implemented in the territory by local authorities, are given by the AgID Three-Year Plan.[3] We can find other references in Forum PA debates[4] as well as another sector research.[5]

Everyday life, as well as retraceable in the current scientific literature, tells us that "we are" experiencing an epochal moment in the great transformation of communication and interrelationships systems between people, places, and communities through a completely new network of "flows" compared that, in fact non-existent, of "only" 50 years ago.

The first question that arises is just how the network of flows, invisible yet so strong and decisive—high interactive among the digitalization, information technology, and the physical world—may have effects (and what kind of effects) on the physical form of the city. The concept of physical "distance" has been reduced by communication among diverse subjects [8]. Yet it is still to be understood to what extent the Internet makes the location of interconnected subjects/things irrelevant (working at home,

---

[3] AgID—The Agency for Digital Italy. It is a public agency, established in Italy by the Monti government (2011–2013). It pursues the highest level of technological innovation in the organization and development of the Public Administration (P.A.).

[4] The Forum PA is a company of the Digital360 Group. The company has been working for three decades to stimulate the digital growth of the P.A., favouring the meeting between PA, businesses, researchers, citizens.

[5] See at the ENEA Convergence, Smart City and Community initiative, "to move from words to deeds", research aimed at "*the digital reorganization of the management processes of urban and territorial contexts, starting from a conceptual, methodological and technological convergence*". ENEA is the main Italian national agency for new technologies, energy and sustainable economic development. It promotes important research on these items.

remote control of things, social meetings, and social interactions, etc.). Indeed, in seeing the ongoing growth of cities does not appear entirely convincing to demonstrate that the most current ways of communication can make the phenomenon of settlement dispersion possible in real. Furthermore, as we well know, human, as "social" animal, needs to participate in a real, physical community to live. The city cannot exist only in cyberspace and it must not deny its relational potential: it must, instead, be a combination of "virtual and physical places". "Invisible networks" and "real places" combine giving rise to physically and functionally changeable urban "forms", but "real".

Here, then, is the new city, the intelligent city. That Smart City where the "invisible networks" allow the use of the city in a different way from half a century ago. "humans and things" become, in fact, real-time "sensors" of the progressive evolution of a new urban reality (ubiquitous computing) or, at least, of some/many of its fundamental components. These components include "public and private physical open spaces". Therefore, from the analogic to the digital era, the city-container assumes changing forms, not only physical but "different over time", even short and flexible in functions". And the concept of "form" of the city changes and acquires a more complex and variable meaning over time [1].

We also fully agree with some authors such as Carlo Ratti and Matthew Claudel, who instead of giving space to scenarios-visions of poor reliability within urban planning conventional processes, say it may be more crucial to design the present by inspecting it and admitting a possible/probable transformation and hypothesizing its next/subsequent scenario (the "future craft" by C. Ratti). In this meaning, it appears convincing to analyze the "information flows" with "data-driver" models, using the enormous amounts of data that can now be collected on the micro and macro themes of each urban/social reality [9].

The new city is transformed quickly supported by data-driven processes: then, which kind of Governance is eligible? Thanks to big data we can understand the actual functioning of cities, the "consistency/resilience" of its forms and its "quality", the political-social coherence of the Governance taken from these data, and the entire data-driven process.

**Governance and Technocracy**. In Western democracies, from the second post-war period to today, Governance understood as the ability to interpret the evolution of the territory toward future scenarios and to "manage" (preferably jointly with the citizens) the necessary transformations, has not had the strength to prevail over traditional applications. The ruling classes preferred to rely on actions that could guarantee mediations between political parties. While the professionals remained faithful to the schools of thought without being able to make an innovative leap in both tools and procedures.

Parag Khanna,[6] XX, in his latest text, contrasts the substantial fragility of a Governance based on a predominantly formal democratic structure with a Governance

[6] Parag Khanna, an internationally renowned geopolitical strategist. In Italy, for Fazi Editore he published the trilogy consisting of The three Empires (2009), How to govern the world (2011) and Connectography (2016).

based on innovative technologies. "In fact, it proposes a decision-making system that makes use of a close relationship between political-representative bodies and technocratic bodies; the latter appointed to provide qualitative and quantitative assessments of the proposed development scenarios. The overall objective, therefore, is the start of an evolution, in a technocratic sense, of the political system" [1].

For Khanna, the term Technocracy is the input for innovation in city governance.[7]

Representative democracy, direct or not, ideological leadership, dictatorships, etc., are forms of government that do not allow transformations to be managed by considering expert analyzes and long-term planning as is the case in "technocratic" management.

The data, according to Khanna, can give awareness of the services that citizens need in a democracy, using medium-long-term scenarios and overcoming the criterion of "representativeness" of many governments. Giving equal weight to "numbers and democracy", Switzerland (European "direct" democracy, bottom-up) and Singapore (Asian technocracy, top-down), are the two states given as an example by the author for the greater balance put in place between democracy and knowledge: the administrators implement the "Governance" by having historical, current, renewable, and confirmable data. Only this type of governance is therefore able to plan and manage smart cities.

In urban/territorial planning—characterized by long—term procedures which, then, inevitably intercept ineluctable transformations and consequent "revisions" of urban planning tools—one wonders who should, and how it is possible, manage the databases in order to use them consistent with the social demand. And here is one of the central points of the topic: the connection between the methods of management processes (Governance) and innovative technologies for data collection and management. The matter of the Governance of the whole process, and above all in relation to the two aspects of "acquisition and management" of data, enters fully (difficult to ignore) as a crucial "frame" for an urban development that is to be maintained within scenarios really "democratic". A Governance that, with the maximum use of new innovative ICT technologies, can redefine, in shared modalities, in time and space, the redesign, even functional, of the city.

## 3 Conclusions and Future Opening

As regards this contribution, we are very interested in understanding "how" public space can be planned and implemented within a form of democratic governance.

In the smart city—an urban structure where the most innovative technologies make the growth of urban welfare and inhabitants faster[8]—it is possible to evaluate how relations between citizens and rulers can and should change.

[7] The author obviously takes as a reference the public administration in the USA.

[8] And in an accelerated way, based on the control and use of an impressive amount of data and the consequent knowledge of social phenomena.

In fact, in the smart city the rulers increase their driving skills and at the same time citizens may seem, for this reason, less and less able to carry out effective control over the public and private management of those enormous amounts of data on the basis of which they propose economic and social development political strategies and territorial infrastructure decisions [1].

Some concluding remarks on the state of Governance and on the current political and administrative system in Italy. The scenario in Italy is changing. Some local administrations, differently located on the national territory, have become aware of the phenomenon and its inevitability, and experiment virtuous processes, in the sense of understanding that the multiple technical definitions between smart, intelligent, eco-sustainable, digital cities, etc. An example, the municipality of Reggio Emilia, while aiming to make the city smart, has achieved the awareness that the "democratic" side of the process implies that "efficiency" must be related to shared objectives with "citizens". The hypotheses that are privileged insert within the forms of politics and administration, still selected with traditional methods, other forms of selection that appreciate some "official" forms of technocracy.

The actions do not only see the use of the term "smart city" but the attribution of an additional content not only limited to the efficiency of the current settlement model but also management, precisely in view of the complex process that can be glimpsed today in the collection, processing and use of big data.

A Governance that must be fully verified in its legal structure (issues of law) and in its ability to produce a city on a human scale (welfare issues).

## References

1. Bellone C, Ranucci P, Geropanta V (2019) The 'Governance' for Smart City Strategies and Territorial Planning. In: Vasant P, Zelinka I, Weber GW (eds) Intelligent computing and optimization. Advances in intelligent systems and computing, pp 76–86, Springer, Berlin
2. Gibelli MC (1999) Dal modello gerarchico alla governance: nuovi approcci alla pianificazione e gestione delle are metropolitane. In: Camagni R, Lombardo S (a cura di), La città metropolitana: strategie per il governo e la pianificazione Ed. Alinea
3. Beguinot C (1999) L'architettura è intelligente, se è capace di (inter) connettere. In: Telèma n. 15, 1998/99
4. Martinotti G (1999) Il vero centro si è spostato, non è più "dentro" ma in periferia, in Telèma n. 15, inverno 1998/99
5. Hannerz U (2001) Esplorare la città. Antropologia della vita urbana, il Mulino, Bologna
6. Camagni R (1998) Agire metropolitano: verso forme e strumenti di governo a geometria variabile. In: Convegno DPTU-DAU pensare ed agire metropolitano: verso una nuova visione istituzionale e funzionale, Roma, 23 April 1998
7. Benveniste G (1999) La pianificazione come gestione a matrice di reti decentrate. In: "Pianificazione strategica e gestione dello sviluppo urbano" a cura di Curti Fausto e Gibelli, Maria Cristina, Ed. Alinea
8. Naselli F (2012) Nuove forme della qualità urbana, FrancoAngeli, Milano
9. Naselli F and Albanese A (2014) La città in comune 3.0, Urbanistica Informazioni Vol. 257, INU edizioni, Roma
10. Beckmann MJ (1958) City hierarchies and the distribution of city size. Economic Development and Cultural Change, Ed Marcel Fafchamps, volume 6, Number 3

11. Camagni R (1999) La pianificazione sostenibile delle aree periurbane, Bologna ed. Il Mulino
12. Caragliu A, Nijkamp P (2008) The impact of regional absorptive capacity on spatial knowledge spillovers. Tinbergen Institute Discussion Papers 08–119/3, Tinbergen Institute, Amsterdam
13. Komninos N (2015) The age of intelligent cities: smart environments and innovation-for-all strategies. Routledge, London and New York
14. Engel JS (2016) Global clusters of innovation: entrepreneurial engines of economic growth around the world. Edward Elgar Publishing, Cheltenham
15. Bolívar MPR (2017) Governance models for the delivery of public services through the web 2.0 technologies: a political view in large Spanish municipalities. Soc. Sci. Computer Rev. 35
16. Khanna P (2017) La rinascita delle città-stato, Fazi Editore
17. Ratti C, with Claudel M, (2017) La città di domani. Einaudi, Passaggi
18. Mochi Sismondi C (2018) http://www.forumpa.it/speaker/2381-carlo-mochi-sismondi

# 3D P-Torus: Wavelength-Based Switching in Packet Granularity for Edge Data Center Networks

Charidimos Chaintoutis, Adonis Bogris, and Dimitris Syvridis

**Abstract** In this paper, we introduce 3D P-Torus, a novel intra-DC architecture, based on the 3-dimensional torus topology, that utilizes wavelength switching at the packet level to achieve improved latency and throughput performance compared to earlier variants of the P-Torus architecture and other wavelength switching intra-DC architectures. By a single-hop communication design between On-Rack Switches, 3D P-Torus achieves great latency and throughput performance, as calculated through detailed simulations. All buffering needs are covered through electronic buffering, avoiding hard to implement and expensive optical buffering. 3D P-Torus can support, with no need for amplification, medium-sized-DCs up to 10k servers, making it an excellent candidate for Edge Data Centers, taking into account its excellent low end-to-end latency.

**Keywords** Wavelength switching · Optical switching · Edge data center · Packet switching · Data center network

C. Chaintoutis (✉) · D. Syvridis
Department of Informatics and Telecommunications, National and Kapodistrian University of Athens, 15784 Panepistimiopolis, Ilissia, Athens, Greece
e-mail: cchaintoutis@di.uoa.gr

D. Syvridis
e-mail: dsyvridi@di.uoa.gr

A. Bogris
Department of Informatics and Computer Engineering, University of West Attica, 12243 Aghiou Spiridonos, Egaleo, Greece
e-mail: abogris@uniwa.gr



X.-S. Yang et al. (eds.), *Proceedings of Sixth International Congress on Information and Communication Technology*, Lecture Notes in Networks and Systems 216,
https://doi.org/10.1007/978-981-16-1781-2_25

# 1 Introduction

## *1.1 General*

Data Centers (DCs) are a core component of our life since everyday online operations produce huge amounts of data. Our dependence on DCs can be seen as the global internet traffic surged by ~40% during the COVID-19 containment measures, between February and April 2020 [1]. This growth comes after a previous rise; since 2010, global internet traffic grew at a rate of ~30% per year [2] and is expected to double by 2022 reaching 4.2 ZB per year [2]. These trends result in a tremendous augmentation in the total traffic inside the DCs: for every bit of data that travels the network from DCs to end-users, another five bits of data are transmitted within and among DCs [3]. The rise of Machine Learning applications also increases traffic and computation needs for the Data Center Networks (DCNs) expecting to deal with "big data" and greater computational requirements [4]. Among all DC operations, networking is becoming the infrastructures' major "pain point", while in the past this was identified in the processing side [5], since workloads in DCs are becoming more distributed, which increases dependence on the networking interconnects. Thus, DCNs should scale at the same rate or the network side will not follow the processing side and the overall system performance will not be able to achieve balanced bandwidth performance [6]. Optics-based DCNs are a promising solution to overcome the problems of electronic switches that are becoming more prominent by scaling to larger topologies. The idea behind optical switching is that the traffic is handled in the optical domain thus avoiding the power-consuming Opto-Electronic conversions achieving better efficiency. Moreover, the switching operation is independent of the bit-rate; and scalability to higher bandwidth can be achieved through Wavelength Division Multiplexing (WDM) schemes. In practice though, it is not easy to achieve a cost-performance equilibrium. In multi-tier architectures, only the communication operation is implemented optically while switching is performed in the electronic domain. Optical Circuit Switching (OCS) architectures use slow optical switches on top of the electronic switching layer [7–10]. Long-living flows travel through the OCSs, and short ones through the electronic switching fabric, thus limiting the OEO conversions [11–13]. In other proposals Arrayed Waveguide Grating Router-based switches (AWGRs) are used to implement wavelength switching operations [14–17].

## *1.2 Edge Data Centers (EDCs)*

In recent years with the IoT devices at the edge producing massive amounts of data, the edge computing paradigm is growing; so, the transmission and compute at DCs doesn't always make sense. For certain applications, traditional DCs cannot guarantee acceptable transfer rates and response times. Edge computing moves the computation away from the DCs toward the network's edge resulting in better response times, leveraging physical proximity to end-users. This need has given birth to the EDC paradigm and the COVID-19 pandemic has brought EDCs further into focus; the centralized approach is no longer enough to satisfy user demand, and creates challenges around capacity, latency, and cost [18]. EDCs can offer a solution as geographic proximity and network connectivity can work together to deliver low latency, enabling peering at the edge and acting as local gateways for high-speed connections to the core and other edges.

### 1.3 Motivation, Novelty, and Contribution

In [19] we introduced P-Torus, a new all-optical DCN architecture that exploits packet-based wavelength switching and is built on the 2D torus topology. P-Torus is a reduced-tier ("flat") architecture that does not employ over-subscription and provides communications with significant benefits over other wavelength-based switching architectures in latency, throughput, cost, and power consumption. Here, we expand our previous work exploring the 3D torus topology. 3D P-Torus offers enhanced wavelength reuse capabilities, supporting all-to-all interconnection between more nodes. This characteristic leads to improved performance for throughput and latency, especially under heavier loads. In Sect. 2 we present the 3D P-Torus architecture, offering details about its operation and the main components. In Sect. 3 we present performance simulation results and comparison with the 2D P-Torus variant.

## 2 Architecture Analysis

### 2.1 Overview

3D P-Torus utilizes the $C_4^3$ topology and is built upon 2 layers: the "edge layer" and the "Inter-cluster Passive Aggregation layer" (Fig. 1). $C_n^d$ stands for the $n$-ary d-cube (d-D cube of size $n$). So, $C_4^3$ denotes the 3D torus with radix 4. The edge layer contains 16 3D torus clusters, each one with 16 racks, with every rack holding 40 servers. Instead of a single Top-of-Rack switch (as in [19]), we use 4 ORS (On-Rack Switches) on each rack. Each ORS is responsible for 10 servers. So, a $C_4^3$ torus is formed in each 3D torus cluster (Fig. 1b). ORSs all-optically switch traffic between every ORS, as well as from/to ORSs and IPA switches. Each ORS is connected to its neighbor ORS and to one IPA switch with one optical link. IPAs are passive, optical components and form a $C_4^2$ torus. Servers in the same cluster but on different racks communicate using the ORSs, with no involvement from IPAs. ORSs from separate clusters use the IPAs. All wavelengths arriving from the ORSs are multiplexed into wavebands (WBs) and are guided to the target IPA for demultiplexing and are

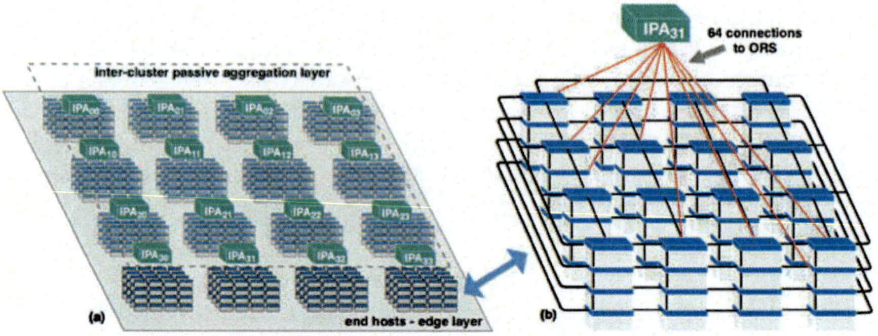


**Fig. 1** The 3D P-Torus topology comprises 2 layers, the edge and the IPA. The edge layer hosts 16 clusters with each cluster having 16 racks with 40 servers each. Each rack holds 4 On-Rack Switches (ORS) with each aggregating traffic from 10 servers. The 64 ORSs in each cluster form a $C_4^3$ torus. The IPA layer hosts 16 IPA switches that are organized in a $C_4^2$ torus. Each IPA switch aggregates communication from a specific cluster to enable inter-cluster communication

then guided to the proper ORS. 3D P-Torus needs only one hop for communication between ORSs, and three hops for communication between any two servers. Moreover, switching is executed optically. On top of the edge and IPA layers we construct all-to-all light-path topologies for intra- and inter-cluster communication that is fulfilled with pre-determined wavelengths and routes under a time-slotted packet transmission operation (see Sect. 3).

## 2.2 *Anatomy of the Light-Path Topology*

3D P-Torus is constructed to enable all-to-all communication between all ORSs, so all-to-all light paths are set up for both the IPA and edge layers. At the same time, we would like to use as few wavelengths as possible. So, for a known topology (torus), for all-to-all communication, and for a metric to minimize (wavelengths) we solve the Routing and Wavelength Assignment (RWA) problem. For d-dimensional tori of even radix $n$ ($C_n^d$) this problem is solved in [20]: for all-to-all communication, the min number of wavelengths is $W_{min} = \frac{n^{d+1}}{8}$, over a set of specific shortest path routes. For the $C_4^2$ torus topology the optimal RWA solution is calculated in [19], and shown in Fig. 2 (different colors depict distinct wavelengths). So, with 8 wavelengths we can achieve non-blocking, all-to-all communication in a $C_4^2$ torus; as, in our case between all the IPA switches in the IPA layer. Using the same formula, we need 32 wavelengths for communication among all nodes in a $C_4^3$ torus; as in our case between all ORSs in the edge layer. The depiction of the optimal light-paths in the 3D space is challenging but the rationale is the same as in the 2D case. For our architecture, this analysis results in 32 wavelengths for intra-cluster communication among ORSs. For a $C_4^2$ torus (as in the IPA layer) one needs only 8 wavelengths for all-to-all communication, but

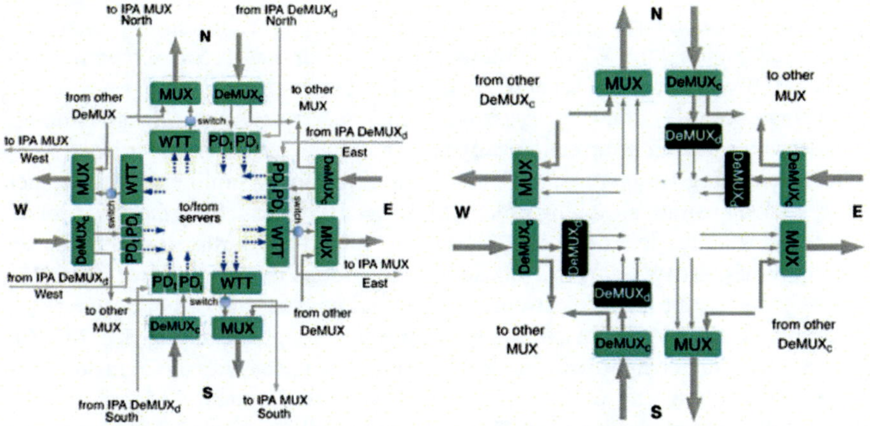


**Fig. 2** (Left) ORS includes one MUX/DeMUX$_c$ pair, one WTT, and four PDs per direction. The WTT is connected to a switch that guides the wavelength to the ORS's MUX or to the IPA's MUX. Connections to/from the servers are shown in blue. All other connections are optical and shown in grey; (Right) IPA overview. Elements with dense channel spacing are dark colored

in our case, we also need to be able to enable communication between all ORSs, so we use 8 WBs, each one containing 64 wavelengths to support the 64 ORSs in each cluster of the edge layer. So, in all clusters *the same* 32 wavelengths are used for intra-cluster communication among ORSs, namely $\lambda_{m1}$, $m \in \{1 \ldots 32\}$ when for inter-cluster communication 8 WBs of 64 wavelengths each are used, with $WB_m = \{\lambda_{mn}\}$, $m \in \{1 \ldots 8\}$, $n \in \{1 \ldots 64\}$.

## 2.3 On-Rack-Switch Operation

ORSs include a Wavelength Tunable Transmitter (WTT), a pair of WDM MUX/DeMUX components, four photodiodes (PDs), and a switch, in each direction (Fig. 2, left). Two PDs ($PD_t$) receive packets from the ORSs on different racks while the $PD_i$s receive packets from the IPAs. MUXes are built with WDM couplers while DeMUXes can be realized using AWG demultiplexers. Multiple DeMUX outputs end up in the same $PD_t$, but each PD receives a single packet at a time using a contention resolution mechanism (Sect. 3) with the help of a 2:1 coupler that lies before every PD. As in most DCs, the amplitude and the phase of the packets vary at the reception, due to the chromatic dispersion and the differentiation of imposed losses in different light paths. Burst-Mode Receivers (BM-RXs) can help in overcoming such challenges since they can adapt to phase alignment and power deviations [21]. For communication among ORSs known wavelengths are transmitted (from a known direction) and are then multiplexed in the WDM mux and guided to the end ORS. At the same time, WTTs are connected to the IPA MUXes (in the same direction), to guide traffic from the rack to the IPA that is responsible for this cluster. To provide both functionalities, the WTT is followed by a switch that directs the wavelength to the MUX component of ORS's/IPA's, for intra/inter-cluster traffic, respectively. These switches (1:2) don't incorporate control headers as the wavelength's choice includes all switching information and can be built with Mach–Zehnder Interferometers, Micro Ring Resonators, or Semiconductor Optical Amplifiers. Off-the-shelf switches demonstrate switching time in the order 10 ns (matching the WTT tuning speed) and insertion loss of ~5 dB [22]. Lower values for insertion loss (<2 dB) and consumption <50 mW (<1 pJ/b at 40 Gbps) can be found in the literature [23]. ORSs, receive server traffic in the electronic domain and transmit packets optically. When the destination is in the same cluster, traffic travels through the appropriate MUX to the next ORSs on the light-path, and arrives to the end ORSs where is being detected in the $PD_t$ (without any OEO conversion). When the destination is in a different cluster, traffic travels from the WTT to the IPA through the IPA's MUX in the same direction. Similarly, traffic reaching the ORS from the IPA is detected in the $PD_i$.

## 2.4 Inter-Cluster Passive Aggregation Switch Operation

IPA switches comprise three MUXes/DeMUXes, similarly to the ORS paradigm (Fig. 2, right), per direction and switching is carried out passively in WB granularity. MUXes can be built with WDM couplers. ORS WTTs send traffic to the MUX in the same direction, where packets are multiplexed in a single WB together with packets coming from other servers and target the same cluster. This WB is optically switched to the IPA that is responsible for the destination cluster. The WB reaches the IPA from the proper $DeMUX_c$ where is demultiplexed in wavelength granularity ($DeMUX_d$). Wavelengths are directed to the linked ORS, and are detected in the proper $PD_i$.

## 2.5 Power Budget

With the diameter of a $C_n^k$ torus being equal to $k\lfloor n/2 \rfloor$, the max shortest path has $n - 1$ steps (odd $n$) or $n$ steps (even $n$). For intra-cluster communication, packets must pass 6 ORSs to reach their destination as opposed to 4 IPAs for the inter-cluster case. For the intra-cluster case, packets traverse: 1*switch, 6*(De)MUX elements, and a 1:2 coupler before detection. For the inter-cluster case: 1*switch, 4*(De)MUX elements, 1*final IPA DeMUX, and a 1:2 coupler before detection. Losses in can be neglected (losses < 0.5 dB/km for all considered wavelengths). DeMUXes can be built with AWG demultiplexers while MUXes can be built with five WDM couplers (2:1). Off-the-shelf a thermal AWG demux elements have losses ~3 dB [24, 25], or even <1 dB [26]. Hence, we can accept losses ~3 dB for off-the-shelf and ~1 dB for State-of-the-Art solutions. Off-the-shelf WDM couplers have insertion loss ~0.2 dB, resulting in ~1 dB for the full MUX [27, 28]. Switches have insertion loss ~2 dB

[23]. So, in the worst-case scenario, losses are ~27 dB for the paths of maximum loss. Accepting WTT output power +10 dBm, and min detection power –18 dBm the packets can withstand losses up to 28 dB, making amplification across the path redundant.

## 3 3D P-Torus Performance

Optical buffering is the greatest obstacle in building OPS architectures. In both P-Torus variants (the 2D P-Torus [19] and the 3D P-Torus) this need is eliminated. Buffering happens before the transmission begins, at the source ORS and using traditional electronic buffering technologies. Packet contention might appear during the packet transmission or reception. PDs shouldn't receive multiple packets at any given time while WTTs can only transmit a single packet at a time. To avoid contention, our architecture utilizes operation in discrete time-slots. Each time slot includes the packet transmission and the WTT tuning duration. To support the time-slotted operation, each WTT uses a separate buffer. When some server generates a packet, knowing the destination, it is directed to the proper WTT buffer. For every timeslot, we compare the destination ORS (and PD) for the packets that are next to be transmitted. If the destination is the same for multiple packets (rack, ORS, and PD) only one is transmitted (arbitrarily chosen); the rest remain in their buffers. In the next time slot the destination of all packets is compared again. New packets are generated according to a Poisson distribution. WTTs' transmission rate is 10 Gbps. Each rack holds 40 servers with 1 Gbps rate each (4 Gbps from the servers to ORS, 40 Gbps for the whole rack). The tuning time of WTTs is 12 ns [28, 29]. Latency includes the transmission time, the queuing delay, and the tuning time. Destination racks are generated according to a uniform distribution. The Overhead imposed by guard time, and clock recovery, is not considered; BM-RXs with preamble and guard time in the order of ns [30, 31] are considered. In Fig. 3 we present the latency for packets of various sizes, under differing load. Latency increases as the packet size or the load increases (upper left). For load up to 75%, latency is almost constant regardless of the packet size. For heavier loads, latency increases, due to the extra time packets spend in the buffer. Regardless, even with the load near 100%, latency remains under 3 $\mu$s; an even better value than the one achieved by P-Torus. Both the single-hop design and the time-slotted operation aid for this performance. In Fig. 3 (upper right, lower left) we see hot-spot scenarios (rack or cluster). Hot-spots receive part of the total traffic. For both cluster and rack hot-spots 3D P-Torus achieves good performance. Latency remains below 6 $\mu$s even for a very hot cluster, even with total traffic exceeding 50%. 3D P-Torus retains high-throughput performance and does not saturate even at very high load; this is attributed to the lack of over-subscription in any layer of the 3D P-Torus network architecture (Fig. 4).

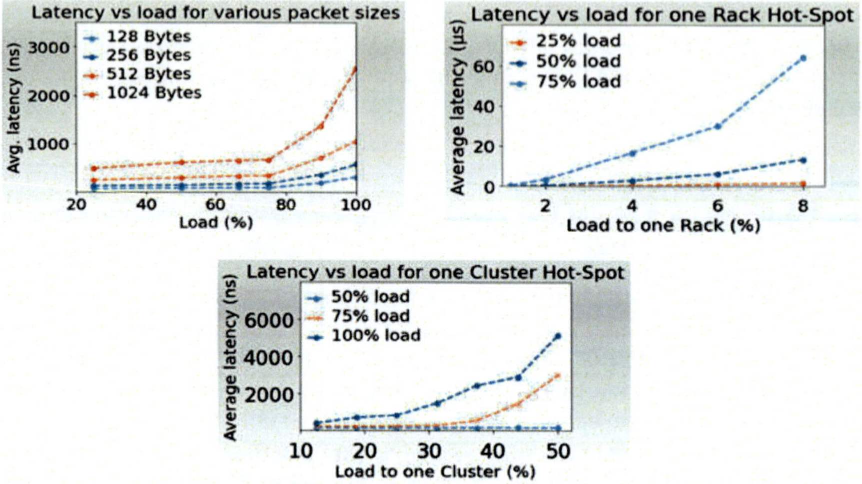


**Fig. 3** Upper left: average latency for various packet sizes under varying load; Lower left: average end-to-end latency for Hot-Spot traffic in a cluster; Upper right: average latency for Hot-Spot traffic in a rack. Latency includes transmission and tuning time, and queuing delay

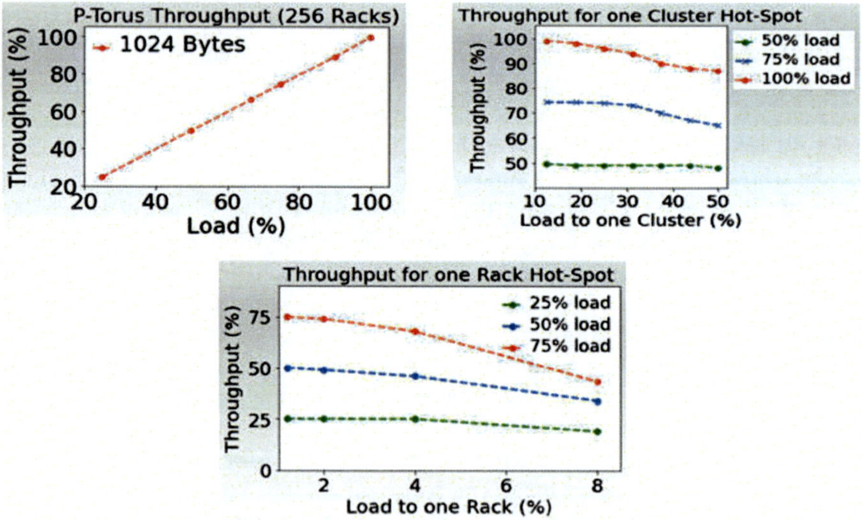


**Fig. 4** Upper left: 3D P-torus throughput. Upper right: throughput for cluster hot-spot scenario. Lower left: throughput for rack hot-spot scenario

## *3.1 Comparison with P-Torus (2D)*

In order to better understand the performance improvements of 3D P-Torus over the 2D variant P-Torus we repeat the most stringent simulation scenarios, now comparing the 2 variants. As we can see in Fig. 5. 3 DP-Torus experiences lower latency in all cases, with the difference becoming more important for heavier loads situations. The same stands for the throughput as we can see in Fig. 6.

It is also worth pointing out that the performance amelioration becomes possible since the 3D variant requires less wavelengths to enable all-to-all communication between the same number of nodes. Thus, we are able to introduce more than one switch on every rack (4 ORS instead of a single Top-of-Rack switch as in [20]). This advantage becomes even more evident comparing the wavelengths that would be needed by a solution utilizing a single AWGR. As can be seen from Fig. 7 in order to

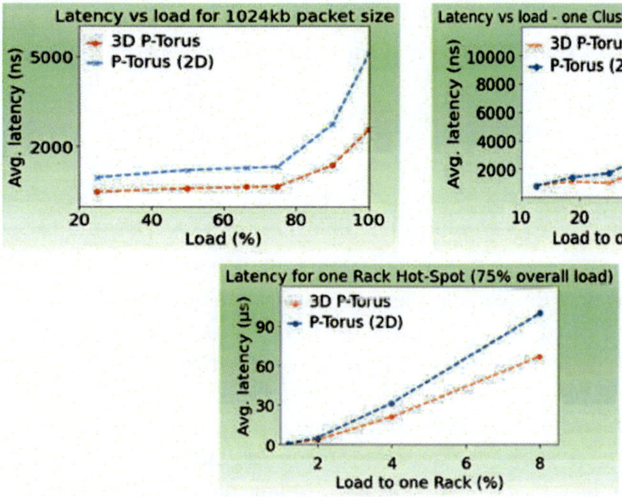


**Fig. 5** Latency comparison between 3 and 2D P-Torus architectures. The 3D variant exhibits lower latency in all scenarios, especially under heavier loads

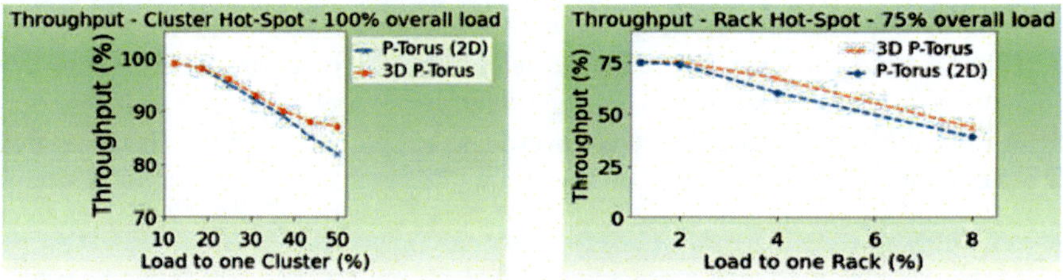


**Fig. 6** Throughput comparison between 3 and 2D P-Torus. The 3D variant retains better the effective throughput, especially under heavier loads, in two hot-spot scenarios (cluster and rack)

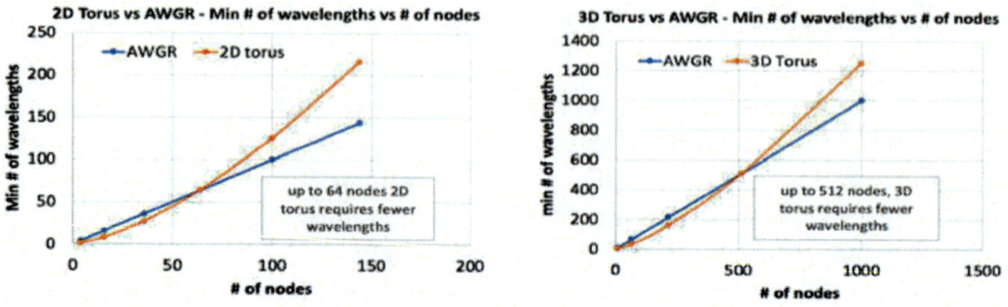


**Fig. 7** The min number of wavelengths needed for all-to-all communication among varying number of nodes in 2D Torus, 3D Torus, and AWGR-based architectures. In 3D P-Torus we construct $C_4^3$ clusters to exploit the best wavelength reuse ratio compared to an AWGR-based solution. AWGR-based solutions for the full 1024 nodes would need at least 1024 wavelengths

achieve all-to-all communication among 1024 nodes (4 ORS/rack, 16 racks/cluster, 16 clusters) we would need at least 1024 wavelengths. So, the efficient wavelength reuse in the clusters and in the IPA layer results in a 50% decrease in the number of the wavelengths needed.

## 4 Conclusion

In this work we propose a new intra-DC architecture, built on the 3D torus topology. 3D P-Torus utilizes wavelength switching at a packet level, leveraging the routing capacity that WDM can offer, to realize all-to-all intra-DC communication, with improved latency and throughput performance compared to earlier P-Torus variants. Through single-hop light-paths between On-Rack Switches it satisfies the requirements of modern Data Centers in terms of latency and throughput, even under very heavy load in hot-spot traffic schemes. Traditional electronic buffering is used since packet buffering takes place at the source rack, avoiding hard-to-implement and expensive optical buffering. 3D P-Torus, employing wavelength reuse schemes, uses 50% less wavelengths compared to other wavelength switching architectures, for same scale DCNs, resulting in lower Capital Expenditure for both tunable sources and WDM components. 3D P-Torus can support, with no need for intermediate amplification, medium-sized-DCs up to 10k servers, making it an excellent candidate to cover the Edge Data Center needs, especially taking into account its excellent low end-to-end latency characteristics. Imminent future research directions include modifications in the 3D P-Torus architecture in order to be able to support the disaggregated DC paradigm.

## References


1. The Global Internet Phenomena Report (2020) Sandvine
2. Data Centres and Data Transmission Networks (2020) IEA, Paris
3. Global Cloud Index Forecast (2020) Cisco
4. Hazelwood K, Bird S, Brooks D, Chintala S, Diril U, Dzhulgakov D, Fawzy M, Jia B, Jia Y, Kalro A, Law J, Lee K, Lu J, Noordhuis P, Smelyanskiy M, Xiong L, Wang X (2018) Applied machine learning at facebook: a datacenter infrastructure perspective. In: 2018 IEEE International symposium on high performance computer architecture (HPCA), pp 620–629. IEEE, Vienna
5. Greenberg A, Lahiri P, Maltz DA, Patel P, Sengupta S (2008) Towards a next generation data center architecture: scalability and commoditization. In: Proceedings of the ACM workshop on programmable routers for extensible services of tomorrow—PRESTO '08, p 57. ACM Press, Seattle, WA, USA
6. Cohen D, Petrini F, Day MD, Ben-Yehuda M, Hunter SW, Cummings U (2009) Applying Amdahl's other law to the data center. IBM J Res Dev 53(5):1–5:12
7. Wang G, Andersen DG, Kaminsky M, Papagiannaki K, Ng TSE, Kozuch M, Ryan M (2010) c-Through: part-time optics in data centers. SIGCOMM Comput Commun Rev 40:327–338
8. Chen K, Singla A, Singh A, Ramachandran K, Xu L, Zhang Y, Wen X, Chen Y (2014) OSA: an optical switching architecture for data center networks with unprecedented flexibility. IEEE/ACM Trans Netw 22:498–511
9. Farrington N, Porter G, Radhakrishnan S, Bazzaz HH, Subramanya V, Fainman Y, Papen G, Vahdat A (2010) Helios: a hybrid electrical/optical switch architecture for modular data centers. In: Proceedings of the ACM SIGCOMM 2010 conference on SIGCOMM—SIGCOMM '10, p 339. ACM Press, New Delhi, India
10. Porter G, Strong R, Farrington N, Forencich A, Chen-Sun P, Rosing T, Fainman Y, Papen G, Vahdat A (2013) Integrating microsecond circuit switching into the data center. SIGCOMM Comput Commun Rev 43:447–458
11. Miao W, Luo J, Di Lucente S, Dorren H, Calabretta N (2014) Novel flat datacenter network architecture based on scalable and flow-controlled optical switch system. Opt Express 22:2465
12. Ye X, Yin Y, Yoo SJB, Mejia P, Proietti R, Akella V (2010) DOS: a scalable optical switch for datacenters. In: Proceedings of the 6th ACM/IEEE symposium on architectures for networking and communications systems—ANCS '10, p 1. ACM Press, La Jolla, California
13. Chao HJ, Kung-Li D, Zhigang J (2003) Petastar: a petabit photonic packet switch. IEEE J Select Areas Commun 21:1096–1112
14. Yawei Y, Proietti R, Xiaohui Y, Nitta CJ, Akella V, Yoo SJB (2013) LIONS: an AWGR-based low-latency optical switch for high-performance computing and data centers. IEEE J Select Topics Quantum Electron 19:3600409–3600409
15. Sato K, Hasegawa H, Niwa T, Watanabe T (2013) A large-scale wavelength routing optical switch for data center networks. IEEE Commun Mag 51:46–52
16. Xu M, Liu C, Subramaniam S (2016) PODCA: A passive optical data center architecture. In: 2016 IEEE international conference on communications (ICC), pp 1–6. IEEE, Kuala Lumpur, Malaysia
17. Ueda K, Mori Y, Hasegawa H, Sato K (2017) Large-scale optical switch utilizing multistage cyclic arrayed-waveguide gratings for intra-datacenter interconnection. IEEE Photon J 9:1–12
18. White Paper (2017) Making the case for edge data centers. Edgeconnex
19. Chaintoutis C, Bogris A, Syvridis D (2019) P-Torus: wavelength-based switching in packet granularity for intra-data-center networks. J Opt Commun Netw 11:491
20. Beauquier B (1998) All-to-all communication for some wavelength-routed all-optical networks
21. Qiu X-Z, Yin X, Verbrugghe J, Moeneclaey B, Vyncke A, Van Praet C, Torfs G, Bauwelinck J, Vandewege J (2014) Fast synchronization 3R burst-mode receivers for passive optical networks. J Lightwave Technol 32:644–659
22. http://epiphotonics.com/products1.htm

23. Testa F, Pavesi L, SpringerLink (Online service) (2018) Optical switching in next generation data centers
24. http://www.alnair.com.my
25. http://www.amstechnologies.com/fileadmin/amsmedia/downloads/100GHzAthermalAWGMultiplexers.pdf.
26. Piels M, Bauters JF, Davenport ML, Heck MJR, Bowers JE (2014) Low-loss silicon nitride AWG demultiplexer heterogeneously integrated with hybrid III–V/silicon photodetectors. J Lightwave Technol 32:817–823
27. https://www.thorlabs.com/newgrouppage9.cfm?objectgroupid=91524
28. https://www.newport.com/p/F-WDM-S11315
29. Zhang X, He J-J, Liu N, Dubowski JJ (2015) Carrier-induced fast wavelength switching in tunable V-cavity laser with quantum well intermixed tuning section. Opt Express 23:26336
30. Miao W, Yin X, Bauwelinck J, Dorren H, Calabretta N (2014) Performance assessment of optical packet switching system with burst-mode receivers for intra-data center networks. In: 2014 the european conference on optical communication (ECOC), pp 1–3. IEEE, Cannes, France
31. Hossain AKMD, Aurangozeb HM (2018) Burst mode optical receiver with 10 ns lock time based on concurrent DC offset and timing recovery technique. J Opt Commun Netw 10:65

# An Image Strategy Based on Saliency Detection Using Luminance Contrast for Artificial Vision with Retinal Prosthesis

**Jing Wang, Jianyun Liu, Yuting Zhang, Haiyi Zhu, Yanling Han, Yun Zhang, Ruyan Zhou, Zhonghua Hong, and Shuhu Yang**

**Abstract** Due to the limited number of implantable microelectrodes, subjects worn by retinal prosthesis received only limited discrete light spot (called artificial vision). Thus, the external abundant information is largely lost. Whereas there is little room for significant improvement in the electrodes number, it would be a feasible way to optimize the low-resolution information in the prosthetic vision by effective image processing. Based on luminance contrast, a real-time saliency detection strategy was presented to enhance the artificial vision, which included color space conversion and a visual attention simulation processing method. The strategy was evaluated by two benchmark databases, and its superiority was validated in extracting image foreground compared with the basic luminance contrast and other algorithms. Meanwhile, simulated experiments were conducted to verify the proposed strategy in the retinal prosthesis.

**Keywords** Retinal prosthesis · Artificial vision · Image processing · Saliency detection

## 1 Introduction

Retinal prosthesis has been viable to the visual function recovery for some blindness like AMD and RP. It is an implantable micro-device encapsulating certain electrical electrodes, along with one micro-camera and vision processing module as external part to collect and convert the image into corresponding electrical stimulation mode. All the information and energy needed are transmitted via wireless transmission unit

J. Wang · J. Liu · H. Zhu · Y. Han (✉) · Y. Zhang · R. Zhou · Z. Hong · S. Yang
College of Information Technology, Shanghai Ocean University, Shanghai 201306, China
e-mail: ylhan@shou.edu.cn

J. Wang · Y. Zhang
Key Laboratory of Fisheries Information, Ministry of Agriculture, Shanghai 201306, China

Y. Zhang
China Mobiles, Shanghai 310000, China



X.-S. Yang et al. (eds.), *Proceedings of Sixth International Congress on Information and Communication Technology*, Lecture Notes in Networks and Systems 216,
https://doi.org/10.1007/978-981-16-1781-2_26

to the microelectrode array in the body, which stimulate the visual nervous system and generate phosphene in the visual center, so as to repair the visual function. However, at present stage and for a long period of time, the artificial vision provided by retinal prosthesis is a kind of perception composed of discrete phosphene, characterized by much low resolution, limited visual field, small dynamic range of expressible brightness, and irregular induction of color information.

It has become important that studying and implementing certain image processing strategies for the vision processing unit in retinal prosthesis, to extract and enhance the most salient information from the captured scene, further to optimize that the implant subjects can perceive. As a pioneer, Parikh and Itti et al. [1] revised Itti' model [2] and applied to a simulated retinal prosthesis system. Focusing on recognizing moving objects, Wang et al. [3] proposed a real-time image strategy using background subtraction. Then Han et al. [4] also presented a foreground extraction to detect moving objects, whereas it cannot process in real-time. Combined with Itti' model, clustering analysis, and grabCut, Wang et al. [5] proposed an object extraction and enhancement algorithm, which processed a mid-size image for a few seconds. Li et al. [6] put forward a real-time saliency detection strategy based on Lab spatial features to optimize prosthesis vision. Later, Li et al. [7] proposed to extract and enhance foreground by graph-based visual saliency and a grabCut-based self-adaptive-iterative optimization model. Guo et al. [8, 9] applied the manifold ranking and proposed an unsupervised moving target segmentation by combining it with foreground and background cues. Experiment results for static images of the strategy were encouraging but its non-real-time processing was an important issue.

Current models were devoted to better performance in extracting salient object, but most had disadvantage in the processing time because of their complexity. For viable image processing strategy in retinal prosthesis for useful information enhancement, this study only selected luminance contrast as the main feature to reduce the time cost. Experimental results showed the proposed algorithm could improve recognition efficiency under simulated artificial vision.

# 2 Methods and Materials

## *2.1 Image Processing Strategy*

In the paper, RGB and YUV color systems were utilized to present image information. The color image was then grayed and computed to generate three saliency maps by luminance contrast (LC) [10]. After equalization, a strategy called "visual attention simulation processing" was proposed for further adaptive refining the salient maps. Under different weights, they were integrated. At last, pixelization was conducted to simulate artificial prosthetic vision. The specific processing is shown in Fig. 1.

**Image saliency computation**. In terms of the luminance, saliency [10] of a pixel is the sum of the Euclidean distance from all other pixels in the image, which generated

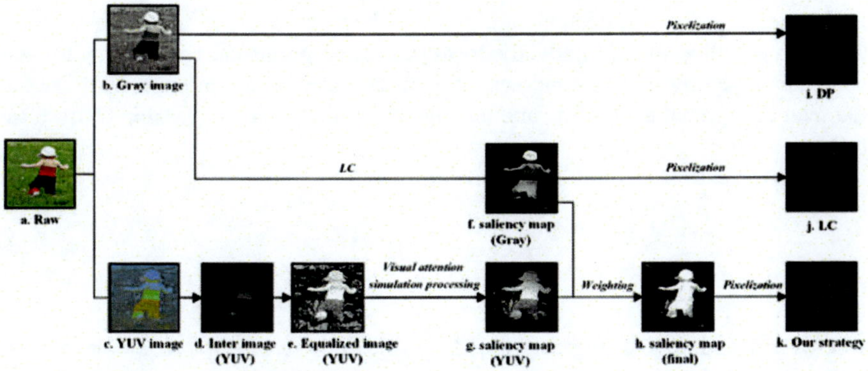


**Fig. 1** A schematic diagram of image processing strategy based on LC

the saliency map $S$. The computation equation is,

$$S(I_k) = \sum_{\forall I_k \in I} \|I_k - I_i\| \quad (1)$$

where $I_i$ is the gray value, $I_i \in [0, 255]$.

**Visual attention simulation processing**. Visual attention simulation processing was proposed to adaptively extract the most attended region. First, the equalized image was processed to get the gravity of the image. Then attended point detection and region expansion were applied to get the most attended region. Based on these results, an adaptive constraint-weighted item $G_{YUV}$ could be obtained. Finally, a YUV saliency map was processed by weighting the equalized image and $G_{YUV}$. Figure 2 illustrated the specific processing.

In the biological visual perception system, the center-peripheral receptive field is a concentric circle [11]. In the paper, inner diameter of the difference of Gaussian (DOG) was used to simulate the human visual field.The gravity of image is used

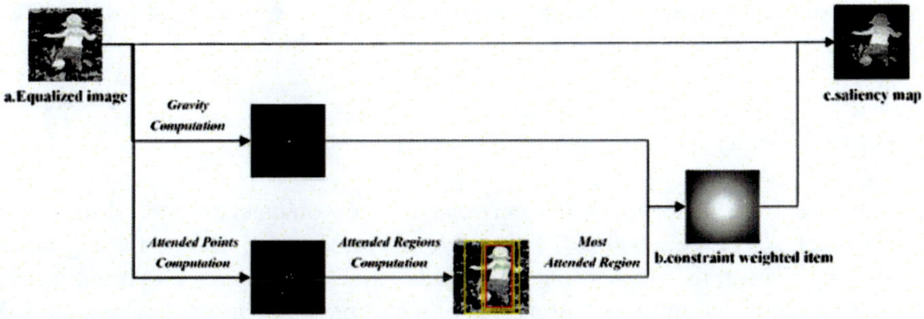


**Fig. 2** Flowchart of visual attention simulation processing

as the center of the Gaussian kernel. $\sigma$ is standard deviation of normal distribution and is an adaptive value. Firstly, the local maximum points called the salient points were calculated by sliding windows. Subsequently, each salient point was taken as the seed of regional expansion, and the algorithm of regional expansion [10] was as follows:

step1, the extraneous potential value of edge $i$ is

$$EP(i) = \frac{\sqrt{E(s_i)E(s_{i'})}}{l_i} \tag{2}$$

where $i$ is the edge of the region, $i \in \{1, 2, 3, 4\}$. $E(s_i)$ ***and*** $E(s_{i'})$ are energy of edge $S_i$ inside and edge $S_{i'}$ outside.

step2, if $EP(i) > Th$, edge $i$ expands one pixel outward, $Th$ is the condition of stopping expansion.

step3, repeat step1 and step2 until the expansion condition is not satisfied.

In regional expansion, the region with high overlapping was salient region. $\sigma$ was computed as follows,

$$\sigma = mean(d_1 + d_2 + d_3 + d_4) \tag{3}$$

where $d_i$ is the edge distance between the salient region and the image, $i \in \{1, 2, 3, 4\}$.

**Saliency map generation**. The final saliency map was computed as follows,

$$S = S_{Gray} + S_{YUV} * G_{YUV} \tag{4}$$

where $S_{Gray}$ is saliency map after gray conversion. $S_{YUV}$ is saliency map in YUV color space. $G_{YUV}$ is the Gaussian map which is $S_{YUV}$ of saliency map to filter noise and enhance the foreground of the image.

**Prosthetic vision simulation**. This study simulated the prosthetic vision by first dividing the saliency map into blocks, then calculating the average gray value of all pixel points in each block, and finally fixing the gray value as the corresponding position of the phosphene in phosphene array. A phosphene spot is defined as in [12].

## 2.2 Subjects

24 subjects from Shanghai ocean university, were volunteer to participate in the experiment. All of them had 20/20 vision or corrected vision. Before the beginning of the experiment, the subjects had fully understood the purpose, and procedure of the experiment, and signed the informed consent. They were ensured to have enough rest before each testing.

### 2.3 Apparatus

The experiment was carried out in a quiet and dark professional lab room to reduce the interference of light and noise. Algorithms were implemented in Visual Studio 2015 with C++. Experimental data were recorded and images were successively displayed on a Windows 7 OS PC (Inter(R) Core(TM) i5-6500 CPU @3.2 GHz, 16.0 GB RAM, Lenovo, China) which connected to the head mounted device (HMD, HTC vive, resolution 480 × 480, Angle of view 40°) and presented to the subjects. Psychtoolbox was used for controlling the test procedure.

### 2.4 Experiment Design

Subject in the experiment wore HMD mounted with a camera to capture the display. Images were processed by algorithms in real time and transmitted to HMD. The subject described the displayed image. Moderate guessing was allowed. Two experimenters judged and recorded the subjects' description to guarantee the objectivity of the results. The recognition time was recorded for test image intervals.

### 2.5 Data Analysis

The recognition accuracy and reaction time were used to represent the experimental result. Meanwhile, Identification Index [13] was used to comprehensively evaluate three image processing algorithms from accuracy and response time. Turkey and LSD were conducted after ANOVA in SPSS to analyze the effect of the algorithm on the recognition. Used Bonferroni to post-check analysis.

$$Identification\ Index = \frac{\%correct - 5.56\%}{tcorrect} \tag{5}$$

## 3 Results

### 3.1 Evaluation of the Proposed Strategy

Aiming to prove the validity of our algorithm, we tested it with the original LC and two real-time algorithms including SR (Spectral Residual) [14], SIM (Saliency Estimation) [15] in two benchmarks: MSRA-10K and ECSSD-1K. Precision-Recall (PR), F-measure and receive operation curves (ROC) were chosen for evaluation.

Results showed in Fig. 3 indicated our strategy had better saliency detection results. PR curves in left column showed our proposed strategy was superior to others. Middle column showed F-measure of our strategy was higher and flatter than others under most segmentation threshold. ROC showed our results were higher in the MSRA-10K. While from the ECSSD-1K, the result was slightly worse than that of SR.

We tested the proposed strategy in ECSSD-1K to get the processing time per image with cropped resolution of 173 × 173. Table 1 gives the average time of each step. The average time per image was 27 ms (~35 frames/s). From statistics analysis,

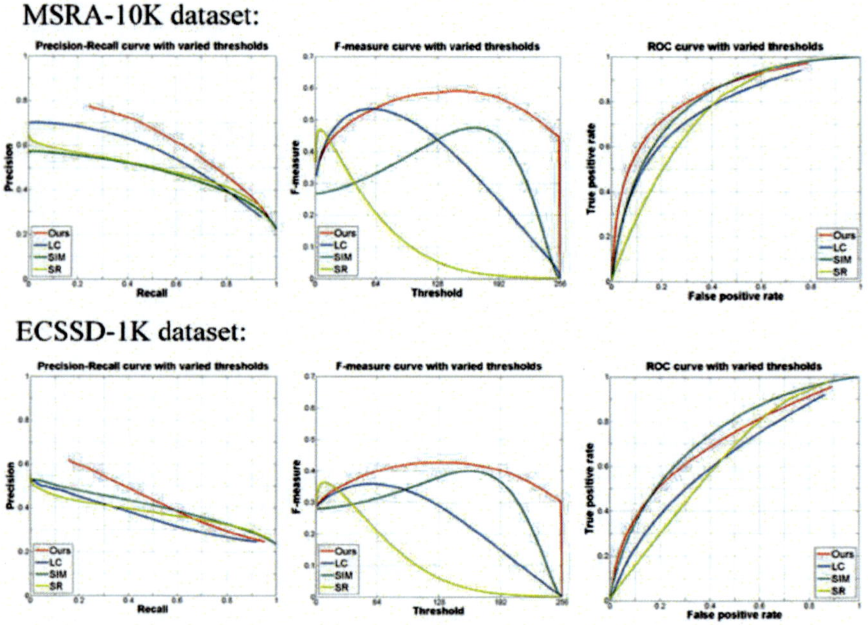


**Fig. 3** Performance comparison of four algorithms

**Table 1** Processing time of each steps

| Processing steps | Strategy implementation (time in ms) |
|---|---|
| RGB → gray | 0.06 |
| RGB → YUV color space | 0.48 |
| Luminance contrast computation | 0.45 |
| Equalization | 0.28 |
| Visual attention simulation processing | 18.63 |
| Frames per second | ~35 (max:54, > = 24 frames/s) |

99.9% of images could be processed above 25 frames/s, which indicated the real-time application of the proposed strategy.

### 3.2 Results of Simulation Experiments

Artificial vision was simulated in experiments. According to epidemiological data, a dataset including 70 objects for blind people was adopted. Results in Fig. 4 indicated our strategy obtain the salient object more precisely, and made it more conspicuous.

Algorithms (DP, LC and Ours) were evaluated by recognition accuracy, reaction time, and identification index. As shown in Fig. 5, it could be seen that the proposed algorithm was more accurate than others. Moreover, accuracy of LC was lower than that of DP. Compared with the results of 10.829 ± 0.355 s in LC and 9.646 ± 0.352 s in DP, our algorithm had the shortest reaction time, 6.317 ± 0.419 s. The identification index of ours was much higher than that of LC and DP. The analytic results showed

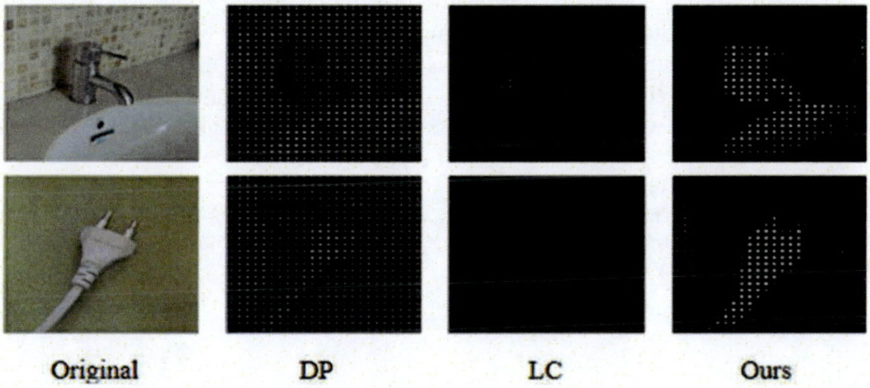


**Fig. 4** Results under different strategies by pixelization

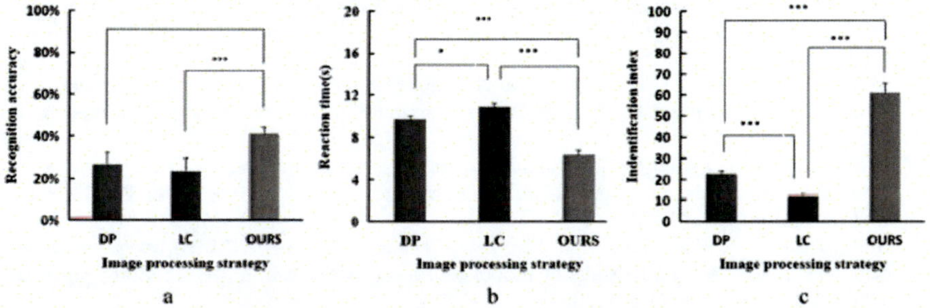


**Fig. 5** Results of the recognition accuracy (**a**), reaction time (**b**) and identification index (**c**) of different image processing algorithms (*$p < 0.05$, **$p < 0.01$, ***$p < 0.001$)

there were significant differences between ours and LC and DP ($p < 0.001$). The above evaluation indexes and the results of Two-way analysis of variance indicated our algorithm is superior to LC and no image processing for retinal prosthesis.

## 4 Discussion

In order to optimize the visual information, the paper presented a method based on LC to obtain the image saliency. "Visual attention simulation processing" was proposed to adaptively enhance foreground and filter other information, which mainly included the region expansion and DOG processing. The experimental results had shown that this processing not only filtered out the background noise caused by foreground enhancement, but also made up the missing foreground information compared to the original LC, thus realized the background suppression while gave better foreground extraction under the requirement of real-time processing. Meanwhile, based on the frequently used objects for blind people, the study showed that the proposed strategy can optimize the salient object information in artificial prosthetic vision.

The paper took into account the luminance of the image, but did not combine other features in visual attention, such as shape, motion, etc. On the other hand, the paper only considered the features under YUV color space. Combined with features in other color space would be feasible. These aspects will be considered in our future works.

**Acknowledgements** This research is supported by the National Natural Science Foundation of China (61806123, 41871325); Shanghai Sailing Program (16YF1415700); National Key R&D Program of China (2019YFD0900805).

## References

1. Parikh N, Itti L, Humayun M, Weiland J (2013) Performance of visually guided tasks using simulated prosthetic vision and saliency-based cues. J Neural Eng 10(2)
2. Parikh N, Itti L, Weiland J (2010) Saliency-based image processing for retinal prostheses. J Neural Eng 7(1)
3. Wang J, Lu YY, Gu LJ, Zhou CQ, Chai XY (2014) Moving object recognition under simulated prosthetic vision using background-subtraction-based image processing strategies. Inf Sci 277(2):512–524
4. Han T, Li H, Lyu Q, Zeng Y, Chai XY (2015) Object recognition based on a foreground extraction method under simulated prosthetic vision. In: International Symposium on Bioelectronics and Bioinformatics (ISBB)
5. Wang J, Li H, Fu WZ, Chen Y, Li LM, Lyu Q, Han TT, Chai XY (2016) Image processing strategies based on a visual saliency model for object recognition under simulated prosthetic vision. Artif Organs 40(1):94–100
6. Li H, Han TT, Wang J, Lu ZF, Cao XF, Chen YL, Li M, Zhou CQ, Chai XY (2017) A real-time image optimization strategy based on global saliency detection for artificial retinal prostheses. Inf Sci s415–416:1–18

7. Li H, Su XF, Wang J, Kan H, Han TT, Zeng YJ, Chai XY (2018) Image processing strategies based on saliency segmentation for object recognition under simulated prosthetic vision. Artif Intell Med 84:64–78
8. Guo F, Yang Y, Gao Y (2018) Optimization of visual information presentation for visual prosthesis. Int J Biomed Imaging 2018(5):1–12
9. Guo F, Yang Y, Xiao Y, Gao Y, Yu NM (2019) Recognition of moving object in high dynamic scene for visual prosthesis. IEICE Trans Inf Syst E102-D(7):1321–1331
10. Zhai Y, Shah M (2006) Visual attention detection in video sequences using spatiotemporal cues. In: 14th ACM International Conference on Multimedia
11. Kuffler SW (1953) Discharge patterns and functional organization of mammalian Retina. J Neurophysiol 16(1):37–68
12. Lu YY, Kan H, Liu J, Wang J, Tao C, Chen Y, Ren QS, Hu J, Chai XY (2013) Optimizing Chinese character displays improves recognition and reading performance of simulated irregular phosphene maps. Invest Ophthalmol Vis 54(4):2918–2926
13. Zhou DD, Dorn JD, Greenberg RJ (2013) The Argus® II retinal prosthesis system: an overview. In: 2013 IEEE international conference on multimedia and expo workshops (ICMEW)
14. Hou X, Zhang L (2007) Saliency detection: a spectral residual approach. In: 2007 IEEE conference on computer vision and pattern recognition
15. Murray N, Vanrell M, Otazu X, Parraga CA (2011) Saliency estimation using a nonparametric low-level vision model. CVPR

# Challenges in Smart Health Applications Using Wearable Medical Internet-of-Things—A Review

**Benedikt Schnell, Patrick Moder, Hans Ehm, Marcel Konstantinov, and Mahmoud Ismail**

**Abstract** The spread of Internet-of-Things (IoT) in domains like health monitoring and the success of cloud services have pushed the horizon of new computing paradigms, such as wearable computing, with health monitoring systems being found in smart homes, hospitals or sports devices. The deployment of off-the-shelf sensors to support monitoring of health functionality, physiology, and activity, can yield to identify health risks. Still, there are several technological challenges to be addressed before biosensor systems with smart healthcare services can be widely adopted. This work addresses latest developments in Smart Health applications, focusing on technical challenges from both system and service perspectives in the context of wearable and medical IoT. We classify relevant publications from the last five years by parameters in the IoT paradigm, including the wearable technology, vital signs, service identification, and IoT computing models such as edge, fog, and cloud. Our findings show that the development faces a lack of system standardization and that data processing moves from servers or cloud locations toward the devices. Second, we examine the challenges present in this domain, the main ones identified are security, privacy, and energy efficiency.

**Keywords** Smart health · Medical IoT · Wearable IoT · Edge computing · Privacy preservation

B. Schnell (✉)
Technical University of Munich, Theresienstr. 90, 80333 Munich, Germany
e-mail: Benedikt.Schnell@infineon.com

B. Schnell · P. Moder · H. Ehm · M. Konstantinov · M. Ismail
Infineon Technologies AG, Am Campeon 1-15, 85579 Neubiberg, Germany
e-mail: Patrick.Moder@infineon.com

H. Ehm
e-mail: Hans.Ehm@infineon.com

M. Konstantinov
e-mail: Marcel.Konstantinov@infineon.com

M. Ismail
e-mail: Mahmoud.Ismail@infineon.com

 
X.-S. Yang et al. (eds.), *Proceedings of Sixth International Congress on Information and Communication Technology*, Lecture Notes in Networks and Systems 216,
https://doi.org/10.1007/978-981-16-1781-2_27

## 1 Introduction

Widespread disease patterns, such as high blood pressure, arrhythmia or diabetes, demand for increased healthcare accessibility, remote monitoring, and early disease detection, at lower overall costs. Smart Health systems in the Medical Internet-of-Things (MIoT) allow researchers and health scientists to develop application-based health solutions, specified to the connection between the physical world, e.g., humans, and the Internet. Varieties of sensors in watches, patches, wristbands, e-textiles are attached to a user's body. Vital signs measurements are the essential signals and are used to detect heart rate, blood pressure, glycosylation, skin temperature, etc. Measurements on motion tracking, like accelerometers and gyroscopes, are additionally useful in studying a patient's habits as well as fall risk assessment in case of an emergency. The strong growth in the number of IoT components in general puts a high downward pressure on the cost of hardware and software systems in fields like MIoT. For the systems, first, the communication and processing of the continuously collected data need to be energy efficient. Sensors must perform sufficiently for prolonging battery life, show physical characteristics that provide unobtrusiveness as well as robustness to minimize maintenance. Second, the system edge components must have wireless connectivity to facilitate networking using widely accepted standards and perform data pre-processing to reduce computational load on gateways and servers. The correctness of data, from retrieval to data security, is a decisive factor for both the medical importance and the acceptance by users. Regarding the underlying services, expectations lie on the exchangeability and isolation of system components as well as on the differentiation and prioritization of health-related critical actions. From a computing point-of-view, Smart Health is an impressive producer of data, with quickly overloading a system. A promising solution is a decentralized allocation of processing units—from the server to the gateway and the sensors nodes. In the IoT, the trend goes from cloud to fog or edge computing, that is wearable computing for the MIoT. Moving the processing toward the devices is creating a link between the various deployed components and the operating cloud backbone.

The remainder of this publication is structured as follows. After having introduced the background and motivation in the domain of interest in this chapter, we present the review methodology in Sect. 2. We describe the classification of reviewed literature, summarize highlighted publications of major interest and discuss our findings as well as implications in Sect. 3. We conclude this paper with a discussion on the conducted work and provide an outlook on potential future research in Sect. 4.

## 2 Methodology

We conduct a literature study to investigate the trend in research that focuses on both latest MioT-related topics and Smart Health applications. The aim is to compare the research with respect to the above-mentioned challenges. The digital library *IEEEXplore* serves as the base for investigation. The library's search engine is fitted with keywords using the "Command Search" option. The structure follows a Boolean logic, which is exploited to find publications with overlapping research fields. The result in Table 1 is a keyword structure that is derived from the index words of contemporary publications. Each content of the cells in the left column is connected with AND operators and all lines are connected with OR operators, respectively. The "Internet-of-Things" or "IoT" serves as an overspanning topic, with "edge computing", as one of the IoT paradigms, focusing on at least the developments within the device level and the underlying computing framework. The category of "Medical IoT" with synonymous terms steers the search in the direction of medical applications. "Wearable, body sensor network" with similar terms and "body area network, ban" serve to further set the focus on interconnectivity and separation from other domains, e.g., WAN and LAN. The intention is to review publications that address the Smart Health industry in an IoT context, hence containing both system- and service-oriented approaches. "Healthcare, service" with similar terms includes the Smart Health-related services accordingly.

The search period for investigation is limited from 2015 to 2020, hence only focusing on the latest findings and innovations. Due to the interconnected research fields in the IoT domain, the publications identified are divided into different categories. Sorting by the publication year can help to derive a trend in the research domain. Further, this allows separation of the papers according to requirements of deeper investigation, as described in the following. After filtering the acquired papers using above-mentioned criteria scheme, we conduct a side-by-side comparison. This assessment contains a qualitative and a quantitative measure on the approach and the methods used. Further, the categorization provides insights into similar research interests from the latest publications.

**Table 1** Keyword structure used in literature search with description

| Keywords | Description |
|---|---|
| Internet of things, iot | General context |
| Medical internet of things, medical iot, miot, wearable internet of things, wearable iot, wiot | Subsections that share systems at a user's body and health services |
| Edge computing | Subsystem at the device level |
| Wearable, body sensor network, bsn, medical cyberphysical system | Interconnected subsystem at device level and user's body |
| Body area network, ban | Subsystem connecting the device level |
| Healthcare, service, tele-care, tele-treatment, telecare, teletreatment | Smart Health service with the help of wireless and remote technology |

# 3 Review

The first part of this chapter presents the classification of the investigated publications from 2015 until the first half of 2020. We then explain the criteria for a deeper investigation and side-by-side comparison of selected papers. Finally, we summarize both the joint research gaps along with the most promising findings within all publications and the identified challenges. First, a categorization for edge, fog, and cloud serves to identify the research field within the various IoT computing approaches. The various paradigms help to distinguish the areas in the IoT, with changing terminology throughout the research history and dependent on the research field, as well [1]. The respective IoT computing paradigms, i.e., edge, fog, and cloud, and the respective research fields are likewise interrelated. Our intention is to find publications that focus on interconnected research topics. Since we identify an emphasis on edge computing in the reviewed publications, the remaining literature on fog and cloud computing are neglected. As depicted in Fig. 1, we observe an increase in general research interest. We can also derive a statistical trend with an increase in edge computing over the last years, while research on fog and cloud computing has seen a lower increase. New system technologies like stronger processing power conjunct with an overall increase in rollout of edge devices serve as an interpretation for this trend [1]. Second, we distribute all publications according to their architectural approach, respectively. Issues such as system component replacement or data migration can be either addressed from a service point-of-view, with a system approach or a joint attempt. System realizations intend to fulfill certain services, while services experience system-independent abstraction. In practice, research interests can be distinguished in this criterium, as depicted in Fig. 2. We can observe a stronger focus on system design in the reviewed research. Beside the overall lack of standardization in system innovations, we find this trend to have a statistical reason, that is, the intentional focus on edge computing in our research, which leads to less findings in services. We moreover identify a trend toward hybrid approaches, hence a combination of systems and services.

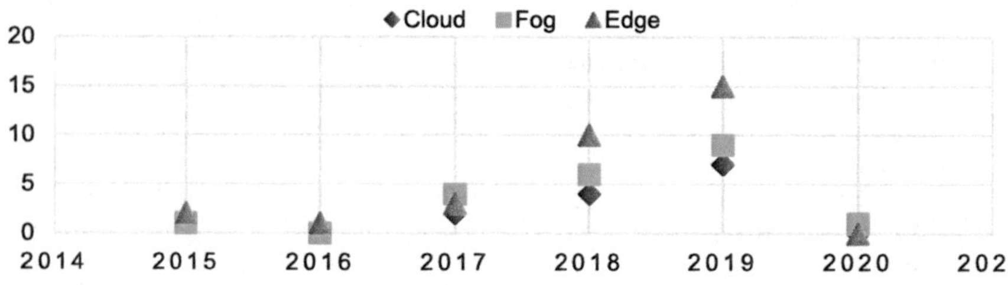


**Fig. 1** Number of research with different computing paradigms and Smart Health applications by year

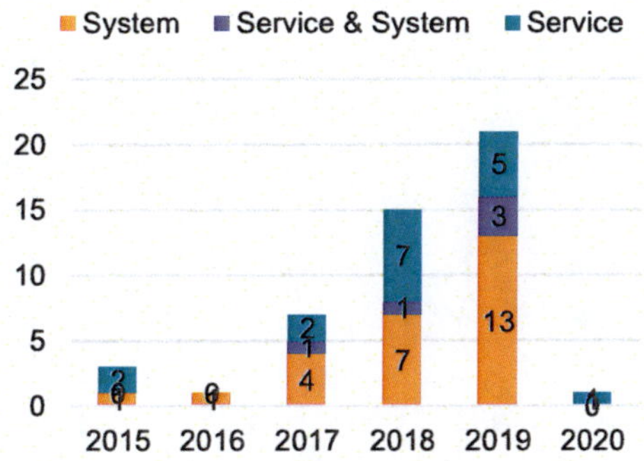


**Fig. 2** Architecture approach by year

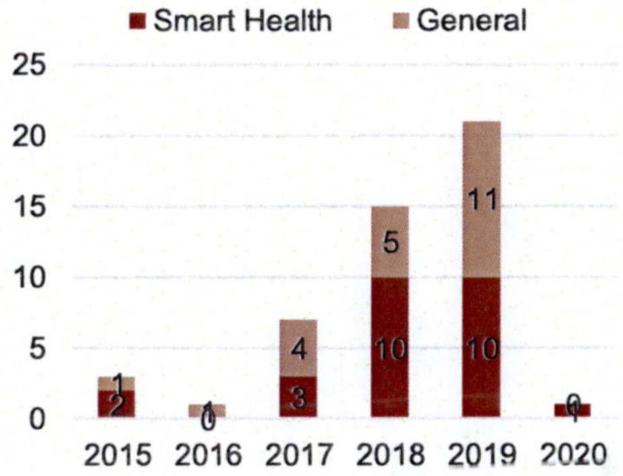


**Fig. 3** Application by year

## 3.1 *Classification*

Lastly, applications in literature are not merely Smart Health specific, but also span across other fields, i.e., mentioning Smart Health as one application field among others. As depicted in Fig. 3, we find that the increasing number of publications in the general context indicates an overall increase of the research interest in other fields, e.g., smart home, smart city, among others, which is set in relation to Smart Health research efforts. We neglect any publications not strictly focusing on Smart Health applications and, as a further criterion, we focus on the ones involving at least edge computing. The selection of papers for in-depth investigation is given in Table 2, together with the addressed problem area, the architecture and the computing paradigm.

## 3.2 *Summary of Reviewed Literature*

In the following, we shortly summarize the literature we select for further comparison, as depicted in Table 2. We structure this part according to the clustering we

**Table 2** Architecture, computing paradigm, and problem area addressed by reviewed literature

| Research | System | Service | Edge | Fog | Cloud | (Big) data analysis, prediction, intelligent inference | architecture, infrastructure, connectivity | Data/workload, offloading, energy efficiency | Service availability, modularity, scalability | Security (privacy, misuse, ownership) | Latency, real-time capability, execution time |
|---|---|---|---|---|---|---|---|---|---|---|---|
| Wang et al. [5] | x | | x | | | | x | | | | |
| Basu et al. [10] | | x | x | | x | x | | x | | | |
| Wang et al. [11] | | x | x | | | | x | | | x | |
| Ritrovato et al. [7] | x | | x | | | x | | | x | | x |
| Prabakaran et al. [12] | x | | x | | | x | x | x | | | |
| Firouzi et al. [8] | | x | x | x | x | | x | | x | | |
| Aileni Raluca et al. [9] | | x | x | | x | x | x | | x | | |
| da Silva Lisboa Tigre et al. [6] | | x | x | x | | x | x | | x | | |
| Jin Kang et al. [13] | x | | x | | | x | | x | | x | x |
| Bohé et al. [2] | | x | x | | | | x | | x | | |
| Hong et al. [14] | x | | x | | x | x | | | | x | |
| Alblooshi et al. [15] | x | | x | x | | x | x | | | | x |
| Pathinarupothi et al. [4] | x | x | x | | | x | x | | x | | x |

(continued)

**Table 2** (continued)

| Research | System | Service | Edge | Fog | Cloud | (Big) data analysis, prediction, intelligent inference | architecture, infrastructure, connectivity | Data/workload, offloading, energy efficiency | Service availability, modularity, scalability | Security (privacy, misuse, ownership) | Latency, real-time capability, execution time |
|---|---|---|---|---|---|---|---|---|---|---|---|
| Sun et al. [16] | x | | x | x | x | x | | | | x | |
| Peddoju et al. [17] | x | | x | | | | x | x | x | | x |
| Chen [3] | x | x | x | x | x | | x | | x | | x |
| Balestrieri et al. [18] | x | | x | | | x | | x | | | |

defined after classification. The remainder of this section follows this order: we present (1) general publications in the selected research field, and then focus on (2) approaches addressing data security, (3) approaches addressing energy efficiency, and (4) approaches addressing both data security and energy efficiency. Table 3 depicts a summary of achieved and future work within the reviewed literature.

**General publications.** The focus in the work of Bohé et al. [2] is on a software system realization to help software integrators in building flexible and open IoT ecosystems on the device level. The approach is complementary to other widely existing approaches, such as service-oriented architectures, but also supports IoT device replacement and maintenance purposes. The work presents a proof-of-concept in Android and does not yet focus on multi-stakeholder support. For a decentralized ubiquitous sensing framework for monitoring and supporting people, Chen [3] proposes a system and service architecture to fuse the plurality of similar sensors and form a virtual sensor. The detailed design and full implementation as well as an experimental evaluation remain for future work. Pathinarupothi et al. [4] present their work about a system to convert vital signals into clinical relevant data, addressing the issue of affordability, accessibility, and availability of remote health care. However, data security and user or service authenticity are not addressed. In order to tackle the challenges of scalability and connectivity of a pervasive healthcare network, Wang et al. [5] propose mobile devices that serve as edge routers. The major advantage of this network is that information exchange happens seamlessly among sensors and the Internet, which in turn offers increased service availability. The authors propose improvement approaches regarding enhanced energy efficiency as future research. Tigre et al. [6] present a system architecture for health monitoring and focus specifically on its availability. The authors analyze multiple scenarios and conclude that based on a sensitivity analysis, both devices in the fog layer and sensors have the most significant impact on system availability. Their model architecture includes the data flow from sensor device layers up to cloud applications. As future work, the model shall be transformed to the prototyping stage with special focus on network connectivity and scalability. For approaching the Internet of Medical Things with Big Data technologies, Ritrovato et al. [7] present a modular, expandable and hybrid technology platform that enables anomaly detection in real-time. Patient data generated by sensors undergo logging on the edge using a staging platform. To follow the approach for semantic data streams, the data gets transformed in triples according to the Resource Description Framework (RDF) as a pre-processing step. A cluster computing part then collects and processes the streams and a database stores the data. The hybrid approach is evaluated with regards to memory usage and performance. The authors state that real-time capabilities need infrastructures that are more powerful. Discussing opportunities and challenges that arise due to the transmission to IoT eHealth design, Firouzi et al. [8] present both general directions of future research as well as a proposed ecosystem architecture. Opportunities include availability, accessibility, and personalization. The main challenges identified are manifold and include scalability, data management, and security, among others. The proposed framework contains a device layer for sensing, a communication layer, edge and fog layers for analysis or data management, as well as a cloud layer for

**Table 3** Method and approach addressed by reviewed literature

| | System realization | Data security | Power saving |
|---|---|---|---|
| Balestrieri et al. [5] | Design validation hierarchical network architecture | | Only addressed as future work |
| Basu et al. [10] | Problem statement and concept proposal | Only addressed as future research | Context inference |
| Wang et al. [11] | General framework with multi-stakeholder view | Ethereum Blockchain smart contracts | |
| Ritrovato et al. [7] | Prototype architectural platform with modular layers | | |
| Prabakaran et al. [12] | Experimental setup, heterogeneous edge computing | | Multicore, inexact, CGRA accelerated |
| Firouzi et al. [8] | 1st case study multilayer cloud architecture | Only addressed as future work | |
| Aileni Raluca et al. [9] | Experimental validation of HW prototype | | Only addressed as future research |
| da Silva Lisboa Tigre et al. [6] | Model for analysis of availability of fog/cloud scenarios | | |

(continued)

**Table 3** (continued)

| | System realization | Data security | Power saving |
|---|---|---|---|
| Jin Kang et al. [13] | | Signal reconstruction quality | Heterogeneous multicore edge platform |
| Bohé et al. [2] | Flexible and open ecosystems for software integrators | | |
| Hong al. [14] | | Scheme proposal for privacy preservation | |
| Alblooshi et al. [15] | Design principles and development concept | | Communication efficiency as future research |
| Pathinarupothi et al. [4] | Convert vital signals into clinical relevant data | | |
| Sun et al. [16] | | Challenges, requirements, threats | |
| Peddoju et al. [17] | System for implementing personalized health services | | Only addressed as future research |
| Chen [3] | Implementation, experiments as future research | | |
| Balestrieri et al. [18] | | | Approximate computing and quality constraints |

integration and data warehousing. The authors point out that future challenges in this domain cover safety, consistency, communication, and power-efficiency, among others. Aileni Raluca et al. [9] study general aspects of wearables in the domain of monitoring health in IoT environments. The authors model the monitoring, analysis and storage of health-related parameters as cloud services and focus particularly on their real-time capabilities.

**Data security**. Privacy preservation is becoming increasingly noteworthy in the edge and cloud hybrid computing-based healthcare applications. Hong et al. [14] propose a scheme that satisfies security properties, data confidentiality, integrity, mutual authentication, forward security, and privacy preservation. However, the implementation approach and tools to be used are not addressed. Sun et al. [16] assess the domain of Internet of Medical Things regarding the security and privacy challenges, the requirements, the threats, as well as future research directions. They find that the resources of devices are constrained with respect to security schemes applied, which requires new solutions that span across the design space of human, cyber and physical elements. Blockchain can be a promising approach to maintain security and trust in Medical IoT without the need of additional parties that serve as authorities for guaranteeing origin or authenticity of devices. Rather than traditional top to bottom approaches to authorization handling, blockchain allows for a distributed or peer-to-peer transaction of events in a trustworthy and tamper-resistant fashion between software applications [11]. Based on the security requirements of highly distributed IoT devices, Alblooshi et al. [15] propose a detailed architecture with distributed ownership that secures trust and tracing of true ownership. The system makes use of smart contracts in order to manage the ownership based on roles and their access rights. The concluding analysis identifies a high level of security. However, confidentiality is not fully addressed. Future work includes real-world applications as well as improvements for both access control and authentication.

**Energy Efficiency**. Peddoju et al. [17] outline the design principles and the development concept for anomaly detection using low power IoT devices. Further, communication efficiency is planned as future research. The invention of a smart shirt by Balestrieri et al. [18] includes a concept to acquire data from several sensors by adopting low power technologies as well as distributed Artificial Intelligence (AI). Their future work contains the selection of electronic components, implementation, and deployment of the distributed AI framework as well as energy consumption analysis. Prabakaran et al. [12] implement an approximate computing stage and achieve energy reductions in energy-constrained, sensory, IoT edge devices, and wearable electronics. While evaluating the quality constraint at two stages, the proposed algorithm reduces the energy consumption by a factor of 20 with 0 percent quality loss and by a factor of 22 with less than 1 percent quality loss. For increasing power-efficiency, Basu et al. [10] present an edge-computing architecture that is used for health monitoring sensors. The IoT system enables operating at ultra-low voltage by the introduction of a multicore system and accelerators together with inexact computing. One major problem concerning energy efficiency is due to overloaded sensor nodes, when real-time forecasting and data analysis are required. Kang et al. [13] present an inference approach that is capable of reducing both frequency and data volume. Data analysis is therefore performed more efficiently and health information may be communicated much faster. However, security and data management concerns persist since infrastructures need further improvement.

**Data security and energy efficiency.** Presenting an edge computing health monitoring system, Al Disi et al. [19] address issues of data security as well as energy efficiency. The presented approach is gateway-centric and employs local edge devices for processing and analysis. It therefore enables real-time capability with limiting resources. The authors moreover present investigations on optimal allocation of resources to increase energy efficiency. Privacy issues are addressed by mobile heterogeneous systems with multicore design.

### *3.3 Results and Discussion*

Resulting from the conducted literature research, we derive the following similarities in the publications. First, there remains a lack of system standardization. The respective IoT computing paradigms, such as edge or fog computing, open up various possibilities for the system architecture. Still, similar to other IoT domains, e.g., Industrial IoT, we see a trend of the computing stage moving toward the devices to reduce both latency and communicational load.

Second, even though both processing power and transmission channel capacity are steadily increasing, data security and privacy in all computational layers remain highly important. We see that latest research has made an effort in finding remaining security issues on the device level and interconnected layers, while successfully focusing on adapting existing security mechanisms from other domains. As depicted in Table 2, we can observe that the publications that focus on data security also set a strong interest on the system design and implementation. Security issues or privacy concerns from a service point-of-view are only partially regarded, however. As data privacy, such as personal identity handling, mostly lies in the service regime, we expect research potential in this field. While Smart Health applications imply handling highly confidential data, data privacy remains an issue as long as a sophisticated service is included in the architecture, i.e., data sharing, mitigation, and identity validation. Last, for one of the most critical bottlenecks for wearable devices, that is the battery consumption, a high computational efficiency is required. The latest publications indicate that the tradeoff between computational efficiency and the quality of services remains an open issue. Worth mentioning, the work of Prabakaran et al. [12], with the implementation including a two-stage assessment of services, achieves low power consumption and high computational accuracy. Workload on the sensor and actuator devices has a major influence on energy efficiency, hence it requires intelligent strategies to shift data processing services to dedicated tools that are part of the system. Investigation in intelligent pre-processing tools might address this. The boundaries between different computing paradigms are fluid and it is also beneficial to stress the services in the design process. This in turn may support a baseline for developing an appropriate computing framework that matches the requirements for both the system, hence low power sensors and actuators, and services, hence privacy handling.

## 4 Conclusion

In this paper, we present a literature review that focuses on Smart Health applications in wearable IoT environments. We present recent trends in literature and classify it based on incorporated architecture and applications. The main findings of our review are a lack of system standardization in the respective domain, and data processing that is moving from server or cloud locations toward the devices. The major challenges we identify in the reviewed approaches are security, privacy, and energy efficiency, or battery consumption, respectively. However, improvements in the respective areas are tackled in parallel, often independently, and lack clear interrelations as to our review. Our findings in Table 3 also show that only half of the researchers propose solutions to tackle security and energy efficiency issues when addressing these challenges, respectively. This may prove that these are indeed real challenges in the domain of interest. The main limitations of this research lie within the methodology, where Smart Health-specific publications have been investigated only, even though the challenges derived apply to other IoT domains, as well.

With more consumer products undergoing efforts to become certified as medical products, the results of our review may serve as an easy access point for integrating novel sensor and actuator devices in a secure and energy-efficient fashion. Our findings indicate potentials of developing a system framework with combined computing paradigms, and hence fulfillment of the essential services. Beyond the advantages of combining system and service architectures, it occurs beneficial to develop customizable frameworks for applying the principles. Considering this, we aim to spend more effort on designing a respective framework and implementing it for specific use cases in order to evaluate its performance. Therefore, we plan to use existing hardware products for vital sensoring and develop solutions for an integration to a service-oriented architecture and implement it in a prototype.

**Acknowledgements** The research has received funding from the EU ECSEL JU under the H2020 Framework Programme, JU grant no. 783163 (project iDev40), no. 826452 (project Arrowhead Tools), and from the partner's national programs/funding authorities.

## References

1. Escamilla-Ambrosio PJ, Rodríguez-Mota A, Aguirre-Anaya E, Acosta-Bermejo R, Salinas-Rosales M (2018) Distributing computing in the internet of things: cloud, fog and edge computing overview. In: Maldonado Y, Trujillo L, Schütze O, Riccardi A, Vasile M (eds) NEO 2016, 731. Springer International Publishing, Cham, pp 87–115

2. Bohé I, Willocx M, Naessens V (2019) An extensible approach for integrating health and activity wearables in mobile IoT apps. In: 2019 IEEE international congress on Internet of Things (ICIOT), pp 69–75. IEEE
3. Chen H (2019) Ubi-Care: a decentralized ubiquitous sensing healthcare system for the elderly living support. In: 2019 IEEE international conference on dependable, autonomic and secure computing (DASC), pp 543–547. IEEE
4. Pathinarupothi RK, Durga P, Rangan ES (2019) IoT-based smart edge for global health: remote monitoring with severity detection and alerts transmission. IEEE Internet Things J 6:2449–2462
5. Wang KI-K, Dubey S, Rajamohan A, Salcic Z (2015) An android-based mobile 6LoWPAN network architecture for pervasive healthcare. In: 2015 international conference on intelligent environments, pp 49–56. IEEE
6. da Silva Lisboa Tigre, MFF, Santos GL, Lynn T, Sadok D, Kelner J, Endo PT (2018) Modeling the availability of an e-health system integrated with edge, fog and cloud infrastructures. In: 2018 IEEE symposium on computers and communications (ISCC), pp 416–421. IEEE
7. Ritrovato P, Xhafa F, Giordano A (2018) Edge and cluster computing as enabling infrastructure for Internet of medical things. In: 2018 IEEE 32nd international conference on advanced information networking and applications (AINA), pp 717–723. IEEE
8. Firouzi F, Farahani B, Ibrahim M, Chakrabarty K (2018) From EDA to IoT eHealth: promises, challenges, and solutions. IEEE Trans Comput Aided Des. Integr Circuits Syst 37:2965–2978
9. Aileni Raluca M, Pasca S, Suciu G (2018) MIoT applications for wearable technologies used for health monitoring. In: 2018 10th international conference on electronics, computers and artificial intelligence (ECAI), pp 1–4. IEEE
10. Basu S, Duch L, Peon-Quiros M, Atienza D, Ansaloni G, Pozzi L (2018) Heterogeneous and inexact: maximizing power efficiency of edge computing sensors for health monitoring applications. In: 2018 IEEE International symposium on circuits and systems (ISCAS), pp 1–5. IEEE
11. Wang X, Zha X, Ni W, Liu RP, Jay Guo Y, Niu X, Zheng K (2019) Survey on blockchain for Internet of Things. Comput Commun 136:10–29
12. Prabakaran BS, Rehman S, Shafique M (2019) XBioSiP: a methodology for approximate bio-signal processing at the edge. In: Proceedings of the 56th annual design automation conference 2019, pp 1–6. ACM, New York, NY, USA
13. Jin Kang J, Adibi S, Larkin H, Luan T (2015) Predictive data mining for converged Internet of Things: a mobile health perspective. In: 2015 International telecommunication networks and applications conference (ITNAC), pp 5–10. IEEE
14. Hong L, Xuanxia Y, Tao Y, Huansheng N (2019) Cooperative privacy preservation for wearable devices in hybrid computing-based smart health. IEEE Internet Things J 6:1352–1362
15. Alblooshi M, Salah K, Alhammadi Y (2018) Blockchain-based ownership management for medical IoT (MIoT) devices. In: 2018 International conference on innovations in information technology (IIT), pp 151–156. IEEE
16. Sun Y, Lo FP-W, Lo B (2019) Security and privacy for the internet of medical things enabled healthcare systems: a survey. IEEE Access 7:183339–183355
17. Peddoju SK, Upadhyay H, Bhansali S (2019) Health monitoring with low power IoT devices using anomaly detection algorithm. In: 2019 Fourth international conference on fog and mobile edge computing (FMEC), pp 278–282. IEEE
18. Balestrieri E, Boldi F, Colavita AR, Vito L de, Laudato G, Oliveto R, Picariello F, Rivaldi S, Scalabrino S, Torchitti P et al (2019) The architecture of an innovative smart T-shirt based on the Internet of Medical Things paradigm. In: 2019 IEEE International symposium on medical measurements and applications (MeMeA), pp 1–6. IEEE
19. Al Disi M, Djelouat H, Amira A, Bensaali F (2018) The accuracy and efficacy of real-time compressed ECG signal reconstruction on a heterogeneous multicore edge-device. In: 2018 21st Euromicro conference on digital system design (DSD), pp 458–463. IEEE

# Development of an Adaptive Fuzzy Algorithm for Identifying Technical Channels of Information Leakage

**Pavel V. Razumov, Alexander A. Zelensky, Larissa V. Cherckesova, Olga A. Safaryan, Dennis A. Korochentsev, Nikolay V. Boldyrikhin, Irina A. Pilipenko, Veronica O. Kravchenko, Vitaly M. Porksheyan, and Nikita G. Lyashenko**

**Abstract** This article is devoted to the problem of identifying technical channels of information leakage using a decision support subsystem and a neural network. Research is aimed at identifying technical channels of information leakage. The article describes a model for ensuring information security, which is based on an adaptive neuro-fuzzy system and a decision support subsystem that ensures the timely identification of technical channels of information leakage, the system also has the functionality to calculate sound insulation, vibration isolation, line-of-sight range, and acoustoelectric conversion. The system specified in the article makes it possible

P. V. Razumov (✉) · A. A. Zelensky · L. V. Cherckesova · O. A. Safaryan · D. A. Korochentsev · N. V. Boldyrikhin · I. A. Pilipenko · I. A. Pilipenko · V. O. Kravchenko · V. O. Kravchenko · V. M. Porksheyan · N. G. Lyashenko
Don State Technical University, Gagarina Square, 1, Rostov-on-Don 344000, Russia

A. A. Zelensky
e-mail: sashaz1696@yandex.ru

L. V. Cherckesova
e-mail: chia2002@inbox.ru

O. A. Safaryan
e-mail: safari_2006@mail.ru

D. A. Korochentsev
e-mail: mytelefon@mail.ru

N. V. Boldyrikhin
e-mail: boldyrikhin@mail.ru

I. A. Pilipenko
e-mail: irenphil@yandex.ru

V. O. Kravchenko
e-mail: olenikr@yandex.ru

V. M. Porksheyan
e-mail: spu-46@donstu.ru

N. G. Lyashenko
e-mail: lyashenko-n.g@yandex.ru



X.-S. Yang et al. (eds.), *Proceedings of Sixth International Congress on Information and Communication Technology*, Lecture Notes in Networks and Systems 216,
https://doi.org/10.1007/978-981-16-1781-2_28

to identify technical channels of information leakage using the description of the premises, and also generates a report on the results of the data obtained with the ability to send such reports via e-mail to a specified address. The practical application of the proposed system is the possibility of its use in the initial research of the protected premises, which allows to build an effective protection system in the organization.

**Keywords** Technical channels of information leakage · Neural network · Fuzzy logic · Database · Backpropagation algorithm · Information leakage · Information leakage channel · Methods of information leakage paths

## 1 Introduction

With the development of technologies and the improvement of technical intelligence tools, the timely identification of technical channels of information leakage, as well as protection against leakage of confidential information through technical channels, becomes the most relevant [1]. A large number of organizations, to one degree or another, are trying to make their systems for finding channels of information leakage effective, or to improve existing systems so that they are much more efficient. The use of fuzzy data can significantly increase the efficiency of the search system for information leakage channels. However, their use in the field of information security is currently not fully understood. Thus, the solution to this issue will be the development of a system for identifying technical channels of information leakage [2].

Motivation of the research is development of an adaptive fuzzy algorithm for identifying information channels for the dissemination of information in the information system of an organization.

The scientific novelty of this work lies in the fact that the developed system for identifying technical channels of information leakage is able to identify technical channels in the absence of human intervention, who attests the protected premises.

The theoretical contribution of the research is that methods of the adaptive model have been investigated and developed, using fuzzy data to identify technical channels of information leakage. Practical value of the obtained scientific results is the use of an adaptive fuzzy-algorithm for identifying technical channels of information leakage for the initial investigation of the premises and recommendations against leakage of confidential information for medium and large organizations [3–5].

## 2 Classification and Main Characteristics of Technical Channels of Information Leakage

The classification of information leakage channels (ILC) is shown in Fig. 1.

Methods of information leakage paths (MILP) from the information system due to technical means are characterized by indicators that make it possible to assess the risk of uncontrolled dissemination of protected data [6–9].

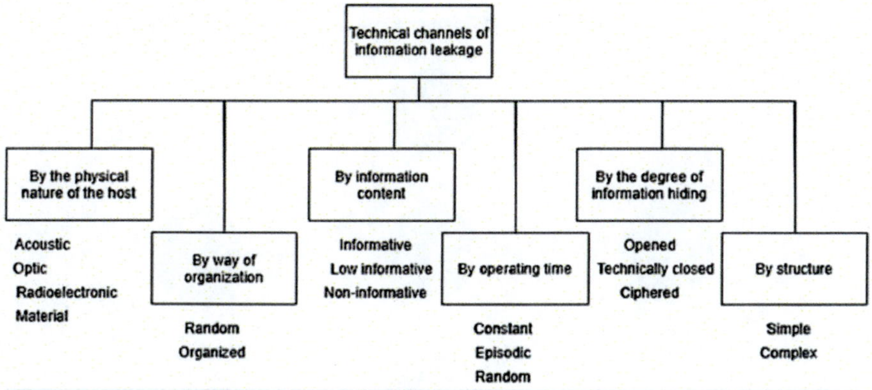


**Fig. 1** Classification of ILC

These indicators are: technical ILC (TILC) throughput, TILC length, conditional information content. Conditional informativeness ranges from zero to one. The transmission medium, the source of the data carrier, and the signal reception are the main components of the information leakage channel (ILC) [10, 11].

The set of physical characteristics that determine the transmission conditions of the data carrier is characterized by the transmission medium. The main ones are: the coefficient of attenuation or energy transfer of the carrier per unit length, the power and type of interference to the signal, the transfer rate of the carrier in the medium, the amplitude-frequency characteristic (if the transfer coefficient depends on the signal frequency).

The parameters of the MILP from the information system at the expense of technical means generally depend on the characteristics of the signal receiver [12–15].

For a signal receiver, the leading characteristics are: sensitivity, range of received frequencies, dynamic range of the data carrier, spatial selectivity of the receiving antenna, the degree and type of distortion of the data carrier.

## 3 Algorithm Description

The program for calculations uses a neural network, consisting of three layers, shown in Fig. 2: an input neuron, 3 hidden neurons, an output neuron. The first hidden neuron receives data on the parameters that form the acoustic channel of information leakage, the second hidden neuron receives data on the parameters that form the optical channel of information leakage, the third hidden layer receives data on the parameters that form the electronic channel of information leakage, the output neuron—the sum of three hidden neurons.

The system actively interacts with the base of this access.

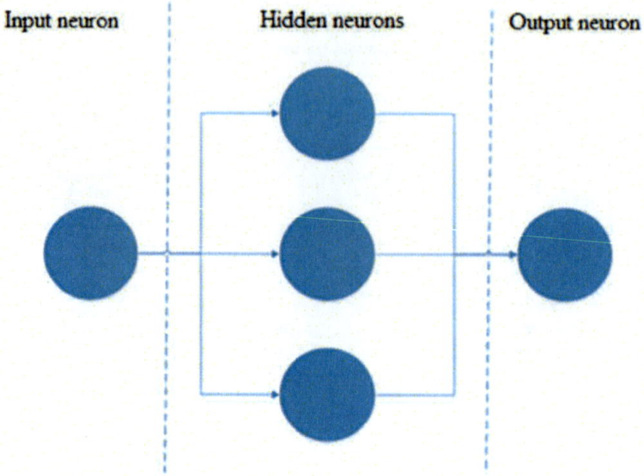


**Fig. 2** Neural network layer diagram

This database stores: synapse coefficients, room component coefficients, logging, data for training a neural network.

For the program to work, you need to train a neural network. Learning is carried out according to the "learning with a teacher" scheme.

(1) The number of room components is randomly determined by formula (1).

$$N = \text{Random}(p), \tag{1}$$

where p denotes the number of room components.
Random is a function that randomly determines a number in the range from 1 to p.

(2) At this stage, a specific component of the room is determined, which is placed in an array consisting of N elements according to formula (2).

$$\text{mas}(i) = \text{Random}(N) \tag{2}$$

where mas (i) denotes an array that contains room components.
N represents the total sampled number of room components.

(3) The coefficients are calculated for each channel of information leakage (formula 3), and the overall coefficient is calculated using the formula (4). Subsequently, the data is compared with the test data stored in the Access database, similar or approximate data is found using the formula (5). Synapse data is set as standard 0.33 for each channel of information leakage.

$$K = \text{sum}(\text{mas}(i)), \tag{3}$$

where K denotes the coefficient of information content, calculated for each of the channels of information leakage.

$$M = k_1 + k_2 + k_3, \tag{4}$$

where M is the total of the odds.

$$K(i) = Kt(i), \tag{5}$$

where M = Mt, where Kt is the test coefficient of information content for technical channels of information leakage; Mt is the test total sum of the information content coefficients for all information leakage channels.

Determination of error—the difference between the test coefficient and the calculated one, and the decrease in the synapse ratio by the difference in the determination of the error. In order to determine the magnitude of the error in the learning process, it is necessary to use the classical backpropagation algorithm according to formula (6). This algorithm works well on two-layer and three-layer neural networks:

$$E_p = 0.5 \sum_j \left(t_{pj} - y_{pj}\right)^2, \tag{6}$$

where $E_p$ is the magnitude of the error function;
$t_{pj}$ is the desired output of the neuron;
$y_{pj}$ is the activated output of the neuron.
After training, you can proceed to work with the program.

## 4 The Scheme of Work in the Subsystem of Decision Support

Decision support subsystems have the following four main features:

- it uses both models and data;
- designed to aid in decision making for unstructured and semi-structured tasks;
- supports, but does not replace, decision making for users;
- improving the efficiency of decisions made is the main goal.

List of characteristics of the "ideal" decision support subsystem:

- operates with semi-structured tasks;
- is intended for persons who make decisions at various levels;

- can be adapted for individual and group use;
- is flexible. Can adapt to changes, including both the internal environment of the organization and its environment;
- easy to modify and use;
- supports both sequential and interdependent solutions;
- supports three phases of the solution process: selection, intelligence, and design;
- supports various methods of solving, which improves the quality of the decision-making process and can be very beneficial when solving the problem by the team that makes the decision;
- a person independently controls the decision-making process using a computer, and not vice versa;
- adapts to changing requirements and supports evolutionary use;
- can be easily modeled if the design logic of the decision support subsystem can be defined;
- supports the modeling process;
- allows you to apply the acquired knowledge.

The decision support subsystem consists of two main subsystems—an information system (IS) and a person who makes decisions. The function of decision makers, as a component of the decision support subsystem, is to enter data and make decisions based on their intuition and knowledge.

In the simplest case, the coefficient technique can be described by formula (7):

$$X = \sum_{i=1}^{g} W_i * X_i, \tag{7}$$

where X is the output indicator,

$X_i$—i-th input indicator (i = 1,…, g),

g is the total number of input indicators,

$W_i \geq 0$—weighting coefficient of the i-th input indicator.

The normalization condition for $W_i$ is presented in formula (8):

$$\sum_{i=1}^{g} W_i = 1, \tag{8}$$

The value $W_i = 0$ indicates the lack of information content of the ILC. The more $W_i$ tends to unity, the more informative this ILC is.

The final result is the overall assessment of the TILC—an integral indicator of the informativeness of the ILC, which is calculated by summing the particular indicators and their informativeness according to formula (8).

Some $X_i$ values are calculated by the $X_j$ indicators, therefore, for them, the $X_i$ calculation will be shown by formula (9):

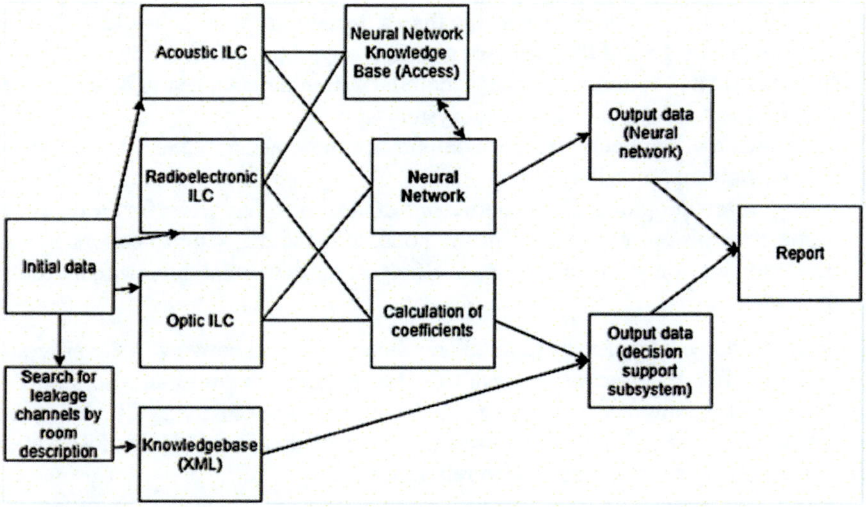


**Fig. 3** General scheme of fuzzy identification system TILC

$$X_i = \sum X_{j,} \tag{9}$$

The input parameters for calculating the information content of the ILC are the normalized values $X_i$ of each ILC. Expert judgment is the source of information about $W_i$ values.

The general scheme of the system is shown in Fig. 3.

Figure shows 2 ways of calculating technical channel identification:

- decision support system;
- adaptive fuzzy algorithm.

After all the calculations, the end user will see the data for each identified channel of information leakage. It will receive the overall coefficient, and the program will also write it to a separate file and provide the user with recommendations for protecting data from leakage for each technical channel identified as a result of the check. The program will be able to get acquainted with this file.

## *4.1 Comparison of the Operation of the Decision Support Subsystem and Neural Network*

Comparing the performance of the same task by both the decision support subsystem and the neural network, we can say the following:

- a neural network is more changeable than a decision support subsystem. This is because the weights of the synapses can be changed by changing the test data;
- for the current program, less code is required to implement a neural network than to implement a decision support subsystem;
- the results that the neural network outputs are more accurate than the results of the decision support subsystem;
- for the current program, all the coefficients of the room parameters for the decision support subsystem are located in the code, and for the neural network in the database, this allows you to edit the coefficients more conveniently, without using the code.

From the above, we can say that the presence of a neural network in the program made it possible to improve the result of the program, but we can also say that neural networks cannot be considered as a universal means of solving all intellectual problems. Their use is justified in those areas in which there are a significant number of similar examples reflecting hidden relationships.

## 5 Discussions of the Investigation Results

The results obtained in the research course were discussed in detail at the scientific seminars of "Cybersecurity of information systems" department of "Informatics and computer engineering" faculty of Don State Technical University. All the comments made by colleagues were taken into account in the final report presented here.

## 6 Conclusion

The developed system makes it easy and convenient to identify technical channels using two methods: a decision support subsystem and a neural network.

Upon completion of the work, a report is generated with recommendations for protection against information leakage, identification of channels according to the description of the room. Also, the presented system has the ability to conduct primary studies of vibration isolation and sound insulation. This system can be used to assess the security of premises from information leakage through technical channels for both large and medium-sized commercial organizations.

The advantages of the developed system are the ability to make informed decisions based on extensive analyzes of available information, as well as the ability to adapt the system and its response to constant external influences.

With the help of the developed system, it is possible to assess the informatively possible technical channels of information leakage, evaluate the most informative technical channel of information leakage, and obtain recommendations for reducing the information content of technical information leakage channels. Also, with the help of this system, conduct primary research in the presence of an acoustic or optical channel of information leakage.

The implemented system allows information security specialists to quickly form a list of technical information leakage channels present in the room and can help with a preliminary examination of an informatization object.

## References


1. Clark L, Algaier W (2007) Surveillance detection the art of prevention, p 197. Cradle Press LLC
2. Keen P (1987) Decision support systems: the next decades. Decis Support Syst 3:253–265
3. Little J (1970) Models and managers: the concept of a decision calculus. Manage Sci 16(8):466–485
4. Pengyu L, Avraham A (2020) Extracting boolean and probabilistic rules from trained neural networks. Neural Netw 126:300–311
5. Yong G, Jian C, Qing D, Anton V, Den H, Qinfeng S, Mingkui T (2020) Multi-way backpropagation for training compact deep neural networks. Neural Netw 126:250–261
6. Zhengxin C (1992) A conceptual framework for expert system description. Comput Edu 18:259–266
7. Yee L (2020) Artificial intelligence and expert systems. International encyclopedia of human geography, 2nd edn, pp 209–215
8. Kushal A, Arun M (2018) Theoretical framework of quantitative analysis based information leakage warning system. Karbala Int J Mod Sci 4:151–163
9. Tao Z, Yangyang S, Yaqin W, Yanzhang G, Long F (2020) An overview of speech endpoint detection algorithms. Appl Acoust 160:107–133
10. Yu F, Tai A (1975) Information leakage through a passive optical channel. Opt Commun 14:51–55
11. Khuat T, Gabrys B (2020) Accelerated learning algorithms of general fuzzy min-max neural network using a novel hyperbox selection rule. Inf Sci
12. Vugar E (2020) Value of fuzzy logic for data mining and machine learning: a case study. Expert Syst Appl 162
13. Mordechai G, Ofer H, Gabi K, Yuval E (2016) An optical covert-channel to leak data through an air-gap. In: 14th Annual Conference on Privacy, Security and Trust (PST)
14. Zhentao G, Yuanyuan C, Zhang Y (2020) A novel method to compute the weights of neural networks. Neurocomputing 407:409–427
15. Mourad G (2020) Hybrid deep neural networks for recommender systems. Neurocomputing 413:23–30

# Towards an Effective Decision Support System for Diabetic Foot Ulcers Diagnostic and Treatment Assessment

**Ricardo Vardasca and Domingos Martinho**

**Abstract** Diabetes mellitus (DM) is a fast-growing metabolic condition that threatens human population quality of living in the overcoming decades. One of its severe consequences is diabetic foot ulcers (DFU), which affect up to a quarter of the DM patients in their lifetime. This consequence leads to high health costs and significant decrease of the patients' quality of life and self-esteem. In order to cope with the rising demands of heath resources and shortage in clinical human assets intelligent computational tools are required to aid in the decision where a patient is in an early stage of a DFU development and on the appraisal of a DFU treatment. It is aim of this research to provide a critical overview of the existing decision support systems (DSS) and publicly available research datasets for diabetic foot ulcers early diagnosis and treatment assessment, and thus proposing a new infrastructure system to deal with it overcoming the past attempts. The existing DFU DSS failed in being introduced in clinical practice due to total discrepancy with current daily clinical practice with DFU and the publicly available DM research datasets are shorter in data for feeding a new DSS. This research presents the actual and promising future data required for effective decisions and discloses a proposed architecture for a DSS applicable to DFU early diagnosis and treatment evaluation. Implementing the proposed system will take time but it will definitely contribute to cope with the patient demands, associated cost reduction and promotion of patients care.

**Keywords** Big data · Clinical datasets · Decision support systems · Diabetic foot ulcers · Diagnostic tools · Multiple data sources · Treatment assessment

R. Vardasca (✉) · D. Martinho
ISLA Santarem, Largo Candido Dos Reis, 2000-241 Santarem, Portugal
e-mail: ricardo.vardasca@islasantarem.pt

 
X.-S. Yang et al. (eds.), *Proceedings of Sixth International Congress on Information and Communication Technology*, Lecture Notes in Networks and Systems 216,
https://doi.org/10.1007/978-981-16-1781-2_29

# 1 Introduction

## *1.1 The Problem of Diabetic Foot Ulcers (DFU)*

Diabetes mellitus (DM) is the most affecting metabolic disease, which majorly threatens the population quality of life, being this aggravated with aging and life expectancy evolution. It can present itself in three forms: type 1 (autoimmune reaction caused, no or low insulin is produced), type 2 (caused by aging and lifestyle habits, characterized by insulin resistance), and gestational (temporary, high blood glucose levels during pregnancy). The estimated worldwide incidence of DM according to the International Diabetes Federation (IDF) is 9.3% for adults aged 20–79 years old, which means one in eleven adults is affected, being the projected rates of 11% and 13.5% for the years of 2030 and 2050, respectively, meaning a 51% increase of the condition [1].

One of the known consequences of DM evolution is the Diabetic Foot Ulcers (DFU), which may affect up to 25% of the DM patients' population, being the risk of recurrence rate of 40% within the first year and up to 65% after five years of healing [2]. This resulting condition can lead in the most severe cases to amputations and consequent death and contributes to the major spending among DM patients' care [3]. With the demand of care caused by the rising of DFU and with limited resources in terms of clinical personal, new intelligent technological instruments are required to facilitate the early identification of DFU and to help on the assessment of its treatments to aid on act promptly to provide the best care possible.

The International Working Group on the Diabetic Foot (IWGDF) in 2019 has defined the guidelines for the classification of diabetic foot ulcers [4], which is an important source of the required data to early detect and characterize a healing DFU.

Since the introduction of Decision Support Systems in medicine in the mid 1950's with the MYCIN that technology proved to be a precious aid to the clinicians daily practice helping them cope with growing demands in clinical care decision, eliminating human error, reducing costs and contributing to provide better overall care to patients [5]. The first known DSS devoted to DFU diagnosis appeared in 2009 for predicting amputations [6].

## *1.2 Current DFU Diagnostic and Treatment Assessment Methods*

The current practical guidelines of the IWGDF established in 2019 [4] determined that to deal with DFU the following procedures have to be implemented:

- Prevention of DFU in subjects with DM
- Offloading DFU in subjects with DM

- Diagnosis, prognosis and management of peripheral artery disease (PAD) in patients with a DFU and DM
- Diagnosis and treatment of DFU infection in subjects with DM
- Interventions to enhance healing of DFU in subjects with DM
- Classification of DFU.

The characterization of DFU patients has a wide spectrum, it involves different geographies, social classes, ages, not specific of a single gender, although it is consensual that it results from a subject with DM and simultaneously having two or more risk factors. The diabetic peripheral neuropathy and PAD usually playing a central role. The first leads to an insensitive and sometimes deformed foot, often causing abnormal loading of the foot. The PAD is characterized by poor circulation in the extremity.

In order to prevent DFU it is very important to Identify the foot at-risk, to regularly inspect and examine the foot at-risk, to educate the patient, his family, and the other healthcare professionals, to ensure routine wearing of appropriate footwear and at least but not less important to treat the risk factors for ulceration.

The current diagnosis is based in the patient history, medical examination of the foot appearance, the vascular status of the limb (which can have its levels of oxygen—$O_2$—monitored) and assessment of the Loss of protective sensation (LOPS). The LOPS can be monitored through the patient pressure perception using the Semmes–Weinstein 10 g monofilament test and/or the vibration perception: 128 Hz tuning fork test.

Based in this assessment the risk of the DFU can be established as presented in Table 1.

Through the history and clinical examination, and DFU can be classified as: neuropathic, neuro-ischemic or ischemic. The measurement of the ankle-brachial index (ABI) using a Doppler instrument is very important for identifying the presence of PAD or not, if it is not present then the DFU is only neuropathic.

For correctly classify the DFU it is important to know the type, cause, site and depth, signs of infection, and patient-related factors.

The severity of infection can be obtained from the IWGDF/ISDA classification criteria (Table 2) if the patient has no PAD but if he has the Wifi system (Table 3)

**Table 1** The DFU Risk with the corresponding foot screening frequency by an expert [4]

| Risk | Characteristics of the examination | Frequency of screening |
|---|---|---|
| Very low | No PAD and no LOPS | Once a year |
| Low | PAD and/or LOPS but no other signs in the foot | Once every 6/12 months |
| Medium | LOPS+PAD, or LOPS+foot deformity or PAD+foot deformity | Once every 3/6 months |
| High | LOPS or PAD, and one or more of the following:<br>– history of a foot ulcer<br>– a lower-extremity amputation (minor or major)<br>– end-stage renal disease | Once every 1/2 months |

**Table 2** IWGDF/ISDA classification criteria [4]

| Clinical manifestations | Infection severity |
|---|---|
| Wound lacking purulence or any manifestations of inflammation | Uninfected |
| Presence of ≥2 manifestations of inflammation (purulence, or erythema, tenderness, warmth, or induration), but any cellulitis/erythema extends ≤2 cm around the ulcer, and infection is limited to the skin or superficial subcutaneous tissues; no other local complications or systemic illness | Mild |
| Infection (as above) in a patient who is systemically well and metabolically stable, but which has ≥1 of the following characteristics: cellulitis extending >2 cm, lymphangitic streaking, spread beneath the superficial fascia, deep-tissue abscess, gangrene, and involvement of muscle, tendon, joint or bone | Moderate |
| Infection in a patient with systemic toxicity or metabolic instability (e.g., fever, chills, tachycardia, hypotension, confusion, vomiting, leukocytosis, acidosis, severe hyperglycemia, or azotemia) | Severe |

**Table 3** Wifi system

| Wound grade | DFU | Gangrene |
|---|---|---|
| 0 | No Ulcer | No Gangrene |
| | Clinical description: minor tissue loss. Salvageable with simple digital amputation (1 or 2 digits) or skin coverage | |
| 1 | Small, shallow ulcer(s) on distal leg or foot; no exposed bone, unless limited to distal phalanx | No gangrene |
| | Clinical description: minor tissue loss. Salvageable with simple digital amputation (1 or 2 digits) or skin coverage | |
| 2 | Deeper ulcer with exposed bone, joint or tendon; generally, not involving the heel; shallow heel ulcer, without calcaneal involvement | Gangrenous changes limited to digits |
| | Clinical description: major tissue loss salvageable with multiple (≥3) digital amputations or standard trans metatarsal amputation (TMA)±skin coverage | |
| 3 | Extensive, deep ulcer involving forefoot and/or midfoot; deep, full thickness heel ulcer±calcaneal involvement | Extensive gangrene involving forefoot and/or midfoot; full thickness heal necrosis with calcaneal involvment |
| | Clinical description: extensive tissue loss salvageable only with a complex foot reconstruction or non-traditional TMA (Chopart or Lisfranc); flap coverage or complex wound management needed for large soft tissue defect | |

should be used for stratifying the possible amputation risk and revascularization benefit [4].

The DFU can be treated according with the status of the wound, it can be healed through pressure offloading and ulcer protection, restoration of tissue perfusion (surgery), infection treatment (clearance and debride of all necrotic tissue and surrounding, oral antibiotic therapy, surgery), metabolic control and treatment of comorbidities (insulin and/or nutrition), local ulcer care (clearance, debride and/or dressings), and education of patient and relatives.

The DFU can be objectively assessed by using the Wagner scale [7, 8] or the University of Texas Diabetic Wound Classification scale [9]. When comparing the two, the second is more descriptive and complete, showing greater association with higher risk of amputation and likelihood of the healing process of ulcers, due to the combination of grade and stage characterization, whether the Wagner scale is simpler and easier to memorize as a visual method, although it does not consider the size, PAD, and LOPS.

## *1.3 Promising Technological Instruments for DFU Diagnostic and Treatment Assessment Methods*

Over the recent years, there were some technological advancements that proved their applicability, although those instruments are not yet fully available in daily clinical setting. Foot pressure loading monitoring is possible through the usage of a sensorial system and it aids to identify the point in which the patient may out more pressure and are more likely to generate a DFU [4] given the load and time of exposure, it can be recorded in rest of exercising. Like pressure load, monitoring the gait can also give important indications on the loss of equilibrium and neurological balance in DM patients [10], it can be performed using sensors such as 3 axis accelerometers and gyroscopes. These two information sources could enter as inputs for a DFU diagnostic SSD.

A recent research also proposed a set of biomarkers that can be of great importance for early detection of DFU [11].

Other important information is foot skin temperature, which is influence by both peripheral vascular and autonomic nervous system, it is known that an asymmetrical or variation to the close surroundings over 2.2 °C is an indicator of possibility of a DFU development one or two weeks before being visible [12]. This type of assessment can be performed using low-cost sensors that require contact [13, 14] or remote monitoring through infrared radiation, which are more expensive. There is also the possibility of using infrared thermal (IRT) imaging, which can measure the temperature of large areas of skin and provides a permanent record that can be used for further analysis, this technique has proved to be of importance either in aiding diagnosis through identifying risky areas (of neuropathic and vascular origin) [15] or through assess the outcome of a treatment [16].

In order to objectively assess DFU healing imaging techniques in visual spectrum can offer real estimation of area, volume, shape, and color of the wound [17, 18] providing a better characterization, along with the Pressure Ulcer Scale for Healing (PUSH) [19–21], already used for chronic wounds in clinical setting and accommodates the ulcer volume, the type of existing tissue and the amount of exudate, it can aid and guide clinicians in a better treatment assessment and in case of poor outcome help to choose a different healing procedure.

Given this and with the recent developments in terms of data analysis and DSSs, it is of major importance to understand where the gaps and opportunities are for its implementation, which is the motivation for this research. The novelty and contribution of this piece of research is to identify the weaknesses of the existing DFU DSS and to disclose the required steps for the implementation of an effective DFU DSS system to give an adequate answer to the growing demand of DM patients care.

It is aim of this research to provide a critical overview of the existing decision support systems and publicly available research datasets for diabetic foot ulcers early diagnosis and treatment assessment, and thus proposing a new infrastructure system to deal with it overcoming the past attempts.

This manuscript is organized in five sections. The first, the current one, introduces the problem, the existing DFU clinical diagnostic and treatments assessment processes and the promising complementary emerging methods. It is followed by a characterization of the existing supportive technology in terms of freely available datasets and DSS implemented and mitigated. In section three a proposal of a new effective DFU DSS is made, it is followed by a discussion between what is proposed and the existing implementations, which is remarked with the conclusions.

## 2 Existing Supportive Technology

In this section, the existing available DM datasets are characterized and the mitigated DFU decision support systems are object of critical assessment.

### *2.1 Existing DM Datasets Available for Research*

In Table 4 are described the DM open available datasets that can be used for research, the purpose of its storage, the number of records, and the attributes of each are presented.

It can be easily observed that mostly insulin and blood glucose levels are the only common attributes to all the datasets, this could be explained by the different nature of the original purpose of the dataset, but it is also possible to note that even the more detailed have insufficient data to classify or predict a DFU.

**Table 4** Characterization of the open available DM datasets

| Size | Purpose | Attributes | References |
|---|---|---|---|
| 442 | Group different DM patients per attribute | Age (years), Sex, body mass index, bp (average blood pressure), s1 tc (T-Cells—a type of white blood cells), s2 ldl (low-density lipoproteins), s3 hdl (high-density lipoproteins), s4 tch (thyroid stimulating hormone), s5 ltg (lamotrigine), s6 glu (blood sugar level) | [22] |
| 6742 | Progression of diabetic kidney disease and trajectory of kidney function decline | BMI, End-stage renal disease in type 2 diabetics, Fasting insulin, Chronic kidney disease in type 2 diabetics, Coronary artery disease in type 2 diabetics, Coronary heart disease or stroke or peripheral vascular disease in type 2 diabetics, eGFR-creat (serum creatinine), HDL cholesterol, Height, HOMA-B, Insulinogenic index, LDL cholesterol, Microalbuminuria, Macroalbuminuria vs. controls, Total cholesterol, Triglycerides, Type 2 diabetes, Urinary albumin-to-creatinine ratio, Waist circumference | Jiang et al. [23] |
| 768 | Verify whether patient shows signs of diabetes according to WHO criteria | Number of times pregnant, Plasma glucose concentration 2 h in an oral glucose tolerance test, Diastolic blood pressure, Triceps skin fold thickness, 2 h serum insulin, Body mass index, Diabetes pedigree function, Age (years), Class variable | [24] |
| 11,830 | General patient data for characterization | ID,Description,Affected Status, Product, Source, Gene, Mutations, Sex, Age at Sampling, Race, Ethnicity | [25] |
| 70 | General DM data of country/region/world overtime | Country/Region/World, ISO, Sex, Year, Crude diabetes prevalence, Lower 95% uncertainty interval, Upper 95% uncertainty interval | [26] |
| 45,000 | 70 sets of data recorded on diabetes patients | Date, Time, Code, Value, Regular insulin dose, NPH insulin dose, UltraLente insulin dose, Unspecified blood glucose measurement, Unspecified blood glucose measurement, Pre-breakfast blood glucose measurement, Post-breakfast blood glucose measurement, Pre-lunch blood glucose measurement, Post-lunch blood glucose measurement, Pre-supper blood glucose measurement, Post-supper blood glucose measurement, Pre-snack blood glucose measurement, Hypoglycemic symptoms, Typical meal ingestion, More-than-usual meal ingestion, Less-than-usual meal ingestion, Typical exercise activity, More-than-usual exercise activity, Less-than-usual exercise activity, Unspecified special event | [27] |

(continued)

**Table 4** (continued)

| Size | Purpose | Attributes | References |
|---|---|---|---|
| 1688 | Demographic information, diabetes status, health indicators, health behaviors, and Problem | class, class_language, age, year, gender, insurance_category, medical_home_category, race_ethnicity, education_level, diabetes_status_yes_no, heart_disease_yes_no, high_blood_pressure_yes_no, tobacco_use_yes_no, previous_diabetes_education_yes_no, diabetes_knowledge, fruits_vegetable_consumption, sugar_sweetened_beverage_consumption, food_measurement, carbohydrate_counting, exercise, problem_area_in_diabetes_paid_scale_score, zip_code_address, zip_code_city, zip_code_state, zip_code_zip | [28] |
| 225 | Verify whether the use of continuous glucose monitoring (CGM) without blood glucose monitoring (BGM) measurements is as safe and effective | Blood Glucose, Glycated hemoglobin level, Insulin, Medications, Patient Demographics | Aleppo [29] |
| 451 | Test CGM as a technology to assist in diabetes care | Glycated hemoglobin level, Insulin, Patient Demographics | Tamborlane [30] |
| 200 | Identify factors associated with severe hypoglycemia in older adults (60+) with type 1 diabetes | Insulin, Medications, Patient Demographics | Weinstock [31] |

## 2.2 *Existing DSS for DFU Diagnostic and Treatment Assessment Methods*

In Table 5 are present the existing decision support systems for DFU, divided by type of DSS, informing the purpose, the used data, and the achieved assessment results. It is important to note that ANN refers to Artificial Neural Networks, acc to accuracy, CNN to Conversional Neural Networks, LDA to Local Discriminant Analysis, SVM to Support Vector Machines, k-NN to k-Nearest Neighbor, GA to Genetic Algorithms, and AUC to Area Under the Curve.

From Table 5, it can be seen that the majority of the DFU DSS are concerned with the classification of the data, only one was applied to risk identification [32] and another to the prediction of amputation [33]. Apart from [44] most of the samples used are very small (<2500). Only four DSS assess the treatment outcome [35, 36,

**Table 5** Characterization of the existing DFU for diagnostic and treatment assessment

| Type | Purpose | Used data | Assessment result | References |
|---|---|---|---|---|
| Risk analysis | Predicting the risk of DFU | 255 samples of genotype data | ANN: 83% acc | Singh et al. [32] |
| Prediction | Predicting the amputation in DM patients with record of DFU | 237 samples of amputation data | Logistic regression and Bayesian network AUC 0.765 | Hüsers et al. [33] |
| Classification | Identify the extent of risk factors for major and minor amputations in patients with DFU | 2321 samples of historical records of DM patient's medical history | Decision tree, validated the international guidelines | Widatalla et al. [6] |
| Classification | Feature extraction and consequent classification into ischemic and non-ischemic or infection and non-infection | 1459 DFU images | Ensemble CNN, 90.3% acc ischemic 72.7% acc infection | McInnes et al. [34] |
| Classification | Classification of wound status | 1000 samples of historical records of DM patient's medical history | logistic regression classifier, error of prediction of 2.8% | Yuan et al. [35] |
| Classification | Classification of microbial species from DFU | 1750 samples | LDA+SVM 99.8% acc | Yusuf et al. [36] |
| Classification | Classify correctly the locations of the DFU in the thermal images | 39 dynamic infrared images of DFU patients | SVM 87.5% acc | Vardasca et al. [37] |
| Classification | Automatic classification of the DFU according to their nature | 54 infrared images of DFU patients | k-NN 93.4% acc | Vardasca et al. [38] |
| Classification | Early detection of DFU considering asymmetry | 100 thermal images of DM patients | Otsu's method and GA, misclassification error of 0.034 | Kaabouch et al. [39] |
| Classification | Classify into healthy and abnormal skin | 754 foot images | SVM 94.5% acc | Alzubaidi et al. [40] |
| Classification | Classify pressure into healthy controls, DM controls and DM neuropathic | 84 samples of dynamic plantar pressure | SVM 96.4% acc | Botros et al. [41] |

(continued)

**Table 5** (continued)

| Type | Purpose | Used data | Assessment result | References |
|---|---|---|---|---|
| Classification | Classify into normal and abnormal skin | 397 foot images | CNN 94.5% acc | Goyal et al. [42] |
| Classification | Localization of the DFU in the images | 1880 foot images | CNN 91.8% | Goyal et al. [43] |
| Classification | Classifying into healing and non-healing wounds | 53,354 samples of demographic and wound information | Gradient Boost Tree models, AUC 0.842 | Jung et al. [44] |
| Classification | Wound classification into tissue types | 50 wound images | SVM 88% acc | Wannous et al. [45] |
| Classification | Separate background from feet and identify correctly the DFU | 76 thermal images of DM patient foot | Clustering k-means, 99.1% acc | Liu et al. [46] |
| Classification | Separate background from feet and identify correctly the DFU | 26 visual spectrum and thermal images | Clustering k-means, 98.25% | Niri et al. [47] |
| Classification | Classify the risk of non-healing in DFUs | Hyperspectral images | PCA 0.66 AUC | Yang et al. [48] |

44, 45], most are concerned with the diagnosis [6, 32–34, 37–42, 48] and other three with the correct location of the DFU in images [43, 46, 47]. The classification method that presents most application is the SVM, good accuracy (acc > 92%) is reported by most of the studies [36, 38, 40–42, 46, 47].

## 3 Proposed Intelligent Effective DSS for Diagnostic and Treatment Assessment Methods

From the previous section, it can be seen that the available datasets are insufficient for effectively aiding with DFU early diagnosis and treatment assessment, there are relevant data missing. Also, the existing DSS do not take into account the current IWGDF guidelines, which is the current clinical daily practice. Most of the used data by the previously implemented DSS was generated and read for the application, being small and dedicated.

In order to address the problematic, it is important to reformulate the data sources and make sure that is possible to achieve all the required data. The DM patient historic data is present on his electronic patient record (EPR), every time a patient undergoes a expert examination because of a suspicious DFU, all the collected data on that consult should be stored in a regular consult DB, which can be linked with the EPR or not. Most patients nowadays use gadgets to monitor their health parameters, in special DM patients must verify their blood glucose level twice a day and they can record it along other parameters in an app in their smartphones with the possibility of having a cloud database to store that data. When they have a DFU that requires treatment, the general practitioner should document the intervention on a database. If any of the promising methods that were described in Sect. 1.3 are available, they can also be documented and stored in a database, it is important that the biomarkers due to its nature to be stored in a different data source. All this data stored in different sources is a challenge to a DSS implementation, it requires remote access to it and a lot of ETL and data preparation, in order to this be transformed and loaded into different data marts that will constitute the DFU data warehouse. Then intelligence will be added through Online Analytic Processing (OLAP) and Data Mining methods (which can be object of further research) to finally make knowledge visible in a dashboard through intelligent and simplified charts. This is the basis for the architecture proposed and present in Fig. 1.

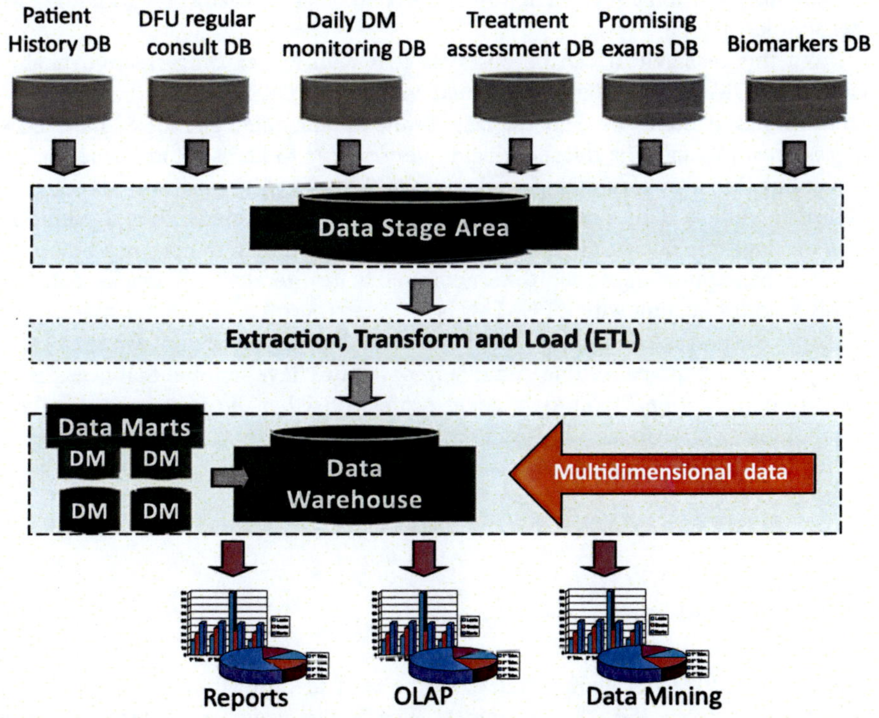


**Fig. 1** The architecture of the proposed effective DSS for DFU diagnosis and treatment assessment

It is important to bear in mind that in the beginning there are two situations: we can be looking for the unknown and then unsupervised methods such as clustering, or aggregation rules should be used to implement the data mining models. On the other hand, if we have a hint and previous knowledge about the classification that we intend, depending on the data (continuous or discrete) and on how many classes are expected, SVM, ANN, k-NN, GA among other methods should be used. When trying to establish a prediction linear or logistic regression should be considered. It is not the intention of this research to propose combinations of data mining methods or suggesting models, or even suggesting implementation technologies, it would only make sense when the data sources were adequately available.

## 4 Discussion

The increase of DM and consequent DFU is a growing problem and burden to every health system, soon the demand will be so high that the health resource will find difficulties to cope with the care requirements, intelligent systems are needed to ease that scenario and to help reducing the associated costs. Preventing will always be cheaper than healing.

From this research, it can be observed that with the existing freely available datasets on DM it cannot be implemented an effective DSS for DFU diagnosis and treatment assessment. Also, the majority of the implemented and tested DFU DSS only addressed a specific question, being very simple to be of future value for the demanding. Most are resultant of academic research and will be difficult to be implemented in clinical daily practice, so the proposed open architecture is dynamic in allowing multiple questions, which through OLAP and data mining is possible to adjust to the required question. It was not found in the literature any implementation of this kind for dealing with DFU. This is first step toward a technological need. To date from the presented existing DSS it seems that developers are not aware of DFU daily practice in clinical setting and it is important to have in mind that is easier to adapt the computational tools to the health professional than the opposite, due to their afraid of losing a job to a smart system the show resistance in this systems adoption, so they have to feel that these tools are useful and only will act as facilitator in their practice.

## 5 Conclusion

The existing DFU DSS are insufficient for the early identification of feet at risk and to assess objectively the DFU treatments. The public available datasets are not in agreement with the current clinical information process required to make decisions nor with the emerging technologies and their objective data. A redefinition of new data sources in line with the IWGDF protocol is required, this along with new emerging related data sources would constitute DM to form a larger DW are essential for more effective Data mining, data analysis, OLAP and Visual dashboards to aid effective DFU decision making.

Only with the implementation of a system as suggested it will definitely contribute to cope with the patient demands, associated cost reduction and promotion of patients care.

## References


1. International Diabetes Federation: IDF diabetes atlas, 9th edn. International Diabetes Federation, Brussels
2. Chatwin KE, Abbott CA, Boulton AJ, Bowling FL, Reeves ND (2020) The role of foot pressure measurement in the prediction and prevention of diabetic foot ulceration—a comprehensive review. Diabet Metabol Res Rev 36(4):e3258
3. Schreml S, Berneburg M (2017) The global burden of diabetic wounds. Br J Dermatol 176(4):845–846
4. Monteiro-Soares M, Russell D, Boyko EJ, Jeffcoate W, Mills JL, Morbach S et al (2020) International working group on the diabetic foot (IWGDF) Guidelines on the classification of diabetic foot ulcers (IWGDF 2019). Diabet Metabol Res Rev 36:e3273
5. Zhou L, Sordo M (2021) Expert systems in medicine. In: Artificial intelligence in medicine, pp 75–100. Academic Press
6. Widatalla AH, Mahadi SEI, Shawer MA, Elsayem HA, Ahmed ME (2009) Implementation of diabetic foot ulcer classification system for research purposes to predict lower extremity amputation. Int J Diabet Dev Countr 29(1):1
7. Wagner FW Jr (1981) The dysvascular foot: a system for diagnosis and treatment. Foot Ankle 2(2):64–122
8. Wagner FW Jr (1986) The diabetic foot and amputation of the foot. In: Mann RA (ed) Surgery of the Foot, 5th edn, pp 421–455. Mosby, St Louis
9. Lavery LA, Armstrong DG, Harkless LB (1996) Classification of diabetic foot wounds. J Foot Ankle Surg 35(6):528–531
10. Scarton A, Guiotto A, Malaquias T, Spolaor F, Sinigaglia G, Cobelli C, Sawacha Z (2018) A methodological framework for detecting ulcers' risk in diabetic foot subjects by combining gait analysis, a new musculoskeletal foot model and a foot finite element model. Gait Posture 60:279–285
11. Wang Y, Shao T, Wang J, Huang X, Deng X, Cao Y, Zhao C et al (2020) An update on potential biomarkers for diagnosing diabetic foot ulcer at early stage. Biomed Pharmacother 133:110991
12. Wijlens AM, Holloway S, Bus SA, van Netten JJ (2017) An explorative study on the validity of various definitions of a 2· 2 C temperature threshold as warning signal for impending diabetic foot ulceration. Int Wound J 14(6):1346–1351
13. Bus SA (2016) Innovations in plantar pressure and foot temperature measurements in diabetes. Diabetes Metab Res Rev 32:221–226
14. Martín-Vaquero J, Hernández Encinas A, Queiruga-Dios A, José Bullón J, Martínez-Nova A, Torreblanca González J, Bullón-Carbajo C (2019) Review on wearables to monitor foot temperature in diabetic patients. Sensors 19(4):776
15. Gatt A, Falzon O, Cassar K, Ellul C, Camilleri KP, Gauci J, Chockalingam N (2018) Establishing differences in thermographic patterns between the various complications in diabetic foot disease. Int J Endocrinol
16. Petrova NL, Donaldson NK, Tang W, MacDonald A, Allen J, Lomas C, Kluwe B (2020) Infrared thermography and ulcer prevention in the high-risk diabetic foot: data from a single-blind multicentre controlled clinical trial. Diabet Med 37(1):95–104

17. Mukherjee R, Tewary S, Routray A (2017) Diagnostic and prognostic utility of non-invasive multimodal imaging in chronic wound monitoring: a systematic review. J Med Syst 41(3):46
18. Frade RA, Vardasca R, Carvalho R, Mendes J (2017) Automatic classification of ulcers through visual spectrum image. In: European congress on computational methods in applied sciences and engineering, pp 297–305. Springer, Cham
19. Cuddigan J (1997) Pressure ulcer classification: What do we have? What do we need? Adv Wound Care 10(5):13–15
20. Stotts NA, Rodeheaver GT, Thomas DR, Frantz RA, Bartolucci AA, Sussman C, MaklebustAn J (2001) Instrument to measure healing in pressure ulcers: development and validation of the pressure ulcer scale for healing (PUSH). J Gerontol A Biol Sci Med Sci 56(12):795–799
21. Thomas DR, Rodeheaver GT, Bartolucci AA, Franz RA, Sussman C, Ferrell BA, Maklebust J (1997) Pressure ulcer scale for healing: derivation and validation of the PUSH tool. Adv Wound Care 10(5):96–101
22. Trevor Hastie's Home Page. https://www4.stat.ncsu.edu/~boos/var.select/diabetes.html. Accessed 29 Nov 2020
23. Jiang G, Luk AOY, Tam CHT, Xie F, Carstensen B, Lau ESH, Ozaki R (2019) Progression of diabetic kidney disease and trajectory of kidney function decline in Chinese patients with type 2 diabetes. Kidney Int 95(1):178–187
24. Pima Indians Diabetes Database. https://datahub.io/machine-learning/diabetes#readme. Accessed 29 Nov 2020
25. Coriell Institute for Medical Research databank. https://www.coriell.org/Search?q=DIABETES. Accessed 29 Nov 2020
26. NCD Risk Factor Collaboration (NCD-RisC) evolution of Diabetes overtime. https://ncdrisc.org/data-downloads-diabetes.html. Accessed 29 Nov 2020
27. University of California Machine Learning Repository—Diabetes dataset, https://archive.ics.uci.edu/ml/datasets/diabetes. Accessed 29 Nov 2020
28. Austin Public Health Diabetes Self-Management Education Participant Demographics 2015–2017. https://data.world/cityofaustin/48iy-4sbg. Accessed 29 Nov 2020
29. Aleppo (2017) Diabetes dataset. https://github.com/irinagain/Awesome-CGM/wiki/Aleppo. Accessed 29 Nov 2020
30. Tamborlane (2008) diabetes dataset. https://github.com/irinagain/Awesome-CGM/wiki/Tamborlane. Accessed 29 Nov 2020
31. Weinstock (2016) diabetes database. https://github.com/irinagain/Awesome-CGM/wiki/Weinstock. Accessed 29 Nov 2020
32. Singh K, Singh VK, Agrawal NK, Gupta SK, Singh K (2013) Association of Toll-like receptor 4 polymorphisms with diabetic foot ulcers and application of artificial neural network in DFU risk assessment in type 2 diabetes patients. BioMed Res Int
33. Hüsers J, Hafer G, Heggemann J, Wiemeyer S, John SM, Hübner UH (2020) Predicting the amputation risk for patients with diabetic foot ulceration–a bayesian decision support tool. BMC Med Inform Decis Making20
34. McInnes MD, Moher D, Thombs BD, McGrath TA, Bossuyt PM, Clifford T, Hunt HA et al (2018) Preferred reporting items for a systematic review and meta-analysis of diagnostic test accuracy studies: the PRISMA-DTA statement. Jama 319(4):388–396
35. Yuan Z, Huang J, Zhao Z, Zahid A, Heidari H, Ghannam R, Abbasi QH (2018) A compact wearable system for detection and estimation of open wound status in diabetic patient. In: 2018 IEEE Asia Pacific conference on postgraduate research in microelectronics and electronics, pp 60–63
36. Yusuf N, Zakaria A, Omar MI, Shakaff AYM, Masnan MJ, Kamarudin LM, Yasin MS (2015) In-vitro diagnosis of single and poly microbial species targeted for diabetic foot infection using e-nose technology. BMC Bioinform 16(1):158
37. Vardasca R, Magalhaes C, Seixas A, Carvalho R, Mendes J (2019) Diabetic foot monitoring using dynamic thermography and AI classifiers. In: Proceedings of the 3rd quantitative infrared thermography Asia conference (QIRT Asia 2019), Tokyo, Japan, pp 1–5

38. Vardasca R, Vaz L, Magalhães C, Seixas A, Mendes J (2018) Towards the diabetic foot ulcers classification with infrared thermal images. In: 14th Quantitative infrared thermography conference, Berlin, Germany
39. Kaabouch N, Hu WC, Chen Y, Anderson JW, Ames F, Paulson R (2010) Predicting neuropathic ulceration: analysis of static temperature distributions in thermal images. J Biomed Opt 15(6):061715
40. Alzubaidi L, Fadhel MA, Oleiwi SR, Al-Shamma O, Zhang J (2020) DFU_QUTNet: diabetic foot ulcer classification using novel deep convolutional neural network. Multimedia Tools Appl 79(21):15655–15677
41. Botros FS, Taher MF, ElSayed NM, Fahmy AS (2016) Prediction of diabetic foot ulceration using spatial and temporal dynamic plantar pressure. In: 2016 8th Cairo international biomedical engineering conference (CIBEC), pp 43–47
42. Goyal M, Reeves ND, Davison AK, Rajbhandari S, Spragg J, Yap MH (2018) Dfunet: Convolutional neural networks for diabetic foot ulcer classification. IEEE Trans Emerg Topics Comput Intell
43. Goyal M, Reeves ND, Rajbhandari S, Yap MH (2018) Robust methods for real-time diabetic foot ulcer detection and localization on mobile devices. IEEE J Biomed Health Inform 23(4):1730–1741
44. Jung K, Covington S, Sen CK, Januszyk M, Kirsner RS, Gurtner GC, Shah NH (2016) Rapid identification of slow healing wounds. Wound Repair Regener 24(1):181–188
45. Wannous H, Lucas Y, Treuillet S (2010) Enhanced assessment of the wound-healing process by accurate multiview tissue classification. IEEE Trans Med Imaging 30(2):315–326
46. Liu C, van Netten JJ, Van Baal JG, Bus SA, van Der Heijden F (2015) Automatic detection of diabetic foot complications with infrared thermography by asymmetric analysis. J Biomed Opt 20(2):026003
47. Niri R, Lucas Y, Treuillet S, Douzi H (2019) Smartphone-based thermal imaging system for diabetic foot ulcer assessment. Sorbonne Universités, Paris, France, Journées d'Etude sur la TéléSanté
48. Yang Q, Sun S, Jeffcoate WJ, Clark DJ, Musgove A, Game FL, Morgan SP (2018) Investigation of the performance of hyperspectral imaging by principal component analysis in the prediction of healing of diabetic foot ulcers. J Imaging 4(12):144

# Open OnDemand as a Platform for Virtual Learning in Higher Education

**Robert E. Settlage, Alan Chalker, Jeff Ohrstrom, Eric Franz, Doug Johnson, and David Hudak**

**Abstract** The need for distance and virtual learning platforms has been emphasized by the COVID-19 pandemic. With the closure of campuses in Spring of 2020 and many classes moving to online only in Fall of 2020, platforms for facilitating computationally oriented curriculum have had to be quickly adopted. Open OnDemand offers a familiar web-based portal to computational resources such as high-performance computing and cloud. Through OnDemands customizable dashboard, students can be offered an interface tailored to the course schedule giving them a just what I need view. Advantages to instructors include a web accessible, platform agnostic interface leading to less time for troubleshooting local student platforms and more time for discussion of the core course curriculum, a fully customizable course page, access controls, and more. Here we present Open OnDemand as a platform for developing, deploying, and presenting software and course material to software-oriented classes as used at Ohio Supercomputer Center and Virginia Tech.

**Keywords** Open OnDemand · Education · HPC · Virtual learning · Distance learning

## 1 Introduction: Virtual Learning Challenges and Opportunities

Education in the Arts and Sciences has been dominated by traditional teacher led in-classroom education in higher education institutions. Through the needs dictated by mitigating the COVID-19 global pandemic, i.e., social and physical distancing, classes have transitioned from in-class to online. This transition has highlighted gaps and opportunities in our approaches, physical infrastructure, and even our curricu-

R. E. Settlage (✉)
Virginia Tech, Blacksburg, VA 24061, USA
e-mail: rsettlag@vt.edu

A. Chalker · J. Ohrstrom · E. Franz · D. Johnson · D. Hudak
Ohio Supercomputer Center, Columbus, OH 43212, USA

 
X.-S. Yang et al. (eds.), *Proceedings of Sixth International Congress on Information and Communication Technology*, Lecture Notes in Networks and Systems 216,
https://doi.org/10.1007/978-981-16-1781-2_30

lum. This is a discussion of some of the issues in offering virtual learning labs and successes we have found in our approach of using Open OnDemand to ease the transition in classes involving computational/software resources. From a educators point of view, many challenges revolve around loss (or dampening) of the interaction between people: instructor–student and student–student. While we can't remove these issues, we can positively affect the instructor, student, and administrators experiences as we describe here.

## 2 Open OnDemand as a Virtual Learning Portal

Open OnDemand [1] provides familiar, site configurable, and extensible web-based access to high-performance computing (HPC) and cloud resources. The use of web-based tools and web forms to request computing and software resources makes traditionally difficult to use computing environments fully accessible to even the most novice users. While providing familiar tools to use HPC/cloud resources, Open OnDemand also provides methods to customize the landing page, dashboard, to include site branding, announcements or messages of the day, etc., to allow creation of fit for purpose portals to HPC and cloud. Additional customizations can include both the tools users have access to and the amount/type of access they have (see [2] for the discussion of the state of the project). In combination, institutions and instructors are able to create class based portals to HPC/cloud resources customized to instruction at the site as a whole and to individual classes.

### *2.1 Center Branded Course Site*

Customizing the branding and dashboard landing page is a core feature of Open OnDemand. Through this feature, sites are able to display announcements, messages of the day, system utilization metrics, and more. An extension of this feature is that, sites can create instruction oriented Open OnDemand portals. Figure 1 shows the Virtual Learning Dashboard at the Ohio Supercomputer Center (OSC). Through this dashboard, students are able to access course materials and apps specific to their study.

### *2.2 Extensibility*

Open OnDemand has a plug-in based architecture allowing for creation of *custom* applications. These custom applications are, at their core, form based HPC/cloud job submission requests. These can be as simple as a form to start RStudio or Jupyter, or more complex to start a compute pipeline. For our purposes in Virtual Learning

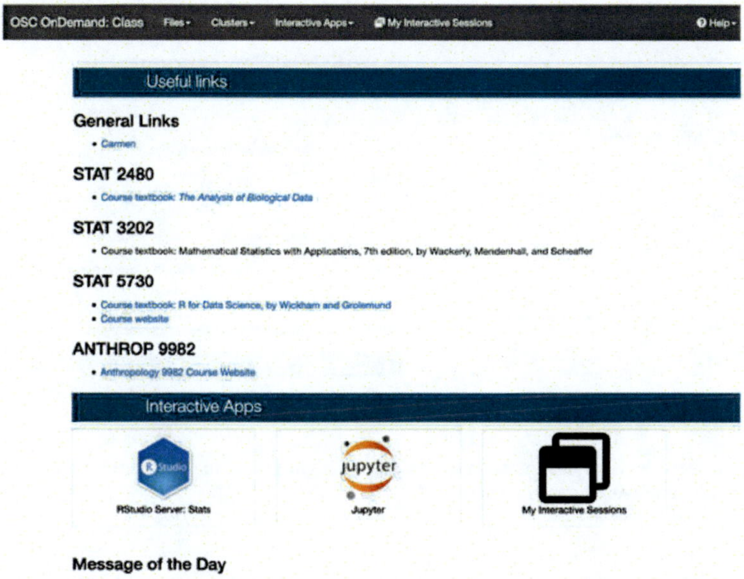


**Fig. 1** Course oriented dashboard at OSC

Labs, we are creating forms to start instances of software specific to a class with job characteristics suitable to that class. As an example, see Fig. 2. For this introductory bio statistics class at Virginia Tech (VT), we have predetermined what a student will receive in terms of memory and cores, limited the length of an interactive session, and only exposed the form fields necessary to start a session. This fine control allows for more flexible instruction. For sessions where parallel computing may be used, we could change the resource request to allow for multi-core functions. For sessions where memory will be a component, again, we can size the backend resources as appropriate. Note, we could have exposed these fields in the web form or we could have simplified the submission form further limiting the account and partition students have access to, but chose to leave these as topics for discussion of computing on large heterogeneous systems as a teaser. These are choices instructors can make to suit the needs of the class.

## *2.3 App Access Control*

"A key feature of OnDemand over a traditional web service is the per-user web server model where all the web applications are run as the Linux user" more fully described in [3]. In contrast to many web services, OnDemand leverages the Linux kernel for security and accounting which means server side security models and access control

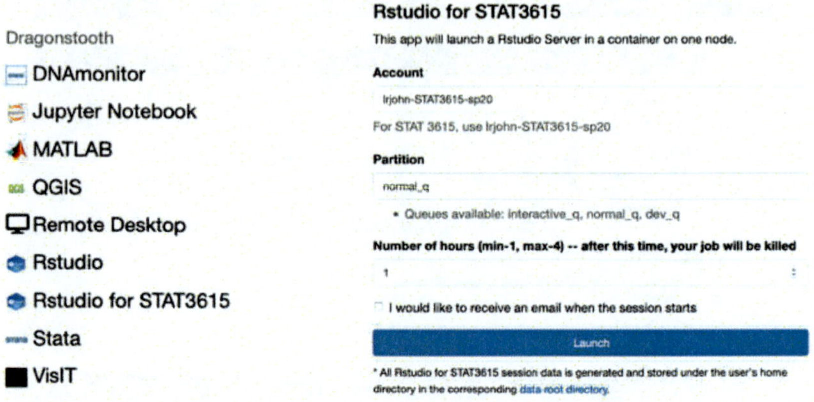


**Fig. 2** Cluster specific OnDemand interactive apps highlighting the presence of a class app and the simplicity a job request form can take

are honored. Operationally, access to specific apps is controlled through normal Linux group membership because the web service is running as the user. Individual apps are deployed on the server within a folder specific to that app. Permissions can be applied to that folder to limit access via group memberships. Users only see apps they have permissions to access because the web service accessing files within the various app folders using the user permissions. This creates an easy environment to maintain; all access is controlled through group memberships in the normal way! In Fig. 2, above, the menu item for the class app is only visible to members of the stat3615 group.

## 3 Open OnDemand Use Cases

Instructional use of Open OnDemand had started prior to COVID-19. However, COVID-19 has stressed the need for stable, consistent, and scalable infrastructure for teaching. Below, we highlight the impacts at Virginia Tech and across Ohio via OSC.

### *3.1 Intro to HPC Computing*

The Advanced Research Computing group at Virginia Tech is often asked to give an introductory "Intro to HPC" lecture in classes such as statistics, math, and engineering. Before Open OnDemand, a typical "Intro to University HPC Resources" lecture would be consumed by platform driven connection issues. "From a Mac, use the terminal, SSH to ..., from a Windows machine, download and install an SSH client (e.g., Putty, MobaXterm, etc.), SSH to ... . Now, if you are configured for 2FA with your phone, have the Duo app ready, and go ... ." For a class of approximately 40 students, getting the students logged in to the cluster was a frustrating exercise often consuming the majority of a 1 h lecture. Open OnDemand changes the discussion to: "using a browser, navigate to *ondemand.<site>.edu*, log in as you would any other university resource, now click the pulldown for cluster access". In most cases, students are logged into the cluster in 5 minutes, leaving the majority of the hour to discuss topics of interest to the class subject.

### 3.2 *Enabling Instruction Across Ohio*

The Ohio Supercomputer Center is a statewide resource that provides super computing services and computational science expertise to Ohio university researchers as well as Ohio industries. With a goal of leading science and engineering research efforts, education and training potential users is an important component of the OSC mission. Students across Ohio can gain exposure to HPC through a variety of opportunities including summer workshops and research events and now class instruction via Open OnDemand. Figure 3 and Table 1 show the reach of the OSC Open OnDemand education of students across Ohio with nearly 1900 students across 14 universities.

Examples of courses utilizing OSC's OnDemand include: Kent State's College of Architecture and Environmental Design ARCH 60103 provided students access to Maya; OSU Physics 6820 provided students a custom Python programming environment using Jupyter through OnDemand; College of Public Health PUBHTLH 5015

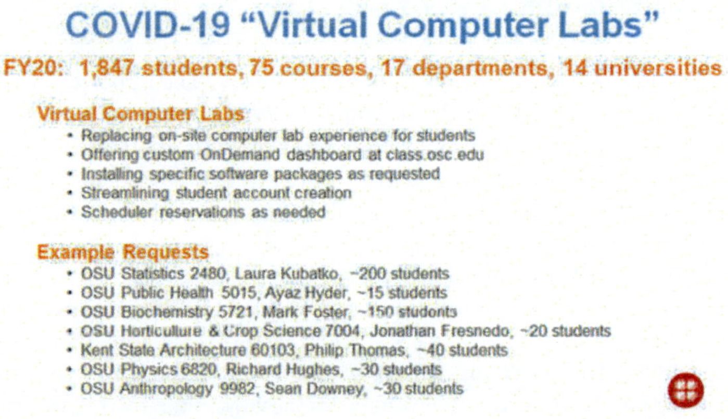


**Fig. 3** Distribution of students accessing OSC resources via OnDemand across Ohio

**Table 1** Snapshot of student distribution. FY2020: 1,847 students, 75 courses, 17 departments, 14 universities

| Institution | Students | Courses | Departments |
|---|---|---|---|
| Bowling Green State University | 10 | 2 | 2 |
| Denison University | 3 | 1 | 1 |
| Kent State University | 23 | 6 | 2 |
| Kenyon College | 18 | 2 | 1 |
| Miami University | 33 | 2 | 2 |
| Mount Union | 8 | 2 | 1 |
| Ohio State University | 1,422 | 46 | 14 |
| Ohio University | 3 | 1 | 1 |
| Stark State College | 17 | 1 | 1 |
| University of Cincinnati | 285 | 8 | 3 |
| University of Dayton | 10 | 1 | 1 |
| Urbana University | 2 | 1 | 1 |
| Wittenberg University | 7 | 1 | 1 |
| Xavier University | 6 | 1 | 1 |

utilized a custom R environment through RStudio via OnDemand; OSU Department of Horticulture and Crop Science's HCS 7004 Genome Analytics course utilized a a custom R environment through RStudio via OnDemand. There are many other courses across many disciplines, including Business Analytics, Bio Chemistry, Statistics, and Anthropology.

## 4 Opportunities and Lessons Learned

While challenging, virtual learning offers opportunities. The opportunities span the gamut from cost savings, potentials to improve the student/instructor experience, modernize the instruction for methods more aligned with real-world problems, and opportunities to scale the classroom to meet increases in demands for education. We highlight several of these below:

**Remote Troubleshooting** One of the hallmarks of instruction in classes involving software and computing are issues surrounding platform issues. As an example, consider RStudio/R for instruction in Statistics classes. Inevitably, immediately following the lecture demonstrating RStudio/R installation and use, student need for troubleshooting platform installation issues effectively monopolizes the instructors office hours for the next 1–2 weeks. Issues include everything from conflicts with previous installs of $LaTeX$, OS language problems (underlying OS was installed as

Chinese version for instance), or simply poorly optimized installs missing links to BLAS or other libraries designed to improve performance. As an instructor, these issues are difficult to troubleshoot in person. Moving this to a remote troubleshooting session tests the ability of both student and instructor in both accurately describing the solution and succinctly describing instructions to fix the problem.

Using a web accessible University resource such as Open OnDemand backed by HPC or Cloud removes most, if not all platform issues. Plus or minus the browser, the software is the same for all students in the class. At Virginia Tech, on our HPC systems, instructors and students are equivalent in terms of access rights and permissions. This means troubleshooting a student issue is reduced to repeating the problem in an identical environment.

**Cost Savings** Moving instruction to virtual modes offers potential savings. Many Universities have computer laboratories. The labs are populated with physical computer workstations, often using some sort of virtualization and networked software licensing, hosting software in common courses such as STATA, R, Matlab, and more. To make sure there are enough workstations, the labs are slightly oversized, but there are still workstation availability bottlenecks at key points throughout a semester. Moving these workstations to a virtual lab giving students access to the software and computing resources through web-based access obviates the need for monitors and even a physical lab. Further, depending on how the virtualization is done and what hardware backs the virtual workstations, there could be opportunities to share hardware that were not possible when a physical keyboard was a limiting device. Frequently, the software use within an introductory course is very minimal from a computational standpoint suggesting high over-subscription rates are possible and use of the physical hardware can continue will past the depreciation period without degradation in user experience [4]. In many cases, both hardware and software can be shared across different uses, such as research and education.

**Improve the Student Experience** Many classes use scientific software to enhance the learning experience. However, the computer is more of a means to an end rather than an end in and of itself. For example, Statistics in Biology may have as the primary goal foundational topics around probability, testing, and prediction. The computer is used to reinforce the learning objectives and begin discussions about data. Local platform issues are really a distraction to the learner. Removing these barriers, at least in introductory courses, removes the distraction, and returns the focus of the student to the class topic. Importantly, virtualization should create a unified student experience.

**Improve the Instructor Experience** While errors and other installation/configu ration issues are often the bane of the first couple of weeks of a course using computational tools, there are other platform oddities that present hurdles to instructors. Consider Microsoft Office. Differences in the interface on Mac and PC versions cause constant frustrations during follow-along lectures. Menu items, tool bars, etc., are different even in the same tool! In some instances, platform differences are desirable, in others, they are frustrating obstacles.

**Reduce Administrative Load** In any computing oriented class, support is a key resource often delivered through a combination of instructor and system admin/computational scientist office hours. Open OnDemand reduces administrative support through pinning workflows to successful job submissions. Additionally, the additional users flowing through classes are administrated the same as normal users in terms of access controls, account allocations, etc. In effect, students are normal HPC users with an intuitive interface. Through interface design, instructors are able to provide contextual help and guide students through use and problems where they would otherwise have been left with a blinking cursor.

**Scaling** Modern instruction in just about every discipline is impacted by the current data deluge. In fact, current estimates by the Visual Capitalist suggest we will generate 463 Exabytes of data everyday [5]. To make sense of this data, we need to instruct students in modern data analysis methods. Open OnDemand facilitates the transition from novice to big data analyst by providing a common platform for both computing paradigms. The underlying infrastructure is the same, the interface can telescope based on user needs and access privileges.

**Limitations of OnDemand for Classroom Support** Providing a large icon launch interface for RStudio and Jupyter on the OSC course oriented dashboard, as shown in Fig. 1, required technical expertise in adding custom HTML to the localization files for the dashboard to include these links. OnDemand 2.0 plans to make this type of customized user or group tailored launch interface easier to provide.

While OnDemand provides a files app for uploading Jupyter notebooks from a student's laptop, professors asked if they could provide students the ability to save and load Jupyter notebooks from internet accessible storage environments such as OneDrive, Google Drive, or GitHub. Without these capabilities, students using tablets for the class may have a difficult time participating if requisite files are not uploaded before hand by the professor to a shared location on the file system. Furthermore, professors and TAs still need access to students completed assignments in the form of Jupyter notebooks in order to grade them. Students need to be able to submit completed assignments to their professor without inadvertently sharing those completed assignments with other students. Ideally, Students would be able to submit the completed assignment through the web interface, completely bypassing the need to first download the notebook to their device. Integrating OnDemand with GitHub Classroom or Canvas LMS may provide an optimal solution to these challenges.

## 5 The Future of Digital Learning

The future of learning is digital. As a society, nothing has more impact on our lives than the computer has now. To make sense of the unimaginable amount of data we are generating everyday, we have a need for computational resources. Educating our workforce in tools necessary to transform, analyze, and extract knowledge from the information we are generating will require digital proficiency. Open OnDemand, as a portal tool set, enables students and instructors alike, provides a familiar framework for accessing computational power necessary for today's problems, and reduces administrative loads while allowing scaling for class demands and problem sets.

**Acknowledgements** National Science Foundation of the United States Awards NSF SI2-SSE-1534949 and CSSI-Frameworks-1835725.

## References

1. Hudak D, Johnson D, Chalker A, Nicklas J, Franz E, Dockendorf T, McMichael B (2018) Open OnDemand: a web-based client portal for HPC centers. J Open Source Softw. https://doi.org/10.21105/joss.00622
2. Chalker A, Franz E, Rodgers M, Dockendorf T, Johnson D, Sajdak D, White JP, Plessinger BD, Zia M, Gallo SM, Settlage RE (2020) Concurrency and computation: practice and experience. https://doi.org/10.1002/cpe.6114
3. Hudak D, Johnson D, Franz E, Baer T, Dockendorf T, Cahill K (2018) http://orcid.org/10.6084/m9.figshare.7069691.v1. https://figshare.com/articles/Open_OnDemand_Access_Clusters_Gateways_and_Interactive_Apps/7069691
4. Franz E, Ozer HG, Dockendorf T, Gadepalli VS, Webb A, Chalker A, Pietrzak M, Rodgers M, Johnson D, Hudak DE (2019) In: Proceedings of the practice and experience in advanced research computing on rise of the machines. https://doi.org/10.1145/3332186.3332211
5. Ronconteur R (2019) Data. https://www.visualcapitalist.com/wp-content/uploads/2019/04/data-generated-each-day-wide.html

# Performance Evaluation of Hybrid Queuing Algorithms for QoS Provision Based on DiffServ Architecture

**Ismael S. Pesántez-Romero, Gabriel E. Pulla-Lojano, Luis F. Guerrero-Vásquez, Edwin J. Coronel-González, Jorge O. Ordoñez-Ordoñez, and Javier E. Martinez-Ledesma**

**Abstract** Based on the Differentiated Services architecture (DiffServ), a network is implemented in a controlled environment for Quality of Service (QoS) tests. A traffic generator is used to generate and transmit test traffic that simulates services and applications. Each service is classified and marked in such a way that the network layer manages the functions of providing a specific QoS treatment, prioritizing services in real time, and allowing non-priority applications to access the network. Subsequently, the network is subjected to congestion situations with overload in different scenarios with QoS policies and the Ethernet Service Activation Methodology (eSAM) performance test is applied. The mechanism used to manage bandwidth is Hierarchical Token Bucket (HTB) and it is combined with queue-based algorithms such as Per Connection Queueing (PCQ), Stochastic Fairness Queueing (SFQ), Random Early Drop (RED), and First-In First-Out (FIFO). The results are contrasted with the performance objectives agreed by the Service Level Agreement (SLA), and finally the feasibility of DiffServ model combined with hybrid queuing algorithms HTB-based to guarantee QoS requirements at the IP level is discussed.

---

I. S. Pesántez-Romero (✉) · G. E. Pulla-Lojano · L. F. Guerrero-Vásquez · E. J. Coronel-González · J. O. Ordoñez-Ordoñez
Universidad Politécnica Salesiana, Cuenca, Ecuador
e-mail: ipesantezr@est.ups.edu.ec

G. E. Pulla-Lojano
e-mail: gpulla@ups.edu.ec

L. F. Guerrero-Vásquez
e-mail: lguerrero@ups.edu.ec

E. J. Coronel-González
e-mail: ecoronel@ups.edu.ec

J. O. Ordoñez-Ordoñez
e-mail: jordonezo@ups.edu.ec

J. E. Martinez-Ledesma
TelecomAustro, Cuenca, Ecuador
e-mail: jmartinez@telecomaustro.com



X.-S. Yang et al. (eds.), *Proceedings of Sixth International Congress on Information and Communication Technology*, Lecture Notes in Networks and Systems 216,
https://doi.org/10.1007/978-981-16-1781-2_31

**Keywords** DiffServ model · QoS · eSAM · Bandwidth management · HTB · Queue algorithms · FIFO · RED · SFQ · PCQ · SLA

## 1 Introduction

Currently, the expansion of Internet access services has made a large percentage of the world's population connected to the large network. However, this growth in the number of users generates technological challenges to Internet Service Providers (ISP), demanding levels of Quality of Service (QoS), which improve the Quality of Experience (QoE) in users. Traffic congestion on the network is one of the biggest problems facing ISPs, due to the increased demand for different heavy services such as real-time and high priority applications [1, 2]. Network quality levels that satisfactorily support the operating requirements for services and applications that use the Internet (IP) protocol are required [3]. IP, as a universal deployment, is not free to present difficulties in the aspect of performance. The problem lies in its packet-based network architecture that cannot guarantee a predetermined capacity or quality permanently, which represents a serious problem when transporting time-sensitive signals [4, 5].

All IP-based networks need the help of additional technologies, protocols, and architectures to establish admission policies, congestion control, and manage different traffic priorities. The ultimate goal is to guarantee QoS and QoE, supporting real-time voice and video applications, and meeting the needs of the service user. On the one hand, QoS consists in providing a special treatment for some type of traffic improving the performance of a given service [6], while QoE refers to the user satisfaction of the network services. Therefore, the QoS parameters such as bandwidth, packet loss, latency, and jitter directly affect the QoE that subscribers perceive of the application offered [7].

Internet Engineering Task Force (IETF) defines two basic models to offer QoS. First model consists of Integrated Services (IntServ) architecture, which is based on the reservation of network resources based on traffic flows. This makes the model not scalable because it implies a high consumption of resources to process data flows [8, 9]. Second model is characterized by prioritization of a certain type of traffic and consists of Differentiated Services (DiffServ) architecture, which improves scalability based on sharing network resources among a set of Traffic Classes (TC) previously defined. Although these QoS models have shown great performance in their tests within laboratories, they have not yet been massively implemented in functional ISPs [10].

A related work is observed in [11], Xu, develops a DiffServ model, mentioning smart queue scheduling for packet forwarding plays an important role in this architecture. In a network with QoS, various queuing techniques are used to control the transmission of traffic. Queuing system for network management consists of different algorithms such as Hierarchical Token Bucket (HTB), Per Connection Queue (PCQ), Stochastic Fairness Queuing (SFQ), Random Early Drop (RED), and First-In Firts-

Out (FIFO); in RouterOS, the queuing system can be set to HTB, PCQ, SFQ, RED, and FIFO. PCQ is a type of non-priority queue and the bandwidth is shared equally among active users. SFQ uses the Hash algorithm to classify the traffic and divide it into a limited number of queues and Round-Robin to distribute the bandwidth. RED is a queuing mechanism that tries to avoid network congestion by controlling the average value of the queue. FIFO runs by creating a simple buffer to hold incoming packets, first packet to arrive is the first packet to be transmitted [12]. The authors [13] detail a study of PCQ and HTB queuing algorithms for packet switching, however, the results demonstrate the advantage of HTB because it achieves to obtain the lowest values of delay, jitter, and packet loss, therefore, HTB is the most appropriate method for networks that have priority levels, allows operate different types of traffic, and also provides bandwidth guarantees.

In this work, the performance evaluation of DiffServ architecture combined with hybrid queue algorithms HTB-PCQ, HTB-SFQ, HTB-RED, and HTB-FIFO is carried out, as mechanisms that allow networks to manage bandwidth, guaranteeing Service Level Agreement (SLA), which specify the performance characteristics of the network in terms of availability, delay, jitter, packet loss, and defines a predetermined level of QoS based on the class of service. Below in this document, in Sect. 2, we present a relevant review of the performance objectives at the IP level, SLA, and QoS over DiffServ used as the design center of our methodology. Later, in Sect. 3, the implemented topology, algorithms for bandwidth management and the evaluation conditions for laboratory tests are shown. And finally, in Sect. 4, the results obtained are discussed and the relevant conclusions are generated.

## 2 Quality of Service

### *2.1 IP Performance Objectives*

The International Telecommunications Union (ITU), in the ITU-T rec. Y.1541 specifies eight application-based QoS classes to define performance objectives in IP networks. ITU-T determines the characteristics of QoS in terms of packet delays and errors for each class. The set of parameters that define QoS at the IP level are:

- IP packet Transfer Delay (IPTD), correspond to end to end latency.
- IP packet Delay Variation (IPDV), correspond to packet delay variation.
- IP packet Loss Ratio (IPLR), correspond to relationship between sent frames and lost response frames.
- IP packet Error Ratio (IPER), correspond to bit error ratio at reception.
- IP packet Reordering Ratio (IPRR), correspond to ratio of incoming packages were out of order.

**Table 1** ITU-T rec. Y.1541: QoS requirements for IP level

| Class | IPTD (ms) | IPDV (ms) | IPLR | IPER | IPRR |
|---|---|---|---|---|---|
| 0 | 100 | 50 | 1 × 10E-3 | 1 × 10E-4 | – |
| 1 | 400 | 50 | 1 × 10E-3 | 1 × 10E-4 | – |
| 2 | 100 | U | 1 × 10E-3 | 1 × 10E-4 | – |
| 3 | 100 | U | 1 × 10E-3 | 1 × 10E-4 | – |
| 4 | 1000 | U | 1 × 10E-3 | 1 × 10E-4 | – |
| 5 | U | U | U | U | – |
| 6 | 100 | 50 | 1 × 10E-5 | 1 × 10E-6 | 1 × 10E-6 |
| 7 | 400 | 50 | 1 × 10E-5 | 1 × 10E-6 | 1 × 10E-6 |

The information shown in the table serves as a reference for QoS levels. However, ISPs may choose to offer performance commitments better than the objectives mentioned in Table 1 [14].

## 2.2 *Service Level Agreement (SLA)*

All traffic on the network can be classified into three types: best effort, real time, and high priority. Each type of traffic is affected differently by the characteristics of the network that should be prepared to meet the minimum performance objectives.

To ensure QoS, ISPs must properly configure their networks to define how internal traffic will be treated. This is achieved by assigning different levels of priority to each type of service and configuring intelligent network congestion algorithms.

SLA is a contract between the service provider and a customer, which guarantees the minimum performance that must be given for the services provided. SLA specifies the forwarding characteristics based on Key Performance Indicators (KPI) based on bandwidth (CIR and EIR), latency, packet loss, and jitter [15, 16].

Compromised Information Rate (CIR), refers to the bandwidth that is guaranteed at all times for a specific service meeting the minimum performance objectives. Excess Information Rate (EIR) refers to the bandwidth that may be available depending on the load and use of the network, without guaranteeing that the minimum performance objectives are met.

## 2.3 *Quality of Service Over DiffServ*

DiffServ architecture is a multiple service model that can meet different QoS requirements. With DiffServ the network tries to deliver a particular type of service based on the specific QoS for each packet. This can occur in different ways, for example,

**Table 2** DSCP values for classes and traffic type

| Traffic type | Traffic class | Service example | DSCP value | PHB |
|---|---|---|---|---|
| Real Time | Premium | Voice/Video | 46 | EF |
| Data | Gold | SNMP | 10 | AF11 |
| | Silver | Telnet | 18 | AF21 |
| | | SMTP | 20 | AF22 |
| | Bronze | HTTP | 26 | AF31 |
| | | FTP | 22 | AF23 |

using 6-bit Differentiated Service Code Point (DSCP), from the IP packet structure, protocols or by source and destination addresses. Thus, the network can take functions of classification and conditioning of traffic such as measurement, marking, and vigilance to provide the appropriate QoS treatment [17].

With the system ability to mark packets, Peer Hop Behavior (PHB) decisions can be made. PHB refers to the programming of packets, intelligent queues, vigilance, and behavior of a node to offer a specific SLA according to the TC [18].

PHB is associated with a PHB Class Selector, which defines the forwarding behavior on all nodes in the DiffServ domain. There are two types of PHB forwarding, the first, Assured Forwarding (AF), which allocates a certain amount of bandwidth, according to the SLA. And the second, Expedited Forwarding (EF), which is used for end to end services of low loss, low latency, low jitter, and guaranteed bandwidth [17].

To provision QoS in DiffServ domain, traffic must be marked with an appropriate DSCP value, in this way, it is possible to classify the type of traffic as shown in Table 2.

Premium class must be forwarded with the lowest possible latency. Gold class should be treated preferentially over the silver class, which, in turn, must have priority over bronze class.

## 3 Implementation

A model is proposed to generate traffic to stress the network and measure performance criteria, monitoring and detecting violation of SLA parameters. DiffServ architecture is applied to establish QoS requirements and hybrid queuing algorithms based on HTB for bandwidth management.

Figure 1 illustrates the proposed model consisting of four basic components for the implementation of Differentiated Services. Classification and marking of packets provide priority levels by service classes that are used to apply QoS. Configuring traffic conditioning (policies and modeling) and Queue Schedule, ensures that packets are adjusted at a specific speed by delaying them and sending them whenever possible, also allows congestion management avoiding network bottlenecks through the use

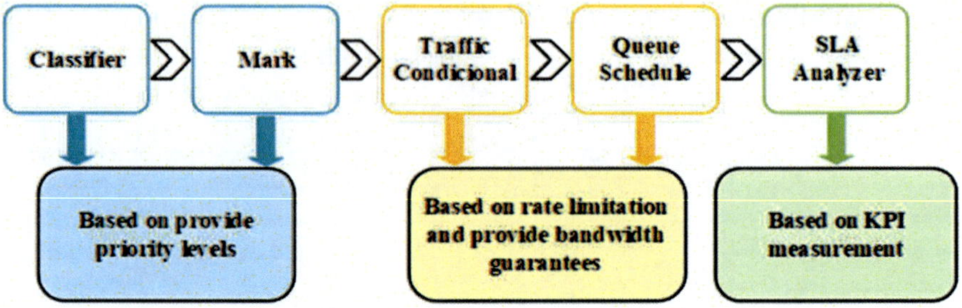


**Fig. 1** Block diagram of proposed model

of queues, which provide different guarantees of bandwidth for different types of traffic.

## 3.1 Network Topology

Figure 2 shows the implementation of QoS with DiffServ model with three routers. In this example, we want to give QoS end to end. RB1 and RB2 represent the edge routers responsible for marking and classifying traffic based on DSCP values.

Traffic is generated that simulates applications Premium, Gold, Silver and Bronze class. Color markers (DSCP) are used to manage different levels of performance in applications, which classify traffic into three sets: green, yellow, and red. Green traffic is transmitted within the limits of bandwidth, delay and loss of frames guaranteed by the SLA agreement, yellow traffic is transmitted, but SLA performance parameters do not apply, finally red traffic is discarded.

## 3.2 Proposed Algorithm

Once traffic classes are marked with the appropriate DSCP values, RC1 uses traffic conditioning and congestion management policies based on HTB.

Figure 3 indicates the HTB structure that is organized in a tree distribution. Limit at and Max limit represent the CIR and EIR respectively. HTB is a queuing method based on priority classes, it allows to create a hierarchical architecture to handle bandwidth in different types of traffic. HTB rule is based on the lending mechanism, sharing bandwidth, the queue with the highest priority can access the additional available bandwidth before the queue with the lowest priority (8 is the lowest priority, 1 is the highest). Table 3 indicates the classification, marking, and prioritization of each traffic class.

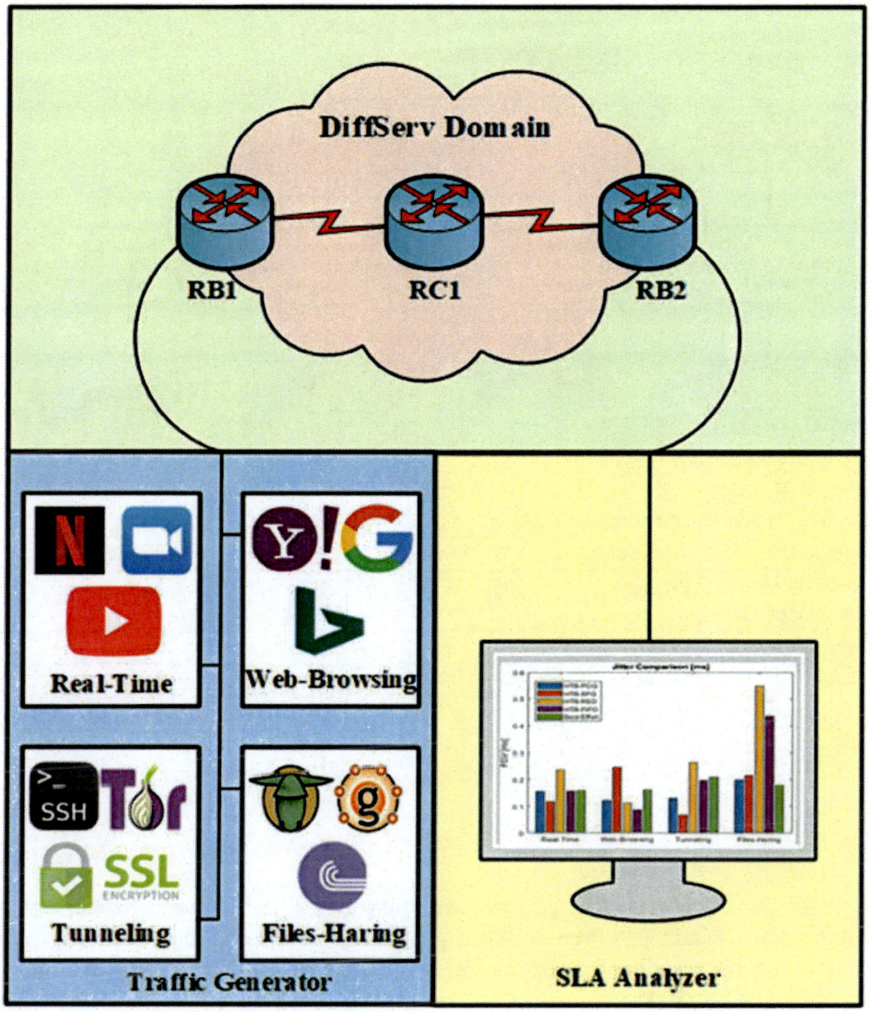


**Fig. 2** Implementation of a DiffServ network

Aouini et al. [19] have conducted an extensive study to characterize traffic based on usage patterns, application distribution, and features. The analysis of traffic services shows that web browsing applications dominate the total volume of traffic with 33.4%, followed by Real-Time entertainment applications with 25.3%, Tunneling and Files-Haring correspond to 13.9% and 11.7%, respectively. Therefore, in this work, the traffic classification to assign priorities is based on this study.

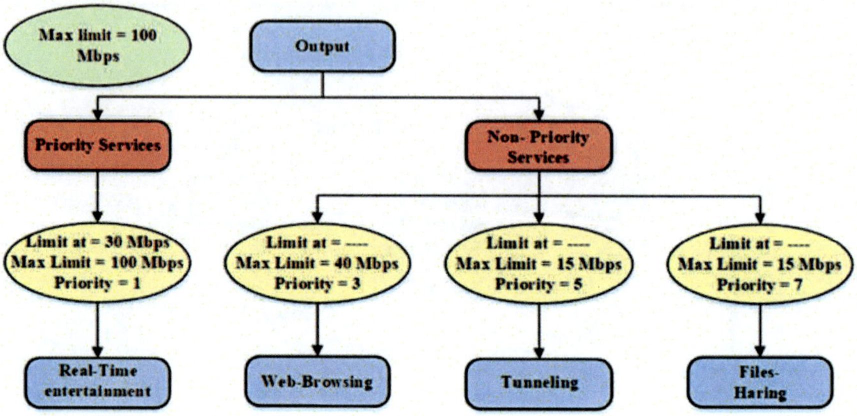


**Fig. 3** HTB structure

**Table 3** Traffic classification and marking

| Application | Traffic class | DSCP value | PHB | Priority |
|---|---|---|---|---|
| Real-Time entertainment | Premium | 46 | EF | 1 |
| Web-Browsing | Gold | 34 | AF41 | 3 |
| Tunneling | Silver | 26 | AF31 | 5 |
| Files-Haring | Bronze | 18 | AF21 | 7 |

### *3.3 Evaluation Conditions*

A traffic generator is used to generate and transmit test traffic over a network with deterministic bandwidth profiles. Data generated is synthetic, that is, it does not contain real information, for this reason, generated frames are replaced by payloads that simulate user traffic, this function is used to force the network, simulating loads that can be configured with specific parameters to measure performance criteria.

Performance test applied is eSAM (Ethernet Service Activation Methodology) is focused on determining the levels of QoS in the network with converged services that depend on various parameters such as architecture, transmission media, resource management, traffic signaling, and other factors.

Test is performed in RouterOS over 100 Mbps link. Bandwidth occupied by each service (CIR) corresponds to 30 Mbps for Real-Time applications, 40 Mbps for Web-Browsing services, Tunneling and Files-Haring obtain 15 Mbps for each one.

RouterOS supports several queuing types (PCQ, SFQ, RED, and FIFO) and it is possible combined with HTB achieving an hybrid queuing algorithm. The purpose of this study is analyze the advantages and weaknesses of these hybrid algorithms (HTB-PCQ, HTB-SFQ, HTB-RED, and HTB-FIFO) and verify which can provide traffic

**Table 4** Performance objectives

| Services | IR (Mbps) | FLR (%) | FTD (ms) | FDV (ms) | Avail (%) |
|---|---|---|---|---|---|
| Real-Time | 35 | 5 | <5 | <1 | 100 |
| Web-Browsing | 45 | 15 | <150 | <75 | 100 |
| Tunneling | 20 | 15 | 150–300 | 75–125 | 100 |
| Files-Haring | 20 | 15 | 150–300 | 75–125 | 100 |

performance that satisfying priority services while allowing non-priority applications access the network. Queue size for algorithms is specified in bytes and are configured with 204800 bytes or 200 KiB.

For the verification of QoS provision, the network is subjected to an overload stress test, exceeding the link capacity with 20 Mbps, all services are loaded simultaneously at a rate higher than CIR. The purpose is to verify that traffic are transported with the required performance in terms of latency (Frame Transfer Delay), jitter (Frame Delay Variation), and frame loss (Frame Loss Ratio) shown in Table 4, that is, KPIs must remain within an acceptable range.

Performance objectives of this study are based on [20], in addition Telecommunications and Internet Protocol Harmonization Over Networks (TIPHON) by European Telecommunication Standards Institute (ETSI) is also used [21].

It is important to establish an adequate frame dimension for the evaluation, it is convenient to generate a series of different sizes because many flows have a considerable length variation. Since the amount of audio/video information per IP packet will vary depending on the coding and compression technique used, the test is simplified by establishing fixed frame size of 256 bytes.

## 4 Results and Discussion

Network is evaluated in two congestion scenarios. In the first, traffic is forwarded using the Best-Effort mechanism, without conditioning and prioritization policies, while in the second, hybrid algorithms based on HTB is applied for the provision of QoS. Performance results based on KPIs parameters are shown in Fig. 4.

Information rate comparison (see Fig. 4a) shows that the network uses a queue-based admission control mechanism to limit the bandwidth to 100 Mbps, queuing the extra 20 Mbps overload and occupying 100% of the bandwidth available. Best-Effort mode fails to meet the CIR rate for Real-Time applications and Web services, these reach 28.317 Mbps and 37.235 Mbps, respectively, while applying QoS (HTB-PCQ, HTB-SFQ, HTB-RED, and HTB-FIFO), all services manage to meet the guarantee requirements for the information rate (Real-Time 30 Mbps, Web-Browsing 40 Mbps, Tunneling 15 Mbps and Files-Haring 15 Mbps).

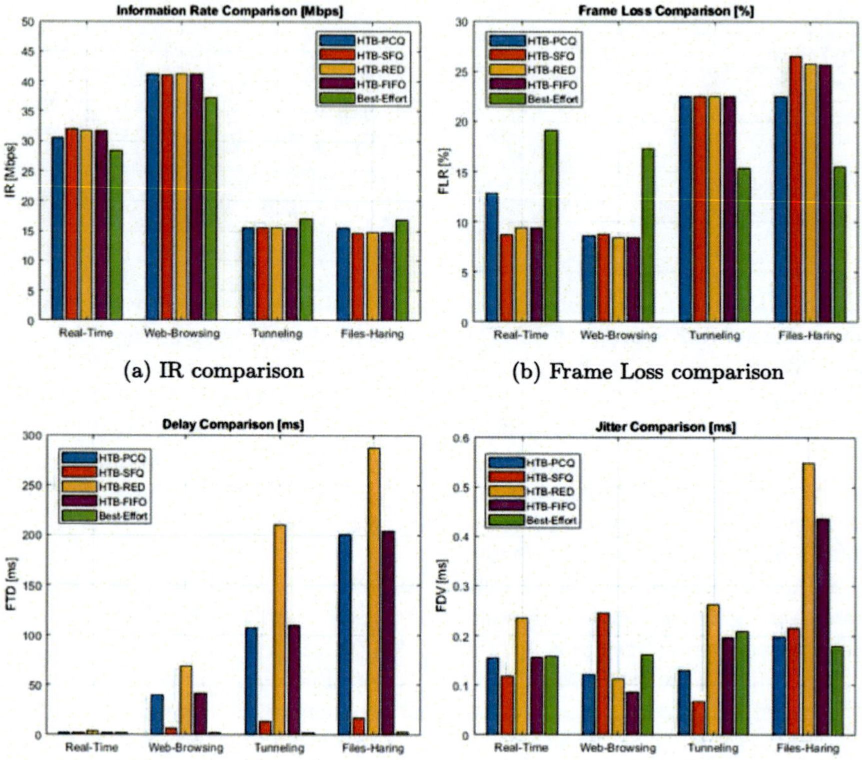


(a) IR comparison (b) Frame Loss comparison

(c) Delay comparison (d) Jitter comparison

**Fig. 4** Performance results

Frame Loss Ratio comparison (Fig. 4b) shows the difference using QoS mechanisms. In Best-Effort, Real-Time services are not priority and FLR reaches almost 20%, which is unacceptable for these types of applications, Web-Browsing, Tunneling, and Files-Haring show FLR of 17.26%, 15.32%, and 15.52%, respectively, categorizing them as a medium degradation. HTB-SFQ indicates the best performance by priority, although it does not reach the 5% target for Real-Time entertainment, set in the performance target, it has the lowest FLR value of 8,729%, Web-Browsing, Tunneling, and Files-Haring show FLR of 8.806%, 22.56%, and 26.51%, respectively, categorizing them into a medium to poor degradation, however, HTB-SFQ, in a scenario under stress with overload, providing performance that satisfies priority services, and allow non-priority applications access to the network.

Figure 4c shows delay comparison, indicates that HTB-PCQ and HTB-FIFO behave similarly. HTB-RED obtains the highest end to end delay values in all services, reaching 3.714 ms in Real-Time applications, 68.756 ms for Web services, 210.007 ms in Tunneling, and 287.243 ms for Files-Haring. HTB-RED, HTB-PCQ,

**Table 5** Performance results under stress without overload

| Services | IR (Mbps) | FLR (%) | FTD (ms) | FDV (ms) | Avail (%) |
|---|---|---|---|---|---|
| Real-Time | 30 | 0 | 0.175 | 0.022 | 100 |
| Web-Browsing | 40 | 0 | 1.47 | 0.091 | 100 |
| Tunneling | 15 | 0 | 1.418 | 0.097 | 100 |
| Files-Haring | 15 | 0 | 0.176 | 0.024 | 100 |

and HTB-FIFO are within the performance objectives, HTB-SFQ improves and obtains the lowest delay, 1.867 ms for Real-Time applications, 6.746 ms for Web services, Tunneling and Files-Haring with 13.732 ms and 17.398 ms, respectively, these values are classified as perfect quality according to TIPHON.

Jitter comparison (see Fig. 4d) indicates that both all QoS mechanisms (HTB-PCQ, HTB-SFQ, HTB-RED, and HTB-FIFO) and Best-Effort meet the performance objective in packet delay variation since all are below 1 ms.

HTB-SFQ hybrid algorithm indicate the best alternative for bandwidth management, obtain the lowest value of FLR, FTD, and FDV, in addition, high priority services do not tend to be dominant over lower priority ones. A second performance test is performed applying HTB-SFQ under stress without overload, all services are charged at the CIR rate. Results in HTB-SFQ scenario (shown in Table 5) indicate that network achieve performance objectives established, categorizing all services with high quality.

## 5 Conclusions

In this article, we have proposed the DiffServ architecture and hybrid algorithms based on HTB (HTB-PCQ, HTB-SFQ, HTB-RED, and HTB-FIFO) for the evaluation of performance in a network under congestion with overload considering the loss of packets, latency and jitter. In addition, we have analyzed the advantages and weaknesses of each one. The HTB-SFQ algorithm is able to provide the FLR, FTD, and FDV closest to the desired value, since the objective quality score is affected by network overload, it provides a performance that satisfies the priority services while allowing the access of non-priority applications to the network. Therefore, we believe that the proposed HTB-SFQ algorithm combined with DiffServ has the potential to provide a desired level of service in terms of FLR, FTD, and FDV and the corresponding QoS/QoE experiences for users using the hybrid relay system.

FLR, FTD, and FDV are a critical measure to evaluate QoS in a network used to provide services with time-sensitive applications. FLR, FTD, and FDV are susceptible to the effects of congestion and other deficiencies that affect end to end performance, therefore, they are important parameters to characterize.

## References


1. Frnda J, Nedoma J, Vanus J, Martinek R (2019) A hybrid QoS-QoE estimation system for IPTV service. Electronics (Switzerland) 8(5)
2. Dumka A, Mandoria HL, Fore V, Dumka K (2015) Implementation of QoS algorithm in integrated services (IntServ) MPLS network. In: 2015 2nd international conference on computing for sustainable global development (INDIACom), pp 1048–1050
3. Abouseda MI, Bozed KA, Zerek AR (2015) Comparative study of QoS and performance of VoIP using MPLS-TE network. In: 2015 16th international conference on sciences and techniques of automatic control and computer engineering (STA), pp 783–788
4. Hu Z, Yan H, Yan T, Geng H, Liu G (2020) Evaluating QoE in VoIP networks with QoS mapping and machine learning algorithms. Neurocomputing 386:63–83
5. Barry MA, Tamgno JK, Lishou C, Cissé MB (2018) Qos impact on multimedia traffic load (IPTV, RoIP, VoIP) in best effort mode. In: 2018 20th international conference on advanced communication technology (ICACT), pp 694–700
6. Mushtaq A, Patterh MS (2017) QoS parameter comparison of DiffServ-aware MPLS network using IPv4 and IPv6. In: 2017 international conference on recent innovations in signal processing and embedded systems (RISE), pp 113–118
7. Chervenets V, Romanchuk V, Beshley H, Khudyy A (2016) QoS/QoE correlation modified model for QoE evaluation on video service. In: 2016 13th international conference on modern problems of radio engineering, telecommunications and computer science (TCSET), pp 664–666
8. Zhigalov K, Skorikova NA, Daudov IM (2020) Interaction of models and methods of providing QoS in networks. J. Phys.: Conf. Ser. 1582:012095
9. Gertsiy A, Rudyk S (2016) Analysis of quality of service parameters in IP-networks. In: 2016 third international scientific-practical conference problems of infocommunications science and technology (PIC S T), pp 75–77
10. Aureli D, Cianfrani A, Diamanti A, Sanchez Vilchez JM, Secci S (2020) Going beyond DiffServ in IP traffic classification. In: NOMS 2020 - 2020 IEEE/IFIP network operations and management symposium, pp 1–6
11. Xu Y (2018) A DiffServ model for video stream. In: 2018 4th annual international conference on network and information systems for computers (ICNISC), pp 30–33
12. Fadjri K, Ritzkal R, Hendrawan H (2020) Computer network analysis using the queue system In Mikrotik: computer network analysis using the queue system In Mikrotik. J Mantik 4(2):483–488
13. Iswadi D, Adriman R, Munadi R (2019) Adaptive switching PCQ-HTB algorithms for bandwidth management in routerOS. In: 2019 IEEE international conference on cybernetics and computational intelligence (CyberneticsCom), pp 61–65
14. ITU-T Y.1541 (2011) Y.1541 network performance objectives for IP-based services. Serie y: Global information infrastructure, internet protocol aspects and next-generation networks internet protocol aspects – quality of service and network performance, p 66
15. Diallo T, Dorais M (2011) EtherSAM: the new standard in ethernet service testing. EXFO assessing next-gen networks, pp 1–12
16. Costantini L, Mammi E, Teodori M, Attanasio V (2017) Polynomial regression models to explain the relationship between network and service key performance indicators. IET Netw 6(5):125–132
17. CISCO (2020) Quality of Service (QoS) configuration guide, pp 1–74
18. Tabassum M, Tikoicina KM, Huda E (2018) Comparative analysis of queuing algorithms and GoS effects on the IoT networks traffic. In: 2018 8th IEEE international conference on control system, computing and engineering (ICCSCE), pp 88–92

19. Aouini Z, Kortebi A, Ghamri-Doudane Y (2016) Towards understanding residential internet traffic: from packets to services. In: 2016 7th international conference on the network of the future (NOF), pp 1–7
20. Lee S, Kim J (2017) Adaptation for the retransmission in VoIP applications. J Commun Netw 19(3):298–305
21. ETSI (2001) Telecommunications and Internet Protocol Harmonization Over Networks (TIPHON); General Aspects of Quality of Service (QoS), pp 1–35

# Performance Analysis in Hybrid OTDM-OCDMA System Using Narrowband Bragg Filter

**Gabriel E. Pulla-Lojano, Ismael S. Pesántez-Romero, Luis F. Guerrero-Vásquez, Edwin J. Coronel-González, Jorge O. Ordoñez-Ordoñez, and German V. Arevalo Bermeo**

**Abstract** In this paper, we analyze the performance of a hybrid optical time division multiplexing (OTDM) system with optical code division multiple access for spectral amplitude encoding (OCDMA), using very narrow filters with fiber Bragg gratings to the system as encoders and decoders in the transmitter and receiver. The performance of the system depends on variables such as: the length on kilometers of the fiber, the number of users, the bandwidth of Fiber Bragg Gratings, and the application of the Walsh Hadamard (WH) codes. The performance of the system is evaluated through the Quality (Q) factor and Bit Error Rate (BER) with different simulations, a comparison of the WH codes and the impact with the number of users is demonstrated.

**Keywords** OTDM · OCDMA · FBG · Walsh Hadamard

G. E. Pulla-Lojano (✉) · I. S. Pesántez-Romero · L. F. Guerrero-Vásquez · E. J. Coronel-González · J. O. Ordoñez-Ordoñez
Universidad Politécnica Salesiana, Cuenca, Ecuador
e-mail: gpulla@ups.edu.ec

I. S. Pesántez-Romero
e-mail: ipesantezr@ups.edu.ec

L. F. Guerrero-Vásquez
e-mail: lguerrero@ups.edu.ec

E. J. Coronel-González
e-mail: ecoronel@ups.edu.ec

J. O. Ordoñez-Ordoñez
e-mail: jordonezo@ups.edu.ec

G. V. A. Bermeo
Universidad Politécnica Salesiana, Quito, Ecuador
e-mail: garevalo@ups.edu.ec



X.-S. Yang et al. (eds.), *Proceedings of Sixth International Congress on Information and Communication Technology*, Lecture Notes in Networks and Systems 216,
https://doi.org/10.1007/978-981-16-1781-2_32

# 1 Introduction

The rapid development of fiber optic communications has revolutionized the telecommunications industry, increasing the demand for greater distances, more bandwidth, and high data transmission speeds; with increasingly diverse requirements and the need for faster and safer systems [1]. Bandwidth sharing and performance are related to physical limitations in transmission systems and the technologies implemented to share resources with users [2]. Consequently, various solutions have been implemented by deploying various multiplexing techniques, mainly Optical Time Division Multiplexing (OTDM) and Optical Code Division Multiple Access (OCDMA), to meet the growing need for bit rate improvement [3, 4]. Despite this, it is still a challenge to implement high data transmission systems, therefore, there is still a need for fast and highly efficient networks that can exploit current fiber optic links.

Regarding multiplexing techniques, Time Division Multiplexing (TDM) is a digital multiplexing technique limited by existing electronic technology that can only multiplex up to approximately 10 Gbps [5]. On the one hand, OTDM overcomes the electronic bottleneck and achieves a single channel high bit rate system [6]. The principle of this technology is to multiplex several low bit rate optical channels in the time domain. OTDM is a method that delivers an extensive capacity to generate high-speed data up to 1 TBps beyond the bandwidth limit [7, 8]. On the other hand, OCDMA has been used as one of the spread spectrum techniques that allows several users to share the same transmission medium, provides asynchronous access to networks, provides high system security, and improves spectral efficiency [9]. This multiplexing technique consists of the spread of the spectrum that allows the dispersion of power in a wide frequency band, and in the form of noise, it gives the transmitted signal that is difficult to detect by disinterested receivers, so the transmission is less sensitive to fluctuations [10, 11]. An advantage of this technique is to take advantage of all the bandwidth simultaneously, it allows a better management of resources. Also, the code sequences are orthogonal in such a way as to reduce user interference that affects system performance [12, 13].

In this context, it is observed that there are notable advantages over the use of different multiplexing techniques, it is even possible to combine various techniques obtaining interesting results. Some examples of combination of multiplexing techniques are observed in WDM (Wavelength Division Multiplexing) -OTDM that when combined increases the bit rate and the transmission distance, it is applied for multicast systems [14, 15], also another system hybrid is Subcarrier Multiplexing (SCM) with Optical Spectrum Code Division Multiplexing (OSCDM), in which they combine the advantages of both techniques, which are high transmission security and increase the capacity of simultaneous users [13].

In this paper, a hybrid system with OTDM and OCDMA is made, where OTDM overcomes the bandwidth limitation of the optoelectric components and increases the bit rate to the system and, with OCDMA, the Spectrum Amplitude Coding (SAC) is applied , due to its ability to eliminate multiple access interference. Walsh Hadamard (WH) codes are implemented with the sequences of 4, 8, and 16. These codes are orthogonal and zero cross-correlation. The implementation uses Fiber Bragg Grating (FBG), which are reflective filters with a very narrow band that allow selecting very specific wavelengths [16, 17].

Next, in what follows this document, Sect. 2 describes the implemented system considering the delay time in the case of OTDM and the WH codes for the OCDMA in the transmission and reception system. Then, in Sect. 3, the simulation configuration is shown with details of the equipment used to apply the OTDM-OCDMA system. Subsequently, in Sect. 4, the results of the simulations are analyzed based on the Q factor and BER parameters, comparisons are made with the Walsh codes, the filter bandwidth and number of users. Finally, in Sect. 5, the conclusions related to the performance of the system are presented.

## 2 System Description

The architecture of the hybrid OTDM-OCDMA system is shown in Fig. 1, it consists of 3 blocks: a transmitter, a transmission channel, and the receiver. The transmitter performs the conversion of the electrical signal to the optical signal and encodes it, the transmission channel carries the encoded signal, and the receiver extracts the information contained in the carrier and decodes the transmitted encoded signal.

The transmitter is composed of a "White light source" with a wavelength of 1550.75 nm at $-100$ dBm, then goes through a multiplexer for 3 users. At the output, the signal is generated by a"Pseudo-Random Bit Sequence Generator" for each of them, which is encoded by means of the "Non-Return to Zero Code" and goes through a "Mach Zehnder Modulator". This modulated signal is delayed in time for time domain multiplexing, the time delays used are 0, $1/(Bitrate) * 0.25$ and $1/(Bitrate) * 0.5$, respectively.

Fiber Bragg Gratings filters are used for coding, they are assigned different wavelength values for each user and have the same bandwidth.

The coded signals converge through the optical combiner, they are connected to the optical fiber, after the fiber has traveled in km, the output passes through an optical splitter assigned to the number of users. Then a 1 to 2 optical splitter is used, the upper part of the splitter is filtered by a decoder using the same code in the transmitter, while the lower part of the splitter is filtered by the complementary decoder. Then 2 photodetectors are used to reveal each component of the optical signal at the output, subtracting both signals from each other with the "Electrical Subtractor". Finally, a low pass filter and an amplifier of the photodetectors are used to obtain the resulting signal, with which the BER can be analyzed.

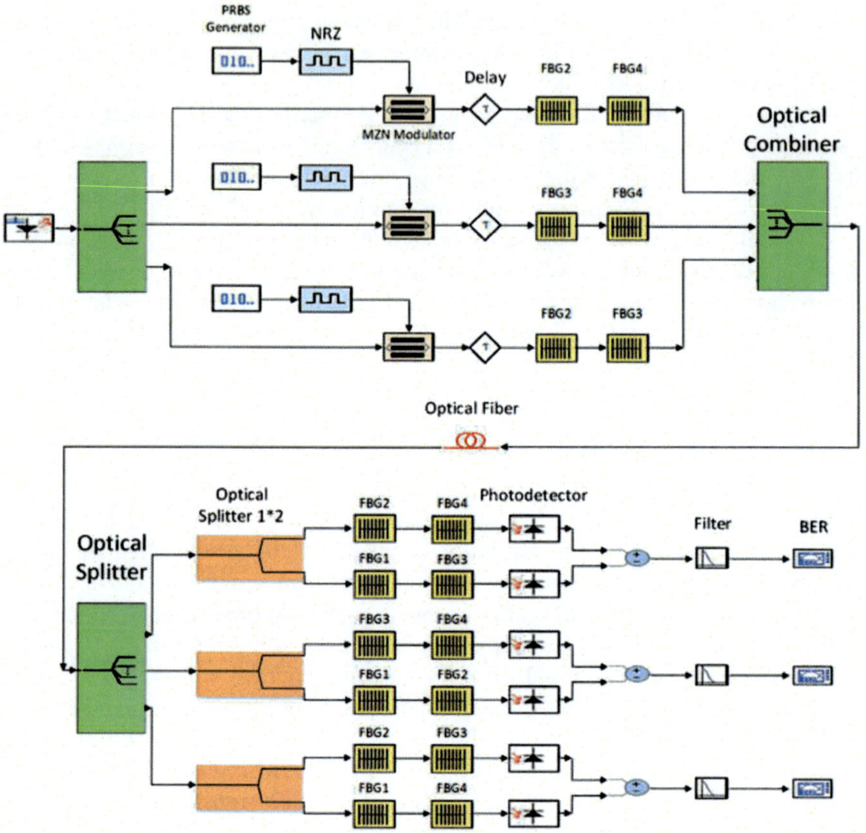


**Fig. 1** BER—number users

## 2.1 Walsh Hadamard Code

The WH code is defined in an orthogonal transformation technique, this technique decomposes each signal into a group of square pulses of levels +1 or −1. The construction of the matrix WH of order $n$ is written in a matrix $n \times n$ of ones and zeros in $HH^T = nI_n$ where $I_n$ is the identity matrix $n \times n$ [10].

$W$ is a Walsh Hadamard matrix of order $n$ that can be written as follows by replacing the value from −1 to 0:

$$W(n) = \begin{bmatrix} W(n/2) & W(n/2) \\ W(n/2) & W'(n/2) \end{bmatrix} \quad (1)$$

For the implementation, the evaluated WH codes 4 and 8 are used, the matrices obtained are:

$$W(4) = \begin{bmatrix} 1\,1\,1\,1 \\ 1\,0\,1\,0 \\ 1\,1\,0\,0 \\ 1\,0\,0\,1 \end{bmatrix} \tag{2}$$

$$W(8) = \begin{bmatrix} 1\,1\,1\,1\,1\,1\,1\,1 \\ 1\,0\,1\,0\,1\,0\,1\,0 \\ 1\,1\,0\,0\,1\,1\,0\,0 \\ 1\,0\,0\,1\,1\,0\,0\,1 \\ 1\,1\,1\,1\,0\,0\,0\,0 \\ 1\,0\,1\,0\,0\,1\,0\,1 \\ 1\,1\,0\,0\,0\,0\,1\,1 \\ 1\,0\,0\,1\,0\,1\,1\,0 \end{bmatrix} \tag{3}$$

## 3 The Simulation Setup

In the simulation, the 3-user OTDM-OCDMA system is performed with the codes Walsh (4), Walsh (8), Walsh (16), and without OTDM. Also, in the same system the number of users is increased for the analysis of the system capacity.

Simulations were carried out in the OTDM-OCDMA system of 3 users with a transmission speed of 200 Mbps, to evaluate the transmission performance, 2 factors are varied: fiber length and FBG bandwidth. Also, for the OTDM-OCDMA system, up to 9 users are simulated. It is interesting to evaluate the effect of the number of users with the BER and the Q factor for different fiber distances.

### 3.1 Simulation of the Performance of OTDM-OCDMA System with Code WH(4)

Based on the diagram in Fig. 1, the values of the different parameters used in the simulation are represented in Table 1.

The FBG values are based on the WH(4) matrix; at the output of FBG encoder, a code is assigned to each user that is unique and different from the other users.

**Table 1** Parameters values used in the simulation

| Parameters | | Values |
|---|---|---|
| White light source | Operating wavelength region | [1546–1553.5] nm |
| | Central wavelength | 1550.75 nm |
| Signal format | | (NRZ) |
| FBGs central wavelength | FBG-1 | 1550.0 nm |
| | FBG-2 | 1550.5 nm |
| | FBG-3 | 1551.0 nm |
| | FBG-4 | 1551.5 nm |
| Optical modulator extinction ratio | | 30 dB |
| Optical fiber | Ref. wavelength | 1550 nm |
| | Dispersion | 16.75 ps/nm/Km |
| | Loss | 0.2 dB/Km |
| Receiver filter bandwidth | | $0.65 * R_b$ |

### 3.2 *Simulation of the Performance of OTDM-OCDMA System with Code WH(8) Y WH(16)*

To measure the performance in the OTDM-OCDMA system when increasing the number of users greater than 3, the sequence code WH(8) and WH(16) described in Table 2 are used. For WH(8), the first 8 lambda lengths are used and the filters are built according to the scheme in Fig. 1. For WH(16), all wavelength values are used for the construction of the filters.

**Table 2** Values of FBG for WH(16)

| FBG | Value (nm) | FBG | Value (nm) |
|---|---|---|---|
| λ1 | 1546.0 | λ9 | 1550.0 |
| λ2 | 1546.5 | λ10 | 1550.5 |
| λ3 | 1547.0 | λ11 | 1551.0 |
| λ4 | 1547.5 | λ12 | 1551.5 |
| λ5 | 1548.0 | λ13 | 1552.0 |
| λ6 | 1548.5 | λ14 | 1552.5 |
| λ7 | 1549.0 | λ15 | 1553.0 |
| λ8 | 1549.5 | λ16 | 1553.5 |

## 4 Analysis and Results

In this section, the simulation results of the OTDM-OCDMA hybrid system are discussed. In the first instance, the performance evaluation is based on the effect of different numbers of users, Q factor, and BER; BER performance parameters are acceptable for values close to below $10^{-9}$, and Q factor around to above 6. Then, the performance is analyzed by varying the bandwidth of the FBG for the WH codes.

### *4.1 Effect of Number of Users on Q Factor and BER for Different Fiber Lengths*

Simulation uses the WH(16) code for the ability to place up to 15 theoretical users. Quality factor analysis is carried out by varying the number of users (3–9). Figure 2 shows the performance of the hybrid OTDM-OCDMA system in terms of distance, quality factor, and BER.

Figure 2a indicates the variation of the Q factor as a function of the number of users. With 3 users, the Q factor is 8,033 at a distance of 10 km, at 50 km it has a value of 6.51. At the number of 4 and 5 users, the quality factor values are 5.68 and 5.24, respectively, at a distance of 10 km, at 50 km they have a quality factor of 4.89 and 4.76, respectively. For the number of 8 and 9 users, the quality factor is low, with a value of 2.24 and 0.97, respectively; at a distance of 10 km, therefore, it is not suitable for this number of users.

Figure 2b shows the variation of BER with respect to the number of users. In the case of 3 users, at 10 km it has $2.19589 \times 10^{-9}$, at 30 km it remains with a value of BER $1.58842 \times 10^{-8}$ while with the number of 4 to 6 users they are between a value of $2.50284 \times 10^{-6}$ in BER; in the 50 km with the 3 users it has a value of

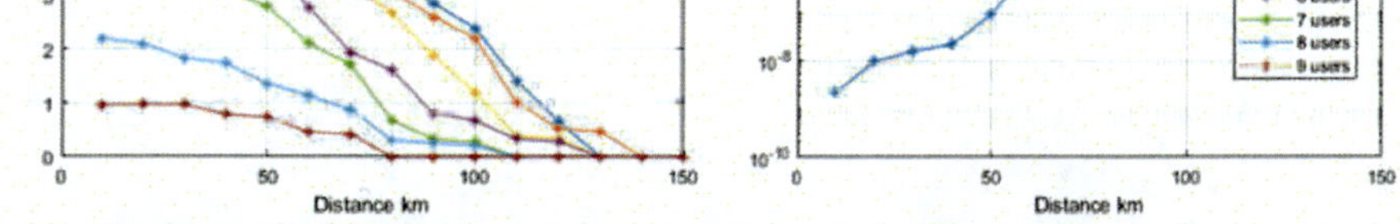


(a) Quality factor - number of users (b) BER - number of users

**Fig. 2** Performance results

$8.91582 \times 10^{-8}$; with the rest of the number of users from 7 to 9, there is a high BER starting at $6.2288 \times 10^{-4}$ for the first 10 km. Each user's BER increases as the total number of users increases.

## 4.2 *Performance of FBG Bandwidth*

The simulation is carried out with 2 fixed parameters: the fiber length of 50 km and the transmission rate of 200 Mbps, and the bandwidth of the FBG is varied between the values 0.1–0.6 nm. The Fig. 3 shows the performance of the Q factor and BER.

The results corresponding to Fig. 3a indicate that Q factor for WH(16) reaches its maximum value of 8.23 for a 0.6 nm FBG. On the other hand, for the WH(4) code described in Fig. 3b, it reaches its maximum value of 6.17 of the corresponding

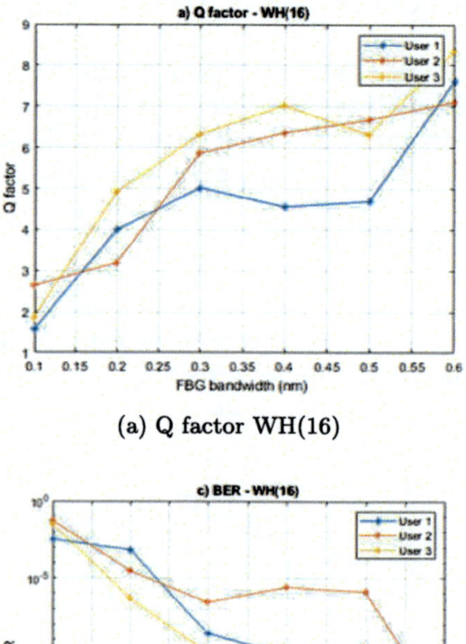


(a) Q factor WH(16)

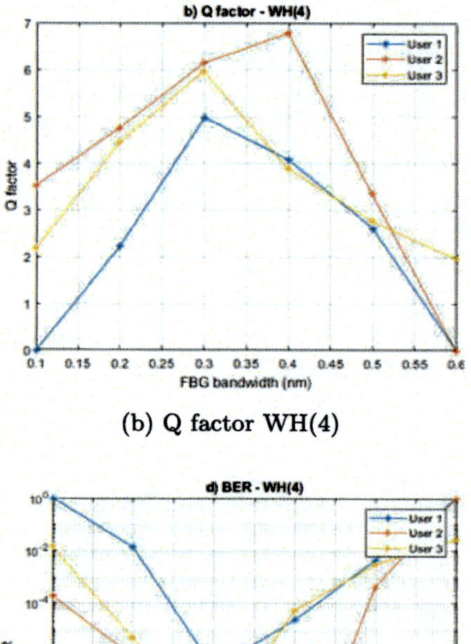


(b) Q factor WH(4)

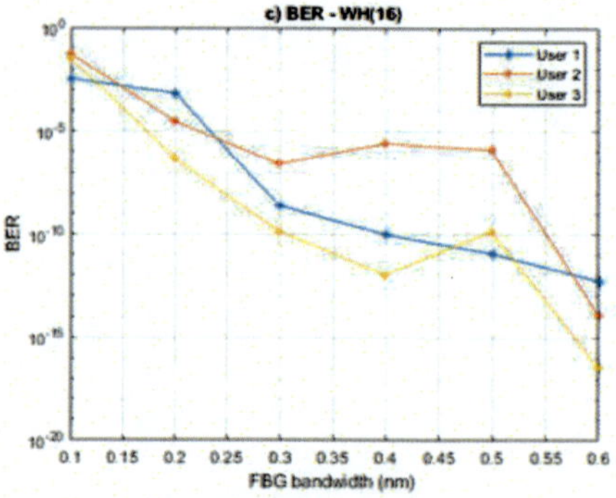


(c) BER WH(16)

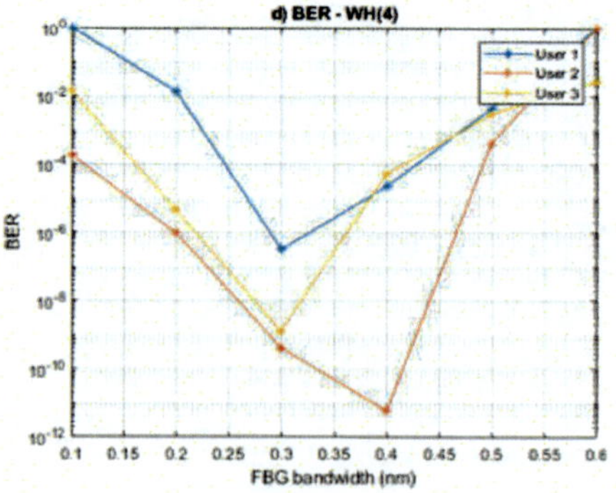


(d) BER WH(4)

**Fig. 3** Performance results

quality factor FBG of 0.3 nm. Figure 3c and 3d indicate the highest value of the BER of WH(16) with $3.89959 \times 10^{-17}$ and of WH(4) with $4.19412 \times 10^{-10}$, respectively.

### 4.3 *Performance of the Q Factor and BER of the OTDM-OCDMA System with Coding WH(4), WH(8), and WH(16)*

Simulation is realized to analyze the performance of the system, applying the best results found from FBG of the previous section. In the case of WH(4) and WH(16), FBG is applied at 0.3 nm and 0.6 nm, respectively. The results are shown in the following Fig. 4.

Figure 4a shows Q factor with respect to distance, WH(16) coding obtain the highest performance, reaching a value of 9.78 at 10 km, with WH(4) having a Q factor of 5.94 at 10 km and can reach a distance of 100 km with a Q factor of 4.15 compared to the other codes, the WH(8) has a low Q factor with a value of 2.13 in the first 10 km , and is maintained up to the distance of 100 km with a Q factor of 1.35 which is a considerably low value. Without the application of OTDM, Q factor is with a value of 7.61 at the first 10 km, as the distance increases, Q factor decreases and the link loses at the distance of 70 km.

Figure 4b shows BER performance of the different Walsh codes, WH(16) obtain BER value of $2.24816 \times 10^{-18}$ at 10 km, it is the code with the lowest BER value in the first 70 km categorizing in good performance. With WH(4) she has a $7.164972 \times 10^{-7}$ BER which is acceptable and is maintained at a distance of 70 km. On the other hand, WH(8) has a high BER value not optimal for transmission with a value of $2.24816 \times 10^{-2}$ categorizing in poor performance.

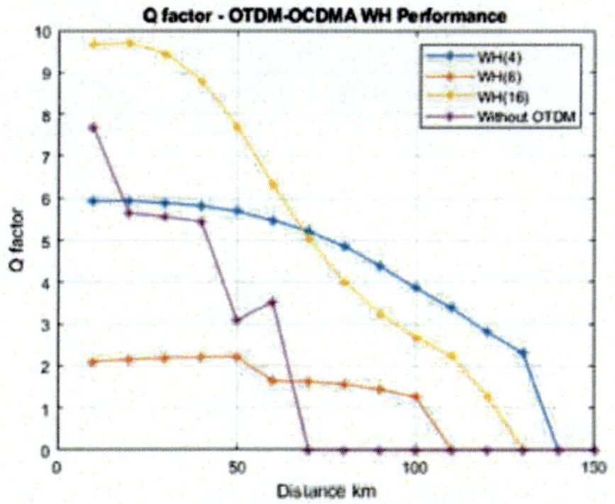


(a) Q factor - OTDM OCDMA

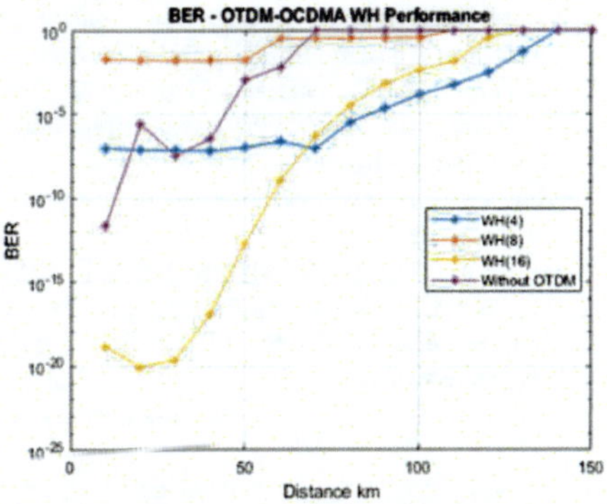


(b) BER - OTDM OCDMA

**Fig. 4** Performance results

## 5 Conclusion

To evaluate the performance of the OTDM-OCDMA system, several simulations have been carried out, and they are related by varying factors such as the length of the fiber, the bandwidth of the FBG filter and the number of users. The results achieved indicate that the optimal transmission efficiency of an OTDM-OCDMA system is the use of the WH (16) code because it can reach a distance of 70 km. On the other hand, without the use of the OTDM system, it does not arrive with optical conditions from BER at 30 km. It also showed that this system works up to a number of 6 users for a distance of 20 km in optical fiber, as the length of the fiber increases, the capacity of the number of users decreases.

We can indicate that multiplexing systems allow users of telecommunications services to obtain faster service, security with the help of FBG filters, and obtain higher data speeds.

## References


1. Mostafa S, Mohamed AENA, El-Samie FEA, Rashed, ANZ (2019) Cyclic shift code for SAC-OCDMA using fiber bragg-grating. arXiv:1904.00373
2. Boukricha S, Ghoumid K, Mekaoui S, Ar-Reyouchi E, Bourouina H, Yahiaoui R (2020) SAC-OCDMA system performance using narrowband bragg filter encoders and decoders. SN Appl Sci 2(6):1–9
3. Panda S, Bhanja U (2016) Design and performance analysis of one dimensional zero cross-correlation coding technique for a fixed wavelength hopping SAC-OCDMA. Int J Electron Commun Eng 9(11):1330–1338
4. Ong CJ (2011) Optical code division multiple access. PhD thesis, UTAR
5. Gupta A (2015) Others: performance analysis of OTDM system at the different data rate. Int J Eng Comput Sci 4(06)
6. Singh S (2018) Performance investigation of 1.68 Tb/s spectral efficient hybrid OTDM–WDM system using orthogonal modulation format. Natl Acad Sci Lett 41(3):151–154
7. Bogoni A, Poti L, Ghelfi P, Scaffardi M, Porzi C, Ponzini F, Meloni G, Berrettini G, Malacarne A, Prati G (2007) OTDM-based optical communications networks at 160 Gbit/s and beyond. Opt Fiber Technol 13(1):1–12
8. Santhiya LP, Jaya NV Simulation and performance evaluation of OTDM in optical communication system
9. Islam M, Ahmed N, Aljunid SA, Ali S, Sayeed S, Sabri N (2016) Performance analysis of DPSK-OCDMA system for optical access network. J Opt Commun 37(1):51–54
10. Majeed MH, Ahmed RK (2020) Performance enhancement of encoding–decoding multidiagonal and Walsh Hadamard codes for spectral amplitude coding-optical code division multiple access (SAC-OCDMA) utilizing dispersion compensated fiber. J Opt Commun 1. (ahead-of-print)
11. Hanane D (2018) Evaluation des performances de la technique cdma dans la transmission optique. PhD thesis, Université de Batna
12. Palushani E, Mulvad HCH, Galili M, Hu H, Oxenlowe LK, Clausen AT, Jeppesen P (2011) OTDM-to-WDM conversion based on time-to-frequency mapping by time-domain optical Fourier transformation. IEEE J Sel Top Quantum Electron 18(2):681–688

13. Sahbudin RKZ, Abdullah MK, Mokhtar M (2009) Performance improvement of hybrid subcarrier multiplexing optical spectrum code division multiplexing system using spectral direct decoding detection technique. Opt Fiber Technol 15(3):266–273
14. Singh S, Singh S (2017) Performance analysis of hybrid WDM-OTDM optical multicast overlay system employing 120 Gbps polarization and subcarrier multiplexed unicast signal with 40 Gbps multicast signal. Opt Commun 385:36–42
15. Kumari M, Sharma R, Sheetal A (2019) Comparative analysis of high speed 20/20 Gbps OTDM-PON, WDM-PON and TWDM-PON for long-reach NG-PON2. J Opt Commun 1. (ahead-of-print)
16. El-Mottaleb A, Somia A, Fayed HA, El-Aziz A, Metawee MA, Aly MH (2018) Others: enhanced spectral amplitude coding OCDMA system utilizing a single photodiode detection. Appl Sci 8(10):1861
17. Majeed MH, Ahmed RK, AbdlJabbar IL (2019) Gb/s broadband spectral amplitude coding-optical code division multiple access (SAC-OCDMA) based on multidiagonal and Walsh Hadamard codes. J Commun 14:802–812

# LDA Algorithm for the Identification of Topics: A Case of Study in the Most Influential Twitter Accounts in Ecuador

**Jorge O. Ordoñez-Ordoñez, Luis F. Guerrero-Vásquez, Paul A. Chasi-Pesántez, David P. Barros-Piedra, Edwin J. Coronel-González, and Katherine C. Bustamante-Cacao**

**Abstract** In the last years, the environment in which we develop has had various changes, most of them due to major technological changes, and the constant development of Information and Communication Technologies (ICTs). As a result of the advances in ICTs, today it is common to interact through social networks and make constant use of them. For this reason, this paper presents an analysis of the 10 most influential Twitter accounts in Ecuador; the objective of this analysis is to detect what the topics or topics addressed in these accounts are. The Latent Dirichlet Allocation (LDA) algorithm using bag of words (BoW) model and also the Term Frequency–Inverse Document Frequency (TF-IDF) model were used for the analysis, finally finding, if the topics provided by both models are similar.

**Keywords** Social networks · ICTs · LDA · Data mining · Machine learning

J. O. Ordoñez-Ordoñez (✉) · L. F. Guerrero-Vásquez · P. A. Chasi-Pesántez · D. P. Barros-Piedra · E. J. Coronel-González
Universidad Politécnica Salesiana, Cuenca, Ecuador
e-mail: jordonezo@ups.edu.ec

L. F. Guerrero-Vásquez
e-mail: lguerrero@ups.edu.ec

P. A. Chasi-Pesántez
e-mail: pchasi@ups.edu.ec

D. P. Barros-Piedra
e-mail: dbarros@est.ups.edu.ec

E. J. Coronel-González
e-mail: ecoronel@ups.edu.ec

K. C. Bustamante-Cacao
IEEE, Madrid, Spain
e-mail: cristinabustamante@ieee.org



X.-S. Yang et al. (eds.), *Proceedings of Sixth International Congress on Information and Communication Technology*, Lecture Notes in Networks and Systems 216,
https://doi.org/10.1007/978-981-16-1781-2_33

# 1 Introduction

During the last decade, the system in which we develop has undergone great and numerous changes, all of which are guided by the set of transformations resulting from technological innovation, and above all by the development of information and communication technologies (ICTs) [1]. The rise of ICTs has transformed our way of living, thinking, and even the way we relate. Additionally, the development of the so-called Web 2.0 has made social media today have a great impact [2]. Within these media, social networks stand out, which act as a meeting point where millions of users can publish and share opinions, information, or simply trivialities on all kinds of topics, thus generating a large amount of information that can be analyzed. Being able to analyze and understand all that information is becoming one of the main objectives for companies and organizations since they see social networks as an excellent area to analyze what is said regarding a specific topic [3].

For content analysis within social networks, organizations have begun to focus their interest in the study of short texts, specifically in the comments that are made, to know the valuations that are given on certain products or services. Of course, this will depend on the social network used, in our case, all the analysis will be done on the social network Twitter.[1]

Twitter, more than a social network, is classified as a two-way communication platform, which limits its messages to 280 characters. However, despite this limitation, today, it is among the most used platforms worldwide with an approximate of 330 million monthly active users and an average of 500 million Tweets per day. Besides, this platform allows you to directly follow famous people, also known as influencers, and observe what their publications are, and how they can directly interact with their fans. On the other hand, it has achieved such a level of popularity since it is widely used to transmit news, and in some events, which have shocked the world, it has been the number one social network in transmitting information; for all this, it is common that today most people have an account on this network [4].

In this context, the purpose of this paper is to carry out an analysis of the topics frequently used by the most influential Twitter accounts in Ecuador. This analysis will be carried out with the use of the Latent Dirichlet Allocation (LDA) algorithm. In this way, the structure of the paper will be as follows. In Sect. 2, some works related to information search and analysis of topics in social networks will be presented. In Sect. 3, we present how the database to be analyzed was made. Section 4 shows the process for obtaining the most used topics from the data set. Finally, in Sect. 5 we present the conclusions of the work.

[1] https://twitter.com.

## 2 Related Works

Twitter has become one of the most widely used platforms worldwide, having as its main feature its messages limited to 280 characters; it allows to carry out various analyses using Machine Learning techniques. Being such an important social network, various studies have examined it from different perspectives, for example, in [5], answer the question whether it is seen as a social network, or as a news center. On the other hand, in [6], they analyze tweets as social sensors of real-time events; in [7], they study the tweets emitted to forecast the box office of new films, among many other studies.

Going into our research, in the literature, we find several works related to the extraction and analysis of tweets through different techniques. Thus, in [8] they propose a technique to measure the influence of users around sensitive topics; for this analysis, they use the LDA algorithm, obtaining, as a result, the detection of the topics that the user posts the most.

In [9], they have created an Android application that extracts trends from Twitter according to location; to meet this objective they use the LDA algorithm. In this research, the authors state that the problem is that existing methods only provide a trend for a country or city and not for a specific location as a tourist spot, a restaurant, among others. For this reason, with their project, they hope to reach business sectors to promote a certain place through a trend in a social network.

For their part, the authors of [10] have created a system to detect the emotions expressed in a tweet, using the LDA algorithm and symbol conversion. With the help of these tools, the researchers try to predict a total of 5 emotions, happy, angry, scared, sad, and shocked. In tests carried out, they can infer that the detection of emotions is better using the conversion of expression symbols, improving concerning for to not using this tool by 14.096%.

On the other hand, in [11] they propose to analyze the content published by Twitter users to determine if it can be considered a news platform. For this, in the analysis carried out they propose a comparison between the content published on the social network, compared to that published in a general digital newspaper such as the New York Times. For comparison of topics, they use unsupervised modeling based on the LDA algorithm.

Similarly, in [12] they characterize Ecuador's political leaders using unsupervised learning. To achieve this objective, they first choose the politicians with the most followers in the country, after which they make a comparison of the topics issued by these people and the headlines of the press, resulting in the issues that politicians deal with some of the headlines in the Ecuadorian press.

As can be seen in the research shown above, several use the LDA algorithm for different purposes; topic detection is a well-known and studied issue in the Big Data world. For this reason, our work differs from the rest, mainly due to the data set with which the analysis will be carried out, and due to the comparison between the techniques of Bag of Words (BoW) and the Term Frequency Model (TMF).

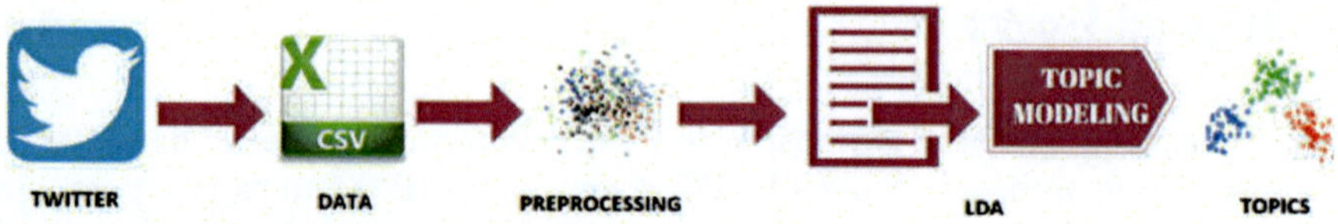


**Fig. 1** Process to follow for the analysis of topics

**Table 1** Ecuador's most influential Twitter accounts

| Twitter account | Number of followers |
|---|---|
| @MashiRafael | 3578870 |
| @miabuelasabia | 2899020 |
| @ecuavisa | 2355123 |
| @teleamazonasec | 2071355 |
| @eluniversocom | 2013454 |
| @Presidencia_Ec | 1567713 |
| @elcomerciocom | 1553791 |
| @enchufetv | 1440081 |
| @flacaguerrerog | 1436404 |
| @EcuavisaInforma | 1308695 |

To carry out our research, we will first choose Twitter accounts with the largest number of followers in Ecuador. The last 3,000 tweets will be extracted from each of the accounts, concerning the date of analysis, and a file with a .csv extension will be created. For the final part, with the help of software made in Python, the preprocessing of the information will be carried out, and the application of the LDA algorithm to obtain the topics; this process can be seen in Fig. 1.

## 3 Creation of the Database

One of the most complicated tasks when performing data analysis is creating the database with which you are going to work. In this way, the first thing defined for our analysis is that the 10 Twitter accounts with the highest number of followers in Ecuador will be chosen.[2] Thus, in Table 1, the most influential Twitter accounts of Ecuador are presented, with their respective number of followers.

Once the accounts from which the information will be extracted are defined, the Application Programming Interface (API) of Twitter will be entered; for this, it will be necessary to log in as a developer; in this way we will be provided with the credentials to be able to enter from an external program. The external program with

[2] https://www.socialbakers.com/statistics/twitter/profiles/ecuador.

which it will work is Python; with this, we will access Twitter using the Tweepy library.[3]

Following the methodology developed for this work, the following will be to extract the tweets emitted by the accounts with which the investigation will be carried out. Due to an API restriction, it only allows us to retrieve the last 3,000 tweets posted by each of the accounts. All this information will be stored in JavaScript Object Notation (JSON) format, and will finally be transformed into a file with a .csv extension to be able to perform analysis.

## 4 Obtaining Topics

This section presents the process carried out to obtain the topics. Remember that the data set to be analyzed is in a file with a .csv extension. Given this, this section will be divided into two parts, the first one analyzes the preprocessing of the data set, and in the second we apply the LDA algorithm using the bag of words model and the term frequency model.

It must be emphasized that the Gensim library was used for this purpose, because it uses a variation of the LDA algorithm that is trained online using variational Bayes [13]. In Algorithm 1, the online variational Bayes for LDA pseudocode is displayed.

**Algorithm 1:** Online variational Bayes for LDA

Define $\rho_t(\tau_0 + t)^{-k}$
Initialize $\lambda$ randomly.
**for** $t = 0$ to $\infty$ **do**
  E step:
  Initialize $\gamma_{tk} = 1$. (The constant 1 is arbitrary.)
  **repeat**
    Set $\phi_{twk} \alpha \, exp \left\{ \mathbb{E}_q \left[ log \theta_{tk} \right] + \mathbb{E}_q \left[ log \beta_{kw} \right] \right\}$
    Set $\gamma_{tk} = \alpha + \sum_w \phi_{twk} n_{tw}$
  **until** $\frac{1}{k} \sum_k |change \ in \ \gamma_{tk}| < 0.00001$
  M step:
  Compute $\tilde{\lambda}_{kw} = \eta + D n_{tw} \phi_{twk}$
  Set $\lambda = (1 - \rho_t) \lambda + \rho_t \tilde{\lambda}$
**end for**

### 4.1 Data Set Preprocessing

When we refer to the preprocessing of data, we are talking about the process used to clean the documents, that is, eliminate punctuation marks, eliminate words that do

[3] https://www.tweepy.org/.

not give more meaning to the sentences, and carry out the process of tokenization and stemming of each one of existing documents.

Our case was different, since when working with tweets, these came with a lot of junk information such as links to web pages, mentions to other Twitter accounts in cases where a tweet is answered, emoticons, emojis, and others information that does not contribute anything to our analysis. For this reason, the first thing that was done was to delete all this information; for this, within our algorithm a function was developed that meets this objective.

Following this, the stopword removal process was performed. Words are known as stopwords that do not have a meaning by themselves, but rather modify or accompany others; this group is usually made up of articles, pronouns, prepositions, adverbs, and even some verbs. To remove this group of words from each document, the Gensim[4] library was used, which already includes a list with stopwords in Spanish.

Finally, the tokenization and lemmatization process was accomplished; for this, the NLTK[5] library was used. The term tokenize refers to dividing the text into sentences, the sentences into words, and each of these words is known by the name of token; this process helps us to analyze which words go together with others and which of these are repeated with more frequency. After this, the stemming process was executed, which is a linguistic process that consists of finding the corresponding motto. The motto is the form that by agreement is accepted as representative of all the flexed forms of the same word. As previously stated, this process was realized with the help of the NLTK library, since it brings these functions to various languages, including Spanish.

In this way, the preprocessing of the data set is fulfilled; once you have the data with all the elimination of the aforementioned, the algorithm for modeling topics is applied.

## 4.2 Topic Modeling

The topic modeling process consists of identifying topics or themes in texts. This is generally accomplished by detecting patterns in a collection of documents called a "corpus" and grouping the words that appear in those documents into topics. Topics are generally defined as a density function pi (w) where w is a word and pi is the density function for topic i. Therefore, pi (w) is the probability of the word w given the topic i.

To achieve modeling, various techniques can be used; in our case, the LDA algorithm will be used. This is an unsupervised learning algorithm that attempts to describe a set of observations as a combination of different categories. These categories are in themselves a probability distribution on the characteristics. LDA is a generative probability model, which means that it tries to provide a model for

[4] https://pypi.org/project/gensim/.

[5] https://www.nltk.org/.

(a) Words assigned using BoW. (b) Words assigned using TF-IDF.

**Fig. 2** Words assigned using different techniques

the distribution of outputs and inputs based on latent variables. This is opposed to discriminatory models, which try to learn how to map inputs to outputs. This algorithm can be used in various tasks, however, in our case it will be used specifically to analyze the created data set.

For the algorithm to work, it is necessary to specify the number of topics of interest. Words are randomly assigned to a topic, receiving a score based on the probability that word belongs to it in the document set. That word is assigned to another topic and the same score is calculated. After this iterative process, a list of words is obtained in each of the topics with associated probabilities. Thus, we can select the k words with the highest probability of belonging to a particular topic and, the combination of these words allows us to describe the topic.

In our case, we are going to apply the LDA algorithm using two techniques, the first will be a word bag (BoW), and the second will be using Term Frequency–Inverse Document Frequency (TF-IDF). Given this, it is necessary to indicate that this process was carried out previously in the program, and then the LDA algorithm was applied.

Assigning the number of topics is a difficult task, since there is no established rule. However, the most recommended, and what was done for our analysis is to run the algorithm with various numbers of topics until finding the greatest amount of coherence. For this process, different numbers were used, going in descending order from 10, thus finding that with number 5 there is more coherence. In Fig. 2a, the words that will determine each of the topics are presented using the word bag model. Similarly, in Fig. 2b, the words to determine the topics are presented, but this time using the TF-IDF technique.

With this set of words, the topics are generated both for the case of the bag of words and for the TF-IDF case used. It should be emphasized that the topics are placed according to the perception of the interpreter. Table 2 presents the topics for each of the cases.

**Table 2** Definition of topics using BoW and TF-IDF

| Topic | BoW | TF-IDF |
|---|---|---|
| 1 | Vida | Justicia |
| 2 | Actualidades | Actualidades |
| 3 | Gobierno | Noticias |
| 4 | Noticias | Politica |
| 5 | Varios | Variedades |

Therefore, using both techniques we obtain consistent topics as a result, and even in some cases the same topics are presented.

## 5 Conclusions and Future Works

In this paper, the topics used by the most influential Twitter accounts in Ecuador were analyzed. To achieve this goal, the last 3000 tweets were extracted from the 10 accounts with the most followers in the country, the information was preprocessed and finally the LDA algorithm was applied. To apply the algorithm, two techniques were used, the first using a bag of words and the second using TF-IDF. The results obtained showed that, for the analyzed data set, it is better to use the TF-IDF technique, since with this a greater coherence of words is obtained to form the topic. However, using both techniques gave related results, so it is concluded that the use of these techniques to test the LDA algorithm is positive. The analysis of tweets using the LDA algorithm has several applications. In this research, the topics shared by the most influential accounts or those with the largest Twitter followers in Ecuador were obtained. Thanks to the results, it is possible to generate marketing-related recommendations to present the main options for product promotion. In the future, those companies or people looking for influencers to promote their products may choose them based on the affinity of Twitter accounts with their products. In this way, it is proposed that future works can analyze the trend of information shared by specific accounts, and the press headlines of the country's news media; in this way, the influence of certain opinions on the news could be quantitatively measured. Similarly, future work can use the algorithm to generate statistics on the reaction of Twitter users due to the health emergency caused by the COVID-19 pandemic.

## References

1. Ng JJ, Li K (2003) Implications of ict for knowledge management in globalization. Inf Manag Comput Secur 11(4):167–174
2. Vilares Calvo D (2014) Análisis de contenidos en twitter: clasificación de mensajes e identificación de la tendencia política de los usuarios. Ferrol: Universidade da Coruña

3. Lage García L (2014) Herramienta para el análisis de la opinión en tweets periodísticos
4. Carballar JA (2011) Twitter. Marketing personal y profesional, RC Libros
5. Kwak H, Lee C, Park H, Moon S (2010) What is twitter, a social network or a news media? In: Proceedings of the 19th international conference on world wide web, pp 591–600
6. Sakaki T, Okazaki M, Matsuo Y (2010) Earthquake shakes twitter users: real-time event detection by social sensors. In: Proceedings of the 19th international conference on world wide web, pp 851–860
7. Asur S, Huberman BA (2010) Predicting the future with social media. In: IEEE/WIC/ACM international conference on web intelligence and intelligent agent technology, vol. 1. IEEE 2010, pp 492–499
8. Weng J, Lim EP, Jiang J, He Q (2010) Twitterrank: finding topic-sensitive influential twitterers. In: Proceedings of the third ACM international conference on Web search and data mining, pp 261–270
9. Jayadharshini J, Sivapriya R, Abirami S (2018) Trend square: an android application for extracting twitter trends based on location. In: 2018 international conference on current trends towards converging technologies (ICCTCT). IEEE 2018, pp 1–5
10. Cahyaningtyas RM, Kusumaningrum R, Riyanto DE et al (2017) Emotion detection of tweets in indonesian language using lda and expression symbol conversion. In: 2017 1st international conference on informatics and computational sciences (ICICoS). IEEE 2017, pp 253–258
11. Zhao WX, Jiang J, Weng J, He J, Lim E-P, Yan H, Li X (2011) Comparing twitter and traditional media using topic models. In: European conference on information retrieval. Springer, Berlin, pp 338–349
12. Torres J, Baquerizo G, Vaca C (2015) Caracterización de líderes políticos de ecuador en twitter usando aprendizaje de maquina no supervisado. Revista Tecnológica-ESPOL 28(5)
13. Hoffman M, Bach FR, Blei DM (2010) Online learning for latent dirichlet allocation. In: Advances in neural information processing systems, pp 856–864

# Access Control in Cyber-Physical Spaces Using Virtual Perimeters

**Brian Greaves, Marijke Coetzee, and Wai Sze Leung**

**Abstract** In recent years, physical access control systems have not developed to the same extent as their cyber counterparts. Cyber access control prevents unwarranted access to sensitive information. When a cyber document is displayed on a monitor, it loses its cyber protection as a passer-by can view the document on the monitor. A disparity thus exists between the level of security provided to a document in the cyber compared to the physical realm. Physically rendered documents can be better protected by establishing dynamic security perimeters, called virtual perimeters, around them in the physical realm. Indoor positioning systems can be used to dynamically locate people to determine if they violate the virtual perimeter around an object. When an event such as someone crossing the virtual perimeter occurs, access control can be performed. This research proposes a dynamic, event-driven, access control framework to provide security and privacy in physical spaces using virtual perimeters and camera-based indoor positioning systems. The Virtual Perimeter Access Control framework aims to prevent physical access to objects by monitoring the locations of entities within the space and prompting access control to be performed once a virtual perimeter is violated.

**Keywords** Virtual perimeters · Cyber-physical · Access control · Privacy · Indoor positioning systems · Framework

B. Greaves (✉) · M. Coetzee · W. S. Leung
University of Johannesburg, Auckland Park 2006, South Africa
e-mail: bgreaves@uj.ac.za

M. Coetzee
e-mail: marijkec@uj.ac.za

W. S. Leung
e-mail: wsleung@uj.ac.za



X.-S. Yang et al. (eds.), *Proceedings of Sixth International Congress on Information and Communication Technology*, Lecture Notes in Networks and Systems 216,
https://doi.org/10.1007/978-981-16-1781-2_34

# 1 Introduction

Modern access control systems can be classified as either physical or cyber in nature [1]. Most modern physical access control systems lack granularity and only protect assets in the physical realm, at the perimeter of the space using doors and locks [1].

Smart technology has brought about an unprecedented rate of development in physical access control systems, with touchless access control becoming a new requirement post COVID-19. Unfortunately, these systems are limited in the level of protection they provide. Suppose a person moves past the door and enters a room. This person may be able to move into a physical location within the room containing sensitive assets such as documents displayed on monitors [2]. Documents are protected by a cyber access control system but once physically displayed on a monitor after access control is performed, the document loses its protection as anyone passing the monitor may view the document. The protection offered at the perimeter of the room is therefore insufficient. A need exists to protect the security and privacy of the information displayed in the physical realm by defining a perimeter around it. GeoFences [3] define a perimeter around assets so that actions can be taken once crossed. If someone stands close enough to a monitor to read private information, the document can be hidden from view. In previous research [4], the researchers proposed virtual perimeters to protect cyber assets within the perimeter of a space using modern indoor positioning systems.

In this paper, a Virtual Perimeter Access Control (VPAC) framework is proposed to protect assets within a virtual perimeter. The VPAC framework supports the monitoring of the locations of people within a space. Changing locations results in events that invoke the access control system to make decisions in both the cyber and physical realms.

The remainder of this paper is laid out as follows: Sect. 2 introduces the reader to the research scenario. Section 3 discusses access control and virtual perimeters. Section 4 presents the Virtual Perimeter Access Control framework. Section 5 discusses the framework and its limitations and the research is concluded in Sect. 6.

The following section discusses the problem domain by means of a scenario.

# 2 Motivating Scenario

Consider an office environment, shown in Fig. 1, where John and Mary are colleagues who deal with sensitive client information that must be kept strictly confidential.

John and Mary use their respective laptops to work. Mary can view documents shown on John's laptop monitor if she stands close enough, in the area labelled *A*. Mary can violate the confidentiality of the document by viewing his monitor. The document has therefore lost its cyber realm protection when displayed in the physical realm [1]. John may also use specialized equipment in area *B* that has no authentication service. To prevent Mary from interfering with it, she must be deterred

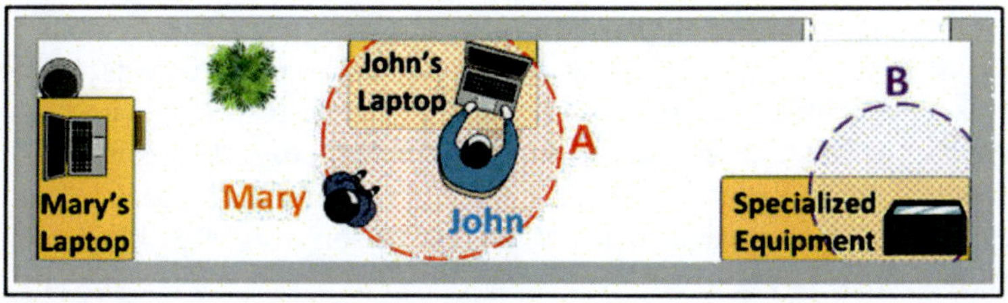


Fig. 1 John and Mary's open-plan office environment

from entering area *B*. Furthermore, John can move his laptop to attend meetings and the documents displayed on his monitor should still be protected. Therefore, there are three identified concerns: First, documents must have the same level of protection in both the physical and cyber realms. Second, equipment must not be used by unauthorized people. Third, the system must handle events as topology changes, for example, when John goes to a meeting.

The next section discusses virtual perimeter access control to address these concerns.

## 3 Virtual Perimeter Access Control

Access control is defined as the security service that limits access to resources for legitimate users of a system [5]. Access control falls into two broad domains: Physical access control provides the ability to control access to a physical space [5], often using doors and locks. Cyber access control limits access to cyber assets such as files [5].

In access control, a *subject* is a person or a user process that requests access to an *object* such as a cyber file or physical door [5]. The general architecture of an access control system is shown in Fig. 2a. The *Policy Enforcement Point (PEP)* intercepts the request, formats it, and sends it to the *Policy Decision Point (PDP)* to make an access decision. The *PDP* uses a *Policy Information Point (PIP)* to access information relevant to the access request. If the request is denied by the *PDP*, the *PEP* enforces

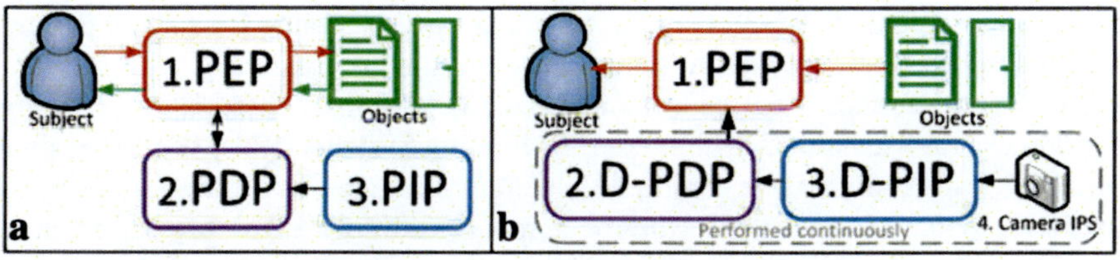


Fig. 2 a Subject- and b Event-driven cyber access control processes

the decision and denies the *subject* access to the *object,* otherwise access is granted. Access control decisions are made as subject request access to objects.

The access control system, however, must be able to perform access control dynamically as Mary enters the vicinity of John's laptop. A sensory mechanism is therefore required to track her location to determine if violations take place. Physical access control thus needs to be enforced in unison with cyber access control to prevent misbehaviours.

Access control using indoor localization is a diverse field of study as discussed in Sect. 3.1. Section 3.2 covers indoor positioning systems and virtual perimeters.

## *3.1 Cyber-Physical Access Control*

Access control for spaces with frequent topological changes is an active field of study. In a model by Jensen et al. [2], access control is performed for subjects entering the line of sight of sensors attached to monitors, preventing access control circumvention.

Yamanaka et al. [6] present an access control model that uses camera-based indoor GeoFences [3] to turn on Wi-Fi as subjects enter the area to reduce power consumption.

Cao et al. [7] propose a Topology-Aware Cyber Physical Access Control (TA-CPAC) model designed to dynamically assign permissions as subjects move about within the building. Each of these models provides possible avenues for virtual perimeter access control. The approach by [2] provides protection by securing the immediate vicinity of the object. The work of [6] highlights the usefulness of camera-based sensors and [7] stresses the importance of real-time topology information. The model from Yamanaka et al. [6] tracks users as they move about and supports highly granular, albeit static, GeoFences. An alternative approach [3] allows for movable GeoFences.

The next section discusses GeoFences using camera indoor positioning systems.

## *3.2 Indoor Positioning Systems for Virtual Perimeters*

GeoFence [3] technology allows for events to trigger when a perimeter is crossed. Outdoor GeoFences use GPS coordinates or 3G triangulation to form static perimeters. These do not adapt well to indoor use due to high inaccuracy or poor signal indoors and are therefore are seldom used [3]. Approaches such as [3] adapt GeoFences for use indoors with Indoor Positioning Systems (IPSs) [8]. In previous work [9], various IPSs were explored to identify a suitable candidate for use in the scenario. Systems such as RFID cards or cellphones are not ideal as they can be intentionally removed or unintentionally lost by those carrying them [8, 9]. The subjects thus become invisible to the system that is tracking their locations. Furthermore, a sensory inaccuracy of at most 30 cm is required to prevent false positives or negatives [1]. This eliminates most forms of triangulation as the expected accuracy is approximately 2 m [8], resulting in the subject being able to interact with equipment but the system does not know to perform access control.

Computer-vision-based pattern recognition systems were identified as the most viable candidates for an IPS [9] as they are accurate to a baseline of 20–30 cm [8] and provide abundant context information [6]. Furthermore, cameras that are installed in the environment are less likely to be tampered with as tokens or cellphones can be.

Previous research [4] presented virtual perimeters as an indoor analogue to GeoFence technology that uses IPSs to establish dynamic security perimeters around physical objects indoors. A *virtual perimeter* is established by specifying minimum proximity that a subject must be relative to the object before access control must be computed. If a subject is within the specified proximity, an access decision must be made and enforced dynamically. If a subject is not within the virtual perimeter, nothing needs to be done. The circles *A* and *B* in Fig. 1 are the virtual perimeters for John's laptop, and the equipment, respectively. An access control policy must specify the virtual perimeter proximity for reference by a PEP, similar to [2]. A policy for equipment should specify the minimum proximity for an alarm to be sounded, similar to [6]. Therefore, a camera-based IPS such as the Stereolabs ZED [10] can be used to determine proximity using depth maps. For this to be possible, the following requirements must be satisfied:

1. *Sensors should accurately identify and locate subjects and objects*: IPS systems must identify, and track subjects and objects dynamically for proximity calculations [9].
2. *Dynamic event-driven access control enabled through real-time IPS monitoring*: The system must report subjects and object proximity to an access control model in real time so that access decisions can be made dynamically [9].
3. *Cyber and physical access control must be enforced in unison*: If 1 and 2 are satisfied, the system must make access decisions when subjects enter virtual perimeters of objects. The decision must be enforced in a cyber capacity and/or physically [2].

The following section presents the VPAC access control framework.

## 4 Virtual Perimeter Access Control Framework

The Virtual Perimeter Access Control (VPAC) framework is presented by describing the access control process in Sect. 4.1. Section 4.2 discusses the framework.

### 4.1 Virtual Perimeter Access Control Process

In subject-driven cyber access control, as depicted in Fig. 2a, a subject requests access to an object, and access is either granted or denied. In the scenario, access control must be continuously computed as subjects move about the space. A coupling between the topology of the physical space reported by the camera IPS and the access control system in the cyber realm is required. When a virtual perimeter is crossed by a subject, an event is triggered to force the access control system to make a decision for any documents visible. If the subject is granted access, nothing needs to be done. Should access be denied, viewing of the documents must be prevented. To address these requirements, this research proposes an event-based access control system similar to [11]: Events trigger the PDP to make access control decisions dynamically. Event-driven decisions cause the PEP to enforce them in real time. Figure 2b shows the changes to the process.

The *Camera-based IPS* system (*4*) to the right provides the *D-PIP* (*3*) with images and depth information that identify *subjects* and *objects* such as laptops. The relative distance between them is computed. The *D-PIP* reports the relative distance of *subjects* and *objects* to the *D-PDP* (2) that computes access decisions for all *subjects* and *objects*. Once the *D-PDP* makes an access decision, it is reported to the *PEP* (*1*) to enforce. The monitor can be dimmed or the application displaying the sensitive document can be temporarily hidden. For specialized equipment, the *PEP* can sound an alarm.

### 4.2 Virtual Perimeter Access Control Framework

The physical and cyber spaces of the scenario must be considered together as each influences the other. The *physical space* contains subjects, their devices, IPSs, and countermeasures. The *cyber space* consists of the logical tools on the laptops and security systems. These are the *user* and *system spaces,* respectively. Each is networked environment within cyber space such that the *system space* is composed of security servers that limit access to files, and the *user space* that is the host environment of each laptop. For a physical subject to work on a cyber document, they must gain access to their laptop's *user space*. Once logged in, they indirectly use the *system space* security services via *user space* software. John is both a physical subject that moves around the physical space and a cyber subject that operates his laptop. He must authenticate to the security server to correlate his physical and cyber identities.

Figure 3 provides the overall structure of the VPAC framework and outlines the interactions of its components among the *physical*, *system*, and *user spaces*. Physical space components are connected to the *system space* components via a network. Prior to framework operation, security administrators register all subjects, objects, and set virtual perimeter proximities in the policy. This is stored in a central database.

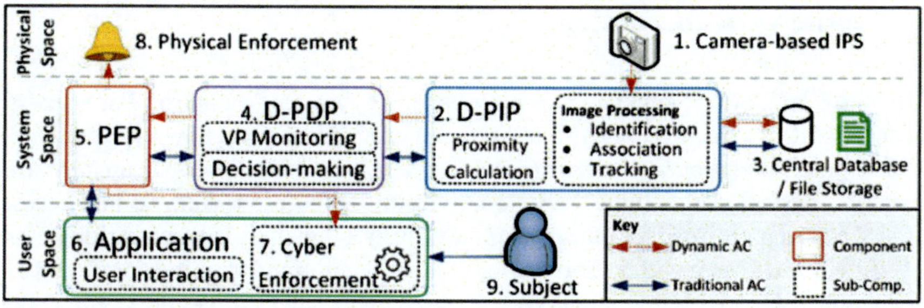


**Fig. 3** The VPAC framework for dynamic virtual perimeter access control

For *cyber access control*: A *subject* (*9*) authenticates at a laptop to use the *application* (*6*) to access files via the *user interaction* subcomponent. (*6*) sends an access request to the *PEP* (*5*) which forwards the request to the *D-PDP* (*4*). (*4*) requests all relevant information from the *D-PIP* (2). The framework is designed with common access control models such as RBAC [5] in mind. However, any model such as TA-CPAC [7] can be used provided that the *D-PDP* can handle the policies and rules. The *D-PIP* gets this information from the *central database* (*3*) before sending it to (*4*). The *D-PDP* makes a decision and sends it to (*5*) to be enforced by sending instructions to the *application.*

For *cyber-physical event-driven access control*: The system knows which permissions have been granted to users in the cyber realm. When a subject enters the room, the system tracks their location. *Camera-based IPSs* (*1*) capture images of the space and its occupants to report to (2). (2) provides policy information to (*4*) as well as processes images using the *image processing* subcomponent. *Identification*: People and objects are identified in images. The identification can be done using various image processing techniques. *Association*: The system associates a physical entity with a cyber entity. TensorFlow with a COCO dataset [12, 13] can be used for this purpose. *Tracking*: The system tracks entities as they move about the space to maintain the association with their cyber identities. DeepSort [14] can be of use here.

Once each entity has been *identified*, *associated* with their cyber identity, and is being *tracked*, the system computes the *proximity* of each person to each physical object by the *proximity calculation* subcomponent to report to (*4*). The *D-PDP* uses the *decision-making* subcomponent to correlate the proximity and tracking information with policy to determine if any subject is violating a virtual perimeter and sends a decision to the *PEP* (*5*). (*5*) instructs the *application* (*6*) to deploy *cyber enforcement* (*7*) measures. For physical equipment, the same process is followed except the resultant enforcement will take place in the physical realm by having (*5*) issue instructions to *physical enforcement* (*8*) to deter misuse via an alarm or deploy security personnel.

The next section discusses the framework and its access control requirements.

# 5 Framework Discussion

The VPAC framework amalgamates physical and cyber access control. For success, it must satisfy the requirements from Sect. 3.2, each of which is now discussed.

*Sensors should accurately identify and locate subjects and objects*: In the physical realm, a physical subject must be uniquely identified by the system so that access control can be performed for objects in their vicinity. Human and object tracking is an active field of research with various forms of identification and tracking possible [13]. The work of [1] highlights a major area where cyber-physical systems fail: There must be a tight coupling between the state of the physical realm and its representation in the cyber realm. The system must be able to correlate the authentication of cyber credentials with a physical person in an image feed from an IPS. This can be done using known static locations such as authentication points [15] to maintaining a tight coupling between physical and cyber identities. For object identification, the system must identify objects such as laptops or specialized equipment. This can be done by identifying equipment on the image feed by a security administrator, or through pattern recognition. Laptops can be moved, so manual administration is difficult and should be automated. Trained datasets can be used to identify laptops. Through experimentation, the researchers have identified that the use of TensorFlow [12] with the Common Object in Context dataset [13] is ideal for identifying people and laptops in images from IPSs. DeepSort [14] can be used to track subjects from the authenticator to the laptop upon which they log in. A physical subject can therefore be associated with a cyber subject. Thus, it is possible to satisfy this requirement using existing technologies.

*Dynamic event-driven access control enabled through real-time IPS monitoring*: Virtual perimeters must be continuously monitored as any lapse in time may allow for misuse of equipment or viewing of documents. Graphical Processing Units (GPUs) and cloud image processing [12] make it possible to process the high number of images produced by IPSs in real time [14]. Through experimentation, stereo camera systems [10] deliver sufficient images and depth maps to identify and track subjects and objects and compute proximity with sufficiently powerful GPUs. The researchers therefore believe it is possible to satisfy this requirement using available technology to enable events to be forwarded to the D-PDP for dynamic enforcement.

*Cyber and physical access control must be enforced in unison*: The system must be able to hide documents in a similar fashion to [2] by removing them from view and also physically deter misuse of equipment. It must also be considered if the system itself is misused to deny other employees from working. Repeated offences could constitute denial of service. In such a case, the reports from the *D-PIP* can be used to discipline repeat offenders. It can therefore be concluded that the VPAC framework is capable of securing John and Mary's office space so that physical objects are not misused, and cyber objects have the same level of protection in the physical realm as they do in the cyber realm. The next section concludes the research.

## 6 Conclusion and Future Work

This research has built upon virtual perimeter research to address problems that arise when access control is viewed from a purely cyber or physical perspective. Under certain conditions, objects that are protected in the cyber realm by cyber access control do not have the same protection when rendered in the physical realm.

A scenario was presented that details possible situations under which physical and cyber access control should be integrated into a single unit to reduce misuse or lapses in privacy. Access control requirements were extracted and a Virtual Perimeter Access Control (VPAC) framework was presented. There is, to the best of the researchers' knowledge, currently no real solution that utilizes virtual perimeters to perform access control to address the privacy or misuse concerns presented in the scenario.

Future work involves the implementation of the VPAC framework in the form of an access control model and prototype to test its real-world viability.

## References

1. Tsigkanos C, Pasquale L, Ghezzi C, Nuseibeh B (2016) On the interplay between cyber and physical spaces for adaptive security. 15:466–480
2. Jensen CD, Geneser K, Willemoes-Wissing IC (2013) Sensor enhanced access control: extending traditional access control models with context-awareness. In: Presented at the IFIP international conference on trust management
3. Young-Hyun E, Young-Keun C, Cho S, Jeon B (2017) FloGeo: a floatable three-dimensional geofence with mobility for the internet of things, pp 114–120
4. Greaves B, Coetzee M, Leung WS (2018) Access control requirements for physical spaces protected by virtual perimeters. In: International conference on trust and privacy in digital business, pp 182–197
5. Sandhu RS, Samarati P (1994) Access control: principle and practice. 32:40–48
6. Yamanaka G, Nishio T, Morikura M, Yamamoto K, Maki Y, Eitoku S, Indo T (2019) Geo-fencing in wireless LANs with camera for location-based access control. In: 16th IEEE annual consumer communications & networking conference (CCNC)
7. Cao Y, Huang Z, Kan S, Fan D, Yang Y (2019) Specification and verification of a topology-aware access control model for cyber-physical space. 24:497–519
8. Deak G, Curran K, Condell J (2012) A survey of active and passive indoor localisation systems. 35:1939–1954
9. ZED Stereo Camera. https://www.stereolabs.com/zed/. Last Accessed 16 April 2020
10. Greaves B, Coetzee M, Leung WS (2020) A comparison of indoor positioning systems for access control using virtual perimeters. In: Fourth international congress on information and communication technology, pp 293–302
11. Hernández JL, Moreno MV, Jara AJ, Skarmeta AF (2014) A soft computing based location-aware access control for smart buildings. 18:1659–1674
12. Bonatti P, Galdi C, Torres D (2013) ERBAC: event-driven RBAC. In: Proceedings of the 18th ACM symposium on access control models and technologies, pp 125–136
13. Unknown: Tensorflow Object Detection API. https://github.com/tensorflow/models/tree/master/research/object_daetection. Last Accessed 16 April 2020

14. Lin TY, Maire M, Belongie S, Hays J, Perona P, Ramanan D, Dollár P, Zitnick CL (2014) Microsoft coco: common objects in context. In: European conference on computer vision, pp 740–755
15. Wojke N, Bewley A (2018) Deep cosine metric learning for person re-identification. In: Proceeding of the winter conference on applications of computer vision (WACV), pp 748–756
16. Yap RH, Sim T, Kwang GX, Ramnath R (2008) Physical access protection using continuous authentication. In: IEEE conference on technologies for homeland security

# COVID-19-Related Communication on Twitter: Analysis of the Croatian and Polish Attitudes

**Karlo Babić, Milan Petrović, Slobodan Beliga, Sanda Martinčić-Ipšić, Andrzej Jarynowski, and Ana Meštrović**

**Abstract** In this paper, we analyze and compare Croatian and Polish Twitter datasets. After collecting tweets related to COVID-19 in the period from 20.01.2020 until 01.07.2020, we automatically annotated positive, negative, and neutral tweets with a simple method, and then used a classifier to annotate the dataset again. To interpret the data, the total number as well as the number of positive and negative tweets are plotted through time for Croatian and Polish tweets. The positive/negative fluctuations in the visualizations are explained in the context of certain events, such as the lockdowns, Easter, and parliamentary elections. In the last step, we analyze tokens by extracting the most frequently occurring tokens in positive or negative tweets and calculating the positive to negative (and reverse) ratios.

**Keywords** COVID-19 · Twitter · Social media · Sentiment analysis · NLP

K. Babić · M. Petrović · S. Beliga · S. Martinčić-Ipšić · A. Meštrović (✉)
University of Rijeka, Department of Informatics, 51000 Rijeka, Croatia
e-mail: amestrovic@uniri.hr

K. Babić
e-mail: karlo.babic@uniri.hr
URL: https://infocov.uniri.hr/

M. Petrović
e-mail: milan.petrovic@uniri.hr

S. Beliga
e-mail: sbeliga@uniri.hr

S. Martinčić-Ipšić
e-mail: smarti@uniri.hr

K. Babić · S. Beliga · S. Martinčić-Ipšić · A. Meštrović
Center for Artificial Intelligence and Cybersecurity, University of Rijeka, Rijeka, Croatia

A. Jarynowski (✉)
Interdisciplinary Research Institute, Wroclaw, Poland
e-mail: ajarynowski@interdisciplinary-research.eu



X.-S. Yang et al. (eds.), *Proceedings of Sixth International Congress on Information and Communication Technology*, Lecture Notes in Networks and Systems 216,
https://doi.org/10.1007/978-981-16-1781-2_35

## 1 Introduction

Social media accelerates information spreading and may cause an infodemic, especially during a crisis. As stated by WHO, the COVID-19 outbreak culminated with a massive infodemic, which is potentially dangerous because it makes it difficult for individuals to find reliable sources of information when they need it. In this light, automatic classification of positive, neutral, and negative attitudes in social media and automatic detection and prediction of fake news spreading play an important role and may improve various aspects of crisis communication.

Sentiment analysis refers to the set of NLP-based techniques used to identify, extract, or characterize opinions, emotions, and subjective information expressed using text [1]. Its main purpose is to classify attitudes related to various topics into positive, negative, or neutral. There are many possible applications of sentiment analysis [4, 12]. One very important application could be during natural disasters and emergencies when knowledge about positive and negative attitudes and sentiments, in general, could be used to improve situational awareness and crisis management. In the field of NLP, there have already been defined numerous approaches and techniques for sentiment analysis and opinion mining. However, the COVID-19 crisis brings a new set of challenges in terms of large communication volumes (massive datasets, new terminology, new aspects, and new specific topics that have come into focus).

As one of the most popular and used social networks, Twitter is one of the most studied networks in the domain of natural language processing (NLP) and social network analysis (SNA). There is a variety of tasks that ranges from tweet classification [17], hate speech detection [9] to link prediction [10, 11], etc.

Our work aims to deploy a multidisciplinary approach with a focus on quantitative analysis of empirical social media data from Twitter, allowing practical applications. Thus, the methods of Digital (computational) Epidemiology [14] make it possible to analyze a huge amount of data (Big Data) at a low cost which allows for the optimization of decision-making processes in public health in the context of COVID-19. We choose a comparative approach [15] between Poland and Croatia (both are post-communistic European countries), as political transformation and second demographic transition have common patterns. Moreover, the sociolinguistics perspective is more or less similar too (e.g., Polish and Croatian are in the same Slavic language group). In both countries, there was an election campaign in June 2020, which has impacted communications on Tweetosphere.

In this research, we deployed techniques from the field of NLP to analyze and compare Croatian and Polish COVID-19 tweets. First, we collected COVID-19-related tweets in the Croatian and Polish languages and performed automatic annotation of tweets using sentiment lexicons and a classifier. Next, we analyzed the distribution of positive and negative tweets over time. We found that there are some peaks of positive and negative tweets and try to explain them in the context of important events in Croatia and Poland. After that, we analyzed the most frequent tokens in positive and negative tweets in both datasets.

## 2 Background and Related Work

There are plenty of attempts around the challenge of media research in the COVID-19 era.

One of the most detailed and relevant studies about tweets' sentiment in the era of the pandemic is described in [19]. Xue et al. perform sentiment analysis of COVID-19 tweets by classifying each tweet into eight pairwise emotions: joy-sadness, trust-disgust, fear-anger, and surprise-anticipation. Their research is based on the 1.9 million tweets related to coronavirus tweeted at the beginning of the pandemic. Moreover, the authors identify 11 topics using the LDA method and show that fear is the dominant emotion in all tweets. In [8], the authors aim to examine trends of four emotions: fear, anger, sadness, and joy during the COVID-19 pandemic. Their results show that public emotions shifted from fear to anger during the pandemic. In [2], the authors examine key themes and topics of COVID-19-related tweets in English at the beginning of the pandemic. All of these studies are focused mostly on the English language domain. Some research examines other languages and states as well, such as [6], a study which analyzes the social dimension of the public health in Poland; [18] which examines irony on Twitter as the first response to the pandemic in Italy; [13] which analyzes sentiment of the tweets in Nepal.

## 3 Methodology

### 3.1 Datasets

We collect 3,945 tweets in the Croatian language and 989,004 tweets in the Polish language during the period of the first five months of the pandemic, January 20–July 1, 2020. We filter relevant Croatian tweets by selecting the tweets that include COVID-19 or other related terms. For Polish tweets [5], we filter the relevant tweets with a selection criteria: tag #Koronawirus and language Polish.

The annotation method we use follows two main steps for both Croatian and Polish datasets: automatic annotation and trained annotation.

Automatic annotation uses a sentiment lexicon [3] (which contains positive and negative words) to score each tweet by calculating the number of positive words minus the number of negative words in a tweet. By doing so a tweet with more positive words than negative will have a positive score, and vice versa. A tweet is negative if the calculated score is smaller than $-5$, positive if the score is larger than 3 for Croatian tweets and larger than 8 for Polish tweets, and neutral otherwise. Words in the lexicons are first shortened by removing the last 20% of letters in words that have more than 4 letters. By doing so, those words have a higher chance of matching the words in tweets.

Annotations produced with the automatic annotator are rough and the number of negative and positive tweets is very low. Because of the strict automatic annotator,

**Table 1** Annotated dataset statistics

| | Total tweets | Negative (%) | Neutral (%) | Positive (%) |
|---|---|---|---|---|
| Croatian | 3,945 | 32.2 | 12.1 | 55.7 |
| Polish | 989,004 | 51.9 | 31.4 | 16.6 |

tweets annotated as negative or positive are very likely in the right class. All tweets for which the automatic annotator is uncertain (a majority) are annotated as neutral.

Trained annotation is an iterative process of training a classifier:

i Initially, the classifier is trained on the annotated data produced with the automatic annotator.
ii After the initial training, it classifies the neutral tweets (as annotated by the automatic annotator) into positive, negative, and neutral.
iii After that step, the total number of positive and negative tweets is higher, and the classifier is trained again on the newly annotated data.
iv All tweets are then again annotated by the classifier.

For the classifier in the trained annotation step, we use Naive Bayes. We vectorized the data by using a bag-of-words approach, where singular words and word-pairs (two-grams) are used as units. After the two steps of annotation, the resulting dataset is described in Table 1.

The annotation method is evaluated on datasets that are manually annotated. They have 99 tweets and 618 tweets for the Croatian and Polish datasets, respectively. The number of negative, neutral, and positive tweets in those datasets is equal. Tweets with the same content as some of the tweets in the evaluation datasets are removed from the training datasets. The Croatian annotator achieves 50.5% accuracy, and the Polish annotator achieves 37.4% accuracy.

## 3.2 *Data Analysis*

We first analyze the number of COVID-19-related tweets per day during the period of 163 days for the Croatian and Polish languages separately.

Next, we analyze the sentiment over time by using mean values and the normalized ratio of positive and negative tweets for both the Croatian and Polish datasets. We explain positive and negative peaks for certain days in the context of the social and political events related to the pandemic.

In the last step, we represent each tweet as a list of tokens. Tokens are sequences of characters separated by spaces or punctuation marks. As lemmatization or stemming did not improve annotations, tokens are not further processed. Then we analyze the most frequent COVID-19-related tokens that appear in positive and negative tweets separately.

# 4 Results

First, by showing the number of tweets per day, we can see when tweets about COVID-19 were trending (Fig. 1). The first peak in the number of tweets about COVID-19 happened at the same time in Croatia and Poland, on the 35th day (25.02.2020). The biggest peak happened first in Poland, on the 50th day (11.03.2020), and Croatia followed on the 60th day (21.03.2020). Interest in coronavirus in Poland is similar to pan-European peak around 11–16.03.2020 [7], however, Croatia is an exceptional case. In Poland, there was a sharp and steep increase in interest, while in Croatia attention was built week by week until it reached its peak. The number of tweets about COVID-19 gradually got smaller, until the 150th day (19.06.2020) when in Croatia the number of COVID-19-related tweets begins to climb again. All the mentioned peaks correlate with the COVID-19 infection waves.

## *4.1 Trends of Positive and Negative Attitudes in Tweets*

By setting the values for negative, neutral, and positive sentiment to −1, 0, and 1, it is possible to calculate and visualize sentiment behavior through time by using different methods.

As is visible in Fig. 1, the number of tweets at the beginning of the pandemic and the end (except the very end which correlates with the second wave) is very low. Because of that, sentiment data is noisy at those periods.

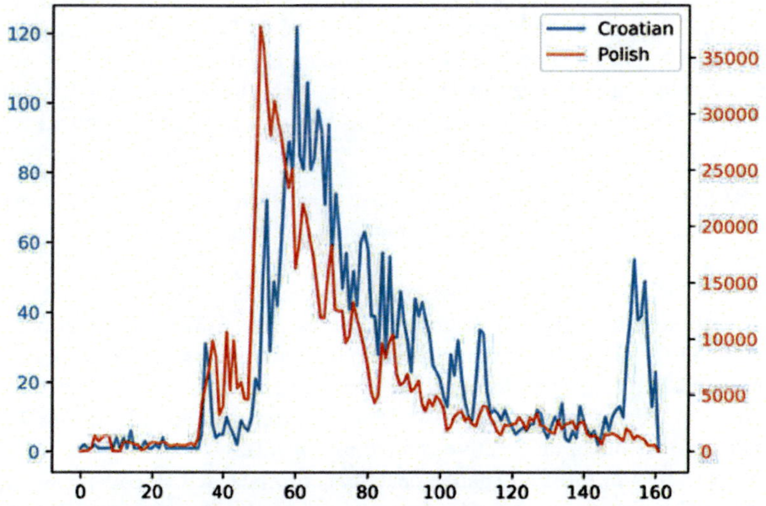


**Fig. 1** The number of Croatian and Polish tweets about COVID-19 (blue and red, respectively) per day calculated from 20.01.2020 until 01.07.2020

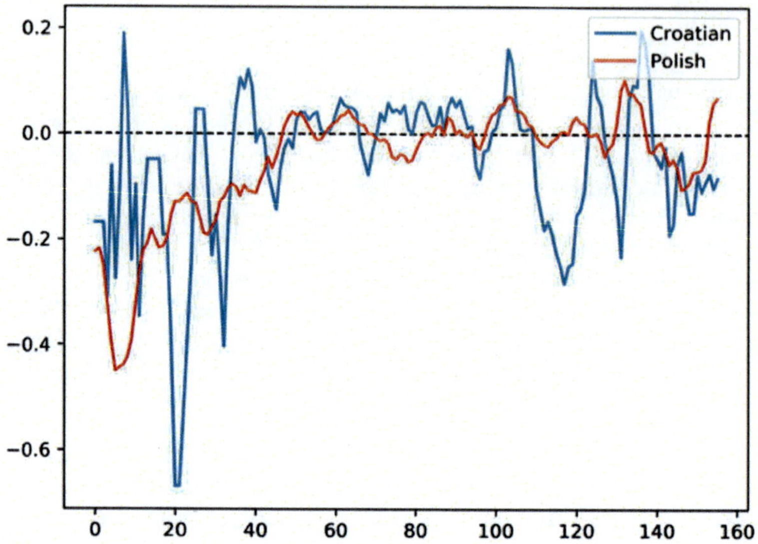


**Fig. 2** Here we plotted the mean value for Croatian and Polish tweets (blue and red, respectively) through time, aggregated by 7 days (sliding window with length equal to 7 days)

For Fig. 2, sentiment values are first averaged per day. That step normalizes the dataset so the differences in the number of tweets per day do not affect the visualizations. The values that are plotted are calculated by taking the average value of the sliding window (which is 7 days long). In this plot correlation of trend in sentiment, change is visible. Both in Croatia and Poland, sentiment was at first at its lowest, after which it begins to climb until the 45th day (06.03.2020), where it stabilizes.

Figure 3 plots the ratio between the normalized number of positive and negative tweets (the values are first averaged with a 7-day long sliding window). By doing so, we can more clearly see the local correlations between Croatia and Poland.

The visible features in Fig. 3 are

- Day 33 (23.02.–01.03. average): A positive peak, especially in Croatian tweets. At that time first COVID-19 cases were detected in Croatia, but it seems that people are ironic and in denial.
- Day 41 (02.03.–09.03.): A negative peak. Croatians began to realize that COVID-19 is a serious threat and events started to cancel. First COVID-19 cases in Poland.
- Day 50 (11.03.–18.03.): After the initial reaction, this positive peak in sentiment corresponds with the acceptance of the current status in both countries. Probably due to consolidation of the society and forming support networks among Internet users (more visible in the relative change in Poland).
- Day 57 (18.03.–25.03.): A negative peak that correlates with the worst day in Italy so far. In Croatia, the number of infections rises and a destructive earthquake hits the capital of Croatia.

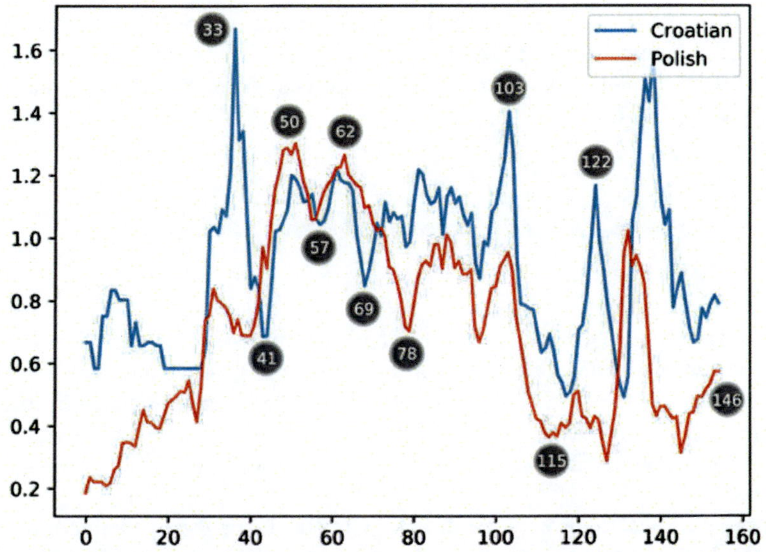


**Fig. 3** Normalized ratio of number of positive to number of negative tweets for Croatian and Polish tweets (blue and red, respectively) through time, aggregated by 7 days (mean value from the sliding window with length equal to 7 days)

- Day 62 (23.03.–30.03.): Sentiment has a positive peak as people started to post tweets supporting the healthcare system, at-risk groups, and respect measures.
- Day 69 (30.03.–06.04.): A negative peak in Croatia that could be explained by the depression caused by isolation and the government closing the borders. Polish tweets show a downward trend due to the lockdown restrictions being tightened and the consolidation effect has probably ended.
- Day 78 (08.04.–15.04.): A negative peak, probably due to isolation at the time of Easter.
- Day 96 (26.04.–03.05.): A negative peak in Poland potentially corresponding to decisions in schooling.
- Day 103 (03.05.–10.05.): A positive peak correlating with almost no new COVID-19 cases in Croatia and restrictions being released (e.g., borders being opened in Poland).
- Day 115 (15.05.–22.05.): A negative peak that was probably due to the Croatian parliamentary election and Polish presidential election campaigns.
- Day 122 (22.05.–29.05.): A positive peak in Croatia as the number of COVID-19 cases is extremely low and it seems as if the pandemic ended.
- Day 129 (29.05.–05.06.): A negative peak after a short period of optimism due to restriction releasing in Poland.
- Day 135 (04.06.–11.06.): A positive peak due to another branch of restriction releasing in Poland.

- Day 146 (15.06.–22.06.): A negative trend correlating with the second wave in Croatia. It could be related to the political campaign and polarization of societies before the Polish presidential elections on 28.06. and Croatian parliamentary election on 05.07.

### *4.2 COVID-19-Related Tokens in Positive and Negative Tweets*

We analyze token frequencies in negative and positive tweets. In Table 2 and Table 3, tokens are sorted by their frequency of occurrence in positive and negative tweets for Croatian and Polish tweets, respectively. Along with the frequency of tokens in positive and negative tweets, ratios are shown as well (e.g., token "masks" has a ratio of 1.43:1, which means that it occurs in 1.43 times more positive than negative tweets).

In Croatia, tweets directly mentioning COVID-19 or the pandemic are mostly negative. Political tweets are often negative as well. Tweets mentioning measures such as quarantine and masks are mostly positive.

In Poland, negative tweets are frequently associated with reporting cases, deficiency in the healthcare system, economic crisis, or elections. Positive tweets are frequently associated with anti-crisis, social mobilization, and support.

## 5 Discussion and Conclusion

This is a comparative approach to present Twitter discourse on COVID-19, in Poland and Croatia, in the first wave of COVID-19 (from 20.01. to 01.07.).

The interest in the COVID-19 topic on Twitter in Europe is primarily of a social, not medical, dimension and the peaks of interest and changes in sentiment are mainly driven by social or political issues. Tweets about COVID-19 in Poland and Croatia have the same intensity patterns (Fig. 1) as in many other countries [7], however, peak of interest in Croatia is a week later than in most of the European countries. Our study is confirming that interest in COVID-19 on Twitter is unrelated to the actual physical risk of acquiring COVID-19 in Poland (official disease prevalence) which could be mostly the case corresponding to Google Trend Studies [7, 16], but it is slightly correlated with the incidence in Croatia. Sentiment analysis (Fig. 3) revealed positive attitudes during Stay at Home consolidation and the support phase during the second part of March and negative attitudes during political campaigns in late June. Positive/negative tweet's ratio (Fig. 3) is a more informative factor than average sentiment (Fig. 2).

**Table 2** The most frequent tokens in positive (left) and negative (right) Croatian tweets translated to English. Ratios of positive to negative (left) and ratios of negative to positive (right) token occurrences. Ratios are normalized with respect to the total number of positive and negative tweets

| N | Ratio | Token | N | Ratio | Token |
|---|---|---|---|---|---|
| 1028 | 0.73 | Coronavirus | 818 | 1.38 | Coronavirus |
| 294 | 0.78 | Virus | 231 | 1.43 | Virus |
| 283 | 1.1 | Corona | 189 | 2.14 | Covid19 |
| 270 | 1.2 | Stayhome | 168 | 1.94 | Corona |
| 255 | 0.86 | Covid19 | 149 | 1.5 | Self-isolation |
| 209 | 1.04 | Measures | 130 | 0.83 | Stayhome |
| 208 | 1.81 | Quarantine | 118 | 1.37 | Viliberos |
| 182 | 2.38 | Croatia | 112 | 1.9 | Croatia |
| 178 | 0.72 | Self-isolation | 107 | 0.99 | Measures |
| 153 | 2.18 | Isolation | 105 | 1.28 | Ravnateljstvocz |
| 149 | 0.73 | Viliberos | 93 | 1.4 | People |
| 142 | 0.78 | Ravnateljstvocz | 80 | 1.28 | Vladarh |
| 108 | 0.78 | Vladarh | 69 | 0.6 | Quarantine |
| 94 | 1.43 | Masks | 61 | 1.53 | Thanks |
| 85 | 0.95 | Headquarters | 60 | 1.12 | Time |
| 83 | 1.1 | Zagreb | 59 | 9.11 | Pandemic |
| 79 | 6.95 | Work | 53 | 13.74 | County |
| 69 | 0.65 | Thanks | 52 | 1.06 | Headquarters |
| 65 | 1.04 | Home | 47 | 1.52 | Zagreb |
| 57 | 1.27 | Help | 46 | 1.66 | Mup |
| 54 | 0.89 | Pandemic | 45 | 3.71 | Hdz |
| 53 | 1.8 | Protection | 44 | 1.58 | Question |
| 52 | 0.51 | Infected | 42 | 1.51 | Davorbozinovic |
| 49 | 1.35 | Crisis | 42 | 9.1 | Test |
| 48 | 0.6 | Mup | 40 | 1.82 | Dnevnikhr |

For future work, we plan to expand the datasets by using a longer range of dates, and by training a classifier that can filter more precisely tweets related to COVID-19. Comparison of sentiment trends for COVID-19-related tweets and tweets that are not related to COVID-19 could provide us with a new understanding. Further, we can examine semantic content by analyzing the distribution of topics through time.

**Table 3** The most frequent tokens in positive (left) and negative (right) Polish tweets translated to English. Ratios of positive to negative (left) and ratios of negative to positive (right) token occurrences. Ratios are normalized with respect to the total number of positive and negative tweets

| N | Ratio | Token |
|---|---|---|
| 162037 | 1.01 | Coronavirus |
| 42291 | 5.47 | People |
| 39266 | 235.18 | Voivodeship |
| 36822 | 16.18 | Cases |
| 32346 | 30752.57 | Confirmed |
| 31776 | 5.35 | Tests |
| 28351 | 1.38 | Government |
| 23936 | 19.21 | Infections |
| 23757 | 8.94 | New |
| 21487 | 67179.23 | Lab |
| 18524 | 57915.4 | Result |
| 18491 | 57812.22 | Positive |
| 15424 | 1.89 | Health |
| 13739 | 1.09 | Law & justice |
| 11438 | 0.93 | Just |
| 9830 | 1.13 | Day |
| 9390 | 13.66 | Issue |
| 8860 | 27700.84 | Related |
| 16370 | 1.09 | Poland |
| 7645 | 0.99 | Stay at home |
| 7236 | 1.24 | Epidemic |
| 6480 | 20259.76 | Lower silesia |
| 6422 | 1.54 | Poles |
| 6420 | 3.7 | Life |
| 6401 | 23.41 | Safety |

| N | Ratio | Token |
|---|---|---|
| 56784 | 0.86 | Government |
| 52595 | 1.77 | Coronavirus in Poland |
| 502491 | 0.99 | Coronavirus |
| 47429 | 0.94 | Poland |
| 39473 | 0.9 | Epidemic |
| 39363 | 0.92 | Law & justice |
| 38549 | 1.74 | Covid19 |
| 30077 | 1.39 | Minister |
| 27182 | 0.88 | Today |
| 25536 | 0.53 | Health |
| 24163 | 0.18 | People |
| 22308 | 1.43 | Time |
| 18556 | 0.19 | Tests |
| 18340 | 1.83 | Polish |
| 18152 | 2.43 | Know better |
| 17820 | 0.9 | May |
| 17795 | 1.16 | Morawiecki |
| 16433 | 1.13 | Lukasz Szumowski |
| 15751 | 1.25 | Pandemic |
| 15428 | 1.64 | Million |
| 14844 | 1.44 | Days |
| 14586 | 0.86 | Relationship |
| 14266 | 2.0 | Thousand |
| 27644 | 0.81 | Fight |
| 13696 | 1.26 | Andrzej duda |

**Acknowledgements** This work has been supported in part by the COST Action CA15109 COST-NET and by the Croatian Science Foundation under the project IP-CORONA-04-2061, "Multilayer Framework for the Information Spreading Characterization in Social Media during the COVID-19 Crisis" (InfoCoV).

## References


1. Beigi G, Hu X, Maciejewski R, Liu H (2016) An overview of sentiment analysis in social media and its applications in disaster relief. In: Sentiment analysis and ontology engineering. Springer, Berlin, pp 313–340. https://doi.org/10.1007/978-3-319-30319-2_13
2. Chandrasekaran R, Mehta V, Valkunde T, Moustakas E (2020) Topics, trends, and sentiments of tweets about the covid-19 pandemic: temporal infoveillance study. J Med Internet Res 22(10):e22–624. https://doi.org/10.2196/22624
3. Chen Y. Skiena S (2014) Building sentiment lexicons for all major languages. In: Proceedings of the 52nd annual meeting of the association for computational linguistics (volume 2: short papers), pp 383–389. https://doi.org/10.3115/v1/P14-2063
4. Jakopović H, Mikelić Preradović N (2016) Identifikacija online imidža organizacija temeljem analize sentimenata korisnički generiranog sadržaja na hrvatskim portalima. Medijska istraživanja: znanstveno-stručni časopis za novinarstvo i medije 22(2):63–82. https://doi.org/10.22572/mi.22.2.4
5. Jarynowski A (2020) A dataset of media releases (Twitter, News and Comments, Youtube, Facebook) form Poland related to COVID-19 for open research. Zenodo. https://doi.org/10.5281/zenodo.4319813
6. Jarynowski A, Wójta-Kempa M, Płatek D, Czopek K (2020) Attempt to understand public health relevant social dimensions of covid-19 outbreak in Poland. Available at SSRN 3570609. https://doi.org/10.2139/ssrn.3570609
7. Lampos V, Moura S, Yom-Tov E, Cox IJ, McKendry R, Edelstein M (2020) Tracking covid-19 using online search. arXiv:2003.08086
8. Lwin MO, Lu J, Sheldenkar A, Schulz PJ, Shin W, Gupta R, Yang Y (2020) Global sentiments surrounding the covid-19 pandemic on twitter: analysis of twitter trends. JMIR Public Health Surveill 6(2):e19–447. https://doi.org/10.2196/19447
9. Markoski F, Zdravevski E, Ljubešić N, Gievska S (2020) Evaluation of recurrent neural network architectures for abusive language detection in cyberbullying contexts. In: Proceedings of the 17th international conference on informatics and information technologies-CIIT 2020. http://hdl.handle.net/20.500.12188/8269
10. Martinčić-Ipšić S, Močibob E, Meštrović A (2016) Link prediction on tweets' content. In: International cconference on information and software technologies. Springer, Berlin, Germany, pp 559–567. https://doi.org/10.1007/978-3-319-46254-7_45
11. Martinčić-Ipšić S, Močibob E, Perc M (2017) Link prediction on twitter. PLoS One 12(7):e0181–079. https://doi.org/10.1371/journal.pone.0181079
12. Načinović L, Perak B, Meštrović A, Martinčić-Ipšić S (2012) Identifying fear related content in croatian texts. In: Proceedings of the eighth language technologies conference, pp 153–156
13. Pokharel BP (2020) Twitter sentiment analysis during covid-19 outbreak in nepal. Available at SSRN 3624719. https://doi.org/10.2139/ssrn.3624719
14. Salathé M (2018) Digital epidemiology: what is it, and where is it going? Life sciences, society and policy 14(1):1. https://doi.org/10.1186/s40504-017-0065-7
15. Strzelecki A, Azevedo A, Albuquerque A (2020) Correlation between the spread of covid-19 and the interest in personal protective measures in Poland and Portugal. In: Healthcare. Multidisciplinary Digital Publishing Institute, p 203. https://doi.org/10.3390/healthcare8030203
16. Szmuda T, Ali S, Hetzger TV, Rosvall P, Słoniewski P (2020) Are online searches for the novel coronavirus (covid-19) related to media or epidemiology? A cross-sectional study. Int J Infect Dis. https://doi.org/10.1016/j.ijid.2020.06.028
17. Tutek M, Sekulić I, Gombar P, Paljak I, Čulinović F, Boltužić F, Karan M, Alagić D, Šnajder J (2016) Takelab at semeval-2016 task 6: stance classification in tweets using a genetic algorithm based ensemble. In: Proceedings of the 10th international workshop on semantic evaluation (SemEval-2016), pp 464–468. https://doi.org/10.18653/v1/S16-1075
18. Vicari S, Murru MF (202) One platform, a thousand worlds: on twitter irony in the early response to the covid-19 pandemic in Italy. Soc Media + Soc 6(3):2056305120948–254. https://doi.org/10.1177/2056305120948254

19. Xue J, Chen J, Chen C, Zheng C, Li S, Zhu T (2020) Public discourse and sentiment during the covid 19 pandemic: using latent dirichlet allocation for topic modeling on twitter. PLoS One 15(9):e0239–441. https://doi.org/10.1371/journal.pone.0239441

# Development of a Prototype of an Intelligent System for the Diagnosis and Treatment Recommendations for Coffee Crop Diseases

**Esteban González Peña, David Alejandro Vergara Arango, and Jorge E. Giraldo Plaza**

**Abstract** Coffee is one of the most important sectors for the Colombian economy, nonetheless, it suffers a lot of complications: from difficulties to acquire resources to climate change and diseases that attack the crops. This is why it is important to provide mechanisms so the farmers can get decisive information about the health state of their crops and get suggestions about possible treatments to resolve their current state. Because of this, we suggest the use of an intelligent system, capable of recognizing coffee diseases using image processing and being able to recommend treatments based on the current environment of the crop. Our proposed system uses a convolutional neural network for image recognition and a collaborative recommendation system for the generation of treatments.

**Keywords** Coffee disease · Identification system · Recommendation system · Intelligent system · Convolutional neural networks · Coffee production

## 1 Introduction

Coffee crops are affected by different nutritional deficiencies, such as nitrogen deficiency in the soils, which is the cause of premature fruit drop. Low phosphorus levels cause, in the worst-case scenario, a partial or total fall of the leaves of the plant thus causing loss of fruits in the growth process. A lack of iron is reflected in the pale color of the leaves [1]. Adding to the deficiencies of vital nutrients, coffee crops are attacked by diseases, among which is the coffee rust or rust, caused by the fungus Hemileia vastatrix which is one of the main threats to coffee production; this fungus mainly

E. G. Peña (✉) · D. A. V. Arango · J. E. G. Plaza
Politécnico Colombiano Jaime Isaza Cadavid, Medellín, Colombia
e-mail: esteban_gonzalez82151@elpoli.edu.co

D. A. V. Arango
e-mail: david_vergara82141@elpoli.edu.co

J. E. G. Plaza
e-mail: jegiraldo@elpoli.edu.co



X.-S. Yang et al. (eds.), *Proceedings of Sixth International Congress on Information and Communication Technology*, Lecture Notes in Networks and Systems 216,
https://doi.org/10.1007/978-981-16-1781-2_36

attacks grains grown in Colombia that come from the Coffea Arabica family; infected crops develop haustoria, which are responsible for depriving the plant of nutrients, turning the leaves with a characteristic yellowish color; once the disease progresses, the infected plant loses the ability to photosynthesize, thus causing defoliation of the plant [2]. On the other hand, the iron stain caused by the fungus Cercospora coffeicola also attacks the leaves of coffee crops, causing spots in which the conidiophores and conidia of the fungus are found, thus generating white-colored spots surrounded by a brown color in halo shape; at the peak of infection, the plant can lose up to 90% of its foliage, and can reach the point of death [3]. In addition, the disease of coffee cherries caused by the fungus Colletotrichum kahawae, characterized by causing anthracnose on the leaves of coffee flowers, causes spots on the edges and tips of the leaves; at later phases it can cause dryness in the branches for their subsequent death [4, 5].

Colombia is ranked as the third country in terms of exports of this product worldwide. [6]. Due to this, it is of vital importance for the sector to have tools with which to combat threats, such as diseases that attack crops that can have consequences such as a decrease in productive capacity.

It is estimated that at least 70% of the diseases that occur in crops are caused by external agents, such as fungi [2], according to data from studies by Rivillas et al. [7]. According to Mahlein et al. [8] at present, the work of identifying the different pathologies can be a tedious, time-consuming, and expensive work, in addition to suffering great difficulties due to the limited qualities of the human being, thus creating a wide range of errors when performing a diagnosis.

## 2 Materials and Methods

For the preparation of this work, different activities were carried out. First, the variables of interest are defined, then the modeling was carried out of the system making use of the information obtained previously. As a next step, the model with its corresponding characteristics was implemented, and finally tests were carried out on the different modules to verify their correct operation. Below is the development of the activities previously listed.

### *2.1 Defining of the Variables Involved in the System*

In Table 1 can be seen the variables that were of interest, because it is considered that those variables have an impact on coffee crops, some of them, influencing the apparition of diseases.

These variables are treated as categorical to guarantee the correct functioning of the recommendation system.

**Table 1** System variables

| Name | Identifier | Type | Range |
|---|---|---|---|
| Precipitation | PCN | INT | 1–6 |
| Relative humidity | HUM | INT | 1–6 |
| Solar time | TSL | INT | 1–6 |
| Solar intensity | ISL | INT | 1–6 |
| Crop age | AGE | INT | 1–6 |
| Past record | ANT | INT | 1–6 |
| Fertilization strategy | EST | INT | 1–6 |
| Crop density | DNS | INT | 1–6 |

## 2.2 *Modeling of the Identification and Recommendation System Using Image Processing*

**Identification System**

The proposed system takes as a starting point the image that was provided by the user, then, processes it and sends it to the convolutional neural network used to provide the identifications to the system.

**Neural Network Design**

The proposed model can be seen in Fig. 1, and is made up of the following components.

**Data Set Used for Network Training**

For the training of the model, the BRACOL [9] data set was used, which has 3892 images of coffee leaves which are distributed among 5 different classes: health, miner, rust, phoma, and cerscospora. The images are distributed as follows (Table 2).

Of the data set, 60% of it was allocated for training, 20% for validation, and the last 20% for testing.

First, the base model used was previously trained with the ImageNet data set; this base model is used as a feature extractor, and its convolutional base was frozen, thus preventing its layer weights from being updated during the first part of the training.

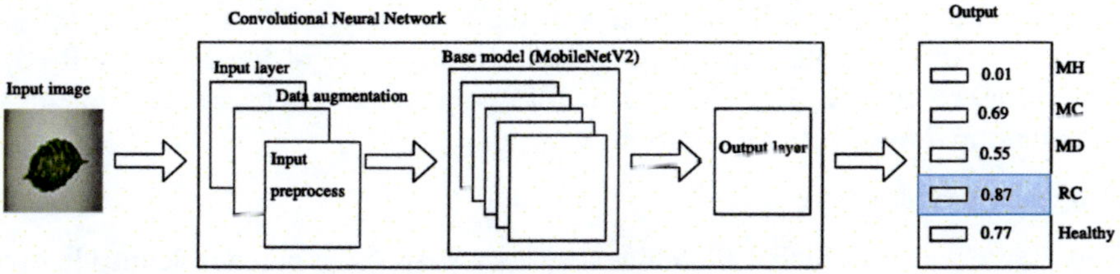


**Fig. 1** Diagram of the proposed convolutional neural network model

**Table 2** Distribution of the classes in the BRACOL data set

| Bracol | |
|---|---|
| Classes | Number of images |
| Health | 533 |
| Miner | 927 |
| Rust | 1152 |
| Phoma | 812 |
| Cercospora | 468 |

Afterwards, the convolutional neural network model proposed for the classification of diseases in coffee was trained with the BRACOL data set. After having been trained for 10 epochs, a fine adjustment was made on the said model; after the fine-tuning was done, the model was trained 20 more times.

**Recommendation System**

The type of recommendation system to be applied is collaborative, since it bases its operation on similar characteristics among users to make recommendations; however, these systems suffer from data deficiencies at the time they come into operation, thus leading to inappropriate recommendations [10].

The proposed recommendation system has four recommendation tables, one for each pathology with which the system works. In these tables, the values of the characteristics entered by the users are stored; with this information, the users who have characteristics more similar to those entered by a new user are searched to obtain a treatment that adjusts to the characteristics entered by this new user; to find the most similar users, the cosine ratio metric is used; before applying this metric, the data will be centered to obtain the similarity patterns between the users; the formula used is described in the following equation:

$$\bar{x} = \bar{x} - mean(\bar{x})$$
$$\bar{y} = \bar{y} - mean(\bar{y})$$
$$sim(x, y) = \frac{\bar{x} * \bar{y}}{\|\bar{x}\| * \|\bar{y}\|} \tag{1}$$

where $\bar{x}$ is the vector of characteristics of the new user, and $\bar{y}$ it is the vector of characteristics of an old user of the system; in this order of ideas, the similarity between user X (new user) and all users $y_n$ (old users) of the system, these calculations must be stored from a vector of size n.

**Proposed Architecture**

The architecture on which the intelligent system prototype is based is defined in two main components which are identification component and recommendation component; there are also other components in charge of controlling the input and output

```
# Cargamos el mode base pre-entrenado MobileNetV2
IMG_SHAPE = IMG_SIZE + (3,)
base_model = tf.keras.applications.MobileNetV2(input_shape=IMG_SHAPE,
                                               include_top=False,
                                               weights='imagenet')
```

**Fig. 2** Mobilenet V2 model load

of information, all of which are, since the components are communicated through controllers, in charge of validating data and transferring information between the information store and the different components of the system; Fig. 2 generally describes the design applied to the system.

## *2.3 Elaboration of the Prototype of the Identification and Recommendation System Making Use of Programming and Training Techniques for the Implementation of the Intelligent System*

For the elaboration of the intelligent system prototype, different development tools were used like *TensorFlow* and *Django framework*. These tools were used for the implementation of the different modules that make up the intelligent system.

**Implementation of the Identification System**

The identification system implemented is based on the image that was provided by the user in the data entry form, then this image must be preprocessed to extract the background, with the aim of minimizing the noise that the convolutional neural network on which the identification system is supported must deal with. This was achieved using the *OpenCV* library, applying the HSV color model (Hue, Saturation, Value).

**Implementation of the Convolutional Neural Network**

For the preprocessing of the data, a data augmentation was carried out, making use of the *Keras* API. It applies random changes to the data such as flips, and rotations, aiming to reduce overfitting and increase the base knowledge of the model.

It was also necessary to normalize the inputs by using a function that is responsible for adjusting the images to values with which the base model Mobilenet_V2 could work.

Then the base model Mobilenet_V2 was loaded, together with the weights of the trained model with the ImageNet data set as shown in Fig. 2.

```
prediction_layer = tf.keras.layers.Dense(5, name = 'Output', activation
                   = 'softmax', kernel_regularizer = regularizers.l2(0.0004))
```

**Fig. 3** Output layer configuration

Finally, an output layer was added to the upper part of the model, which is responsible for the classification task through a softmax function as shown in Fig. 3.

Finally, the prediction is passed to the recommendation system through the controller, which queries the database to obtain all known treatments for the disease.

**Implementation of the Recommendation System**

Once all the similarity calculations have been carried out using formula (1), data must be sorted descendingly so the first three values are returned to the *Django* controller to make the necessary queries to the database to obtain the information of the corresponding recommendations, so that they are shown to the user in the final form of the diagnosis, where the details are of the disease identified, in addition to the recommendations to treat the disease.

However, if at least ten treatments are not known for the disease identified by the identification system, the above is omitted, and generic treatments provided by CENICAFÉ literature are recommended; this is done due to the cold start problem that collaborative recommendation systems have.

## 2.4 Validation of the Prototype Obtained Through Practical Verification with Experts for the Detection and Correction of Errors

**Identification System**

After using de designed part of the data set meant for testing, the model archived the following results, where the following table describes the performance obtained for the model (Table 3).

**Recommendation System**

The results of applying Eq. (2) vary between [0, 1], with 0.5 being the perfect value, as it means that the recommendations associated with the diagnosis are completely

**Table 3** General model metrics

| | |
|---|---|
| Accuracy | 0.90 |
| Precision | 0.91 |
| Sensibility | 0.90 |
| Specificity | 0.90 |
| F-Value | 0.91 |

**Table 4** Values and results of applying precision formula

| | Input #1 | Input #2 | Input #3 | Input #4 | Input #5 | Mean |
|---|---|---|---|---|---|---|
| Input values | 2.90 | 3.44 | 3.78 | 4.11 | 3.33 | 3.51 |
| Recommendation values | 2.68 | 3.59 | 2.44 | 2.81 | 2.85 | 2.88 |
| Precision | 0.48 | 0.51 | 0.39 | 0.41 | 0.46 | 0.48 |

ideal; the average result of precision, being 0.45, shows that the recommendations made by the system do not conform enough to the diagnoses made by the users (Table 4).

$$1 - \frac{input\ mean}{recommendation\ mean + input\ mean} \quad (2)$$

The recommendation system finds users with very similar characteristics in the case of data entry number one and two. Likewise, there are cases in which the difference is significant; this may be due to the limited diversity of recommendations that the system has, showing that a minimum of ten treatments known to the system may not be the best number to start making personalized recommendations.

## 3 Conclusions and Future Work

The identification mechanisms supported by image processing that currently exists are not effective enough, either because they require expensive equipment or because they are capable of identifying a single type of disease; in addition, they do not have mechanisms capable of performing support to users, which allows them to know possible solutions to possible diseases that their crops present.

That is why the proposed system, by offering, not only the ability to identify various coffee pathologies, it also makes available to users control measures for the identified pathology, this by providing treatments that are adapted to the circumstances in which it is currently being cultivated; this provides a powerful tool for the care of coffee crops.

For future work is considered the possibility of implementing a batch identification system that can receive a large number of images and based on these, makes identifications for each one of the images provided, leading to faster diagnosis on big crops.

## References


1. CENICAFÉ Disturbios fisiológicos y nutricionales del café. (No 17). https://www.cenicafe.org/es/ [consultado el 25 de agosto de 2020]
2. Castro W, Oblitas J, Maicelo J, Avila-George H (2018) Evaluation of expert systems techniques for classifying different stages of coffee rust infection in hyperspectral images. Int J Comput Intell Syst 11(1):86–100. https://doi.org/10.2991/ijcis.11.1.8
3. Guzmán Piedrahita OA, Rivillas Osorio CA (2008) Relación de Glomus manihotis y G. fasciculatum con el crecimiento de plantas de café y la severidad de la mancha de hierro. Cenicafe 58(3):236–257
4. Giraldo LA, Flórez CP, Maldonado CE (2020) Resistencia Genética A La Enfermedad De La Cereza Del Café En Variedades Cultivadas En Colombia 71(1):68–90, 2020
5. CENICAFÉ (2005) Manejo Integrado de Enfermedades y corrección de disturbios nutricionales del café. Servicio Nacional de Aprendizaje
6. Ensayos sobre economia cafeteria (2014) Fed Nac Café Colomb 30:5
7. Rivillas CA, Serna CA, Cristancho MA, Gaitán AL (2011) La roya del cafeto en Colombia. Impacto, manejo y costos del control. Resultados de investigación. https://www.cenicafe.org/es/publications/bot036.pdf
8. Mahlein A-K, Kuska MT, Behmann J, Polder G, Walter A (2018) Hyperspectral sensors and imaging technologies in phytopathology: state of the art. Annu Rev Phytopathol 56(1):535–558. https://doi.org/10.1146/annurev-phyto-080417-050100
9. Krohling R, Esgario J, Ventura J (2019) BRACOL—a Brazilian Arabica coffee leaf images dataset to identification and quantification of coffee diseases and pests. Mendeley Data V1. https://doi.org/10.17632/yy2k5y8mxg.1
10. Zhu Y et al (2020) Addressing the item cold-start problem by attribute-driven active learning. IEEE Trans Knowl Data Eng 32(4):631–644. https://doi.org/10.1109/TKDE.2019.2891530

# Maturity Model for Automatization of Service Provision and Decision-Making Processes in Municipalities

**Ingmar Pappel, Teona Gelashvili, and Ingrid Pappel**

**Abstract** Citizens expect local governments to provide high-quality public services quickly and without hidden additional costs. Today, public service provision takes place mainly through e-services, using various document and information management systems. In this paper, we present the five-stage maturity model describing the automatization of service provision and decision-making processes in the back offices of municipalities. The model is developed after 20 years of continuous work in the field of e-governance. The process has been backed up by a scientific research method called Action Design Research (ADR) together with existing maturity model development methods. This model makes it possible to assess the level of maturity of local governments based on the carrier and format of data, as well as introducing other criteria and various easy-to-apply KPIs.

**Keywords** Maturity model · e-Governance · Action design research · Service provision · e-Service · Artificial intelligence · Local government · EDRMS

## 1 Introduction

This paper is based on the authors' experience in implementing digital administration and developing document management software (Amphora) in Estonia over the past 20 years. Today, digital records management and digital workflows are used in most organizations; nevertheless, the information processed in digital records management systems is still not machine-readable. The extreme importance of document management at the municipal level in connection with the state's e-governance level has been

I. Pappel (✉)
Amphora Information Management LLC, 12915 Tallinn, Estonia

T. Gelashvili (✉) · I. Pappel
Tallinn University of Technology, 12616 Tallinn, Estonia
e-mail: teona.gelashvili@taltech.ee

I. Pappel
e-mail: ingrid.pappel@taltech.ee



X.-S. Yang et al. (eds.), *Proceedings of Sixth International Congress on Information and Communication Technology*, Lecture Notes in Networks and Systems 216,
https://doi.org/10.1007/978-981-16-1781-2_37

highlighted numerously [1, 2]. Therefore, having the model which enables measurement of maturity levels of advanced service provisioning and decision-making in the local municipalities is crucial for Estonia, as one of the leading countries in e-governance [3, 4].

The maturity model presented in this paper has been created based on a five-step maturity model development framework [5], which consists of the following phases: scope, design, populate, test, and deploy and proposes three levels for models: descriptive, prescriptive, and comparative. The general principles of the Action Design Research (ADR) and Agile methodology have been followed during the creation process. The first cycle of model development was a theoretical and unvalidated prototype based on the authors' experience, and it is covered by this paper. The second cycle is planned as the further advancement of the model, promoting it to the prescriptive level and piloting in five municipalities. The third cycle is the development of the model to the highest, comparative level and the compilation of a comparison of Estonian local governments.

### *1.1 Development of Amphora*

In 2003, when Amphora was created, document management in Estonia was handled on paper, and there was almost no usage of a digital signature. One of the biggest challenges in municipalities was the loss of citizens' applications and letters. Development of the TWAIN interface in 2005 enabled Amphora to work with document scanners and transition to the use of digital copies in the workflows; to a small extent, letters were signed digitally. When Amphora was introduced outside the clerical offices, support for digital signatures was developed, and the first digitally signed documents began to emerge. In 2008, measurement of an organizational development process in municipalities became vital, thus the first Key Performance Indicators (KPIs) were developed and monitored.

Administrative processes in local governments to be efficient, the digitization of the input of public services started with a creation of e-forms on the portal eesti.ee (Estonian gateway to government information and e-services). In 80 local governments (of 237 in total at this time), the first two new e-services have been implemented. Later, a package of additional 25 new e-services (e-forms) was developed. Even though municipalities increased introducing e-forms and implementing e-services, a broader use remained low. In 2010 et seq., the transition to digitally signed documents was significant, and Amphora was increasingly used for digital workflows.

By 2018, signing the letters and documents digitally was predominant, but the information was still managed in the "A4" format. Soon a self-service portal and the machine-readable XML legislation system in Amphora started to develop. Interfaces for other popular self-service portal providers (SPOKU, KOVMEN), business rule-based automatic registration and distribution of input documents from self-service portals, and data exchange with council information system VOLIS were introduced.

During 2020, several of these systems have been implemented by approximately 30 percent of local governments, but none of them have reached fully automatic processes.

Prediction is that over the next 5 years, the transition to machine-processed information will take place, and most activities will be automated. It is also expected that in the next 5 years, the first municipalities in Estonia will begin to implement rules for making automated decisions that are approved by humans. From 2025 onwards, it is likely that AI (Artificial Intelligence) will be introduced for simpler (discretionary) decisions, and the role of the individual will be a random check of accuracy. By 2030, AI could make an initial (discretionary) decision that will be validated by humans. Further steps would encompass giving AI greater legal power.

### *1.2 Digital Records Management in Municipalities*

Three leading digital records management processes can be distinguished in municipalities: the correspondence handling process, the decision-making process, and the rule-making process. The correspondence process consists of registering an incoming application, forwarding it to a person responsible for preparing a draft of decision or answer, formulating a draft of the decision or answer, signing or passing (on town government meeting), and lastly, sending the answer back to the applicant. The decision-making process consists of preparing a draft of the decision, adding a draft to the agenda of the meeting, approving the agenda before the meeting, presenting the draft decision at the meeting, formal or substantive decision-making at the meeting, post-meeting formalization, digital signing of the decision, and sending the decision to the applicant. The rule-making process consists of drafting the terms of regulation, drafting the legislation (regulation) in the government, submitting the draft to the government meeting, passing the regulation on the meeting, submitting the draft to the council meeting, adopting (or rejecting) the regulation on the meeting, post-meeting formulation, and sending the formulated regulation to the Riigi Teataja (Estonian state portal for legal acts) for publishing and entering into force.

During the development of the model, careful attention has been paid to the integration of these three processes. The internal and personnel management processes have been eliminated from the model as nonsignificant in the context of service provisioning and decision-making.

## 2 Literature Review

Maturity models have been widely used in various disciplines, and their definition varies accordingly [6, 7]. In the Information System (IS) discipline, the term "maturity" is defined as "a measure to evaluate the capabilities of an organization" [8]. Nevertheless, on a large scale, the idea behind it is similar; they define the evolution path of a certain artifact or process within the organization and it can be displayed either through a top-down or bottom-up approach [5, 9]. Maturity models are grouped into two categories: fixed-level maturity models and focus area maturity models [10]. Irrespective of what the scheme when describing the evolution path is, Utterback and Abernathy argue that the progress of an innovation follows an S-curve [11, 12] and can be represented in four stages: emerging, pacing, disruptive, and mature.

*Maturity Model in Information Governance.* The development and usage of maturity models have been significant in Information Governance (IG) discipline. Proenca and Borbinha develop how to create an IG maturity model through Design Science Research [13]. The following three maturity models belong to five fixed-level category schemes and focus on different areas: Information Governance Maturity Model through which organizations are able to grasp IG characteristics and advance across the levels [14]; Asset Management Maturity Model developed in the Netherlands for the asset managers, helping them to make better investment decisions [15]; Digital Asset Management (DAM) Maturity Model with a rather broad focus and including people, systems, information, and processes [16].

*Maturity Model in IT Service Delivery.* IT Service delivery maturity models presented below consist of five levels [17] and assess various aspects. The Control Objects for Information and related Technology (COBIT) represents a framework for IT management and identifies IT processes; Service Management Process Maturity Framework (PMF) focuses on assessing the maturity of each of the Service Management processes; IT Service Capability Maturity Model (IT Service CMM) is oriented to assess the maturity of IT service processes and identify a direction for improvement, and its target is to help service organizations improve service quality [17].

*Enterprise Content Management (ECM) Maturity Model.* The concept of ECM is widely spread in various disciplines and has been used interchangeably with other related terms such as Electronic Document Management Systems (EDMS), Electronic Records Management Systems (ERMS), and Electronic Document and Records Management Systems (EDRMS) [18]. The core idea of ECM software applications and strategies is to allow the organization to manage its information effectively [19, 20]. The Enterprise Content Management Maturity Model (ECM3) was launched in March 2009 [19] and has five levels of maturity.

*Maturity Model for IT Management.* Improvement of IT performance in connection with economic efficiency is one of the main goals of introducing a maturity model in IT management. The role of the maturity model here is to assess the as-is position of the organization, identify the existing maturity state, and assign improvement measures to be implemented for further progress [21].

*Business Process Management (BPM) Maturity Models.* The Software Engineering Institute at Carnegie Mellon University has developed the Capability Maturity Model (CMM) which is the foundation for most of the Business Process Management maturity measurement models [22]. Among others, Harmon developed a Business Process Management Maturity (BPMM) model based on the Capability Maturity Model [23, 24]. Similarly, Fisher [25] combines five "levers of change" with five states of maturity. And lastly, the Infrastructure Management Maturity Matrix

($IM^3$) has a stronger focus on organizational communication reflected in the columns internal and external communication [26].

## 3 Maturity Model of Service Provisioning and Decision-Making Processes

### 3.1 Methodology

The authors of the presented maturity model have been active participants in the projects of the document exchange center of Estonia, contributing to conceptualizing a paperless management approach at the national level and involving in the state digitization processes. Therefore, the development of the suggested maturity model has been in close cooperation with state entities and ongoing digital advancements at the national level. The very first iteration of the model started in 2004 with the creation of Amphora. From then on, the process has been continued, with Amphora being an excellent platform where the implementation of all the existing hypotheses became possible together with testing their validity and applicability. The development process included a plethora of workshops in which the participants have been experts in the fields together with the actual users (from various municipalities, governmental entities, and citizens). Therefore, the methodology through which the model has been developed is based mainly on the qualitative approach. For building the model offered in this paper, together with their own practical experience, authors have conducted a thorough review of the existing maturity models in different fields and the methodologies which have enabled their development. Action Design Research, according to which "IT artifacts are ensembles shaped by the organizational context during development and use" [27, 28], has been chosen to apply the development process of the model. Application of ADR has enabled to deal with two challenges: firstly, it allowed to address the problem in the organizational setting and provided evaluation possibility and secondly, after identifying the existing problem, respective IT artifact could have been constructed which mitigated typified difficulty [27]. Consequently, the immense value offered by this maturity model has been in the fact that it has enabled authors to involve the dimension beyond the technological and incorporate all the feedback, reviews, experience, and opinions received from the users in the form of data and build an artifact based on the contextual framework. The active engagement with the stakeholders has been the foundation for the co-creation aspect [29, 30], which has also guided through the model formation process.

Maturity Model for Automatization of Service Provision and Decision-making Processes in Municipalities is represented in Table 1 below, followed by the description of each phase.

**First Phase.** In the first phase, the original application is copied several times on paper and all correspondence is delivered using postal service or hand-to-hand. There are

**Table 1** Maturity model of service provisioning and decision-making processes

| No. | Phase name | Information mode | Information carrier | Decision-making | Time for deciding |
|---|---|---|---|---|---|
| 1 | Paper original/paper copy | Unstructured | Paper/paper | Human | Weeks or months |
| 2 | Paper original/digital copy | Unstructured | Paper/digital | Human | Days |
| 3 | Digital free text + metadata/legalized paper copy | Unstructured | Digital/paper | Human | Days/hours |
| 4 | Data/(legalized) digital or paper copy | Structured | Digital | Human/AI | Nearly real-time |
| 5 | Data and AI | Structured | Digital | AI/human | Realtime/proactive |

two main problems identified at this level: documents are getting lost when processed, and since the documents must be moved physically, workflows are time-consuming. The decision-making act is formalized and signed on paper. This the first and most basic level, and there is no transitioning path to this level. The most distinctive difference between this and the next phase is that at this level, there is no digital information used on any step of the processes and paper is only a "data carrier". The main flaw at this level is that documents are lost when processed, and the circulation of original documents is not controlled.

Maturity at this level can be detected through observation, interviews, and questionnaires. Exit criteria to the next are every paper document to be scanned, digital copies to be used for processing, and digital records management system to be implemented. Detecting this phase can be done by observations or using questionnaires and interviews.

**Second Phase**. At this stage, applications are scanned and archived, digital copies are used in workflows, EDRMS is handling digital copies, and digital channels are being used for digital copies. Decisions are formulated and signed still on paper, and then scanned. Archiving and preserving are happening on paper. There are three significant problems identified at this level: scanning a document is time-consuming and costly, only paper documents are accepted and sent because only paper documents are trusted outside of ERDMS, and formulation of the decision on a paper is time-consuming and cannot be automated. However, there is an advantage compared to the previous stage: digital copies in ERDMS can be trusted, workflows are faster than in the previous phase, and most importantly, because of no physical delivery of documents, paper originals are not lost anymore. The transitioning path from level 1 is to follow the rule: everything must be digitized.

The list of the challenges to be mitigated includes technical capabilities (ERDMS must be implemented, scanners must be in places where papers appear), human capabilities, and change management (how to work with digital copies). The distinctive characteristic of this phase from others is that the data carrier is now a hybrid of paper/digital with two main criteria of progress: the percentage of digital copies in ERDMS and the percentage of digital workflows.

In order to initiate the transition from the previous phase, every document must be scanned, digital copies must be used in workflows, and paper originals must be stored in a controlled place. For exiting from this phase, digitally signed documents and e-forms over a trusted channel (self-service portals) should be accepted, as well as decision-making acts should be signed digitally or be delivered to the self-service portal. Maturity at this level can be detected through data acquired from EDRMS, observation, interviews, and questionnaires when there is no direct access to data.

**Third Phase**. During this phase, applications are accepted in digitally signed form or from trusted channels like self-service portals, decisions are formulated and signed digitally or delivered to trusted channels, and all documents are handled and processed in ERDMS. Archiving is also done in digital format. Because of the usage of unstructured free-text, it is hard to implement automated processing and decision-making. Moreover, human work is needed because of the unstructured and non-machine-readable format of data, which makes it time-consuming and error-prone. Identified advantages compared to the previous level are that some processes can be automated based on metadata or based on the processing of free-text, delivery of digital originals is nearly real-time, and every "copy" of the digitally signed document has the same legal power. There is no more distinction between copy and original. To transition from level 2, every signature must be digital, no more signing of paper documents.

The list of the main challenges occurring at this level consists of technical capability—servers, digital signing, document formats, archiving of digital documents, tendency of creating more "copies", and lastly, human capability and change management. A crucial aspect at this stage is the implementation of a digital archive. A distinctive feature of this phase is the fact that the data carrier is now digital, but the format is still unstructured, with the two main progress criteria being the percent of digitally signed documents and the percent of digitally signed decisions in ERDMS.

To transition from the previous phase, everything must be digitally signed, digitally signed documents or e-forms from self-service portals should be accepted (and preferred), and digital documents should be processed faster than paper. For exiting from this phase, IS working with structured and machine-readable data, registries, and databases should be implemented, or ERDMS must have this functionality. Maturity at this level is measured through data from ERDMS, observation, interviews, and questionnaires only when there is no direct access to data.

**Fourth Phase**. At this phase, applications are accepted from self-service portals using e-forms or inserted directly into IS, decisions are formulated in IS, and generated, signed digitally, or stamped as an extract of the state of data. Delivered over trusted channels, all documents are handled and processed in ERDMS or IS. Changes in data are archived when a change in data has different legal power and consequences in comparison with the previous state of the data. There are no papers preserved anymore except receiving the paper original from the external party. The list of the main problems existing at this level include the development of IS capable of working with unstructured data to be more expensive than implementing EDRMS, changes in the structure of data, business rules and workflows may be costly to implement, decisions are still made by humans, time-consuming, error-prone, and open to corruption.

Advantages over the previous level include the workflows to be nearly real-time and automated, some decisions can be automated, delivery of digital data happens in real-time, the system is interoperable with other IS, registries, and databases, and data can be re-used. The biggest challenge during the transition from level 3 lies within the technical capability—development and implementation of specialized IS. Distinctive characteristics of this phase are that the data carrier is now digital and format is unstructured with the new main criterion of progress—percent of decisions done in IS.

To initiate the transition from the previous phase, data should be in the re-usable format, and there should be interfaces to other interconnected information systems directly or over data exchange layers like X-road (Estonian official data exchange layer). For exiting from this phase, implementation of the AI-assisted decision-making system is crucial, and AI should be used for decision making; there must be strict rules and procedures to prove and validate decisions made by AI. Maturity at this level is measured through data from IS, observation, interviews, and questionnaires when there is no direct access to data.

**Fifth Phase**. In the fifth phase, only structured and machine-readable data is used; most of the decision-making is done by AI, and humans must be able to validate AI decisions. The main problems at this level include trust in AI, decisions are not "humanly soft" anymore, AI has no compassion, changing business rules may be costly, laws and regulations are not machine-understandable, and AI may be non-ethical and discriminating. Advantages over the previous level are that decisions are automated and real-time, proactive service can be provided.

The transitioning path from level 4 every decision is prepared or supported by AI. Main challenges include technical capability—development and implementation of AI as well as trust in AI. A distinctive characteristic from others is that decisions are made by AI and the main progress criterion is the percent of decisions done by AI.

Transition factors from the previous phase are AI does not make mistakes (if rules are correct), and AI is honest and non-corruptible. Since this is the highest level on this maturity model, there is no exit. Maturity at this level is measured through the percentage of automated decisions.

## 4 Discussion and Shortcomings

Firstly, understanding how convenient the implementation of the proposed maturity model is, and secondly, whether it is more beneficial than those which already exist will become apparent after the testing stage followed shortly after the current development phase. Nevertheless, it is still possible to identify a few expected shortcomings, even without having tangible post-implementation results. The challenges can be divided into several categories, such as technical, financial, legal, and human-related. When it comes to technical limitations, the main issue could be archiving problems of hybrid case files as well as archiving problems of structured data. Transitioning between the phases includes the advancement of technology, and thus, the development of EDRMS and IS capable of working with unstructured data might be costly. When it comes to the challenges of the legal aspects, continuous translation of legal acts to machine-readable rules will also be something to consider beforehand and think about possible mitigations. One of the very important aspects of the transition is the human factor, and the challenges related to change management are expected to be identified at each transitional phase. Furthermore, at the levels when decision-making is supported by AI, another important factor to consider is trust—trust in AI.

## 5 Future Work

The current paper where the development of the model has been presented belongs to the first cycle of model development. The result of this cycle is a theoretical and unvalidated prototype, which is based on the authors' experience, consultations (with experts in the field as well as customers and users), and existing literature. The second development cycle is planned as the further advancement of the model, promoting it to the prescriptive level and piloting in five municipalities. The third cycle is the additional improvement of the model to the highest, comparative level and the compilation of a comparison of Estonian local governments on the basis of the model. The maturity model is scalable and can be applied to different cases. It should be highlighted that the transition to each cycle is a result of thorough research, design, and consultation with all of the stakeholders. As already mentioned, an integral part of this development process is a collaborative aspect, which enables authors to keep the right track and progress.

## References

1. Pappel I, Pappel I (2011) Methodology for measuring the digital capability of local governments. In: 5th international conference on theory and practice of electronic governance, Tallinn

2. Pappel I, Pappel I, Tampere T, Draheim D (2017) Implementation of e-invoicing principles in Estonian local governments. In: Proceedings of the 17th European conference on digital government (ECDG). Military Academy, Lisbon
3. Pappel I, Tsap V, Draheim D (2020) The e-LocGov model for introducing e-Governance into local governments: an Estonian case study. In: IEEE transactions on emerging topics in computing
4. Pappel I, Tsap V, Pappel I, Draheim D (2018) Exploring e-services development in local government authorities by means of electronic document management systems. In: 5th international conference on electronic governance and open society challenges in Eurasia, St. Petersburg
5. De Bruin T, Freeze R, Kulkarni U, Rosemann M (2005) Understanding the main phases of developing a maturity assessment model. In: Proceedings of the ACIS 2005
6. Mettler T (2011) Maturity assessment models: a design science research approach. Int J Soc Syst Sci 81–98
7. Peffers K, Tuunanen T, Rothenberger M, Chatterjee S (2007) A design science research methodology for information systems research. J Manag Inf Syst 45–78
8. Rosemann M, De Bruin T (2005) Towards a business process management maturity model.In: Proceedings of the thirteenth European conference on information systems ECIS 2005
9. Proença D, Vieira R, Borbinha J (2016) A maturity model for information governance. In: 2016 11th Iberian conference on information systems and technologies (CISTI), Lisbon
10. Utterback J, Abernathy W (1975) A dynamic model of process and product innovation. Omega, p 639–656
11. Van Steenbergen M, Bos R, Brinkkemper S, Van de Weerd I (2010) The design of focus area maturity models. In: International conference on design science research in information systems
12. Lahrmann G, Marx FMT, Winter R, Wortmann F (2011) Inductive design of maturity models: applying the Rasch algorithm for design science research. In: International conference on design science research in information systems
13. Proença D, Borbinha J (2018) Maturity models for data and information management. In: 22nd international conference on theory and practice of digital libraries
14. ARMA international. https://www.arma.org/page/principles
15. Lei T, Volker L, Herder P, Ligtvoet A (2011) Evaluating asset management maturity in the Netherlands: a compact benchmark of eight different asset management organizations.In: 6th world congress of engineering asset management
16. DAM foundation. What is the DAM maturity model? https://www.dammaturitymodel.org/.
17. Flores J, Rusu L, Johannesson P (2011) A maturity model of IT service delivery. In: Proceedings of the CONF-IRM 2011
18. Katuu S (2016) Assessing the functionality of the enterprise content management maturity. Records Manag J
19. Pelz-Sharpe A, Durga A, Smigiel D, Hartman E, Byrne T, Gingras J (2010) ECM3 maturity model
20. Milenković DS (2016) Determining the maturity level of business organizations for ICT implementation in the case of electronic content management. Military Technical Courier, p 1119–1131
21. Becker J, Knackstedt R, Pöppelbuß J (2009) Developing maturity models for IT management—a procedure model and its application. Business & Information Systems Engineering, pp 213–222
22. Paulk M, Curtis B, Chrissis M (1993) Capability maturity model for software, version 1.1. Software Engineering Institute, CMU/SEI-93-TR-24
23. Harmon P (2003) Business process change: a manager's guide to improving, redesigning, and automating processes (The Morgan Kaufmann series in data management systems), 1st edn. Morgan Kaufmann
24. Harmon P (2004) Evaluating an organization's business process maturity
25. Fisher D (2004) The business process maturity model: a practical approach for identifying opportunities for optimization. Business Process Trends

26. Volker L, Van der Lei T, Van den Boomen M, Van der Velde J (2012) Continued learning in asset management for the Dutch transport network. In: Third international engineering systems symposium CESUN 2012
27. Sein M, Henfridsson O, Purao S, Rossi M, Lindgren R (2011) Action design research. MIS Q 35(1):37–56
28. Petersson ALJ (2016) Applying action design research (ADR) to develop concept generation and selection methods. Elsevier, p 222 – 227
29. Ind N (2013) The meanings of co-creation. Eur Bus Rev 25(1)
30. Bhalla G (2011) Collaboration and co-creation. Springer-Verlag, New York

# Particle-Based Physics for Interactive Applications

**Yalmar Ponce Atencio, Manuel J. Ibarra, Juan José Oré Cerrón, Roberto Quispe Quispe, Richard Flores Condori, Julio Huanca Marín, and Mary Huaman Carrión**

**Abstract** Physics-based simulation is one of the most intensively studied fields in recent decades, and currently there are many wonderful techniques, especially to simulate the dynamics of solids, fluids, and the interaction between them. In order to simulate rigid solids, classical methods based on the application of forces or impulses are used. On the other hand, response to detected contacts or collisions requires using fast and efficient techniques, and position-based methods have become popular. These research fields have become very popular in the Computer Graphics community. Currently, the methods have evolved to be more accurate, fast, stable, and controllable, but more complex to implement, which makes them adequate for use in interactive environments like computer games, virtual reality applications, and to generate special effects in movies. In this research work, we show some position-based methods which are well-suited for the simulation of solids (rigid and deformable) and fluids. We will use particles as small spheres, and the concept of

Y. P. Atencio (✉) · J. J. O. Cerrón · R. Q. Quispe · R. F. Condori · J. H. Marín · M. H. Carrión
Escuela Profesional de Ingeniería de Sistemas, Universidad Nacional José María Arguedas, Andahuaylas, Apurímac, Peru
e-mail: yalmar@unajama.edu.pe

J. J. O. Cerrón
e-mail: jore@unajama.edu.pe

R. Q. Quispe
e-mail: rquispe@unajama.edu.pe

R. F. Condori
e-mail: rflores@unajama.edu.pe

J. H. Marín
e-mail: jhuanca@unajama.edu.pe

M. H. Carrión
e-mail: mhuaman@unajama.edu.pe

M. J. Ibarra
Escuela Profesional de Ingeniería Informática y Sistemas,
Universidad Nacional Micaela Bastidas de Apurímac, Abancay, Apurímac, Peru

 
X.-S. Yang et al. (eds.), *Proceedings of Sixth International Congress on Information and Communication Technology*, Lecture Notes in Networks and Systems 216,
https://doi.org/10.1007/978-981-16-1781-2_38

position-based dynamics is entirely referred to them. Also we show the implementation of several techniques proposed in the last decade, for instance, simulation based on shape matching, Smooth Particle Hydrodynamics (SPH) approach for liquids, Impulse-based dynamics, and the interaction between them. In addition, we will go into some detail in the implementation of these methods with practical examples that demonstrate how useful they are for interactive applications.

**Keywords** Particle-based simulation · Rigid and soft bodies · SPH · Interactive applications

## 1 Introduction

Physically based simulation has been an active and important research field in computer graphics at least over the last three decades. This field was introduced to the researchers community by Terzopoulos [1, 2] before the nineties, and from there, a large list of works has been published. There exists a variety of survey papers [3–7] which details this development process. All the topics related with physically based simulation are exciting and challenging, and there are many methods for achieving the proposed goals, some more expensive than others and some easier to implement than others. In this research work, we propose to focus on particle-based techniques, i.e., methods where bodies are represented as collections of particles which behave in a physically plausible way by imposing certain rules. Many approaches in this class have been proposed and shown to be flexible, powerful, and especially amenable to implementation using parallel architectures in general, and GPUs in particular. We plan on covering several of these techniques, showing how they can be used to simulate the interaction between rigid bodies, between rigid and soft (deformable) bodies such as cloth or ropes, and between these and fluids using SPH. In the research work, we will also discuss related issues such as collision detection and response and rendering techniques. Our main emphasis will be on applications where physics must be computed at interactive rates, such as virtual reality, computer games, and movie effects, among others. In computer graphics, for simulating dynamic systems, the most used approaches are based on forces, where internal and external forces are accumulated in order to compute accelerations using Newton's second law for motion. Later, a time step integration method is used in order to update the velocities, and finally the positions of the objects are updated. On the other hand, some simulation methods (most frequently for rigid bodies) use impulse-based dynamics, manipulating directly the velocities. In addition, these methods are used with point-based or mesh-based approaches to represent the objects. The main advantage of the point-based approach is its controllability, surpassing problems existing in explicit integration schemes, when force based systems are used. Within force-based approaches, one of the simplest methods is to represent and simulate solids represented by mass-spring networks. A mass-spring system consists of a set of point masses which are connected by springs. The physics for such a system is straightforward and easy to implement a simulator. However, there are some significant drawbacks for that simple method. Thus, much of the object behavior depends on the way the spring network is set up. Furthermore, it can be difficult to choose the adequate spring constants to get a desired behavior. Mass-spring networks are not capable to capture volumetric effects directly such as volume conservation or to prevent volume inversions. But, for overcome these problems, a good idea was represent solids with particles (small spheres) and use the shape matching approach for compute plausible deformations; although this method is not physically correct, the results look real enough compared to other approaches like the finite elements method. Finally, collisions can be handled by computing repulsion forces on interpenetration particles, which are applied for each collided object.

## 2 Solids as a Collection of Chunks

Consider a body $B$. It can be broken up into chunks (or elements), and each chunk can be treated as a single particle. Then, the mechanics and dynamics of any kind of body could be modeled in an easy way, since each body is composed of a collection of chunks and the formulae are essentially the sum of their parts. The representation of objects by these chunks could easily be obtained by a voxelization process.

### *2.1 Voxelization*

The word Voxel is similar to the word Pixel, where a pixel is in 2D and Voxel stands for a volume in 3D. So, if a 3D object is taken to voxelize, it is possible to discretize it with the desired resolution. In general, for a voxelization process, a Voxel doesn't know anything about the "shape" within its volume, since for each voxel only is known if its center is inside the shape or not (0 or 1). The value could also be a gradient between 0 and X, for instance, a gas cloud that slowly fills a volume. All this to say that a voxel is a value V at a point C in a grid with dimensions Wx, Wy, and Wz. Then, it is only necessary to define the density of volumes using grids, which means that it is necessary to decide how big the voxels are. If we have a grid $1 \times 1 \times 1$, it would basically only be known if an object is in the volume or not; if we have a grid of $10 \times 10 \times 10$ (1,000 voxels), it would have a very rough definition of the shape; if we have $1000 \times 1000 \times 1000$ (1 000 000 000 voxels), that is quite a bit of data and the shape would be decently represented. Ultimately, imagine grid volumes as small as an atom; you could potentially have an almost perfect replication of your object, but the amount of data would be enormous and practically impossible to render on today's computers.

## 3 Rigid Body Simulation

For the motion, the technique presented in [8] computes the movement of a rigid body in two parts: the translation and the rotation. The translation describes the center mass movement, while the rotation describes how the rigid body turns around its center mass. A detailed description can be found in [9].

**Translation** The center mass movement is trivial, and can be considered similar to a particle movement. Thus, when a force $F$ acts over a rigid body ($RB$), this force could cause a variation of the $RB$'s linear momentum $P$. There is a relation between $P$ and $F$, then, the velocity $v$ can be got dividing $P$ by the rigid body mass $M$, and finally, the velocity $v$ is the derivative of $x$ (position of the mass center) related to time $t$ ($\frac{dP}{dt} = F$ , $v = \frac{P}{M}$ and then $\frac{dx}{dt} = v$).

**Rotation** The force $F$ which acts over any surface point on the rigid body generates a change of angular momentum $L$. Then, this variation depends of $RB$'s relative position $r$. Thus, the torque is the derivative of $L$ related to time, and in turn is the cross product between $r$ and $F$ as $\frac{dL}{dt} = r \times F$

The angular velocity $\omega$ is the derivative of angular momentum $L$, represents the length of the rotation velocity, and is computed by $\omega = I(t)^{-1}L$, where the $3 \times 3$ matrix $I(t)$ represents the inertia tensor of the $RB$, at time $t$, and $I(t)^{-1}$ is the matrix inverse. The inertia tensor is a physical term, which indicates the body resistance to the rotational movement. The inertia tensor, for every time step, changes in relation to the rigid body orientation, then it must be recomputed for every time step. The matrix $I(t)^{-1}$ is computed using $I(t)^{-1} = R(t)I(0)^{-1}R(t)^T$, where $R(t)$ is the rotation matrix which describes the orientation of the rigid body at time $t$. Having the angular velocity $\omega$, the follow step is to compute the rigid body orientation related to time, using the angular velocity $\omega$. In order to save the the orientation, typically two representations are used: rotation matrices and quaternions. Quaternions are better over matrices, for instance, matrices tend to accumulate non-rotational components over the time during the simulation. Therefore, numerical errors, which will be added during repetitive combinations of rotational matrices, can lead to get a distorted rigid body. On the other hand, in order to avoid degrees of freedom for non-rotational movement, quaternions are preferable to use. A quaternion, usually denoted by a quad-vector $q = [\theta, v_x, v_y, v_z]$, represents a rotation of $\theta$ radians in turn of $v = (v_x, v_y, v_z)$ axis. For simplicity, a quaternion can be represented as $q = [\theta, v]$. An object is represented by $q$, and suffers a change on its orientation given by a velocity $\omega$ during an interval of time $dt$; its orientation will be changed by the equation $dq = (\cos(\frac{\theta}{2}), v\sin(\frac{\theta}{2}))$, where $v = \omega/\|\omega\|$ is a unit vector and represents the axis of rotation and $\theta = \|\omega dt\|$ is the angle of rotation. Later, the quaternion at time $t + dt$ is updated as $d(q + dt) = q \times dq$.

### 3.1 Collision Detection

Since a rigid body $RB$ is represented by a set of particles, a good technique for this kind of representation is presented in [10]. This technique is adequate to be implemented on GPUs as presented in [8]. Typically, collision detection schemes use a two-step process *broad phase* and *narrow phase*; in this approach, only *narrow phase* is used, since the objects are represented by sets of small particles. Later, the collision detection process is improved by the use of a uniform three-dimensional grid, where all the particles (from all the bodies) are mapped. Then, the collision detection method is summarized on the collision test of each particle, stored on a grid cell, with all the other particles stored on the same grid cell and the 26 neighbor grid cells again. Clearly this technique reduces dramatically the computational cost compared to $N^2$ brute force method, and its implementation in parallel architectures is relatively simple and yields a very high performance. Note that GPU implementations use textures or GPU arrays to represent the objects so that there is no need to transfer data between CPU and GPU during the simulation. The main drawback is that for every time step, all the particles must be mapped onto the grid again.

The advantage of the particle representation is the controllability between precision and performance. Since more particles is more accurate simulation, but less performance and less particles represent a roughly approximation but better performance. Take a look at Fig. 1.

### 3.2 Collision Response

The collision detection process verifies if a particle collides with its local neighbors. If there is collision, repulsion forces are computed by applying the discrete elements method (DEM is a method for simulating granular materials [11]). The repulsion force $f_{i,s}$ is modeled by a linear spring, and the dumping force, $f_{i,d}$, is modeled by a dump which dissipates energy between collided particles. Thus, for the par-

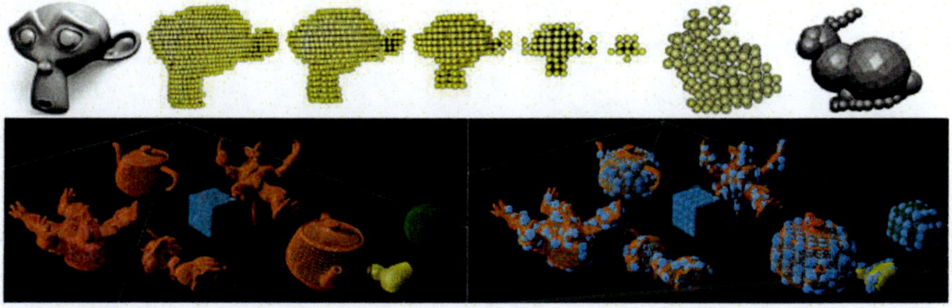

**Fig. 1** Different chunk sizes for representing a body (top). Different bodies represented by particles (bottom)

ticle $i$ which collides with the particle $j$, the forces are computed by the equation $f_{i,s} = -\kappa \left(d - |r_{ij}|\right) \frac{r_{ij}}{|r_{ij}|}$, $f_{i,d} = \eta v_{ij}$, where $\kappa$, $d$, $r_{ij}$, $v_{ij}$, and $\eta$ are the restitution coefficient of the spring, particle diameter, the relative position, and the relative velocity of the particle $p_j$ related to $p_i$, and the dumping coefficient, respectively. Next, the shear force $f_{i,t}$ is computed too, as a force proportional to the relative tangential velocity $v_{ij,t}$, by the equation $f_{i,t} = \kappa_t v_{ij,t}$, where the $v_{ij,t}$ is computed by $v_{ij,t} = v_{ij} - \left(v_{ij} \frac{r_{ij}}{|r_{ij}|}\right) \frac{r_{ij}}{|r_{ij}|}$. Then, the total force $F$ and the total torque $T$, applied to a rigid body $RB$, are the sum of all the forces and torques applied over all the particles which belong to the rigid body $RB$ by the equations $F = \sum_i \left(f_{i,s} + f_{i,d} + f_{i,t}\right)$ and $T = \sum_i \left(r_i \times \left(f_{i,s} + f_{i,d} + f_{i,t}\right)\right)$, where $r_i$ is the current relative position of the particle $i$ related to the center mass.

Because a simulation of a large number of rigid solids is computationally expensive, real-time simulation of such a number of rigid bodies is still a challenging topic. In this work, we present the implementation of a point-based method for simulating a huge quantity of rigid solids in real time, in a similar way to [8], where we can take advantage of the high-performance parallel architecture of GPUs, which can be used to accelerate the simulation drastically.

The results are fast enough, however, the behavior looks a little bit springy, interpenetrating between objects and objects with the obstacles (walls). A more robust approach is to use impulses instead of forces, which is more adequate because impulse are applied on contact points over the surface of the objects. Figure 2, top right, shows an example about this method.

Another extension is to consider variable sized objects. In this case, for avoiding a huge quantity of particles, we use variable sized particles. In Fig. 1, on the top right, a bunny model is represented with fixed and variable sized particles, and a more complex example is shown in Fig. 1, on the top center part.

Although the performance decreases, it is better than using a lot of particles, since the collision handling algorithms will be required to check all the particles in grid cell, and for huge objects it will be unnecessary to check their large quantity of particles. Moreover, objects are very different than spheres, so that not many particles are placed on grid cells.

## 4 Soft Body Simulation

In the field of soft bodies, the meshless shape matching method presented by [12] is a good choice to use in interactive applications (see Fig. 2, on the middle left). In this work, we present an implementation for simulating deformable objects using particles. The underlying model handles point-based objects and is geometrically motivated, furthermore it does not need information about connectivity. This approach does not require any previous processing, is relatively simple to implement, and is able to generate stable dynamic simulations [13]. The underlying idea is that the deformable model replaces forces by distances and energies by geometric constraints; all of them are computed by the goal positions from the current positions. The goal positions are computed by using a generalized shape matching technique which depends from an undeformed rest state of the object with the current deformed state of the object (the point cloud) which represents the deformable body. Since the points are always toward well-defined locations, then, the common problem, of the explicit integration schemes, is surpassed. Later, the versatility of this approach, in terms of object representations that can be handled, is the efficiency in terms of memory and saves computational effort, and the stability of the dynamic simulation allows the approach particularly interesting for interactive applications, such as interactive games, movie effects, and virtual reality, among others.

Different from rigid bodies, to take advantage of the point-based approach, it's necessary to put small particles over the object surface. It is in order to maintain compatibility with the original approach and for collision handling purposes.

## 4.1 Integration Method

As in [12], for the dynamics part, a semi-implicit Euler method is used. Which includes an explicit part that updates the velocity, and an implicit part which updates the position. Later, the used integration method is $\mathbf{v}_i(t+dt) = \mathbf{v}_i(t) + \alpha\frac{\mathbf{g}_i(t)-\mathbf{x}_i(t)}{dt} + \frac{dt}{m_i}f_{ext}(t)$ and $\mathbf{x}_i(t+dt) = \mathbf{x}_i(t) + dt\mathbf{v}_i(t+dt)$, where $dt$ is the time step, $m_i$ the mass, $\mathbf{v}_i(t+dt)$ the velocity after a time step, $\mathbf{v}_i(t)$ the current velocity, $\alpha \in [0, 1]$ is a parameter of control used to set up the rigidity of the objects, $\mathbf{g}_i(t)$ is the goal position for the vertex $\mathbf{x}_i(t)$, and finally $\mathbf{f}_{ext}(t)$ are external forces like gravity, wind, among others.

## 4.2 Deformation Types

**Rigid Deformation** The following method to be described can be used to imitate a rigid body, but it is not exactly a rigid body, although neither in real world there are rigid body, then, in an implementation, a rigid body behavior could be achieved by setting $\alpha = 1$. Thus, the points must be moved to the goal positions $\mathbf{g}_i$, for every time step, by using the following method.

The technique has two principal components: first find the best rigid transformation which approximates the next orientation and position for the simulated object, and that is done by moving the original vertices to the goal position applying a model of linear deformation type. Considering the weights of the particles (vertices), it is a matching problem and consists in find translation vectors $\mathbf{t}$ and $\mathbf{t}_0$ and also the rotation matrix $\mathbf{R}$ that are computed minimizing $\sum_i w_i\left(\mathbf{R}\left(x_i^0 - \mathbf{t}_0\right) + \mathbf{t} - \mathbf{x}_i\right)^2$.

In order to simplify, $w_i = m_i$ is used, where $m_i$ is the mass for the particle $i$. The best translation vectors, between the mass centers, from the object in their initial position and the current position, are computed using the equations $\mathbf{t}_0 = \mathbf{x}_{cm}^0 = \frac{\sum_i m_i \mathbf{x}_i^0}{\sum_i m_i}$ and $\mathbf{t} = \mathbf{x}_{cm} = \frac{\sum_i m_i \mathbf{x}_i}{\sum_i m_i}$.

The best rotation **R** could be found by computing a linear transformation matrix **A**. In this case, **A** is not necessarily orthonormal. Later, considering the relative positions of the particles, the current positions are estimated by $\mathbf{q}_i = \mathbf{x}_i^0 - \mathbf{x}_{cm}^0$ and $\mathbf{p}_i = \mathbf{x}_i - \mathbf{x}_{cm}$; a good choice for solve the matching problem could be using the least square technique. Thus, the matrix **A** is found by minimizing the term $\sum_i m_i (\mathbf{A}\mathbf{q}_i - \mathbf{p}_i)^2$.

Setting to zero the first derivatives of this expression related to the coefficients of **A**, we can get $\mathbf{A} = \left(\sum m_i \mathbf{p}_i \mathbf{q}_i^T\right)\left(\sum m_i \mathbf{q}_i \mathbf{q}_i^T\right)^{-1} = \mathbf{A}_{pq}\mathbf{A}_{qq}$, where $\mathbf{A}_{qq}$ is a symmetric matrix which only contains scale information and $\mathbf{A}_{pq}$ is a correlation matrix. Furthermore, in order to find **R**, first, it is necessary to find the rotational component $\mathbf{A}_{pq}$. According to [12], a polar decomposition method is a good choice in order to compute $textbf R = \mathbf{A}_{pq}\mathbf{S}^{-1}$, where $\mathbf{S} = \sqrt{\mathbf{A}_{pq}^T \mathbf{A}_{pq}}$ is a symmetric matrix. In order to get $\mathbf{S}^{-1}$, and the matrix $\mathbf{A}_{pq}^T \mathbf{A}_{pq}$ is necessary to diagonalize by using less than 10 Jacobi rotations.

Last, all the vertices (particles) are displaced to their goal position, which are computed by $\mathbf{g}_i = \mathbf{R}\left(\mathbf{x}_i^0 - \mathbf{x}_{cm}^0\right) + \mathbf{x}_{cm}$. This process is repeated for every time step.

**Linear Deformation** In order to have more degrees on deformation movement, a matrix for linear deformation is used **A**. This matrix describes the best linear transformation which matches the current shape from the initial shape in less square terms. Instead of using **R** to compute $\mathbf{g}_i$, the combination $\beta\mathbf{A} + (1 - \beta)\mathbf{R}$ is used, where $\beta$ is an additional control parameter. The presence of **R** in the sum ensures that the current shape have a tendency to back the original shape. In order to ensure that the volume be conserved, the matrix **A** is divided by $\sqrt[3]{det(\mathbf{A})}$ ensuring that $det(\mathbf{A}) = 1$. Different from rigid deformation, where only computing $\mathbf{A}_{pq}$ is necessary, here it is necessary to compute a new matrix $\mathbf{A}_{qq} = \left(\sum_i m_i \mathbf{q}_i \mathbf{q}_i^T\right)^{-1}$. But this matrix is symmetric and can be pre-computed.

**Quadratic Deformation** Notice that the deformation model presented before achieves only linear deformations, namely shear and stretch deformations. To extend the range of deformations like torsion and bending, a redefinition of the deformation matrices is done $\mathbf{g}_i = \beta[\mathbf{AQM}] + (1 - \beta)\left[\mathbf{R}\,\tilde{\mathbf{0}}\,\mathbf{0}\right]\tilde{\mathbf{q}}_i + \mathbf{x}_{cm}$, where $\mathbf{g}_i \in \Re^3$, $\tilde{\mathbf{q}} = [q_x, q_y, q_z, q_x^2, q_y^2, q_z^2, q_x q_y, q_y q_z, q_z q_x]^T \in \Re^9$. $\mathbf{A} \in \Re^{3\times3}$ contains the linear term coefficients, $\mathbf{Q} \in \Re^{3\times3}$ contains the coefficients of terms purely quadratic, and $\mathbf{M} \in \Re^{3\times3}$ contains the mixed term coefficients. With $\tilde{\mathbf{A}} = [\mathbf{AQM}] \in \Re^{3\times9}$ must be minimize $\sum_i m_i \left(\tilde{\mathbf{A}}\tilde{\mathbf{q}}_i - \mathbf{p}_i\right)^2$. Then, the optimal quadratic transformation becomes $\tilde{\mathbf{A}} = \left(\sum_i m_i \mathbf{p}_i \tilde{\mathbf{q}}_i^T\right)\left(\sum_i i m_i \tilde{\mathbf{q}}_i \tilde{\mathbf{q}}_i^T\right)^{-1} = \tilde{\mathbf{A}}_{pq}\tilde{\mathbf{A}}_{qq}$. Here, the $\tilde{\mathbf{A}}_{qq} \in \Re^{9\times9}$ is a symmetric matrix and the $\tilde{\mathbf{q}}_i$ terms can be pre-computed too. In an analogous way to the linear deformation model, the flexibility can be varied using $\beta\tilde{\mathbf{A}} + (1 - \beta)\tilde{\mathbf{R}}$ on goal points computing process, where $\tilde{\mathbf{R}} \in \Re^{3\times 9} = [\mathbf{R}\ \mathbf{0}\ \mathbf{0}]$. Figure 2, on the middle-left part, shows a scene where objects are suffering quadratic deformations.

## 5 Liquids with SPH

The smoothed-particle hydrodynamics (SPH) is a computational method which was developed by Gingold and Monaghan [14] and is used for simulating fluid flows. In the last years, this method has become into a very important research topic since it was presented by Harada [15]. Currently, it is used in many research fields, including oceanography, astrophysics, ballistics, volcanology, and other applications which are based on granular materials. In a nutshell, it is a mesh-free Lagrangian method (where the particles move according the fluid dynamics). Another important advantage is that the resolution of the method can be adjusted easily with respect to variables such as pressure, density, viscosity, and even surface tension. Although there are many advanced methods for simulating fluids [16–21], the classical technique presented by Harada [15] is enough for use in interactive applications like games, mainly because the real-time condition must be achieved.

The SPH method works by dividing the fluid volume into a set of discrete elements, which are referred to as particles. These particles have a spatial distance of influence (known as the "smoothing length", where, in equations, typically they are represented by $h$), where their properties are "smoothed" by using a kernel function. Later, for each time step, the physical quantity for any particle can be obtained by summing the relevant properties of all the particles which lie within the range of influence of the used kernel function. For example, we can use Monaghan's popular cubic spline kernel in order to compute the force at position $\mathbf{r}$, which depends on the forces of all the particles within a radial distance $2h$ of $\mathbf{r}$.

### *5.1 Discretization of the Space Occupied by the Fluid*

SPH is an interpolation method for particle systems, where field quantities, which are defined at the particle positions, can be evaluated anywhere in the space occupied by the fluid. For this purpose, the SPH method distributes quantities in a local neighborhood of each particle using a symmetric radial smoothing function. Thus, a scalar quantity $\phi$ is interpolated at a location $x$ by a weighted sum of contributions from all particles in a neighborhood of radius $h$ is computed by $\phi(x) = \sum_j m_j \frac{\phi_j}{\rho_j} W(x - x_j, h)$, where $j$ means it checks for all particles in the neighborhood and $m_j, \rho_j, x_j, \phi_j$ are the mass, density, position, and a field measure for particle $j$, respectively. The function $W(x - x_j, h)$ is a smoothing function on the radius $h$. The terms of particle mass and density appear in the equation since a particle $i$ represents a volume given by $m_i/\rho_i$. While the mass $m_i$ is constant, and in our case it is the same for all particles, the density $\rho_i$ varies and needs to be calculated at each step of time. Thus, we can calculate the density of the particle at position $x$ by substituting $\phi_j$ for $\rho_j$ obtaining $\rho(x) = \sum_j m_j W(x - x_j)$.

The fluid pressure is computed using the equation $p = p_0 + \kappa(\rho - \rho_0)$, where $p_0$, $\rho_0$ are the resting pressure and density, respectively, and $\kappa$ is a constant.

Then, to calculate momentum conservation, the Gradient and Laplacian operators must be modeled, since they are used to solve the pressure and also the viscosity forces for all the particles. Thus, the pressure force $f_i^{press}$ and the viscosity force $f_i^{vis}$, for each particle $i$, are computed as $f_i^{press} = -\sum_j m_j \frac{p_i + p_j}{2\rho_j} \nabla W_p(r_{ij})$ and $f_i^{vis} = \upsilon \sum_j m_j \frac{v_j - v_i}{\rho_j} \nabla^2 W_\upsilon(r_{ij})$, where $r_{ij} = r_j - r_i$ is the relative position vector, likewise $r_i$ and $r_j$ represent the positions of the particles $i$ and $j$.

The weighted functions for pressure, viscosity, and other conditions are $\nabla W_p(r) = \frac{45}{\pi h^6}(h - |r|)^3 \frac{r}{|r|}$, $\nabla W_\upsilon(r) = \frac{45}{\pi h^6}(h - |r|)$ and $W(r) = \frac{315}{64\pi h^9}(h^2 - |r|^2)^3$. In these functions, the value out of the radius $h$ is 0.

### 5.2 Boundary Condition

**Pressure** The pressure force brings the liquid to a constant density, when calculating the incompressible flow equations, keeping the particles at a distance $d$, which is the rest distance. At the boundary, it is assumed that a particle $i$ is at a distance $\|r_{iw}\|$, where $\|r_{iw}\| < h$. The pressure force pushes the particle $i$ back for a distance $d$, in the direction of $n_w(r_i)$, which is the normal vector on the surface of the boundary. Later, the pressure force $f_i^{press\_wall}$ is computed by $f_i^{press\_wall} = m_i \frac{\Delta x_i}{dt^2} = m_i \frac{(d - |r_{iw}|)n(r_i)}{dt^2}$.

**Density** When a particle is close to the surface of the boundary, at a distance less than $h$, the contribution of the boundary to the density of the particle is estimated assuming that there are particles inside the surface, and is calculated by $\rho_i(r_i) = \sum_{j \in fluid} m_j W(r_{ij}) + \sum_{j \in wall} m_j W(r_{ij})$.

The particle distribution at the boundary is determined assuming that they are uniformly placed perpendicular to the surface and the mean curvature is 0. Thus, the contribution of the boundary particles is computed by using a distance function $\|r_{iw}\|$ as $\rho_i(r_i) = \sum_{j \in fluid} m_j W(r_{ij}) + Z_{wall}^{\rho}(|r_{ij}|)$.

The weighting function $Z_{wall}^{\rho}(\|r_i w\|)$ of the boundary surface depends from the distance $\|r_i w\|$ that can be pre-computed and used directly when it is needed in the simulation. Then the value of the function, for any arbitrary position, is computed by a linear interpolation. Similarly, in order to obtain the distance of each particle to the boundary surface, a distance function is used that is pre-computed and stored in memory to be used during the simulation.

## 5.3 Additional Considerations

The contribution of each particle to the closest others is weighted according to its distance from the particle of interest, and its density. Mathematically, this is governed by the kernel function ($W$). Commonly, the kernel functions are used including a Gaussian function and a cubic spline. The latter function returns zero for particles farther from the smoothing action radius (unlike Gaussian, where, at any finite distance, there is a small contribution). This has the advantage of save computational effort since it is not including the minor contributions from far particles. Thus, the computation of a quantity $A$, with respect to any point $\mathbf{r}$, is given by the $A(\mathbf{r}) = \sum_j m_j \frac{A_j}{\rho_j} W(|\mathbf{r} - \mathbf{r}_j|, h)$, where $m_j$ is the mass of particle $j$, , $\rho_j$ is the density associated with the particle $j$, $A_j$ is the value of the quantity $A$ for the particle $j$, $\mathbf{r}$ denotes the position, and $W$ is the kernel function mentioned in the above section. Later, the density of the particle $i$ ($\rho_i$) can be expressed as $\rho_i = \rho(\mathbf{r}_i) = \sum_j m_j \frac{\rho_j}{\rho_j} W(|\mathbf{r}_i - \mathbf{r}_j|, h) = \sum_j m_j W(\mathbf{r}_i - \mathbf{r}_j, h)$, where the summation around the particle $i$, denoted by all the $j$ particles, includes all particles which can influence in the simulation. In the same way, the spatial derivative of a quantity can be easily obtained considering the linearity of the derivative $\nabla A(\mathbf{r}) = \sum_j m_j \frac{A_j}{\rho_j} \nabla W(|\mathbf{r} - \mathbf{r}_j|, h)$.

To compute the force on a particle, we use the same technique described on the rigid body section. The density is computed in the same manner, and finally the particles' positions are updated for the next time step. In Fig. 2, in the bottom-left part, we show the interaction of a SPH fluid in a cubic container.

## 5.4 Interaction Between Liquid and Solids

In the previous Sects. (3 and 5) were described how simulations of liquids and rigid bodies, respectively, have been implemented. Considering both simulation schemes, liquid and solid bodies, the same collision detection system has been used, mainly based on the use of a uniform grid. So, in the same way, we can simply compute forces between particles and change the physical properties for the liquid or for the solid bodies that interact in the simulation, achieving a two-way coupling.

In the collision detection process, solid body particles are considered as fluid particles, and the force computation process consists essentially of the computation of $f_i^{fluid}$ and $f_i^{solid}$, where $f_i^{fluid} = f_i^{press} + f_i^{vis} + f_i^{solid}$. Then, each solid body $j$ receives a total force $F_j$ and a torque $T_j$ which is the sum of forces and torques of all the particles that compose it as $F_j = \sum_{i \in RB_j} f_i^{fluid}$ and $T_j = \sum_{i \in RB_j} r_i \times f_i^{fluid}$. Note that the forces $f_i^{press}$, $f_i^{vis}$, and $f_i^{solid}$ are calculated and applied whenever necessary, for example, particles inside a solid body do not need to apply $f_i^{press}$, $f_i^{vis}$, but is needed for the particles on the bodies' surfaces. In addition, other external forces $f_i^{ext}$ might also need to be added on the force $f_i^{fluid}$ computation.

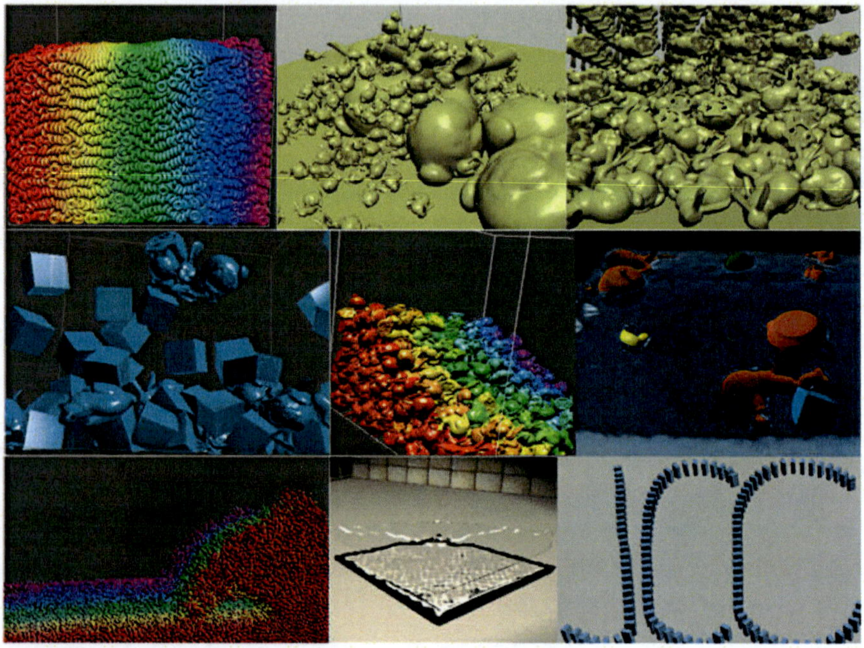

**Fig. 2** Different simulations using all the described methods

Finally, after computing these forces and torques, the positions and orientations of the objects are updated as described in Sect. 3, and the particles are also updated to their new positions. In the case of liquid particles, the physical properties update is carried out using the SPH method described in Sect. 5.

## 6 Results

We have implemented a group of techniques for physically based simulations, where the objects are represented by particles. Meshes are only used for rendering. Some results are shown in Fig. 2. Our implementations also could be applied in real-time games. In order to prove that, we have implemented a simple dominoes simulator game, which consists of stroking a path with the mouse, where the dominoes are placed; later, by pressing a key the animation starts, and the domino pieces drop colliding each other in a realistic way.

## 7 Discussion

In this work, techniques that use particles to represent objects were presented, as well as the SPH technique to model fluids. Practical guidelines for physics-based simulation implementations were presented, for fluids and solids, and several other concepts that are particularly relevant for graphical applications. We can say that with recent improvements, SPH models have matured and finally emerged as a competitive alternative to Eulerian fluid simulations or hybrid approaches. Particular SPH challenges have been overcome related to neighborhood search algorithms, pressure solvers, or versatile fluid-solid interaction techniques. With improved robustness and efficiency, millions of particles can now be simulated on a single desktop computer. Consequently, the Lagrangian SPH method achieves an unprecedented level of visual quality, where fine-scale surface effects and flow details are reliably captured. The SPH community, both in graphics and other research disciplines, is very active and the field is rapidly advancing [22]. Each community contributes to different aspects of the SPH simulations and the research often finds applications across disciplines. For graphics applications, it was especially important to efficiently enforce incompressibility in unstructured particles and thus to remove the severe time-step restrictions of standard SPH techniques. A current difficulty is that these approaches make the passage of time more challenging, as the largest possible passage of time does not necessarily result in better overall performance. On the other hand, the use of large amounts of particles is one of the key components to obtain high-quality visual and production-level results [23]. These high-resolution simulations pose new challenges and existing concepts may need to be revised. Especially speed, flexibility, and controllability are central aspects, for which solutions are still largely lacking. This problem, however, affects not only the field of SPH but also all kinds of applications that use fluid in graphics, and has sparked research on pre- and post-processing methods or data-driven approaches. To reinforce all of the concepts covered in this article, there are excellent reference implementations, such as SPlisHSPlasH (https://github.com/InteractiveComputerGraphics/SPlisHSPlasH). This implementation is an excellent tool and may serve as a starting point for students, researchers, and practitioners for future research.

## 8 Conclusion

Currently, there are many remarkable techniques based on points or particles, however, they have grow up to be very complex to understand and implement, and others are executed not in real time. Our presented approach is in the sense it is simplified and possible to be implement physically interaction of objects for interactive applications like games, since there are all the necessary techniques even some of them are very simple to implement. This allow rigid bodies, deformable bodies, and fluids to be simulated in a plausibly way in real time. On the other hand, there are some inconveniences, for example, since particles are spheres they cannot accurately represent flat surfaces and sharp edges. But, mixing several techniques it is possible to represent even these kinds of objects, but reducing the efficiency. In order to overcome efficiency issues, there are parallelization techniques for all these algorithms and data structures are also being implemented and nowadays there are many excellent implementations, however, for serious applications, it is necessary to organize all the components carefully.

**Acknowledgements** The authors would like to specially thank NVIDIA and its grant programs.

## References

1. Terzopoulos D, Witkin A (1988) Physically based models with rigid and deformable components. IEEE Comput Graph Appl 8(6):41–51. https://doi.org/10.1109/38.20317
2. Terzopoulos D, Platt J, Barr A, Fleischer K (1987) Elastically deformable models. In: SIGGRAPH '87: proceedings of the 14th annual conference on Computer graphics and interactive techniques. ACM, New York, NY, USA, pp 205–214
3. Gibson SFF, Mirtich B (1997) A survey of deformable modeling in computer graphics. Technical Report
4. Magnenat-Thalmann N, Volino P (2005) From early draping to haute couture models: 20 years of research. Vis Comput 21(8):506–519 October
5. Nealen A, Müller M, Keiser R, Boxerman E, Carlson M, Ageia N (2005) Eurographics 2005 star—state of the art report physically based deformable models in computer graphics
6. Bender J, Erleben K, Trinkle J, Coumans E (2012) Interactive simulation of rigid body dynamics in computer graphics. In: EUROGRAPHICS 2012 state of the art reports. Eurographics Association
7. Bender J, Müller M, Otaduy MA, Teschner M (2013) Position-based methods for the simulation of solid objects in computer graphics. In: EUROGRAPHICS 2013 state of the art reports. Eurographics Association
8. Harada T (2007) GPU Gems 3, chapter Rigid Bodies on GPU, pp 611–632. Addison Wesley. http://www.iii.u-tokyo.ac.jp/~takahiroharada/
9. Baraff D (1997) An introduction to physically based modeling: Rigid Body simulation i-unconstrained rigid body dynamics. In: In An introduction to physically based modelling, SIGGRAPH '97 course notes, p 97
10. Green S (2007) Particle simulation using cuda. Report, NVIDIA, In
11. Mishra BK (2003) A review of computer simulation of tumbling mills by the discrete element method: part ii–practical applications. Int J Miner Process 71(1–4):95–112. http://www.sciencedirect.com/science/article/B6VBN-48J45N7-1/2/c584739799c6b54f2d73e85e5e167718
12. Müller M, Heidelberger B, Teschner M, Gross M (2005) Meshless deformations based on shape matching. In: ACM SIGGRAPH 2005 papers, ser. SIGGRAPH '05. ACM, New York, NY, USA, pp 471–478. https://doi.org/10.1145/1186822.1073216
13. Atencio Y, Alzamora G, Esperança C (2010) Enhanced physically-based animation of deformable bodies using shape-matching. Comput Entertain 7:52:1–52:19. https://doi.org/10.1145/1658866.1658871
14. Gingold RA, Monaghan JJ (1977) Smoothed particle hydrodynamics-theory and application to non-spherical stars, vol 181, pp 375–389
15. Harada T, Koshizuka S, Kawaguchi Y (2007) Smoothed particle hydrodynamics on GPUs, pp 63–70

16. Koschier D, Bender J, Solenthaler B, Teschner M (2019) Smoothed particle hydrodynamics for physically-based simulation of fluids and solids. In: EUROGRAPHICS '19 tutorials. Eurographics Association
17. Bender J, Koschier D, Kugelstadt T, Weiler M (2019) Turbulent micropolar sph fluids with foam. IEEE Trans Vis Comput Graph 25(6):2284–2295 June
18. Weiler M, Koschier D, Brand M, Bender J (2018) A physically consistent implicit viscosity solver for sph fluids. Comput Graph Forum (Eurographics) 37(2)
19. Hu Y, Fang Y, Ge Z, Qu Z, Zhu Y, Pradhana A, Jiang C (2018) A moving least squares material point method with displacement discontinuity and two-way rigid body coupling. ACM Trans Graph (TOG) 37(4):150
20. Peer A, Gissler C, Band S, Teschner M (2017) An implicit sph formulation for incompressible linearly elastic solids: implicit elastic sph solids. Comput Graph Forum 37:12
21. Macklin M, Müller M, Chentanez N, Kim T (2014) Unified particle physics for real-time applications. ACM Trans Graph 33:1–12
22. Koschier D, Bender J, Solenthaler B, Teschner M (2019) Smoothed particle hydrodynamics techniques for the physics based simulation of fluids and solids. In: Jakob W, Puppo E (eds) Eurographics 2019 tutorials. The Eurographics Association
23. Macklin M, Müller M, Chentanez N, Kim T-Y (2014) Unified particle physics for real-time applications. ACM Trans Graph 33(4):153:1–153:12. https://doi.org/10.1145/2601097.2601152

# Study of Electric Vehicle Charging Impact

**Salma Sraidi and Mohamed Maaroufi**

**Abstract** The technology of electric vehicles is well known. Dozens of vehicles have already been marketed. Nowadays, several models are emerging, whether they are city vehicles or luxury roadsters. Car manufacturers all over the world are developing cars with electric motors allowing lightning accelerations, which require high energy. However, the impact of this new energy demand is still ambiguous with regard to the electricity grid. Since there are no large-scale application examples of electric vehicles, the existing research basically analyzes the traffic laws of ordinary vehicles to infer the possible future charging load characteristics of electric vehicles, and then conduct an impact assessment. According to the evolution of electric vehicles, today, it's necessary to investigate the behavior of electric vehicles in practice and obtain a more realistic charging load model, in addition, to study charging consequence on the distribution network in order to obtain more accurate results.

**Keywords** Electric vehicle charging · Grid impact · Optimal dispatching · Distribution grid

## 1 Introduction

Driven by national policies, the world's electric vehicle (EV) industry is developing vigorously. The national electric vehicle charging power will account for up to 25% of the national installed capacity in 2030. Electric vehicles need to be fed from the power grid. They are "large users" of random electricity consumption, and in turn have the ability to feedback electrical energy to the grid, and can be used as distributed energy storage (ES) for the electrical network if necessary [1]. Thus, as more and more EVs are connected to the grid, it will have a serious impact on the scheduling and operation of the transmission grid in addition to the distribution grid. As a kind of bidirectional energy interactive device with partial controllability in the

S. Sraidi (✉) · M. Maaroufi
Research in Electric Energy and Command, Mohammadia Engineering School, Mohamed V University, Rabat, Morocco
e-mail: salmasraidi@research.emi.ac.ma



X.-S. Yang et al. (eds.), *Proceedings of Sixth International Congress on Information and Communication Technology*, Lecture Notes in Networks and Systems 216,
https://doi.org/10.1007/978-981-16-1781-2_39

dispatching operation of the transmission and distribution network, electric vehicles have complex characteristics and are worthy of in-depth study.

This article will review the research results of electric vehicles from the perspective of the impact of EVs on the operation of transmission and distribution grids and the optimal scheduling of charging.

## 2 Impact of Electric Vehicle Charging

Evaluating the impact of electric vehicle charging from the electrical network is currently the most concerning issue for many power grid companies, and it is also the basis for the subsequent optimization and scheduling of EV charging and discharging. The review will start from two aspects: charging load modeling and impact assessment.

### *2.1 Charging Load Modeling*

Power flow analysis model

For the majority of ongoing researches, a charging load of electric vehicles could be modeled by PQ-type load. The literature [2, 3] considered the operating characteristics of the charging interface inverter when modeling the electric vehicle load, and established a more refined power flow analysis model. The literature [2] established a steady-state power flow model for a voltage source inverter-type electric vehicle charging interface, equivalent to a series of internal potential and leakage reactance inside the inverter, and based on this, proposed an extended Jacobian matrix. The power flow algorithm is used to resolve the technique of the power flow equation after the EV is connected to the network. The literature [3] further studies the power flow algorithm for EV to connect to the three-phase distribution grid on the basis of the literature [2]. In the calculation, the internal potential inside the charging interface of the EV is considered as the PV node, besides, the network point of the charging interface is regarded similarly to the PQ node, and the reactive power injected by the electric vehicle into the distribution network is corrected by analyzing the internal potential and the node voltage. For the EV field, the PQ-type load modeling idea regards the voltage support effect of the electric potential in the charging interface, in this way, under the same alternative conditions, the electric vehicle solved by the latter is located the node voltage is usually higher.

Uncertainty analysis model

The timing for EV to enter the grid for charging is non-predictable; the demand, the requested power linked to each type of vehicle, is not expected either compared to a conventional load. Therefore, when evaluating the impact of EV charging, it is necessary to model these uncertain factors to improve the accuracy and comprehensiveness of the evaluation results. The modeling methods of charging uncertainty mainly include queuing theory modeling and statistical modeling.

In the queuing theory modeling method, the three-parameter x/Y/Z model is mainly used, where x represents the distribution of car arrival time interval; Y represents the distribution of charging time; z is the number of charging parking spaces. The literature [4] adopts the MIMIn-type queuing theory model (where x and l are all obeyed exponential distribution, and l is the number of charging parking spaces connected by the node) which established the number of n vehicles in the same medium-voltage distribution network node Charging model. The literature [5] adopted a similar model and combined it with the expressway dynamic traffic model to establish a random charging load model for electric vehicles in expressway fast-charging stations. The literature [6] first analyzed the statistical law of the driving mode of a single EV, and therefore established the charging load model of EV clusters in charging stations and populated space to establish the platoon theory.

The literature [7] regards different levels of electric vehicle charging interfaces as different queues, combined with the influence of electricity prices on the number of EVs connected to the grid, and obtained the charging load expectation model of EV clusters.

The statistical modeling method is to extract the distribution function of limitations as the charging time, off-grid time, and driving distance of EV from the traffic history database and the transition probability between different charging locations through statistical methods, and then establish the random charging of electric vehicles Load model. The literature [8, 9] analyzed the data in the 2001 U.S. Department of Transportation's Household Vehicle Survey Database (NHTS2001), and obtained the statistical laws of electric vehicle travel, end time, and travel distance, and compiled the statistics of the EV based on the Random Charging Charging Model. Reference [10] used the 2009 NHTS database to conduct a similar study. The literature [11] carefully considered the power "uphill" and "downhill" processes in the charging of EV, and used convolution to show the charging power of EV clusters at different times during the charging process. It is worth noting that most of the current studies assume that parameters such as network access time, offline time, and travel distance are independent of each other. The literature [12] analyzed the actual charging process of electric vehicles and pointed out that these parameters are actually related. It is recommended to use conditional probability to establish the charging load model of electric vehicles built with the sequence of events in the charging process. The literature [13, 14] uses discrete Markov chain to establish the distribution probability matrix of the transfer of a single electric vehicle between different charging locations, and studies the temporal and spatial distribution characteristics of the charging load and its influence on the distribution grid. The literature [15] uses a fuzzy arrangement making approach to simulate the irregular time of charging and location decision case. The literature [16] established the spatial distribution of EV load that depends on the revised rate of picking up vehicles from charging stations. The literature [17] simulates the uncertainty in the charging behavior of EV based on the traffic simulation system, and also generates a charging load based on this.

## 2.2 Evaluation of the Impact of Electric Vehicles

The disorganized charging of EVs usually uses a boost in the peak-to-valley difference of electric load, which affects the quality power of the electrical network. For example, when the charging load is large, problems of voltage offset and three-phase imbalance will occur, which will endanger the stability and stability of the power grid Economics operation. Researchers in many countries have analyzed the influence of the charging load according to different EV infiltration rates to the grid charge curve, network loss, and other indicators by selecting the typical charging mode of electric vehicles.

The literature [18] studied the effect of EV charging load on the distribution network of Washington State in the United States, and indicated that when the penetration rate of EV reaches a certain percentage, the sum of its charging load and conventional load will exceed the amount of power on the existing distribution network. Therefore, a charging load must be optimally controlled. The literature [19] selected 400 V and 10 kV distribution networks in the Swedish city of Gothenburg to study the charging effects of electric vehicles in commercial and residential areas, and pointed out that if electric vehicles are charged at peak load, the medium-voltage distribution network will have line thermal overload, and low-voltage power distribution will have transformer thermal overload. The literature [13] studied the charging effects of electric vehicles at 25% and 50% penetration rates on the 15 kV medium-voltage distribution network on Flores Island, Portugal, and found that there will be increased network loss, line overload, etc., and demonstrated the necessity of optimization of the charging control. The literature [20] selected a low-voltage feeder in the suburbs of Dublin, Ireland, and studied the effect of the charging load on the three-phase node voltage of the distribution network under different electric vehicle penetration rates. Research has found that in some cases, the charging load has different effects on different phases, and a three-phase power distribution power flow model is needed for analyzing this. The literature [21] is based on the seasonal load curves of College Station in East Texas, USA, and simulates the influence of the charging load on the thermal effect of the transformer under the temperature of each season for different vehicle types and different usage conditions of daytime vehicles.

Studies have pointed out that the rise in the penetration rate of EV will heat up the transformer, and the service life of the transformer will be shortened (especially in summer). Changing the decentralized charging to a relatively concentrated charging will alleviate this impact.

The literature [22] directly used a 25kVA transformer to carry out the thermal effect experiment, and obtained the heating condition of the transformer under different working conditions. The literature [23] selects typical residential curves in Belgium in summer and winter to study the effect of EV charging load on the daily activity of the distribution network, and finds that the natural charging behavior of car owners (called disordered charging mode) is usually the same as the evening peak of residential load. Partial overlap results in a large deviation of the node voltage on the low-voltage distribution grid side and an increase in network loss. The literature [24] and [16] studied the impact of EV charging on China's distribution network. Reference [25] studies the impact of wireless charging on the electrical grid and how the integration of supercapacitor on the grid side smoothened the pic of charging of EV.

It is worth noting that, unlike the existing literature, the literature [26] evaluated the effect of EV charging load on the global power system under different penetration rates from the perspective of the global transmission and distribution system, and the impact of different voltage control modes was considered in the evaluation. Studies have shown that the impact of EV charging cannot be ignored even at a medium scale of development.

In addition, the non-linear characteristics of EV charging piles could produce certain harmonic pollution to the grid, which will lead to increased losses and faster loss of power distribution equipment, which could affect the economy of the distribution grid. The literature [27] considered the harmonic effects of EV charging equipment, and evaluated the impact of electric vehicle charging on the power quality of the power network (like the total harmonic distortion of voltage and current) through decoupling harmonic power flow calculations. When the infiltration rate of electric vehicles is huge, the power quality could be seriously affected. The literature [28] studied the impact of fast charging stations on voltage flicker.

On the other hand, as EVs are related to the power grid on a large scale, operators that uniformly manage the charging and discharging of EVs will expect a certain impact on the electricity market 29. The presence of electric vehicles in the electricity market in the form of aggregators will change the supply and demand of the electricity market, thereby affecting the equilibrium price of the market. The literature [30] conducted a simulation analysis on the known data to study the relevant factors of the local electricity market price and charging load curve affected by the access of electric vehicles. It shows that the transaction price and charging load curve in the electricity market are greatly affected by the size of the vehicle cluster. The literature [31] uses IEEE-30-node power grid as an example, introduces real traffic data and two different charging power modes, and studies the impact of electric vehicle cluster access to power grid charging on grid load and electricity market-clearing prices.

# 3 Research on the Participation of Electric Vehicles in the Optimal Dispatch of the Transmission Grid

The considerable integration of EVs into the grid not only affects the distribution grid, but also affects the transmission grid. Assuming that the charging and discharging attitude of EVs are effectively regulated, they can be used as a form of load response to coordinate and interact with current energy sources, for instance, wind or solar power on the power generation and transmission side to increase the efficiency of grid operation. In view of this, many researchers have studied the coordinated dispatch of electric vehicles and new energy represented by wind power from the perspective of the transmission grid.

## 3.1 *Electric Vehicle Clusters Participate in the Economic Dispatch of Power Transmission*

The literature [32, 33] is an earlier paper on the collaborative economic dispatch of electric vehicles and wind power. The literature [32] present the probability density distribution of EV charging and discharging through simulation, and combined them with the expectation density distribution of solar power, using electric vehicle discharge power, wind power output, and thermal power unit output as the optimization variables, the stochastic economy of the transmission grid was established. The scheduling model is solved by the stochastic optimization method.

The literature [33] uses a deterministic model for research. The literature [34] extracted the probability distribution function of user traffic behavior established with the NHTS database and established a dynamic probability example of the charging load based on it. The semi-invariant approach was used to count the transmission grid of electric vehicles and wind power in different charging modes. The literature [35] focuses on the deterministic scheduling mode of electric vehicles with the goal of peak shaving and valley filling in the Chinese grid. Different from the literature [32] in modeling, the power grid not only optimizes the discharge power of electric vehicles but also optimizes the charging power of vehicles. The literature [36] further quantitatively calculated the average carbon emission level of the grid and the carbon emission level of vehicle unit under different charging modes on the basis of the literature [35]. The literature [37, 38] further studied the issue of windmill coordinated economic dispatch that takes into account carbon emissions on the basis of the literature [33]. The literature [37] analyzed in detail the benefits of windmill cooperative dispatch under different electric vehicle penetration rates, charging and discharging modes, transmission section constraints, and charging and discharging power price ratios. The literature [39] operate with game theory and robust optimization methods to give a robust coordinated scheduling strategy to the windmill coordinated scheduling problem considering the uncertainty of wind power available power. However, the study did not consider the uncertainty of electric vehicles, and there is still room for further improvement of the charging and discharging model of EVs.

## 3.2 Electric Vehicle Clusters Participate in the Combination of Transmission Units

The literature [40] studied the problem of EV cooperating in the unit commitment of wind power systems (but did not consider the adjustability of electric vehicle charging power), using particle swarm optimization algorithm to resolve the approach, and evaluated the energy saving and reduction of EV in the transmission grid after participating in the unit combination.

The literature [41] studied the influence of EV charging mode on the random unit commitment plan of the transmission grid. During the modeling, for different wind power output scenarios, a model of EV adjustable charging power on the load side to provide backup for the system was established, and finally a two-stage stochastic optimization algorithm was used to solve it. The article points out that due to the controllability of EV charging, the load curve of the system becomes flatter, and the total running cost of the system will be decreased. The literature [42] further considered the spatial movement characteristics of electric vehicle charging load and the characteristics of random disconnection of transmission grid lines when studying the problem of electric vehicle participation in unit commitment, and established a safety constrained unit commitment (SCUC) model. Studies have shown that if congestion occurs in the transmission grid, the spatial mobility characteristics of electric vehicles are more effective in alleviating network congestion. The literature [43] models the traffic behavior of electric vehicles more elaborately than the literature [42]. On the one hand, it uses transition probability to model the transfer of vehicle groups between different nodes. Differently, it models the transfer behavior of EV under the same bus. Cars have been divided into more detail, making the charging model of the car group closer to the real scene. The literature [44] also studied the problem of the joint SCUC of EV and wind power but didn't recognize the operating costs and benefits of electric vehicles. Under this assumption, the conditions under which the complementary constraints of charging and discharging of electric vehicles can be relaxed are derived, which can decrease the amount of mixed-integer variables in the optimization problem.

## 3.3 Cooperative Mode of Electric Vehicles on the Transmission System and Distribution Side

Since electric vehicles are connected from the distribution network, cluster agents are usually required to contribute to the optimal dispatch of the transmission grid, that is, the so-called aggregator is used as a middleware to implement cluster dispatching for a large amount of EV to improve the distributed access from the distribution network. The schedule ability of the car can then improve the system operating characteristics [45].

The literature [46] pointed out that in terms of communication reliability, this hierarchical dispatching model through an aggregator is more reliable than the dispatching architecture in which the grid directly manages each electric vehicle. The literature [47] constructed a two-way interactive system and related communication mechanism of “electric vehicle-aggregator-grid”. On the one hand, the aggregator directly interacts with the dispatching agency for charging demand and metering fee settlement, and receives the dispatching instructions from the dispatching agency, and on the other hand, scheduling the charging and discharging process of each car.

The literature [37] proposed a three-level dispatch architecture for every single electric vehicle from the transmission grid to the distribution grid. The aggregator centralizes the charging demand and other information of the electric vehicles under its jurisdiction to form an optimized model and upload it to the transmission grid dispatching center. After that, the transmission grid dispatching center issues a plan to the aggregator, and the aggregator decomposes the instructions into the charging plan of each car according to the specific charging demand of each car under its jurisdiction, and then distributes it to each car. The two-level optimal scheduling schemes in [48, 49] are similar to those in [37]. The literature [50, 51] defines the internal decision-making problem of the aggregator as an “optimal tracking problem”, that is, by optimizing the charging and discharging energy of each vehicle, so that the charging and discharging power of the electric vehicle group matches the cluster optimization issued by the superior control center as much as possible Charge and discharge plan curve. This document proposes an efficient distributed algorithm for this problem. For the multi-period optimal tracking problem of thousands of electric vehicles, it only takes about 20 ms to converge to the optimal resolution of the approach.

Shortly, the future large-scale EV participating in the dispatching action of the transmission network in the form of clusters has shown good prospects and the cluster agent is the key link in it. Future research can further carry out the analysis of cluster dispatch ability, dispatch model parameter calculation, and cluster distributed management and economic incentive system for electric vehicles within the country.

## 4 Conclusion

A review of existing research shows that in terms of the impact of massive electric vehicle charging on the electrical network, this impact has a great impact on the operation of the transmission grid and the distribution network. To end this paper, it is necessary to design an effective EV charging guidance scheme based on the characteristics of the distribution market; it is necessary to further explore EV and the potential of automobile clusters to participate in power system dispatching to transform the challenge of charging EV from the grid into an opportunity for the grid as a new situation, and better accomplish the effect of grid peak and valley filling and promote the consumption of new energy on the power generation and transmission side. In addition, the future large-scale electric vehicles in the form of clusters to participate in the dispatch operation of the transmission grid have shown good prospects.

## References

1. Luo Z, Hu Z (2012) Coordinated charging and discharging of large scale plug in electric vehicles with cost and capacity benefit analysis. Autom Electr Power Syst 36(10):19–26
2. Jimenez A, Garcla N (2011) Power flow modeling and analysis of voltage source converter based plug-in electric vehicles. In: IEEE PES general meeting, San Diego, CA
3. Jimenez A, Garcla N (2012) Unbalanced three-phase power flow studies of distribution systems with plug-in electric vehicles. In: North American power symposium (NAPS). Champaign, IL, pp 1–6
4. Garcia R, Vlachogiannis J (2009) Electric vehicle demand model for load flow studies. Electr Power Compon Syst 37(5):577–582
5. Bae S, Kwasinsi A (2012) Spatial and temporal model of electric vehicle charging demand. IEEE Trans Smart Grid 3(1):394–403
6. Li G, Zhang X (2012) Modeling of plug-in hybrid electric vehicle charging demand in probabilistic power flow calculations. IEEE Trans Smart Grid 492–499
7. Alizadeh M, Scaglionea M (201) A scalable stochastic model for the electricity demand of electric and Plug-in hybrid vehicles. IEEE Trans Smart Grid 5(2):848–860
8. Tian L, Shi S, Jia Z (2010) A statistical model for charging power demand of electric vehicles. Power Syst Technol 4(11):126–130
9. Darabi Z, Ferdowsi M (2011) Aggregated impact of plug-in hybrid electric vehicles on electricity demand profile. IEEE Trans Sustain Energy 2(4):501–508
10. Shafiee S, Rastegar M (2013) Investigating the impacts of plug-in hybrid electric vehicles on power distribution systems. IEEE Trans Smart Grid 4(3):1351–1360
11. Qian K, Zhou C, Allan M (2011) Modeling of load demand due to EV battery charging in distribution systems. IEEE Trans Power Syst 26:802–810
12. Ashtari A, Bibeau E, Shahijad S (2012) PEV charging profile prediction and analysis based on vehicle usage data. IEEE Trans Smart Grid 3(1):341–350
13. Soares FJ, Peaslopes JA (2010) Monte Carlo method to evaluate electric vehicles impacts in distribution networks. In: IEEE conference on innovative technologies for an efficient and reliable electricity supply (CITRES). Waltham, MA, pp 365–372
14. Grahn P, Alvehag K, Soder L (2014) PHEV utilization model considering type of trip and recharging flexibility. IEEE Trans Smart Grid 5(1):139–148
15. Ghiasnezhad O, Filizadeh S (2014) Location based forecasting of vehicular charging load on the distribution system. IEEE Trans Smart Grid 5(2):632–641
16. Zhang H, Hu Z, Song Y (2014) A prediction method for electric vehicle charging and considering spatial and temporal distribution. Autom Electr Power Syst 38(1):13–20
17. Xin S, Guo Q, Sun H (2013) A hybrid simulation method for EVs operation considering power grid and traffic information. In: IEEE power and energy society general meeting. Vancouver, BC, pp 1–5
18. Schneider K, Gerkensmeyer C, Kintner Meyer M (2008) Impact assessment of plug—in hybrid vehicles on pacific North West distribution systems. In: IEEE power and energy society general meeting conversion and delivery of electrical energy in the 21st Century. Pittsburgh, PA, pp 1–6

19. Babaei S, Steen D, Tuan LA (2010) Effects of plug-in electric vehicles on distribution systems: A real case of Gothen burg. In: IEEE power and energy society innovative smart grid technologies conference Europe (ISGT Europe). Gothenburg, Sweden, pp 1–8
20. Richardson P, Flynn D, Keane A (2010) Impact assessment of varying penetrations of electric vehicles on low voltage distribution systems. In: IEEE power and energy society general meeting. Minneapolis, MN, pp 1–6
21. Qin Y, Kezunovic M (2012) Impact analysis of electric vehicle charging on distribution system. In: North American power symposium, pp 1–6
22. Gong Q, Midlam-Mohlers S, Serra E (2015) PEV charging control considering transformer life and experimental validation of a 25kvA distribution transformer. IEEE Trans Smart Grid 6(2):648–656
23. Clement-Nynsk K, Haesen E, Driesen J (2010) Impact of charging plug-in hybrid electric vehicles on a residential distribution. IEEE Trans Power Syst 25
24. Li H, Bai X (2011) Impacts of electric vehicles charging on distribution grid. Autom Electr Power Syst 35(17):38–43
25. Sraidi S, Maaroufi M (2019) Ultracapacitor high power booster for dynamic wireless charging. In: IEEE transaction, international symposium on advanced electrical and communication technologies (ISAECT)
26. Li Z, Sun H, Guo Q (2012) GPF based method for evaluating EVs free charging impacts in distribution system. In: IEEE power and energy society general meeting. San Diego, CA, pp 1–7
27. Moses PS, Deilami S, Masoum AS (2010) Power quality of smart grids with plug-in electric vehicles considering battery charging profile. In: IEEE power and energy society innovative smart grid technologies conference Europe (ISGT Europe). Gothenburg, Sweden, pp 1–7
28. Etezadi M, Choma K, Stefani J (2010) Rapid-charge electric vehicle stations. IEEE Trans Power Deliv 25(3):1883–1887
29. Wang X, Xie M, Zhang S (2014) Analysis on equilibrium of electricity market under grid integration of electric vehicles. Power Sys Technol 38(11):2993–2998
30. Goncalves R, Saraiva JT (2013) Impact of electric vehicles on the electricity prices and on the load curves of the iberian electricity market. In: 10th International conference on the European energy market. IEEE Stockholm, pp 1–8
31. Jain P, Jain T (2014) Assessment of electric vehicle charging load and its impact on electricity market price. In: 2014 International conference on connected vehicles and expo (ICCVE)
32. Zhao J, Wen F, Xue Y (2010) Power system stochastic economic dispatch considering uncertain out-puts from plug-in electric vehicles and wind generators. Autom Electr Power Syst 34(20):22–29
33. Li Z, Sun H, Guo Q (2011) Study on wind-EV complementation in transmission grid side. In: IEEE power and energy society general meeting. Detroit, MI, pp 1–6
34. Cai D, Qian B, Chen J (2013) Analysis on dynamic probabilistic characteristic of power grid connected with electric vehicle charging load and wind power. Power Syst Technol 37(3):590–596
35. Yu D, Song S, Zhang B (2011) Synergistic dispatch of PEVs charging and wind powering Chinese regional power grids. Autom Electr Power Syst 35(14):24–29
36. Yu D, Huang H, Lei M (2012) $CO_2$ reduction benefit by coordinated dispatch of electric vehicle charging and wind power. Autom Electr Power Syst 36(10):14–18
37. Li Z, Guo Q, Sun H (2013) Emission-concemed wind EV coordination on the transmission grid side with network constraints: concept and case study. IEEE Trans Smart Grid 4(3):1692–1704
38. Li Z, Sun H, Guo Q (2012) Study on wind EV complementation on the transmission grid side considering carbon emission. Proc CSEE '12 32(10):41–48_
39. Mei S, Guo W (2013) A game model for robust optimization of power systems and its application Proc CSEE '13 33(19):47–56
40. Saber AY, Venayag GK (2011) Plug-in vehicles and renewable energy sources for cost and emission reductions. IEEE Trans Industr Electron 58(4):1229–1238

41. Liu C, Wang J, Botterud A (2012) Assessment of impacts of PHEV charging patterns on wind-thermal scheduling by stochastic unit commitment. IEEE Trans Smart Grid 3(2):675–683.
42. Khodayar ME, Wu L, Shahidehpour M (2012) Hourly coordination of electric vehicle operation and volatile wind power generation in SCUC. IEEE Trans Smart Grid 3(3):1271–1279
43. Momber I, Morales G, Ramos A (2014) PEV storage in multi-bus scheduling problems. IEEE Trans Smart Grid 5(2):1079–1087
44. Shao C, Wang X (2014) Cooperative dispatch of wind generation and electric vehicles with battery storage capacity constraints in SCUC. IEEE Trans Smart Grid 5(5):2219–2226
45. Kempton W, Tomic J, Letendre S (2001) Vehicle to grid power: battery, hybrid and fuel cell vehicles as resources for distributed electric power in California. California: Institute of Transportation Studies, UC Davis
46. Quinn C, Zimmerle D, Bradley TH (2010) The effect of communication architecture on the availability, reliability, and economics of plug-in hybrid electric vehicle-to-grid ancillary services. J Power Sources 195(5):1500–1509
47. Lu L, Wen F, Xue Y (2012) An intelligent network control system for plug-in electric vehicles and the associate communication mechanism. Autom Electr Power Syste (20):44–49
48. Khodayar ME, Wu L (2013) Electric vehicle mobility in transmission constrained hourly power generation scheduling. IEEE Trans Smart Grid 4(2):779–788
49. Yao W, Zhao J, Wen F (2013) A hierarchical decomposition approach for coordinated dispatch of plug-in electric vehicles. IEEE Trans Power Syst 28(3):2768–2778
50. Li Z, Guo Q, Sun H (2014) A decentralized optimization method to track electric vehicle aggregator's optimal charging plan. In: IEEE power and energy society general meeting. National Harbor, MD, pp 1–5
51. Li Z, Guo Q, Sun H (2015) Study on a decentralized optimal charging algorithm for electric vehicles inside an aggregator. Electr Power Constr 36(7):114–119

# Cloud Accountability Method: Towards Accountable Cloud Service-Level Agreements

**David Ebo Adjepon-Yamoah**

**Abstract** The paper expands on our position paper Adjepon Yamoah and Wen (Assuring Dependable Cloud-Based System Engineering: A Cloud Accountability Method. Vasteras, Sweden, pp. 181, 2016), [2], to make two contributions: demonstrate cloud accountability by applying a novel method for assuring cloud dependability to a Reactive Architecture, Adjepon-Yamoah et al. (A Reactive Architecture for Cloud-based System Engineering. ACM, New York, USA, pp. 77, 2015), [3], and conduct an evidence-based stochastic (Markov chains) analysis of the Reactive Architecture with a Cloud Accountability System (CAS). This CAS facilitates the novel Cloud Accountability Method (CAM), which provides dependability assurance evidence for the Reactive Architecture. This is achieved using a virtual machine introspection (VMI) technique that counter-checks the dependability metrics (related to availability and reliability) provided by the cloud service providers (CSPs) as digital evidence. This work provides a means of addressing the perceived lack of trust for cloud-based services towards cloud accountability. **Method**: We evaluate CAM by implementing a prototype of the cloud accountability system (CAS) to facilitate the active monitoring of the Reactive Architecture. Our evaluation is performed using a cloud-based test-bed in the Amazon Web Service (AWS) Elastic Cloud Compute (EC2) environment comprising 5 Linux Virtual Machines (VMs) over a period of 3 h. The 5 instances are classified into two groups: the target VMs of the RA components and the monitoring VM of the CAS. The CAM facilitates the collection, examination, analysis, and reporting of dependability metrics as digital evidence to assure dependability of the AWS EC2 environment. **Result**: Introspecting our 5 VMs for 3 h, we identified some relevant but unreported dependability incidents on AWS, which cost the user about $65. We then conclude that the CAM is sufficiently capable of meaningfully assuring the dependability of deployed systems in the cloud.

D. E. Adjepon-Yamoah (✉)
Ashesi University, 1 University Avenue, Berekuso, Ghana
e-mail: dadjepon@ashesi.edu.gh
URL: https://www.ashesi.edu.gh/

PMB 3 CT Cantonment, Accra 00233, Ghana

 
X.-S. Yang et al. (eds.), *Proceedings of Sixth International Congress on Information and Communication Technology*, Lecture Notes in Networks and Systems 216,
https://doi.org/10.1007/978-981-16-1781-2_40

**Keywords** Cloud accountability · SLA · CSP · Trust assurance · Digital forensics · Markov chains

# 1 Introduction

Cloud computing promises to be an effective business model for staging varying sizes of enterprise systems to provide services to intended users. It is a model for enabling convenient, on-demand network access to a shared pool of configurable computing resources (e.g. servers and storage) that can be rapidly provisioned and released with minimal management effort or service provider interaction. Cloud computing offers some benefits such as low cost, unlimited smart storage, flexibility, improved performance, and increased data reliability [10]. That said, cloud computing faces challenges such as lack of customer trust [17], weak service level agreements (SLAs) [14], perceived lack of reliability [4], and absence of independent quality assurance body [9].

With the mentioned challenges of cloud computing, we expand on our position paper [2] by applying our novel methodology for assuring cloud accountability in terms of dependability (i.e. availability and reliability) to analyse the Reactive Architecture [3] test-bed on AWS. A digital forensic model is adopted to guide the data collection, examination, evidence analysis, and reporting of digital evidence. To achieve this, we present our method to perform virtual machine introspection for the purpose of assuring the dependability of the Reactive Architecture. Here, some objectives are identified:

- *01*: Demonstrate the accountability method by applying it to a cloud-based test-bed of the Reactive Architecture;
- *02*: Conduct an evidence-based trust analysis supported by stochastic continuous discrete time analysis for the purpose of dependability assurance of the cloud-based Reactive Architecture;
- *03*: Quantify the SLA/dependability issues identified on the cloud platform that devalues stakeholder investments using Net Present Value (NPV).

# 2 Background

We briefly recall two relevant concepts of this work from our position paper [2]: cloud dependability assurance and cloud accountability analysis.

For cloud dependability assurance, some notable approaches are through the cloud service level agreements (SLAs) and benchmarks. In this work, we focus on the SLA provided by Amazon Web Services (AWS). We then assessed AWS SLA for their Elastic Cloud Compute (EC2) service [5]. Our findings were that it guarantees 99.9% availability but does not specifically cover the reliability of the AWS EC2 instances. Though it provides some assurances for availability, we found them to be very ambiguous. It is, however, possible to define some boundaries for the assurance of system availability. In terms of benchmarks, we consider the Cloud IaaS 2016 Benchmark [23] provided by Standard Performance Evaluation Corporation (SPEC) to measure cloud performance. The benchmark addresses the performance of public or private Infrastructure-as-a-Service (IaaS) cloud platforms, which are designed to stress provisioning as well as runtime aspects of a cloud using input-output (I/O) and CPU-intensive cloud computing workloads. In addition, the "Nines" of Availability [27] is also used to help reduce the mean time to repair (MTTR) of high-demand systems and redundant system components.

Cloud accountability can be explained as a clear disclosure of service obligations: faithfully honouring of disclosed obligations, or otherwise assuming the liability for the unsatisfactory performance of the obligations [28] by cloud agents (i.e. service providers and software engineers). Typically, accountability in service is achieved through the enforcement of a legal and paper-based contract such as the SLA. It is important for cloud computing services to provide *assurance* based on detailed trust-focused auditing to enable forensic conclusions to be drawn for the purpose of accountability. It is relevant to note that the indiscriminate addition of auditing to a run-time environment introduces performance overheads. With cloud accountability, the cloud users can investigate whether the provider is meeting their expectations according to the service-level agreement; if violations are reported, the cloud users can provide evidence to verify who is responsible, and if there is a dispute, the cloud users can present proof to a third party (i.e. regulator). Digital forensic is about obtaining, preserving, analysing, and documenting digital evidence from digital devices [6]. Digital forensic has been widely accepted and largely used in investigating cloud security issues [20, 22]. This activity is called cloud forensic, which is defined broadly by [21] as an application of digital forensic science in cloud computing environments. Technically, it consists of a hybrid forensic approach (e.g. remote, virtual, network, live, large-scale, thin-client, and thick-client) towards the generation of digital evidence. Generally, cloud forensic is based on the well-established and widely accepted standard of NIST SP800-86 [15].

The process of virtual machine introspection (VMI) [18] facilitates the access to digital forensic evidence on a running virtual machine from another detached running VM which is co-located on the same hardware. The introspecting virtual machine must acquire privileges to access the hypervisor layer. Some open-source programming libraries such as LibVMI [25] are able to read from or write to RAMs of virtual machines. Therefore, monitoring applications can access the memory state, CPU registers, and disc activity of target operating systems in a safe and efficient manner [16]. Digital evidence is collected from multiple sources across a cloud system: Hypervisor, and the Cloud Management System (CMS) like the AWS CloudWatch. Based on the digital evidence, any breach in the contract (i.e. SLA) will have to be quantified using net present value (NPV)[19] towards monetary compensation of affected cloud agent.

## 3 Related Work

We identify some related works that identify some concepts in trust assurance. In our assessment from the literature, we identify that cloud computing essentially requires trust. We gather that a mere perception of reputation as well as CSP self-assessment crucially affects trust. Here, we briefly narrow down on evidence-based trust mechanism out of the generally classified mechanisms of (a) reputation-based trust [1, 7, 13], (b) SLA verification-based trust [8, 11, 24], and (c) evidence-based trust that have been introduced in the literature to assure trust for cloud services.

Evidence-based trust is tightly linked to QoS monitoring and SLA verification. Here, the actual performance attributes from the cloud service are compared or analysed relative to the contract (i.e. SLA) provided by the CSP. In our first considered work [12], the authors argue that a trustor's belief in the expected behaviour of the trustee is based on the evidence about the trustee's attributes of competency, goodwill, and integrity, with respect to that expectation. Other researchers, [26] propose a model dynamic trust-level scheduling (DLS) for cloud computing, based on the trust relationship models of sociology as well as the Bayesian cognitive model. Their approach was two-pronged: a novel Bayesian method-based cognitive trust model, and a trust dynamic-level scheduling algorithm via mixing the existing DLS algorithm. They used simulation and theoretic analysis to demonstrate that their algorithm securely and proficiently meets the cloud computing workloads in trust and performance requirements. However, its safety, confidentiality, and integrity were low.

## 4 Cloud Accountability System

The Cloud Accountability System is a component of the Reactive Architecture (see Fig. 1a), and it is constituted by four main components (see Fig. 1b). The Reactive Middleware, Evidence Collector, Logging System, and Auditor are its components. A brief overview of the mentioned components is given below.

1° **Evidence Collector (EC)**: The Evidence Collector system collects data for the availability and reliability metrics from the required set of the RA's VM hypervisor through two means: Virtual Machine Introspection (VMI) technique and the Cloud Management System (CMS). Refer to steps *2* and *3* of Fig. 1b. The VMI uses a VMI library developed using the C language, and also Python scripts to collect metrics data. Secondly, the CMS uses Amazon Web Services' (AWS) API known as the AwsSdkMetrics. Figure 2 shows a high-level view of the EC system. To achieve data integrity, a reconciliation of these two-sourced data is undertaking by the Arbitrator. All relevant activities in steps *2* and *3* are saved as logs (see step *5* in Fig. 1b).

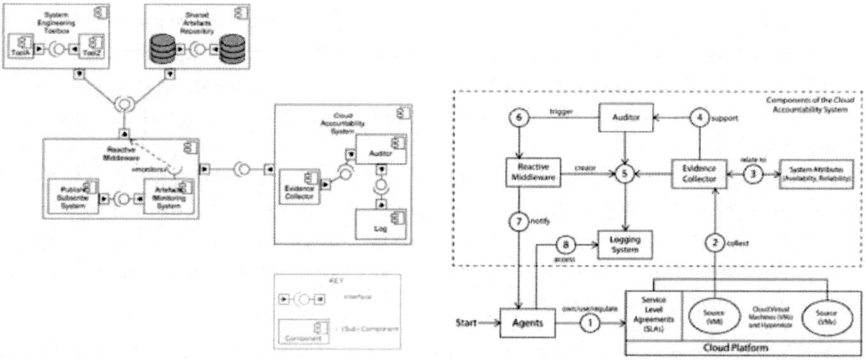


**Fig. 1** **a** Overview of reactive architecture, and **b** Operational model of the cloud accountability system

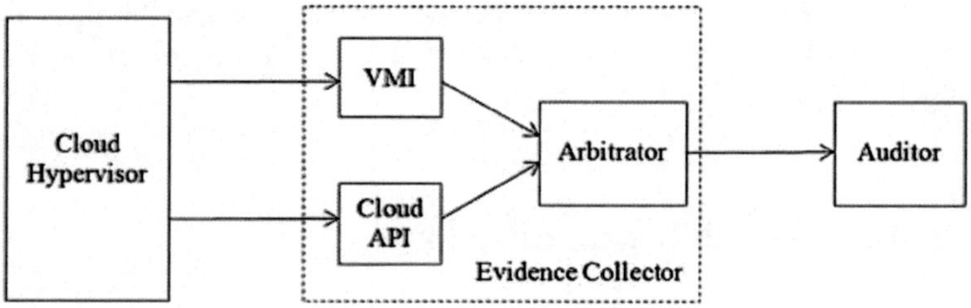


**Fig. 2** Evidence collector

2° **Auditor**: The Auditor conducts an evidence-based trust analysis. Here, a simplistic yet relatable and effective weighted trust value approach is used. All systems are initialised to zero, the absence of violations gets one, however, the presence of it gets negative one. Generally, the Auditor works by collating the set of metrics data relating to availability and reliability from the Evidence Collector (see step *4* in Fig. 1b). Then the availability and reliability are computed for each system being monitored. At this point, the availability and reliability values from each monitored system are obtained from the CSP API, and then compared with those from the VMI. Violations from the two data sources are checked in the context of the stipulated SLA. If there are violations (see step *6* in Fig. 1b), the Reactive Middleware is initiated to handle notifications. Then, to understand the impact of such violation on the cloud agent's investment, the Auditor computes the NPV. show a financial perspective to the cloud agents (e.g. the system developers). Log are used to save every activity so far (see step *5* in Fig. 1b). These logs are periodically updated since cloud service performance varies due to the dynamic quality attributes which affects the metrics over time.

3° **Reactive Middleware (RM)**: The RM manages notifications in the RA. In situations where there are SLA violations, cloud agents are provided with appropriate options to handle them (see step *7* in Fig. 1b). All activities involved here are all documented as *logs*. Also, the cloud agents can access the logs of their systems (see step *8* in Fig. 1b).

4° **Logging System (LS)**: This is where all the generated *evidence* and associated cloud *events* are stored as logs. In situations where claim compensations are expected, such logs are provided as evidence. Again, logs can be used as datasets for deep learning to, say, predict frequency of SLA violations on different cloud platforms using machine learning.

The CAS is implemented to collect data from the RA's VM in run-time as per the metrics that are related to the architecture reliability, and availability. This work focuses on metrics such as system operation time, mean time to failure (MTTF) or mean time between/before failures (MTBF), and mean time to repair (MTTR). Such metric data are then compared with those from the CSP and the VMI. If there are identified violations, the cloud agents are then notified. Figure 1b provides an operational overview of this CAS.

## 5 Cloud Accountability Methodology

The Cloud Accountability Methodology (CAM) is facilitated by the CAS. The CAM is guided by the widely accepted NIST SP800-86 digital forensics model, for supporting incident handling. The forensic process shown in Fig. 3 involves the collection, examination, analysis, and reporting digital evidence. The steps of the CAM as well as its algorithm (see Algorithm 1) are recalled below.

(1) The EC is assigned to the composing systems of the Reactive Architecture (RA), deployed on virtual machines (VMs) in the cloud.
(2) The EC collects metrics data for availability, [A], and reliability, [R], including processing times in a synchronised manner from their respective VMs at a defined constant time duration, $[t_i, t_{i+1}]$ (*i* is initialised to zero).
(3) The gathered metrics data are sorted and examined for data integrity by the Arbitrator.

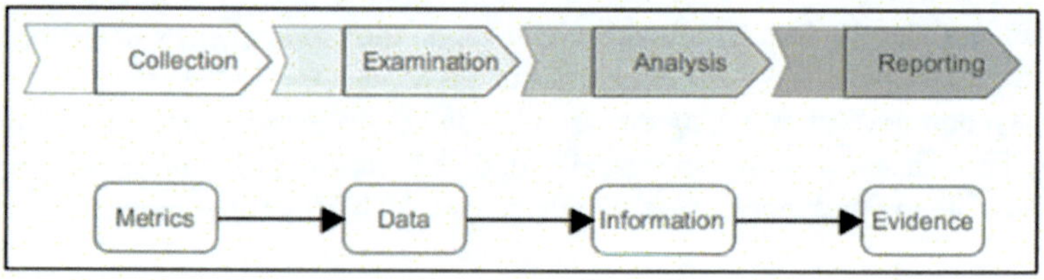


**Fig. 3** Forensic process

(4) The set $[A, R]_N$ is sent to the Auditor using synchronous procedure calls (SPC).
(5) The Auditor initialises the trust values of the VMs to *0*.
(6) The Auditor classifies $[A_N]$ and $[R_N]$ for the VMs.
(7) The Auditor compares the sets of metrics data from the cloud evidence sources (i.e. VMI, CMS) to understand their relationship based on the SLA (i.e. acceptable metric range, $[AMR]_c^m$) of the cloud platform.
(8) If any metric data of the two evidence sources violates the SLA's $[AMR]_c^m$, the Auditor assigns a value of *−1* to the trust value of the related VM, and then initiates a call to the Reactive Middleware (RM).
(9) All activities including VM users, duration of VMs observation, date, details of failed VMs, analysis of failure, etc. are *logged.*
(10) The RM then sends a notification to the cloud agents when all VMs are "unavailable".
(11) The RM provides options such as requesting for "Service Credits" from the CSP using the *log* from (9).
(12) The Auditor also computes the Net Present Value (NPV) of the set of cloud resources with respect to the processing times from (7), and/or violations from (8).
(13) If there was no violation from (8), the Auditor assigns *1* to the trust value of the related VM, an analysis of the comparison in (7), and the NPV from (12) are saved as a *log*, $[l_N]$.
(14) This log, $[l_N]$, is sent as a notification to the cloud agents as periodic reports (e.g. monthly and yearly).
(15) The method continues in a loop at (2) for time duration $[t_i, t_{i+1}]$, if none of the assigned VMs in (1) failed.
(16) If there is any recorded violation, the method will continue in a loop at (1) for the given time duration, $[t_i, t_{i+1}]$.

## 6 Evaluating the Cloud Accountability System

Here, we attempt to provide support for cloud dependability assurance focusing on availability and reliability with a cloud accountability method. We use the SLA's assured values for cloud availability as the primary basis for such assurance. To achieve our objectives (i.e. O1, O2, and O3), a research question is posed: "can a cloud accountability method be used to meaningfully assure cloud dependability of deployed systems?".

**Algorithm 1** Cloud Accountability Methodology

**Require:** $VM_{RM}(A, R), VM_{SET}(A, R)$
**Require:** $VM_{SAR}(A, R), VM_{FDS}(A, R)$
**Ensure:** notify $(l_N)$
$EC_{RM} \leftarrow VM_{RM}(A, R)$
$EC_{SET} \leftarrow VM_{SET}(A, R)$
$EC_{SAR} \leftarrow VM_{SAR}(A, R)$
$EC_{FDS} \leftarrow VM_{FDS}(A, R)$
$duration \leftarrow 0$
$i \leftarrow 0$
$trust_{VM_N} \leftarrow 0$

*top*:
**while** $t_i \leq duration \leq t_{i+1}$ **do**
  $arbitrate_{time}(EC_{RM}, EC_{SET}, EC_{SAR}, EC_{FDS})$
  $audit_{SPC}(EC_{RM}, EC_{SET}, EC_{SAR}, EC_{FDS})$
  $audit_{classify}(EC_{RM}, EC_{SET}, EC_{SAR}, EC_{FDS})_R^A$
  $computeA_N, R_N$
  $audit_{metdata}(EC_{RM}, EC_{SET}, EC_{SAR})_{CMS}^{VMI}$

  **if** $(EC_{RM}, EC_{SET}, EC_{SAR}, EC_{FDS})_{CMS}^{VMI}$ violates $SLA's[AMR]_c^m$ **then**
    $trust_{VM_N} \leftarrow -1$
    $l_N(violation_{VM}, trust)$
    notify $(l_N)$
    $request_{ServiceCredit}()$

    **if** $VM$ is unavailable **then**
      notify $(l_N)$
    **else** $\{l_N(audit_{metdata}(), audit_{NPV}()\}$

    **end if**

  **else** $\{l_N(audit_{NPV}(), audit_{metdata}(), violation_{VM}())\}$
    $trust_{VM_N} \leftarrow 1$
    $l_N(violation_{VM}, trust)$
  **end if**

**end while**
**goto** *top*

We evaluate our work by answering our research question and draw conclusions. We begin by recalling how we apply the cloud accountability method (CAM) (see Sect. 5) to the Reactive Architecture (RA). To achieve objective 1 (O1), we implement a prototype of the cloud accountability system (CAS) to facilitate the active monitoring of the RA. Crucially, our evaluation is performed using a cloud-based test-bed in the Amazon Web Service (AWS) Elastic Cloud Compute (EC2) environment. The AWS/EC2 test-bed is run in the *eu-west-1b* and *eu-west-1c availability zones* (i.e. geographic locations of AWS cloud infrastructure/service) in Ireland. Such a decision is necessary since these availability zones provide the least network latency to Ashesi University where the evaluation is conducted. It also allows for the monitor-

**Fig. 4** AWS/EC2 CAS test-bed instances

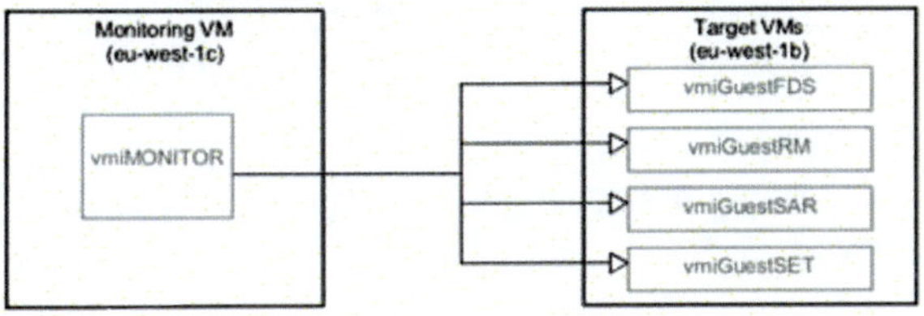


**Fig. 5** Test-bed's VMs introspection in two AWS availability zones

ing or introspection of the group of VMs for the set of RA components from another availability zone. For the test-bed, we create a set of 5 Linux virtual machines (i.e. Ubuntu 16.04 × 64 t2.micro with 1 GB of memory, 1 vCPU, SSD Volume Type and variable ECU—see Fig. 4).

The 5 instances are classified into two main VMs: the *target VMs* and the *monitoring VM* (see Fig. 5). The *target VMs* are constituted with *vmiGuestFDS* (i.e. Formal Decomposition System), *vmiGuestRM* (i.e. Reactive Middleware), *vmiGuestSAR* (i.e. Shared Artefacts Repository), and *vmiGuestSET* (i.e. System Engineering Toolbox) VM instances. The *target VMs* are located at the *eu-west-1b* availability zone. Also, the *monitoring VM* is the *vmiMONITOR* VM instance located at the *eu-west-1c* availability zone. The separation of the availability zones allows for the effective introspection of the *target VMs* by the *vmiMONITOR* VM, without being subjected to the same dependability situation at the *eu-west-1b* availability zone.

**Cloud Accountability Method to Assure Dependability**
We classify the steps of the presented cloud accountability method according to the phases of the NIST SP800-86 guide (see Table 1). The operational model (i.e. Fig. 1b) of the cloud accountability system is also used to support this classification.

## 6.1 Collection

To implement the CAS, we look at the source of digital evidence collection. In our work, we identify two sources for accessing data related to dependability metrics from the *hypervisor* of the RA virtual machines. The choice for the two sources of data below is because the data from the CSPs are perceived to be untrustworthy, so a

**Table 1** Classified steps of cloud accountability method with the forensic process phases (NIST SP800-86 guide)

| The forensic process phases (NIST SP800-86 guide) | Cloud accountability method steps |
|---|---|
| Collection | 1, 2, 4, 5, 15, and 16 |
| Examination | 3 and 6 |
| Analysis | 7, 8, 11, and 12 |
| Reporting | 9, 10, 13, and 14 |

second data source using a reliable digital forensic approach is considered to check the former data source.

**AWS CloudWatch** The API used here is the *AmazonCloudWatchClient* API, which provides data related to metric statistics. The *metrics data* provided by the AWSCloudWatchClient API also provide information for CPU utilisation, status checks for VMs and system, network packets, network state, disc write operations, disc read operations, CPU credit usage, and CPU credit balance.

Status check metrics are available at 1- and 5-minute frequencies. However, we focus exclusively on the *CPU utilisation*, *status checks for VMs*, and *status check for system* metrics data collected at a frequency of 5 minutes. This time frequency is appropriate since it is long enough to accommodate our 1-minute data collection call from the target VMs, and network latency which can be unpredictable. We believe that this set of metrics are representative for providing a reliable picture of the dependability of a set of cloud-based VMs. All the mentioned metrics are also provided to users of AWS as graphs in AWS CloudWatch relative to users' AWS resources (e.g. EC2). In this work, we consider metrics from the AWS CloudWatch API, as well as the generated graphs from AWS CloudWatch. In Fig. 6, we provide a brief description of some of the focal metrics we consider in our evidence collection. This same approach is used to collect the other three focal metrics.

Initially, we took a look at the focal metrics as presented by the AWS CloudWatch for the test-bed. From Figs. 7, 8, 9, and 10, we observed a set of line graphs that

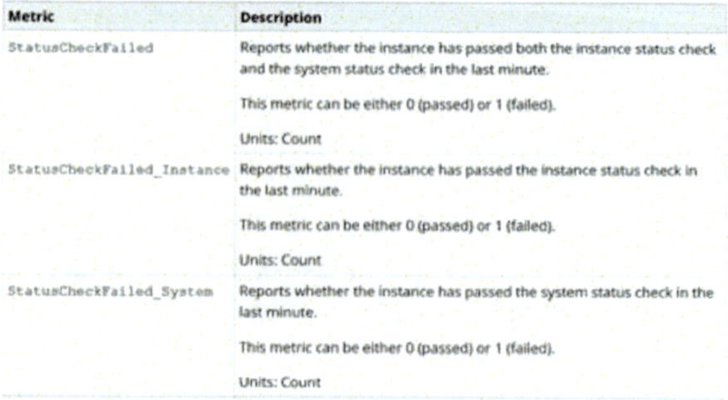

| Metric | Description |
|---|---|
| StatusCheckFailed | Reports whether the instance has passed both the instance status check and the system status check in the last minute.<br>This metric can be either 0 (passed) or 1 (failed).<br>Units: Count |
| StatusCheckFailed_Instance | Reports whether the instance has passed the instance status check in the last minute.<br>This metric can be either 0 (passed) or 1 (failed).<br>Units: Count |
| StatusCheckFailed_System | Reports whether the instance has passed the system status check in the last minute.<br>This metric can be either 0 (passed) or 1 (failed).<br>Units: Count |

**Fig. 6** Some AWS/EC2 instance and system metrics

**Fig. 7** vmiGuestFDS: AWS CloudWatch line graphs for some metrics

**Fig. 8** vmiGuestRM: AWS CloudWatch line graphs for some metrics

indicate that the instances are "healthy" per the set of metrics. Here, the data for the *status checks* metrics indicate results of "0", implying that there has not been any recorded failures for the target VMs in the 3 h observation period. That said, our work takes a step further to collect our own set of evidence relative to the said set of metrics. We consider a well-used digital forensic process known as *virtual machine introspection* as our reliable means of accessing such data. This process is discussed below. Here, with the aid of the metrics provided by the APIs, we aim to deduce the state of dependability of the CAS target VM instances by obtaining the mean time to failure (*MTTF*) or mean time between failures (*MTBF*), mean time to recover (*MTTR*), and *system operation time*.

**Fig. 9** vmiGuestSAR: AWS CloudWatch line graphs for some metrics

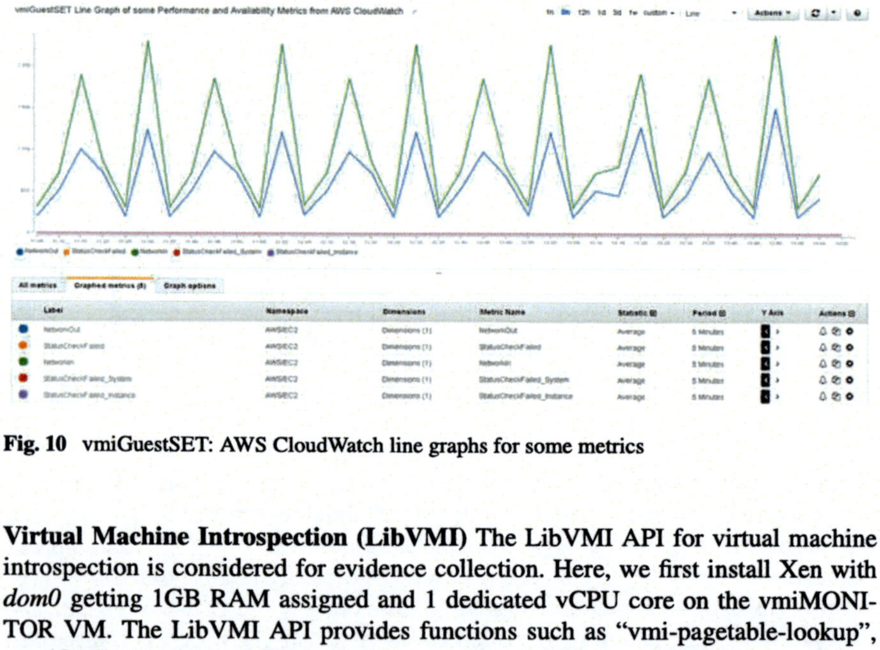


**Fig. 10** vmiGuestSET: AWS CloudWatch line graphs for some metrics

**Virtual Machine Introspection (LibVMI)** The LibVMI API for virtual machine introspection is considered for evidence collection. Here, we first install Xen with *dom0* getting 1GB RAM assigned and 1 dedicated vCPU core on the vmiMONITOR VM. The LibVMI API provides functions such as "vmi-pagetable-lookup", specifically "vmi-pagetable-lookup-extended" to check the performance status of VMs. Importantly, it returns VMI_SUCCESS for an "active" and "healthy" VM, or VMI_FAILURE if the VM is "invalid" or "inactive". To further check for VM introspection failure, we consider "vmi-get-linux-sysmap" to show a Linux system path to an active VM.

It must be said that other equally effective LibVMI API functions can be used, but we are also cautious since the performance of both the monitoring VM (i.e.

vmiMONITOR) and the target VMs is sensitive to the size of request parameters and the resulting return values. The API functions called by "module-list" (such as "vmi-pagetable-lookup-extended()") have fewer parameters and small return values (i.e. primitive variable type) compared to "process-list" such as "vmi_read_pa( )", which returns specified size of physical memory in binary which will introduce more significant overhead. This identified API function call is used to help in observing the times between VMs fail (i.e. obtaining MTBF or MTTF), times for VMs to recover (i.e. obtaining MTTR), and system operation time. That said, we proceed to apply our defined CAM to the CAS test-bed.

With the provided overview for our evidence collection approach supported by **Steps 1**, **2**, **4**, and **5**, we synchronise the collection of *operation times* for the respective VMs from both the AWS CloudWatch and LibVMI APIs evidence sources. This activity is undertaken between 11:00 GMT and 14:00 GMT which falls within the range of time considered to be AWS/EC2 peak time in the Ireland availability zone. Such a time period sees a higher level of user requests, processing and scaling of AWS/EC2 resources. We presume that it will be relatively easier to identify dependability issues at this time period. Also, the "trust values" for the VMs are initialised to 0.

During the 3 h period, the evidence collection process for both evidence sources is repeated every 5 minutes in line with **Steps 15** and **16** of the cloud accountability methodology.

## *6.2 Examination*

The two sets of collected processing times are first assessed by the Arbitrator for *outliers* in relation to **Step 3**. Here, we consider processing times that span a period of *100 ms* to *650 ms* as the acceptable processing time, and hence any other processing time is classified as an "outlier". However, we identify processing times less than 100 ms as a "failure" of the VM. This failure classification is informed by an earlier assessment of the minimum latency to each VM, and it was observed to be an average of 119 ms. Also, this decision is complemented by at least one evidence source API check of the VM in consideration.

Following from this, an "outlier" was observed for the vmiGuestRM instance at 12:15 GMT (see Fig. 11). The LibVMI and AWS CloudWatch API recorded processing times of about *93.23* ms and *84.53* ms, respectively. Since the recorded processing times are below 100 ms, the *Arbitrator* indicates a VM failure. However, the next sets of processing times recorded after 5 minutes shows values of an average of 300 ms. This shows the *vmiGuestRM* instance "recovered" during the said time. Such recovery is made possible by cloud virtualisation features such as snapshot, autoscaling, and load balancing. All other processing times are considered to be "acceptable" by the Arbitrator, and hence passed on to the Auditor.

The computation of system availability and reliability is often a laborious and complex process requiring system observation of many months to a year, or even

| Results After 3 (hr) | | | | |
|---|---|---|---|---|
| VMI Time (HH:mm) | Introspection Time for Reactive Architecture instances (ms) | | | |
| | vmiGuestFDS | vmiGuestRM | vmiGuestSAR | vmiGuestSET |
| 11:00 | 358.9100474 | 370.4532001 | 362.3901615 | 368.9284697 |
| 11:05 | 402.4200017 | 421.3621982 | 409.8791003 | 413.6350998 |
| 11:10 | 301.4189565 | 313.8665493 | 306.0723782 | 308.1937303 |
| 11:15 | 297.8903279 | 317.1003756 | 308.9685978 | 312.7643812 |
| 11:20 | 313.9864337 | 325.7511986 | 322.1276639 | 324.7610075 |
| 11:25 | 349.3487084 | 409.1083746 | 403.4632698 | 405.6768527 |
| 11:30 | 419.6762789 | 479.7667729 | 452.9875453 | 467.2397629 |
| 11:35 | 591.4008266 | 607.9875236 | 602.6296398 | 605.7553753 |
| 11:40 | 550.2486899 | 581.9764272 | 575.8276272 | 580.1098338 |
| 11:45 | 512.8007165 | 529.8762282 | 522.4200909 | 527.1009313 |
| 11:50 | 420.9743678 | 444.1098747 | 430.8891071 | 433.8600201 |
| 11:55 | 357.1238735 | 483.6572761 | 360.6826887 | 362.9761101 |
| 12:00 | 373.6189474 | 469.2702891 | 371.3915638 | 375.6569127 |
| 12:05 | 307.2290375 | 319.3899188 | 311.9553051 | 313.9010798 |
| 12:10 | 311.1829513 | 320.5449531 | 313.3771778 | 317.5432301 |
| 12:15 | 499.1203273 | 93.2335726 | 518.1200923 | 511.6776723 |
| 12:20 | 313.6174435 | 321.5113181 | 320.1036738 | 321.1010207 |
| 12:25 | 379.8417021 | 399.2103796 | 389.4632142 | 393.4234735 |
| 12:30 | 357.7100278 | 388.1646221 | 359.5232673 | 371.8884201 |
| 12:35 | 589.1456006 | 601.9875236 | 591.3109129 | 600.3001502 |
| 12:40 | 470.2486899 | 487.7212702 | 476.1011893 | 483.9638371 |
| 12:45 | 412.2973143 | 430.2900421 | 425.7507005 | 497.1998203 |
| 12:50 | 367.4004711 | 381.9011253 | 330.1754662 | 330.6617236 |
| 12:55 | 371.8015578 | 380.5567106 | 374.6131891 | 380.1391556 |
| 13:00 | 334.0648725 | 351.2902506 | 347.0417628 | 348.2182318 |
| 13:05 | 352.2037276 | 368.6743919 | 366.9710383 | 367.6350193 |
| 13:10 | 371.1080564 | 373.8165412 | 372.0723787 | 373.1007302 |
| 13:15 | 300.8907252 | 311.1003756 | 303.1600748 | 307.4321847 |
| 13:20 | 303.8304332 | 306.2300107 | 304.2766301 | 305.6947523 |
| 13:25 | 351.4235882 | 372.2010567 | 364.4637121 | 369.9768528 |
| 13:30 | 357.7100278 | 368.1646227 | 359.5238673 | 361.8904203 |
| 13:35 | 391.1296064 | 394.9875236 | 392.7132122 | 393.7098532 |
| 13:40 | 372.2426843 | 387.7702764 | 376.9986124 | 383.9638323 |
| 13:45 | 382.7983186 | 364.1764042 | 362.5103251 | 367.1020317 |
| 13:50 | 427.7913875 | 371.1127638 | 370.7219792 | 371.6203974 |
| 13:55 | 419.1689932 | 381.1698247 | 376.8909574 | 384.1750054 |

| Results After 3 (hr) | | | | |
|---|---|---|---|---|
| AWS Time (HH:mm) | AWS/EC2 API Metrics Collection Time for Reactive Architecture Instances (ms) | | | |
| | awsec2FDS | awsec2RM | awsec2SAR | awsec2SET |
| 11:00 | 308.8097518 | 321.7528 | 311.89006 | 318.828360 |
| 11:05 | 352.3197061 | 372.6617981 | 359.3789988 | 363.5349981 |
| 11:10 | 251.3186609 | 265.1661492 | 255.5722767 | 258.0936286 |
| 11:15 | 247.7900323 | 268.3999755 | 258.4684963 | 262.6642795 |
| 11:20 | 263.8861381 | 277.0507985 | 271.6275624 | 274.6609058 |
| 11:25 | 299.2484128 | 360.4079745 | 352.9631683 | 355.576751 |
| 11:30 | 379.5759833 | 431.0663728 | 402.4874438 | 417.1396612 |
| 11:35 | 541.300531 | 559.2871235 | 552.1295383 | 555.6552736 |
| 11:40 | 504.1483943 | 533.2760271 | 525.3275257 | 530.0097321 |
| 11:45 | 462.7004209 | 481.1758281 | 471.9199894 | 477.0008296 |
| 11:50 | 370.8740722 | 395.4094746 | 380.3890056 | 383.7599184 |
| 11:55 | 337.0235779 | 434.956876 | 310.1825872 | 312.8760084 |
| 12:00 | 351.5186518 | 420.569889 | 320.8914623 | 295.556811 |
| 12:05 | 257.1287419 | 270.6895187 | 261.4550036 | 263.8009781 |
| 12:10 | 261.0826557 | 271.844553 | 262.8770763 | 267.4431284 |
| 12:15 | 466.0200317 | 84.5331725 | 467.6199908 | 461.5775706 |
| 12:20 | 263.5171479 | 272.810918 | 269.6035723 | 271.000919 |
| 12:25 | 329.7414065 | 350.5099795 | 338.9631127 | 343.3233718 |
| 12:30 | 307.6097322 | 339.464222 | 309.0231658 | 321.7883184 |
| 12:35 | 539.045305 | 483.2871235 | 540.8108114 | 550.2000485 |
| 12:40 | 420.1483943 | 439.0208701 | 425.6010878 | 433.8637354 |
| 12:45 | 362.1970187 | 381.589642 | 375.250599 | 447.0997186 |
| 12:50 | 317.3001755 | 333.2007252 | 279.6753647 | 280.5616219 |
| 12:55 | 321.7012622 | 331.8563105 | 324.1130876 | 330.0390539 |
| 13:00 | 283.9645769 | 302.5898505 | 295.5416613 | 298.1181301 |
| 13:05 | 302.895432 | 319.9739918 | 316.4709368 | 317.5349176 |
| 13:10 | 361.1077608 | 325.1161411 | 321.5722772 | 323.0006285 |
| 13:15 | 250.7904296 | 262.3999755 | 252.6599733 | 257.332083 |
| 13:20 | 253.7301376 | 257.5296106 | 253.7765286 | 255.5946506 |
| 13:25 | 301.3232926 | 323.5006566 | 313.9636106 | 319.8767511 |
| 13:30 | 307.6097322 | 319.4642226 | 309.0237658 | 311.7903186 |
| 13:35 | 341.0293108 | 346.2871235 | 342.2131107 | 343.6097515 |
| 13:40 | 322.1423887 | 339.0698763 | 328.4985109 | 333.8637306 |
| 13:45 | 362.698023 | 355.4760041 | 312.0102236 | 347.00193 |
| 13:50 | 418.7910954 | 322.4123637 | 320.2218777 | 321.5202957 |
| 13:55 | 389.0686976 | 332.4694246 | 326.3908559 | 334.6749037 |

**Fig. 11** Processing times (ms) from the evidence sources

longer. This period usually records several incidents of failure and recovery from failure, which facilitates a more easily acceptable analysis of system availability and reliability. We are, however, limited by time to conduct such a detailed process; the amount of data to analyse from the two evidence sources; and by the cost of using the cloud platform over a prolonged period. Our aim for this work is to provide information towards the assurance of cloud platform's availability and reliability. That said, we base our analysis on the data collected during 3 h of the peak time for the AWS cloud platform, to mainly demonstrate that our methodology can provide relatively more information towards the assurance of cloud dependability to cloud *agents*.

In this context, we proceed to **Step 6**, where the availability, $[A_N]$ and reliability, $[R_N]$ of the Reactive Architecture VMs are derived and classified. A classification of $[A_N]$ and $[R_N]$ values with respect to their corresponding VMs is shown in Table 2. Using the *autoscaling* feature for AWS VMs, a *stop-start cycle* for recovering from failure is provided as less than 3 minutes. So in computing the availability, $[A_{RM}]$ of *vmiGuestRM* after its failure at 12:15 GMT, we consider the *MTTR* as *3 minutes* instead of 5 minutes for our evidence collection frequency.

**Table 2** Classification of availability and reliability for the test-bed VMs

| Instances $[S_N]$ | Availability | | | Reliability | | |
|---|---|---|---|---|---|---|
| | MTBF/ MTTF (hr) | MTTR (hr) | $[A_N]$ | MTBF/ MTTF (hr) | TIME (hr) | $[R_N]$ |
| vmiGuestFDS | 3 | 0 | 1 | 3 | 3 | 0.36788 |
| vmiGuestRM | 1.15 | 0.30 | 0.7931 | 1.15 | 3 | 0.07363 |
| vmiGuestSAR | 3 | 0 | 1 | 3 | 3 | 0.36788 |
| vmiGuestSET | 3 | 0 | 1 | 3 | 3 | 0.36788 |

## 6.3 Analysis

We analyse the examined processing times from the *Evidence Collector* towards meaningfully assuring availability and reliability of deployed cloud-based systems using the presented cloud accountability method. This meets objective 2 (O2). The processing times for the set of VMs from the two evidence sources are compared to identify significant variance(s) that may exist between the two sets of processing times.

We also observed that the AWS CloudWatch API collected evidence at an average of *48.25* ms faster than that of the LibVMI API. We believe that the implementation of the LibVMI API introduces some overhead cost which affects its processing time, and/or it is as a result of efficiency on the part of the AWS CloudWatch API. That said, comparing the failure incident of vmiGuestRM at 12:15 GMT recorded by both the AWS CloudWatch API and LibVMI APIs, with the AWS CloudWatch generated line graph (refer to Fig. 8) shows a supportive recording. Here, the AWS CloudWatch line graph shows a drop in network data for both *NetworkIn* and *NetworkOut* metrics. However, this drop in network data does not seem out of place. The oscillation of the network data in the line graph seems to follow a regular pattern. Furthermore, the *StatusCheckFailed*, *StatusCheckFailed_System*, and *StatusCheckFailed_Instance* metrics from the AWS CloudWatch generated line graph indicates no failure for the *vmiGuestRM* VM.

Here, for our evidence-based trust analysis which has been assigning trust value of 1 to VMs with no violations to the SLA, and $-1$ to VMs with violations, a resulting graph is generated as shown in Fig. 12. With the recorded failure incident of vmiGuestRM at 12:15 GMT, a trust value of $-1$ is assigned to this VM. A history of the trust values are made available to cloud agent for their consideration. As mentioned earlier, trust values are assigned to all VMs for every 5-minute cycle to assure real-time monitoring. For a much more complex analysis, we conduct a stochastic analysis, involving continuous discrete time Markov chains (dtmc). In Fig. 13, we show five states of the vmGuestRM VM and its state transitions (based on a right-stochastic transition matrix of this stochastic process). We simulate a 36-step (i.e. for each 5 minutes in the 3 h) random walk that starts from a random state.

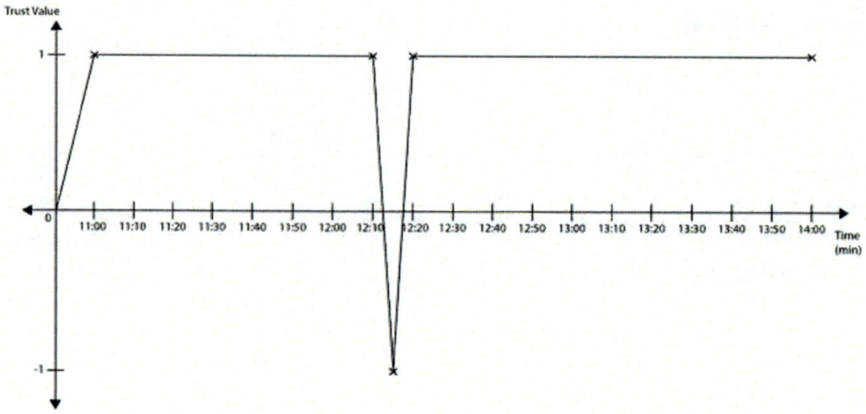


**Fig. 12** Evidence-based trust analysis of reactive middleware's VM

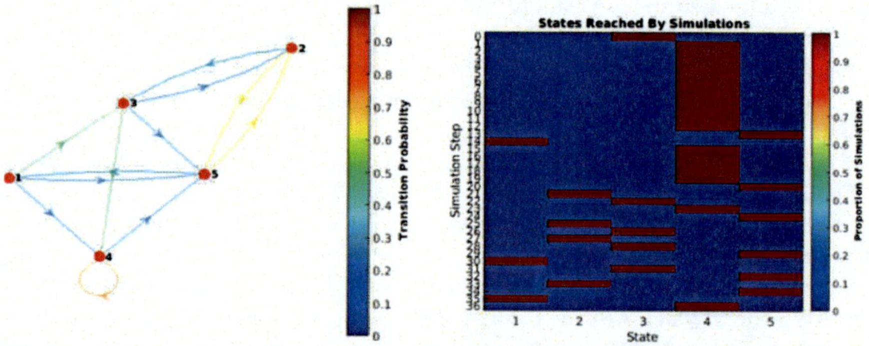


**Fig. 13** Markov chain simulations of the vmiGuestRM state transitions—*Initiated* (1), *Starting* (2), *Running* (3), *Waiting* (4), *Failed* (5) [Tool: The MathWorks, Inc.]

The reported failure is evident between transitions 2 and 5 and also at transition 4 (refer to Transition Probability in Fig. 13). Also, steps 1–12 and 15–19 for state 4 (refer to Simulation Steps in Fig. 13) show the simulated impact of the failure of vmGuestRM.

Since there is no justified reason according to the AWS SLA to request for "Service Credit" compensation, **Step 11** is not undertaken. At this point, the Auditor computes the net present value (NPV) of the Reactive Architecture's VMs as in **Step 12** to meet objective 3 (O3). The formula for NPV varies slightly depending on the consistency with which returns are generated, and if the investment was a one-time event. Also, since each period generates returns in equal amounts and the observation time is relatively short, the mathematical expression for the NPV in our case is

$$NPV = \frac{C}{(1+r)^i} - C_0 \tag{1}$$

where
**C**—the expected cash flow per period (i.e. \$ 96.82 per 3 h for 5 AWS t2.micro VMs including data transfer cost, storage cost, and VAT.);
**r**—the required rate of return. The recommended Ghana public service discount rate is 14.4196% at 25 Nov 2020;
**T**—the number of periods over which the project is expected to generate income (i.e. 3 h);
$C_0$—the initial investment (\$ 129.27 fee for AWS EC2 Reserved Instance).

$$NPV = \frac{96.82}{(1 + 0.144196)^3} - 129.27 \tag{2}$$

$$NPV = -\$64.637 \tag{3}$$

The negative NPV value from Eq. (3) indicates that the investment in the 5 AWS virtual machines is *undesirable* subject to its prevailing *discount rate* over the 3-hour period. Since there was no violation to the SLA, the NPV value indicates the value of the cloud investment after the said time.

## *6.4 Reporting*

Generally, our examination and analysis of evidence using a wide range of analysis options (i.e. availability and reliability classification, evidence-based trust analysis, and line graphs from two evidence sources), combined with a widely accepted forensic process model, make our analysis considerably trustworthy. Here, our analysis provides relatively more detailed information than the state of the art towards the assurance of dependability (i.e. availability and reliability) of the cloud-based Reactive Architecture.

The cloud agents are provided with a *log* of the analysis (i.e. comparison of the processing times, and the NPV). Refer to Fig. 14 for an XML presentation of the mentioned log. This log, together with the figures (including the evidence-based trust graph—Fig. 12) and tables generated in the forensic process, is provided to the cloud agents for assuring dependability of AWS resources. This activity is in line with **Steps 9, 10, 13**, and **14**.

To compare our analysis to a relevant benchmarking of the performance of cloud infrastructure, we consider the Cloud IaaS 2016 Benchmark. Here, our performance evaluation is conducted over a period of "46 minutes". This evaluation period mentioned justifies our 3 h period as a suitable time interval to observe cloud-based resources for analysis. Based on the "downtime per week" availability from the "Nines of Availability", our 3 h AWS EC2 peak time recording which was 94.828% (refer to Fig. 2) has 2.8 h of weekly downtime. This indicates about 99.0% availability (i.e. 2 nines) with a potential 3.65 days downtime per year. The reliability during the said time is 0.2943175.

```xml
<cas_log>
    <log_item id="user01">
        <user>System Engineer</user>
        <privilidge>Administrator</privilidge>
        <publish_date>2020-11-23</publish_date>
        <duration>3 hours</duration>
        <description>Dependability Assurance of RA's AWS VMs</description>
    </log_item>
    <log_item id="fail_vm_i-6d16b7ad">
        <name>vmiGuestRM</name>
        <fail_time>12:15</fail_time>
        <fail_value_vmi>93.23</fail_value_vmi>
        <fail_value_vmi>84.53</fail_value_vmi>
        <availability>0.7931</availability>
        <reliability>0.07363</reliability>
        <description>Reactive Middleware VM fails ones</description>
    </log_item>
    <log_item id="sys_analysis">
        <ci>95%</ci>
        <alpha>0.05</alpha>
        <pop_variance>reject</pop_variance>
        <sla_violation>no</sla_violation>
        <cash_flow>$96.82</cash_flow>
        <disc_rate>14.419%</disc_rate>
        <npv>$64.637</npv>
        <npv_decision>desirable</npv_decision>
        <description>No AWS SLA violation but one failure</description>
    </log_item>
</cas_log>
```


**Fig. 14** CAS XML log for cloud agents

## 7 Summary

We organise our paper to begin with an introduction which highlights our three research objectives, followed by the contextual research background. Other related and relevant research works are reviewed to identify gaps that then led to the proposal of the cloud accountability system and methodology. Our paper ends by evaluating our proposal relative to the three research objectives and a research question.

In this work, we make two contributions: demonstrate cloud accountability by applying a novel method for assuring cloud dependability to a Reactive Architecture, and conduct an evidence-based stochastic (Markov chains) analysis of the Reactive Architecture with a Cloud Accountability System (CAS). These contributions meet our three objectives (O1, O2, and O3). Our work aims to provide a meaningful level of assurance of cloud dependability in terms of availability and reliability of cloud-based architectures. This assurance is enabled by the proposed cloud accountability methodology. Our approach for VMI metrics collection, data examination, information analysis, and evidence reporting is guided by the NIST SP800-86 digital forensic model to assure cloud agents of cloud dependability in terms of availability and reliability. Here, cloud agents are provided with NPV computations and logs that give concise information about cloud resource activities. The information from our work aims to provide more detailed assessment of the cloud agents' cloud-based resources than what is provided by the CSPs. Also, the SLA from the CSP was identified to be vague, and provides virtually no basis for assuring the dependability of cloud resources to system engineers. Here, we counter-check the digital evidence based on the SLA from the CSPs, with those collected from the digital forensic process. An evidence-based stochastic trust analysis is then conducted to provide a perspective to the state of the clients' cloud-based VMs. After introspecting our set of VMs for 3 hours, we identified some relevant but unreported dependability incidents on AWS relating to *vmGuestRM*, which cost the user about $65. We then conclude that the CAM is sufficiently capable of meaningfully assuring the dependability of deployed systems in the cloud.

## References


1. Abawajy J (2011) Establishing trust in hybrid cloud computing environments. In: 2011 IEEE 10th international conference on trust, security and privacy in computing and communications, pp 118–125. https://doi.org/10.1109/TrustCom.2011.18
2. Adjepon Yamoah D, Wen Z (2016) Assuring dependable cloud-based system engineering: a cloud accountability method. In: 12th European dependable computing conference (EDCC 2016). IEEE, Vasteras, Sweden, pp 181–184. https://doi.org/10.1109/EDCC.2016.20
3. Adjepon-Yamoah D, Romanovsky A, Iliasov A (2015) A reactive aarchitecture for cloud-based system engineering. In: Proceedings of the 2015 international conference on software and system process, ICSSP 2015. ACM, New York, NY, USA, pp 77–81. https://doi.org/10.1145/2785592.2785611. http://doi.acm.org/10.1145/2785592.2785611
4. Armbrust M, Fox A, Griffith R, Joseph AD, Katz RH, Konwinski A, Lee G, Patterson DA, Rabkin A, Zaharia M (2009) Above the clouds: a berkeley view of cloud computing. Technical Report, Electrical Engineering and Computer Sciences; University of California at Berkeley; UC Berkeley Reliable Adaptive Distributed Systems Laboratory
5. AWS (2020) Amazon EC2 service level agreement. http://aws.amazon.com/ec2/sla/
6. Chen Z, Han F, Cao J, Jiang X, Chen S (2013) Cloud computing-based forensic analysis for collaborative network security management system. Tsinghua Sci Technol 18(1):40–50
7. Chiregi M, Navimipour NJ (2016) A new method for trust and reputation evaluation in the cloud environments using the recommendations of opinion leaders' entities and removing the effect of troll entities. Comput Hum Behav 60:280–292. https://doi.org/10.1016/j.chb.2016.02.029. http://www.sciencedirect.com/science/article/pii/S0747563216300814
8. Fan W, Perros H (2014) A novel trust management framework for multi-cloud environments based on trust service providers. Knowl Based Syst 70:392–406. https://doi.org/10.1016/j.knosys.2014.07.018. http://www.sciencedirect.com/science/article/pii/S0950705114002755
9. Habib SM, Ries S, Muhlhauser M (2010) Cloud computing landscape and research challenges regarding trust and reputation. In: Symposia and workshops on ubiquitous, autonomic and trusted computing, UIC-ATC '10. IEEE Computer Society, Washington, DC, USA, pp 410–415. https://doi.org/10.1109/UIC-ATC.2010.48
10. Hande S, Mane S (2015) An analysis on data accountability and security in cloud. In: 2015 International conference on industrial instrumentation and control (ICIC), Pune, India, pp 713–717. https://doi.org/10.1109/IIC.2015.7150835
11. Haq IU, Alnemr R, Paschke A, Schikuta E, Boley H, Meinel C (2010) Distributed trust management for validating sla choreographies. In: Wieder P, Yahyapour R, Ziegler W (eds) Grids and service-oriented architectures for service level agreements. Springer, US, Boston, MA, pp 45–55

12. Huang J, Nicol DM (2013) Trust mechanisms for cloud computing. J Cloud Comput Adv Syst Appl 2(1):9. https://doi.org/10.1186/2192-113X-2-9
13. Hwang K, Kulkareni S, Hu Y (2009) Cloud security with virtualized defense and reputation-based trust mangement. In: 2009 Eighth IEEE international conference on dependable, autonomic and secure computing, pp 717–722. https://doi.org/10.1109/DASC.2009.149
14. Jayathilaka H, Krintz C, Wolski R (2015) Response time service level agreements for cloud-hosted web applications. In: Proceedings of the sixth ACM symposium on cloud computing, SoCC '15. ACM, New York, NY, USA, pp 315–328. https://doi.org/10.1145/2806777.2806842. http://doi.acm.org/10.1145/2806777.2806842
15. Kent K, Chevalier S, Grance T, Dang H (2006) Guide to integrating forensic techniques into incident response. Technical Report, SP 800-86, Gaithersburg, MD, United States
16. Kourai K, Nakamura K (2014) Efficient VM introspection in KVM and performance comparison with Xen. In: Proceedings of the 2014 IEEE 20th pacific rim international symposium on dependable computing, PRDC '14. IEEE Computer Society, Washington, DC, USA, pp 192–202. https://doi.org/10.1109/PRDC.2014.33
17. Lansing J, Sunyaev A (2016) Trust in cloud computing: conceptual typology and trust-building antecedents. SIGMIS Database 47(2):58–96. https://doi.org/10.1145/2963175.2963179. http://doi.acm.org/10.1145/2963175.2963179
18. Lee M, Park S, Song Y, Eom YI (2019) Introspection of virtual machine memory resource in the virtualized systems. In: 2019 IEEE international conference on big data and smart computing (BigComp), pp 1–4. https://doi.org/10.1109/BIGCOMP.2019.8679210
19. Martínez SD, Fernández FAC, Collado JV (2019) Generation expansion planning based on positive net present value. In: 2019 16th international conference on the European energy market (EEM), pp 1–6. https://doi.org/10.1109/EEM.2019.8916509
20. Mohite M, Ardhapurkar S (2015) Overcast: developing digital forensic tool in cloud computing environment. In: 2015 International conference on innovations in information, embedded and communication systems (ICIIECS). Coimbatore, India, pp 1–4. https://doi.org/10.1109/ICIIECS.2015.7193220
21. Ruan K, Carthy J, Kechadi T, Baggili I (2013) Cloud forensics definitions and critical criteria for cloud forensic capability: an overview of survey results. Digit Investig 10(1):34–43. https://doi.org/10.1016/j.diin.2013.02.004
22. Sharevski F (2013) Digital forensic investigation in cloud computing environment: impact on privacy. In: Eighth international workshop on systematic approaches to digital forensic engineering (SADFE). Pok Fu Lam, Hong Kong, pp 1–6. https://doi.org/10.1109/SADFE.2013.6911542
23. Standard Performance Evaluation Corporation (2016) SPEC™ Cloud IaaS 2016. http://www.spec.org/cloud_iaas2016/index.html
24. Tang M, Dai X, Liu J, Chen J (2017) Towards a trust evaluation middleware for cloud service selection. Future Gener Comput Syst 74:302–312. https://doi.org/10.1016/j.future.2016.01.009. http://www.sciencedirect.com/science/article/pii/S0167739X1600011X
25. Utomo AP, Winarno I, Syarif I, (2019) Detecting hang on the virtual machine using LibVMI. In: 2019 International electronics symposium (IES), pp 618–621. https://doi.org/10.1109/ELECSYM.2019.8901677
26. Wang W, Zeng G, Tang D, Yao J (2012) Cloud-DLS: dynamic trusted scheduling for cloud computing. Expert Syst Appl 39(3):2321–2329. https://doi.org/10.1016/j.eswa.2011.08.048
27. Weygant PS (2001) Clusters for high availability: a primer of HP solutions, 2nd edn. Prentice Hall PTR, Upper Saddle River, NJ, USA
28. Zou J, Wang Y, Lin KJ (2010) A formal service contract model for accountable saas and cloud services. In: IEEE international conference on services computing, Miami, Florida, USA, pp 73–80. https://doi.org/10.1109/SCC.2010.85

# FOREX Movement Direction Predictions with Similar Wavelet Patterns

**Milton Saulo Raimundo**

**Abstract** The recent increase in trading volume registered on the world stock exchanges market has led to a growing interest in algorithms and techniques capable of predicting future prices of financial securities. The usage of time series combined with graphical analysis tools, application of indicators, and trend techniques represents an important approach to this research question. This article presents the SWVP (Similar Wavelet Patterns) indicator, a forecasted approach developed to make predictions as a trend indicator on the future prices of currencies using as a pattern of decomposed ripples through Wavelets as a trend indicator on the future prices of currencies, conceived by translations and dilations of a Wavelet named as *mother-Wavelet*, formulated over the financial series, highlighting the ripples with better correlations with the original series. The objective of this research is to develop this new trend indicator, for the future prices of currencies, particularly FOREX security type, based on patterns with one of the periodical functions, composed of harmonically related sine and cosines, through a series of Wavelets family, generating points of interest for selling or buying financial securities integrated positions as a strategy to start trading.

**Keywords** Trend indicator · Wavelets · Patterns · Graphical analysis · Forecast · FOREX

## 1 Introduction

The investment data investigation is a potty task due to its diversity, considerable volume, and intricacy and has been conducted to rapid development and the new prospect in visualizations and visual analytics methods [1], including trend indicators.

Added to this is the fact that goals and direction are perhaps the most important when it comes to the lifeblood of trading. The yield of large capital returns relics on the type of security and the indicator being used [2].

M. S. Raimundo (✉)
Universidade de São Paulo, São Paulo, SP 05508-010, Brazil
e-mail: miltonsr@alumni.usp.br



X.-S. Yang et al. (eds.), *Proceedings of Sixth International Congress on Information and Communication Technology*, Lecture Notes in Networks and Systems 216,
https://doi.org/10.1007/978-981-16-1781-2_41

The FOREX market, Currency FOREX Market, Foreign Exchange Market, or only Foreign Exchange (FX), is worldwide commerce for off-centered currency market [3]. In terms of market volume, it is the biggest worldwide market [4]. The principal dealers in this sector are the global banks, accountable for setting the relative prices of the distinct coins.

With the globalization of the world economy, corporations operating in this fair are naturally subject to the threats of the floating ratio in both directions, negative and positive. The ability of a company to compete internationally depends on its ability to adequately and effectively manage the risks arising from multinational operations. Consequently, forecasting this type of financial asset has an important significance for multinational companies and is a crucial factor for the success of financial fund managers [5].

Many economic, political, and even psychological factors affect the movements of exchange rates, which makes predicting this type of series a task that is neither easy nor trivial. The historical series of this type of financial asset generally has non-stationary characteristics, which means that traditional statistical methods are neither adequate nor efficient in their prediction, requiring the application of more advanced forecasting techniques and methods.

In this context, the Currency Market is a tricky and demanding task for forecasting due to the unsure activity of the currency exchange. As knowing as FX historical data, these time series over distinct time-frames proffers to be repeatable trend patterns. Features pulled from trend patterns could be used and feasible to model and forecast the next trend [6].

Several indicators were developed to predict the trends of securities or assets, which conduct discussions on modern perspectives in the forecasting time series of securities or assets [7]. Therefore, despite the good results found in several recent studies, the challenge of finding models with better capacity for data forecasting is still open, whether for predicting market assets or trends for financial indicators.

The objective of this paper is to develop a new trend indicator, SWVP (Similar Wavelet Patterns) indicator, for future currency prices, in particular the type of FOREX securities, which integrates Wavelets family of functions, based on patterns with one of the periodical functions, conceived by translations and dilations of a Wavelet named as *mother-Wavelet*, highlighting the ripples with better correlations with the original series, generating points of interest for selling or buying financial securities, for forecasting financial trends.

Wavelet patterns are applied to learn the historical time series trend models and employed to forecast the next movement trends, which is compared with different indicators, such as Moving Average, Crossover, and Bollinger Band®, proving the capacity and efficiency of the SWVP indicator in predicting trends in financial time series of FOREX securities.

This paper is arranged as follows: Sect. 2, Wavelet Analysis, presents a review of Wavelets' formulations and their theory, relating it to time series. Section 3, SWVP Indicator, specifies and defines the new SWVP indicator to forecast trends in financial series, particularly for the FOREX security type, by applying decomposed ripple patterns from Wavelets formulations and transformations, conceived by translations and dilations of a Wavelet named as *mother-Wavelet*, emphasizing the ripples with better correlations with the original series. Section 4, Traditional Indicators, exposes a brief description of the Moving Average, Crossover, and Bollinger Band® indicators, subjective to the experimental tests and final analysis. Section 5, FOREX data source, discusses the FOREX type time series, obtained from an open knowledge base. Section 6, Methodology, sets out the method for applying the SWVP trend indicator over the main series analyzed and the process for obtaining the trends, as well as the steps for obtaining the relevant points of selling or buying positions. Section 7, Experimental Results, exposures an analysis against different indicators, such as Moving Average, Crossover, and Bollinger Band®, in forecasting trends in financial time series of FOREX securities. Finally, Sect. 8, Conclusion, cites the general analysis and contributions of this article, as well as suggestions for future articles.

## 2 Wavelet Decompositions

Wavelet decompositions are a very useful mathematical technics for numerical studies and manipulation of discrete multi or unidimensional signals. As wavelets stretch data ranges, parting them into distinct elements of scale and frequency, allowing an inspection of each element in its appropriated scale [8].

The idea of estimation, athwart the overlap of functions, was initialized with the studies of Joseph Fourier who detected, in the nineteenth century, that could be possible to apply cosines and sines to state different functions [8]. It can also be demonstrated that the Fourier approximation can be stated as [9]:

$$f_n(x) = \frac{a_0}{\sqrt{2}} + \sum_{k=1}^{n} a_k \cos(kx) + \sum_{k=1}^{n} b_k \sin(kx) \tag{1}$$

Figure 1 presents some estimates for a continuous even function [9].

The novelty, concerning Fourier, is associated with the basis of Fourier functions that are independent of time and subordinated on frequency. It means that insignificant changes in the frequency domain generate changes in the time domain. Wavelets are subjected in both domains, in frequency by dilatations and in time by translations, which is an advantage in several cases [10].

Wavelets and their properties are used in the study of multi-fractal properties, in functional analysis, pattern recognition, the image and sound compression, among others [11].

Thereby, both in Fourier Analysis and in Wavelet Analysis, the idea is to approximate a function by the linear combination of sines and cosines or wavelets, respectively [12].

Extending the concept of Fourier analysis to $L^2(\Re)$ space, it is possible to deduce the $\psi$ function, called the *mother-Wavelet*, generated by $\psi$ expansions and

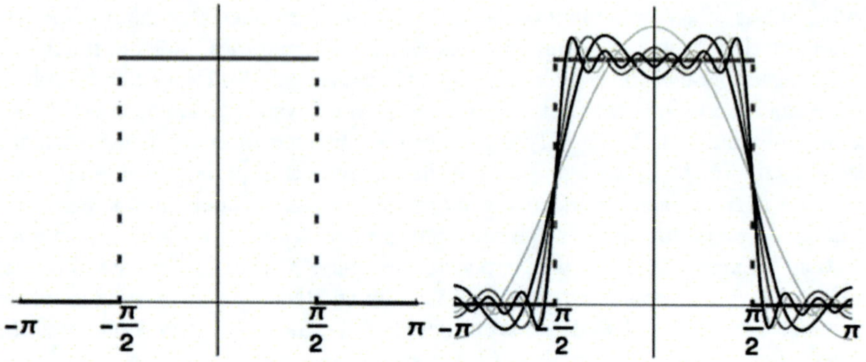


**Fig. 1** Representation of some estimates for a continuous even function [9]

translations, producing a family of $\psi_{a,b}$ functions, such as [13]:

$$\psi_{a,b}(t) = {}^{1}\big/_{\sqrt{|a|}}\psi\left({}^{t-b}\big/_{a}\right) \tag{2}$$

where $a,\ b \in \mathbb{R}$ and $a \neq 0$.

Called dilation equations [14] $h_k$, a low-pass filter coefficient, and $g_k$, a high-pass filters coefficient, is used to calculate the Discrete Wavelet Transform (DWT), as follows:

$$h_k = \sqrt{2}\int_{-\infty}^{\infty} \phi(t)\phi(2t-k)dt \tag{3}$$

$$g_k = \sqrt{2}\int_{-\infty}^{\infty} \psi(t)\phi(2t-k)dt \tag{4}$$

Consequently, given the orthonormal system $\{\phi_{j,k}(t),\ \psi_{j,k}(t),\ j,k \in Z\}$, can be formulated $f(t) \in L^2(R)$ as:

$$f(t) = \sum_k c_{J,k}\phi_{J,k}(t) + \sum_{j\leq J}\sum_k d_{j,k}\psi_{j,k}(t) \tag{5}$$

wherein

$$c_{J,k} = f(t), \phi_{J,k}(t) = \int_{-\infty}^{\infty} f(t)\phi_{J,k}(t)dt \tag{6}$$

and

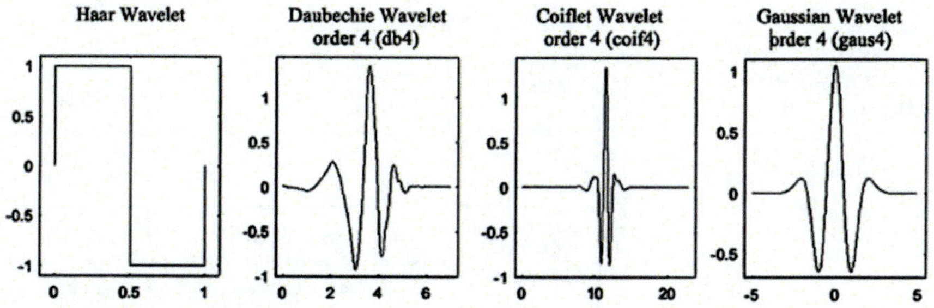


**Fig. 2** Illustration of the Haar, Daubechies order 4 (db4), Coiflet order 4 (coif4), and Gaussian order 4 (gaus4) Wavelet families obtained using MATLAB® software

$$d_{j,k} = f(t), \psi_{j,k}(t) = \int_{-\infty}^{\infty} f(t)\psi_{j,k}(t)dt \quad (7)$$

The wavelet decomposition basis is to exteriorize a signal or function by a series of functions engender by expansions and translations [15].

The present paper emphasizes the use of the Symlets wavelet, order 4, for the implementation of the Wavelet transform. However, other Wavelet families such as Haar, Daubechies, Coiflets, and Gaussian can be used and analyzed to see which one is more suitable and adherent to the financial time series.

Figure 2 illustrates the Haar, Daubechies order 4 (db4), Coiflet order 4 (coif4), and Gaussian order 4 (gaus4) Wavelets families using MATLAB® software.

## 3 SWVP Indicator

The SWVP (Similar Wavelet Patterns) indicator is formulated to forecast trends in financial series, by applying decomposed ripples through Wavelets family of functions, conceived by translations and dilations of a Wavelet named as *mother-Wavelet*, formulated over the financial series, highlighting the ripples with better correlations with the original series, generating points of interest for selling or buying financial securities integrated positions as a strategy to start trading, as described in Fig. 3.

The goal is to predict the trend of the next movement in the series, $X = \{X_1, X_2, X_3, \ldots\}$. First, using the DWT, the original series is sliced into $n$ Wavelets using one of the Wavelets families.

A reduction is performed on the set of input data for level $J_0$. The primal series is hereafter exhibited as the summation of the oncoming clements $\tilde{N}_{J0}$ and of the detail wavelets $\tilde{a}_J$, $J = 1, \ldots, J0$.

Then, it is verified which wavelet is most correlated with the original series. Originally, it was found that the wavelet most correlated with the original series could not be the most suitable for overlapping with the original series, opting for the

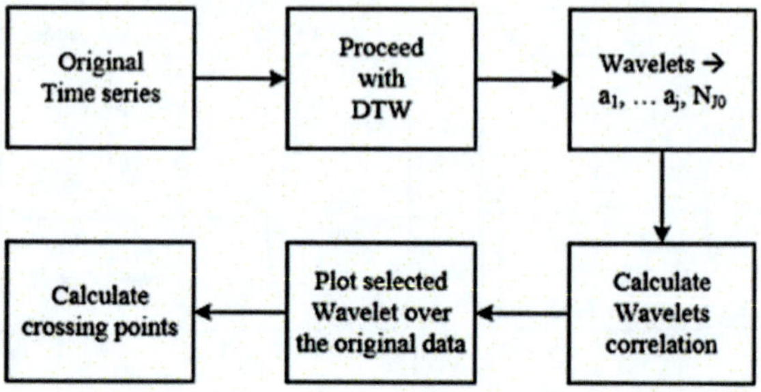


**Fig. 3** Flowchart with the steps taken to select the wavelet for further trend analysis

second or third wavelet, with less correlation with the original series. A further study is necessary to segregate possible wavelet candidates to carry out the trend analysis.

Finally, the selected wavelet is plotted over the original series, inspecting the crossing points between the selected wavelet and the original series.

Figure 3 briefly illustrates the steps taken to select the wavelet for further trend analysis.

# 4 Traditional Indicators

To verify the efficiency of the SWVP indicator, this paper uses three traditional indicators of the Financial Market: Moving Average, Crossover, and Bollinger Band® indicators.

## *4.1 Moving Average*

The moving average (MA) is generally referred to in technical analysis as a stock indicator, aiding to flat the price by generating a continually up-to-date average price. [16]. An uprising MA denotes that the asset is in an upward trend, while a down rising MA designates that it is in a downward trend [16].

A Simple Moving Average (SMA) is computed by evaluating the arithmetic mean of a series of prices through a particular quantity of past days. The formula for estimating the Simple Moving Average of asset is as go after [16]:

$$SMA = {A_1 + A_2 + \ldots + A_n}/{n} \tag{8}$$

where $A$ is the average in period $n$, and $n$ is the number of periods.

Exponential moving average (EMA) is a weighted average that provides more significance to the price of an asset in more latter days. The formulation for estimating the exponential moving average of an asset is as come after [16]:

$$EMA_t = \left[V_t * \left({}^{s}/{1+d}\right)\right] + EMA_y * \left[1 - \left({}^{s}/{1+d}\right)\right] \tag{9}$$

where $EMA_t$ is the EMA value for today, $V_t$ is the value today, $EMA_y$ is the EMA value for yesterday, $s$ is the smoothing factor, and $d$ is the number of days.

## 4.2 Crossover

The crossover is a point on the trading graph where two indicators cross or whenever a line of technical indicator and the price of an asset cross [16].

It is used to estimate and forecast future changes in trends, such as reversions and disruptions [16].

Moving average (MA) crossovers are ordinary, which includes the Golden-Cross, a candlestick template, downward indicator in which a fairly short-term MA passes through a long-term MA, and the Death-Cross, the converse of a Golden-Cross, used in technical analysis to confirm trends creating signals of a buy or a sell [16].

## 4.3 Bollinger Band®

A Bollinger Band® is determinate by a set of trend lines plotted two standard deviations from an SMA of a security price [16].

As developed by John Bollinger, it was projected to detect chances that give investors a greater likelihood of correctly recognizing when an asset is oversold. There are composed of three lines: an upper band (UBB), a lower band (LBB), and an SMA, a middle band. The formula for evaluating the Bollinger Band® of an asset is as follows [16]:

$$ABB = SMA(CTP, n) + m * \sigma[CTP, n] \tag{10}$$

$$BBB = SMA(CTP, n) - m * \sigma[CTP, n] \tag{11}$$

where $ABB$ is the Atop Bollinger Band, $BBB$ is the Bottom Bollinger Band, S $MA$ is the Simple Moving Average, $CTP$ is the Close Typical Price, i.e. $(High + Low + Close\ prices)/3$, $n$ is the smoothing term, $m$ is the standard deviations, and $\sigma[CTP, n]$ is the standard deviation through the last $n$ periods of $CTP$ [16].

## 5 FOREX Data Source

Among the various security types of the Financial Market, such as Common Stocks, Commodities, Futures, Options, Fixed Income, Swaps, among others, this paper emphasized and uses the type FOREX, Foreign Exchange Market, because it is the largest worldwide market [4] and both historical and intraday series are easily retrieved from some open or public knowledge base [17].

The Currency FOREX Market is a worldwide market for decentralized currency market [17]. The main dealers in this sector are the largest global banks, accountable for setting the relative prices of the distinct coins.

The main centers for FOREX are London, New York, Tokyo, Hong Kong, and Singapore. These markets trade 5 days a week non-stop and are influenced by many geopolitical and economic factors, such as the balance of payments in each country, the trade balance, fiscal and monetary policies, political stability, and the credit strength of governments [18].

The current structure of the FOREX market has evolved from fluctuating exchange rates. The most traded currency pairs on the FOREX spot market today are 12% GBP-USD [18], 13% USD-JPY, and 27% EUR-USD [18].

This paper selected the FOREX financial time series for the parity on the American Dollar (USD) and the Brazilian Reais (BRL) coins.

The selected FOREX USD-BRL asset is a time series compounded of the intraday currency prices for the date of 11/27/2020 between the period of 07:00 am to 04:30 pm, with 1-min intervals, obtained from the open knowledge base [19].

A Python 3.7.6 (08/01/2020) script is created, developed in Jupyter Notebook, 64-bit, macOS Big Sur© environment, Version 11.0.1, 3.3 GHz Core i7 processor, with 16 GB of memory, for selecting the USD-BRL series for the date of 11/27/2020, totaling 571 points in the series.

Figure 4 presents a chart with the intraday series of FOREX USD-BRL asset prices with 1-min intervals. 571 points in the series are sampled.

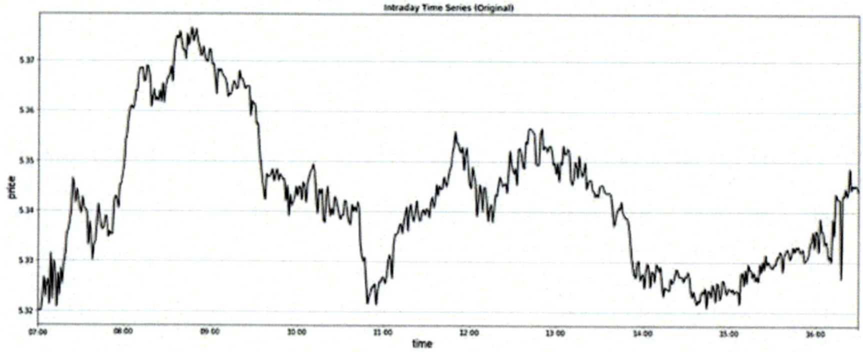


**Fig. 4** Chart with the intraday series of FOREX USD-BRL asset prices with 1-min intervals. 571 points in the series are sampled

## 6 Methodology

The development of the SWVP trend indicator for future currency prices, in particular the type of FOREX securities, using Wavelets family functions, attempts to generate points of interest for selling or buying financial securities, for forecasting financial trends.

This is possible due to the perspective of using patterns with one of the periodical functions of the Wavelets family, conceived by translations and dilations of a Wavelet named as *mother-Wavelet*, and by the highlighted and selected ripples with better correlations that are applied to the analyzed original time series.

Figure 5 clarifies the suggested model for forecasting time series trends using the SWVP trend indicator.

First, there is the instantiation of the open knowledge base, applying an API from the library of APIs available by the provider of the open knowledge base on its API Documentation pages [19].

Additionally, it is necessary to provide a given credential to access the data, as well as to configure the time interval to be analyzed, as shown in the source code below:

```
from alpha_vantage.timeseries import TimeSeries
ts=TimeSeries(key='personal-credentials')
data,meta_data=ts.get_intraday(symbol='USDBRL')
```

Next, with the series already persisted, either in JSON format or even in DataFrame format on pandas, the series is plotted by verifying its shape and vertices contained in the generated curve, as well as their nuances, such as trends or even seasonality.

Afterward, using the DWT, the original series is sliced into *n* Wavelets using one of the Wavelets families, evidenced by the source code down.

```
import pywt
wavelets=pywt.wavedec(data=df,wavelet='sym2',level=4)
```

This paper opted to use the Wavelet Symlets (sym) family, because some tests were carried out with other families, such as Daubechies (db), Haar (haar), and Gaussian wavelets (gaus), there was no evidence of best fit or performance. However, a more in-depth study is necessary to better highlight this point.

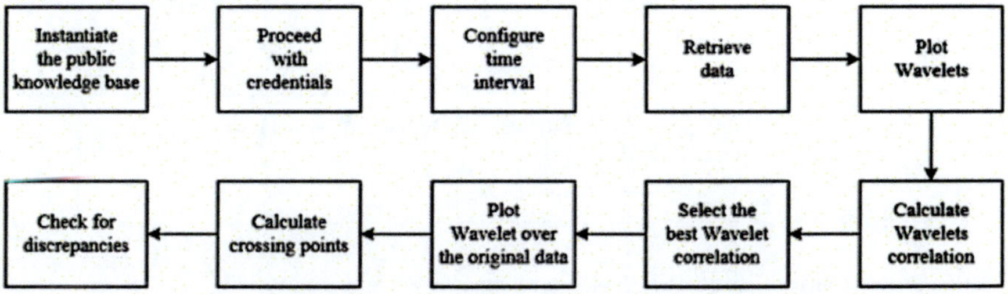


**Fig. 5** Flowchart with the steps to use SWVP indicator to forecast trends in financial series

Figure 6 presents the original series sliced into *n* Wavelets using the sym Wavelet family. The primal series is hereafter exhibited as the summation of the oncoming elements $\tilde{N}_{J0}$ and of the detail wavelets $\tilde{a}_J$, $J = 1, \ldots, J0$.

Then, it is verified which wavelet is most correlated with the original series, as explained by the partial source code ahead.

Originally, it was found that the wavelet most correlated with the original series could not be the most suitable for overlapping with the original series, opting for the second or third wavelet, with less correlation with the original series. A further study is necessary to segregate possible wavelet candidates to carry out the trend analysis.

```
import numpy as np
for i, wavelet in enumerate(wavelets):
    ...
    aclose.append(np.allclose(y, y1))
    aMSE.append(np.sum((y-y1)**2)/len(x))
    acorrcoef.append(np.corrcoef(df, y2))
```

Finally, the selected wavelet is plotted over the original series, inspecting the crossing points between the selected wavelet and the original series, generating points of interest for selling or buying financial securities, for forecasting financial trends.

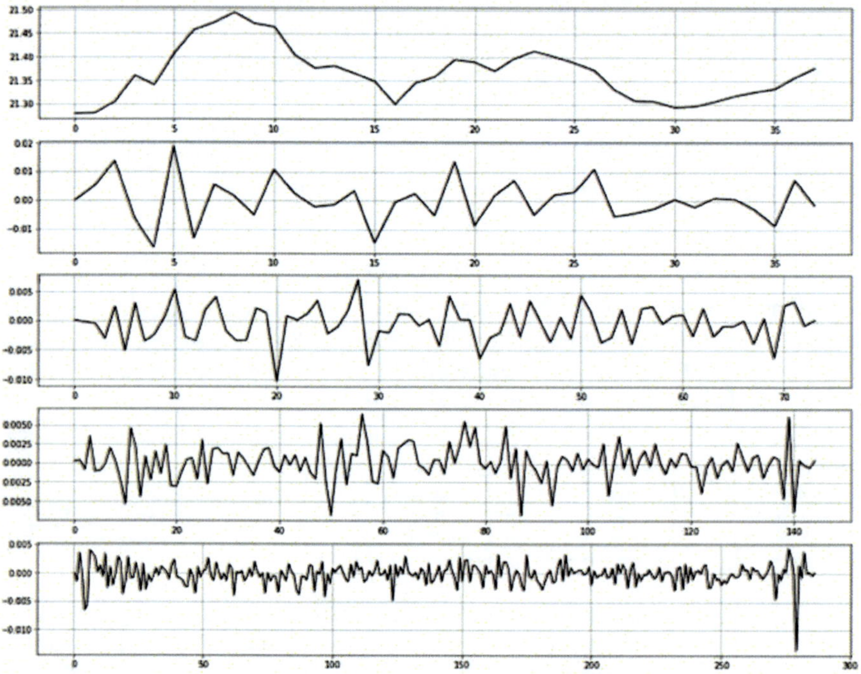

**Fig. 6** Original series sliced into *n* Wavelets using sym Wavelet family

**Table 1** Summary of the correlations found between the Wavelets and the original time series

| Wavelets components | | | | | |
|---|---|---|---|---|---|
| | 1st detail | 2nd detail | 3rd detail | 4th detail | Approximation |
| MSE | 1.295e−29 | 5.533e−36 | 5.979e−37 | 2.522e−37 | 1.772e−37 |
| Correlation | 0.885089 | 0.214275 | 0. 085,099 | 0. 009,233 | 0. 001,932 |

Table 1 summarizes the correlations found between the Wavelets and the original time series, where the third row is the mean square error (MSE) and the fourth one is the correlation coefficient.

## 7 Experimental Results

After the Wavelet patterns are applied to learn the historical trend models, used to predict the next movement trends, an analysis against different indicators, such as Moving Average (SMA), Crossover (CO), and Bollinger Band® (BB), is to verify the capacity and efficiency of the SWVP indicator in predicting trends in financial time series of FOREX securities.

For the execution of the tests, some steps were adopted according to what is shown in Fig. 7 with the macro flowchart of the main tasks in the process of execution of the tests and the final analysis.

First, using the original time series, SMA and its crossing points are calculated. Afterward, the same way has calculated the CO and its crossing points. Then, applying the original time series is considered the BB indicator, and its crossing points are demonstrated by the source code below.

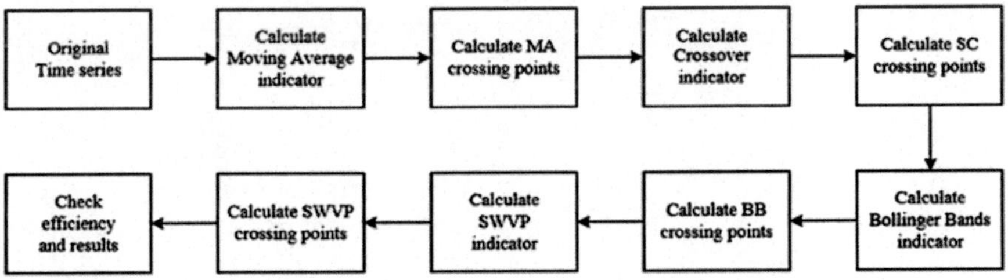


**Fig. 7** Main tasks in the process of execution of the tests and the final analysis

```
from alpha_vantage.techindicators import TechIndicators
ti = TechIndicators(key='personal-credentials', out-
put_format='pandas')
data_sma, meta_data_sma = ti.get_sma(symbol='USDBRL',
interval='1min'time_period=30,series_type='close')
data_co, meta_data_co = ti.get_ema(symbol='USDBRL',
interval='1min', time_period=40,series_type='close')
data_bb, meta_data_bb = ti.get_bbands(symbol='USDBRL',
interval='1min', time_period=30,series_type='close')
```

Finally, following the steps detailed in Sect. 6, Methodology, is computed the SWVP indicator and also its crossing points.

Figure 8 portrays the extracted graphs and generated by the SMA (a), CO (b), and BB (c) indicators, drew over the original series, as well as the new SWVP (d) indicator.

Analyzing the points of the graphs in Fig. 8, evidenced by small grey arrows, it is possible to recognize that the SWVP indicator, in (d), exceeds in numbers of correct answers when compared with the SMA, CO, and BB indicators, respectively, in (a), (b), and (c).

These last three indicators have shown themselves to be weak in hits at some points, which can generate false alerts, entries, and exits of positions, or unreal buys or sells securities.

Furthermore, in (d), it is possible to use an algorithm to filter some points even if they are in a little excessive number, which is neither found in (a), nor in (b), and (c).

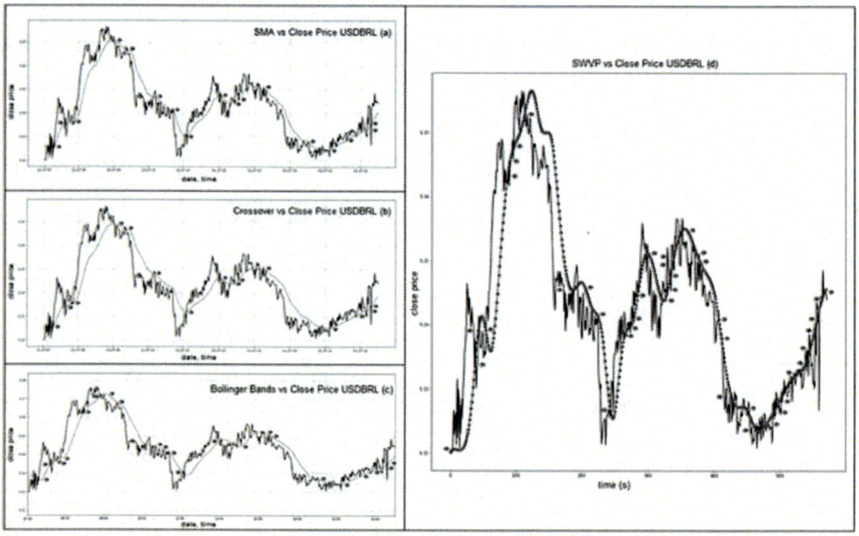


**Fig. 8** Extracted graphs and generated by the SMA (a), CO (b), and BB (c) and SWVP (d) indicators drew over the original series

The points, for selling or buying financial securities, can be retrieved using a small algorithm salient in the source code below. The algorithm generates an array of the crossover points where 1 is a crossover from below to above and −1 is a crossover from above to below.

```
crossover = np.pad(np.diff(np.array(data >
dataNorm).astype(int)), (1,0), 'constant', constant_val-
ues = (0,))
```

## 8 Conclusion

This paper emphasized some significant trend indicators used for the predictions of trends of financial securities, such as Moving Average, Crossover, and Bollinger Band®, and presented a new and suitable indicator, SWVP, for forecasting trends, particularly for FOREX securities, which integrates Wavelets family of functions, based on patterns with one of the periodical functions, conceived by translations and dilations of a Wavelet named as *mother-Wavelet*, highlighting the ripples with better correlations with the original series, generating points of interest for selling or buying financial securities.

SWVP indicator has presented to be efficient capable in forecasting trends in the financial time series, notedly FOREX assets, which can also be confirmed by the content of Table 1 that includes the Mean Square Error (MSE) and the series correlation coefficient generated by the Wavelet Symlet wavelet, used to calculate the crossing points with the original series. Their operations, in contrast to other conventional indicators, are remarkably better and overcome the traditional indicators in forecasting the trending of non-linear and non-stationary time series, such as the FOREX assets.

Although, time series trends' forecasting still demands substantial academic studies to guarantee that it stretches out its plenitude and can perform its full competence.

Some studies may enlarge the endowment of this paper, such as more in-deep studies among other Wavelet families, for instance, Daubechies and Gaussian, and the number of sliced ripples or the order of the Wavelet family employed could be also explored.

Also, originally it was found that the wavelet most correlated with the original series could not be the most suitable for overlapping with the original series, opting for the second or third wavelet, with less correlation with the original series. A further study could be interesting to discriminate possible wavelet candidates to carry out the trend analysis.

Additionally, other securities could be part of a new study to verify if the new trend indicator, SWVP, could also be used as and a good option for trend indicators in the Financial Market.

## References


1. Ko S, Cho I, Afzal S, Yau C, Chae J, Malik A, Beck K, Jang Y, Ribarsky W, Ebert DS (2016) A survey on visual analysis approaches for financial data. Comput Graph Forum 35:599–617
2. Rosillo R, de la Fuente D, Brugos JAL (2013) Technical analysis and the Spanish stock exchange: testing the RSI, MACD, momentum and stochastic rules using Spanish market companies. Appl Econ 45:1541–1550
3. Foreign exchange turnover in April 2019 (2019) https://www.bis.org/statistics/rpfx19_fx.htm. Last Accessed 24 Nov 2020
4. Settlements BI (2013) Triennial central bank survey, foreign exchange turnover in April, 2013: preliminary global results. Bank for International Settlements, p 24
5. Ristovska K, Ristovska A (2014) The impact of globalization on the business. Econ Anal 47(3–4):83–89
6. Lee Y, Ow LTC, Ling DNC (2014) Hidden markov models for forex trends prediction. In: International conference on information science & applications (ICISA), pp 1–4
7. Morettin PA (2017) Econometria financeira: um curso em séries temporais financeiras. 3rd edn. Blucher, São Paulo, Brazil
8. Misiti M, Misiti Y, Oppenheim G, Poggi JM (2009) Wavelet toolbox: for use with MATLAB®. The Math Works, Inc.
9. Knill O (2010) Fourier series, lecture 29. Math Dept. Harvard College
10. Tsay RS (2010) Analysis of financial time series 3. Wiley, Chicago
11. Berg JC (2010) Wavelets in physics, pp I–Iv. Cambridge University Press, Cambridge
12. Morettin PA (2014) Ondas e ondaletas: da análise de Fourier à análise de ondaletas. Ciências Exatas edn. Coleção Academia, vol. 23, Edusp, São Paulo, Brazil
13. Addison PS (2020) The illustrated wavelet transform handbook: introductory theory and applications in science, engineering, medicine and finance, 2nd edn. CRC Press, Florida
14. Gupta K, Goyal K (2019) Daubechies wavelets: theory and applications. School of Mathematics of Thapar Institute of Engineering and Technology
15. Ryan Ø (2019) Linear algebra, signal processing, and wavelets—a unified approach: python version 1. Springer, Norway
16. Investopedia. https://www.investopedia.com/financial-term-dictionary-4769738. Last Accessed 24 Nov 2020
17. Free forex historical data, public knowledge base, HistData.com. Last Accessed 24 Nov 2020
18. Premanode B (2013) Prediction of nonlinear nonstationary time series data using a new digital filter and support vector regression. Imperial College London Department of Electrical and Electronic Engineering, High Wycombe, UK
19. Alpha vantage | Free stocks API in JSON & Excel, open knowledge base, https://www.alphavantage.co/. Last Accessed 24 Nov 2020

# Performance Study of Zigbee Networks in an Apartment-Based Indoor Environment

**Biswajit Kumar Dash and Jun Peng**

**Abstract** The Internet of Things (IoT) networks using Zigbee are very popular in smart homes. However, Zigbee networks are vulnerable to the interference of Wi-Fi networks because they share the same 2.4 GHz Industrial, Scientific, and Medical (ISM) radio frequency band. Studies have shown that weaker Zigbee signals might be significantly interfered by stronger Wi-Fi signals. This type of interference may cause severe problems when these types of networks coexist in an indoor environment such as in a smart home. In this paper, the performance of a Zigbee network with and without the presence of Wi-Fi interference traffic has been evaluated in an apartment-based indoor environment mimicking a smart home. The experimental results were obtained and analyzed in terms of packet drop rate (PDR), received signal strength indicator (RSSI), and throughput under various operating channels, distances between Zigbee and Wi-Fi devices, transmission intervals of Zigbee packets, transmit power of Zigbee transmitter, and Zigbee packet lengths.

**Keywords** Internet of Things · Smart Home · Zigbee · Wi-Fi · Interference · Indoor Environment · Coexistence

## 1 Introduction

The concept of the Internet of Things (IoT) is gaining popularity day by day in smart homes. The vital part of an IoT system is to communicate and collaborate wirelessly among different types of devices connected to it. Popular wireless communication technology, ZigBee (IEEE 802.15.4) serves this purpose because of its numerous

B. K. Dash · J. Peng (✉)
Department of Electrical and Computer Engineering, University of Texas Rio Grande Valley, Edinburg, TX 78539, USA
e-mail: jun.peng@utrgv.edu

B. K. Dash
e-mail: biswajitkumar.dash01@utrgv.edu



X.-S. Yang et al. (eds.), *Proceedings of Sixth International Congress on Information and Communication Technology*, Lecture Notes in Networks and Systems 216,
https://doi.org/10.1007/978-981-16-1781-2_42

advantages, for example, low power consumption, large scaling capability, and flexible topology [1]. Besides Zigbee, Wi-Fi (IEEE 802.11) technology is also popular in indoor environments, mainly for internet access.

The coexistence of Zigbee and Wi-Fi networks in indoor environments generates interference from each other, but Zigbee networks suffer more from interference due to their low transmission power (1mW) in the 2.4 GHz frequency band. Therefore, the coexistence of Wi-Fi and Zigbee in proximity creates a great challenge in smart home environments. Taking account of this challenge, this paper attempts to present a comprehensive experimental study of Zigbee's performance in terms of packet drop rate (PDR), received signal strength indicator (RSSI), and throughput. To get to the real feeling of a smart home environment, we conducted our experiments in a home apartment by changing operating channels, distances between Wi-Fi and Zigbee devices, Zigbee transmit power, Zigbee packet lengths, and transmission intervals of Zigbee packets.

Starting from the introduction in Section 1, the rest of the paper is organized as follows: Section 2 offers a brief literature review on the coexistence of Wi-Fi and Zigbee networks, whereas Section 3 illustrates the experimental testbed. Experiments and experimental results are reported in Section 4. Finally, Section 5 concludes the paper.

## 2 Related Works

Due to the popularity of Zigbee and Wi-Fi technologies in low power and short-distance communications, the study of the coexistence of these two networks has earned a great point of interest among researchers. So far, different studies and mechanisms have been accommodated to improve the coexistence of Zigbee and Wi-Fi networks, which are discussed in the following subsections.

### *2.1 Interference of Wi-Fi Traffic to Zigbee Networks*

Studies have shown that stronger Wi-Fi networks interrupt Zigbee communications badly when they coexist in proximity [2]. Zigbee's performance in terms of packet loss ratio (PLR) and packet error rate (PER) has been evaluated in [3] with the presence of 802.11b/g interference. In paper [4], the authors have analyzed the interference level of IEEE 802.15.4 standardized wireless sensor networks by varying channel spacings. The impact of interference on Zigbee networks with respect to PLR and average round trip time (RTT) using overlapped and non-overlapped channels has been observed in [5].

The effect of interference in the case of adjacent and alternate channels has been investigated in [6] with respect to packet drop ratio. In paper [7], the authors have conducted a vast survey to evaluate the impact of continuously changing communication environments on various parameters (e.g., RSSI and latency).

## 2.2 *Ways to Facilitate the Coexistence of Wi-Fi and Zigbee Networks*

Besides investigating the interference, some ways have been suggested to ensure the coexistence of Zigbee and Wi-Fi networks. For example, the insertion of a transmission time interval between two consecutive packets of Wi-Fi traffic has been proposed in [8] to mitigate interference between Zigbee and Wi-Fi. Mechanisms like WiseBee and BuzzBuzz have been proposed in [9] and [10], respectively, to help the coexistence of Zigbee and Wi-Fi in IoT systems. A heterogeneous network integrating multiple wireless technologies has been proposed in [11] to facilitate Wi-Fi networks to access Zigbee. The extended network allocation mechanism has been demonstrated in [12], which reduces a maximum of 50% IEEE 802.15.4 frame losses in the case of strong interference. An adaptive scheme based on multiple radio channels has been designed and evaluated in [13] for large-scale sensor network applications.

Though some suggestions mentioned in [8–13] have been made case by case, still the nature of interference between Zigbee and Wi-Fi networks is an interesting topic among researchers because of the growing applications of Zigbee.

# 3 Overview of the Testbed

The testbed consisting of Zigbee and Wi-Fi networks is based on a home apartment to mimic a smart home. Digi International's XBee Zigbee modules and Configuration & Test Utility (XCTU) software are used to deploy and configure the Zigbee devices. The coordinator and end devices are configured as a Zigbee receiver (RX) and a transmitter (TX), respectively. A Wi-Fi network is deployed using a home Wi-Fi router, acting as a Wi-Fi AP, along with two smartphones running Iperf3 on them to exchange Wi-Fi traffic. Figure 1 shows our experimental testbed where $d_z$, $d_{zw}$, and $d_w$ denote the distance between Zigbee TX and Zigbee RX, Zigbee RX and Wi-Fi TX, and Wi-Fi TX and Wi-Fi RX, respectively.

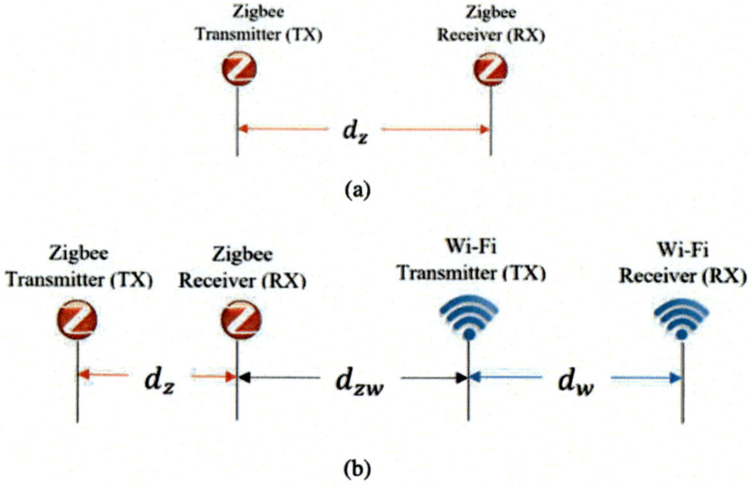


**Fig. 1** Testbed topology

# 4 Experiments and Results

The topology in Fig. 1a is considered where the experiments involve no Wi-Fi traffic, whereas Fig. 1b is used where the experiments consider Wi-Fi traffic. The following experiments are carried out to evaluate the performance of a Zigbee network under various testbed settings and performance metrics. Moreover, each data are taken three times with the same network settings to get a reliable result.

## *4.1 Packet Drop Rate (PDR) Measurement*

This experiment measures and analyzes the adverse effects of Wi-Fi interference traffic on Zigbee communication in terms of PDR. Before going to the main experiment, we first conduct a baseline study without Wi-Fi traffic to obtain a modest Zigbee transmission interval at which the PDR is minimal. Considering the Fig. 1a and parameters in Table 1, the experiment is conducted with two transmission intervals (100 and 500 ms) at different values of $d_z$ (0–5 m). In each case, the Zigbee TX sends 200 Zigbee packets to Zigbee RX that calculates the number of successfully received packets to generate the PDR using the following formula.

$$\text{Packet Drop Rate, PDR}(\%) = \frac{P_{Tx} - P_{Rx}}{P_{Tx}} * 100 \quad (1)$$

where, $P_{Tx}$ = Number of packets sent by Zigbee TX.

**Table 1** Experimental parameters for experiment 4.1

| Parameters | | Zigbee | Wi-Fi |
|---|---|---|---|
| Operating channel | Case 1 | 12 | 1 |
| | Case 2 | 14 | 1 |
| | Case 3 | 12 | 11 |
| Transmit power, $P_t$ | | 8 dBm | – |
| Packet size | | 64 Bytes | – |
| Average bandwidth | | – | 18.8 Mbits/s |
| Average data rate | | – | 17.84 Mbits/s |
| Distances between the devices | | $d_z$= 1 m, $d_w$ = 1 m, $d_{zw}$ = 0.5 m – 4.5 m (varied) | |

$P_{Rx}$= Number of successfully received packets by Zigbee RX.

We found 500 ms transmission interval generates less PDR than that of 100 ms at every measurement as shown in Fig. 2. Hence, the 500 ms is used as the Zigbee packet transmission interval for the rest of the experiments. Next, we conduct the same experiment according to the topology in Fig. 1b for three Wi-Fi interference cases:

- Case 1: Overlapping channel (Zigbee channel 12 and Wi-Fi channel 1)
- Case 2: Adjacent channel (Zigbee channel 14 and Wi-Fi channel 1), and
- Case 3: Non-overlapping channel (Zigbee channel 12 and Wi-Fi channel 11).

Table 1 provides the parameters used in this experiment. It is worth mentioning that to generate Wi-Fi traffic we didn't assign any specific bandwidth for the Iperf TCP session. So, Iperf assigns an ideal bandwidth for every TCP session and averages all values to generate the average bandwidth. On the other hand, the average data rate is the average value of actual data flow per second during these sessions.

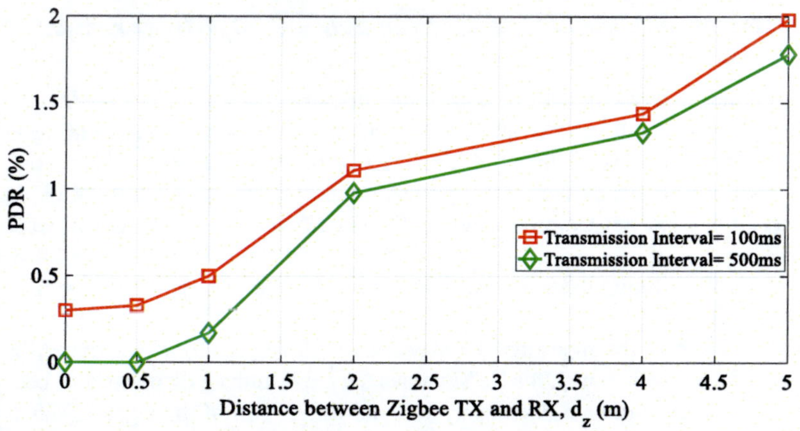


**Fig. 2** PDR with respect to various values of $d_z$ and transmission intervals

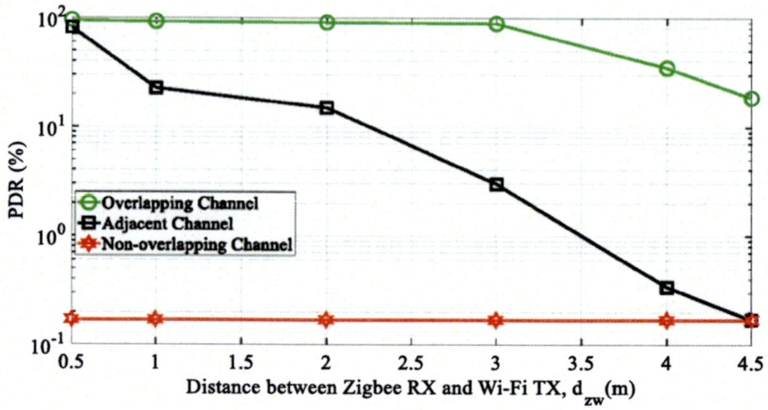


**Fig. 3** PDR versus $d_{zw}$ for three interference cases

The experimental results are shown in Fig. 3. For the non-overlapping channel case, the Wi-Fi network has nearly zero impact on Zigbee communication (PDR = 0.17%), whereas the effect of overlapping channel and adjacent channel cases are obvious. The worst impact is noticed in the case of the overlapping channel, especially for smaller values of $d_{zw}$. According to our observations, this is because.

1. Zigbee devices compete with Wi-Fi devices to get channel access but fail due to the continuous flow of stronger Wi-Fi traffic. Even if the Zigbee devices take channel access and send the packets, the packets lose in the air due to strong interruption by Wi-Fi traffic.
2. Additionally, the Zigbee TX and RX devices sometimes get disconnected or disassociated due to heavy Wi-Fi traffic during the experiment.

## *4.2 Received Signal Strength Indicator (RSSI) Measurement*

The goal of this measurement is to examine the performance of the Zigbee network in terms of RSSI with various values of $d_z$ and transmit power, $P_t$. According to the topology in Fig. 1a, the Zigbee TX transmits 50 packets to Zigbee RX and the RX sends back the echoes of successfully received packets to calculate the RSSI value of each Zigbee packet from the Zigbee TX side. Five different values of transmit power, $P_t$ are studied in this case: (i) 8 dBm, (ii) 5 dBm, (iii) 1 dBm, (iv) –1 dBm, and (v) –5 dBm. Experimental specifications used in this measurement are listed in Table 2.

Figure 4 shows the measured RSSI values with respect to $d_z$ for five different values of $P_t$. As anticipated, the RSSI values decrease with an increase in $d_z$ because of networking phenomena like absorption and diffraction of the signals. When the Zigbee TX and RX devices stay at close contact (say, $d_z = 0$ m), the RSSI value

**Table 2** Experimental parameters for the experiment 4.2

| Parameters | Zigbee | Wi-Fi |
|---|---|---|
| Operating channel | 12 | No Wi-Fi traffic |
| RF payload size | 50 Bytes | |
| Distances between devices | $d_z$= 0 m – 5.5 m (varied) | |

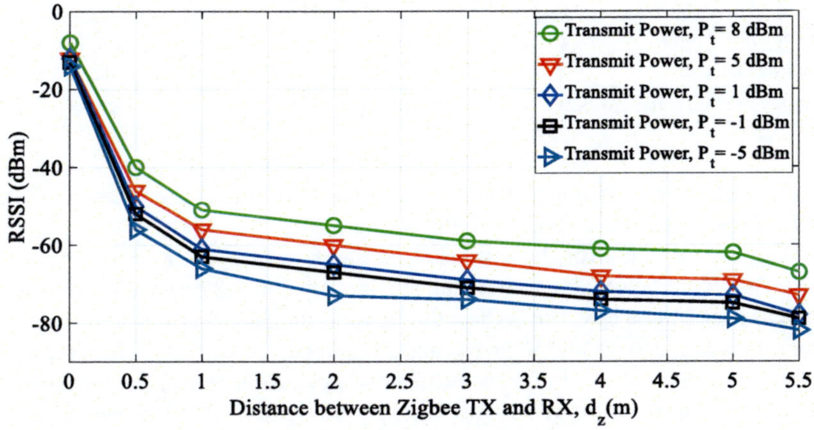


**Fig. 4** RSSI values versus $d_z$ for different levels of transmit power

doesn't degrade much as there is no scope to affect the signal by any external networking factors.

## 4.3 Throughput Measurement

This experiment measures the throughput of Zigbee modules with the presence of Wi-Fi traffic considering Fig. 1b. The throughput allows us to measure the data transfer ratio between two radio modules located in the same network and can be calculated from the packet size and total transmission time using the below formula.

$$\text{Throughput} = \frac{8 * \text{Number of bytes}}{\text{Total transmission time (s)}} (\text{bit/s}) \quad (2)$$

The specifications used in this experiment are shown in Table 3. First, we conduct a baseline experiment without Wi-Fi traffic and get a throughput of 3,287.04 bit/s (bps). During the experiment, the value of $d_{zw}$ is varied maintaining a fixed value of $d_z$ and $d_w$. Each time 50 Zigbee packets are exchanged between the Zigbee devices

**Table 3** Experimental parameters for the experiment 4.3

| Parameters | | Zigbee | Wi-Fi |
|---|---|---|---|
| Operating channel | Case 1 | 12 | 1 |
| | Case 2 | 14 | 1 |
| | Case 3 | 12 | 11 |
| Transmit power, $P_t$ | | 8 dBm | – |
| Packet size | | 64 Bytes | – |
| Average bandwidth | | – | 13.3 Mbits/s |
| Average data rate | | – | 12.72 Mbits/s |
| Distances between the devices | | $d_z$= 1 m, $d_w$ = 1 m, $d_{zw}$ = 0.5 m – 4.5 m (varied) | |

and calculated the throughput using Eq. (2). Then, we repeat the same experiment for the three interference cases stated in Section 4.1.

Figure 5 shows a throughput graph as a function of $d_{zw}$ for three interference cases. The worst impact observes for the overlapping channel case having the highest throughput of 1484.8 bps. The highest throughputs obtained for non-overlapping and adjacent channel cases are 3266.56 bps and 2713.6 bps, respectively, which are almost close to the without Wi-Fi traffic case (3,287.04 bps). As expected, the throughput of the Zigbee network is improved with distances between Zigbee RX and Wi-Fi TX.

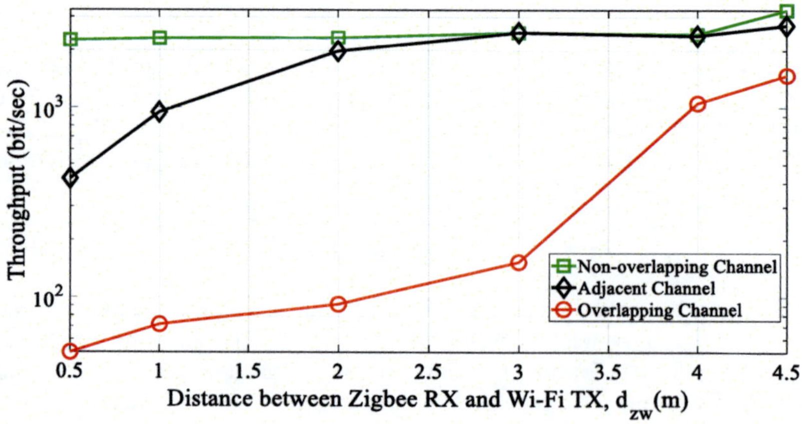


**Fig. 5** Throughput versus $d_{zw}$ for different interference cases

## 5 Conclusion

Zigbee networks are very popular for smart home applications, but the interference from other IoT networks is the main obstacle to get their full performance. This paper provides the performance study of a Zigbee network with and without the presence of Wi-Fi traffic. Our studies broadly show that the communication performance of a Zigbee network degrades significantly in the proximity of an interference network (a Wi-Fi network in this case) with overlapping channel selection. Therefore, one must ensure proper channel selections between Zigbee and Wi-Fi networks when they coexist, particularly in indoor environments. While the current research is on Zigbee and Wi-Fi networks, the performance of a Zigbee network with the presence of both Wi-Fi and Bluetooth traffic is the topic of our future study.

## References

1. De Nardis L, Di Benedetto M (2007) Overview of the IEEE 802.15.4/4a standards for low data rate wireless personal data networks. In: 2007 4th workshop on positioning, navigation and communication, Hannover, pp 285–289. https://doi.org/10.1109/WPNC.2007.353647
2. Thonet G, Allard-Jacquin P, Colle P (2008) ZigBee - WiFi coexistence: white paper and test report. Technical report, Schneider Electrics
3. Yang G, Yu Y (2009) ZigBee networks performance under WLAN 802.11b/g interference. In: 2009 4th international symposium on wireless pervasive computing, Melbourne, VIC, pp 1–4. https://doi.org/10.1109/ISWPC.2009.4800615
4. Incel OD, Dulman S, Jansen P, Mullender S (2006) Multi-channel interference measurements for wireless sensor networks. In: Proceedings. 2006 31st IEEE conference on local computer networks, Tampa, FL, pp 694–701. https://doi.org/10.1109/LCN.2006.322179.
5. Abrignani MD, Buratti C, Frost L, Verdone R (2014) Testing the impact of Wi-Fi interference on Zigbee networks. In: 2014 Euro Med Telco Conference (EMTC), Naples, pp 1–6. https://doi.org/10.1109/EMTC.2014.6996634
6. Khan S, Krishnan TSS, Kothari S, Ebenezer J, Madhusoodanan K, Murty SAVS (2016) Interference study in adjacent and alternate channels of IEEE 802.15.4 spectrum. In: 2016 international conference on wireless communications, signal processing and networking (WiSPNET), Chennai, pp 391–395. https://doi.org/10.1109/WiSPNET.2016.7566162
7. Sherazi HHR, Iqbal R, Ul Hassan S, Chaudary MH, Gilani SA (2016) ZigBee's received signal strength and latency evaluation under varying environments. J Comput Netw Commun 5. https://doi.org/10.1155/2016/9409402
8. Nomura K, Sato F (2014) A performance study of ZigBee network under Wi-Fi interference. In: 2014 17th International Conference on Network-Based Information Systems, Salerno, pp 201–207. https://doi.org/10.1109/NBiS.2014.86
9. Jacob S, Ravi P (2015) Enabling coexistence of ZigBee and WiFi. Commun Appl Electron 2(6):28–34. Published by Foundation of Computer Science (FCS), NY, USA. https://doi.org/10.5120/cae2015651788
10. Liang C-JM, Priyantha NB, Liu J, Terzis A (2010). Surviving wi-fi interference in low power ZigBee networks. In: Proceedings of the 8th ACM conference on embedded networked sensor systems - SenSys '10. https://doi.org/10.1145/1869983.1870014
11. Wang X, Yang K (2017) A real-life experimental investigation of cross interference between WiFi and ZigBee in indoor environment. In: 2017 IEEE international conference on Internet of Things (iThings) and IEEE green computing and communications (GreenCom) and IEEE

cyber, physical and social computing (CPSCom) and IEEE smart data (SmartData), Exeter, pp 598–603. https://doi.org/10.1109/iThings-GreenCom-CPSCom-SmartData.2017.94
12. Leugner S, Hellbrück H (2019) eNAV - Enhanced Co-Existence of IEEE 802.15.4 and IEEE 802.11. In: 2019 IEEE international symposium on local and metropolitan area networks (LANMAN), Paris, France, pp 1–6. https://doi.org/10.1109/LANMAN.2019.8847147
13. Won C, Youn J-H, Ali H, Sharif H, Deogun J (2005) Adaptive radio channel allocation for supporting coexistence of 802.15.4 and 802.11b. In: VTC-2005-Fall. 2005 IEEE 62nd vehicular technology conference, 2005, Dallas, TX, USA, pp 2522–2526. https://doi.org/10.1109/VETECF.2005.1559004

# Analysis of the Impact of ICT on Virtual Teaching and Learning: A Review of the Scientific Literature 2011–2020

Josep Reyes-Rojas and Michael Cabanillas-Carbonell

**Abstract** The evolution of technology has generated significant changes in education, providing new forms of teaching–learning through the use of ICT. For this purpose, a systematic review of the scientific literature was carried out, in which articles were searched and selected based on the subject of the research using the EBSCO, ScienceDirect, ProQuest, Redalyc.org and Dialnet databases, resulting in 50 sources of information. As a conclusion of the investigation, it was determined that the ICT in virtual education has granted great advantages and benefits by the tools that offer, in addition, that different types from learning are applied in special the collaborative one that turns out to be very essential in the education of the students.

**Keywords** Online teaching and learning · Virtual education · ICT · Systematic review

## 1 Introduction

The spread of the infectious disease COVID-19 has come to affect the lifestyle of all people in the world since it not only has a devastating impact on health but it has also threatened the continuity of education, causing the closure of universities and schools in 138 countries, harming 1.37 billion students, which represents more than three out of four children and young people worldwide [1].

This crisis has made it possible to see with greater importance the opportunities provided by technology since they allow the transformation of the modality of face-to-face education into virtual education by making use of information and communication technologies (ICT) [2], which provide a lot of flexibility with respect to space–time and favor institutions to develop their functions more efficiently [3].

J. Reyes-Rojas (✉) · M. Cabanillas-Carbonell
Universidad Privada del Norte, Lima, Peru
e-mail: n00132321@upn.pe

M. Cabanillas-Carbonell
e-mail: mcabanillas@ieee.org

 
X.-S. Yang et al. (eds.), *Proceedings of Sixth International Congress on Information and Communication Technology*, Lecture Notes in Networks and Systems 216,
https://doi.org/10.1007/978-981-16-1781-2_43

The incorporation of ICTs has managed to offer great possibilities in student learning because they provide various tools that help in the development of problems, promote participation and teamwork, facilitate the search for information and contribute to increasing their knowledge and intellectual skills [4].

The use of ICTs has allowed the development of different ways of carrying out the learning process, among which it is associated with the collaborative, meaningful, cooperative and autonomous, which are used effectively with the continuous guidance of teachers [5]. UNESCO states that education and ICTs are more than talking about equipment, computers, devices and/or programs, it is the opportunity to reflect on how we are thinking about education and how students and teachers learn and teach [6].

## 2 Methodology

### *2.1 Research Question*

**RQ1**. What impact do ICTs have on virtual teaching and learning between 2011 and 2020?

**RQ2.** What is the type of learning mostly used in virtual education through ICT?

### *2.2 Search Strategies*

To answer the research questions, articles were searched in the EBSCO, ScienceDirect, ProQuest, Redalyc.org and Dialnet databases, resulting in 106 articles with publication years between 2011 and 2020.

For this, we used the keywords: "ICT", "virtual education", "virtual teaching AND ICT", "ICT AND virtual education", "virtual learning AND ICT", "online teaching AND learning", "e-learning".

### *2.3 Inclusion and Exclusion Criteria*

See Table 1.

**Table 1** Inclusion and exclusion criteria

| | Criterion | |
|---|---|---|
| Inclusion | I01 | Articles related to virtual education and ICT |
| | I02 | Articles published between 2011 and 2020 |
| | I03 | Articles in Spanish and English |
| Exclusion | E01 | Articles not related to virtual education and ICT |
| | E02 | Articles published more than 10 years ago |
| | E03 | Closed access articles |

## 3 Results

In this review, 106 scientific articles were retrieved, of which 10 were eliminated because they were repeated. Likewise, 23 articles were discarded, since some were in Portuguese and others were thesis and book researches. Then, with the 73 articles selected, 23 had to be excluded because several had issues different from the education sector, and others did not focus entirely on virtual teaching and learning. Finally, there were 50 articles that would help answer the questions established in the research (Fig. 1).

Of the articles collected, it was possible to identify that 35 articles have been applied in three educational levels where ICT for virtual teaching and learning would

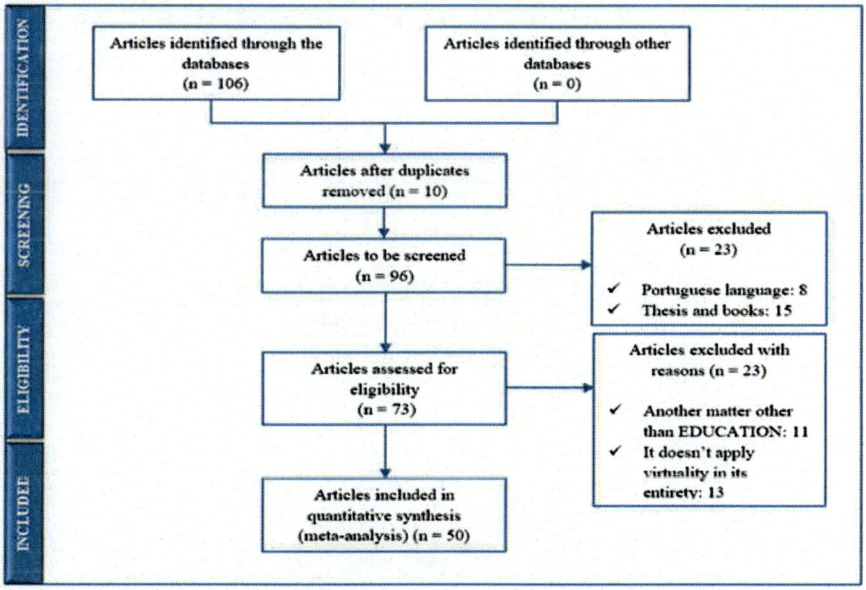


**Fig. 1** Item selection flowchart

have been evaluated and applied, arriving to observe that higher education is the one that had more attachment to the way in which it is taught and learned in a virtual way due to the fact that 22 articles were found, followed by secondary education where 8 articles were found and finally primary education with 5 articles as shown in Fig. 2.

Likewise, it was possible to identify the advantages that ICTs have given both teachers and students for the improvement of education as shown in Table 2.

On the other hand, in Table 3, the articles were classified with respect to the benefits that produce the use of ICT in virtual education.

Additionally, the different types of tools that were applied as support in virtual teaching and learning were identified, distributed as shown in Table 4.

On the other hand, it was identified that collaborative learning had a greater use in virtual education with 53% of the articles, then it was followed by the significant learning that was found in 21% of the articles used, followed by the cooperative learning, which covered 16% of the articles and finally the autonomous learning that had 10% of the articles.

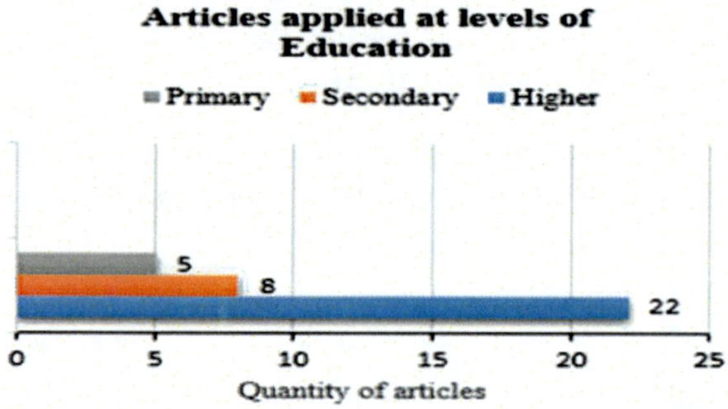


**Fig. 2** Distribution of articles that have been applied in primary, secondary and higher

**Table 2** Advantages of ICT in virtual education

| Advantages of ICT | Articles |
|---|---|
| Keeps the teacher constantly updated | [7–13] |
| They provide teaching resources | [4, 14–17] |
| They promote innovation in pedagogy | [3, 18, 19] |
| Provides flexibility in distance and time | [20, 21] |
| Facilitates the learning process | [22] |

**Table 3** Benefits of ICT in virtual education

| ICT benefits | Articles |
|---|---|
| Improves academic performance | [15, 23–27] |
| They generate good attitudes | [19, 28] |
| They motivate the student | [23, 29] |
| Generate greater autonomy | [30, 31] |

**Table 4** Tools used in virtual teaching and learning

| Types of virtual tools | Articles |
|---|---|
| Exchange of ideas and experiences | [25, 27, 32–44] |
| Information distribution | [26, 45, 46] |
| Application and experimentation | [22, 47] |
| Evaluation | [29, 48] |

## 4 Discussion

With the present review, it was possible to answer the research question: **What impact do ICTs have on virtual teaching and learning between 2011 and 2020?** Emphasizing the importance of ICT in virtual teaching and learning, since it was possible to obtain through the collected articles, the advantages and benefits that they give to students and teachers.

In Table 2, it was highlighted that the use of ICTs provides great flexibility in education, since they have allowed to break the barriers of space and time in which classes are held, and have also facilitated learning in students due to the teaching resources they provide. On the other hand, the use of ICTs requires that teachers are constantly updated in the use of digital tools because this will allow them to adapt to new teaching methodologies.

Table 3 shows that learning through ICTs has provided great benefits in academic performance due to the use of various tools that generate good attitudes and motivate students to continue learning.

In Table 4, it was identified that the tools for exchanging ideas and experiences were mostly used with 68% since they allowed communication and collaborative work between students and teachers using wikis, blogs, social networks, videoconferences and forums.

On the other hand, in Fig. 2, it was shown that the highest rate of use of ICT is reflected in the higher education level with 63%, due to the fact that teachers are more prepared and have a greater infrastructure in technology.

In addition, it was possible to answer the second research question: **What is the type of learning mostly used in virtual education through ICT?**

In Fig. 3, it was shown that collaborative learning stood out among the other learning with a 53% preference, since it encouraged the participation of all the members of a group for the development of a certain work by performing different tasks, but with a common objective.

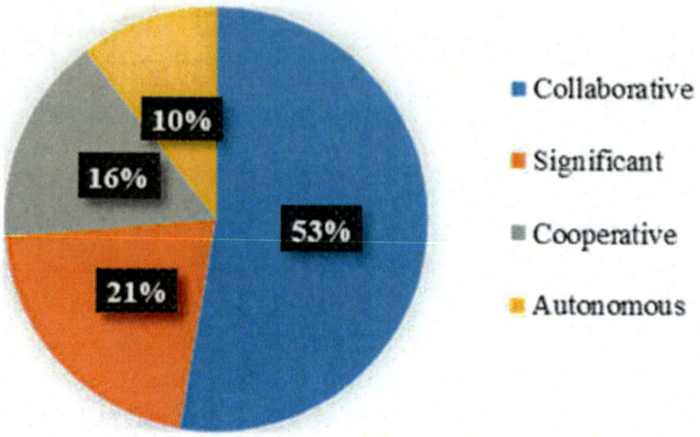


**Fig. 3** Types of learning used in virtual education

## 5 Conclusion

From the results, it was demonstrated that ICTs provide great advantages in virtual education, among which the flexibility in which education is provided and the variety of teaching resources that they provide, as well as generating a good attitude and greater motivation on the part of students, strengthening their autonomy and academic performance. In spite of this, the limitation that was found to carry out the use of the ICT was that the teachers must be in constant update so that they can face the diverse challenges that entail the use of these tools since they will be the people who will use it for the improvement of their pedagogy as well as guide the students in the effective use of these tools.

On the other hand, it was possible to identify the types of learning associated with virtual education among which the most outstanding was the use of collaborative work among students.

Finally, for future research work, it is recommended to analyze in more detail the factors that prevent the change from traditional to virtual education in developed and underdeveloped countries of the world. Likewise, it is to identify the reasons why educational institutions do not make use of virtual education in its totality in the present twenty-first century.

## References

1. UNESCO (2020) 1.370 millions students are already home with COVID-19 school closings, ministers expand multimedia approaches to ensure continuity of learning. Accessed: Nov. 29, 2020. Available: https://es.unesco.org/news/1370-millones-estudiantes-ya-estan-casa-cierre-escuelas-covid-19-ministros-amplian-enfoques
2. GESTIÓN (2020) Importance of technology in education. Accessed: Nov. 29, 2020. Available: https://gestion.pe/tecnologia/importancia-de-la-tecnologia-en-la-educacion-noticia/

3. Sánchez MRG, Reyes Añorve J, Godínez Alarcón G (2017) ICT in higher education, innovations and challenges. RICSH 6(12):1–18. doi:https://doi.org/10.23913/ricsh.v6i12.135
4. Aguilera Almaguer O, Aguilera Borjas M, Peña Cruz E (2011) Integration of ICT teaching in education. Cuad Educ y Desarro 3(29):1–15
5. Claro M, Sunkel G, Trucco D (2010) Impact of ICT on student learning. CEPAL 43(3):190–192
6. UNESCO (2013) Strategic approach to ICT in education in Latin America and the Caribbean
7. González Garcete CC (2017) Use of ICTs in continuous teacher training. Rev Científica Int 4(1):1–28. Available: http://www.utic.edu.py/revista.ojs/revistas/4/07-Gonzalez_ARANDUUTIC_VOL_IV.pdf
8. Davis Castro D, Palomeque Solano ME, Crespo Ayala PE (2016) The use of Information and communication technologies as a didactic resource in higher education. Didasc@lia 7(4):259–272. doi:https://doi.org/10.1017/CBO9781107415324.004
9. García Aretio L (2020) Teaching knowledge and skills in distance and digital education. A reflection for training. RIED 23(2):9–30. doi:https://doi.org/10.5944/ried.23.2.26540
10. Orrego Cumpa R, Quiñones Rodríguez S, Hernández RM (2018) New ways of learning: teacher training versus the use of ICT. Propósitos y Represent 6(2):671–701
11. Méndez MRR, Aguirre Aguilar G (2013) Teaching work, ICT and virtual or distance education. Apertura 5(2):108–123. Accessed: Sep. 08, 2020. Available: https://www.redalyc.org/articulo.oa?id=68830444010.
12. Turpo Gebera O (2014) Pedagogical usability of web resources in the continuous training of teachers. 3C TIC 3(3):133–155
13. Ocaña-fernández Y, Valenzuela-fernández L, Morillo-flores J (2020) Digital competence in university teachers. Propósitos y Represent 8(1)
14. Anciani JCR (2020) Guiding guidelines for ICT as a didactic resource in Primary Education. Rev Sci 5(15):150–167. doi:https://doi.org/10.29394/scientific.issn.2542-2987.2020.5.15.7.150-167
15. Mirete Ruiz AB, García Sánchez FA (2014) Academic performance and ICT. An experience with didactic websites at the University of Murcia. Pixel-Bit 44:169–183
16. Pérez Cardoso CN, Suárez Mella RP, Rosillo Suárez NA (2018) Interactive virtual education, the paradigm of the future. Rev Atenas 4(44). Accessed: Sep. 11, 2020. Available: https://www.redalyc.org/jatsRepo/4780/478055154009/html/index.html
17. Torres Chávez TE, García Martínez A (2019) Reflections on adaptive virtual teaching materials. Rev Cuba Educ Super 38(3)
18. Arruda R, Prata-Linhares M, Paredes J (2018) Pedagogical innovation with the use of ICT: perceptions of faculty members from Brazil, Spain and Mexico about their practices. Acta Sci Educ 40(3). doi:https://doi.org/10.4025/actascieduc.v40i3.36217
19. Gámiz-Sánchez V-M (2017) ICT-based active methodologies. Procedia Soc Behav Sci 237:606–612. https://doi.org/10.1016/j.sbspro.2017.02.018
20. Amador YA (2014) Tutor-student interaction in distance education settings. Rev Lenguas Mod 20:285–294
21. Bergamin PB, Werlen E, Siegenthaler E, Ziska S (2012) The relationship between flexible and self-regulated learning in open and distance universities. Int Rev Res Open Distance Learn 13(2):101–123. https://doi.org/10.19173/irrodl.v13i2.1124
22. Arriassecq I, Santos G (2017) New information technologies as facilitators of meaningful learning. Arch Ciencias la Educ 11(12)
23. Huertas Montes A, Pantoja Vallejo A (2016) Effects of an educational program based on the use of ICT on student academic performance and motivation in the secondary education technology subject. Educ XX1 19(2):229–250. doi:https://doi.org/10.5944/educXX1.14224
24. Loza Arenas JE, Salinas Urbina V, Glasserman Morales LD (2017) Academic performance of secondary school students participating in the digital classroom program. 6(2):60–80
25. García-Martín S, Cantón-Mayo I (2019) Use of technologies and academic performance in adolescent students. Comunicar 27(59):73–81. https://doi.org/10.3916/C59-2019-07

26. de la Fuente Sánchez D, Hernández Solis M, Pra Martos I (2018) Educational video and academic performance in distance higher education. Rev Iberoam Educ a Distancia 21(1):323–341
27. Tejada Garitano E, Garay Ruiz U, Romero Andoneg A (2017) The incidence of cooperative participation in virtual environments on academic performance. EDUTEC (60):1–11
28. Centeno Moreno G, Cubo Delgado S (2013) Evaluation of digital competence and attitudes towards ict of university students. Rev Investig Educ 31(2):517–536. doi:https://doi.org/10.6018/rie.31.2.169271
29. de Mingo-López DV, Vidal-Meliá L (2019) Kahoot Activities! in the classroom and student satisfaction. 3C TIC 8(1):96–115
30. Sierra Varón CA (2011) Virtual education as a facilitator of autonomous learning. Panorama 5(9)
31. Pérez Lasprilla MA (2020) Autonomous learning in higher education, virtual modality: a reading from anthropotechnics. Rev Acad y Virtualidad 13(1):80–92
32. Alaslani K, Alandejani M (2020) Identifying factors that influence students performance through social networking sites : an exploratory case study. Heliyon 6. doi:https://doi.org/10.1016/j.heliyon.2020.e03686
33. Santos H, Batista J, Marques RP (2019) Digital transformation in higher education: the use of communication technologies by students. Procedia Comput Sci 164:123–130. https://doi.org/10.1016/j.procs.2019.12.163
34. Buil I, Hernández B, Sesé FJ, Urquizu P (2012) Discussion forums and their benefits in virtual teaching: recommendations for efficient use. INNOVAR 22(43):131–143
35. Duta N, Martínez-rivera O (2015) Between theory and practice : the importance of ICT in Higher Education as a tool for collaborative learning. Procedia Soc Behav Sci 180:1466–1473. https://doi.org/10.1016/j.sbspro.2015.02.294
36. Pinheiro MM, Simões D (2012) Constructing Knowledge: an experience of active and collaborative learning in ICT classrooms. Procedia Soc Behav Sci 64:392–401. https://doi.org/10.1016/j.sbspro.2012.11.046
37. García C (2016) Project-based learning in virtual groups - collaboration and learning outcomes in a virtual training course for teachers. Procedia Soc Behav Sci 228:100–105. https://doi.org/10.1016/j.sbspro.2016.07.015
38. Sangrá LM (2015) Learning history with wiki in secondary education. 14:39–47. doi:https://doi.org/10.1344/ECCSS2015.14.4
39. Molina P (2019) Innovation and implementation of ICT in Primary Education. DIM 37
40. Daud MY, Zakaria E (2012) Web 2. 0 application to cultivate creativity in ICT literacy. Procedia Soc Behav Sci 59:459–466. https://doi.org/10.1016/j.sbspro.2012.09.301
41. Casteleyn J, Mottart A, Van Vooren V (2013) Web 2. 0 presentation tools as an alternative to real-life performance. Procedia Soc Behav Sci 93:221–224. https://doi.org/10.1016/j.sbspro.2013.09.181
42. Uribe Zapata A, Ramírez Salazar DA, Henao Álvarez O (2017) Exploration of an online collaborative writing exercise for a group of elementary school students. Rev Lasallista Investig 14(1): 29–41. doi:https://doi.org/10.22507/rli.v14n1a2
43. Benítez Lima MG, Barajas Villarruel JI, Noyola Rivera R (2016) The usefulness of the virtual forum for collaborative learning, from the opinion of the students. Campus Virtuales 5(2):122–133
44. Tobar-esparza AP, Lozada-Yánez RM, Maldonado-Gavilánez CE (2017) Asynchronous collaborative tools in the academic learning teaching process. Dominio las Ciencias 3:439–453
45. Troche-Isalgué NA, Valdés-López M (2019) Particularidades del video tutorial como medio didáctico digitalizado. Santiago 148:19–27. Available: http://search.ebscohost.com/login.aspx?direct=true&db=fua&AN=135810875&lang=es&site=eds-live
46. Etchegaray MC, Guzmán MD, Duarte AM (2017) Diseño de un recurso multimedia on line basado en Inteligencias Múltiples. Campus Virtuales 6(1):51–65

47. Magal-royo T, Jorda-Albiñana B, Gonzalez del Rio J, Ampuero O, Gimenez-López J (2012) Online collaborative environments in the creative process of product development for engineering students. Procedia Soc Behav Sci 51:677–681. https://doi.org/10.1016/j.sbspro.2012.08.223
48. Kyriazi T (2015) Using technology to introduce frequent assessments for effective learning: registering student perceptions. Procedia Soc Behav Sci 197:570–576. https://doi.org/10.1016/j.sbspro.2015.07.195

# Mapping of TOGAF ADM and TMForum Frameworx in the Telecommunication Industry

**Muharman Lubis, Firdaus Adi Nugroho, Lyvia Winiyanti Lumingkewas, and Arif Ridho Lubis**

**Abstract** Nowadays, the telecommunication market is developing actively and dynamically. Flexibility and agility are required, in which the alignment between business process and IT strategies is a mandatory attribute to be able to sustain in the hard competition for the open market. To allow the alignment, several companies adopt the use of an Enterprise Architecture (EA), which in the process of designing, it is necessary to use specific framework such as TOGAF based on certain best practice. In the telecommunication industry, a collaboration framework is needed due to several factors such as availability, novelty and technicality, in which one of the telecommunication associations, known as TMForum, provides frameworks, namely, eTOM. To identify the mapping between these frameworks, interviews and observations are used in this study. The secondary data source is the result of a review of publications on the implementation of EA and the use of a broader framework in the telecommunication industry such as the utilization of subcomponents of eTOM such as SID and TAM. Therefore, in this study, TOGAF will be mapped to eTOM.

**Keywords** Enterprise architecture · Telecommunication · SID · TOGAF · eTOM · TAM

M. Lubis (✉) · F. A. Nugroho · L. W. Lumingkewas · A. R. Lubis
School of Industrial Engineering, Telkom University, Telekomunikasi No. 1, Bandung 40257, Indonesia
e-mail: muharmanlubis@telkomuniversity.ac.id

Department of Computer Engineering and Informatics, Politeknik Negeri Medan, Medan City 20155, Indonesia

A. R. Lubis
e-mail: arifridho@polmed.ac.id

 
X.-S. Yang et al. (eds.), *Proceedings of Sixth International Congress on Information and Communication Technology*, Lecture Notes in Networks and Systems 216,
https://doi.org/10.1007/978-981-16-1781-2_44

# 1 Introduction

Today's market especially within the telecommunication industry is developing tremendously and dynamically with the primary customer segments, which are related to telephone communication, mobile services, satellite data transmission and Internet usage [1]. Therefore, the digital age in an actual sense has a huge impact on business operations to stay competitive despite rapid innovations and market demands [2]. It requires flexibility, agility and collaborative modeling concepts coupled with increased stakeholder participation and direct implementation of process models [3]. To gain a competitive and agile advantage, the resource and strategy have critical roles to allow sustainability and growth in modern organizations. It should be followed by the alignment of business process with IT application in the primary project based on strategic plans and organization resources [4] to keep being innovative and creative in the open market. In order to ensure this kind of process has been done considerably, the corporate structure should put their business strategy and planning in accordance with the technology and information system architecture. On the other hand, the EA is a hierarchical approach to align the fields of business and information technology by integrating business processes, information systems [5], organizational functions and stakeholders within an organization. It has the ultimate goals that the existing activities not limited only to indicate the urgent needs but also to develop the ability to create an organization that can compete, grow and satisfy the customer concurrently because its core competence has been embedded with its competitive advantage to create a sustainable organization [6].

Actually, in the process of designing an EA, the best practice is using the consortium-based due to the widely used to allow flexibility and transparency such as TOGAF. In order to develop the EA for the companies, especially the telecommunications industry, a cooperative framework is also required due to certain issues that must be covered in a careful manner such as technicality and novelty. There is a global non-profit industry union that promotes collective problem-solving and joint initiative in terms of cooperative work to optimize the performance and benefit in the telecom industry to provide the digital service and ecosystem known as ForumTM. It is dedicated to telecommunication sector as the best practices for the adaptation within the companies around the world by providing a framework called Business Process Framework (eTOM), which has other two subcomponents, which are Shared Information and Data Model Framework (SID) and the Application Framework (TAM) [7]. Therefore, in this study, the mapping process will be conducted to allow the TOGAF be assigned into the Forum FrameworxTM.

# 2 Literature Review

## *2.1 The Open Group Architecture Framework (TOGAF)*

TOGAF involves the approach, way and tools to assist the production receipt, coherent use and continuous maintenance within EA. Actually, it has four main domains called technology architecture, data architecture, application architecture and business architecture, which applies an iterative process model that allows a company to continue to change and adjust to the company's business targets to be achieved [8]. There are several advantages of TOGAF, namely, it is flexible and open source, then systematic, focus on the implementation cycle (ADM) and process, richful in architectural technical areas and resource based provides lots of reference material. Today, various EA approaches have been introduced to generate roadmap, blueprint, rules, consideration, capabilities, initiatives, scope, etc. for creating the information systems' (IS) attributes, namely, personalization, customization, integration and automation. These IS must pinpoint the issues and challenges of the current organization to adjust the process of modernizing the commercial structure and unit of the company through the proper attributes and characteristics. There is some criticism of the standard of the agreement for explaining EA to the public with the term esoteric or remote execution [9–13]. The results will be in the form of separate documents called artifacts that present critical instrument and informational content for description, implementation and decision-making.

## *2.2 Business Process Framework (eTOM)*

The eTOM is the analysis and perspective that emphasizes the functional area to explain and investigate the multiple stages of business to the importance and necessity of the enterprise to manage and directing the operation into sustainability within telecom industry. In the conceptual level, it can be viewed as the platform for drawing the three main process areas, which are business operations that are organization area, operations area and strategy, infrastructure, product area, while the technical model consists of four levels of abstraction that allow the view of processes and tasks in various details. Organization indicates the interaction between processes while operation depicts the fulfillment of service delivery life cycle. Meanwhile, the strategy, infrastructure and product (SIP) offer iterative and incremental planning to manage the activity and operations that cover the primary routine and customer support [1]. Actually, the description in the artifact should be presented in the level periodically, so it can determine from which perspective to be seen while at the same time, it provides the proper requirement to allow the integration process as the open framework [2, 14]. The strategies are acquired and assessed through the companys' operating model based on the suggested platform according to the system design and engineering knowledge [15, 16]. On the other hand, the complexity and various entities within the organization together with their different type of rules, standards and protocols require diverse and specific components that are used to perform every critical monitoring and maintenance activities and process. By default, eTOM presents the perspective of each level from lowest level 0 (general) to highest level 4 (more detail) with overall benefit as the expectation to be achieved, which of course required the functional support namely:

1. As a blueprint model for executing the business process in every units, systems, department, partners, vendors, suppliers, investors and stakeholders to reduce several project times, costs and risks of system acquisition and plan.
2. Adopting standardized schemes, categorization and idioms for implementing the plan into executing processes in order to paraphrase and rationalizing internal and external activity by exploiting the opportunities and chance of partnerships within and between industries.
3. Execute the development of comprehensive and continuous business processes through technology resources within the company to be reused later on and consistently.
4. Understand the service design and IT management of proper use of every application to support the requirements and specification based on user demand by aligning with the needs and work culture.
5. Develop a sustainable environment with high-quality process flow from start to finish by terminating or shortening the gaps, redudancy, missing, invalid and duplication within the relationship and interaction.
6. Identify the chances by exploiting the opportunities to increase the performance by re-using the potential and stable component systems [17].

## 3 Research Method

EA applies architecture principles and practices to guide organizations through changes in business processes, information, processes and technology needed to implement their strategies. These practices use various aspects of the company to identify, motivate and make these changes. TOGAF ADM is a form of iteration cycle in each process in each phase, regardless of whether new decisions must be made with each iteration [18–20]. On the other hand, eTOM is a specific framework for the telecommunications industry that has multiple domains, so that it is possible to map TOGAF on eTOM. To identify the mapping between these frameworks, interviews and observations related to develop of EA in the telecommunication industry are used. The secondary data source is the result of a review of publications on the implementation of EA and the use of a broader framework in the telecommunications industry. After the information is collected, the analysis is carried out, and the TOGAF is mapped to eTOM and TAM.

# 4 Result and Discussions

## *4.1 Positioning of TOGAF and Frameworx*

The initial stage is not explicitly available in Frameworx, which should be considered complementary when applying the EA approach, which both can be identified and adopted as an architectural frame of reference for addressing EA development. Phase A—Vision Architecture also represents a new pattern for Frameworx, being covered only partly in the form of definitions for business strategy formulation. In this stage, TOGAF describes the basis and scope of target architectures for all dimensions of the Frameworx architecture. The three Frameworx solutions were assigned to Phases A, B and C of TOGAF ADM. The business process framework of the Enhanced Communications Process Map (eTOM) describes the core business processes of Communications Service Provider (CSP) and is designated for TOGAF ADM Phase B. The SID provides a comprehensive common data and information model for Telecom SP and is mapped to the TOGAF ADM Phase C. In addition, the SID framework relates to eTOM business processes in different domains, which are made up of different groups of entities Multilayer Combined Commercial (ABE). The TAM provides a framework for service provider-related applications in the telecommunication industry that can be implemented and used to classify their complex application landscape, which usually speaks of BSS/O (Business/Process Support Systems). Meanwhile, it can refer to the TOGAF ADM Phase C application architecture that the integration framework will be part of the layout of the second document. Stage D refers to Information Technology Engineering. TOGAF does not have a specific integration stage in ADM. Stage D refers to Information Technology Engineering. TOGAF does not include a specific integration stage into ADM while the digital transformation of the company often the discussion related to the concept use of the contextual and strategic plan for the control and management, particularly through the collaboration between the various telecommunications and utility sectors [21, 22].

## *4.2 The Mapping of TOGAF and TM Forum Frameworx Architecture Principles*

There are some additional Frameworx principles that can be used specifically by the telecom industry. The Frameworx principles mentioned in the TAM of the Frameworx solution are related to the different principles of TOGAF engineering. Certainly, these principles are closely related to Frameworx's solutions and telecom operator needs. TOGAF brings an additional benefit to this part of Frameworx by providing methods, models, standards and features for developing other good principles in TOGAF engineering.

## 4.3 TOGAF Metamodel Mapping to Frameworx

The intersection of customer power with evolving trends to enable the technologies rapidly drives the time of unprecedented opportunity and risk for organizations [18–20]. Thus, creating value and benefits for customers should shape digital transformation, while mature standards can help make operations more efficient and effective by enabling creative thinking [14, 23, 24]. The following table shows the objects in the TOGAF Content Metamodel and their mapping to the Frameworx eTOM, SID and TAM.

## 4.4 Frameworx Solution and TOGAF Architectural Artifacts and Deliverables

The mapping is done by mapping architectural principles, metamodels, artifacts and the results of TOGAF and TMForum Frameworx submissions. It can be seen that three TMForum solutions Frameworx can be mapped in the phases of vision architecture, business architecture and information system architecture. Table 1 also shows that not all of the principles of Frameworx can be covered by TOGAF, so that

**Table 1** The mapping principles

| TOGAF principles | Frameworx principles |
|---|---|
| Architecture principles | • Provides complete company-wide operational solutions for the fixed, mobile, cable and centralized industries |
| Business principles | • Transferring business to operators |
| | • Allows business processes to be modified easily without changing the schedule by separating control of the business process flow from the application process |
| Data principles | • Allows full sharing of company data across the company and, if necessary, with business partners |
| Application principles | • Reduce software development costs and risks based on industry best practices and current standards |
| | • Enables operator organizations to thrive using distributed systems that are loosely coupled without system failure |
| Technology principles | • Reduce IT costs and lead times by using ready-to-use software components (COTS) at scale |
| | • Allow clear pathways for migration through integration and further development of legacy systems |
| | • Enables simplified system integration (Plug and Play) via a clear node interface between applications |
| | • Allows simplified system integration using a common communication bus between applications |

Frameworx is able to provide specific architectural principles for the telecommunications industry. In Table 2, it can be seen that not all metamodel in TOGAF is in Frameworx but it can be adapted to the three Frameworx solutions in its implementation. In addition, there are also artifacts and deliverables that indicate the absence of specific domains in the TOGAF, which is not the weaknesses, but the TOGAF broadly explains the design of the EA and produces several artifacts in general while Frameworx provides more specific domains to be applied in the development of EA. With the mapping of TOGAF to Frameworx, the design and implementation of EA have become more comprehensive and in accordance with the needs of the telecommunications industry.

**Table 2** TOGAF metamodel mapping to Frameworx

| TOGAF content metamodel | Frameworx |
|---|---|
| Organizational unit | • Not included in Frameworx, as part of the business architecture, it maps to the Frameworx solution eTOM, Within eTOM, it can be considered as the conceptual view of eTOM and represents the horizontal and vertical process groupings, which represent the line management responsibilities, goals, objectives and measures |
| Actor/role | • Not included in Frameworx, as part of the business architecture, it maps to the Frameworx solution eTOM. An actor can be a person, organization or system that has a role that interacts with activities or functions as part of functional process grouping. This actor can have different roles within a process or rather function |
| Business function | • As part of the business architecture, it maps to the eTOM. A function or activity delivers a business capability aligned to an organization |
| Process | • As part of the business architecture, it maps to the eTOM. A process is a flow/sequence of steps that produce a specified outcome. It can be decomposed into sub-processes. TOGAF process definition refers to a logical level representing the HOW? and Frameworx refers to the conceptual level as the WHAT? |
| Data entity | • As part of the data architecture, it maps to the Frameworx SID. Data entity is an encapsulation of data that is classified within a business domain as a thing of interest to the business. In SID terms, these are business entities that map to data entity in TOGAF |
| Logical data component | • As part of the data architecture, it maps to the Frameworx solution SID as an information model. This is a boundary zone that encapsulates related data entities to form a logical location to be held |
| Physical data component | • Does not map to the SID definition, because it reflects the implementation of a data entity, and in architecture no implementations exist |

(continued)

**Table 2** (continued)

| TOGAF content metamodel | Frameworx |
|---|---|
| Information system service | • This is an automated element of a business service and may support business services. Both, business and information system services are described in the integration framework in Frameworx |
| Logical application component | • As part of the application architecture, it can map to the TAM Frameworx. In the TAM, applications are not described as computer systems, but as logical groups of capabilities that manage the data objects and support the business function in business architecture. The applications and their capabilities are defined without reference to particular technologies |
| Physical application component | • Does not map to the TAM definition, because it is a conceptual architecture framework, not an implementable solution |
| Business services | • A business service supports capabilities through an explicity defined interface and is governed by an organization. The integration Frameworx focuses on business services, which represent a specification of system capability to achieve a stated business purpose |

## 5 Conclusion

There are several principles, metamodel, artifacts and deliverables that are not provided by TOGAF for the telecommunications industry specifically, TOGAF can be used as a base framework in the design of EA, which is then mapped to Frameworx, which provides domain-specific domains and details for the telecommunications industry (Tables 3 and 4). *With the mapping of TOGAF to Frameworx, the design and implementation of EA have become more in accordance with the needs of the telecommunications industry. Finally, it can help companies be more flexible and agile although certain circumstances and environments should be considered first such as the current business process and the strategic planning of the company.*

**Table 3** Frameworx solution and TOGAF architectural artifacts mapping

| TOGAF principles | Frameworx artifacts |
|---|---|
| Artifacts (catalogs, matrices and diagrams) of preliminary | • Frameworx application solution TAM identifies specific principles |
| Artifacts (catalogs, matrices and diagrams) of architecture vision—Phase A | • Not covered directly. However, Frameworx mentions the different type of stakeholders, which support the development of the stakeholder map |

(continued)

**Table 3** (continued)

| TOGAF principles | Frameworx artifacts |
|---|---|
| Artifacts (catalogs, matrices and diagrams) of business architecture—Phase B | • Frameworx process eTOM supports the development of the artifacts, in which the result will be a RACI matrix equal to actor/role matrix |
| Artifacts (catalogs, matrices and diagrams) of data architecture—Phase C | • Frameworx data SID and TAM support the development of these artifacts |
| Artifacts (catalogs, matrices and diagrams) of application architecture—Phase D | • Not covered, system architecture contains a set of concepts that are similar or equal to contracts in Frameworx |

**Table 4** Mapping TOGAF deliverables into Frameworx

| TOGAF content metamodel | Frameworx |
|---|---|
| Architecture definition document | • Not covered |
| Architecture principles | • A number of architecture principles are defined as part of Frameworx documentation, particularly within the application framework TAM |
| Architecture repository | • Not covered |
| Architecture requirements specification | • Architecture requirements are taken into account and described as use cases. Use cases are a key artifact for gathering business requirements and confirm how processes and information actually interact, eTOM processes are a good starting point for modeling and defining use cases |
| Architecture roadmap | • Frameworx's goal is to provide an integrated business architecture roadmap that helps achieve business process automation along with dramatically increased business flexibility and agility |
| Architecture vision | • Not covered |
| Business principles, business goals, business drivers | • A number of architecture principles are defined as part of Frameworx documentation |
| Capability assessment | • The Frameworx solution eTOM process maturity assessment and team supports the development of this deliverable |
| Change request | • Not covered |
| Communications plan | • Not covered |

(continued)

**Table 4** (continued)

| TOGAF content metamodel | Frameworx |
|---|---|
| Compliance assessment | • This assessment is relevant to Frameworx as it relates to the SID and eTOM conformance and compliance assessment principles. In compliance assessment methodology, it can be very helpful in the assessment of conformance or compliance |
| Implementation and migration plan | • Not covered and Frameworx can benefit from TOGAF |
| Implementation governance model | • Not covered and Frameworx can benefit from TOGAF |
| Organizational model for enterprise architecture | • Not covered and Frameworx can benefit from TOGAF |
| Request for architecture work | • Not covered and Frameworx can benefit from TOGAF |
| Requirements impact assessment | • Not covered and Frameworx can benefit from TOGAF |
| Solution building block | • There is a notion of application components defined in the TAM |
| Statement of architecture work | • Not covered and Frameworx can benefit from TOGAF |
| Tailored architecture framework | • Tailored Frameworx solutions eTOM, SID, TAM |
| Transition architecture | • Not covered and Frameworx can benefit from TOGAF |

# References


1. Ahlan AR, Lubis M, Lubis AR (2015) Information security awareness at the knowledge-based institution: its antecedents and measures. Procedia Comput Sci 72:361–373
2. Tcukanova O, Torosyan E, Morozova M, Shekhovtsova E (2019) Standardization of eTOM model for solving problems of the Russian telecommunications market. In: E3S web of conferences, 110, pp 1–7
3. vom Brocke J, Mendling J (2016) Business process management cases – learning from real-world experience. BPTrends Column, Class Notes, pp. 1–10
4. Nursinski-Stolberg A, Gangatharan K, Czarnecki C (2016) Development of a subject-oriented reference process model for the telecommunications industry. In: Lecture notes in informatics (LNI), series of the Gesellschaft fur Informatik (GI), P-259, pp. 699–712.
5. Olsen DH, Trelsgård K (2016) Enterprise architecture adoption challenges: an exploratory case study of the Norwegian higher education sector. Procedia Comp Sci 100(1877):804–811
6. Karouw S, Wowor H (2013) Analysis and design application for quantifying IT business value. In: Information systems international conference, pp. 7–12
7. Murti DN, Prasetyo YA, Fajrillah AAN (2017) Perancangan Enterprise Architecture Pada Fungsi Sumber Daya Manusia (SDM) Di Universitas Telkom Menggunakan Togaf ADM. J Rekayasa Sist Inf 4(01):47
8. Milham D (2016) TM Forum and MEF collaboration. MEF seminars, building dynamic third network service

9. Seraoui Y, Elmostafa B, Bellafkih M, Raouyane B (2017) eTOM mapping onto NFV framework: IMS use case. In: International conference on computing and wireless communication
10. Prasetyo YA, Lubis M (2020) Smart city architecture development methodology (SCADM): a meta analysis using SOA-EA and SoS approach. SAGE Open 10(2)
11. Jannah AM, Lubis M, Saedudin RR, Dwiputri FS (2020) Designing the Smart Health Function towards Puskesmas (Citizen Health Centre) based on Smart City Concept. In: IOP conference series: materials science and engineering 847(1)
12. Wardani AE, Fajrillah AAN, Gumilang SFS, Lubis M (2018) Project evaluation for business and IT alignment with enterprise architecture for water distribution company. In: Proceedings of international conference on computer science and information technology
13. Saffanah TNA, Fajrillah AAN, Lubis M (2018) IT roadmap to improve business strategy using TOGAF ADM: a case study of government-owned electricity company. In: Proceedings of international conference on computer science and information technology
14. Benhima M, Himi A, Ameyao C, Adou EA (2013) Telco business process transformation using agile lean six sigma and Frameworx components: focus on the core engineering aspects with a case study. Int J Comp Sci Issues 10(1/1)
15. Dachyar M, Zagloel TYM, Saragih LR (2020) Enterprise architecture breakthrough for telecommunications transformation: a reconciliation model to solve bankruptcy. Heliyon 6:e05273
16. Raouyane B, Bellafkih M, Mohammed E, Leghroudi D, Ranc D, Ramdani M (2011) eTOM business processes conception in NGN monitoring. CESM-CCIS Springer 176:133–143
17. Rouhani BD, Mahrin MN, Nikpay F, Nikfard P (2013) A comparison enterprise architecture implementation methodologies. In: Proceesings of the International conference on informatics and creative multimedia (ICICM), pp. 1–6
18. Andi STN, Fajrillah AAN, Lubis M (2019) IT roadmap to improve business strategy using TOGAF ADM: a case study of government-owned electricity company. J Phys Conf Ser 1361:012017
19. Lubis M, Azizah AH (2018) Towards achieving the efficiency in Zakat management system: interaction design for optimization in Indonesia. In: International conference on user science and engineering, User Science and Engineering, pp 289–301
20. Lubis M, Kartiwi M, Zulhuda S (2016) Election fraud and privacy related issues: addressing electoral integrity. In: Proceedings of the IEEE ICIC
21. Armour F, Kaisler SH (2018) Introduction to the minitrack on enterprise architecture and business processes analysis. In: Hawaii International Conference on System Sciences
22. Vakaloudis A, O'Leary (2019) A framework for rapid integration of IoT systems with industrial environments. In: IEEE World Forum on Internet of Things
23. Gagne D (2017) Alignment of strategies to operations using TM Forum Frameworx. TMForumLive! Driving the roadmap to digital success
24. Andriole S (2020) Why no one understands enterprise architecture & why technology abstractions always fail. Forbers Article
25. Formigo SDD (2018) Information systems investment decision making using return of investment: application of CRUD framework with eTOM business process framework in Portuguese Telecommunication Industry. Master disertation, NOVA Inf. Man. School. Universidade Nova de Lisboa

# A Feasibility Study Towards E-processing Supply Chain Management Transactions in South African Municipalities

**Silma Koekemoer and Rossouw Von Solms**

**Abstract** The continued regressing in audit outcomes relating to supply chain management practices of South African municipalities, together with the new ways of working brought about by COVID-19 restrictions, has highlighted the need for e-processing of supply chain management transactions, to ensure uninterrupted service delivery to the community. The purpose of this research paper is, therefore, to present a feasibility study based on primary data collection and a literature review, which informs the development of a model for the implementation of e-processing of supply chain management transactions in South African municipalities. It was found that one single implementation model will not suffice, so a system-based hybrid model is recommended to be implemented in phases, involving municipalities, which are grouped together based on their capacity and contribution to the total municipal expenditure. The implementation should be supported by regulating the minimum IT system specifications, supporting business processes and operating procedures.

**Keywords** E-processing · Supply chain management · Supply chain value chain · Feasibility study · Implementation models · South African municipalities · Auditor general of Southern Africa · Local government audit outcomes · Regulatory compliance · Supply chain management oversight · Supply chain management controls · Contract management · Procurement processes · Fruitless and wasteful expenditure

S. Koekemoer (✉) · R. Von Solms
School of IT, Nelson Mandela University, PO Box 77000 NMU, Gqeberha 6031, South Africa

R. Von Solms
e-mail: rossouw@mandela.ac.za



X.-S. Yang et al. (eds.), *Proceedings of Sixth International Congress on Information and Communication Technology*, Lecture Notes in Networks and Systems 216,
https://doi.org/10.1007/978-981-16-1781-2_45

## 1 Introduction

The General Report of the Auditor General of South Africa (AGSA) on the Local Government Audit Outcomes [1] laments the current state of financial affairs at municipalities across the country, indicating that compliance with supply chain management (SCM)-related legislation has steadily regressed since 2016–2017. The report continues that oversight and controls are lacking, and most municipalities do not have contract management adequately institutionalized, with fruitless and wasteful expenditure increasing annually. Projects displayed widespread underspending of budget allocations, were found to be refuting SCM legislation, and were impeded by high levels of irregular expenditure [1]. Together with these regressing audit outcomes, the advent of the COVID-19 pandemic has renewed the interest in implementing the e-processing of SCM transactions in government institutions to enable and continue service delivery to the community under new remote working conditions.

The municipal SCM value chain is complex, with multiple integrated components, and is underpinned by a legislative and regulatory framework. It is also a source of continued adverse audit findings and therefore, it is clear that new ways of performing SCM transactions in municipalities could potentially make a significant difference. Thus, this research paper provides a South African case study, investigating the feasibility of e-processing of SCM transactions in municipalities.

## 2 Purpose

The research investigates the feasibility of enabling the e-processing of SCM transactions in South African municipalities. Therefore, a review of the SCM legislative requirements and system specifications is presented. The research further investigates what IT functionality is currently available and used for the e-processing of SCM transactions in municipalities. Thus, the purpose of this research paper is to present a feasibility study based on primary data collection and a literature review, which informs the development of a model for the implementation of e-processing of SCM transactions in municipalities.

## 3 Research Methodology

This research employed a literature review and the collection of primary data from municipalities and financial IT system vendors, using questionnaires and unstructured telephonic interviews.

The practical research and data collection were based on questionnaires created with Google forms. The questionnaires for financial IT system vendors and municipalities were similar, making the comparison of the findings possible.

The sample of financial IT system vendors included all the vendors, which provide financial IT systems in the current municipal environment. A total population of 11 vendors were invited to respond to the questionnaire, of which only 10 vendors have participated. The municipal sample was nominated by the financial IT system vendors and, therefore, can be regarded as a sample of convenience. There are 257 municipalities in South Africa, of which 32 were nominated and invited to participate (12%), but only 16 of these responded (6%). The municipal participants represent 10 of the 11 financial IT system vendors, as follows in Table 1.

While the sample adequately represents the available IT functionality, Table 1 indicates that category B and C municipalities were not as well represented as those from category A. However, the overall findings correlate with the findings of the AGSA, and the views of the IT system vendors in general, and therefore are accepted as representative of the current situation across municipalities.

**Table 1** Research sample segmentation

| Category of municipality | Population size | Sample size | Percentage of category size |
|---|---|---|---|
| A: Metro municipalities | 8 | 2 | 25% |
| B: Local municipalities | 205 | 12 | 6% |
| C: District municipalities | 44 | 2 | 5% |
| **Total number of municipalities:** | **257** | **16** | **6%** |
| Financial IT system vendor | | | |
| Vendor 1 | 49 | 3 | 6% |
| Vendor 2 | 4 | 0 | NA |
| Vendor 3 | 21 | 2 | 10% |
| Vendor 4 | 47 | 1 | 2% |
| Vendor 5 | 57 | 5 | 8% |
| Vendor 6 | 1 | 0 | NA |
| Vendor 7 | 6 | 0 | NA |
| Vendor 8 | 6 | 1 | 17% |
| Vendor 9 | 32 | 1 | 3% |
| Vendor 10 | 16 | 1 | 6% |
| Vendor 11 | 17 | 1 | 6% |
| Other—in house custom developed | 1 | 1 | 100% |
| **Total number of municipalities** | **257** | **16** | **6%** |

*Source* Researcher

## 4 Literature Review

The objective of the MFMA is to "secure sound and sustainable management of the fiscal and financial affairs of municipalities ..." [2]. Article 115 addresses the implementation of a SCM system. The term system does not refer to e-processing per se, but a system consisting of best practice business processes, enabled by appropriate technology, and managed by skilled people. This research paper mainly focuses on the software component of "appropriate technology", and the feasibility of deploying e-processing of SCM transactions in municipalities.

The Municipal SCM Regulations, issued in 2005 [3], informed the development of minimum system specifications applicable to SCM processing in municipalities. These regulations and the minimum system specifications were used as a framework for the research questions, which were prepared for the practical research and data collection.

National Treasury published MFMA Circular 80 [4] and Annexure B of this circular details the minimum financial IT system specifications applicable to municipalities. A content analysis of the specifications related to SCM processes only [4] clearly indicates the intent to entrench e-processing of SCM transactions.

Since the publication of MFMA Circular 80 [4], several additional circulars have been published, relating to SCM in municipalities. MFMA Circular No. 81 [5] introduces the Web-based CSD and is applicable from 1 July 2016. MFMA Circular 83 [6] "provides information to municipalities and municipal entities regarding the advertisement of bids and the publication of notices" on the e-Tender Publication Portal and aligns with the principles of OCDS [7]. The use of the e-Tender portal has been a requirement for tenders above ZAR200 000 since 1 July 2015.

Schatteman [8] states that internationally municipalities are experiencing problems with the procurement of goods and services, and e-procurement. Their findings are supported by the AGSA [1], reporting that of the 16 participating municipalities, only 1 achieved a financially unqualified audit report with no findings. All of the participating municipalities received either findings or material findings, in the area of SCM.

Public procurement should be fair, equitable, transparent, competitive, cost-effective, simple to access and technology driven [9]. However, Ambe [10] explains the complexity of SCM in local government and the comprehensive regulatory framework. An article by Bizana et al. [11] further highlights sourcing challenges that negatively affect municipal service delivery. The authors recommend that cross-functional sourcing teams (shared services) should be formed. Kramer [9] acknowledges the progress made towards introducing e-processing of public procurement, with Habtemichael and Cloete [12] adding that the use of IT in the procurement process can contribute towards controlling corruption. Introducing IT should improve transparency in SCM, but on the downside, could empower technology experts to engage more effectively in corruption.

The literature review revealed that the e-processing of SCM transactions in municipalities, the people required to perform the e-processing, as well as the mechanisms to enforce this, should be considered when conducting this research and making recommendations. It became apparent that providing e-processing capabilities alone will not sufficiently address the current state of SCM in municipalities.

## 5 Research Analysis and Findings

Eleven financial IT system vendors were invited to participate, with 10 responding. The non-responsive financial system vendor serves only 16 municipal clients. However, one of their clients did participate and the responses from this municipality were used to gauge the functionality provided by the vendor.

The research questions were modelled on the framework created by the minimum system specifications [4] and the responses provided by the respective IT system vendors indicated that significant investments have been made to ensure that the IT system functionality enables and supports these requirements. Only one vendor indicated that their core financial IT system does not include SCM-enabling functionality. This is not considered a challenge, as the vendor only serves one municipality. Four vendors reported that they provide optional SCM functionality, which should be purchased with the core financial system.

Sixteen municipalities participated in the research. The responses were compared with each other, as well as with the responses of the respective financial IT system vendors. The comparison identified many discrepancies in what vendors' state as being available, versus municipalities' perception of what IT systems can provide. It became evident that municipal employees need SCM business process and system-specific training to understand the functionality, which is provided to them. The municipal responses contained many blank fields and it appeared that "not available" was used as a default response. Only two municipalities indicated that their respective vendors do not provide SCM functionality. One of these municipalities uses SCM functionality from a third-party party vendor and provided comprehensive feedback regarding this functionality, while the other did not indicate anything.

Functionality for costing and pricing analysis of demand can only be provided by three vendors, collectively serving 117 municipalities. The remaining IT system vendors indicated that the functionality could be provided as custom developments. Eight municipalities indicated that their IT system vendors do not provide functionality for the planning of SCM and procurement activities. Another five selected "not available" to all the questions, however, all the system vendors claim to provide SCM planning functionality.

All of the system vendors indicated that they provide the functionality, which enables preferential procurement. Nine municipalities consistently responded to this group of questions that the required IT functionality is "not available". Two municipalities indicated that preferential procurement functionality is a "feature of the core financial system", and another two municipalities, that it required "custom development". Academic research conducted by Barahona and Elizondo [13], Mkele [14] and Duma [15] revealed that procurement at municipalities is not pro-actively utilized to address the goals of a developmental state, or to support preferential procurement.

System vendors stated that IT functionality for SCM e-processing, with a few exceptions, is readily available. In comparison, nine municipalities responded that the functionality is "not available", and one municipality left all the questions blank. These responses lead to the conclusion that the municipalities do not know the extent of the available IT system functionality, do not use the available functionality, or need in depth training regarding SCM processes and the available e-processing capabilities.

System vendors answered the questions relating to logistics, materials, inventory, and disposal management, that the required functionality is available, except for two of the vendors, representing less than 10 municipalities. The municipalities confirmed this position, but several do not use it. The reason for this was not investigated.

All the IT system vendors provide functionality to manage vendors and contracts, and for integration with the CSD. Seven municipalities confirmed that information, and five stated that the functionality for vendor and contract management is either "not available" or "not being used".

Data extraction and manual analysis thereof are available to all municipalities, although the skills to perform the analysis and interpret the findings may not be readily available at municipalities. However, automation of the analysis, strategic reporting and dashboards to display the results, can only be made available through custom developments. Only three system vendors, serving 71 municipalities in total, provide all of the functionality mentioned in these questions. The municipal responses correlate with the answers provided by the respective IT system vendors. The value of strategic reporting and monitoring dashboards cannot be emphasized enough when addressing oversight and adverse audit findings, and providing this functionality should be prioritized.

Municipalities were required to publish tenders and procurement information on the e-Tender Portal since 1 July 2015 [6] and integrate with the CSD from 1 April 2016 [5]. Only one system vendor provides the functionality to enable these requirements. The other vendors can provide the functionality through custom developments, third-party vendors, or not at all. Nine municipalities reported that integration with the CSD is not available, seven indicated that the principles of OCDS are not applied, and in most cases, the e-tender portal is not utilized. In lieu of regulatory compliance, all municipalities should have access to, and use this functionality.

The last two groups of questions addressed legislative, regulatory and policy compliance, as well as end-to-end procurement processing, and can be accommodated by all of the vendors, with the exception of a few items here and there. The AGSA is mandated to review and report on the legislative compliance of auditees, and the annual report confirms these findings [1].

When analyzing the responses from the financial IT system vendors regarding the SCM enabling functionality that they provide, it becomes clear from their responses that the functionality required for e-processing of SCM transactions is readily available. However, the responses from the participating municipalities do not correlate with these findings. It seems that municipalities either do not know what their respective vendors offer or are not able to use the functionality as it is intended. The result of this is continued adverse audit findings and regression in audit outcomes [1].

## 6 Implementation Models

The research findings informed the development of implementation models for consideration towards implementing e-processing of SCM transactions at the municipal level. The models are compared in Table 2, briefly indicating the advantages, impediments, complexity and risks associated with each.

## 7 Recommendation—A System-Based Implementation Model for E-processing of SCM in Municipalities

The recommended model for the implementation of e-processing of SCM transactions is a system-based model, which can be summarized as follows:

- Circulars and practise notes should be regulated. Regulatory compliance is reviewed and reported by the AGSA, with consequences for non-compliance;
- The minimum IT system specifications, supporting business processes and operating procedures to enable e-processing of SCM transactions, should be regulated;
- Extensive upskilling of municipal employees is required;
- A phased approach is recommended for implementation, based on municipal capacity, financial position and contribution to overall municipal spend;
- Cooperation with IT system vendors, line departments and provincial treasuries will improve the success rate of the respective municipal implementations; and
- Local economic development agencies, NGOs and education/training organisations should participate to maximise the intended benefits to the community.

## 8 Conclusion

The purpose of this research paper was to present a feasibility study based on primary data collection and a literature review, which informs the development of an implementation model for e-processing of SCM transactions in South African municipalities. Five implementation models were proposed, of which not one single model was considered suitable by itself, resulting in a system-based hybrid model being recommended.

**Table 2** Feasibility of the proposed models

| Model | Advantages | Impediments | Risks and Complexity |
|---|---|---|---|
| **Regulatory model**—regulate the minimum system specifications and articulate, enable and enforce municipal compliance | Familiar approach;<br>CSD, OCDS and *m*SCOA already in place;<br>High-level specifications available;<br>Regulation of processes already communicated;<br>Structures exist for this approach | Costly;<br>Time to regulate;<br>Change fatigue; No enforcement and consequence management;<br>On-going municipal decline | E-processing of SCM requires a structured and guiding regulatory framework; Regulations follow an extended process before publication, phased implementation and limited enforcement thereafter |
| **People-focused model**—intensive focus on training and capacity building; SCM value chain, business processes, IT systems and functionality | Improved skills and ability to manage SCM processes;<br>Direct and indirect participation in developmental initiatives; Motivation to get involved;<br>Improved service delivery | Key positions remaining vacant;<br>High staff turnover;<br>Demotivated staff;<br>Widespread corruption and mismanagement;<br>Lack of consequence management; Selective participation in reforms and projects;<br>Limited opportunities for growth | Varied skills levels across the municipalities;<br>Limited knowledge and understanding of enabling technology;<br>People intensive, time consuming and costly to reach everyone. The impact of unionisation should be considered for this alternative |
| **Centralized processing model**—shared service for processing of SCM transactions | Reduced cost due to shared infrastructure and overheads;<br>No maintenance and support required at local level; Sufficient infrastructure centrally available | Reducing economic opportunities from where it is most needed; Anti-competitive; Not supported by legislation; Need affordable and reliable power and communication technology | Few working examples of successful shared service<br>Deficient infrastructure, administrative arrangements and political mandates increase the complexity |
| **Vendor-driven model**—implementation and use of the SCM processing functionality provided | Keen to maintain market share;<br>Access to the required skills;<br>One development to benefit multiple clients;<br>Replicate lessons learnt | Costly, not affordable for all municipalities;<br>Can only implement at a number of clients at a time; Limited skills transfer to protect IP;<br>Limited integration across systems | Municipalities locked into extended contracts due to high exit barriers; Resist reforms that are externally driven; Do not necessarily select the best available vendors/functionality |
| **System-based model**—comprehensive approach from a system perspective: people, processes and technology | Comprehensive approach to entrench the initiative; Underpinned by regulatory framework and transfer of skills; Supported by technology; Small successes regularly;<br>Gradual improvements | Extended time to roll out to all municipalities;<br>Costly—not affordable by all municipalities;<br>People intensive to succeed; Requires management support and external enforcement to succeed | Reduced effort after the first initial success, resulting in partial implementation only<br>This proposal is costly, people-intensive and requires an extended period of time to implement |

*Source* Researcher

The municipal sample was proposed by the participating IT system vendors, resulting in the municipal responses not sufficiently representing the current municipal environment. However, this would not necessarily change the recommendations. The responses from IT system vendors were positive and helpful. It would be beneficial to involve the IT system vendors in this reform, possibly even to allow them to drive the implementation at their respective municipal clients.

Recommended areas for further study include an assessment to determine how e-processing of SCM transactions can be organized in districts or provinces, to accommodate smaller and financially constrained municipalities, and how the impact of digitization, particularly in the rural and economically challenged areas of the country, can be embraced to stimulate new opportunities for economic development.

## References

1. Auditor General of South Africa (2019) General report on the local government audit outcomes 2018/19. Auditor General of South Africa, Pretoria
2. South Africa (2003) Municipal Finance Management Act, Act 56 of 2003. National Treasury, Pretoria
3. South Africa (2005) Municipal supply chain management regulations. National Treasury, Pretoria
4. South Africa (2016) MFMA circular 80: municipal finance systems and processes requirements in support of mSCOA. National Treasury, Pretoria
5. South Africa (2016) MFMA circular 81: Web based central supplier database. National Treasury, Pretoria
6. South Africa (2016) MFMA circular 83: eTender portal. National Treasury, Pretoria
7. Hayman G, Frauscher K (2020) Open contracting. https://www.standard.open-contracting.org. Accessed 17 Oct 2020
8. Schatteman A (2010) The state of Ontario's municipal performance reports: a crytical analysis. Can Public Adm 53(4)
9. Kramer G (2016) Electronic public processing as a tool of reform in South African public procurement. Afr Public Procure Law J 3(1)
10. Ambe I (2016) Insight into supply chain management in a municipal context. Public Munic Financ 5(2)
11. Bizana N, Naude M, Ambe I (2015) Supply chain management as a contributing factor to local government service delivery in South Africa. J Contemp Manag 12
12. Habtemichael F, Cloete F (2009) The role of ICT's in reducing administrative corruption in the South African public sector. Adm Publica 17(4)
13. Barahona J, Elizondo A (2012) The disruptive innovation theory applied to national implementations of E-processing. Electron J E-Gov 10(2)
14. Mkele M (2019) The challenges facing small, medium and microsize enterprise in public procurement: a case of Sundays river valley municipality. Nelson Mandela University, Port Elizabeth
15. Duma P (2018) The adoption of the E-procurement system by the Ethekwini municipality for SMME owners: the views of small medium and micro enterprise (SMME) owners. University of Kwazulu Natal, Durban

# Video Recording and Archiving of Surgical Operations

**Yaroslav Yurchyshyn and Vasyl Yurchyshyn**

**Abstract** The development of medical informatics, the development of approaches and methods of digital medicine, the basic existing medical information systems, and information technologies at rendering of medical services to patients are considered. Modern possibilities for information technology to support the storage and processing of digital medical information are presented. The possibility of using and applying changes in the registered information at all stages of the lifecycle is emphasized, which is acceptable in business projects but unacceptable in quality control systems for the provision of medical services by health professionals to the average patient. It is noted that it is impossible to use unquestionable control functions when using the existing technologies due to the possibility of editing medical information at all stages of the lifecycle. The necessity of creating a new fundamentally different system of medical information registration by video recording of a medical operation with a complete ban on software editing and making changes to the registered information at all stages of the lifecycle is shown.

**Keywords** Information medical technologies · Long-term storage of digital medical images · Digital medicine · Image transmission · Telemedicine networks · Video recording of surgery · Medical video recording · Storage of medical video recordings · Editing medical video · Video recording and archiving of surgical operations

## 1 Introduction

Currently, the health care system is intensively implementing electronic technologies including the widespread use of telemedicine. Information and data circulating in medical networks require new approaches to their processing and analysis; medical images must be not only transmitted but also stored, simultaneously processed and analyzed. Ensuring large-scale flows of transmitted information implies a global

Y. Yurchyshyn (✉) · V. Yurchyshyn
Department of Computer Systems Software, National Technical University of Ukraine "Igor Sikorsky Kyiv Polytechnic Institute", Kyiv 03056, Ukraine

 
X.-S. Yang et al. (eds.), *Proceedings of Sixth International Congress on Information and Communication Technology*, Lecture Notes in Networks and Systems 216,
https://doi.org/10.1007/978-981-16-1781-2_46

rethinking and updating of the existing medical systems. The work of telemedicine systems is currently aimed at providing services to health professionals in the course of their professional activities or providing services to patients to meet their needs in the treatment or diagnosis of disease. This work is supported by a number of communication standards. The latest approaches to visualization are also becoming widespread: virtual endoscopy, multiplanar reconstruction of organs, 3D-virtual energy echoangiography, X-ray osteodensitometry, radiothermography, electroimpedance tomography, digital a posteriori algorithmic processing, etc. In this regard, the world has done a great job. We have no technical comments on these decisions. There are, however, comments on their functioning; they allow changes in the registered medical information at all stages of the lifecycle.

We consider this work as debatable.

The purpose of the article is to draw the attention of the scientific community to the need to create an electronic system to support any surgical operations under video surveillance and the mandatory registration of information from the video surveillance system in an electronic database with a complete prohibition of any change in the recorded information from the video system at both recording and storage stages.

The paper briefly considers the current state of digital medicine, algorithms for creating and saving medical videos and images. The work also emphasizes the need to create the above-mentioned electronic system in the form of a Unified State Electronic Database of Surgical Operations and Childbirth.

## 2 Information Technologies of Digital Medicine

The field of information technology implementation in healthcare has become large and continues to grow. Many well-known organizations are involved in the projects of creating a system for processing and storing medical images. Thousands of medical facilities use Brainlab's surgical technology. One of the top priorities in digital surgery systems is to optimize and consolidate the data flow. Brainlab Digital O.R. solutions support many configurations to ensure smooth execution of workflows and provide modern approaches to documenting medical operations [1]. During the operation, the surgeon can record video or take screenshots from any connected video sources (endoscope, microscope, camera on a lighting lamp) as well as from applications running in the Buzz system. Progress tracking information is automatically entered into the patient record in the HIS system. Data can be accessed immediately after surgery. After surgery, the documentation is automatically or manually transferred to the selected image archive or storage location (PACS, HIS, VNA, USB storage, Quentry cloud service). It can then be accessed from any client computer through the hospital's HIS system for viewing or consultation. Surgical records can be stored on a temporary scalable repository server before *editing* in Brainlab Video Editor. This ensures that only the most important data will be stored in the patient's card in the HIS archive or available for download in MP4 format. Brainlab solutions integrate data from HIS, PACS and VNA systems into the operating system using a simple data-processing node. Brainlab platforms for visual-controlled surgery and intraoperative imaging as well as robotic support technologies increase the efficiency of surgical procedures and allow surgeons to achieve maximum efficiency in the provision of medical services.

Various scalable technologies are designed to solve the problems of visualization, navigation, data acquisition and transmission. These works are standardized [2]. Medical images are represented by huge amounts of data; one image can take from a few megabytes to a gigabyte or more. Legal issues related to the archiving of medical data are resolved differently in different countries, but there is a general European trend—to ensure long-term storage for at least 20 years.

Today, distributed medical imaging systems based on cloud solutions have been created. Such systems include a DICOM format parser, which allows reading information from DICOM fields for further processing; data storage for files; database server; authorization module; patient depersonalization module; doctor's web interface; and administrator's interface. Data from medical equipment are stored in DICOM format [2]. A unique random identifier and QR code of the analysis are formed, which are provided to patients for access to their own analyses in the future. Data are processed, and images and meta-information are separated. Images are stored in the repository as files, and the meta-information enters the database. The DICOM server forwards the received data to the parser, which distinguishes data for entering into a database. The data then pass through the depersonalization module. The module filters DICOM fields and deletes personal information. Such fields are not stored in the database. In addition, the depersonalization module removes fields with personal data from the structure of the DICOM file. After saving, the new file is added to the research repository.

A medical DICOM image, according to the standard, is an instance of a class called a service object pair (SOP). An SOP, in turn, contains data blocks:

1. patient—metadata that identifies a particular patient
2. research—metadata that identifies the doctor, the time of the study, its parameters
3. series—metadata that identifies the modality, equipment, manufacturer, organization
4. image—metadata that is a digital description of the content of the image: type (can be a picture or frames of a video recording), storage, pixels, placement, playback, frames, configuration, photometric interpretation and more.

Thus, DICOM files contain both the image itself and additional information about the patient. Patient information and research cannot be separated from the image itself.

A typical solution for storing and archiving digital medical images based on PACS systems allows storing medical images, controlling access to them, visualizing them, and editing metadata. This requires mandatory consideration of all stages of the lifecycle. At the same time, the main condition for using the system to solve medical problems is observed—ensuring the confidentiality of medical information when it leaves the medical institution and protection of personal data.

The process of transferring medical images to the repository is as follows:

1. ordering of patient metadata—image sources in accordance with the card index or database (database) of patients
2. depersonalization of images intended for transfer to the repository—a procedure that includes:
   a. the formation of a substitute identifier (ID) of the patient in the metadata
   b. storing patient-identifying metadata and a substitute patient ID in a depersonalization database
   c. deletion in the medical image of metadata that identifies the patient
3. uploading depersonalized images to the appropriate directory.

## 3 With What We Do not Agree and What is Offered

We have deliberately described the Brainlab Digital O.R. and existing solutions to draw attention to the fact that the received information can be modified after its receipt and during storage. We draw special attention to this. This is what we are not satisfied with in the existing solutions. All these developments are aimed at providing services to health professionals in the course of their professional activities or to patients to meet their needs in terms of treatment or diagnosis of the disease. In Ukraine, many institutions have also created a strong enough base to perform such work [3]. But it should be noted that the above reflects only a small share of work performed in the field of electronic medicine. Their list can be continually extended.

What we offer is not related to these works; rather, it is related but concerns a completely different issue. The above work is the provision or improvement of medical services to both healthcare professionals and patients. These are business projects. They are well developed and fairly complete. These solutions are extremely relevant, but they allow changes in the registered medical information at all stages of its lifecycle.

We offer something else. We offer a control system—control over the provision of medical services by medical workers to the population during surgical operations.

What is it? Any surgical operation, any penetration into the internal organs of a person in a medical institution, regardless of ownership, must be carried out under video surveillance with subsequent mandatory documentation in an electronic database, access to which should be equated to access to state electronic databases with limited access. This also includes the birth of a child.

## 4 Discussion

The electronic database of surgical operations should be an automated system for collecting, registering, storing and protecting information and data obtained during surgical operations and childbirth in all medical institutions of the country, regardless of their subordination and ownership. Such an electronic database must be state-owned and has the status of a Unified State Electronic Database. The Unified State Electronic Database of Surgical Operations and Childbirth should be owned by the state and managed by the Ministry of Health, and the technical administrator should be a state-owned enterprise designated by the government. The Unified State Electronic Database of Surgical Operations must be used when performing any surgical operation on any person and when giving birth throughout the country or when performing any intervention or penetration into the internal organs of a person.

It should be noted first that this is a control system; all surgical interventions (surgical operations) in the country should be carried out under video surveillance with the subsequent obligatory documentation in the electronic database in which nobody has the right to edit and enter any changes to the information that is recorded. The Unified State Electronic Database of Surgical Operations and Childbirth should be an integrated information and telecommunication system. The technical means of such a system must consist of a set of automated workstations, integrated into a single information system by means of communication using remote access technology, and be connected to public communication networks with the delimitation of access rights, providing protection against violations of information integrity and ensuring different levels of availability of open information and information with limited access, the protection of which is established by law. All information that will be contained in the Unified State Electronic Database of Surgical Operations and Childbirth, except for personal data and information with limited access, must be available in the format of open data. Everyone must have full access to all information about themselves entered in the Unified State Electronic Database of Surgical Operations and Childbirth. The issue of the rights and forms of access of patients or healthcare professionals to the data in which they personally appear should be discussed separately when creating the system. What is fundamental is the registration system, and no one ever, under any circumstances, has the right to change the visual information entered into the database by the video surveillance system.

The different stages in the lifecycle when recording the progress of a medical operation should also be indicated. With existing diagnostic information technologies, the lifecycle describes the process of creating, sequentially transforming and subsequently storing a medical image. In our case, the lifecycle includes the process of creating an image, recording this image and long-term (decades) storage with a complete ban on editing at all stages.

The fundamental difference between the existing technologies of documenting medical images and the proposed one is a complete ban on editing information from the video system at all stages of its lifecycle. Software must be created that will make it impossible to make any changes, at any time and by any person, to the information received from the medical support video system from its beginning to its end.

This should be a very complex, costly and strictly standardized state system of quality control of medical services provided to the population. At the stage of formation, it will be accompanied by considerable opposition from state and non-state bodies, including the medical staff of medical institutions. Its functioning should be regulated by more than one state standard and rely on appropriate legislative state support.

As already mentioned, there are Grid and cloud structures that specialize in providing medical services to both the public and medical staff, but all these are either business projects or services provided to medical staff, providing services to the driver of the car rather than monitoring the actions while driving. One can only imagine the voluntary nature of drivers to perceive such control.

Is it worth doing? This is extremely necessary despite all resistance, obstacles and hindrances. There are many examples of this. It is known that foreign objects are left inside the bodies of many women during childbirth, such as gauze and cotton swabs. For some, it costs their life; others managed to survive. The wife of our friend had a cotton swab left inside her during childbirth. Then, for more than a week, she fought for her life. The body itself pushed out the unfortunate swab, and it saved her life. If this had not happened, she would not have survived, because the doctors were looking for anything anywhere but not for a swab left behind. In another example, a student studying at the Sikorsky Kyiv Polytechnic Institute had scissors sewn into his abdomen during appendicitis surgery.

These patients survived. How many did not? And in what circumstances? And what is the reason for this? At the same time, there are accusations, both grounded and baseless, from both guilty and innocent people.

Is it too early or too late to do this now? The time is just right. The world is more technically and methodologically ready for this than ever. It is a difficult job, but there is an experience in creating and operating such systems. In the future, all countries of the world will go through this. It is only a matter of time. The country that does this first will have significant advantages. We are convinced that, in the future, state electronic databases of surgical operations will be created in many countries of the world, after which a global electronic database will be created, which will be regulated by relevant international standards.

As already mentioned, the opposition to the implementation of such a system will be huge, because the greatest benefit in its creation will be to ordinary citizens, not medical workers. For the latter, it will be a remarkable "headache."

Access to the information to be contained in the Unified State Electronic Database of Surgical Operations and Childbirth should be through the official website of the central executive body in the medical field. One of the features of this system should be the inability to edit the recorded information in the process of observation, transmission and registration and in the process of its storage in any place (both temporary and long-term storage).

There are many developments in the world that can be used to implement this project, and the world is ready to create such a project and has the necessary experience. Existing developments can be used (Zoom; databases, for example, Mongo DB; global networks; police body cams). It is only a matter of financial costs, and these are not so much to create a system as to maintain it. All this will be expensive and will require the utmost interest of each person in the implementation of this project.

It is easy to calculate what kind of system this will be. There are now about 1,700 hospitals in Ukraine. This is somewhere around 0.63 per 10 thousand people; in the EU, it is 0.31. For Ukraine, the file storage should be hundreds of TB of data. Shelf life is at least 20 years.

We must look for other approaches to the practical implementation of such a system, which can only be done by attracting a wide range of concerned enthusiasts for whom finances are not an issue. Pilot projects based on individual hospitals will not yield the desired results and may delay the implementation of such a system indefinitely. This may be something like trying to involve car thieves in voluntarily sponsoring the creation of a database of stolen cars. We must start by searching for cheaper implementation of the system as a whole with something like WhatsApp, Viber or Telegram.

It is also important to involve concerned government officials in addressing this issue to provide state support for such work. Enthusiasm alone will not be enough.

## 5 Conclusion

It is necessary to create an electronic system of video recordings of surgical operations and childbirth with software providing a complete, absolute ban on making changes to the video information recorded, both at the stage of monitoring the progress of surgery or childbirth and in the process of transmission, registration and storage (temporary and long-term). Such an electronic database must be state-owned and has the status of a Unified State Electronic Database of Surgical Operations and Childbirth. It must be a restricted access system. Its functioning should be regulated by state regulations and rely on appropriate legislation.

It is necessary to point out the important value of such a system not only in the presence of special control functions; it will also be a unique archive and scientific and practical material for the training of medical staff and the development of medical science in general. Without exaggeration, such an archive would be a great practical asset for the global medical community.

## References

1. Sinsky C et al (2016) Allocation of physician time in ambulatory practice: a time and motion study in 4 specialties. Ann Inern Med 165(11):753–760
2. ISO 12052:2017(en): Health informatics - digital imaging and communication in medicine (DICOM) including workflow and data management. https://www.iso.org/obp/ui/#iso:std:iso:12052:ed-2:v1:en
3. Kovalenko AS, Kozak LM, Romanyuk OA (2017) Information technology of digital medicine. Cybern Comput 1(187):67–80. Kyiv

# Design of Capital Distribution and Settlement Platform of Citizen Card Among Cities Based on Unified Standard

**Ren Ying, Lina Wang, Du Jing, and Yanhong Zhang**

**Abstract** Build a bridge among cities by combining with the construction of citizen card projects, so as to make between cities and heterogeneous system of citizen card consumption funds distribution and settlement become to reality. Cross-regional consumption can be distributed and settled in time, greatly enriched the scope of citizen cards, reduced the consumption processing and fund transfer in transit, and promoted the communication and exchange between city areas. Analyzing the process of distribution and settlement on consumption funds of citizen cards between cities focuses on the analysis and design, suggests the realization by using WebService and message middleware. It proposes a set of solutions about citizen card consumption funds' distribution and settlement between cities. In the provinces, cities, districts, and counties of the unified key management system, it achieves IC card payment, which crosses regionals and cities, achieves the maximum compatibility for the application differences in data format and other aspects in each region. The system differences between regions will be well avoided, making the crossing cities' application of cards transparent. The platform of funds' distribution and settlement between cities not only promotes the communication and exchange between cities but also has certain social and economic benefits.

**Keywords** Between cities · Citizen card · IC card · Funds distribution and settlement

## 1 Introduction

With the continuous deepening of the application level of IC cards in various cities and the vigorous development of various industry application systems, especially the gradual improvement and implementation of the application plan of IC cards in various cities and cities, the scope and field of the application of IC cards in

R. Ying (✉) · L. Wang · D. Jing · Y. Zhang
Naval Aviation University Aviation Foundation, Yantai 264000, China
e-mail: ren.ying@163.com



X.-S. Yang et al. (eds.), *Proceedings of Sixth International Congress on Information and Communication Technology*, Lecture Notes in Networks and Systems 216,
https://doi.org/10.1007/978-981-16-1781-2_47

construction undertakings have been continuously expanded. “One card multi-use” and “interconnection” will become an important development trend [1].

Intercity IC card application interconnection can effectively integrate existing resources, convenient for citizens to travel, consumption, and in the elimination of counterfeit money, reduce the tedious process of zero-fixing will also have a considerable role. The application of the inter-city card clearing and settlement platform (hereinafter referred to as the platform) will build a bridge between the existing one-card cities, making the consumption of cross-regional and heterogeneous systems of citizen cards a reality [2], cross-regional consumption can be cleared in time, greatly enrich the scope of use of citizen cards, reduce consumption processing and fund transfer in transit, and promote communication and exchange between metropolitan areas.

Within the province, city, district, and county of the unified key management system, the small payment of IC card is realized across regions and cities, and the functions of data combing, transaction flow clearing and clearing are provided for IC card application. The platform will be responsible for the access to the various urban areas, cross-city data collection, sorting, and other processing behavior, as well as to provide relevant banks with an effective basis for accounting, reconciliation, so as to achieve real connectivity [3]. The value of this core function is that the platform will achieve maximum compatibility for possible application differences in data formats between cities, and between cities, and the system differences between cities will be well avoided to make the cross-city application of cards transparent [4].

## 2 Overall Architecture Design

### *2.1 Network Architecture*

The functional core of the platform consists of the database server and an application server. The database realizes high availability in the form of dual computer backup and provides data access and storage services for the platform. The application server uses the form of two server clusters to balance the dynamic access pressure and ensure a stable access service supply.

The application server uses the form of two server clusters to balance the dynamic access pressure and ensure the stable access service supply. Encryption machines and key management realize the management function of the platform-related key. The digital authentication server carries on the real-time digital authentication for the access city card identity [5]. The digital authentication server carries on the real-time digital authentication for the access city card identity and guarantees the data and the service security. The transaction front and application server cluster provide the platform with data exchange service, data clearing service, business statistics service, and settlement fund transfer with the bank. Each city card center and bank

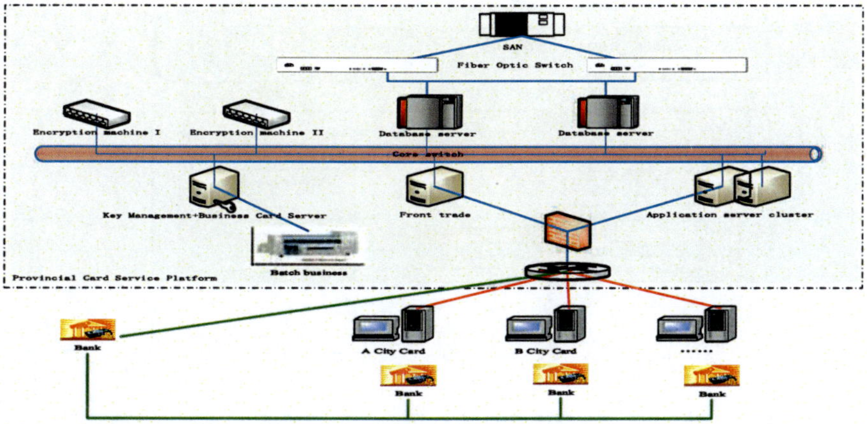


**Fig. 1** One-card fund clearing network architecture

access units use a special network and provincial card platform. The overall network structure of the platform is detailed in Fig. 1.

## *2.2 Functional Architecture*

The main functions of the platform include: digital certificate identity authentication, data exchange, transaction clearing. The data center of one-card city can exchange data and business through integrated service access and data exchange system and platform. The clearing and clearing system carries out the clearing, clearing and reconciliation of transaction data through the data exchange system, as well as the transfer and liquidation of account funds [6]. The specific functional architecture is shown in Fig. 2.

# 3 Functional Design

## *3.1 Digital Certificate System*

In the common information system, the authentication method of user name/password is generally adopted. With the development of information technology, the limitation of this authentication method is gradually manifested.

In order to ensure the effective authentication of user identity, digital certificate is issued for each user. The digital certificate system is shown in Fig. 3.

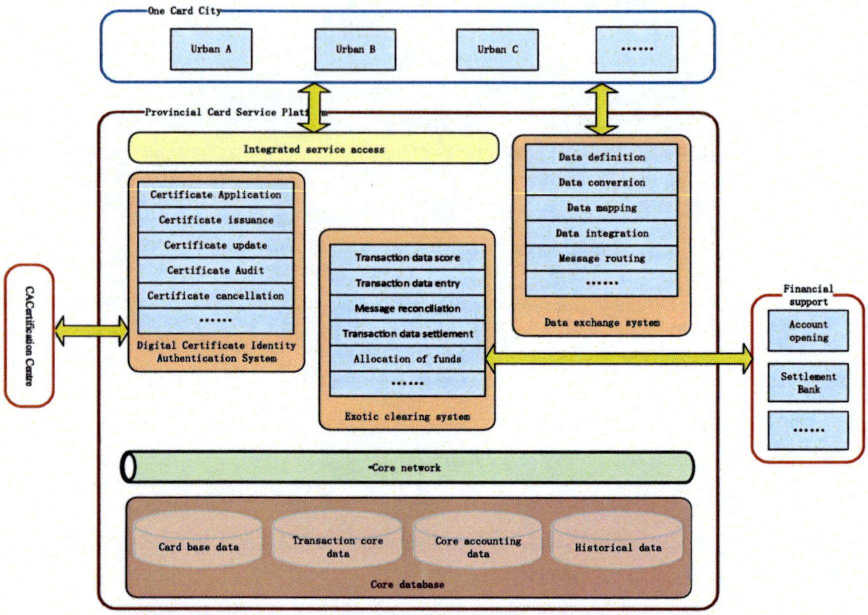


Fig. 2 Functional framework of the one-card clearing platform

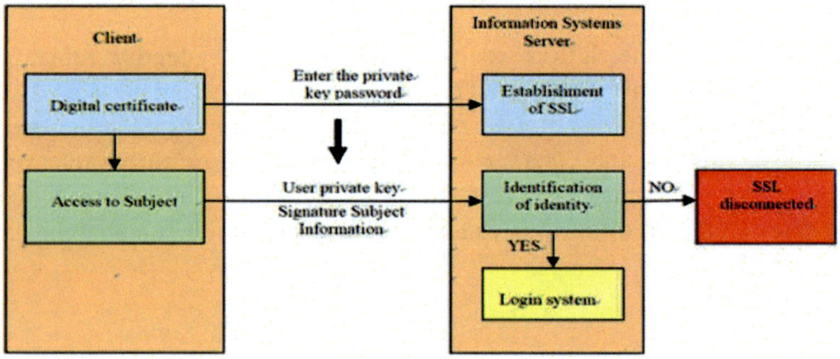


Fig. 3 Digital certificate system

The user uses the certificate to log on to the information system, the system after the identity authentication, if the user is legal then logs in to the system, otherwise does not allow to log in, the information system data encryption transmission, prevents the tampering [7].

When the user sends the data to the information system, the key is used to encrypt the transmitted information and send it to the server, which can ensure the security of the information being submitted.

Realize the identity authentication and responsibility identification of the end user and ensure the timely and accurate information submitted by the end user. When sending information, the user first signs the information electronically using his own certificate and then sends the signature value and information to the information system together. It can effectively prevent illegal users from transmitting data and can also ensure that the data submitted are checked.

For the users of each information, the authority given by the system is different. The strict identity authentication mechanism is implemented to ensure the correctness of the system information and to strictly prevent information leakage and information tampering caused by identity counterfeiting.

## 3.2 Data Exchange System

### 3.2.1 Mode of Communication

The data exchange system is the hub of data exchange between the platform and various urban institutions. The exchange of real-time and non-real-time business information between cities or institutions and platforms through data exchange system is carried out in order to complete the exchange of data between metropolitan areas [9].

On the platform, the data exchange system uses WebService forms to provide data query and data acquisition services to visitors outside the platform; for system data clearing and settlement [10], the data exchange system adopts TCP/IP communication protocol between the data exchange system and various institutions, and transfers transaction instructions and data files by short SOCKET and FTP communication according to business requirements. Some access units with software conditions can transmit messages in the form of message middleware [11].

### 3.2.2 Data Constraints

As the intermediate hub of information exchange, the data exchange module needs to exchange data with the information systems of various platforms of various relevant institutions and units. As a result, the data exchange and interface of the data exchange module follow the specifications of XML, WEB service, ensure the standard and openness of the data, and realize the seamless connection with the information system based on various operating system platforms [12].

### 3.2.3 Exchange of Content

(1) Consumer transaction data

As the clearing center of cross-city consumption, the platform uploads the data of each city card system to the platform through the data exchange system, which also plays the role of data backup. In data exchange, the platform requires the transmission of data by checking and verifying the transaction summary data and the transaction detail data to ensure the integrity and unification of the data [13]. The details are as follows (Table 1).

Details of cross-city transaction data are shown in Table 2.

(2) Accounts and funds transfer data

The liquidation results of cross-city consumption data are sent to the settlement bank and each city through the data exchange system, and the settlement bank transfers funds between the cities according to the data. Prefectural and municipal account information is shown in Table 3. Information on cross-city accounts transfers is shown in Table 4.

(3) Blacklist data

The blacklist generated by each city is summarized, processed, and distributed through the data exchange system. The blacklist adopts the dynamic updating mechanism to ensure the timely updating of the effective blacklist. Blacklist information is shown in Table 5.

**Table 1** Transaction summary data between cities

| Serial number | Data item name | Type | Length | Dictionary code | Constraints | Reference standards |
|---|---|---|---|---|---|---|
| 1 | Transaction current number | N | 15 | JYLSH | NOT NULL, UNIQUE | |
| 2 | Place of transaction | C | 2 | JYFSDS | NOT NULL | GB/T 22 2260-2002 |
| 3 , | Cities to which the card belongs | C | 2 | KSSDS | NOT NULL | GB/T 22 2260-2002 |
| 4 | Trading time- | T | | JYSJ | NOT NULL | YYYY-MM-DD hh:mm:ss |
| 5 | Amount of transaction | N | 12, 2 | JVJE | NOT NULL | |
| 6 | Check identification | C | 2 | HDBS | | |

*Note* character data items are represented by C; date data items are represented by D; a time-based data item is represented by a T; digital data items are represented by N; binary type data items are represented by B. in bytes. "m," for the length of digital data items n "or" "m" means, n of which represents decimals, m minus n is an integer partial digit. The following is the same

**Table 2** Transaction details data between cities

| Serial number | Data item name | Type | Length | Dictionary code | Constraints | Reference standards |
|---|---|---|---|---|---|---|
| 1 | Citizen card | N | 15 | SMKH | NOT NULL, UNIQUE | |
| 2 | Cities to which the card belongs | C | 2 | KSSDS | NOT NULL | GB/T 22 2260-2002 |
| 3 | Consumption in cities | C | 2 | XFFSDS | NOT NULL | GB/T 22 2260-2002 |
| 4 | Terminal identification number | C | 15 | ZDJSBH | NOT NULL | |
| 5 | Consumer network name | C | 50 | XFWDMC | NOT NULL | |

**Table 3** Account information of cities

| Serial number | Data item name | Type | Length | Dictionary code | Constraints | Reference standards |
|---|---|---|---|---|---|---|
| 1 | Account opening account | C | 15 | KHZH | NOT NULL, UNIQUE HIE | |
| 2 | Account opening bank | C | 50 | ZHKHH | NOT NULL | |
| 3 | Balance of accounts | N | 12, 2 | ZHYE | NOT NULL | |
| 4 | City of the Account | C | 2 | ZHSSDS | NOT NULL | GB/T 2260–2002 |

## *3.3 Transaction Clearing System*

### 3.3.1 Process Design

Each city card center liquidates the city's data, uploads the cross-city consumption data to the platform, and after the platform unified liquidation produces the transfer account and the reconciliation information.

The cross-city consumption data clearing system reads the consumption transaction record one by one to carry on the legitimacy check; for the transaction which passes the legitimacy check, carries on the liquidation processing according to the liquidation rule, statistics the city fund settlement account. The function is processed regularly every day and the results are sent back to card issuing and consumer cities [14].

**Table 4** Account transfer information between cities

| Serial number | Data item name | Type | Length | Dictionary code | Constraints | Reference standards |
|---|---|---|---|---|---|---|
| 1 | Date | D | | RQ | NOT NULL | |
| 2 | Underline account number | C | 15 | HCZHBH | NOT NULL | |
| 3 | Municipalities to which accounts are assigned | C | 2 | HCZHSSDS | NOT NULL<br>NOT NULL | GB/T 2260–2002 |
| 4 | Assigned to account number | C | 15 | HRZHBH | NOT NULL | |
| 5 | Municipalities assigned to accounts | C | 2, | HRZHSSDS | NOT NULL<br>NOT NULL | GB/T 2260–2002 |
| 6 | Amount allocated | N | 12,2 | HBJE | NOT NULL | |
| 7 | Transaction current number | N | 15 | JYLSH | NOT NULL | |
| 8 | Operator number | C | 15 | CZYH | NOT NULL | |

**Table 5** Blacklist information

| Serial number | Data item name | Type | Length | Dictionary code | Constraints | Reference standards |
|---|---|---|---|---|---|---|
| 1 | Citizen card | C | 15 | SMKH | NOT NULL, UNIQUE | – |
| 2 | Cities to which the card belongs | C | 2 | KSSDS | NOT NULL | GB/T 2260–2002 |
| 3 | New blacklist identification time | T | – | XZHMDSJ | NOT NULL | – |
| 4 | Delete blacklist identification time | T | – | SCHMDSJ | NOT NULL | – |

After receiving this kind of data, the card-issuing city and the consumer city shall carry out the liquidation of the city according to the liquidation rules, and then reconcile with the platform. Reconciliation information includes total transaction amount, transaction details, city code, city industry code, etc.

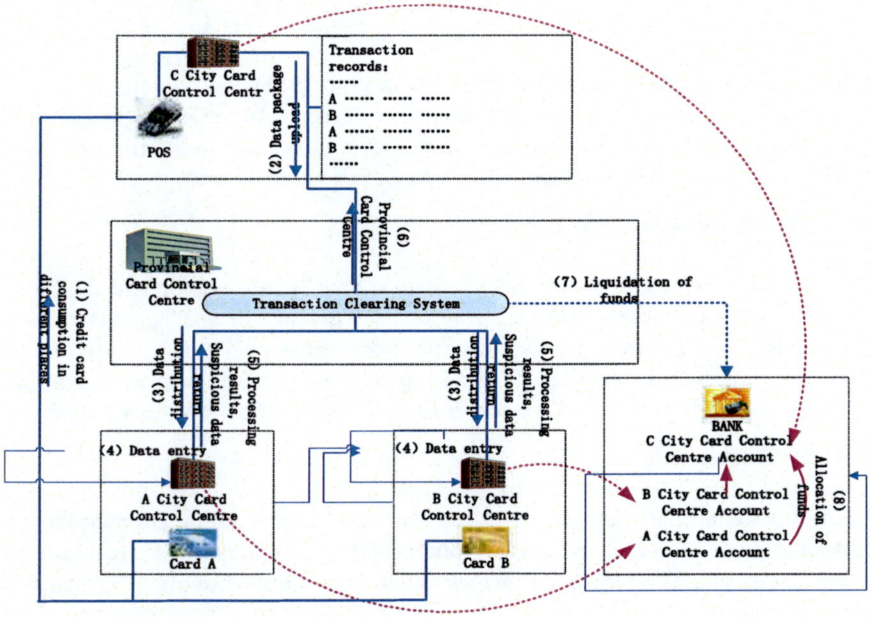


Fig. 4 One-card transaction clearing process

Reconciliation is correct, the platform will transfer data to the settlement bank, fund liquidation, and transfer. The flow diagram is shown in Fig. 4.

### 3.3.2 Functional Design

(1) Transaction data score

If there is a city card center where cards are consumed in different places, the consumption data will be uploaded to the platform. The clearing system will divide the data according to the city code, and the data will be checked for legality, including the MAC, TAC code of the data, and then classified according to the application code and recorded in the transaction database [15].

(2) Transaction data entry by the card issuer

After sorting the data, according to the city code for regional data packaging, and through the data exchange system sent to the card-issuing city card center. After the card issuer records the data, the reconciliation results are uploaded to the platform through the data exchange system. The clearing system carries on the specific transaction detail check and the status modification according to the reconciliation result.

(3) Consumer reconciliation

After all the card-issuing city card centers have processed the consumption of their own remote consumption cards, the transaction clearing and settlement system will then package all the consumption data of the batch and send it to the consumer city card center for summary reconciliation and detailed reconciliation.

(4) Transaction data settlement

Through the checking of the transaction data summary information and the transaction detail information by the clearing and settlement system, the consumption amount that the consumer city should settle at present as well as the corresponding amount that each card issuing city should pay is confirmed. The data are exchanged through the clearing system to the corresponding clearing bank for fund settlement.

(5) Settlement cycle

According to the characteristics of citizen card off-machine trading, refer to the use of trading day T 1 for data upload. The platform divides and distributes to each city, T 2 each city carries on the entry, the result data return the write-off, T 3 carries on the fund transfer settlement.

(6) Handling of anomalies and disputes

- Consumption Data Transmission

After the data correction, the consumer city relaunched the settlement process to process.

- TAC code error

After analyzing the number of wallet transactions and the continuity of the transaction amount, the issuing party makes an opinion, including the unconventional operation methods such as compulsory entry of the transaction.

- Overdue upload

Because the non-contact card is offline electronic wallet consumption, if the consumer city cannot upload the consumption data in time because of abnormal equipment or long-term data acquisition, the loss caused by the consumption will be borne by the city where the consumption is located.

- Repeat transactions

Two transactions have the same card transaction serial number or transaction equipment serial number, the repeat transaction issuer does not confirm the settlement.

- Lack of continuity between transactions

The system assumes that the transaction can be uploaded to the center within a certain period of time. If the transaction of the card lacks continuity, that is, the two recent transactions owned by the system are not continuous, and the latter transaction time exceeds the time specified by the system.

- Exception of the transaction amount

The transaction content and its amount do not conform to the system's transaction audit rules. For example, in the recharge transaction, the system stipulates that the recharge amount is greater than 0 and less than the maximum recharge amount. When the recharge value exceeds this range, it is an abnormal transaction, and the issuer of the abnormal transaction does not confirm the settlement.

- Transaction Details

After the end of the whole clearing and settlement process, the transaction clearing and settlement system can also query the transaction details and can query the transaction details for specific citizen cards, card issuing cities, consumer cities, dates, etc.

- Account Enquiry Statistics

The transaction clearing and settlement system have the function of account inquiry, which can query the current and historical transfer ledger, transaction detail information, summary information, and so on.

The contents of inquiry and statistics mainly include the accumulative processing of the accounts of the service platform, the settlement and accumulative processing of the accounts of the one-card sub-centers in each city, the settlement and accumulative business handling fee and other service charges, and the accumulative handling of the abnormal situation.

## 4 Concluding Remarks

This paper makes an in-depth study and analysis on the settlement of one-card funds between metropolitan areas and puts forward the process of clearing settlement, which lays a foundation for the consumption of one-card between metropolitan areas. The application of the platform will build a bridge between the existing one-card cities, make the consumption of citizen cards across regions and heterogeneous systems become a reality, and promote communication and communication between metropolitan areas.

## References


1. Fang Q, Xie Z, Chang Z (2013) Analysis of operation characteristics of traffic card in Guangdong Province. China Traffic Inf 3:129–130
2. Zhou Y, Yuan G, Jincheng W et al (2019) The challenge and opportunity of national transportation card interconnection to urban transportation card. Obs Reflect 01(005):14–16
3. Fang Q, Zhang J, Feng X et al (2016) Design and implementation of operation and management platform of traffic card city. Inf Technol Inf 5:53–55
4. Yang X, Hongling Y (2017) Study on the transformation and development model of transportation card enterprises. China Manag Inf 20(19):101–102
5. Crandall DJ, Backstrom L, Dan C et al (2010) Inferring social ties from geographic coincidences. Proc Natl Acad Sci USA 107(52):22436–22441
6. Blondel VD, Decuyper A, Krings G (2015) A surver of results on mobile phone datasets analysis. EPJ Data Sci 4(1):1–55
7. Zhang L, Liu H (2014) Mining of students' breakfast habits and results association rules based on data analysis of card. J Fuyang Norm Univ (Nat Sci Ed) 31(04):92–95
8. Xu Y, Shuai W (2017) The further exploration of the application of campus card in colleges and universities-- the transformation of one-card to one-call under the intelligent campus environment. J Cent China Norm Univ (Nat Sci Ed) S1:142–145
9. Haowei S, Wen X, Yang Y (2018) Application design and realization of urban public transport card based on financial IC card. Traffic Technol 01(023):66–68
10. Wang Y, Zhu J, Yanchao et al (2017) A study on the design of card system for smart city. Comput Appl Softw 34(04):93–98
11. Dou X, Guo X, Zhao K (2017) Singapore public transport service quality improvement strategy. Mod Urban Stud 03:58–63
12. Cheng Q (2016) Development of intelligent transportation card function -taking the "travel card" in Taiwan as an example. Urban Manag Sci Technol 1:76–78
13. Xie Z, Yu H, Fang Q (2014) Analysis of development model and experience of Japanese traffic card. Transp Ports 06(3):59–62
14. Liu Qiang X, Feng JH et al (2018) A study on the method of trusted recharge of urban traffic card. Urban Public Transp 10:32–36
15. Li T (2015) Campus card construction under the intelligent campus environment. J East China Norm Univ (Nat Sci Ed) S1:530–535
16. Lim JB, Hurson AR (2002) Maintaining consistency of data in mobile distributed environments. IEEE Trans Knowl Data Eng 14(6):1330–1346.

# The Security Concerns and Countermeasures Towards V2V and Autonomous Cars

**Huu Phuoc Dai Nguyen and Rajnai Zoltán**

**Abstract** Autonomous vehicles (AVs) refer to ones that are capable of perceiving the surrounding environment and directing without human assistance. With the boom of this technology, these driverless vehicles can detect the traffic, environment and pedestrians in the street, and then they make decisions about a safe and efficient way to get to the destination. Though this technology brings advantages, the security problems of such vehicles should be taken into consideration. This paper, therefore, presents an overview of several security concerns towards vehicle-to-vehicle communication and autonomous vehicles. Moreover, this paper also provides insights into some countermeasures against cyber threats for V2V and autonomous cars.

**Keywords** Autonomous vehicle · Countermeasure · Threats · Security concern · V2V security

## 1 Introduction

Traffic accidents can cause massive traffic congestion, which may lead to huge delays in the street; family sufferings and social difficulties. Autonomous vehicles mainly aim to improve road safety and give a possible solution to these problems. Some researchers believe that autonomous cars can offer independent mobility for drivers; reduce stress and tedium; and minimize congestion, accident, and pollution problems [1]. However, this new technology also presents new challenges to users. In recent times, cybersecurity for other aspects such as transportation, home alliances, smart devices, utilities, especially in the autonomous vehicle has become a serious social concern [2]. This paper provides an overview of security concerns towards V2V

H. P. D. Nguyen (✉) · R. Zoltán
Doctoral School on Safety and Security Science, Óbuda University, Budapest, Hungary
e-mail: phuoc.daitt@bgk.uni-obuda.hu

R. Zoltán
e-mail: rajnai.zoltan@bgk.uni-obuda.hu



X.-S. Yang et al. (eds.), *Proceedings of Sixth International Congress on Information and Communication Technology*, Lecture Notes in Networks and Systems 216,
https://doi.org/10.1007/978-981-16-1781-2_48

communication, and autonomous vehicles, as well as countermeasures against cyber threats for them.

This article's structure is as follows: Section 1 illustrates the introduction of the autonomous car. Section 2 expresses the literature reviews of V2V and autonomous cars. Section 3 shows cyber-attacks towards V2V. Section 4 describes possible threats and attacks for the VANET system and AVs. Section 5 indicates several countermeasures for V2V and AVs. Finally, Section 6 discusses the conclusion of this article.

## 2 Theoretical Background

### *2.1 Vehicle to Vehicle Communication (V2V)*

Vehicular ad hoc networks (VANETs) refer to a special mobile network communication for exchanging data between one vehicle and the others. VANET has two types like vehicle to vehicle (V2V) and vehicle to infrastructure (V2I) communication [3]. V2V communication uses two types of communication such as vehicle to vehicle by using dedicated short-range communication standard (DSRC) protocol [4] and other technologies [5] with the nearby vehicle, and long-distance communication. V2V technology enhances the level of safety of drivers by reducing vehicle crashes and the number of injuries each year [6].

### *2.2 Autonomous Vehicle*

An autonomous vehicle refers to a car that can determine the best route to get to the destination and automatically drive itself without the driver's control [7]. Besides, the autonomous vehicle is related to self-driving, driverless, or robotic vehicles [8]. Moreover, according to the National Highway Traffic Safety Administration (NHTSA), they classified autonomous vehicles into five categories such as no automation, function-specific automation, combined function automation, limited self-driving automation and full self-driving automation [9].

## 3 Cyber-Attacks in V2V

### *3.1 Controller Area Network (CAN)*

Bosch developed CAN in 1985 for in-vehicle networks to replace the electronic wiring systems inside the vehicles [10]. It brought several advantages like reducing cost, complexity and weight; integrating a serial bus system for smart devices' networking; and corresponding with the standard in-vehicle network. However, Hoppe et al. [11] expressed that attackers can take advantage of CAN's security holes by using special software—namely CarShark. There are several security mechanisms based on authenticating messages with Message Authentication Codes (MACs) like [12] to protect CAN protocol.

## *3.2 In-Car Wireless Network*

The tire pressure monitoring system (TPMS) was the first wireless network and was integrated by carmakers in the USA [13]. It provides road safety, enhances traction, saves fuel because of preventing tire failure by giving a warning signal to the drivers. Nonetheless, Rouf et al. [14] showed that an attacker could attack a nearby vehicle's TPMS in 40 m. Luckily, Seifert and Obermaisser [15] built the anomalies and failure detection system on the gateway to secure in-vehicle networks.

## *3.3 Electronic Control Units (ECUs)*

Modern cars are integrated with many embedded electronic components, which are interconnected, managed and executed by ECUs in the car network [5]. However, Mastakar [16] illustrated that one attacker can control an ECU to execute some control modules in the body, engine and electronic brake. In contrast, Wolf and Gendrullis [17] provided a security solution that can protect the ECUs and their communication.

## *3.4 Cooperative Adaptive Cruise Control (CACC) Vehicle Stream*

Adaptive cruise control (ACC) refers to a radar cruise control system for vehicles to keep a safe zone among vehicles by managing their speed automatically [18]. CACC is the improvement of ACC to give more preview information around the vehicles for the drivers based on the front vehicles by radar or LiDAR [19]. Besides, it seems to be a straight single-lane vehicle stream that moves during vehicle operation. Therefore, this stream can be recognized by the attackers as the application layer, network layer, system-level and privacy leakage [20–24] (Fig. 1).

| Level | Type of attack | Description and impact | Countermeasure |
|---|---|---|---|
| Application layer | Message falsification | The adversary exploit the message during transmission and manipulate beacon signals to change the acceleration field. | Using state of the art security architectures with a strong cryptographic system (digital signatures and data integrity) in order to support authentication and non- repudiation services for the messages. |
| | Spoofing attack | The attacker pretends as another vehicle and send fake information to victim's vehicle to change, delay, or drop messages to CACC. | |
| | Replay attack | The hacker captures and keeps beacon signals, then tries to replay them at another time with malicious purpose. It leads to collision. | |
| Network layer | Denial of Service(DoS) or Distributed denial of service (DDoS) | The adversary use a vehicular botnet to attack. This attack can disable hardware security module (HSM). | Using public key infrastructure for VANET communication |
| | Radio jamming | Likewise DoS attack, this attack can interrupt communication | Using several techniques like channel switching, frequency hopping, technology switching, or multiple radio transceivers |
| System level | Hardware or software | Malicious hardware or software can be done by insiders in the car manufacturers or outsiders in maintenance process. | Using tamper-roof sensors, tamper-roof hardware (tamper resistance, tamper evidence, tamper detection and tamper response) or misbehavior detection techniques solution. |
| Privacy leakage | Eavesdropping | Hackers can extract valuable information to clarify | Using encryption to achieve data privacy or anonymity techniques to get identity and location privacy |

**Fig. 1** Types of attack and its countermeasures in CACC stream

# 4 Possible Threats and Attacks

## *4.1 The Threats*

*Carry-in device*

Nowadays, the usage of carrying devices (USB, music player, and so on) in the vehicle is more popular. Virus, malware or Trojans are embedded with the applications, music and video files stored in carrying in devices. When these devices connect to the vehicle, viruses and malware can target vehicle electronic systems via their entertainment systems or terminals [25], (Fig. 2).

*Hardware and software malfunction*

Any device can occur failure even hardware or software. Particularly, the complex electronic system in the vehicle can have several failures such as operating system failure, false sensor, software error and distorted signal, which cause tragic accidents [8].

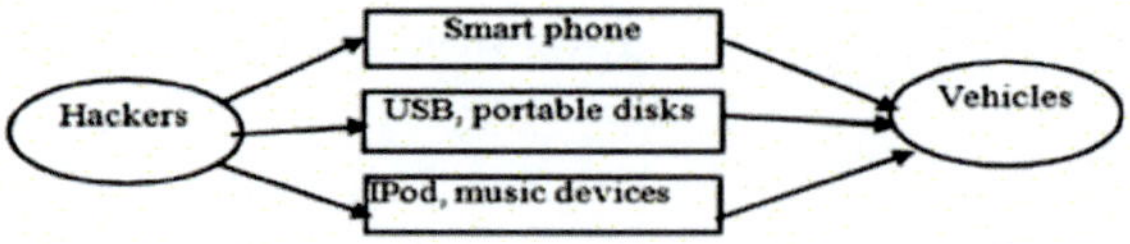


**Fig. 2** Cyber risks from carry-in device

*Malicious hacking*

In 2015, there was an illustration to hack a Jeep Cherokee remotely based on malicious software and wireless connections [26]. In addition, jailbreaking, rooting firmware or an operating system from the users may be created security holes for applications and chances for the attackers to exploit them [27].

*Sybil attack*

Douceur [28] described a Sybil attack in which one vehicle node can send many messages to the others with a different and fake source's identity. The major purpose is to make confusion to other vehicle nodes with wrong messages to inform other ones on the route to get out the street for the benefit of the attackers [29].

*Node impersonation*

When a car accident happens, the police can track the vehicle identifier as well as the driver's identity easily. An impersonation attack is that a node sends a message with wrong information to forge the originator with some other car's identity; therefore, the police can't figure out the driver's identity [29].

*Sending false information*

An attacker uses this attack for sending wrong or false information of actual situation by a node to other nodes in the street; as a result, the users will leave the road and make the road free for the attacker's purposes [23].

*ID disclosure or car tracking*

In a VANET, the car's identity and current position tracking are broadcasted in the network. Thus, it is easy for the attackers to take these data of the victim's node by sending a virus. If data are collected, it can be used by a third party with a commercial goal [29].

## 4.2 The Possible Attacks

*DoS and DDoS Attack*

Soomro et al. [30, 31] demonstrated several types of DoS and DDoS attack in VANET system relied on three main categories such as "the communication channel jamming, network overloading, and packets dropping" [5, 13, 32], as follow: basic level overwhelms the node resources, extended level/jamming the channel and DDoS.

*Hybrid DoS/DDoS*

Moreover, Rawat et al. [5, 33] classified all possible DoS and DDoS attacks in the VANET environment based on availability, authentication and confidentiality (Fig. 3).

| Type of attacks | Effect of attacks | Impact |
|---|---|---|
| Spamming | Using up the bandwidth of the network | Low |
| Black hole | Data lost, creating the first step for man in the middle attack | High |
| Malware | Reducing VANET operation | High |
| Sybil attack | Making the confusion for many vehicles to force the vehicles get off the road for attacker's goal | High |
| Node Impersonation | Modifying driver's identity and car's identity to deceive the police when accident happens | Medium |
| Message suppression | Opposing the authorities and RSU to know about the collision | Medium |
| Alteration | Shifting or altering exist data to deceive the users | Medium |
| Relay | Taking and replaying the packets to puzzle the authorities and prevent vehicle's identity in any accident | High |
| GPS spoofing | Getting the identity and geographic location of vehicles on the network to fool them. | High |
| Tunneling | Controlling a traffic analysis or preparing for forwarding attack | High |
| Timing attack | Altering the actual content time to create delay time for message | High |
| Home attack | Getting the control of user vehicle | High |
| Man in the middle attack | Conducting all communication between the sender and receiver to inject or modify message between vehicles | Medium |
| Traffic analysis | Taking the packets to analyze the important information of vehicle like ID, location | High |
| Social attack | Distracting the victim to make driver disturb | Low |
| Brute force | Breaking the cryptography key | High |
| ID disclosure | Getting vehicle's ID and monitoring the victim s' vehicle route | High |
| Bogus information | Sending false information to affect the decision of other vehicles | Medium |

**Fig. 3** Classification of DoS/DDoS attack by availability, confidentiality and authentication

*Light imaging Detection and ranging (LiDAR) attack*

LiDAR is a sensor that releases the light vibration and calculates its time reflection to identify the objects in the street in any weather condition [13]. Nevertheless, the attackers use several attacks such as [13]; as a result, it can cause difficulties for this sensor to observe the objects in the street and it may lead to distraction for the collision avoidance system.

*Camera attack*

A camera is essential digital equipment to help autonomous vehicles collect outside objects [34] by using several detection techniques; for example, "lane detection [35], horizon/vanishing point [36], object detection and tracking [37], and so on" [13]. The adversary can attack the camera with the main goal-blinding the camera incompletely or fully in order to hide the vehicle's surroundings objects by radiating the light into it [34].

# 5 Countermeasures

## 5.1 *Possible VANET Attack's Solutions*

In the VANET system, the transferring messages (traffic information) do not consist of any sensitive data. It only requires authentication without encryption and no need for confidentiality; therefore, digital signatures are the best option [20, 38–40] (Fig. 4).

| Solution | Description |
|---|---|
| Digital signatures without certificates | - Using asymmetric cryptography with a public-private key infrastructure.<br>- Providing a non-repudiation method to prevent impersonation attacks |
| Digital signature with certificates | - Supports certificates from a trusted certificate authority system to valid certificate for trusted nodes.<br>- Using the public key infrastructure to revoke the certificates from attackers, misbehaving or malfunctioning nodes, as a result, they can't use these certificates to do attack again |
| Data correlation | - Regarding the correlating of the received information, a node can identify the messages are legal or not according to the credibility, consistency, and relevance of the information |
| IEEE 1609. 2 standard | - Ensuring the safety and security for vehicular communication messages in order to fight against cyber threats and attacks like eavesdropping, masquerading, replay, and other attacks<br>- Providing revoking vehicle's certificates based on public-key cryptography |

**Fig. 4** Solutions in VANET attacks

## 5.2 *DoS/DDoS Countermeasures*

According to [30, 31, 33], they demonstrated some possible solutions to counter against DDoS and DDoS attacks such as channel switching, technology switching, frequency hopping spread spectrum and multiple radio transceivers.

## 5.3 *Camera Attack's Prevention*

There are several countermeasures to ensure the safety of the cameras from tampering, sensitivity, image quality and the like; however, they require that the cameras should be modified. First of all, a removable near-infrared-cut filter tool is used to separate the light between day and night to bring a better image but only in the day time. Another option to improve the camera's attack is by using photochromic lenses to change the color for clarifying the lights [41].

# 6 Conclusions

Autonomous cars are becoming more and more popular. Researchers proposed that such vehicles will be convenient, safe transportation means. Moreover, they can replace human beings to reduce stress, tedium, traffic jams, delays and fatal accidents. Although this technology benefits users in several ways, the security challenges of autonomous vehicles should be considered. This article has clarified several cyber-security threats in the VANET system, particularly in V2V and autonomous car communication. Moreover, the authors figured out the feasible impacts of security threats and possible countermeasures for reducing or handling them. Besides, this paper also aims to increase awareness of these security threats toward autonomous vehicles and keep the users informed of potential countermeasures.

**Acknowledgements** The research presented in this paper was carried out as part of the EFOP-3.6.2-16-2017-00016 project in the framework of the New Széchenyi Plan. The completion of this project was funded by the European Union and co-financed by the European Social Fund.

## References

1. Arbib J, Seba T (2017) Rethinking transportation 2020–2030, p 70. https://static1.squarespace.com/static/585c3439be65942f022bbf9b/t/591a2e4be6f2e1c13df930c5/1494888038959/RethinkX+Report_051517.pdf
2. Neil M, Pedersen J, Spring S (2011) Security and critical infrastructure protection progress 275
3. Kim S, Lee I (2014) A secure and efficient vehicle to vehicle communication scheme using bloom filter in VANETs 8(2):9–24
4. Watson T, Maple C (2017) Cyber security standards and issues in V2X communications 1 Ivan Ivanov, 2 Sang-Woo 9:2–7. http://www.worldresearchlibrary.org/up_proc/pdf/722-14918904992-7.pdf
5. Nguyen HPD, Zoltán R (2018) The current security challenges of vehicle communication in the future transportation system. In: SISY 2018 - IEEE 16th International Symposium on Intelligent System and Informatics, Proceedings, pp 161–165. https://doi.org/10.1109/SISY.2018.8524773
6. NHTSA (2014) What are the advantages of V2V?, pp 1–4
7. Yeomans G (2014) Autonomous vehicles - handing over control: opportunities and risks for insurance, Lloyd's, pp 1–27
8. Litman T (2014) Autonomous vehicle implementation predictions: implications for transport planning. Transp Res Board Annu Meet 2014:36–42. https://doi.org/10.1613/jair.301
9. Anderson J, Kalra N, Stanley K, Sorensen P, Samaras C, Oluwatola O (2016) Autonomous vehicle technology: a guide for policymakers
10. Corrigan S (2016) Introduction to the Controller Area Network (CAN). In: Texas instruments, August 2002, pp 1–17
11. Hoppe T, Kiltz S, Dittmann J (2011) Security threats to automotive CAN networks—practical examples and selected short-term countermeasures. Reliab Eng Syst Saf 96(1):11–25. https://doi.org/10.1016/j.ress.2010.06.026
12. Van Herrewege A, Singelee D, Verbauwhede I (2011) CANAuth - a simple, backward compatible broadcast authentication protocol for CAN bus. In: ECRYPT workshop on lightweight cryptography, November 2011, pp 299–235. http://www.uclouvain.be/crypto/ecrypt_lc11/static/post_proceedings.pdf
13. Dai NHP, Zoltán R (2012) The obstacles of autonomous cars before taking the wheel in the future 91(5)
14. Rouf I et al (2010) Security and privacy vulnerabilities of in-car wireless networks: a tire pressure monitoring system case study. Proc. USENIX Secur. Symp. 39(4):11–13. https://doi.org/10.1177/004057368303900411
15. Seifert S, Obermaisser R (2014) Secure automotive gateway - secure communication for future cars. In: Proc. - 2014 12th IEEE international conference on industrial informatics, INDIN 2014, pp 213–220. https://doi.org/10.1109/INDIN.2014.6945510
16. Mastakar G (2012) Experimental security analysis of a modern automobile. In: IEEE symposium on security and privacy, pp 1–16. https://doi.org/10.1109/SP.2010.34

17. Wolf M, Gendrullis T (2012) Design, implementation, and evaluation of a vehicular hardware security module. In: Lecture notes in computer science (including Subseries Lecture notes in artificial intelligence and Lecture notes in bioinformatics). LNCS, vol 7259, pp 302–318. https://doi.org/10.1007/978-3-642-31912-9_20
18. Adaptive cruise control system. https://en.wikipedia.org/wiki/Autonomous_cruise_control_system
19. Amoozadeh M et al (2015) Security vulnerabilities of connected vehicle streams and their impact on cooperative driving. IEEE Commun Mag 53(6):126–132. https://doi.org/10.1109/MCOM.2015.7120028
20. Rao A, Sangwan A, Kherani AA, Varghese A, Bellur B, Shorey R (2007) Secure V2V communication with certificate revocations. In: 2007 mobile networking for vehicular environments movE, pp 127–132. https://doi.org/10.1109/MOVE.2007.4300817
21. Hasbullah H, Soomro IA, Ab Manan JL (2010) Denial of service (DOS) attack and its possible solutions in VANET. World Acad Sci Eng Technol 65(5):411–415. https://doi.org/10.5281/zenodo.1085740
22. Lin X, Member S, Sun X, Ho P, Shen XS, Member S, GSIS : a secure and privacy preserving protocol for vehicular communications XX(Xx):1–14
23. Lipiński B, Mazurczyk W, Szczypiorski K, Śmietanka P (2015) Towards effective security framework for vehicular ad-hoc networks. J Adv Comput Netw 3(2):134–140. https://doi.org/10.7763/JACN.2015.V3.155
24. Garip MT, Gursoy ME, Reiher P, Gerla M (2015) Congestion attacks to autonomous cars using vehicular botnets. In: NDSS Symposium 2015, pp 1–9. https://doi.org/10.14722/sent.2015.23001
25. Onishi H (2012) Paradigm change of vehicle cyber security. In: 2012 4th International conference on cyber conflict (CYCON), pp 1–11
26. Hackers hitch cars. http://www.latimes.com/business/autos/la-fi-hy-car-hacking-20150914-story.html
27. Bécsi T, Aradi S, Gáspár P (2015) Security issues and vulnerabilities in connected car systems. In: 2015 international conference on models and technologies for intelligent transportation system MT-ITS 2015, April 2016, pp 477–482. https://doi.org/10.1109/MTITS.2015.7223297
28. Douceur JR (2002) The Sybil attack, pp 251–260. https://doi.org/10.1007/3-540-45748-8_24
29. Guette G, Bryce C (2008) Using TPMs to secure Vehicular Ad-Hoc Networks (VANETs). In: Lecture notes in computer science (including Subseries Lecture notes in artificial intelligence and Lecture notes in bioinformatics). LNCS, vol 5019, pp 106–116. https://doi.org/10.1007/978-3-540-79966-5_8
30. Halabi Hasbullah JAM, Soomro A (2010) Denial of Service (DOS) attack and its possible solution in VANET 4(5):813–817
31. Raghuwanshi V, Jain S (2015) Denial of service attack in VANET: a survey. Int J Eng Trends Technol 28(1):15–20. http://www.ijettjournal.org
32. Blum J, Eskandarian A (2004) The threat of intelligent collisions. IT Prof 6(1):24–29. https://doi.org/10.1109/MITP.2004.1265539
33. Rawat A, Sharma S, Sushil R (2014) Vanet: security attacks and its possible solutions. J Inf Oper Manag 3(1):301–304. https://doi.org/10.1007/s13398-014-0173-7.2
34. Petit J, Stottelaar B, Feiri M, Kargl F (2015) Remote attacks on automated vehicles sensors: experiments on camera and LiDAR, pp 1–13, pp. 1–13. Blackhat.com. https://www.blackhat.com/docs/eu-15/materials/eu-15-Petit-Self-Driving-And-Connected-Cars-Fooling-Sensors-And-Tracking-Drivers-wp1.pdf.
35. Bai L, Wang Y (2010) A sensor fusion framework using multiple particle filters for video-based navigation. IEEE Trans Intell Transp Syst 11(2):348–358. https://doi.org/10.1109/TITS.2010.2043431
36. Kong H, Akakin HC, Sarma SE (2013) A generalized laplacian of gaussian filter for blob detection and its applications. IEEE Trans. Cybern. 43(6):1719–1733. https://doi.org/10.1109/TSMCB.2012.2228639

37. Keller CG, Enzweiler M, Rohrbach M, Llorca DF, Schnörr C, Gavrila DM (2011) The benefits of dense stereo for pedestrian detection. IEEE Trans Intell Transp Syst 12(4):1096–1106
38. Leinmuller T, Schoch E, Maihofer C (2007) Security requirements and solution concepts in vehicular ad hoc networks. In: 2007 fourth annual conference on wireless demand network systems and services, pp 84–91. https://doi.org/10.1109/WONS.2007.340489
39. Raya M, Raya M, Hubaux J, Hubaux J (2007) Securing vehicular ad hoc networks. J Comput Secur 15:39–68. https://doi.org/10.1111/j.1523-1755.2002.kid554.x
40. Raya M, Hubaux JP (2005) Security aspects of inter-vehicle communications. Swiss Transp. Res. Conf. http://en.scientificcommons.org/22427989
41. Ziari M, Steier WH, Ranon PM, Trivedi S, Klein MB (1992) Photorefractivity in vanadium-doped ZnTe. Appl Phys Lett 60(9):1052–1054. https://doi.org/10.1063/1.106441

# Emotion Detection from Text Using LSTM

**Keerthi Reddy Alla, Nikhil Kandibanda, Priyanka Katta, Abhinav Muthavarapu, and Swarna Kuchibhotla**

**Abstract** Emotion can be expressed in several forms, which can be facial expressions, voice and text. This paper implements a methodology based on the LSTM (long-short-term memory) theory that includes principles to identify the text emotions from NLP (Natural Language Processing) domain. It proposes an approach based on LSTM, which is defined to identify feelings in text using pre-trained words from Glove Word Embeddings.

**Keywords** NLP · LSTM · Glove word embeddings · PyTorch

## 1 Introduction

The Deep Learning Architectures and NLP tasks are discussed in several scientific papers. These papers use technical terminology, and interpretation can often be difficult. In recent years, these have accomplished great success since feature engineering is not necessary. Multiple DL architectures have been proposed in Natural Language Processing (NLP) to solve several tasks [1], from speech recognition to analysis. Detecting a person's emotional state by evaluating a textual document written by one seems to be complicated, and moreover significant because text-based emotions are most often used by direct emotional language, and also it results in the understanding the meaning of concepts and the interaction of concepts mentioned in the textual document. In human–computer interaction (HCI), the identification of the statement's emotions plays a very important role. Awareness of these textual emotions plays an important role in the relationship between humans and robots. Emotions may be conveyed in a person's written word. The facial expression of the human being and the human being's vocabulary is seen as text emotions, facial emotions and voice emotions [2]. Enough work on voice emotion and Image emotion recognition is already performed, however, researchers. The text-based must always be

K. R. Alla (✉) · N. Kandibanda · P. Katta · A. Muthavarapu · S. Kuchibhotla
Koneru Lakshmaiah Education Foundation, Guntur, Andhra Pradesh, India



X.-S. Yang et al. (eds.), *Proceedings of Sixth International Congress on Information and Communication Technology*, Lecture Notes in Networks and Systems 216,
https://doi.org/10.1007/978-981-16-1781-2_49

drawn to the emotion recognition system. In machine linguistics, from an applicative point of view, the recognition of the text of human feelings is now becoming highly relevant [3]. In addition to a variety of symbols, Cowie et al. identified feelings, thought movements, speech movements and images. Although emotional detection technology can recognize individual emotional signs from various gestures, many more people have preferred to communicate their emotions in order to create their social networks through the exchange of few emotional texts on the social network. Thus the perception of emotions has become increasingly important to many textual expression tools of emotions. The purpose of this research is to establish an automatic system for the recognition of emotions in the client's everyday text [4]. Only a fraction of the thoughts exchanged by a person by email may be registered. As follows, there are only two methods of emotional expression:

1. Emotional vocabulary: words that directly refer to human emotion terms, such as happy, angry, sad, etc.
2. Evaluative items: emotional repertoire words are also not used in the presentation of emotions, i.e. expressions of feelings based on the context that do not have a clear relationship [5]. The most popular emotional models in the literature, the approaches to the identification of text emotion, the prominent statistics, and the output measures used to validate the developed works. To assess feelings, a number of sources can be used [6].

There are applications available that can respond to feelings and emotions, but rather to several environmental stimulations. A range of techniques can also be used for analyzing attitudes. Email, social talk, voice are only a couple of the major changes occurring and affecting the mental state of the person, and the machine will recognize and process them.

## 2 Methodology

Using LSTM and fully linked layers, we have created our Deep Neural Network. We created an LSTM model that takes word sequences into account as input word order. To represent words, we have used 50-dimensional pre-trained GloVe word embeddings. We then add them to an LSTM as an entry, which predicts the most fitting text emotion. The initial LSTM model comprises a single hidden LSTM layer, preceded by a layer of LSTM regular feed-forward output layer. The stacked LSTM is a model with multiple secret LSTM layers. In each layer, there are several memory cells with an extension of this model.

These are the libraries that we have used:

PyTorch is a deep learning open-source platform that is used to implement network architectures such as RNN, CNN, LSTM, etc. and other high-level algorithms. It is used by ML AI academics, companies and societies. It is used for applications including the development of natural languages. It is mainly developed by the artificial intelligence research group of Facebook, and it is based on Uber's "Pyro" probabilistic programming language software.

PyTorch provides two high-level features:

Tensor computing (like NumPy) with powerful acceleration via processing units for graphics (GPU).

Deep neural networks are built on a tap-based automatic differentiate system. Matplotlib is a Python programming language library for plotting, which is an extension of NumPy.

Pandas is an open source python package, it is used for machine learning tasks. It is built on top of another package named Numpy it supports the multidimensional arrays.

Numpy stands for numerical python where it mostly used to do scientific calculations based on the array forms.

## *2.1 Word Embeddings*

A distributed collection of words might be one of the main advancements and for the incredible success of deep learning methods in the processing of natural language challenges. Valid vector representation in a similar image, also known as word embedding, is a type of word certification with the same meaning.

Word embedding is a word group composed of methods in which individual words in a predefined vector space are represented as real-value vectors. It is necessary for the approach to use a dense, distributed representation. Each description is based on the single vector function and the vector characteristics are trained in such a way that the technique is integrated with the machine learning field. Any term, sometimes in tens or hundreds of dimensions, has a real value vector. This contrasts tens or hundreds of thousands of parameters used for fluorescent word vectors, such as one-hot encoding.

Based on the use of words, the distributed representation is learned. This enables terms they are used in similar ways to lead to specific representations, capturing their context naturally. This may be as a crisp and delicate portrayal of a word bag style in which different words, regardless of how they are used, have distinct depictions. Word vector value techniques, for a predefined fixed-sized vocabulary, learn factual feature vectors from a corpus of text. In certain activities, such as document classification [7], that language learning has been either combined mostly with neural-network model or is an unsupervised algorithm with textual statistics. The most famous are GloVe, Word2vec.

GloVe, the new global one is a model of log bi-linear regression unmentioned learning for the depictions of words that outperformed. Other term reference templates, word resemblance, recognition and operations called objects. A WordPress dump in 2010 of 1 billion tokens, a WordPress dump in 2014 of 1.6 billion tokens, Gigaword 5 of 4.3 billion tokens, the best model utilizes the 50-dimensional publicly accessible word embedding released by five companies of various sizes. Combination of 6 billion tokens from Gigaword5 and network awareness of 42 billion tokens. Which were the models in trained on the smaller companies on Gigaword5. Furthermore, word-based word embedding was found to work better actually using the GloVe [8] program that is publicly available. Gigaword5 dataset re-implementation 5 of the word2vec program to train word embedding as well.

## *2.2 LSTM*

LSTM (Long–Short-Term Memory) network is a special form of (RNN), able to learn dependencies in long term, typically only called "LSTM". LSTM is deliberately designed to avoid the issue of long-term dependence. It is basically their default behavior to remember details over long periods of time. The error they can be back-propagated through period and layers is preserved by LSTM. They enable recurring networks to continue to learn by maintaining a more constant error over many stages (over 1000), thus opening up a channel to connect causes and consequences remotely. Data are contained by LSTM in a gated cell beyond the recurrent network's regular flow. Data from a cell may be stored, written or read. The neuron makes choices about what to contain and makes it possible to, through gates that open and close, to read, write.

As follows, the layers are:

1. Tokenization: This isn't an LSTM network layer, but a compulsory move transformation from our phrases into tokens (integers).
2. Embedding Layer: The conversion of our term tokens (integers) relevant size for embedding.
3. LSTM (Long–Short-Term Memory) Layer: Unknown state and number of layers found
4. Fully Connected Layer: Mapping the LSTM layer output to the desired output size.
5. Sigmoid Activation Layer: Both output values will be translated to values in the range between 0 and 1.
6. Output: The final time stage of Sigmoid production is considered to be the final network's production.

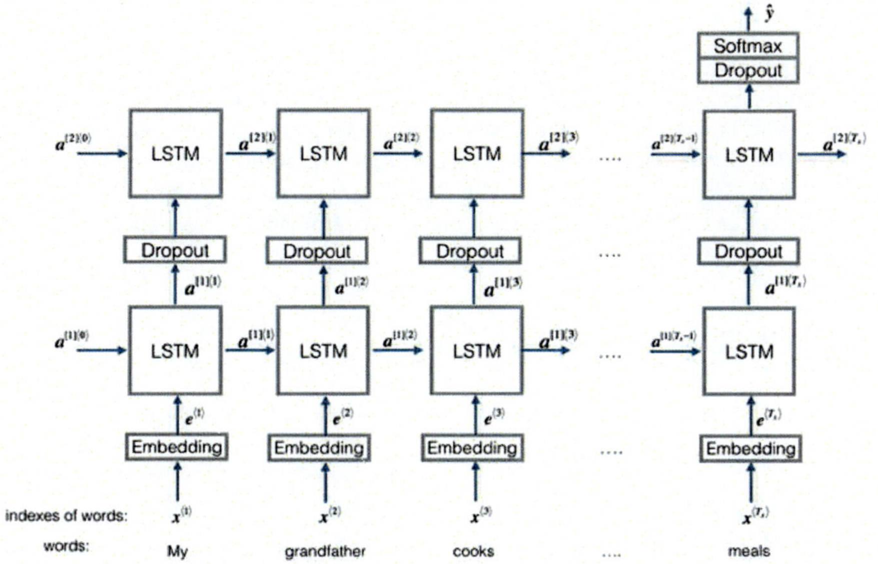


Long–short-term memory architecture

NLP is a subfield of computer science and artificial intelligence that deals with computer-to-computer interactions (natural human languages). It is used for implementing text and speech machine learning algorithms. To create a model for detecting text emotions, we have implemented an NLP mission. Word Embedding in the NLP is a policy used to map vocabulary words or phrases to a connected number vector used to locate word predictions and word comparisons.

## 2.3 *Memory Cell*

For memorizing long-term dependencies, this is a special neuron. In LSTM, the vector passing from one cell to the other in its inner state is used. LSTM is wise enough to determine how much time it takes to keep the old data in mind and to skip how to link the new information to the old memory.

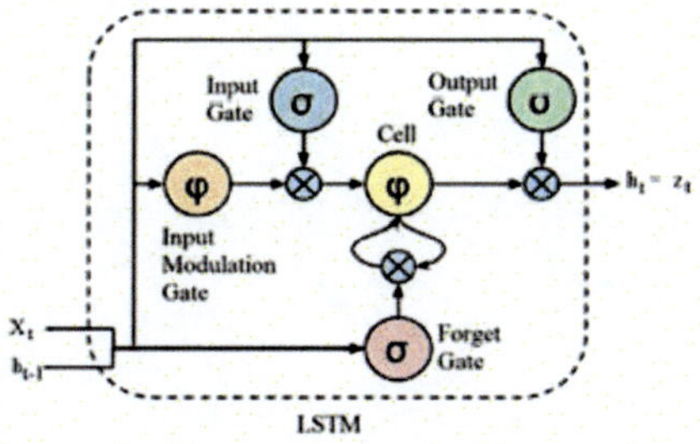

Architecture of memory cell in LSTM

We will generate an LSTM model that takes as input word sequences, which will take word order into account. We will use 50-dimensional pre-trained word integrations from GloVe in order to describe words. Then we feed the word embeddings into an LSTM as an input that predicts the most acceptable emotion for the text. The LSTM takes embedded words as inputs and hidden layers as outputs from architecture of memory cell in LSTM. The linear layer maps space from hidden state space to output.

## 3 Datasets

In this portion, we are responsible for the data used to train or validate the architecture.

Our dataset consists of two phases:

1. Training phase.
2. Testing phase.

Training phase has 132 instances whereas the testing phase has 56 instances. Each sentence contains maximum 10 words. We created our own deep neural network using GloVe (global vectors for word representation) Word Embeddings, which is a large matrix of co-occurrence information and it is constructed so that we will count each "word" (rows) and how frequently we see this word in some "columns" in a large corpus.

For the following five emotions, we will create an emotion detection:

1. Loving
2. Playful
3. Happy
4. Annoyed
5. Foodie.

We had labeled the emotions between 0 and 4 where "Loving" is labeled as [0], "Playful" is labeled as [1], "Happy" is labeled as [2], "Annoyed" is labeled as [3], "Foodie" is labeled as [4] and we append these emotions with emojis.

| Sentence | Emotion |
|---|---|
| food is life | 🍽 Foodie |
| I love you mum | ❤️ Loving |
| Stop saying bullshit | 😒 Annoyed |
| congratulations on your acceptance | 😄 Happy |
| The assignment is too long | 😒 Annoyed |
| I want to go play | ⚽ Playful |
| she did not answer my text | 😒 Annoyed |
| Your stupidity has no limit | 😒 Annoyed |
| how many points did he score | ⚽ Playful |
| my algorithm performs poorly | 😒 Annoyed |
| I got approved | 😄 Happy |

Example of dataset with sentences and emotions.

In the training phase, we erase previous gradients, make a prediction, calculate how much we missed, figure out which weights caused us to miss and change those weights accordingly. While testing phase contains labels for each sentence between [0, 4] corresponding to five emotions.

Finally, we test our model with any sentence and all the words are in GloVe Embeddings only.

The text file is named as glove.6B.50d file, which contains a statement, followed by N numbers contained in the text file row. The feature is about to be read from a guidance file in the glove file with pre-trained terms vector values. The N number is the vector of the term's location. N is 50 for this project, which relies on the vector, since we use glove.6B.50d.

## 4 Experiment Results

This study suggests an LSTM-based strategy to identify texts in the user's emotional language. The textual feature values from each of them have derived the words in the text entry document using a word2vec technique. In order to achieve the emotional feature vectors, an evaluation language is applied to the textual words in the speech space. The characteristics derived are then combined with the semantic term vector to construct a map of the textual function. Finally, a description of time dynamics of attitude is used in each input text using the LSTM algorithm [9].

We can construct a model that achieves 84.4% precision quickly. They were able to use our user input language model in our preparation package and were able to generate predicted outputs of emotions.

The following impacts are emotional predictions of user input phrases:

```
Input Text:    I hate you
Emotion:       😒 Annoyed

Input Text:    I want a pizza
Emotion:       🍽 Foodie

Input Text:    Lets see the game
Emotion:       ⚽ Playful

Input Text:    I love you Lisa
Emotion:       ❤ Loving

Input Text:    This is the best day of my life
Emotion:       😄 Happy
```

Output

Validation loss should usually be close to but slightly greater than, training loss. As long as the loss of validation is less than or even equal to the loss of training, one should continue doing more training. If training loss decreases without an improvement in validation loss, then do more training again. If the loss of validity begins to increase, then it is time to stop.

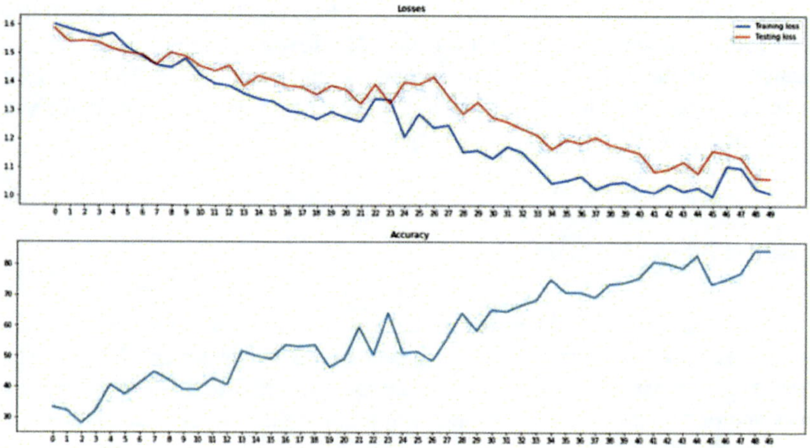

## 5 Conclusion and Future Work

Emotion detection in human–computer interaction can be seen as an important area of study. While it is still a recent and hot field of study to classify emotions from textual data, researchers have done enough work to detect emotions from facial and audio information. This study explores methods that are currently being used to discern emotions from a textual and distinct layout of computers, indicating that this paper would work successfully. There could be more options for data collection in the future contemplated. It is also necessary to take into account the personality of the user, such as when creating device models, in order to make the experience more unique. Protection and privacy are what bring our model a solid life, and this is where federated learning comes in. Not only does federated learning allow us to train the model at the user end or at the neutral end but also serves as a framework for managing the aggregation of data from multiple teaching devices. This concept of emotion detection from the text can be useful and productive for modern day keyboard prediction and emoji prediction on the fly by being able to train on distributed text data (imitating the concept of having and deploying trained models or description from teachers in this case devices such as smartphone and computers).

## References


1. Shivhare SN, Khethawat (2012) Emotion detection from text
2. Sebe N, Cohen I, Gevers T, Huang TS (2005) Multimodal approaches for emotion recognition. In: Proceedings of SPIE - the international society for optical engineering
3. Tripathi V, Joshi A, Bhattacharyya P (1749) Emotion analysis from text: a survey
4. Cowie R, Douglas-Cowie E, Tsapatsoulis N, Votsis G, Kollias S, Fellenz W, Taylor JG (2001) Emotion recognition in human-computer interaction
5. Al-Saaqa S, Abdel-Nabi H, Awajan A (2018) Efficient estimation of word representations in vector space
6. Pennington J, Socher R, Manning CD (2014) Glove: global vectors for word representation
7. Lechner M, Hasani R (2011) Learning structured embeddings of knowledge bases
8. Su M-H, Wu C-H, Huang K-Y, Hong Q-B (2018) LSTM-based text emotion recognition using semantic and emotional word vectors
9. Gosai DD, Gohil HJ, Jayswal HS (2018) A review on a emotion detection and recognization from text using natural language process ing

# Design and Implementation of an Efficient IIR Filter Architecture Using Merged Delay Transformation

**P. Bujjibabu, M. Kamaraju, and K. Babulu**

**Abstract** Signal transformation is such an essential function in modern applications. The filter is a vital element in signal processing for conversion and is designed with effective functioning and efficient naturally. Indeed, a filter design must ensure less area and the least power consumption. One can also reduce the hardware complexity and achieve a much better critical path delay by using infinite impulsive response (IIR) instead of finite impulsive response (FIR) filters. IIR filters are superior to FIR filters to a keep flat frequency phase response for all sample rates. From an area point of view, the existing IIR decimation filter implementation method is not up to the mark. Merged delay transformation (MDT) is being used for filter design to reduce the target architecture's hardware and power consumption. The current architectures directly produce the new value of output without working the intermediate results. The new samples are dependent only on $M$th old output sample and the $M$ input samples. The performance of the modified design is compared with the existing IIR decimation filter. The stability is increased and is more effective in areas with reduced sub-model modules and better power consumption. The number of multiplications involved in computing would be less as compared with traditional IIR filters. The efficiency of the modified structure is evaluated by using MATLAB Simulink, and then power dissipation is observed in Xilinx XPower Analyzer.

**Keywords** Hardware complexity · Infinite Impulsive Response (IIR) · Merged Delay Transformation (MDT) · MATLAB Simulink

P. Bujjibabu (✉)
Aditya Engineering College (A), Aditya Nagar, ADB Road, Surampalem, Andhra Pradesh 533437, India
e-mail: bujjibabu_penumuchi@aec.edu.in

M. Kamaraju
Gudlavalleru Engineering College, Seshadri Rao Knowledge Village, Gudlavalleru, Krishna (dt), Andhra Pradesh 521356, India

K. Babulu
JNTUV, Kakinada, Dwarapudi, Vizianagaram, Andhra Pradesh 535 003, India

 
X.-S. Yang et al. (eds.), *Proceedings of Sixth International Congress on Information and Communication Technology*, Lecture Notes in Networks and Systems 216,
https://doi.org/10.1007/978-981-16-1781-2_50

## 1 Introduction

The present era of mobile computing and multimedia technology demands complex and high-performance digital signal processing (DSP) processors. The crucial elements used for signal processing are digital filters. These are of different types and designed using distinct realization techniques toward optimized design as per the interest application. IIR filter [1] is one of the commonly used filters in DSP applications. The efficient filter [2, 14] design in terms of power, area, and speed is the dominant issue for all signal processing applications. To improve performance, MDT [3] is applied. The superiority of IIR filters over FIR filters is fewer coefficients to execute similar filtering operations [4], with high computation speed, with improved sensitivity, and occupy less space so that potential saving in memory space. The use of a lower sampling rate is usually cost-effective. Hence, decimation [5] is applied to reduce the processing cost, the calculation, and the memory required to implement a DSP system. Filter design is designing a signal processing module that satisfies a set of requirements to transform input, which passed through. A digital filter is either known by its transfer function or, only, its difference equation. For complete design and analysis, one would have recommended considering the problem statement with a set of specifications, and then the architecture is realized from the derived transfer function. Finally, we compared the architectures for optimized parameters. Then the architecture with a smaller area is subjected to a method called MDT for the betterment. Suitable architecture identification is always a typical task in designing a module that suits a set of specifications. The purpose is to find a type of realization for the filter that meets each requirement with the allowable computation cost. Ample methods materialized to analyze the characteristics of a given digital filter. Typically, one can rate a filter by knowing its response to a simple input such as an impulse, then extend this information to the next level. Designing a filter consists of developing the right architecture for the specified constraints and its functional response with a much-reduced difference between the actual and standard responses. The most common technique for designing IIR digital filters, known as the indirect method, involves creating an analog prototype filter and then transforming it into a digital filter. For the given specifications of a digital filter, the transfer function's derivation requires three steps: 1. It was mapping the required digital filter characteristics with the equivalent analog set. 2. Deriving the analog function prototype. 3. Transform that into an equivalent digital model. The filters used to derive the analog transfer function for given specifications are Butterworth filter: produces a flat response in the pass band and is also known as a maximally flat magnitude filter; Chebyshev Type-I: Offers Equi-ripple behavior in the pass band and with much better linearity in actual and expected responses; Chebyshev type-II: Known as inverse Chebyshev filter and is not preferred to use because of its role of inability.

This paper's main objective is to identify the type of IIR filter architecture, which is efficient in terms of hardware and power dissipation to a remarkable level by incorporating the MDT. The design can then be modeled in MATLAB/Simulink [6] using Xilinx System Generator, configured with MATLAB, and constraints verified in XILINX Vivado tool through HDL description. The other part of the paper is organized as follows: discussion on IIR filter architecture is elaborated in Sect. 2. The implementation is depicted in Sect. 3. Simulation results and validations are presented in Sect. 4, and the conclusion, with the design's future scope, is shown in Sect. 5.

## 2 IIR Filter Architectures

In this section, the basic design procedure with the help of a standard problem statement is described.

### 2.1 Design of IIR Filters from Analog Filters

IIR filter offers infinite response with a pulse excitation, designed in many ways, but the lesser difference in the behavior cum constraints will top up the methods. Transformation Methods for handling the intended task are:

***Bilinear Transformation (BT) Method***: BT is a mathematical mapping of variables. It transforms the Ωj axis into the unit circle in the z-plane. The mapping between S-plane into Z-plane is by $H[Z] = H[S]$; where $S = \frac{2}{T}\left(\frac{1-Z^{-1}}{1+Z^{-1}}\right)$.

***Impulse Invariance (II) Method***: This technique is used to design discrete-time IIR filters from continuous-time filters. Express the analog filter transfer function as the sum of single-pole filters to develop a digital filter by using this method, and is represented below. If $H[s] = \sum_{k=1}^{N} \frac{c^k}{s-p^k}$;then $H[S] = \sum_{k=1}^{N} \frac{c_k}{1-e^{P_k}z^{-1}}$.

***Approximation of Derivatives (AD) Method***: This method of digitizing an analog filter is by approximating an equivalent difference equation. The system function for IIR filter obtained as a result of the approximation of derivatives and is given by; $H[Z] = H[S]; S = \left(\frac{1-Z^{-1}}{T}\right)$.

***Matched Z-Transformation-(MZT) Method***: In this method, conversion is achieved by directly mapping the poles and zeros in the z-plane. If $H[s] = \Pi_{k=1}^{M} \frac{s-Z_k}{s-p_k}$; then $H[S] = \Pi_{k=1}^{N} \frac{1-e^{z^k T} Z^k}{1-e^{p^k T} Z^k}$.

### 2.2 Realizations of IIR Digital Filters

Definition of a digital filter model is possible in two fundamental ways [7] named; recursive and non-recursive. In recursive realization, the output sample $y[n]$ is a function of past outputs, past with present inputs and producing infinite response with an impulse, known as IIR filter, which is in the following forms. Viz. Direct Form-I, Direct form-II, Cascaded form and Parallel forms.

*Direct Form-I:* Consider a recursive system by the difference equation,

$$y[n] = -\sum_{k=1}^{M} a_k y[n-k] + \sum_{k=0}^{M} b_k x[n-k] \tag{1}$$

*Direct Form-II:* Consider the difference equation,

$$y(n) = -\sum_{k=1}^{M} a_k y(n-k) + \sum_{k=0}^{N} b_k x(n-k) \tag{2}$$

The system function of the above equation can be expressed as;

$$H(z) = \frac{Y(z)}{X(z)} = \frac{\sum_{k=0}^{N} b_k z^{-k}}{1 + \sum_{k=1}^{M} a_k z^{-k}} \tag{3}$$

Let $\frac{Y(z)}{X(z)} = \frac{W(z)}{X(z)} \frac{Y(z)}{W(z)}$; where $\frac{W(z)}{X(z)} = \frac{1}{1+\sum_{k=1}^{M} a_k z^{-k}}$ that gives

$$W(z) = X(z) - a_1 z^{-1} W(z) + a_2 z^{-2} W(z) \ldots - a_M z^{-M} W(z) \tag{4}$$

Applying Inverse Z-transform on both sides;

$$w[n] = x[n] - a_1 w[n-1] - a_2 w[n-2] - \cdots - a_M w[n-M] \tag{5}$$

And $\frac{Y(z)}{W(z)} = \sum_{k=0}^{N} b_k z^{-k}$ from which $Y(z) = b_0 W(z) - b_1 z^{-1} W(z) + b_2 z^{-2} W(z) \ldots - b_N z^{-N} W(z)$

Applying Inverse Z-transform on both sides;

$$y[n] = b_0 w[n] + b_1 w[n-1] + \cdots + b_N x[n-N] \tag{6}$$

*Cascaded form:* An IIR filter can be represented with function; $H(z) = H_1(z) H_2(z) \ldots H_k(z)$.

*Parallel form:* A parallel form of realization for an IIR system can be achieved by performing a partial expansion of $H(z) = c + \sum_{k=1}^{N} \frac{c_k}{1-p_k z^{-1}}$ the same equation can be expanded as;

$$H(z) = c + \frac{c_1}{1 - p_1 z^{-1}} + \frac{c_2}{1 - p_2 z^{-1}} + \cdots + \frac{c_N}{1 - p_N z^{-1}} \tag{7}$$

By considering an example, detailed architectures drew. At this point in the stage, in each architecture, it is worth noting a variable number of submodules required to develop the architecture for the same functional response.

## 2.3 Design Example with Specifications

For example, a low pass filter is being designed with set of specifications that are given below. Pass band frequency: $Wp = 0.1\pi$, stop band frequency: $Ws = 0.35\pi$ pass band ripple: $-3$ dB $\leq |H(e^{jw})| \leq 0$ dB, $w \leq Wp$ which is $0.707 \leq |H(e^{jw})| \leq 1$, stop band attenuation: $|H(e^{jw})| \leq -20\text{dB}ws \leq |w| \leq Wp$ that is $|H(e^{jw})| \leq 0.1$.

The filter has drawn using MATLAB/FDA tool, which requires the filter specifications, as per these characteristics, the possible architectures are theoretically realized upon observing a satisfactory level of response, elaborated in the coming session.

*Butterworth Filter*: The analog transfer function of the Butterworth low pass filter can be obtained by the following procedure. The order of filter is given by the formula $N \geq \frac{\log\sqrt{\left[\frac{10^{0.1\alpha_s}-1}{10^{0.1\alpha_p}-1}\right]}}{\log\left[\frac{\Omega_s}{\Omega_p}\right]}$ and the value of filter order $N$ is 2.449, which is rounded to 3, at which the transfer function could be defined as $H(s) = \frac{1}{(s+1)(s^2+s+1)}$. Replacing $s$ with $S/\Omega_c$ where $\Omega_c = \frac{\Omega_p}{\varepsilon^{1/N}} = \frac{0.15\pi}{1} = 0.471$ and hence the modified analog transfer function would be $H(s) = \frac{0.1044}{(s+0.471)(s^2+0.471s+0.2218)}$.

The transformations that are applied to these analog transfer functions to obtain the digital IIR function (as shown in Fig. 1) are as follows:

Bilinear transformation: To use the bilinear transformation for the analog transfer function $H(s)$, replace $s$ with $S = \frac{2}{T}(\frac{1-z^{-1}}{1+z^{-1}})$(with $T$ unit delay). Then the resulting transfer function is as follows;

$$H(z) = \frac{0.1044z^{-3} + 0.3132z^{-2} + 0.3132z^{-1} + 0.1044}{-5.014z^{-3} + 19.65z^{-2} - 26.56z^{-1} + 12.75} \tag{8}$$

The realizations for this transfer function in various forms are as follows:

*Architecture in Direct Form-I*: Now consider $H(Z)$ equation and finding the inverse Z-transform for its final solution, which is given by

$$\begin{aligned} y[n] &= 0.008x[n] + 0.024x[n-1] + 0.024x[n-2] + 0.008x[n-3] \\ &\quad + 2.083y[n-1] - 1.54y[n-2] + 0.039y[n-3] \end{aligned} \tag{9}$$

*Architecture in Direct form-II:* Solution for the $H(Z)$ in direct form II is (shown in Fig. 2) given by

$$y[n] = 0.10w[n-3] + 0.31w[n-2] + 0.31w[n-1] + 0.10w[n] \tag{10}$$

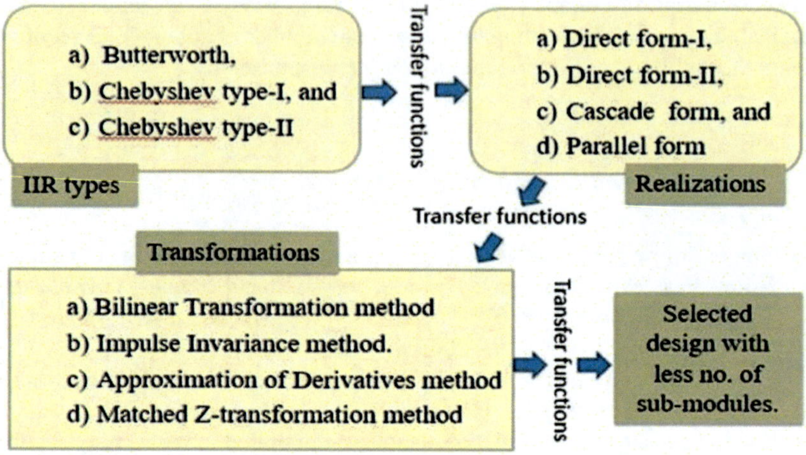


**Fig. 1** Algorithmic approach for digital IIR filter architecture evaluation

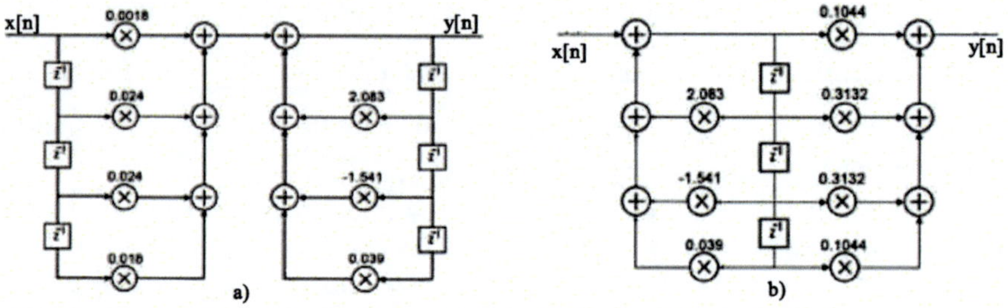


**Fig. 2** Architecture of Butterworth in Bilinear Transformation **a** Direct Form-I **b** Direct form-II

where $w[n] = x[n] + 2.083x[n-1] - 1.54x[n-2] + 0.39x[n-3]$.

Architecture in Cascaded form: Solution for the $H(Z)$ in Cascaded form is (shown in Fig. 3) given by

$$H(z) = 0.0185\frac{(1+z^{-1})}{(1-0.6187z^{-1})}\frac{(1+z^{-1})}{(1-0.868z^{-1})}\frac{(1+z^{-1})}{(1-0.868z^{-1})} \quad (11)$$

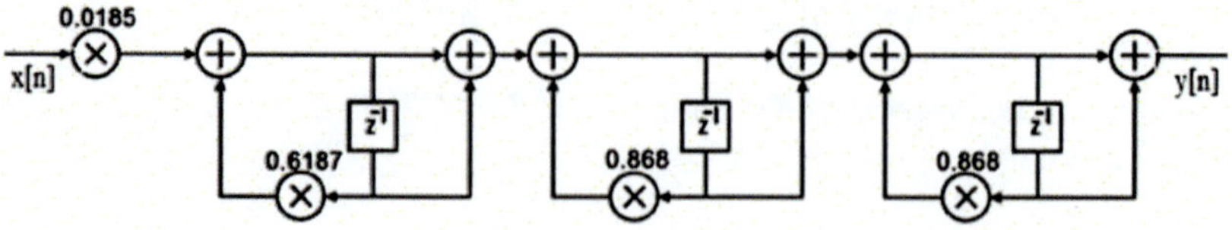


**Fig. 3** Cascaded form realization of Butterworth-with Bilinear Transformation

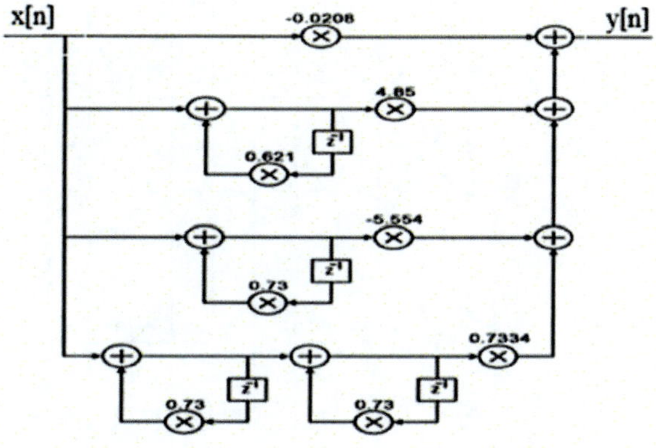


**Fig. 4** Parallel form realization of Butterworth-with Bilinear Transformation

Architecture in parallel form: Solution for the H(Z) in parallel form is (shown in Fig. 4) given by

$$H(z) = -0.0208 + \frac{4.85}{(1 - 0.621z^{-1})} + \frac{-5.554}{(1 - 0.73z^{-1})} + \frac{0.7334}{(1 - 0.73z^{-1})^2} \quad (12)$$

The architecture is said to be good if it is realized with less number of sub-modules required. Among all possible architectures, we will identify effective and efficient ones. So, for $H(Z)$, the architectures are also realized by using the impulse invariance method, Approximation of derivatives method, and matched Z-Transformation method, by incorporating Chebyshev Type-I and Chebyshev Type-II approaches, to observe the number of adders, multipliers, and delay elements required for each realization.

From Table1, the type of realization, which is taking less hardware, is selected for the next round of design challenges toward delay optimization.

## 2.4 *Merged Delay Transformation Application to Filter*

To face the competition between different architectures or the systems for the same function, the designer or vendor must search for the best possible style to design optimized architecture [8] that fits for digital IIR filter implementation. One technique is the MDT method, which directly computes the new value of the output without adding the intermediate results. In other words, it means that the current output sample becomes dependent only on *M*th old output and *M* input samples. Filter design [9,10] using the MDT is featuring the stability, besides the low power and better area. In this method, *M* delay elements are integrated into the recursive track to decrease the output data rate. The sampling rate is the same at the input and output sides. The transformed form of the IIR equation is suitable for conversion into a downsampling filter.

**Table 1** Comparison and discussion

| Transformation technique | Filter type | Butterworth | | | | Chebyshev type-I | | | | Chebyshev type-II | | | |
|---|---|---|---|---|---|---|---|---|---|---|---|---|---|
| | realization | Mult's | Add's | Delay elements | Total modules | Mult's | Add's | Delay elements | Total modules | Mult's | Add's | Delay elements | Total modules |
| Bilinear | DF-I | 7 | 6 | 6 | 19 | 7 | 6 | 6 | 19 | 7 | 6 | 6 | 19 |
| Transformation | DF-II | 7 | 6 | 3 | 16 | 7 | 6 | 3 | 16 | 7 | 6 | 3 | 16 |
| Method | CAS | 4 | 6 | 3 | 13 | 4 | 6 | 3 | 13 | 6 | 6 | 3 | 15 |
| | PAR | 8 | 7 | 4 | 19 | 5 | 4 | 5 | 16 | 7 | 7 | 4 | 18 |
| Impulse | DF-I | 6 | 5 | 5 | 16 | 6 | 4 | 5 | 15 | 6 | 5 | 6 | 17 |
| Invariance | DF-II | 6 | 5 | 3 | 14 | 6 | 4 | 3 | 13 | 7 | 5 | 3 | 15 |
| Method | CAS | 5 | 5 | 3 | 13 | 5 | 4 | 3 | 12 | 6 | 5 | 3 | 14 |
| | PAR | 7 | 6 | 4 | 17 | 6 | 6 | 4 | 16 | 7 | 7 | 4 | 18 |
| Approximation of derivatives | DF-I | 4 | 3 | 3 | 10 | 6 | 3 | 3 | 12 | 5 | 5 | 6 | 16 |
| | DF-II | 4 | 3 | 3 | 10 | 4 | 3 | 3 | 10 | 5 | 5 | 3 | 13 |
| Method | CAS | 4 | 3 | 3 | 10 | 4 | 3 | 3 | 10 | 6 | 5 | 3 | 14 |
| | PAR | 7 | 6 | 4 | 17 | 6 | 6 | 4 | 16 | 6 | 6 | 4 | 16 |
| Matched | DF-I | 4 | 3 | 3 | 10 | 5 | 4 | 5 | 14 | 6 | 5 | 6 | 17 |
| Z-Transform | DF-II | 4 | 3 | 3 | 10 | 5 | 6 | 3 | 14 | 6 | 5 | 3 | 14 |
| Method | CAS | 4 | 3 | 3 | 10 | 3 | 3 | 3 | 09 | 5 | 5 | 3 | 13 |
| | PAR | 7 | 6 | 4 | 17 | 6 | 6 | 4 | 16 | 7 | 7 | 4 | 18 |

DF-I: Direct Form-I, DF-II: Direct Form-II, CAS: Cascaded Form, PAR: Parallel Form Mult's: Number. of Multipliers, Add's: Number of Adders

# 3 Design and Implementation of IIR Filters

The functional responses of all realized architectures are verified using MATLAB/FDA tool. Its performance is in Xilinx Vivado Tool to know the impact of specifications on designing the filter. The next stage of filter design will be followed up on successful verification of its response as per the specifications.

## *3.1 Designing a Simulink Model for IIR Decimation Filters*

Xilinx System generator configured with MATLAB Simulink to create a Simulink model for the IIR filters' architectures realized in the previous sections.

Then create the corresponding model using the communication toolbox; the coefficients to every multiplier are supplied as per the one generated in the FDA Tool (or from the manually designed architectures) and verify its expected functional response with a specified set of input samples, as shown in Fig. 5. Then the digital filter response is observed. The overall Simulink model of transformed third-order MDT design in two sub-blocks connects the input–output blocks, as shown in Fig. 6. The standard procedure is used to arrive at the HDL code from a Simulink block set. The generated code can be run/synthesis on any formal FPGA Board [11–13]. Once the code is generated and then parameters like; load in terms of capacitance, switching activity, and the supply voltage for a selected range of frequencies varied so as to observe the performance/design constraints of the model.

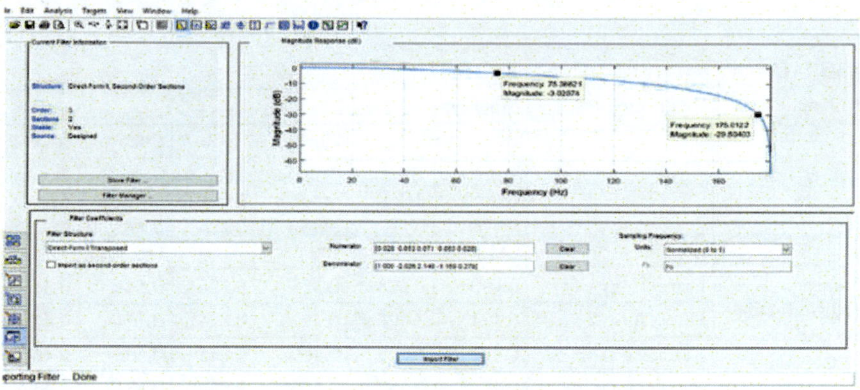

**Fig. 5** Magnitude response of low pass filter on MATLAB FDA tool

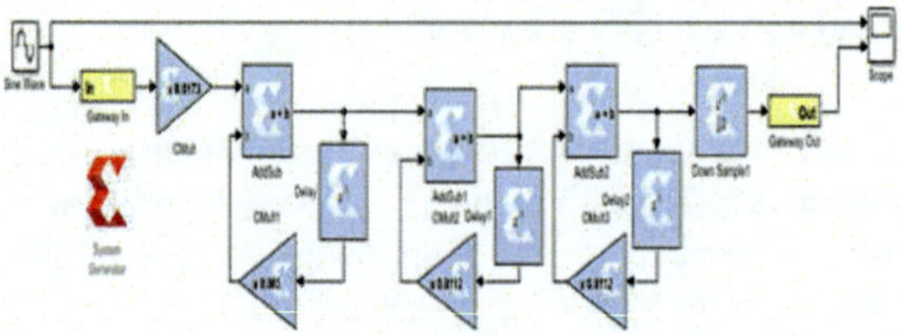

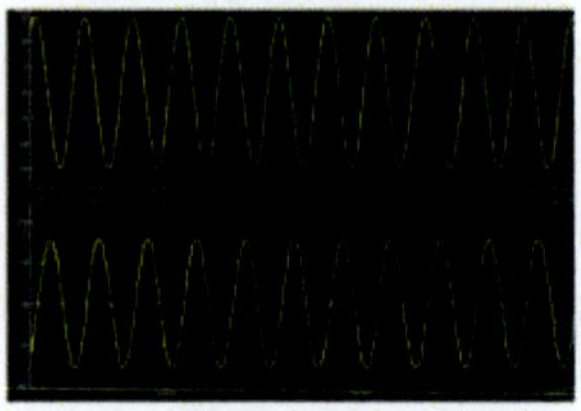

**Fig. 6** Simulink model and the frequency response of IIR filter

## 4 Results and Validations

Initially, the number of modules [15] required to implement the IIR filter for a given specification is identified. Table 1 shows that the Chebyshev Type-I filter realized in cascaded form using Matched Z-transformation takes a total of 9 sub-modules for the same function rather than 19 in Direct Form-I with Bilinear Transformation method, will be used in implementation using MDT so that the filter can be with high speed and low power dissipation.

From the magnitude response plot, it is observed that the frequencies less than $0.15\pi$ rad/s, i.e.,75 Hz have a magnitude between 0 dB and –3 dB, and the frequencies above $0.35\pi$ rad/s, i.e., 175 Hz have the magnitude less than –20 dB. Therefore, the magnitude response of the low pass filter met the specifications. Table 2 illustrates the amount of power and delay characteristics with different possible architectures realized using bilinear transformation, impulse invariance method, approximation of derivatives method, and matched Z-transform methods. From Fig. 7, also observed the drastic difference in total power dissipation between the architectures realized

**Table 2** On-chip power and delay comparison

| Technique | Realization | Filter type | | | | | |
|---|---|---|---|---|---|---|---|
| | | Butterworth | | Chebyshev type-1 | | Chebyshev type-2 | |
| | | Total on-chip power (w) | Total delay (ns) | Total on-chip power (w) | Total delay (ns) | Total on-chip power (w) | Total delay (ns) |
| Bilinear Transformation method | Direct form 1 | 0.490 | 0.428 | 0.456 | 0.896 | 0.548 | 0.315 |
| | Direct form 2 | 0.424 | 1.408 | 0.412 | 0.814 | 0.454 | 0.405 |
| | Cascaded form | 0.421 | 1.025 | 0.409 | 0.845 | 0.408 | 0.460 |
| | Parallel form | 0.498 | 0.657 | 0.486 | 0.856 | 0.471 | 0.457 |
| Impulse Invariance method | Direct form 1 | 0.466 | 0.742 | 0.434 | 0.862 | 0.366 | 0.455 |
| | Direct form 2 | 0.491 | 0.728 | 0.418 | 0.748 | 0.540 | 0.554 |
| | Cascaded form | 0.448 | 0.769 | 0.424 | 0.882 | 0.466 | 1.019 |
| | Parallel form | 0.565 | 0.562 | 0.552 | 0. 756 | 0.562 | 1.255 |

(continued)

**Table 2** (continued)

| Technique | Realization | Filter type | | | | | |
|---|---|---|---|---|---|---|---|
| | | Butterworth | | Chebyshev type-1 | | Chebyshev type-2 | |
| | | Total on-chip power (w) | Total delay (ns) | Total on-chip power (w) | Total delay (ns) | Total on-chip power (w) | Total delay (ns) |
| Approximation of derivatives method | Direct form 1 | 0.389 | 0.456 | 0.365 | 0.256 | 0.422 | 0.466 |
| | Direct form 2 | 0.378 | 0.477 | 0.369 | 0.376 | 0.458 | 0.455 |
| | Cascaded form | 0.399 | 0.489 | **0.363** | **0.232** | 0.399 | 0.378 |
| | Parallel form | 0.425 | 0.469 | 0.491 | 0.993 | 0.451 | 0.478 |
| Matched z-transform method | Direct form 1 | 0.385 | 0.489 | 0.456 | 0.463 | 0.478 | 0.378 |
| | Direct form 2 | 0.369 | 0.437 | 0.417 | 0.423 | 0.415 | 0.421 |
| | Cascaded form | 0.367 | 0.389 | 0.399 | 0.421 | 0.452 | 0.416 |
| | Parallel form | 0.451 | 0.462 | 0.475 | 0.475 | 0.467 | 0.455 |

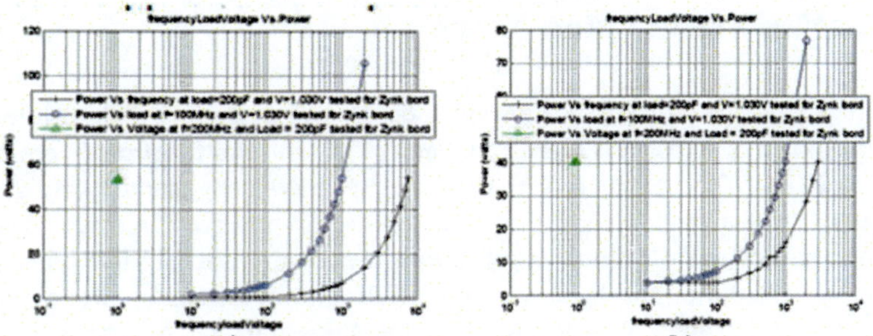

a) b)

**Fig. 7** Total on-chip power at different frequencies **a** without MDT **b** with MDT

with and without incorporating the MDT techniques under the same test environment w.r.t operating frequency. The architecture without MDT reaches a maximum power of 100 mW, whereas the MDT application will scale down the power to a value of nearly 80 mW. In Fig. 8, it is also noticed that there is an impact on power concerning load, frequency, and supply voltages. The interesting point is that the architecture shows its reluctance at or above a load of 500 pF. At this stage, the device current and junction ambient temperature are beyond the threshold value. In this condition, the prototype is not suggestible to go for the final design. Operating the structure with an even lower level of supply voltages can compensate for this condition.

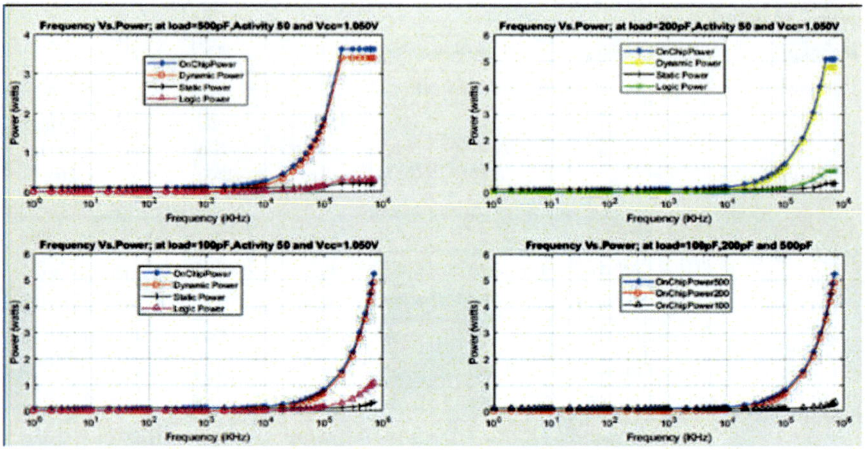


**Fig. 8** MDT power performance w.r.t fan-out or load, frequency and switching activity

## 5 Conclusion and Future Scope

In this paper, the procedure to design power-efficient and effective architectures from a given problem statement is discussed using FDA tool. For functional verification, Simulink Xilinx Toolbox is used. With the help of a system generator configured to MATLAB, the design is verified for its constraints. It is observed that the architecture is taking a significantly less number of submodules to arrive at the same function and is a method of solution for the processors suffering from area and power dissipation problems. Besides the MDT, the IIR filter is required to realize by incorporating the order reduction techniques (like balanced truncation model order reduction, optimal model order reduction in the Hankel Norm, and SVD) improvement in the hardware requirement. One tries for additional development, in the delay reduction, in a type of MDT-based filter architecture by incorporating the order minimization techniques.

**Acknowledgements** I want to thank the scholars for their potential information and inspiration for designing efficient filter architectures. I thank the management of Aditya Engineering College (Autonomous), Surampalem, Department of ECE, UCEK, JNTUK Kakinada, Gudlavalleru Engineering College, Gudlavalleru, ApplyVolt, Vijayawada for their timely help and permission to utilize the Tools to carry out the experimentation with MATLAB Simulink/System Generator, in concluding these outcomes.

## References

1. Naqvi SZ, Hassan SZ, Kamal T (2016) A power consumption and area improved design of IIR decimation filters via MDT. In: Intelligent system engineering (ICISE), 15–17 Jan 2016

2. Smitha GL (2017) System generator based area reduction of IIR decimation filter. Int J Innov Res Comput Commun Eng (IJIRCCE) 5(6)
3. Rashid U, Siddiq F, Muhammad T, Jamal H (2013) Area efficient decimation filter based on MDT for wireless applications. The Nucleus 50(4):301
4. Patni A (2015) Merged delay transformed IIR down sampler filter. In: International conference on industrial instrumentation and control (ICIC), College of Engineering pune, India, 28–30 May 2015
5. Farooq U (2008) Efficient architectural transformation of multi rate recursive filters. PhD thesis, University of Engineering and Technology Texila, Pakistan
6. Chenliang Z (2012) IIR digital filter design research and simulation on MATLAB. In: International conference on signal processing systems (ICSPS 2012), vol 58
7. Maurya AK (2011) Multi-rate signal processing. Int J Electron Commun Eng 4(4):443–452. ISSN 0974-2166
8. Moon GG (2014) Design approach for decimation filter for ADC applications. Int J Eng Trends Technol (IJETT) 10(12)
9. Wheeler D (2013) High performance IIR filters for interpolation and decimation, Ensilica, July 2013
10. Yeary MB (2006) Theory and implementation of a computationally efficient decimation filter for power-aware embedded systems. IEEE Trans Instrum Meas 55(5):1839–1849
11. Abed KH (2003) Low power and hardware efficient decimation filter. In: Wireless communications and networking. WCNC 2003, March 2003, vol 1, pp 16–20, 454–459. IEEE
12. Rushell AI (2000) Efficient rational sampling rate alteration using IIR filters. IEEE Signal Process Lett 7(1): 6–7
13. Hwang C, Shin Y-P Routh approximations for reducing order of discrete systems. J Dyn Syst Meas Control 104(1):107, 1
14. Coffey MW (2013) Optimizing multistage decimation and interpolation processing, IEEE Signal Process Lett 10(4):107–110
15. Bujjibabu P, Sravani N (2017) Architecture based performance evaluation of IIR digital filters for DSP applications. In: International conference on big data analytics and computational intelligence (IEEE-ICBDAC, year=2017), pp 224–229

# Prediction Model for Malaria: An Ensemble of Machine Learning and Hydrological Drought Indices

**Paulina Phoobane, Muthoni Masinde, and Joel Botai**

**Abstract** Literature is awash with evidence that points to climate change as one of the reasons for the upward increase in infectious disease outbreaks around the globe. With the advancement in artificial intelligence (AI) and big climate data, accurate analysis of medical and climate data allows the prediction and early detection of infectious diseases associated with climate change. The effectiveness of AI and drought indices in predicting oncoming droughts has been demonstrated elsewhere in research, this has, however, not been explored in the prediction of infectious diseases. Consequently, in this paper, a model that uses machine learning to predict the outbreak of infectious diseases using two drought indices and historical incidents of malaria in Limpopo (South Africa) is presented. Having achieved up to 99% prediction accuracies, we demonstrate that such a model equips stakeholders with a new perspective when it comes to the prediction of infectious diseases' outbreaks that are associated with extreme climate variations.

**Keywords** Big climate data · Effective drought index (EDI) · Available water resource index (AWRI) · Infectious diseases prediction · Machine learning · Malaria · Limpopo · Southern Africa

## 1 Introduction

Infectious diseases claim the lives of many people each year globally and Africa continues to bear the burden of morbidity. For instance, the 2019 World Malaria Report confirms that in 2018, 228 million cases of malaria were reported worldwide

P. Phoobane (✉) · M. Masinde · J. Botai
Department of Technology, Central University of Technology, Free State, Bloemfontein, South Africa

J. Botai
South Africa Weather Service, Pretoria, South Africa



X.-S. Yang et al. (eds.), *Proceedings of Sixth International Congress on Information and Communication Technology*, Lecture Notes in Networks and Systems 216,
https://doi.org/10.1007/978-981-16-1781-2_51

with 93% of these cases occurring in Africa [1]. Infectious diseases such as malaria, diarrhoea, tuberculosis are among the top 10 causes of death in low-income countries [2].

Most of the existing health systems that are developed to address infectious diseases and other health issues, in general, face multiple challenges including the rising burden of illness and greater demand for health services [3]. Additionally, most of the health systems are not effective because of the poor and reactive management starting interventions too late in the epidemics [4]. Machine learning conjoined with drought indices holds the promise of achieving more with less to address some of these challenges. However, the nature and extent of this promise have not been systematically assessed [3, 5].

During the last three decades, the development of geographical information systems and satellites for earth observation has made important progress in the monitoring of the weather, climate, environmental and anthropogenic factors that influence the reduction or the re-emergence of infectious diseases [6]. Consequently, to prevent and control disease outbreaks, several infectious prediction tools were developed. However, most of these tools emphasize the international or regional scope of information with minimal reference to information about the disease at the local level [7]. Several methods, ranging from traditional methods to mathematical models and recently machine learning-based approaches, have also been used to explore hidden patterns in big climate data. The aim is to solve among many, health issues that arise from climate change [8]. Many of these methods assist in diagnosis, classification and risk forecasting and, few predict future disease outbreaks [9]. Moreover, most of the predicting models are based on mathematical models and very few are based on machine learning [10]. Predicting infectious disease outbreaks is a complex task and incorporating climate data makes it even more complex because of the dynamic complex nature of the climate and as such, it calls for effective prediction tools such as machine learning [6, 11, 12].

Other recent advancements in computer science and related fields have resulted in interesting tools and technologies that can be leveraged in countering the effects of climate change [13, 14]. For example, the user-as-a-sensor concept, a concept already successfully demonstrated in many projects (WeSenseIt, Ground Truth 2.0, Landsense, GROW, SCENT, AfriAlliance), takes advantage of the capabilities of mobile smartphones for acquiring observations via crowdsourcing and citizen science, mainly referring to the current weather conditions, pest, disease and crop yield reporting. In this way, an increasing amount of data can now be obtained from a huge variety of non-traditional sources—from smartphone sensors to amateur weather stations to canvassing members of the public. Although a very powerful "engine" of data, citizen science comes with some challenges, such as a lack of anonymity, security, transparency, ethics, data manipulation cases, data quality concerns, and, not least, incentives for continued citizen engagement and provision of data [15]. On the positive side, however, this rich and diverse big climate data can be utilized in developing all different kinds of early warning systems.

The Effective Drought Index (EDI), by Byun and Wilhite [16] is a very effective index compared with other drought indices. EDI computes the intensity of droughts by using cumulative precipitation as a weighting function of time and also gives the available water resources Index (AWRI); the latter is a measure of hydrological drought and can be used to assess the quantity of soil moisture. To the knowledge of the authors, the correlation between AWRI and the ability of malaria-carrying vector to thrive has not been investigated.

Given the above gap analysis, the following research objectives (in the context of Africa) are pursued in this paper:

(1) What is the correlation between hydrological drought indices and outbreaks of malaria?
(2) What machine learning algorithms are capable of efficiently and effectively monitoring and predicting the correlation in question 1 above?

To answer the above research questions, the main objective of this research was demarcated as: to develop a machine learning algorithm for predicting outbreaks of malaria using drought indices and historical outbreaks of malaria. A case study of the malaria outbreaks in Limpopo Province in South Africa is selected based on the availability and reliability of data.

## 2 Related Literature Review

### *2.1 Big Climate Data*

In [17], big data is described as having a basic characteristic that can be defined by 3Vs: the volume of data, variety of types of data and the velocity at which the data grow. Big data can be sufficiently analyzed to get useful insight that can solve problems such as climate change, disease diagnosis, disease outbreak prediction and drug recommendation [18, 19]. Big data has been experiencing exponential expansion recently. This proliferation of big data is due to the tremendous amount of data in various formats that are being generated continuously, shared and used in different aspects of life [19].

In climate science which is the study of climate [20], a lot of data that are mostly structured have been generated through sensor networks, pervasive computing and artificial intelligence. Consequently, the amount of climate data and the advancement in climate processing technology have grown enormously giving rise to big climate data. The main use of these big climate data is to address the issues of climate change by recording and pointing out the world climate extremes and variability [17].

Some research has been made in the past decades on big climate data to mine valuable insight. For example, ref. [21] developed a framework called Information Technology and Indigenous Knowledge with Intelligence (ITIKI) on big climate data to predict drought in the Sub Sahara Africa to enhance agriculture, especially for small farmers, ref. [19] developed a big climate data-based surveillance system that continuously monitors the correlation between climate change and dengue in

India. The proposed system models the correlation between the climate parameters and dengue incidences. In the uMkhanyakude district in KwaZulu-Natal province of SA, ref. [22] developed a model using big climate data from Google Earth Engine and International Research Institute (IRI) Data library. The model indicates that slow-moving surface water with extreme temperature is the best predictor of the snails, which transmit a disease called schistosomiasis.

## 2.2 Infectious Diseases Prediction

With the growth in big data, accurate analysis of medical and climate data allows prediction and early detection of infectious diseases associated with climate change [14, 23]. There is ample evidence that links most of the world's infectious diseases to climatic variations [8, 24, 25]. It is indeed indisputable that climate change is threatening both fauna and flora [26, 27]. In the health sector, the resulting risks of climate change for human populations depend on where and how people live. The rise of global temperatures makes urban populations, particularly those in tropical mega-cities, vulnerable to heat stress, infectious diseases and air pollution, while populations in rural areas are vulnerable to infection agents and their associated vector organisms. This happens because infection agents' reproduction and survival rates are strongly affected by temperature fluctuations [28]. Climate change is, therefore, a fundamental threat to global good health. Increased temperature variability, variations in amounts and frequency of precipitation, a greater occurrence of dry spells and droughts, along with the intensification of extreme weather events, the salinization of arable land and freshwater and disease outbreaks are among the most important challenges Africa, and the world at large must address [28].

## 2.3 Correlation Between Malaria and Climate

Malaria is a vector-borne disease that is spread by the bite of the infected Anopheles mosquito. The World Health Organization [1] classifies malaria as a communicable disease, among the likes of water-borne diseases like cholera and tuberculosis. Transmission is more intense in places where the mosquito lifespan is longer and where it prefers to bite humans rather than other animals. The long lifespan and strong human-biting habit of the African vector species are the main reasons why approximately 90% of the world's malaria cases are in Africa [29]. Transmission of malaria also depends on the climatic conditions such as rainfall patterns, temperature and humidity. Currently, many countries with a high burden of malaria have weak surveillance systems and are not in a position to assess disease distribution trends [24]. Malaria control is based on the scientific principle and ongoing success requires continual research input, government commitment to control of the disease and appropriately skilled and trained personnel. Malaria is endemic in the low-altitude areas of the northern and eastern parts of South Africa along the border with Mozambique and Zimbabwe, with the transmission taking place mainly in Limpopo, Mpumalanga and Kwazulu-Natal provinces, and occasionally in the North West Province.

Even though there is ample evidence that links malaria incidence to climate variations [24, 25], there has not been considerable reports on the relationship between malaria and rainfall or water [30, 31]. Githeko and Ndegwa [30] assert that an abundance amount of rainfall leads to more malaria incidence. Receded water after the rainfall gives favourable conditions for mosquitos to breed. This, in turn, increases the number of adult-biting mosquitos. In the study carried by the duo (Githeko and Ndegwa), it was discovered that increased rainfall doubled the number of man-biting mosquitos. A malaria predictive model reported in [32] made use of rainfall to predict malaria incidences. Even though this model showed a good performance, the authors point out that for optimal prediction, modelling of malaria incidence should be done at a micro-level such as a village or district level.

## 2.4 *Drought as a Climatic Variable*

Drought is a condition of moisture deficit sufficient to have an adverse effect on vegetation, animal and man over a sizeable area [33]. From the viewpoint of the frequency, duration, intensity and severity, drought can be categorized as meteorological, agricultural, hydrological and socioeconomic. All the drought categories are interlinked: meteorological drought can degenerate into agricultural drought, hydrological drought and socioeconomic drought, also the longer the duration of drought the more spatially extensive it becomes [34]. Drought indices quantify drought conditions through the integration of, for example, precipitation, rainfall, streamflow and snowpack data into a logical big picture. Many indices have thus far been developed over the years, two of these are Palmer Drought Severity Index (PDSI) [35] and the Effective Drought Index (EDI) [16]. Developed to address the shortcomings of Standard Precipitation Index, EDI considers daily water accumulation with a weighting function for time passage. In this case, the Available Water Resources Index (AWRI) is used, it expresses the actual amount of available water [36]. When it comes to malaria transmission, therefore, we hypothesise that the latter could play a role.

## 2.5 *Machine Learning in Predicting Infectious Diseases Associated with Climate Change*

Machine learning is the new era of AI. AI solves complex problems by developing computer systems with the capacity to perform tasks that require human intelligence such as learning and visual perception [3]. Different IA techniques are normally applied to analyze and model vast amounts of relevant data, it is then learnt which conditions are most likely to result in disease outbreak [37]. For example, through the use of deep learning, Masinde [38] discovered that malaria in Africa tends to occur along latitudes of 0°–15°.

Some of the popular techniques for machine learning algorithms include artificial neural networks (ANNs) and deep learning. ANN has been adopted to diagnose and predict different types of diseases from infectious diseases to chronic diseases such as diabetics and heart diseases [8, 39, 40]. Projects in which ANNs have been used for predicting infectious diseases include the ANN-based prediction of malaria abundances using big data done in India [41]. ANNs, when used to predict infectious diseases, are capable of taking into consideration multiple factors such as health data, climate data. On the other hand, the capabilities of deep learning to add more hidden layers to ANN allow exploring more complex non-linear patterns in the data [37]. Chae et al. [42] developed models based on deep neural networks and long-short term memory (LSTM) to predict chicken pox, malaria and scarlet fever using climate data (temperature, humidity), data on infectious disease occurrence and internet search query data. The authors found that deep neural networks were very accurate and hence outperformed the regression models. Indeed, machine learning can be used to build robust prediction models that handle both linear and nonlinear data and produce results with a high degree of accuracy. Consequently, this makes machine learning suitable for the prediction of certain disease outbreaks that are associated with climate factors.

# 3 Research Methodology

## *3.1 Datasets*

This study uses two datasets that include: (1) Monthly malaria incidence data per district; this was sourced from Limpopo province of South Africa and it is for the period of 1998–2019 (21 years period). The administrative districts consist of Capricorn, Greater Sekhukhune, Mopani, Vhembe and Waterberg. (2) Monthly weather data consisting of effective drought index (EDI), available water resources index (AWRI) and monthly rainfall/precipitation. These data are grouped by climatic districts as classified by the South Africa Weather Services (SAWS). Malaria data were sourced from the Limpopo department of health while weather data were acquired from (SAWS). Figure 1 below depicts both climate and administrative districts of Limpopo province. There are five administrative districts and 13 climate districts. The climate districts are labelled using numbers. Figure 2 below depicts malaria incidence distribution in Limpopo province. Vhembe and Mopani show the highest malaria incidence while Great Sekhukhune and Waterberg show the least malaria incidence. It can also be observed that Vhembe and Mopane experienced the highest malaria incidences between 2026 and 2018. As indicated earlier, Malaria

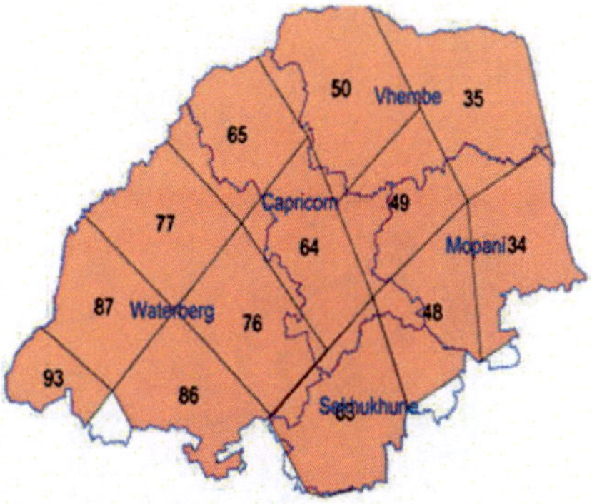


**Fig. 1** Climate and administrative districts of Limpopo

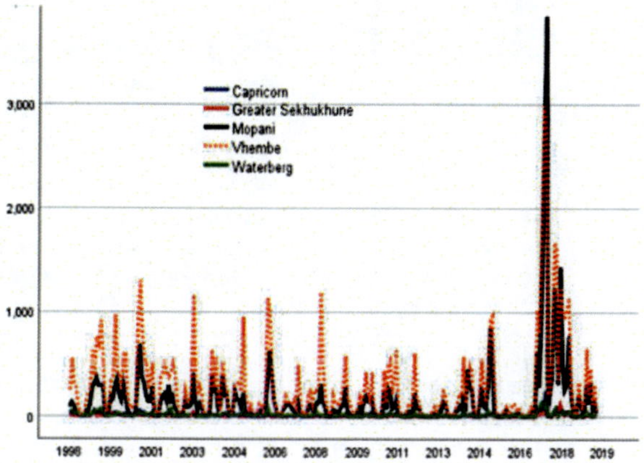


**Fig. 2** Malaria incidence distribution in five districts of Limpopo

incidence and weather dataset considered in this study were for a period of 21 years. Dataset for 20 years was used to train the model while dataset for 1 year was selected to test the model. The selection was done randomly.

## *3.2 Data Processing*

The entire dataset consisted of 258 cases of monthly data for January 1998–June 2019. Furthermore, there were 37 variables consisting of monthly AWRI, EDI and precipitation values for 10 climate districts (total of 20) climatic districts, monthly malaria incidence values for the 5 administrative districts, a column for the total incidence for the province and finally, the date column for the date (month and year). To clean and prepare the data for building, the predictive model, data pre-processing and feature reduction processes were done. Data reliability on malaria incidence dataset and weather dataset was determined using Cronbach's alpha test based on standardized items. The items returned a Cronbach's alpha value of 0.811

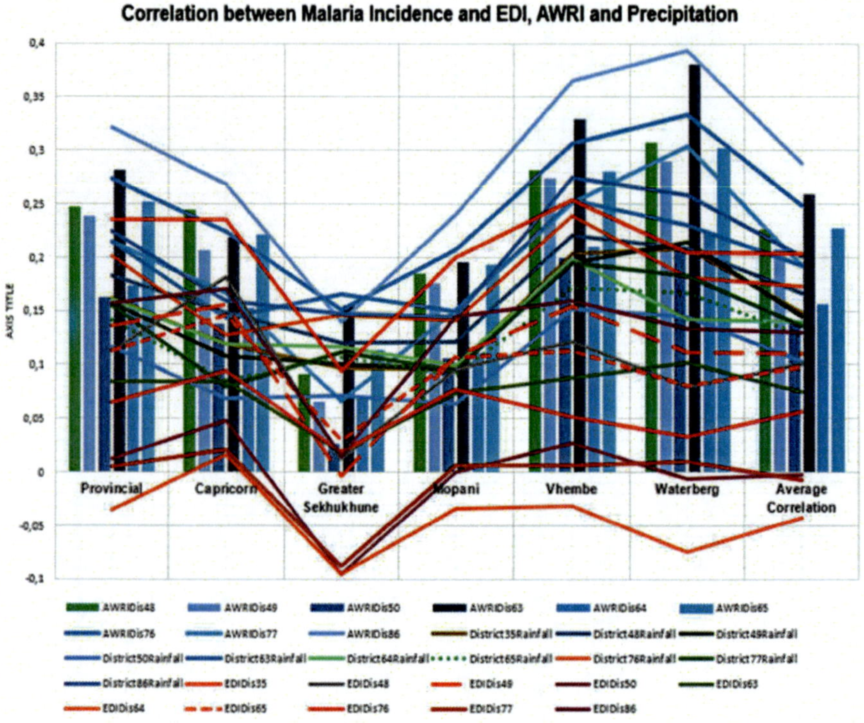


**Fig. 3** Correlation between malaria incidence and EDI, AWRI and precipitation

and a Cronbach's alpha based on standardized items' value of 0.962—hence the internal consistencey was good. The results of correlation measure on the individual measures returned high levels of correlation. The correlation between the IDE, AWRI, precipitation and malaria is shown in Fig. 3. On the other hand, Table 1 depicts the total rankings of the variables based on their correlation to malaria incidence.

The rankings in Table 1 were computed based on the 30 climatic variables' correlation strength for each of the regions (districts and provincial level)—the highest score being 30 and lowest being 1. For instance, for Vhembe district, the strongest correlation (with malaria incidence reported in the district) was the AWRI value for rainfall station number 86 while the lowest correlation was recorded for EDI value for rainfall station number 64; these are similar to the correlations for total incidence recorded in the entire province, as well as for Capricorn, Mopani and Waterberg districts. The average of the individual correlation of the variables was then calculated to determine the overall ranking. The lowest-ranked variables are the EDI values for districts 64, 77, 50, 76 and 63 in ascending order. The Pearson correlation between these and the district's malaria cases is insignificant and was therefore discarded. On the other hand, the highest-ranked variables are the AWRI for districts 86, 63,76, 65, 48 and 49.

**Table 1** Correlation between malaria incidence and EDI, AWRI and precipitation. Green: Correlation is significant at the 0.01 level (two-tailed); Red: Correlation is significant at the 0.05 level (two-tailed)

| Dependent Variable | Provincial | Capricorn | Greater Sekhukhune | Mopani | Vhembe | Waterberg | Average Correlation |
|---|---|---|---|---|---|---|---|
| EDIDis64 | -0,035 | 0,016 | -0,095 | -0,034 | -0,032 | -0,074 | -0,043 |
| EDIDis77 | 0,005 | 0,021 | -0,088 | 0,007 | 0,006 | 0,009 | -0,007 |
| EDIDis50 | 0,013 | 0,048 | -0,095 | 0,000 | 0,026 | -0,006 | -0,002 |
| EDIDis76 | 0,065 | 0,095 | 0,018 | 0,077 | 0,051 | 0,033 | 0,057 |
| EDIDis63 | 0,085 | 0,084 | 0,019 | 0,074 | 0,088 | 0,102 | 0,075 |
| EDIDis65 | 0,113 | 0,146 | 0,030 | 0,107 | 0,113 | 0,081 | 0,098 |
| EDIDis48 | 0,113 | 0,181 | 0,012 | 0,096 | 0,121 | 0,080 | 0,100 |
| District50Rainfall | 0,117 | 0,069 | 0,071 | 0,064 | 0,152 | 0,145 | 0,103 |
| EDIDis49 | 0,136 | 0,156 | -0,003 | 0,108 | 0,155 | 0,112 | 0,111 |
| District65Rainfall | 0,141 | 0,086 | 0,106 | 0,091 | 0,172 | 0,167 | 0,127 |
| EDIDis86 | 0,157 | 0,171 | 0,014 | 0,145 | 0,159 | 0,134 | 0,130 |
| District77Rainfall | 0,155 | 0,080 | 0,112 | 0,094 | 0,195 | 0,183 | 0,136 |
| District64Rainfall | 0,16 | 0,119 | 0,117 | 0,100 | 0,199 | 0,142 | 0,139 |
| AWRIDis50 | 0,163 | 0,138 | 0,027 | 0,106 | 0,199 | 0,207 | 0,140 |
| District49Rainfall | 0,157 | 0,107 | 0,101 | 0,096 | 0,195 | 0,214 | 0,145 |
| District35Rainfall | 0,162 | 0,118 | 0,097 | 0,097 | 0,204 | 0,209 | 0,148 |
| AWRIDis64 | 0,173 | 0,153 | 0,068 | 0,111 | 0,21 | 0,224 | 0,157 |
| District48Rainfall | 0,183 | 0,148 | 0,120 | 0,121 | 0,221 | 0,209 | 0,167 |
| District76Rainfall | 0,201 | 0,128 | 0,146 | 0,141 | 0,239 | 0,182 | 0,173 |
| AWRIDis77 | 0,214 | 0,167 | 0,066 | 0,151 | 0,251 | 0,304 | 0,192 |
| District63Rainfall | 0,215 | 0,143 | 0,166 | 0,149 | 0,254 | 0,23 | 0,193 |
| District86Rainfall | 0,224 | 0,159 | 0,147 | 0,145 | 0,274 | 0,259 | 0,201 |
| EDIDis35 | 0,235 | 0,235 | 0,094 | 0,2 | 0,254 | 0,205 | 0,204 |
| AWRIDis49 | 0,238 | 0,206 | 0,065 | 0,176 | 0,274 | 0,29 | 0,208 |
| AWRIDis48 | 0,247 | 0,244 | 0,091 | 0,184 | 0,281 | 0,307 | 0,226 |
| AWRIDis65 | 0,251 | 0,221 | 0,113 | 0,193 | 0,28 | 0,301 | 0,226 |
| AWRIDis76 | 0,273 | 0,224 | 0,151 | 0,207 | 0,306 | 0,334 | 0,249 |
| AWRIDis63 | 0,281 | 0,219 | 0,155 | 0,195 | 0,329 | 0,38 | 0,260 |
| AWRIDis86 | 0,321 | 0,269 | 0,14 | 0,24 | 0,365 | 0,394 | 0,288 |

## 3.3 *Building the Model*

After the data processing presented in the section above, RapidMiner was used to simulate data and build the following predictor models: Deep Learning, Decision Trees, Random Forest, Gradient Boosted Trees, Support Vector Machine, Generalized Linear Model, Naive Bayes, Logistic Regression and Fast Large Margin. RapidMiner is open-source software that is both commercially and freely available and it is very popular for its friendly user interface. Additionally, it is platform-independent as it works on cross-platform operating systems. It is also widely used for quick prototyping and validation of models [38, 43]. The variables shown in Table 2 were

**Table 2** Variables with the highest correlation to malaria incidence in Vhembe and Provincial

| Vhembe District | | Provincial Level | |
|---|---|---|---|
| **Variable** | **Correlation** | **AWRIDis86** | **Correlation** |
| AWRIDis86 | 0,365 | AWRIDis63 | 0,321 |
| AWRIDis63 | 0,329 | AWRIDis76 | 0,281 |
| AWRIDis76 | 0,306 | AWRIDis65 | 0,273 |
| AWRIDis48 | 0,281 | AWRIDis48 | 0,251 |
| AWRIDis65 | 0,28 | AWRIDis49 | 0,247 |
| District86Rainfall | 0,274 | EDIDis35 | 0,238 |
| AWRIDis49 | 0,274 | District86Rainfall | 0,235 |
| EDIDis35 | 0,254 | District63Rainfall | 0,224 |
| District63Rainfall | 0,254 | AWRIDis77 | 0,215 |
| AWRIDis77 | 0,251 | District76Rainfall | 0,214 |
| District76Rainfall | 0,239 | District48Rainfall | 0,201 |
| District48Rainfall | 0,221 | AWRIDis64 | 0,183 |
| AWRIDis64 | 0,21 | AWRIDis50 | 0,173 |
| District35Rainfall | 0,204 | District35Rainfall | 0,163 |
| District64Rainfall | 0,199 | AWRIDis86 | 0,162 |
| AWRIDis50 | 0,199 | | |
| District77Rainfall | 0,195 | | |
| District49Rainfall | 0,195 | | |
| District65Rainfall | 0,172 | | |

used to build the models owing to their higher correlations with the reported malaria incidences. Vhembe was chosen because of its high malaria incidence compared with the other three districts: Great Sekhukhune, Capricorn and Waterberg.

### *3.4 Machine Learning Algorithm Performance.*

After running nine models through the Auto model of the RapidMiner, the performance of the algorithms was compared to choose the best model that can be used to develop a malaria predictive model. The algorithms' performance comparison was done based on (1) classification performance and (2) computation time performance [43]. Tables 3 and 4 depict the classification performance of each algorithm in terms of accuracy, area under curve (AUC), classification error, FMeasure, precision, recall and sensitivity at Vhembe district level and provincial level. On the other hand, Table 6 depicts the computation time performance and overall rankings of the algorithms using scoring, training and total computation times.

As seen in Tables 3 and 4, all the algorithms show very high and equal performance. However, Fast Large Margin showed errors and was, therefore, excluded in further analysis.

**Table 3** Machine learning algorithm classification performance (Provincial data)

| Model | Accuracy (%) | AUC (%) | Classification error (%) | F Measure (%) | Precision (%) | Recall (%) | Sensitivity (%) |
|---|---|---|---|---|---|---|---|
| Decision tree | 98.7 | 0 | 1.3 | 99.3 | 98.7 | 100 | 100 |
| Deep learning | 98.7 | 0 | 1.3 | 99.3 | 98.7 | 100 | 100 |
| Fast large margin | 98.7 | 0 | 1.3 | 99.3 | 98.7 | 100 | 100 |
| Generalized linear model | 98.7 | 0 | 1.3 | 99.3 | 98.7 | 100 | 100 |
| Gradient boosted trees | 98.7 | 0 | 1.3 | 99.3 | 98.7 | 100 | 100 |
| Logistic regression | 98.7 | 0 | 1.3 | 99.3 | 98.7 | 100 | 100 |
| Naive Bayes | 98.7 | 0 | 1.3 | 99.3 | 98.7 | 100 | 100 |
| Random forest | 98.7 | 0 | 1.3 | 99.3 | 98.7 | 100 | 100 |
| Support vector machine | 98.7 | 0 | 1.3 | 99.3 | 98.7 | 100 | 100 |

Table 5 depicts the total, training and scoring times of each machine learning algorithm used in this study. The algorithms have different rankings for the total, training, scoring, computation times. For example, Random Forest has the best training time, first position, but came sixth in total time. On the other hand, even though Logistic Regression was among the worst-performing algorithm based on training time, it has a high ranking on both total and scoring times as it came in first position for both times. To compute subjectively the overall performance of the algorithms, the sum of total, training and scoring times of each algorithm were calculated and used for ranking. In the final ranking (column "Overall Rank"), Decision Tree, Logistic Regression, Random Forest, Naive Bayes, and Deep Learning were the top four best-performing algorithms. While Gradient Boosted Trees, Support Vector Machine and Generalized Linear Model had the worst-ranking performance. Since all the algorithms have equal classification performance, when choosing the best algorithm, we used the computation time ranking factor (Table 6).

## 3.5 *Malaria Prediction System Development*

Based on the performance rankings done in the previous section, Decision Tree was selected to build a malaria prediction system. Using the variables identified earlier in

**Table 4** Machine learning algorithm classification performance (Vhembe data)

| Model | Accuracy | AUC | Classification Error | F Measure | Precision | Recall | Sensitivity |
|---|---|---|---|---|---|---|---|
| Decision tree (%) | 100 | 0 | 0 | 100 | 100 | 100 | 100 |
| Deep learning (%) | 100 | 0 | 0 | 100 | 100 | 100 | 100 |
| Fast large margin | Error | Error | Error | Error | Error | Error | Error |
| Generalized linear model (%) | 100 | 0 | 0 | 100 | 100 | 100 | 100 |
| Gradient boosted trees (%) | 100 | 0 | 0 | 100 | 100 | 100 | 100 |
| Logistic regression (%) | 100 | 0 | 0 | 100 | 100 | 100 | 100 |
| Naive Bayes (%) | 100 | 0 | 0 | 100 | 100 | 100 | 100 |
| Random forest (%) | 100 | 0 | 0 | 100 | 100 | 100 | 100 |
| Support vector machine (%) | 100 | 0 | 0 | 100 | 100 | 100 | 100 |

**Table 5** Machine learning algorithm computation time performance ranking

| Model | Total time | Training time (1,000 Rows) | Scoring time (1,000 Rows) | Rank (Total) | Rank (Training) | Rank (Scoring) | Overall rank |
|---|---|---|---|---|---|---|---|
| Naive Bayes | 32,731,0 | 3135,7 | 951,5 | 5 | 3 | 6 | (4)14 |
| Generalized linear model | 20,311,0 | 10,182,2 | 757,3 | 3 | 8 | 4 | (5)15 |
| Logistic regression | 15,004,0 | 7639,5 | 417,5 | 1 | 6 | 1 | (2)8 |
| Deep learning | 22,747,0 | 9984,5 | 679,6 | 4 | 7 | 3 | (4)14 |
| Decision tree | 17,840,0 | 2934,1 | 466,0 | 2 | 2 | 2 | (1)6 |
| Random forest | 41,618,0 | 651,2 | 825,2 | 6 | 1 | 5 | (3)12 |
| Gradient boosted trees | 383,287,0 | 3581,4 | 2019,4 | 8 | 4 | 8 | (7)20 |
| Support vector machine | 70,159,0 | 4391,5 | 1349,5 | 7 | 5 | 7 | (6)19 |

**Table 6** Decision tree simulator for the prediction of malaria incidence

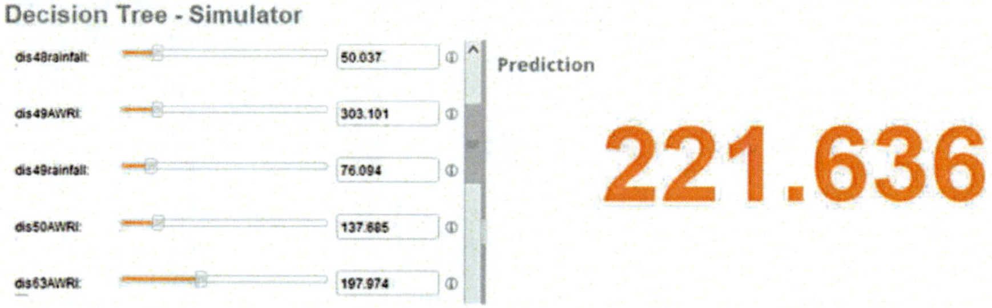


Sect. 3.4, RapidMiner was used to predict malaria incidence and the highest weights contributed by the variables are as follows: AWRIDis86 = 0.3, AWRIDis63 = 0.2, AWRIDis76 = 0.2, AWRIDis48 = 0.1, AWRIDis65 = 0.1. The weights of other predictors were very close to zero. This indicates that AWRI is the main indicator for malaria prediction in Vhembe district and Limpopo province. Figure 7 depicts the Decision Tree simulator. The dependent variables are input into the model and the output is malaria incidence predicted based on the dependent variables.

## 4 Evaluation, Discussion and Conclusion

### *4.1 Evaluation*

As indicated in Sect. 3.1, Malaria incidence and weather dataset for 20 years were used to train the model while dataset for one year was selected randomly to test the model. Using the developed model, the prediction of the occurrence of malaria was done based on the prediction factors identified earlier. The Simulator tool in RapidMiner was used to predict malaria. The results of this evaluation are depicted in Table 7.

### *4.2 Discussion and Conclusion*

AWRI is an excellent measure of conduciveness of malaria-causing vector. The fact that the whole EDI is a bottom-up approach, it is able to capture micro-details of hydrological droughts and is, therefore, a good predictor for malaria incidence. The eight machine learning algorithms used in this study showed very good performance during the training process; 100% for accuracy, precision, FMeasure, sensitivity and recall and 0% for AUC and classification error at the district level, Vhembe district. Less accurate but still very good performance of the same algorithms was seen when

**Table 7** Predictor for Mopani malaria incidence—evaluation results

| Month | Actual value | Predicted value | Relative error (%) |
|---|---|---|---|
| 1 | 626 | 620,2 | 1.0 |
| 2 | 289 | 291,1 | 0.7 |
| 3 | 1197 | 1172,4 | 2.1 |
| 4 | 2499 | 2458,0 | 1.7 |
| 5 | 3845 | 1258,3 | 1.6 |
| 6 | 285 | 290,9 | 2.1 |
| 7 | 177 | 180,4 | 1.9 |
| 8 | 373 | 380,7 | 2.1 |
| 9 | 1238 | 1209,9 | 2.3 |
| 10 | 1042 | 1051,9 | 1.0 |
| 11 | 313 | 311,6 | 0.4 |
| 12 | 1437 | 1450,6 | 0.9 |

using provincial data. The algorithms obtained 98.7% for accuracy, precision, and 99.3% for FMeasure, 100% for sensitivity and recall, 0% for AUC, and 1.3% for classification error. Modelling of malaria incidence being done at the micro-level, district and provincial level explains this high performance of the algorithms. This implies that as the scope of coverage increases, the performance of the algorithms is likely to decrease. For example, if this model can be extended to the national level, South Africa, or even to large scale, South Sahara Africa, less accuracy of the algorithms could be expected.

As per our research objective 1, according to the results presented in this paper, it is evident that there is a very strong correlation between hydrological drought indices and malaria incidence reported in Limpopo. More scientific explanations for this association have also been provided. As per our research objective 2, the objective was obtained to develop Decision Tree Malaria Predictor System using in RapidMiner. Even though Decision Tree and other machine learning algorithms used in this study, had a very high accuracy during the training process as explained previously, during the testing phase, the relative error of the developed malaria prediction system, is as high as 23% even though it goes as low as 4%. This is a good performance; however, better results were expected considering that the modelling is done at a micro-level (district and provincial level). The authors acknowledge that this due to the completeness and accuracy of most weather datasets—there were few missing data. Nevertheless, the prediction capacity of this malaria prediction system an excellent decision tool for stakeholders, especially in the health sector—it aids in preparing ahead of malaria outbreaks.

In its current form, our malaria prediction model can operate at the provincial level health centres of Limpopo and can be deployed in getting early alerts that can aid in preventive action before malaria outbreaks occur. Further work in terms of incorporating other variables such as temperature, humidity and vegetation indices is proposed. The authors also recommend the training of our models using datasets from other demographic areas for purpose of further evaluating models and ascertaining the model's scalability to the other parts of South Africa, Southern Africa and Africa at large.

## References

1. WHO (2019) World malaria Report 2019, Geneva
2. WHO (2015) Global technical strategy for malaria 2016–2030
3. Panch T, Szolovits P, Atun R (2018) Artificial intelligence, machine learning and health systems. J Glob Health 8(2):1–8. https://doi.org/10.7189/jogh.08.020303
4. Ramadona AL, Lazuardi L, Hii YL, Holmner Å, Kusnanto H, Rocklöv J (2016) Prediction of dengue outbreaks based on disease surveillance and meteorological data. PLoS ONE 11(3):1–18. https://doi.org/10.1371/journal.pone.0152688
5. Macherera M, Chimbari MJ (2016) A review of studies on community based early warning systems. Jamba J Disaster Risk Stud 8(1):1–10. https://doi.org/10.4102/jamba.v8i1.206
6. Ceccato P, Ramirez B, Manyangadze T, Gwakisa P, Thomson MC (2018) Data and tools to integrate climate and environmental information into public health. Infect Dis Poverty 7(1):1–11. https://doi.org/10.1186/s40249-018-0501-9
7. Bansal S, Chowell G, Simonsen L, Vespignani A, Viboud C (2016) Big data for infectious disease surveillance and modeling. J Infect Dis 214(Suppl 4):S375–S379. https://doi.org/10.1093/infdis/jiw400
8. Sharma V, Kumar A, Panat L, Karajkhede G (2015) Malaria outbreak prediction model using ML 4(12):4415–4419
9. Sarkar BK, Sana SS (2019) An e-healthcare system for disease prediction using hybrid data mining technique. J Model Manag 14(3):628–661. https://doi.org/10.1108/JM2-05-2018-0069
10. Modu B, Polovina N, Lan Y, Konur S, Asyhari AT, Peng Y (2017) Towards a predictive analytics-based intelligent malaria outbreak warning system. Appl Sci 7(8):1–20, 2017. https://doi.org/10.3390/app7080836
11. Wu J, Guo S, Li J, Zeng D (2016) Big data meet green challenges: greening big data. IEEE Syst J 10(3):873–887. https://doi.org/10.1109/JSYST.2016.2550538
12. Hassani H, Huang X, Silva E (2019) Big data and climate change. Big Data Cogn Comput 3(1):12. https://doi.org/10.3390/bdcc3010012
13. Cella W et al (2019) Do climate changes alter the distribution and transmission of malaria? Evidence assessment and recommendations for future studies. Rev Soc Bras Med Trop 52:0–2. https://doi.org/10.1590/0037-8682-0308-2019
14. Immergluck LC et al (2019) Geographic surveillance of community associated MRSA infections in children using electronic health record data. BMC Infect Dis 19(1):1–12. https://doi.org/10.1186/s12879-019-3682-3
15. Shirk R, Bonney (2020) Handbook of Citizen science in ecology and conservatio. University of California Press, California
16. Byun HR, Wilhite DA (1999) Objective quantification of drought severity and duration. J Clim 12:2747–2756
17. Manogaran G, Lopez D (2017) Disease surveillance system for big climate data processing and Dengue transmission. Int J Ambient Comput Intell 8(2):88–105. https://doi.org/10.4018/IJACI.2017040106
18. Li Z, Huang Q, Carbone GJ, Hu F (2017) A high performance query analytical framework for supporting data-intensive climate studies. Comput Environ Urban Syst 62(2017):210–221. https://doi.org/10.1016/j.compenvurbsys.2016.12.003
19. Manogaran G, Lopez D, Chilamkurti N (2018) In-mapper combiner based MapReduce algorithm for processing of big climate data. Futur Gener Comput Syst 86:433–445. https://doi.org/10.1016/j.future.2018.02.048

20. JLSA et al (2014) MERRA analytic services: meeting the big data challenges of climate science through cloud-enabled climate analytics-as-a-service. Comput Environ Urban Syst
21. Masinde M (2014) An effective drought early warning system for sub-Saharan Africa: integrating modern and indigenous approaches. ACM Int Conf Proceed Ser 60–69. https://doi.org/10.1145/2664591.2664629
22. Manyangadze T, Chimbari MJ, Gebreslasie M, Ceccato P, Mukaratirwa S (2016) Modelling the spatial and seasonal distribution of suitable habitats of schistosomiasis intermediate host snails using Maxent in Ndumo area, KwaZulu-Natal Province, South Africa. Parasit Vectors 9(1):1–10. https://doi.org/10.1186/s13071-016-1834-5
23. Abisoye O, Gbenga JR (2015) A hybrid intelligent forecasting model to determine malaria
24. Amuakwa-Mensah F, Marbuah G, Mubanga M (2017) Climate variability and infectious diseases nexus: evidence from Sweden. Infect Dis Model 2(2):203–217. https://doi.org/10.1016/j.idm.2017.03.003
25. Altizer S, Ostfeld RS, Johnson PTJ, Kutz S, Harvell CD (2013) Climate change and infectious diseases: from evidence to a predictive framework. Science 341(6145):514–519. https://doi.org/10.1126/science.1239401
26. Bellard C, Bertelsmeier C, Leadley P, Thuiller W, Courchamp F (2012) Impacts of climate change on the future of biodiversity. Ecol Lett 15(4):365–377
27. Wu X, Lu Y, Zhou S, Chen L, Xu B (Jan. 2016) Impact of climate change on human infectious diseases: empirical evidence and human adaptation. Environ Int 86:14–23. https://doi.org/10.1016/j.envint.2015.09.007
28. Thomson MC, Muñoz ÁG, Cousin R, Shumake-Guillemot J (2018) Climate drivers of vector-borne diseases in Africa and their relevance to control programmes. Infect Dis Poverty 7(1):1–22. https://doi.org/10.1186/s40249-018-0460-1
29. World Health Organisation (2019) Guideline for malaria vector control, Geneva
30. Githeko AK, Ndegwa W (2001) Githeko_and_Ndegwa_Epidemic_prediction2(1)
31. Sharma G (2017) Impact factor: 5.2 Pros and cons of different sampling techniques. Int J Appl Res 3(7):749–752. www.allresearchjournal.com
32. Kifle MM, Teklemariam TT, Teweldeberhan AM, Tesfamariam EH, Andegiorgish AK, Kidane EA (2019) Malaria risk stratification and modeling the effect of rainfall on malaria incidence in Eritrea. J Environ Public Health 2019. https://doi.org/10.1155/2019/7314129
33. Warwick RA (1975) Drought hazard in the United States: A research assessment. NSF/RA/E-75/004 edn. c
34. Mishra VP, Singh AK (2011) Drought modeling—A review. J. Hydrol 403:157–175
35. PWC (1968) Keeping track of crop moisture conditions, nationwide: the new crop moisture index. Weatherwise 21:156–161
36. Wilhite H, Glantz M (1985) Understanding the drought phenomenon: the role of definitions. Water Int 10(3):111–120
37. Jiang F et al (2017) Artificial intelligence in healthcare: past, present and future. Stroke Vasc Neurol 2(4):230–243. https://doi.org/10.1136/svn-2017-000101
38. Masinde M (2020) Machine learning—application of machine learning on malaria incidence and climate data
39. Sadek RM et al (2019) Parkinson's disease prediction using artificial neural network 3(1). http://dstore.alazhar.edu.ps/xmlui/handle/123456789/302
40. Pradhan N, Rani G, Dhaka VS, Poonia RC (2020) Diabetes prediction using artificial neural network. Deep Learn Tech Biomed Heal Inform 121:327–339. https://doi.org/10.1016/b978-0-12-819061-6.00014-8
41. Thakur S, Dharavath R (2019) Artificial neural network based prediction of malaria abundances using big data: a knowledge capturing approach. Clin Epidemiol Glob Heal 7(1):121–126. https://doi.org/10.1016/j.cegh.2018.03.001
42. Chae S, Kwon S, Lee D (2018) Predicting infectious disease using deep learning and big data. Int J Environ Res Public Health 15(8). https://doi.org/10.3390/ijerph15081596
43. Wieland M, Pittore M (2014) Performance evaluation of machine learning algorithms for urban pattern recognition from multi-spectral satellite images. Remote Sens 6(4):2912–2939. https://doi.org/10.3390/rs6042912

# Arabic Speech Recognition Independent of Vocabulary for Isolated Words

**Ahmed Boumehdi and Abdellah Yousfi**

**Abstract** This paper presents a vocabulary-independent speech recognition model of isolated Arabic words. This approach uses a restricted dictionary of syllables and phonemes, and it is based on the hidden Markov model (HMM). The main objective of this contribution is to remedy the problem of insufficient vocabularies in automatic speech recognition systems. This new model has made it possible to recognize any given word even if it has never been pronounced. The model gives a very considerable recognition rate comparable to the limited vocabulary speech recognition system of isolated words.

**Keywords** Automatic speech recognition (ASR) · Arabic language · Hidden Markov Models (HMMs) · Semi-syllable · Phoneme · Isolated words · Vocabulary independent

## 1 Introduction

Nowadays, many intensive works have focused on automatic Arabic speech recognition. For any language, using a hidden Markov model-based system gives satisfactory results. However, the fact that the Arabic language has a huge vocabulary made the learning and the recognition phase delicate.

In fact, a traditional speech recognition system of a given language is based, during the learning phase, on a dictionary containing acoustic references of each word. Each word must be pronounced by several speakers. However, the Arabic language is very rich in vocabulary, which requires the use of thousands of records in order to learn only a very limited number of words.

---

A. Boumehdi (✉)
National School of Computer Science and Systems Analysis (ENSIAS), Information, Communication and Embedded Systems (ICES), Rabat, Morocco

A. Yousfi
Faculty of Law, Economics and Social Sciences, Souissi-Rabat Mohammed V University, Rabat, Morocco

 
X.-S. Yang et al. (eds.), *Proceedings of Sixth International Congress on Information and Communication Technology*, Lecture Notes in Networks and Systems 216,
https://doi.org/10.1007/978-981-16-1781-2_52

Several research studies have been carried out in order to develop a speech recognition system whose capabilities are close to those of humans.

In 1952, Davis et al. [1] built an automatic speech recognition system of isolated numbers for a single speaker. This system was able to recognize 98% of these numbers.

In 1970, Itakura and Saito formulated the fundamental principles of the LPC method (linear predictive coding). This method was used in the extraction of voice characteristics.

In 2002, 2006, Yousfi and Meziane introduced a prosodic parameter vector associated with each phonetic unit, in a speaker-independent speech recognition system of isolated French words. The obtained recognition rate showed the importance of taking into account the prosodic parameters at the phonetic level [2, 3].

In 2016, Dzmitry [4] used a hybrid speech recognition system. This system uses a combination of neural networks with statistical models. The training and test data were made from a corpus containing 81 h of recordings. The error rate varied from 9.3 to 18.6% depending on the adopted language model (bi gram, trigram …).

In 2019, Zulkarnaen implemented a speech recognition system for small vocabulary composed of the first five numbers in the Indonesian language (0, 1, 2, 3, 4). This system was based mainly on HTK and resulted in a recognition rate of 100% [5].

## 2 HMM-Based Automatic Speech Recognition

In general, a speech recognition system is built according to the following three stages: the features extraction stage (acoustic analysis), the model construction (learning) stage and the recognition or decoding stage (Fig. 1).

### 2.1 *Acoustic Analysis*

The speech signal is a signal that contains a lot of redundancy, and this sets a cost problem (in terms of time and storage) when it is used in a raw way in computer speech processing applications.

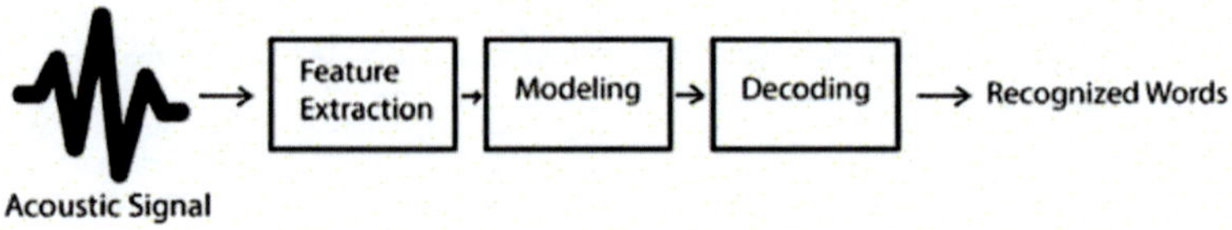


**Fig. 1** ASR principal stages

The speech signal is an almost stationary signal, that is why when it is divided into small frames, the signal is considered stationary. In general, the duration of these frames is between 16 and 40 ms.

Then, and once the signal has been cut into small frames, it only needs to extract relevant information from the speech signal and to eliminate any redundancy. Several acoustic analysis techniques exist, which makes it possible to recover and extract several types of parameters, among which we quote: MFCC coefficients (Mel-frequency Cepstral coefficients), LPC coefficients (linear predictive coding), LPCC coefficients, etc.

In this work, we used the 13 MFCC coefficients and their first and second derivatives. The MFCC vector calculation was done using the HTK HCopy function.

## *2.2 The Modeling and Learning Phase*

This phase consists of building models allowing the recognition of words spoken during a verbal speech.

This phase is made up of two main stages:

- The construction and the proposal of adequate architectures for the models used in recognition, such as, for example, the number of layers and the number of nodes when using the connectionist model, or the number of states, and the types of transition between states in the case of HMM models.
- The learning of the used model: after initializing the used model's parameters, we pass onto the step of re-estimating these parameters using one of the same algorithms used in the learning process.

## *2.3 The Recognition Phase (Decoding)*

Starting with information derived out of the acoustic analysis of the voice signal, this step makes it possible to extract the most suitable linguistic message associated with that signal.

To decode the information derived out of the acoustic analysis, one of the following approaches is often used: the analytical approach, the global approach, the statistical approach or the hybrid approach, which is a combination of these approaches.

1. Analytical approach: it consists of segmenting the signal into phonetic-sized units. Each segment is identified independently of the others, only to be then regrouped together in order to recognize the spoken words.

2. Global approach: it consists of identifying a word or a sentence as elementary entities by comparing them to recorded references. This comparison is carried out by using the principle of dynamic time warping (DTW), which is executed using the principle of optimality of Bellman [6].
3. Statistical approach: statistical models are used to decode the voice signal and to associate the most likely words with this signal. Among the most prevalent statistical models are the hidden Markov models (HMM), which have become the most popular statistical models in the past two decades.

In this work, we have used a hybrid approach, which combines the statistical approach based on the hidden Markov models (HMM) and the analytical approach.

## 3 HMM-Based Recognition

A hidden Markov model is a statistical model, defined by two stochastic processes [7, 8].

- $X_t$ a Markov stochastic process of the order of 1 with value in the set of states $Q = (q_1, q_2, \ldots, q_n)$. This process verifies:

$$\Pr \Pr(X_{t-1} = q_i, \ldots, X_1 = q_{i1}) = \Pr \Pr(X_{t-1} = q_i) = a_{ij}$$

- $Y_t$ an observable stochastic process with value in the set Y. $Y_t$ verifies:

$$\Pr \Pr(X_t = q_i, \ldots, X_1 = q_{i1}, Y_{t-1} = y_{t-1}, \ldots, Y_1 = y_1) = \Pr \Pr(X_t = q_i) = b_i(y_t)$$

$b_i(y_t)$ is the probability of getting the observation $y_t$ starting from the state $q_i$. In this case, each phoneme is defined by a set of states plus an initial state $q_I$ and a final state $q_F$. Below is an HMM model, figure showing five states of the consonant "ل" (L) (Fig. 2).

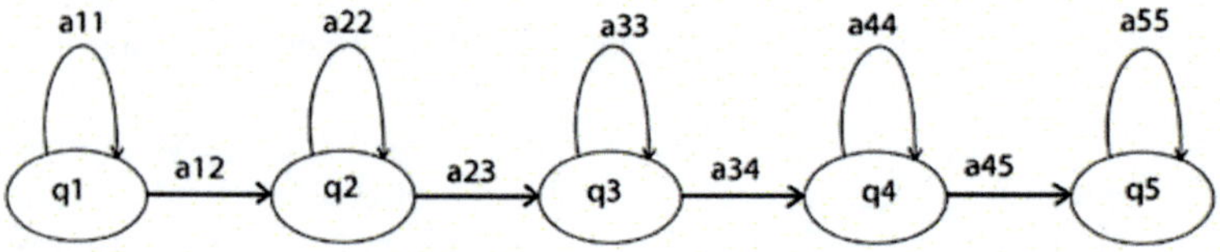


**Fig. 2** Example of a five-state HMM

# 4 Proposal of a Vocabulary Independent Recognition Model of Isolated Words

Currently, most automatic speech recognition systems are based on vocabularies with a limited number of words. In the recognition phase, this system always relies on this vocabulary to compare the voice signal and the words in this vocabulary. Several works have been carried out concerning the Arabic language in recent years, we cite for example:

- Satori and other authors have implemented a speech recognition system for the first 10 Arabic numbers (0–9) on Sphinx. The average recognition rate obtained is around 85.33% [9].
- Abubaker and others have developed a system for recognizing isolated Arabic words using neural networks. They tested their system on 14 Arabic words using a learning set made of several pronunciations of these words by 14 speakers [10].
- Other works on the Arabic language or dialects have been made, we cite for example [11–15].
- Khalifa and Yousfi have carried out a set of works on the recognition of Koranic sounds by introducing duration [14–17].

This kind of system has several drawbacks, for example:

- The system always gives recognition results belonging to vocabulary, even if the word to be recognized does not exist in the system's vocabulary.
- The capacity is very limited compared with the number of words to be recognized, whatever the size of the vocabulary, this system remains always incapable of correctly recognizing words outside of the vocabulary.

To remedy these problems, we propose in this article a new model for isolated word recognition, independent from the vocabulary. This system relies only on units of the same order as syllables (which we note syllable-phoneme or semi-syllable) to recognize any given single word.

## 4.1 The Semi-Syllable Unit

In this work, we tried to deal with the problem of insufficient lexicons in ASR systems, by defining a new dictionary made only of phonetic units, which are less influenced by the rules of coarticulation between these units. Each unit consists of one or two consonants followed or not by a vowel. For example, the semi-syllable units that can be generated starting from the letter "ب" (b) are given in Table 1.

Taking that into account, we defined 20 semi-syllables units for every single Arabic letter, and since the Arabic language is composed of 28 letters, it resulted in a dictionary containing a total of 28 × 20 = 532 units. Therefore, instead of setting

**Table 1** An extract from the list of semi-syllables related to letter "ب" (b)

| Segment | Segment code | Composition | Composition code |
|---|---|---|---|
| بْ | BS01 | ب | Bs |
| بَ | BS02 | ◌َ + ب | bs as |
| بِي | BS07 | ب + ي | bs + is+ it |
| بٌ | BS09 | ◌ٌ + ب | bs + us + nn |
| بَّ | BS12 | ◌َّ + ب | bs + bs + ad |
| بُّو | BS16 | ب + ◌ّ + و | bs + bs + uw |
| بًّا | BS18 | ب + ◌ّ + ا | bs + bs + ad + nn |

up a system with a dictionary containing words, we can have a system with only 532 semi-syllables, yet it will be capable of recognizing any word of Arabic.

## 4.2 *Global HMM Model*

In the hereby work, we present three levels of the proposed model:

- Semi-syllable level: it represents the semi-syllable units.
- Phoneme level: each semi-syllable is broken down into a set of phonemes.
- Acoustic level: it represents the HMM model associated with each phoneme.

Example:

After the construction of all the models for each syllable/phoneme, a global network can be built. It is represented by the (Fig. 3).

This global network consists of 532 semi-syllables, which are linked together by an initial state qI and a final state qF, plus a closed link, which directly ties the state qI to the state QF, and which makes it possible to connect between the semi-syllables in order to form words. In this global network, each semi-syllable is broken down into phonemes, which are modeled with HMMs.

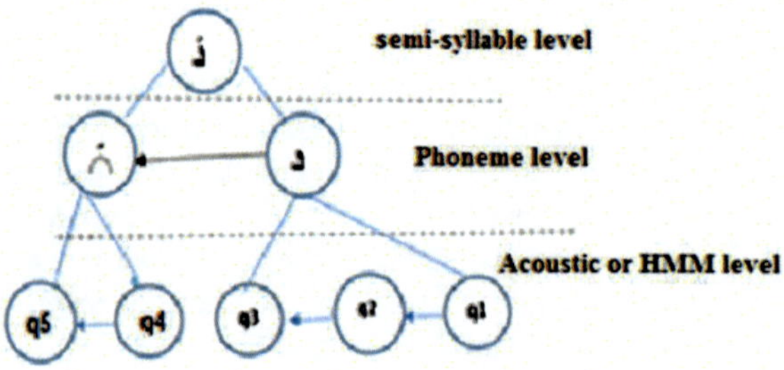


**Fig. 3** Three levels of the proposed model

### 4.3 Arabic Speech Recognition Systems Based on Semi-Syllables

This work's objective is to build a speech recognition system that does not depend on the system's vocabulary, which means, a system that is able to recognize any word of the spoken language. In order to reach this goal, we have introduced what we call semi-syllable units, and instead of recognizing words directly, we adapted the HMM-based system in such a way as to recognize semi-syllables, and to connect between these semi-syllables in order to form the pronounced word.

To build our model, we went through the following steps:

(1) The recording frequency is fe = 48000 Hz, and each sample was made in 32 bits.
(2) Each spoken word was recorded and then was cut into semi-syllabic units.
(3) The construction of a global HMM model (Fig. 4).
(4) The recognition phase: This phase allows to firstly recognize all the semi-syllables composing the spoken word, then concatenate these units to form the pronounced word.

The recognition of isolated words based on semi-syllables is done by searching for the optimal paths in the global HMM model. This research is done by the Viterbi algorithm [18]:

Let $w = y_1, y_2 \ldots y_n$ be a pronounced word composed of n observations. The recognition of this word based on the approach that we propose in this paper, and to the Viterbi algorithm is as follows:

$$\delta_t(q_j) = \max_{q_i \to q_j} \delta_{t-1}(q_i) \times a_{ij} \times b_j(y_t) \times f(\Pr(\phi_k(q_i)|\phi_{k-1}))$$

Fig. 4 Global network of semi-syllables model

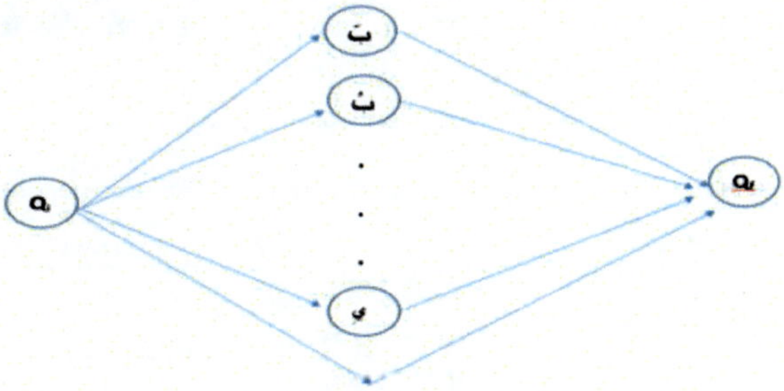

With:

- $\phi_k(q_i)$ is the semi-syllable containing $q_i$.

$$f(\Pr(\phi_k(q_i)|\phi_{k-1})) = \begin{cases} \Pr(\phi_k(q_i)|\phi_{k-1}) & si\ \phi_k(q_i) \neq \phi_k(q_j) \\ 1 & si\ \phi_k(q_i) = \phi_k(q_j) \end{cases}$$

- $\phi_{k-1}$ is the semi-syllable which is just before $\phi_k(q_i)$.

The HMM models were built using the HTK function HRest.

# 5 Tests and Results

## 5.1 Training Corpus

In this work, we have used a training corpus composed of 1500 pronunciations of all the semi-syllables and 2457 pronunciations of the 38 phonemes in the Arabic language, which means an average of 2.8 pronunciations per semi-syllable, and an average of 64 pronunciations per phoneme. A single speaker spoke these pronunciations.

For the estimation of the b-grams language model parameters, we used the set of all Koranic words (77,760 words). The training on this set allowed estimating 11,561 bi-gram probabilities between two semi-syllables.

## 5.2 Test and Results

The test is carried out on types of word sets spoken by the same speaker. This set contains 211 words, which are different from the words included in the training phase. It is called TestSet (Table 2).

**Table 2** The recognition rates of semi-syllables in the training data

| Semi-syllable | Code | Count | Recognition rate (%) |
|---|---|---|---|
| ... سّ - صْ | SS09-SB01 ... | 10 | 0–40 |
| ... ب - صِ | BS14-SB04 ... | 2 | 40–50 |
| ... بّا - سّا | BS08-SS08 ... | 5 | 50–60 |
| ... مُ - ذّا | MS03-DS08 ... | 22 | 60–70 |
| ... جو - قّي | JB06-QS07 ... | 14 | 70–80 |
| ... صو - را | SB06-RS05 ... | 31 | 80–90 |
| ... جَ - لَ | JB02-LS02 ... | 161 | 90–100 |

**Table 3** The recognition rates of semi-syllables in the test data

| Semi-syllable | Code | Number | Recognition rate (%) |
|---|---|---|---|
| ذْ - بٰ ... | SS01-BS12 … | 4 | 0–40 |
| – | – | 0 | 40–50 |
| لْ - لو ... | LS06-LS01 … | 5 | 50–60 |
| ءُ - جا ... | JB05-HZ03 … | 3 | 60–70 |
| تٌّ - قَ ... | QS02-TS09 … | 3 | 70–80 |
| نَ - بُ ... | BS03-NS02 … | 8 | 80–90 |
| فِ - رِ ... | FS04-RS04 … | 90 | 90–100 |

**Table 4** The recognition rates of phonemes in the training data

| Phoneme | Recognition rate (%) | Phoneme | Recognition rate (%) |
|---|---|---|---|
| ب | 69.57 | ي | 100.0 |
| ن | 70.0 | ج | 100.0 |
| ز | 75.0 | ك | 100.0 |
| س | 78.95 | ذ | 100.0 |
| ت | 80.0 | غ | 100.0 |
| ق | 83.33 | ز | 100.0 |
| ض | 87.5 | ص | 100.0 |
| د | 88.89 | ش | 100.0 |
| ع | 91.67 | ل | 100.0 |
| م | 93.94 | خ | 100.0 |
| ر | 100.0 | ح | 100.0 |
| أ | 100.0 | و | 100.0 |
| ف | 100.0 | ث | 100.0 |
| هـ | 100.0 | ظ | 100.0 |

The table below shows the lowest recognition rates and the highest recognition rates that were recorded and compared with the training data (Table 3).

The table below shows the lowest recognition rates and the highest recognition rates that were recorded and compared with the test data (Table 4).

Concerning the test data, the table below summarizes the recognition rate of each phoneme in Arabic (Table 5).

Regarding the learning data, the table below summarizes the recognition rate of each phoneme of the Arabic language (Table 5).

For the recognition rate on the 211 words that are never pronounced in this system, this rate is 82.46% for the words and 93.40% for the semi-syllables.

In general, the results obtained are very satisfactory since most of the phonemes or syllables are recognized with an almost 100% percentage. The errors observed are mainly due to the fact that the phonemes or semi-syllables are very similar in terms of their pronunciation, for example س and ص and also ض and د.

**Table 5** The recognition rates of phonemes in test data

| Phoneme | Recognition rate (%) | Phoneme | Recognition rate (%) |
|---|---|---|---|
| ض | 70.0 | ت | 96.67 |
| ن | 82.35 | ط | 100.0 |
| ص | 86.67 | و | 100.0 |
| ز | 87.5 | ش | 100.0 |
| س | 90.32 | ر | 100.0 |
| ل | 90.48 | ي | 100.0 |
| ك | 92.31 | خ | 100.0 |
| ب | 92.98 | ذ | 100.0 |
| ء | 93.33 | ز | 100.0 |
| ق | 95.45 | ه | 100.0 |
| ع | 95.65 | غ | 100.0 |
| د | 95.83 | ح | 100.0 |
| ف | 96.55 | ج | 100.0 |
| م | 96.67 | | |

## 6 Conclusion

This paper presents a new approach that aims to improve the Arabic language automatic recognition by setting up a vocabulary independent model. This model is indeed not based on a vocabulary of words carrying out the recognition, although it is able to recognize any word of the Arabic language.

This model uses a pronunciation set of Arabic words containing all the semi-syllables of the language. The model, therefore, needs a few thousand words to cover all the 530 semi-syllables of the Arabic language.

With a learning corpus containing only 1,400 pronunciations, a reasonable success rate of 85.19 was achieved. The obtained test results are clearly very satisfactory and encouraging and show the value of the approach we have proposed. Following this work, we will extend this system so that it is multi-speaker in order to attain recognition of both isolated words and continuous speech.

## References

1. Davis KH, Biddulph R, Balashek S (1952) Automatic recognition of spoken digits. J Acoust Soc Am 24:637
2. Yousfi A, Meziane A (2002) Introduction of the speaking rate in the model of speech recognition. Int J Math Math Sci IJMMS 29(2):121–124
3. Yousfi A, Meziane A (2006) The centisecond two levels hidden semi markov model (CTLHSMM). In: Fifth international conference on parallel computing in electrical engineering (PARELEC 2006), Bialystok, Poland

4. Bahdanau D, Chorowski J, Serdyuk D, Brakel P, Bengio Y (2016) End-to-end attention-based large vocabulary speech recognition. In: International conference on acoustics, speech and signal processing (ICASSP), Shanghai, China
5. Hatala Z, Elektro JT, Ambon PN, Practical speech recognition with HTK
6. Bellman R (1954) The theory of dynamic programming. Bull Am Math Soc 60(6):503–515
7. Baker JK (1975) The DRAGON system—an overview. IEEE Trans Acoustics Speech Signal Process 23(1):24–29
8. Jelinek F (1976) Continuous recognition by statistical methods. In: Proceedings IEEE, vol 64, no A, pp 532–555
9. Satori H, Harti M, Chenfour N (2007) Système de Reconnaissance Automatique de l'arabe basé sur CMUSphinx
10. Abushofa A, Hmad NFM (2010) Arabic speech recognition (ASR). In: The Libyan Arab international conference on electrical and electronic engineering LAICEEE At, Tripoli, Libya
11. Dammak AM (2016) Approche hybride pour la reconnaissance automatique de la parole en langue arabe. Université du Maine, France, Thèse de doctorat
12. Menacer MA, Mella O, Fohr D, Jouvet D, Langlois D, Smaili K (2017) An enhanced automatic speech recognition system for Arabic. In: Proceedings of the third Arabic natural language processing workshop, Valencia, Spain
13. Menacer MA, Mella O, Fohr D, Jouvet D, Langlois D, Smaili K (2017) Development of the Arabic Loria automatic speech recognition system (ALASR) and its evaluation for Algerian dialect. In: 3rd International conference on Arabic computational linguistics, ACLing, Dubai, United Arab Emirates
14. Khelifa MOM, Elhadj YM, Yousfi A, Belkasmi M (2017) Constructing accurate and robust HMM/GMM models for an Arabic speech recognition system. Int J Speech Technol 20(3)
15. Khelifa M, Elhadj YOM, Yousfi A, Belkasmi M (February 2017) Helpful statistics in recognizing basic Arabic phonemes. Int J Adv Comput Sci Appl 8(2):238–244
16. Khelifa M, Yousfi A, Elhadj YOM, Belkasmi M (2017) An accurate HSMM-based system for Arabic phoneme recognition. In: The ninth international conference on advanced computational intelligence at, Doha, Qatar
17. Elhadj YOM, Khelifa M, Yousfi A, Belkasmi M (2016) An accurate recognizer for basic Arabic sounds. ARPN J Eng Appl Sci 11(5)
18. Robert CP, Celeux G, Diebolt J (1993) Bayesian estimation of hidden Markov chains: a stochastic implementation. Stat Probab Lett 16:77–83

# Development of Design of a Low-Cost Air Quality Monitoring System in Binh Duong Province

**Thuyet That Vo Nguyen, Nguyen Duc Thinh, Hien Vo, Tran Quang Nhu, and Nguyen Binh Minh**

**Abstract** This paper represents the development of the design of a low-cost air quality monitoring system in the local area in Binh Duong, Vietnam. The system monitors air quality at a position based on the humidity, temperature, PM. 10, PM. 2.5, and PM 1.0 index in real time for a month. Then the system was developed to monitor air quality at the Vietnamese-German University campus in Binh Duong, Vietnam. The first-generation model was designed by using Raspberry Pi as the central module for collecting data and then sent them to the cloud. The model is small and compact, it can withstand in Vietnamese environment, no different temperature inside and outside of the case (self-cooling), minimum resistance of airflow through the case from any direction. The second-generation model was designed by using STM32L151 microcontroller-based low-cost modules that can be integrated with a variety of low-cost sensors.

**Keywords** IoT · Air monitoring · Optical sensor · Microcontroller · PM1 · PM2.5 · PM10

## 1 Introduction

Air quality in Vietnam: According to the report of Environmental Performance Index (EPI), Vietnam was at 161/180 of the rank [1]. According to the Vietnam National Environmental Performance Assessment Report, Vietnam did not approach the expected requirement [2]. Binh Duong province is one of the most developed industrial areas in Vietnam, having 28 industrial parks with a total area of more than 10,000 hectares with a population of 1.7 million [3] but monitoring air quality around these industrial zones is not considered seriously. Urban health takes this effect obviously even the number of people and diseases related to lung and breathing problems are not listed. The authority had organized a Hackathon competition and scientific

T. T. V. Nguyen · N. D. Thinh (✉) · H. Vo · T. Q. Nhu · N. B. Minh
Vietnamese-German University, Binh Duong, Vietnam
e-mail: thinh.nd@vgu.edu.vn

 
X.-S. Yang et al. (eds.), *Proceedings of Sixth International Congress on Information and Communication Technology*, Lecture Notes in Networks and Systems 216,
https://doi.org/10.1007/978-981-16-1781-2_53

workshop about smart solutions for Binh Duong New City in 2017 and monitoring air quality project was the first price.

This paper presents the development of a low-cost Air Quality Index (AQI) system at Vietnamese-German University. The first AQI system was designed by using Raspberry Pi as the central module for collecting data but the cost was high and it was a problem when exposing the number of AQI in many nodes in Binh Duong. The second AQI was designed by using a low-cost microcontroller (STM32L151) that can be integrated with a variety of low-cost sensors for 10 nodes in the VGU campus.

The rest of this paper is organized as follows. Section 2 presents some related work. Section 3 explains the proposed methodology. Section 4 compares two designs of the first model and the second model and how the models have designed. Section 5 is about the results of five measured parameters of two designs. Section 6 gives the conclusion and future work.

## 2 Literature Review

There are researches related to air quality monitoring in the world. OpenSense integrated the air quality measurements from heterogeneous mobile and crowd-sensing data sources, to understand the health impacts of air pollution exposure and to provide high-resolution urban air quality maps in Zürich [4]. The Array of Things, an intelligent urban measurement project in Chicago aimed to construct a sensor network for monitoring the air quality of the city [5]. The sensors used in this project are the ones used in our project. Xinghua Li investigated the relative differences (RD) between estimated PM2.5, 1.0 sampler concentrations and actual PM2.5, 1.0 concentrations associated with the interaction between performance characteristics for PM2.5, 1.0 size separator and measured typical urban PSDs in four seasons covering all pollution levels [6]. Georgia Miskell examined the effect of local sitting on data quality to address the overall inquiry on the use of low-cost data. Data from a network of instruments (mounted on a variety of sitting options, such as on regulatory stations or walls of private houses) were analyzed for their intra-(within a site) and inter-(between sites) variability. Differences within a site were compared with their surroundings using regression and standard statistical diagnostics to ascertain whether certain factors were related to large intra-site differences [7]. Florentin M. L Bulot evaluated the long-term capacity of the sensors to produce hourly data, informing about trends in pollution, a number of low-cost sensors needed at the same place, and the capacity of these sensors to detect short-lived events not detected by Automatic Urban and Rural Network (AURN) stations [8]. Jen-Hao Liu proposed a wireless sensor network air quality monitoring system integrated with the global system for mobile communications in the Taipei city to monitor the carbon monoxide (CO) concentration. Sensor nodes, a gateway, and a back-end platform are controlled by the LabVIEW program through which sensing data can be stored in a database [9]. William Lassman compared the total measured PM concentrations to Global Burden of Disease (GBD) estimates, as well as presented speciated-PM results, put the PM

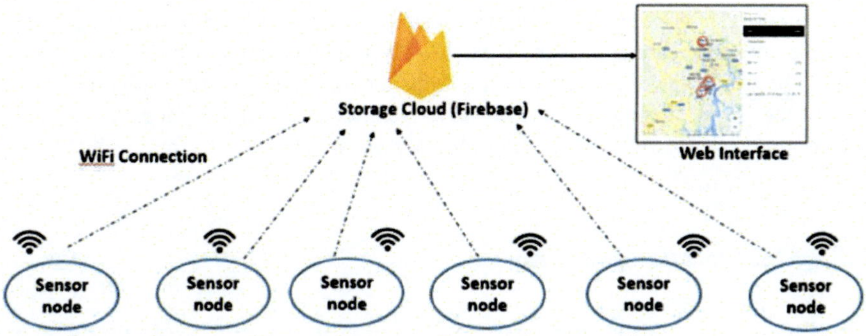


**Fig. 1** Air monitoring system with many nodes sending data to the cloud

measurements in the context of aerosol optical depth (AOD) measured by the Aerosol Mass and Optical Depth (AMOD) and satellites, and briefly discussed the important sources for PM2.5 by Plantower PMS 5003 in Palapye, Botswana [10] (Fig. 1).

# 3 Methodology

## *3.1 Working Principle of the System*

Working principle of the optical sensor: it allows air into an open scattering chamber with an induction fan. Light emitted from a laser source passes through a stream of sample air then scatted. All of the scattered light from the air stream that strikes the mirror is focused onto the photodiode, which produces a voltage as shown in Fig. 2.

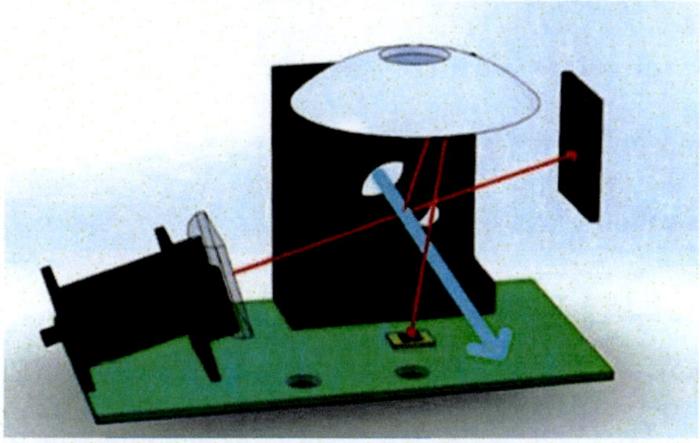

**Fig. 2** The interior of the OPC-N2 sensor, showing the relative directions of the laser beam (red), airstream (blue), and reflected light rays (red) reflecting off the elliptical mirror. Images created by Nicholas Wallace, 2018 [11]

The voltage range is dependent on the scattered light intensity, which is related to the size and concentration of the particle. The calculation is based on the look-up table. The advantage of using the optical sensor counting compared with gravimetric consulting is that it shows the real-time data collection of the concentration during the combustion. However, its results vary widely in price, design, and accuracy.

## 3.2 Mathematic Modeling

The output of the sensor represents the amount of uncertainty passing through the chamber by measuring a voltage produced by light scattering. The unit of particle measurement (PM) is in [$\mu g/m^3$] and it is defined by the following equation.

$$PM\left(M, \dot{V}_a, T\right) = \frac{M}{\dot{V}_a T} \tag{1}$$

where $\dot{V}_a\left(m^3/s\right)$ is the volumetric flow rate of air over the sample period $T(s)$.

$M(\mu g)$ is the mass of the particle sensed during one period,

$T$ is constantly based on the programming data acquisition system,

$\dot{V}_a$ is a constant, depending on the assumption that air is incompressible as it passes through the fan.

$M$ is the sum of all particle masses in relevant size bins multiplied by correction factors as the equation below.

$$M\left(V_p, \rho_p, \omega, ER, P\right) = \sum V_{p,i} \cdot \rho_{p,i} \cdot \omega_i \cdot ER_i \cdot P_i \tag{2}$$

where $\rho_{p,i}\left(kg/m^3\right)$ is the average density of particles in bin $i$,

$\omega_i$ is dimensionless sample volume weighting for bin $i$,

$ER_i$ (%) is a percentage of corresponding air quality standards of the country of origin [12].

$P_i$ is the number of particles in bin $i$,

$V_{p,i}$ $\left(m^3\right)$ is the volume of particles in the bin $i$ for $b$ bins, and be calculated in the following equation.

$$V_{p,i}\left(d_{\text{upper}}, d_{\text{lower}}\right) = \frac{3}{4}\pi\left(\frac{d_{\text{upper},i} + d_{\text{lower},i}}{4}\right)^3 \tag{3}$$

where:$d_{\text{upper},i}$, $d_{\text{lower},i}$ are the diameters of the upper and lower bin boundaries.

The particles being measured are assumed in the same size, the summation is eliminated and $V_p$ is calculated as:

$$V_p(d) = \frac{4}{3}\pi\left(\frac{d}{2}\right)^3 = \frac{\pi d^3}{6} \tag{4}$$

For non-spherical particles, the convention is to use aerodynamic diameter $d_a$ [13]

$$d_a = \sqrt{\frac{\mu u_s}{\rho_0 g}} \tag{5}$$

where $\mu$ is the viscosity coefficient of air,
$u_s$ is the particle settling velocity,
$\rho_0 = 1\ (\text{g/cm}^3)$ is the unit density or density of water,
$g = 9.81\ (\text{m/s}^2)$ is the gravitational acceleration.
$PM_1$ depends only on particles that are smaller than 1 $\mu$m,
$PM_{2.5}$ depends only on particles that are smaller than 2.5 $\mu$m [14].

# 4 Design

## *4.1 Hardware Design*

- *The first-generation model* (Fig. 3).

Design criteria:

- The air monitoring system is small and compact
- Can withstand in the Vietnamese environment

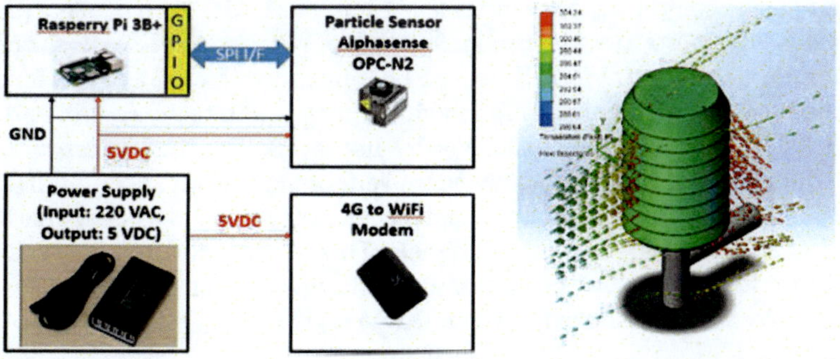


**Fig. 3** Block diagram of the first model using Raspberry Pi and outer case of the air monitoring system simulated in software

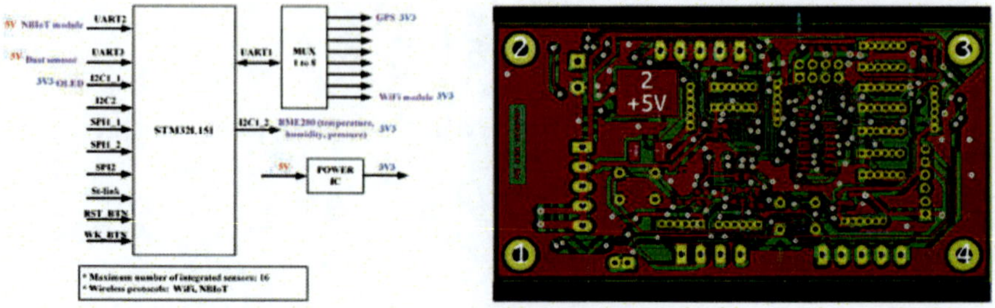

**Fig. 4** Block diagram and PCB layout of the second generation model

- No different temperature inside and outside of the case (self-cooling) to minimize the hot air effect in the measurement result.
- Minimum resistance of airflow through the case from any direction

The design case is taken in an airflow simulation environment to verify the design criteria. The result of the simulation shows the small difference in temperature inside and outside the case at an airflow velocity of 50 m/s. There is almost no different temperature between inside the case and outside the case.

- *The second-generation model.*

Design criteria: to build low-cost devices that can be integrated with a variety of low-cost sensors for AQI monitoring and wireless communication protocols (Wifi, NBIoT, etc.) for sending data to the cloud. Furthermore, a wireless sensor network for monitoring the AQI of a wide area can be built and deployed based on these low-cost devices (Fig. 4).

The central part is an ARM Cortex-M3 microcontroller (STM32L151) [15] playing the role of the main control unit. Sensors and wireless modules are also connected to the control unit for data transmission. The UART1 protocol is expanded from one to eight ports to achieve more numbers of sensors connected to the control unit. Besides, the other UARTs, SPIs, and I2Cs interfaces are used to maximize the number of sensors or communication modules integrated into the control unit up to 16. The 5VDC input is passed to a 5 V–3V3 linear regulator IC. The capacitors are used to filter out noises and make the output voltage more stable. The IC 74HC4051 is used as a MUX/DEMUX 1 to 8 to expand UART1 from one to 8 ports. This IC can be used with both analog and digital signals. This is a low-cost solution to expand micro-controller interfaces. The cost of a sensor node is given in Table 2. The PCB is designed by an open-source EDA tool, KiCAD [16].

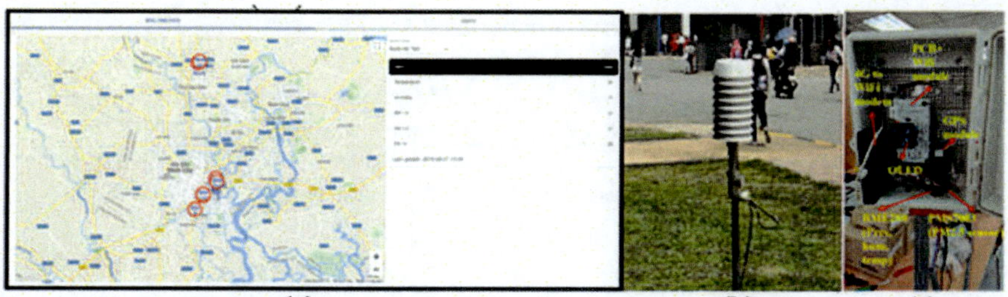

(a) (b) (c)

**Fig. 5** **a** Real-time data shown on Google map. **b** The first working prototype at the Vietnamese-German University campus. **c** The second working prototype with an OLED display at the Vietnamese-German University campus

## *4.2 Firmware*

Written by Python.

Libraries: spidev, firebase admin, opc.

https://py-opc.readthedocs.io/en/latest/

Python script starts automatically.

after Raspberry Pi finishes booting.

Web Guided User Interface (GUI) (Fig. 5).

**Experiment setup**: the node is fixed on a pole at 1.5-m high outdoor campus. The sensors collect data every 3 min then send them to the cloud. The value of data collected every 3 min is the average value that sensors measure. Users can get the data from the cloud in file.json. For the second model, the measurement values can be observed on the OLED display.

## 5 Measurement Result of Two Models

See (Table 1).

## 6 Conclusion

The first model was designed and run successfully with Raspberry Pi. The mechanical design showed the advanced when the different temperature inside and outside the case is not much. All data were collected then sent to the cloud. The second model is low cost, compact devices. It is very convenient to create wireless sensor networks

**Table 1** Measurement result of the first generation model in August 2019; measurement result of the second generation model in July 2020

| *Measurement result of the first generation model in August 2019* | *Measurement result of the second generation model in Jully 2020* |
|---|---|
| 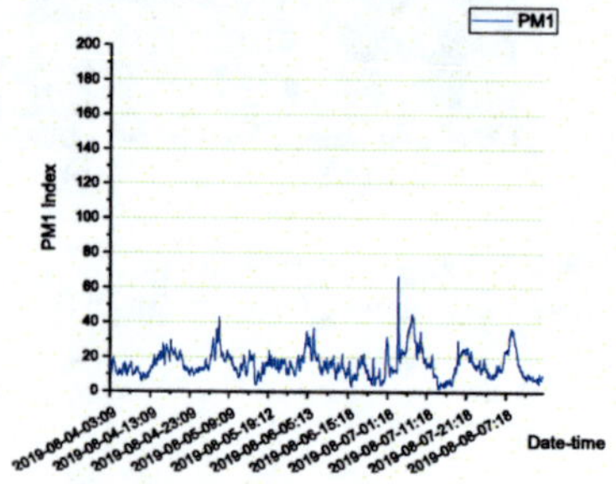 | 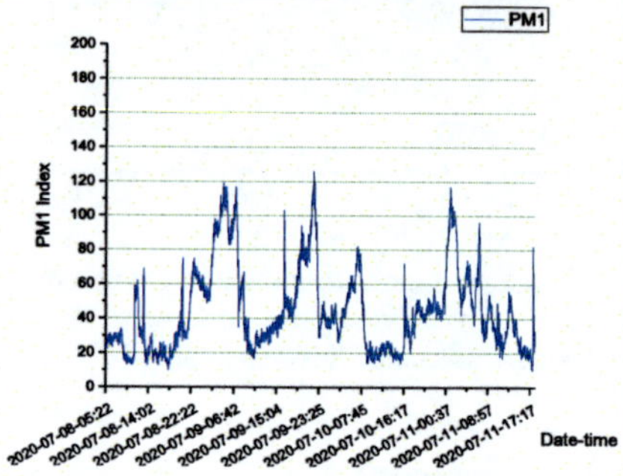 |
| 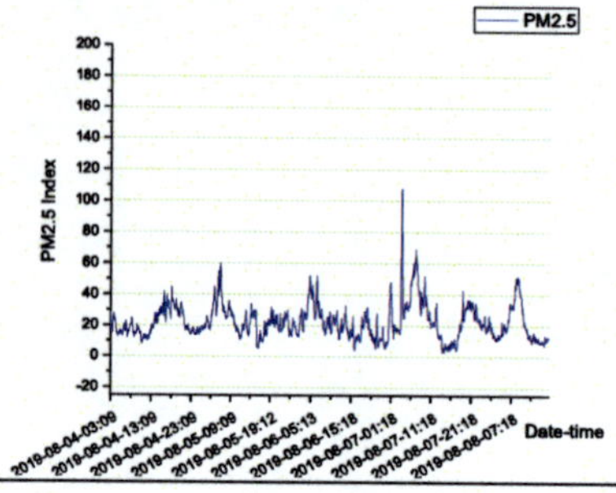 | 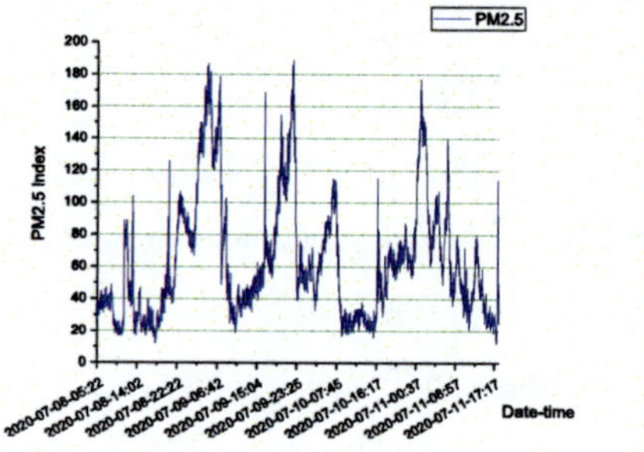 |

(continued)

from these devices. The components of the device are easy to find in the local market that definitely supports mass production. The second model has an OLED display to show the measured value and its design is more stable with RTOS and ARM-based microcontrollers. Besides, the advantages of NBIoT technology help create products with low-power consumption and high coverage signal.

The results of PM1, PM2.5, and PM10 values show that the second model has sensitivity better than the first model, however, the noise error of the second model is higher than the first model. Both results show a similar trend that the index value is high in day time and low at night time when the temperature is dropped. The humidity range measured in the first model and the second model is almost the same while the temperature range of the first model is a bit higher than the temperature range of the second model.

The limitation of the sensor is dependent on humidity level based on sensor manufacture specification. Long-term data need to be collected to remove errors from the trend. The collected data are from only one node in the same season but different years.

**Table 1** (continued)

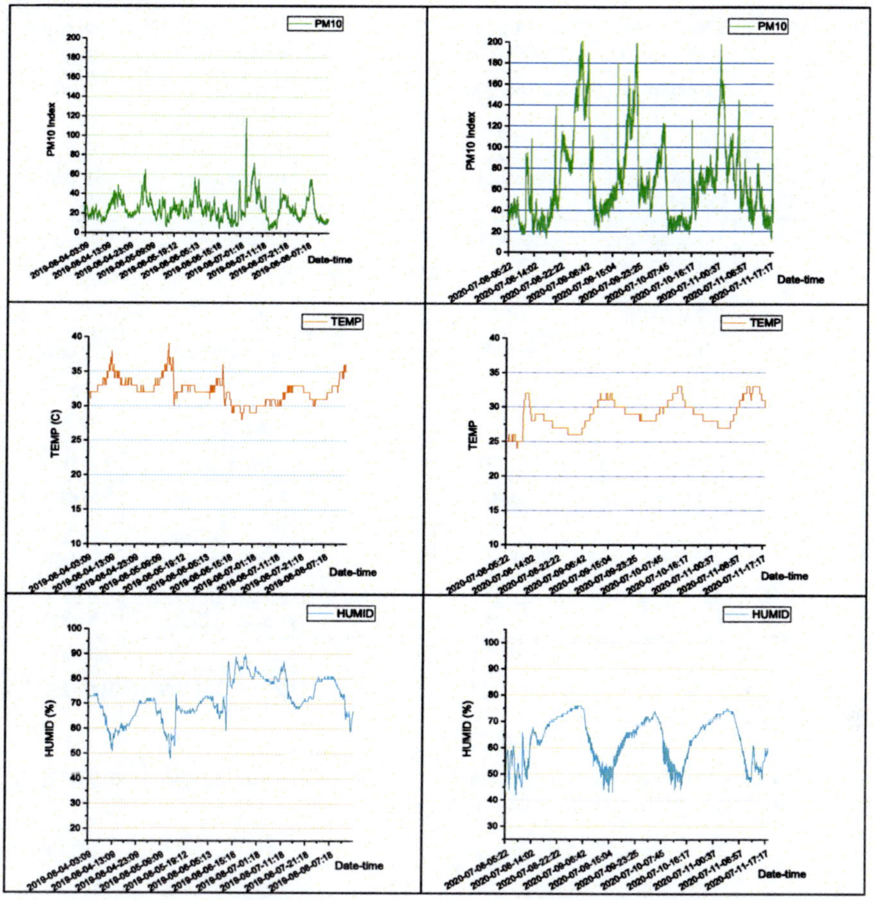


Future work of this research is that we will investigate by setting up three nodes at the same position but different attitudes to see how the behavior of air quality in different layers is high. The nodes will be set up by close groups to check the reliability of the result and stability of the second model system (Table 2).

**Table 2** Cost of one sensor node

| Number | The first AQI system (in 2019) | Unit price (VND) | Number | The second AQI system (in 2020) | Unit price (VND) |
|---|---|---|---|---|---|
| 1 | Raspberry Pi & SD card | 1 146 000 | 1 | PCB + electronics components | 200 000 |
| 2 | Alphasense OPC-N2 | Donated | 2 | PM sensor (PMS7003) | 485 000 |
| 3 | Temperature, humidity, and pressure sensor (BME280) | 150 000 | 3 | OLED | 95 000 |
| 4 | 4G LTE-Advanced TP-Link M7350 V5 150Mbps | 1 298 900 | 4 | Temperature, humidity, and pressure sensor (BME280) | 150 000 |
| 5 | Power adapter | 120 000 | 5 | GPS module | 290 000 |
| 6 | 3D printed cover | 2 000 000 | 6 | Wifi module (6 or 7) | 70 000 |
| 7 | Acrylic templates | 70 000 | 7 | NBIoT module (6 or 7) | 650 000 |
| | | | 8 | Power adapter | 120 000 |
| | | | 9 | Box | 100 000 |
| Total (max) | 4 784 900 VND (>200 USD) | | Total (max) | 2 090 000 VND (<100 USD) | |

**Acknowledgements** This research is funded by MoET under grant number B2019-VGU-03. Vietnamese-German University, Binh Duong, Vietnam.

## References


1. Wendling Z, et al (2018) Environmental performance index (EPI). Yale University & Columbia University. Accessed 12 Dec 2020
2. Vietnam National Environmental Performance Assessment Report (2008). Accessed 22 Nov 2019
3. https://en.wikipedia.org/wiki/Binh_Duong_Province. Accessed 14 Dec 2020
4. http://opensense.epfl.ch/wiki/index.php/OpenSense_2.html. Accessed 26 Nov 2019
5. https://arrayofthings.github.io/. Accessed 12 Dec 2019
6. Li X, Ruan B, Hopke PK, Mehmood T (2019) On the performance parameters of PM2.5 and PM1.0 size separators for ambient aerosol monitoring. Aerosol Air Qual Res 19:2173–2184
7. Miskell G, Salmond J, Williams DE (2016) Low-cost sensors and crowd-sourced data: observation of siting impacts on a network of air-quality instruments. Sci Total Environ 575:1119–1129 (Elsevier)
8. Bulot FMJ, Johnston SJ, Basford (2019) Long-term field comparison of multiple low-cost particulate matter sensors in an outdood urban environment. Nat Sci Reports 9

9. Liu J-H, Chen Y-F, Lin T-S (2011) Developed Urban air quality monitoring system based on wireless sensor networks. In: Fifth international conference on sensing technology. IEEE, New Zealand
10. Lassman W, Pierce JR, Bangs EJ (2020) Using low-cost measurement systems to investigate air quallity: a case study in palapye, botswana. Atmosphere 11(6)
11. Blanchard JT (2018) Uncertainty in optical particulate counting sensors. Undergraduate Honors Theses 12. https://scholarsarchive.byu.edu/studentpub_uht/12. Accessed 16 Dec 2020
12. E.C.f.S (CEN) (1993) Workplace atmospheres Size fraction definition for measurement of airbond particles, Bristish standard (481):1–15
13. Okubo M, Kuwahara T (2020) Principle and design of emission control systems. In: Chapter 3 in new technologies for emission control in marine diesel engines. Elsevier, pp 53–143
14. I.O for standardization (ISO) Air quality particle size fraction definition for health related sampling ISO publication (7708). Accessed 14 Dec 2020
15. Ultra-low-power 32-bit MCU ARM®-based Cortex®-M3 with 512KB
16. Flash, 80KB SRAM, 16KB EEPROM, LCD, USB, ADC, DAC. Accessed 16 Dec 2020

# Using Big Data to Compare Classification Models for Household Credit Rating in Kuwait

**Najla Albarrak, Hessa Alsanousi, Irene Moulitsas, and Salvatore Filippone**

**Abstract** Credit rating risks have become the backbone of bank performance. They are the reflection of the current status of the bank and the milestone for future planning. A good credit assessment can better anticipate expected losses and will minimize unexpected losses from accumulating. Given advancements in technology as well as the big data available within banks about customers in an oil country such as Kuwait, a built-in model to help in-household credit scoring is at management's decision. Compared with the current 'black box' rating models, we did a comparison between different classification models for two types of banking: conventional and Islamic. The classification models are as follows: Logistic Regression, Fine Decision Tree, Linear Support Vector Machines, Kernel Naïve Bayes, and RUSBoosted. Sufficiently, the last could be used to classify banks' household customers and determine their default cases.

**Keywords** Credit rating model · Credit risk · Technology · Conventional banking · Islamic banking · Classification models · Household customers · Machine learning · Logistic regression · Fine decision tree · Linear support vector machines · Kernel Naïve Bayes · RUSBoosted

N. Albarrak (✉) · H. Alsanousi
Central Bank of Kuwait, P. O. Box: 526, 13006 Safat, Kuwait
e-mail: n.albarrak@cranfield.ac.uk

H. Alsanousi
e-mail: h.alsanousi@cranfield.ac.uk

N. Albarrak · H. Alsanousi · I. Moulitsas · S. Filippone
Cranfield University, Cranfield MK43 0AL, Bedfordshire, UK
e-mail: i.moulitsas@cranfield.ac.uk

S. Filippone
e-mail: salvatore.filippone@cranfield.ac.uk

S. Filippone
University of Rome Tor Vergata, Via Cracovia, 50, 00133 Roma, RM, Italy



X.-S. Yang et al. (eds.), *Proceedings of Sixth International Congress on Information and Communication Technology*, Lecture Notes in Networks and Systems 216,
https://doi.org/10.1007/978-981-16-1781-2_54

# 1 Introduction

Financial crisis is not a new phenomenon or term. Rather, it is an ongoing bubble from the early stages of the financial world. The early stages left us with lessons in order to avoid problems before the next crisis appear. The main obstacle is how much this lesson costs and who is in charge of paying the bill. Banks in Kuwait have gone through several crises including the 2008 crisis and the oil prices drop. Kuwait stood strong without the need for any support from regulators' buddy for backup. This is due to the stringent regulations in place and the timely adoption of international regulations with additional safe buffers than internationally accepted benchmarks. Although we satisfactory overcame the crisis, there is always space for improvement to ensure a better climate for the next crisis arrival. Crises cannot be avoided given the globalization in the business world; however, we could be better prepared for it. From the international 2008 financial crisis, Basel committee has emphasized on the minimum capital available to cover the riskiness of bank's assets and investment decisions. Banking industry mainly operates by utilizing raised capital and borrowed funds to lend money and profit from the different rates.

The remainder of our paper is organized as follows. Section 2 provides a review of relevant literature. Section 3 describes our methodology. Section 4 describes our data collection. Our results are presented and analyzed in Section 5. Section 6 concludes the article.

# 2 Literature Review

Recent events shed importance on credit risks, which eventually drew the attention of bankers and regulators with regards to managing the credit portfolio efficiently. There are several machine learning and deep learning options to examine the credit probability of default [1]. In a very recent International Monetary Fund working paper [2], the advantages in financial technology are discussed and how machine learning solutions could reduce the cost of credit and provide much clearer solutions than the 'black box' templates for nontechnical audiences. From corporate to household customers, the recent literature recognized the importance of programming in loan granting process. Advancements in technology have created a collaboration between fields, computer science and finance, in order to benefit from the witnessed technological efficiency. In the context of using classification models for calculating the probability of customers loan default, there is a wide range of options to be implemented. Providing a credit rating model is fruitful for regulators and banks' decision-making evenly due to the advancement in the technology and the outperformance of the models developed and tested [3]. A study compared 17 different methods for credit classes showing that it is suitable to conduct classification methods for credit rating [4]. However, they did not use heavily imbalanced data as the case in their study. Imbalanced data in our context are data that are not well distributed between the two classes. In our data, almost 85% of the data are considered as good customers and the remaining 15% are prone to default. Given that when comparing machine learning methods to classify credit default customers, several studies showed that RUSBoost is the most significant method [5]. In research, it is evidenced that RUSBoost performance is significantly better for imbalanced data [6].

This paper aims to cover the gap of small-size samples for studies with long-big data samples [1]. As recommended in the literature, ensemble learning and gradient boosting decision trees are a solution for solving the disadvantages of decision trees especially if the data are large and have a long history, which is the case of our research [2]. The research done on similar work was on a period of 3 years while our aim is to expand it up to 11 years [3]. Another important aspect is that most studies have relayed on the same set of data gathered from the customer disregarding the data available with banks. Our study relied on data available with the banks currently and made use of it to estimate default customers [7]. The paper will also provide a multi-period observation along with historical events due to that the type of loan picked which has a term of 15 years. For the parameters, we chose customer characteristics already used along with new variables that have not been used before (monthly number of transactions done in bank accounts and monthly average cash flow in bank accounts). Moreover, in order to evaluate our results properly, we will use the standard measures in the field of credit scoring [1, 8]. Specifically, the standard measures are average accuracy, type I error, and type II error.

## 3 Methodology

The aim of this study is to come up with the most appropriate credit rating model to predict households default rate. In order to do so, we made a comparison between different classification model: Logistic Regression, Decision Tree, Linear Support Vector Machines, Bayesian Network, and RUSBoosted.

### *3.1 Logistic Regression*

In this paper, we will be distinguishing between two classes of creditors, good or bad [9]. For this binary response model, the response variable Y can take one of two set of values Y = 0 if the customer is good, non-defaulter, or Y = 1 if the customer is bad, defaulter. Xs are the columns vector of M explanatory variables, $\pi = P(Y = 1|X)$ is the responses probability and N is the number of observations

$$\text{logit}(\pi) = \log\left(\frac{\pi}{1-\pi}\right) = \alpha + \beta^T x \tag{1}$$

where $\alpha$ is the intercept and $\beta T$ is the coefficients.

## 3.2 Decision Tree

In classification decision trees, it starts with a single node then through a binary differences (1,0) results in the most information about the class [10]. Then we repeat the process with the resulting new node until we reach a position to stop. Usually, the tree is too large, so we back test it through a cross-validation. The dependent variable Y is categorical, so, by using information theory in measuring how much we know about it from knowing the value of another separate variable A

$$I[Y; A] = \sum_{a} \Pr(A = a) I[Y; A = a] \quad (2)$$

$I\ [Y; A = a]$ is the value of the uncertainty about Y decreases from knowing that $A = a$ given that we go from full population to sample where $A = a$. Therefore, $I\ [Y; A]$ is how much our doubt about Y reduces on average from knowing the value of $A$.

## 3.3 Support Vector Machines

From assuming a training set of $N\ \{(Xi,\ Yi)\}Ni = 1$ with input data Xi ε Rn and consistent binary class labels $Yi\ \varepsilon\ \{-1 + 1\}$, the SVM classifier in Vapnik's theory satisfies the following

$$yi\left[W^T \varphi(xi) + b\right] \geq 1, \qquad i = 1, \ldots, N \quad (3)$$

The non-liner function of φ (.) plots the input space to a high-dimensional feature space [4]. In which, the mentioned variations construct a hyperplane WT φ $(X) + b = 0$ discriminating between two classes. In original weighted space, the following equation is used for the classifier

$$y(x) = \text{sign}[w^T \varphi(x) + b] \quad (4)$$

But it is never evaluated in this form where the curved optimization problem could be defined as

$$\min_{w,b,\xi} j(w, b, \xi) = \frac{1}{2} w^T w + C \sum_{i=1}^{N} \xi i \quad (5)$$

Subject to

$$\begin{cases} y_i\left[W^T \varphi(xi) + b\right] \geq 1 - \xi_i, & i = 1, \ldots, N \\ \xi_i \geq 0, & i = 1, \ldots, N \end{cases} \quad (6)$$

The variables used in ξi are loose variables, which are needed to allow the misclassifications to occur in the set of inequalities due to overlying distribution. The first section of the objective function is set to maximize the margin between two classes in the feature space. The second part is set to minimize the misclassification error.

## 3.4 Bayesian Network

Bayesian Network is a simple and high-performance classifier [4]. This classification model works through learning the class condition probability p (*Xi*l*Y*) from each input variable *Xi* I = 1 … *n* given the class label *Y*. Then, a new observation is classified by Bayes' rule to calculate the following probability of each class of *Y* given the vector of observed feature values

$$p(y|x) = \frac{p(x|y)p(y)}{p(x)} \tag{7}$$

To make things easier, an assumption behind the naïve Bayes classifier is that the features are in theory independent given the class label, therefore,

$$p(x|y) = \prod_{i=1}^{n} p(x_i|y) \tag{8}$$

The probabilities *p* (*Xi*l*Y*) are then estimated through using the frequency counts for the discrete features and a normal-based method for the continued features.

## 3.5 Gradient Boosting

The gradient boosting technique, which is an ensemble algorithm founded by [11], is used to calculate the probability of loan default. It relies on incremental minimization of the error term, in which, improves the precision of the prediction function [9]. In trees, after setting the learner base, every tree calculated then is fit to the 'pseudo residuals', which is the deviation from the median and not from the expectation, from the earlier predictions in order to lower the error in general. Therefore, the following model is used

$$F(x) = G_0 + \beta_1 T_1(x) + \beta_2 T_2(x) + \cdots + \beta_n T_n(x) \tag{9}$$

*G*0 is the initial value for the set. $T_1$ … $T_n$ are the trees and $\beta_1$ … $\beta_n$ are the coefficients for particular tree nodes calculated by the algorithm. To conduct the gradient boosting classifier, a maximum branch size needs to be set. We chose 30 learning cycles and 0.1 learn rate.

## 4 Data

Our study is based on household customers in Kuwait with a sample of two banks, one conventional and one Islamic. The period chosen is 11 years, from 2008 to 2018, on monthly basis. There are two types of household loans: consumer loans and installment loans.

Consumer Loans: loans for the purpose of personal needs and durable goods with a limit of 15,000 KWD or 15 times the salary (whichever is less).

Installment Loans: loans for the purpose of maintenance or purchase of private residents with a limit of 70,000 KWD.

We chose installment loans exposures given the higher amount granted, which gives a higher impact on the economy. For bank 1, the conventional bank, we took a sample of 100,000 customer base, out of which 37,488 have loans (installment and consumer). The number of customers with installment loans was 28,033 with 996 default cases. When we calculate them in terms of observations for machine learning classification, we have 347,977 transactions given that each customer could have more than one Installment loan. For bank 2, the Islamic bank, we took a sample of 100,000 customer base, out of which 21,559 have loans (installment and consumer). The number of customers with installment loans is 15,108 with 1,394 default cases. When we calculate them in terms of observations for machine learning classification, we have 249,567 transactions given that each customer could have more than one installment loan.

To calculate the probability of household loans default, we gathered data of the loan's portfolio including outstanding balance, which is the amount left from the loan granted to be paid. The principal amount that is a part of the monthly total payment is to be paid against the principal amount of the loan. The remaining part of the monthly payment is the interest charge of the loan. Defining the default cases is the case when the customer fails to meet his monthly total obligations for 3 consecutive months while there is a remaining outstanding balance. This means that the customer is defaulting (bad customers). Hence, the ongoing payments of monthly obligations are considered non-defaulting (good customers).

The parameters, or independent variables, chosen for this study are:

- Credit card exposure
- Income
- Age
- Gender
- Education
- Nationality
- Relationship duration

- Monthly number of transactions done in bank accounts
- Monthly average cash flow in bank accounts
- Number of loans.

A limitation with bank 2 is that they did not provide us with nationality information.

## 5 Results

After running the classification models to predict the probability of default, a comparison of the area under the curve (AUC) is done to facilitate which model to choose (Table 1).

Also, Tables 2 and 3 are the summary of the confusion matrix output and the performance indicators.

From the illustration provided earlier, the RUSBoosted can be the most efficient model for calculating the probability of default due to, as discussed earlier, that our sample set is very large in bank 1 and runs for long period 11 years. These huge datasets show the lack of existing classification methods and enhance the importance of ensemble models. From an AUC of 0.86 to average accuracy of 79% with

**Table 1** AUC results

| Bank 1 | AUC | Bank 2 | AUC |
|---|---|---|---|
| Logistic regression | 0.71 | Logistic regression | 0.72 |
| Decision tree | 0.7 | Decision tree | 0.69 |
| Linear support Vector machines | 0.45 | Linear support Vector machines | 0.66 |
| Bayesian network | 0.67 | Bayesian network | 0.70 |
| RUSBoosted | 0.86 | RUSBoosted | 0.80 |

**Table 2** Confusion matrix results and model; performance results for bank 1

| Bank 1 | True positive (%) | True negative (%) | False positive (%) | False negative (%) | Average accuracy (%) | Type I error (%) | Type II error (%) |
|---|---|---|---|---|---|---|---|
| Logistic regression | 100 | 0 | 0 | 100 | 50 | 50 | 0 |
| Decision tree | 99 | 0 | 1 | 100 | 50 | 50 | 1 |
| Linear support Vector machines | 100 | 0 | 0 | 100 | 50 | 50 | 0 |
| Bayesian network | 99 | 0 | 1 | 100 | 50 | 50 | 1 |
| RUSBoosted | 84 | 74 | 16 | 26 | 79 | 24 | 16 |

**Table 3** Confusion matrix results and model; performance results for bank 2

| Bank 2 | True positive (%) | True negative (%) | False positive (%) | False negative (%) | Average accuracy (%) | Type I error (%) | Type II error (%) |
|---|---|---|---|---|---|---|---|
| Logistic regression | 99 | 0 | 1 | 100 | 50 | 50 | 1 |
| Decision tree | 99 | 2 | 1 | 98 | 51 | 50 | 1 |
| Linear support Vector machines | 100 | 0 | 0 | 100 | 50 | 50 | 0 |
| Bayesian network | 99 | 7 | 93 | 1 | 53 | 1 | 48 |
| RUSBoosted | 67 | 78 | 22 | 33 | 73 | 33 | 25 |

lowest value in type I error (24%), the RUSBoosted is the ideal solution for big data classifications, especially in imbalanced data. The increase rate in type II error of 16% could be justified by that RUSBoosted is the only model working for the data set on hand indicating false cases while the rest models are overfitted with zero or 1% values in false positive. An important clarification is that this model is used to estimate the credit default for household's customers in order to classify the customers into stages for determining the unexpected loss of good customers. Given that we defined the default rate for three consecutive non-payments, as per the provisioning scheme, the high false-negative rates are covered risk wise through the provision charges.

In bank 2, an AUC of 0.80 and average accuracy of 73% with the lowest value in type II error, 25%, the RUSBoosted is the ideal solution for big data classifications, especially in imbalanced data. The increase rate in type I error of 33% could be justified by, as stated, that RUSBoosted is the only model working for the data set on hand indicating false cases while the rest models are overfitted with zero or 1% values in false positive.

To evaluate our model, a training subsample for bank 1 of 70% (243,584 observations) from total observations of 347,977 has been tested, and the following AUC has been calculated to show the accuracy of the model (see Fig. 1). Nevertheless, a training subsample for bank 2 of 70% (174,697 observations) from total observations of 249,567 has been tested, and the following AUC has been calculated to show the accuracy of the model (see Fig. 2).

The high number of AUC, from the test and training samples' AUC for both figures, indicates that the True-Positive Rate (TPR) is greater than the False-Positive Rate (FPR). The TPR corresponds to the proportion of positive data points that are correctly considered as positive while FPR corresponds to the proportion of negative data points that are mistakenly considered as positive. The higher AUC, the fewer positive data points, we will miss while the less negative data points will be miss classified as positive, which specifies that the model is sufficient.

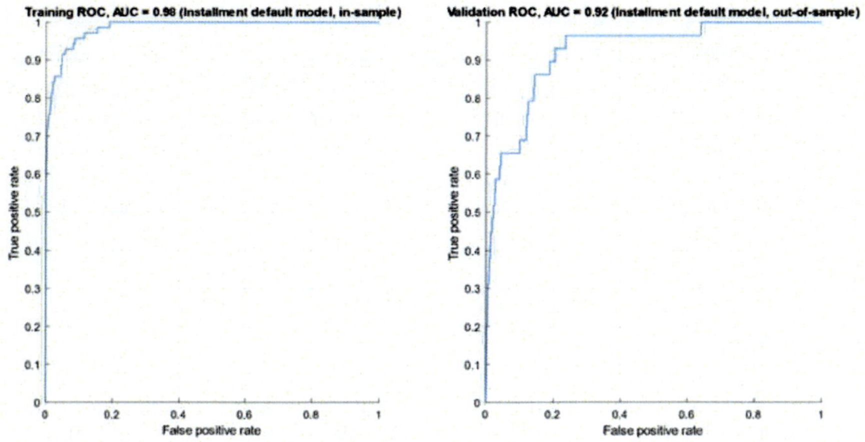


**Fig. 1** AUC results for training and validation bank 1

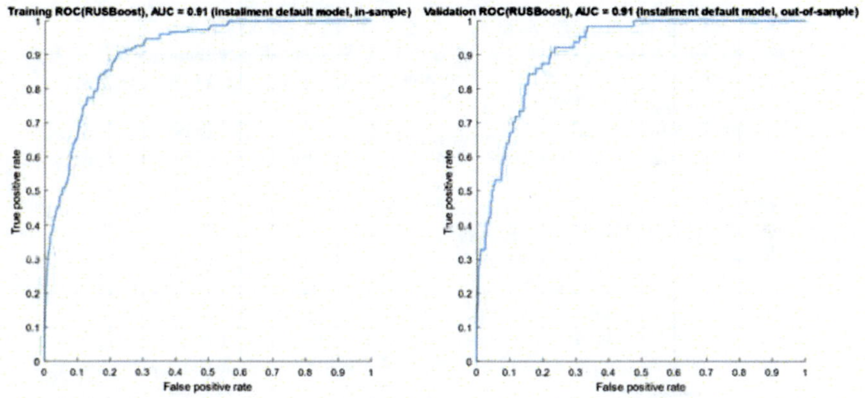


**Fig. 2** AUC results for training and validation bank 2

# 6 Conclusion

In general, classification methods are increasingly implemented in other fields than computer science. The literature review is full of studies evidencing the efficiency of such models in knowing the expected resulting different classes. Nevertheless, classification methods have been used to categorize credit default classes, good or bad, in order to benefit regulators and bankers to better anticipate risks. This paper compared the different methods of classifications (Logistic Regression, Decision Tree, Linear Support Vector Machines, Bayesian Network, and RUSBoosted) in order to examine the credit default cases. Our study relied on big data from Kuwaiti bank for 11 years to tackle the gap of not lengthy data. The parameters also have included new items—more than what has been used currently. Moreover, those data came from banks indicating the importance of data existing within bank databases. From the AUC, average accuracy, type I error, and type II error RUSBoosted were chosen as the outperforming method. A supporting result for training sets has indicated the efficiency of the model selected.

**Acknowledgements** This work is partially sponsored by the Central Bank of Kuwait.

## References

1. Wang G, Ma J, Huang L, Xu K (2012) Two credit scoring models based on dual strategy ensemble trees. Knowl-Based Syst 26:61–68. https://doi.org/10.1016/j.knosys.2011.06.020
2. Bazarbash M (2019) FinTech in financial inclusion: machine learning applications in assessing credit risk. IMF Work Pap 19:1. https://doi.org/10.5089/9781498314428.001
3. Petropoulos A, Siakoulis V, Stavroulakis E, Klamargias A (2018) A robust machine learning approach for credit risk analysis of large loan level datasets using deep learning and extreme gradient boosting. use big data. Anal Artif Intell Cent Bank 50:30–31
4. Baesens B, Van Gestel T, Viaene S, Stepanova M, Suykens J, Vanthienen J (2003) Benchmarking state-of-the-art classification algorithms for credit scoring. J Oper Res Soc 54:627–635. https://doi.org/10.1057/palgrave.jors.2601545
5. Munkhdalai L, Munkhdalai T, Namsrai OE, Lee JY, Ryu KH (2019) An empirical comparison of machine-learning methods on bank client credit assessments. Sustain 11:1–23. https://doi.org/10.3390/su11030699
6. Seiffert C, Khoshgoftaar TM, Van Hulse J, Napolitano A (2010) RUSBoost: a hybrid approach to alleviating class imbalance. IEEE Trans Syst Man Cybern Part A Syst Human 40:185–197. https://doi.org/10.1109/TSMCA.2009.2029559
7. Nyathi K, Ndlovu S, Moyo S, Nyathi T (2014) Optimisation of the linear probability model for credit risk management. Int J Comput Inf Technol 03:1340–1345
8. Samreen A, Zaidi FB, Sarwar A (2013) Design and development of credit scoring model for the commercial banks of Pakistan: forecasting creditworthiness of individual borrowers. Int J Bus Soc Sci 2:1–26
9. Brown I, Mues C (2012) An experimental comparison of classification algorithms for imbalanced credit scoring data sets. Expert Syst Appl 39:3446–3453. https://doi.org/10.1016/j.eswa.2011.09.033
10. Speybroeck N (2012) Classification and regression trees. Int J Public Health 57:243–246. https://doi.org/10.1007/s00038-011-0315-z
11. Friedman JH (2002) Stochastic gradient boosting. Comput Stat Data Anal 38:367–378. https://doi.org/10.1016/S0167-9473(01)00065-2

# Design Thinking in the Improvement of Usability of Physics Mobile Educational Application

**Dewitt Scott Chavez-Ponce, Luis Daniel Villanueva-Montoya, and Jorge Joo-Nagata**

**Abstract** This article describes the creation of a mobile application following the design thinking (DT) approach and its assessment of usability. DT is thought of as a methodology to create innovative solutions to the needs of end-users through well-established five stages. We worked with high school students, a pedagogical team, and a group of undergraduate students in software engineering. In each of the DT stages, diagrams, maps, matrices, groupings, prototypes, and simulations were elaborated that allowed us to empathize with the user, define, create, and test the creative ideas generated throughout the process (development process). To evaluate the usability of the application, the USE questionnaire (Usefulness, Satisfaction, and Ease of use) was applied to a group of students from the education career. The results indicate that it has been possible to develop an educational mobile application with a good level of usability that meets the needs of users.

**Keywords** Design thinking · M-learning · Usability · Mobile computing

The design of software products that give a relevant added value in the development of a course is of vital importance due to the high level of impact that it can produce on students. Using DT provides a better understanding of what people want and need in their lives by looking directly at them. That is, it uses the sensitivity and methods of the designer to meet their needs and provide value to the user [1].

DT is being used by organizations for agile development and requirements elicitation [2, 3]. Whether it is perfecting the functionality of the requirements, relying on

D. S. Chavez-Ponce (✉) · L. D. Villanueva-Montoya
Universidad Nacional de San Agustín de Arequipa, Arequipa, Perú
e-mail: dchavezp@unsa.edu.pe

L. D. Villanueva-Montoya
e-mail: lvillanuevamo@unsa.edu.pe

J. Joo-Nagata
Universidad Metropolitana de Ciencias de La Educación, Santiago, Chile
e-mail: jorge.joo@umce.cl



X.-S. Yang et al. (eds.), *Proceedings of Sixth International Congress on Information and Communication Technology*, Lecture Notes in Networks and Systems 216,
https://doi.org/10.1007/978-981-16-1781-2_55

prototypes to detail the functionalities of a system, and adding agile design and development methodologies such as extreme programming (XP), it allows the creation of higher quality projects that respond to the needs of the final user [10, 13, 15].

The objective of this work is to develop a mobile virtual physics laboratory using the DT approach to meet the needs of the end-users and evaluate the level of usability when using the DT approach.

The document is presented as follows: in Section II, related works are presented in which DT is applied for software development, Section III explains the theoretical framework on DT and usability. Section IV introduces the methodology and explains about the application, how DT has been applied in the mobile laboratory, the design of the evaluation, the participants, and the data collection and analysis method, Section V details the results obtained, and finally, Section VI shows the conclusions.

## 1 Related Work

Dieu Pham et al. conduct a workshop, following DT's approach, to help students generate ideas for a mobile app development project course. Achieving that students can obtain a human-centered perspective toward the generation of innovative solutions and propose new iterations based on our lessons learned [7].

Valentin et al. indicate that DT helps to design an application from two perspectives: personal and collective. The techniques developed in the personal perspective were the map of empathy and people, in the collective perspective, the techniques of brainstorming ideas and cooperative workshops were developed. They conclude that the lessons learned in the teaching of DT: is the time used to teach and apply DT in the classroom for the interaction between the team participants [18].

## 2 Background

### *2.1 Design Thinking*

Design is a process that turns a requirement into a finished product, and creativity is present throughout the process to a large extent [19]. DT involves a series of steps that are performed to obtain information from the end-users and their needs. This approach has some variations in the stages that are proposed in the different models [14]. Stanford University is largely credited with spreading DT and has proposed a tiered iterative process framework that is often described as a sequence of activities that iterate through the process and is the one that has become particularly widespread, this composed of five stages, which are, empathize, devise, prototype, and test [4, 5].

## 2.2 Usability

The decision to adopt the technology depends on the degree to which users perceive that the system is usable and will improve work performance [12, 16].

The usability of mobile applications is defined as the ability of a user to use the application for its intended purpose without getting frustrated. Studies carried out on usability have been recognized as an important tool in the evaluation of mobile applications, despite this, there are very few usability tests that have been developed for mobile learning applications [9, 12].

The usability test that is applied to mobile learning applications is an emerging area in the research since it faces great challenges that significantly affect the usability of the applications, and this is due to mobile devices such as the size of the screen, limited input capacity, and rapidly changing environment [11, 12].

The technical limitation of mobile devices and ensuring that users feel satisfied is important for the success of an application, therefore, to ensure that a mobile learning application is effective and easy to use it is necessary to carry out usability tests [12].

## 2.3 Methodology

**About Fisica M-Lab**. The Fisica M-Lab application is focused on conducting physics experiments using mobile sensors such as the speedometer, accelerometer, gyroscope, and microphone, to this are added the moments that are followed in a practical session according to the national curriculum. Each session contains the moments of beginning and development and closing and they develop activities as follows:

- Start: Initial concepts and questions are developed to motivate the student.
- Development: The theoretical framework and practical exercises are developed
- Closing: the session ends with a question that makes the student reflects after having developed the exercises.

**DT in the development of Fisica M-Lab**. The development of the mobile virtual laboratory is detailed below, shows how each of the DT stages was worked and the tools used in each of the stages [6, 17]:

*Empathize*

- Map of actors: In our proposal, we consider the student and teacher actors. Both actors are related to carrying out their activities with the application.
- Focus Group: The prototype of the mobile application of the virtual laboratory was prepared that integrates a laboratory session based on the measurement. Subsequently, the focus group was carried out to a group of eight students to evaluate their experiences.

*Define*

- Clustering: The information obtained was processed and analyzed to identify and obtain key aspects that must be implemented in the application, then it was grouped into two topics: Content of the application and the Design and interaction.
- Priority diagram: With the key aspects identified, a target-shaped graph was made where each of them was located.

*Ideate*

- Brainstorming workshops: The workshops allowed us to define the following: a) Implement animations in texts and graphics to provide a good user experience. B) Employ a gamification approach for the theoretical part and the practical exercises. For this reason, it was decided to use Unity to develop these points since it provides facilities for the implementation of these points.

*Prototype*

- Prototypes were developed to be validated by the students. Fig. 1 shows the prototypes developed in a class session, and Fig. 2 shows the prototypes for the practical part using sensors.

**Fig. 1** Prototypes of the session class

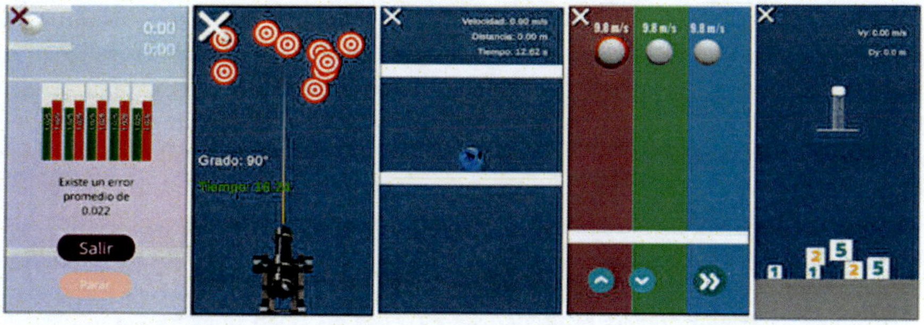


**Fig. 2** Prototypes of the practical exercises

*Test*

- Roleplay: Roles were assigned between the members of the development team and the pedagogical team, then a complete laboratory session was simulated.
- Value Evaluation: First, a survey was prepared based on usability and user experience attributes, then the prototypes were presented to a group of students and teachers. Finally, they were asked to answer the survey.

**Evaluation Design**. The Fisica M-Lab application was developed to support the learning of physics practically, it was planned that the study would be carried out during the educational process of a school to fifth-year high school students, but due to the current situation of COVID-19, it was carried out virtually and with other participants.

The objective of the research was to develop an application for the teaching of physics in a practical way using the sensors of the smartphone, to later carry out a redesign of it and evaluate the interaction with users considering usability aspects for a better understanding and helps in learning.

This research has two different phases, the first data to be obtained through a focus group, the purpose of this activity is to collect data from the perception of the participants to determine the changes to be made in the application.

Therefore, the data collected in this phase would provide a fundamental description to propose the development of the redesign through the application of the DT methodology. After that, a survey based on the USE questionnaire was carried out to obtain quantitative data that helped to analyze and explain the results of the changes made. Subsequently, a comparison was made between the versions of the Fisica M-Lab application to develop a real result of the change for the study [8].

**Participants**. The study was conducted virtually through meetings in the real-time video conferencing application, Meet. The sample group was conformed of eight students of the education career with a mathematical physics specialty in the last year of the Universidad Nacional de San Agustín, Arequipa, Peru. These students were selected for their critical ability in the context as they possess technical knowledge as future teachers in the field of physics.

**Data Collection and Analysis**. In the first stage of the process, the students were given the application so that they can download and test it. Also, the students received a short induction on the use of the application, and some topics that were relevant for the evaluation of the physics learning application were explained, focusing on visual and content aspects. At the end of this stage, a Focus Group was held in which different topics were touched to see the point of view of each of the participants and thus be able to identify the changes to be made in the Fisica M-Lab application, the topics that were discussed were focused on the content, functionality, and design of the application.

In this first stage, it was evaluated using a set of open questions and easy to understand for the participants, the questions were divided into four parts, general or opening questions (three items), transition questions (two items), specific questions (five items), and closing questions (one item). Given that the purpose of this stage was to analyze and define the changes that the Fisica M-Lab application should have, three tools based on the DT methodology were used for data analysis, which are clustering, priority diagram and impact, and effort matrix.

In the final stage of the process, after having made the redesign changes to the application using the DT methodology, the participants proceeded to download and test the application in its new version. The way the evaluation was carried out in this stage was different from the previous stage, the participants evaluated the redesigned application using USE, the evaluation was based on general experiences they had when testing the application [8].

Since the purpose of this final stage is to evaluate the development of the participants and the effectiveness of the application, the application was evaluated after the participants had tried it, it was evaluated through a questionnaire that consists of 30 items based on USE, these were adapted to the context of the application so that the participants have a better understanding of the questions. These items are grouped into four areas, usefulness (8 items), ease of use (11 items), ease of learning (4 items), and satisfaction (6 items) [8]. For the evaluation, a 5-point Likert scale was used for each question (1 = "totally disagree"; 5 = "totally agree").

The analysis of the data obtained from the questionnaire was processed with the help of an Excel sheet and the SPSS tool, a detailed analysis of the data was preceded, for this, the mean and standard deviation were calculated, which are simple descriptive statistical values that will help to understand the opinion of the participants about the usefulness and support in learning physics of the Fisica M-Lab application.

## *2.4 Results*

The quantitative results of the study were obtained after having applied the USE questionnaire in which the answers of eight students of the education career were collected and analyzed to validate the utility, ease of use, and ease of learning of the Fisica M-Lab application.

**Table 1** The usefulness of the Fisica M-Lab application

| Statements | Mean | Standard deviation |
|---|---|---|
| Q1. It helped to be more effective | 4.13 | 0.641 |
| Q2. It helped to be more productive | 3.75 | 0.707 |
| Q3. It is useful | 4.50 | 0.535 |
| Q4. It gives me more control over the activities in my life | 4.38 | 0.518 |
| Q5. It makes the things one would want to accomplish easier to get done | 3.50 | 0.756 |
| Q6. It saves me time when I use it | 4.25 | 0.463 |
| Q7. It meets one's need | 3.88 | 0.354 |
| Q8. It does everything I would expect it to do | 3.38 | 0.518 |

Table 1 shows the statistical results of the Fisica M-Lab Application Utility. As can be seen, most of the participants agreed that the application is useful, with a mean value of 3.971 and a standard deviation of 0.561, which on the 5-point Likert scale applied, is in the range of totally agree. Besides, most participants agree that the application helps them to be more effective, allows them to better tackle physics topics, saves them time, and satisfies their needs to learn physics.

The result of the ease of use of the Fisica M-Lab application is illustrated in Table 2, it is observed that most of the participants were moderately in agreement and in agreement that the application was easy to use. The mean was 3.458 and the standard deviation was 0.732 and the result implies that it was not difficult to use, it was simple, user-friendly, and that it does not require effort to use it.

Descriptive statistics for ease of learning are shown in Table 3 where the results on the management of the application by the participants, in which a mean value

**Table 2** The ease of use of the Fisica M-Lab application

| Statements | Mean | Standard deviation |
|---|---|---|
| Q9. It is easy to use | 4.38 | 0.744 |
| Q10. It is simple to use | 4.13 | 0.641 |
| Q11. It is user friendly | 3.63 | 0.518 |
| Q12. It requires the fewest steps possible to accomplish what I want to do with it | 3.63 | 0.916 |
| Q13. It is flexible | 2.38 | 1,302 |
| Q14. Using it is effortless | 2.50 | 1,414 |
| Q15. I can use it without written instructions | 2.00 | 0.926 |
| Q16. I do not notice any inconsistencies as I use it | 3.38 | 1,598 |
| Q17. Both occasional and regular users would like it | 3.63 | 0.744 |
| Q18. I can recover from mistakes quickly and easily | 3.88 | 0.641 |
| Q19. I can use it successfully every time | 4.50 | 0.756 |

**Table 3** The ease of learning of the Fisica M-Lab application

| Statements | Mean | Standard deviation |
|---|---|---|
| Q20. I learned to use it quickly | 4.63 | 0.744 |
| Q21. I easily remember how to use it | 4.38 | 0.744 |
| Q22. It is easy to learn to use it | 3.75 | 0.707 |
| Q23. I quickly became skillful with it | 4.00 | 0.535 |

**Table 4** The satisfaction of the Fisica M-Lab application

| Statements | Mean | Standard Deviation |
|---|---|---|
| Q24. I am satisfied with it | 4,50 | 0.756 |
| Q25. I would recommend it to a friend | 4,13 | 0.835 |
| Q26. It is fun to use | 3,88 | 0.354 |
| Q27. It works the way one wants it to work | 4,25 | 0.707 |
| Q28. It is wonderful | 3,50 | 0.535 |
| Q29. I feel I need to have it | 4,25 | 0.463 |
| Q30. It is pleasant to use | 3,75 | 0.886 |

of 4.19 and standard deviation of 0.682, were obtained; therefore, it is concluded that most of the participants "agree" or they "totally agree" on the ease of use of the Fisica M-Lab application, and they agree that it is easy to learn to use and remember.

Finally, Table 4 shows the results for the satisfaction of using the Fisica M-Lab application. A mean value of 4.03 and a standard deviation of 0.378 were obtained, with the positive attitude of the participants towards the application was validated, showing that most of them are satisfied and that it works as it should, in addition to being mostly in complete agreement on recommending it.

## 3 Conclusions

The Fisica M-Lab application has been developed following DT's approach to ensuring that this mobile virtual laboratory meets the expectations of end-users. The interested party, both students and teachers, participated in the development of the mobile application, providing opinions for the Empathize phase and the Prototype phase.

The Define and Ideate stages were developed with two teams, the pedagogical team and the Software development team. The workshops developed in these phases allowed the requirements to be clearly defined and accepted by the users, also, the creative ideas were developed in a better way by integrating the knowledge of both teams.

The creative ideas were focused on providing a good user experience, having a Gamification approach to ensure that the student is motivated when carrying out the laboratory activities, and making use of mobile sensors for the experimentation part. Due to this, the good design practices offered by Material Design were followed, a system of achievements and scores was created, animations were implemented in the theoretical part and the practical part, and mobile sensors were used in the experimentation part.

The results obtained have shown that developing an application following the DT approach allows us to implement an application with a good level of usability and that responds to user needs.

**Acknowledgements** Thanks to the Universidad Nacional de San Agustín de Arequipa, for financing the research project according to the NIAI-015-2018-UNSA Contract, and the project team that made it possible to develop the proposal.

## References

1. Brown T et al (2008) Design thinking. Harv Bus Rev 86(6):84
2. Canedo E, Vinhadelli Papadópolis A, Cerqueira A, De Arauja F, Patricia A (2020) Application of design thinking for elicitation requirements in mobile applications. In: Proceedings of the 53rd Hawait international conference on system sciences
3. Canedo ED, da Costa RP (2018) The use of design thinking in agile software require-ments survey: a case study. In: International conference of design, user experience, and usability. Springer, pp 642–657
4. Carlgren L, Elmquist M, Rauth I (2016) The challenges of using design thinking in industry–experiences from five large firms. Creat Innov Manag 25(3):344–362
5. Carlgren L, Rauth I, Elmquist M (2016) Framing design thinking: the concept in idea and enactment. Creat Innov Manag 25(1):38–57
6. Chen P, Huang R (2017) Design thinking in app inventor game design and development: a case study. In: 2017 IEEE 17th international conference on advanced learning technologies (ICALT). IEEE, pp 139–141
7. Dieu Pham Y, Fucci D, Maalej W (2018) A first implementation of a design thinking workshop during a mobile app development project course. arXiv preprintarXiv:1803.10587
8. Gao M, Kortum P, Oswald F (2018) Psychometric evaluation of the use (usefulness, satisfaction, and ease of use) questionnaire for reliability and validity. In: Proceedings of the human factors and ergonomics society annual meeting, vol 62. SAGE Publications Sage CA, Los Angeles, CA, pp1414–1418
9. Goel S, Nagpal R, Mehrotra D (2018) Mobile applications usability parameters: taking an insight view. In: Information and communication technology for sustainable development. Springer, pp 35–43
10. Hehn J, Uebernickel F (2018) The use of design thinking for requirements engineering: an ongoing case study in the field of innovative software-intensive systems. IEEE, pp 400–405. https://doi.org/10.1109/RE.2018.00-18
11. Joyce G, Lilley M, Barker T, Jefferies A (2016) Mobile application usability: heuristic evaluation and evaluation of heuristics. In: Advances in human factors, software, and systems engineering. Springer, pp 77–86

12. Kumar BA, Mohite P (2018) Usability of mobile learning applications: a systematic literature review. J Comput Edu 5(1):1–17
13. Nedeltcheva GN, Shoikova E (2017) Coupling design thinking, user experience design and agile: towards cooperation framework. In: Proceedings of the international conference on big data and internet of thing, pp 225–229
14. Pereira JC, de FSM Russo R (2018) Design thinking integrated in agile software development: a systematic literature review. Procedia Comput Sci138:775–782
15. Piras L, Dellagiacoma D, Perini A, Susi A, Giorgini P, Mylopoulos J (2019) Design thinking and acceptance requirements for designing gamified software. IEEE, pp 1–12. https://doi.org/10.1109/RCIS.2019.8876973
16. Sohaib O, Solanki H, Dhaliwa N, Hussain W, Asif M (2019) Integrating designthinking into extreme programming. J Ambient Intell Human Comput 10(6):2485–2492
17. Steinke GH, Al-Deen MS, LaBrie RC (2018) Innovating information system development methodologies with design thinking. In: Titel: proceedings of the 5th conference in innovations in IT, vol 5. Bibliothek, Hochschule Anhalt
18. Valentim NMC, Silva W, Conte T (2017) The students' perspectives on applying design thinking for the design of mobile applications. In: 2017 IEEE/ACM 39th international conference on software engineering: software engineering education and training track (ICSE-SEET). IEEE, pp 77–86
19. Wolniak R (2017) The design thinking method and its stages. System y Wspomaganiaw in˙zynierii Produkcji 6

# Mono- and Cross-Language Information Retrieval based on Analogical Proportions: A Review

**Bilel Elayeb and Myriam Bounhas**

**Abstract** Since more than a decade, many researchers have been focused on analogical proportions and their applications in different domains. Analogical proportions are known as statements of the form "$x$ is to $y$ as $z$ is to $t$" that we denote "$x : y :: z : t$". This relationship expresses that "$x$ differs from $y$ as $z$ differs from $t$", as well as "$y$ differs from $x$ as $t$ differs from $z$". Making the assumption that items $x$, $y$, $z$, and $t$ form a valid analogical proportion, and given that item $t$ is unknown, this is the basic root for extrapolation enabling to guess the fourth item $t$ based on the values of the triple $(x, y, z)$. This extrapolation principle confirms that analogy may be seen as much a matter of similarity as a matter of dissimilarity. This paper introduces the reader to the main motivation for exploiting analogical proportions in mono- and cross-language information retrieval (IR/CLIR) tasks. The assessment of the existing analogical IR/CLIR approaches is also reviewed. Finally, some challenges and future trends are suggested.

**Keywords** Mono- and cross-language information retrieval · Analogical proportions · Analogical learning

B. Elayeb (✉) · M. Bounhas
Emirates College of Technology, Abu Dhabi, United Arab Emirates
e-mail: Bilel.Elayeb@riadi.rnu.tn

M. Bounhas
e-mail: Myriam_Bounhas@yahoo.fr

B. Elayeb
RIADI Research Laboratory, ENSI, Manouba University, Manouba, Tunisia

M. Bounhas
LARODEC Research Laboratory, ISGT, Tunis University, Tunis, Tunisia



X.-S. Yang et al. (eds.), *Proceedings of Sixth International Congress on Information and Communication Technology*, Lecture Notes in Networks and Systems 216,
https://doi.org/10.1007/978-981-16-1781-2_56

# 1 Introduction

Reasoning by analogy [34], which overlaps the area of linguistics and artificial intelligence (AI), is the foundation of psychological basics of human behavior, in which an inference is exploited to examine and classify new issues by highlighting their likeness to issues already answered.

In the last decade, analogical reasoning has received increasing attention within the community of artificial intelligence and linguistics as a tool for problem-solving, deduction and learning. The main goal of analogical reasoning in AI is to develop analogical learning systems that mimic humans in their ability to learn by analogy. *Analogical proportions*, known to be a formal representation of this kind of reasoning, are efficient in comparing cases stated by differences that are equated to other differences. In other words, they are statement modeled as: "$x$ is to $y$ as $z$ is to $t$", known as "$x : y :: z : t$" that means that "$x$ differs from $y$ as $z$ differs from $t$", as well as "$y$ differs from $x$ as $t$ differs from $z$" [30]. This can be extended to pairs of items by considering the pair $(x,y)$ is analogous to the pair $(z,t)$ [21]. Analogical inference can be stated as follows: given the statement that four items $x$, $y$, $z$, $t$ are creating an analogical proportion on a set $S$ of given variables, it can also remain holding on another feature $l \notin S$ but linked to them. The problem is then to guess feature $l$ for $t$ using the recognized features $l$ for $x$, $y$, and $z$ in case this feature makes an analogical proportion.

Analogical learning has been considered as a two-step inference process: The first one consists of computing a mapping between a new and a memorized situation, while the second step amounts to transform the known to the unknown situation [44]. The main challenge beyond such learning methods is the ability to find and then use this mapping based on analogical proportions. In fact, the definition of these proportions as well as the implementation of their computation are the main goals in any analogical learning approach.

Analogical proportions have attracted the attention of several research fields in the last two decades [27, 30, 36, 44, 47]. They have shown their efficiency in modeling morphological linguistic analysis [43], information retrieval (IR) [5], text summarization [18], classification [3, 8, 9, 19, 29, 41], preference learning [6] and data set expansion [7] and have achieved significant results in terms of complexity and accuracy.

To the best of our knowledge, the existing reviews and surveys related to the domains of analogical proportions and learning are collected in the computational approaches to analogical reasoning book of [37]. In these series of Studies in Computational Intelligence (SCI), the authors selected some reviews and surveys papers such as [38, 39, 45]. More recently [2] have discussed the new perspectives on analogical proportions, while [40] have introduced and discussed the relationship between analogical proportions and analogical reasoning. Unfortunately, the existing reviews and survey works are not dedicated to survey a particular domain of applications. Moreover, there is only few number of existing research works in the domains of analogical IR/CLIR and most of them are relatively outdated if compared to the available literature on other analogical proportions applications, such as classification [22]. Accordingly, we are interested in this paper to review the domains of analogical IR/CLIR given the intuitive relationship between them. We present and discuss in this paper several research works in the fields of IR and CLIR based on analogical proportions, their learning types, their assessment techniques and metrics, as well as their benchmark datasets and experimental results. We also propose some solutions to the current problems and some future trends relevant to enhance research efforts in these areas.

This paper is organized as follows. Section 2 provides a short background of analogical proportions in terms of definitions and general properties. Section 3 presents and discusses the existing IR/CLIR approaches based on analogical proportions. This discussion concludes that analogical proportions are promising for learning especially for areas including large databases of structured data, as it is specifically the case of several IR/CLIR applications. We discuss in Sect. 4 some open problems and propose some directions for future research. Finally, we conclude the review in Sect. 5 and we suggest further future works.

## 2 Background on Analogical Proportions

In this section, we provide a brief overview of analogical proportion between Boolean variables. We recall the basic properties, the solving equation, and the inference process. It is important to note that analogical proportion are instanced for different situations where objects are binary vectors, vectors of real values , or sequence of finite alphabets [29]. In this paper, we only focus on the first instantiation.

### 2.1 *Properties of Analogical Proportions*

Analogical proportion is a specific concept of a more general case known as *logical proportion* [36]. The latter is a Boolean function $F(x, y, z, t)$ involving four Boolean features $x, y, z, t$. It is modeled by a pair of two distinct equivalences, linking a conjunction of elements $x, y$ on the first side, and a conjunction of elements $z, t$ on the other side of $\equiv$, where each element may be negated (further details on logical proportions are available in [34]). For a pair $(x, y)$ of Boolean features, we can generate four Boolean functions associated to such a pair, namely,

- $x \wedge y$ and $\neg x \wedge \neg y$: they represent, respectively, *positive similarity* and *negative similarity* between $x$ and $y$ ; for example, $x \wedge y$ is true iff both $x$ and $y$ are true;
- $x \wedge \neg y$ and $\neg x \wedge y$: they represent *dissimilarity* between variables $x$ and $y$; they are true iff only one of $x$ or $y$ is true and the other is false.

A logical proportion is known as *heterogenuous* if it combines dissimilarities and similarities in its definition, while it is called *homogeneous* in the contrary situation.

**Table 1** Valuations where analogical proportion holds true

| $x$ $y$ $z$ $t$ | $x : y :: z : t$ |
|---|---|
| 0 0 0 0 | 1 |
| 1 1 1 1 | 1 |
| 0 0 1 1 | 1 |
| 1 1 0 0 | 1 |
| 0 1 0 1 | 1 |
| 1 0 1 0 | 1 |

A logical proportion is called *homogeneous* if it does not combine similarities and dissimilarities in its definition and *heterogenuous* in the opposite case. In general, we can generate up to 120 semantically distinct logical proportions. Further details about the set of logical proportions can be find in [36].

*Analogical proportion*, known as $x : y :: z : t$, is a specific case of logical proportion and are defined as follows [35]:

$$x : y :: z : t = (x \wedge \neg y \equiv z \wedge \neg t) \wedge (\neg x \wedge y \equiv \neg z \wedge t)$$

Besides, analogical proportion uses only dissimilarities and can be read: "$x$ **is to** $y$ **as** $z$ **is to** $t$" or more informally as "$x$ differs from $y$ as $z$ differs from $t$ and vice versa". The four elements $x$, $y$, $z$, and $t$ are assumed to be in *analogical proportion*.

Table 1 shows the six situations (among $2^4 = 16$ possible ones), which make true an analogical proportion. We note that the truth table remains the same if we exchange 0 and 1 everywhere. This expresses that analogical proportion are independent regarding the negative or positive encoding of variables. Analogical proportion also fulfills various properties, usually estimated from numerical proportion such as

- *Symmetry*: $x : y :: z : t \rightarrow z : t :: x : y$
- *Central permutation*: $x : y :: z : t \rightarrow x : z :: y : t$.
- Thanks to the central permutation property, a third property requires that the two following implications also hold: $x : y :: x : w \rightarrow w = y$ and $x : x :: y : w \rightarrow w = y$
- *Transitivity*: $x : y :: z : t \wedge z : t :: p : q \rightarrow x : y :: p : q$.

Briefly speaking, given two distinct values $a$ and $b$, an analogical proportion usually holds for the three following patterns: $a : a :: a : a$, $a : a :: b : b$, and $a : b :: a : b$.

## 2.2 *Solving Analogical Equation*

Solving analogical equation consists to find a fourth value of the analogical equation if the first three are being known. Given a triple $(x, y, z)$ of Boolean values, the analogical equation $x : y :: z : w$ yields a solution $t$ if and only if $x : y :: z : t = 1$.

**Table 2** Possible solutions of analogical proportion in case of Boolean vectors

| | | | | | | | | |
|---|---|---|---|---|---|---|---|---|
| $\overrightarrow{x}$ | 0 | 0 | 0 | 0 | 1 | 1 | 1 | 1 |
| $\overrightarrow{y}$ | 0 | 0 | 1 | 1 | 0 | 0 | 1 | 1 |
| $\overrightarrow{z}$ | 0 | 1 | 0 | 1 | 0 | 1 | 0 | 1 |
| $\overrightarrow{t}$ | 0 | 1 | 1 | ? | ? | 0 | 0 | 1 |

In the Boolean case, $x : y :: z : t$ holds true only for the 6 distinct patterns among $2^3 = 8$ given in Table 1. For example, for the two following situations, the equations $1 : 0 :: 0 : w = 1$ and $0 : 1 :: 1 : w = 1$ do not yield any solution. Moreover, the analogical equation $x : y :: z : w = 1$ is solvable iff $(x \equiv y) \vee (x \equiv z)$ is valid. The *unique* solution, in this case, is $w = x \equiv (y \equiv z)$ [30].

### 2.3 Analogical Equation for Boolean Vectors

Analogical proportion can be applied to objects represented as *vectors* of Boolean features. Let us consider a set $\vartheta$ of vectors $\in \{0, 1\}^n$, each vector $\overrightarrow{w} \in \vartheta$ is represented by $n$ attributes as $\overrightarrow{w} = (w_1, \cdots, w_n)$.

Given 4 vectors $\overrightarrow{x}, \overrightarrow{y}, \overrightarrow{z}$, and $\overrightarrow{t} \in \vartheta$. For each attribute $i \in [1, n]$, only 8 possible combination of values of $\overrightarrow{x}$, $\overrightarrow{y}$, and $\overrightarrow{z}$ might be created (see Table 2). As stated before, we can see that in two situations among eight, the equation don't provide any solution.

Let consider the analogical equation $\overrightarrow{x} : \overrightarrow{y} :: \overrightarrow{z} : \overrightarrow{t}$ between 4 Boolean vectors. Applying the following extension of the previous definitions to Boolean vectors in $\{0, 1\}^n$ enables to solve this equation.

$$\overrightarrow{x} : \overrightarrow{y} :: \overrightarrow{z} : \overrightarrow{t} \text{ iff } \forall i \in [1, n],\ x_i : y_i :: z_i : t_i$$

This last extension assumes that analogical proportion between 4 vectors is valid iff the analogical proportion is valid componentwise between *all* their attributes.

## 3 Analogical IR/CLIR Approaches

This section provides a literature review of the existing analogical IR/CLIR approaches. These research work can be divided into two basic categories of analogy-based machine learning approaches: supervised and unsupervised. The supervised approaches apply various machine learning techniques to train the proposed analogical model (e.g., classifier, translator, etc.) using labeled training sets including examples with their suitable correct answer (e.g., class, translation, etc.). The unsupervised approaches don't require any labeled training sets and only a set of unlabeled examples is given as input to the suggested system. On the other hand, an analogical

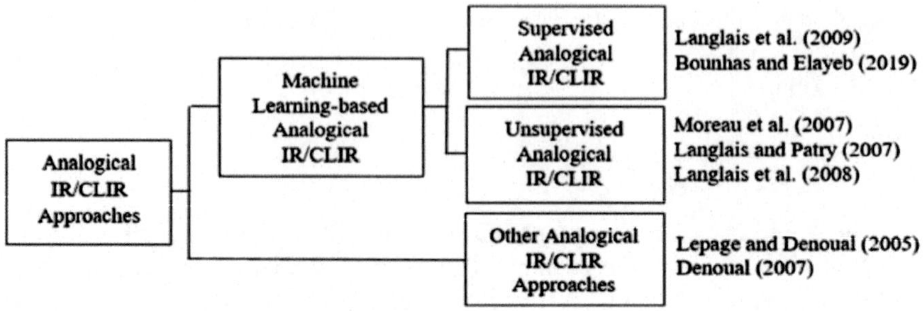


**Fig. 1** Summary of analogical IR/CLIR approaches

approach may involve both labeled and unlabeled data to train the model, in this case it is known as a semi-supervised approach.

Figure 1 summarizes the following literature review of the existing analogical IR/CLIR approaches. To the best of our knowledge, this area is not reviewed yet. This review investigates to study and discuss some of these methods which may help to suggest for the community some challenges and future trends. In this work, we are only limited to review analogical IR/CLIR topics, but readers can refer to [37] in which computational approaches to analogical reasoning have been reviewed. In the following, we explore the IR/CLIR approaches based on analogical proportions. We first summarize the IR/CLIR standard metrics. Then, the existing supervised, unsupervised, and other IR/CLIR approaches-based analogical proportions are presented, discussed, and compared.

## *3.1 Analogical IR/CLIR Standard Metrics*

The evaluation of the existing analogical IR/CLIR approaches has been performed using several standard assessment measures such as Recall and Precision. Indeed, the precision value is the number of relevant documents found divided by the total number of documents proposed for a given query. The recall value is the number of relevant documents found divided by the number of relevant documents in the database. The weighted harmonic mean that combines precision and recall is called F-measure or F-score. It is also known as F1 metric if the precision and the recall are weighted equally.

Besides, the recall-precision curves have been commonly used in the IR/CLIR assessment protocol, and known as interpolated average precision (IAP). Indeed, the Y-axis is presented via the precision values, while the X-axis is presented via 11 points of recall. The non-interpolated average precision is known as the mean average precision (MAP), whereas the *exact precision (R-Precision)* is defined as the precision value at rank $R$. The latter is the size of the set of relevant documents. More formally, the *MAP* and the *R-Precision* are defined as follows:

$$MAP = \frac{1}{|Q|}\sum_{j=1}^{|Q|}\frac{1}{|\text{rel}_j|}\sum_{r=1}^{N_j}P(r) \times \text{isRel}(r)$$

where

$$P(r) = \frac{\text{Number of returned relevant documents at rank r or less}}{r}$$

and

- $|Q|$: The total number of queries.
- $|rel_j|$: The number of relevant documents returned for query $Q_j$.
- $N_j$: The total number of documents returned for query $Q_j$.
- $isRel(r)$: A binary function that returns 1 if the result at rank r is relevant and 0 otherwise.

$$R - \text{Precision} = \frac{1}{|Q|}\sum_{j=1}^{|Q|}\text{Precision}(\{D_{kj}\})$$

where

- $Precision(\{D_{kj}\})$ is the precision of returned $k$-first documents for query $Q_j$.
- $k$ corresponds to the number of returned relevant documents for query $Q_j$.

Besides, the Precision (resp. Recall) values at different top $i$ documents, namely, $P(i)$ (resp. $R(i)$) are also considered as relevant IR/CLIR assessment metrics. For example, the precision at point 10, denoted $P(10)$, is the ratio of relevant documents among the top 10 retrieved documents.

For a comparative study between IR/CLIR models, it is also relevant to compute the improvement percentage of an approach $B$ compared to an approach $A$ is computed by

$$\%\text{improvement} = \frac{B - A}{A} \times 100$$

On the other hand, some analogical IR/CLIR approaches have been assessed using the BLEU-N metric [32]. It is based on a modified form of precision. It assesses the similarity score between the reference documents and the generated ones. The modified precision is calculated through Eq. (1).

$$p_n = \frac{\sum_{C\in\{\text{Candidates}\}}\sum_{n-gram\in C} Count_{clip}(n-gram)}{\sum_{C'\in\{\text{Candidates}\}}\sum_{n-gram\in C'} Count(n-gram)} \tag{1}$$

where

- $Count(n - gram)$: is the n-grams number that candidate in the test set.
- $Count_{clip}(n - gram)$: is the clipped n-grams number for the candidate documents.

The BLEU score is computed via Eq. (2).

$$BLEU = BP * \exp(\sum_{n=1}^{N} w_n \log p_n) \tag{2}$$

where

- $BP$: is the brevity penalty.
- $N$: is the n-grams length.
- $w_n$: is the positive weights.
- $p_n$: is the modified precision score.

The NIST metric has been formulated based on some modifications on the BLEU metric. In fact, the BLEU computes n-gram precision and assigns equal weight to each one, while NIST computes how informative a specific n-gram is. In other words, when a suitable n-gram is established, the rarer the n-gram is, the more weight is given. Besides, there is a difference between NIST and BLEU in computing the brevity penalty, since slight differences in the length of the translation do not have an important effect on the overall score. Formula of calculating NIST metric is available in [16].

The Word-Error-Rate (WER) metric has been founded based on Levenshtein distance. However, the WER operates at the word level, while the Levenshtein distance operates at the character level. Besides, WER has been initially exploited for computing the performance of speech recognition systems in addition to the assessment of MT systems. Indeed, this metric is dedicated to compute the number of words that differ between a translated text through MT and a reference translation. Moreover, the Position-independent word Error Rate (PER) is a derived metric from WER. The PER provides the opportunity to re-order words and sequences of words between a given translated text via MT and a reference translation. The formula of computing WER metric is available in [33].

## 3.2 *Supervised Analogical IR/CLIR Approaches*

In the following, we present two supervised analogical IR/CLIR approaches. We firstly detail the multi-terms translation-based analogical learning approach. Then, an analogical proportion-based matching model for domain-specific IR is exposed. Finally, a summary and a discussion of these two analogical supervised approaches are given at the end of this subsection.

### 3.2.1 Analogical Learning for Terms Translation

Existing approaches of analogical learning in the literature have been enhanced by [25] in different ways. Firstly, the authors have suggested and implemented a more efficient analogical solver. Secondly, they have proposed and tested an effective search engine based on analogical learning. Thirdly, they have explored the ability of a classifier to be trained to identify wrong candidates generated by analogical learning. The assessment of this analogical search engine has been performed on a set of translation multi-terms of a medical domain. In fact, the building of new words, with complex lexical constructions, is more frequent in medical area if compared to other application domains.

The proposed approaches have been evaluated using datasets issued from the Medical Subject Headings (MeSH) thesaurus. The latter has been exploited by the US National Library of Medicine to index the biomedical scientific literature in the MEDLINE database. Five language pairs, with various morphology, have been involved in the experimentations, three of them are known as quite close European languages such as English-Spanish, English-French, and English-Swedish. Moreover, there is one language pair including different scripts (English-Russian) and a more distant one (English-Finnish). Ten translation directions have been considered in the experimentations and were assessed using different assessment metrics such as Coverage, Recall, and Precision. The authors confirmed that their analogical approach is still efficient even when translating into languages with different scripts (such as Russian), or into a morphology rich language (such as Finnish).

On average across all translation directions, the best variant has translated in first position 21% of the words with a precision of 57%. Besides, 30% of the terms have been translated with significant precision. The authors showed that analogical translations outperformed the translations generated by a phrase-based engine trained at the character level, but with a poor recall. Besides, and if compared to SMT alone, a direct combination of the two approaches achieved an improvement of 5.3 BLUE points.

### 3.2.2 Analogical Proportions-Based Matching Model for Domain-Specific IR

Bounhas and Elayeb [5] have suggested, assessed and compared a domain-specific IR matching model, denoted APMM, based on analogical proportions between queries and documents. The proposed model is founded based on the following logic: Let us consider $q_1$ and $q_2$ as two IR queries and $rel(q_1)$ and $rel(q_2)$ their corresponding sets of relevant documents, the *APMM* model supposes that the difference between $q_1$ and $q_2$ can have a significant impact on the differences between $rel(q_1)$ and $rel(q_2)$. This may be modeled as "$q_1$ differs from $q_2$ as $rel(q_1)$" differs from $rel(q_2)$.

Given an new test query, the logic of *APMM* is to expect the irrelevance or relevance of any document in the collection to this new query, based on its analogical proportions with queries in the training set. For this purpose, the authors have defined an analogical inference allowing to assess document relevance. The matching model proposed in [5] has been dedicated to domain-specific IR known as a sub-field of IR, but queries and the set of document collection are built from a particular area such as neuroscience, sport, politic, etc. The authors have exploited a domain-specific IR test collection in order to find and benefit from analogy between a new "test query", "training queries" and their known sets of relevant documents, with the aim at retrieving the unknown set of relevant document for the new given test query.

The basic idea of the *APMM* matching model is the following: given a new test query with a particular indexing terms, the authors have formalized two logical proportions joining queries and their sets of relevant documents for an analogical inference. A disagreement and agreement scores have been defined based on the two suggested logical proportions with the aim at expecting the degree of rejecting or accepting any document from the collection given a new query. The *Agreement* degree between query and document is computed based on the common words between them, while the *Disagreement* degree represents the set of words they differ. These two degrees exploit documents in a different way: the goal of the *Disagreement* degree is to reject the irrelevant documents from the returned list, whereas the *Agreement* degree aims at reinforcing the relevance of the remaining documents not rejected by the *Disagreement*.

This analogical model differs from existing IR matching models on two main features: First, the *APMM* considers two reasons for selecting a document as relevant and for rejecting it as irrelevant by introducing the two previously defined agreement and disagreement scores. The final document relevance is assessed by aggregating the Agreement and the Disagreement scores. Second, to predict document relevance, *APMM* only restricts the search to the sub-set of queries having high similarity. For this purpose, the authors have started their approach by computing similarity between queries using a Cosine function. This helps to order the "training queries" based on their similarity to the "test query" and then identify its $k$-nearest neighbors queries. Then, instead of focusing on the full document collection and computing the relevance score for each document as in classic IR models, the *APMM* has focused only on the sub-collection of candidate documents that corresponds to the set of $k$-nearest neighbors queries with the aim at selecting the set of relevant documents. This filter is beneficial in terms of complexity and may significantly contribute to enhance the precision by reducing the set of noisy documents.

The analogy-based IR matching model has been tested using three IR Glasgow[1] test collections (CRAN: 1400 documents and 225 queries; CACM: 3204 documents and 64 queries; CISI: 1460 documents and 112 queries). The experimental results highlighted the effectiveness of the model, in terms of MAP, R-Precision, the precision values at top document and the improvement percentage, if compared to OKAPI-BM25 matching model. For the CRAN collection, the best achieved improvement percentage in terms of MAP = 53.59%; R-Precision = 73.13% and P(30) = 65.59%. For the CACM collection, the MAP = 33.81%; R-Precision = 16.67% and P(30) = 1.24%, while the CISI collection leads to MAP = -7.8%; R-Precision = 28.11% and

[1] http://ir.dcs.gla.ac.uk/resources/testcollections/.

P(30) = 26.67%. Furthermore, the *APMM* has been compared to the historical work of [20] using the three same test collections. Results have shown an improvement of *APMM* if compared to [20] model, except in P(15) for the CRAN dataset where their approach is slightly better than *APMM*.

### 3.2.3 Synthesis and Discussion of Supervised Analogical IR/CLIR Approaches

Langlais et al. [25] have proposed and implemented a solver for some practical problems in analogical learning. The authors have suggested an efficient search technique based on analogical learning and outperformed the existing search strategy of [24]. Besides, the proposed classifier is trained to generate suitable candidate translations for multi-terms of a medical domain. This classifier outperforms the most frequently generated heuristic involved in many approaches on analogical learning.

Despite the multiple advantages of the proposed analogical translator [25], Bounhas believed that the efficiency of an SMT may be enhanced by using morpheme-based system [46]. In this context, an unsupervised technique is necessary to acquire morphemes. In fact, there is a lack of availability of morpheme lists in the medical domain for different languages pairs. In this case, the comparative study between SMT and analogical approaches seems more objective since both of them work at the character level.

On the other hand, it is also interesting to test the proposed analogical devices using terminologies issued from different domains with diverse sizes of the training sets. Besides, it is also relevant to consider the process of analyzing the segmentation supported by analogical learning. A combination between analogical translation and a front-end statistical translation engine may lead to significant results. Finally, even if the analogical learning has been exploited in translating terminology, it can be extended to translate infrequent entities.

The analogical proportions have shown their efficiency in several research fields such as natural language processing (NLP) and classification, which encourage [5] to propose their successful analogical IR matching model. The authors have been mainly interested to investigate the analogy between new test query, training queries, and their known sets of relevant documents in order to build the unknown set of relevant documents for the new test query. From a logical point of view, one may assume that what differs one query to another (in terms of indexing terms) should also affect their corresponding relevant documents. Then, they have formalized appropriate logical proportions to link queries and their corresponding relevant documents in an analogical inference.

Although its effectiveness, especially for test collections with high similarity between queries, the *APMM* can be enhanced: in fact, this model has achieved high precision on top retrieved documents. Therefore, an iteration of a blind relevance feedback technique on the top returned documents can improve the effectiveness of the achieved results.

**Table 3** Summary of the comparative study between supervised analogical IR/CLIR approaches

| Analogical IR/CLIR approach | Standard metrics | Benchmark datasets | Experimental results (Best ones) |
|---|---|---|---|
| Multi-terms translation-based analogical learning [25] | – Coverage<br>– Precision (P)<br>– Recall (R) | **MeSH datasets**:<br>– Finnish: 19.787 pairs<br>– French: 17.230 pairs<br>– Russian: 21.407 pairs<br>– Spanish: 19.021 pairs<br>– Swedish: 17.090 pairs | – Translation (FI-EN): P = 53.6%; R = 61.3%<br>– Translation (FR-EN): P = 57.5%; R = 68.4%<br>– Translation (RU-EN): P = 61.9%; R = 66.7%<br>– Translation (SP-EN): P = 64.3%; R = 70.0%<br>– Translation (SW-EN): P = 53.1%; R = 64.4% |
| Analogical proportions-based matching model for domain-specific IR [5] | – MAP<br>– R-Precision<br>– P(i)<br>– %improvement | – **CRAN**:<br>1400 documents<br>225 queries<br>– **CACM**:<br>3204 documents<br>64 queries<br>– **CISI**:<br>1460 documents<br>112 queries | – **CRAN (%imp)**:<br>MAP = 53.59%<br>R-Precision = 73.13%<br>P(30) = 65.59%<br>– **CACM (%imp)**:<br>MAP = 33.81%<br>R-Precision = 16.67%<br>P(30) = 1.24%<br>– **CISI (%imp)**:<br>MAP = –7.8%<br>R-Precision = 28.11%<br>P(30) = 26.67% |

Table 3 summarizes and compares the existing supervised analogical IR/CLIR approaches in terms of standard metrics, benchmark datasets, and experimental results.

## *3.3 Unsupervised Analogical IR/CLIR Approaches*

We have find three unsupervised analogical IR/CLIR approaches in the literature. The first one addresses the problem of analogical query expansion. The second one develops an analogical unknown words translation. The last approach applies an analogical translation of medical terms in different languages. Finally, we recapitulate and discuss these three unsupervised IR/CLIR approaches in terms of standard metrics, benchmark datasets and the best achieved experimental results.

### 3.3.1 Analogical Query Expansion

Moreau et al. [31] have suggested an analogy learning-based technique to automatically identify morphological variants within documents that have been exploited to query expansion. Among the different kinds of linguistic analysis that can be applied (syntactic, or semantic morphological), the performance enhancement of a given information retrieval system (IRS) generally depends on the efficiency of the morphological analysis process. This latter consists at identifying the list of words that are forms of the same word but graphically different (i.e., morphological variants) such as reduce, reduced, reducing, and reducer. Consequently, the matching of these graphically different but semantically close forms may enhance the IR efficiency.

The state-of-the-art technics for detecting morphological variants are still suffering from non-portability problem. Indeed, most of them exploit external knowledge (dictionary, list of endings, recoding rules, etc.) and are dedicated for a single given language. Therefore, they cannot be re-exploited out of their designing context. However, the proposed approach [31] for detecting morphological variants is based on an unsupervised machine learning technique characterized by the following advantages: (i) it is totally automatic; (ii) it does't require any external resources or a priori knowledge; and (iii) it can support several languages. Moreover, this approach has been used in a process of query expansion useful to enhance the IR efficiency. The basic statement is to select the terms in a text collection that are in morphological relationships with original query terms. These words are injected in this original query during the expansion process.

The evaluation of the proposed approach has been performed using three IR collections: for English: TIPSTER (from TREC-3: 175.000 documents and 50 queries); French: INIST (163.000 documents and 30 queries) and ELRA (3.511 documents and 30 queries), which is a kind of "questions/answers" collection of the European Commission, available in English, French, Portuguese, German, Italian, and Spanish. Besides, LEMUR[2] is used as IRS and OKAPI-BM25 as IR matching model. The reported results highlight an important enhancement of the IRS efficiency, if compared to some available tools such as a lemmatizer (Treetagger) and a stemmer (Porter's stemmer), according to several IR evaluation metrics (recall, precision, MAP, R-Precision, improvement percentage, etc.) for each language.

### 3.3.2 Analogical Unknown Words Translation

Stroppa and Yvon [43] have proven the success of analogical learning as a morphology analysis tool for several languages. Later [24] have investigated better the effect of analogical learning on the identification of suitable translations of unknown words or phrases. Many literature works have exploited big non-parallel corpora to compute distributional collocational properties useful in the translation process of unknown words. Although non-parallel corpora are more available than bitexts, this

[2] http://www.lemurproject.org.

branch of work is still highly dependent on external resources having huge size. On the contrary, the approach of [24] does not require any additional huge resources (i.e., morphological knowledge) from the source, target, or both languages.

Besides, this technique is not only dedicated to translate unknown words, but also can be useful to enrich existing terminologies in a dictionary. Although it is known that many statistical translation methods still suffer from the translation of low frequency terms, this approach shows its efficiency to identify their suitable translations. Moreover, it is able to find the candidate translation corresponding to a valid sense of the French term not existing in the training set.

Furthermore, this approach is not only limited to translate single terms, since separators can simply be treated as any other character. For this purpose, the authors have exploited analogical reasoning with the aim at translating a set of at most 5 terms in the test set that include an unknown term. Given that there are several more sequences than single terms, the input space of these tests is greatly larger. Thus, a pruning technique has been used to select the roots of the sequences before conducting French-to-English translation.

The experimental process has been performed using 441 unknown words extracted from 200 sentences of the Europarl[3] parallel corpus, in addition to 1064 sentences from the Project Syndicate.[4] Three languages, French, Spanish, and German, have been tested in terms of Recall (R), Precision (P), Word-Error-Rate (WER), and BLEU metric. As introduced before, the BLEU is a precision-based measure that represents the precision modified form. It is based on assessing the similarity's degree between the reference and the generated translations. The registered results for these three languages are as the following: for French (P = 59.4%; R = 65.2%; WER = 61.6% and BLEU = 22.90%), for Spanish (P = 54.0%; R = 66.5%; WER = 53.7% and BLEU = 27.27%) and for German (P = 49.3%; R = 23.1%; WER = 69.7% and BLEU = 18.30%). Moreover, this approach has provided correct English translations for 80% of ordinary unknown French words.

### 3.3.3 Analogical Translation of Medical Words in Different Languages

An analogical translation-based approach of medical terms in various languages has been developed and evaluated by [23]. In this approach, the candidate translations are generated using partial bilingual lexicon useful to solve bilingual analogical equations. The proposed analogical approach is addressed to answer a series of queries such as (i) are medical word-forming devices sufficiently parallel in diverse languages to predict the translation of a given new term? For example, let us consider the knowledge of a given set of medical terms in a source language SL and their corresponding translations in a target language TL, how can we find the translation Tw in TL of an unseen medical term Sw in SL? In fact, there are some "generative" methods in the literature based on machine learning techniques such as in the work

[3] https://www.statmt.org/europarl/.

[4] http://www.project-syndicate.com.

[14] (generating French words from English medical words), or using human expertise such as in [42] (generating Spanish words from Portuguese medical words). Both of these two generative approaches generate new target words corresponding to previously unseen source words.

The proposed work exploits some existing generative approaches based on analogical learning such as [24, 26, 43]. The training step in this approach uses a bilingual lexicon, then it exploits the formal similarity of medical terms in some languages to generate new translations. Unlike external approaches, it can provide translations for unseen words. In case of medical words [23], benefits from the strings of characters formal relations (for example: *aortotomy* : *aortitis* :: *spondylotomy* : *spondylitis*). An analogical proportion including an unknown item ($a : b :: c :?$) is recognized as an analogical solving equation. Moreover, an analogical proportion may consider complex object such as pairs of terms like (*source*, *target*). In fact, source and target exist in the bilingual dictionary. In case an object configures a missing part (for example, missing target), applying the analogical inference helps to guess this part by solving the analogical equation. For this purpose, three steps are usually required: (i) building a set of triples of word pairs, then an analogical proportion is defined using the first items of the pairs, i.e: source items of each pair; (ii) analogical equations between the corresponding second items (i.e: target items) should be solved, this may lead to different solutions; and (iii) the best candidate among these solutions will be selected. Despite the advantages of this approach, especially its ability to identify translations for unseen terms, it has a lack in case of huge number of analogical proportions in the first step, which requires sampling and using of appropriate data structures.

The assessment of the proposed approach is performed using a series of datasets of medical words and compare it to an identification-based, non-generative approach based on edit-distance. The precision is about 81% using the dataset MeSH (Spanish-English), while the recall is about 55% using the dataset MedDRA (Spanish-English). A combination of this approach with an existing edit-distance with a large word list has led to a significant improvement of the results. Besides, this analogical approach still efficient on a series of language pairs such as different scripts (Cyrillic) and distant languages (Finnish).

### 3.3.4 Synthesis and Discussion of Unsupervised Analogical IR/CLIR Approaches

The analogical query expansion approach [31] has been dedicated to detect morphological variant of a given word to be exploited for a query expansion process. Experimental results have confirmed the relevant impact of this technique to enhance the IRS efficiency. Besides, significant results are also shown in case of languages known by their poor morphological level. If compared to some related works, the proposed approach seems less sensitive to query length. This enhancement is still constantly noticed and comparable especially for query lengths between 5 and 15 words.

Although its noticed efficiency, this analogical approach for query terms enrichment still suffers from some drawbacks. In particular, there is a lack to detect some words related to query terms. For example, the relationship between the words *paid* and *pays* or between *hazard* and *hazardous* cannot be detected. In these cases, the results using Porter's stemmer seem more significant, since they avoid these kinds of errors.

In the worst cases, some irrelevant words are added to the original query which decrease the IRS efficiency. Among these cases, we can distinguish: (i) some polysemous terms that are detected as morphologically linked such as *production-reproduction*. These terms are linked in result in case of *reproduction* and *production*, but not in case of *animal reproduction*; and (ii) sometimes there is no semantic relationship between the original terms and the detected words; i.e., the morphological link between them is no longer used as for *exposition-composition*. In order to overcome these kinds of errors, the authors have avoided, in the query expansion process, to involve terms that are themselves related to extensions. Consequently, the expansion approach seems more flexible than the conflation one in which *production* and *reproduction*, as well as their variants, would all be converted into a unique form.

Regarding unknown words translation, the analogical approach [24] has achieved significant results by successfully translating about 80% of ordinary unknown words in French-to-English translation test. Moreover, this approach has been extended to perform suitable translation for unknown phrases. Although its efficiency, this approach still suffers from many drawbacks: (i) it is still silent for some words or phrases. Consequently, it is necessary to enhance the identification technique of word or phrase stems; (ii) the phrase-based SMT can be improved by using an efficient enhancement of a phrase-transfer table; (iii) this approach requires the learning of regularities from the obtained analogies in the training process; therefore, there is no need to any training material during translation task; and (iv) the performance of this approach should be compared to other approaches of unsupervised morphology acquisition.

The analogical translation approach of medical words of [23] has been assessed using various datasets in different languages. However, the authors have a difficulty to compare the efficiency of their approach to the work of [14] since they have used in their experiments quite different test set including pairs of identical words. In case of a random selection of test words [14], achieves a precision of about 75% (containing 10–12% of identical words), while no recall rate is provided. The proposed analogical approach [23] is able to identify a suitable translation for unseen words. The analogical inference consists to form analogical equations combining three known words that make this equation solvable. Then, solving each equation allows to guess/build a new assumed word as a solution for this equation. Generating a huge number of analogical equations and providing solution for each one forms an increasing support for most encouraging assumptions. Besides, selecting the suitable translation is based on a simple frequency ordering method. This selection may also be supported by the use of a suitable classifier. On the other hand, we believe that the proposed analogical approach may be enhanced by using extra knowledge on morphemes or "sub-words". This may be exploited to impose morphemic limitations when providing solutions to the analogical equations and then decrease the number of produced forms, or to implement a posteriori filtering of candidate translations.

Table 4 summarizes and compares the existing unsupervised analogical IR/CLIR approaches in terms of standard metrics, benchmark datasets, and experimental results.

## 3.4 Other Analogical IR/CLIR Approaches

There are two analogical IR/CLIR approaches that aren't known as machine learning-based ones. The first one is the analogical machine translation system (ALEPH), while the second one is focusing on analogical translation of unknown terms in a statistical machine translation (SMT) system. As before, a summary and a discussion of these two approaches is provided at the end of this subsection.

### 3.4.1 ALEPH: Analogical Machine Translation System

Lepage and Denoual (2005) [28] have developed an analogical machine translation system denoted ALEPH that has been tested using the IWSLT-2004 campaign.[5] ALEPH system has some particular characteristics if compared to the state-of-the-art systems that can be summarized as follows: (i) it is a pure example-based MT system applying analogical equations that ignores translation divergences in different manner; (ii) ALEPH system does not require any training, templates or variables; (iii) it does not use any preprocessing of the aligned examples; and (iv) it does not need any explicit transfer component. During the Unrestricted Data track of IWSLT-2004, ALEPH system has achieved third rank in Japanese-English (with the best BLEU), and the second rank in Chinese-English translations. However, during the assessment campaign of IWSLT-2005[6] [17], especially in the C-STAR tracks, ALEPH has shown that it can be applied to any available language pairs.

In IWSLT-2004, ALEPH has been evaluated on both Chinese-English and Japanese-English translations in the Unrestricted Data track. The C-STAR corpus (160K aligned sentences) has been used for both language pairs. According to the BLEU metric, the ALEPH system has achieved the third rank in Japanese-English (0.634) and the second rank in Chinese-English (0.522, first at 0.524). In all IWSLT-2005 C-STAR tracks, ALEPH system assessment has been limited to the use of 160K C-STAR translation pairs. For example, the system has used only 20K supplied translation pairs during the Arabic-English track. Experimental results, using the BLEU metric, for the following language pairs are summarized as follows: Arabic-English (0.382), Korean-English (0.412), Chinese-English (0.477), Japanese-English (0.593), and English-Chinese (0.098).

[5] https://www.isca-speech.org/archive/iswlt_04/.

[6] https://www.isca-speech.org/archive/iwslt_05/.

**Table 4** Summary of the comparative study between unsupervised analogical IR/CLIR approaches

| Analogical IR/CLIR approach | Standard metrics | Benchmark datasets | Experimental results (Best ones) |
|---|---|---|---|
| Analogical query expansion [31] | – MAP<br>– IAP<br>– R-Precision<br>– P(i)<br>– R(i)<br>– %improvement | – **English**:<br>TIPSTER (from TREC-3: 175.000 documents and 50 queries)<br>– **French**:<br>INIST (163.000 documents and 30 queries)<br>– **"Questions/Answers" collection for 6 languages**:<br>ELRA (3.511 documents and 30 queries) | – **English (%imp)**:<br>MAP = 27.18%<br>IAP = 29.44%<br>R-Precision = 31.96%<br>P(10) = 47.0%<br>R(500)= 67.68%<br>– **French (%imp)**:<br>MAP = 18.45%<br>IAP = 19.93%<br>R-Precision = 21.63%<br>P(10) = 38.67%<br>R(5000) = 72.20%<br>– **For the 6 languages (%imp)**:<br>MAP = 17.52% (English)<br>IAP = 16.66% (English)<br>R-Precision = 10.23% (English, Italian)<br>P(1000) = 24.64% (Portuguese)<br>R(500) = 21.68% (English) |
| Analogical unknown words translation [24] | – Recall (R)<br>– Precision (P)<br>– Word-Error-Rate (WER)<br>– BLEU | – **Europarl corpus**:<br>2000 sentences<br>– **Project Syndicate**:<br>1064 sentences<br>– **Total**: 441 unknown words | – **French**:<br>P = 59.4%<br>R = 65.2%<br>WER = 61.6%<br>BLEU = 22.90%<br>– **Spanish**:<br>P = 54.0%<br>R = 66.5%<br>WER = 53.7%<br>BLEU = 27.27%<br>– **German**:<br>P = 49.3%<br>R = 23.1%<br>WER = 69.7%<br>BLEU = 18.30% |
| Analogical translation of medical words [23] | – Coverage<br>– Precision (P)<br>– Recall (R)<br>– Edit-distance | – mesh-SW-EN : 19090 terms<br>– mesh-FR-EN : 19230 terms<br>– mesh-SP-EN : 21021 terms<br>– mesh-FI-EN : 21787 terms<br>– mesh-RU-EN: 23394 terms<br>– meddra-SP-EN : 67523 terms<br>– snomedct-SP-EN : 284255 terms | – P = 81.0% using MeSH (SP-EN)<br>– R = 55.0% using MedDRA (SP-EN) |

The ALEPH system shows its efficiency as a MT system without requiring any NLP tool such as tagger. Besides, no dictionary is used to support the corpus of sentence examples. It is also independent of any tokenization device. Moreover, ALEPH does not use any distance between examples, nor any threshold. For all these reasons, the results from the ALEPH system can be considered as a sort of baseline for all C-STAR tracks. Despite its success, ALEPH system still has some limitations. It has been tested using simple and short sentences of a small domain, a special cases of the BTEC corpus. For instance, there is no guarantee that the ALEPH system will still be sufficiently effective when translating news articles or any other long sentences of a large domain.

### 3.4.2 Analogical Translation of Unknown Words in a SMT

Denoual [15] has proposed and assessed an approach for analogical translation of unknown words in a statistical machine translation (SMT) context. Machine translation-based data-driven is founded on the principle of ignoring, in the output, the translation of unknown terms. In fact, the problem of unknown word translation becomes a serious challenge in case of limited bilingual data. The author has exploited proportional analogy at the character level in order to perform an efficient unknown words translation process. The suggested technique is integrated in a SMT system performing unknown words translation from Japanese to English.

Experimental results have been performed using a test set of 500 sentences used in the IWSLT06 campaign to assess Japanese to English MT systems. The automatic assessment metrics BLEU and NIST are used to evaluate the quality of translations. In fact, to confirm a better correlation with “fluency”, it is recommended to use the BLEU metric [32], while a better correlation with “adequacy” can be confirmed using the NIST metric [16]. The assessment of every translation is achieved via seven reference sentences suggested by human translators. The results when the output considers the unknown words translation are BLEU = 0.1638 and NIST = 5.8054, while the results when the output ignores the unknown words translation are BLEU = 0.1790 and NIST = 5.7627. Objective assessment metrics confirm that the translated sentences have a similar “fluency”, while their “adequacy” is higher than that obtained by a baseline system.

### 3.4.3 Synthesis and Discussion of Other Analogical IR/CLIR Approaches

The analogical MT system ALEPH [28] presents a quadratic complexity which significantly weaken this system. Nevertheless, ALEPH can improve the translation times if it applies efficient heuristics to identify only relevant pairs useful to solve analogical equations. On the other hand, ALEPH is known as a pure example-based MT system, it will be interesting in future to focus more on a better exploitation of data examples. Even if ALEPH has used some heuristics to identify sentences from the original corpus, useful to build and solve analogical equations, these heuristics are efficient only a quarter of the time. Consequently, it is necessary to enhance the efficiency of these heuristics in order to enlarge the system coverage. In addition, the performance of ALEPH can be supported by using other NLP tools such as a tagger or resources such as paraphrases or bilingual lexicons. It should also be beneficial to involve a training step in the translation task in which the translation of a given test phrase can benefit from the transitional translation pairs generated in the translation of previous test sentences.

Analogical proportions have been exploited by [15] to translate unknown words as strings of characters, while classical MT approaches still perform their translation at the level of words. The proposed analogical translation approach seems useful in a statistical machine translation context and contributes to enhance the quality of translating unknown words. Operating at the level of characters rather than words enables to capture more linguistic information by exploring the commutations that occur inside words. The efficiency of this approach is assessed using BLEU and NIST metrics. Supposing that NIST is more related to "adequacy" and BLEU to "fluency", the translation "adequacy" (more information is transmitted) is more increased thanks to the proposed analogical approach whereas the sentence "fluency" is almost similar to that achieved by the baseline. The author has also improved the quality of translation by considering only a part of the unknown words. Indeed, unknown words are sorted according to the number of analogical equations used during the translation task, which enables to slightly enhance "fluency" while consistently enhances "adequacy".

The improvement of the proposed approach can be performed at two levels. Firstly, more effective reorganization techniques should be explored at the higher level of integrating analogical translation into SMT system. The addition of unknown translated words to the original statistical word alignments does not allow the phrase table to be estimated sufficiently. Consequently, there will be a negative effect on objective assessment metrics such as BLUE or NIST, which essentially operate by computing sequences of words. Secondly, better results can be achieved at the lower level of analogical translation of unknown words. This requires the study of more efficient heuristics to improve the ratio of solved analogies over the total number of attempted analogies. Besides, since word segmentation may be erroneous in the first stage, it seems relevant to translate series of character substrings and to investigate more recombination schemes at the character level. Finally, the proposed analogical approach can be suitably incorporated in a phrase-based SMT system, since it is completely language-independent and performs promising results by increasing the "adequacy" of the translations in terms of objective assessment metrics.

Table 5 summarizes and compares existing other analogical IR/CLIR approaches in terms of standard metrics, benchmark datasets, and best achieved experimental results.

**Table 5** Summary of the comparative study between other analogical IR/CLIR approaches

| Analogical IR/CLIR approach | Standard metrics | Benchmark datasets | Experimental results (Best ones) |
|---|---|---|---|
| ALEPH: Analogical MT system [28] | – mWER<br>– mPER<br>– BLEU<br>– NIST<br>– GTM | **C-STAR corpus**:<br>– 160K aligned sentences<br>– For Arabic-English: 20K tranlation pairs | – **Results of IWSLT-2004**:<br>**Chinese-English:**<br>mWER = 0.434; mPER = 0.400; BLEU = 0.522; NIST = 8.42; GTM = 0.687<br>**Japanese-English**:<br>mWER = 0.324; mPER = 0.300; BLEU = 0.634; NIST = 9.19; GTM = 0.731<br>– **Results of IWSLT-2005**:<br>**Korean-English**:<br>mWER = 0.530; mPER = 0.486; BLEU = 0.412; NIST = 7.12; GTM = 0.446<br>**Chinese-English**:<br>mWER = 0.454; mPER = 0.418; BLEU = 0.477; NIST = 7.85; GTM = 0.553<br>**Japanese-English**:<br>mWER = 0.361; mPER = 0.323; BLEU = 0.593; NIST = 9.82; GTM = 0.607<br>**Arabic-English**:<br>mWER = 0.527; mPER = 0.497; BLEU = 0.382; NIST = 6.22; GTM = 0.481<br>**English-Chinese**:<br>mWER = 0.798;e mPER = 0.746;e BLEU = 0.098; NIST = 3.029; GTM = 0.363 |
| Analogical translation of unknown words in a SMT [15] | – BLEU<br>– NIST | 500 sentences used in the IWSLT06 campaign to assess Japanese to English MT systems | – BLEU = 0.1638<br>– NIST = 5.8054 |

# 4 Open Problems and Future Directions

This last section is devoted to propose and discuss some open research areas related to the fields of analogical IR/CLIR. We first focus on analogical query disambiguation, expansion, and translation. Second. analogical proportions can be useful in predicting user's preferences in personalized IR. We detail, in the following, the specificity of each of these open trends and we discuss their expected challenges.

## 4.1 *Analogical Query Disambiguation, Expansion, and Translation*

Existing analogical IR/CLIR approaches may benefit from analogical proportions by extending their use for query disambiguation, expansion, and translation. On the first hand, the goal of analogical query disambiguation task is to select the suitable sense and meaning of an ambiguous query term by considering the full context of the query. This problem can be designed using analogical proportions modeled as "$x : y :: z : t$", where $x$ refers to an ambiguous word and $y$ refers to its correct sense selected from an external linguistic resource such as lexicon, thesauri, corpus.. While $z$ refers to another ambiguous query term and $t$ refers to its unknown correct sense.

The idea is to retrieve all pairs $(a, b)$ in the training set that forms a valid analogical proportion with pairs $(z, t)$. Then, solving analogical equations "$a : b :: z : t$" for all these pairs $(a, b)$ enables to identify the correct sense $t$ of $z$. This is just a preliminary reflection on this problem that requires to be more investigated using other related domains of IR and CLIR such as analogical query expansion and analogical query translation. While the former aims at building analogical relationships between the original query terms and their corresponding neighbors in terms of meaning that helps to enlarge the query context, the latter consists of constructing the analogical relationships between the source query terms and their corresponding suitable translations in order to generate the translated query.

## 4.2 *Predicting User's Preferences in Personalized IR Using Analogical Proportions*

In the context of IR, user's preference enables to specify the relevant document satisfying the user's needs, rather than simply indicating whether a document is relevant or not. Moreover, classic IRS performance has been assessed using some binary relevance judgment metrics such as Recall and Precision, which are not suitable for multi-grade relevance. However, the efficiency of recent IRS is evaluated based on the document user's preference. In fact, it is proven that user's preference may be useful to define and evaluate a multi-grade relevance [48].

In the context of preferences learning [6] have recently proposed and tested a comparative study between analogical approaches for preferences expectation. The authors aim at predicting the preference between a new pair of items, based on a set of preferences between items taken by pairs and defined in terms of nominal or numeric feature values. For this purpose, the authors have exploited analogical proportions to develop two generic preference learning methods: the first approach exploits triples of pairs of elements for which preferences are known and which create analogical proportions, completely with the novel pair. Taken attribute-by-attribute, these proportions assess to what extent the modification in values between the elements of the first two pairs is similar to the last two pairs. Noting that analogical proportions between attributes of the four elements are valid and the proportion between preference relation is solvable, this enables to predict the preference relation linked with the fourth pair, while ensuring that no conflicting trade-off are produced. The second method uses the pairs of elements compared one by one. Indeed, the task consists of looking for another pair of elements whose preference is known in order to predict the preference between two elements in a similar way as in k-NN logic. The difference between the first pair items is equivalent to the difference between the second pair items.

Thanks to its generic nature, we believe that analogical preference learning can be useful in predicting user's preferences in personalized information retrieval (PIR) and may lead to promising enhancement in terms of IR efficiency.

## 5 Conclusion

This paper aims to review the most existing IR/CLIR analogical proportions-based approaches. We have presented and discussed different IR/CLIR approaches existing in the literature during the last two decades (2000–2020). Most of these research works have been proposed in the first decade of 2000, especially during the period 2001–2011. However, we can only find the unique recent work of [5] considered as a supervised analogical IR approach. Besides, most of the existing analogical IR/CLIR approaches are unsupervised. This is because of the lack of labeled training sets, which are not usually available for different languages, domains, and tasks.

Existing works in the fields of analogical IR/CLIR are quite limited in all languages, thus we have recommended to apply analogical learning in the domain of query disambiguation, expansion, and translation. It is also relevant to investigate the domain of analogical document/query indexing in IR and CLIR. Besides, we have also recommended the application of analogical proportions for predicting user's preferences in personalized IR. However, these new research trends may require some domain-specific datasets given that analogical learning-based IR models are mainly based on similar cases in the training set. Recently [10–13], have proposed a domain-specific dataset tested in analogical Arabic text summarization [18] and classification. Besides [1, 4], have suggested a domain-specific standard dataset for Arabic NLP and IR/CLIR. We expect that these domain-specific Arabic linguistic resources can support research efforts in analogical Arabic IR/CLIR.

**Acknowledgements** This work was funded by the Emirates College of Technology in Abu Dhabi (UAE) under research grant IRG-BIT-002-2020.

## References

1. Ayed R, Chouigui A, Elayeb B (2018) A new morphological annotation tool for arabic texts. In: Proceedings of the AICCSA, pp 1–6
2. Barbot N, Miclet L, Prade H, Richard G (2019) A new perspective on analogical proportions. In: Proceedings of the ECSQARU, pp 163–174
3. Bayoudh S, Miclet L, Delhay A (2007) Learning by analogy: a classification rule for binary and nominal data. Proc. IJCAI 2007:678–683
4. BenKhiroun O, Ayed R, Elayeb B, Bounhas I, Saoud NB, Evrard F (2014) Towards a new standard arabic test collection for mono- and cross-language information retrieval. In: Proceedings of the NLDB, pp 168–171
5. Bounhas M, Elayeb B (2019) Analogy-based matching model for domain-specific information retrieval. In: Proceedings of the ICAART, pp 496–505
6. Bounhas M, Pirlot M, Prade H, Sobrie O (2019) Comparison of analogy-based methods for predicting preferences. Proc SUM LNCS 11940:339–354
7. Bounhas M, Prade H (2019) An analogical interpolation method for enlarging a training dataset. Proc SUM LNCS 11940:136–152
8. Bounhas M, Prade H, Richard G (2017) Analogy-based classifiers for nominal or numerical data. Int J Approx Reason 91:36–55
9. Bounhas M, Prade H, Richard G (2018) Oddness-based classification: a new way of exploiting neighbors. Int J Intell Syst 33(12):2379–2401
10. Chouigui A, BenKhiroun O, Elayeb B (2017) ANT corpus: an arabic news text collection for textual classification. In: Proceedings of the AICCSA, pp 135–142
11. Chouigui A, BenKhiroun O, Elayeb B (2018) Related terms extraction from arabic news corpus using word embedding. In: Proceedings of the otm workshops, revised selected papers, pp 230–240
12. Chouigui A, BenKhiroun O, Elayeb B (2018) A TF-IDF and co-occurrence based approach for events extraction from arabic news corpus. In: Proceedings of the NLDB, pp 272–280
13. Chouigui A, BenKhiroun O, Elayeb B (2021) An arabic multi-source news corpus: experimenting on single-document extractive summarization. Arab J Sci Eng (In press)
14. Claveau V, Zweigenbaum P (2005) Translating biomedical terms by inferring transducers. In: Proceedings of the AIME, pp 236–240
15. Denoual E (2007) Analogical translation of unknown words in a statistical machine translation framework. In: Proceedings of the MT Summit XI, pp 135–141
16. Doddington G (2002) Automatic evaluation of machine translation quality using n-gram co-occurrence statistics. In: Proceedings of the HLT, pp 138–145
17. Eck M, Hori C (2005) Overview of the IWSLT 2005 evaluation campaign. In: Proceedings of the IWSLT, pp 1–22
18. Elayeb B, Chouigui A, Bounhas M, BenKhiroun O (2020) Automatic arabic text summarization using analogical proportions. Cogn Comput 12(5):1043–1069
19. Essid M, Bounhas M, Prade H (2020) Continuous analogical proportions-based classifier. In: Proceedings of the IPMU, pp 541–555
20. Fuhr N, Buckley C (1991) A probabilistic learning approach for document indexing. ACM Trans Inf Syst 9(3):223–248

21. Hesse M (1959) On defining analogy. Proc. Aristot Soc 60:79–100
22. Hug N, Prade H, Richard G, Serrurier M (2016) Analogical classifiers: a theoretical perspective. In: Proceedings of the ECAI, pp 689–697
23. Langlais P, Yvon F, Zweigenbaum P (2008) Analogical translation of medical words in different languages. In: Nordström B, Ranta A (eds) Advances in natural language processing, LNCS, vol 5221. Springer, pp 284–295
24. Langlais P, Patry A (2007) Translating unknown words by analogical learning. In: Proceedings of the EMNLP-CoNLL, pp 877–886
25. Langlais P, Yvon F, Zweigenbaum P (2009) Improvements in analogical learning: application to translating multi-terms of the medical domain. In: Proceedings of the EACL, pp. 487–495
26. Lepage Y (1998) Solving analogies on words: an algorithm. Proc ACL-COLING 1:728–734
27. Lepage Y (2001) Analogy and formal languages. Proc FG/MOL 2001:373–378
28. Lepage Y, Denoual E (2005) Purest ever example-based machine translation: detailed presentation and assessment. Mach Trans 19(3–4):251–282
29. Miclet L, Bayoudh S, Delhay A (2008) Analogical dissimilarity: definition, algorithms and two experiments in machine learning. J Artif Intell Res 32:793–824
30. Miclet L, Prade H (2009) Handling analogical proportions in classical logic and fuzzy logics settings. In: Proceedings ECSQARU'09. Springer, LNCS, pp 638–650
31. Moreau F, Claveau V, Sébillot P (2007) Automatic morphological query expansion using analogy-based machine learning. In: Proceedings of the ECIR, pp. 222–233
32. Papineni K, Roukos S, Ward T, Zhu W (2002) Bleu: a method for automatic evaluation of machine translation. In: Proceedings of the ACL, pp 311–318
33. Popovic M, Ney H (2007) Word error rates: decomposition over POS classes and applications for error analysis. In: Proceedings of the WMT@ACL, pp 48–55
34. Prade H, Richard G (2010) Reasoning with logical proportions. Proc KR 2010:545–555
35. Prade H, Richard G (2012) Homogeneous logical proportions: their uniqueness and their role in similarity-based prediction. Proc KR 2012:402–412
36. Prade H, Richard G (2013) From analogical proportion to logical proportions. Logica Universalis 7(4):441–505
37. Prade H, Richard G (eds) (2014) Computational approaches to analogical reasoning: current trends. Stud Comput Intell 548 (Springer)
38. Prade H, Richard G (2014) From analogical proportion to logical proportions: a survey. In: Computational approaches to analogical reasoning: current trends, pp 217–244
39. Prade H, Richard G (2014) A short introduction to computational trends in analogical reasoning. In: Computational approaches to analogical reasoning: current trends, pp 1–22
40. Prade H, Richard G (2017) Analogical proportions and analogical reasoning—an introduction. In: Proceedings of the ICCBR, pp 16–32
41. Prade H, Richard G, Yao B (2012) Enforcing regularity by means of analogy-related proportions-a new approach to classification. Int J Comput Inf Syst Ind Manag Appl 4:648–658
42. Schulz S, Markó KG, Sbrissia E, Nohama P, Hahn U (2004) Cognate mapping—A heuristic strategy for the semi-supervised acquisition of a spanish lexicon from a portuguese seed lexicon. In: Proceedings of the COLING
43. Stroppa N, Yvon F (2005) An analogical learner for morphological analysis. In: Proceedings of the CoNLL, pp 120–127
44. Stroppa N, Yvon F (2005) Analogical learning and formal proportions: Definitions and methodological issues. Tech. rep, ENST (June
45. de la Tour TB, Peltier N (2014) Analogy in automated deduction: a survey. In: Computational approaches to analogical reasoning: current trends, pp 103–130
46. Toutanova K, Suzuki H, Ruopp A (2008) Applying morphology generation models to machine translation. In: Proceedings of the ACL, pp 514–522
47. Yvon F, Stroppa N, Delhay A, Miclet L (2004) Solving analogies on words. Technical Report D005, ENST, Paris, France
48. Zhou B, Yao Y (2010) Evaluating information retrieval system performance based on user preference. J Intell Inf Syst 34(3):227–248

# Maximizing Parallelism in the Cloud: A Hybrid Load Balancing Approach

**Alberto Arteta Albert, Nuria Gómez Blas, and Luis Fernando de Mingo López**

**Abstract** When the number of processes requiring CPU is higher than the number of cores in a system, achieving parallel systems show many challenges. However, different approaches are known to work well for achieving an acceptable degree of parallelism. Load balancing operations are one of the main concerns in virtual systems and, specifically, in Cloud Computing. An optimal load balancing strategy can increase the multiprogramming rate, significantly impacting the parallelism degree of a system. There are many algorithms proposed and designed to carry out client requests at distributed servers in the cloud. All of them target maximizing the performance of the overall system. In this manuscript, various existing load balancing algorithms are discussed and compared with a new hybrid algorithm. This hybrid algorithm has been developed to handle the client requests based on the priority and availability of the cluster nodes, which are part of the virtual system. This algorithm maintains two tables with a list of node id, priority, and availability so that the load can be optimally distributed. That enables selecting an appropriate node to achieve efficient load balancing and therefore s a higher degree of overall parallelism.

**Keywords** Load balancing · Parallel cloud computing · Virtual systems performance

---

A. Arteta Albert (✉)
Department of Computer Science, Troy University, Alabama, USA
e-mail: aarteta@troy.edu

N. G. Blas · L. F. de M. López
Departamento de Sistemas Informáticos, Universidad Politécnica de Madrid, Madrid, Spain
e-mail: nuria.gomez.blas@upm.es

L. F. de M. López
e-mail: fernando.demingo@upm.es

 
X.-S. Yang et al. (eds.), *Proceedings of Sixth International Congress on Information and Communication Technology*, Lecture Notes in Networks and Systems 216,
https://doi.org/10.1007/978-981-16-1781-2_57

# 1 Introduction

Cloud computing refers to providing software through the www network. It exists as a paradigm to enable, network points to a shared area of resources [1]. It is done very efficiently. It works as (IaaS) [2], SaaS, SaaS. Specifically it locates its applications on the cloud infrastructure of service providers and is delivered to users through web interfaces and other software. Cloud computers are considered to have built their zero points from Software as a Service [3]. PaaS indeed is a model supported by cloud service providers [4]. Thus, scalability, availability fast processing, and storage capacities are managed by PaaS. Hence, PaaS run and manage their applications using distal frameworks. However, users do not have absolute control over the core cloud platforms such as servers, operating systems, or storage [5], deploying any kind of software known as Software on the cloud [6]. An increase in the number of users worldwide has resulted in many requests at a rapid rate. Load balancing in clouds distributes the tasks in a balanced manner to all the nodes in place. It is used to achieve good feedback from the final user and to improve the resource utilization ratio, ensuring that no single node is in a bottleneck damaging the overall performance of the system, detecting that issue boosts the overall performance of the system. If a proper load balancing operation is done, the available resources will be handled optimally, thereby minimizing resource consumption. It also helps implement fail-over, enabling scalability, avoiding bottlenecks that decrease response time, and overall latency among other essential metrics. Balancing the load can be done either dynamically or in a static way. Dynamic ones work better for homo and heterogeneous scenarios. The static on the other hand produces less overhead [5].

Cloud computing is a revolutionary innovative technology that can benefit from solutions that target load balancing. Moreover, Cloud computers balance the load out over different nodes dynamically, avoiding bottlenecks, and boosting the resource usage [7].

Depending on the workload, the load balancer software is usually divided into two categories (static and dynamic) load balancing algorithms (Fig. 1).

## *1.1 Static Algorithms (SLB)*

SLB use the average metrics to make the balancing decision operations. The main aim of SLB algorithms is to decrease the running time, so the communication between nodes can flow without issues [16].

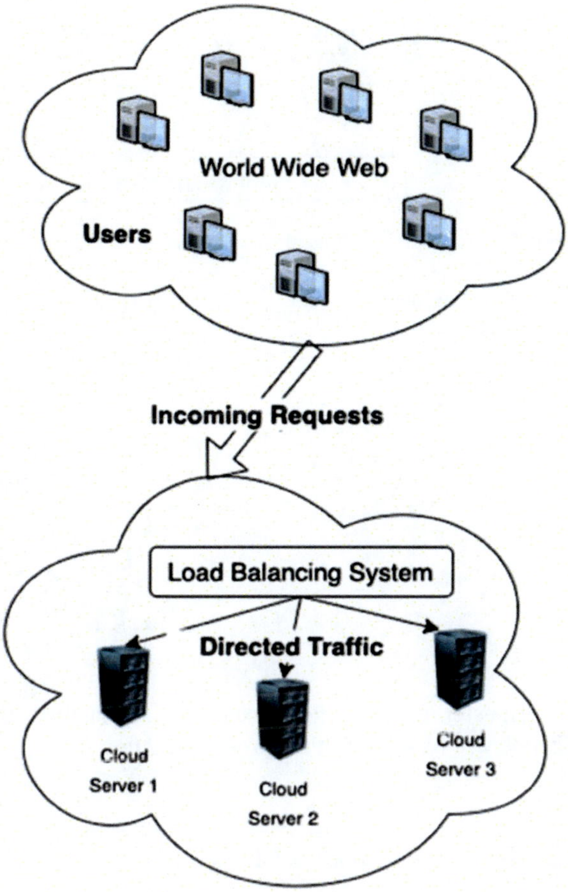


**Fig. 1** Load balancing in the Cloud computing environment

## *1.2 DLB Algorithms*

DLB algorithms distribute the tasks among the processors during execution time. The master node assigns new processes to the slaves based on the new information collected [17, 18]. Also, Unlike SLB algorithms, DLB dynamically addresses when one of the processors becomes underloaded. Instead, they are buffered in the queue on the primary host and distributed dynamically upon requests from remote nodes. DLB continually monitor the processors' workload and redistributes it when the imbalance reaches some predefined level, the redistribution of work occurs at the expense of increasing the CPI (cycles per instructions), so it should not be used without caution as the overhead increases.

LBD has five major components [19].

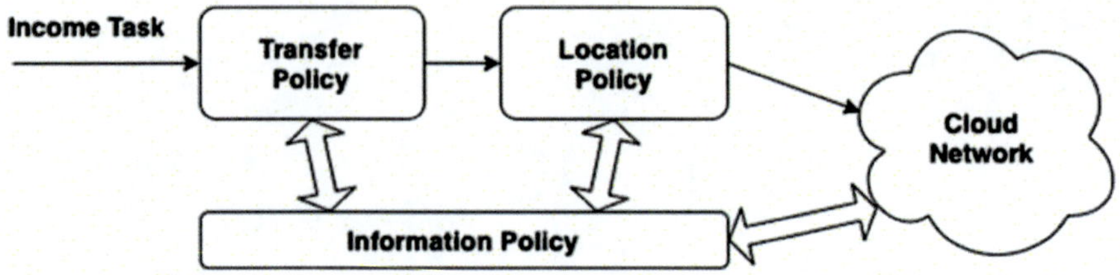


**Fig. 2** Components of dynamic load balancing

**Fig. 3** Round robin mechanism

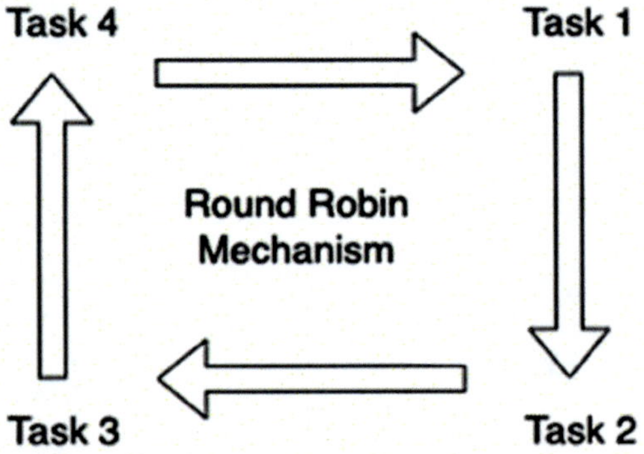


Transfering, Selection, Location, Information, and Load Estimation Policies: These will be named in the proposal section (Fig. 2).

## 2 Related Work and Discussion

Load algorithms balance out the tasks load across all the nodes and also provide a way to reach the proper usage of resources and a better feedback from the final user. Hence, using an appropriate load balancing algorithm is a must to ensure an efficient setup. Hence, using an appropriate load balancing algorithm is a must to ensure an efficient setup. In [8], LB algorithm optimally manages the incoming tasks among VMs. The authors in [9] work with a node with LB model (OPH-LB) [10] introduces a load balancer named (LBaaS) in the cloud (Figs. 3 and 4).

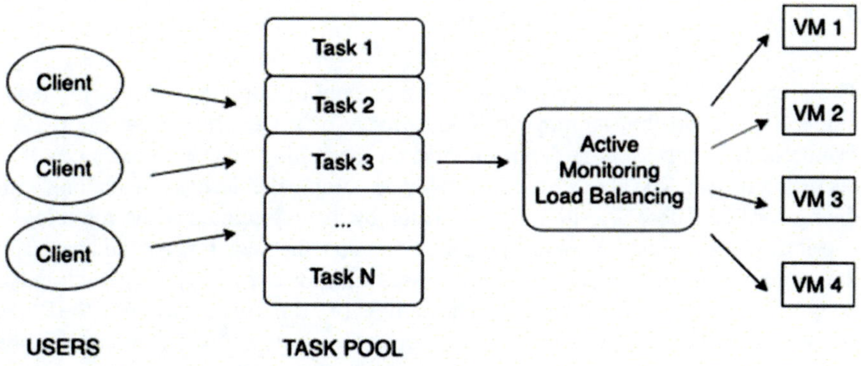


**Fig. 4** Active monitoring-based load balancing technique

# 3 Existing LB Algorithms in Cloud Computing

## *3.1 Round Robin Algorithm*

The Round Robin approach is a time-shared LB algorithm [8] with new improvements in [10].

The WRR( Weighted Round Robin) is an updated version of RR [7].

## *3.2 Active Monitoring-Based Load Balancing Algorithm*

AMBLB [15] is another vital load that identifies the least loaded VM.

## *3.3 Ant Colony Optimization (ACO)*

ACO algorithm [20] is mainly proposed for load balancing of nodes and aims to distribute workload among the nodes [22–24] optimally. Ants move toward the food provider from the top node. Then the ants record their data for future decisions.

## 4 Proposed Hybrid Algorithm

The proposed model is a combination of static and dynamic load balancing techniques. This algorithm requires prior knowledge of system resources; the performance of the processors is determined at the beginning of the execution as it is occurring in [24]. The algorithm uses a table containing the list of servers along with their priorities. These priorities are calculated by using weighted round robin logic. A priority function calculates the weights. After that, the load is distributed based on the weights assigned to each node. Processors with lower cycles per instruction are assigned a more considerable value. The number of processes allocated in the RAM and previous statistics of average responses in the cluster of nodes determines the node weight. That information is enough to build the function of priorities. Hence the highest weighted servers end up receiving more tasks. In a situation where all weights become equal, servers will receive balanced traffic. In that case, our algorithm behaves like a static balancer.

Where $x_1, x_2$ and $x_3$ are the processor's capacity, amount of process admitted to the RAM, and average statistical response per process, respectively. For the hybrid algorithm, the weights assigned to the formula are 0.4 for the processor factor, 0.3 for the number of processes, and 0.3 for the node's average response per process. And, $f(x)$ evaluates these inputs for every node; then it returns the priority, which is used by the Hybrid algorithm to send the load to the targeted node, when available (Fig. 5).

Therefore, the server with higher priority will receive more number of tasks compared to the server with less priority. Numerical values quantify the servers that means the server with higher priority has a more considerable value than the server with lower priority. Once the priority of the nodes is determined, the next section of the algorithm comes into focus, which is determining the nodes' availability for a particular request. Based on the priority table, we found the order of priority of servers. Now the task is to assign that particular client request to the virtual machine of the highest priority. To perform this task, another table named the availability table is maintained, which is dynamically updated based on the availability of the virtual machines at that moment. This algorithm finds which virtual machine of that server is available and assigns the user request to that particular server. The proposed algorithm is given (Fig. 6 shows the flowchart of the proposed algorithm):

```
While number of system process > K
    For each node:
      Calculate priority f
      Calculate availability
      Calculate node with
                max(availability, priority)
    endFor
  Distribute load-K  to the node
  EndWhile
```

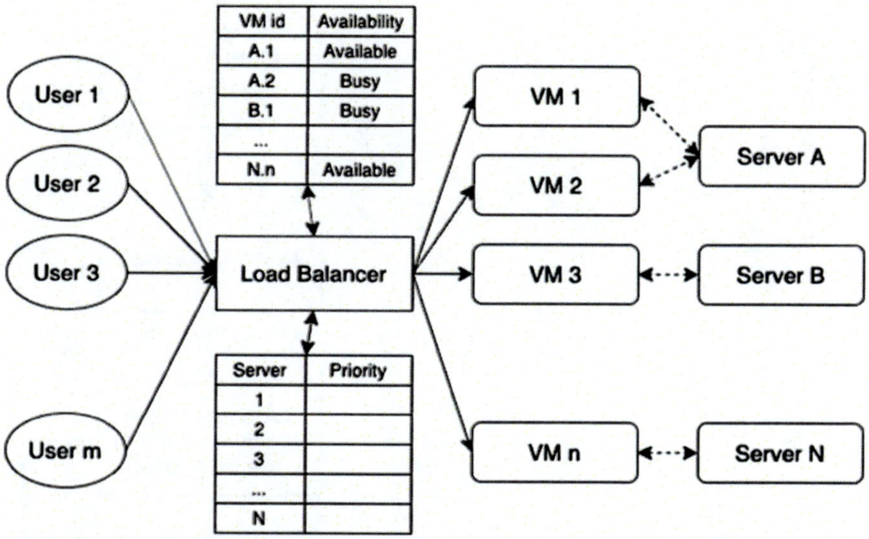


**Fig. 5** Proposed load balancing algorithm

For instance, there is a user/client request, now in this algorithm, all the requests go through the load balancer system by which it checks the priority associated with each server based on its system resources. Here A, B, N are the servers, A.1, A.2, B.1, B.2, and N.n are the virtual machines of server A, server B, server N, respectively. Let us assume that server A has a priority value of 100, B has the priority value of 90. server B has a lower capacity in handling the client requests than server A, so here the load balancer will stabilize the load (requests) by assigning the request to server A at this level. Now, deciding which virtual machine of server A will handle the request. Suppose in node A. Virtual machine A.1 is busy, then the algorithm searches for the next VM available with the highest priority and assigns the request to that particular VM. This state of the virtual machine, whether available or busy, is updated in the second table dynamically that maintains the list of virtual machine ids and their availabilities. If all the virtual machines are found busy in that particular server, it moves to the server with the next highest priority, and the same process is repeated.

The same environment and architecture, so the metrics could not be affected by the system's architecture/topology. All of the load balancing algorithms worked equally and optimally when the number of processes and service requests were stable (see Table 1). The main differences in timing occurred when the overload was continuously forced in different degrees, so switching to different nodes was differently done by the four algorithms (see Table 2). Below there is a table summarizing different normalized metrics extracted from the systems' data collector in processing the same

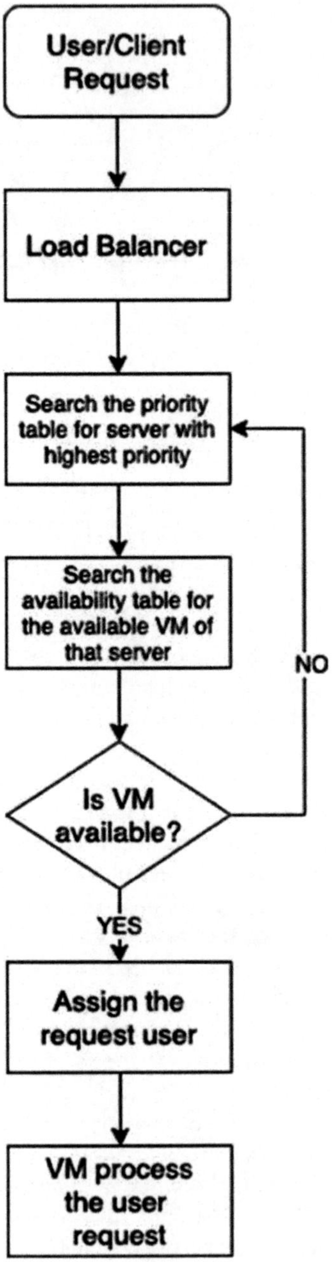


**Fig. 6** Flowchart of the proposed load balancing algorithm

**Table 1** Snapshot of normalized metrics with a stable variation of threads

| | CPUT | ATPC | KFM | PD | AVG |
|---|---|---|---|---|---|
| ACO | 0.2334 | 0.4432 | 0.2340 | 0.7560 | 0.3331 |
| AMB | 0.2454 | 0.4323 | 0.2382 | 0.7498 | 0.3255 |
| WRR | 0.2462 | 0.4413 | 0.2376 | 0.7732 | 0.3439 |
| SLB | 0.2389 | 0.4387 | 0.2325 | 0.7698 | 0.3234 |
| HYB | 0.2443 | 0.4382 | 0.2397 | 0.7588 | 0.3315 |

**Table 2** Snapshot of normalized metrics with a dramatic increase of threads (power three growth)

| | CPUT | ATPC | KFM | PD | AVG |
|---|---|---|---|---|---|
| ACO | 0.4456 | 0.5634 | 0.3740 | 0.7560 | 0.536 |
| AMB | 0.4344 | 0.5536 | 0.3821 | 0.7498 | 0.575 |
| WRR | 0.4662 | 0.5713 | 0.4267 | 0.7732 | 0.639 |
| SLB | 0.4089 | 0.5237 | 0.3601 | 0.7698 | 0.553 |
| HYB | 0.4095 | 0.5568 | 0.3798 | 0.7982 | 0.497 |

threads (randomly created) for the different approaches. Completed processes per unit of time (CPUT), Average Threads per core (ATPC), Karp-flat metric (KFM), the average response time (AVG), and parallelism degree (PD) are defined in Amdhals' law.

## 5 Results

The proposed algorithm and the traditional ones have been compared in an environment with three nodes in a cluster and a distributed OS. The three nodes have been virtualized, simulating the structure of the load balancing module in a cloud system (parameters and specifics described above). All the approaches mentioned in the literature have been tested in the environment, and the data collector has been recording latencies and other performance metrics for the implementation of the known approaches and the proposed one. The virtual cluster was logging the activity of every node in the virtual one. A daemon process was running, generating system calls, and creating processes in every node asking for different resources (randomly fixed). These services were switching from terminal, printer, hard disk, and USB units randomly selected. The goal was overloading the system generating some processes > K, K prefixed threshold symbolizing the number of the overloaded processes. Then, the virtual system could switch the control to other nodes to attend the request as efficiently as possible.

The experiment shows that all the approaches respond well and efficiently balance the processing load when the number of processes continuously or linearly increases.

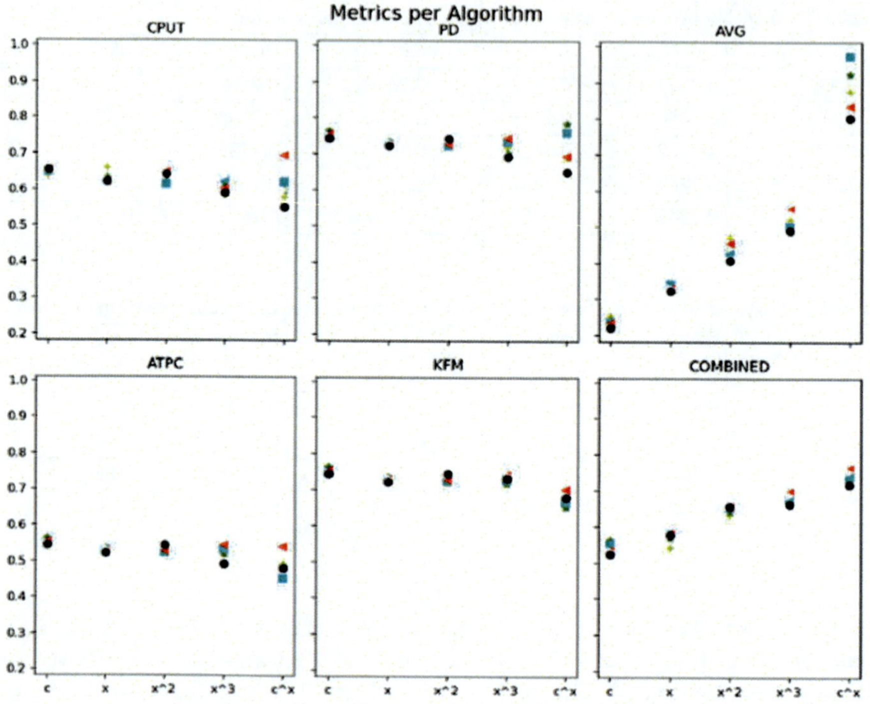


**Fig. 7** Metrics per algorithm

Interestingly the differences start when the number of threads to process in our system increases according to a higher power. The processing order and the chosen nodes differ, affecting the metrics described above.

Figure 7 below depicts the relationship between the variation in the number of threads and the metrics. It shows a noticeable difference when the number of treads increases in a power three order and a clear difference in the metrics where the threads are exponentially generated. Interestingly, all metrics slightly increase their values when the threads grow from linear to quadratic order, suggesting that the architecture of the system was optimally designed for managing this kind of growth. A new metric (COMBINED) is added in the figure, summarizing the rest of the metrics in one (Every metric has the same weight as part of COMBINED). The hybrid approach (black circles in the graph) shows the lowest value in most metrics when compared with ACO (triangles), AMB (+), WRR (*), and SLB (squares), specifically noticeable and significant the AVG and the COMBINED metric, with a clear improvement when the threads are exponentially generated.

## 6 Conclusions

Cloud computing and any other virtualize system with a clustered environment must decide an optimal algorithm or load balancing to distribute dynamic workload to all the nodes in the cloud environment to improve performance and resource utilization. That requires complicated analysis. This paper has focused on different approaches (static, dynamic, and theirs combination). All of them aimed for maximum resource utilization and maximized throughput, and reduced response time. The increase in the user usage of the cloud has put a load on the computational capacity of the VM. It is imperative to distribute the load to the appropriate VM. The proposed algorithm that chooses an appropriate VM considering the system resources and their availability as well, to increase the throughput. There are various factors to be considered for the perfect load balancer. In the same environment, the most popular approaches have been compared with our proposal.

However, the appropriate load balancer always is an open challenge for the future. For future work, there are other factors like memory capacity, performance, etc. that could be considered to prioritize the servers. Also, there are still some issues related to cloud security. Apart from availability, reliability is also one of the critical factors for cloud computing, which would also be considered in future work. The results are obtained in a virtual system with 2 servers, simulating a micro-cloud system. Ideally, this algorithm could be tested in a real cloud-provider environment.

## References

1. Int J Inf Manage (2018). www.elsevier.com/locate/ijinfomgt, 38:128–139, Accessed 25 July 2017. ISSN: 0268-4012
2. Zissis D, Lekkas D (2012) Addressing cloud computing security issues. Future Gener Comput Syst 28(3):583–592. https://doi.org/10.1016/j.future.2010.12.006
3. Mell P, Grance T (2011) The NIST definition of cloud computing
4. Marston S, Li Z, Bandyopadhyay S, Zhang J, Ghalsasi A (2011) Cloud computing—the business perspective. Decis Support Syst 51(1):176–189. https://doi.org/10.1016/j.dss.2010.12.006
5. Sultan N (2010) Cloud computing for education: a new dawn? Int J Inf Manag 30(2):109–116. https://doi.org/10.1016/j.ijinfomgt.2009.09.004
6. Mateescu G, Gentzsch W, Ribbens CJ (2011) Hybrid computing—where HPC meets grid and cloud computing. Future Gener Comput Syst 27(5):440–453. https://doi.org/10.1016/j.future.2010.11.003
7. Bhandari A, Kaur K, An enhanced post-migration algorithm for dynamic load balancing in cloud computing environment. In: Amritsar, K. Kaur. Chakraborty et al (eds) Proceedings of international ethical hacking conference 2018, advances in intelligent systems and computing 811. Department of Computer Engineering and Technology, Guru Nanak Dev University. https://doi.org/10.1007/978-981-13-1544-2_6
8. Dhinesh Babua LD, Venkata Krishna P (2013) Honey bee behavior inspired load balancing of tasks in cloud computing environments. Appl Soft Comput 13(5):2292–2303
9. Chhabra S, Singh AK (2018) A probabilistic model for finding an optimal host framework and load distribution in cloud environment. Procedia Comput Sci 125:683–690

10. Rahman M, Iqbal S, Gao J (2014) Load balancer as a service in cloud computing. In: 2014 IEEE 8th international symposium on service oriented system engineering, pp 204–211
11. Sharma S, Singh S, Meenakshi S (2008) Performance analysis of load balancing algorithms. Int J Civ Environ Eng 2(2):367–370
12. Sran N, Kaur N (2013) Zero proof authentication and efficient load balancing algorithm for dynamic cloud environment. Int J Adv Res Comput Sci Softw Eng 3(7):128–2277
13. Randles M, Lamb D, Taleb-Bendiab A (2010) A comparative study into distributed load balancing algorithms for cloud computing. In: 24th international conference on advanced information networking and applications workshops, pp 551–556
14. Padhy RP, Rao PGP (2011) Load balancing in cloud computing system. In: Department of computer science and engineering, National Institute of Technology, Rourkela, Orissa, India
15. Bryhni H, Klovning E, Kure O (2000) A comparison of load balancing techniques for scalable Web servers. IEEE Netw 14(4):58–64
16. Rajguru AA, Apte S (2012) A Comparative performance analysis of load balancing algorithms in distributed system using qualitative parameters. Int J Recent Technol Eng (IJRTE) 1, 3
17. Malik S (2000) Dynamic load balancing in a network of workstation. 95.515 Research Report
18. Wang Y, Morris R (1985) Load balancing in distributed systems. IEEE Trans Comput C-34(3):204–217
19. Amar M, Anurag K, Rakesh K, Rupesh K, Prashant Y (2011) SLA driven load balancing for web applications in cloud computing environment. Inf Knowl Manag 1(1):5–13
20. Nishant KP, Sharma V, Krishna C, Gupta KP, Singh NN, Rastogi R (2012) Load balancing of nodes in cloud using ant colony optimization. In: Proceedings of the 14th international conference on computer modelling and simulation (UKSim). IEEE, pp 3-8
21. Dam S, Mandal G, Dasgupta K, Dutta P (2014) An ant colony based load balancing strategy in cloud computing. In: Advanced computing, networking and informatics, vol 2. Springer International Publishing, pp 403–413
22. Kumar S, Rana DS, Dimri SC (2015) Fault tolerance and load balancing algorithm in cloud computing: a survey. Int J Adv Res Comput Commun Eng
23. Kashyap D, Viradiya J (2014) A survey of various load balancing algorithms in cloud computing. Int J Sci Technol Res 3(11)
24. Shah N, Farik M (2015) Static load balancing algorithms in cloud computing: challenges and solutions. Int J Sci Technol Res 4(10). ISSN 2277-8616

# Secure Data Transmission in Smart Meters Using Q-Learning in Fog Computing Environment

**Saurabh Shukla, Subhasis Thakur, and John G. Breslin**

**Abstract** With the advancement in technology and connectivity, the concept of smart grids with smart meters has significantly increased the interest of both electric consumers and power suppliers. Nowadays the smart cities are using the technology of smart meters in their household, which can transfer the electrical data in a bi-directional way to consumers and smart grids. It consists of various Internet-connected sensors, software applications, to connect different devices. This makes the smart grid network more complex which is further vulnerable to the cyber-attack. Moreover, it gives rise to various malicious attack such as false data injection (FDI) in smart meters which leads to incorrect decisions due to a large number of open access points in a network; which is accessible to outside intruders. In smart meters, Home Area Network (HAN) and smart meter collectors are more vulnerable for cyber-attacks as they are directly involved in bi-directional communication between the devices such as smart grid, data connectors, and smart meters. Since smart meters work in a distributed environment where fog computing (FC) placed at the edge of the networks can play a major role by integrating with blockchain for secure electric data transmission and to avoid the FDI in smart meters. Here FC nodes will act as miners to makes the decisions on the data transmission while monitoring the states of the system. Hence, we propose a novel solution for the above mention issue which consists of 3-tier FC-based blockchain architecture and Smart Meter-based Q-Learning Encryption (SMQE) algorithm to verify the system in FC environment. The algorithm performs the identification of FDI, and malicious attacks conducted on smart meters during data transmission. It opens a blockchain-based channel for secure data transmission in smart meters. The simulation tools used for this proposed work are iFogSim, SimBlock, and Python (Anaconda) with Geth version 1.9.25.

S. Shukla (✉) · S. Thakur · J. G. Breslin
National University of Ireland Galway, Galway, Ireland
e-mail: saurabh.shukla@nuigalway.ie

S. Thakur
e-mail: subhasis.thakur@nuigalway.ie

J. G. Breslin
e-mail: john.breslin@nuigalway.ie



X.-S. Yang et al. (eds.), *Proceedings of Sixth International Congress on Information and Communication Technology*, Lecture Notes in Networks and Systems 216,
https://doi.org/10.1007/978-981-16-1781-2_58

**Keywords** Smart grid · Smart meter · Fog computing · Cloud computing Cyber-attack · Q-learning · Markov decision process · Reinforcement learning

## 1 Introduction

Traditional gird with conventional meters has been used for a quite long period in our households. These grids connect the meters, power grids, and the consumers in a single directional way. Here meters have a limited role to play for the electric data transmission. They lack the real-time data transmission, connectivity between different electrical appliances and meters which makes them unsuitable for the current existing smart cities[1]. Secure electric data transmission with low latency and high throughput play a central role in smart meters. Lately due to the advancement in the deployment of smart meters in smart cities have opened a large security challenge to avoid cyber-attack on the complex network of smart grid. Here the electric data is usually transferred from Home Area Network (HAN) to Neighborhood Area Network (NAN) and Wide Area Network (WAN) [2]. Current power consumption in cities requires constant monitoring and efficient secure transmission between smart grids, smart meters, and consumers in real-time. These smart meters can be connected and disconnected at the ease of consumers depending upon their requirements and usage in their households. Smart meters are a kind of computers which makes them a part of the cyber world [3]. The network of a smart grid is open and decentralized which further makes it more vulnerable from cyber-attacks such as malicious data inclusion, false data injection (FDI), and erroneous values transmission [4].

The attackers can easily change or manipulate the meter value according to their requirements and needs by monitoring the user power consumption level. With the advancement in technology especially for distributed networks; edge computing and fog computing (FC) plays a major role to safeguard the electric data distribution across a spread decentralized open network. FC works in Local Area Network (LAN) and can connect with HAN, NAN, and WAN in smart meter network [5], whereas blockchain also works in a centralized manner by keeping the transaction records of all the blocks using an attached timestamp [6, 7].

Therefore, we have proposed a Smart Meter-based Q-Learning Encryption (SMQE) algorithm which uses Q-learning technique of Reinforcement Learning (RL) algorithm to makes the decisions in the stochastic environment by solving the problem of Markov Decision Process (MDP) as the system state of smart meter network changes with time.

Recent works mostly focused on the detection of electricity theft in smart meters, classification of malicious data using a machine learning algorithm. They lack real-world implementation. Hence, we propose a novel solution using a machine learning technique in FC-based blockchain environment for secure electric data transmission in the smart meter network.

## 2 Background and Related Work

The six main aspects of cybersecurity concern of smart meters and smarts grid which are open to research work are (1) Integrity (2) Confidentiality (3) Availability (4) Reliability (5) Authentication and (6) Authorization [8–10]. Here the cyber threat in smart grid network is related to all the three networks, i.e., NAN, HAN, and WAN. Attacks such as false data injection (FDI) can easily change the system state which leads to stochastic state and it becomes difficult to monitor the data and identify the malicious attack [1, 11]. This attack can further incorporate the flow of data traffic by misleading the original flow of data.

The FDI can be sub-divided into four major attacks such as (1) Attack on smart meter device (2) Data attack (3) Attack on privacy and (4) Attack on the network [2, 6]. There is the major gap to fulfill the QoS requirement for electric data transmission between smart meters, consumers, prosumers, meter management systems, and utility offices.

In [12], the authors used a fog computing-based model to design a grid topology using MATLAB Simulink for smart grid monitoring. Their proposed model was developed to monitor voltage and power loss during transmission of data in the smart grid network. Their primary focus was on the monitoring of data transmission using Fog-Cloud network. However, no work has been done to avoid the cyber-attack in the smart grid. In [13], the authors proposed a novel solution to identify the FDI using an input observer-based distributed detection system. The system was able to easily detect the transition in states. Moreover, they were able to isolate the FDI using their proposed distributed detection approach. They mostly considered the stealthy characteristics of False Data Injection Algorithm (FDIA). However, it lacks real-world implementation.

Similarly, in [14], the authors suggested blockchain as a tool for security in a smart grid system. They also mentioned how blockchain as a tool can be efficient for a smart grid complex network system by keeping the record of several transactions with an attached timestamp. In [1], the author conducted a complete survey analysis of smart grid and smart meter highlighting the benefits and drawbacks related to the security aspect of electric data transmission and consumers records. Moreover, the author has also highlighted the network topology used in the smart grid network along with communication protocols.

On a similar note, the authors in [4], highlighted the issue of cybersecurity and cyber threat on the smart grid network. Furthermore, they discussed the challenges, threats, and potential measures to overcome these attacks. They conducted an in-depth analysis of several network threats and their effects on national security. In [15], the authors implemented the integrated blockchain model with Ethereum and Hyperledger to measure the performance of the smart grid network. Furthermore, they presented a virtual network of smart gird to monitor the data transaction and transmission by making a slight modification in their smart contracts.

The authors in [16], proposed a detection method for FDI in smart grids. This method was based on the Graphical Signal Processing (GSP) by utilizing the graph

Fourier transform. Their derivation was based on the graph model and the flow of the AC power. In [17], the authors suggested and conducted a detailed survey analysis of blockchain for futuristic smart grids. They further discussed the issues related to the complex network system of the smart grid system and role of blockchain as a decentralized system to minimize the cyber-attacks and keep a record of the previous transaction.

Similarly, the authors in [18] discussed the security and communication in the smart grid network. Moreover, they highlighted the importance of cybersecurity by mentioning its key features and aspects for smart grid network topology. In [19], the authors used machine learning and statistical techniques such as SVM and reinforcement learning to overcome the issue of FDI in the smart grid and smart meters.

## 3 3-Tier FC-Based Blockchain Architecture and Advanced System Model

This section discusses the proposed 3-tier FC-based architecture with blockchain and advanced system model for a smart meter. The section further discusses the importance of FC and blockchain for smart meters deployed at the user's household. See Fig. 1 for 3-tier FC architecture.

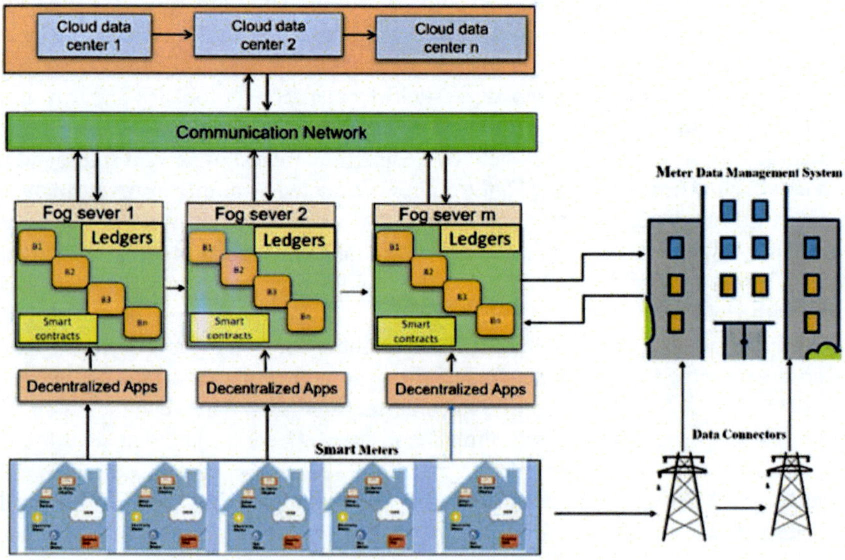


**Fig. 1** 3-tier FC-based blockchain architecture

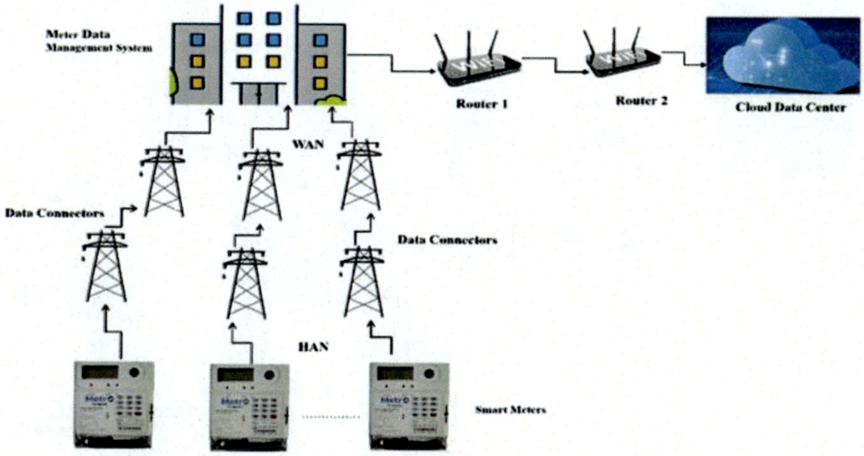


**Fig. 2** Conventional system model for electric data transmission

The proposed 3-tire FC-based architecture consists of three layers. The first layer consists of smart meters placed at the consumer site. These smart meters are connected using HAN and NAN to FC nodes which are deployed at the middleware layer. The smart meters are further connected to the meter data management system through various data connectors using WAN. The electric data is then transferred to the middleware layer which consists of fog servers; these fog servers are divided into various virtual machines (VM). These machines consist of software and application for smart contracts and decentralized apps with ledgers to identify the false data injected during transmission from smart meters to fog servers and meter data management systems.

See Fig. 2 shows the conventional method of electric data transmission.

Figure 2 is a conventional system model which is used nowadays for electric data transmission. Here the smart meters are connected through data connectors. The data is transmitted from HAN to WAN. Furthermore, the meter data management system is connected to a cloud server for the request and retrieval of information for user power consumption. The main drawback with this system is that it is an open network where smart meters are connected to many data connectors with multiple hop counts which makes it vulnerable to cyber-attacks on the electric data. See Fig. 3 shows the advanced system model for secure data transmission.

Figure 3 shows an advance system model for secure electric data transmission for smart meters. The model utilizes the concept of distributed FC deployed at the local household sites. These fog nodes work in HAN and connect through WAN to the meter management system. The smart meters measure the reading and send the data through data connectors to fog nodes. The fog nodes receive the data in the form of ledgers by an attached timestamp for the identification of smart meters and electric data readings. These fog nodes now act as miners using private blockchain model. The fog serves uses Q-learning technique of reinforcement learning (RL) algorithm

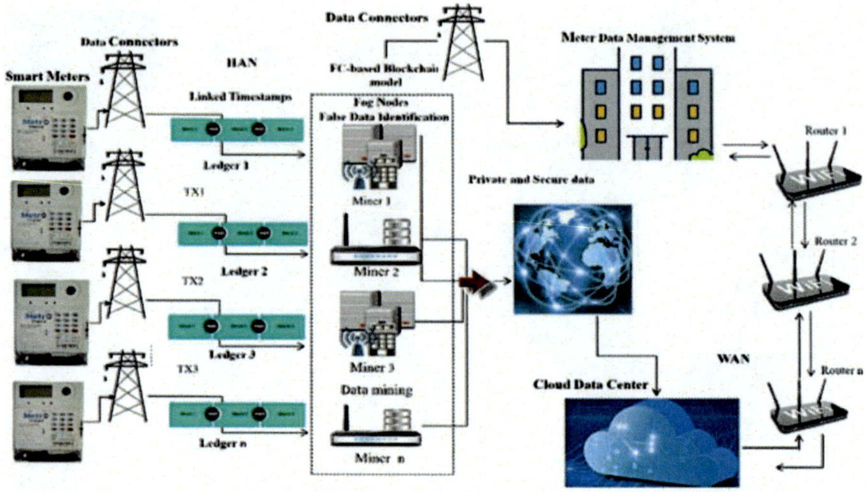


**Fig. 3** Proposed advanced system model

to solve the Markov Decision Process (MDP) using the best quality of action to identify the FDI inside the stochastic system.

## 4 Smart Meter-Based Q-Learning Encryption (SMQE) Algorithm

The principal aim of this research is the development of a novel approach to meet the QoS requirement related to security, identification, and authenticity of data transmission between smart meters, fog nodes, data connectors, consumers, and the cloud servers using FC-based blockchain approach. To accomplish the main aim, we propose a 3-tier FC-based blockchain architecture for smart meters. Next, we designed and developed a novel SMQE algorithm. The algorithm performs the identification of False Data Injection (FDI) by predicting the change in the system state, verification of data sources and targets when data is transmitted data via different smart meters and fog nodes.

The algorithm uses a Q-learning technique to solve MDP in smart meters. The data is encrypted and then decrypted using private blockchain and different cryptographic operations inside fog nodes. FC nodes acting as miners in smart meter network plays a major role by identifying the FDI. Here FC nodes make decisions on the transmission of flow of data while monitoring the states of the system. Miners here consist of encryption algorithm to encrypt the meter data and readings from the outside world.

Asymmetric public–private key pairs are also used in the development of an algorithm.

**SMQE Algorithm Symbol Notations**

$C_m$ : Consumer
$M_s$ : Meter management system
$S_m$ : Smart Meters
$E_d$ : Electric data
$E_{d}\_ID$: ID of the participated smart meters
$F_n$: Fog nodes
L: The list of the configuration $S_m$ devices sent by the $F_n$

$\gamma_i$ : Discount factor
$(\in)$ : Exploration policy
$K_{sym}$ : Symmetric key
$K_{pub}$ : Public key
$F_n K_{pub}$ : Fog node public key
$Encrypt_{Sym}$ : Symmetric encryption
$Encrypt_{Asym}$ : Asymmetric encryption
C: Ciphertext
$C_K$ : Cipher key
$F_n K_{prvt}$ : Fog node private key
$K_{Sym}$ : Symmetric key
$F_s$ : Fog server
$T_s$ : Timestamp
$K$: Key
$H_c$ : Hash code
$Prvt_K$ : Private key
$D_F$ : Data format
$M$ : Miners
$C_s$ : Cloud server
$S_m$ _$D$: Smart meter data
$E_{da}$: Electric data allocation
$Q_L$ : Q-learning

**SMQE Algorithm:**
**Input:** Encrypted $S_m\_D$ C, Encrypted Symmetric key $C_k$
**Output:** Decrypted $S_m\_D$
***START***
*(FC-based blockchain system is created)*
Data classification
**if** ($S_m\_D$ == *Malicoius data*) then
get geo-location and send the data for verification to $F_n$ using *SPARK*
**else if** ($S_m\_D$ == *non − malicious data*)
then
$S_m\_D$ send to $F_n$ to $C_s$
$F_n$ allocates the $E_d$ to $F_s$
**for each** $S_m\_D$ do ($S_m$ <− $C_T$)
$C_T + T_s$ <- $S_m\_D$
**if** $F_s$ == Available
allocate the $E_d$
**else** no allocation
**end if**
**end**
***Set*** $Q(s,a)=0(\forall s\in S_i)(\forall a\in A_i(a))$, *iter* := 0 , *and* $s:=(1,1\{(Q_1 \text{--------} Q_N)\mid Q_i=0\})$
***While*** (*iter* ≤ maximum iteration) ***do***
*Select* $a\in A_i(a)$ *applying* ∈*- greedy algorithm*
*Allocate the data packets conferring to action* $a$ *and examine the next state* $s'$ *and reward* $r$*.*
$Q(s,a)\leftarrow(1-\ _i)Q(s,a)+\alpha_i\left[R_i(s,a)+\gamma\max_{\in A_{i'}}Q(s',a')\right]$
$s\leftarrow s'$
$iter\leftarrow iter+1$
**function** Encryption ($S_m\_D$ )
**if** $S_m$ confirms $E_d$ storage over blockchain then
Generate a $K_{sym}$
$C$ <− $Encrypt_{Sym}$ ($S_m\_D$ , $K_{sym}$)
$C_k$ <− $Encrypt_{Asym}$($K_{sym}$, $F_nK_{pub}$)
**else**
do no operation
**end if**
**end function**
**function** *DECRYPTION* (C, $C_k$, $F_nK_{prvt}$, $K_{sym}$)
$K_{sym}$ <-$Decryption_{Asym}$($C_k$, $F_nK_{prv}$)
$S_m\_D$ <- Decryption (C, $K_{sym}$)
**end function**

# 5 Results and Discussion

This section discusses the result and simulation of the proposed work. The SMQE algorithm is simulated using iFogSim and SimBlock simulator [20]. The simulation of nodes for new state entries, new block headers, and new block receipts is conducted using SimBlock. Ethereum was chosen as the blockchain-based technology for Proof of Work (PoW). See Table 1 for software and hardware specifications.

Figure 4 represents physical topology configuration.

Figure 4 shows the physical topology for configuration built-in the iFogSim simulator. The configuration is solely based on the concept of a proposed system.

See Fig. 5 shows the comparison of packet error in fog and cloud during a simulation.

Figure 5 shows the comparisons of packet error using SMQE algorithm between fog nodes and cloud servers at different time intervals. The figure shows the packet error in fog nodes is much lesser when compared to the cloud during data transmission.

**Table 1** The hardware and software used for the implementation of the proposed algorithm

| Hardware and software | Specification |
|---|---|
| Processor | Inter® Core$^{TM}$ i9-8750H |
| CPU | 5.30 GHz |
| RAM | 32 GB |
| System Type | 64-bit Windows 10 |
| Platform | iFogSim, SimBlock, and Spyder |
| Language | Java and Python |

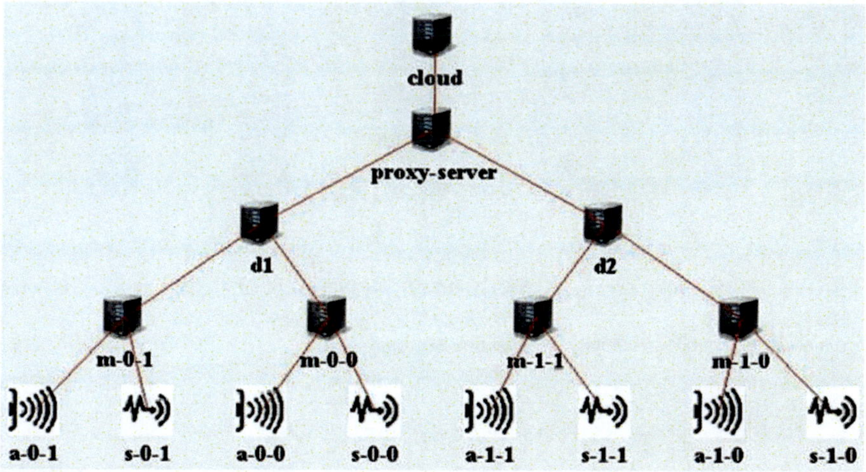


**Fig. 4** GUI configuration

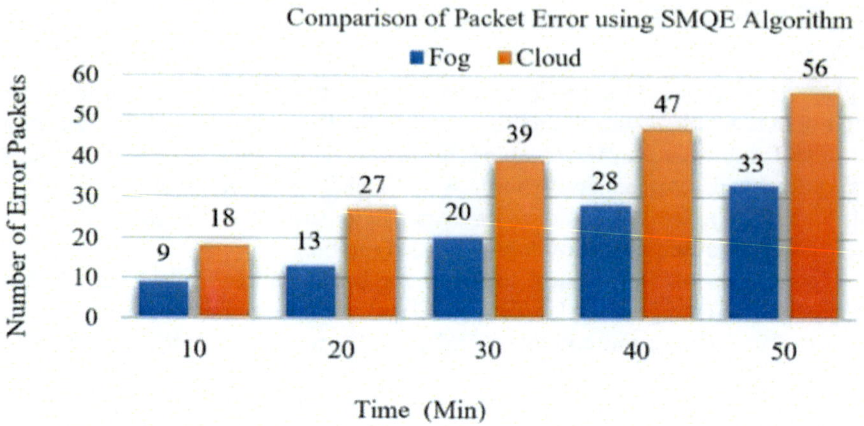


Fig. 5 Comparison of packet error

Figure 6 shows the malicious node percentage in fog and cloud along with the detection accuracy in percentage. The figure shows that the detection accuracy in fog nodes is much greater when compared to the cloud during data transmission.

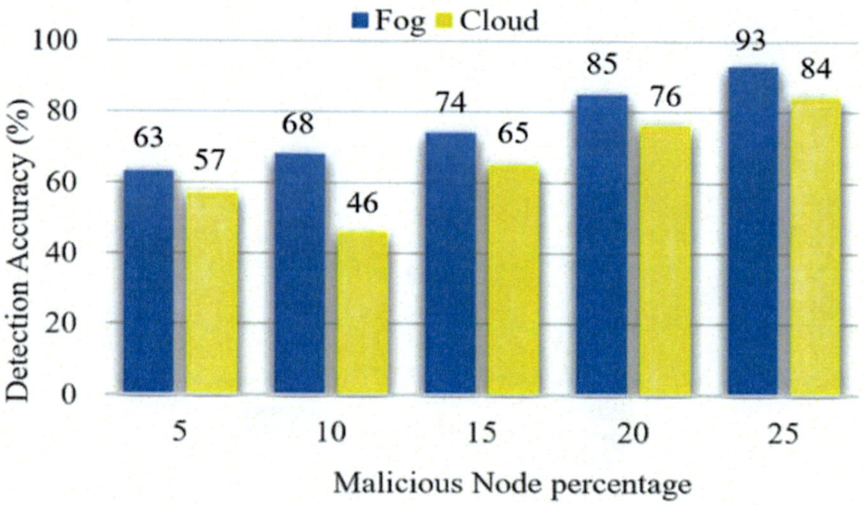


Fig. 6 Malicious node percentage vs detection accuracy

## 6 Conclusion

Smart meters are now used in smart cities for the transmission of electric data with minimum delay. The consumers can access the data through physical means such as using their mobile devices through HAN, NAN, and WAN. Smart meters are a cyber-physical system such as smart computer where information can be access in real time. However, the system is so complex and wide that it makes it vulnerable to cyber-attacks such as FDI where an intruder or an attacker adds some malicious data and changes the configuration where the meter starts responding and generating false or erroneous value, which leads to huge loss to smart city and consumer society.

Therefore to overcome this issue we have proposed a 3-tier FC-based blockchain architecture, and Smart Meter-based Q-Learning Encryption (SMQE) algorithm to identify the FDI and encrypt the data which further helps in a secure data transmission between smart meters, fog nodes acting as miners, consumers, meter management system, and cloud servers. The future work includes the benchmarking and comparison with other existing works in terms of algorithm performance and efficiency.

**Acknowledgements** This publication has emanated from research supported in part by a research grant from Cooperative Energy Trading System (CENTS) under Grant Number REI1633, and also by a research grant from Science Foundation Ireland (SFI) under Grant Number SFI 12/RC/2289_P2 (Insight), co-funded by the European Regional Development Fund.

## References

1. Kabalci Y (2016) A survey on smart metering and smart grid communication. Renew Sustain Energy Rev 57:302–318
2. Khattak AM, Khanji SI, Khan WA (2019) Smart meter security: vulnerabilities, threat impacts, and countermeasures. In: International conference on ubiquitous information management and communication. Springer, pp 554–562
3. Hussain M, Beg M (2019) Fog computing for internet of things (IoT)-aided smart grid architectures. Big Data Cogn Comput 3(1):8
4. Gunduz MZ, Das R (2020) Cyber-security on smart grid: threats and potential solutions. Comput Netw 169:107094
5. Khan S, Parkinson S, Qin Y (2017) Fog computing security: a review of current applications and security solutions. J Cloud Comput 6(1):19
6. Pop C, Cioara T, Antal M, Anghel I, Salomie I, Bertoncini M (2018) Blockchain based decentralized management of demand response programs in smart energy grids. Sensors 18(1):162
7. Li M, Zhang K, Liu J, Gong H, Zhang Z (2020) Blockchain-based anomaly detection of electricity consumption in smart grids. Pattern Recogn Lett 138:476–482
8. Musleh AS, Yao G, Muyeen S (2019) Blockchain applications in smart grid—review and frameworks. IEEE Access 7:86746–86757
9. Gilbert GM, Naiman S, Kimaro H, Bagile B (2019) A critical review of edge and fog computing for smart grid applications. In: International conference on social implications of computers in developing countries. Springer, pp 763–775

10. Nehaï Z, Guerard G (2017) Integration of the blockchain in a smart grid model. In: The 14th international conference of young scientists on energy issues (CYSENI) 2017, pp 127–134
11. Zhang M, Shen C, He N, Han S, Li Q, Wang Q, Guan X (2019) False data injection attacks against smart gird state estimation: construction, detection and defense. Sci China Technol Sci 1–11
12. Forcan M, Maksimović M (2020) Cloud-fog-based approach for smart grid monitoring. Simul Modell Pract Theory 101:101988
13. Wang X, Luo X, Zhang M, Guan X (2019) Distributed detection and isolation of false data injection attacks in smart grids via nonlinear unknown input observers. Int J Electr Power Energy Syst 110:208–222
14. Agung AAG, Handayani R (2020) Blockchain for smart grid. J King Saud Univ-Comput Inf Sci
15. Malik H, Manzoor A, Ylianttila M, Liyanage M (2019) Performance analysis of blockchain based smart grids with Ethereum and hyperledger implementations. In: IEEE international conference on advanced networks and telecommunications systems, pp 1–5
16. Drayer E, Routtenberg T (2019) Detection of false data injection attacks in smart grids based on graph signal processing. IEEE Syst J
17. Mollah MB, Zhao J, Niyato D, Lam K-Y, Zhang X, Ghias AM, Koh LH, Yang L (2020) Blockchain for future smart grid: a comprehensive survey. IEEE Internet Things J
18. Ericsson GN (2010) Cyber security and power system communication—essential parts of a smart grid infrastructure. IEEE Trans Power Delivery 25(3):1501–1507
19. Esmalifalak M, Liu L, Nguyen N, Zheng R, Han Z (2014) Detecting stealthy false data injection using machine learning in smart grid. IEEE Syst J 11(3):1644–1652
20. Shukla S, Hassan MF, Khan MK, Jung LT, Awang A (2019) An analytical model to minimize the latency in healthcare internet-of-things in fog computing environment. PloS one 14(11):e0224934

# Methodology for Random Number Generation Based on FPGA

**Dexuan Kong, Zaiming Fu, Qiang Wang, and Zhong Wu**

**Abstract** Random numbers are playing an increasingly important role in the modern information technology field. The methods for generating random numbers are mainly divided into two categories: analog and digital methods. Due to their portability and simple structure, digital methods have been widely used in communication and other fields. The digital method based on M-sequence is the mainstream method for generating random numbers. The existing M-sequence methods for generating random numbers have disadvantages such as insufficient period and insufficient bit width. In this paper, we propose a random number generation method based on FPGA hardware, which adds a multi-tap structure to M-sequence and synthesizes multiple parallel M-sequence outputs into a new random number generator. It implements a random number generator with arbitrary bit width and ultra-long period. Finally, we analyzed 512 k output sample data of the FPGA chip xcku040 used to implement this design. The experimental results verify the feasibility and superiority of the method.

**Keywords** Random number generator · Pseudo-random sequence · M-sequence · FPGA · Shift register

## 1 Introduction

Random numbers are increasingly used in electronic information fields such as communication, cryptography, navigation, and high-speed circuit testing. Due to the different sources of randomness, random numbers are also divided into true random numbers and pseudo-random numbers [1]. Most of them generated by numerical methods are pseudo-random numbers. Traditional pseudo-random numbers generally refer to M-sequences and pseudo-random sequences derived from M-sequences. M-sequences have excellent pseudo-random sequence properties and are widely used in radar, geophysical measurements, electronic countermeasures, and other fields

D. Kong · Z. Fu (✉) · Q. Wang · Z. Wu
University of Electronic Science and Technology of China, Chengdu 611731, SC, China
e-mail: fuzaimin@163.com



X.-S. Yang et al. (eds.), *Proceedings of Sixth International Congress on Information and Communication Technology*, Lecture Notes in Networks and Systems 216,
https://doi.org/10.1007/978-981-16-1781-2_59

[2–4]. Currently, there are many options for generating M-sequence generators. M-sequence generators based on application-specific integrated circuits can achieve faster data output rates. However, due to the inconvenient modification of hardware circuits, only a single M-sequence can be generated, which cannot meet the requirements of diverse random sequences. Software-based M-sequence generators can use flexible data query methods to obtain arbitrary M-sequence generators, but the implementation process is more complicated and the speed is lower. In high-speed design situations, simpler hardware implementations based on FPGAs are often used. Although existing pseudo-random generation theories have been shown to have some good properties, these schemes have two major drawbacks. On the one hand, their outputs cannot achieve arbitrary bit-width selection, and on the other hand, the short sequence period [5].

In this paper, an FPGA-based pseudo-random number generator is proposed, which takes advantage of the FPGA hardware circuit and combines the flexibility of the design to implement an ultra-long-period pseudo-random number generator with arbitrary bit width. The experimental results show the feasibility and superiority of this design.

The rest of this paper is organized as follows. In Section II, the principle of the M-sequence is explained; in Section III, a new random number generation design is proposed; in Section IV, experimental results will show the effectiveness of the proposed method, and Section V presents the conclusions of this paper.

## 2 Principles of M-sequence

The M-sequence is formed by adding a feedback loop to the n-level linear shift register, as shown in Fig. 1. Its structure is simple and easy to implement. Under clock control, the state of the register is determined by the input information ("0" or "1"). For example, the state of the i shift register is determined by the state of the i-1 shift register after the previous clock pulse.

The $c_i$ in Fig. 1 is the feedback coefficient, which is determined by the primitive polynomial. When $c_i = 1$, the feedback intervenes and when $c_i = 0$, the feedback loop is broken. $a_i$ indicates the state of the $i$th level flip-flop and can only indicate whether the flip-flop is currently storing a value of 0 or 1. When the generator operates, the resulting M-sequence is output from the right end of the shift register, and in each

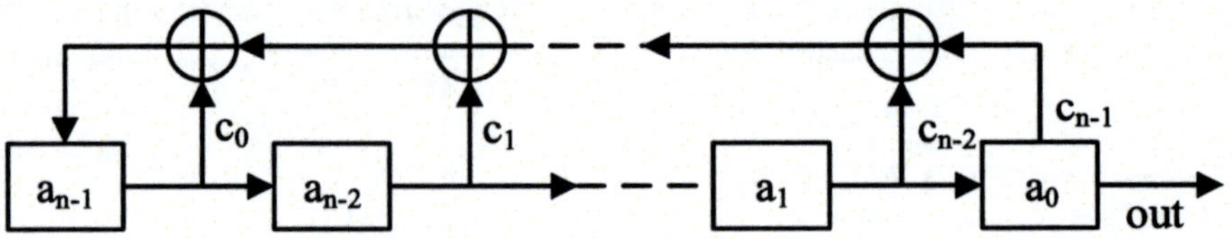


**Fig. 1** N-level M-sequence generator

**Table 1** Some commonly used M-sequences

| M-sequence level | Feedback coefficient (Octal) | Period |
|---|---|---|
| 3 | 13 | 7 |
| 4 | 23 | 15 |
| 5 | 45, 67, 75 | 31 |
| 6 | 103, 147, 155 | 63 |
| 7 | 203, 211, 217, 235, 277, 313, 325 | 127 |
| 8 | 435, 453, 537, 543, 545, 551, 703 | 255 |
| 9 | 1021, 1055, 1131, 1157, 1167 | 511 |
| 10 | 2011, 2033, 2157, 2443, 2745 | 1023 |
| ... | ... | ... |

clock, it produces one bit. Suppose the initial state of each flip-flop is $a_{n-1}$, ..., $a_1$, $a_0$. Then, in each clock cycle, the left end of the shift register will get a new input $\mathrm{a_n}$, whose value can be expressed as following:

$$a_n = c_0 a_{n-1} \oplus c_1 a_{n-2} \oplus \cdots \oplus c_{n-2} a_1 \oplus c_{n-1} a_0 = \sum_{i=0}^{n-1} c_i a_{n-1-i} \quad (1)$$

It can be seen from formula (1) that the output of the M-sequence depends on its initial state and the feedback of each flip-flop ($c_i$), where the value of ci is often represented by the coefficients of the primitive polynomial:

$$f(x) = c_0 + c_1 x^1 + \cdots + c_{n-1} x^{n-1} = \sum_{i=1}^{n-1} c_i x^i \quad (2)$$

The commonly used primitive polynomials of M-sequences and their periods are shown in Table 1.

## 3 Design of Random Number Generator

To adjust the M-sequence, it is necessary to change the primitive polynomial used in the design, but this method is not applicable in some cases. There are only more than 300 primitive polynomials applied, and the M-sequence obtained by changing the primitive polynomial is very limited. In addition, from the feedback process of the M-sequence, in each cycle, the shift register shifts only the highest bit and feeds a value to the lowest bit. Therefore, the adjacent states are not independent, which seriously affects the randomness of the output. In this paper, we propose a random number generation method with multiple sequence taps, which greatly improves the flexibility and periodicity of the M-sequence.

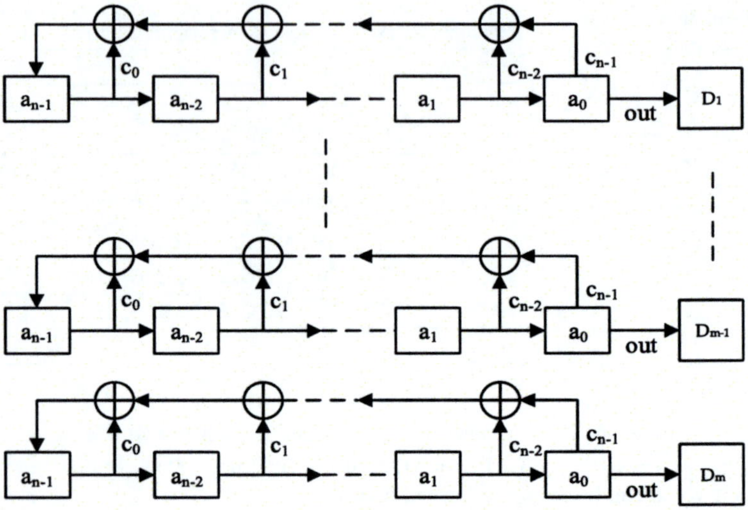


**Fig. 2** Multi-tap structure

As shown in Fig. 2, M-sequences of M of different levels are concatenated, and only one bit is extracted from each M-sequence, and these extracted bits are arranged in order into the random number generator. Since each bit in the random number comes from a different sequence, this scheme can greatly improve the flexibility of the sequence. The bit width can be chosen independently, and the period of the random number is greatly increased. For example, suppose a new random number generator is formed by selecting 19 sequences of M with sequence levels 10–28, the period of the generator is

$$\text{Period} = [2^{10} - 1, 2^{11} - 1, \ldots, 2^{28} - 1] \tag{3}$$

where [6] is the least common multiple. The period of the random generator is as long as decades at the output rate of 2G/s. With this design, the correlation between adjacent bits in the sequence is also greatly reduced, and the randomness gets stronger.

## 4 Experimental Results

We programmed the FPGA chip (Xilinx xcku040) in Verilog using the EDA tool Vivado 2018.3. Afterward the generated random numbers were captured, and 512 k samples were collected using an online logic analyzer tool in Vivado.

The analog signal diagram of the 19-bit random number output is shown in Fig. 3. And the output of each bit is shown in Fig. 4. Such a time domain plot does not accurately represent the statistical characteristics of the generated random numbers.

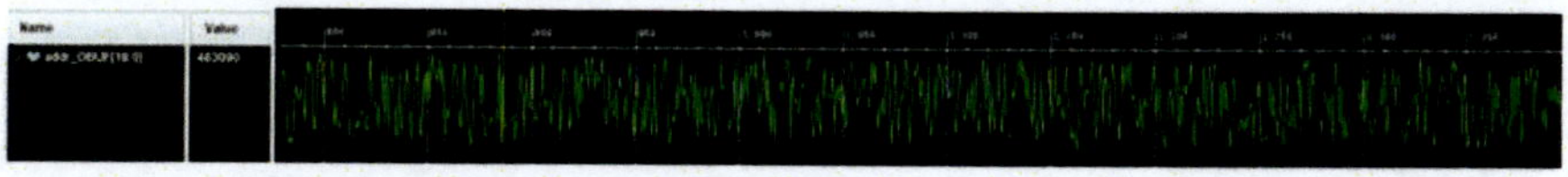

**Fig. 3** 19-bits random number analog signal diagram

**Fig. 4** Each bit of the output

**Fig. 5** Random number sample histogram

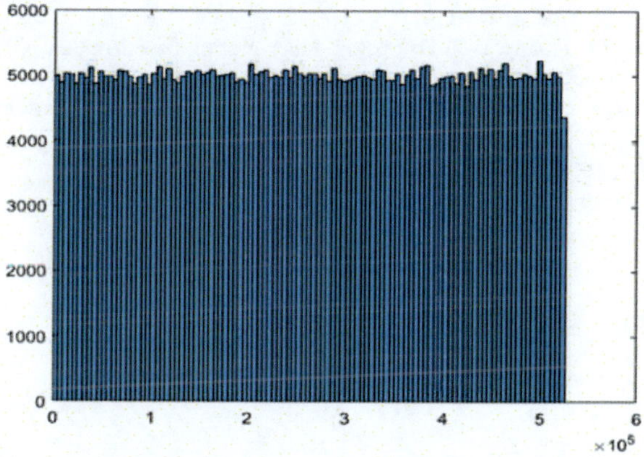


So we draw the collected samples as a histogram, which is Fig. 5. From Fig. 5, we can see that the histogram is very flat, which means that the probability of occurrence of each small interval of values from 0 to 524,288 is essentially the same.

## 5 Conclusion

In this paper, a new random number generation method is proposed to overcome the short period and inflexibility of the traditional method. Random number generation of arbitrary bit width can be achieved, and the period becomes extremely large with the increase of bit width, which can reach several decades. Experimental results show that the generated random numbers are very evenly distributed and have good randomness. Although this method is more flexible and has a longer period than the traditional method, the generated random numbers are still not true random numbers. The disadvantage of predictability still exists when applied to fields such as communication, which should be addressed in future research.

## References

1. SK Park KW Miller 1988 Random number generators: good ones are hard to find Commun ACM 31 1192 1201
2. Shujun L, Xuanqin M, Yuanlong C (2001) Pseudo-random bit generator based on couple chaotic systems and its applications in stream-cipher cryptography. In: International conference on cryptology in India. Springer, pp 316–329
3. Wan-Li Y (2004) A pseudo random (PR) code based ultrasonic navigation system (UNS). In: IEEE Conference on Robotics, Automation and Mechatronics, 2004. pp. 752–757 vol.2 (2004). https://doi.org/10.1109/RAMECH.2004.1438012
4. I Pomeranz SM Reddy 1993 3-weight pseudo-random test generation based on a deterministic test set for combinational and sequential circuits IEEE Trans Comput Aided Des Integr Circuits Syst 12 1050 1058
5. J-F Huang D-Z Hsu 2000 Fiber-grating-based optical CDMA spectral coding with nearly orthogonal M-sequence codes IEEE Photonics Technol Lett 12 1252 1254
6. Golomb SW (1967) Shift register sequences. Aegean Park Press

# An Improvement to FRAGMENT: A Web Application for Database Fragmentation, Allocation, and Replication Over a Cloud Environment

**Felipe Castro-Medina, Lisbeth Rodríguez-Mazahua, Asdrúbal López-Chau, Giner Alor-Hernández, Ulises Juárez-Martínez, and Cuauhtémoc Sánchez-Ramírez**

**Abstract** Fragmentation plays a very important role in databases since this achieves an adequate design in distributed environments such as the cloud, which in terms of execution cost of the read and write operations, benefits are obtained. This work presents an improvement to the design of the overlapping fragmentation algorithm shown in previous works, which is part of a method to also carry out allocation and replication. The improvement consists of ordering the predicates by cost and in this way obtaining each predicate contemplating that the fragment overlap will occur, however, the fragments with a higher cost will be kept more intact than those with a lower cost. Experiments with the TPC-E benchmark adapted to a distributed environment demonstrate that our enhanced approach achieves a significant reduction of query response time.

**Keywords** Horizontal fragmentation · Replication · Distributed environment · Fragment overlap · Cost precedence

F. Castro-Medina (✉) · L. Rodríguez-Mazahua · G. Alor-Hernández · U. Juárez-Martínez · C. Sánchez-Ramírez
Tecnológico Nacional de México, Instituto Tecnológico de Orizaba, Veracruz, México

L. Rodríguez-Mazahua
e-mail: lrodriguezm@orizaba.tecnm.mx

G. Alor-Hernández
e-mail: galorh@orizaba.tecnm.mx

U. Juárez-Martínez
e-mail: ujuarezm@orizaba.tecnm.mx

C. Sánchez-Ramírez
e-mail: csanchezr@orizaba.tecnm.mx

A. López-Chau
Universidad Autónoma del Estado de México, Centro Universitario UAEM Zumpango, Zumpango Estado de México, México
e-mail: alchau@uaemex.mx



X.-S. Yang et al. (eds.), *Proceedings of Sixth International Congress on Information and Communication Technology*, Lecture Notes in Networks and Systems 216,
https://doi.org/10.1007/978-981-16-1781-2_60

# 1 Introduction

The demand for information in distributed systems forces database designers to find suitable alternatives to improve performance in these environments, and horizontal fragmentation is one of those options. Abdel et al. [1] presented a horizontal fragmentation and allocation approach that contemplates the cost of the basic operations performed in a database. The fragmentation scheme is determined using a cost matrix called the MCRUD (Modified Create, Read, Update and Delete) matrix, which shows the cost of each predicate used in the operations performed on the database from different sites. The MCRUD determines the cost of each attribute by producing a table called ALP (Attribute Locality Precedence), which shows the most expensive attribute. This attribute generates the fragmentation scheme using the predicates that are related to it. Each predicate determines a fragment that is allocated to the node where it produced the highest cost and copied to the next node in cost precedence.

Castro-Medina et al. [2] based their work on the algorithm of Abdel et al. [1] analyzing the problem of overlapping fragments, which is presented in the final stage when obtaining the predicates that will be applied. The algorithm was used in the context of a Web application called FRAGMENT. Castro-Medina et al. [2] addressed the problem of overlapping fragments, always keeping the largest ones. In both works, the costs obtained in the MCRUD are not used directly when obtaining the predicates for their application. This work shows an improvement to [2] solving the problem of overlapping fragments by means of cost precedence obtained from MCRUD. The presented algorithm is tested under the FRAGMENT Web application in a distributed environment. This work has the following structure: The second section reviews background, the third section shows the state of the art of the works related to the topics of interest of this research, in Sect. 4 the improvement to the overlapping fragmentation algorithm is presented, the results are discussed in the fifth section. Finally, the sixth section gives the conclusions.

# 2 Background

Fragmentation divides a database table $t$ into a collection of fragments $t_1$, $t_2$,... $t_n$. These have sufficient data to reconstruct the table $t$ [3]. Two basic approaches exist for data fragmentation: vertical and horizontal. However, there is a type of fragmentation that combines the previous two called hybrid fragmentation [4].

Horizontal fragmentation (HF) splits a relation across tuples. Therefore, every fragment is a subset of rows of that table. There are two types of HF: primary and derived. The former is done using predicates that are utilized in that relation. The latter, differently, is the partitioning of a relation as a result of predicates that are defined on another relation [4].

Vertical fragmentation consists of dividing a relation into subsets of attributes and in each fragment placing the primary key of the relation to allow its reconstruction.

The goal of vertical partitioning is to break a relation into a group of smaller relations so that most applications run in a single fragment [4].

In many contexts, a horizontal or vertical fragmentation of a database schema is not enough to fulfill the demands of the applications. In these cases, a vertical fragmentation is carried out after a horizontal one or in reverse order, generating a structured tree of partitions. This technique is called hybrid fragmentation, mixed fragmentation, or nested fragmentation [3].

## 3 Related Work

The study presented in [5] analyzed the improvement of the availability of federated storage services to produce finer service quality to customers with few computational resources. López et al. proposed smart replication and a method of improving storage availability through sentiment analysis of the load file. The methodology was based on a mixture of two kinds of load file analysis, calculation of availability, and performance metrics to choose the optimal subset of federated cloud storage services.

In [6], three protocols for the management of distributed data in the cloud were analyzed: Read-One-Write-All-Available (ROWAA), Majority Quorum (MQ), and Data Partitioning (DP). The authors introduced BEOWULF, a meta-protocol based on a complete cost model that fuses all three protocols and chooses in a dynamic way the protocol with the minimum response time for a given set of queries.

Rodríguez-Mazahua et al. [7] proposed DYMOND (DYnamic Multimedia ONline Distribution), a system that vertically and dynamically fragments a multimedia database based on active rules. DYMOND monitors the queries executed to decide when a new fragmentation is required.

In [8], the authors presented a dynamic DDBS (Distributed Database System) architecture over a cloud environment that permits FAR (fragmentation, allocation, and replication) decisions to be made in a dynamic way over clustered database sites. A horizontal fragmentation technique was proposed, which helps to make fragmentation decisions in the early phases of the distributed database design.

The authors in [9] addressed fuzzy querying by presenting a fragmentation methodology. In addition, several query execution strategies were shown whose objective is to minimize the number of accesses to fragments. Within the query execution strategies, the overlapping predicates were addressed, of which a strategic algorithm is shown to contemplate this characteristic within the work approach.

Wise et al. [10] presented a replication strategy in a DDBS that contemplates multiple fragmentations of a certain data table. The main characteristic of the proposed replication approach is that replication and recovery respect the overlapping fragments that come from other fragmentations.

In [11], an enhanced heuristic process for horizontal fragmentation and data allocation was developed, which is aimed to propose adequate data fragmentation, accurate data allocation, and design a practical algorithm for site clustering.

Moral and Kumar [12] proposed a system that uses fragmentation and single replication (FASR) to improve security and response times instead of using traditional encryption techniques. The main contribution of FASR was to prevent the loss of information in the event of an attack.

Abdel et al. [1] developed an improved dynamic DDBS over a cloud environment, which enables runtime allocation, replication, and fragmentation decisions to be made dynamically. In addition, an improved replication and allocation technique was introduced that is applied early in the design of distributed databases when there is no query execution information. Using the information of the operations carried out on a database, a matrix called MCRUD is created, which relates the types of operations performed with their predicates and the sites where they were executed. Through the MCRUD, the ALP table is obtained, which shows the costs of using each attribute. In this way, a fragmentation scheme is created considering the most expensive attribute.

In [2], the authors showed a study of distinct procedures for the fragmentation, allocation, and replication of databases and a Web application called FRAGMENT that takes into account the working approach that was chosen in the stage of analysis, since it is a method of partitioning and replication, applies to a cloud context, focuses on enhancing the efficiency of operations performed on the database, and is based on a cost model. FRAGMENT observes the operations executed on any database table, the most expensive attribute determines the fragmentation scheme and the system assigns and replicates the scheme that the user chooses in a distributed environment in the cloud. In [2], it was shown a usual problem in fragmentation techniques, overlapping fragments, and provided a strategy to solve it. This proposal obtains the predicates that will define every partition. Evaluations with the TPC-E benchmark revealed that the application reduces response time for queries executed in the distributed database obtained by FRAGMENT in contrast to a centralized database.

Table 1 shows a comparison of the works mentioned in this section. The comparison criteria are that the works include both fragmentation and replication approaches, that they are applied to the cloud, with easy implementation, focused on the performance of queries, that the aforementioned works have everything necessary to be implemented, they are based on a cost model and, finally, that solve the problem of overlapping fragments through cost precedence.

As described in Table 1, [2] presents all the desired characteristics and even proposes a method to solve the problem of overlapping fragments, however, the proposed solution is not directly based on costs. This work is based on the fragmentation approach used in [2] proposing an improvement and adapting it to the Web application already developed by the authors.

**Table 1** Comparison of related works

| Work | A | B | C | D | E | F | G | H |
|---|---|---|---|---|---|---|---|---|
| Abdel et al. [1] | Yes | Yes | Yes | Yes | Yes | Yes | Yes | No |
| Castro-Medina et al. [2] | Yes | Yes | Yes | Yes | Yes | Yes | Yes | No |
| López et al. [5] | No | Yes | Yes | No | Yes | Yes | Yes | No |
| Stiemer et al. [6] | Yes | Yes | Yes | No | Yes | No | Yes | No |
| Rodríguez-Mazahua et al. [7] | Yes | No | No | Yes | Yes | Yes | Yes | No |
| Abdel et al. [8] | Yes | Yes | Yes | No | Yes | No | No | No |
| Drissi et al. [9] | Yes | No | No | Yes | Yes | Yes | No | No |
| Wiese et al. [10] | Yes | Yes | No | Yes | Yes | Yes | Yes | No |
| Ali et al. [11] | Yes | Yes | No | Yes | Yes | No | Yes | No |
| Kumar and Gupta [12] | Yes | Yes | Yes | No | Yes | No | No | No |
| This work | Yes | Yes | Yes | Yes | Yes | Yes | Yes | Yes |

(A) Fragmentation, (B) Replication, (C) Applied to the cloud, (D) Ease of implementation, (E) Performance, (F) Completeness, (G) Cost model, (H) Overlapping fragments with cost precedence

## 4 An Improvement to FRAGMENT

FRAGMENT focuses on the operations performed on a database in the MySQL DBMS (Database Management System) in a distributed environment and uses its log file to obtain the cost of all predicates. Figure 1 shows the process that the Web application uses to obtain and apply the fragmentation, allocation, and replication scheme.

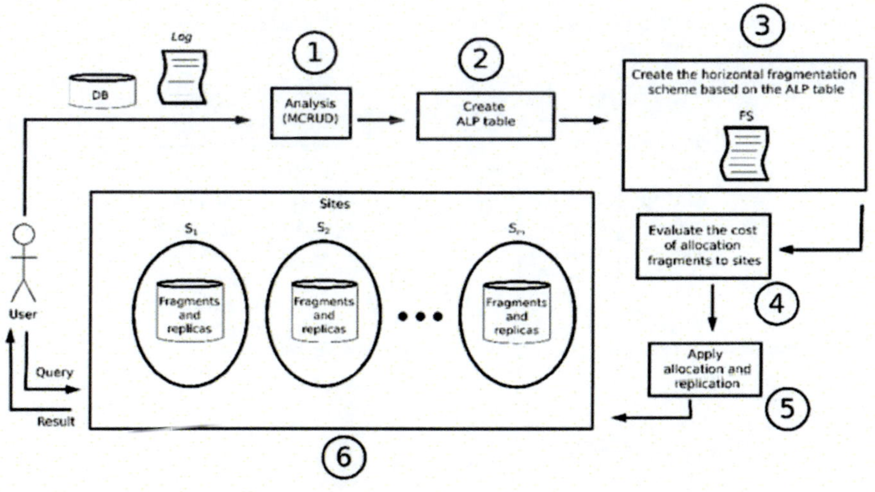


**Fig. 1** Web application workflow

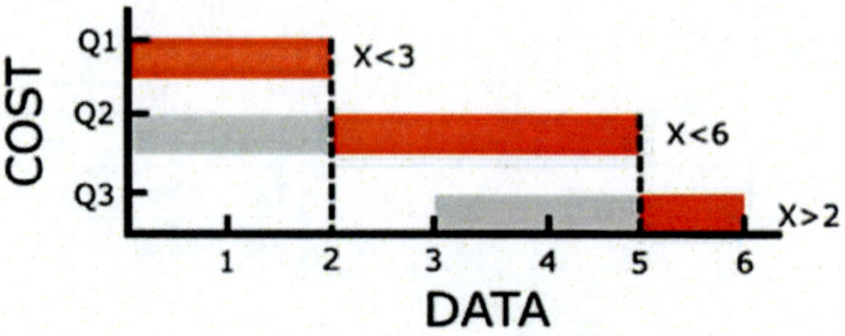


**Fig. 2** Operation of the overlap algorithm with cost precedence

In the Web application workflow, generating the ALP table loses the specification of the costs of each predicate since the costs are calculated by attribute and, in general, obtaining the fragments using the most expensive attribute cannot determine the precedence of the predicates, thus causing more importance to be given to predicates that do not have it.

Figure 2 shows the operation of the proposed approach, solving the overlap by allowing it but through the precedence of costs. Only the red stripes are preserved. It is observed in Fig. 2 that the most expensive fragments are conserved, in this way the fragmentation improves its performance and the scheme that is obtained achieves better results.

Algorithm 1, Fragmentation with overlapping and cost precedence, represents the pseudocode to obtain a matrix of predicates, where the rows are the set of predicates to obtain each fragment. In this way, placing the predicates of each row inside a query, when executed, will return each fragment with its corresponding tuples.

In Algorithm 1, *fragment predicates* is the array of predicates. If a list of input predicates contains id < 30, id < 20, id = 40 of a table named *accounts*, with costs 4, 8, and 1, respectively, having tuples with id from 1 to 50, as a result, the fragment predicates will contain the matrix shown on the right side of Fig. 3

| | |
|---|---|
| (id<30) | !(id=40) |
| !(id<30) | (id=40) |
| !(id<30) | !(id=40) |

| | | |
|---|---|---|
| (id<20) | | |
| !(id<20) | (id<30) | |
| !(id<20) | !(id<30) | (id=40) |
| !(id<20) | !(id<30) | !(id=40) |

**Fig. 3** Results of the example using [2] and the proposed approach

**Algorithm 1**: Fragmentation with overlapping and cost precedence

```
input: Predicate list[], Table name, MCRUD[][]
output: Fragment list
for w from 0 to size of MCRUD[][] do
    maxValue ← getMaximumValueOfPredicate(MCRUD[w][])
    maxPosition ← getPositionOfMaxValue(MCRUD[w][])
    sum ← 0
    for r from 0 to size of MCRUD[w][] do
        if r != maxPosition
            sum ← sum + MCRUD[w][r]
    predicateValues[w][] ← maxValue - sum
for x from 0 to size of Predicate list[] do
    valuePredicateList[x] ← searchValueInPredicateValues(Predicate list[x])
sort Predicate list[] by valuePredicateList[]
define line[] as a list of strings
for y from 0 to size of Predicate list[] do
    negation[size of negation] ← "!(" + Predicate list[y] + ")"
    erase everything in line[]
    line[size of line] ← "(" + Predicate list[y] + ")"
    define previous[] as a list of strings
    for z from 0 to size of previous[] do
        line[size of line] ← previous[z]
    previous[size of previous] ← "!(" + Predicate list[y] + ")"
    fragments predicates[y][] ← line[]
fragments predicates[size of Predicate list][] ← negation[]
for i from 0 to size of fragments predicates[][] do
    query ← "select * from " + Table name + " where "
    for j from 0 to size of fragments predicates[i][] do
        query ← query + fragments predicates[i][j] + " && "
    query ← query - last three characters
    query ← query +" limit 1;"
    result set ← execute query in database
    if result set is empty then
        indexesToRemove[] ← i
remove indexes from indexesToRemove[] in fragment predicates[][]
return fragments predicates[][]
```

Figure 3 shows the comparison of the results of the algorithm proposed in [2] and the algorithm proposed in this work.

As seen in Fig. 3, the work in [2] proposes to eliminate the smaller predicates contained in larger predicates. The id < 20 predicate is removed regardless of cost since the id < 30 predicate contains it. To carry out this elimination, the row and column related to this predicate are removed from the table. In this work, the approach proposes to keep all predicates. It can be seen in Fig. 3 that the predicate table is ordered from highest to lowest cost and the most expensive predicates are applied with fewer restrictions. The id < 20 predicate is the most expensive predicate, for this reason, it is applied without restrictions, id < 30 has a lower cost and is applied respecting the most expensive predicate. The same happens with the predicate id =

40, and finally, all the predicates are restricted to obtain all the tuples that are not contained in them.

## 5 Results

This part of the work presents the success obtained through the modification made to FRAGMENT under the TPC-E benchmark [13] in a distributed environment made up of three computers with Windows as the operating system, 2 GB in RAM, and a 1.3 GHz dual-core inside processor.

Table 2 shows the operations performed on the main computer, which contains the benchmark database. Operations are performed on the CUSTOMER_ACCOUNT table from the main computer and the other two computers connected to a local network. This table contains 5000 tuples produced using the data generator provided from the TPC official page.

The ALP table of the operations performed after having been analyzed in the MCRUD by FRAGMENT is shown in Fig. 4. It is observed that the most expensive attribute is CA_BAL, which represents the economic balance of the clients' accounts in the benchmark database. After obtaining the ALP table, it is observed that the Web application displays the final table, which shows the fragments that will be applied. Figure 5 presents the final table obtained with the new algorithm.

Fragment overlap is observed in this evaluation in many predicates, two of them are CA_BAL > 60,000 and CA_BAL > 0, which are allowed taking into account the order of costs. Fragments that do not contain tuples from the CUSTOMER_ACCOUNT table are not applied. The last column of the final table represents the fragments to be applied, marked with a sign. The negation of the predicates to be applied is not presented since it does not contain tuples.

Figure 6 shows the result obtained with the previous approach, solving the overlap of fragments with the algorithm presented in [2]. Regardless of the execution times that are improved, the scheme by means of the precedence of costs is better, since each fragment has importance in the last stage considering the predicates that were more expensive. However, the improvement in query execution times is also observed and is presented in Fig. 7.

The three most expensive queries executed in the fragments produced by the old and the new approach are seen in Fig. 7. Each operation was carried out 5 times and the average of the response times obtained was determined to generate more accurate results. The time difference is immediately associated with the number of tuples present in each fragment to which the query is executed, however, it is also appropriate to highlight that the cost of the queries is obtained through the type of operation and the number of times it is used, for this reason, the benefit in improving execution times is greater than the one presented, since the most expensive queries will be executed a greater number of times than the least expensive ones.

**Table 2** Operations performed on the main computer

| Query | IP |
|---|---|
| SELECT * FROM 'customer_account' WHERE 'CA_NAME' = 'Joshua Fowle Savings Account' | 127.0.0.1 |
| SELECT * FROM 'customer_account' WHERE 'CA_C_ID' = '4,300,000,020' | 127.0.0.1 |
| SELECT * FROM 'customer_account' WHERE 'CA_B_ID' = '4,300,000,008' | 127.0.0.1 |
| SELECT * FROM 'customer_account' WHERE 'CA_ID' = '43,000,000,244' | 127.0.0.1 |
| SELECT * FROM 'customer_account' WHERE 'CA_TAX_ST' = 0 | 127.0.0.1 |
| UPDATE 'customer_account' SET 'CA_TAX_ST' = 1 WHERE 'CA_TAX_ST' = 2 | 127.0.0.1 |
| SELECT * FROM 'customer_account' WHERE 'CA_C_ID' = '43,000,000,044' | 127.0.0.1 |
| DELETE FROM 'customer_account' WHERE 'CA_ID' = '4,300,000,019' | 127.0.0.1 |
| SELECT * FROM 'customer_account' WHERE 'CA_NAME' = 'Lizzie Nail IRA-SEP' | 127.0.0.1 |
| INSERT INTO customer_account VALUES (1,2,3,'Felipe Castro Medina',0,40.99) | 127.0.0.1 |
| INSERT INTO customer_account VALUES (80,20,33,'Sunny Martinez Guarneros',3,10,080.54) | 192.168.0.12 |
| SELECT * FROM 'customer_account' WHERE 'CA_NAME' = 'Richard Moe Health Savings' | 192.168.0.12 |
| DELETE FROM 'customer_account' WHERE 'CA_ID' = '43,000,000,554' | 192.168.0.12 |
| SELECT * FROM 'customer_account' WHERE 'CA_C_ID' = '4,300,000,010' | 192.168.0.12 |
| UPDATE 'customer_account' SET 'CA_TAX_ST' = 2 WHERE 'CA_TAX_ST' = 1 | 192.168.0.12 |
| SELECT * FROM 'customer_account' WHERE 'CA_TAX_ST' = 1 | 192.168.0.12 |
| SELECT * FROM 'customer_account' WHERE 'CA_BAL' > 60,000 | 192.168.0.12 |
| SELECT * FROM 'customer_account' WHERE 'CA_BAL' > 60,000 | 192.168.0.12 |
| SELECT * FROM 'customer_account' WHERE 'CA_BAL' > 60,000 | 192.168.0.12 |
| SELECT * FROM 'customer_account' WHERE 'CA_NAME' = 'Stella Wilson Emergency Expenses' | 192.168.0.12 |
| SELECT * FROM 'customer_account' WHERE 'CA_BAL' < 0 | 192.168.0.90 |
| SELECT * FROM 'customer_account' WHERE 'CA_BAL' = 0 | 192.168.0.90 |
| SELECT * FROM 'customer_account' WHERE 'CA_BAL' > 0 | 192.168.0.90 |
| SELECT * FROM 'customer_account' WHERE 'CA_BAL' < -10,000 | 192.168.0.90 |
| SELECT * FROM 'customer_account' WHERE 'CA_BAL' > 10,000 | 192.168.0.90 |
| SELECT * FROM 'customer_account' WHERE 'CA_BAL' > 15,000 | 192.168.0.90 |
| SELECT * FROM 'customer_account' WHERE 'CA_BAL' > 80,000 | 192.168.0.90 |
| SELECT * FROM 'customer_account' WHERE 'CA_BAL' < 15,000 | 192.168.0.90 |
| SELECT * FROM 'customer_account' WHERE 'CA_BAL' < 30,000 | 192.168.0.90 |
| SELECT * FROM 'customer_account' WHERE 'CA_BAL' < –5000 | 192.168.0.90 |

| ALP Table | | |
|---|---|---|
| **Attribute** | **ALP Value** | |
| CA_ID | 9 | |
| CA_B_ID | 5 | |
| CA_C_ID | 7 | |
| CA_NAME | 8 | |
| CA_TAX_ST | 6 | |
| CA_BAL | 17 | Select |

**Fig. 4** ALP table of operations on "CUSTOMER_ACCOUNT"

| Final result | | | | |
|---|---|---|---|---|
| **Fragment** | **Assign** | **Replicate** | **Cost** | |
| `CA_BAL`>60000 | 192.168.0.12 | localhost | 3 | ✓ |
| CA_BAL=40.99 | localhost | 192.168.0.12 | 2 | |
| CA_BAL=10080.54 | 192.168.0.12 | localhost | 2 | |
| `CA_BAL`<0 | 192.168.0.90 | localhost | 1 | ✓ |
| `CA_BAL`=0 | 192.168.0.90 | localhost | 1 | |
| `CA_BAL`>0 | 192.168.0.90 | localhost | 1 | ✓ |
| `CA_BAL`<-10000 | 192.168.0.90 | localhost | 1 | |
| `CA_BAL`>10000 | 192.168.0.90 | localhost | 1 | |
| `CA_BAL`>15000 | 192.168.0.90 | localhost | 1 | |
| `CA_BAL`>80000 | 192.168.0.90 | localhost | 1 | |
| `CA_BAL`<15000 | 192.168.0.90 | localhost | 1 | |
| `CA_BAL`<30000 | 192.168.0.90 | localhost | 1 | |
| `CA_BAL`<-5000 | 192.168.0.90 | localhost | 1 | |

**Fig. 5** Proposed allocation and replication scheme

| Final result | | | |
|---|---|---|---|
| **Fragment** | **Assign** | **Replicate** | |
| CA_BAL=40.99 | localhost | 192.168.0.12 | |
| CA_BAL=10080.54 | 192.168.0.12 | localhost | |
| `CA_BAL`>60000 | 192.168.0.12 | localhost | |
| `CA_BAL`<0 | 192.168.0.90 | localhost | |
| `CA_BAL`=0 | 192.168.0.90 | localhost | |
| `CA_BAL`>0 | 192.168.0.90 | localhost | |
| `CA_BAL`<-10000 | 192.168.0.90 | localhost | |
| `CA_BAL`>10000 | 192.168.0.90 | localhost | |
| `CA_BAL`>15000 | 192.168.0.90 | localhost | ✓ |
| `CA_BAL`>80000 | 192.168.0.90 | localhost | |
| `CA_BAL`<15000 | 192.168.0.90 | localhost | |
| `CA_BAL`<30000 | 192.168.0.90 | localhost | ✓ |
| `CA_BAL`<-5000 | 192.168.0.90 | localhost | |

**Fig. 6** Proposed allocation and replication scheme using the algorithm of [2]

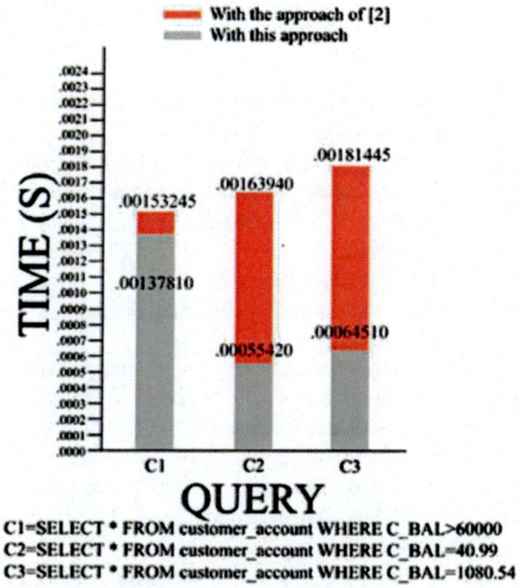


**Fig. 7** Comparison of execution times of the two approaches

# 6 Conclusions

A new approach to solving overlapping fragments was introduced, which is a usual issue in the fragmentation activity. The technique was appropriately adapted to the environment of the FRAGMENT Web application and improved the schema produced by reducing the execution time of the queries related to the most expensive attribute compared to the schema produced with the previous approach.

The major contribution of this research is to use cost precedence when producing the final fragmentation scheme in the workflow used in [1]. Thus, fragment preservation is based on cost precedence and not size, since in many cases the largest fragment is the least used or the smallest fragment is the most used.

In future work, the fragment grouping will be used with the same cost and in this way improve the scheme produced by avoiding fragments with a single element obtained by equality operations. In addition, the suppression of subtraction in obtaining costs will be analyzed, since, in the current process, although a predicate is the most widely used by all sites, the cost of this is not directly proportional to its use.

**Acknowledgements** The authors thank The National Technological of Mexico (TecNM) for its support for this work. Also, the National Council of Science and Technology (CONACYT) has sponsored this research paper.

## References


1. Abdel AE, Badr NL, Tolba MF (2016) Distributed database system (DSS) design over a cloud environment. Multimed Forensics Secur 115:97–116
2. Castro-Medina F, Rodriguez-Mazahua L, Lopez-Chau A, Abud-Figueroa M, Alor-Hernandez G (2020) FRAGMENT: a web application for database fragmentation, allocation and replication over a cloud environment. IEEE Latin Am Trans 18(6):1126–1134
3. Silberschatz A, Korth HF, Sudarshan S (2011) Database systems concepts, 6th edn. Yale University, McGraw-Hill, New Haven, CT
4. Özsu M, Valduriez P (2020) Principles of distributed database systems, 4th edn. Springer International Publishing, Ontario, CA
5. López C, Heinsen R, Huh E (2017) Improving availability applying intelligent replication in federated cloud storage based on log analysis. In: Proceedings of the 2017 international conference on machine learning and soft computing—ICMLSC '17. Association for Computing Machinery, Ho Chi Minh City, Vietnam, pp 148–153
6. Stiemer A, Fetai I, Schuldt H (2016) Analyzing the performance of data replication and data partitioning in the cloud: the BEOWULF approach. In: IEEE international conference on big data (big data). IEEE, Washington, DC, pp 2837–2846
7. Rodríguez-Mazahua L, Alor-Hernández G, Li X, Cervantes J, López-Chau A (2017) Active rule base development for dynamic vertical partitioning of multimedia databases. J Intell Inf Syst 48(2):421–451
8. Abdel AE, Badr NL, Tolba MF (2014) Dynamic distributed database over cloud environment. In: International conference on advanced machine learning technologies and applications. Springer, Cairo, Egypt, pp 67–76
9. Drissi A, Nait-Bahloul S, Benouaret K, Benslimane D (2018) Horizontal fragmentation for fuzzy querying databases. Distrib Parallel Databases 37(3):441–468
10. Wiese L, Waage T, Bollwein F (2016) A replication scheme for multiple fragmentations with overlapping fragments. Comput J 60(3):308–328
11. Ali A, Adel A, Tacha MA (2017) An optimized approach for simultaneous horizontal data fragmentation and allocation in distributed database systems (DDBSs). Heliyon 3(12):1–42
12. Kumar R, Gupta N (2014) An extended approach to non-replicated dynamic fragment allocation in distributed database systems. In: 2014 international conference on issues and challenges in intelligent computing techniques. IEEE, Ghaziabad, India, pp 861–865
13. TPC Homepage, www.tpc.org, last accessed: 2020/08/30

# Vetting the Influence of ISO-9126 Software Characteristics in the Development of Mobile Applications

**Pilita A. Amahan**

**Abstract** This research aimed to revisit the influence of the mobile applications developed by the students taking the Capstone Projects for three consecutive years of 2016–2019. The Capstone Projects were randomly selected having at least 15 manuscripts per year for the total of 45 manuscripts subject for assessment. The six characteristics of ISO-9126 like contents, organization and design, mechanics, comprehensibility, functionality, and maintainability were observed based on the summary of its evaluations. The results found that among the characteristics it is the functionality that gained the least average mean of 4.38. It also reveals that the ease of the software's operation receives the lowest mean of 4.53. With the results of the study, and with the help of the literature cited, the proposed framework is hereby recommended to escalate the functional assessment. Moreover, an interesting work on how a software product be assessed using a pure functional requirement using other metrics aside from ISO-9126.

**Keywords** Capstone project · Mobile applications · SDLC · ISO-9126 · Software products · Off-grid mobile applications · Vetting influence · Software characteristics · Functional assessment · Software development · Software vulnerability

## 1 Introduction

The infrastructures of mobile applications have become dependent on how it is being built by software developers. With that dependence given, it attracts cyber-attacks that cause malicious intents in cyberspace either by exploitation or by using the unique keys of the emerging technologies [1]. According to [2] a software becomes vulnerable when there is a problem in the implementation, specification of a software system and whose execution does not conform to the implicit security policy. Such

P. A. Amahan (✉)
Occidental Mindoro State College, San Jose Occidental Mindoro, Philippines



X.-S. Yang et al. (eds.), *Proceedings of Sixth International Congress on Information and Communication Technology*, Lecture Notes in Networks and Systems 216,
https://doi.org/10.1007/978-981-16-1781-2_61

cases lead the software with varied scenarios like having criticality in the installation of those patches and the possibility of not running it in accordance with its design.

With the increasing growth of off-grid mobile applications, especially for those students whose projects are the initial developer of the software tend to experience such technicalities. In developing a system, students were usually guided by the methodology of Software Development Life Cycle (SDLC) that includes the phases of analysis, design, coding, testing, installing, and maintaining the system [3]. Although there are varieties of models followed by the educators and students in institutions, SDLC has become the most useful model.

Also, the generations of today are not only dependent on the online activities, but rather, their focus is also on the development of off-grid applications because of its ease of accessibility and use. However, despite the presence of SDLC phases, some of the noted issues during the testing phase are still being noted like the crashing and freezing of data, data conflicts, data synchronizations, and such that leads the proponent to vet the functionality issues that lie during the development of these applications. As reiterated by [4] offline applications are very rare to receive updates and therefore, should be reexamined periodically for the intent of improving its tools and techniques.

Thus, this research aims to revisit the assessment of mobile applications developed by the Information Technology (IT) graduating students with their Capstone Project covering the year of 2016–2019. Specifically, the study aimed to identify the degree of influence of mobile applications against the metrics of ISO-9126 and determine the least sub-characteristics of the software product from the evaluation of ISO-9126. Also, this study would like to propose a framework that will benefit the trade-off analysis in the enhancement of the characteristics found from the evaluation.

## 2 Methods Used

### *2.1 Influence of Mobile Applications on ISO-9126 Software Characteristics*

Through random sampling techniques the study randomly selected at least 15 capstone manuscripts having the total of 45 of the academic years of 2016–2019. With the permission given by the research coordinator, the influence of mobile applications was able to assess by getting the average per characteristics of ISO-9126 such as contents, organization and design, mechanics, comprehensibility, functionality, and maintainability. Table 1 shows the recorded assessment per characteristics of ISO-9126.

With the result gathered, the influence of mobile applications against ISO-9126 six characteristics were summarized (see Fig. 1). It shows that the functionality of the system for three (3) consecutive academic years had drastically lowered down the result of 4.38 as to compare among the rest of software characteristics. However,

**Table 1** Influence of the mobile applications on ISO-9126 software characteristics

| ISO-9126 characteristics | 2017 | 2018 | 2019 |
|---|---|---|---|
| Contents | 4.461333 | 4.530667 | 4.670667 |
| Organization and design | 4.46 | 4.478 | 4.675333 |
| Mechanics | 4.426 | 4.416667 | 4.636667 |
| Comprehensibility | 4.392667 | 4.443333 | 4.609333 |
| Functionality | 4.524 | 4.52 | 4.698 |
| Maintainability | 4.436 | 4.262 | 4.440667 |

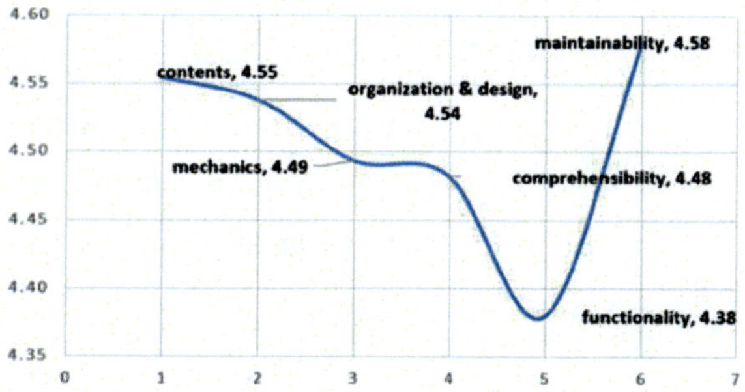


**Fig. 1** Summary of influence of the mobile applications on ISO-9126 software characteristics

it is the maintainability that dominated with the average mean of 4.58. Meaning, that the software is modifiable with the significant information and data needed for the system.

Having the noted concern in functionality, Table 2 shows the mean average for its sub-characteristics such as ease of operation, provision of comfort and convenience, and user friendliness. From the normative of the International Organization for Standardization (ISO), the functionality is the totality of essential functions that the software product provides, and others are just some sort of degree that contributes to the functionality. However, the relationship of the software functionality to the overall process of the business has been already outside the scope of ISO-9126.

Moreover, in vetting the security of the software, it is the functionality that chained the sub-characteristics of security. However, the security here according to the ISO is only defined as the relationship of unauthorized access to the software functions. Added to this, the study investigated the sub-characteristics that may influence the status of functionality and it is the ease of use that affects the most the functionality characteristic (see Fig. 2). To understand the factors affecting the usability of the software product, related researches have also been noted and most of it speaks about the lack of training for the use of it [6–10].

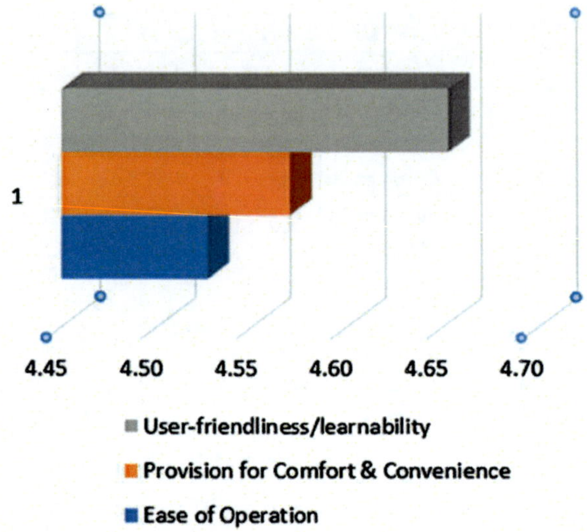


**Fig. 2** Sub-characteristics of functionality

## 3 The Proposed Framework

With the results and in the review of revisiting the influence of mobile applications developed by the graduating students against the ISO-9126 software characteristics the framework is hereby being proposed (see Fig. 3). With the five (5) phases of the SDLC, proper governance must be given with equal attention during the planning phase and on the development phase. Planning project the goals supported with the detailed requirements necessary in building a software. And during the development

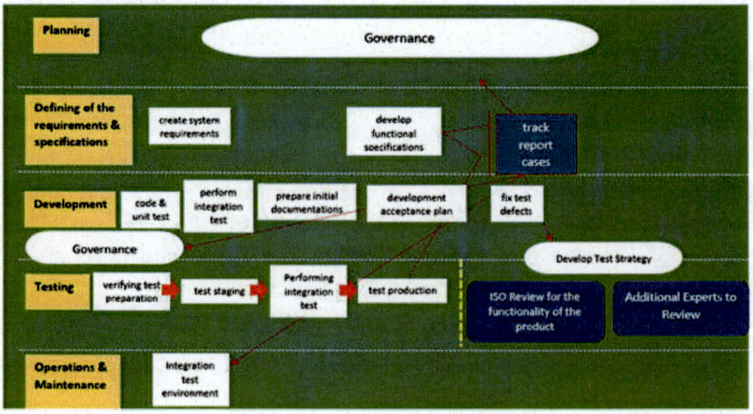


**Fig. 3** Proposed framework for the trade-off analysis of SDLC

phase, it is essential to observe if the specific requirements are being integrated. So, tracking of report cases will be done with the guidance of the experts and team. During the testing phase, a unique formation of a test strategy will be observed by the developers to ensure the quality of the functionality of the system. The functionality matters the most because it represents the totality of the system to serve the very purpose of the product which is to generate service. Moreover, the entire stages were guided by the ISO-9126 and it is too general that can test the specificity of the functionality and thus, proposing to determine a specific ISO metric that will test the inner functionality of the system is being emphasized on the proposed framework.

## 4 Results and Discussion

The history of the records of the three consecutive years of the academic years of 2016–2019 has been revisited for the purpose of identifying the possible strengths and weaknesses of the software projects found in the summary of evaluation from the experts, students, and target audiences. With this regard, it is found out that the functionality of the system fluctuates to the average mean of 4.38. Though it still receives the descriptive rating of "Very Good" the data shows the need for its improvement. As added by the ISO, the functionality is the totality of essential functions that the software product provides so functionality really matters. On the other hand, what may also affect the evaluation of the functionality are the external factors like the people who will be able to test the product. So proper training and orientation before used should also be considered.

## 5 Conclusions

The following are the conclusions derived from the findings of the study. In vetting the influence of ISO-9126 software characteristics against the capstone project of the graduating students, it is found out the functionality gained the lowest general mean average of 4.38. Although it is still on the passing level of a five-point Likert Scale, the metric used in determining the extent of respondents' attitude in asking their perception on the quality of the software, it became the next concern of the students as well as the teachers in developing software applications. With the results of the study, the proposed framework for a better trade-off analysis of SDLC has been proposed. Since the developed capstone project is just using one metric, the proposed framework is suggesting to have other metrics that will purely evaluate the functionality of the system. With this, the functionality of the system is considered the ground among the software characteristics that would help uplift the utilization status of the developed capstone projects of the students.

## 6 Recommendations

The demand of developing mobile applications is becoming rampant nowadays because of the satisfaction, convenience, and other benefits that it could give amongst the end users. However, it is also a must that developers should also look upon the possible way to improve more on how the end users would be satisfied with the service of the product. The tail of the study also implies that developers should also have to review and/or add other metrics that will have the focus on the evaluation of the functionality of the software product. A proper orientation on the usability of the software product must also be considered. Moreover, for future work the study of how a software product conforms with the requirement specifications against the pure functional requirements should also be considered.

## References

1. Jang-Jaccard J, Nepal S (2014) A survey of emerging threats in cybersecurity. J Comput Syst Sci 80(5):973–993. https://doi.org/10.1016/j.jcss.2014.02.005
2. Sadeghi, A., Bagheri, H., Malek, S.: Analysis of android inter-app security vulnerabilities using COVERT. In: 37th IEEE International Conference on Software Engineering, pp. 725–728. IEEE/ACM, Florence (2015). doi: https://doi.org/10.1109/ICSE.2015.233
3. Hoffer, J., George, J., Valacich, J.: Modern system analysis and design. 7th edn. Prentice Hall/Pearson Education (2011).
4. Ogata, M., Franklin., J., Voas, J., Sritapan, V. Quirolgico, S.: Vetting the security of mobile applications. National Institute of Standards and Technology/United States (2019). doi: https://doi.org/10.6028/NIST.SP.800-163rl
5. Tahar, A., Riyadh, HA., Sofyani, H., Purnomo.: Perceived ease of use, perceived usefulness, perceived security and intention to use e-filing: The role of technology readiness. The Journal of Asian, Finance, Economics, and Business 7(9), 537–547. doi: https://doi.org/10.13106/jafeb.2020.vol7.no9.537
6. Goedicke, M., Bosch, J., Olsson, H.: Which factors influence practitioners' usage of build automation tools. In: 3rd International Workshop on Rapid Continuous Software Engineering (RCoSE), pp. 1–20, IEEE/Argentina (2017).
7. Phau I, Ng J (2010) Predictors of usage intentions of pirated software. J Bus Ethics 94:23–37. https://doi.org/10.1007/s10551-009-0247-1
8. Ali A, Anbari F, Honey W (2008) Impact of organizational and project factors on acceptance and usage of project management software and perceived project success. J Bus Ethics 39(2):2–4. https://doi.org/10.1002/pmj.20041
9. Green G, Hevner A, Collins R (2005) The impacts of quality and productivity perceptions on the use of software process improvement innovations. Inf Softw Technol 47(8):543–553. https://doi.org/10.1016/j.infsof.2004.10.004
10. Chiasson M, Lovato C (2001) Factors influencing the formation of a user's perception and use of a DSS software innovation. ACM SIGMIS Database for Advances in Information Systems 32(3):16–35. https://doi.org/10.1145/506724.506727

# An Efficient Approach for Multiple Moving Objects Tracking with Occlusion

**Bayar Shahab and Fattah Alizadeh**

**Abstract** Some of the main issues for the governments and businesses nowadays are security and monitoring. Several critical locations need high security such as military bases, airports, malls, and checkpoints. The ambition and objective of this paper are to detect and track multiple objects while considering the challenges of partial and full occlusion. This paper presents an efficient algorithm for object tracking which can track objects in the presence of occlusion. The results are compared with the original Kalman filter-based approach. The proposed algorithm has an average of 0.51% improvement comparing to the original algorithm according to the following three test scenarios. It has 0.43% improvement for test 1, which is object tracking having full occlusion, 0.57% improvement for the second scenario, which is object tracking having full occlusion having a different dataset and 0.52% improvement for the third scenario in the tracked objects have been partially occluded. The accuracy, efficiency, and analogy of the proposed algorithm with the original algorithm show promising results.

**Keywords** Object tracking · Occlusion · Multiple moving objects tracking · Kalman filter · MOG · GMM

## 1 Introduction

A huge consideration has been given to the video surveillance processing for the purpose of monitoring, controlling, and security [1]. The motivation for this work is to utilize available camera and then applying a background subtraction and an improved tracking algorithm to process the images in order to perform a reliable detection and tracking. No expensive hardware device or machine is required as the proposed algorithms needs to be implemented using a software program on the available camera. As it is clear, object detection is principally locating an object of interest in a sequence of images [2]. On the other hand, object tracking is monitoring

B. Shahab (✉) · F. Alizadeh
University of Kurdistan Hewler, Erbil 44001, Iraq

 
X.-S. Yang et al. (eds.), *Proceedings of Sixth International Congress on Information and Communication Technology*, Lecture Notes in Networks and Systems 216,
https://doi.org/10.1007/978-981-16-1781-2_62

the changes such as shape, location, and size of the object of interest [3]. Despite the benefits and great usages of the available approaches, there are several challenges associated with detection and tracking such as occlusion [4], non-stationary camera [5], appearance and disappearance of objects and shadow, light, noise, and weather conditions [6]. In this work, background subtraction and specifically Mixture of Gaussian (MOG) [7] has been utilized as the detection method. MOG has a strong foreground detection and background maintenance process [8]. An improved method of Kalman-based filter has been utilized for object tracking. Additionally, most of the tracking methods suffer from the challenges stated earlier such as occlusion problem, sudden appearance, and disappearance of the objects. Kalman filter typically suffers from object deviation from the motion [9]. For example, if an object moves and then turns 90 degrees to left or right, the filter will have false tracking. Another challenge for Kalman filter-based approach is predicting the location of the object while it is blocked or occluded. Therefore, the proposed approach and algorithm is an improved Kalman Filter-based approach to have a robust tracking to solve the occlusion problem.

## 2 Related Works

There are several research conducted to give a comprehensive review about the available methodologies as in attempts of [4, 5, 8, 10, 11]. Ultimately, some are setting benchmarks for object detection and tracking [12]. In the sequence of the subsections, a brief review of the utilized methodologies of object detection and tracking will be given.

### *2.1 Background Subtraction—MOG*

Background subtraction can be categorized as recursive/ parametric [10, 13] and non-recursive/non-parametric [13]. Followings are some of the recursive/ parametric approaches; W4 Method, Mixture of Gaussian (MOG), Three Frame Differencing, and Differencing and Summing Technique. Furthermore, some of the non-recursive/ non-parametric methods are KDE, CodeBook, Fuzzy Logic Based, SACON, BNN, SOBS, ViBe, RaBS, and PBAS. In this work, MOG has been used for object detection. For more details on background subtraction using MOG, the readers are referred to a complete survey published by the same authors of the current work [14].

### 2.2 *Object Tracking—Kalman Filter*

There are several classifications for object tracking [10, 11, 15] have categorized and classified object tracking methodologies as point-based, kernel-based, and Silhouette-based tracking. While some authors have other categorizations for tracking techniques. Authors of [16] group tracking methods as matching-based tracking, filtering-based tracking, class-based tracking, and fusion-based tracking. This paper proposes an enhanced Kalman filter-based approach which is a point-based method for tracking. Point-based specifies a point or several points across the frames. These points are normally the motion information of the object such as velocity or location. As Sindhuja and Devi described; point-based tracking is a complex problem and in cases of occlusion it can be difficult to track objects [10]. Some recognized methods are Kalman filter and particle filter. On the other side, Kalman-based filter is a well-known approach for object tracking algorithm is Kalman filter. Welch and Bishop [17] have done a comprehensive introduction to the original Kalman filter. It models the motion of the human (detected objects) using position, velocity, and acceleration. Kalman filter has two stages: prediction and correction.

## 3 Proposed Method

As mentioned beforehand in the introduction, in this paper, an enhanced version of the Kalman filter-based approach has been proposed which is an adaptive Kalman filtering to solve the occlusion problem. Adaptive Kalman filtering means to update the state model, i.e., updating the process noise and measurement noise to obtain a more realistic value.

In addition, every object tracking process starts with a good object detection and as stated MOG has been utilized as background subtraction-based object detection method. The method can simply be described as following:

- Each pixel will be assigned and represented by K-number of Gaussian distributions.
- The current pixel is compared with the background model by matching the Gaussian values of the current pixel and the background model.
- If some match found, according to the distributions of the Gaussian, the pixel is labeled as background or foreground.
- If no match found, the pixel is absolutely considered as foreground.

The probability of representing each pixel value X at frame t with the K-number of Gaussians is shown in Eq. (1)

$$P(Xt) = \sum_{i-1}^{K} W\,i{*}\ \ GPdf(Xt, u, i, t,\ C\,i,\ t) \tag{1}$$

W: Weight, Gpdf: GaussainPdf, u: mean, C: covariance.

After parameter initialization, foreground detection can take place using *Mahalanobis distance's* equation [18]. The method will use a pixel match test to determine the foreground pixels.

Finally, to apply the update equations after the foreground detection, we must consider the matched and unmatched pixels.

On the other hand, Kalman filter is a set of mathematical equations that are used to estimate the location/position and velocity of an object when the measurements used to measure position and velocity are not certain or contain errors [17]. The goal of the Kalman filter is to get closer to the actual value/measurement of the object. Kalman filter is associated with a motion model to predict or estimate the location and velocity. The motion models are a function of velocity or acceleration.

There are two main steps/concepts associated with Kalman filtering which are prediction and correction:

1. **Prediction:**

Prediction is accomplished to predict the location of an object in the current frame [11]. Prediction is using previous frame's location to calculate the current location. Using a motion model/transition matrix such as velocity motion model and multiplying it by the previous state, the current location can be predicted. There is a variable associated with the prediction called process noise or error in the estimate/prediction. The variable is constant and compensates the variations between the actual and predicted location of the object. Equations (2 and 3) are the formulas used for prediction.

$$X_{K_p}{=}AX_{k-1} + Bu_k + W_k \tag{2}$$

$$P_{K_p}{=}AP_{k-1}A^T + Q_k \tag{3}$$

where the parameters are:
$X_{K_p}$: State matrix, new predicted State based on previous state.
$P_{K_p}$: Process Covariance matrix, error in the estimate/process.
u: Control variable matrix, i.e., acceleration.
W: Predicted state noise matrix.
Q: Process noise covariance matrix.
A and B are matrices to convert values to the same matrices format.

2. **Correction**

Correction is a process which follows the prediction phase. Correction is to correct the location of a measured object (detected object) to update the state of the object [11]. There is a variable associated with correction called measurement noise. The measurement noise compensates the differences between the predicted measurement and actual measurement. The predicted value is corrected using the residual value. The residual value is the difference between the measured and predicted value. In order to update or correct the prediction, all the factors, Kalman gain, predicted values, and residuals are used. Once the correction step is done, the process of prediction continues iterating. Equations (4, 5, 6 and 7) are the formulas used for correction.

$$Y = CX_{K_m} + Z_k \tag{4}$$

$$X_{k=}X_{K_p} + KG[Y - HX_{K_p}] \tag{5}$$

$$X_{k=}X_{K_p} + KG[Y - HX_{K_p}] \tag{6}$$

$$P_{k=}(I - KH)P_{K_p} \tag{7}$$

where

Y: Measurement of the state, $X_{K_m}$: Measurement input,$Z_K$: Measurement noise (uncertainty), KG: Kalman gain, R: Sensor noise covariant matrix, measurement error, H is a matrix to convert values to the same matrices format.

Original Kalman-filter-based approach has shortage in its update process and is not efficiently adaptive and accurate. In prediction stage, there might be some challenging factors/errors such as wind, object deviating from the motion model (velocity or acceleration), and full or partial occlusion of the moving objects. These factors or errors can change the predicted values. Here, an enhanced version of the Kalman-filter-based approach is proposed that can solve the occlusion problem specifically. We propose to put more emphasis/trust on the prediction rather than the detection/measurement in the case of occlusion.

We achieve that by making the Kalman gain smaller and smaller in each step/iteration that makes the tracker to put more trust on the predictions. Figure 1 shows the Kalman gain bar and how it affects the tracker to put more trust on either the predictions or actual measurements. From the Kalman gain graph we can observe two scenarios as following:

Scenario 1: Detector or measurement tool is reliable and strong with minimal error in the measurement tool represented as R in Eq. (5). In this case, you can trust the detector measurement and put less emphasis on the predictor tool by making the Kalman Gain value to approach the value 1.

Scenario 2: Detector or measurement tool not reliable and has a lot of error in the measurement as in the case of occlusion. In this case, the predictor tool can see the bigger role in deciding the location of the object. This is achieved by making the Kalman Gain value to approach 0 in each iteration of the process.

**Fig. 1** Kalman Gain Bar

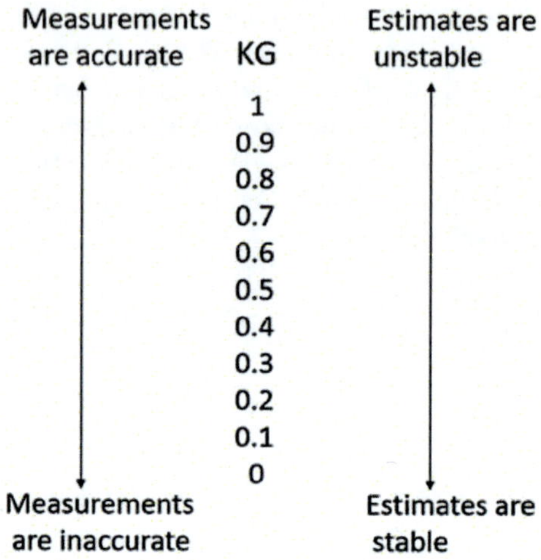


Process noise or process covariance matrix represented as $P_{K_p}$ has a direct effect on the value of Kalman gain as in Eq. (5). So, the smaller the value of $P_{K_p}$ will be, the value of Kalman Gain would be lower. This means the tracker will have more contribution in deciding the location of the object rather than the actual measurements of the detector since process noise is decreasing in each iteration of the process. The reason why we make Kalman Gain smaller is because we cannot trust the detectors as in the case of occlusion.

We update the motion noise in a way in each step to fit the predictions and to compensate the differences between the predicted measurement and actual measurement. Using experimental methods by conducting several tests, we have concluded that for the occlusion-based tracking using Kalman filter, trusting the predictor will give more accurate results comparing the actual measurements.

Detected object has been assigned to existing tracks using Munkres version of the Hungarian algorithm [19]. Figure 2 shows a flowchart of the implemented algorithm using MATLAB software. The diagram shows step by step the overall system process and data flow.

## 4 Results

This section presents the main test scenarios conducted to approve the viability of the proposed algorithm while comparing the results with the original work. It is divided into three subsections following the three test scenarios conducted. Video samples have been taken from datasets as in [20, 21]. MATLAB is utilized as the main software for testing and showing the results. For presenting the data, statistical, and graphical methodologies have been utilized. The performance measurement has taken multi-target tracking into consideration. The evaluation measure is the center error, in which

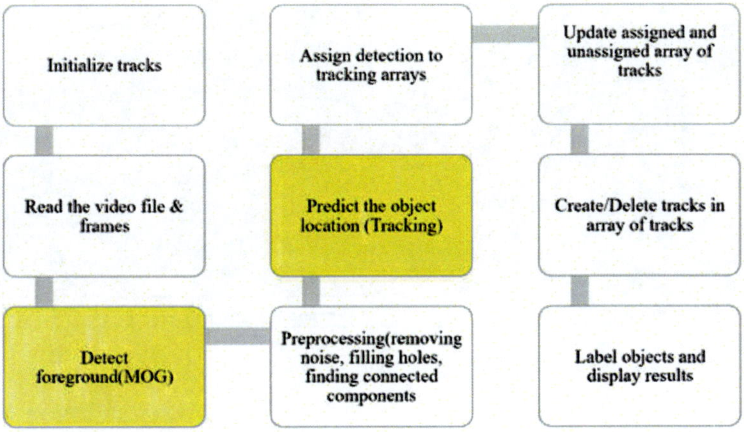


**Fig. 2** Flow chart of the proposed algorithm

the difference between the actual(detected) centroids and the predicted centroids are calculated in each observed frame. Euclidian distance metric [22] has been used to find the center error value as in Eq. (8). After that, Percentage change (improvement) [23] has been used to find the percentage of improvement ratio between both algorithms shown in Eq. (9). Percentage improvement has been calculated between the average of the center error values of both original and proposed methods (Table 1).

$$\text{Euclidian distance} = \sqrt{(X2 - X1)^2 + (Y2 - Y1)^2} \tag{8}$$

$$\text{Percentage change} = \frac{\text{New value} - \text{Old value}}{\text{Old value}} \tag{9}$$

**Table 1** Test scenario 1

| Algorithm | # of tested frames | Sum of center error values (Euclidian distance) | Avg. of center error values | Percentage change |
|---|---|---|---|---|
| Original algorithm | 51 | 1138.06 | 22.31 | 0.43% |
| Proposed Algorithm | | 652.97 | 12.80 | |

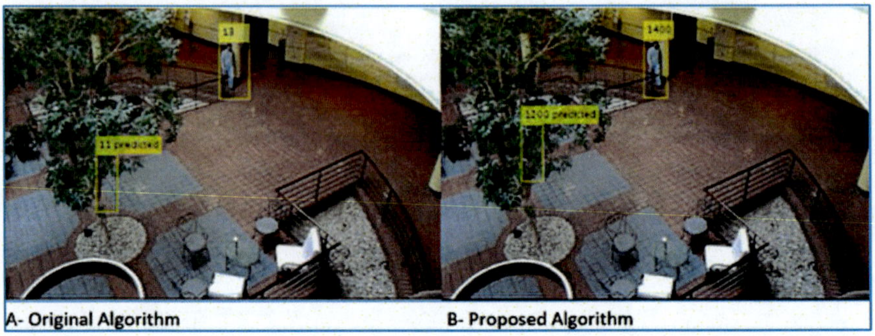


**Fig. 3** Test scenario 1—tracking/prediction result

## 5 Test Scenario 1—Object Tracking Having Full Occlusion

This test scenario is based on the atrium.mp4 video from MATLAB directory used for testing object detection and tracking. The video sequence image size is 640 × 360, and the framerate is 30 fps (frame/second) with the total of 600 frames. The analysis was done for 51 frames starting from frame 500. The reasons of why analysis was done only for 51 frames are as the following. Firstly, the object labeled 1200 in Fig. 3 (This label is 11 in the original algorithm) would be blocked by a tree. Next, the process will be very expensive in terms of time and speed to process all the frames. Additionally, the fact that our image labeling numbers are different than the original algorithm is only due to the extra codes we have introduced to our proposed algorithm. However, the process will remain the same and this labeling does not affect any of the processes. Above are true and generalized for other test scenarios as well.

The proposed algorithm performs 0.43% better than the original work in case of full occlusion. Figure 3 demonstrates a certain frame where the proposed algorithm has predicted the location of a person while it has been blocked by a tree. The original algorithm has predicted the location below the central point.

## 6 Test Scenario 2—Object Tracking Having Full Occlusion

This test scenario is considering full occlusion using i-Lids bag and vehicle detection challenge dataset [43]. The video file is used in this dataset is AVSS_AB_Easy_Dix. Image size is 640 × 460, the framerate is 29 fps (frame/second) and the total of 1131 frames. Analysis was done for 48 frames starting from frame 100. The test was done in this period as the object labeled 200 in Fig. 4 (This label is 2 in the original algorithm) would be blocked by a pillar.

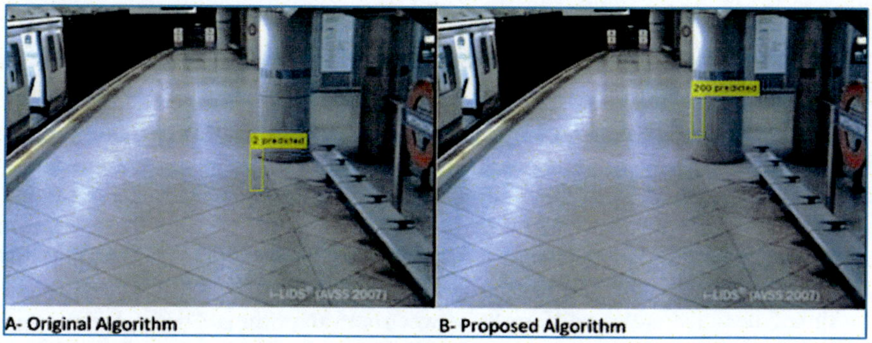


**Fig. 4** Test scenario 2—tracking/prediction result

From Table 2, it can be observed that the proposed algorithm performs 0.57% better than the original work. Figure 4 demonstrates a certain frame where the proposed algorithm has predicted the location of a person while he has been blocked by a Pillar. The original algorithm has a false prediction and does not predict the actual location of the object. Figure 5 shows a full track of predictions for object labeled 2 (this label is 200 in the proposed algorithm) in some successive frames. Starting from frame 248, both predictions are similar as there is no occlusion. In frame 255, the object would be blocked by the wall pillar. In frame 264 this comparison is even more clear as the object has been blocked completely by the wall pillar.

**Table 2** Test scenario 2

| Algorithm | # of tested frames | Sum of center error values (Euclidian distance) | Avg. of center error values | Percentage change |
|---|---|---|---|---|
| Original algorithm | 48 | 1268.13 | 26.42 | 0.57% |
| Proposed Algorithm | | 541.89 | 11.29 | |

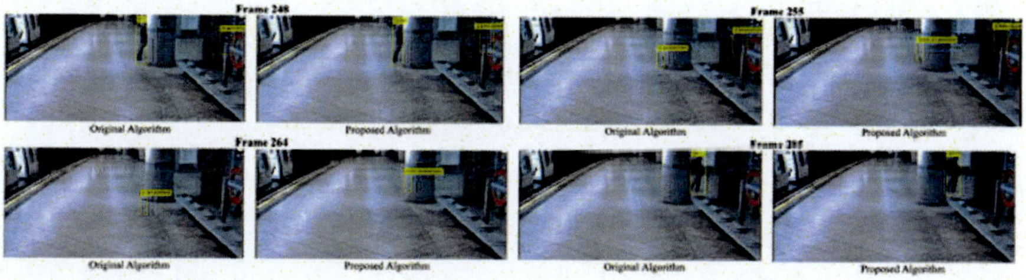

**Fig. 5** Test scenario 2—full track of the object

## 7 Test Scenario 3—Object Tracking Having Partial Occlusion

This test scenario is based on our own recorded video in the UKH (University of Kurdistan Hewler) library. Image sequence size is 640 × 480, and the framerate is 29 fps (frame/second) with the total of 181 frames. The analysis was done for 81 frames starting from frame 100 up to 181 and calculations were performed for 40 frames. The test was done in this period as the object labeled 100 in Fig. 6 (this label is 1 in the original algorithm) will be partially occluded/blocked by a leaflet holder.

Table 3 demonstrates the difference between both algorithms. The final percentage improvement is 0.5%. Figure 6 demonstrates a certain frame where the proposed algorithm has predicted the location of the person while it has been partially blocked by a leaflet. Figure 7 shows a full track of predictions for object labeled 1 (this label is 100 in the proposed algorithm) in some successive frames. As the object is moving toward the leaflet, it would partially be hidden/blocked as shown in Fig. 7. Both frames 160 and 172 are showing similar predictions for both algorithms as the object is not yet blocked or occluded. However, the object in frames 178 and 180 is partially occluded/blocked by the leaflet. The proposed algorithm has better predictions than the original algorithm in both frames 178 and 180.<

**Fig. 6** Test scenario 3—tracking/prediction result

**Table 3** Test scenario 3

| Algorithm | # of tested frames | Sum of center error values (Euclidian distance) | Avg. of center error values | Percentage change |
|---|---|---|---|---|
| Original algorithm | 40 | 546.87 | 13.67 | 0.52% |
| Proposed Algorithm | | 262.83 | 6.57 | |

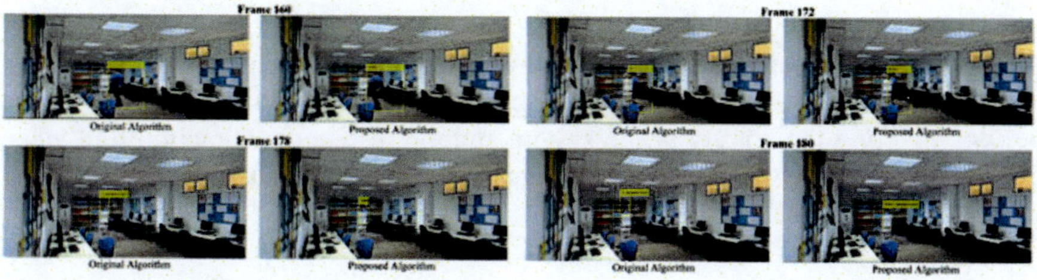

**Fig. 7** Test scenario 3—full track of the object

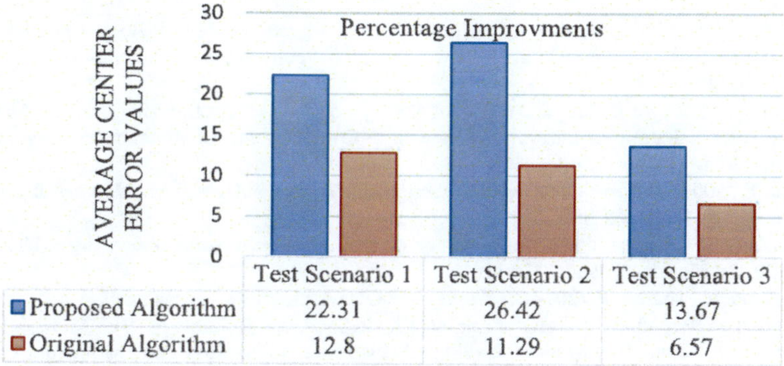


**Fig. 8** Average percentage improvement bar chart

Figure 8 Shows the overall percentage improvement. The proposed algorithm performs 0.51% better than the original algorithm as an average of all the three percentage improvements.

## 8 Conclusion and Future Works

This paper has proposed an algorithm for object tracking which is an enhanced version of the original Kalman filter-based approach. The original Kalman filter works well however, in some cases like occlusion it fails to track and predict the locations of objects. In addition, detection and tracking of multiple objects need to go through several stages to be well-accomplished. Background subtraction/object detection has been implemented using Mixture of Gaussian. It is one of the famous methods for object detection. Object tracking has been accomplished using an enhanced version of Kalman filter-based approach. From the results, it can be observed that the proposed algorithm performs better than the original algorithm in the presence of occlusion. The reason is because the proposed improved Kalman filter-based algorithm puts more emphasize on the predictions rather than the detections. The precision of the results leads to a conclusion to state that the proposed algorithm has satisfactory results comparing to the original algorithm. There were some defects affecting the quality of the system and the results as following. Bad video qualities, availability of camera column to hold the camera for a better recording, and bad weather conditions.

## References

1. IFSECGlobal (2014, January 1) Role of CCTV Cameras : Public, Privacy and Protection. Retrieved from ifsecglobal: https://www.ifsecglobal.com/role-cctv-cameras-public-privacy-protection/.
2. Parekh H, Thakore D, Jaliya U (2014) A Survey on Object Detection and Tracking Methods. International Journal of Innovative Research in Computer and Communication Engineering, 2970–2978
3. Rout R (2008) A Survey on Object Detection and Tracking Algorithms. National Institute of Technology Rourkela, India
4. Shu G (2009) Human detection, tracking and segmentation in surveillance video. University of Central Florida, Orlando, Florida
5. Ragland K, Tharcis P (2014) A Survey on Object Detection, Classification and Tracking Methods. International Journal of Engineering Research & Technology (IJERT), 622–628
6. Jacques JC, Jung CR, Musse SR (2005) Background subtraction and shadow detection in grayscale video sequences. In Computer Graphics and Image Processing (pp. 189–196). Brazil: Brazilian Symposium
7. Stauffer C, Grimson E (1999) Adaptive background mixture models for real-time tracking. Computer Vision and Pattern Recognition
8. Vacavant, A., & Sobral, A. (2014). A comprehensive review of background subtraction algorithms evaluated with synthetic and real videos. Computer Vision and Image Understanding, 4–21
9. mathworks. (2017). *Motion-Based Multiple Object Tracking*. Retrieved from mathworks: https://www.mathworks.com/help/vision/examples/motion-based-multiple-object-tracking.html
10. Sindhuja, G., & Devi, R. (2015). A Survey on Detection and Tracking of Objects in Video Sequence. *International Journal of Engineering Research and General Science* , 418–426.
11. Yilmaz, A., & Javed, O. (2006). Object Tracking: A Survey. *ACM Computing Surveys*, 45 pages.
12. Dhome, Y., Vacavant, A., Chateau, T., & Goyat, Y. (2010). A benchmark for Background Subtraction Algorithms in Monocular Vision : A Comparative Study. *IEEE Xplore*. IEEE.
13. Setitra, I., & Larabi, S. (n.d.). Background subtraction algorithms with post. Algiers, Algeria: University of Science and Technology USTHB.
14. B. Azeez and F. Alizadeh, "Review and Classification of Trending Background Subtraction-Based Object Detection Techniques," 2020 6th International Engineering Conference "Sustainable Technology and Development" (IEC), Erbil, Iraq, 2020, pp. 185-190, doi: https://doi.org/10.1109/IEC49899.2020.9122929
15. Lakshmeeswari, G., & Karthik, K. (2016). Survey on Algorithms for Object Tracking in Video. International Journal of Computer Applications, 0975–8887.
16. Li, M., Cai, Z., Wei, C., & Yuan, Y. (2015). A Survey of Video Object Tracking. International Journal of Control and Automation, 303–312.
17. Welch, G., & Bishop, G. (2006). An Introduction to the Kalman Filter. NC: UNC.

18. Bouwmans, T., El Baf, F., & Vachon, B. (2008). Background Modeling using Mixture of Gaussians for Foreground Detection - A Survey. Recent Patents on, 219–237
19. Munkres, J. (1957). Algorithms for Assignment and Transportation Problems. Journal of the Society for Industrial and Applied Mathematics.
20. AVSS 2007 (2007). 2007 IEEE International Conference on -Advanced Video and Signal based Surveillance. Retrieved from eecs.qmul: http://www.eecs.qmul.ac.uk/~andrea/avss2007_d.html
21. Vacavant, A., Chateua, T., Wilhelm, A., & Lequievr, L. (2012). A Benchmark Dataset for Outdoor Foreground/Background Extraction. ACCV 2012, Workshop: Background Models Challenge.Korea
22. Weisstein E (N.D) Distance. Retrieved from MathWorld--A Wolfram Web Resource. http://mathworld.wolfram.com/Distance.html
23. Math is Fun (N.D) Percentage Change. Retrived from https://www.mathsisfun.com/numbers/percentage-change.html

# Interaction Design of Asthma Inhaler Assistant Device

**Shinyoung Park, Suhyun Choi, Yumi Cha, Dongryun Lee, Seungwon Shin, and Dokshin Lim**

**Abstract** Asthma is an allergic disease that more than 300 million people worldwide suffer from. Recurrent attacks of breathlessness and wheezing are main symptoms of asthma and inhalers are used for the treatment. Surprisingly, 84 percent of patients are using the inhaler incorrectly and 53.7 percent of medical doctors confess the difficulty in educating the inhaler usage due to a lack of time and manpower. It is important for patients of chronic diseases such as asthma to form proper inhaler usage habit from the very beginning when they are in the early stage of the disease. However, they have to rely only on doctors whom they do not meet every day. Otherwise, they can read or watch instruction materials, but it cannot correct their behavior of use in person. Our asthma inhaler assistant device is a working prototype with a purpose of aiding asthma patients in three critical steps: keep the inhaler horizontally, inhale deeply enough, and hold the breath for 10 s. We designed an interactive audio-visual real-time feedback system using LED and sound to deliver not only informative, but also motivational aid to the users of asthma inhalers. Finally, we validated our working prototype with actual patients or people suffering from similar diseases. The results indicate that it has the effect of actually correcting participants' usage behavior judging from the peak inspiratory flow rate and breath-holding time.

**Keywords** Asthma · Inhaler usage · Ellipta · DPI (Dry Powder Inhaler) · Real-time correction · Assistant device · Interaction design · UX

## 1 Introduction

### *1.1 Background*

Asthma is a chronic respiratory disease that causes repeated and paroxysmal symptoms such as wheezing and attacks of breathlessness. Around the world, 339 million

S. Park · S. Choi · Y. Cha · D. Lee · S. Shin · D. Lim (✉)
Department of Mechanical and System Design Engineering, Hongik University, Seoul, Korea
e-mail: doslim@hongik.ac.kr



X.-S. Yang et al. (eds.), *Proceedings of Sixth International Congress on Information and Communication Technology*, Lecture Notes in Networks and Systems 216,
https://doi.org/10.1007/978-981-16-1781-2_63

people are suffering from asthma and the prevalence is increasing [1]. The use of asthma inhalers is the most recommended method for the treatment of asthma [2]. However, since asthma inhalers rely on the patient's inhalation techniques for the effective drug delivery, it is very important for the patient to use it properly. If the inhaler is not used correctly, the dosage reaching to the lungs is about one-third as compared to when the inhaler is used correctly [3]. Surprisingly, 84 percent of asthmatic patients are using inhalers incorrectly [4], and medical staffs are in difficult situations to help it [5]. Our own user interviews with ten asthma inhaler-using patients and one medical doctor revealed that most users did not follow the correct instruction. Moreover, there was a lack of training on how to use it and the training was not reinforced by follow-up checks. It was hard for them to assess by themselves whether they were using it correctly or not.

In order to help asthma patients, we propose an "Asthma inhaler assistant device" that monitors patients' usage behavior and conveys real-time feedback. We built a working prototype that provides information to the user in three steps: keep the inhaler horizontally, inhale deeply enough, and hold the breath for 10 s. Our interactive audio-visual real-time feedback system was implemented in form of an attachable device so that current users do not have to handle another separate device than the main inhaler in such a short usage time. This paper presents an interaction design that effectively corrects the usage habit of asthma patients and demonstrates the development process of a new product category called "Asthma inhaler assistant device".

## 1.2 Problem Definition and Design Goal

Above all, we have to begin with understanding the application status of the developed sensor system. The developed system targets the Dry Powder Inhaler (DPI), RELVAR Ellipta. The DPI is a method of sucking dry powder drug only by the patient's inspiratory effort, and it is commonly used to hold one's breath after inhaling deeply in a certain position [6]. While holding breath, most users show similar behavior pattern of lowering their hands which hold the inhaler. In case of Ellipta, it is recommended to maintain the horizontal posture before and during the inhalation and be administered as 1 inhalation once daily for 30 days [7].

We conducted a series of user interviews with ten asthmatic patients using inhalers and a medical doctor with a purpose of identifying current pain points and unmet needs in using DPI. We interviewed them individually using a mobile chat service in April 2020. Based on the results, we defined our representative persona and the user journey map [8] as is shown in Fig. 1. Major pain point in using DPI is that most of users are not sure if they are using the inhaler correctly, nor they are able to tell how to correct it right. In order to recognize wrong behavior simultaneously during the usage which happens in such a short time, the best way might be to provide immediate feedback on the inhaler itself. Our problem is formulated as "How might we guide users to correct behavior while they are using the DPI?".

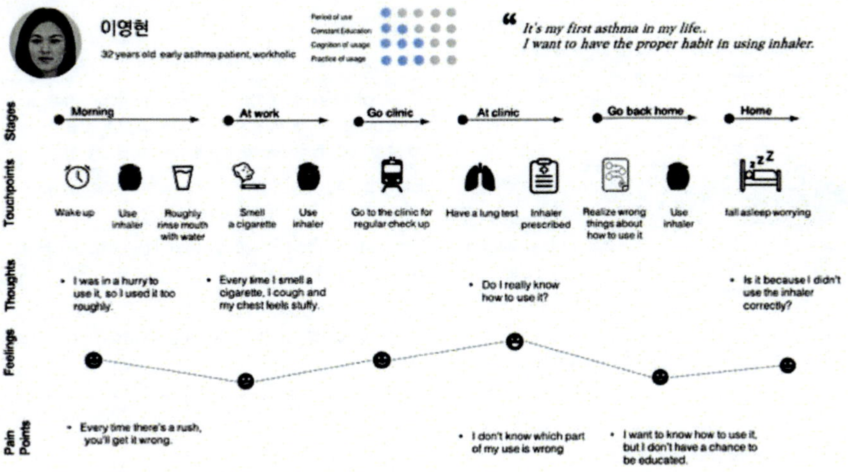


**Fig. 1** User journey map of DPI users

We then considered all senses (see Table 1) to choose best way for users to know how they do with the inhaler, most importantly without obstructing the inhalation process. Taste and smell are excluded because they are not adequate to the inhalation process. The tactile sense has many advantages in communicating without seeing but is excluded due to the risk of affecting the decomposition of powder drugs inside the inhaler. Visual sense is considered as a main channel, but it is not evident that users pay attention to it on the stage of holding the breath. Auditory sense is quicker than the vision and can be transferred easily regardless of the direction the user is facing. Finally, we set a more detail interaction design goal to establish an interaction that gives real-time feedback on LEDs and motivates users to hold the breath long enough (which is 10 s) by using sound.

**Table 1** Comparative analysis of human senses in line with DPI usage steps

| Human senses | Signal | Hold the inhaler horizontally | Inhale deeply enough | Hold the breath for 10 s |
|---|---|---|---|---|
| Vision | Light | O | O | O |
| Tactile | Vibration | X | X | O |
| Hearing | Sound | O | O | O |
| Taste | Flavor | X | X | X |
| Olfaction | Smell | X | X | X |

**Table 2** Existing visual interfaces on the latest digital devices

| Sample UI | Product | User interface description |
|---|---|---|
|  | Homey Pro | Control values of smart home devices such as dimming of light, color of lamp, etc. are expressed in color and pattern of light |
|  | Google Home Mini | Expressing the state of device by motion and color changing of 4 LEDs |
|   | Apple Air pods pro | Presenting the device charge and connect state as color of 1 light |
|   | LG PuriCare mini | Real-time air quality status is expressed as light color |

## 2 Interaction Design

For visual interface design, we rely on existing metaphor of colors. For auditory interface design, we quickly created sound samples and tested with people to see which sound may make people feel motivated and enjoyable when they hold the breath for 10 s.

### 2.1 *Visual Interface Design*

We conducted a benchmarking study on existing visual interfaces that communicate the system status in real-time (see Table 2) and found that they are quite effective even though they dispose only small number of colors in a restricted area. The colors reflect the status of the system constantly, which enhances the user's understanding of the system. Small LED light modules can deliver information intuitively and effectively to the user using color and pattern which became standards these days.

Our asthma inhaler assistant device[1] (in Fig. 2) aims at very intuitive user interface and applies familiar metaphor. Simply with three small circular LED lights, users form natural mapping to three steps to focus upon. At each step corresponding to each LED, the user is informed of their usage whether they are doing right or wrong. To minimize interference during the inhaler usage flow, status feedback for each step appears progressively in real-time using only color changing without any pattern, except for feedback on the last overall result. The colors take the metaphor of traffic lights which helps users forming familiar mental model. Consistently with the semantic meaning that everyone has, red is used for warning and hazard, orange/yellow for caution, and green for positive results (see Table 3).

[1] This asthma inhaler assistant device was awarded at Korea Engineering Education Festa 2020 jointly organized by The Engineering Education Information Center and The Korea Institute for Advancement of Technology.

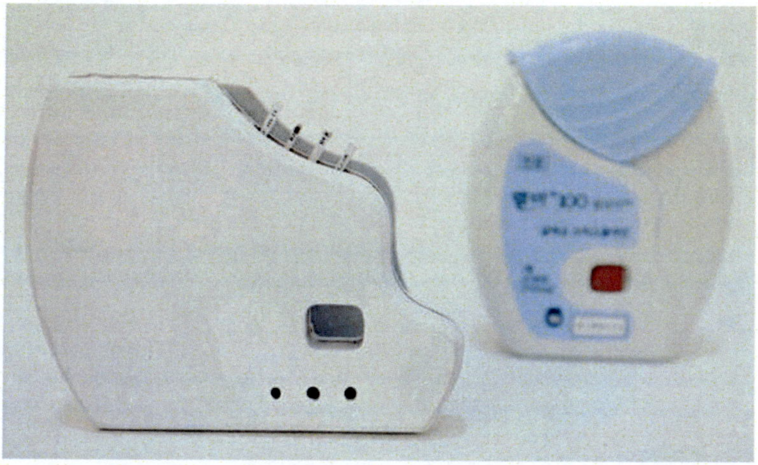

**Fig. 2** Inhaler assistant device (our prototype on the left) and the Ellipta inhaler (on the right)

**Table 3** Stepwise color feedback system design

| Steps | Visual feedback | LED displays |
|---|---|---|
| Hold the inhaler horizontally | – The 1st LED indicates the horizontal balance of the inhaler. If the inhaler is horizontal, it is green | 🟢 ○ ○ |
| | – Otherwise, it is red | 🔴 ○ ○ |
| Inhale deeply enough | – The 2nd LED indicates the PIFR (Peak Inspiratory Flow Rate)[2]. If it is same or higher than 60L/min., it displays green | ○ 🟢 ○ |
| | – If it is between 30 and 60L/min., it is orange | ○ 🟠 ○ |
| | – Else (if it is lower than 30L/min.), it is red | ○ 🔴 ○ |
| Hold the breath for 10 s | – The 3rd LED indicates the success or failure of holding the breath for 10 s. If the user keeps the inhaler lid open for more than 10 s., it displays green | ○ ○ 🟢 |
| | – Otherwise (If the user closes the inhaler lid in less than 10 s.), it displays red | ○ ○ 🔴 |
| Deliver the overall performance | – The overall result of three steps is shown and flashing until the lid of inhaler is closed | e.g. 🟢 🟢 🔴 |

[2] Peak inspiratory flow rate (PIFR), the measure of a patient's inspiratory effort, can be used to assess a patient's ability to generate adequate inspiratory flow rate from DPIs such as the Ellipta. Adequate PIFR for Ellipta is over 60L/min and inadequate PIFR is under 30L/min [9].

**Table 4** Sound Samples (Factorial Design) & Evaluation (Likert Scale-5 Mean & SD)

| No | Factor 1: Each unit | Factor 2: Mid-point | Factor 3: End | Boring: Mean (Standard Deviation) | Tension: Mean (Standard Deviation) | Distraction: Mean (Standard Deviation) |
|---|---|---|---|---|---|---|
| 1 | + | + | + | 2.50 (1.27) | 3.19 (1.11) | 2.38 (1.50) |
| 2 | + | + | – | 2.81 (1.11) | 2.63 (1.15) | 2.88 (1.16) |
| 3 | – | + | + | 3.75 (1.24) | 2.94 (1.06) | 2.63 (1.15) |
| 4 | – | + | – | 3.75 (1.29) | 2.44 (0.81) | 2.25 (1.13) |
| 5 | + | – | + | 2.31 (1.01) | 2.94 (1.29) | 2.38 (1.03) |
| 6 | + | – | – | 2.38 (1.26) | 2.25 (1.07) | 2.31 (0.95) |
| 7 | – | – | + | 4.06 (1.00) | 2.37 (1.03) | 3.13 (1.15) |
| 8 | – | – | – | 4.19 (1.11) | 2.25 (1.48) | 2.00 (1.21) |

In terms of the position to embed LEDs, the bottom of the inhaler assistant device is best to be seen easily and quickly considering the user's viewing angle during inhalation.

## 2.2 *Auditory Interface Design*

Sound has many advantages in interaction design. In ergonomics, it is known that response time is faster by 0.03 s. when using sound than vision. Also, people perceive low pitch sounds less enjoyable and more complex. Moreover, in general, repeating fast-paced sounds make people feel the tension with high attention level [10]. Our inhaler assistant device needs to help users in holding their breath for 10 s. We designed several sound samples with three factors in mind: unit alert, mid-point alert, and end alert. They all aim not only to be informative, but also to be motivational and enjoyable. To understand these three factors' influence on how users feel motivated and enjoyable, we set an experiment of three-way factorial design.

The three-way factorial design has three grouping factors (independent variables: unit alert, mid-point alert, and end alert) and three evaluation criteria (dependent variables: Feel bored, tensed, and distracted). Subjects heard eight sample sounds (composition of three factors, each of which is two levels) for 10 s and evaluated all eight samples (within-subjects). We asked them to evaluate how much they feel bored, tensed, and distracted on Likert Scale-5. All 16 people took part in our test and the order of presenting sound samples was random for each participant. The least boring sounds are No. 5 and No. 6, which alert each unit and not in the mid-point. When there are all types of alerts, people feel most tensed. On the contrary, people feel least distracted when there is no alert (see Table 4).

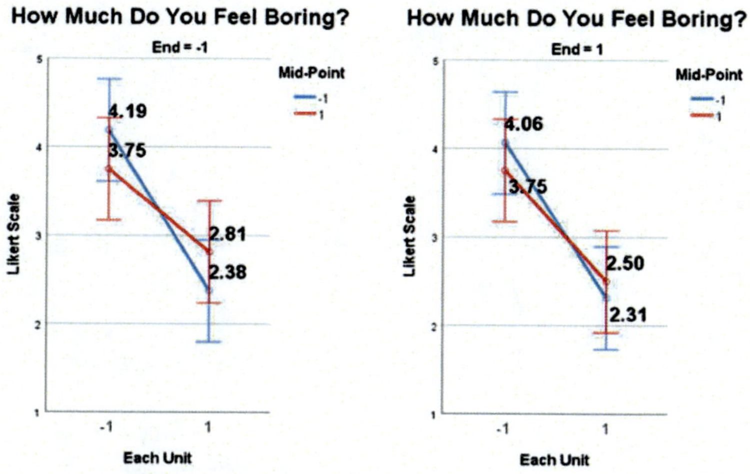


**Fig. 3** Three-way ANOVA on boring feeling

The result shows that only the boring feeling varied depending on sounds (p-value 0.000), not the tension (p-value 0.151) nor the distraction (p-value 0.155). Subsequently, a three-way ANOVA for further analysis (dependent variable = feeling, three factors = Unit alerts, Mid-point alert, end alert) gave the following results.

Figure 3 indicates clearly that people felt less bored when there were unit alerts (p-value 0.000). Slight interaction effect of (unit alert)*(mid-point alert) (p-value 0.098) was also shown, which means that people felt less bored when the sound notified each unit but adding mid-point notification increased slightly the boring feeling. Similarly, a two-way ANOVA on feeling tensed showed that when there was an end alert, people felt more tension (p-value 0.021). In terms of distracted feeling, it showed an interaction effect of (unit alert)*(end alert) (p-value 0.020), which means that people felt more distracted when there was end alert and when there was no unit alert. However, it didn't influence when there were unit alerts (Fig. 4).

Accordingly, motivational and attentive sound was designed. To avoid feeling boring, a beep alert is given every second from 1 to 6 s followed by a beep-beep note from 7 to 9 s to form tension and not to feel distracted. Finally, long beep note assures that they reached the end of 10 s breath-holding. In addition, the success alert sound and failure alert sound for the overall performance were used with intuitive and concise scales to reduce boredom. Thus, the success alert sound was decided as the rising note of C code and the failure alert sound as the falling note of F code. At this point, the note of failure sound was composed of a tone going down so that it felt like it was ending rather than a negative warning because it might let the users to lose the will to use the device.

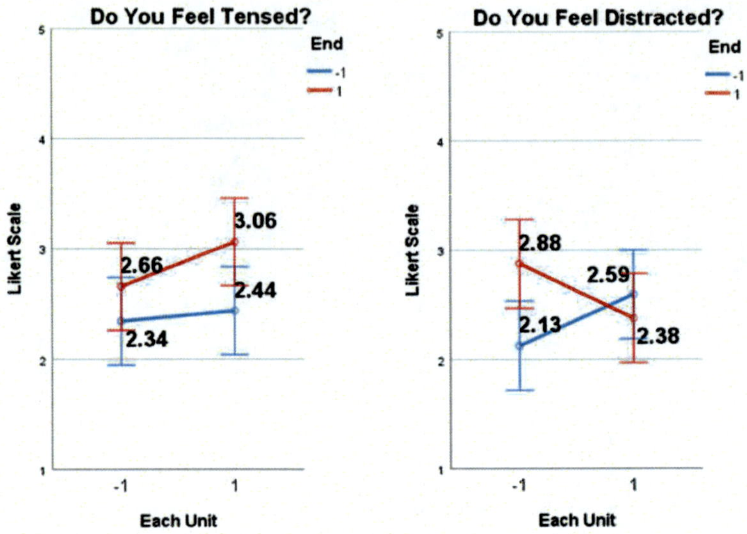


**Fig. 4** Two-way ANOVA on feeling tension and distraction

# 3 Usability Test

## *3.1 Purpose and Procedure*

To validate our concept of the asthma inhaler assistant device, we recruited those who are actual asthma patients or suffer from similar diseases and let them try our working prototype. We wanted to find people as close as possible to our target user persona (patients in early stage of asthma who need to learn correct inhaler usage) but due to COVID-19 it was very difficult to find people who accept to try our working prototype which necessitates inhaling. Finally, five subjects shown in Table 5 participated in the test. Five is not sufficient to draw statistical conclusion but is recommended as number of subjects to find maximum 75% of usability issues [11].

Each session was individual and semi-structured. It took around 60 min each. We began with basic questions about their past and current experience of asthma (their disease) and inhalers (their tools for treatments). Regardless of their prior experience of using the Ellipta inhaler, we began with a brief training on how to use the Ellipta correctly. Then, we introduced the new inhaler assistant device and let them discover by themselves using manual if necessary. After enough exploration, we asked them to try Ellipta, and this time with the inhaler assistant device (our working prototype). We observed them using Don Norman's Seven Stages of Action frame [12]: goal, plan, specify, perform, perceive, interpret, and compare.

**Table 5** Test participants

| No of subject & color | Gender | Age | Asthmatic symptoms | Inhaler user |
|---|---|---|---|---|
| #1 ■ | Male | 29 | N/A | No |
| #2 ■ | Female | 24 | 9 months | Yes |
| #3 ■ | Female | 28 | 10 years | Yes |
| #4 ■ | Female | 20 | 2 years | No |
| #5 ■ | Male | 27 | 25 years | Yes |

## *3.2 Usability Assessments and Issues*

**Understand the Goal, Plan, and Specify**. All users found easily that the two products assemble together for better use of the inhaler. Furthermore, patients with asthma for a long time quickly understood what these kinds of attachable devices do with inhalers. However, an issue arose when they actually tried to assemble. Two subjects (#1 and #3) out of five did it wrong. They tried to assemble in vertical direction. The right way of doing it is horizontal assembly as shown in Fig. 5. The outer curve cut in the same way with the Ellipta (go back to Fig. 2) was not evidently figured out by some users and the shape needs to improve the detail reflection of the original outer lines of the main inhaler to signify how the two parts can be combined (Fig. 6).

**Perform and perceive**. When the two devices are assembled, they are asked to use the inhaler as naturally as they can. They could explore the printed manual to understand more of the assistant device and three times. We measured their performances at three steps and each time both quantitatively and qualitatively.

Firstly, does the inhaler assistant device help users to hold the inhaler horizontally? The proportion of time they kept the horizontal balance over time they tried to hold the inhaler might give a sense of it. The data show dramatic improvement in performance (see Fig. 7). For example, subject #4 achieved 100% horizontal posture every time she tried with the assistant device compared to 10% when she used Ellipta inhaler

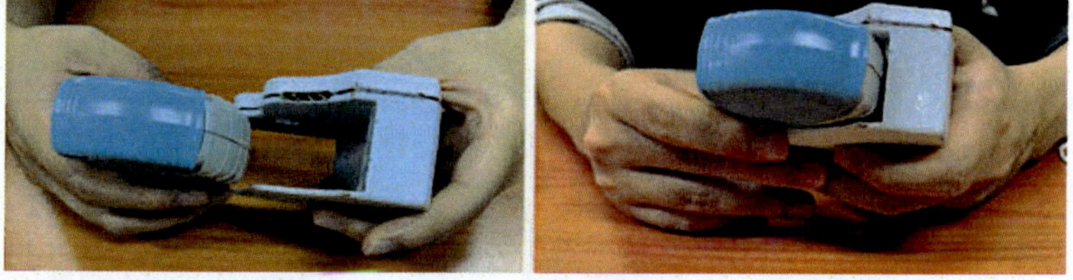

**Fig. 5** Assemble the Ellipta inhaler and the assistant device

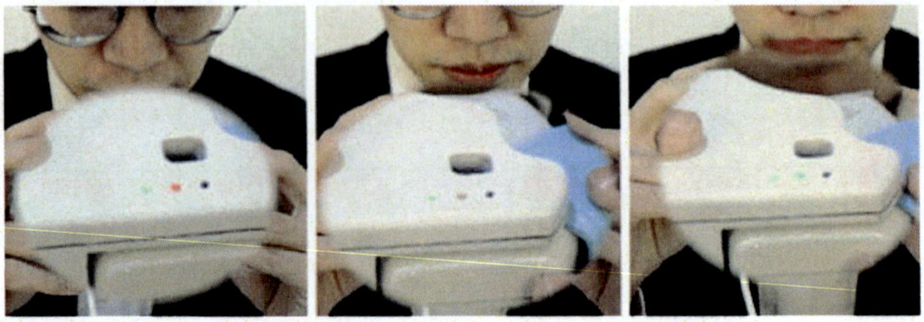

**Fig. 6** Use the Ellipta inhaler with the assistant device

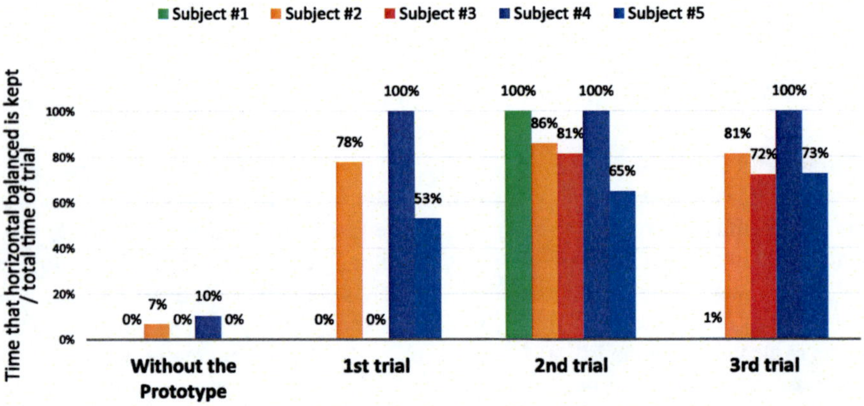


**Fig. 7** The proportion of time they held the inhaler horizontally over time they tried to hold

alone. Subject #2 and #5 also showed much better performance when they used the assistant device. Overall, the proportional time they kept the balanced posture is more than 19 times (65.95% compared to 3.42% in average) when they used the inhaler with the assistant device than when they used the inhaler alone.

Secondly, does the inhaler assistant device help users to inhale deeply enough? Fig. 8 shows the data of three subjects (Subject #1 and #3 did not want to actually inhale due to the caution of COVID-19). They all performed better judging from the PIFR (peak inspiratory flow rate) when they used the inhaler with the assistant device attached to it. On the 3rd trial, they all reached maximum and two of them far exceeded the criterion value (88.1L/min. and 73.93L/min. vs. 60L/min.) and the rest one almost reached this level (57.93L/min.), which is very encouraging. A usability issue was found at this step, though. Some of them misunderstood the meaning of the 2nd LED color as amount of the drug intake while it means the degree of inhalation power.

Finally, does the inhaler assistant device help users holding the breath for more than 10 s? Three out of four reached the goal as is shown in Fig. 9. Subject #2 did not reach the goal but improved after the 2nd trial. We may expect most users understand

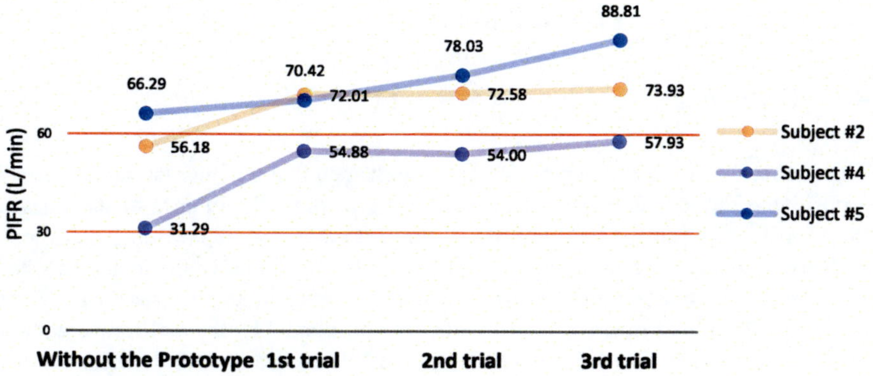


**Fig. 8** Peak inspiratory flow rates

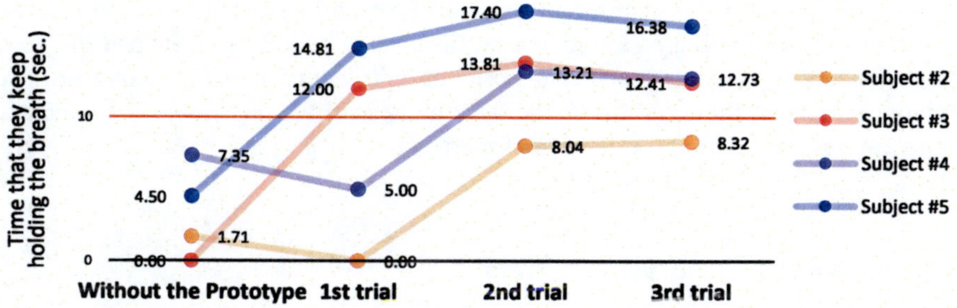


**Fig. 9** Longest time of breath-holding (exceeding 10 s. is recommended)

and reach 10 s when they get accustomed to the device in person. The sound alert was quite intuitive to lead 10 s. Usability issues at this stage lied on the sound quality (too sharp), which can be better with hardware improvement.

**Interpret and Compare with the Goal**. When more than two steps out of three are successful, the overall result is given with positive sound feedback. Otherwise, it is regarded as failed. We asked whether the final sound feedback was communicated as we intended. Subjects were aware of the difference of the designed overall result sound feedback in terms of notes. But they did not really match each sound to the success or failure. It seemed that it was understood differently by person. Furthermore, after the breath-holding sound for 10 s, the overall evaluation sound just followed afterward, which was not clearly communicated as a separate message to the users. Therefore, we decided to change the overall result feedback sound to one note, F code, to convey only the meaning of simple termination of the inhaler use process and device use.

# 4 Conclusion and Limitation

## 4.1 Conclusion

The qualitative and quantitative evaluation confirmed that the interaction design had a positive effect on improving the actual usage of the DPI. In case of the Ellipta, the front face should be held upward to prevent powdery drugs from leaking out. However, most users were not aware of it. With the inhaler assistant device attached to the inhaler, all five test subjects used the inhaler correct direction to see the LEDs. In addition, all subjects made sure the inhaler was horizontal by the real-time feedback of the 1st LED indicator. As they were trying with the assistant device multiple times, the proportion of the time that they held the inhaler horizontally over the time that they held the device for the inhalation improved. Moreover, the PIFR increased compared to when they were using the Ellipta alone. This inhalation technique improvement was also achieved by the real-time feedback of the 2nd LED indicator. Finally, the designed breath-holding sound increased the rate at which they reached the 10-s goal. Thanks to the sound, subjects were naturally counting each second in their minds. Moreover, they said the sound gave a cheerful feeling. Subjects remember this experience as both informative and motivational.

## 4.2 Limitation and Future Work

Our results rely on casual usability testing. We deliberately did not analyze the data statistically in the previous chapter but showed only the brief pattern that we got from the small set of data. Although performance measures were better after two or three times of use, real-time feedback using LED indicators sometimes engendered misunderstanding of the meaning or misleading of the behavior. Since horizontal holding posture is important for guaranteeing about a dosage of inhalation, the color of 1st LED changes instantly during inhalation. Without understanding that the color meant for the balance, some users hesitated. The color of 2nd LED appears when an inhalation is over because it means the PIFR. Therefore, when subjects saw the 2nd LED in orange or red, they tended to repeat inhalation rather than moving to the next step (hold breath). It might make them inhale more than the recommended dose and at the same time may lead to insufficient lung deposition because they do not hold breath.

To improve the understanding of the interaction, we plan to add icons on the front-facing surface of the device. They make sure to use the inhaler in a correct way throughout the three progressive steps (see Fig. 10).

In addition, resolving design issues found on usability test is an important aspect for future study. For instance, there is an insight that red or orange color of 2nd LED misled users to repeat inhalation. Users are inclined to perceive red and orange color as failure of step, which rather motivates users to try again. Instead of just changing

**Fig. 10** Icons to be added on the surface of the product to represent three steps

the current color composition of LED, future study should investigate how to create an interactive environment where users don't make the mistake. With more continuous use flow through software improvement, the interaction environment would prevent the users leave from correct usage path.

**Acknowledgements** This work was supported by University Innovation. (No. 2020 University Innovation-132, University Innovation (Hongik University)).

## References


1. Chronic respiratory diseases: Asthma (n.d.) Retrieved from https://www.who.int/news-room/q-a-detail/chronic-respiratory-diseases-asthma, last accessed 2020/12/21.
2. Asthma pill tested against inhalers (2017, August 21) Nhs.UK. https://www.nhs.uk/news/heart-and-lungs/asthma-pill-tested-against-inhalers/, last accessed 2020/12/21.
3. Newman SP, Weisz AW, Talaee N, Clarke SW (1991) Improvement of drug delivery with a breath actuated pressurised aerosol for patients with poor inhaler technique. Thorax 46(10):712–716. https://doi.org/10.1136/thx.46.10.712
4. Anderson WC, Gondalia R, Hoch HE, Kaye L, Szefler SJ, Stempel DA (2019) Screening for inhalation technique errors with electronic medication monitors. The Journal of Allergy and Clinical Immunology: In Practice, 7(6), 2065–2067. https://doi.org/10.1016/j.jaip.2019.02.006
5. Ltd BPG (2012) Improving inhaler technique – who needs teaching? Drug Ther Bull 50(10):109
6. Levy ML, Carroll W, Izquierdo Alonso JL, Keller C, Lavorini F, Lehtimäki L (2019) Understanding Dry Powder Inhalers: Key Technical and Patient Preference Attributes. Adv Ther 36(10):2547–2557. https://doi.org/10.1007/s12325-019-01066-6
7. How to use an Ellipta inhaler. (n.d.). Asthma UK. Retrieved from https://www.asthma.org.uk/advice/inhaler-videos/ellipta, last accessed 2020/12/21.
8. Gibbsons S (2018) Journey Mapping 101. Nielsen Norman Group. https://www.nngroup.com/articles/journey-mapping-101/ last accessed 2020/12/21
9. Grant AC, Walker R, Hamilton M, Garrill K (2015) The ELLIPTA® Dry Powder Inhaler: Design, Functionality, In Vitro Dosing Performance and Critical Task Compliance by Patients and Caregivers. J Aerosol Med Pulm Drug Deliv 28(6):474–485. https://doi.org/10.1089/jamp.2015.1223
10. Bomin K, Joo-Yeon L, Kyung-Won C (2008) A study on the Suitability of Sound Feedback in Product User Interface. Archives of Design Research 21(5):8
11. Nielsen J (2000) Why You Only Need to Test with 5 Users. Nielsen Norman Group. Retrieved from https://www.nngroup.com/articles/why-you-only-need-to-test-with-5-users/ last accessed 2020/12/21
12. Norman D (2013) The Design of Everyday Things: Revised and Expanded Edition. Basic Books

# Exposure and Evaluation of Different Indoor Localization Systems

**Youssef Ibnatta, Mohammed Khaldoun, and Mohammed Sadik**

**Abstract** Indoor location attracts a lot of attention because people spend their maximum time indoors. As well as the loss of GPS signal power through walls, which requires the integration of other approaches to locate passengers indoors. With the technological growth of telecommunications systems such as 5G and the development of the Internet of Things technique, indoor location is becoming an applicable reality, in order to improve the services offered indoors. In addition, ensuring security in public environments such as airports, train stations, shopping malls, supermarkets… In this paper, we present results acquired during the bibliography phase, before studying more than 78 approaches dominant in the literature. The majority of indoor positioning techniques only use accuracy as a criterion that determines the quality of the approach, while system performance is strongly related to other criteria, such as energy consumption, stability, cost, response time, measurement heterogeneity, environmental change, and others. For this reason, we evaluate the proposed techniques of different categories of indoor positioning systems on these criteria. This document is structured around the following points:

- We expose the different techniques of indoor positioning
- We present the different criteria for indoor locations
- We evaluate indoor positioning techniques

**Keywords** RSS · PDR · IMU · Localization by vision · RFID · WKNN · Kalman filter · RMSE · UWB

Y. Ibnatta (✉) · M. Khaldoun · M. Sadik
Department of Electrical Engineering, NEST Research Group ENSEM, Hassan II University, Casablanca, Morocco
e-mail: youssef.ibnatta@ensem.ac.ma

M. Khaldoun
e-mail: m.khaldoun@ensem.ac.ma

M. Sadik
e-mail: m.sadik@ensem.ac.ma

 
X.-S. Yang et al. (eds.), *Proceedings of Sixth International Congress on Information and Communication Technology*, Lecture Notes in Networks and Systems 216,
https://doi.org/10.1007/978-981-16-1781-2_64

# 1 Introduction

With the technological development of mobile devices, such as smartphones, tablets, etc., the application of a high-quality indoor positioning system has become more reliable and less expensive. These devices contain a wide of sensors, which we can use to deduce the positions of customers, among these sensors, accelerometer, gyroscope, magnetometer, compass, and pedometer.

This set of measurements is called IMU inertial measurements, mobile devices equipped with a network card can use the Wi-Fi protocol to determine the positions of targets, also the Bluetooth low-energy communication technology (BLE) is widely used, moreover the availability of cameras in mobile devices, it will allow us to realize a location by vision. Light is the technology of the future because it ensures data security, simple to use, easy to deploy, powerful… This communication tool can guide us to have a high precision positioning, the laser rangefinder is also used in this technology but remains limited in specific conditions. Communication by radio waves is the most useful tool in indoor positioning, RSS (Received Signal Strength) is a localization technique either Wi-Fi or Bluetooth based; consists in comparing the received powers for each access point, in order to build the radio card offline, the powers received during the online phase are compared with the offline database to determine the customers' positions. RFID radio wave identification can also be used to identify the positions of passengers inside a building by a simple TAG/READER communication.

In this article, we focus on the exposure and evaluation of different indoor positioning systems, for this, we ask the following questions: can we say that a system is efficient because only its accuracy is high or medium? If not, what are the other criteria for choosing a good indoor positioning system?

In the following, Sect. 2 presents the major constraints of indoor positioning systems, Sect. 3 presents the evaluation criteria for indoor positioning systems, Sect. 4 presents the different indoor location classes, Sect. 5 presents the evaluation of indoor positioning systems, Sect. 6 presents the discussion, for the rest, Sect. 7 begins our future work, Sect. 8 presents the conclusion.

# 2 The Major Constraints of Indoor Positioning Systems

In indoor positioning technology, the majority of techniques are influenced by several constraints which break their qualities, among these constraints:

- Multiple path effect (NLOS) and direct path conditions (LOS) not detected.
- The instability effect of the measures.
- The penetration effect of radio signals.
- Signal interference.
- The shading effects.
- The effect of light on vision-based localization systems.

- Card Constraints.

## 3 Evaluation Criteria for Indoor Positioning Systems

The choice of a location system, the most reliable within an environment, depends on the objectives of use. For example, if we want to guide passengers to specific areas or help people to navigate indoors, this does not require high accuracy, however, the approach response time must be taken into account, as a simple delay destroys the localization objective, as well as energy consumption and other criteria must also be taken into account. We are not talking now about localization, but earlier about the goal of localization, because if we lose the money and time to provide a few features that will be unusable in practice, and that will not change the quality of the system, this is not science. In this framework, we quote some criteria that control the performance of the systems, among these criteria:

- Precision
- Energy consumption
- Cost
- Easy to deploy
- Stability
- Response time
- Adaptation to changes in the environment and the heterogeneity of devices.

## 4 Exposure of the Different Indoor Location Classes

Indoor localization is a great challenge in front of the constraints that influence the qualities of the systems. Researchers are proposing several solutions to solve localization problems, we classify these techniques into the following six categories:

- RSS-based approaches (Wi-Fi or Bluetooth)
- Triangulation and trilateration-based approaches
- PDR or IMU-based approaches
- Approaches of identification by communication Tag/Reader
- Localization by Vision
- Multimodal location.

The RSS-based approaches are the most usable in indoor positioning techniques, thanks to the simplicity of deployment, low cost, but they suffer from calibration problems, change of environment, heterogeneity of power values received by different brands of smartphones. In general, RSS-based approaches are divided into two categories, one with site survey and one without site survey. For techniques that use site survey [10], TILOC aiming to improve the robustness and accuracy of indoor fingerprint-based localization, the author applied a robust principal component analysis, to form the fingerprints offline/online to reduce the noise of sparse peaks, TILOC still uses the WKNN algorithm to deduce the client location, the simulation error is less than 3 m.

For approaches without site survey, Walki-Markie [16] uses the data collected by the crowd of sensors, this method automatically generates the floor plan without prior knowledge of the floor plan by using Wi-Fi, the error for 90% of CDF is between 1.65 and 2.9 m. Semantic-SLAM [1] uses SLAM technology to form the database automatically; therefore, the author merged all inertial sensors such as accelerometer, gyroscope, compass with Wi-Fi, to improve the system the author used particle filter, the average error is 0.53 m.

Mathematical games of antenna locations, such as triangulation and trilateration, have attracted the attention of researchers, since the birth of indoor location technology because they can provide low-cost positioning with three or more antennas. In this category, the approaches use communication between antennas to infer the user position, among those methods that use triangulation and trilateration are TOA, TDOA, AOA, ADOA. In [2], the author tried to correct the errors produced by the multipath effect, and the undetected direct path conditions, by using the TOA method merged with UWB, the multipath error decreased to 5.23 m. MOSAIC [22] is a framework for joint localization and mapping, the concept is based on a few steps based on receiver localization, followed by VPN estimation and then mapping and dimensioning of obstacles, the EKF (Extended KALMAN filter) model was adopted to improve the accuracy of mapping of reaching obstacles, triangulation was used to determine the target location, the RMSE is less than 0.7 m.

The use of inertial sensors in indoor location is attracting the attention of researchers. However, these sensors not only suffer from the constraints of measurement instability but also the deviation of heading angle given by the gyroscope, again the problem of calibration of accelerometer measurements; moreover, the magnetometer is influenced by ferromagnetic metals existing inside buildings. In [7], the high-precision localization and calibration of magnetic positioning systems were proposed, by using magnetic coils which certainly found a link between all terrestrial magnetic fields and positions. The author succeeded in creating a mathematical model, the established error of distance between 0.3 and 0.8 m and the angle between 2.8° and 9.7°. SMARTPDR [9] which follows pedestrians through a typical approach to calculating the number of deaths using data from inertial sensors embedded in smartphones, the proposed method is to find a correlation between the gyroscope and the magnetometer, in order to eliminate heading errors, accelerometer measurements are used to calculate the step length, both tools are encapsulated in a mathematical formula, in order to deduce the customer position, the error is less than 2 m. Telecommunications systems suffer from signal penetration constraints, wave interference, and attenuation. In the literature, some approaches use protocols for target identification using simple communication between antennas. In [6], a system investigated the possibility of identifying and locating passive tags by adapting a single millimeter-wave (mm-wave) RFID reader, performing beam steering, the approach evaluated the achievable distance in terms of packet error probability, comparing the results with those achieved by RFID-UWB at microwave frequencies, the RFID technique with UWB based on microwave frequencies requires a pulse count of $4.10^3$ to achieve an identification distance of 6.5 m, whereas the method proposed with mm-wave-based RFID can achieve the same identification distance with $5.10^2$ pulses. Such a request for pulses is also accompanied by a location error.

The proposed approach overrides that of the UWB thanks to the less stringent EIRP requirements, the approach also gives the possibility of locating a label with a single reader. Indeed, in a noiseless scenario, the error range is about 4 cm but with an angle error of about 3.7°. The two types of errors are joined in RMSE of value 50 cm. In brief, with $4.10^4$ impulses, the RMSE can reach a few centimeters. MIMO implementations are of several types that differ in the way the data is sent and the number of antennas used at the transmitter and receiver. For example, the implementation of Wi-Fi with the Beamforming system, which uses two antennas at the transmitter and receiver, in order to improve the quality of the SNR signal-to-noise ratio as well as the transmission rate, which involves reducing the emission level of radio signals in order to reduce the surrounding electromagnetic pollution and also to extend the battery life in the case of a smartphone. In [27], the author focused on active beamforming to reduce user equipment location error for mm-wave MIMO systems. Such beamforming for localization is difficult because of its real user optimization cost function and instantaneous channel states, which are unknown in advance. To meet this challenge, a new SLAB localization and successive beamforming scheme are proposed, the approach achieves an error of less than 1 m.

With the proliferation of mm-wave systems and visible light communications (VLC), indoor location can find multiple applications. When high localization accuracy is required, and triangulation is not possible due to infrastructure and scenario limitations, the computational complexity of the virtual grid searching can become excessive. In [3], the author merged mm-wave-based localization in uplink and VLC in downlink. The author used quantum search algorithms to reduce the computational complexity required to achieve optimal performance based on the full search. For the uplink localization, the author used a single anchor with multiple antennas. Triangulation was used based on VLC for the downlink, there was an average positioning error of 5.6 cm. In [25], VIVID a vision-based inland navigation system for intelligent devices. The system adopts a cloud-assisted edge computing architecture to distribute the network throughput and offload computing tasks from a centralized cloud server to edge nodes. The author also proposed a discrete grid-based map to eliminate the need for domain knowledge about the environment. Also, an automatic map maintenance mechanism based on deep learning to mitigate human involvement in map updates due to environmental changes. The maximum error of VIVID is 60 cm. In [5], an indoor robot VLP visible light localization system based on ROS (Robot Operating System) is presented, aiming to promote the application of VLP in a natural robotic system. The author has innovatively designed a VLP localization package which contains the basic operational control of robots, extraction, and recognition of LED-ID functionalities, the localization accuracy is less than 1 cm, the average response time is 0.08 s (Table 1).

**Table 1** Constraints of different indoor location approaches

| IPS constraints | Multipath | Instability | The Penetration effect | Signal interference | Light effect | Card Constraints |
|---|---|---|---|---|---|---|
| PDR OR IMU | | x | | | | |
| Wi-Fi-based RSS | x | x | x | | | x |
| Bluetooth | x | | x | | | x |
| RFID | | | x | | | |
| Triangulation | x | x | x | | x | x |
| UWB | | | x | x | | |
| Vision and light | | | | | x | x |
| ZigBee | | x | x | | | |
| Ultrasound | x | x | x | | | |

Indoor positioning approaches suffer from the impact of imprecision and measurement instability, which requires opening up to other indoor positioning strategies. Some researchers have proposed to merge the fundamentals of localization techniques in order to take advantage of the performance of each method and reduce their errors. In [4], VINAV detects the trajectories of users and adding to the 3D model. In practice, data from sensors including accelerometer and gyroscope readers are used as a basis to detect the trajectories of users. These trajectories are automatically calibrated by photos taken along the way, Wi-Fi fingerprints collected along user tracks are automatically georeferenced and added to the 3D model for quick location. Barometer readers along user tracks are also collected to detect connection paths between multiple floors. Finally, the POI (Points of Interest) information their corresponding locations are computed based on the visual data from the CROWDSOURCING and the generated 3D model, the author used three algorithms to deduce the positions such as KNN closest neighbor algorithm, DIJSKASTRA algorithm, DUCKHAM algorithm. The method error is less than 2 m, the Wi-Fi fingerprint error is less than 10 m for 90% of the cases. In [13], a new PDR method assisted by VLP visible light positioning on smartphone was studied. The author proposed a new calculation method for the PDR heading angle by calculating the extreme values of the data points which may decrease the probability of positioning error of the traditional PDR method. On the other hand, the author has developed the double-checking method consisting of CRC and range constraint to improve the accuracy of VLP to obtain the absolute position. the approach applies both the heading angle and CRC improvements, the method shows much better performance than traditional PDR and Wi-Fi techniques. The mean error is less than 1 m. In [21], an indoor visual location method has been proposed to achieve continuous and accurate indoor location based on image matching. This method uses smartphones to collect multi-sensor data, including geotagged images in a dataset. Several image search strategies based on spatial constraints are also designed to further reduce computational time. The visual localization error is 0.4 m. The average accuracy is 1.3 m and the computation time is 0.59 s.

## 5 Evaluation of Indoor Positioning Systems

In this section, we evaluate the different categories of indoor positioning systems based on the criteria presented in sections 3. In this framework, we have examined more than 78 approaches classified into 6 categories. Table 2 presents the evaluation of indoor location approaches divided into six categories.

## 6 Discussion

In Table 2, we compare several approaches to the different categories of indoor locations, based on the criteria presented in Sect. 3. In [8], the system has an error of less than 30 cm as well as low response time, but high energy consumption, difficult to deploy, unstable, not adaptable to changing environment. In [9], SMART-PDR has an average error of 1.35 m, with low power consumption, low cost, simple to deploy, low response time, but the approach not adapting to changing environment and heterogeneity of devices, unstable. Reference [26] has high precision, on the other hand, her energy consumption is high, not adapting to the change of environment, not adapting to the heterogeneity of the devices, her cost is high. From the data displayed in Table 2, we can say that the choice of a location system does not only depend on accuracy but also other evaluation tools that determine the performance of the system. For this reason, we cannot speak about localization but rather about the objective of localization, because this is the only way that determines the nature and values of the selection criteria presented in Sect. 3, therefore, we obtain a high-performance system.

## 7 Future Works

In our future perspective, we will study the qualities and shortcomings of the UWB (Ultra-Wide Band) modulation technique, and we will generate a general view on the UWB based indoor localization approaches available in the literature. In the second step, we will elaborate on a fingerprint RSS-based localization system. We will improve different handling criteria (accuracy, stability, simplicity of deployment, response time, cost...) in order to have a high-performance system, for this we will integrate the UWB technique into the system.

**Table 2** Evaluation of indoor positioning systems

| Approach title | Accuracy | Energy consumption | Cost | Simple to deploy | Stable | Response time | Adapting to a Changing environment | Adapt to device heterogeneity |
|---|---|---|---|---|---|---|---|---|
| Multi-phase Fingerprint [30] | The average error is 1,87 m | Low | Low | Yes | unstable | Low | Yes | – |
| WinIPS [29] | The average error is 1,718 m | High | Medium | Yes | Yes | – | Yes | Yes |
| A Map-Free [8] | The size of area is 1335 × 700 $cm^2$ error is less than 30 cm | High | – | No | No | Low | No | – |
| Mosaic [22] | The size of area is 10mx10m the error is centimetric | – | High | No | No | Low | No | – |
| G-LOC [17] | The accuracy is 4,1 m | High | Medium | Yes | Yes | Low | No | No |
| Optimal Target Tracking [20] | The Office size is $324m^2$, 10 Aps for 1000 samples error is 0.4729 m | High | Low or Medium | Yes | No | Response time is 1,094 s | Yes | No |

(continued)

**Table 2** (continued)

| Approach title | Accuracy | Energy consumption | Cost | Simple to deploy | Stable | Response time | Adapting to a Changing environment | Adapt to device heterogeneity |
|---|---|---|---|---|---|---|---|---|
| Signle-and Multiple-Access point [14] | The size of area is 18mx16m for 5 APs and 80% of CDF error is less than 1 m | High | High | No | – ——— | Low | No | – |
| Accurate Doa Estimation with AAPD [28] | The average error is 1,9044° | – | Medium or High | Yes | Medium | t = 45 ms, response time is low | No | – |
| High-precision localization and calibration [7] | The step error is between 0,3 m and 0,8 m, angle error is between 2,8° and 9,7° | – | Medium | Yes | No | – | No | Yes |
| SMART-PDR [9] | The average error is 1,35 m | Low | Low | Yes | No | Low | No | No |
| Millimeter wave Beamsteeing [6] | RMSE = 50 cm | Low | High | No | Yes | Low | Yes | Yes |
| Successive Localization [27] | The error is less than 1 m | – | High | No | – | Low | No | – |
| Vivid [25] | The error is 60 cm | High | High | Medium | – ——— | Low | Yes | – |
| AVF [19] | The average error is 1,32 m | High | High | Medium | No | Medium or High | Yes | – |

(continued)

**Table 2** (continued)

| Approach title | Accuracy | Energy consumption | Cost | Simple to deploy | Stable | Response time | Adapting to a Changing environment | Adapt to device heterogeneity |
|---|---|---|---|---|---|---|---|---|
| Quantum-Assisted IPS [3] | The size of room is 15mx15m, the number of APs is 64, the average error is 30 cm | LED power = 5,5 mW, AP power = 20 W | High | Medium | No | Low | – | – |
| High-Accracy Robot [5] | The average error is 0,82 cm | High | High | Yes | Yes | Low | No | – |
| Improved Smartphone [21] | The error is less than 1 m | Medium or High | Low | Yes | No | Low | No | No |
| Mpiloc [15] | For 37 Aps the average error is 1,82 m | High | Low or Medium | Yes | No | Low | Yes | No |
| Gradient-Based Fingerprint [18] | The error is less than 4 m for 80% of CDF | High | Low or Medium | Yes | No | Low | Yes | Yes |
| Enhancing camera [26] | For 92% of CDF the error is 0,2 m | High | Medium or High | Yes | No | Low | No | No |

## 8 Conclusion

In this paper, we have focused on the exposure and evaluation of the different categories of indoor positioning techniques, which is divided into the following six classes: RSS-based techniques, triangulation and trilateration-based approaches, PDR-based approaches, identification techniques, localization by vision, and other multimodal. We evaluated these techniques on certain criteria which are accuracy, energy consumption, cost, simplicity of deployment, stability, response time, adaptation to changing environment, adaptation to the heterogeneity of devices. Accuracy is not only a criterion for choosing a good positioning system but also other criteria that determine the performance of systems. A strategy for the development of RSS-based positioning systems, using UWB pulse-width modulation, has been proposed in order to correct the problems of instability, accuracy, device heterogeneity, change of environment, and other constraints of this positioning technique.

## References

1. Abdelnasser H, Mohamed R, Elgohary A, Alzantot MF, Wang H, Sen S, Choudhury RR, Youssef M (2015) Semantic-slam: Using environment landmarks for unsupervised indoor localization. IEEE Trans Mob Comput 15(7):1770–1782
2. Alavi B, Pahlavan K (2006) Modeling of the toa-based distance measurement error using uwb indoor radio measurements. IEEE Commun Lett 10(4):275–277
3. Botsinis P, Alanis D, Feng S, Babar Z, Nguyen HV, Chandra D, Ng SX, Zhang R, Hanzo L (2017) Quantum-assisted indoor localization for uplink mm-wave and downlink visible light communication systems. IEEE Access 5:23327–23351
4. Dong J, Noreikis M, Xiao Y, A. Yl¨a-J¨a¨aski. Vinav (2018) A vision based indoor navigation system for smartphones. IEEE Transactions on Mobile Computing, 18(6):1461–1475
5. Guan W, Chen S, Wen S, Tan Z, Song H, Hou W (2020) High-accuracy robot indoor localization scheme based on robot operating system using visible light positioning. IEEE Photonics J 12(2):1–16
6. Guidi F, Decarli N, D. Dardari, F. Mani, Errico RD (2018) Millimeter-wave beamsteering for passive rfid tag localization. IEEE Journal of Radio Frequency Identification, 2(1):9–14
7. Hehn M, Sippel E, Carlowitz C, Vossiek M (2018) High-accuracy localization and calibration for 5-dof indoor magnetic positioning systems. IEEE Trans Instrum Meas 68(10):4135–4145
8. Ji Y, Hejselbæk J, Fan W, Pedersen GF (2018) A map-free indoor localization method using ultrawideband large-scale array systems. IEEE Antennas Wirel Propag Lett 17(9):1682–1686
9. Kang W, Han Y (2014) Smartpdr: Smartphone-based pedestrian dead reckoning for indoor localization. IEEE Sens J 15(5):2906–2916
10. Li H, Qian Z, Tian C, Wang X (2020) Tiloc: Improving the robustness and accuracy for fingerprint-based indoor localization. IEEE Internet Things J 7(4):3053–3066
11. Li P, Yang X, Yin Y, Gao S, Niu Q (2020) Smartphone-based indoor localization with integrated fingerprint signal. IEEE Access 8:33178–33187
12. Liu M, Wang H, Yang Y, Zhang Y, Ma L, Wang N (2018) Rfid 3-d indoor localization for tag and tag-free target based on interference. IEEE Trans Instrum Meas 68(10):3718–3732
13. Luo C, Hong H, Chan MC, Li J, Zhang X, Ming Z (2017) Mpiloc: Self-calibrating multi-floor indoor localization exploiting participatory sensing. IEEE Trans Mob Comput 17(1):141–154
14. Palacios J, Bielsa G, Casari P, Widmer J (2019) Single-and multiple access point indoor localization for millimeter-wave networks. IEEE Trans Wireless Commun 18(3):1927–1942

15. Rezazadeh J, Subramanian R, Sandrasegaran K, Kong X, Moradi M, Khodamoradi F (2018) Novel ibeacon placement for indoor positioning in iot. IEEE Sens J 18(24):10240–10247
16. G. Shen, Z. Chen, P. Zhang, T. Moscibroda, and Y. Zhang. Walkiemarkie: Indoor pathway mapping made easy. In Presented as part of the 10th fUSENIXg Symposium on Networked Systems Design and Implementation (fNSDIg 13), pages 85–98, 2013.
17. Y. Shu, P. Cou´e, Y. Huang, J. Zhang, P. Cheng, and J. Chen. G-loc: Indoor localization leveraging gradient-based fingerprint map. In 2014 IEEE Conference on Computer Communications Workshops (INFOCOM WKSHPS), pages 129–130. IEEE, 2014.
18. Y. Shu, Y. Huang, J. Zhang, P. Cou´e, P. Cheng, J. Chen, and K. G. Shin. Gradient-based fingerprinting for indoor localization and tracking. IEEE Transactions on Industrial Electronics, 63(4):2424–2433, 2015.
19. F. Vedadi and S. Valaee. Automatic visual fingerprinting for indoor image-based localization applications. IEEE Transactions on Systems, Man, and Cybernetics: Systems, 2017.
20. Wang C, Luo J, Zheng Y (2018) Optimal target tracking based on dynamic fingerprint in indoor wireless network. IEEE Access 6:77226–77239
21. Wang Y, Zhao H (2018) Improved smartphone-based indoor pedestrian dead reckoning assisted by visible light positioning. IEEE Sens J 19(8):2902–2908
22. Yassin A, Nasser Y, Al-Dubai AY, Awad M (2018) Mosaic: Simultaneous localization and environment mapping using mmwave without a-priori knowledge. IEEE Access 6:68932–68947
23. Yin Z, Jiang X, Yang Z, Zhao N, Chen Y (2017) Wub-ip: A highprecision uwb positioning scheme for indoor multiuser applications. IEEE Syst J 13(1):279–288
24. Zhang C, Subbu KP, Luo J, Wu J (2014) Groping: Geomagnetism and crowdsensing powered indoor navigation. IEEE Trans Mob Comput 14(2):387–400
25. Zhao W, Xu L, Qi B, Hu J, Wang T, Runge T (2020) Vivid: Augmenting vision-based indoor navigation system with edge computing. IEEE Access 8:42909–42923
26. Zhao Y, Xu J, Wu J, Hao J, Qian H (2019) Enhancing camera-based multimodal indoor localization with device-free movement measurement using wifi. IEEE Internet Things J 7(2):1024–1038
27. Zhou B, Liu A, Lau V (2019) Successive localization and beamforming in 5g mmwave mimo communication systems. IEEE Trans Signal Process 67(6):1620–1635
28. Zhou X, Chen L, Yan J, Chen R (2020) Accurate doa estimation with adjacent angle power difference for indoor localization. IEEE Access 8:44702–44713
29. Zou H, Jin M, Jiang H, Xie L, Spanos CJ (2017) Winips: Wifi-based non-intrusive indoor positioning system with online radio map construction and adaptation. IEEE Trans Wireless Commun 16(12):8118–8130
30. Zuo J, Liu S, Xia H, Qiao Y (2018) Multi-phase fingerprint map based on interpolation for indoor localization using ibeacons. IEEE Sens J 18(8):3351–3359

# Metric Analysis of Big Data of Payment System Based on Kolmogorov–Shannon Coding Methods

**Ivan G. Kamenev and Daria A. Andrianova**

**Abstract** The article describes the problem of research of "raw" socio-technical big data. We explain the features of big data in comparison with other areas of data analysis, bringing the terminology of data science into line with the terminology of social statistics. The article shows some problems that limit the application of algorithmic methods of big data research (machine learning, data mining, etc.) to social data. We explain the necessity of the development of additional preprocessing and exploratory research methods for such data. We propose to use the method of metric data analysis (MMDA), which includes entropy-optimal stochastic encoding and approximation methods for a set of data by metric Kolmogorov-Shannon nets. The proposed approach is illustrated by an example of the real big data flow research: large payment system. It is shown how the metric analysis allows us to detect various errors in the databank, form a general idea about the flow, segment it, and formalize some forecasting problems solvable by machine learning methods.

**Keywords** Multidimensional statistical analysis · General population · Sociological sample · Metric net · Data visualization · Data analysis · Data optimization · Method of metric data analysis · Decisive minorities · Raw data · Data mining

## 1 Introduction

Many industries are increasingly dependent on big data (hereinafter, BD). Created and accumulated by business practice, BD is of great practical interest for this practices' optimization. However, such data usage poses increasing challenges for data

I. G. Kamenev (✉)
Federal Research Center of Computer Science and Control Dorodnitsyn,
Computing Centre of the Russian Academy of Sciences, Moscow, Russia

National research university, Higher school of economics, Moscow, Russia

D. A. Andrianova
Saint Petersburg, Yandex N.V., Russia

 
X.-S. Yang et al. (eds.), *Proceedings of Sixth International Congress on Information and Communication Technology*, Lecture Notes in Networks and Systems 216,
https://doi.org/10.1007/978-981-16-1781-2_65

science. Traditional social statistics methods are only partially applicable here due to the problems with data quality and reliability of conclusions (see Sect. 3). Automation and algorithmization of BD usage created new areas of data science, in which terminology and methodological apparatus are determined by the logic of software development.

The fundamental scientific challenge in this area includes the necessity, on the one hand, to state for algorithms a highly clear formulation of the problems, and, on the other hand, to give a meaningful interpretation of the identified dependencies. Both require a good understanding of the data structure, the optimization problems to be solved, and the format of the forecast to be formulated.

As a result, BD is actively used with mostly technical systems with a small number of characteristics that are massively replicated for stable reproducible tasks (for example, image recognition). It is much more difficult to use BD from systems with a strong social aspect, i.e., data on human behavior. The formulations of the tasks to be solved in this case change rapidly, as does the composition of relevant characteristics. Thus, such researches contain conclusions practically significant for the researcher but with very uncertain limits of their applicability beyond the research itself.

In this work, we use the method of metric data analysis (MMDA [8]), which complements the traditionally more developed methods in both social statistics and algorithmic data science for the measures (probabilities, densities, averages, etc.) with a new metrics (topology, combinations of data characteristics) analysis apparatus. Our research shows that the metric properties of BD allow us to answer many research questions and, more importantly, establish a general understanding of the data. This allows us to formulate specific forecasting and optimization problems that can be solved by machine learning technologies.

This article illustrates the BD analysis with an example of BD of a certain payment system for almost 12 years of its existence. It records payments, including a large number of characteristics of the payer, recipient, and the transfer itself. Research questions for these BD are vaguely formalized and constantly change as the payment system develops. Therefore, a meaningful analysis of such BD required us to use new methods of data analysis.

## 2 Features of BD as a Data Type

There are several approaches to BD definition. Some sources define them only by the amount of information (number of records) criterion [5], others by the complexity of the structure (data that does not fit into the logic of standard databases) [6], and others refer to arrays that analysts are not able to process in traditional ways [7]. Combining these features, we define BD as the data sets obtained from the data flow of the socio-technical system (being continuously updated, supplemented, and modified), so-called "log" of a technical system that is a by-product of the various social systems' activities, which can not be processed nor visualized by standard methods due to the large volume (more than 1 million records) and real multidimensionality (at least 4 data characteristics). Any BD array is a sample from the log, for example, a log on a specific date.

Structurally, big data is a sequence of records (rows), each of them contains a row's ID and recorded information represented as a hard-defined set of characteristics. For example, information about transactions of payment system's clients forms BD. BD usage entails a number of fundamental problems:

- sampling nature of any BD array with unknown characteristics of completeness and reliability;
- collection of many BD characteristics in a format not adapted for statistical analysis; inability to visualize large multidimensional data;
- overlooking of significant (decisive) minorities with specific combinations of characteristics in statistical analysis (while using general characteristics such as sample average value, dispersion, distribution by decile groups, etc.);
- formation of a false majority in two-dimensional data projections and in their statistical analysis (correlation coefficients, etc.);
- presence of a large number of exceptions (errors, atypical row, etc.) among the data, requiring interpretation.

All these circumstances make it fundamentally impossible to apply traditional statistical methods to BD. The solution for these problems is currently known only in special cases and mainly engineeringly (without the measurement of completeness, accuracy, reliability, etc., see Sect. 3). We will propose new methods for big data analysis with references to their mathematical basis, weakening these fundamental problems, and demonstrate the proposed approach using the example of payment data analysis in a payment system.

The researched object is the data flow of payments in the payment system. The flow volume is $10^8$ records annually, and each payment in the system has more than 20 characteristics. This volume of flow makes it impossible to use standard analytical tools for analysis, except for the simplest statistical calculations (min/maximum, average, correlations, etc.). A number of more complicated questions appear, including: Was any data lost or recorded incorrectly? Does the system track effectively the user's identity and his activities' factors? All the questions are vaguely formalized, and it is difficult to set a machine learning task for them. Therefore, we need to form a general understanding of the data structure, their interrelation, and features of their recording and storage.

## 3 Traditional Methods of BD Analysis and Their Applicability Limits

Processing, analysis, and interpretation of BD can be described by the following logical scheme:

1. Forming the BD array (sampling from the BD flow).

2. Automated preprocessing of the BD.
3. Automated statistical analysis of the BD.
4. Automated BD modeling.

Let's look at each of these stages in more detail.

1. Since the BD array is by definition a sample from the general population (the data flow of the socio-technical system), the content of the BD array is initially determined by the sampling method. This sampling can be done by row numbers (by date). Covered period should be well-justified and sufficiently long, but small enough, so it is realistic to combine all the data into a single array. Then the array can be considered census (i.e., observation of the entire General population of $G$). Otherwise, randomized sampling can be applied (including $M$ records from the flow sampled by random numbers), assuming the random nature of $G$ and the independence of the characteristics distributions. There are some risks associated with randomized sampling. Many data flows are non-uniform, which leads to distortions in frequency analysis. However, it allows to cover longer periods of time and check the universality of flow features.

Distributed computing, non-relational databases, and systems like *hadoop* and *ClickHouse* that combine distributed computing and data storage on different clusters are frequently used to create and store BD arrays. These systems are convenient for data accumulation and storage. But the graphical interface is either missing or displays only a small part of the information about the array which leads to the necessity of a large number of queries for pattern recognition.

2. Preprocessing (manual or automated) involves converting the BD array to a technical format and interpreting or correcting any gaps and errors. Some unformatted and distorted characteristics in some rows are almost unavoidable in any real BD flow, so this stage plays a key role in these problems' elimination. Preprocessing is often performed several times as long as new problems are identified at subsequent analysis stages.

3. Statistical analysis of BD is usually performed in the same systems where the array integration was performed. There is a general limitation of the classical apparatus of statistical analysis of BD: automated commands are specialized at generalized characteristics calculation, rather than research of their internal metric features (impossible combinations of characteristics, outliers, etc.). The higher the dimension of BD, the less standard statistical methods are suitable for research (less informative and, more importantly, more resource-intensive). For example, if the computational complexity of commands for a single characteristic is proportional to $N$, where $N$ is the number of records (rows), then the complexity of the clustering is proportional to at least $N^2$. Their application to BD samples with cardinality $10^6$ or more and more than 2–3 significant characteristics is possible only with computing time's manual restriction.

4. The data (in particular, BD) model is a system of equations and inequalities that describes (approximates) the entire data array. This system should reflect the properties of the data essential for the researcher. It consists of a combination of object's characteristics (corresponding to BD dimensions) linked by some coefficients. The

types of equations are set by the researcher. In general, these are linear equations or equations of low degrees, but in great numbers.

A complex computer algorithm identifies the model by setting the coefficients of the equations—that, in fact, is the so-called machine learning. One-time execution of machine learning gives usually disappointingly low-quality predictions, so the researcher has to repeat it, correcting the algorithm manually, or using deep learning to adjust algorithms automatically. Both ways require a clear statement of the problem and an intuitive understanding of BD properties. Thus, 2–3 stages of the mentioned scheme are usually mentioned as data mining, and stage 4 as machine learning.

Social data are usually multidimensional, so the machine learning tasks are harder, while algorithm testing on smaller samples is incorrect due to the heterogeneous intensity of the initial BD flows. Preliminary multidimensional metric data analysis allows us to solve this problem. For this purpose, we suggest to apply the method of metric data analysis (MMDA) [8]. It should complement the existing methods of preprocessing and statistical analysis of BD. Thus, using data science terminology, it can be qualified as a method of data mining.

## 4 Method of Metric Data Analysis

The metric analysis is a methodological approach which includes the research of the data's topological and metric properties (the distribution of observations in the metric space of characteristics). This approach is mostly developed in the (Engineering) Sciences, where it is used to solve standardized problems in small-dimensional space (for example, in image recognition). Multidimensional metric analysis, on the contrary, remains the area where the application of existing theoretical concepts is limitly developed, especially for Social Sciences.

In this study, we will demonstrate the method of Metric data analysis (MMDA), which, in turn, is an integration of the family of specialized methods, developed by our scientific school over the past three decades. Let us briefly consider the general logic of the method. The mathematical basis of the method consists of constructing of $(\varepsilon, \delta)$-nets and $(\varepsilon, \delta)$-coverings, introduced by Claude Shannon [2], for the general population (transactions in the data flow). The universal method for "coding" or "tabulating" nonconvex sets is the construction of metric $\varepsilon$-nets [3]. Geometrically, this means that the set is approximated by a collection ($\varepsilon$-covering) of simple figures (balls in the Euclidean metric, or cubes in the Chebyshev's metric (as in this research), or any other relevant metrics), whose diameters are taken smaller to achieve a higher approximation accuracy. A metric $\varepsilon$-nets of any set is its subset with $N$ points for which it is true that any point of the set lies at the distance of at most $\varepsilon$ from the subset.

Bounded sets can be approximated by applying adaptive stochastic techniques based on the Deep Holes method (DHM) [10]. We omit the mathematical details, but it is essential that these methods allow us to obtain not just some net, but the net with known characteristics of completeness $\eta$ and precision $\varepsilon$ and reliability. The completeness of the approximation is characterized by the fact that only a small part (a part with a small measure) of the approximated set lies outside the constructed collection of balls. The collection of balls with accuracy $\varepsilon$ and completeness $\eta$ forms so-called $(\varepsilon, \delta)$-coverings, $\delta = 1 - \eta$. The optimal complexity of $\varepsilon$-nets constructing is characterized using the concepts of metric $\varepsilon$-entropy and in small by metric (fractal) dimension [3]. Metric (fractal) dimension also characterizes the reliability of conclusions about the $G$ from the $(\varepsilon, \delta)$-covering as multidimensional sample.

For visualization purposes we use Interactive Decision Maps (IDM, [9]): 2D and 3D maps and atlases of multidimensional explicitly given sets. The full algorithm of the Metric data analysis method includes ([11]) 6 stages:

1. Collection of data about characteristics.
2. Selection of relevant characteristics.
3. Creation of the multidimensional space (metrics).
4. General population reconstruction by $(\varepsilon, \delta)$-covering approximation: determining the reliability, accuracy, and completeness of the representation of the General population by the sample.
5. Structural and visual analysis of $(\varepsilon, \delta)$-covering approximation: identification of types, classes, and separate groups based on metrical proximity.
6. Generalization of characteristics and study of the behavior of selected classes.

## 5 The Construction of the BD Array of Payments

In this research, we analyze the BD flow of some payment systems, which is a collection of records about characteristics of payments. The flow capacity is $10^8$ annually, and each payment in the system has several dozen characteristics. This publication presents the main results (including applied results as well as our method fully adapted to the specifics of BD) of the research, the first approaches to which were described in ([12]) and some other reports. The BD flow is recorded on several servers, coming from different sources. It is caused by different interfaces of operations (website, mobile application, plastic cards, contactless payments by phones, etc.), as well as by the usage of different payment gates for acquiring. Operation times are brought to a single standard: the time of logging on the server. Logs store information about payments, transfers, users and their accounts, cash register data, properties of accounts, and cards, however, for our research, only payment data and some non-personalized account data are used.

Each payment in the system has up to 17 key characteristics: transaction identifier, date and time of the transaction, unixtime of the transaction, where the payment came from, where payment goes, payment purpose, payer type, payee type, payment method, payment sub-method, payer's account type, payment status, operation interface, operation value, operation (net) income, accrued bonuses for operation, (partner's) commission from operation, payer's identifier, wallet number (id, date/time, unixtime, payer, receiver, purpose, payer type, receiver type, method, submethod, status, interface, sum, profit, bonus, commission, all payment, payer entity id, account; these shorter variables' names are highlighted in the article below).

Based on the available computing power, we decided to form a randomized by payment ID sample $M \sim 5 * 10^6$ ($\sim 4 * 10^6$, taking into account the drop-down empty id ranges). It was carried out by the *Python* module "random", using *random.seed()function* [13]. Before data analysis we made some adjustments. Confidential data (user id, store id etc.) where removed or substituted. *date/time*, for the convenience of quantitative analysis, was converted into 5 different indicators: *year*, *month*, *day*, *hour*, *dayoftheweek*. 5 calculated characteristics were added to the array: the number of user's active months, the number of months from the first payment, the user's turnover for all time, the total user's income for all time, the total number of user operations: *activity age*, *month from first payment*, *all sum*, *all profit*, *all trans*.

The metric dimension test [12] has shown that this sample size is not sufficient to make informed judgments ($\varepsilon \sim 0.2$) about the combinations of all characteristics (as dimensions of a multidimensional space), but it is convenient to operate with 5–10 characteristics with $\varepsilon \sim 0.0001$. (Note that each combination of characteristics had to be tested independently).

## 6 Preprocessing and Statistical Analysis of the Payment BD Array

All quantitative indicators during preprocessing were normalized to the maximum (i.e., converted to unit fractions) to protect commercial secrets. Years were renumbered from 1 to 12. For all characteristics of the array M, simple statistical tests were performed: minimum, maximum, frequency analysis.

Since big data of natural origin can contain any kind of technical errors, failures, unique operations, data analysis can be meaningful only after eliminating all these outliers. Therefore, all dysfunctional or abnormal records were moved from the general array to a separate array. Sometimes it was possible to restore the missing values (for example, due to the fact that the dating characteristics are redundant). The total number of excluded entries was 11304. For such outliers, the BD logs were checked for their prevalence and a meaningful interpretation was given. Further analysis was performed without outliers.

The metric analysis also showed that in some cases, logs may not contain reliable information about the payer's account. For example, if a payment was made via a browser, the payer's account information is completely lost if the payer used incognito mode. Therefore, we decided to put in a separate array and research independently those payments that were made by the payer for whom only one payment was registered. There were 756275 such samples.

There are many variables with nominal scales among the array characteristics. For them, we had to check if all their ranks actually occur in the array, so they

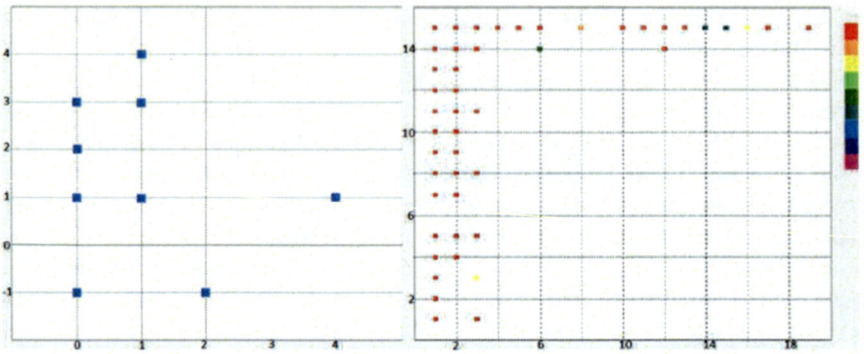


**Fig. 1** On the left: X-axis: payer, Y-axis: payer's account type. On the right: X-axis: payment method, Y-axis: payment interface, the color indicates the day of the week (from above, i.e.. non-red color means that there are no such payments on Sunday)

do not distort the metric. Accordingly, ranks not found in the array were checked against logs and excluded. Another feature of characteristics with natural scales is their artificial assignment by overlapping/interrelated classifications. To reduce the dimension of the metric, such related characteristics were combined. Let's illustrate the possibilities of metric analysis for combining interrelated characteristics using the example of combining payer and *payertype* (see Fig. 1 on the left). It can be seen that 1 and 2 categories of payer are divided into subcategories by type of payer, and 3 and 5 have an unambiguous interpretation of the type. Most significantly, the comparison of these two classifications eliminated the uncertainty that existed in the BD flow. For example, the value Missing for payer type could have a different interpretation, but in combination with Security and Support in payer, it is clearly interpreted as payments from the account of the payment system administration to the user as some compensation (see Table 1). Let's also clarify that the term payment gate in the payment system refers to payments received not directly from the payer (for example, en route through another payment system).

## 7 Metric Properties of the Payments BD

It is technically possible to visualize the entire sample M, but building such a visualization in a three-dimensional or larger projection and/or visualizing a separate section of M consumes too much computing power and time. A full visualization of a two-dimensional or three-dimensional projection of the M array can take 5–10 mins (this time has to be spent every time the projection axes and/or metric constraints are changed). It is possible to visualize already suspected patterns on M patterns, but not to detect them (which requires testing many hypotheses). Thus, we built exploratory

**Table 1** Re-coding of combined characteristics *payer* and *payer type* (account)

| *Payer* New | *Payer* Source | *Payertype* |
|---|---|---|
| Unknown payment gate | Payment gate | Unknown |
| Credit card through payment gate | Payment gate | Main current account |
| Technical account for a payment of unauthorized user from payment gate | Payment gate | Quick-account (technical account for a single operation) |
| Temporary corporate account for funds placement | Payment gate | Light-account (temporary account for funds placement) |
| Main current account | User | Main current account |
| Temporary user account for funds placement | User | Light-account |
| Temporary bearer call account | User | Light-call account |
| Another | Another | Unknown |
| Security and Support account | Security and Support | Unknown |

approximations for several key combinations of attributes, aiming to identify significant patterns in the data using limited time before overall visualization.

Exploratory approximations were made by metric nets with a capacity of up to $N = 10^5$ for groups of 3–5 characteristics. Such net size does not give a small enough $\varepsilon$ to make reasonable conclusions about the metric properties of $M$ or $G$ ($\varepsilon \sim 0.2$). However, most of the metric properties are already visible with this net quality, which allowed us to identify significant properties of the metric. Then, for relevant combinations of characteristics, we constructed approximations with $\varepsilon \sim 0.001$ or less ($N$ up to 500000), if necessary with restrictions on other characteristics (for more detailed study of local sections of space).

During the exploratory approximations, some features of the database were revealed. They can be divided into four types:

1. Misrepresentation of metrics and measures caused by key clients

Metric analysis has shown that several large corporate clients (unique ID in $G$, qualified in $M$ without ID by a unique combination of total payment, total profit, and total number of payments) strongly distort the metric and measure of BD. Obviously, such large clients are served personalizingly, under special contracts and tariffs. However, they are included in all statistics (averages, variances, etc.) calculated by analysts and marketing services of the payment system. Thus, clearing statistics of their contribution significantly change any commercial conclusions. For example, in Fig. 3 on the right, most of the observations outside the lower-left square [0.1, 0.1] disappear. Details of this effect will be omitted in this publication, as they are the subject of a separate study.

2. Technical errors in BD flow:

For example, the interface of one of the mobile applications (see *interface* 6 in Fig. 1 on the right) turns out to not identify the user, although the specifics of

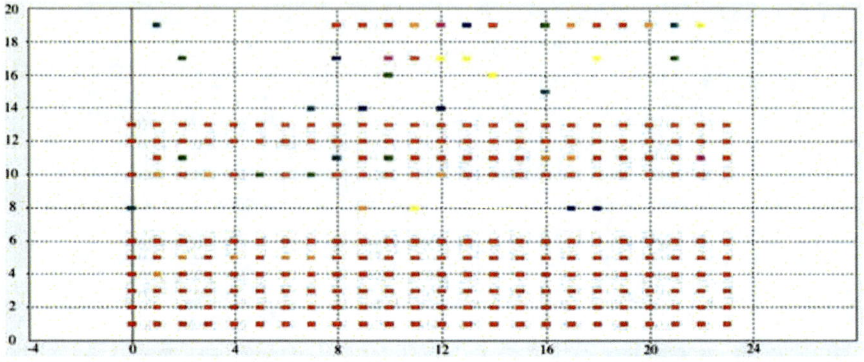


**Fig. 2** X-axis: hour when the payment was made, Y-axis: payment method and the color indicates the day of the week (from above)

this platform require unambiguous identification. As a result, all operations on this interface were among single payments. Some other problems with identifying unique users were also identified: some payment methods and purposes only have unique payers. In Fig. 2, *methods* 7, 9, 18, and 20 − 22 are missing—they are "single" (not associated with the user profile) in $M$ and $G$. Metric analysis of other characteristics shows that the user accounting system remains stable: if some methods disappear from the array of "single" payments in later years, it is only due to obsolescence and disappearance in the $G$ flow, and not due to the transition to "multiple" ones.

3. Some trends were revealed and a general understanding of the interrelation between payment characteristics was achieved. This includes changes in the structure of revenue and turnover.

Note a methodological problem that requires a particularly careful interpretation. The BD array $M$ is a random sample from the BD flow $G$, and constructed metric nets are a probabilistic construction on $M$. Therefore, any conclusions (from any methods of BD analysis including MMDA) about the properties of the BD array cannot be extended to the original BD stream. Theoretically, MMDA allows you to get around this problem by dividing an array into a test and a control ones. Then the characteristics of the metric net can be checked directly on the $G$ (data flow). In this study, we instead checked our conclusions directly on the source flow $G$ which is also methodically correct, since it is simple enough to make an automated statistical query from the opposite.

Ignoring this type of verification completely may lead to incorrect conclusions. Let's see an example on Fig. 3 on the left. Firstly, you can note the constant presence of small payments that are not monetized. With new services introduction, as long as the payment system develops, the values and profits become more diverse, but the figure shows that this process is heterogeneous (for example, in years NoNo6 − 8, small variety is approximated). A check on $G$ shows that the first two conclusions are correct, but the third is invalid. Since the intensity of the payment flow increased

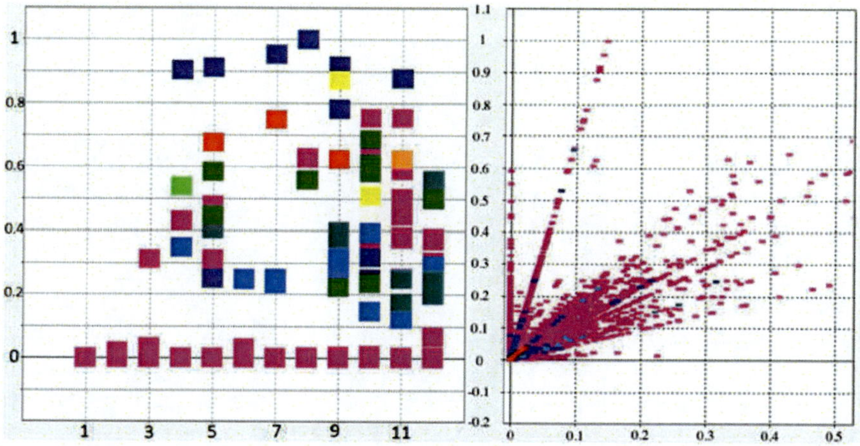

**Fig. 3** On the left: X-axis: year (number), Y-axis: payment value, the color indicates the scale of net income from the payment (from above); On the right: X-axis: net revenue from the payment, Y-axis: payment value and the color indicates the turnover for the entire user's history (from above)

dramatically (for example, for years 6 and 11, they differ by more than an order of magnitude), the amount of payments in $M$ is smaller for earlier years, missing some of the characteristics' variety. The drop in diversity in years $6-8$ turned out to be a sampling artifact.

Visualization of $M$ also allows us to see the profitability coefficients of operations that have been executed over the entire history and at the same time learn the characteristics of users who performed these operations. For example, users who bring the largest total revenue to the payment system make small payments with moderate commissions (see the red line in Fig. 3 on the right). Clearly visible lines in the figure correspond to the commission rates. We omit some trivial properties, known long before and confirmed in the results of metric analysis.

4. Features of the distribution of payments over time:

Working hours differ from non-working hours and weekends in several ways: payments by some payers and for some purposes are made only on weekdays during business hours; the maximum payment value on weekdays during business hours is significantly higher (see Fig. 4); a number of interfaces are only used during business hours. Note that these conclusions are not trivial, because the payment system operates in territories in different time zones, while the payment time is recorded as server time, not user time. A meaningful interpretation of this phenomenon is possible, but cannot be presented in an open publication.

Due to the limited possibilities of visualization in print, we present visualization in the form of projections. However, as stated in the examples above, metric properties are usually identified by a combination of 4–6 characteristics.

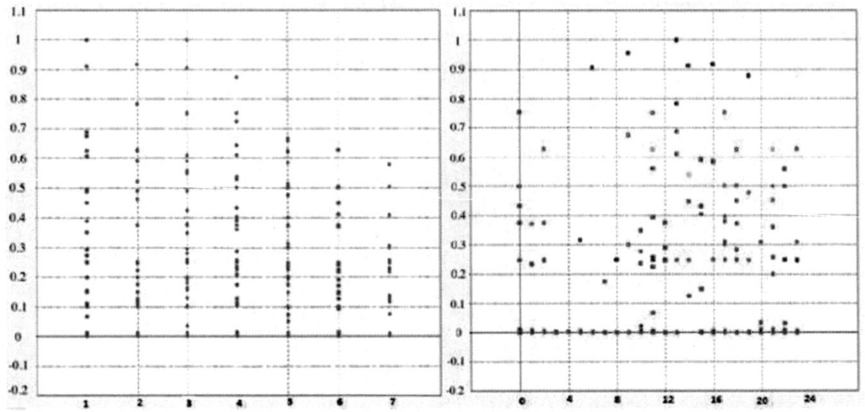

**Fig. 4** On the left: X-axis: day of the week, Y-axis: value of payments. On the right: X-axis: time (payment hour), Y-axis: payment value and the color indicates the net income (from above)

# 8 Conclusions and Available Problem Statements for Automated BD Modeling

The presented study shows the advantages of metric analysis (in particular, the Method of Metric data analysis, MMDA) usage for socio-technical systems' Data Mining. The method, including logical and software algorithms, allows to solve applied problems inaccessible for conventional data mining methods, as well as to measure the completeness and reliability of the conclusions. In particular, it is shown that this method allows us to process the data with non-uniform intensity, to identify compact minorities, and to eliminate many consequences of an inefficient data structure. The algorithms are optimal in terms of the calculation speed, which allows us to test a large number of hypotheses without making special requests to the data flow.

The search analysis of data presented in this publication allowed us to identify a number of problems in the payment registration system and pointed out the fallacy of some marketing hypotheses. Additionally, it also allowed us to form a general understanding of the data structure sufficient for machine learning tasks setting. Here are highlighted highly formalized questions, the answer to which can potentially be obtained from this BD source using algorithmic methods:

- is it possible to reliably determine whether the payment is the first one for the payer, having all the information about the payment, but not about the payer?
- is it possible, having all the information about the payment, to determine the erroneous nature of the recording of its individual characteristics?
- is it possible to reliably restore various information about individual payment's characteristics based on other characteristics when the information is lost?

- is it possible to reliably determine whether a profitable or unprofitable client makes a payment with all the information about the payment, but not about the payer?
- is it possible to reliably determine whether the payer is a person or a corporation (including financial control purposes), having all the information about the payment, but not about the payer?
- is it possible to predict the dynamics of significant characteristics of particular groups of payers (values, number of payments, profitability) for future periods?

The application of metric analysis for measure analysis and clustering tasks remains outside the scope of this publication and should be presented separately.

**Acknowledgements** The reported study was co-funded by RFBR, project number: 18-01-00465 a: Development of social data multidimensional metric analysis methods.

## References

1. Lotov AV, Bushenkov VA, Kamenev GK (1999) Feasible Goals Method: Mathematical Foundations and Environmental Applications. The Edwin Mellen Press, Lewiston, NY USA
2. Shannon C.: The Mathematical Theory of Communication. The Bell System Technical Journal, **27–28**, pp. 379–423, 623–656 (1948)
3. Kolmogorov AN, Tikhomirov VM (1959) Epsilon-entropy and Epsilon-capacity of sets in functional spaces. Advances in mathematical Sciences 25(2):3–86
4. Kamenev GK (2001) Approximation of Completely Bounded Sets by the Deep Holes Method. Computational Mathematics and Mathematical Physics 41(11):1667–1675
5. Leetaru K. How Do We Define Big Data And Just What Counts As A Big Data Analysis? https://www.forbes.com/sites/kalevleetaru/2019/01/09/how-do-we-define-big-data-and-just-what-counts-as-a-big-data-analysis/#13d886081b66
6. Ward J. S., Barker A. Undefined by data: a survey of big data definitions //arXiv preprint arXiv:1309.5821. (2013)
7. Davenport Thomas H, Barth Paul, Bean Randy (2012) How Big Data Is Different. MIT Sloan Management Review 54(1)
8. Kamenev, G.K., Kamenev, I.G. Multidimensional statistical sets and their metric analysis, Proceedings of the Institute for System Analysis of the Russian Academy of Sciences CC FRC of CSC of the RAS, **68**(2), 30–33. https://doi.org/10.14357/20790279180207 (2018)
9. Lotov, A.V., Bushenkov, V.A., Kamenev, G.K. Interactive Decision Maps. Approximation and Visualization of Pareto Frontier, Applied Optimization **89** Kluwer Academic Publishers. Boston / Dordrecht / New York / London (2004)
10. Kamenev GK (2001) Approximation of Completely Bounded Sets by the Deep Holes Method. Computational Mathematics and Mathematical Physics 41(11):1667–1675
11. Kamenev, G.K., Kamenev I.G. Metric analysis of multidimensional sociological samples [Metricheskij analiz mnogomernyh sociologicheskih vyborok] // Pospelov, I.G. et al. (edditors). Proceedings of the conference "Modeling of coevolution of nature and society: problems and experience. To the 100-th anniversary from the birthday of academician N.N. Moiseev". [Modelirovanie koevolyucii prirody i obshchestva: problemy i opyt. Trudy Vserossijskoj nauchnoj konferencii] M.: FITZ of RAS, 2017. pp. 198–209

12. Kamenev I.G., Andrianova D.A. Mathematical and statistical methods of pre-processing and exploratory analysis of Big social Data on the example of payments stream analysis [Matematicheskie i statisticheskie metody predobrabotki i poiskovogo analiza bol'shih social'nyh dannyh na primere analiza potoka platezhej]. Proceedings of the 62nd all-Russian scientific conference of MIPT, Applied mathematics and computer science [Trudy 62-j Vserossijskoj nauchnoj konferencii MFTI] (2019). Moscow: MIPT (2019). 40–42
13. Python 3.8.3 documentation: random – generate pseudo-random numbers https://docs.python.org/3/library/random.html

# Dates Fruit Classification Using Convolution Neural Networks

**Haya Alaskar, Shaikah Alhewaidi, Bayan Obaid, Ghadah Alzahrani, Aisha Abdulahi, Zohra Sbai, and Thavavel Vaiyapuri**

**Abstract** Date is a fruit rich in dietary fiber with global commercial value due to its nutritional and medicinal value in curing and preventing various diseases. The grading and sorting process of date fruits is still a complex task in the industry requiring a great deal of human expertise. The breakthrough of deep learning in computer vision has opened up new path for various industrial applications. Therefore, the objective of this paper is to employ convolutional deep learning network and develop an improved automatic sorting system for classifying the images of three main varieties of date fruits, namely, Ekhlas, Nbute Sultan, and Shayshi. To this end, a convolutional neural network (CNN) was designed and trained from scratch to efficiently learn and discern the different varieties of date fruits with high accuracy. The generalization ability of the developed CNN architecture was improved with a combination of convolutional and max-pooling layers. Added to, it was trained from scratch with nearly 3165 date fruit images which include 1055 images of each three varieties. A comparative analysis with different gradient descent optimization methods on the developed CNN architecture is presented. The obtained results testified the efficacy and generalization ability of the developed CNN architecture to discriminate date cultivars with higher accuracy. Therein, the integration of the developed CNN architecture in a suitable affordable framework, will make it possible to reduce the inconsistency and lag of human labor by raising the accuracy and speed of the process.

**Keywords** Dates Processing for Industry 4.0 · Dates fruits Classification · Image processing · Gradient optimization algorithm · Deep learning · Convolutional neural network

H. Alaskar (✉) · S. Alhewaidi · B. Obaid · G. Alzahrani · A. Abdulahi · Z. Sbai · T. Vaiyapuri
College of Computer Engineering and Sciences, Prince Sattam bin Abdulaziz University, Alkharj, Kingdom of Saudi Arabia
e-mail: h.alaskar@psau.edu.sa

Z. Sbai
National Engineering School of Tunis, Tunis El Manar University, Tunis, Tunisia

 
X.-S. Yang et al. (eds.), *Proceedings of Sixth International Congress on Information and Communication Technology*, Lecture Notes in Networks and Systems 216,
https://doi.org/10.1007/978-981-16-1781-2_66

# 1 Introduction

Date palm trees are not only the major fruit crops cultivated in the regions of the Arabian Peninsula, North Africa, and the Middle East but also are considered as the most valuable natural resources in these regions [1]. They have also been introduced over the last three centuries to new production areas in many countries across the world such as the United States, South America, Australia, Southern Africa, India/Pakistan, and Mexico [2, 3]. Further, in many countries, dates are regarded as the primary source of income and staple food in the main market. As a result, it plays significant roles in those countries' economies, communities, and environments. The annual global dates production was documented at 7.5 million tons in 2008 [4–6].

The date fruit is a prehistoric fruit providing nutritional value with symbolic significance. In modern world, it is defined as a dietary food of many nations and is also transformed into several other food products with strong economic significance into global market [7, 8]. Hundreds of varieties with different textures, colors, nutritional values, and flavors are available for use in food processing. It has been recorded that worldwide greater than 3000 date varieties are grown [9–12]. Moreover, there are several value-added products that can be produced from date fruit such as jelly, jam, confectionery products, chutney, pickles, chutney, date butter, date-in-syrup, date bars, candy, and date flour [13–16]. Yet, the community is more interested to investigate and discover more products from dates varieties.

The name of the date palm tree was believed to be derived from the Greek word, "phoenix" that means red or purple fruit, and "dactylifera" which means the fruit bunch appears like fingers [17]. The ripening of data fruits passes through four distinct stages which in Arabic are termed as, "Kimri," "Khalal," "Rutab," and "Tamar," These four distinct stages describe the immature orange, mature full color, soft brown, and hard-raisin texture, respectively. Most of the date varities are harvested at Rutab, Khalal, and Tamr stages considering as the best time for harvesting irrespective of its variety [18]. For date fruits, the process of post-harvesting involves fumigating, washing, sorting, grading, glazing, weighing, and packaging [19, 20]. The sorting process is known in the agriculture industry to be a repetitive procedure [21]. The recognition procedure of date variety is still done empirically based on the experience and knowledge of the examiners. It is manually done by human labor, who visually inspect dates one by one. Such a procedure is sluggish and expensive and needs especially skilled workers. Therefore, this manual classification system is laborious and is error-prone. In tandem to this, recently there is a common global trend to replace manual systems of classification with automatic and digitized ones [2]. Automated classification not only speeds up the sorting process but also leads to more efficient and more consistent results which could drastically improve the productivity level and economic growth [22]. In light of this, the industrial application of computer vision and sensor technology is growing at a rapid pace, receiving the attention of researchers to a greater extent to develop automated solutions, that can optimize and improve the industrial processes.

Inspired by the recent development in automated sorting solutions, this paper investigates the application of deep learning on dates fruit images with no prior feature engineering or feature selection step. This will allow the neural network to analyze and find the optimal parameters to differentiate between the three types of dates fruit. An efficient framework for dates fruit classification using CNN is proposed. Also, the optimization of the proposed network structure with five different methods, namely: Adaptive Moment Estimation (Adam), Stochastic Gradient Decent (SGD), Adamax, Adagrad, and Root Mean Squared Propagation (RMSProp) is presented.

The remainder content of the paper is arranged as follows. Section 2 presents the literature review for the state-of-the-art sorting systems. Section 3 introduces our methodology, proposed system, and the dataset. Section 4 presents and discusses the obtained experimental results. We conclude with Sect. 5 summarizing our findings and final remarks.

## 2 Literature Review

This section explores the state-of-the-art research done in the field of date fruits classification for grading or sorting purposes.

In recent years, several studies have been conducted to create automated date classification systems. These systems either categorizes the dates into different grades within the same variety or recognized the variety to which a given date belongs. In [23–25], the systems measured the quality and ripeness of the date fruits through the detection of cracks, moisture, or grading dates based on their colors. For instance, the study in [26] attempted to develop an automatic system to grade Berhee variety with three levels of maturity.

However, the literature includes very few publications dealing with the issue of classifying dates into different varieties. In [27], the researchers developed date fruit classification system using artificial neural networks (ANN) to sort seven date cultivars. They extracted physical features like moisture content and water movement and integrated hem with color characteristics. However, this method is hard to implement because it requires physical measurements. In [28], probabilistic neural networks with color descriptors in RGB space were used to classify five date varieties. The study in [29] developed date fruit grading system based on computer vision. In this research, the author proposed a real-life sorting system prototype to sort the dates through a computer-based system. Here, the system comprises two key modules, namely, image processing and pattern recognition. In the image processing modules, a number of features were used to identify the quality of the dates. These features were the flabbiness, size, shape, intensity, and defects. ANNs were used for the classification and the results showed 80% accuracy for sorting dates.

In [30], color, shape, and texture features were extracted from seven varieties of date images. Three types of classifiers were used, namely, linear discriminant analysis, K-nearest neighbor, and ANN. Here, ANN achieves the best classification accuracy of 98.6%. A multi-class support vector machine (SVM) was used to develop a system for automatic date classification [31], where mean color, calibre, and homogeneity features were extracted from dates images.

In [32], the researcher proposed another approach to classify dates fruit utilizing different features such as shape, size, and texture. Here, the textures are extracted from images in the form of local binary pattern (LBP) and Weber local descriptor (WLD) histograms. SVM has been used to classify dates images into four varieties of dates such as Sagai, Ajwah, Sukkary, and Sellaj. They used a dataset of 200 images for each type and a total number of 800 distinct images. A number of experiments were carried away based on different sets of features. The highest result was 98.1% accuracy using the selected WLD features with size and shape features from chrominance (Cb and Cr) and luminance (Y) color space.

The authors in [33] have used a dataset with 5000 date images for training process and have classified single date images based on their types. Here, the authors have used gray-level co-occurrence matrix, RGB color histogram, and four shape features such as eccentricity and area along with Gaussian mixture model.

The handcrafted feature-based methodologies implemented in earlier studies are not suitable for constructing results. Several feature sets can be extracted from images; however, determining the optimal features to represent the properties of images needs time for examination and inspection [34]. Besides, the feature sets search space is considerably large which renders any search method used, manually or algorithmically, suboptimal at best.

Furthermore, since these methods are intended to be integrated into the date industry, they must be tested on a large dataset simulating the reality of manufacturing environments. Moreover, dates received by manufacturers could vary in terms of the level of maturity, degree of hardness, and form. They could even include some obsolete dates, like those deformed due to transportation conditions. The limited handcrafted features might not cover the diversity of dates at the production stage. Hence, developing a more efficient and accurate method to classify dates regardless of the large variations within certain varieties and the small inter-variation between certain varieties. Developing such a system would be very useful for dates production industry. To overcome this issue, we propose an automated intelligent system that can learn directly from the variety of date fruit maturity, shape, or form that are sent for production using deep learning.

Deep learning is a new area of machine learning research; and one of the most influential and fastest-growing fields in artificial intelligence. Once the ImageNet image classification champion was awarded to Hinton's team, the focus shifted towards deep learning and continued to develop ever since [35]. Hence, deep learning received major demand compared with other machine learning algorithms. The main advantage of deep learning is its ability to learn local and global image features and to extract related information, without the need for the traditional handcrafted feature methodologies. It saves time and effort related to pre-processing and feature engineering tasks. Deep learning has shown the ability to deal with computer vision challenges and real-world issues. Deep learning models, benefiting from their high level of professional information processing, can deliver more powerful representations of complex high-dimensional data.

Therefore, it could be very efficient for fruit classification and recognition. However, the research in fruit classification employing this method is less current. Deep neural networks have been used on a fruit detection system by [34]. The deep learning model is trained to classify seven fruits it was used to automatically extracts several features. The study in [36] suggested an approach to enable analysis on packaged lettuce. They used CNN to classify vegetables with minimum color distortions. In terms of dates classification, the most recent approach is proposed in [37], where the authors gained an improved accuracy of 99.2% applying deep learning network to classify images of dates. Here, the authors built a dataset with the images of all four varieties of dates from the Google search engine. In another study [38], two types of pre-trained deep CNN are utilized named AlexNet and VGGNet. The highest result was 99.01% for VGGNet

## 3 Proposed System Architecture

This section describes the architecture of the proposed system which includes the four key modules: image acquisition, preprocessing, deep learning, gradient descent optimization, and evaluation metrics.

### *3.1 Date Fruit Image Acquisition*

In this paper, we introduce three popular date fruit cultivars in Saudi Arabia, namely Ekhlas, NbuteSultan, and Shayshi. The Ekhlas and Shayshi varieties are generally oval and cylindrical in shape but Ekhlas is golden in color, whereas Shayshi is dark brown and wrinkled. On other hand, the NbuteSultan variety is round, glossy, and brown in color. Images of these date fruit cultivars were acquired during the "Rutab" stage when it is fully ripe, using iPhone 7 mobile phones. More importantly, a dataset containing 1055 different images of each of the three types of the date fruit with a total number of 3165 images was created for this study purpose. Each of these images contained only one date fruit with a uniform background as showed in Fig. 1a.

### *3.2 Date Fruit Image Preprocessing*

To keep the date fruit in frame, images were captured varying the camera distance. As result, the size of date fruits in the obtained images was not apparent as illustrated in Fig. 1. Further, it can be observed that the images placed on a white background had a varied effect of lighting across different images. Due to this effect, shadows can be observed in some date images and, in few other images, the white background does not cover the entire edge, showing dark triangles on edges and corners [39].

(a) Captured Input Date Images (b) Preprocessed Input Dates Images

**Fig. 1** Date fruit cultivars (Ekhlas, Nebute Sultan, and Sheeshee)

Therefore, to remove these background noise and shadow artifacts from the images, as a first step, adaptive thresholding and Sobel operator were employed in sequel to extract the foreground image map and edge information respectively. Later, utilizing these two extracted information, the bounding box of the image was found out to crop and eliminate shadows in the captured image as shown in Fig. 1b.

### *3.3 CNN Architecture for Date Fruit Image Classification*

In this paper, we propose a smart classification system for date fruits using the state-of-the art CNN model. CNN is a special type of deep neural network commonly used in computer vision systems. The structure of CNN enables it to interpret image data by learning the features through different layers of abstraction. It starts from colors and edges in the first layer to using them to recognize corners and contours in the subsequent layer, to more abstract features that a fully connected neural network can easily classify. CNNs have three major components; convolutional layers, pooling layers, and fully connected layers. These layers are intricately connected in a particular order with the fully connected layer used towards the classification end of the network.

(a) Convolutional layers: generally applies convolution kernel on input data to create feature maps which holds the output from previous layers. These layers are regarded as feature extractor of the network. Each unit in this layer is associated with a set of weights called filter bank as local patches in the feature maps of the previous layer. The summation of these weights is passed into activation functions. Here we apply ReLU, the non-linearity activation function.
(b) Pooling layers: Pooling layer's task is to merge semantically similar features following a convolutional layer. A standard pooling unit selects a representative

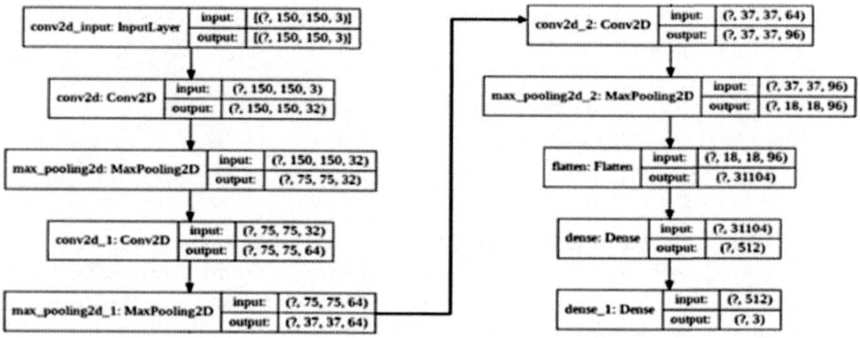


**Fig. 2** The developed CNN

value from the local area of a unit, such as the maximum value in max pooling. Neighboring pooling units receive inputs from areas moved by more than one row or column, thereby reducing the features number and creating an invariance to minor modifications and distortions.

(c) Fully connected layers: After the data passes through multiple convolutional and pooling layers, the size of the output feature maps is decreased due to the increasingly high-level features extracted and the downsampling of the pooling layer. The fully connected layer task maps the extracted features to final output (e.g. classification).

Our developed CNN contains four convolutional and pooling layers followed by two fully connected layers as illustrated in Fig. 2. We trained our CNN to learn all its parameters from scratch using five different types of optimization algorithms.

## *3.4 Gradient Decent Optimization*

Training CNNs is a high-dimensional, non-convex optimization problem. Backpropagation and gradient descent algorithms are used to find the derivatives of the weights and to update the parameters, respectively. The parameters' updates follow the derivatives of the weights to minimize the loss function and reach the optimal weights for the given task. Different optimization algorithms have been designed by researchers to improve the standard gradient descent optimization algorithm to enhance the performance of CNNs. We introduce here some of the most popular optimization algorithm used in deep learning networks that we applied on our CNN.

(a) Stochastic gradient descent (SGD) : is regarded as the extended version of gradient descent. Here, at each iteration, the gradient of the cost function is calculated for each sample in the training set despite of calculating the gradient sum of the cost function for the given training set. It is known to be faster than the typical gradient descent and less prone to get stuck in a local minimum [40]. Below are the weights update's for SGD:

$$g_{t,i,j} = \nabla\theta J(\theta t, j) \quad (1)$$

$$\theta_i = \theta_i - \alpha g_{t,i,j} \quad (2)$$

(b) AdaGrad: This optimization technique adopts adaptive learning rate during the training process to learn all the involved parameters with smaller rates for regular feature parameters and with greater updates for uncommon featured parameters. For the network parameters updates see Eq. (3) below.

$$\theta_t = \theta_{t-1} - \frac{\alpha}{\sqrt{G_t + \epsilon}} g_t \quad (3)$$

(c) Root Mean Square Propagation (RMSprop): This optimization adjusts the learning rate automatically to stabilize the oscillation of gradients. Here, the learning rate is divided with average of squared gradients which decays exponentially. In addition, this technique selects different learning rate for each parameter. It was developed by Geoffrey Hinton [41]. The Eqs. 4 and 5 describe each update. These update are done separately for each parameter.

$$E[g^2]_t = 0.9[g^2]_{t-1} + 0.1[g^2]_{t_t} \quad (4)$$

$$\theta_t = \theta_{t-1} - \frac{\alpha}{\sqrt{E[g^2]_t + \epsilon}} g_t \quad (5)$$

(d) Adaptive Moment Estimation (Adam): Adam is a gradient descent optimization algorithm which learns model parameters updating the learning rate for each parameter [42]. In Adam algorithm, two types of optimization algorithms were integrated which are RMSprop and SGD with momentum. The network parameters are updated using the Eqs. (6)–(10).

$$m_t = \beta_1 m_{t-1} + (1 - \beta_1) g_t \quad (6)$$

$$v_t = \beta_2 v_{t-1} + (1 - \beta_2) g_t^2 \quad (7)$$

$$\widehat{m_t} = \frac{m_t}{1 - \beta_1^t} \quad (8)$$

$$\widehat{v_t} = \frac{v_t}{1 - \beta_2^t} \quad (9)$$

$$\theta t = \theta t - 1 - \frac{\alpha}{\sqrt{\widehat{v_t} + \epsilon}} \widehat{m_t} \quad (10)$$

The authors suggested using the default values of 0.9, 0.999 and $10^{-8}$ for $\beta 1, \beta 2 and \epsilon$ respectively. Also, it is proved empirically that the practical performance of Adam is better compared to other algorithms of the adaptive learning process. It is so popular and applied in wide range of application of deep learning. The main advantages of this type of optimizations is that they are computationally efficient, and needs little memory. Accordingly, they considered better choice and suitable technique for problems which are large in regard to parameters or data.

(e) AdaMax: It's a variant of Adam optimization method based on infinity norm. LU generally exhibits stable behavior, and hence this variant of Adam optimizer was proposed in [42] where the authors show that it does indeed converges to a more stable value.

## 4 System Implementation

For this work, we used Colaboratory of Google (Colab) which provides the preconfigured runtime environment of Python with rich and essential libraries for statistical, artificial intelligence, and deep learning. It also freely provides GPU and TPU accelerated runtimes with few limitations on time and availability.

$$\begin{aligned} v_t &= \beta_2^{\infty} v_{t-1} + (1 - \beta_2^{\infty})|g_t^{\infty}| \\ &= \max(\beta_2 \cdot v_{t-1}, |g_t^{\infty}|) \end{aligned} \quad (11)$$

Our convolutional neural network models were developed using the available libraries using Python 3.7 including TensorFlow 2.0 and Keras 2.3 (TensorFlow implementation) both run on a Colab cloud TPU accelerator. For our implementation, we used CoLab Pro for less restricted experimentation, which afforded us a TPU with 8 cores and more than 28 GB RAM at all times. However, to use the API available for the training with TPU, it is required to use Google Cloud (GC) storage rather than local or google drive storage. Our google cloud bucket, the basic containers in GC storage, is organized into training and testing folders each containing the designated files for the date fruits cultivars.

We used tf.data TensorFlow library for efficiently building asynchronous, highly optimized data pipeline, it loads the images from memory, applies optimized transformations, creates batches, and sends them to the processing unit, TPU in our case. It also prevents the TPU's from data starvation by prefetching the data while training. For reproducibility purposes, we share the work done in the paper in the opensource at [43]

## 5 System Evaluation Metrics

We ran two experiments to evaluate the performance of the developed CNN models in classifying the provided types of dates. The first experiment used the dates images directly to train the developed CNN, while the second experiment used the cropped images instead. The experiments were conducted to investigate the automatic solution of the classification task without requiring multi-step machine learning methods. This section describes the experiments carried out to evaluate the developed CNN models. The developed CNN learns directly from dates images. The performance of each CNN was estimated using three types of evaluation measurements: accuracy, loss, and training time. These measures were calculated by Eqs. 12 and 13, respectively.

$$\text{Accuracy} = (\text{TP} + \text{TN})/(\text{TP} + \text{TN} + \text{FP} + \text{FN}) \quad (12)$$

$$\text{Loss} = -\Sigma\, yj \quad \log(y) \quad (13)$$

## 6 Result and Discussion

### *6.1 Results*

For both experiments, the last fully connected layer of the CNN was set to three neurons to represent the date type classes. The training dataset contained 2404 images of almost equal proportions for all classes, however, for training they were divided into mini-batches each containing 128 images. The CNN networks were trained for 540 iterations, but updated their weights 30 epochs only and accumulated the changes from 18 samples from the mini-batches. On the other hand, the models were tested on a dataset containing 756 images.

**First experiment** In this experiment, each image was sent directly to CNN models. The performance of each CNNs is presented in Table 1. The optimizers achieved accuracies of 84.44%, 94.44%, 94.92%, 96.22%, and 98.89% for the developed CNN+SGD, CNN+Adamax, CNN+Adagrad, CNN+RMSprop, and CNN+Adam models, respectively. The CNN+Adam model achieved better accuracy than others CNNs in this experiment. The training and validation accuracies and losses are presented in Fig. 3. As it can be observed in Fig. 3, the losses of the SGD are considerably higher than all the other algorithms even though the accuracy is seemingly high, the training has not finalized yet for this algorithm. In addition, while the progress for all algorithms is relatively consistent, RMSprop progress is highly unstable throughout the training process. The confusion matrix for Adam optimizer is shown in Figs. 4 and 5.

**Table 1** Performance of each optimization algorithm on test set with the original images)

| Optimizer | Accuracy (%) | Loss | Wall Time (s) | System time (ms) | User time (s) |
|---|---|---|---|---|---|
| SGD | 52.34 | 0.969082 | 2 mins 53 | 669 | 3.47 |
| RMSprop | 94.53 | 0.196288 | 1 min 26 | 654 | 3.89 |
| Adagrad | 95.31 | 0.149 | 1 min 25 | 640 | 3.87 |
| Adam | 99.22 | 0.102978 | 1 min 29 | 676 | 4.36 |
| Adamax | 97.66 | 0.106877 | 1 min 28 | 692 | 4.25 |

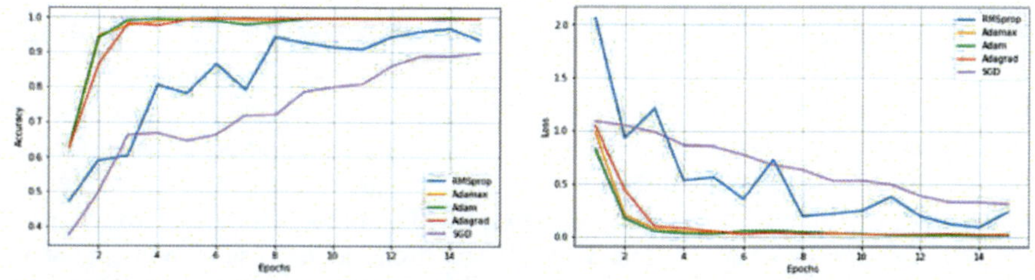


**Fig. 3** The training progress for each epoch

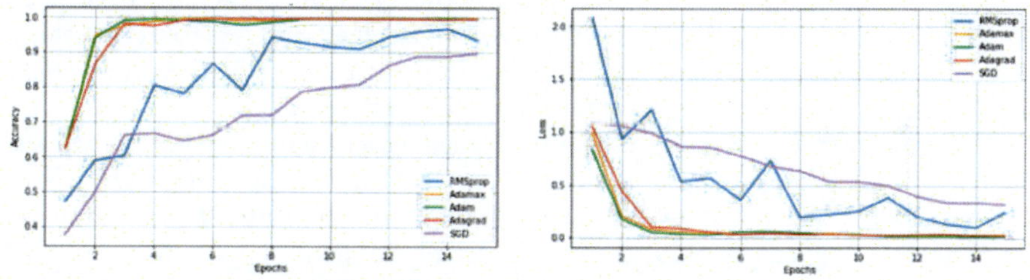


**Fig. 4** The training progress of the model with the cropped images

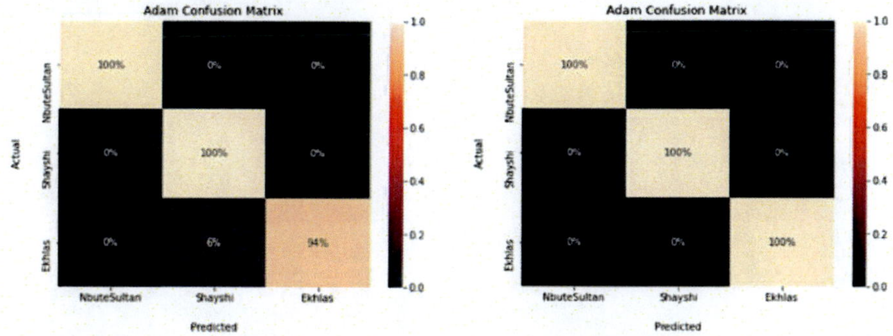


(a) First Experiment (b) Second Experiment

**Fig. 5** Confusion matrix results

**Table 2** The performance of each optimization algorithm on test set with the cropping images)

| Optimizer | Accuracy (%) | Loss | Wall time (s) | System time (ms) |
|---|---|---|---|---|
| SGD | 78.91 | 0.460064 | 3 mins 35 | 513 |
| RMSprop | 98.44 | 0.076828 | 1 min 27 | 677 |
| Adagrad | 99.22 | 0.044421 | 1 min 26 | 605 |
| Adam | 100.00 | 0.008167 | 1 min 30 | 764 |
| Adamax | 96.09 | 0.134929 | 1 min 27 | 826 |

**Second experiment** In this experiment, the cropped images were sent to the developed CNN model. The performance of CNNs with the various optimization algorithms is presented in Table 2. The highest accuracy of 100% was obtained by CNN+Adam model. The confusion matrix in Fig. 5 shows the accurate model of CNNs correctly classifying all the varieties of dates provided in the test images. It can be observed that the general performance of CNN models increased in comparison with the first experiment. The date segmentation method has indeed enhanced the performance of CNNs. The accuracy and loss during the training stage for the various optimization algorithms are presented in Fig. 6.

As it can be observed from Fig. 6, the accuracy of all optimizers have been increasing asymptotically at different rates, Adam, Adamax, and Adagrad converging at around the eighth epoch or earlier. RMSprop training results still fluctuating while SGD exhibiting steady slow convergence for all experiments.

To obtain the results of the developed CNN model for date fruit images, we present the feature maps applied on a date image from the Nbute Sultan variety, see Fig. 7. This is usually done to know which image features are preserved or detected in the CNN model. The feature maps that are closer to the input images are capable of detecting fine and small details whereas other feature maps are involved in capturing the most generic features that are closer to model output. In the first convolutional

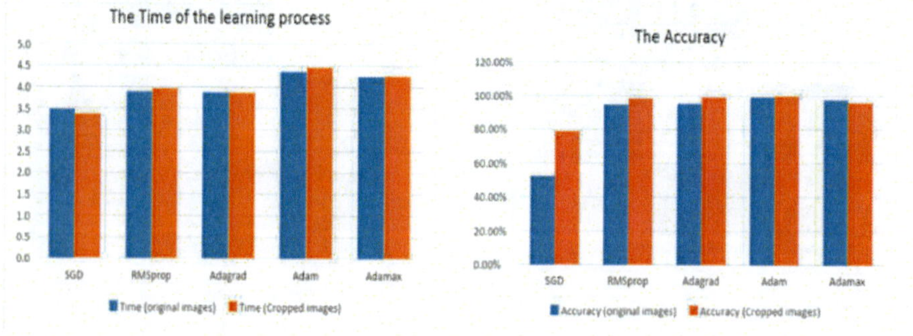


(a) Computational Time (b) Accuracy

**Fig. 6** Comparison graph for different optimization algorithms

**Fig. 7** Feature maps for the developed CNN layers. **a**, **b**, **c**, and **d** are for the first, second, third, and fourth layer, respectively

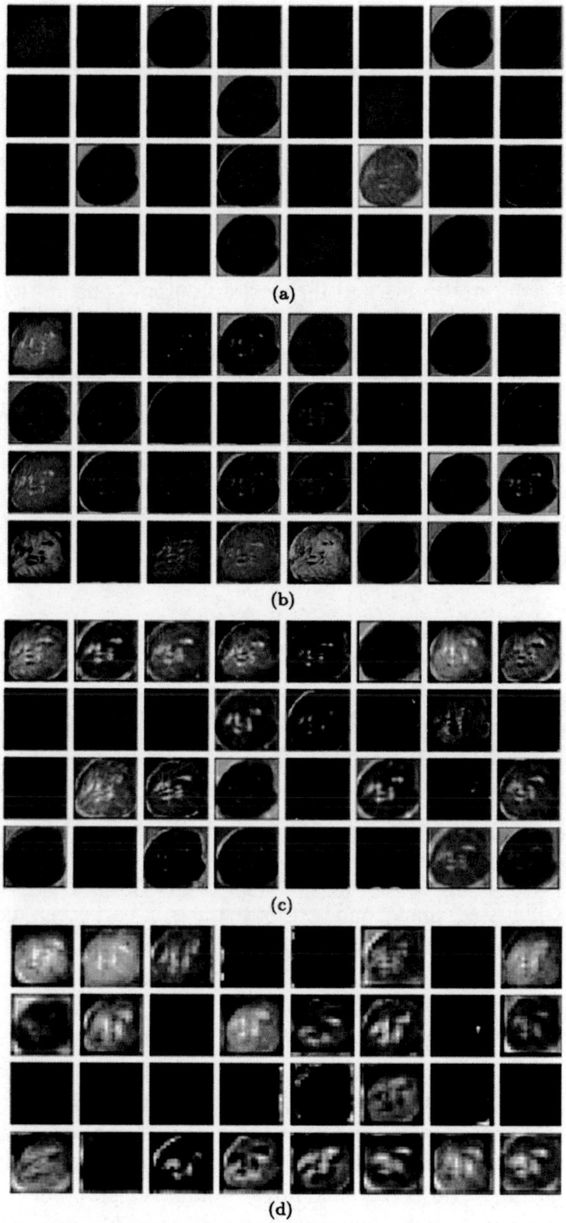

layer, the filters are applied to display the results. Fig. 7a, highlights different features of the date image in various ways. However, as the filters get closer to the output our model extracts the features from the image and form the more general concepts that can be used to make a classification. Even though our model is of four layers only, it is obvious from the feature maps at output layer that the model is looking at a date image, Fig.7d, generally, it loses the capability to interpret those most deep feature maps.

We also visualize our CNN filters using the activation maximization technique, which aims to maximize the activation of certain neurons. So in this technique, we keep the weights and the desired output constant while modifying the input such that it maximizes the output of certain neurons. In Fig. 8, a selected number of filters from the different convolutional layers are presented. We can see that the network has learned a variety of date surface textures both vertical and diagonal resembling the wrinkles shown in Shayshi and Nbute Sultan date images. As well as other formations pertaining to the different colors and shapes present in dates images.

## 6.2 Discussion

Deep CNNs models were trained to classify three different types of dates. The experiments explored five different optimization algorithms applied to update CNN parameters. Tables 1 and 2 show that using a CNN with Adam optimization algorithm achieved the best result when compared to the others. The performance of the developed CNN compared with the methods used in other studies of date fruit classification is shown in Table 3.

The authors of [32] extracted size, shape, LBP, and WLD local texture features. They used SVM to classify different varieties of date fruits and were able to achieve a reasonable accuracy of 98.1%. In [29], ANN is employed 15 features involving color, size, shape, and texture features to sort seven dates varieties. The authors used a number of feature extraction and feature reduction methods in order to achieve the best performance. The study achieves a good and better accuracy of 98.6%. The authors of [33], extracted a number of features such as RGB color histogram, co-occurrence matrix of gray level, and four shape features such as eccentricity and area. Also, for classification, the authors have applied the Gaussian mixture model to classify ten different types of dates. They recorded 97.5% accuracy.

Other studies investigated the ability to use CNN in data classification. In [44] used deep learning models to classify the four types of dates. They achieved 99.24% accuracy; however, each image contains a group of date in same variety. In addition, another study has employed pre-trained CNN named AlexNet and VGG-16 models [38]. They used same type of image, which contains a group of dates, on same type. The study reported 99.01% accuracy obtained by VGG-16.

Nonetheless, our study shows that CNN has potential to extract the most robust and compact set of features automatically. Further, the study aims to investigate how the highly complex nature of CNN architectures can be utilized for data classification

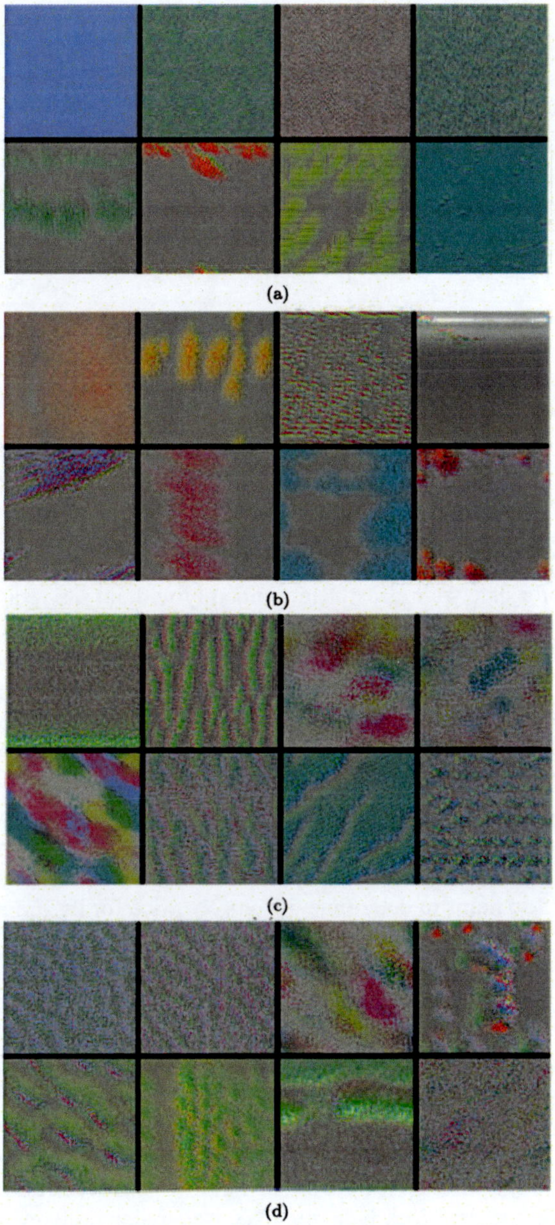

**Fig. 8** The developed CNN layers visualization using activation maximization technique. **a**, **b**, **c**, and **d** are for the first, second, third, and fourth layers, respectively

**Table 3** Accuracy comparison with related works

| Related works | Features | Classifiers | Accuracy |
|---|---|---|---|
| [29] | Color, shape, size and texture features | ANN | 98.6 |
| [32] | LBP and WLD local texture features | SVM | 98.1 |
| [33] | Color, shape, co-occurence matrix | Gaussian mixture model | 97.5 |
| [38] | – | VCG-16 and AlexNet | 99.01 |
| [44] | – | DNN | 99.24 |

and also has focused to design tests to identify the characteristics that can help to differentiate between the three types of dates. Though some dates images for these three types of dates show similar characteristics as been illustrated in Fig. 1. The high accuracy results show that CNN can be used to achieve high performance. It is also demonstrated that CNN can be utilized to achieve a greater performance for date fruit classification notably without the requirement for pre-processing and feature extraction process. Hence, the present study accomplished outstanding results compared to that achieved by the state-of-the-art methods.

Despite the number of research studies have attempted to address the requirement and the importance of automated sorting system, our developed CNN model can provide a suitable approach for sorting dates varieties and the results obtained ensure that it can be implemented on the dates industry, especially when dates varieties need for packaging. In addition, the models can be used to build a mobile application to identify the varieties of dates. Such application can help supermarkets employees to find not only the most appropriate variety of the fruit bought by the customer rather also helps to determine the correct price for its variety.

## 7 Conclusion

Date fruits which are regarded to have high medicinal value in treating a wide range of diseases are top grown agricultural product in the kingdom of Saudi Arabia. The need for good quality date fruits in meeting the customer satisfaction and target market requirements have led to pressing need for automated grading and sorting system in the date fruit industry. This was major impetus to develop the proposed system employing deep learning network to learn and discriminate the date cultivars namely, namely Ekhlas, Nbute Sultan, and Shayshi. The comparative study with other published works reveals the superiority of the proposed system with improved accuracy and clearly claims that the developed system can replace the conventional methods used for sorting process with increased marketability of date fruit. Expanding the model for other varieties of date fruits will be our future study. Also, the future work, taking the advancement of 5G technology, will attempt to integrate the developed system with smart mobile applications which the date fruit interested consumers can use as a shopping guide in purchasing the high-quality medicinal value dates.

## References


1. Anuja Bhargava and Atul Bansal. Fruits and vegetables quality evaluation using computer vision: A review. *Journal of King Saud University-Computer and Information Sciences*, 2018
2. Alavi N et al (2012) Date grading using rule-based fuzzy inference system. Journal of Agricultural Technology 8(4):1243–1254
3. John B Njoroge, Kazunori Ninomiya, Naoshi Kondo, and Hideki Toita. Automated fruit grading system using image processing. In *Proceedings of the 41st SICE Annual Conference. SICE 2002.*, volume 2, pages 1346–1351. IEEE, 2002
4. Shebiah Arivazhagan, R Newlin Shebiah, S Selva Nidhyanandhan, and L Ganesan. Fruit recognition using color and texture features. *Journal of Emerging Trends in Computing and Information Sciences*, 1(2):90–94, 2010
5. An underutilized food seeking industrial valorization (2017) Sami Ghnimi, Syed Umer, Azharul Karim, and Afaf Kamal-Eldin. Date fruit (phoenix dactylifera l.). NFS journal 6:1–10
6. Hajira Khan and Sameen Ahmed Khan (2016) Date palm revisited. Research Journal of Pharmaceutical Biological and Chemical Sciences 7(3):2010–2019
7. Yun Jeong Hong, FA Tomas-Barberan, Adel A Kader, and Alyson E Mitchell. The flavonoid glycosides and procyanidin composition of deglet noor dates (phoenix dactylifera). *Journal of agricultural and food chemistry*, 54 (6):2405–2411, 2006
8. Boubaker Dhehibi, Mohamed Ben Salah, and Aymen Frija. Date palm value chain analysis and marketing opportunities for the gulf cooperation council (gcc) countries. In *Agricultural Economics-Current Issues*. IntechOpen, 2018
9. Mohammed I Yasawy. The unexpected truth about dates and hypoglycemia. *Journal of family & community medicine*, 23 (2):115, 2016
10. CJ Miller, EV Dunn, and IB Hashim. The glycaemic index of dates and date/yoghurt mixed meals. are dates 'the candy that grows on trees'? *European Journal of Clinical Nutrition*, 57 (3):427–430, 2003
11. Campbell J Miller, Earl V Dunn, and Isameldin B Hashim. Glycemic index of 3 varieties of dates. *Saudi medical journal*, 23(5):536–538, 2002
12. Juma M Alkaabi, Bayan Al-Dabbagh, Shakeel Ahmad, Hussein F Saadi, Salah Gariballa, and Mustafa Al Ghazali. Glycemic indices of five varieties of dates in healthy and diabetic subjects. *Nutrition journal*, 10(1):59, 2011
13. A Manickavasagan, M Mohamed Essa, and Ethirajan Sukumar. *Dates: production, processing, food, and medicinal values*. CRC Press, 2012
14. Jameel M Al-Khayri, Shri Mohan Jain, and Dennis V Johnson. *Date palm genetic resources and utilization*, volume 1. Springer, 2015
15. Ghnimi Sami, Kamal-Eldin Afaf (2015) Fruit nutraceuticals seeking clinical evidence and technological exploitations. Journal of Bioequivalence & Bioavailability 7(3):1
16. Mohamed Ali Al-Farsi* and Chang Yong Lee. Nutritional and functional properties of dates: a review. *Critical reviews in food science and nutrition*, 48 (10):877–887, 2008
17. Nawal Nasrallah. *Dates: A global history*. Reaktion Books, 2011
18. Wulfsohn D, Sarig Y, Algazi RV (1993) Defect sorting of dry dates by image analysis. Canadian Agricultural Engineering 35(2):133–139

19. Hans Muller, Rudolf Gayler, and Nikolas Rapsomanikis. Process for the extraction of date fruits, 31 1984. US Patent 4,428,969
20. Shay Cohen, Gili Haberman, Rom Kshuk, and Limor Gendler. Date kernel preparation, 16 2011. US Patent App. 13/058,298
21. Joseph Kanner, Shlomo Navarro, and Avinoam Daos. Process of treating dates, 30 2004. US Patent 6,824,812
22. Efthimia Mavridou, Eleni Vrochidou, George A Papakostas, Theodore Pachidis, and Vassilis G Kaburlasos. Machine vision systems in precision agriculture for crop farming. *Journal of Imaging*, 5(12):89, 2019
23. AHMED Roukhe, Bouzid Abdenabi, Noureddine El Barbri, et al. Sorting dates fruit bunches based on their maturity using camera sensor system. *Journal of Theoretical and Applied Information Technology*, 56(3):324–337, 2013
24. Al-Rahbi Sawsana, Manickavasagan Annamalai, Thomas Gabriel (2015) Back propagation neural network (bpnn) to detect surface crack on dates using rgb images. Journal of Medical and bioengineering 4(1)
25. Ismail KM, Al-Gaadi KA (2009) Development of an electronic sensor for date sorting based on moisture content. Misr Journal of Agricultural Engineering 26(4):1923–1932
26. Razieh Pourdarbani, Hamid Reza Ghassemzadeh, Hadi Seyedarabi, Fariborz Zaare Nahandi, and Mohammad Moghaddam Vahed. Study on an automatic sorting system for date fruits. *Journal of the Saudi Society of Agricultural Sciences*, 14(1):83–90, 2015
27. Ali I Hobani, AmarNishad M Thottam, and Khaled AM Ahmed. Development of a neural network classifier for date fruit varieties using some physical attributes. *King Saud University-Agricultural Research Center*, 2003
28. Fadel M (2007) Date fruits classification using probabilistic neural networks. CIGR Journal, Agricultural Engineering International
29. Yousef Al Ohali (2011) Computer vision based date fruit grading system: Design and implementation. Journal of King Saud University-Computer and Information Sciences 23(1):29–36
30. Abdulhamid Haidar, Haiwei Dong, and Nikolaos Mavridis. Image-based date fruit classification. In *2012 IV International Congress on Ultra Modern Telecommunications and Control Systems*, pages 357–363. IEEE, 2012
31. Djeffal A, Rais HE, Babahenini MC (2012) Un système de tri automatique des dattes par svm 1vsr. Revue des Sciences et de la Technologie 3(1):106–115
32. Ghulam Muhammad. Automatic date fruit classification by using local texture descriptors and shape-size features. In *2014 European Modelling Symposium*, pages 174–179. IEEE, 2014
33. Oussama Aiadi and Mohammed Lamine Kherfi (2017) A new method for automatic date fruit classification. International Journal of Computational Vision and Robotics 7(6):692–711
34. Sa Inkyu, Ge Zongyuan, Dayoub Feras, Upcroft Ben, Perez Tristan, McCool Chris (2016) Deepfruits: A fruit detection system using deep neural networks. Sensors 16(8):1222
35. Alex Krizhevsky, Ilya Sutskever, and Geoffrey E Hinton. Imagenet classification with deep convolutional neural networks. *Communications of the ACM*, 60(6):84–90, 2017
36. Dario Pietro Cavallo, Maria Cefola, Bernardo Pace, Antonio Francesco Logrieco, and Giovanni Attolico. Non-destructive automatic quality evaluation of fresh-cut iceberg lettuce through packaging material. *Journal of Food Engineering*, 223:46–52, 2018
37. M Shamim Hossain, Ghulam Muhammad, and Syed Umar Amin. Improving consumer satisfaction in smart cities using edge computing and caching: A case study of date fruits classification. *Future Generation Computer Systems*, 88:333–341, 2018
38. Muhammad Ghulam (2015) Date fruits classification using texture descriptors and shape-size features. Engineering Applications of Artificial Intelligence 37:361–367
39. Altaheri Hamdi, Alsulaiman Mansour, Muhammad Ghulam (2019) Date fruit classification for robotic harvesting in a natural environment using deep learning. IEEE Access 7:117115–117133
40. Duchi John, Hazan Elad, Singer Yoram (2011) Adaptive subgradient methods for online learning and stochastic optimization. Journal of machine learning research 12(7)

41. Geoffrey Hinton, Nitish Srivastava, and Kevin Swersky. Lecture 6a overview of mini-batch gradient descent (2012). *Coursera Lecture slides https://class. coursera. org/neuralnets-2012-001/lecture*, 2012
42. Diederik P Kingma and Jimmy Ba. Adam: A method for stochastic optimization. *arXiv preprint* arXiv:1412.6980, 2014
43. Ekaba Bisong. Google colaboratory. In *Building Machine Learning and Deep Learning Models on Google Cloud Platform*, pages 59–64. Springer, 2019
44. Nasiri Amin, Taheri-Garavand Amin, Zhang Yu-Dong (2019) Image-based deep learning automated sorting of date fruit. Postharvest Biology and Technology 153:133–141

# Low-Cost Embedded System for Customer Loyalty

**Mauro Lima, José F. Morgado, and Rui P. Duarte**

**Abstract** Nowadays, digital marketing has become an indispensable tool for companies to promote and sell high-quality products to cope with increasingly competitive markets. For these to be shaped according to the customer's interests and preferences, technology has been pushed to evolve every day trying to keep up with high global demand. However, there is a particular problem in identifying and locating people in indoor spaces. Many applications were developed leveraging beacon technology that ranges from user's localization and indoor navigation to personalized assistants that would assist in the decision-making process. This, however, is still an open problem due to the difficulty in creating scalable indoor navigation systems, that are independent of the physical plan, and lack of precision in the detection of the customer indoor position. This paper will disclose the possibility to develop an efficient, low-cost indoor navigation system that solves both problems. It proposes an improvement on the accuracy of customer-beacon proximity distance, and a plan-free navigation method is proposed which allows customers to travel from a position to a specific beacon. Moreover, we integrate this with the concept of proximity marketing to embed the system in a physical context, focused on product promotions to increase customer loyalty.

**Keywords** Indoor navigation · Indoor location · Bluetooth low energy · Positional information · Customer experience · Signal strength

M. Lima · J. F. Morgado · R. P. Duarte (✉)
Polytechnic Institute of Viseu, Viseu, Portugal
e-mail: pduarte@estgv.ipv.pt

J. F. Morgado
e-mail: fmorgado@estgv.ipv.pt

R. P. Duarte
CISeD - Research Centre in Digital Services, Viseu, Portugal

 
X.-S. Yang et al. (eds.), *Proceedings of Sixth International Congress on Information and Communication Technology*, Lecture Notes in Networks and Systems 216,
https://doi.org/10.1007/978-981-16-1781-2_67

## 1 Introduction

With the technological advances in mobile devices, the location of people or devices has become a reality. In this context, the most common technologies are geolocation (GPS) and wireless networks (WiFi). GPS, fully operational in 1993 [1] is one of the leading technologies based on satellite triangulation and wave emission at a given frequency. This frequency presents two significant problems: it moves slowly through solid objects and can be hampered by the noise inside buildings [2]. At the WiFi level, there are several constraints to take into account, such as the service provider contracted speed, number of users accessing the network, and signal quality. Moreover, to deploy an infrastructure that is able to support thousands of users, it would be necessary to spend thousands of euros in proportion. To overcome these constraints, beacon Bluetooth Low-Energy (BLE) technology was developed, being a low-cost, scalable technology, and it works as an indoor GPS.

A commercial space that contains retail stores is a typical context of indoor buildings, where two major players arise: the customer and the retailer. While the first wants to buy products, the second wants to sell them, and a win–win situation occurs when this relation is well established. Thereof, retail stores can use beacon technology to notify clients of promotions, which can be viewed as products of interest by customers. The major contributions of this paper are device location and proximity marketing. The system will be able to execute one-to-one relations between beacons and customers, and the system will consider the customer's location so that beacons can make available information for them. To achieve this, the accuracy of the customer–beacon proximity distance is improved and a plan-free navigation system is proposed to better assist a customer in the navigation to a store. Moreover, the collected data is used for proximity marketing with customized promotions based on the interests of customers.

This paper is organized in the following manner. Section 2 reviews methods for indoor device location and associated technologies; Sect. 3 details the approach by presenting the authentication, navigation, and communication methods. In Sect. 4, a mobile application is presented that takes advantage of the embedded system targeting promotion coupons for customer loyalty. Section 5 concludes the paper, providing some hints to future work.

## 2 Related Work

BLE systems are a well-established reality focusing on location algorithms and technologies. A complete overview of methods and technologies is presented in [3]. Location algorithms have gained popularity in the last years, with focus on Lateration [4–7], Fingerprinting [8, 9], and Angulation [4, 10].

Lateration is used to determine the location of a smart device in a closed space. It considers the distance between a device and three anchor points which define the position of the beacons. This distance is provided by the Received Signal Strength Indication (RSSI) of each beacon and is considered as the circumference radius around the beacon location. This results in a system of equations whose solution represents the location of the mobile device in the indoor space. Wang et al. [5] determined the distance between two Bluetooth devices and results show that they achieved good performance with a reduced error rate. Oguejiofor et al. [6] developed an algorithm to determine the position of the nodes in a WiFi network. In order to track people, Mahiddin et al. [7] used this technique and the signal emitted by each Access Point (AP) so that the mobile device can evaluate the signal strength of the AP and convert it into a distance.

Fingerprinting is an algorithm based on the subdivision of a map into segments. The starting point is the online training stage where each segment of the map is assigned the average of RSSI and the position of the segment, which are stored in a Database (DB). Once the training stage is completed, the offline stage starts, and the device location is tracked. As in the online stage, the same parameters are collected for the location system and are compared to those obtained in the first stage. The best match between the data defines the position of the user in a certain fingerprint. In this context, Kaemarungsi et al. [8] proposed a new modular system that aims not only to optimize the implementation time but also to quantify an improvement in accuracy and speed to the method.

Angulation is a geometry-based technique [4, 10]. In two dimensions, this tracking method requires two points of communication with the mobile device. The location of the device is carried out by considering the intersection of one or more pairs of angular direction lines. To determine this intersection, the angle between beacon $A$ and user position is determined as well as the angle between beacon $B$ and the user. The point where these angular direction lines intersect provides the user position. In three dimensions, three beacons are required [4]. The advantage of this method is the accurate estimation of the device position using few communication points. Subhan et al. [11] presented a method that focuses on mitigating typical problems associated with the quality of the accuracy (such as differences in temperature and noise) and improved in about 45% the distance accuracy. This technique is used in this paper for the navigation system.

With the rapid technological evolution experienced in recent years, several technologies used to locate smart devices inside buildings have emerged, with special focus on WiFi, Bluetooth, and communication networks. Other technologies like magnetometer [12, 13], image sensor [12, 14, 15], gyroscope [16], and accelerometer [17], are device related.

WiFi and Bluetooth are present in all smartphones and tablets, making them desirable for use in the context of locating smart devices inside buildings. Bluetooth is a short-range wireless communication system created with the need to replace the cable connection between fixed and mobile devices. The application that handles the data received by the signal emitters and defines the user location should take into account the signal variation problem [18]. Another problem of these technologies is the noise caused by other objects between the emitter and the receiver [19]. In this paper, we develop a method that is optimized over time by considering more samples. However, considering all these less positive points, it is imperative to say that they are quite viable technologies for locating devices inside buildings [20]. Communication networks work through the multi-directional exchange of magnetic wave information between the mobile phone and the signal antennas. Since each antenna has a limited number of pairings, for a very densely populated area, several must be installed, ensuring the best possible communication. Otsason et al. [21] proposed that this technology should be used to estimate the user's position based on the fingerprinting algorithm. They mapped six antennas at a distance of 1.5 m from each other to obtain a user's location in a closed building. Tian [22] revealed that it is ultimately possible to use this technology for indoor location, by synchronizing this technology with the fingerprinting location algorithm, but with the particularity of using smartphone sensors that allow the required adjustments.

## 3 Approach

Indoor location is related to three major components: the location of the customer, the communication with a beacon, and the navigation to the beacon.

### *3.1 Authentication, Navigation, and Communication Modules*

This paper relates to the interaction of a mobile application with beacons associated to stores that contain promotions. When a customer enters a commercial space, the mobile application starts to monitor which beacons are valid within its reach. A valid beacon contains preferences defined by customers. Modules that allow this interaction are specified in this section.

The authentication of customers in mobile applications can be carried out in several ways. The most common are those that require an username and a password. In this paper, a different approach is used, based on a single user ID. This unique ID is generated and is associated to a customer, which is validated every time the customer enters the application. This ID is stored in the memory of the mobile device and the ID is validated, in the DB, through a web-service. If so, the web-service returns the promotions associated to that customer.

The navigation module allows the evaluation of customer's position and provides a view of notifications of several categories of interests, as shown in Fig. 1a). The detection of beacons is carried out by obtaining their $(x, y)$ positions from the DB. The center of the $(x, y)$ axis corresponds to the fixed position of the customer, represented in the mobile application by an arrow. The rotation angle, $\alpha$, is determined by considering the center of the axis related the position of the selected beacon. The center of the coordinate axes is determined given by considering the average of the $(x, y)$ val-

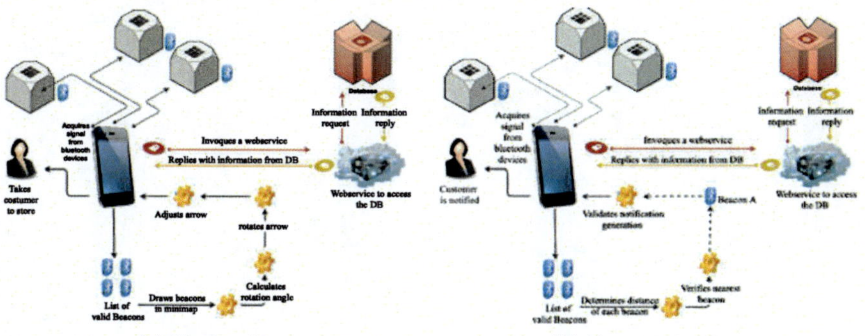


(a) Navigation (b) Communication

**Fig. 1** Customer–Beacon architecture

ues and is given by $(x_c, y_c) = ((\max(x) + \min(x))/2, (\max(y) + \min(y))/2)$. The tangent between the value of the axis center, $(x_c, y_c)$, and the position of the beacon, $(P_x, P_y)$, is given by $\tan(\theta) = \sin(P_y - y_c)/\cos(P_x - x_c)$. Finally, the rotation angle, $\alpha$, is given by

$$\alpha = \frac{\tan^{-1}(\frac{P_y - y_c}{P_x - x_c}) \times 180}{\pi} \tag{1}$$

and the arrow rotates $\alpha$ degrees indicating the user the direction to reach the desired beacon.

The last component of this system portrays the communication one-to-one between the customer and a beacon and is presented in Fig. 1b. In this module, the application scans, at every second, for beacons that are in the proximity of the mobile device. After detection, the distance, $d$, between the beacons and the customer is determined, and the information regarding the nearest beacon is sent to the customer.

The underlying problem in this relation corresponds to the accuracy of the distance measured between the beacon and the mobile device versus the real distance between a customer and a store. To determine the nearest beacon, two parameters are used: the Received Signal Strength (*RSSI*) and the Power of Transmission (*Tpower*). While the *RSSI* provides information about the strength signal that a receiver device can read (typically $RSSI \in [-26, -100]$), the value of *Tpower* is related to the amount of energy of an emitter ($Tpower \in [-40\ dBm, +4\ dBm]$).

With these two variables, the distance, $d$, between a beacon and a smart device is typically given by

$$d = 10^{\frac{Tpower - RSSI}{10 \times N_{[2,4]}}} \tag{2}$$

where $N_{[2,4]}$ is an environmental variable related to the building where the beacon is installed. However, given constraints associated to closed spaces, the relation presented in Eq. (2) is not efficient. To overcome this problem, an adjustment parameter,

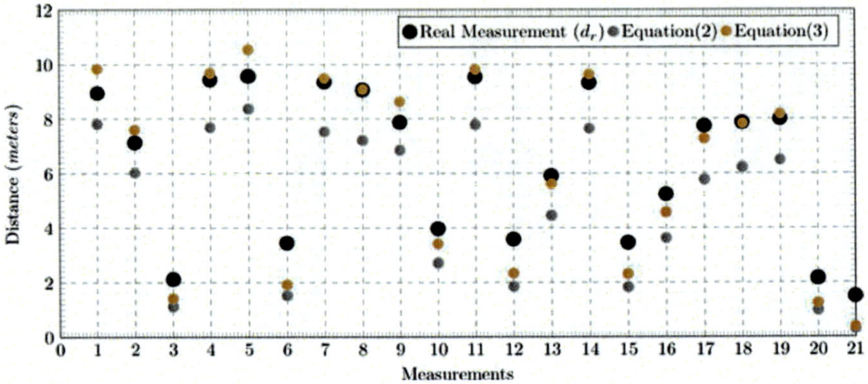


**Fig. 2** Distance measurements comparison between real values and approximations ($N_{[2,4]} = 2$, in Eq. (3))

$\gamma$, is introduced in Eq. (2). It is a three steps procedure: first, we determine the error between the real measurement of distance, $d_r$, and the distance value $d$ given by Eq. (2). Next, the average coefficient of the error obtained in the first stage is determined, $\gamma = \sum_{i=1}^{N}(d_r - d)/N$ ($N$ represents the number of samples stored in the DB). The value of $\gamma$ is continuously refined, since every time the customer accesses the mobile application this value is updated with new samples from the DB. Finally, we introduce $\gamma$ in Eq. (2) and obtain the following relation:

$$d = 10^{\frac{Tpower - RSSI}{10 \times N_{[2,4]}}} \times \gamma \quad (3)$$

To validate this approach, 21 test scenarios were considered. First, the real distance between the customer and the beacon was measured, and then Eq. (2) and Eq. (3)

where used. Figure 2 shows that $\gamma$ introduces more accuracy in the overall predicted value of the distance given in Eq. (3), when compared to the values obtained using Eq. (2). Given the physical positioning of retail stores, the improvement on the accuracy of the distance function increases the effectiveness of proximity marketing.

## 4 Experimental Design

Considering the growth potential of the beacon technology, a scalable mobile application was also developed as a proof of concept (Fig. 3). When a customer first uses the application, he is asked to define categories of interest for products. After that, beacons associated to stores that have these categories allow the customer to generate promotion coupons (Fig. 3a).

When the user wins a prize, he can choose between going to the store, or to keep the coupon and reuse it in another time (Fig. 3b). By choosing to go to claim a prize,

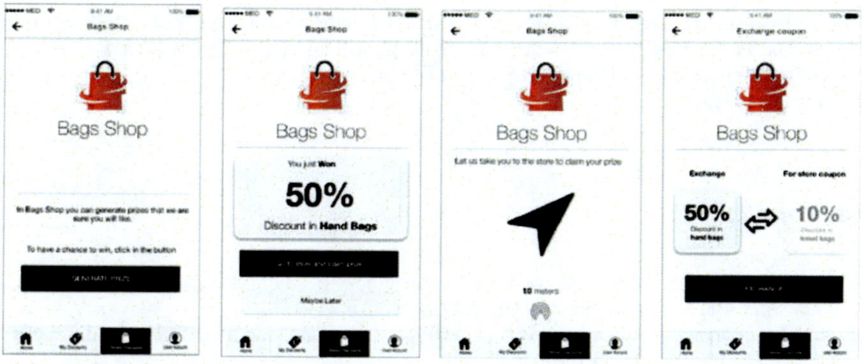


(a) Generate prize (b) Promotion (c) Navigation (d) Exchange

**Fig. 3** Customer mobile application for store promotions

the navigation module is activated, and the user is guided to the store (Fig. 3c) to claim the promotion. Finally, if the customer does not want to use the promotion he can store it to use in another occasion, or exchange it with another promotion. If he chooses to exchange the promotion, he can select another from the coupons available in the store (made available by other customers) that are of his interest, as presented in Fig. 3d).

## 5 Conclusions

This paper presents a scalable method based on beacon technology. It targets both the navigation between a user and a beacon and uses this technology for proximity marketing. At the navigation level, the beacon was considered a magnetic pole to where a user will navigate. To overcome errors on the distance accuracy between devices and the beacon, an improvement on the distance function was implemented, thus reducing the error in proximity measurements. Moreover, the method takes advantage of beacon technology for proximity marketing, by developing a customer-centered system, based on the generation of promotion coupons. These are generated from customer interests and are provided by active beacons that represent retail stores of the commercial space. To provide flexibility to the customer, several types of promotions are generated.

Future work will follow two directions. The $z$ direction is to be implemented in the navigation module to allow a better navigation between floors. At the coupon generation level, a recommendation system based on machine learning can be implemented to generate promotions according to the customer profile.

**Acknowledgements** This work is funded by National Funds through the FCT—Foundation for Science and Technology, I.P., within the scope of the project Ref. UIDB/05583/2020. Furthermore, we would like to thank the Research Centre in Digital Services (CISeD) and the Polytechnic of Viseu for their support.

## References

1. Getting IA (1993) Perspective/navigation-the global positioning system. IEEE Spectrum 30(12):36–38. https://doi.org/10.1109/6.272176
2. Leick A, Rapoport L, Tatarnikov D (2015) GPS satellite surveying. John Wiley & Sons. https://doi.org/10.1002/9781119018612
3. Ranasinghe C, Kray C (2018) Location information quality: A review. Sensors 18(11):3999. https://doi.org/10.3390/s18113999
4. Hightower J, Borriello G (2001) Location sensing techniques. IEEE. Computer 34(8):57–66
5. Wang, Y., Yang, X., Zhao, Y., Liu, Y., Cuthbert, L.: Bluetooth positioning using rssi and triangulation methods. In: 2013 IEEE 10th Consumer Communications and Networking Conference (CCNC), pp. 837–842. IEEE (2013). https://doi.org/10.1109/CCNC.2013.6488558
6. Oguejiofor O, Aniedu A, Ejiofor H, Okolibe A (2013) Trilateration based localization algorithm for wireless sensor network. International Journal of Science and Modern Engineering (IJISME) 1(10):2319–6386
7. Mahiddin, N.A., Safie, N., Nadia, E., Safei, S., Fadzli, E.: Indoor position detection using wifi and trilateration technique. In: The International Conference on Informatics and Applications (ICIA2012), pp. 362–366 (2012)
8. Kaemarungsi, K.: Efficient design of indoor positioning systems based on location fingerprinting. In: 2005 International Conference on Wireless Networks, Communications and Mobile Computing, vol. 1, pp. 181–186. IEEE (2005). https://doi.org/10.1109/WIRLES.2005.1549406
9. Alhmiedat, T., Samara, G., Salem, A.O.A.: An indoor fingerprinting localization approach for zigbee wireless sensor networks. arXiv preprint arXiv:1308.1809 (2013)
10. Goswami, S.: Indoor location technologies. Springer Science & Business Media (2012)
11. Subhan, F., Hasbullah, H., Rozyyev, A., Bakhsh, S.T.: Indoor positioning in bluetooth networks using fingerprinting and lateration approach. In: 2011 International Conference on Information Science and Applications, pp. 1–9. IEEE (2011). https://doi.org/10.1109/ICISA.2011.5772436
12. Navarro, D., Benet, G.: Magnetic map building for mobile robot localization purpose. In: 2009 IEEE Conference on Emerging Technologies & Factory Automation, pp. 1–4. IEEE (2009). https://doi.org/10.1109/ETFA.2009.5347181
13. Chung, J., Donahoe, M., Schmandt, C., Kim, I.J., Razavai, P., Wiseman, M.: Indoor location sensing using geo-magnetism. In: Proceedings of the 9th international conference on Mobile systems, applications, and services, pp. 141–154. ACM (2011). https://doi.org/10.1145/1999995.2000010
14. Yoshino, M., Haruyama, S., Nakagawa, M.: High-accuracy positioning system using visible led lights and image sensor. In: 2008 IEEE Radio and Wireless Symposium, pp. 439–442. IEEE (2008). https://doi.org/10.1109/RWS.2008.4463523
15. Galván-Tejada C, García-Vázquez J, Galván-Tejada J, Delgado-Contreras J, Brena R (2015) Infrastructure-less indoor localization using the microphone, magnetometer and light sensor of a smartphone. Sensors 15(8):20355–20372
16. Trein, G., Singh, N., Maddila, P.: Simple approach for indoor mapping using low-cost accelerometer and gyroscope sensors. DOCPLAYER (2013)
17. Ladetto, Q., Merminod, B.: Digital magnetic compass and gyroscope integration for pedestrian navigation. In: 9th Saint Petersburg international conference on integrated navigation systems, Saint Petersburg, Russia, CONF (2002)

18. Ababneh, N.: Radio irregularity problem in wireless sensor networks: New experimental results. In: 2009 IEEE Sarnoff Symposium, pp. 1–5. IEEE (2009). https://doi.org/10.1109/SARNOF.2009.4850343
19. Boes, B.S.: System and method for reducing signal interference between bluetooth and wlan communications (2014). US Patent 8,824,966
20. Villanueva, F.J., Gazzano, J.D., Villa, D., Vallejo, D., Mora, C., Morcillo, C.G., López, J.C.: Distributed architecture for efficient indoor localization and orientation. In: 2013 IEEE international conference on consumer electronics (ICCE), pp. 57–58. IEEE (2013). https://doi.org/10.1109/ICCE.2013.6486793
21. Otsason, V., Varshavsky, A., LaMarca, A., De Lara, E.: Accurate gsm indoor localization. In: International conference on ubiquitous computing, pp. 141–158. Springer (2005). https://doi.org/10.1007/11551201_9
22. Tian, Y.: Practical indoor localization system using gsm fingerprints and embedded sensors. Ph.D. thesis (2015)

# Using Narrative-Based Video on Gaining Safety Driving: Focusing on Career Resilience and Learning Style in Automated Driving Level 3

**Maki Arame, Junko Handa, Yoshiko Goda, Masashi Toda, Ryuichi Matsuba, Huiping Zhou, Makoto Itoh, and Satoshi Kitazaki**

**Abstract** Before automated driving vehicles are put to practical use, people should know how those vehicles run and how a driver should operate them for safety. Some drivers are interested in learning about automated driving vehicles, others are less interested in it. There are individual differences in learning styles as well as a risk reduction method such as resilience. Thus, two types of motivational video materials were developed and investigated which type of video was effective. As a result, narrative-based motivational video was most effective. In short, this study especially focused on resilience and learning style as personal attributes and verified the effectiveness of narrative-based motivational materials.

**Keywords** Motivational video contents · Career resilience · Narrative · Conditional driving automation

## 1 Introduction

Before Level 3 and 4 self-driving cars are put to practical use, people should gain knowledge about them for safety [1]. When one learns about operation of automated driving vehicles, it is thought that individual differences will occur depending on the driving experience and the learning experience on safe driving [2]. Motivation is needed for more people to learn effectively [3].

M. Arame (✉) · J. Handa
Polytechnic University of Japan, Kodaira, Tokyo, Japan
e-mail: arame@uitec.ac.jp

M. Arame · J. Handa · Y. Goda · M. Toda · R. Matsuba
Kumamoto University, Kumamoto-city, Kumamoto, Japan

H. Zhou · M. Itoh
University of Tsukuba, Tsukuba-city, Ibaraki, Japan

S. Kitazaki
National Institute of Advanced Industrial Science and Technology, Tsukuba-city, Ibaraki, Japan



X.-S. Yang et al. (eds.), *Proceedings of Sixth International Congress on Information and Communication Technology*, Lecture Notes in Networks and Systems 216,
https://doi.org/10.1007/978-981-16-1781-2_68

In the verification of the effectiveness of safe driving education content, the effectiveness of video materials had been suggested [4], and it had been verified that learning materials of video format were effective, because it absorbed the differences of one's career resilience and learning style [5]. As a next step, the effective contents of video format as a learning material should be considered. However, there has not been investigated whether or not a certain type of motivational content can be effective regardless of personal attributes.

In this study, the effectiveness of motivational learning materials developed using the narrative approach and fact-based information including data using the graphs were examined if either one shows better outcomes.

## 2 Literature Review

### *2.1 Hierarchical Approach Theory of Driving Behavior*

In order to prevent accidents caused by human error, there is a method that classifies the elements of the driver's safe driving behavior into hierarchies and examines countermeasures. Keskinen's hierarchical approach theory of driving behavior [6] consists of four layers. The first layer includes basic driving skills and knowledge of laws and regulations, and the second layer includes dealing with other vehicles and pedestrians and predicting danger. The third layer is the skill related to acknowledge the purpose of driving and to make the operation plan (setting of driving time and route), and the fourth layer encompass the self-control skill such as awareness and responsibility to drive safely and not cause an accident. The current traffic safety education mainly targets up to the contents of the second layer but in order to realize safety education in society as a whole, it is necessary to motivate them to focus on higher layers [3]. In this study, we developed motivational teaching/learning materials on safe driving of automated driving levels 2 and 3 using Keller's motivational model (ARCS) [7] which is used when developing learning materials in instructional design and verify the effectiveness of the materials.

### *2.2 Narrative Approach*

Resilience is considered one of the methods to reduce the risk for complex systems.

The word "resilience" can be explained "elasticity" and "restoring force." Hollnagel described resilience as a function of its own, before changing the condition or occurring a disturbance, in order for the system to continue the required operation under expected or unexpected conditions. It is defined as an essential ability that can be adjusted during or after the outbreak [8].

In current society, which presupposes the perception that change is normal, it is impossible to make decisions with complete information that takes all changes into account. Decision-making in social and technical systems will be made under conditions where only incomplete information is available. There is a limit to human awareness under incomplete information. Resilience supplementary requirements include support methods that overcome the constraints of awareness. It recommends the introduction of heuristics and the use of narrative approaches to support temporal or empirical reference reflections, "Have similar things happened in the past?".

Cognitive psychology studies the subdivision of narrow research that relies on computational theory of information processing, deviating from the original human research and being isolated from other human sciences. On the other hand, the narrative approach is an approach that has emerged as a method that criticizes such cognitive science and focuses on the "meaning" that cognitive science has abstracted [9].

In the narrative approach, one thinks that one has a time structure and conceptual framework composed of the languages acquired so far and the languages acquired through learning in community life, and "talking about things" has the effect of transforming "the way of looking at things."

Ando [10], explained two important points of a structural feature of narrative, based on the previous research of Bruner [11], Gergen and Gergen [12], Escalas [13]: (1) the story has a time axis and is connected (2) there is a causal relationship between the elements that are inferred.

Furthermore, she provides five elements as a measure of narrative: (1) it is not a general theory or an abstract theory, (2) the characters are clear, (3) the temporal transition is clear, (4) there is a dramatic structure including introduction, development, turn and conclusion, and (5) the causal relationship is clear. By introducing narratives related to driving automation as motivation, it may be possible to support learning beyond the restrictions of learners' ability of awareness.

## 3 Purpose of This Study

At the public training site in traffic safety education, the lesson time is assumed to be about 5 min. The hierarchical model of driving behavior points out the need for not only skill-based education but also education of higher levels such as risk prediction ability and driving planning, and motivation for learning should be considered. In safety education, motivational learning materials may be important.

In this research, the learning materials that can learn the difference in driving automation levels 2–5, and the role of the driver, and fosters safety awareness were developed based on the ARCS model. There was no difference between the two types of motivational learning materials developed in the preliminary evaluation using the ARCS evaluation sheet [14]. That is, it has the same motivational effect on the ARCS model. According to narrative theory, the use of narratives may overcome the constraints of learners' awareness. Thus, this study focused on the contents of the motivational learning material with video format and investigated the effectiveness of two types of content developed based on narrative and fact-based information.

# 4 Method

## 4.1 Data Collection

The survey was conducted in October 2020 on the Internet. The data were collected from the survey. 930 of the 2,790 surveys were used in this study. As Fig. 1 indicates, the participants watch a pattern of motivational video learning material and then, they learned the same contents of automated driving. In the survey, they were asked about demographic information such as age and gender, as well as questions about learning styles and career resilience. The test questions were not about driving skills but driving attitudes and the operation plan which were the higher levels of Keskinen's hierarchical approach theory. In this survey, when learning about automated driving levels 2 to 5, there were three patterns of learning materials and conducted tests before and after learning. The tests were conducted before and after the learning about automated level 3driving. Using the pre- and post-test scores, two groups were

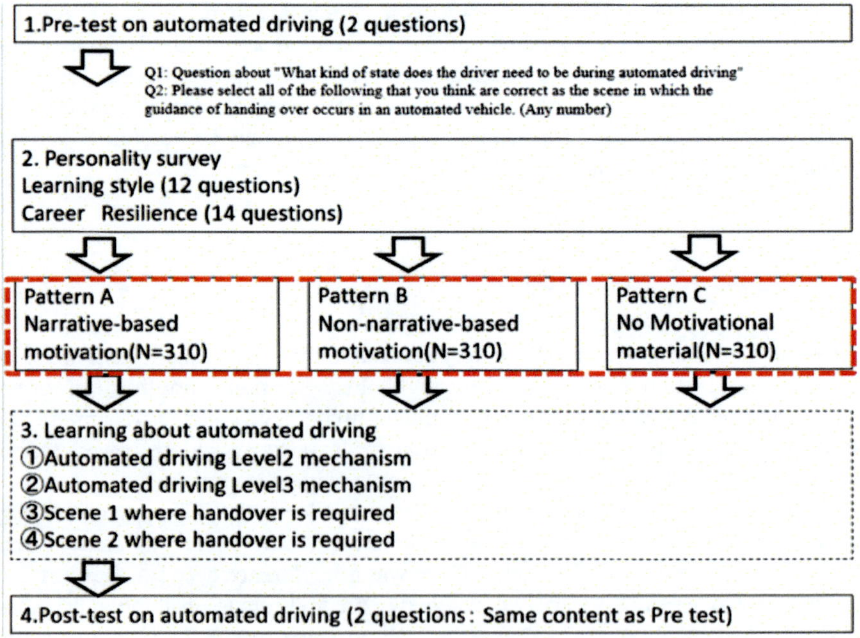


**Fig. 1** Web survey structure 2020

made. One is up-group that went up the score. The other is non-up-group that did not increase the score.

The assumption is that if there is no difference, it is conceived that the learning material works for anyone regardless of the different personal attributes.

Moreover, this study examined the factors that can reduce the differences by driving frequency. In order to find the factors, the personal attributes of the up-group were investigated for each motivational material type.

## 4.2 *Overview of Motivational Materials*

Pattern A of motivational video has a story of going for a drive using an automated driving vehicle with a family. While they are enjoying driving, a request to intervene (RtI) gave warning to take over driving. The father, driver, as well as family members were very surprised. The story made them realize the gap between level 5 and level 3 and familiarize the issue of themselves that may face near future.

Pattern B of the motivational video includes the need for safe driving as a general theory and an abstract theory based on the results of data analysis on automobile accidents. The two patterns of motivational video materials have been verified to have the same effect by pre-evaluation using the ARCS evaluation items.

## 4.3 *Measurement*

There is still no consensus among researchers on how to measure resilience at the individual level [15, 16]. Since there are many common factors, the Kodama scale [17] was used in this study. Career resilience scale is surveyed with 14 questions ("Factor 1: Challenge," "Factor 2: Novelty / Diversity," "Factor 3: Future-oriented," "Factor 4: Assistance-oriented"). Answers are obtained on a 5-point scale from "I agree very much (5 points)" to "I do not agree at all (1 point)" for each item.

The learning style was investigated with 16 questions of 4 factors with reference to the model [18] proposed by Felder. Felder–Silverman model shows their characteristics in four pairs of items. ACT_REF: Active types prefer discussions and explorations with others, while reflective types prefer organizing information and summarizing important points. SEN_INT: Sensing types prefer learning with facts, while intuitive types prefer a change in reference to learning. VIS_VIR: Visual types prefer learning with visual images, such as photos, diagrams, flowcharts, timelines, and videos, while verbal types prefer learning through words (i.e., written or oral explanations). SEQ_GLO: Sequential types prefer a linear thinking process, while global types prefer a holistic thinking process.

### 4.4 Data Analyses

It is assumed that the understanding of autonomous driving depends on individual attributes. Logistic regression analysis was performed by dividing two groups based on the test scores before and after learning. A group with increased scores was determined an up-group and a group with non-increased scores was determined non-up-group. This calculated the probability that multiple independent variables affected the dependent variable. Those with a small difference in the calculated numerical values (propensity scores) were selected from the up-group and the non-up-group, and the attributes and characteristics of the up-group and the non-up-group were selected. In this study, after matching propensity scores, the effectiveness of the motivational learning materials was examined.

## 5 Results

### 5.1 Results of Pre- and Post-tests by Learning Material Type

Table 1 shows the basic attributes of respondents.

Table 2 compared the results of the pre–post-tests for each pattern. Question 1 is a multiple choice question and has four alternatives. Question 2 consists of 10

**Table 1** Respondent attributes

| | | | Pattern A | | Pattern B | | Pattern C | |
|---|---|---|---|---|---|---|---|---|
| Age | Male | Female | Male | Female | Male | Female | Male | Female |
| ~ 39 | 93 | 93 | 31 | 31 | 31 | 31 | 31 | 31 |
| 40–49 | 93 | 93 | 31 | 31 | 31 | 31 | 31 | 31 |
| 50–59 | 93 | 93 | 31 | 31 | 31 | 31 | 31 | 31 |
| 60–69 | 93 | 93 | 31 | 31 | 31 | 31 | 31 | 31 |
| 70 ~ | 93 | 93 | 31 | 31 | 31 | 31 | 31 | 31 |
| Total | 930 | | 310 | | 310 | | 310 | |

**Table 2** Pre–post-test score comparison

| | | Pattern A<br>N = 310 | Pattern B<br>N = 310 | Pattern C<br>N = 310 |
|---|---|---|---|---|
| Pre Test | Mean | 7.05 | 7.07 | 7.31 |
| | SD | 2.14 | 2.14 | 2.13 |
| Post Test | Mean | 8.32 | 8.11 | 8.44 |
| | SD | 2.24 | 2.5 | 2.23 |

**Table 3** Pre–post-test score comparison by pattern

| | | Pattern A N = 310 | Pattern B N = 310 | Pattern C N = 310 |
|---|---|---|---|---|
| Factor 1:challenge | Mean | 2.65 | 2.69 | 2.69 |
| | SD | 0.55 | 0.53 | 0.53 |
| Factor 2:novelty/diversity | Mean | 2.79 | 2.73 | 2.75 |
| | SD | 0.63 | 0.56 | 0.62 |
| Factor 3:future-oriented | Mean | 2.32 | 2.46 | 2.39 |
| | SD | 0.69 | 0.69 | 0.69 |
| Factor 4:help-seeking | Mean | 2.81 | 2.83 | 2.82 |
| | SD | 0.62 | 0.61 | 0.60 |

questions and is an alternative question. Therefore, the perfect score for this test is 11 points. Test results after learning were significantly higher, regardless of the pattern. The score increase rate of the pre–post-test for each pattern was 1.18 times for pattern A, 1.15 times for pattern B, and 1.15 times for pattern C.

## 5.2 *Results of Career Resilience by Pattern*

The calculation results of the factor confidence coefficient are Factor 1: α (challenge = 0.80), factor 2: novelty/diversity (α = 0.78), factor 3: future-oriented (α = 0.88) Factor 5: Help-Seeking Preferences (α = 0.78). There was no problem with internal consistency. Table 3 compares the average scores of the learner's four factors of career resilience. There was no significant difference between the three patterns. As a result of the T-test for each pattern, there was a significant difference in factor 3: future orientation of A and B at the 5% level, and pattern A was significantly lower.

## 5.3 *Results of Learning Style by Pattern*

Table 4 and Fig. 2 show the results of comparing the average scores of the four learning style factors by each motivational video pattern. The ratio of pairs in each category is high for the same items regardless of pattern. In the ACT-REF pair, REF scored high, similarly, SEN_INT has higher SEN, VIS_VIR has higher VIS, and SEQ_GLO has higher SEQ. It is considered effective to use learning materials that use not only words but also images and to use the facts.

**Table 4** Learning style score comparison

| | | Pattern A<br>N = 310 | Pattern B<br>N = 310 | Pattern C<br>N = 310 |
|---|---|---|---|---|
| LS1_ACT_REF | M | 0.40 | 0.40 | 0.35 |
| | SD | 0.49 | 0.49 | 0.48 |
| LS2_SEN_INT | M | 0.69 | 0.74 | 0.67 |
| | SD | 0.46 | 0.44 | 0.47 |
| LS3_VIS_VIR | M | 0.69 | 0.68 | 0.65 |
| | SD | 0.46 | 0.47 | 0.48 |
| LS4_SEQ_GLO | M | 0.65 | 0.65 | 0.64 |
| | SD | 0.48 | 0.48 | 0.48 |

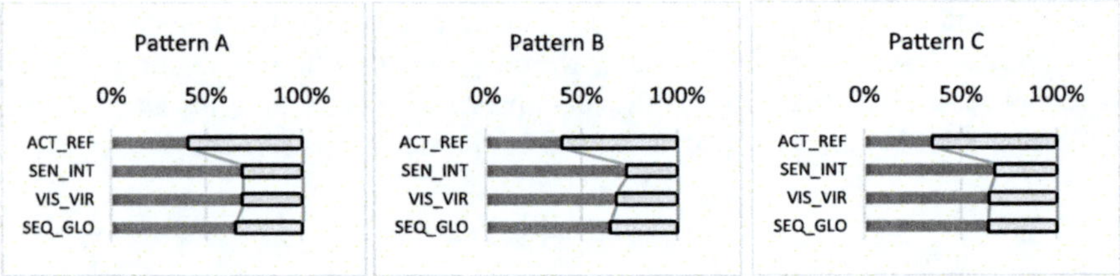


**Fig. 2** Learning style score comparison

## 5.4 *Results of Analysis of Variance*

Figure 3 are graphs that divide the total resilience score into four stages and summarize the pre- and post-scoring scores by motivational video patterns.

It can be seen that the higher the resilience score, the higher the post-score, regardless of the pattern. This result is consistent with previous studies on resilience. Pattern B and pattern C show interactions depending on the level of resilience.

## 5.5 *Logistic Regression Analysis Results*

The method used for analysis was logistic regression for each motivational video pattern to investigate how much one's career resilience and learning style affects one's learning. Tables 5, 6, and 7 show the results of logistic regression analysis of each pattern.

$p < 0.05$ $^{**}p < 0.01$ Cox-Snell $R^2$ 0.12 $N = 310$.

$p < 0.05$ $^{**}p < 0.01$ Cox-Snell $R^2$ 0.11 $N = 310$.

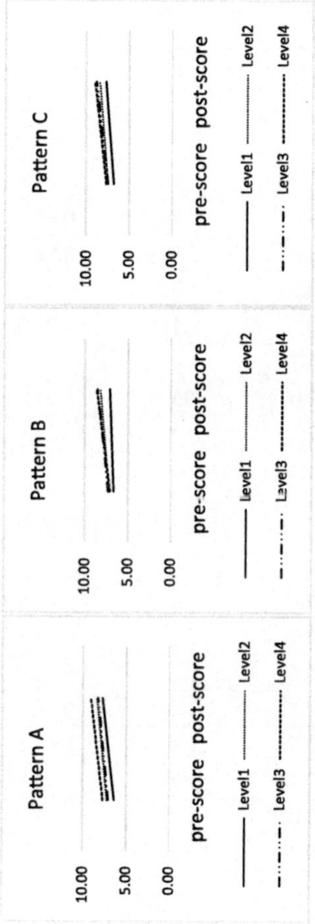


**Fig. 3** Comparison by pattern resilience level

**Table 5** Pattern A analysis results

| | B | Sig | Exp(B) | Lower limit | Upper limit |
|---|---|---|---|---|---|
| Pre Test Score | −0.33 | 0.00** | 0.72 | 0.63 | 0.82 |
| LS1_ACT_REF | −0.43 | 0.10 | 0.65 | 0.39 | 1.09 |
| LS2_SEN_INT | 0.91 | 0.00** | 2.49 | 1.44 | 4.33 |
| LS3_VIS_VIR | 0.14 | 0.62 | 1.15 | 0.66 | 2.00 |
| LS4_SEQ_GLO | 0.14 | 0.61 | 1.15 | 0.67 | 1.98 |
| Resilience score | 0.02 | 0.89 | 1.02 | 0.79 | 1.29 |
| Constant | 2.42 | 0.00 | 11.26 | | |

**Table 6** Pattern B analysis results

| | B | Sig | Exp(B) | Lower limit | Upper limit |
|---|---|---|---|---|---|
| Pre Test Score | −0.12 | 0.05* | 0.89 | 0.80 | 0.99 |
| LS1_ACT_REF | −0.12 | 0.63 | 0.88 | 0.52 | 1.39 |
| LS2_SEN_INT | 0.08 | 0.79 | 1.08 | 0.66 | 1.97 |
| LS3_VIS_VIR | 0.12 | 0.64 | 1.13 | 0.70 | 1.89 |
| LS4_SEQ_GLO | 0.08 | 0.75 | 1.09 | 0.65 | 1.83 |
| Resilience score | 0.37 | 0.00** | 1.53 | 1.21 | 1.94 |
| Constant | −0.26 | 0.60 | 0.20 | | |

*$p < 0.05$ **$p < 0.01$ Cox-Snell $R^2$ 0.09 $N = 310$

**Table 7** Pattern C analysis results

| | B | Sig | Exp(B) | Lower limit | Upper limit |
|---|---|---|---|---|---|
| Pre Test Score | −0.29 | 0.00** | 0.75 | 0.66 | 0.85 |
| LS1_ACT_REF | −0.13 | 0.63 | 0.89 | 0.52 | 1.48 |
| LS2_SEN_INT | 0.42 | 0.13 | 1.52 | 0.89 | 2.59 |
| LS3_VIS_VIR | 0.10 | 0.69 | 1.11 | 0.67 | 1.84 |
| LS4_SEQ_GLO | 0.19 | 0.48 | 1.21 | 0.71 | 2.07 |
| Resilience score | 0.32 | 0.01* | 1.37 | 1.08 | 1.74 |
| Constant | 1.51 | 0.01 | 4.51 | | |

As a result of the analysis, it can be seen that in pattern A, the learning style influences the increase in the score rather than the resilience. Patterns B and C are more affected by resilience than learning styles, and the odds ratios are 1.44 for pattern B and 1.32 for pattern C, indicating that pattern B is more affected.

**Table 8** Propensity score matching results

| | | Pattern A N = 214 | Pattern B N = 247 | Pattern C N = 215 |
|---|---|---|---|---|
| Pre Test | Mean | 7.62 | 7.44 | 7.54 |
| | SD | 1.86 | 1.99 | 1.90 |
| Post Test | Mean | 8.37 | 8.02 | 8.32 |
| | SD | 2.31 | 2.62 | 2.37 |

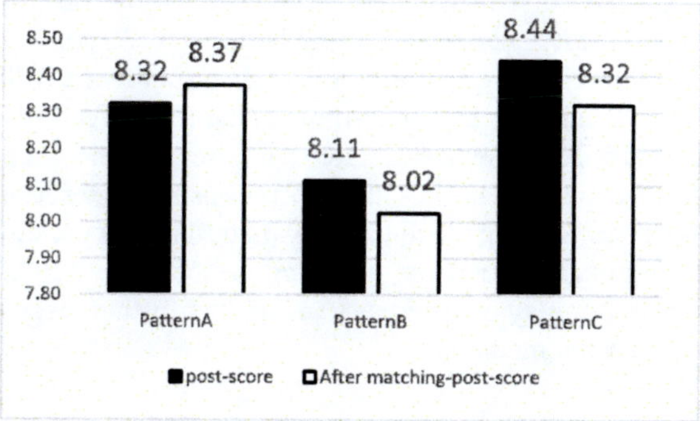


**Fig. 4** Comparison by pattern resilience level

## 6 Discussion

In order to confirm the motivational effect of pattern A, matching was performed by propensity score which consists of pre-score and learning style (SEN). In propensity score matching, they were divided into two groups. One was the propensity scores of the non-rising group and the other was the rising group those who have the same or extremely small difference in propensity scores.

Table 8 compares the pre- and post-scoring again using only the matching data. The pre-scoring increased in all patterns, but the post-scoring increased only in pattern A (Fig. 4). These test questions were not asked about driving skills but driving attitudes and operation planning which were the higher levels of Keskinen's hierarchical approach theory. Thus, pattern A, which is a narrative-based material was the most effective.

## 7 Conclusion

This study examined if motivational video materials increased the learning effectiveness of the higher levels, which were described by Keskinen's hierarchical approach theory. The result of propensity score matching indicates that the learning effect of pattern A using the narrative method was the highest and that it absorbed individual resilience factors.

Case-based pattern B and no motivational video were found to be impacted by resilience. Narrative-based motivational materials may have the effect of absorbing individual resilience.

The survey found two points:

1. Narrative-based motivational materials may have the effect of absorbing individual resilience.

2. On the other hand, it can be influenced by learning style. When using narrative materials, it is necessary to find better ways to convey the facts.

When conducting safety education on automated driving for drivers, it is important to make them empathize with the character's feelings. Although the pre-evaluation using the ARCS evaluation items was verified the same effect in two patterns of motivational video materials, the analyses using career resilience showed the differences. When developing video learning materials, it is possible to absorb the characteristics of personal attributes by utilizing not only the ARCS model but also the narrative method. It may be necessary to conduct further research and keep seeking effective educational methods.

## 8 Limitations and Recommendations

There are some limitations to this study. In this survey, gender, marital status, age, etc. were examined, but self-efficacy and expected results after learning were not investigated. In addition, in this study, the learning effect was tested by a selective test, and in order to investigate the degree of comprehension more accurately, it is necessary to consider the test questioning method. As a further study, it is necessary to investigate the learning effect when the learning materials presented are changed according to the pre-score, learning style, and resilience score.

As a further study, the relationships between ARCS model evaluation items and career resilience should be investigated more.

**Acknowledgements** This work was supported by Council for Science, Technology and Innovation (CSTI), Cross-ministerial Strategic Innovation Promotion Program (SIP), entitled "Human Factors and HMI Research for Automated Driving " (funding agency: NEDO).

## References

1. Zhou, H., Itoh, M., Kitazaki, S.: How Does Knowledge about System Limitations Contribute to Interventions into Partial Automation Among Elderly Drivers? Proceedings of The 2018

IEEE international Conference on Systems, Man, and Cybernetics (SMC2018), pp. 851–820.Miyazaki, Japan(2018).
2. Zhou, H., Itoh, M., Kitazaki, S.: Long-term Effect of Experiencing System Malfunction on Driver Take-over Control in Conditional Driving Automation. Proceedings of The 2019 IEEE international Conference on Systems, Man, and Cybernetics (SMC2019), pp. 1950–1955. Bari, Italy (2019).
3. Okamoto, M. , Nakahira, K. , Kitajima, M.: The Measure towards the Improvement in Motivation of Safety Education using "Visualization" of Traffic Accident Damage. FIT2012, pp. 659–660, (2012).
4. Arame, M., Hideshima, E., Kanda, K. .: Study on Scaffolding Effect of Video Learning materials in Participatory Learning for Improving Preparedness. JSCE D3, 68 (5), pp. 35–43, (2012).
5. Arame, M., Handa, J., Goda, Y., Toda, M., Matsuba, R., Zhou, H., Itoh, M., Kitazaki, S.: Learning Effects of Different Learning Materials about Automated Driving Level 3: Evidence from a Propensity Score Matching Estimator. Proceedings of Fifth International Congress on Information and Communication Technology, ICICT 2020, London, Volume2, pp. 387–394, (2020).
6. Keskinen E (1994) Why do young drivers have more accidents? Junge Fahrer und Fahrerinnen, Referate der Esten Interdiziplinären Fachkonferenz, pp 12–14
7. Suzuki, K. : On the Framework of Designing and Developing "appealing instruction": The ARCS Motivation Model. Journal of Educational Media Research Vol(1) Issue1, pp. 50–61, (1995).
8. Nemeth, C. P. , Hollnagel, E. (eds.), Kitamura, M. (translator) .: Guidelines for Application of Resilience Engineering: To Become a Resilient Organization. Nikka Giren Publisher, (2017).
9. Imai, Y.: A Study on the Narrative Theory on Bruner, Bulletin of the Graduate School of Education, Hiroshima University. Part 1, Learning and curriculum development (59), pp. 51–57, (2010).
10. Ando, K.: Narrative Transportation Effects on Blog Readers. Advances in Consumer Studies Vol (21), Issue 1_2, 1_2_25–1_2_46, (2015).
11. Bruner J (1990) Acts of Meaning. Harvard University Press, Cambridge, MA
12. Gergen, K, J. , Gergen, M. M.: Narrative and the Self as Relationship. Advances in Experimental Social Psychology, 21, ed. Leonard Berkowitz, New York: Academic Press, pp. 17–56, (1988).
13. Escalas JE (1998) Advertising Narratives: What Are They and How do They Work? In: Barbara B (ed) Representing Consumers: Voices, Views and Visions. Routledge Press, Stern, New York, pp 267–289
14. Kogo, C., Suzuki, K.: An analysis of the structure of course evaluation items based on ARCS motivation model, II. Proceedings of ICCE 2000, 8th International Conference on Computers in Education, Taipei, Taiwan. 2. pp. 1577–1578, (2000).
15. Grotberg, E. H. (Ed.) : Tapping your inner strength: How to find the resilience to deal with anything. New Harbinger Pubns, (1999).
16. Hartmann S, Weiss M, Newman A, Hoegl M (2019) Resilience in the workplace: A multilevel review and synthesis. Appl Psychol 69(3):913–959
17. Kodama M (2015) Examination of Construct of Career Resilience and Development of Measurement Scale. Psychology Research 86(2):150–159
18. Felder RM, Silverman LK (1988) Learning Styles and Teaching Styles in Engineering Education. Engr. Education 78(7):674–681

# New Way to Generating and Simulation QKD

**Hasan Abbas Al-Mohammed and Elias Yaacoub**

**Abstract** Due to the necessity for using the security to the end-users, the conventional methods for security are not sufficient for the highly equipped attacker and new ways to violate security. Moreover, the quantum age is coming and brings more vulnerability to classical security techniques. On the other hand, Quantum security relies on the natural physical phenomena (quantum mechanics) and gives a suitable and strong way for security. One of the quantum security protocols is BB84 that depends on the polarization of photons. This paper proposes a new method for the simulation of quantum key distribution (QKD) in the BB84 protocol. The method is practical to use; moreover, it is flexible and can implement any required number of photons as quantum bits, also, the paper evaluates the technique with other techniques that generate and simulate QKD.

**Keywords** Quantum security · BB84 protocol · QKD · Quantum · Photon · Quantum key generator

## 1 Introduction

Cryptography is the art of preventing unauthorized access to private records, maintaining the confidentiality and security of files, and other activities [1]. Classical cryptographic technology in use today is based on the hardness of certain numerical methods, such as integer factoring or the problem of discrete logarithms. However, because these problems are usually not known to be challenging to a malicious entity with quantum computing capability, the resulting cryptosystems are theoretically insecure.

H. A. Al-Mohammed (✉) · E. Yaacoub
Computer Science and Engineering Department, Qatar University, Doha, Qatar
e-mail: ha1800217@qu.edu.qa

E. Yaacoub
e-mail: eliasy@ieee.org

 
X.-S. Yang et al. (eds.), *Proceedings of Sixth International Congress on Information and Communication Technology*, Lecture Notes in Networks and Systems 216,
https://doi.org/10.1007/978-981-16-1781-2_69

One of the most basic cryptographic primitives is the establishment of symmetric keys among two distant parties via an insecure network, and it underlies many cryptographic schemes that are used nowadays. Public-key encryption is widely used to achieve this. No other approach can be found which does not use out-of-band communication, quantum communication, or third parties that are trustworthy [2].

Quantum key distribution (QKD) facilitates the key establishment via an untrusted network like classical public-key cryptography. This approach is known as the distribution of quantum keys, thus the naming QKD. The protection of QKD relies on the natural phenomena of quantum mechanics, not on the complexity of mathematical problems, and can be proved even against an eavesdropper, Eve, who has the infinite computational capacity [2].

If the key is produced as long as the message is to be sent and is used once only (one-time pad), the encrypted message cannot be decrypted for any computing capacity, including the most efficient equipment. The theoretical security of information is known as this kind of security. Since the advent of the original protocol, BB84 [3, 4], other quantum security protocols have also been developed and introduced such as the E91 protocol, etc., for distribution of the key that was generated to ensure secrecy between two devices.

QKD produces a secret key of random photons (Qubit) in an end-to-end relation. A transmitter, a receiver, and two communicating networks compose QKD communication. One is a quantum channel connected to the transmission of quantum random-bit signals by the encoder side and the decoder side, another is a conventional channel linking the main distillers. For the filtration of the main, the primary distillers perform error correction and privacy amplification on exchanged random pieces.

An attacker, who has infinite computing capacity, is able to enter the quantum and classical networks, but cannot access the encoder (polarization filter), decoder (polarization filter at the receiver side), and key distiller's internal bits [5], thus attacker (Eve) cannot detect the key.

It is important to authenticate the classical channel, called the public channel. By checking the changing in the threshold that is calculated before the transmission, the sender and the receiver will detect it as Eve eavesdrops on the quantum channel in every way allowed by the laws of physics. If the error goes above a certain threshold and leads to discarding random bits along with those limitations, the sender and the receiver stop producing the secret key [3].

In this paper, we use BB84 QKD generation to describe the photon polarization states used for adopting optical fiber to relay high-security telecommunication information. We also use easy and functional code in this article to measure the duration of a key for a certain number of photons (quantum bit) protocol BB84. Quantum cryptography uses a quantum channel to confidentially share passwords, barring private information from unauthorized parties or eavesdroppers. QKD offers a special way of exchanging a random sequence of bits between users with a degree of anonymity that cannot be accomplished by encoding the data in quantum states with any other classical cryptography methods to share a random secret key.

## 2 Related Work

By utilizing the states (four polarizations) of the photons describing the quantum key shared in the quantum link, the BB84 protocol [6] was implemented in [7]. Optical fiber is a standard function well known in OptSim (simulating models according to System Dynamics methodology) for its grouping of channels. This technique provides an easy way to overcome the issue of the PBS (Polarization beam splitter).

As a PBS and a random collection of the incoming photons, the part called "pick" can be used [16]. In QKD experiments, to deliver the photons to the recipient, random polarization choices are made. For the incoming photon measurement, the receiver often selects random polarization. Lastly, on the basis of polarization, detectors may be triggered. The public channel discusses all photon values reported by Alice and Bob, which illustrates the basic operation of the QKD scenario [7].

Moreover, QKD is implemented using Python packages such as QuTip in [8], where the authors have used BB84 protocol and the package makes use of different science libraries. The software was designed in such a way that, as in the actual quantum system, it can replicate the quantum world and maintain the randomness of the system. Furthermore, modeling Fuzzy Logic QKD using Simulink is simulated in [9].

## 3 Quantum Key Distribution

QKD is a technology that requires parties (Alice and Bob) to be involved to share a secret symmetric key. If an eavesdropper (Eve) attempts to snatch the key in a QKD protocol, communicators may use applicable quantum laws to observe it (e.g., the well-known Heisenberg uncertainty theory) [10]. In quantum mechanics, QKD usually relies on the inability to analyze a structure without disturbing it. She would definitely leave any signs that can be detected, as Eve needs to eavesdrop on Alice and Bob's quantum correspondence. The QKD protocol thus achieves security. On the other hand, the protocol randomly produces long keys as long as Eve is passive.

In addition, the protocol identifies the attack and terminates the generation of keys as Eve tampers with quantum networks. Also, as long as Eve is passive, the protocol will not be suspended. The likelihood that the protocol does not stop and an attacker copies the generated keys is so tiny for any attack on quantum networks [11].

## 4 BB84 Protocol

Bennet and Brassard proposed in 1984 the first QKD protocol, It was referred to it as BB84. At the moment, BB84 is the most general and powerful quantum cryptography protocol that transmits data using photon polarization states [10].

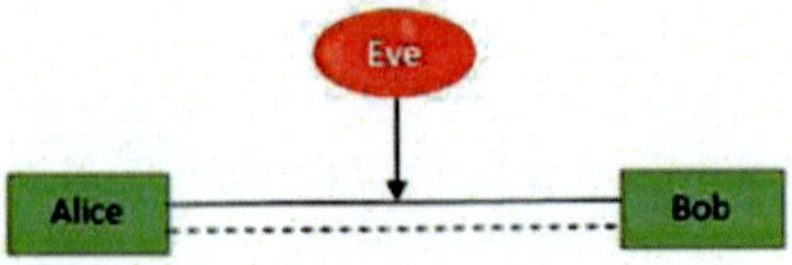


**Fig. 1** A QKD protocol generally consists of a quantum channel (the strong line) and an authenticated classical public channel (the dashed)

In quantum cryptography, as a universal convention, Alice transfers quantum states through a quantum channel to Bob. An eavesdropper (Eve) is accused of listening to the quantum channel, as illustrated in Fig. 1. Alice and Bob can come up with a shared key that they can then use with a symmetrical encryption method, such as a one-time pad, Alice has four polarized filters, with each stream having one of four different polarizations, each two of polarizations represent "1" or "0," and can be shared with Bob, Photons are used as qubits in the initial formulation of the protocol, the information being stored in their polarization [12, 13] (see Fig. 1).

## 5 Proposed Method for Generating QKD

To build photon streams, Alice has four polarized filters where each stream has one of four distinct polarizations, rectilinear polarizations of 0°, 90° and diagonal polarizations of 45° and 135°. For the binary state "0," Alice and Bob settle on 0° or 90° to represent it, whereas 45° or 135° will represent the binary state "1," as shown in Table 1.

Alice produces the first step of the main source, Alice generates a polarized photon stream, choosing their polarization at random. Alice transmits these photons one polarization at a time through the quantum tube, Alice makes a note of the unpredictability polarization chain that Bob has no way of guessing what polarization any of the photons has. Bob has two detectors:

A rectilinear filter (+): Photons with a rectilinear position move unchanged into this filter, the rectilinear filter switches to a random rectilinear state (0° or 90°) the state of a diagonally polarized photon.

**Table 1** The polarization states and corresponding bit represented

| Polarization/bit | 0 | 1 |
|---|---|---|
| Rectilinear + | ↕ | ↔ |
| Diagonal x | ⤢ | ⤡ |

Photons with a diagonal state pass unchanged across a diagonal filter (x). The diagonal filter alters the state to a random diagonal state (45° or 135°) of a rectilinearly polarized photon, as described in Table 1.

When each photon arrives, Bob arbitrarily guides it to one of his two detectors that Alice and Bob interact over the normal contact channel to address Bob's detector preference, discarding all the wrong bits that Bob acquired from an incorrect filter range. To extract a shared key stream, the remaining bits are then used. Eve follows the same approach used by Bob, sometimes using the right detector and sometimes the wrong detector. If her decision is correct, the photon will continue with its prior polarization on its path. If her decision is false, she is going to change the polarization of the photon that she passes on to Bob. The resulting bit stream would not be right in any way due to Eve's interference at the end of Bob's detection, and after consulting with Alice to discard wrongly observed photons, Alice and Bob execute a single search before the mutual key stream is used comfortably. If they discover a bunch of mismatches, they publicly pick and compare a number of bits picked at random from their key streams, and the error rate crosses a negotiated threshold, they delete the whole key stream and produce a new one.

Algorithm 1: Algorithm for simulation and generating QKD

```
1: A[i] = 0;//declaration an array for the sender
2: B[j] = 0;//declaration an array for the receiver
3: key[k] = 0;
4: for i: = N;//N = number of photons (Qubits)
5: for m: = long number;//for increasing the randomness
6: A[i] = rand(1,4)//prefer to be true random number that depend on natural phenomena such as
   noise of semiconductor
7: for j: = N;//N = number of photons (Qubits)
8: for l: = long number;//for increasing the randomness
9: B[j] = rand(1,4)//prefer to be true random number that depend on natural phenomena such as
   noise of semiconductor
10: for m: = N;
11: if A[i] = = B[j] then
12: key[i] = A[i];//key length;
```

Algorithm 1 describes the approach. At the beginning, we declare arrays containing the random numbers for sender and receiver, also the last array is called key for containing the key values that were generated and reached the receiver. Line 6 generates a random number from 1 up to 4, each number representing a polarization of photon (Qubit) such as 1 for 0°, 2 for 45°, 3 for 90°, and 4 for 135°. Line 11 compares the two arrays (representing the polarization of the sender and receiver photons), takes each cell in the Alice array, and compares with each cell in the Bob array (corresponding cell, such as cell A[1] compared to B[1], and guarantees that the polarization at each end is equal when the polarization is the same, then the polarization of the photon will be stored in the corresponding cell of the key array (e.g., when A[1] equals B[1] then A[1] stored in the key[1]), the loop will continue until the end of the arrays.

This algorithm is dynamic as it can adjust the number of the sent photons to be equal to the number required for achieving the requested key length.

## 6 Results and Discussion

The algorithm is implemented by using MATLAB. To evaluate the work of this paper, first, we have taken other work and compared it with our work, such as the work at Luxembourg university presented on the website in [15]. This work is online and simple but the simulator is limited in the number of photons. Therefore, we took 50 cases from 510 photons up to 1000 photons. Also, we took the number of sent photons (Qubits) and the final key length and analyzed them by using SPSS to find the correlation, we found the correlation illustrated in Table 2. Moreover, we plot the sent photons with respect to the key length in Fig. 2. This figure illustrates the correlation and links each number of sent photons to the corresponding key length. on the other hand, as part of our results, we simulate 7010 up to 7500 photons (Qubit). We used these numbers of photons to show that our approach is flexible and can be used with any photons, moreover, we can compare our work easily with other scenarios such as in [14], where the authors have used (IDQ-Clavis2) device to generate and implement a QKD.

Table 2 shows the relationship between the number of initial photons sent by Alice and the number of final key lengths that produce secrecy at each end (Alice and Bob), these results are regenerated from the simulator in [15], whereas the proposed simulation results show the same correlations, as shown by the Pearson Correlation (which is a statistic that tests the linear correlation between the two variables) in Table

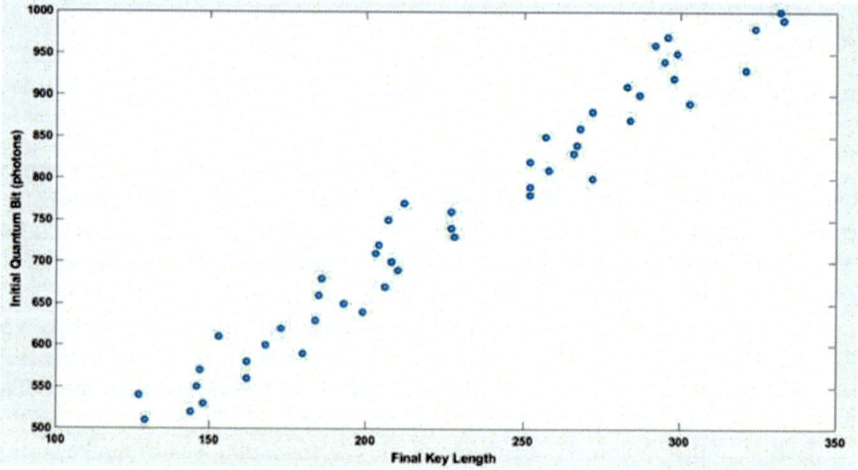


**Fig. 2** Initial photons (Qubit) with respect to the key length

**Table 2** The correlation of the Initial number of qubits (photons) to the key length

Algorithm 1: Algorithm for simulation and generating QKD

```
A[i] = 0;//declaration an array for the sender
B[j] = 0;//declaration an array for the receiver
key[k] = 0;
for i: = N;//N = number of photons (Qubits)
for m: = long number;//for increasing the randomness
A[i] = rand(1,4)//prefer to be true random number that depend on natural phenomena such as noise of semiconductor
for j: = N;//N = number of photons (Qubits)
for l: = long number;//for increasing the randomness
B[j] = rand(1,4)//prefer to be true random number that depend on natural phenomena such as noise of semiconductor
for m: = N;
if A[i] = = B[j] then
key[i] = A[i];//key length;
```

**Table 3** The correlation of the Initial number of qubits (photons) to the key

Algorithm 1: Algorithm for simulation and generating QKD

```
A[i] = 0;//declaration an array for the sender
B[j] = 0;//declaration an array for the receiver
key[k] = 0;
for i: = N;//N = number of photons (Qubits)
for m: = long number;//for increasing the randomness
A[i] = rand(1,4)//prefer to be true random number that depend on natural phenomena such as noise of semiconductor
for j: = N;//N = number of photons (Qubits)
for l: = long number;//for increasing the randomness
B[j] = rand(1,4)//prefer to be true random number that depend on natural phenomena such as noise of semiconductor
for m: = N;
if A[i] = = B[j] then
key[i] = A[i];//key length;
```

3. This correlation also demonstrates the precision of the new approach relative to other simulations.

In addition, Fig. 2 shows the number of initial photons versus the key length; for example, when 510 photons are sent they produce 144 bits as the key length and 322 is the key length for 1000 photons when the proposed method is equivalent to these results ( because the quantum mechanics is truly random, therefore, the results did not exact the same at the two approaches).

The main benefit of the proposed method is more realistic than the other methods since each number in the algorithm (the random number between 1 and 4) represents the polarization state of the photon, the other methods simulate the final decision "0" or "1" the proposed method actually indicates the randomness of the generation of the quantum key, and the method can also predict the number of each polarization

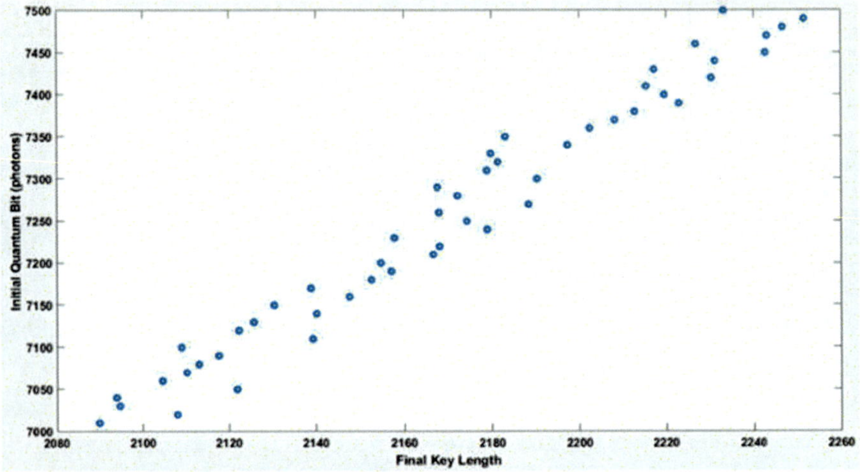


**Fig. 3** Initial photons (Qubit) versus the key length (the proposed method)

state (such as how many photons with 0, 45, 90, and 135 can be used for other applications in pure physics science), therefore, in the algorithm, it is preferable to use the true random number generator that depends on the physical phenomena.

Figure 3 shows that the new method can give the corresponding number of key lengths to the photons sent, such as for (N = 7500) photons the key length is 2089, and so on. For other methods, such as [15], there is a restriction, as the key length is given for up to 1000 initial photons transmitted. Finally, the proposed simulation method is less complicated than the ones using fuzzy logic [7], since the algorithm is simple (almost 11 lines of code).

## 7 Conclusion

The paper proposed a new practical method for simulating a BB84 protocol for generating a QKD. The new technique is easy and can implement any required number of photons (Qubit). Furthermore, the paper evaluates the new method compared to other works, and the results show that this method is promising. We can extend this work to show the thresholds of the highest and lowest numbers of the key length, and then it can be used to easily detect an eavesdropper in the middle (man in the middle). We leave this investigation for future work.

## References


1. Bruss D, Erdélyi G, Meyer T, Riege T, Rothe J (2007) Quantum cryptography: a survey ACM Comput Surv 39(2)
2. Lohachab A, Lohachab A, Jangra A (2020) A comprehensive survey of prominent cryptographic aspects for securing communication in post-quantum IoT networks. Internet of Things 9, p. 100174
3. Sasaki M (2018) Quantum key distribution and its applications. IEEE Secur Priv 16(5):42–48
4. Editors F et al (2010) Applied Quantum Cryptography (Lecture Notes in Physics, 797). Springer Berlin Heidelberg
5. Muruganantham B, Shamili P, Ganesh Kumar S, Murugan A (2020) Quantum cryptography for secured communication networks. Int J Electr Comput Eng **10**(1): 407–414
6. Shor PW, Preskill J (2000) Simple proof of security of the BB84 quantum key distribution protocol. Phys Rev Lett 85(2):441–444
7. M. E. System, "using OptSim," IEEE sponsere, Int. Conf. Electron. Commun. Syst. (ICECS 2015) Implement., no. Icecs, pp. 457–460, 2015.
8. Prithvik HC, Charan KR, Sachin BS, Rakesh KR (2020) "Implementation of Quantum Key Distribution using Python", Wutan Huatan Jisuan Jishu, vol. XVI 190:190–196
9. Quantum cryptography, Modelling Fuzzy Logic Quantum Key Distribution using Simulink, https://www.mathworks.com/content/dam/mathworks/mathworks-dot-com/solutions/automotive/files/in-expo-2015/modelling-fuzzy-logic-quantum-key-distribution-using-simulink.pdf, last accessed 2021/1/19.
10. Zhang H, Ji Z, Wang H, Wu W (2019) Survey on quantum information security. China Commun. 16(10):1–36
11. El-Latif AAA et al (2020) Providing End-to-End Security Using Quantum Walks in IoT Networks. IEEE Access 8:92687–92696
12. M. Mina and E. Simion, "A Scalable Simulation of the BB84 Protocol Involving Eavesdropping," *Cryptol. ePrint Arch.*, vol. 1, no. Report 2020/1074, pp. 1–19, 2020.
13. Singh H, Gupta DL, Singh A (2014) Quantum Key Distribution Protocols: A Review. IOSR J. Comput. Eng. 16(2):01–09
14. Mavromatis A, Ntavou F, Salas EH, Kanellos GT, Nejabati R, Simeonidou D (2018) Experimental Demonstration of Quantum Key Distribution (QKD) for Energy-Efficient Software-Defined Internet of Things. Eur. Conf. Opt. Commun. ECOC, vol. 2018-Septe, no. 1, pp. 1–3
15. QKD simulator, Homepage. https://www.qkdsimulator.com/. Last accessed 19 Jan 2019
16. Viacheslav B (2015) Security Issues of quantum cryptographic systems with imperfect detectors. Doctoral Diss., vol 151,September 2016, pp. 1–133

# Arabic Handwriting Recognition with Wavelets

**Zahia Tamen, Dalila Boughaci, and Habiba Drias**

**Abstract** In this paper, we explore wavelet analysis for recognizing Arabic handwriting. Thus, two main studies are undertaken. First, we use some wavelets namely (db4, sym4 and coif1) and related techniques of feature extraction for the purpose of recognition of Arabic handwritten words' images coming from the IFN/ENIT database. The multi-layer perceptron is used in the classification step. Further, after deciding on the parameters' combination that worked the best, the latter is fed in an SVM classifier with optimized parameters using the K-Fold cross-validation technique. In this second study, a large number of wavelet families and members are tested. The results compete with the state-of-the-art ones.

**Keywords** Wavelets · Arabic handwriting recognition · MLP · SVM · Pattern recognition · IFN/ENIT database

## 1 Introduction

The recognition of the manuscript in general and the Arabic manuscript, in particular, remains an area with great challenges. Although research continues to yield important results for Latin and Chinese manuscripts, there is as yet no a commercialized recognition system for the Arabic manuscript. The interest of such a field leaves no doubt in the daily tasks such as postal sorting, the verification and validation of postal check amounts, the verification and authentication of signatures, etc. This interest increases when we know the enormous cultural heritage contained in handwritten documents which are saved in digital versions from scanners and others. These documents are only waiting for the appearance of reliable systems to transform them into symbolic representations where documentary and information retrieval will be

Supported by organization DGRSDT.

Z. Tamen (✉) · D. Boughaci · H. Drias
USTHB, Computer Science Department, LRIA Laboratory, University of Science and Technology Houari Boumediene, Houari, Algeria



X.-S. Yang et al. (eds.), *Proceedings of Sixth International Congress on Information and Communication Technology*, Lecture Notes in Networks and Systems 216,
https://doi.org/10.1007/978-981-16-1781-2_70

facilitated. The use of wavelet analysis in the field of pattern recognition and in particular in the recognition of the manuscript has already been tested and has already given satisfactory results. But til now, we do not yet know which parameters are able to optimize this recognition. The goal of this paper is to make an in-depth study of the wavelet analysis for the purpose of Arabic handwritten recognition. Most of the parameters which derive from the wavelet analysis are undertaken in an intensive experiment to deduce which is the best parameter combination able to optimize this recognition using an MLP classifier. The resulting parameters are used as the input of an SVM classifier with optimized parameters using the cross-validation technique. For this, we used for the validation of the results the IFN/ENIT database of Tunisian town and village names. The rest of this paper is structured as follows: in the first section, a study of the state of the art of the field is reviewed. The second section is devoted to wavelet analysis in particular. The different parameters which can be taken into consideration in this field are defined and the works in which they have been tested are explained. A review of the classifiers used is also undertaken. Section three will explain in detail the experimentation that will be undertaken with all the pre-processing and post-processing. The implementation of the two systems, as well as the experiments undertaken, will be relegated to the fourth section. A detailed analysis of the results is also described there. A conclusion and some perspectives will end this paper.

## 2 State of the Art

In this section, we first give an overview of the Arabic handwriting recognition problem. Then we discuss how wavelets have been used in such systems.

### *2.1 Arabic Handwriting*

More than 260 million people speak Arabic as their native language, and over one billion people use the Arabic language in several religion-related activities [4]. Further, the Persian, Urdu and Jawi languages use Arabic characters in their transcriptions. Thus, the importance of systems performing a good recognition of such script is without doubt.

The recognition of Arabic handwriting is still very challenging due to some particularities and difficulties that are specific to this language.

The Arabic script is written from right to left in a cursive way even in the printed form. Automatic recognition of handwritten remains a challenging task even though the latest improvements of recognition methods and systems are very promising. Especially for the automatic recognition of Arabic handwritten words a lot of work is still to be done [6]. The recognition of handwriting can be achieved with or without segmentation of words or PAWs (Parts of Arabic Words) into their characters constituents [6]. This segmentation can be explicit [11, 16] or not explicit meaning that in the latest case, the recognition of the PAW is based upon the recognition of its constituents without actually segmenting it [6]. In this paper, we're concerned with segmentation-free recognition of Arabic handwritten words. Whole words are considered as the units to be recognized. In this topic, moments have been the features of predilection that were used to characterize the images [3, 16]. The fractal theory has been also used for this purpose [13]. Wavelets have been tested using Gabor filter in [8, 9] and in [17]. For classification, HMM (Hidden Markovian Models) have been used in [5, 12] besides with Neural Networks in [4, 17], SVM and random forests RF in [1, 8]. Combination of classifiers have been used [1]. Deep architecture can be found in [6] where CNN are combined to SVM classifiers and in [7]. We're concerned with recognizing Arabic handwritten words' images using wavelet analysis, at the features extraction step and neural networks and SVM at the recognition step. Wavelet analysis, when applied to digitized images is described below.

## 2.2 *Wavelet Analysis and Parameters*

Discrete signal wavelet decomposition by wavelet transform (DWT) provides an alternative to the discrete Fourier transform (DFT) for signal analysis allowing decomposition into two-dimensional functions of time and scale. The main benefit of DWT over DFT is in its multi-resolution time-scale analysis ability [17]. Wavelet functions used for signal analysis are derived from the initial function W(t) forming a basis for the set of functions for discrete parameters of dilation $a = 2^m$ and $b = k2^m$ translation. Wavelet dilation, which is closely related to spectrum compression, enables local and global signal analysis [10]. From an intuitive point of view, the wavelet decomposition consists of calculating a 'resemblance index' between the signal and the wavelet. If the index is large, the resemblance is strong, otherwise, it is slight. The indices are the wavelet coefficients [2]. For many signals, the low-frequency content is the most important part. It is what gives the signal its identity. The high-frequency content, on the other hand, imparts flavor or nuance. The different wavelets taken in this study are Daubechies wavelets (db4), Symlet (sym4) and Coiflets (coif1). Figure 1 shows the results of a level 1 wavelet decomposition using 'db1' for an image of the IFN/ENIT database.

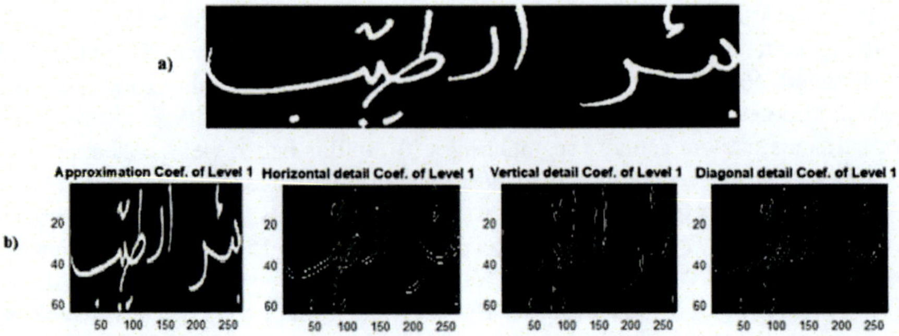


**Fig. 1** Image decomposition using the ‘db1’ wavelet **a** the original image, **b** the approximation and detail images

## 3 Description of the Approach

The aim here is to find out the best parameters discriminating the words’ images when using wavelet analysis. The approach consists of four main steps given as follows:

1. **Preprocessing:** As the images in the IFN/ENIT database are already clean and binarized, processing concerned the image resizing. In order to normalize the size of the words’ images and due to previous experiments [16], we normalized the images to sizes, which are: $32 * 128$ and $64 * 256$. Figure 2 shows the results for an image of the dataset.
2. **Sliding windows:** In order to have some local information that will prevent to confuse words with the globally same shape and different such as in the Fig. 3, a sliding window with some pixels overlapping has been applied to words’ images. Figure 4 shows an example of this process.
3. **Wavelet decomposition:** in this step, we decompose the resized images to the $n^{th}$ level. In each level, we have four images representing the ’approximation image’ and three detail images (see Fig. 1). For these $4 * n$ images, we applied these operators:

**Fig. 2** The size normalization of an image of the dataset

**Fig. 3** Words with same overall shapes

**Fig. 4** 5 Sliding windows with an overlapping of 3 pixels of the images in Fig. 3

- **Mean**: We estimated the mean of the wavelet decomposition at each level and for all approximation and detail matrices as:

$$me_d^k = \frac{1}{n*m}\sum_{i=1}^{n}\sum_{j=1}^{m} x_{ij}^k$$

where $n*m$ represents the number of coefficients in each decomposition image at the level $d$, and $k = 1, \ldots, 4$ represents the decomposition image (k=1, approximation, k=2, horizontal details, k=3, vertical details and k=4 diagonal details).

- **Standard deviation**: The standard deviation gives an idea about how the data fluctuate in regard to the mean value. It is calculated as

$$\sigma_{X^k} = \sqrt{E[(X^k - E[X^k])^2]}$$

**Algorithm 1** Energy(image)

```
Energie ← 0
[Lig, Col] ← size(image);
for i = 1 to Lig do
  for j = 1 to Col do
    if image(i, j) ≠ 0 then
      Energie ← Energie + image(i, j)^2
    end if
  end for
end for
```

- **Energy**: This parameter gives the information about the energy contained in each of the decomposition images. It is calculated as given in Algorithm 1.
- **Eigenvalues**: Eigenvalues of a matrix A are the values for which we have:

$$det(A - \lambda I) = 0$$

where I is the identity matrix of the same rank as the matrix A. Initially used to study the main axes of movement of rigid bodies in rotation, eigenvalues and eigenvectors have a wide range of applications today, for example in the analysis of the stability of systems, the study and analysis vibrations, study of atomic orbits, facial recognition, etc.

4. **Parameters Combination** We have taken each of the statistical parameters individually and then concatenated them with each other by taking into account all the possible combinations without omitting the order of concatenation. The number of combinations is calculated as $C = \Sigma_{s=1}^{p}(C_p^s)$. For $p = 4$, (we have four statistical parameters), $C = 15$.

# 4 Numerical Results

Table 1 gives the results with the MLP classifier. We give the classification achieved with the two dimensions [32, 128] et [64, 256] with the parameters [Mean, Standard deviation, Energy] with the $Coif1$ wavelet with an overlapping of the sliding windows of 2 pixels. We notice that the dimension [32, 128] gave the best results. Also, Table 1 shows the impact of the overlapping of the sliding windows with the same parameters that helped to fix the image dimension. As the classification rates were not significantly different, we fixed the overlapping to 2 pixels. Further, we give the results of the Statistical parameters, wavelet and decomposition level. As we can see from Table 1, the parameter 'mean' gave the worst results with all the wavelets families. $Coiflet$ gave the best results with the decomposition level 6. For the Standard deviation, this parameter gave better results than those given by the 'mean' parameter. $Coiflet$ gave the best results with the level 6. For the Energy parameter: the results are better than the two first parameters with $Coiflet$ and $Symlet$ with the level 7. For the Eigenvalues parameter : this parameter gave decidedly the best results. (See Table 1).

Table 2 shows the best results achieved with the best configuration of the MLP. $Coiflet$ worked the best with decomposition levels of 6 and 7. The MLP best architecture is as follows: Layers number =1, Neurons number = 27, Trainlm as a Training function and a Sigmoid Activation function.

In the second study, we deal with the SVM classifier parameters such as the kernel, the Gamma and the C parameters. Table 3 shows the best values obtained with the combination 'mean', 'standard deviation', 'energy' and 10 eigenvalues of the 'coif1' wavelet in terms of mean results using the K-Fold cross-validation technique with K=10.

Figure 5 summarizes the best results obtained with each wavelet family and we can see that the 'coif4' member of the Coiflet wavelet family gave the best results with a rate of 94.03 with the RBF kernel.

The RBF kernel gave decidedly the best results with all the wavelet families.

## 4.1 Comparison with Some State of the Art Works

Table 4 summarizes the comparison of our system with some of the state of the art works described in [14] and shows that our results are quiet better.

**Table 1** Classification rates with MLP

| Decomposition level | With dimensions [32 128] and [64 256] | | | Depending on the overlapping | | |
|---|---|---|---|---|---|---|
| | [32 128] | [64 256] | | 1 | 2 | 3 |
| 4 | 89.6261 | 68.6095 | | 90.5337 | 91.1045 | 89.6261 |
| 5 | 82.6850 | 92.0813 | | 93.0800 | 91.6523 | 82.6850 |
| 6 | 91.8556 | 89.9220 | | 92.7177 | 89.9170 | 91.8556 |
| 7 | 90.9587 | 75.0575 | | 94.6565 | 93.6741 | 90.9587 |
| Decomposition level | Wavelet families and the parameter mean | | | With the standard deviation | | |
| | coif1 | db4 | sym4 | coif1 | db4 | sym4 |
| 4 | 59.3642 | 57.1955 | 42.7205 | 71.6437 | 75.2509 | 70.1319 |
| 5 | 71.2154 | 53.9863 | 70.2679 | 81.5238 | 81.0306 | 81.5924 |
| 6 | 71.9901 | 53.1721 | 75.8937 | 86.2324 | 79.9851 | 74.7747 |
| 7 | 66.4198 | 51.2603 | 63.5876 | 77.1827 | 67.6913 | 20.9228 |
| Decomposition level | With the parameter 'energy' | | | With the eigenvalues | | |
| | coif1 | db4 | sym4 | coif1 | db4 | sym4 |
| 4 | 82.8600 | 71.0365 | 83.2073 | 0.854382 | 0.810724 | 0.844451 |
| 5 | 85.4000 | 66.2599 | 84.8026 | 0.814653 | 0.847282 | 0.841546 |
| 6 | 81.5426 | 28.9546 | 86.9283 | 0.858517 | 0.819668 | 0.670947 |
| 7 | 86.2116 | 86.1304 | 87.2141 | 0.862429 | 0.842973 | 0.869875 |

**Table 2** The best classification results with MLP

| Statistical Parameters | Size | Overlapping | Wavelet | Dec. level | K | Recognition rate |
|---|---|---|---|---|---|---|
| 1 2 3 4 | [32, 128] | 2 | coif1 | 6 | 10 | **93.7981** |
| 1 2 3 | [32, 128] | 2 | coif1 | 7 | 0 | 93.6741 |
| 1 2 3 4 | [32, 128] | 2 | coif1 | 6 | 5 | 93.5858 |
| 2 3 | [64, 256] | 1 | coif1 | 7 | 0 | 92.7057 |
| 1 2 | [64, 256] | 1 | coif1 | 7 | 0 | 91.9200 |
| 1 3 | [32, 128] | 2 | coif1 | 5 | 0 | 90.3371 |
| 3 | [64, 256] | 1 | coif1 | 6 | 0 | 89.5360 |
| 4 | [32, 128] | 1 | coif1 | 7 | 5 | 88.9385 |
| 2 | [32, 128] | 2 | coif1 | 7 | 0 | 86.2360 |
| 1 | [64, 256] | 2 | sym4 | 6 | 0 | 75.8937 |

**Table 3** SVM optimized parameters

| Kernel | C | Gamma | Coef0 | Degree |
|---|---|---|---|---|
| Linear | 0.5 | – | – | – |
| RBF | 100 | Default | – | – |
| Sigmoid | 100 | 0 | Defaults | – |
| Polynomial | 1000 | 0.9 | 0 | 3 |

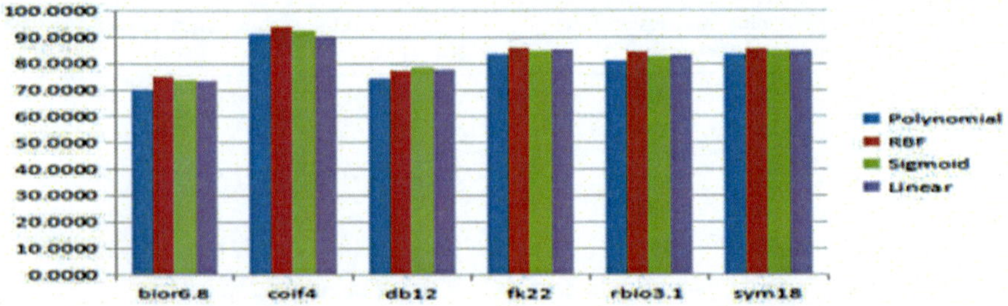


**Fig. 5** The best results with all wavelet families and SVM

**Table 4** Comparison with some of the state of the art results [14]

| Author(s) | Classifier | Data | Accuracy (%) |
|---|---|---|---|
| Pechwitz and Margner 2006 | Semi-continuous 1-D HMM | IFN/ENIT Train: a.b,c Test: d | 89.1 |
| Khorsheed 2007 | HMM (one for each word) | 32,000 words | 85% |
| Ben Cheikh et al. 2008 | Transparent Neural Networks | 2,250 words | 91 |
| Haboubi et al. 2009 | ANN | 16,107 images from IFN/ENIT | 87.1 |
| Azizi et al. 2010 | Multi-classifiers(SVM,K-NN,ANN,HMM) | 10,000 words | 93.96% |
| Chen et al. 2010 | SVM | 7,346 PAWs | 82.7 |
| Khémiriet al. 2014 | PGM | 5,279 words | 92.19 |
| Our system | MLP, SVM | 7560 words | **94.03** |

## 5 Conclusion

This paper describes a synthesis of a work that lasted for years with the wavelet families in order to achieve an Arabic handwriting recognition system. The study and experiments can be divided into two categories: In the first category, we tried to define the best statistical parameters to use when working with wavelet analysis for the purpose of recognizing Arabic handwriting. Thus, we used the 'mean', 'standard deviation', 'energy' and 'eigenvalues' as the statistical parameters and the MLP networks as the classifier tool. In this step of the study, we used the wavelets 'db4', 'sym8' and 'coif1' at the feature extraction step. The best results were achieved by the 'coif1' wavelet with the combination of all the statistical parameters and ten eigenvalues.

In the second step, we used the features resulted in the first step as the input of an SVM classifier with optimized parameters with the K-Fold cross-validation (K=10). In this part of the study, we worked with each of the 'Daubechies', the 'Coiflet', the 'Biorthogonal', the 'Reverse Biorthogonal', the 'Fejer-Korovkin' wavelet families. The results show that decidedly the 'Coiflet' family overpasses all the other wavelet families and that the 'coif4' (see Fig. 5) works the best in recognizing the Arabic handwriting from the IFN/ENIT database.

## References


1. Ali A. A. A., Sureshi M. (2019), A novel features and classifiers fusion technique for recognition of Arabic handwritten character script, SN Applied Sciences (2019) 1:1286
2. Aujol J. F., Aubert G., and Blanc-Féraud L. (2003), Wavelet-Based Level Set Evolution for Classification of Textured Images, IEEE Transactions on Image processing, VOL. 12, NO. 12
3. Broumandnia A, Shanbehzadeh J (2007) Fast Zernike wavelet moments for Farsi character recognition. Image and Vision Computing 25:717–726
4. Chergui L. and Kef M. (2014), A Serial Combination of Neural Network for Arabic OCR, A. Elmoataz et al. (Eds.): ICISP 2014, LNCS 8509, pp. 297–303, 2014. Springer International Publishing Switzerland 2014
5. Dehghan M, Faez K, Ahmadi M, Shridhar M (2001) Handwritten Farsi (Arabic) word recognition: a holistic approach using discrete HMM. Pattern Recognition 34:1057–1065
6. Elleuch M., Maalej R. and Kherallah M.(2016), A New Design Based-SVM of the CNN Classifier Architecture with Dropout for Offline Arabic Handwritten Recognition,Procedia Computer Science, Volume 80, 2016, Pages 1712–1723, The International Conference on Computational Science ICCS 2016
7. Eltay M., Zidouri A., and AHMAD I., Exploring Deep Learning Approaches to Recognize Handwritten Arabic Texts, Digital Object Identifier 10.1109/IEEE. ACCESS.2020.2994248
8. Elzobi M., Al-Hamadi A., Saeed A. and Dings L., Arabic handwriting recognition using Gabor wavelet transform and SVM, 2012 IEEE 11th International Conference on Signal Processing, Beijing, 2154-2158
9. Hamida S., Cherradi B., Terrada O., Raihani A., Ouajji H. and Laghmati S., A Novel Feature Extraction System for Cursive Word Vocabulary Recognition using Local Features Descriptors and Gabor Filter, 2020 3rd International Conference on Advanced Communication Technologies and Networking (CommNet), Marrakech, Morocco, 2020, pp. 1-7
10. Huang K, Aviyente S (2008) Wavelet Feature Selection for Image Classification. IEEE Transactions on Image Processing 17(9):1709–1720
11. Kechad C (2016) Impact des coefficients d'ondelettes dans la reconnaissance du manuscrit arabe. Master, USTHB
12. Khorsheed MS (2007) HMM-Based System for Recognizing Words in Historical Arabic Manuscript. International Journal of Robotics and Automation
13. Mozaffari S., Faez K., Kanan H. R. (2004), Feature Comparison between Fractal Codes and Wavelet Transform in Handwritten Alphanumeric Recognition Using SVM Classifier, Proceedings of the 17th International Conference on Pattern Recognition (ICPR'04)
14. Parvez MT, Mahmoud SA (2013) Offline arabic handwritten text recognition: A survey. ACM Comput, Surv, p 45

15. Pechwitz M., Snoussi Maddouri S., Märgner V., Ellouze N., Amiri H. (2002), IFN/ENIT database of handwritten Arabic words. Colloque International Francophone sur l'Ecrit et le Document, CIFED, Oct. 21-23, 2002, Hammamet, Tunis
16. Tamen. Z, Drias.H, and Boughaci.D, An efficient multiple classifier system for Arabic handwritten words recognition. in Pattern Recognition Letters , 93 (2017), 123132. 2017
17. Yadav A., Gurav J. J., Yadav U., Image Classification by Combining Wavelet Transform and Neural Network, National Conference on Technological Advancement and Automatization in Engineering, January 2016
18. Zou W. and Li Y., (2007), Image Classification Using Wavelet Coefficients in Low-pass Bands, Proceedings of International Joint Conference on Neural Networks, Orlando, Florida, USA, August 12-17

# Development and Pilot Testing of GoTrace: A Web Application for Contact Tracing Using Quick Response (QR) Codes

**Gee Jay C. Bartolome, Mary Ann E. Ignaco, Rhuzel Leinster Dean J. Julian, and John Patrick D. Buco**

**Abstract** It is known that the COVID-19 spread mainly from having close contact with people who have histories of exposure to the novel Corona virus (2019-nCoV or SARS-CoV-2). Though some countries have succeeded in developing the vaccine, its arrival remains a challenge in most countries like the Philippines. For this purpose, strict health protocols and contact tracing remain to be the principal public health interventions and preventive measures to limit the spread of COVID-19. With the number of available contact tracing applications that use Bluetooth and GPS technologies, this study focused on the development and pilot testing of GoTrace, a web-based contact tracing application that uses Quick Response (QR) codes. GoTrace integrates the collection of daily health surveys and location tracking data of the end-users on a digital and web platform to generate real-time, accurate, and reliable contact tracing reports. With the positive feedback of the participants during the pilot testing, GoTrace will offer a promising mobile contact tracing technology that can be adopted in various environments through its dynamic, flexible, and user-friendly architecture.

**Keywords** Health · Web application · Contact tracing · Electronic survey · Quick Response (QR) code · COVID-19 in the Philippines

G. J. C. Bartolome (✉)
Department of Engineering, Cavite State University-CCAT Campus, Cavite, Philippines
e-mail: gjcbartolome@cvsu.edu.ph

M. A. E. Ignaco
Department of Computer Studies, Cavite State University-CCAT Campus, Cavite, Philippines

R. L. D. J. Julian · J. P. D. Buco
Research and Extension, Cavite State University-CCAT Campus, Cavite, Philippines

 
X.-S. Yang et al. (eds.), *Proceedings of Sixth International Congress on Information and Communication Technology*, Lecture Notes in Networks and Systems 216,
https://doi.org/10.1007/978-981-16-1781-2_71

# 1 Introduction

## *1.1 Background of the Study*

COVID-19, an infectious disease caused by a newly discovered Corona virus, spreads primarily through droplets of saliva or discharge from the nose when an infected person coughs or sneeze [1]. As a highly transmittable disease and the third pandemic in the twenty-first century, COVID-19 threatens the human population worldwide [2]. COVID-19 has clear potential for a long-lasting global pandemic, high fatality rates, and incapacitated health systems.

The National Capital Region (NCR) was announced to be placed under quarantine on March 13, 2020. By March 17, the quarantine was expanded into the whole island of Luzon due to the rising number of cases. Under Enhanced Community Quarantine (ECQ), all modes of transportation were banned, except medical personnel and goods, sanitation operations, and emergency cases. Work from home and online communications are encouraged to avoid gatherings. The ECQ was extended until April 30, 2020 [3]. Several versions of the quarantine with variations in leniency such as Modified Enhanced Community Quarantine (MECQ), General Community Quarantine (GCQ), and Modified General Community Quarantine (MGCQ) have also been implemented throughout the country to aid in economic recovery. As policies and guidelines grow more lenient, establishments and government offices resume operations to help the economy bounce back. The DOH reported inflation of confirmed cases to 514,996 including 475,422 recoveries and 10,292 deaths in the Philippines by January 25, 2021.

Isolation and quarantine, contact tracing, physical distancing, decontamination, and hygiene measures were proven effective preventive approaches against the infection [4]. Contact tracing was considered by the WHO as COVID-19's principal public health intervention [5] and a preventive measure on the spread of the virus by classifying, evaluating, and managing personalities with symptoms or close contact. However, the efficiency of contact tracing relies on the community's collective involvement to protect their community through daily reporting of perceived COVID-19 symptoms and willingness to be isolated when symptoms occur [1]. DOH also encouraged the use of personal protective equipment, sanitation, physical distancing, and contact tracing as preventive measures. Coupled with immediate treatment and isolation, contact tracing is proven highly efficient as a small-scale control measure against infectious diseases [6]. Most of the establishments and government offices resorted to pen-and-paper method for contact tracing. This practice requires individuals to list down contacts within the day. Aside from the infection risk with the shared pen and paper, this method also poses high inaccuracy as it is prone to human errors. Moreover, printed documents can be easily damaged or misplaced. DOH [5] said that modern alternatives like digital logbook applications should replace the pen-and-paper setup.

With the existence of several contact tracing technologies that use Bluetooth and GPS in mobile devices, this study focused on the development and pilot testing of a web application for contact tracing, called GoTrace, within the premises of a state university in Cavite. It primarily aimed to digitize contact tracing using Quick Response.

QR codes and direct information to the cloud storage. The GoTrace features a contactless logging in and out of the campus facilities tagged with a unique QR code as well as a secured and live updating of health survey for a quicker and accurate contact tracing. Compared with conventional barcode technologies, QR code offers many advantages. It enables the storage of 100× more information while a QR code-based system can easily be implemented on a mobile device making the technology easy to build, yet reliable.

## 2 Literature Review

Sars-CoV-2 or 2019-nCoV is transmitted through respiratory droplets, and until a vaccine emerges, the key factor to limit the harm caused by the virus is to identify and isolate infected individuals as quickly as possible [7]. The process of contract tracing that emphasizes quarantine and isolation of an index case (with confirmed infection) and its close contacts has shown evidences of mitigating the spread of the disease in countries such as China, Singapore, and South Korea [8]. During contact tracing, the index case and a close contact which is defined by the WHO [1] as "*any person who had one-meter proximity with a confirmed case during their symptomatic period including 4 days before symptom onset*" are challenged to provide information such as the personal health status and time course of contact between one another [9]. However, manual contact tracing can be time-consuming and inefficient, hence, technology-based solutions like smartphone applications have been proposed to automate the process [10]. A modeling study showed that the transmission of the disease can be reduced up to 80% with optimized testing and tracing covered and minimized delays in contact tracing with an app-based technology [11]. In Taiwan for instance, a 22.5–61.5% increase in the responses in contact tracing was recorded with the complement of a web application system [12]. Furthermore, contact tracing apps were also proved to substantially contribute to reduction in infection rates when accompanied by prioritized testing of symptomatic cases [10].

In the Philippines, contract tracing activities mostly rely on the memory of the index patient when submitting a list of close contacts and in fact, the recent data showed a very low contact tracing ratio of 1:7 when the ideal ratio should be 1:37 and 1:30 for urban and rural areas, respectively [13]. In an event that a COVID-19 positive fails to remember his/her close contacts, technology-based contact tracing applications enabled by GPS, Wi-Fi, Bluetooth, network-based API, mobile tracking, and other smartphone technologies can help identify and notify contact persons, or with possible exposure to the index patient [14]. Several contact tracing applications have been developed and deployed in some countries. These contact tracing apps include TraceTogether (Singapore), CoronApp (Colombia), COVIDSafe (Australia), Stopp Corona (Austria), HaNagen (Israel), and StayHomeSafe (China), among others [15].

With the rapid increase in COVID-19 cases and deaths in all countries worldwide, there in inevitable need to integrate emerging technologies in the conduct of contact tracing activities.

One of the common things with countries that have flattened their COVID-19 curves and maintained low mortality rates is their response to integrating digital technology into policies. These comprehensive responses of countries that have been successful at containment and mitigation quickly deployed digital technologies to facilitate planning, surveillance, testing, contact tracing, quarantine, and clinical management to limit the efforts required from their frontliners [16]. With the weak contact tracing in the Philippines due to the absence of system and proper encoding, the Philippines' COVID-19 task force mandated the national government and local government units to use its preferred contact tracing application called StaySafe [13]. However, the insecurity and trust issues of the target users resulted in very low achievement of target end-users [17]. This triggered other government agencies to develop their own contact tracing applications like the Tanod COVID by the Department of Science and Technology (DOST), TRAZE App by the Philippine Airlines, and Trace by the Department of Transportation (DOTr). These technologies exhibit the following common features: (a) end-users are required to manually input their information using digital forms; (b) the software is designed for the specific environment and user types; and (c) the software is implemented in a controlled environment and monitored and maintained by local system administrators.

## 3 Methodology

### *3.1 System Development Model*

The agile methodology was used in the design and development of GoTrace. Agile encourages rapid responses to change with the continuous iteration while it employs continual planning, design, development, and testing, deployment, and system review to deliver a software solution. With the scale and gravity of the problem of transmission of COVID-19, a novel software application must be rapidly developed to provide a contactless and efficient replacement to the conventional contact tracing. Agile will enable modifications even after the implementation of the system where software problems are resolved, and it allows integration of new features while the system adapts to changes in health protocols. Though agile is among the simplest software development models, it was found to be very effective and efficient in turning software plans into realistic and implementable IT solution (Fig. 1).

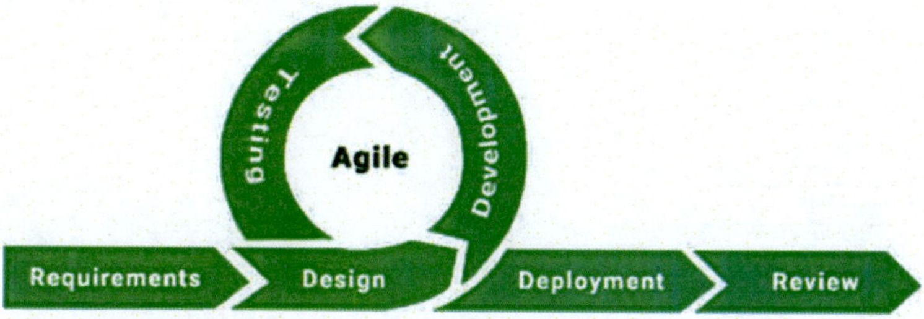


**Fig. 1** The agile software development model

## 3.2 Requirements

From the unstructured interviews and focus group discussions with health representatives and selected participants of contact tracing activities in Cavite State University-CCAT Campus, common problems encountered were determined and analyzed. Contact tracing in the University is done by requiring the employees to fill-out a standard health declaration form and this form is submitted to the campus infirmary on a daily basis. The responses are encoded in spreadsheets either by the University physician or the nurses for manual contact tracing. In addition, the Human Resource Management Office (HRMO) requires the university staff to fill out a separate log sheet for staff monitoring while the University security officers implement locator sheets for visitors/guests. A unified digital health declaration form and computer-assisted contact tracing will overcome the challenges on the physical delays and inaccuracy of the reports, cross-referenced with existing health policies and guidelines and data privacy considerations.

From the analysis of the requirements, the following outcomes were perceived: (a) on the frontend, the types of users who shall respond to the health declaration form should accommodate the faculty, non-teaching staff, students, and visitors; and (b) on the backend, three levels of system administrators will be accommodated consisting of the technical support with full access to the system (except the health survey data), the medical staff with exclusive access to the medical records, and the human resource officer with access only to the daily log-in and log-out records.

Furthermore, the final specifications of the web app must ensure that it can support a large number of users (>50,000) with varying models and specifications of mobile devices. There must also be provisions for quick installation, data interchange, and secured data storage.

## 3.3 Design

After determining the system requirements, the architecture of the software was designed. The flow charts, use-case diagrams, and wireframes were drawn according

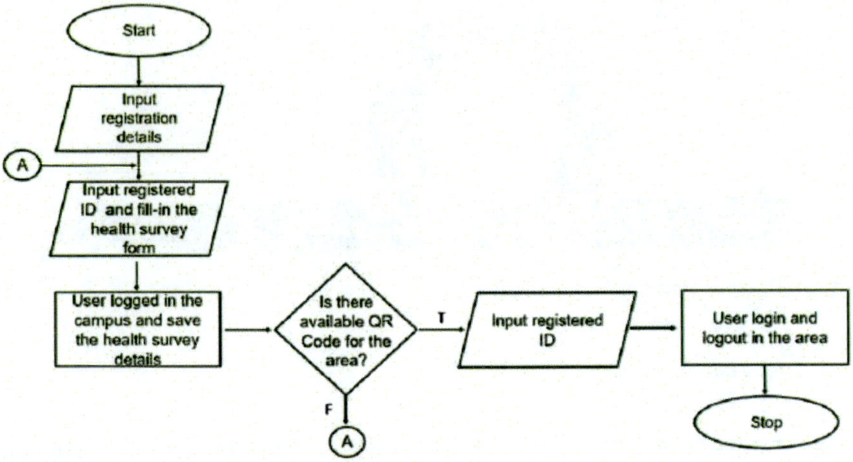


**Fig. 2** Flow chart for encoding the log-in and log-out details and the responses to the health survey form of the end-user to assist in contact tracing

to specifications. The software requirements, hardware requirements, and the database were also designed according to the data to be stored and information to be generated. Flow charts were used as base structure for the development of modules.

**Contact Tracing Mechanism.** Contact tracing with GoTrace becomes possible in three easy steps (Fig. 2). To begin with, the end-user must register to create an account. During registration, end-users are required to submit basic information that includes the user type, ID number, full name, age, gender, address, contact number, e-mail, and workstation/office. Next, when the account has been confirmed, he/she will have access to the daily health survey form. A QR code is assigned to access the link for the health survey which will require the ID number of the end-user. Lastly, once the responses to the health survey are recorded, the end-user can now access the other QR codes in various locations in the campus or site. Scanning the QR codes will record the date and time when the end-user visited a specific location.

All the responses to the health survey and attached log-in and log-out data can be accessed by the health representatives through the system's backend (Fig. 3). Data can also be imported in spreadsheets for processing.

## *3.4 Development*

**Development tools.** The software was developed on a web-based platform to minimize the download and installation procedures while ensuring compatibility to (but not limited to) the following standard devices: desktop computers (1440 × 900 or 1600 × 900), iPhone 5 or lower versions (320 × 568px), iPhone 6/7/8 (414 × 736px),

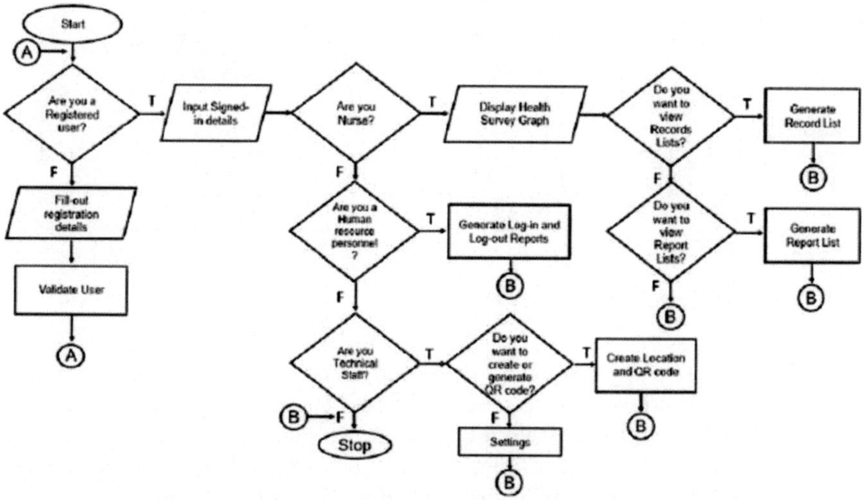


**Fig. 3** A flow chart showing how the system admin generates contact tracing reports

iPhone X (375 × 812), and iPad/Tablets (768 × 1024) using Bootstrap. The graphics were created using Adobe Photoshop and Adobe Illustrator. The database was developed using MySQL while programming was done using PHP and JavaScript. The web application was developed using a computer with specifications: Intel Core i7 8th Gen processor, 8 GB RAM, 120 GB SSD, and 1 TB HDD.

## 3.5 Testing

GoTrace was subjected to alpha and beta testing to ensure the system components are functional, error-free, and ready for deployment. Alpha testing was performed to identify bugs before the software product is released to real users. In doing the alpha tests, the developers reviewed the design specifications and functional requirements, developed a comprehensive test plan and test cases, then executed the test plan to identify log defects and fix the defects. Re-tests were done for validation. The devised test case tackled the following: test case ID which is a unique number that identifies the test case; use-case reference which refers to the reference number on the requirement document that served as the basis for the test case; objective and purpose of the test case to be performed; actions that indicate the steps needed to run the test case; expected results, and the status to show whether the application has passed or failed the test case.

**Table 1** Likert scale

| Numerical scale | Interpretation |
|---|---|
| 4.51–5.00 | Excellent |
| 3.51–4.50 | Very Good |
| 2.51–3.50 | Good |
| 1.51–2.50 | Fair |
| 1.00–1.50 | Poor |

### *3.6 Deployment and Review*

The system was deployed on a third-party web hosting site for pilot testing. During the deployment, end-users were required to register in the application through a link. To solicit the feedback of the end-users, 30 participants consisting of 24 faculty members, a head nurse, two IT instructors, and three web developers were purposively selected to evaluate the web application. The adopted ISO 25,010 software evaluation instrument helped in the assessment of the web application in terms of functional suitability, efficiency, effectiveness, context coverage, satisfaction, usability, security, maintainability, and portability. The scores during the evaluation were obtained using a five-point Likert scale as described below (Table 1).

## 4 Results and Discussion

### *4.1 GoTrace Versus Traditional Contact Tracing*

Transcriptions from an interview during the requirement analysis stage are summarized to highlight the advantages of using the web-based contact tracing app (Table 2).

GoTrace has the potential to replace the traditional health survey and contact tracing and overcome the difficulty in manual distribution and retrieval of survey forms. The web app is also advantageous in terms of the scope of contact tracing and completeness of contact reports with the quick submission of responses to the health survey and real-time analysis of data.

### *4.2 Test Case Results*

The results of the functionality testing of the software conducted by the developers revealed the compliance of the application to the specified functionality scenarios (see Table 3). GoTrace was designed such that any mobile device with Internet access can also access the web application when inside the campus premises. Registered

**Table 2** Comparison between GoTrace and traditional contact tracing methods

| Description | Traditional | GoTrace |
|---|---|---|
| Survey forms | | |
| Percentage retrieval of daily health survey of participants | 50–60% | 85–95% |
| Identification of suspected COVID-19 case based on symptoms | Undetermined | Real time |
| Number of forms filled out daily | 2 (daily log sheet and health declaration form with contact tracing) | 1 |
| Average time required to fill out a form | 60–180 s | 30–60 s |
| Accuracy and completeness of data | 65–75% | 100% |
| Data security | Secured | Highly secured |
| Contact tracing | | |
| Percentage of close contacts declared per participant | 30–35% (based on the average contact tracing ratio of 1:7) | 100% |
| Scope of the contact tracing | 60–70% (of the site of implementation) | 100% |
| Possibility of prediction of possible COVID-19 hotspots | No | Yes |

accounts are secured in the database. The unique ID number provided by the end-user is counterchecked with the registration record prior to the acceptance of the accomplished health survey data and other transactions.

Initial findings of the alpha testing through the test cases revealed that the GoTrace can facilitate a faster and more accurate generation of contact tracing reports when compared with paper-based system. The mobile app eliminates the time required to receive the accomplished daily health survey forms and the need for manual recording and analysis of data. Since GoTrace is web-based, the University physician and the nurses are given the ability to simultaneously monitor a large population of end-users in real time, thus, enabling them to provide prompt actions in the case of data anomalies. Furthermore, the embedded QR code generator also allows the easier creation of new QR codes for additional locations. All these features make the web app a dynamic, flexible, and implementable contact tracing application.

One interesting feature of the web application is the visual representation of the health survey and the data counters. These elements make the information easier to read and interpret. Various reports are easily generated using filter options. These reports include the locations visited by each employee, log sheet records, most visited locations, and visitors per location (Fig. 4). These are used as decision support tools for monitoring of hot spots for transmission of COVID-19 (Fig. 5).

**Table 3** Test case results

| Functionality | Test Case ID | Status | Details |
|---|---|---|---|
| Frontend | | | |
| Registration | TC-001 | Passed | Accomplished the account registration form |
| Submission of health survey information | TC-002 | Passed | Uploaded the health survey information through the digital health survey form |
| QR code scanning to log in and log out of tagged locations | TC-003 | Passed | Successful log in and log out by scanning the QR codes |
| Backend | | | |
| Adding new location and QR code generation | TC-004 | Passed | Created new locations with unique QR code; the QR code generator was embedded in the web app |
| Account registration and validation | TC-005 | Passed | Sign-up accounts were accepted, validated, and counted |
| Real-time viewing of health survey count | TC-006 | Passed | Health survey results and counter were display in the dashboard in real time |
| Report generation | TC-007 | Passed | Various reports are generated accurately including the most visited locations, log records, visitors per location, etc. |

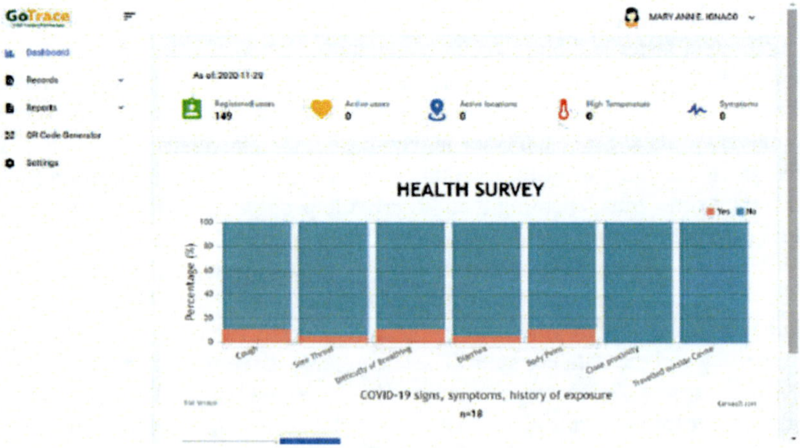


**Fig. 4** Bar chart of the health survey information showing the incidence of COVID-19 symptoms

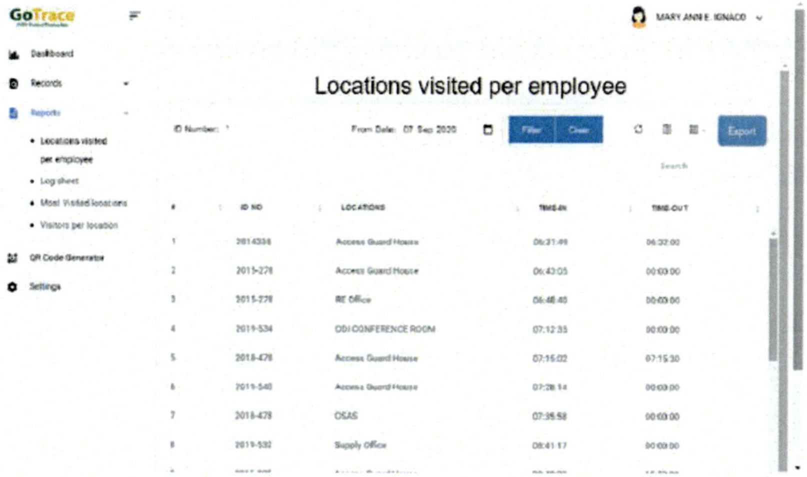


**Fig. 5** Dashboard options for generating various location tracking and contact tracing reports

## 4.3 *End-User Evaluation*

As a result of the pilot testing, the perception of the end-users on the features of the web app were summarized (see Table 4). Contact tracing using mobile technology remains an effective tool in limiting the COVID 19 outbreak especially that the vaccine in countries like the Philippines is still unavailable. Based on the indicators specified, the web app developers ($n = 5$) perceived GoTrace as a very good contact app (*mean rating* $= 4.4 \pm 0.16$) while the other end-users ($n = 25$) perceived it as excellent (*mean rating* $= 4.61 \pm 0.08$).

Compared with the manual contact tracing and health survey, the manual approach is not only slow but also exhibits a late response since it requires the contact tracers to identify persons who were in close contact with persons suspected of exposure to COVID-19. In addition, it is highly vulnerable to common errors of missing and inaccurate data due to manual encoding of the health survey. All the physical delays combined with issues on data storage and archiving negatively impact the efficiency and effectiveness of contact tracing and the quality of contact tracing information.

The use of the smartphone for contact tracing with GoTrace showed promising results among varied types of end-users as evident on the pilot testing. Focusing on the efficiency, effectiveness, and overall satisfaction, with the user experience, the participants agreed that the web app was able to deliver its design function with a high accuracy and completeness. It helps overcome the issues on the manual contact tracing with the ability to quickly generate complete, real-time, accurate reports while maintaining high interaction with the end-users due to its simple and easy-to-understand user interface. Furthermore, the embedded features in the web app's backend also enable the adoption of the technology in various environments, thus, helping in easy, practical, and more economical implementation.

**Table 4** End-user evaluation of the web application

| Indicator | Rating | Description |
|---|---|---|
| Web application developer | | |
| Functional suitability | 4.45 | Excellent |
| Performance efficiency | 4.65 | Excellent |
| Usability | 4.24 | Very good |
| Security | 4.25 | Very good |
| Maintainability | 4.62 | Excellent |
| Portability | 4.46 | Very good |
| **Mean** | **4.46** | **Very good** |
| **Standard deviation** | **0.16** | |
| Other end-users | | |
| Usability | 4.63 | Excellent |
| Effectiveness | 4.76 | Excellent |
| Efficiency | 4.56 | Excellent |
| Satisfaction | 4.56 | Excellent |
| Context coverage | 4.56 | Excellent |
| **Mean** | **4.61** | **Excellent** |
| **Standard deviation** | **0.08** | |

However, while other contact tracing apps rely on the use of Bluetooth technology and GPS functions in smartphones for automated tracking, GoTrace uses the QR code which is largely dependent on the number of people using the application. This dependence limits the usage of the app and requires that the target population and technology adopters clearly understand the context and importance of contact tracing. Since it also web-based, Internet connectivity with better network coverage remains a major challenge.

## 5 Conclusions and Recommendation

Unlike other contact tracing technologies, the developed GoTrace web app uses the QR codes to effect location tracking and collection and storage of health survey data as inputs to contact tracing. This contact tracing web app provides an alternative to manual contact tracing, resolving the issues on physical delays and reliability and validity of contact tracing reports. The successful integration of the digital health survey and tagging of the locations in the campus using the QR codes allowed a faster data retrieval and even real-time generation of various reports. The perception of the end-users on the efficiency, effectiveness, and satisfaction of the contact tracing with GoTrace, along with other features of the technology, revealed promising results as bases for the full adoption and implementation of the technology. The dynamic design of the web app also supports customization for other environments. However, along with its adoption, it is a must that the end-users fully understand the significance and context of contact tracing as a strict measure to limit the spread of contagious diseases such as COVID-19. The end-user's perspective toward the use of the technology will significantly impact the extent of the scope of contact tracing and the quality of contact tracing information.

## References


1. World Health Organization, Report of the WHO-China joint mission on Coronavirus disease 2019 (COVID-19), February 28, 2020, https://www.who-.int/publicationsi/item/report-of-the-who-china-joint-mission-oncoronavirus-disease-2019-(covid-19)
2. Perlman S (2020) Another decade, another corona virus. N Engl J Med. https://doi.org/10.1056/NEJMe2001126
3. Vallejo BM, Ong RAC (2020) Policy responses and government science advice for the COVID 19 pandemic in the Philippines: January to April 2020. Prog Disaster Sci 7(2020). https://doi.org/10.1016/j.pdisas.2020.100115
4. Ferretti L, Wymant C, Kendall M, Quantifying SARS-CoV-2 transmission suggests epidemic control with digital contact tracing. Science 368(6491), https://science.sciencemag.org/content/368/6491/eabb6936.abstract
5. Department of Health (2020) Updated guidelines on contact tracing of close contacts of confirmed coronavirus disease (COVID-19) cases, https://doh.gov.ph/sites/default/files/health-update/dm2020-0189.pdf?fbclid=IwAR210xI3CGgfeU7T-4mHXXpbc0Wgm_DKAxBNOkB9gn1FPLkbV0wEiM0WLWQ
6. Eames KTD, Keeling MJ (2003) Contact tracing and disease control. R Soc doi:https://doi.org/10.1098/rspb.2003.2554
7. O'Callaghan ME, Buckley J, Fitzgerald B, Johnson K, Laffey J, McNicholas B, Nuseibeh B, O'Keeffe D, O'Keeffe I, Razzaq A, Rekanar K, Richardson I, Simpkin A, Abedin J, Storni C, Tsvyatkova D, Walsh J, Welsh T, Glynn L (2020) A national survey of attitudes to COVID-19 digital contact tracing in the Republic of Ireland, https://link.springer.com/article/https://doi.org/10.1007/s11845-020-02389-y
8. Koetter P, Pelton M, Gonzalo J, Du P, Exten C, Bogale K, Buzzelli L, Connolly M, Edel K, Hoffman A, Legro NR, Medina D, Sood N, Blaker J, Kearcher K, Sciamanna C (2020) Implementation and process of a COVID-19 contact tracing initiative: leveraging health professional students to extend the workforce during a pandemic, https://linkinghub.elsevier.com/retrieve/pii/S0196655320307793
9. Müllera J, Kretzschmarc M, Contact tracing – old models and new challenges, released January 2021, doi:https://doi.org/10.1016/j.idm.2020.12.005
10. Almagor J, Picascia S (2020) Exploring the effectiveness of a COVID-19 contact tracing app using an agent-based model, https://www.nature.com/articles/s41598-020-79000-y
11. Kretzschmar ME, Rozhnova G, Bootsma MCJ, van Boven M, van de Wijgert JHHM, Bonten MJM (2020) Impact of delays on effectiveness of contact tracing strategies for COVID-19: a modelling study. 10.1016
12. Jian SW, Cheng HY, Huang XT, Liu DP (2020) Int J Infect Dis 101:348–352. https://doi.org/10.1016/j.ijid.2020.09.1483
13. CNN Philippines official website, Gov't mandates use of StaySafe contact tracing app in local, national level, November 27, 2020, https://cnnphilippines.com/news/-2020/11/27/StaySafe-contact-tracing-app-COVID-19.html
14. Lalmuanawma S, Hussain J, Chhakchhuak L (2020) Applications of machine learning and artificial intelligence for Covid-19 (SARS-CoV-2) pandemic: a review. https://doi.org/10.1016/j.chaos.2020.110059

15. Mbunge E (2020) Integrating emerging technologies into COVID-19 contact tracing: opportunities, challenges and pitfalls from diabetes & metabolic syndrome: clinical research & reviews, https://www.sciencedirect.com/science/article/abs/pii/S1871402120303325?via%3Dihub
16. Mamas WSP, Philcd MD, Harriette ET, Van GC, Spall D (2020) Applications of digital technology in COVID-19 pandemic planning and response. doi:https://doi.org/10.1016/S2589-7500(20)30142-4
17. Inquirer Philippines official website, Use of staysafe.ph app now a must for LGUs, nat'l agencies, national level, November 28, 2020, https://newsinfo.inquirer.net/1365781/use-of-staysafe-ph-app-now-a-must-for-lgus-natl-agencies
18. Maccari L, Cagno V (2020) Do we need a contact tracing app?, Comput Commun. doi:https://doi.org/10.1016/j.comcom.2020.11.007
19. Cho H, Ippolito D, Yu YW (2020) Contact tracing mobile apps for COVID-19: privacy considerations and related trade-offs, Collonel University, https://arxiv.org/abs/2003.11511
20. Lunn PD, Motivating social distancing during the COVID-19 pandemic: an online experiment, https://ideas.repec.org/p/esr/wpaper/wp658.html
21. Auqib ABD, Saniya HL, Afshan Z, Naazb AKR (2020) Applicability of mobile contact tracing in fighting pandemic (COVID-19): issues, challenges and solutions, Applicability of mobile contact tracing in fighting pandemic (COVID-19): issues, challenges and solutions. doi: https://doi.org/10.1016/j.cosrev.2020.100307
22. Hao SWJ, Xiang YC, Ding TH, Liuac P (2020) Contact tracing with digital assistance in Taiwan's COVID-19 outbreak response. Int J Infect Dis. doi:https://doi.org/10.1016/j.ijid.2020.09.1483

# Sense Prediction for Explicit Discourse Relations with BERT

Jiří Mírovský and Lucie Poláková

**Abstract** We present experiments in predicting a discourse sense for explicit intersentential discourse relations in Czech and English, using embedding and deep learning (fine-tuned BERT) to predict the senses, and annotation projection from English to Czech to increase the size of training data for Czech. We show that strengthening information about the explicit connective improves the results and, on the other hand, that annotation projection from an annotation system with a different set of senses may not help if used in a straightforward way, which we discuss in detail.

**Keywords** Discourse · Explicit relations · Sense prediction · BERT · Annotation projection

## 1 Introduction

Shallow discourse parsing consists of several subtasks, from argument labeling (identification of text spans that represent two arguments of a discourse relation) to connective identification (applicable for explicit discourse relations: identification of a text span that represents a connective), to classification of the relation (prediction of a discourse sense for the relation by choosing from a predefined set of senses). End-to-end shallow discourse parsers need to address all these tasks; see, for example, [5].

The present contribution focuses on one subtask of shallow discourse parsing—discourse sense classification for explicit discourse relations. Later in the text, for the purposes of the present experiments, we refine and even further narrow the definition of the task. The paper is organized as follows: in the rest of Introduction, we put our

J. Mírovský (✉) · L. Poláková
Faculty of Mathematics and Physics, Institute of Formal and Applied Linguistics, Charles University, Prague, Czech Republic
e-mail: mirovsky@ufal.mff.cuni.cz

L. Poláková
e-mail: polakova@ufal.mff.cuni.cz

 
X.-S. Yang et al. (eds.), *Proceedings of Sixth International Congress on Information and Communication Technology*, Lecture Notes in Networks and Systems 216,
https://doi.org/10.1007/978-981-16-1781-2_72

contribution in context with related works and state our research questions. Section 2 gives an overview of all data resources used and presents our methodology. Section 3 describes in detail our experiments' setting and offers their results. The results are further discussed in Sect. 4, which also hints at two future research directions.

Discourse sense classification has been studied extensively in recent years, although exact settings of the experiments differ significantly in various aspects: language (most frequently English and Chinese), implicit versus explicit versus both types of relations (implicit relations are in focus recently), number of predicted classes (often only four main discourse classes are distinguished). Xue et al. [14] report on CoNLL 2016 shared task on multilingual shallow discourse parsing, where (as one of the subtasks) both arguments and the connective for explicit relations were given as the input; they used a 10-way classification for Chinese, and 21 labels for English. Varia et al. [12] use convolutional networks for classifying both implicit and explicit discourse relations in English (a four-way classification only). Weiss and Bajec [13] designed a language-independent end-to-end trainable system for both explicit and implicit discourse relations based on focused recurrent neural networks, following the experimental setting of the CoNLL 2016 shared task. Shi and Demberg [11] use BERT for sense prediction for implicit discourse relations in English (PDTB, an 11-way classification task).

Inspired mostly by the last-mentioned work, we have formulated the following three research questions:

1. Can BERT (as a well-known system for embedding and deep learning, already successfully used for implicit relations) be used with good results to classify explicit discourse relations, i.e. relations between two text spans with an explicit connective present in the text?
2. Can resources such as lexicons of discourse connectives, here represented merely by the additional information about "what is the connective", help in this scenario?
3. If we have suitable parallel data, can annotation projection provide beneficial additional training data even with two slightly different discourse sense taxonomies?

## 2 Data, Methods and Tools

Our study deals with discourse sense prediction in Czech and English and with annotation projection from English to Czech. The following three data resources were used in the experiments:

- **The Prague Discourse Treebank—PDT** [8] is a manually annotated corpus of Czech newspaper texts, first published as [7]. We use the newest version of the discourse annotation of the data, published in the Prague Dependency Treebank 3.5 [4]. The corpus contains approximately 50 thousand sentences annotated in detail on the levels of morphology, surface and deep syntax, and discourse phenomena (explicit discourse relations, coreference and bridging anaphora).

- **The Penn Discourse Treebank—PDTB** [9] is a corpus of English newspaper texts, annotated manually with discourse relations. The corpus consists of approximately 50 thousand sentences of the Wall Street Journal section of the Penn Treebank (Treebank 2, [6]). We use the most recent version of the discourse annotation of the data, the Penn Discourse Treebank 3.0 [10], which contains approximately 40 thousand annotated discourse relations.

  The PDT and the PDTB are comparable in genre (journalism), size (50k sentences) and also the annotation scenario, although there are differences in the sense hierarchies and the extent of the annotated relations (e.g. implicit relations were not annotated in the PDT).
- **The Prague Czech–English Dependency Treebank—PCEDT** [2] is a corpus of English–Czech parallel texts and their mostly manual annotations on several layers of language description, following the annotation scenario of the PDT. Importantly, the English part of the PCEDT contains the same texts as the PDTB, i.e. the Wall Street Journal section of the Penn Treebank. The Czech part is based on human translations of the English texts to Czech, 1:1 sentence-aligned, with an additional automatic alignment on the word/node level on all annotation layers. We use the newest version of the PCEDT, the Prague Czech–English Dependency Treebank 2.0 [3].

## 2.1 *Annotation Projection*

The fact that the PDTB and the English part of the PCEDT are based on the same texts allowed us to perform annotation projection of the discourse annotation from the PDTB to the Czech part of the PCEDT, with the following intermediate steps:

1. transformation of the PDTB discourse annotation from the raw-text-based annotation to the phrase structure trees of the Penn Treebank,
2. further transformation from the phrase-structure trees to the dependency trees of the deep syntax (tectogrammatical) layer of the English part of the PCEDT,
3. using the (automatic) alignment of words/nodes, annotation projection of the discourse relations (argument spans, connectives, discourse senses, additional info) from the English part of the PCEDT to the Czech part of the PCEDT,
4. and last, transformation of the PDTB hierarchy of discourse senses to the Czech (PDT) hierarchy of discourse senses; an N:1 function transforming a single PDTB discourse sense to a single PDT discourse sense (see Table 1) was devised and used for each discourse relation without any other consideration (such as taking a context into account); for the purposes of the present experiments, this is the most problematic step, which is further discussed in the conclusion.

The projected data were then used as additional training data in the experiments with Czech; see below.

**Table 1** The PDTB—PDT sense transformation table. With the exception of the relation Expansion.Level-of-detail, which maps to two different PDT discourse senses, we abstract from the argument semantics (i.e. the direction of the relation)

| The PDTB sense | The PDT sense |
|---|---|
| Comparison.Concession | Concession |
| Comparison.Concession + SpeechAct | Confrontation |
| Comparison.Contrast | Opposition |
| Comparison.Similarity | Conjunctive alternative |
| Contingency.Cause | Reason–result |
| Contingency.Cause + Belief | Pragmatic reason–result |
| Contingency.Cause + SpeechAct | Pragmatic reason–result |
| Contingency.Condition | Condition |
| Contingency.Condition + SpeechAct | Pragmatic condition |
| Contingency.Negative-cause | Reason–result |
| Contingency.Negative-condition | Condition |
| Contingency.Purpose | Purpose |
| Expansion.Conjunction | Conjunction |
| Expansion.Disjunction | Disjunctive alternative |
| Expansion.Equivalence | Equivalence |
| Expansion.Exception | Restrictive opposition |
| Expansion.Instantiation | Instantiation |
| Expansion.Level-of-detail.Arg1-as-detail | Generalization |
| Expansion.Level-of-detail.Arg2-as-detail | Specification |
| Expansion.Manner | Explication |
| Expansion.Substitution | Correction |
| Temporal.Asynchronous | Precedence–succession |
| Temporal.Synchronous | Synchrony |

## *2.2 BERT*

BERT, the Bidirectional Encoder Representation from Transformers, proposed by Devlin et al. [1], is a deep learning system pre-trained using a combination of masked language modeling objective and next sentence prediction, which can be fine-tuned for a specific NLP task with just one additional output layer. We use its implementation in the HuggingFace library[1] with the multilingual uncased pre-trained model (trained on lower-cased Wikipedia texts in 102 languages) for Czech, and the base uncased pre-trained model for English.[2] For the fine-tuning, we use the BertForSe-

[1] https://huggingface.co/transformers/.

[2] For Czech, we also experimented with the DeepPavlov model trained for four Slavic languages (https://huggingface.co/DeepPavlov/bert-base-bg-cs-pl-ru-cased), with the same results.

**Table 2** Data sizes in the experiments

| Experiment | Training data | Additional training data | Test data | Number of labels |
|---|---|---|---|---|
| Czech | 5 064 | 4 781 | 531 | 23 |
| English | 5 066 | – | 563 | 27 |

quenceClassification transformer with a sequence classification/regression head on top and three training epochs.

## 3 Experiment Settings and Results

From the available data resources, we have first selected only explicit discourse relations holding between whole consequent sentences, i.e. not intra-sentential relations, neither inter-sentential relations with an argument formed, e.g. only by one clause of a compound sentence. It is a technical (rather than linguistic) condition which simply guarantees that the connective is in most cases present in the right-hand argument.

Table 2 gives an overview of the size of datasets obtained this way from the available resources. For English, we use the complete PDTB 3.0 sense taxonomy, and for Czech, we use the Czech discourse sense taxonomy without a distinction of argument semantics (e.g. for the precedence–succession relation, we do not distinguish which argument is the first in time). The table shows how many different labels (discourse senses) appeared in the selected data.

Following the research questions formulated in the Introduction, we have compiled the following experiment settings:

- **Baseline 1**: For each discourse relation, we predict the most frequent discourse sense that was found in the basic training data,[3] i.e. we predict the same discourse sense for all test cases.
- **Baseline 2**: For each discourse relation, we predict the most frequent discourse sense that was found in the basic training data at the given connective, i.e. we predict the same discourse sense for all relations marked with the same connective.
- **BERT**: We fine-tune the pre-trained BERT model with training data with instances consisting of the correct label (discourse sense) and two consequent sentences with an explicit discourse relation[4] annotated between these two whole sentences. As BERT offers (and requires), the two sentences are separated by a special token.
- **BERT + connective prepended**: As "BERT" above but at the beginning of each second sentence, the connective is redundantly added. It means that it appears

[3] Meaning "without the additional training data".

[4] We use the term "explicit discourse relation" in a broad sense, i.e. a discourse relation explicitly marked by a connective or an alternative lexicalization in the terms used in the PDTB, or by both primary and secondary connectives in the terms used in the PDT.

**Table 3** Results—accuracy of the sense classification

| Setting | Czech | English |
|---|---|---|
| Baseline 1 | 0.29 | 0.36 |
| Baseline 2 | 0.76 | 0.86 |
| BERT | 0.77 | 0.89 |
| BERT + connective prepended | 0.79 | 0.90 |
| BERT + conn. prep. + add. data | 0.76 | – |

twice in each training/test instance—once in its original position, the second time as the new beginning of the second sentence.

- **BERT + connective prepended + additional training data**: As "BERT + connective prepended" above but additional training data from the annotation projection are used.

Table 3 gives an overview of the results of the various settings of the experiment. For Czech, there has been no previous work to compare to, for English, our results are similar with the best systems of the CoNLL 2016 shared task, as reported in [14]. However, our experiment setting was narrower (only inter-sentential relations between whole sentences) and evaluated on a slightly different part and newer version of the PDTB data, including the new PDTB 3.0 discourse sense taxonomy.

## 4 Discussion

The results show that in the task of predicting a discourse sense for inter-sentential explicit relations, it is quite difficult to significantly outperform a relatively simple yet strong baseline (Baseline 2) that chooses the most common discourse sense for each connective, as observed in the training data. Knowing the connective and pinpointing it for the sense-predicting system (here accomplished by redundantly prepending the connective before the second argument) in the training/test instances improves the results, which justifies the incorporation of lexicons of discourse connectives in end-to-end discourse parsers.

The system performs much better in English than in Czech. It might be partially attributed to a better representation of English in pre-trained BERT models but the lower baselines in Czech suggest that the Czech hierarchy of discourse senses and annotation strategies present a greater challenge for the classifier than the PDTB 3.0 hierarchy, although they are of a similar size (in our experiment, the Czech hierarchy is even smaller than the English one; see Table 2). Indeed, if we look at the sense ambiguity of discourse connectives in the Czech and English datasets used in our experiments, we can see that for Czech, 85 out of total 349 different connectives[5] show sense ambiguity, the most ambiguous ones being *tedy* [*so, therefore*] (10 different senses), *přitom* [*at the same time*] (10 senses), *ale* [*but*] (10 senses), *však* [*however*] (8 senses) and *tak* [*so*] (8 senses), while for English, only 32 out of total 456 connectives show sense ambiguity, the most ambiguous ones being *indeed* (8 different senses), *or* (6 senses), *in fact* (5 senses), *but* (5 senses) and *then* (4 senses). Further analysis of the results would be needed to identify the most problematic cases, i.e. (pairs of) senses that are difficult to distinguish (with or without respect to connectives) and that is the direction in which our research will continue.

Additional training data obtained by the annotation projection and discourse sense transformation from English to Czech surprisingly worsened the results. This might be partly caused by different distributions of sense labels in the original (PDT) and the additional (projected PDTB) Czech training datasets—compare, for example, 25% of the conjunction sense in the PDT dataset versus 38% in the projected PDTB, 17% of the reason–result label (PDT) versus 7% (PDTB). Some senses occurring in the PDT dataset have not appeared in the projected and selected PDTB data at all (confrontation, gradation, explication, pragmatic reason–result, conjunctive alternative), which is in some cases caused by the too simplistic N:1 sense transformation function. It seems that for a completely successful sense transformation, manual work would be required, to distinguish, for example, Czech senses opposition and correction (both Comparison.Contrast in the PDTB 3.0 taxonomy), pragmatic contrast and explication (both Contingency.Cause + Belief or + SpeechAct), or conjunctive alternative and disjunctive alternative (both Expansion.Disjunction). Some of the disproportions result quite naturally from the mismatches in the label mapping but others can be most likely attributed also to different annotation guidelines and different definitions of the senses (30% of concession as the second most frequent sense in the projected PDTB data manifests different annotation strategies for *but, however* and other connectives of the Comparison group than in Czech for their translation equivalents (4% concession, 27% opposition in the PDT dataset)). This also calls for a further analysis and leaves room for improvement.

**Acknowledgements** The authors gratefully acknowledge support from the Grant Agency of the Czech Republic (project GA19-03490S). The research reported in the present contribution has been using language resources developed, stored and distributed by the LINDAT/CLARIAH-CZ project of the Ministry of Education, Youth and Sports of the Czech Republic (LM2018101).

[5] Please note again that we understand a connective here as a broad class that includes alternative lexicalizations.

## References

1. Devlin J, Chang MW, Lee K, Toutanova K (2019) Bert: pre-training of deep bidirectional transformers for language understanding. In: Proceedings of the 2019 conference of the North American chapter of the association for computational linguistics: human language technologies, pp 4171–4186
2. Hajič J, Hajicová E, Panevová J, Sgall P, Bojar O, Cinková S, Fucíková E, Mikulová M, Pajas P, Popelka J et al (2012) Announcing prague Czech–English dependency treebank 2.0. In: LREC, pp 3153–3160
3. Hajič J, Hajiová E, Panevová J, Sgall P, Cinková S, Fucíková E, Mikulová M, Pajas P, Popelka J, Semecký J, Šindlerová J, Štěpánek J, Toman J, Urešová Z, Žabokrtský Z (2012) Prague Czech-English dependency treebank 2.0. data/software, linguistic data consortium. University of Pennsylvania, Philadelphia. LDC2012T08
4. Hajič J, Bejček E, Bémová A, Buráňová E, Hajičová E, Havelka J, Homola P, Kárník J, Kettnerová V, Klyueva N, Kolářová V, Kučová L, Lopatková M, Mikulová M, Mírovský J, Nedoluzhko A, Pajas P, Panevová J, Poláková L, Rysová M, Sgall P, Spoustová J, Straňák P, Synková P, Ševčíková M, Štěpánek J, Urešová Z, Hladká BV, Zeman D, Zikánová Š, Žabokrtský Z (2018) Prague dependency treebank 3.5. data/software, Univerzita Karlova, MFF, ÚFAL, Prague, Czech Republic
5. Lin Z, Ng HT, Kan MY (2014) A PDTB-styled end-to-end discourse parser. Nat Lang Eng 20(2):151–184
6. Marcus MP, Santorini B, Marcinkiewicz MA (1995) Treebank-2. data/software, linguistic data consortium. University of Pennsylvania, Philadelphia. LDC95T7
7. Poláková L, Jínová P, Zikánová Š, Hajičová E, Mírovský J, Nedoluzhko A, Rysová M, Pavlíková V, Zdeňková J, Pergler J, Ocelák R (2012) Prague discourse treebank 1.0. data/software, ÚFAL MFF UK, Prague, Czech Republic
8. Poláková L, Mírovský J, Nedoluzhko A, Jínová P, Zikánová Š, Hajičová E (2013) Introducing the prague discourse treebank 1.0. In: Proceedings of the 6th international joint conference on natural language processing. Asian Federation of Natural Language Processing, Asian Federation of Natural Language Processing, Nagoya, Japan, pp 91–99
9. Prasad R, Lee A, Dinesh N, Miltsakaki E, Campion G, Joshi A, Webber B (2008) Penn discourse treebank version 2.0. data/software, linguistic data consortium. University of Pennsylvania, Philadelphia. LDC2008T05
10. Prasad R, Webber B, Lee A, Joshi A (2019) Penn discourse treebank version 3.0. data/software, linguistic data consortium. University of Pennsylvania, Philadelphia. LDC2019T05
11. Shi W, Demberg V (2019) Next sentence prediction helps implicit discourse relation classification within and across domains. In: Proceedings of the 2019 conference on empirical methods in natural language processing and the 9th international joint conference on natural language processing (EMNLP-IJCNLP), pp 5794–5800
12. Varia S, Hidey C, Chakrabarty T (2019) Discourse relation prediction: revisiting word pairs with convolutional networks. In: Proceedings of the 20th annual SIGdial meeting on discourse and dialogue, pp 442–452
13. Weiss G, Bajec M (2018) Sense classification of shallow discourse relations with focused rnns. PloS One 13(10):e0206,057
14. Xue N, Ng HT, Pradhan S, Rutherford A, Webber B, Wang C, Wang H (2016) CoNLL 2016 shared task on multilingual shallow discourse parsing. In: Proceedings of the CoNLL-16 shared task, pp 1–19

# Hour-Ahead Electric Load Forecasting Using Artificial Neural Networks

**Lemuel Clark P. Velasco, Karl Anthony S. Arnejo, Justine Shane S. Macarat, and Mia Amor C. Tinam-isan**

**Abstract** To ensure an efficient supply of electricity, power utility companies make use of a combination of either short-term, medium-term, and long-term forecasting techniques. This paper presents a strategy for implementing hour-ahead electricity load forecasting using artificial neural networks (ANN). Data preparation through data selection, cleaning, partitioning, and transformation was performed to the dataset provided by a power utility in Southern Philippines. ANN models using a feedforward architecture having 9 input neurons, 6 hidden neurons, and 1 output neuron with a learning rate set to 0.00001, momentum set to 0.7 with an epoch value of 20,000, and a maximum error of 0.001 was implemented using a Java-based open-source library. Results showed that the ANN model with a quick propagation training algorithm and sigmoid activation function had a Mean Absolute Percentage Error (MAPE) of 2.91%, and the ANN model with resilient propagation training algorithm and sigmoid activation function had a MAPE of 3.49%. This study shows that with appropriate data preparation and machine learning implementation, ANN has the potential to aid the decision-making of power utility companies through short-term forecasting in terms of hour-ahead electric load consumption prediction.

**Keywords** Artificial neural networks · Hour-ahead electric load forecasting · Short-term electric load forecasting

## 1 Introduction

Load forecasting is categorized by electricity distributors into short-term, medium-term, and long-term forecasts depending on the granularity of their prediction along with the intent of what decisions should be undertaken for the efficiency of the power system. While long-term load forecasting varies from 1 year to several years of projection, medium-term load forecasting generates from 1 month to several months

L. C. P. Velasco (✉) · K. A. S. Arnejo · J. S. S. Macarat · M. A. C. Tinam-isan
Premier Research Institute of Science and Mathematics, Mindanao State University-Iligan Institute of Technology, 9200 Iligan City, Philippines
e-mail: lemuelclark.velasco@g.msuiit.edu.ph



X.-S. Yang et al. (eds.), *Proceedings of Sixth International Congress on Information and Communication Technology*, Lecture Notes in Networks and Systems 216,
https://doi.org/10.1007/978-981-16-1781-2_73

ahead of the forecast while short-term load forecasting varies from 1 hour to several days ahead of prediction [1–3]. With each granular predicted outputs serving their own purpose in the operations of a power utility, precise prediction of electricity load is important for accurate bidding of electricity to maximize the profit and expenditure for both the consumers and producers. Serving as the electricity distributor, power utility companies in a competitive power market tend to maintain their generation reserve close to the minimum required in case there is a fluctuation of electricity that day-ahead and week-ahead load forecasts cannot detect [2, 4]. With spot prices variedly changing in very short time intervals among electricity spot markets, there is an emphasis among power utilities to augment the existing long-term and medium-term forecasts with very short-term predictions that will coincide with a geography's power consumption as well as the market's spot price. Electricity spot prices are generally higher compared to bilateral contracts which are long-term in nature, that is why an independent forecasting tool that generates short-term predictions to augment immediate consumption margin should be used vis-à-vis the highly granular outputs [3, 5, 6]. An accurate and instantaneous load forecast such as an hour ahead of granular prediction is needed over and on top of the days, weeks, and months ahead of analysis will be needed to ensure there will be avoidance of underestimation which cause failure to deliver enough electricity and overestimation that can lead to additional operating costs.

There are various data modeling tools that can be used as statistical analysis and machine learning approaches in predicting the hour-ahead load consumption of a given locality such as time series analysis, exponential smoothing, support vector machines, fuzzy logic, and artificial neural network (ANN) techniques [2, 3, 6, 7]. Researchers have found out that ANN has been proven effective in week-ahead and day-ahead electric load forecast but has less promising performance in hour-ahead predictions due to the nature of its learning capability [2, 8]. Being a data modeling tool that is composed of hundreds and thousands of computing elements usually called neurons in several layers that can be used in analyzing large numbers of inputs, ANN is starting to replace traditional methods in many application offerings like that of electric load forecasting due to its capability to map out nonlinear relationships based from historical electric load records [1, 4, 9]. The three-layer fully connected feedforward neural network composed of an input layer, a hidden layer, and an output layer is generally used for medium-term and long-term electric load forecasts but has not been much explored for short-term predictions that will generate hour-ahead consumption [6, 8]. Multilayer perceptron (MLP), an ANN architecture commonly used for the analysis of largely granular data due to its ability to operate fast, ease of implementation, smaller training set requirements, and the capability to solve function approximation problems, is yet to be explored and validated if it also shares promising predictive power [10]. Challenges in the applicability of the ANN model for very short-term forecast will involve appropriate data cleaning as well as determining and implementing various configurations of the network's layers that will function efficiently in its training and testing phases. Just like any endeavor in predictive analytics, an attempt to validate ANN's ability for very short-term predictive tasks can easily be measured by scientific comparison of the actual and the forecasted results of the network along with patient fine-tuning of the model in order for it to deliver promising results.

A power utility company located in Southern Philippines has sufficient data that can be used for an ANN model to examine its predictive ability on the hour-ahead forecast. This power utility that distributes electricity to a highly urbanized city already has day- and week-ahead electric load prediction models but has no hour-ahead electric load prediction model that it can use for instantaneous load forecast in order to maintain their generation reserve to the required minimum in case there is power supply fluctuation [4, 6]. Existing studies show the effectiveness of augmented analysis of incorporating short-term load forecast with medium- and long-term forecast so that electric distributors can optimally perform efficient economic dispatching, system planning, reliability evaluation, and prevention overutilization as well as underutilization of resources [3, 9, 11]. This research aims to implement and evaluate an hour-ahead electric load forecasting model using ANN for a power utility. Specifically, this research will attempt to conduct electric load data preparation as well as implement a designed ANN model that will generate a close to accurate hour-ahead forecast of a given geographic locality's electric consumption. With the evaluation of the implemented ANN model's predictive capability, this study hopes to aid power utility companies by providing a tool in their decision-making, electric load planning, and load utilization while at the same time exhibiting the ability of ANN for an accurate hour-ahead forecast.

## 2 Methodology

### 2.1 Electric Load Data Preparation

Being the crucial preliminary step in any data modeling task, this study will employ data selection, data cleaning, data partitioning, and data transformation in preparing the electric load data to ensure pertinent input sets to the ANN model. As an initial step, data selection was conducted to finding the most relevant attributes of the dataset for model training for the task of load forecasting [12]. Raw monthly electric load data of the power utility company from three metering points for the years 2012–2018 was provided to the researchers but data for the years 2015–2018 were not included due to the unavailability and data inconsistency in some metering points with the second metering point not being able to capture most of the electric load consumption. The dataset came in a *.xls* format and contained a 15-minute load consumption for the entire month of the load consumption of a city's power system. As shown in Table 1, each sheet contained the date, time, kilowatt delivered (KW_DEL), kilowatt per hour delivered (KWH_DEL), kilovolt amps reactive hours delivered (KVARH_DEL), and kilowatt hour delivered (KWH_DEL). The attributes referring to date, time, and the kilowatt delivered reflecting the consumed electric load were selected to proceed for

**Table 1** Sample of the raw electric load data

| Date | Time | KW DEL | KWH_DEL | KVARH_DEL |
|---|---|---|---|---|
| NN/NN/NNNN | 00:15 | NN | NN | NN |
| NN/NN/NNNN | 00:30 | NN | NN | NN |

the data cleaning and data normalized while all the other attributes from the raw electric load dataset were eliminated.

Data cleaning was then conducted to ensure the integrity of the dataset and eliminate the data errors caused by either human intervention or computer interaction arising in every step of the power utility's data collection processes [7, 10]. Outliers within the electric load consumption dataset may come in terms of human errors, power outages, and power interruptions which might be present in the load data obtained, therefore, data cleaning must be made to produce accurate results [4, 13]. Since recording the load data does not involve human intervention, human errors were excluded as noise, and only out-of-range values such as power interruptions that output zeroes and null values were considered as noise. Data cleaning was performed manually from the spreadsheet file before importing the data into the database to prevent out-of-range values. After data cleaning was performed, the three metering points were then aggregated to get the total load demand. The cleaned data that has undergone the necessary data corrections were then converted into a *.csv* file so it can be imported into the database.

The cleaned dataset has undergone data partitioning which divided the electric load consumption dataset into training, testing, and validation sets to evaluate how the ANN model would behave when applied to data on various phases. Studies strongly suggest that the training set should comprise the largest part of the dataset while the testing and validation sets are assigned to what remained from the original dataset [7–9, 14]. The training set comprises the data that was used for adjusting and training the weights on the neural network while the testing set was a separate independent subset used for testing the design of the network to confirm its actual predictive power as well as to track errors during testing [11]. Validation set as used in this study is another subset of independent data which acted as the actual everyday one-hour load data from the power utility company which was used to evaluate the performance of the final neural network as it would be used in the operations of the power utility company. A transformation phase was conducted to the dataset in order for the data to be consolidated into forms appropriate for the process because neural networks cannot just receive any data and yield accurate results. The variables needed to transform the data are queried in the database. Data transformation in this study includes data normalization which is significantly important to make the training of the network faster, memory-efficient, and to ensure the good quality of data. This was conducted to ensure that the dataset is compliant with certain ranges in such a way that the dataset became uniform and that no value overpowered another value to prevent the instability of data [10]. The projected neural network will only work within the range of 0.0–1 since negative values were considered outliers and can affect the accuracy of the load forecasting [15]. Thus, the dataset was transformed to comply with this specified range employing the Min–Max normalization technique as shown in Eq. 1 where $\min(x)$ is the minimum value of an attribute and $\max(x)$ is the maximum value of the attribute. Min–Max normalization technique was used as suggested by researchers showing that Min–Max normalization yields positive values which suits the electric load data best and performs efficient design in training the electric load datasets producing the lowest tree growing time exhibiting optimality compared to other normalization methods such as Z-score and decimal scaling normalization [9, 13]. With this, the electric load data were transformed into uniformly distributed values for network input that was then fed for the ANN training. After the training, testing, and validation sets were normalized and fed into the neural network, the outputs from the ANN model underwent a denormalization phase through a derived formula so that the forecasted values were revealed and compared to the actual load values:

$$z = \frac{x - \min(x)}{\max(x) - \min(x)} \tag{1}$$

## 2.2 *ANN Model Implementation*

The researchers implemented a feedforward neural network simplified in a multilayer perceptron architecture embedded in a forecasting system that performed the processes from data importation up to forecasting the next-hour electric load. As shown in Fig. 1, the multilayer perceptron neural network independently used resilient propagation and quick propagation training algorithms along with sigmoid activation function among the individual layers of the ANN's input, hidden, and output layers with its respective 9 input neurons, 6 hidden neurons, and 1 output neuron. The network's input neurons include the 15-minute historical electric load consumption, the day of the week which determines the data to be between Monday–Sunday, an indicator on what type the day falls, either as a weekend or weekday, an indicator if the day is a regular day or a holiday, the week number where the electric load was taken, the time in 15-minute interval, the date of the day, and the month where the date falls under and the year of the captured electric load data [4, 7, 8, 14]. The learning rate was set to 0.00001 while the momentum was set to 0.7 with an epoch value of 20,000 and a maximum error of 0.001. The network training facilitated the signals of the neural networks to be multiplied with bias weights and they were mapped into only one output neuron that predicted the electric load consumption of the next 15-minute output to be used in the evaluation of its predictive capability.

Using the Java-based Encog library machine learning framework, a desktop-based forecasting system was developed to implement the ANN model. The functions of the forecasting system were grouped into classes such as the database connection, normalization, and the class for training, evaluation, and prediction with the

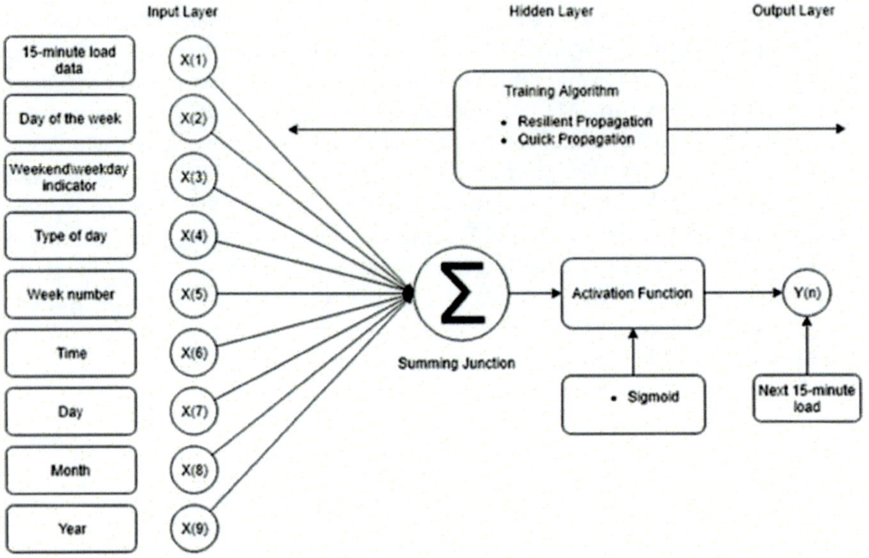


**Fig. 1** Block diagram of the implemented neural network

system's use cases as data importation of the cleaned dataset, data normalization, network training, network evaluation, and generation of the predicted results. There will be two phases in the implementation of the neural network. The first phase of the implementation was to set up the environment along with the development of the forecasting system that will embed the ANN model for implementation. In this phase, the researchers implemented the designed neural network that performed the process from data importation up to data normalization, network training, and network testing up until the generation of the prediction results that will forecast the next-hour electric load. The second phase will involve the performance evaluation of the ANN model's predictive capability for the hour-ahead electric load forecasting data along with the measurement of forecast accuracy and comparison of the actual load with the forecasted load.

Error computation was conducted with the help of the Mean Absolute Percentage Error (MAPE) as a measurement of the amount of error in percentage that arises in the forecasted electricity load by determining how close the forecasted electric load is compared to the actual electric load [9, 11, 14]. Studies highly recommend MAPE in electric load forecasting since it measures the size of the error in terms of percentage making it easier to be analyzed with the 5% globally acceptable forecasting error in electric load forecasting [2, 4, 6]. The MAPE formula is shown in Eq. 2 where $P_A$ is the actual load, $P_F$ is the forecasted load, and $N$ is the number of data. After evaluating the MAPE, the forecasted values generated by ANN's output layer were denormalized and were compared to the actual next-hour date in terms of four 15 min ahead rows from the validation subset. A graphical representation of the computations is generated with the purpose of ease of analysis in illustrating the comparison between the actual and forecasted load values:

$$\text{MAPE}(\%) = \frac{1}{N}\sum_{i=1}^{N}\frac{|P_A^i - P_F^i|}{P_A^i} \times 100 \quad (2)$$

# 3 Results and Discussion

## *3.1 Electric Load Data Preparation Results*

The data selection process that initialized data preparation of the electric load data eliminated unnecessary attributes from the raw electric load dataset leaving only the attributes of Date, Time, and Kilowatt Delivered (KW_DEL). The aggregation of the cleaned data was imported into a database design using PostgreSQL. Figure 2 shows the Entity Relationship Diagram for the hour-ahead electricity load forecasting system dividing into four tables: raw, normalized, user, and logs. The raw table was where the raw data were stored, the normalized table was where the normalized data were stored correspondingly, the user table contains the user details, and the logs table contains the log records of the current user using the system. The raw data stored in the raw table consists of 87,648 rows with the corresponding 87,648 rows, in the normalized table. While the user table stored user details from the power utility company, the logs table is where the actions of the currently logged-in user were stored for security and to trace which user to call in case their actions caused some misconfiguration on the hour-ahead load forecasting system.

Outliers and zero values caused by power outages and power interruptions were then found in the historical data making the dataset inconsistent and inaccurate. It was found out that the number of rows that output zeroes did not imply that the consumers of the power utility chose to not use electricity but is caused by power interruptions very common in Southern Philippines. There were 2263 rows or 2.6% of the whole dataset that contain zeroes implying power outages that could potentially throw off the results of the network. With the number of rows that needed to be cleaned being significantly less compared to the total amount of data, data cleaning as shown in Table 2 was performed by correcting the zeroes caused by the power interruptions with the average of load consumption from the preceding and succeeding dates of the respective time. This method was supported by studies which recommended that the method of data correction to replace outliers and zero values is by computing the average of both preceding and succeeding values of the same time [13]. Aside from the outliers brought about by the scheduled and unscheduled power interruptions, no other anomaly was identified from the dataset that will warrant further corrections.

In order to be fed into the ANN model, the resulting electric load dataset after the data cleaning was then optimally divided into training, testing, and validation sets.

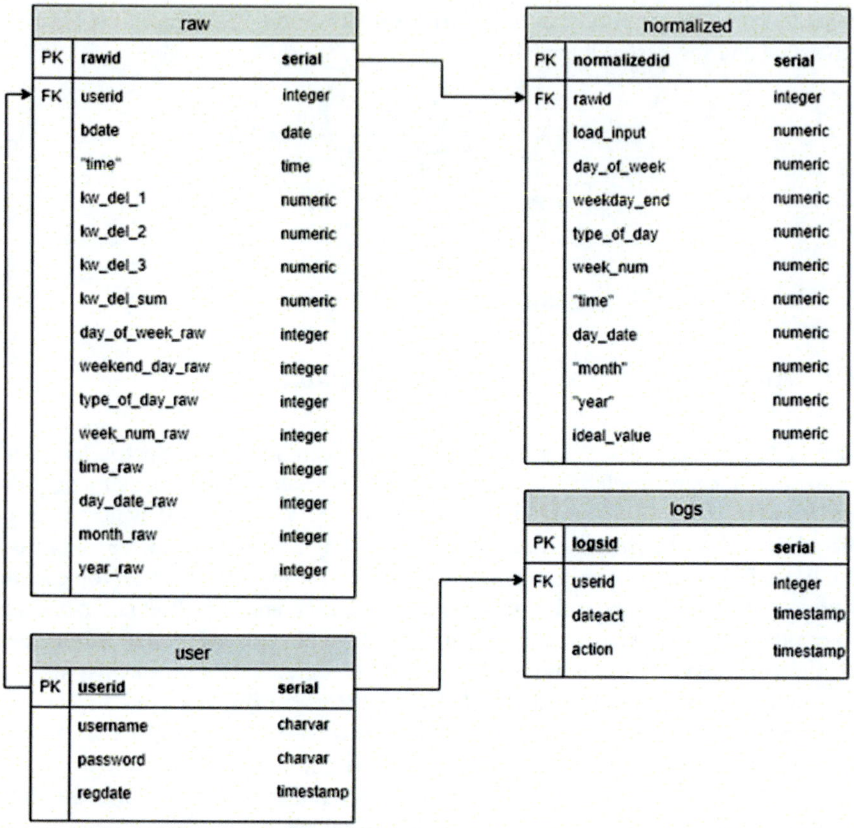


**Fig. 2** The formulated entity relationship diagram

83,267 rows (95%) of the data were allotted into the training subset which comprise the load record from 00:00:00 of April 01, 2012, to 07:45:00 of August 16, 2014. The remaining 5% of the data was partitioned for the testing subset with 2629 rows of data from 08:00:00 of August 16, 2014, to 17:15:00 of September 12, 2014, while the validation subset comprised 1753 rows of the data from 17:30:00 of September 12, 2014, to 23:45:00 of September 30, 2014. These partition results were based on similar studies which suggested that the training set must contain the largest amount of data for the efficient performance of the ANN model [3, 16, 17]. To consolidate the dataset into forms appropriate for the ANN process, the data was then normalized using the Min–Max normalization that produced transformed output within the range of 0–1. It was found out that normalizing the electric load alone using Min–Max normalization produced non-converging results during training, evaluation, and actual prediction during validation. Thus, the factors affecting the electricity load were also transformed with the day of the week represented from 1 to 7, the weekend/weekday indicator as 0 or 1, the daily 15-minute time capture from 1 to

**Table 2** Data correction of zero values caused by power interruption

| Original data | | | After data correction | | | Data correction sources | | |
|---|---|---|---|---|---|---|---|---|
| Date | Time | KW_DEL | Date | Time | KW_DEL | Time | KW_DEL of preceding date | KW_DEL of succeeding date |
| NN | TT | 0 | NN | TT | YY | TT | XX | ZZ |
| NN | TT | 0 | NN | TT | YY | TT | XX | ZZ |

96, the date from 1 to 31, the week number within the year from 1 to 52, the months from 1 to 12, and the years from 2012 to 2014.

## 3.2 ANN Model Implementation Results

The hour-ahead electricity load forecasting system that implemented the ANN models was programmed on Eclipse Mars IDE, a Java platform along with its frontend using the Eclipse WindowBuilder and was designed to run on the Windows operating system. As shown in Fig. 3, the forecasting system had five use cases which allowed the user from the power utility to log in, import data, train the network, evaluate the network, and generate the prediction. The import data use case is where the user will import their data in the database through a *.csv* file for training, testing, or prediction. The forecasting system then normalizes the imported data and will be ready to be used for training, testing, or prediction. The train network use case is where the power utility's user will train the customized network indicating the number of hidden neurons, the training algorithm, the activation function, the learning rate and momentum if applicable to their chosen training algorithm, and the range of data needed for training. The evaluated network use case is where the user picks and evaluates the trained network in its predictive capability using the desired testing data. Finally, the forecast use case is where the user will forecast the next-hour load in a 15-min interval using the latest four rows stored in the database as inputs.

The evaluated models were then tested using the inputs from the validation set that acted as the actual everyday operations of one-hour load data from the power utility company. The four sets of inputs from the validation set in 15-min intervals are from September 30, 2014, 23:00–23:45. The forecasted outputs from these inputs were then compared to the next 1 hour in 15-min interval extra data outside of the datasets mentioned in data partitioning from October 1, 2014, 00:00–00:45. Table 3 shows Quick Propagation having a MAPE of 2.91% and Resilient Propagation having a MAPE of 3.49%. The researchers can infer that not all models that perform well on model evaluation can perform well with the validation data or the actual everyday

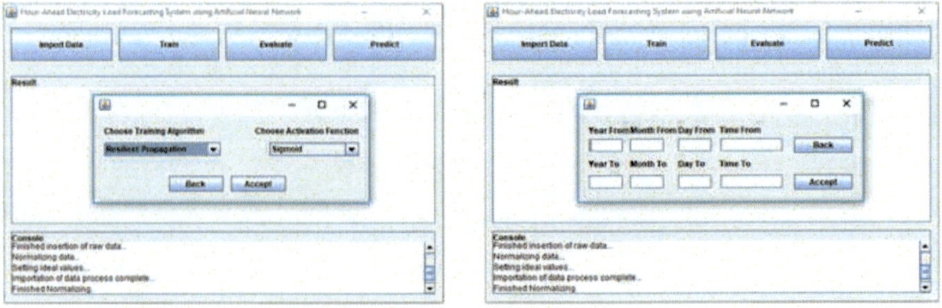


**Fig. 3** Sample user interfaces of the forecasting system

**Table 3** MAPE of the ANN models

| ANN model | MAPE (%) |
|---|---|
| Quick propagation training algorithm and sigmoid activation function | 2.91 |
| Resilient propagation training algorithm and sigmoid activation function | 3.49 |

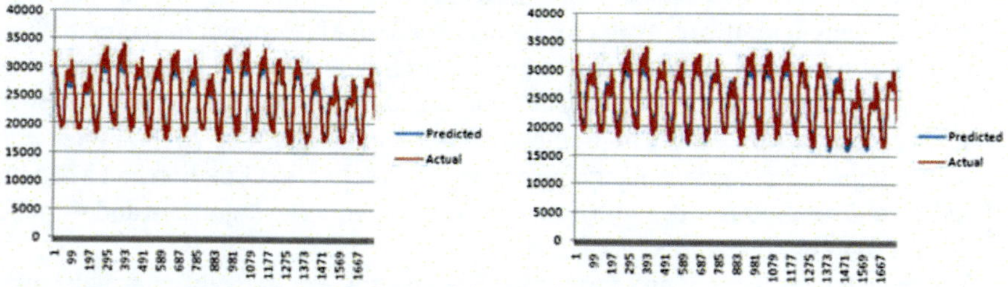


**Fig. 4** Comparison of the predicted and actual electric load consumption per 15-min data observations

data. The reason behind this could be that there are no expected values or ideal values in the actual prediction while there are in the model evaluation. Having an expected value can affect how the network scales the output data compared to not having it [9, 11]. Thus, the networks on model evaluation with the testing set performed well compared to the actual prediction of the next 1 hour in a 15-min interval using the validation set.

The predicted outputs of the ANN models were then denormalized resulting in Fig. 4 which compared the predicted and the actual electric load within the operations of the power utility from September 12 to 29, 2014. The researchers can observe that the predicted output of the ANN model which utilized resilient propagation as training algorithm and sigmoid as activation function sticks closer to the actual output compared to ANN model which utilized quick propagation as training algorithm and sigmoid as activation function. Moreover, the ANN model with resilient propagation also produced rather consistent close to accurate results compared to the other model with the actual and the predicted hour-ahead electric load values closer to each other. Contrary to their recorded MAPE values, a study has confirmed this stating that one cannot determine the accuracy of the result by network testing alone even if they yield lower MAPE results [14]. Furthermore, a confirmatory test that will show the accuracy of the predictive model using the validation set being the real everyday data can exhibit the actual and the forecasted coinciding with each other. Nonetheless, these ANN models with their error computations and their accurate predictions produce a MAPE of less than 5% of the internationally accepted error for load prediction for power utility companies. With this, the researchers have shown that the hour-ahead electricity load forecasting in a 15-minute interval using artificial neural networks is feasible and optimal for power utility companies.

# 4 Conclusion and Recommendations

Hour-ahead electric load forecasting is essential for power utility companies as it plays a major role in maintaining their generation reserve close to the minimum required in case there is a fluctuation of electricity that day-ahead and week-ahead load forecasts cannot detect. In this study, a machine learning technique using ANN was implemented to forecast the hour-ahead electric load. To achieve this, necessary procedures were undertaken such as data preparation, ANN model implementation, and prediction evaluation. It was found out that after data cleaning, data partitioning, and data transformation, only certain attributes of the dataset provided by the power utility company are necessary to predict the hour-ahead load. The ANN model with resilient propagation training algorithm and sigmoid activation function had a MAPE of 3.49%, and the ANN model with quick propagation training algorithm and sigmoid activation function had a MAPE of 2.91%.

The researchers recommend further studies that will explore other normalization methods, training algorithms, activation functions, and ANN architectures. Additional resources such as the acquisition of the latest electricity load data as well as determining more factors that can possibly affect the locality's electricity consumption such as temperature and weather data can be considered to improve ANN's performance. With the provision from power utility companies, further studies could also produce more accurate results if the granularity of the data is much lesser than the 15-min interval.

**Acknowledgements** The authors would like to thank the support of the Mindanao State University-Iligan Institute of Technology (MSUIIT) Office of the Vice Chancellor for Research and Extension through the Premiere Research Institute in Science and Mathematics (PRISM) for their assistance in this study. This work is supported by MSU-IIT as an internally funded research under the PRISM-Applied Mathematics and Statistics Group.

## References

1. Jammanakatti V, Kusagur A (2015) Week-ahead load forecasting by ANN approach. Int J Innov Res Electr Elecron Instrum Cont Eng 3(6):105–109
2. Laouafi A, Mordjaoui M, Dib D (2015) One-hour ahead electric load forecasting using neuro-fuzzy system in a parallel approach. In: Computational intelligence applications in modeling and control, Springer International Publishing, pp 95–121
3. Benlembarek K, Khadir MT, Benabbas F (2010) A web based system for short-term forecasting of algerian electricity load using artificial neural network. J Autom Syst Eng
4. Al-Shareef AJ, Mohamed EA, Al-Judaibi E (2008) One hour ahead load forecasting using artificial neural network for the western area of Saudi Arabia. Int J Electr Syst Sci Eng 1(1):35–40
5. Baharudin Z, Kamel N. (2007) One week ahead short term load forecasting. In: Proceedings of the 7th IASTED international conference on power and energy systems, Spain, pp 91–96
6. Senjyu T, Takara H, Uezato K, Funabashi T (2002) One-hour-ahead load forecasting using neural network. IEEE Trans Power Syst 17(1):113–118. https://doi.org/10.1109/59.982201

7. Chen Y, Luh B, Zhao Y, Michel L, Coolbeth M, Friedland P, Rourke S (2010) Short-term load forecasting: similar day-based wavelet neural networks. IEEE Trans Power Syst 25(1):322–330. https://doi.org/10.1109/TPWRS.2009.2030426
8. Guan C (2013) Wavelet neural network based very short-term load forecasting and prediction interval estimation. Doctoral dissertations, University of Connecticut
9. Hayati M, Shirvany Y (2007) Artificial neural network approach for short term load forecasting for Illam region. World Acad Sci Eng Technol 28:280–284
10. Barzamini R, Menhaj MB, Khosravi A, Kamalvand SH (2005) Short term load forecasting for Iran national power system and its regions using multi layer perceptron and fuzzy inference systems. Procee IEEE Int Joint Conf Neural Netw 4:2619–2624
11. Mandal P, Senjyu T, Urusaki N, Funabashi T (2006) A neural network based several-hour-ahead electric load forecasting using similar days approach. Int J Electr Power Energy Syst 28(6):367–373
12. Axelrod A, Resnik P, He X, Ostendorf M (2015) Data selection with fewer words. In: Proceedings of the tenth workshop on statistical machine translation, Lisboa, Portugal, pp. 58–65
13. Adepoju GA, Ogunjuyigbe SOA, Alawode KO (2007) Application of neural network to load forecasting in Nigerian electrical power system. Pac J Sci Technol 8(1):68–72
14. Karlik B, Olgac AV (2011) Performance analysis of various activation functions in generalized MLP architectures of neural networks. Int J Artif Intell Exp Syst 1(4):111–122
15. Vafaei N, Almeida Ribeiro R, Camarinha-Matos L (2015) Importance of data normalization in decision making: case study with TOPSIS method. In: Proceedings of the 1st International Conference on Decision Support Systems Technologies, Belgrade, Serbia
16. Adamowski JF (2008) Development of a short-term river flood forecasting method for snowmelt driven floods based on wavelet and cross-wavelet analysis. J Hydrol 353(3–4):247–266
17. Bantugon MJT, Gallano RJC (2016) Short-and long-term electricity load forecasting using classical and neural network based approach: a case study for the Philippines. In: Region 10 Conference (TENCON), IEEE, pp. 3822–3825

# Prevent the Occurrence of False Signals in the Power Transformer Internal Protection in Substation to Improve Reliability with Differential Operational Amplifier Safety

**Soni Asmaul Fuadi and Ario Dwi Prabowo**

**Abstract** Substation transformer disturbances often occur because of false signals caused by the presence of water as a connection media and the presence of conductive dust that causes trip and results in decreased reliability or TROD KPI is not achieved. To overcome this, a detection and prevention device for power transformers was made at the substation due to this false signal by using a differential operational amplifier security. The working principle of this tool is to provide an indicator of the voltage difference that is read between the 2 sensor terminals in the Auxiliary Relay with four criteria, namely normal for voltage 0–25 V DC-, Warning for voltage 25–35 V DC and Trip false signal for voltage 35–78 V DC and Original Siyal Trip with 110 V DC voltage. This electronic circuit is also equipped with a relay driver that prevents the emergence of false signals to work so that it is ensured that this security can solve the problem of the false signal with the prediction of Saving Energy Not Served (ENS) 63,066,400 IDR.

**Keywords** Internal transformer safety · False signals · Differential operational amplifier safety

## 1 Introduction

Disturbances in the Power Transformer at the Main Station, especially at the Central Kalimantan UPT in 2019 which caused the KPI TROD not to be achieved due to a false signal that caused the trip of the GI power transformer are as shown in the graph below (Fig. 1).

The motivation for this study was carried out because of the frequent occurrence of false signals in transformer protection at substations where there is currently no early detection. This research has never existed and is a new thing that is applied in preventing the trip transformer which results in power outages and harms the company. The advantage obtained from the research and application of this journal

S. A. Fuadi (✉) · A. D. Prabowo
PT PLN (Persero) Corporate University, UPDL Banjarbaru, Banjarbaru 70711, South Borneo, Indonesia

 
X.-S. Yang et al. (eds.), *Proceedings of Sixth International Congress on Information and Communication Technology*, Lecture Notes in Networks and Systems 216,
https://doi.org/10.1007/978-981-16-1781-2_74

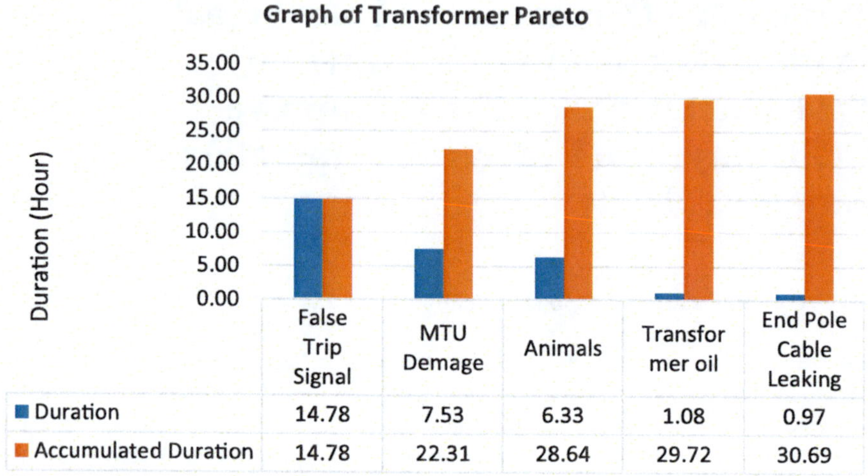


**Fig. 1** Graph of pareto transformer disruption in 2019

is that it eliminates the potential for ENS that causes companies to lose and satisfies consumers with the reliability of the electrical power supply.

Of the disturbances, the most dominant one was caused by a false signal so that the KPI TROD was not reached from the target of 0.4 while the realization was 0.58. This false signal arises from the presence of rainwater or dirt attached to the cable terminals from the internal sensor of the substation power transformer where the water and dirt are as conductors which causes the relay to work to plug the power transformer. The substation transformer internal sensor consists of Sudden Pressure, Bucholz, Jansen Relay and Thermal Relay. This electronic circuit is also equipped with a relay driver that prevents the emergence of false signals to work so that it is ensured that this security can solve the problem of the false signal with the prediction of Saving Energy Not Served (ENS) 63,066,400 IDR.

To prevent the emergence of this false signal, it is necessary to make a detection device and prevent the trip of the power transformer because of the false signal by using a differential operational amplifier circuit. The block diagram of the circuit can be illustrated as shown in Fig. 2.

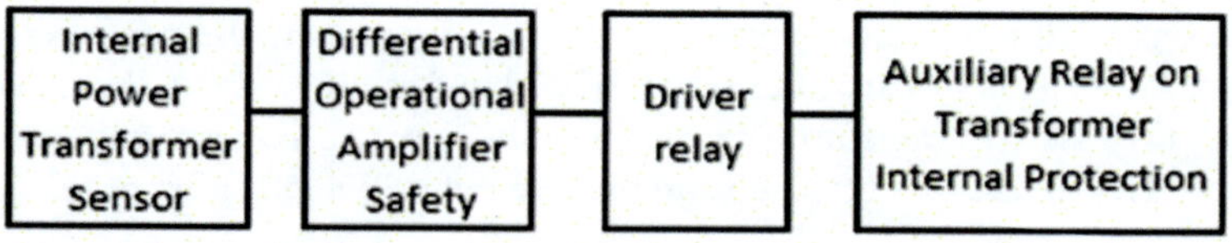


**Fig. 2** Block diagram equipment

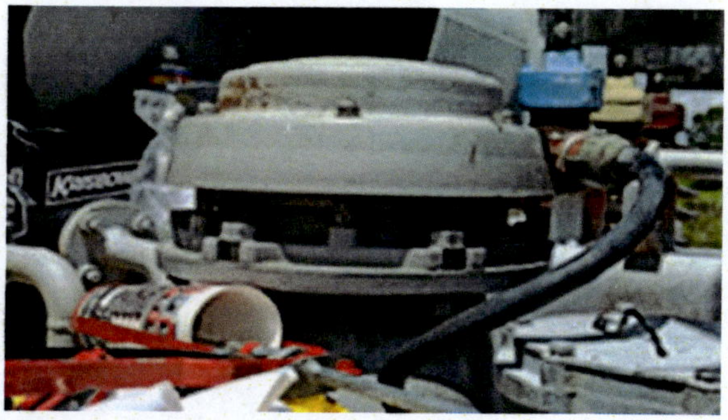

**Fig. 3** Sudden pressure

# 2 The Block Diagram Part and How It Works

## 2.1 *Internal Power Transformer Sensor*

The part of the power transformer is that there are four sensors installed for the internal safety of the transformer, namely

- Sudden Pressure,
- Bucholz,
- Jansen Relay and
- Thermal Relay.

### 2.1.1 How It Works: Sudden Pressure

How to protect the transformer from excessive pressure disturbances caused by disturbances in the Main Tank of the transformer by using a type valve that is pressed by a spring when the pressure in the transformer exceeds the spring pressure then it will open and throw pressure. Next Figure Sudden Pressure (Fig. 3).

### 2.1.2 How It Works: Bucholz

It serves to secure the transformer from internal disturbances that cause gas due to a short circuit or due to arcs in the transformer OLTC with alarm and trip contacts. The picture of Bucholz is shown in the (Fig. 4).

### 2.1.3 How It Works: Jansen Relay

Jansen relays are used to secure the transformer from interference in OLTC. It works with the same principle as a Bucholz relay but has only one tripping contact. The picture of Jansen relay is shown in the (Fig. 5).

**Fig. 4** Bucholz

**Fig. 5** Jansen relay

### 2.1.4 How It Works: Thermal Relay

It serves to protect the transformer from excessive temperature proportional to the loading factor and the outside air temperature of the transformer. It can be divided into two types, namely winding temperature relays (windings) and Thermal Relay as shown in the (Fig. 6).

**Fig. 6** Thermal relay

## 2.2 *Differential Operational Amplifier Safety*

- Quad Operational Amplifier,
- Resistor and
- Lamp LED.

### 2.2.1 Quad Operational Amplifier

- Operational Amplifier—Operational Amplifier or better known as the Op-Amp is one form of Linear IC that functions as an electric signal amplifier. An Op-Amp consists of several Transistors, Diodes, Resistors and Capacitors that are interconnected and integrated to enable it to produce high gain over a wide frequency range where Quad Op-Amp in one IC has 4 Op-Amps, one of which is LM 324 IC.
- In Fig. 7, it can be seen that there are two inputs, namely input inverting and non-inverting; there are also two input sources as the power source of the Op-Amp, namely positive voltage (+Vcc) and negative voltage (−Vee).

- To be able to understand the working system of the Op-Amp, it is necessary to know in advance some of the properties of an ideal Op-Amp, namely
- Difference in input voltage (Vdm) = 0 b Input current Op-Amp (ia) = 0;
- Infinite open loop amplifier (AVOL);
- The open ring output resistance (Ro, ol) is zero;
- Infinite open loop (Ri, ol) input barriers;
- Infinite bandwidth (bandwidth) or Δf infinite;
- Infinite Common Mode Rejection Ratio (CMRR).

**Op-Amp as a differential amplifier**

A differential amplifier is to strengthen the voltage difference between inverting and non-inverting inputs. So far we have only used one of the operational amplifier inputs

Fig. 7 Op-Amp symbol in the circuit

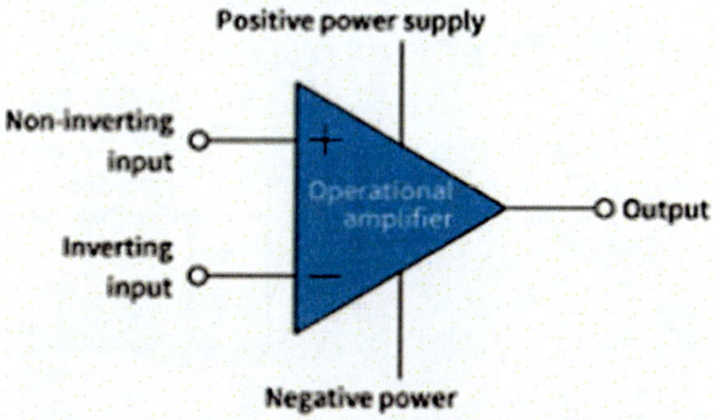

(Op-Amps) to connect to the amplifier, using either the "inverting Op-Amp" or "non-inverting Op-Amp" input terminals to amplify a single input signal with other inputs connected to the ground.

But because the standard operational amplifier (Op-Amp) has two inputs, inverting and non-inverting, we can also connect signals to these two inputs at the same time to produce a general type of Operational Amplifier circuit called a Differential Amplifier or Differential Amplifier. The configuration connects one voltage signal to one input terminal and another voltage signal to another input terminal, the resulting output voltage will be proportional to the "difference/difference" between the two input voltage signals V1 and V2. This type of op-amp circuit is commonly known as a Differential Amplifier configuration and is shown in Fig. 8.

The formula of the differential equation is as follows:

$$V_{out} = -V_1\frac{R_3}{R_1} + V_2\left[\frac{R_4}{R_2 + R_4}\right]\left[\frac{R_1 + R_3}{R_1}\right] \quad (1)$$

where when resistors, R1 = R2 and R3 = R4, the formula becomes

$$V_{out} = \frac{R_3}{R_1}[V_2 - V_1] \quad (2)$$

If all resistors have the same ohm value, namely R1 = R2 = R3 = R4, the circuit will be a Gain Unity Differential Amplifier and the amplifier's voltage gain will be exactly one or a unit. Then the output expression will only be $V_{out}$ = V2−V1. The workings of this Op-Amp differential safety device for safety and prevention of false signal interference caused by dirt and water is to read the voltage to be measured at the auxiliary relay coil terminals that cause a trip. The types of auxiliary relays used by internal transformer protection can be seen in Table 1.

From Table 2, there are 4 conditions that can be made, namely

- Normal Conditions,
- Warning Conditions,
- False Signal Trip Conditions and

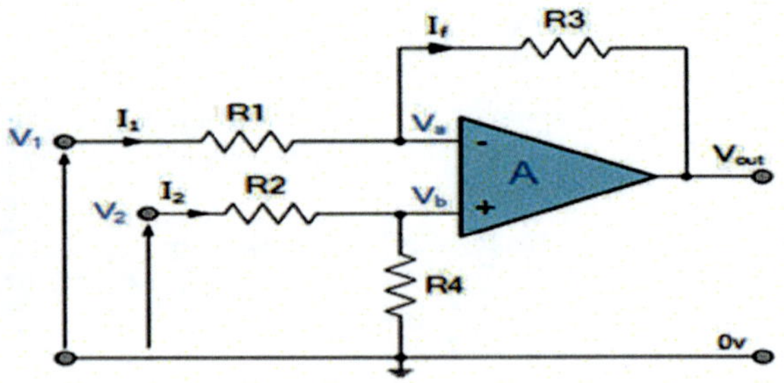


**Fig. 8** Differential op-amp in the circuit

**Table 1** Results of auxiliary relay pickup tests

| No. | Auxiliary relay | Pickup relay | Reading results of Aux relay voltage | | | Information |
|---|---|---|---|---|---|---|
| | | | Normal | Warning | False signal trip | |
| 1 | RELPOL | 78 V DC | 0–25 V DC | 25–35 V DC | 35–110 V DC | When sensors work then the resistance of content sensor 0 Ω so that Aux relay voltage with source voltage reduced drop in cable |
| 2 | SMITT | 70 V DC | | | | |
| 3 | ALEN BREADLY | 68 V DC | | | | |
| 4 | ARTCHE | 74 V DC | | | | |
| 5 | OMRON | 35 V DC | | | | |

- Original Signal Trip Condition (V source = V Aux) Transformer internal sensor circuit diagram can be explained in the (Fig. 9).

This auxiliary relay can work in a minimum voltage range of 35–78 V DC depending on the brand of Auxiliary Relay used. In the condition of the internal sensor, the transformer does not work (contact open), no current flows in the circuit, the voltage reading at the auxiliary relay coil terminal will be equal to 0 V or the Auxillary Relay does not work. If there is dirt and water in the sensor terminal, resistance will occur which causes current to flow in the circuit. The more dirty and wet the sensor terminal is, the smaller is the resistance so that it will enlarge the current flowing in the circuit which causes the voltage at the auxiliary relay terminal also to rise which can cause a transformer trip if the pickup voltage has been reached. In designing this tool, 4 different differential safety op-amps are used to determine the status condition of the voltage that is read at the Auxiliary Relay terminal with the following voltage:

- Normal Conditions (Voltage 0–25 V DC);
- Warning Conditions (Voltage 25–35 V DC);
- False Signal Trip Conditions (Voltage 35–110 V DC);
- Original Signal Trip Condition (V source = V Aux).

**Drivers**:

The output signal from the differential Op-Amp is included in the driver circuit. The figure for driver circuit can be seen in (Fig. 10).

Relay overcome consists of a coil and contact. This coil creates a magnetic field when electric current flows so that it mechanically pulls the contact on the relay which contacts NO contacts to NC and vice versa. Relays also have a section named common that is connected to NC under normal circumstances. Auxiliary Relay on Transformer Internal Protections:

**Table 2** Internal transformers disturbance in 2019

| Years | Month | Date | Transfomer | Capacity (MVA) | Condition | | | | Duration | | Indication | Cause | OFF (MW) | ENS (kWh) |
|---|---|---|---|---|---|---|---|---|---|---|---|---|---|---|
| | | | | | Trip | | Close | | Minute | Hour | | | | |
| 2019 | February | 4 | Trafo #1 GI Seberang Barlto | 20 | 22 | 2 | 23 | 0 | 58 | 0,967 | Master trip | Cable impulse leaks | 3,94 | 4463 |
| | February | 5 | Trafo #1 GI Pu lang Plsau | 10 | 6 | 5 | 6 | 58 | 53 | 0,883 | Differential | Animal | 3,13 | 2396 |
| | March | 3 | Trafo #1 GI Pu lang Plsau | 10 | 2 | 52 | 3 | 47 | 55 | 0,917 | Differential | Animal | 3,43 | 3669 |
| | March | 7 | Trafo #1 GI Seberang Barlto | 20 | 1 | 54 | 2 | 59 | 65 | 1,083 | Differential | Quality of transformer oil | 4,86 | 6804 |
| | March | 17 | Trafo #1 GI Ulin | 30 | 9 | : 52 | 16 | 46 | 414 | 6,900 | Differential | CT 66 kV phasa S bay Trafo #1 GI Ulin Break Down | 28,72 | 44,999 |

(continued)

**Table 2** (continued)

| Years | Month | Date | Transfomer | Capacity (MVA) | Condition | | | | Duration | | Indication | Cause | OFF (MW) | ENS (kWh) |
|---|---|---|---|---|---|---|---|---|---|---|---|---|---|---|
| | | | | | Trip | | Close | | Minute | Hour | | | | |
| | May | 23 | Trafo #2 GI Selat | 30 | 8 | : 55 | 9 | 33 | 38 | 0,633 | SBEF | Anomali NGR & CT | 5,81 | 2279 |
| | July | 23 | Trafo #1 GI Kayutangl | 30 | 21 | :4 | 1 | 36 | 272 | 4,533 | Differential | Animal | 19,3 | 73,348 |
| | October | 31 | Trafo #3 GI Cempaka | 10 | 15 | : 15 | 17 | 30 | 135 | 2,250 | Master Trip | False signal (Bucholz) | 4,05 | 9112,5 |
| | November | 14 | Trafo #1 GI Bagendang | 20 | 11 | 52 | 23 | 39 | 707 | 11,783 | Master Trip | False signal (Oil temperature indication) | 3,57 | 17,662 |
| | December | 10 | Trafo #5 GI Cempaka | 30 | 15: 29 | | 16: 14 | | 45 | 0,750 | Master Trip | False signal (Winding temperature indication) | 9.66 | 4403 |

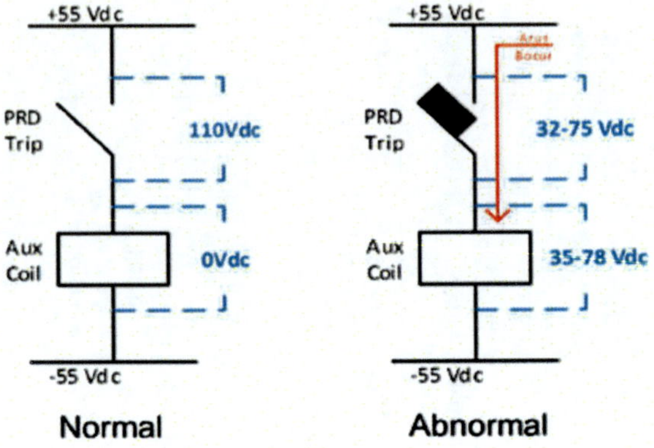


**Fig. 9** Transformer internal sensor circuit

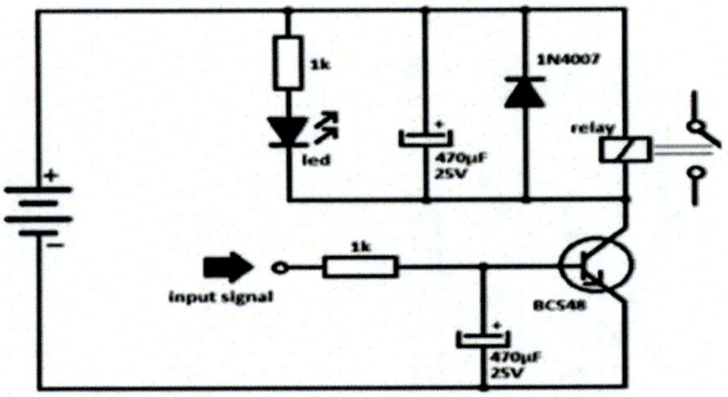


**Fig. 10** Driver relay circuit

The series of Internal Protections Transformer diagrams on each power transformer is different, one of which is shown in the (Fig. 11).

Trip Data and ENS Calculations

Disturbances Transformer in 2019 at PLN South and Central Kalimantan UPT can be seen as below.

## 3 Results

The prediction of saving calculation from the installation of the transformer internal protection to prevent the trip from the transformer due to false signals at PLN UPT South and Central Kalimantan is estimated at Rp 63,066,000—the TROD prediction is down from 0.59 in 2019 to 0.30 in 2020. ENS and Saving can be seen in Table 3.

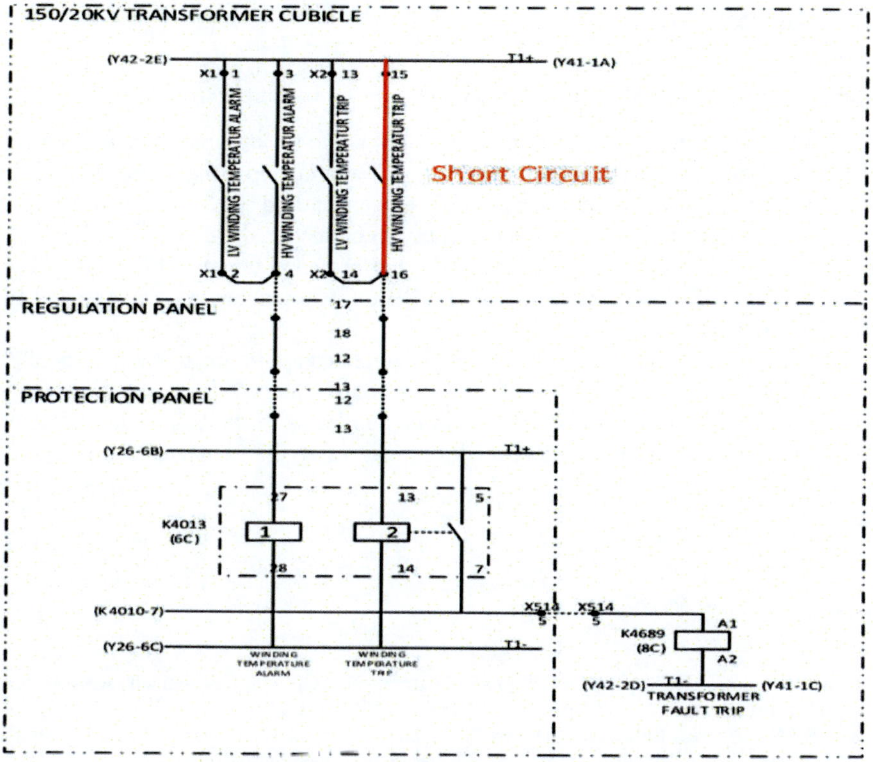


**Fig. 11** Schematic internal transformer protection diagram

**Table 3** ENS and saving

| Causes of transformer disorders | Off (MW) | Time (hour) | Energy not served (kWh) | Equipment making costs (Rp) | Saving ( Rp) |
|---|---|---|---|---|---|
| False signal (Bucholz) | 4,05 | 2,25 | 9.112,50 | 400.000,00 | 9.623.750,00 |
| False signal (oil temperature indication) | 3,57 | 11,78 | 42.066,50 | 400.000,00 | 45.873.150,00 |
| False signal (winding temperature indication) | 9,66 | 0,75 | 7.245,00 | 400.000,00 | 7.569.500,00 |
| Total | | | 58.424,00 | 1.200.000,00 | 63.066.400,00 |

## 4 Conclusions

The conclusions that can be drawn from making this tool are as follows:

1. Trip Transformer generator and substation due to false signals can be detected early and can be prevented by using differential Op-Amp protection. This research is limited to false signals only that are caused by the presence of water as a connection media and the presence of conductive dust.
2. As a consideration in determining the performance target (KPI) of PT PLN (Persero) UPT South Kalimantan and Central Kalimantan that can be reduced from 0.59 to 0.30.
3. Saving Energy Not Served (ENS) can be obtained approximately as 63,066,400 IDR due to false signals from 52 existing transformers.
4. For further research, this system can be applied with the ESP32 IoT system; then the monitoring and data recording can easily be managed online by the user.

## References

1. Klumperink EAM, Seevinck E (1989) MOS current gain cells with electronically variable gain and constant bandwidth. IEEE J Solid-State Circ 24:1465–1467
2. ABB Power Grids Italy SpA COMEM Operating Unit Strada Regionale 11, Signolo 22 36054 Montebello Vicentino (VI)–Italy www.abb.com/transformercomponents
3. Toumazou C, Lidgey FJ, Haigh DG (1990) Analogue IC design: the current mode approach, Peter Peregrinus, London, U.K.
4. Bruun E (1991) A differential-input, differential-output current mode operational amplifier. Int J Electr 71:1048–1056
5. Bruun E (1991) Constant-bandwidth current mode operational amplifier. Electron Lett 27:1673–1674
6. Temes GC, Ki WH (1987) Fast CMOS current gain amplifier and buffer stage. Electron Lett 23:696–697
7. Bowers DF (1988) A precision dual current feedback operational amplifier. In: Proceedings of the IEEE 1988 bipolar circuits and technology meeting, Minneapolis, pp 68–70
8. Bruun E (1993) Design considerations for a high speed CMOS current op amp. In: Proceedings of the Norchip seminar, Trontheim, Norway, pp 24–31
9. Zele RH, Lee SS, Allstot DJ (1992) A high gain current-mode operational amplifier. In: Proceedings of the IEEE international symposium on circuits and systems, San Diego, U.S.A., pp 2852–2855
10. Zele RH, Allstot DJ, Fiez TS (1991) Fully-differential CMOS current-mode circuits. In: Proceedings of the IEEE custom integrated circuits conference, San Diego, U.S.A, pp 24.1.1–24.1.4
11. https://www.ablic.com/en/semicon/products/analog/opamp/intro/
12. https://www.electroschematics.com/relay-driver/

# Risk Factors in the Appearance of Breast Cancer, Tools, Models and Current Issues

**Ramón Santana Fernández, José Manuel Valencia Moreno, and Everardo Gutiérrez López**

**Abstract** Breast cancer is one of the diseases that mostly affects the female population worldwide. Not only because of the high mortality rate and the difficulty of detection, but also because of the physical and psychological sequelae, it causes in women. Currently, there is a set of tools and models to assist the early detection of breast cancer for the U.S. population, however, its application to the Latino and Caribbean population has not had the expected results. The models and tools use risk factors predefined in 1989 as part of the Gail model for the North American white population, without the possibility of introducing new risk factors or evaluating the incidence factor of each one of them in the appearance of the disease. Other works have been proposed in the bibliography, however, they are calibration of the original model, having the same limitations. In the present work, an analysis of the models and tools is carried out with the aim of contrasting the results obtained and presenting in the immediate future new techniques for the calculation of the factor or index of incidence of risk factors in the manifestation of the disease.

**Keywords** Breast cancer · Risk factors · Risk factor incidence

## 1 Introduction

Breast cancer is the second most common type of cancer affecting the female population worldwide [1]. Currently, hundreds of medical data are collected from patient interviews and medical examinations which can be analyzed. The use of new data analysis techniques could reveal relevant information about factors that trigger the

R. S. Fernández (✉) · J. M. V. Moreno · E. G. López
Autonomous University of Baja California, 22890 Ensenada, BC, Mexico
e-mail: ramon.santana@uabc.edu.mx

J. M. V. Moreno
e-mail: jova@uabc.edu.mx

E. G. López
e-mail: everardo.gutierrez@uabc.edu.mx



X.-S. Yang et al. (eds.), *Proceedings of Sixth International Congress on Information and Communication Technology*, Lecture Notes in Networks and Systems 216,
https://doi.org/10.1007/978-981-16-1781-2_75

disease. Currently in Mexico, no analysis of this data is performed; only statistics are captured and processed for reports, with the greatest weight given to the performance of tests. The multivariate analysis of the data from different points of view could support the specialists and patients in the early detection of the disease, improving this way their life expectancy. The medical databases by the nature of their data are private, but in them resides the necessary knowledge to fight the disease using different techniques of data analysis and knowledge discovery.

Early detection of breast cancer is one of the most important success factors in reducing patient mortality. This decreases the possibility of severe physical and psychological damage among others. For the detection of the disease, numerous laboratory tests and medical examinations have been developed which are expensive and not within the reach of everyone. In order to make diagnosis more accessible and more economical, both for the government and for individuals, a probabilistic model was developed that estimates the probability of developing breast cancer based on the incidence of various risk factors, called the Gail Model [2].

This model presents excellent results in Caucasian patients who were born in the United States or who have descent from this country. Other models have been proposed subsequently, however, none of them have presented good results when analyzed for Latin American populations. This poses a problem that has not yet been resolved, highlighting the possible applications of risk factor data analysis to detect signs of patterns that could lead to the development of breast cancer in a patient. In the present work, risk factors are analyzed as a tool to support the early detection of the disease.

## 2 Previous Work

Among the first works for the analysis of risk factors found in the literature is the model for estimating the probability of suffering from breast cancer [2]. This model consists of estimating the probability of developing breast cancer based on risk factors. The importance of each risk factor is taken into account and the calculation is made using the statistical method of pseudostratified logistic regression. The risk factors used are

- Age;
- Age of first menarche;
- Age at which you had your first child;
- Number of biopsies;
- Number of blood relatives with breast cancer.

This allows the calculation of a woman's probability of developing breast cancer at a specific age, so the woman's age is taken into account. To address this issue, the authors of the paper [2] subdivide the problem into the following:

- To determine which risk factors are important and to estimate the relative risk of a subject with a certain constellation of risk factors at a certain age compared to a subject without the identified risk factors.
- To estimate the age-specific risk rate for breast cancer, which is the rate for a woman with no identified risk factors.
- Project the long-term probability of developing breast cancer based on a consideration of competing risks and outcomes.

To solve the problem, the group of authors proposed the estimation of a set of numerical values which are multiplied by each risk factor during the estimation process. The calculation of the incidence rates of each risk factor was initially carried out with two techniques, the logistic regression and the technique of analysis by pairs, which obtained similar results. In view of this situation, in [3] the logistic regression technique was selected, which performs the analysis of a dependent variable with an independent variable from the data, with 5 variables to be analyzed.

For the validation of this model, 2852 cases of white women and 3146 control cases were used. As a result, it was concluded that the model presents good results for North American Caucasian populations and has been demonstrated practically and theoretically in the United States [4]. It is used in that country to calculate the probability of those suffering from breast cancer and to predict its development at both 5 and 10 years.

When applying the model to other populations, it has not had the expected results and therefore different adaptations have been made for specific populations.

## 2.1 *Other Models Proposed Based on Gail's Model*

In 2012, Banegas and a group of authors performed a calibration of Gail's model for a specific population [5]. In this work, the model was recalibrated using the observed/expected (O/E) method for Hispanic women. The Hispanic women in this research were conceptualized as being of Hispanic descent living on the San Francisco coast. The observed/expected method was applied for model recalibration and another regression method called the Cox method was used for validation. This method performs the single dependent variable analysis with several independent variables, which can still leave hidden knowledge in the data to improve the efficiency of the process.

In 2017, this same group of authors publishes a new study in which the Hispanic population is divided into two groups, 533 born in the U.S. and 533 born outside the U.S. The same method used in [6] is used to calculate the incidence rate of each risk factor. This method groups the risk factors in a manner determined by the authors. In the publication of [1], the use of the observed and expected method for model calibration is described, obtaining good results in the selected population.

In 2016, a group of authors proposed a model for estimating breast cancer disease based on risk factors for Iranian women [7]. In this case, Gail's model was modified in order to adapt it to the specific conditions of the Iranian population.

## 2.2 *Risk Factors*

A risk factor is defined by the World Health Organization as any trait, characteristic or exposure of an individual that increases his or her likelihood of suffering an illness or injury [8]. In [9], different risk factors that may be involved in the appearance of the disease are analyzed. These are

1. Being a woman,
2. Age,
3. Family history,
4. Genetics,
5. Personal history of breast cancer,
6. Radiation to the chest and face before age 30,
7. Certain changes in the breast,
8. Race or ethnicity,
9. Overweight,
10. Pregnancy history,
11. Lactation history,
12. Menstruation history,
13. Use of hormone replacement therapy,
14. Ingestion of alcoholic beverages,
15. Having dense breasts,
16. Do not exercise,
17. Smoking,
18. Low levels of vitamin D,
19. Exposure to light at night (night workers),
20. Exposure to diethylstilbestrol while in their mother's womb,
21. Junk food or unhealthy foods,
22. Exposure to cosmetic chemicals,
23. Exposure to chemicals in food,
24. Exposure to chemicals in gardens or planting sites,
25. Exposure to chemicals in plastics,
26. Exposure to chemicals in sunscreen,
27. Exposure to chemicals in water and
28. Exposure to chemicals in grilled or prepared foods.

This work shows that there are multiple risk factors in the appearance of breast cancer and that they are present in modern life. In Gail's model, only 5 of them are analyzed, leaving a vast field of analysis of risk factors that have been appearing and will continue to appear with the advance of modern life. Most studies focus on breast cancer survivors, leaving a niche of knowledge that is not being analyzed.

In [10], a set of risk factors which have an incidence in the appearance of breast cancer are analyzed. These factors are documented later in [9] which asserts their importance despite the passage of time. The tools for prediction and analysis of risk factors developed from several models proposed in the literature only take into account some of these risk factors, which does not allow a greater and more detailed observation of the phenomenon. The analysis of the incidence of each factor on the disease is carried out with techniques such as logistic regression, which only takes into account one de-pendent variable and one independent variable. This type of analysis limits the potential for the discovery of knowledge in the data.

These models are implemented in the Breast Cancer Risk Assessment Tool. It is a free solution that can be consulted in [6] and provides the estimation of the probability of developing the disease based on the risk factors described by several authors.

The Mexican breast cancer standard [11] establishes the risk factors that are used in Mexico and states that in 70% of confirmed cases they present only 1 or none of the described risk factors. This states that the risk factors vary and are not yet definitively described.

## 3 Discussion

The models and solutions analyzed above allow us to estimate the probability of a person to develop breast cancer and make a prediction at 5 and 10 years. The results published by different authors at various times have been very good, however, there are still elements that are not contemplated among which are the following:

- The tool does not present a way to add or change the risk factors proposed by the studies that are implemented in it.
- It does not allow the calculation of the incidence factor that has a risk factor in the appearance of the disease. In this sense, the analysis is made with the same incidence factors for each factor that was proposed in 1989, which evidences the need to re-evaluate both the incidence factors and the risk factors.
- The model is not fed with historical data nor is it fed back with data that is incorporated with the purpose of evaluating the change in risk factors over time, which places it as a static model with very good results for a part of the population but which could be improved for continuous learning from the new cases that are introduced into it daily.

As part of this work, the situation of a Latin American population where the tool to evaluate the risk of breast cancer has not presented the expected results published in other works is analyzed. In Baja California, Mexico, it was tested with real cases of which both their diagnosis and the risk factors proposed by Gail are known, and the results are shown in (Table 1) [12]. As can be seen, a question emerges

**Table 1** Risk factors and models published in the literature. All models were implemented in the Breast Cancer Risk Assessment Tool

| Article | Statistical techniques | Main risk factors used | Target population |
|---|---|---|---|
| Gail et al. [3] | Stratified logistic regression | Age, age of first menarche, age at which you had your first child, number of biopsies, number of relatives of first degree with breast cancer | North American population |
| Banegas et al. [5] | Stratified logistic regression; and observed and expected for model calibration | Age, age of first menarche, age at which you had your first child, number of biopsies, number of relatives of first degree with breast cancer | North American population of Latin-American descent |
| Mirghafourvand et al. [7] | ANOVA, one-way test, T-Test and Pearson correlation | Demographic factors and reproductive factors (both groups of factors can be found in the original publication) | Iranian population |
| Banegas et al. [1] | Conditional logistic regression | Age, age of first menarche, age at which you had your first child, number of biopsies, number of relatives of first degree with breast cancer, benign breast tumor biopsy | North American population of Latin-American descent |

for populations different from those in North America: Will the risk factors be the same for these populations and for the populations in different Latin American and Caribbean states? Will the incidence factor be the same for each risk factor in the populations analyzed?

## 4 Conclusions and Future Work

In the present work, an analysis of the methods for calculating the probability of developing breast cancer based on risk factors was carried out in order to determine their basis and operation. Both the risk factors proposed in 1989 with the first tool developed for this purpose, as well as the risk factors proposed later by other authors, were evidenced, so that it is possible to extend the analysis of risk factors in the calculation of the factor or incidence rate of these in the occurrence of the disease.

As a continuation of this work, a tool will be developed for the calculation of the importance factor of risk factors using pattern recognition methods that allow the multivariate analysis of risk factors. The purpose of the development of the tool is to determine which risk factors have an impact on the appearance of breast cancer and thus promote differentiated care for patients in specialized consultations.

It is planned to implement the tool as part of the state breast cancer program in Baja California. With this, we intend to determine if the risk factors of these populations are the same as the risk factors exposed by other authors.

## References

1. Banegas MP et al (2017) Projecting individualized absolute invasive breast cancer risk in US hispanic women. J Natl Cancer Inst 109(2):1–8. https://doi.org/10.1093/jnci/djw215
2. Gail MH, et al (1989) Projecting individualized probabilities of developing breast cancer for white females who are being examined annually. JNCI J Natl Cancer Inst 81(24):1879–1886. Available: http://jnci.oxfordjournals.org/
3. Gail MH et al (1989) Projecting individualized probabilities of developing breast cancer for white females who are being examined annually. J Natl Cancer Inst 81(24):1879–1886. https://doi.org/10.1093/jnci/81.24.1879
4. Rockhill B, Spiegelman D, Byrne C, Hunter DJ, Colditz GA (2001) Validation of the Gail et al. Model of breast cancer risk prediction and implications for chemoprevention. J Natl Cancer Inst 93(5):358–366. doi:https://doi.org/10.1093/jnci/93.5.358
5. Banegas MP et al (2012) Evaluating breast cancer risk projections for Hispanic women. Breast Cancer Res Treat 132(1):347–353. https://doi.org/10.1007/s10549-011-1900-9
6. N. C. Institute (2018) Breast Cancer Risk Assessment Tool, https://bcrisktool.cancer.gov/calculator.html#demographics-section.
7. Mirghafourvand M, Mohammad-Alizadeh-Charandabi S, Ahmadpour P, Rahi P (2016) Breast cancer risk based on the Gail model and its predictors in Iranian women. Asian Pac J Cancer Prev 17(8):3741–3745. doi:https://doi.org/10.14456/apjcp.2016.163
8. OMS | Factores de riesgo, WHO, 2011
9. Chaurasia V, Pal S, Tiwari BB (2018) Prediction of benign and malignant breast cancer using data mining techniques. J Algorithms Comput Technol 12(2):119–126. https://doi.org/10.1177/1748301818756225
10. Singletary SE (2003) Rating the risk factors for breast cancer. Ann Surg 237(4):474–482. https://doi.org/10.1097/01.SLA.0000059969.64262.87
11. Salud S (2011) Norma Oficial Mexicana NOM-041-SSA2-2011, Para la prevención, diagnóstico, tratamiento, control y vigilancia epidemiológica del cáncer de mama. D. Of. la Fed., 1–36
12. Valencia-Moreno JM, López EG, Pérez JFR, Rodríguez JPF, Xochihua OÁ (2020) Exploring breast cancer prediction for cuban women. In: Advances in intelligent systems and computing, vol 1137 AISC, pp 480–489, doi:https://doi.org/10.1007/978-3-030-40690-5_47

# A Study on the Comparison of Feature Extraction Methods for Classification of Patent Litigation

**Youngho Kim, Junseok Lee, Jiho Kang, Juhyun Lee, Dongsik Jang, and Sangsung Park**

**Abstract** Patents contain detailed information about the developed technology. In addition, patents grant exclusive rights to the developed technology. For this reason, many companies use patents for technology protection. Patent litigation occurs when the operating activities of one company infringe on the scope of the patent rights of another company. When patent litigation occurs, a lot of time and money are consumed. Therefore, it is necessary to prevent patent litigation in advance. In this paper, an appropriate feature extraction method is sought when constructing a model for classifying patent litigation. Principal component analysis and Autoencoder are used to perform the proposed research. The experimental data are those registered with the USPTO as patents related to artificial intelligence. Feature extraction is performed on the quantitative indicators of the collected patents. In addition, performance is measured with various classification algorithms. As a result of the experiment, the classification performance of the method combining Autoencoder and K-Nearest neighbor was good.

Y. Kim (✉) · J. Lee · D. Jang
Department of Industrial Management Engineering, Korea University, Seoul, Republic of Korea
e-mail: Youngho0928@korea.ac.kr

J. Lee
e-mail: leeju@korea.ac.kr

D. Jang
e-mail: jang@korea.ac.kr

J. Lee
MICube Solution, Seoul, Republic of Korea
e-mail: js.lee@micube.co.kr

J. Kang
Machine Learning Big Data Institute, Korea University, Seoul, Republic of Korea
e-mail: kangmae@korea.ac.kr

S. Park
Department of Big Data Statistics, CheongJu University, Cheong Ju, Republic of Korea
e-mail: hanyul@cju.ac.kr

 
X.-S. Yang et al. (eds.), *Proceedings of Sixth International Congress on Information and Communication Technology*, Lecture Notes in Networks and Systems 216,
https://doi.org/10.1007/978-981-16-1781-2_76

**Keywords** Patent litigation · Feature extraction · Patent big data analysis · Classification

# 1 Introduction

Patents contain detailed information on the developed technology and give exclusive rights [1]. For this reason, many companies use patents to protect their technology. Patent litigation occurs when the operating activities of a specific company infringe on the scope of the patent rights of another company. When patent litigation occurs, companies have to resolve disputes by consuming a lot of time and money [2]. Therefore, patent litigation causes a decline in the competitiveness of a company. To solve this problem, various studies have been conducted to pre-classify patents for which patent litigation may occur [3–6]. There are qualitative and quantitative methods to classify patent litigation. Qualitative methods are carried out manually by some experts. Therefore, the accuracy of classification may be high, but a lot of time is consumed. Quantitative methods are efficient because they use statistical or machine learning algorithms. Recently, patents are recognized as big data, and algorithm learning data is abundant [7]. For this reason, the classification performance of the quantitative method can be improved than before.

Patent litigation classification using a quantitative method uses various classification algorithms for patent indicators. In this paper, we obtain new features for learning classification algorithms from patent indicators using various feature extraction methods. In addition, the patent litigation classification performance is measured by applying various classification algorithms. As experimental data, a patent registered in the United States Patent and Trademark Office (USPTO) is used as a patent related to artificial intelligence (AI). In general, there are very few patents in which litigation has occurred compared to those that do not. Therefore, non-litigation patents are sampled and used through Propensity Scoring Matching (PSM). Principal component analysis (PCA) and Autoencoder (AE) are used for feature extraction. As classification algorithms, logistic regression (LR), K-nearest neighbor (KNN), support vector machine (SVM), decision tree (DT), random forest (RF), and AdaBoost (AB) are used.

# 2 Related Works

## *2.1 Classification of Patent Litigation*

The classification of patent litigation using a quantitative method uses various indicators and classification algorithms of the patent. Kim et al. [3] built a model that predicts patents that may cause patent disputes. Kim et al. used not only quantitative indicators of patents but also applicants' sales, as input variables. In addition, feature selection was carried out using the Stepwise method. In this paper, new features are extracted from patent indicators through PCA and AE. Chien [4] studied patent indicators and company information to LR to predict patent disputes. Bessen and Meurer [5] conducted a study to explore the risk of patent litigation using corporate and financial information. Chien, Bessen, and Meurer used indicator values as input variables without feature selection or extraction [4, 5]. Lanjouw and Schankerman [6] analyzed the main factors of patents in which patent litigation occurs. In previous studies for prediction or classification of patent litigation, the patent index was used as it is or feature selection was performed. In this study, new features are extracted from patent indicators using PCA and SAE and used as input variables for patent litigation classification. In addition, the performance of various classification algorithms is compared.

## 2.2 Feature Extraction

Feature extraction is to project original high-dimensional features into low-dimensional new features [8]. Feature extraction is different from feature selection because it creates a new variable by combining the original variables. PCA is a representative method of feature extraction. PCA creates new variables that are orthogonal to each other by converting the original variables linearly while preserving the variance of the data as much as possible [9, 10]. PCA creates the same number of principal components (PCs) as the original variable. PC is derived as in Eq. (1) below.

$$
\begin{aligned}
PC_1 &= a_{11}x_1 + \ldots + a_{1n}x_n \\
PC_2 &= a_{21}x_1 + \ldots + a_{2n}x_n \\
PC_m &= a_{m1}x_1 + \ldots + a_{mn}x_n
\end{aligned} \quad (1)
$$

In Eq. (1), $PC_m$ represents the mth PC, $x_n$ is the original data, and $a_{mn}$ represents factor loadings. AE extracts features using neural networks [11, 12]. AE creates a bottleneck by making the hidden layer smaller than the input layer in the neural network structure. Due to this unique structure of AE, features of the data can be extracted. AE can learn nonlinear relationships in data using the nonlinear activation function. These features show that AE is different from PCA. In this study, we extract new features from patent indicators using PCA and AE. In addition, the extracted features are trained on a classification model and performance is compared.

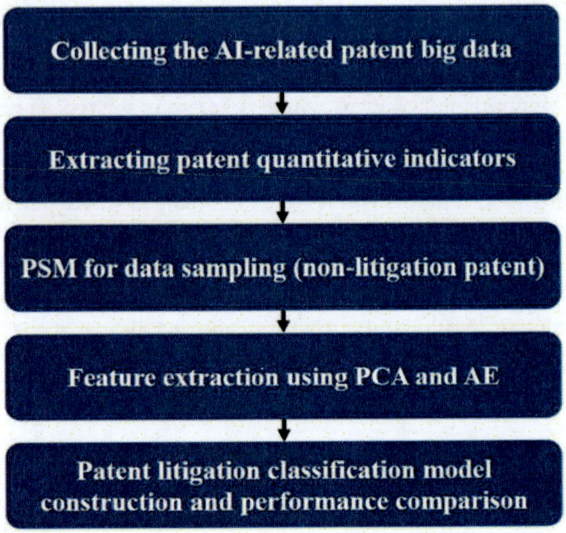


**Fig. 1** Proposed methodology

## 3 Proposed Methodology

In this paper, an appropriate feature extraction method is sought when constructing a model for classifying patent litigation. The proposed method creates a new feature using the feature extraction method for patent quantitative indicators. In addition, a classification model is constructed to compare the patent litigation classification performance. The proposed method is shown in Fig. 1.

In Fig. 1, first, patent big data related to AI technology is collected. Next, the patent quantitative indicators are extracted. In general, there are very few patents in which litigation occurs. For this reason, if the original data is learned in the model, the characteristics of the litigation patent cannot be reflected well. To solve this problem, this paper samples non-litigation patents with propensity scores similar to litigation patents through PSM. A new feature is extracted by performing PCA and AE on the sampled patent data, respectively. The extracted features are learned by various classification algorithms and their performance is compared.

## 4 Experimental Data

Actual patent big data is collected and used to perform the method proposed in this paper. The collected patents are related to AI technology and are registered with the USPTO. Among them, 222 patents were litigation. Non-litigation patents are matched at a ratio of 1:5 through PSM. Table 1 shows the patent quantitative indicators extracted for PSM.

**Table 1** Patent indicators for PSM

| Indicator | Description | Variable type |
|---|---|---|
| all_IPC_count | Number of IPC codes | Numerical |
| citation_count | Number of patent backward citations | Numerical |
| forward_citation_count | Number of patent forward citations | Numerical |
| family_nation_count | Number of family patent nations | Numerical |
| family_doc_count | Number of family patents | Numerical |
| all_claim_count | Number of claims | Numerical |
| inventor_count | Number of inventors | Numerical |
| app_to_regi | Time required from application to registration | Numerical |
| app_to_trans | Time required from application to technology transfer | Numerical |
| transfer_yn | Technology transfer status | Dummy |

Non-litigation patents are sampled so that the propensity score is similar using patent quantitative indicators extracted as in Table 1. There are 1110 non-litigation patents sampled. Therefore, an experiment is performed with a total of 1332 patents. Figure 2 below shows the PSM results.

Figure 2 shows the density propensity score. As shown in Fig. 2, the litigation patent group and the non-litigation patent group were matched to have similar propensity scores.

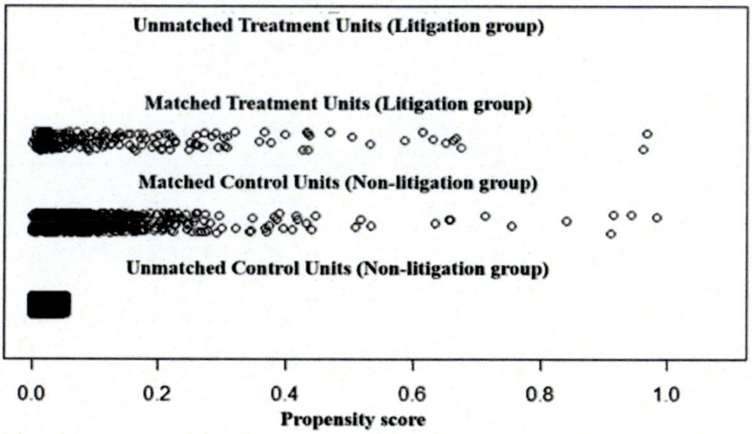


**Fig. 2** PSM results

## 5 Experimental Results

New features were extracted by inputting the patent quantitative indicators of Table 1 into PCA and AE, respectively. Table 2 below shows the description of PCA and AE that performed feature extraction.

As shown in Table 2, PCA selects a PC that can explain more than 95% of the variance of the original data. Therefore, it can be reduced to a total of five dimensions. AE is a structure with three hidden layers. In addition, it can be reduced to a total of seven dimensions. The extracted new features are used as input variables for various algorithms. Table 3 below shows the parameters for each classification algorithm.

Classification performance was measured by inputting features extracted from PCA and AE into a model as shown in Table 3. Models performed tenfold cross validation. Table 4 below shows the classification performance of models.

When the accuracy is highest in Table 4 above, AE is used and classified into KNN and DT. However, F1-score was higher than DT as KNN showed 0.26. Therefore, it is best to extract features using AE and classify patent litigation with KNN.

**Table 2** Parameters of PCA and AE

| Methods | Parameter |
|---|---|
| PCA | PCs with more than 95% explanation of variance are extracted<br>Number of dimensions: 5 |
| AE | Number of hidden layers: 3<br>Activation function: ReLu<br>Number of epochs: 200<br>Number of dimensions: 7 |

**Table 3** Parameters of classification algorithms

| Classification algorithm | Parameter |
|---|---|
| LR | L2 penalty |
| KNN | K = 6, Distance = Minkowski |
| SVM | Kernel = RBF, C = 0.001, Gamma = 0.001 |
| DT | Criterion = Gini, Max_leaf_nodes = 6 |
| RF | Criterion = Gini, Number of features = 2 |
| AB | Criterion = Gini, Number of features = 2 |

**Table 4** Classification performance

| Method | Measure | LR | KNN | SVM | DT | RF | AB |
|---|---|---|---|---|---|---|---|
| PCA | Accuracy | 0.83 | 0.82 | 0.83 | 0.82 | 0.81 | 0.80 |
| | F1-score | 0.00 | 0.13 | 0.00 | 0.10 | 0.22 | 0.20 |
| AE | Accuracy | 0.83 | 0.84 | 0.83 | 0.84 | 0.83 | 0.82 |
| | F1-score | 0.00 | 0.26 | 0.00 | 0.23 | 0.25 | 0.16 |

## 6 Conclusions and Future Work

In this paper, feature extraction methods for effective patent litigation classification are compared. For this, quantitative indicators were extracted from the patent and PCA and AE were used. In addition, performance was measured using various classification algorithms. In the experiment, a patent registered in the USPTO was actually used, and non-litigation patents were sampled through PSM. As a result of the experiment, the highest performance was obtained by extracting features using AE and inputting them into KNN.

In this paper, only quantitative indicators of patents were extracted and used as input variables. However, the patent includes detailed information on the developed technology in text form. For this reason, future studies will need to build a classification model by reflecting the text information of the patent.

**Acknowledgements** This work was supported by the National Research Foundation of Korea (NRF) grant funded by the Republic of Korea government (MSIT) (No. NRF-2020R1A2C1005918). This research was supported by the MOTIE (Ministry of Trade, Industry, and Energy) in Korea, under the Fostering Global Talents for Innovative Growth Program (P0008749) supervised by the Korea Institute for Advancement of Technology (KIAT).

## References

1. Eisenberg RS (1989) Patents and the progress of science: exclusive rights and experimental use. U Chi L Rev 56(3):1017–1086
2. Casey KR (1993) Alternate dispute resolution and patent law. Fed Cir B J 3(1):1–14
3. Kim A, Partee NW, Reynolds T, Santamaria MA, Advisor F, Beling P, Ferron DV, Winer DS, Woolley DF (2002) Patent litigation risk-scoring model. In: 2002 IEEE systems and information design symposium, IEEE Press, New York, pp 13–17
4. Chien CV (2011) Predicting patent litigation. Tex L Rev 90(2):283–330
5. Bessen J, Meurer MJ (2013) The patent litigation explosion. Loy U Chi L J 45(2):401–440
6. Lanjouw JO, Schankerman M (2001) Characteristics of patent litigation: a window on competition. RAND J Econ 32(1):129–151
7. World Intellectual Property Organization (2019) World intellectual property indicators 2019. WIOP, Gen
8. Li J, Cheng K, Wang S, Morstatter F, Trevino RP, Tang J, Liu H (2016) Feature selection: a data perspective. arXiv. doi: https://doi.org/10.1145/3136625

9. Wold S, Esbensen K, Geladi P (1987) Principal component analysis. Chemometr Intell Lab Syst 2(1–3):37–52
10. Abdi H, Williams LJ (2010) Principal component analysis. WIREs Comp Stats 2(4):433–459
11. Hinton GE, Salakhutdinov RR (2006) Reducing the dimensionality of data with neural networks. Science 313(5786):504–507
12. Meng Q, Catchpoole D, Skillicom D, Kennedy PJ (2017) Relational autoencoder for feature extraction. In: 2017 international joint conference on neural networks, IEEE Press, New York, pp 364–371

# Performance Analysis and Comparison of Classification Algorithms for EEG-Based BCI System

**Navtej Singh, Balkrishan Jindal, and Mandeep Kaur Ghumman**

**Abstract** The signals emanating from the human scalp due to motor imagery activity happening in the brain can be fetched using various techniques. Electroencephalogram (EEG) is a popular signal fetching technique for providing input to the brain-computer interface system. This field of signal processing can be used for the rehabilitation of patients suffering from neurological disorders. It can also be used in a variety of other applications including environmental control, device communication, and mobility devices. Artificial intelligence techniques can be used to implement various phases of brain-computer interface. In this paper, we have discussed, evaluated, and compared the performance of many such popular techniques used for the classification of the EEG signals.

**Keywords** Brain-computer interface · Feature extraction · Classification · Motor imagery · EEG · AI

## 1 Introduction

Brain-computer interface (BCI) is a technique of establishing an alternative communication channel from a human brain to some external device. It involves fetching of signals emanating from the scalp of the human brain, during motor imagery activity. These signals are further processed and classified into appropriate class. This technique has found applications in various fields including rehabilitation of patients. The

N. Singh (✉)
Department of CSE, Punjabi University, Patiala, Punjab, India
e-mail: navtejghumman@sbsstc.ac.in

B. Jindal
Department of CE, Yadavindra College of Engineering, Talwandi Sabo, Punjab, India
e-mail: balkrishan@pbi.ac.in

M. K. Ghumman
Department of ECE, Chandigarh University, NH-95, Mohali, Punjab, India
e-mail: mandeep.e10105@cumail.in



X.-S. Yang et al. (eds.), *Proceedings of Sixth International Congress on Information and Communication Technology*, Lecture Notes in Networks and Systems 216,
https://doi.org/10.1007/978-981-16-1781-2_77

BCI transforms the imaging of intended task directly into machine-implementable commands by bypassing the conventional neural muscular conduction pathway in the human body [1]. It is basically a communication mechanism which does not require any movement of peripheral muscles. Applications of BCI are found in areas of robotics, mobility devices, environmental control, and device communication. It has the capability to provide an alternative communication as well as control technology, to patients suffering from severe neuromuscular disorders [2]. Most commonly used BCI systems use the technology of electroencephalogram (EEG) because of its affordable recording equipment of non-invasive nature which is conducive for its real-time operation [3]. EEG is a test that involves the measurement of electrical signals emanated to the scalp by the brain, by placing electrodes on the outer surface of scalp. The EEG-based BCIs, are preferred over other methods of data fetching from brain waves, as they provide an affordable and easy-to-use recording equipment, which can be used to provide real-time operation [3]. EEG signals are fetched from the scalp of the brain, which are generated as an associated result of motor imagery (MI). MI is a mere mental intention of a motor act, without its actual execution by the corresponding muscle. MI is a commonly performed mental task in which subjects are asked to imagine themselves executing a particular movement such as that of a hand or foot but without actually performing it physically. It is observed by electrical recordings from the scalp surface of the subject person, as it is accompanied by a continuous activity of the brain waves in various parts of the brain [4]. A BCI system requires to be vigorous, portable, safe, affordable, and easy to use, for communication or device control. The equipment required for signal acquisition should be minimalist and easy to set up so that it can be used without extensive setup and training [2]. The brain waves lack regularity, and thus it is difficult to establish a specific pattern in the EEG [5]. The signals emanating from motor imagery activity have to be recognized by the BCI system. It is a daunting task as the EEG signal is non-stationary in nature and is adversely affected by undesirable electrical interference, involuntary muscular movement artifacts, and is also subject specific. The precise and accurate classification of MI tasks requires that the most distinguishing features are extracted from the brain signals acquired in the data fetching phase of BCI. Many algorithms, which are collectively called classifiers, are used for this purpose. EEG is a typical non-invasive method which records brain's electrical activity along the scalp surface, however, invasive electrodes are sometimes used in specific applications. In such a BCI implementation, electrical signals are recorded from the scalp surface of the head, as there is a continuous electrical activity in the brain, which is directly in correspondence to intended tasks [4]. The EEG signals are further categorized into sub-categories of EEG sub-bands—delta, theta, alpha, beta, and gamma based on frequency of signal. The standardized 10–20 electrode placement system of EEG demarcates the position of electrodes on various parts of the scalp. It is used to record signals emanating as a byproduct of a mental activity, from somatosensory and the motor areas of the brain. The signals recorded from these electrodes reflect the motor activity of the person like hand movements, foot movements, tongue movements, etc. [6]. The EEG-based BCIs can manipulate MI of users, and thus can cater to the requirement of robust communication performance between the brain and a computer. There are many other techniques of acquiring such signals, such as that of magnetic resonance imaging (MRI) and positrons emission tomography (PET). However, EEG has emerged as preferred choice for the implementation of non-invasive BCI system, as it is inexpensive, has higher temporal resolution (approximately 0.05 s), and is easy to use [7]. The foremost application of BCI is to aid patients suffering from neurological disabilities, by providing them with an alternative method of communication with some external devices. The other BCI applications include transmitting instructions in entertainment applications besides communicating with personal computers by use of thoughts' translation into appropriate corresponding control signals. BCI is a integrative research field, which encompasses domains of psychology, neurology, human-computer interaction (HCI), rehabilitation engineering, signal processing, and machine learning. The main objective of BCI applications is to meticulously translate the brain wave patterns automatically extracted from emanating EEG signals into desired machine control commands. The aim of many researches is to improve the accuracy of interpretation of the harvested EEG signals [8]. Many BCI systems, which are based on EEG for signal fetching, were recently developed. In these systems, patterns of EEG emanating from different motor imagery tasks were analyzed and classified by using different algorithms in multiple stages of a BCI system. It has been concluded in the researches that efficiency and effectiveness of a BCI system depend on the quality of algorithms used for signal acquisition, preprocessing, feature selection, and classification phase [9]. A BCI is implemented using a particular pattern recognition approach which is characterized by extracting features from EEG signals and then using an appropriate algorithm for classifying it into a mental task (Figs. 1 and 2).

### *1.1 Properties of EEG Signals*

The fetched EEG signals can be analyzed to extract various features which can be further classified for implementation of BCI system, These features include amplitude of EEG signal, value of power spectral density (PSD), band powers (BP), autoregressive parameter, besides some time-frequency features [10]. Some of the properties of EEG signals are as follows.

- Noise and Outliers: EEG signals fetched from the scalp of a human brain are highly prone to noise, due to sheer complexity of the brain. The human body is a continuous host to various activities of physiological and neural nature, which are closely intertwined, and are often undistinguishable from the analysis of the waves emanating from the skin.
- High Dimensionality: In the implementation of a BCI, the feature vector extracted from EEG signals is multi-dimensional, as multiple features are fetched from several channels, and can span to multiple time segments. These multiple features

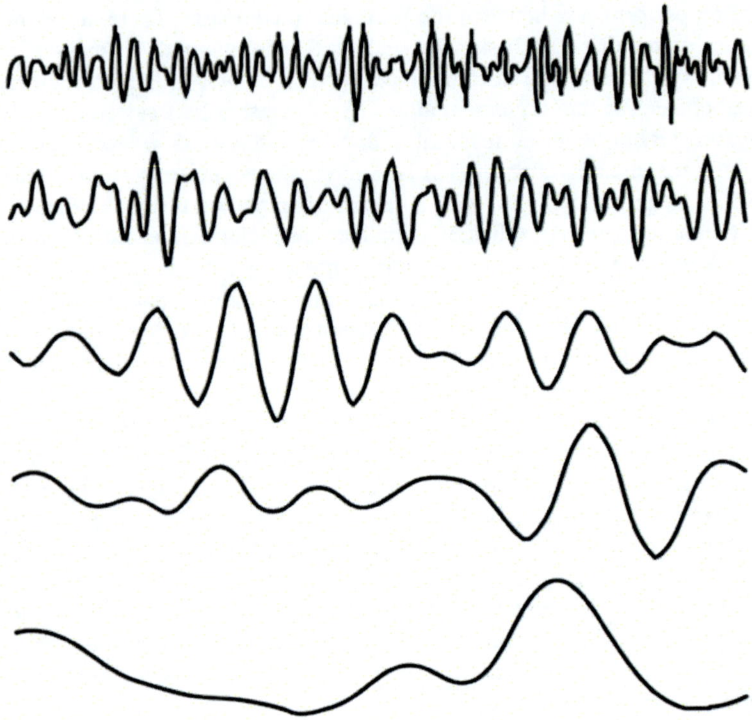

**Fig. 1** EEG sub-bands

**Fig. 2** BCI model

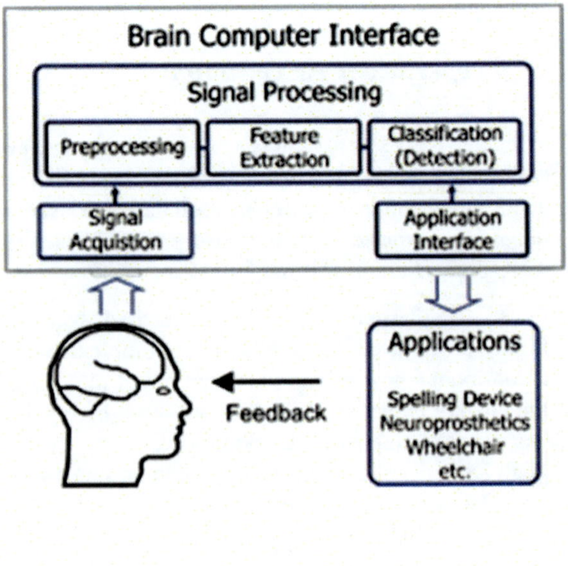

which are extracted are preprocessed and then combined and concatenated to a single feature vector.

- Temporal Information: The BCI features represented by EEG signals should have particular time-stamped information engrossed in them, to represent the particular time variations of the signals.
- Volatility: The BCI features are highly volatile in nature, and are stationary for a very short time and vary at a great rate. Hence, the signals have to be fetched at a very high speed, which leads to an increase in the complexity. The short life span of the signals complicates the signal fetching stage of the BCI.
- Small Training Set: The BCI system has to be trained, which is a time-consuming exercise. This also has adverse an effect on subjects, who have to spend considerable time. Usually, the subjects are not very cooperative while training the proposed BCI.

## 2 Data Acquisition

The motor imagery data is fetched by measurement of brain activity, using many methods, which are broadly classified as invasive and non-invasive. However, the EEG has gained most acceptance, mainly because of its non-invasive character, low setup cost, and relatively higher resolution. The electric potential in a single dipole EEG system a function of value of its charge, the distance separating the poles and distance to the poles and is represented as following:

$$V_P = \frac{qd.r}{4\pi\varepsilon_0 r^3} = \frac{qd\cos\Theta}{4\pi\varepsilon_0 r^2}. \quad (1)$$

Many BCI systems have also been developed using other alternative methods like functional magnetic resonance imaging (fMRI). It detects variation in local cerebral blood volume, its rate of flow besides levels of oxygenation. These parameters change during neural activation by means of electromagnetic fields and magnetoencephalogram (MEG). It registers the brains magnetic activity emanating from the scalp, by measuring magnetic induction. The data fetching in BCI systems can also be performed by invasive interfacing like electrocorticograms (ECoG). It requires surgical procedure, so as to place the electrode directly underneath the scalp, to measure patterns of electrical activity taking place in the area of cerebral cortex. The surgically installed electrodes are directly over the surface of the brain, which requires a surgical operation of craniotomy to create an electrode grid [11]. This can be a serious health hazard, and thus non-invasive BCIs are preferred over it. The BCI systems are usually based on data fetching by recording of the EEG signals, associated with brain activity to recognize associated patterns of various mental tasks. It has been established that various motor imagery tasks like episode of only imagination of moving of the left and right and hands are correlated with the characteristic pattern of EEG activity observed from the electrodes placed on the scalp of respective right and left side portions of the motor cortex. Such characteristic patterns are associated with various variations in EEG activity. For instance, it has been established that during movement or even preparation for the movement, the beta rhythms (18–26 Hz) and mu rhythms (8–12 Hz) decrease which is called event-related desynchronizing. They have been observed to increase immediately following the execution of this movement, which is called event-related synchronization [2]. Many researchers have made use of the data prepared and provided by BCI competition IV, which is a continuation in the series of previous BCI competitions, which are organized to provide neuroscientific data, which has a high quality and can be openly accessed by the scientific community [12]. The data set has signals recorded from the scalps of 9 human subjects, while they perform motor imagery tasks. The EEG recordings include a total of 25 channels that mostly were distributed over and around sensory motor areas of the scalp. Each participating subject contributed in a total of two sessions consisting of a calibration session and an evaluation session. A calibration session includes the performance of four of the motor imagery tasks, by each of the subject. These are categorized on the basis of imagining movement of the left hand as class-1, right hand as class-2, both of the feet as class-3, and the tongue as class-4. A visual cue is shown on the computer screen as a signal to the subject, for performing a specific motor imagery task for 4 s. Data set was recorded from 9 subjects, labeled as A01, A02, A03,…A09), while they performed four-class motor imagery tasks. The 9 subjects were seated in front of a computer screen in an arm chair. A signal appears on the screen which is a fixation cross appearing on black screen, which signals start of a trial (t = 0s). Another cue appears on the screen after two seconds (t = 2s), in the form of an arrow pointing to the right, left right, up or down. Each of this is corresponding to four classes of the right hand, left hand, tongue or foot movement, which appears on the screen for 1.25 s. These arrows displayed in one of the four directions give signals to the subjects to undertake the intended motor imagery task, without providing any feedback. The participating subjects were instructed to execute the motor imagery task as long as the fixation cross continues to appear on the screen (t = 6s). A brief break of blank screen was incorporated between each cycle of the above data recording. Two such recording sessions were executed for each of the protracting subject. Each of the above session comprised six executions, which were separated by short breaks in between. Thus each of the execution consisted of 48 such recording trials, which made up of 12 runs for each of the four possible MI classes, generating data for a total of 228 different trials.

## 3 Preprocessing and Feature Extraction

This is the first phase of signal processing, which as the role of preparing the recorded electric signal by enhancing it, so as to enhance the features for their improved distinction for detection. Preprocessing techniques further improve the ratio of signal to noise (SNR), by rejecting unwanted artifacts. The efficiency of the preprocessing phase has a great role in the efficiency attainment of overall BCI system [13]. Technique of independent component analysis (ICA) is usually applied for the removal of power line noise as well as EOG, EMG, and ECG artifacts. Once the brain signals have been converted into digital format, they are then processed by one or more feature extraction methods. The feature extraction methods extract specific characteristics from the fetched signals, which contain messages or commands emanating from subject's scalp of the brain by evoked or spontaneous inputs. Preprocessing and feature extraction are two important steps in EEG signal processing. Preprocessing techniques are employed to isolate unintended artifacts from the EEG signal. This leads to improvement in the signal-to-noise ratio, which leads to a better performance of the system. Subsequently, a feature extraction process retrieves the relevant features from the input signal. These features are required for decision-making mechanism in generating the required output. EEG signals are very volatile in nature, and sustain for a very short time. There is a change in spectrum with time hence different feature extraction methods are required for this variation. There are many feature extraction methods, which are available. However, methods of autoregressive modeling, wavelet transform, and fast Fourier transform are popularly employed to accomplish feature extraction in BCI. Autoregressive model (AR) is a popular feature extraction methodology for executing classification stage of BCI, which involves random task-based EEG signal processing. This model establishes that the output variable is dependent on its own prior value. This model of feature extraction is suitable in application which performs conversion of imagined letters into appropriate classes. The AR model is extensively used in neural network-based approaches. The EEG signal experience variations in its parameters of voltage and frequency, while a variety of mental tasks are performed. Henceforth, these features can thus be captured and then be extracted by use of the AR model [14]. This model is based on ICA and prior EEG, which are used in combination with current EEG samples. Wavelet transforms (WT) performs analysis of signals emanating from various biomedical events in human body. Compared with other spectral analyzing methods, WT is relatively more suitable and is appropriate for temporary non-stationary biomedical signals. The representation of wavelet is effective for time frequency of a signal. WT successfully represents real frequency information at the low frequencies, and also accurate time information at high frequencies. As such, biomedical applications inherently contain components with higher frequency, which have ashort time period and also have some of the low-frequency components with a long time period, making WT a preferred method for feature extraction as non-stationary signals in particular wavelet transforms provide multi-resolution analysis.

Fast Fourier transform (FFT) is another popular feature extraction algorithm which converts time domain of EEG signals to frequency domain. It is an implementation of discrete Fourier transform, but it is faster. It is an efficient method for performing feature extraction and is used in a variety of mathematical operations which have time dependency [15]. In contrast, the procedure in discrete Fourier transform (DFT) has higher complexity and thus is time consuming. Hence, to gain reduction in operation time and improve the speed, FFT is a preferred choice. FFT is used in many applications which range from digital signal processing, solving partial differential equations to that of implementation of large integer numbers multiplication.

## 4 Feature Selection

Measurement of brain activity, by analysis of EEG, generates a huge quantity of data. The EEG signals can be recorded with a sampling frequency which is in the range 100–1000 Hz. These signals have to be converted into a fewer number of values, which characterize specific features of signals. For example, these features could be the representation on the scale of power of the EEG signals belonging to various frequency bands. Such features then are normally further combined into a form of vector, which is called a feature vector. Feature extraction thus may be defined as a process of transformation of one or multiple signals to a single feature vector. However, the selection and extraction of good features improves the performance of the system, which eases the task of further stage of classification algorithm [16]. The designer of BCI system can choose from a variety of many available feature selection algorithms. The performance can be validated, by categorizing the input data into a set of training data, and another into a set of validation data. Feature selection is performed on a set of training data, by selection of the discriminative CSP features, which is based on the relation between each of the features and the respective motor imagery class [5]. Next stage in FBCSP is to select a combination of discriminative CSP features from the subject task. It uses mutual information-based best individual feature algorithm for feature selection. For verifying the efficiency attained in the BCI, cross-validation is executed, wherein input data is randomly categorized into two sets of the training data and the validation data. This method performs operation of feature selection only upon the training data by selecting the discriminative CSP features, which calculates the degree of correct correlation established between each of the occurring feature, with the appropriate intended motor imagery classes.

## 5 Classification

The classification stage of BCI executes a classification algorithm to model the selected CSP features and classifies it into MI classes. The purpose of the classification stage is to perform automatic association of a MI class to an appropriate feature vector, which is extracted in the previous stage. This class identifies the type of mental task performed by the BCI user. To accomplish the above task of correlating various EEG signals, the classification of the selected features obtained from feature extraction, the following classification algorithms are often used.

### *5.1 Linear Discriminant Analysis (LDA)*

The classification in many BCI applications is executed by using linear discriminant analysis (LDA) as it offers higher efficiency. LDA classifies a recorded set of obser-

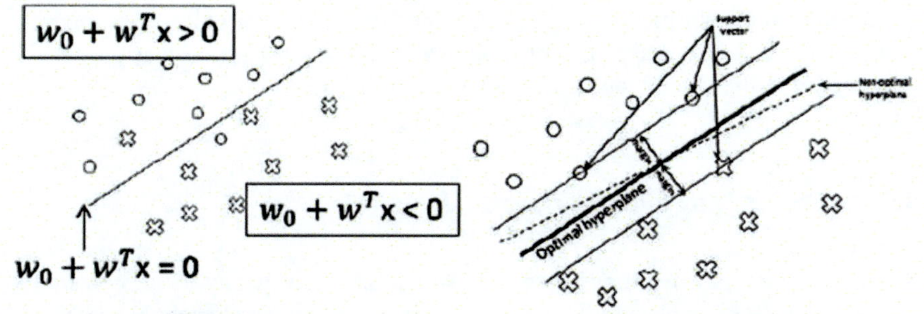


(a) LDA hyperplane (b) SVM hyperplane

**Fig. 3** LDA and SVM hyperplanes

vations into pre-configured classes. It executes linear transformation of data from that of a high dimensionality space to that in a lower dimensional space. Subsequently, it decides the classification based on a set of observations in a low dimensional space, in conjunction with a defined decision boundary. Hence, the choice of deciding an appropriate boundary has a direct influence in the outcome efficiency of classification. The distributions of class include some degree of finite variance that are included in the selected space during this process. Hence, the value of mean and the standard variance of the targeted data are taken into consideration, to establish dimensions of the above decision boundary. LDA is used to establish a hyperplane and linear combination of features, which can successfully demarcate multiple classes. LDA is applicable, when there is a normal distribution of raw data, which has equal covariance matrices for each of the classes. The separating hyperplane is established by calculating the value of projection that produces a maximum distance between means of the two classes [17]. It also has to take into account that the interclass variance is also minimum. However, multiple hyperplanes are required, if the number of classes is more than two, and these hyperplanes are required to be calculated. LDA has lower computational requirements, which makes it as a proffered choice, whenever the available hardware is small, and for establishing an online BCI systems. This classifier is also preferred due to its ease of use, and offers higher success for classification. Hence, the LDA has been successfully employed in many motor imagery-based BCIs [16] (Fig. 3).

## 5.2 *Support Vector Machine (SVM)*

Support vector machine (SVM) is another classification algorithm, which uses a discriminant hyperplane for performing distinction between many classes [10]. In SVM, the selected hyperplane is such that it establishes maximum margin, which is calculated as the distance from the closest training points. This also increases the generalization capabilities, lending greater performance, in a multi-class environment [10]. SVM also makes use of a parameter for regularization that provides it tolerance to outliers and inbuilt errors on the training set.

## *5.3 Nearest Neighbor Classifier (NNC)*

KNN is a supervised learning algorithm, which has the capacity to store available case scenarios of signals and their classes. It is an easy to implement algorithm, in which new cases are classified to a particular class, on the basis of a parameter called similarity measure. NNC makes a comparison of training data with that of featured values of testing data. It performs analysis of featured values by using the distance function like Euclidean distance of the given numeric parameter values. It classifies by using the concept of majority voting of neighbors of the object to be classified. The object is allocated a class, which it shares with its k-neighbors [18]. The number of nearest values under consideration for such class allocation is represented by K.

## *5.4 Naive Bayes Probability*

The Bayesian classifier allocates the class to the selected features based upon its description by feature vector. It is a simple method to establish classifiers, which assume independence between different features. It uses maximum likelihood training to infer the probability of selected feature to belong to a finite set of classes [19]. It establishes probability of classification, by using heuristic data to ascertain the probability.

## *5.5 Neural Networks*

The neural networks (NN) are inspired by biological nervous system, which has strong learning capability to learn from examples. This method has internal parameters which are redesigned if the classification is not of the desired class. Hence, this classification algorithm is not dependent upon the knowledge of the classification strategy. It extensively makes use of multilayered perceptron (MLP), which is a collection of many layers of neurons, which propagate the message to one another based on the weights they are assigned [20].

**Table 1** Comparison of classification accuracy [21]

| | CSP | | | AAR | | |
|---|---|---|---|---|---|---|
| Subject | LDA (%) | NNC (%) | SVM (%) | LDA (%) | NNC (%) | SVM (%) |
| Highly Experienced—K3 | 91.1 | 86.3 | 89.0 | 69.4 | 49.8 | 74.5 |
| Less Experienced—L1 | 69.3 | 67.7 | 69.2 | 42.7 | 30.9 | 49.4 |
| Beginner—K6 | 56.6 | 52.4 | 55.3 | 43.2 | 30.6 | 53.2 |

## 5.6 Performance Comparison of LDA, NNC, and SVM Classification Algorithms

Many researchers have used various combinations of algorithms in different phases of implementation of BCI. The researchers in [21] have compared the performance of LDA, NNC, and SVM by applying to EEG data which is preprocessed by the common spatial pattern (CSP) method and adaptive autoregressive parameters (AAP). The method of CSP is used to preprocess two-class EEG signals by finding their linear subsets, and then by increasing the variance of the first to a maximum, and reducing the variance of the second to a minimum value. The adaptive autoregressive (AAR) parameter describes the EEG data by consideration of the variation of the EEG spectrum with time, and estimates the AAR parameters by adaptive process. Both of these methods were used in the preprocessing of the data in this implementation of this BCI. The authors have then applied the various classifiers to the EEG data sets recorded from 3 subjects, named as K3, K6, and L1 having varying degree of experience in BCI training. The data set, K3, was recorded in 9 training runs from subjects who were experienced in BCI training, whereas K6 was recorded 6 times from subjects who had little experience in BCI training, and L1 was recorded 6 times from the subject who was a beginner in BCI training. Within each run, each of the four movements (left hand, right hand, foot or tongue movement) was trained 10 times. Subsequently, in each of the two preprocessing methods of CSP and AAR, they have applied LDA, NNC, and SVM classifiers to evaluate their performance. Table 1 shows the percentage accuracy attained, which depends on the experience level of the subject, the preprocessing method employed, and the classification method used. The bar charts in Figs. 5 and 4 show the percentage of accuracy attained in successful classification of motor imagery task, when respectively CSP and AAR methods are applied for preprocessing. This shows that a successful BCI system heavily depends upon the suitable choice of algorithms employed in various phases of its implementation.

The comparison of the percentage age accuracy attained and various factors affecting it is clearly seen from the graph in Fig. 5.

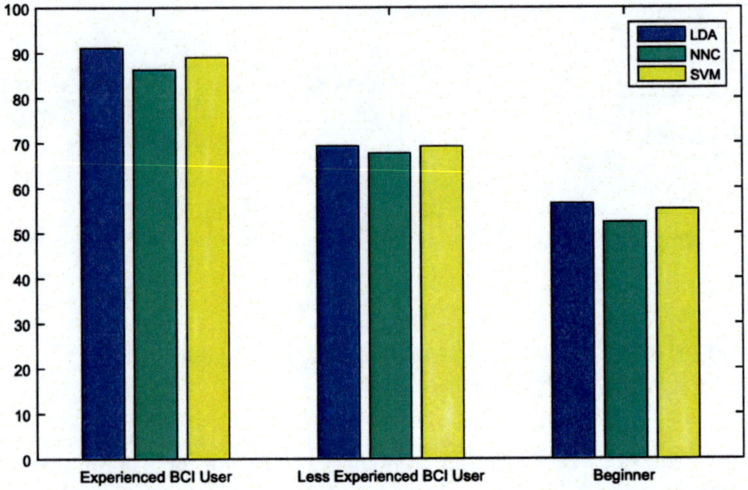


**Fig. 4** Classification accuracy of LDA, NNC, and SVM with CSP method

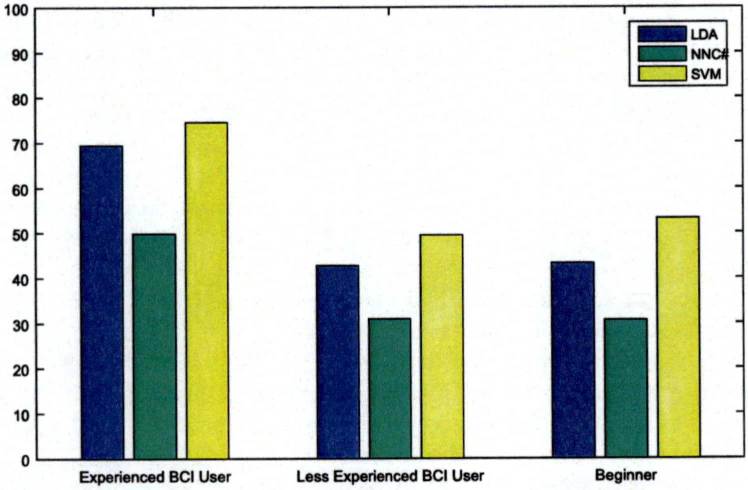


**Fig. 5** Classification accuracy of LDA, NNC, and SVM with AAR method

# 6 Applications

The BCI has a potential to offer a wide range of applications ranging from entertainment to that of medical rehabilitation. One of the interesting areas of BCI research is design and development of peripheral devices that can be controlled by thoughts, without relying on muscle activity. To accomplish this, the basic mechanism of converting thoughts into computerized or robotic action is evolving, thus offering limitless potential uses of this technology. The primary objective of the BCI is to provide alternatives to the functionality of muscles, to accomplish the intended task, for rehabilitation of people suffering from severe disabilities such as locked-in syndrome. Such patients lose partial or full capability to use their muscles or limbs and may be completely paralyzed and are unable to perform daily chores or even speak. The BCI system has the potential to offer alternative channels of communication between the human brain and the external devices, without relying upon neural pathway. The BCI researchers have proposed and experimented with several alternative methods of assisted communication, which widely vary from simpleton simple binary (yes/no) capabilities [22] to state-of-the-art applications such as TalkAssist [23]. They have also designed virtual keyboards that support spelling, with partial input from user. Several alternative approaches to spelling have been proposed, which have been used by a patient suffering from severe neurological disorder, to compose words in a realistic home environment, for effective communication with outside world. Other applications of BCI include the field of creation of virtual reality, which is used for gaming, entertainment, and environmental control. The researchers have described a virtual driving environment that has evaluated motor imagery signal responses when subjects saw a stoplight [24]. Subsequent works includes design of a virtual apartment which has allowed the residents to interact with virtual people and devices, by establishing alternative communication channels directly with their thoughts process. The researchers have proposed a BCI system that allows users to transcend a complex puzzle of maze, by taking turn decisions at crossroads [25]. Virtual reality has a potential to offer a safe environment and opportunity for training and configuring the BCI controlled interfaces to practically used devices, such as a powered wheelchair for physically challenged people.

## 7 Conclusion

This paper gives a glimpse of various methods used in different phases of BCI applications. Researchers are employing different techniques to improve the robustness, efficiency, and accuracy by choosing the appropriate methods in the implementation of different phases of BCI. The signals emanating from the scalp are identified by EEG data acquisition, which are first preprocessed to reduce unwanted artifacts from the EEG signal. These are then processed by feature extraction and classification methods. Appropriate selection of feature extraction, feature selection, and the classification methods are important building blocks in the development of a robust and efficient EEG-based BCI system. The improved accuracy attained and lower error rate observed are a guiding objective to choose appropriate methods for the implementation of a BCI system.

## 8 Future Scope

The improving understanding of human brain and its cognitive skills, along with rapid advances in hardware and software technologies, offers huge scope for BCI in a wide range of applications. In future, the proposed study can be verified in real-life scenario, to ascertain its efficacy and reliability.

## References


1. Huang D, Qian K, Fei D-Y, Jia W, Chen X, Bai O (2012) Electroencephalography (eeg)-based brain-computer interface (bci): A 2-d virtual wheelchair control based on event-related desynchronization/synchronization and state control. IEEE Trans Neural Syst Rehabil Eng 20(3):379–388
2. Wolpow J, Birbaumer N, McFarland DJ, Pfurtscheller G, Vaughan T (2002) Brain-computer interfaces for communication and control. Clinical Neurophysiol 113(6):767–791
3. Park S-A, Hwang H-J, Lim J-H, Choi J-H, Jung H-K, Im C-H (2013) Evaluation of feature extraction methods for eeg-based brain-computer interfaces in terms of robustness to slight changes in electrode locations. Med Biol Eng Comput 51(5):571–579
4. Jessy P (2009) Analysis of eeg signals for eeg-based brain-computer interface
5. Sarma P, Tripathi P, Sarma MP, Sarma KK (2016) Pre-processing and feature extraction techniques for eegbci applications-a review of recent research. ADBU J Eng Technol 5(1)
6. Lee JC, Tan DS (2006) Using a low-cost electroencephalograph for task classification in hci research. In: Proceedings of the 19th annual ACM symposium on User interface software and technology. ACM, pp 81–90
7. Siuly S, Li Y (2012) Improving the separability of motor imagery eeg signals using a cross correlation-based least square support vector machine for brain-computer interface. IEEE Trans Neural Syst Rehabil Eng 20(4):526–538
8. Alomari MH, Samaha A, AlKamha K (2013) Automated classification of l/r hand movement eeg signals using advanced feature extraction and machine learning. arXiv:1312.2877
9. Li K, Sun G, Zhang B, Wu S, Wu G (2009) Correlation between forehead eeg and sensorimotor area eeg in motor imagery task. In: 2009 Eighth IEEE international conference on dependable, autonomic and secure computing. IEEE, pp 430–435
10. Burges CJ (1998) A tutorial on support vector machines for pattern recognition. Data Mining Knowl Discov 2(2):121–167
11. Abdulkader SN, Atia A, Mostafa M-SM (2015) Brain computer interfacing: Applications and challenges. Egyptian Inf J 16(2):213–230
12. Tangermann M, Müller K-R, Aertsen A, Birbaumer N, Braun C, Brunner C, Leeb R, Mehring C, Miller KJ, Mueller-Putz G et al (2012) Review of the bci competition iv. Frontiers Neurosci 6:55
13. Mason SG, Birch GE (2003) A general framework for brain-computer interface design. IEEE Trans Neural Syst Rehabil Eng 11(1):70–85
14. McFarland DJ, Wolpaw JR (2011) Brain-computer interfaces for communication and control. Commun ACM 54(5):60–66
15. Zhang D, Yao L, Zhang X, Wang S, Chen W, Boots R, Benatallah B (2018) Cascade and parallel convolutional recurrent neural networks on eeg-based intention recognition for brain computer interface. In: Thirty-second AAAI conference on artificial intelligence
16. Pfurtscheller G, Da Silva FL (1999) Event-related eeg/meg synchronization and desynchronization: basic principles. Clinical Neurophysiol 110(11):1842–1857

17. Brunner C, Naeem M, Leeb R, Graimann B, Pfurtscheller G (2007) Spatial filtering and selection of optimized components in four class motor imagery eeg data using independent components analysis. Pattern Recognit Lett 28(8):957–964
18. Yazdani A, Ebrahimi T, Hoffmann U (2009) Classification of eeg signals using dempster shafer theory and a k-nearest neighbor classifier. In: 2009 4th international IEEE/EMBS conference on neural engineering. IEEE, pp 327–330
19. Wang H, Zhang Y et al (2016) Detection of motor imagery eeg signals employing naïve bayes based learning process. Measurement 86:148–158
20. Coyle D, McGinnity TM, Prasad G (2010) Improving the separability of multiple eeg features for a bci by neural-time-series-prediction-preprocessing. Biomed Signal Process Control 5(3):196–204
21. Lee F, Scherer R, Leeb R, Neuper C, Bischof H, Pfurtscheller G (2005) A comparative analysis of multi-class eeg classification for brain computer interface. In: Proceedings of the 10th computer vision winter workshop, pp 195–204
22. Wolpaw JR, McFarland DJ, Vaughan TM (2000) Brain-computer interface research at the wadsworth center. IEEE Trans Rehabil Eng 8(2):222–226
23. Kennedy PR, Bakay RA, Moore MM, Adams K, Goldwaithe J (2000) Direct control of a computer from the human central nervous system. IEEE Trans Rehabil Eng 8(2):198–202
24. Bayliss JD, Ballard DH (2000) A virtual reality testbed for brain-computer interface research. IEEE Trans Rehabil Eng 8(2):188–190
25. Birch GE, Mason SG (2000) Brain-computer interface research at the neil squire foundation. IEEE Trans Rehabil Eng 8(2):193–195

# Correlation Analysis of Node and Edge Centrality Measures in Artificial Complex Networks

**Annamaria Ficara, Giacomo Fiumara, Pasquale De Meo, and Antonio Liotta**

**Abstract** The role of an actor in a social network is identified through a set of measures called centrality. Degree centrality, betweenness centrality, closeness centrality, and clustering coefficient are the most frequently used metrics to compute the node centrality. Their computational complexity in some cases makes unfeasible, when not practically impossible, their computations. For this reason, we focused on two alternative measures, WERW-Kpath and Game of Thieves, which are at the same time highly descriptive and computationally affordable. Our experiments show that a strong correlation exists between WERW-Kpath and Game of Thieves and the classical centrality measures. This may suggest the possibility of using them as useful and more economic replacements of the classical centrality measures.

**Keywords** Complex networks · Social network analysis · Centrality measures · Correlation coefficients · K-path

## 1 Introduction

Social network analysis (SNA) uses graph theory and networks to understand social structures and has diverse fields of application such us psychology, economics, criminology [1–4], organizational studies, and information science.

A. Ficara (✉)
University of Palermo, Palermo, Italy
e-mail: aficara@unime.it

A. Ficara · G. Fiumara · P. De Meo
University of Messina, Messina, Italy
e-mail: gfiumara@unime.it

P. De Meo
e-mail: pdemeo@unime.it

A. Liotta
University of Bozen-Bolzano, Bozen-Bolzano, Italy
e-mail: Antonio.Liotta@unibz.it

 
X.-S. Yang et al. (eds.), *Proceedings of Sixth International Congress on Information and Communication Technology*, Lecture Notes in Networks and Systems 216,
https://doi.org/10.1007/978-981-16-1781-2_78

One of the most effective SNA tools to measure social interactions has been a simple graph which consists of a set of nodes (i.e. individuals or organizations), often called actors in the SNA tradition, with edges between them, also called links or connections (i.e. relational ties, e.g. friendship relationships).

The term node centrality indicates a family of measures aimed at characterizing the importance or the position of an actor in a network [5]. An actor might be important for being connected to a large number of nodes or for his power in the control over the information flowing through the network. His absence could also result in a social network made of many isolated components.

Despite an abundance of methods for measuring centrality of individual nodes, there are by now only a few metrics to measure centrality of individual edges. De Meo et al. [6–8] presented a novel measure called the $K$-path to compute link centrality. The advantage of using this metric is that it can be computed with a near-linear time algorithm called WERW-Kpath. Most recently, Mocanu et al. [9] developed an algorithm called Game of Thieves which is able to compute actor and link centrality in a polylogarithmic time.

An often asked, but rarely answered, question is: are these centrality measures correlated? [10] If there exists a high correlation between the centrality metrics, they will have a similar behavior in statistical analyses and for this reason the development of multiple measures could be redundant. If there is no high correlation, these measures are unique and they can be associated with different outcomes. Many researchers carried out studies on the correlations between centrality measures. Valente et al. [10] investigated the correlation among the most commonly used centrality measures and they identified degree as the measure with the strongest overall correlations. Shao et al. [11] use degree to approximate closeness, betweenness, and eigenvector in artificial and real networks. Oldham et al. [12] used 212 different real networks and calculated correlations between 17 different centrality measures showing a positive correlation.

In this work, as in a previous study [13], we try to answer to a different question: are these centrality measures correlated with Game of Thieves? If so, we can use Game of Thieves thus considerably reducing the execution time in the computation of node and edge centrality in very large networks.

We have done a correlation analysis using the most common types of coefficients which are Pearson, Spearman, and Kendall and three kinds of artificial complex networks: scale-free, small-world, and Erdös-Rényi random networks.

In a previous work [13], we analyzed the correlation coefficients among node centrality metrics when the number of nodes in both artificial and real networks increased using GoT only to measure an actor centrality. In this study, we have made an analysis on artificial networks taking into account the increase of the number of links when the network size does not change. We also used GoT to compute edge centrality comparing it with the WERW-Kpath.

The paper is planned as follows. Section 2 presents a brief description of the main theoretical definitions of the considered centrality metrics and correlation coefficients. Section 3 contains a discussion of our methods and the obtained results. Section 4 shows the conclusions and future developments of our study.

# 2 Background

## 2.1 Basic Definitions

A graph $G$ is an ordered pair $G = (N, E)$ where $N$ and $E$ are the sets of nodes and edges, respectively. A graph is called undirected if edges are bidirectional. The main node centrality measures are

- *degree centrality* (DC) [14], which is defined as

$$DC_i = \frac{d_i}{n-1}, \tag{1}$$

  where $d_i$ is the number of edges of a node $i$ (i.e. the degree of $i$) and $n$ is the network size.
- *betweenness centrality* (BC) [15], which is given by the following formula:

$$BC_i = \sum_{h,k} \frac{v^i_{hk}}{g_{hk}}, \tag{2}$$

  where $v^i_{hk}$ is the number of shortest paths from a node $h$ to a node $k$ by passing through $i$ and $g_{hk}$ is the total number of shortest paths from $h$ to $k$.
- *closeness centrality* (CL) [14], which is defined as

$$CL_i = \frac{n}{\sum_j d_{ij}}, \tag{3}$$

  where $d_{ij}$ is the distance between $i$ and $j$ and $n$ is the network size.
- *clustering coefficient* (CC) [16], which is defined as the number of complete triangles $T_i$ in which a node $i$ participates divided by the maximum possible number of these triangles:

$$CC_i = \frac{2T_i}{d_i(d_i-1)}. \tag{4}$$

A more recent centrality measure is called *Game of Thieves* (GoT) [9]. The game proceeds in epochs. When it begins, each node has a certain number of thieves and virtual diamonds or vdiamonds. At each epoch $e$, a thief located on a node $i$ randomly picks a neighbor of $i$. He moves to this new node and, if he finds a vdiamond, he fetches it. Then, he brings the vdiamond back to his home node. After $T$ epochs, the centrality of each node $i$ is computed as the average number of vdiamonds present on $i$:

$$\Phi^i_T = \frac{1}{T}\sum_{e=0}^{T} \Phi^i_e. \tag{5}$$

**Table 1 Centrality measures.** $n$ is the cardinality of $N$, and $m$ is the cardinality of $E$ in a graph $G(N, E)$

| Measure | Centrality | Computational complexity |
|---|---|---|
| DC | Nodes | $O(m)$ |
| BC | Nodes | $O(mn)$ or $O(n^3)$ |
| CL | Nodes | $O(n^3)$ |
| CC | Nodes | $O(n^2)$ |
| GoT | Nodes and edges | $O(log^2 n)$ or $O(log^3 n)$ |
| WERW-Kpath | Edges | $O(km)$ |

The centrality of each link $l$ is computed as the average number of thieves who carry a vdiamond passing through $l$:

$$\Psi_T^l = \frac{1}{T}\sum_{e=0}^{T} \Psi_e^l. \tag{6}$$

Another recent measure of link centrality for social networks is the *K-path* which is defined as

$$L^k(l) = \sum_i \frac{\sigma_i^k(l)}{\sigma_i^k}, \tag{7}$$

where $\sigma_i^k(l)$ is the number of $k$-paths originating from a node $i$ and traversing the link $l$ and $\sigma_i^k$ is the number of $k$-paths which originate from $i$. A near-linear time algorithm called WERW-Kpath (Weighted Edge Random Walks—$K$ Path) [6–8] is able to compute this centrality index.

Table 1 shows how GoT represents a great step forward in terms of computational complexity with respect to classical algorithms of node centrality such us BC, CL, and CC which have a quadratic computational complexity. The DC has a linear time complexity but GoT still remains a better option because it is also able to compute the centrality of edges.

## 2.2 Correlation Coefficients

Given two random variables $a$ and $b$, *Pearson's r correlation coefficient* [17] is defined as

$$r = \frac{cov(a, b)}{\sigma_a \sigma_b}, \tag{8}$$

where $cov(a, b)$ is the covariance of $a$ and $b$, and $\sigma_a \sigma_b$ is the product of their standard deviations.

*Spearman's ρ rank correlation coefficient* [18] is defined as Pearson's $r$ between the rank values of the two variables. For a sample of size $s$, the $s$ raw scores $a$ and $b$ are converted to ranks $rg_a$ and $rg_b$, and $\rho$ is computed as

$$\rho = \frac{cov(rg_a, rg_b)}{\sigma_{rg_a}\sigma_{rg_b}}, \tag{9}$$

where $cov(rg_a, rg_b)$ is the covariance of the rank variables and $\sigma_{rg_a}\sigma_{rg_b}$ are the standard deviations of the rank variables.

Given two samples $a$ and $b$, where each sample size is $s$, *Kendall's τ rank correlation coefficient* [19] is defined according to the following formula:

$$\tau = \frac{s_c - s_d}{\frac{1}{2}s(s-1)}, \tag{10}$$

where $s_c$ and $s_d$ are the numbers of concordant and discordant pairs, respectively, and $\frac{1}{2}s(s-1)$ is the total number of pairings with $a$ and $b$.

## 3 Methodology and Results

We investigated the correlations among the node and edge centrality measures, described in Sect. 2.1, in different kinds of artificial networks. The network models include the scale-free (SF) [20], the small-world (SW) [21], and the Erdös-Rényi (ER) [22] random networks. For each class, we randomly generated undirected and unweighted networks with 10,000 nodes and between 5,000 and 50,000 edges.

We used the model proposed by Holme and Kim [23] to generate the SF networks. In each experiment, we chose the number of nodes $n = 10{,}000$, we added $m = \{5, 15, 25, 35, 50\}$ random edges for each new node $i$, and we picked a probability $p = 0.3$ of adding a triangle after adding a random edge.

SW graphs are generated using Newman–Watts–Strogatz (NWS) small-world model [21]. In each experiment, we chose the number of nodes $n = 10{,}000$, $k = \{6, 18, 32, 64\}$ neighbors with which we connect each node $i$ in the ring topology, and a probability $p = 0.6$ of rewiring each edge.

ER networks are generated choosing the network size $n = 10{,}000$, and for edge creation the probability values $p = \{0.001, 0.003, 0.005, 0.010\}$.

For the implementation of the artificial networks and the centrality metrics such as DC, BC, CL, and CC, we used Python and NetworkX module [24]. To run the WERW-Kpath[1] [6–8], we set the value of the random walk to $k = 10$. For GoT[2], we set 1 thief per node and the initial amount of vdiamonds per node equal to the network size. We let GoT to run for $T = \log^3 n$ epochs.

[1] Available at http://www.emilio.ferrara.name/code/werw-kpath/.

[2] Available at http://github.com/dcmocanu/centrality-metrics-complex-networks.

In Fig. 1 are shown the results of Pearson's $r$ correlation coefficient (left column), Spearman's $\rho$ (middle column), and Kendall's $\tau$ rank correlation coefficients (right column) for SF networks (upper row), SW networks (middle row), and ER random networks (lower row).

In SF networks (see Fig. 1 upper row), GoT and DC have the strongest negative correlation. GoT exhibits a strong negative correlation with the BC and slightly weaker with the CL. There is a positive correlation between GoT and the CC which becomes stronger for the rank correlation coefficients. In particular, we can observe that Spearman's $\rho$ is always larger than Kendall's $\tau$. Except for the DC, we can observe a small deviation of the rank correlation coefficients when the number of

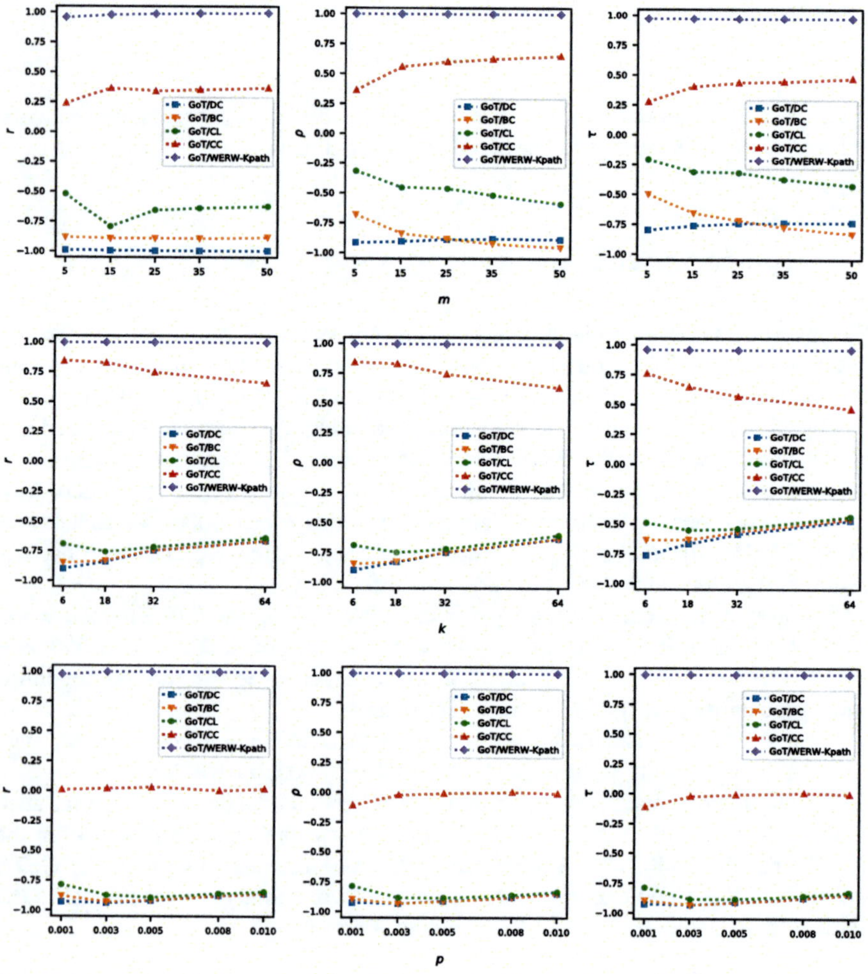


**Fig. 1** Pearson's $r$, Spearman's $\rho$, and Kendall's $\tau$ correlation coefficients

edges in the analyzed networks is smaller which is not visible anymore when edges grow big enough.

In SW networks (see Fig. 1 middle row), there is an almost identical strong negative correlation among GoT and DC, BC, and CL. We can also still observe a strong positive correlation between GoT and the CC. All of these correlations become weaker when the number of edges grows.

In ER networks (see Fig. 1 lower row), the correlation among GoT and DC, BC, and CL is the same as in SW networks. But, in this kind of networks, there is no correlation between GoT and the CC even if we can observe a very little deviation when the number of edges is small.

In all networks, we can observe a strong positive correlation among GoT and WERW-Kpath which remains constant when the number of edges grows.

## 4 Conclusions

In this work, we used artificial networks, i.e. SF, SW, and ER networks, to examine the correlation between well-known and more recently proposed centrality measures. A strong correlation implies the possibility of approximating the metric with the highest computational complexity using the other. We have done an analysis observing the Pearson, Spearman, and Kendall correlation coefficients on artificial networks by changing the amount of links given a fixed network size. Our results show a strong correlation among GoT and the most known centrality algorithms. For this reason, it can substitute them in the computation of node centrality in large networks. GoT can also replace the CC as a property of SW and SF networks.

A future development of our work will be to focus more on the application of WERW-Kpath, changing the values of $k$, and GoT to the case of edge centrality. The two measures seem to be equivalent and it could be also interesting to make a deeper analysis on their computational costs on large networks.

## References

1. Ficara A, Cavallaro L, De Meo P, Fiumara G, Catanese S, Bagdasar O, Liotta A (2020) Social network analysis of Sicilian Mafia interconnections. In: Cherifi H, Gaito S, Mendes JF, Moro E, Rocha LM (eds) Complex networks and their applications VIII. Springer International Publishing, Cham, pp 440–450. https://doi.org/10.1007/978-3-030-36683-4_36
2. Calderoni F, Catanese S, De Meo P, Ficara A, Fiumara G (2020) Robust link prediction in criminal networks: a case study of the Sicilian Mafia. Expert Syst Appl 161:113,666. https://doi.org/10.1016/j.eswa.2020.113666
3. Cavallaro L, Ficara A, De Meo P, Fiumara G, Catanese S, Bagdasar O, Song W, Liotta A (2020) Disrupting resilient criminal networks through data analysis: The case of Sicilian Mafia. PLOS ONE 15(8):1–22. https://doi.org/10.1371/journal.pone.0236476
4. Cavallaro L, Ficara A, Curreri F, Fiumara G, De Meo P, Bagdasar O, Liotta A (2021) Graph comparison and artificial models for simulating real criminal networks. In: Benito RM, Cherifi

C, Cherifi H, Moro E, Rocha LM, Sales-Pardo M (eds) Complex networks and their applications IX. Springer International Publishing, Cham, pp 286–297. https://doi.org/10.1007/978-3-030-65351-4_23
5. Wasserman S, Faust K, Granovetter M, Iacobucci D (1994) Social network analysis: methods and applications. Structural analysis in the social sciences. Cambridge University Press
6. De Meo P, Ferrara E, Fiumara G, Ricciardello A (2012) A novel measure of edge centrality in social networks. Knowledge Based Systems 30:136–150. https://doi.org/10.1016/j.knosys.2012.01.007
7. De Meo P, Ferrara E, Fiumara G, Provetti A (2013) Enhancing community detection using a network weighting strategy. Information Sciences 222:648–668. https://doi.org/10.1016/j.ins.2012.08.001
8. De Meo P, Ferrara E, Fiumara G, Provetti A (2014) Mixing local and global information for community detection in large networks. Journal of Computer and System Sciences 80(1):72–87. https://doi.org/10.1016/j.jcss.2013.03.012
9. Mocanu DC, Exarchakos G, Liotta A (2018) Decentralized dynamic understanding of hidden relations in complex networks. Scientific Reports 8(1):1571. https://doi.org/10.1038/s41598-018-19356-4
10. Valente TW, Coronges K, Lakon C, Costenbader E (2008) How correlated are network centrality measures? Connections (Toronto, Ont.) 28(1):16–26
11. Shao C, Cui P, Xun P, Peng Y, Jiang X (2018) Rank correlation between centrality metrics in complex networks: An empirical study. Open Physics 16(1):1009–1023. https://doi.org/10.1515/phys-2018-0122
12. Oldham, S., Fulcher, B., Parkes, L., Arnatkeviciute, A., Suo, C., Fornito, A.: Consistency and differences between centrality measures across distinct classes of networks. PLOS ONE 14(7), 1–23 (2019). https://doi.org/10.1371/journal.pone.0220061
13. Ficara A, Fiumara G, De Meo P, Liotta A (2021) Correlations among Game of Thieves and other centrality measures in complex networks. In: Fortino G, Liotta A, Gravina R, Longheu A (eds) Data science and Internet of Things. Springer International Publishing. https://doi.org/10.1007/978-3-030-67197-6_3
14. Freeman LC (1978) Centrality in social networks conceptual clarification. Social Networks 1(3):215–239. https://doi.org/10.1016/0378-8733(78)90021-7
15. Brandes U (2008) On variants of shortest-path betweenness centrality and their generic computation. Social Networks 30(2):136–145. https://doi.org/10.1016/j.socnet.2007.11.001
16. Saramäki J, Kivelä M, Onnela JP, Kaski K, Kertész J (2007) Generalizations of the clustering coefficient to weighted complex networks. Phys Rev E 75:027,105. https://doi.org/10.1103/PhysRevE.75.027105
17. Chen P, Popovich P (2002) Correlation: parametric and nonparametric measures. Sage university papers series. No. 07-139. Sage Publications
18. Spearman C (1904) General intelligence, objectively determined and measured. The American Journal of Psychology 15(2):201–292. https://doi.org/10.2307/1412107
19. Kendall M, Gibbons J (1990) Rank correlation methods. Charles Griffin Book. E, Arnold
20. Barabási, A.L., Albert, R.: Emergence of scaling in random networks. Science 286(5439), 509–512 (1999). https://doi.org/10.1126/science.286.5439.509
21. Newman M, Watts D (1999) Renormalization group analysis of the small-world network model. Physics Letters A 263(4):341–346. https://doi.org/10.1016/S0375-9601(99)00757-4
22. Erdös P, Rényi A (1959) On Random Graphs I. Publicationes Mathematicae Debrecen 6:290
23. Holme P, Kim BJ (2002) Growing scale-free networks with tunable clustering. Phys Rev E 65:026,107. https://doi.org/10.1103/PhysRevE.65.026107
24. Hagberg AA, Schult DA, Swart PJ (2008) Exploring network structure, dynamics, and function using NetworkX. In: Varoquaux G, Vaught T, Millman J (eds) Proceedings of the 7th Python in science conference. Pasadena, CA USA, pp 11–15

# Helping People with Visual Impairments to Avoid Obstacles Using Deep Learning

**Mostafa Elgendy and Cecilia Sik Lanyi**

**Abstract** Doing activities such as navigation is a big problem for people with visual impairment. It makes them inactive and isolates them from communicating with the people around them. A lot of technological interventions have been proposed to solve and overcome these problems. This paper proposes a solution to identify popular objects and avoid obstacles around them. YOLOv3 and Tiny-YOLO3 deep learning models are trained with multiple images containing obstacles that the visually impaired person will face indoors. The results show an average accuracy of 94.6% for object detection while using the YOLOv3 model, and 97.91% recognition accuracy is achieved for using the same model.

**Keywords** YOLOv3 · Tiny-YOLOv3 · Deep learning · People with visual impairment · Obstacle detecting · Indoor navigation

## 1 Introduction

Currently, there are over 253 million people all over the world living with visual impairment [1]. People with Visual Impairment (PVI) have problems with the functions of their vision system. These problems prevent them from seeing and communicating with the environment or doing daily activities such as walking indoors and outdoors [2]. One of the most significant problems that PVI face is to detect objects and avoid obstacles which lead to parlous situations. PVI uses assistive technology to detect and avoid obstacles which provide safe navigation. So, researchers tried to develop solutions to solve this problem by using sensors like ultrasonic, radar, and laser sensors. Other solutions used computer vision and Deep Learning (DL) [3].

M. Elgendy (✉) · C. S. Lanyi
University of Pannonia, Veszprem, Hungary
e-mail: mostafa.elgendy@virt.uni-pannon.hu

C. S. Lanyi
e-mail: lanyi@almos.uni-pannon.hu



X.-S. Yang et al. (eds.), *Proceedings of Sixth International Congress on Information and Communication Technology*, Lecture Notes in Networks and Systems 216,
https://doi.org/10.1007/978-981-16-1781-2_79

Dias et al. provided a system which used a white cane with an ultrasonic sensor and Inertial Measurement Unit (IMU). It used IMU to report orientation which helps PVI walking on a straight path. The ultrasonic sensor allowed them to detect and avoid obstacles. The system used an algorithm to detect obstacle size and the distance from PVI. It gave the required direction as audio feedback using earphones [4]. Also, smart walker used laser scanners for the perception of obstacles during navigation. It used a notebook for data processing and a vibration motor attached to each handle for tactile feedback. The system is connected to a vibrotactile belt around the waist to give haptic feedback. Furthermore, it incorporates a microcontroller to receive data from PVI and controls the vibration motors [5].

Mohamed proposed an obstacle avoidance and recognition system using computer vision technologies. It used a laser sensor to calculate the distance from PVI to obstacles and used markers to determine the PVI location. The system used a path planning module to calculate a safe route from PVI current position to the destination. It used a recognition system to get information about objects present around PVI. The system includes a speech recognition module to receive instructions and give voice feedback to the user [6]. Aladren et al. developed a navigation system to improve PVI abilities for better interaction with the environment. The system used RGB images and range information to recognize the elements in the view around PVI. It extracted the floor from images to detect obstacles easily. This system sent navigation commands via sound map and voice feedback. This map was created using stereo beeps and used their frequency to indicate the distance to obstacles. The system is able to detect near and far obstacles [7].

Caldini et al. proposed a novel system using a smartphone inertial gyroscope and camera to detect obstacles near the PVI. It used a modified version of structure from the motion algorithm to process captured images. This algorithm used information from the gyroscope in the smartphone. Then, it detected obstacles and their distance from PVI [8]. Mocanu et al. proposed a system to improve the cognition of PVI by recognizing persons using computer vision and deep learning algorithms. It used deep convolutional neural networks to help PVI distinguish between known and unknown face identities. It used acoustic warning messages through headphones to give semantic information to PVI [9].

This article proposes a navigation system to help PVI avoid obstacles during navigating inside the building using DL. We used the images dataset to train YOLOv3 and Tiny-YOLOv3 models, then we compare both of them to select the model which gives better accuracy. The paper is structured in the following way: Sect. 2 explains our system design, while Sect. 3 shows our experiments and results. Finally, Sect. 4 outlines the conclusion.

## 2 Proposed Architecture

Precise and fast indoor object detection and recognition is an important task to help the PVI interacting with the external world. DL models are useful and proved their big performances in object detection and recognition tasks. We propose a system to help PVI avoiding obstacles using a deep learning model. It starts by opening the camera and asking the PVI to move to reach their destination. While walking, a real stream of images from the smartphone camera is given to our system, then these images are converted to grayscale ones and sent to our deep learning model to detect obstacles. If any obstacle is detected, our system returns feedback to PVI to avoid it. However, if it fails to do so, it decides that no obstacles are available and continues processing the next image. Figure 1 shows the flowchart for this process.

### 2.1 *Dataset*

We have collected images for five different indoor obstacles for simplification. We have used 2000 images for each class, so a total of 10,000 images for training. For validation, we used 400 images per class which is a total of 2000. For testing, we used 100 images per class which is a total of 500 images. Table 1 shows examples of these images. We annotate all images manually and use text annotation files to store information about bounding boxes. Then, we use the final dataset to train the YOLOv3 and Tiny-YOLOv3 models and select the best-trained model for our system.

### 2.2 *Deep Learning Model*

DL algorithms for object detection are divided into one stage and two stages. Two-stages algorithms use a region proposal network in the first stage to generate regions

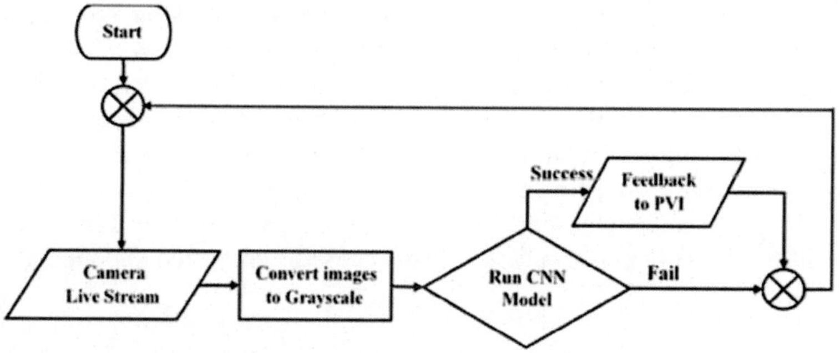


**Fig. 1** Flowchart of the detection process

**Table 1** Samples from indoor obstacles collected from dataset

| Class Name | Chair | Desk |
|---|---|---|
| Class Image |  |  |
| Class Name | Door | Stairs |
| Class Image |  |  |

of interest and send them to the second stage for object classification. While these models give high accuracy, they need more resources and processing time. One-stage models such as YOLO handle detection as a simple regression problem [10]. They take images and learn the class probabilities and bounding box coordinates. While they are faster than two stages, they achieve lower accuracy [11]. Several versions of the YOLO model were proposed to improve the accuracy. We choose YOLOv3 because it can balance the performance on accuracy and processing time well [12]. Figure 2 shows the architecture of the YOLOv3 model.

Also, Tiny-YOLOv3 is a simplified version of the YOLOv3 model that accepts images as input. It consists of two main blocks: feature extractor and detector. The feature extractor uses the input images to extract features embedding at different scales. They feed these features into two detectors to get bounding boxes and class information Figure 3 shows the architecture of the Tiny-YOLOv3 model. We used both models to compare and balance the accuracy and the execution time. Given an image of this model, the final output is a list of bounding boxes along with the recognized classes. At first, the dimension of input images is reduced, and the features are extracted by going through several convolutional layers which makes the detection classifications. Finally, the output is given as a feature map which represents the network class prediction. This output is converted to bounding boxes with class id.

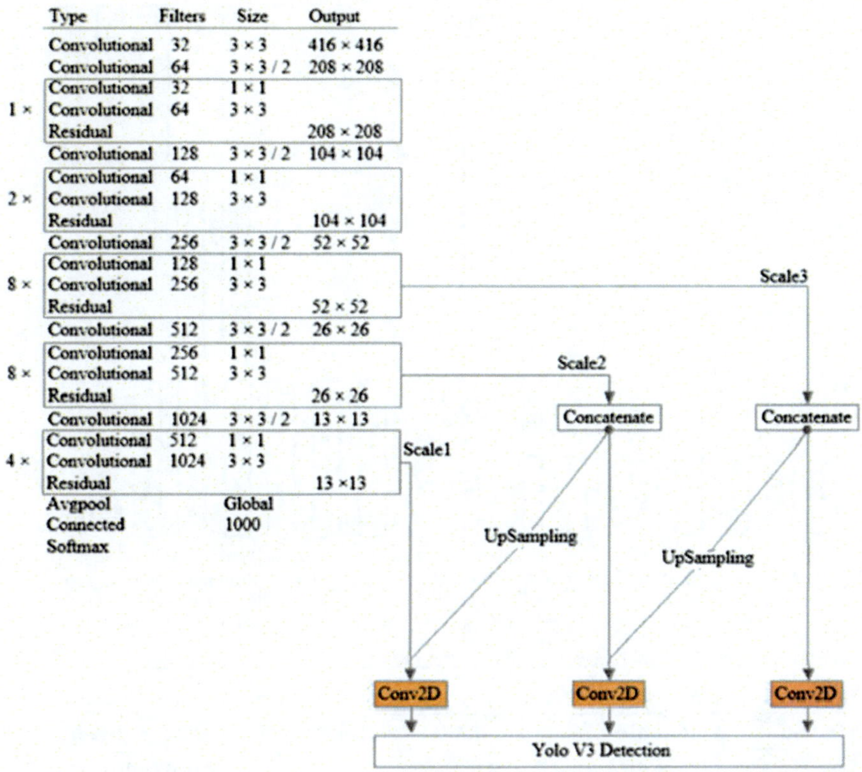


**Fig. 2** The architecture of YOLOv3 model using input images with dimension 416 × 416 [13]

## 3 Experiments

Training our marker detection model requires a lot of resources. Therefore, we used Google Colab to leverage the power of free GPU for training our dataset quite easily. The models were evaluated using videos on a Dell Inspiron N5110 computer with Intel Core i7-2630 QM 2.00 GHz CPU, 6 MB cache, quad-core, and 8 GB RAM. The models were uploaded to the mobile device with the following specifications. HTC Desire 826 smartphone with 2 GB RAM, octa-core CPU, and Adreno 405 GPU. As discussed, we trained the YOLOv3 and Tiny-YOLOv3 models using our dataset. The models are trained with the dataset and initialized parameters as shown in Table 2. The size of the input images is adjusted to 416 × 416 pixels. The training steps are set to 10,000 to give better analysis, and 16 is used as a batch size. After completing the training step, 100 test images for each class are used to verify the algorithm performance. Table 3 shows the performance analysis of the YOLOv3 and Tiny-YOLOv3 models using the test set.

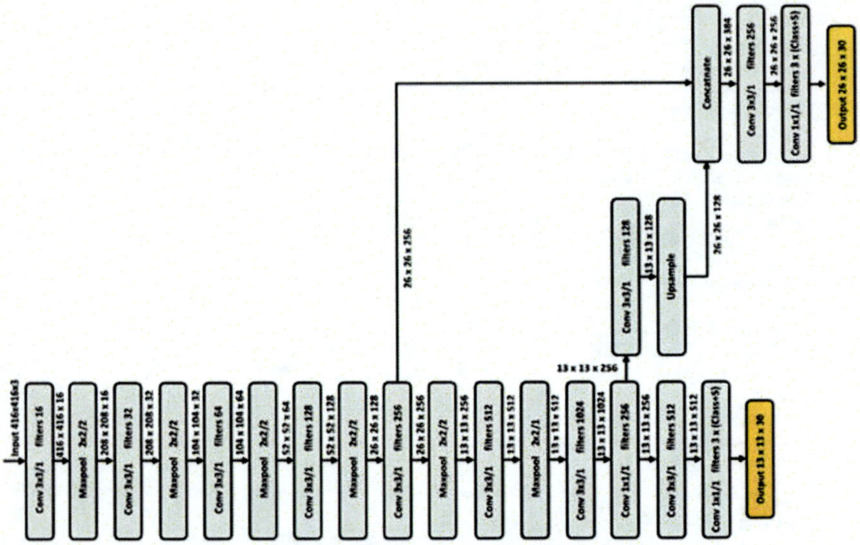


**Fig. 3** The architecture of Tiny-YOLOv3 model using images with dimension 416 × 416

**Table 2** The initialization parameters of the YOLOv3 and Tiny-YOLOv3 models

| Model | Input size | Batch size | Momentum | Learning rate | Decay | Training steps |
|---|---|---|---|---|---|---|
| YOLOv3 | 416 × 416 | 16 | 0.9 | 0.001 | 0.0005 | 10,000 |
| Tiny-YOLOv3 | 416 × 416 | 16 | 0.9 | 0.001 | 0.0005 | 10,000 |

As shown, an average accuracy of 94.6% is achieved for object detection using the YOLOv3 model while 52.6% for object detection using the Tiny-YOLOv3 model. Also, 97.91% for recognition accuracy using YOLOv3 model while 95.12% for Tiny-YOLOv3 model was achieved. The results showed that YOLOv3 gives better accuracy for detecting and recognition. Furthermore, once the object is detected, the two models properly classify it among the list of object classes.

From these results, we found that the YOLOv3 model gave better accuracy for detection and used it for our mobile application. The average detection time for both models is shown in Table 4. Results for detecting different object classes are shown below in Fig. 4. The YOLOv3 model can detect and recognize the object correctly.

## 4 Conclusion

We proposed an assistive system for PVI to detect and avoid obstacles independently where all feedback is provided to PVI in the form of audio. Images of obstacles are collected and manually labeled to create the dataset which is used to train the

**Table 3** Performance analysis of the YOLOv3 and Tiny-YOLOv3 models on the five classes

| Objects | Total images | Correctly detected | | Correctly recognized | | Detection accuracy (%) | | Recognition accuracy (%) | |
|---|---|---|---|---|---|---|---|---|---|
| | | YOLOv3 | Tiny-YOLOv3 | YOLOv3 | Tiny-YOLOv3 | YOLOv3 | Tiny-YOLOv3 | YOLOv3 | Tiny-YOLOv3 |
| Chair | 100 | 95 | 53 | 93 | 51 | 95 | 53 | 97.9 | 96.23 |
| Table | 100 | 94 | 51 | 92 | 49 | 94 | 51 | 97.9 | 96.08 |
| Door | 100 | 96 | 54 | 92 | 49 | 96 | 54 | 95.83 | 90.74 |
| Person | 100 | 91 | 50 | 91 | 49 | 91 | 50 | 100 | 98 |
| Stairs | 100 | 97 | 55 | 95 | 52 | 97 | 55 | 97.94 | 94.55 |
| **Average** | | | | | | **94.6** | **52.6** | **97.91** | **95.12** |

**Table 4** The average detection time of the two models in milliseconds

| Model | GPU | CPU |
|---|---|---|
| YOLOv3 | 9.8 | 2000 |
| Tiny-YOLOv3 | 3.16 | 250 |

**Fig. 4** Results after object detection and recognition

DL model. In this system, we have selected and compared YOLOv3 and Tiny-YOLOv3 models for detecting obstacles. The results showed that YOLOv3 gives better accuracy for detecting and recognition as the accuracy of 94.6% for object detection is achieved using the YOLOv3 model while 52.6% is achieved for object detection for the Tiny-YOLOv3 model. Also, 97.91% is achieved for recognition accuracy for the YOLOv3 model while 95.12% for the Tiny-YOLOv3 model. The results showed that the Tiny-YOLOv3 model is faster than the YOLOv3 model as it took an average of 3.16 ms for recognition while YOLOv3 took 9.8 ms using GPU. Tiny-YOLOv3 model took an average of 250 ms for recognition while YOLOv3 took 2 s using CPU. So, we have selected YOLOv3 for our system. Future work will focus on minimizing recognition time for YOLOv3 and adding more objects for obstacles in the dataset to make it more useful for PVI.

**Acknowledgements** The authors would like to acknowledge the financial support of Széchenyi 2020 under the EFOP-3.6.1-16-2016-00015.

## References


1. Ackland P, Resnikoff S, Bourne R (2018) World blindness and visual impairment: despite many successes, the problem is growing. Commun Eye Heal J 30:71–73
2. Elgendy M, Sik-Lanyi C, Kelemen A (2019) Making shopping easy for people with visual impairment using mobile assistive technologies. Appl Sci 9:1061
3. Manjari K, Verma M, Singal G (2020) A survey on assistive technology for visually impaired. Internet Things. 11
4. De Silva SA, Dias D (2016) A sensor platform for the visually impaired to walk straight avoiding obstacles. In: Proceedings of the international conference on sensing technology (ICST), pp 838–843
5. Wachaja A, Agarwal P, Zink M, Adame MR, Möller K, Burgard W (2017) Navigating blind people with walking impairments using a smart walker. Autono Rob 41:555–573
6. Mekhalfi ML, Melgani F, Zeggada A, De Natale FGB, Salem MA-M, Khamis A (2016) Recovering the sight to blind people in indoor environments with smart technologies. Expert Syst Appl 46:129–138
7. Aladren A, Lopez-Nicolas G, Puig L, Guerrero JJ (2016) Navigation assistance for the visually impaired using RGB-D sensor with range expansion. IEEE Syst 10:922–932
8. Liao C-F (2016) An integrated assistive system to support wayfinding and situation awareness for people with vision impairment. ProQuest Dissertion Theses, 291
9. Mocanu B, Tapu R, Zaharia T (2018) Deep-see face: a mobile face recognition system dedicated to visually impaired people. IEEE Access 6:51975–51985
10. Redmon J, Divvala S, Girshick R, Farhadi A (2016) You only look once: unified, real-time object detection. In: IEEE conference on computer vision and pattern recognition (CVPR), pp 779–788
11. Soviany P, Ionescu RT (2018) Optimizing the trade-off between single-stage and two-stage deep object detectors using image difficulty prediction. In: IEEE 20th international symposium on symbolic and numeric algorithms for scientific computing (SYNASC), pp. 209–214
12. Redmon J, Farhadi A (2017) YOLO9000: better, faster, stronger. In: IEEE conference on computer vision and pattern recognition (CVPR), pp 6517–6525
13. Redmon J, Farhadi A (2018) YOLOv3: an incremental improvement

# Chatbot Accessibility Guidance: A Review and Way Forward

**Jeff Stanley, Ronna ten Brink, Alexandra Valiton, Trevor Bostic, and Becca Scollan**

**Abstract** Chatbots are an increasingly popular way to deliver and support online services. These services must be accessible for people with disabilities; however, it is not yet clear how to extend existing guidance on web content accessibility to chatbots. Researchers and developers from different sectors are producing recommendations by a variety of methods. This paper surveys existing chatbot accessibility guidance. We found seventeen different sources yielding 157 unique recommendations for creating an accessible chatbot experience, which we organized into five categories: content, user interface, integration with other web content, developer process and training, and testing. We identify and discuss trends, themes, and opportunities to refine these recommendations. We believe that there is a need to further refine and expand the recommendations presented in this paper to facilitate the creation of accessible chatbots. Through our analysis, we intend to guide future research, inform the development of materials for designing, developing, and evaluating chatbots; and promote a shared understanding of how chatbots can deliver a successful experience for all users.

**Keywords** Chatbot · Accessibility · Human–computer interaction

J. Stanley (✉) · R. Brink · A. Valiton · T. Bostic · B. Scollan
The MITRE Corporation, McLean, VA 22152, USA
e-mail: jstanley@mitre.org

R. Brink
e-mail: rtenbrink@mitre.org

A. Valiton
e-mail: avaliton@mitre.org

T. Bostic
e-mail: tbostic@mitre.org

B. Scollan
e-mail: rscollan@mitre.org



X.-S. Yang et al. (eds.), *Proceedings of Sixth International Congress on Information and Communication Technology*, Lecture Notes in Networks and Systems 216,
https://doi.org/10.1007/978-981-16-1781-2_80

## 1 Introduction

As chatbot technology is becoming easier to acquire, customize, and deploy, organizations are increasingly relying on chatbots to deliver web content and services while reducing costs and staff workload [1, 2]. These organizations include industry, academia, and government, delivering services as varied as ordering food [3]; providing information to university students and staff [4]; answering general tax filing questions [5]; and answering questions about coronavirus and healthcare resources for veterans [6].

Since the onset of the COVID-19 pandemic, more people than ever need remote access to public and private web-based services. These services are often unequally available due to differences in users' access to and familiarity with technology. In fact, the users with the least access are likely to already be at risk from chronic health conditions [7, 8]. Even when users have familiarity with and access to technological resources, they can be cut off from digital service due to gaps in interface design and implementation [9, 10]. While all users benefit from well-designed applications, people with physical, cognitive, and other disabilities depend on designers and developers to create experiences that meet a range of needs and abilities.

The World Wide Web Consortium (W3C) publishes Web Content Accessibility Guidelines (WCAG) and associated success criteria that mindful designers and developers apply to ensure equal access for people with disabilities [11]. However, there is not yet consensus in the literature on accessibility guidelines specifically for chatbots. To fill the gap, various entities have released their own recommendations for producing chatbots that offer access to users with disabilities and provide satisfying experiences for all users. The landscape of chatbot accessibility guidance is fragmented and encompasses several points of view. In this literature survey and analysis, we integrate and find patterns across these disparate sources. We examine trends and organize a variety of recommendations for producing accessible chatbots. For the purpose of this work, a *chatbot* is defined as a text-based conversational interaction interface with at least one automated and at least one human conversational partner [12] (though we also touch on some relevant research on voice-based human–machine conversational interfaces). *Accessibility* on the web means giving people with disabilities "access to information comparable to the access available to others" [13].

## 2 Methodology

Keyword searches were performed in Google Scholar and Google Search to find sources. Keywords used were variations of "chatbot" or "conversational" combined with variations of "accessibility" or "disability." Sources' reference lists were combed for further results. Results were selected according to the following criteria:

- The source must acknowledge that it is attempting to address a user's disability, accessibility concern, or accessibility in general.
- The source must address chatbots. Sources focusing on automated voice interfaces were included (e.g., [14, 15]). Accessibility recommendations for voice interfaces may apply to textual chat as well, particularly regarding conversation structure and content.
- The source must offer some kind of *recommendation*; that is, the author(s) must include some proposed way to improve chatbot accessibility.

Keyword searches were performed in English. In three cases, English-language sources referenced relevant non-English-language sources that passed criteria [12, 16, 17]. These were translated into English using Google Translate and included as sources.

We identified and extracted any recommendations (as defined above) from each source. We clustered similar recommendations to arrive at a categorization of recommendations across all sources. In some cases, a single recommendation was relevant to and included in more than one category.

## 3 Results

We found 17 sources offering guidance for chatbot accessibility (Table 1), yielding a total of 157 unique recommendations. The survey reveals trends including which sectors are producing guidance, how guidance is produced, and which user needs are most frequently addressed. In this section, we present these trends and discuss our organization of the recommendations into categories and themes.

Except for one source from 2008, all sources were from 2017 and later. (Two online sources were not explicitly dated but copyrighted through the present date.) This finding is supported by a study conducted a few years ago which found "no publications about chatbot accessibility" before 2017 [29]. This trend follows the more general trend of increasing interest in chatbots. "Chatbot" as a search term has been used worldwide in the Google search engine since at least 2004 (earliest data available from Google Trends) but started steadily rising in search frequency in 2016–2017 [30].

The 17 sources were generated by 26 different entities. Fourteen of the entities are universities. Four are accessibility-focused firms or organizations (Bureau of Internet Accessibility [20], Fundación Sidar [27], CANAXESS [22], Funka [17]). Four are technology, telecom, or design companies (Microsoft Research [25], Rogerwilco [19], Orange [26], AB Stelacon [12]). Three sources involved government funding or participation. The remaining source [21] was published independently by an industry expert.

We inferred which users' needs are addressed in the literature by examining which standards were cited and what participants were recruited for usability tests. Most sources addressed at least three different kinds of disabilities or disabilities in

**Table 1** Sources of recommendations for chatbot accessibility

| Citation | Type of source | Originating entity | Country |
|---|---|---|---|
| Abdolrahmani et al [14] | Conference paper | University of Maryland, Baltimore County; University of California, Irvine | USA |
| Baldauf [15] | Conference paper | University of Applied Sciences St. Gallen | Switzerland |
| Bigham et al. [18] | Conference paper | University of Washington | USA |
| Boshoff [19] | Blog or web page | Rogerwilco | South Africa |
| Bur. of Internet Accessibility [20] | Blog or web page | Bureau of Internet Accessibility | USA |
| Campbell [21] | Blog or web page | UX Collective | Virtual |
| CANAXESS [22] | Blog or web page | CANAXESS | Australia |
| De Filippis et al. [23] | Conference paper | University of Perugia; University of Twente; Imperial College; Hertfordshire University; Italian Ministry of Economic Development | Italy; Netherlands; UK |
| Furini et al. [24] | Conference paper | University of Modena and Reggio Emilia; University of Bologna | Italy |
| Lilleby and Marstein [16] | Thesis | University of Oslo | Norway |
| Lister et al. [10] | Conference paper | The Open University | UK |
| Microsoft [25] | Blog or web page | Microsoft Research | USA |
| Orange [26] | Blog or web page | Orange | France |
| PTS [12] | Report | Swedish Post and Telecom Authority; AB Stelacon | Sweden |
| Restrepo et al. [27] | Conference paper | National University of Distance Education; Fundación Sidar | Spain |
| Simonsen [17] | Report | Funka; Directorate for Children, Youth, and Families | Norway |
| State [28] | Thesis | Kungliga Tekniska Högskolan | Sweden |

general. The most commonly addressed disability is visual impairment, and the most commonly addressed assistive technologies (AT) are screen readers and magnifiers. Multiple papers addressed keyboard control for users preferring keyboard to mouse, across a variety of disabilities. Fewer papers focused explicitly on other accessibility concerns such as dyslexia, cognitive impairments, mental health, hearing loss, and mobility and motor impairments. Most papers focused on users having a single disability.

Our survey found some recommendations that referred to specific features, functionality, or measurable quantities. These recommendations can be helpful at the time those features are being developed and tested. For example, "do not include flashing content" [10] and "use adequate color contrast" [22] are specific recommendations that can be implemented in the development phase and can be specifically addressed in testing (e.g., asking users whether the content was overstimulating, and measuring color contrast against the WCAG standard).

Other recommendations are less specific and should be addressed earlier in the design process. These recommendations should be interpreted to fit the specific chatbot use case. For instance, "support context-aware conversations" [15] and "minimize distractions" [10] do not provide specific criteria for development or testing. In order to act on these recommendations, the chatbot design team would need to determine what context the chatbot needs to be aware of and which distractions can be eliminated. Based on the needs determined by the design team, these recommendations may be addressed with appropriate chatbot platform selection or website design, among other ways. However, methods for designing and measuring chatbot context awareness or user distraction depend on the project's specific needs and goals. In order to evaluate whether these recommendations have been acceptably implemented, the design team may need to perform task-based user testing [12, 16, 17, 27, 31] or other general usability assessment.

## 3.1 *How is Chatbot Accessibility Guidance Produced?*

From our sources, we identified three ways of producing chatbot accessibility guidance: adapting established accessibility standards to chatbots; conducting empirical studies such as interviews, questionnaires, and usability tests; and reasoning about the user experience by leveraging authors' expertise. Some sources employed more than one method. We describe each of these methods in greater detail below.

**Adapting existing standards.** Recommendations can be explicitly grounded in existing standards, and authors might discuss how existing web accessibility standards (e.g., WCAG) can be interpreted to apply to chatbots. In some cases, standards like WCAG success criteria apply in a straightforward manner, such as 1.4.3: "providing adequate color contrast for users with varying degrees of vision". Since chatbots have a visual interface, there are opportunities to manage color contrast [22]. Other success criteria warrant discussion and interpretation, such as those falling under WCAG guideline 3.2: "Predictability". While WCAG gives specific criteria for assessing predictability of web interfaces, these criteria do not apply directly to chatbots. What does it mean for a chatbot to be predictable in its interactions [10]?

We marked a recommendation as adapting existing standards if it explicitly refers to a standard from a recognized standards organization (such as W3C, ANSI, or NIST) or governing body and recommended its application to a chatbot. In practice, all examples we found were from WCAG.

**Empirical studies.** Recommendations may come directly from user-based research, including interviews and testing with potential chatbot users. For example, Lilleby and Marstein [16] observed users with dyslexia while they performed assigned tasks using two Norwegian government chatbots and interviewed them afterwards; design recommendations were derived from accessibility gaps as perceived by the participants. The authors of that study distinguished between studies that focus on standards and those that focus on data collection with users and identified their work as an example of the latter. On the other hand, other sources acknowledge no such contrast and complement user-centered research with relevant WCAG standards [12].

We marked a recommendation as an empirical study if the authors used data obtained through direct engagements with potential users in their research.

**Reasoning about the user experience.** Recommendations may be presented without accompanying experimental support or agreed-upon standards. These recommendations are typically based on the authors' experience in accessible, inclusive or universal design, on findings and suggestions from other papers, or on a proposed solution to challenges they witnessed in their own work but have not tested. This way of producing chatbot guidance includes literature surveys as well as expert evaluation. For example, The Bureau of Internet Accessibility [20], an internet accessibility firm and W3C member, presents a set of best practices for designing chatbots, and Lister [10] presents a survey of literature on accessibility challenges for a diverse range of users and associated recommendations. While neither of these sources presents their own empirical findings or adapts existing web accessibility standards to chatbots, they aggregate and organize existing work in a useful way.

We marked a recommendation as reasoning about the user experience if the authors do not adapt an accessibility standard to chatbots or present empirical evidence from user testing or interviews to support the recommendation.

## 3.2 Recommendations

We organized the 157 unique recommendations into five categories based on whether we interpreted them as addressing content, user interface, integration into the website, development process, or testing. Some recommendations were included in more than one category. For the most part, each category is oriented toward a particular role or step in the chatbot lifecycle. Within each category, we grouped similar recommendations into themes. This section briefly describes how recommendations were created, introduces each category, and provides the identified recommendation categories and themes. The Appendix contains the full list of recommendations by category and theme. Categories and themes are further addressed in the discussion.

In the listed recommendations, "A" denotes an Adapted source, "E" denotes a source we consider Empirical, and "R" denotes a Reasoned source (terms described in Sect. 3.3). As shown in Table 2, the majority of recommendations were from Reasoned sources.

Recommendations were clustered into five categories: content, interface, integration, developer process and training, and testing. Recommendations for chatbot *content* pertain to the language and content of the conversation, regardless of the interaction medium (text, voice, etc.). Chatbot *interface* recommendations cover the mechanics of inputting into the chatbot and perceiving the output, regardless of the content. Chatbot *integration* recommendations are related to the mechanics of navigating to the chatbot, notifications of chatbot activity, and navigating to content linked or attached by the chatbot. Recommendations for chatbot *developer process and training* suggest activities throughout the ideation, design, and development process which may lead to a more accessible chatbot. Chatbot *testing* recommendations pertain to activities evaluating the chatbot accessibility, usability, or efficacy.

# 4 Discussion

The goal of this discussion is to highlight patterns, gaps, and discrepancies that appear in the recommendation list, explore other potentially relevant sources, and suggest research directions.

## *4.1 Overall Trends*

The variety of entities producing chatbot accessibility guidance and the high proportion of these entities that are academic suggest that researchers are still exploring what accessibility means for chatbots and how to derive accessibility guidance for chatbots. At the same time, government involvement in literature reviews [23] and user studies [12, 17] indicates a desire for standard criteria that can be used to evaluate and remediate chatbot accessibility. Most of the recommendations we found were produced by reasoning (see Sect. 3.3), meaning they do not reference user studies or existing guidelines. We see this as an indication that the field needs more empirically validated recommendations.

Chatbots have been a topic of conversation at W3C Accessibility Guidelines Working Group meetings [32], and chatbots are mentioned in the working draft of the forthcoming WCAG 2.2 standard under success criterion 3.2.6 "Findable Help." However, the proposed standard only mandates that the chatbot should "meet other WCAG success criteria for your conformance level." In addition, in order to "work

**Table 2** Summary of how many recommendations per category and theme are adapted (A), empirical (E), or reasoned (R) (terms described in Sect. 3.3). Some recommendations are included in more than one category and theme

| Category | Theme | A | E | R |
|---|---|---|---|---|
| Content | Identify the bot appropriately | 0 | 2 | 1 |
| Content | Use suitable personality | 0 | 0 | 2 |
| Content | Manage conversation length and complexity | 1 | 7 | 10 |
| Content | Be aware of conversational context | 0 | 1 | 2 |
| Content | Use both image and text output | 0 | 2 | 0 |
| Content | Offer choices appropriately | 0 | 1 | 2 |
| Content | Set user expectations for the conversation | 0 | 1 | 6 |
| Content | Manage stimulating content | 0 | 0 | 3 |
| Interface | Support multiple modalities for input and output | 0 | 5 | 9 |
| Interface | Support interaction with AT | 6 | 2 | 5 |
| Interface | Use visually clear text and interface | 3 | 4 | 9 |
| Interface | Promote user awareness of interface elements | 0 | 1 | 4 |
| Interface | Tolerate user errors and input variation | 0 | 4 | 5 |
| Interface | Support personalization and configuration | 0 | 0 | 4 |
| Interface | Keep the interface simple | 0 | 1 | 6 |
| Interface | Manage conversation length and pacing | 0 | 2 | 4 |
| Interface | Offer choices appropriately | 0 | 1 | 4 |
| Integration | Facilitate navigating to chatbot | 5 | 4 | 7 |
| Integration | Manage conversation end and transitions to external resources | 0 | 3 | 3 |
| Integration | Support AT | 1 | 0 | 5 |
| Integration | Manage interruptions and notifications | 3 | 1 | 4 |
| Dev process/training | Include people with disabilities in the process | 0 | 2 | 2 |
| Dev process/training | Follow WCAG | 0 | 0 | 3 |
| Dev process/training | Choose an accessible platform and plan on customizing it | 1 | 1 | 2 |
| Testing | Test early and continuously | 0 | 2 | 1 |
| Testing | Test with realistic technology scenarios | 1 | 1 | 0 |
| Testing | Use validated, standardized testing methodologies | 0 | 2 | 3 |
| Testing | Test with diverse users and with target users | 0 | 3 | 2 |
| All | All | 21 | 53 | 108 |

more effectively for everyone, and particularly for people with cognitive disabilities," the chatbot should "recognize misspelled words, provide human contact details if the chatbot is unable to provide a satisfactory response after 3 attempts, and be dismissed with a single interaction, and recalled using a link or button" [33].

We expect that as recommendations for chatbot accessibility are consolidated and validated by academia and other research entities, this relatively sparse guidance will mature into a more complete set of standards that developers can use to assess the accessibility of their chatbots. In particular, further research can make general recommendations more precise and actionable as discussed in the following sections.

All but one source in our analysis was from 2017 or later (see Table 1), following increasing worldwide interest in chatbots since 2016. It stands to reason that interest in chatbot accessibility would rise along with general interest in chatbots. As chatbots continue to proliferate, chatbot accessibility becomes more urgent, and we are hopeful that more attention will be given to creating and validating actionable guidelines for accessible chatbots.

Most papers did not address chatbot accessibility for users with multiple disabilities. 9.7 million working-age adults in the USA had two or more disabilities as of 2014 [34]. Users with more than one disability may have an intersection of needs different from users with one disability. Future research should include users with multiple disabilities to create needed guidance.

Few sources are focused on accessibility concerns other than visual impairment. While designing for blind and visually impaired users, can increase accessibility for other user groups—for example, ensuring good keyboard navigation is a benefit for people with limited mobility as well as for blind and low vision users—other disabilities pose their own unique challenges. Future research should further consider chatbot accessibility for under-researched user populations, including but not limited to dyslexia, cognitive disabilities, mental health, mobility disabilities, and hearing loss.

## 4.2 *Content*

When it comes to the bot's communication style, most sources recommend the chatbot use straightforward, literal language, and a single-minded focus on the end-user's request. However, two sources suggest injecting the conversation with humor or emotion. Experimentation with digital assistants that exhibit empathy has affirmed that humanlike personality can improve the user's liking, trust, and engagement with the software [35–37]; so deliberately designing chatbot personality could be a way to increase user satisfaction. Humanlike personality is not necessarily at odds with straightforward communication, though humor and emotion may make conversations longer than needed to effectively respond to the user's request. The sources recommending humor and emotion were produced through market analysis, expert interviews, and literature surveys and present no specific empirical evidence that these features improve outcomes for users with disabilities, only that they may be viewed as important by some individuals. Future research should determine experimentally how chatbot personality traits affect outcomes for users with various disabilities.

Where sources recommend short simple messages, we suggest that simplicity can be assessed by existing WCAG success criteria under guideline 3.1 "Readable." These criteria include defining abbreviations and unusual words and writing content for a lower-secondary education reading level. Several of the sources make recommendations closely in line with these success criteria.

Three sources recommend a short simple conversation structure. While the intent of these recommendations is clear (to help the user track and navigate the conversation), it is not clear how to achieve a simple conversation structure. Further literature reviews focusing on conversation design for users in general might identify ways to measure and minimize conversation complexity.

Across sources, we found slightly different uses of the word "context," but all refer to how various parts of the conversation and situation relate to one another. Therefore, a chatbot that is aware of context [15, 24] is able to offer responses that take into account conversation history and knowledge of the end-user and use case. A chatbot that provides context [21] is also able to offer this information to the user to fill potential gaps in the user's understanding or memory, which is particularly helpful for users with disabilities that affect situational awareness. What is not clear is how to evaluate whether a chatbot's use of context is sufficient or appropriate. Future research should explore techniques for capturing, conveying, and assessing context.

Recommendations for minimizing conversation complexity, providing context, and promoting user awareness all have the goal of creating an experience that is easy for users to understand and navigate over the course of the conversation. Therefore, tools for measuring situational awareness such as the Situation Awareness Rating Technique [38] may be helpful for research to refine these recommendations.

## *4.3 Interface*

An interface challenge unique to chatbots is conversation pacing. Since chatbots present information sequentially, the question arises of how to manage message size and timing to accommodate a variety of user needs. Lister et al [10] point out that WCAG guideline 2.2 "Enough Time" needs additional interpretation to be applied to chatbots. Future research should investigate different conversation rhythms or mechanisms for adjusting the timing of chatbot messages, as well as appropriate timeouts for user inputs, to understand the impacts on populations of users with disabilities.

Many web interface accessibility guidelines are relevant to a chatbot interface, such as those addressing color, fonts, focus, typed inputs, keyboard control, and screen readers; these are already covered by existing WCAG guidelines. However, some interface recommendations must be considered jointly with others. Notably, the solution to ensuring high-enough color contrast [20, 22] is not simply to use the highest color contrast for a design theme, as that may not meet the recommendation to avoid stimulating colors [10]. As another example, designers and developers will need to balance recommendations to use a large interface and large text [12, 16] with those to avoid requiring too much scrolling [10]. These may come into conflict particularly in chatbots that deliver large amounts of information. A related interface design challenge is to allow for the user to zoom (e.g., to 200%) in the browser and still be able to interpret the conversation and limit scrolling.

Content recommendations to supplement or complement text with images [28] must also ensure that those pictures are accessible to users with vision impairment [10, 12, 20, 22, 28] in a way that fits the flow of chatbot conversation. Future work should determine whether a typical website picture alt-text writing style is appropriate for images that are part of a chatbot conversation.

Finally, a novel, chatbot-specific interface challenge is to distinguish between chatbot and user messages in the conversation history both visually and audibly [12, 26]. This is a future research opportunity.

## 4.4 Integration

Some recommendations focus on making sure notifications are communicated clearly with both audible and visible alerts, such as making a sound, changing the window title, and interfacing with screen readers [19, 22, 26].

At the same time, several sources recommend subtlety in notifications to avoid distracting or alarming the user [10, 14]. Alerts may be delayed, suppressed, or handled in an unintrusive way to avoid interrupting important tasks. These recommendations are supported by experimentation with app notifications (not involving chatbots) among older adults: perceived disruption and inappropriateness of a notification was greater when the information was non-critical or when it interrupted a cognitively demanding task [39].

In a study of chat widgets (supporting user-to-user conversations) for users with vision impairments, users reported a significantly better experience when chat notifications were delivered by an entirely different voice from that used by the screen reader than when using the same voice; this solution allows clear and timely notification while minimizing interference with other tasks [31]. To facilitate this solution, screen reader software developers could consider offering chatbot and other web content developers access to a range of different voices.

## 4.5 Developer Process and Training

There were relatively few developer process and training recommendations. "Follow WCAG" recommendations [10, 17, 25] were specific and immediately actionable. However, all other recommendations tended to be less specific. We see a need for more specific guidelines in the themes "Include people with disabilities in the process" and "Choose an accessible platform and plan on customizing it." Developers would benefit from aids that are ready-to-deploy or easily adaptable, such as pre-made diverse personas, design and evaluation methodologies that include diverse users, a list of chatbot platforms known to be accessible, or customization-specific accessibility checks to keep in mind when customizing a chatbot platform. Restrepo et al [27] have referenced personas for addressing accessibility and inclusiveness in chatbot design, but we were unable to locate those resources.

## 4.6 Testing

Testing recommendations align well with existing industry design and development practices and processes such as agile testing [40], user-centered design [41], and design thinking [42]. Although multiple sources recommend using "validated, standardized testing methodologies", they list disparate testing tools, from the empirical Usability Metric for User Experience (UMUX) satisfaction survey [43] to WCAG 2.0 checklists, both of which can be adapted to chatbot use cases but neither of which are specifically suited to them. Many industry-standard usability surveys, such as UMUX, do not delve into features specific to chatbots (such as chat history) nor do they account for accessibility. Current WCAG guidelines and testing tools do not cover features specific to chatbots [10]. We did not find validated methodologies for evaluating the impact of chatbot verbal responses on users with disabilities or standardized guidance for accessible conversation design. We believe a compilation of testing methodologies and resources specific to chatbots would significantly help developers test chatbots for accessibility.

## 4.7 Future Research Directions

Through our analysis, we identified many gaps in chatbot accessibility guidance. We summarize future research questions from our work here:

- Overall Trends:
  - How can general recommendations be made more precise and actionable?
  - What chatbot accessibility considerations support users with multiple disabilities and other under-researched user populations?
- Content:
  - How does chatbot personality affect outcomes for users with disabilities?
  - How might chatbot designers achieve a simple conversation structure?

  - How can a chatbot's use of context be assessed to determine whether it is sufficient or appropriate?

- Interface:
  - How does conversation pace impact users with disabilities?
  - What are appropriate user input timeouts for users with disabilities?
  - How can chatbot designers balance large interfaces with minimal scrolling?
  - How much scrolling in chatbot interactions is too much?
  - How might chatbot design support browser magnification without sacrificing usability?
  - Is typical alt-text writing appropriate for images in a chatbot conversation?
  - What design techniques help users distinguish between message authors?

- Integration:
  - What specific design techniques clearly communicate notifications while not distracting or alarming the user?
  - How might a range of voices support chatbot accessibility?

- Developer process and training:
  - What aids are available to support accessible chatbot development?

- Testing:
  - What chatbot testing methods and resources help developers test accessibility?

## 5 Conclusion

Chatbots are increasingly used to offer support for government and commercial services. However, existing standards for web accessibility are not sufficient for ensuring equal access for all chatbot users. To fill this gap, multiple research efforts in industry, academia, and government have attempted to recommend best practices for chatbot developers to improve the accessibility of chatbots. We have compiled these recommendations, organized them into categories and themes, and analyzed them for trends, gaps, and discrepancies. In particular we note:

- research on chatbot accessibility is often limited to addressing users with visual impairment, and should be expanded to include users with diverse abilities;
- chatbot accessibility recommendations are often not tested with groups of target users, which limits their validity—in sum, the field needs more empirically validated recommendations;
- research in accessibility of similar applications, like user-to-user chat applications and applications that use notifications, may offer experimentally validated accessibility guidelines that are appropriate for chatbots;

- many accessibility recommendations are non-specific or only loosely defined and future work should strive to offer chatbot developers more specific and implementable recommendations, and to augment them with specific examples and methods.

This literature survey offers an organized summary of the state of the art in chatbot accessibility and highlights key areas of future research that will make the greatest impact on accessible chatbot design. Through our analysis, we intend to facilitate future research, inform the development of materials for designing, developing, and evaluating chatbots; and promote a shared understanding of how chatbots can deliver a successful experience for all users.

**Acknowledgements** This research was funded by the MITRE Innovation Program. 

## Appendix

This section contains the full list of chatbot accessibility recommendations identified in the literature review, organized by category and theme.

**Content.** Recommendations for chatbot content pertain to the language and content of the conversation, regardless of the interaction medium (text, voice, etc.).

A. **Identify the bot appropriately**

(1) Clearly state the bot's nature and don't try to impersonate a human [10] (R)
(2) Introduce the chatbot as a bot and state its purpose [12] (E)
(3) Clearly communicate that users are talking to a computer and not with a customer service representative [17] (E)

B. **Use suitable personality**

(1) Consider integrating emotional intelligence into bots [15] (R)
(2) Add humor to engage the user and understand user's irony/sarcasm [12] (R)

C. **Manage conversation length and complexity**

(1) Allow the user to adjust the level of detail in chatbot's messages [14] (E)
(2) Avoid jargon and acronyms, use easy-to-understand conversational patterns [19] (R)
(3) Use simple language [20] (R)

(4) Provide context to user in conversation, especially in questions [21] (R)
(5) Use a linear, time-efficient architecture, even in voice interfaces [21] (R)
(6) Use literal language, short and simple sentences, and full-length words [21] (R)
(7) Keep messages short, or allow the screen reader to skip longer messages [21] (R)
(8) Use a suitable reading level and language style for the chatbot's intended audience [22] (A)
(9) Reduce unnecessary messages [16] (E)
(10) Explain complex concepts and difficult words [16] (E)
(11) Keep messages short and simple [10] (R)
(12) Use clear, easily understandable speech [10] (R)
(13) Avoid figures of speech and idioms, exaggeration, ambiguous language, or turns of phrase [10] (R)
(14) Limit the number of steps in a conversation-based task [10] (R)
(15) Divide long messages into paragraphs [12] (E)
(16) Reduce amount of text and use symbols and colors to enhance meaning [28] (E)
(17) Use simple, clear, and understandable language [28] (E)
(18) Use simple questions with layman's terms [28] (E)

D. **Be aware of conversational context**

(1) Support context-aware conversations [15] (E)
(2) Provide context to user in conversation, especially in questions [21] (R)
(3) Be context aware [24] (R)

E. **Use both image and text output**

(1) Combine icons with a description to avoid confusion [28] (E)
(2) Reduce amount of text and use symbols and colors to enhance meaning [28] (E)

F. **Offer choices appropriately**

(1) Put the most important information or most common answer first or last [21] (R)
(2) Offer choices one by one [10] (R)
(3) If the chatbot cannot understand the user's input, provide input suggestions [12] (E)

G. **Set user expectations for the conversation**

(1) Provide context to user in conversation, especially in questions [21] (R)
(2) Let the user know what commands they can issue [24] (R)
(3) Clearly indicate next steps and expectations [10] (R)
(4) Clearly manage expectations about privacy and outcomes [10] (R)

(5) Provide hints and tips to guide user queries [10] (R)
(6) Clearly identify conversational errors or miscomprehension by agent [10] (R)
(7) Suggest to the user ways to interact or how to get started [12] (E)

H. **Manage stimulating content**

(1) Reduce panic triggers [10] (R)
(2) Do not include flashing content [10] (R)
(3) Present information in a calming way [10] (R)

**Interface.** Recommendations for chatbot interface pertain to the mechanics of inputting into the chatbot and perceiving the output, regardless of the content.

A. **Support multiple modalities for input and output**

(1) Use equivalent visual and non-visual cues/affordances [14] (E)
(2) Use both text and audio for all content and notifications [19] (R)
(3) Support voice-to-text [19] (R)
(4) Offer multi-modal experiences [21] (R)
(5) Allow volume control [21] (R)
(6) Make sure user knows about potential for voice interaction [21] (R)
(7) Build in speech-to-text and text-to-speech [16] (E)
(8) Offer a visual option for voice conversational interfaces [10] (R)
(9) Offer an audio version of a chatbot or ensure that the interface is accessible with screen reader [10] (R)
(10) Allow combinations of input/output modalities [10] (R)
(11) Consider using a touch screen interface [10] (R)
(12) Allow multiple means of interfacing with the chatbot [12] (E)
(13) Allow multiple forms of interaction [28] (E)
(14) Provide text-to-speech and speech-to-text [28] (E)

B. **Support interaction with AT**

(1) Make sure response option buttons are tagged for screen readers [19] (R)
(2) Support navigation with keyboard or screen reader [20] (R)
(3) Ensure that rich media is accessible [20] (R)
(4) Ensure conversation history can be navigated with a keyboard [22] (A)
(5) Ensure that rich media is accessible [22] (A)
(6) Make the conversation history identifiable with ARIA-labelling [22] (A)
(7) Offer an audio version of a chatbot or ensure that the interface is accessible with screen reader [10] (R)
(8) Make sure all elements including chat history window can be navigated/triggered/scrolled with keyboard [26] (R)
(9) Add ARIA labels to all interactive elements [26] (A)
(10) Include the sender before each message in aspan [26] (A)

(11) Make sure interface works with screen reader [12] (E)
(12) Use correct WAI-ARIA tags [17] (A)
(13) Make sure interface works with screen reader [28] (E)

C. Use visually clear text and interface

(1) Use accessible color contrast and fonts [20] (R)
(2) Make sure focus shows the most recent message [22] (A)
(3) Use appropriate and flexible font size [22] (A)
(4) Use adequate color contrast [22] (A)
(5) Employ a larger interface when possible [16] (E)
(6) Give the user the ability to change font size, magnify the chat interface up to 200%, and change the interface color/contrast [10] (R)
(7) Design buttons that are inclusively sized, or adaptable in settings [10] (R)
(8) Use a clear visual focus [10] (R)
(9) Avoid periods of sustained scrolling [10] (R)
(10) Minimize distractions [10] (R)
(11) Clearly label buttons [10] (R)
(12) Use low arousal colors [10] (R)
(13) Avoid italics, block capitals, underlining, serif fonts, justified alignment, animation and small sizes in text formatting [10] (R)
(14) Limit message length so that messages do not exceed the size of the message window [12] (E)
(15) Align chatbot messages to the left, and user messages to the right, and use different colors [12] (E)
(16) Use high contrast and large text, and offer a zoomable view [12] (E)

D. **Promote user awareness of interface elements**

(1) Make sure the user knows how to start dialog with voice interaction [21] (R)
(2) Clearly label buttons [10] (R)
(3) Use a clear visual focus [10] (R)
(4) Make sure focus is visible [26] (R)
(5) Align chatbot messages to the left, and user messages to the right, and use different colors [12] (E)

E. **Tolerate user errors and input variation**

(1) Support complex user commands [14] (E)
(2) Use robust speech-to-text or typo-tolerance [15] (E)
(3) Accept pauses [21] (R)
(4) Tolerate shaky/ broken speech [21] (R)
(5) Tolerate poor English [21] (R)
(6) Use a slow or adaptable time-out [10] (R)
(7) Tolerate typos [10] (R)

(8) Tolerate typos [12] (E)
(9) Tolerate simple word choice [12] (E)

F. **Support personalization and configuration**

(1) Allow volume control [21] (R)
(2) Allow speech rate control [21] (R)
(3) Allow users to customize the interaction style/settings [24] (R)
(4) Give the user the ability to change font size, magnify the chat interface up to 200%, and change the interface color/contrast [10] (R)

G. **Keep the interface simple**

(1) Limit the number of voice-commands needed [21] (R)
(2) Avoid periods of sustained scrolling [10] (R)
(3) Use a clean design [10] (R)
(4) Minimize distractions [10] (R)
(5) Use low arousal colors [10] (R)
(6) Limit choices [10] (R)
(7) Employ consistent, clean interfaces [28] (E)

H. **Manage conversation length and pacing**

(1) Allow speech rate control [21] (R)
(2) Avoid periods of sustained scrolling [10] (R)
(3) Use slow or adaptable time-outs for the chatbot re-prompting for input or ending the conversation [10] (R)
(4) Reduce friction in interactions (except when offering a chance to correct mistakes) [10] (R)
(5) Limit message length so that messages do not exceed the size of the message window [12] (E)
(6) Send one message at a time [12] (E)

I. **Offer choices appropriately**

(1) Make sure option buttons are tagged for screen readers [19] (R)
(2) Use a slow or adaptable time-out [10] (R)
(3) Limit choices [10] (R)
(4) When displaying options, move focus to before the options and make sure the options are selectable [26] (R)
(5) Place buttons in a vertical list [12] (E)

**Integration.** Recommendations for chatbot integration pertain to the mechanics of navigating to the chatbot, notifications of chatbot activity, and navigating to content linked or attached by the chatbot.

A. **Facilitate navigating to chatbot**

(1) Make sure the chatbot is discoverable [21] (R)
(2) Make sure the button for activating the chatbot is clear, visible, selectable, etc. [19] (R)
(3) Use skip links to navigate directly to the chatbot [22] (A)
(4) Use responsive web design techniques [22] (A)
(5) Ensure that navigation is accessible [16](E)
(6) Ensure that the chatbot is accessible in both desktop and mobile [10] (R)
(7) Avoid unexpected or involuntary activation [10] (R)
(8) Ensure consistent navigation [10] (R)
(9) Give the button for bringing up the chatbot a clear label [26] (A)
(10) Make the button for bringing up the chatbot accessible via the keyboard using a element [26] (A)
(11) Keep the chatbot button away from the edge of the window so that it can be magnified [26] (R)
(12) Make sure the chatbot button is not covered by nearby tooltips [26] (R)
(13) Use skip links, ARIA landmarks, and/or a header tag so the user is aware of the chatbot button [26] (A)
(14) Ensure correct tab order [12] (E)
(15) Make sure that the chatbot window is easily accessible and visible on the website [17] (E)
(16) Allow interaction without login (users may not be able to access login) [28] (E)

B. Manage conversation end and transitions to external resources

(1) Offer longer interaction timeouts [14] (E)
(2) Prepare resources that can be easily converted into an accessible form, like electronic format [18] (E)
(3) Use slow or adaptable time-outs for the chatbot re-prompting for input or ending the conversation [10] (R)
(4) Carry accessibility settings forward to linked resources [10] (R)
(5) Use only linked or embedded content that is accessible by AT [10] (R)
(6) Ensure users have the option of interacting with a human customer service representative (not solely a chatbot) [17] (E)

C. **Support AT**

(1) Consider conventions of transmitting to hearing devices before playing sounds [21] (R)
(2) Use responsive web design techniques [22] (A)
(3) Support users' AT of choice [10] (R)
(4) Support use of touch screens [10] (R)
(5) Linked or embedded content should be accessible by AT [10] (R)

(6) Make sure the chat window width supports magnification [26] (R)

D. **Manage Interruptions and Notifications**

(1) Offer whisper or discreet mode for audio content [14] (E)
(2) Use both text and audio for all content and notifications [19] (R)
(3) Ensure messages are announced with ARIA-live [22] (A)
(4) Employ subtle notifications instead of intrusive and demanding [10] (R)
(5) Minimize distractions [10] (R)
(6) Use ARIA-live to notify of new chat content [26] (A)
(7) Make a sound when sending and receiving messages [26] (R)
(8) Change the window title when receiving a message [26] (A)

**Developer process and training.** Recommendations for chatbot developer process and training suggest activities throughout the ideation, design, and development process which may lead to a more accessible chatbot.

A. **Include people with disabilities in the process**

(1) Design bots to respect full range of human ability [25] (R)
(2) Maintain an open dialogue with representatives of target user groups to involve them in development [12] (E)
(3) Introduce diverse personas to developers including ability, experience, and demographics [27] (R)
(4) Include diverse users in design process [28] (E)

B. **Follow WCAG**

(1) Follow WCAG 2.0 [10] (R)
(2) Follow WCAG accessibility guidelines [25] (R)
(3) Ensure at least one developer is familiar with WCAG requirements [17] (R)

C. **Choose an accessible platform and plan on customizing it**

(1) Consider technical requirements and accessibility when selecting a platform [18] (E)
(2) Plan on customizing a chatbot platform [20] (R)
(3) Choose a customizable platform [22] (A)
(4) Request accessibility certifications from vendors if purchasing a chatbot platform [17] (R)

**Testing.** Recommendations for chatbot testing pertain to activities evaluating the chatbot accessibility, usability, or efficacy.

A. **Test early and continuously**

(1) Test early and extensively with members of the target population (or at least the tools the users will use) [18] (E)
(2) Conduct user testing continuously [12] (E)
(3) Conduct user tests with people with different disabilities, preferably early and continuously throughout the development process [17] (R)

B. Test with realistic technology scenarios

(1) Test early and extensively with members of the target population (or at least the tools the users will use) [18] (E)
(2) Test with different browsers and versions of AT [22] (A)

C. Use validated, standardized testing methodologies

(1) Use validated, standardized methodologies to measure satisfaction, acceptability, and safety (managing risks in use) [23] (R)
(2) Lack of complaint is not a sufficient measure to assess chatbot adoption or safety [23] (R)
(3) Measure satisfaction and acceptability separately [23] (R)
(4) Validate accessibility with WCAG 2.0 [12] (E)
(5) Test with real users using a cognitive task analysis [28] (E)

D. **Test with diverse users and with target users**

(1) Test early and extensively with members of the target population (or at least the tools the users will use) [18] (E)
(2) Perform user testing [16] (E)
(3) Involve target users in testing [25] (R)
(4) Conduct user tests with people with different disabilities, preferably early and continuously throughout the development process [17] (R)
(5) Test with real users using a cognitive task analysis [28] (E)

## References

1. Reddy T (2017) How chatbots can help reduce customer service costs by 30%. IBM Watson Blog. https://www.ibm.com/blogs/watson/2017/10/how-chatbots-reduce-customer-service-costs-by-30-percent/. Accessed 22 December 2020
2. Sweezey M (2018) Council post: the value of chatbots for today's consumers. Forbes Communications Council. https://www.forbes.com/sites/forbescommunicationscouncil/2018/02/13/the-value-of-chatbots-for-todays-consumers. Accessed 22 December 2020
3. Domino's: Order with Dom, Domino's interactive chatbot (2021) https://www.dominos.com/chat-pizza-order/. Accessed 22 December 2020
4. Brereton E (2019) Universities use AI chatbots to improve student services. EdTech Magazine. https://edtechmagazine.com/higher/article/2019/10/universities-use-ai-chatbots-improve-student-services. Accessed 22 December 2020

5. Canada Revenue Agency: What's new for the 2020 tax-filing season. CRA Newsroom (2020) https://www.canada.ca/en/revenue-agency/news/newsroom/tax-tips/tax-tips-2020/tax-filing-season.html. Accessed 22 December 2020
6. U.S. Department of Veterans Affairs: VA coronavirus chatbot (2020) https://www.va.gov/coronavirus-chatbot/. Accessed 22 December 2020
7. Beaunoyer E, Dupéré S, Guitton MJ (2020) COVID-19 and digital inequalities: reciprocal impacts and mitigation strategies. Comput Human Behav. 111. doi:https://doi.org/10.1016/j.chb.2020.106424
8. Guitton MJ (2020) Cyberpsychology research and COVID-19. Comput Hum Behav 111. doi:https://doi.org/10.1016/j.chb.2020.106357
9. Calado C (2017) We need to talk about accessibility on chatbots. UX Design. https://uxdesign.cc/we-need-to-talk-about-accessibility-on-chatbots-98cf93c54963. Accessed 17 December 2020
10. Lister K, Coughlan T, Iniesto F, Freear N, Devine P (2020) Accessible conversational user interfaces: considerations for design. In: Proceedings of the 17th international web for all conference (W4A '20), vol 111, Taipei, Taiwan, pp 1–11, April 2020. Association for Computing Machinery, New York, NY, USA. doi:https://doi.org/10.1145/3371300.3383343
11. W3C: Web Content Accessibility Guidelines (WCAG) 2.2: W3C Working Draft 11 August 2020 (2020) https://www.w3.org/TR/2020/WD-WCAG22-20200811/. Accessed 22 December 2020
12. PTS: Utvärdering av tillgängligheten i chattbotar för personer med funktionsnedsättningar och äldre [Evaluation of the accessibility of chatbots for people with disabilities and the elderly]. Swedish Post and Telecom Agency (PTS) (2017) https://www.pts.se/globalassets/startpage/dokument/icke-legala-dokument/rapporter/2018/ovrigt/chattbotstudie/huvudrapport/utvardering-av-tillganglighet-i-chattbotar-huvudrapport.pdf. Accessed 22 December 2020
13. U.S. General Services Administration: IT Accessibility/Section 508 (2020) https://www.gsa.gov/policy-regulations/policy/information-integrity-and-access/it-accessibility-section-508. Accessed 22 December 2020
14. Abdolrahmani A, Kuber R, Branham SM (2018) Siri Talks at You: an empirical investigation of voice-activated personal assistant (VAPA) usage by individuals who are blind. In: Proceedings of the 20th international ACM SIGACCESS conference on computers and accessibility, Galway, Ireland, October 2018. Association for Computing Machinery, New York, NY, USA, pp 249–258. doi:https://doi.org/10.1145/3234695.3236344
15. Baldauf M, Bösch R, Frei C, Hautle F, Jenny M (2018) Exploring requirements and opportunities of conversational user interfaces for the cognitively impaired. In: Proceedings of the 20th international conference on human-computer interaction with mobile devices and services adjunct, Barcelona, Spain, September 2018. Association for Computing Machinery, New York, NY, USA, pp 119–126. doi:https://doi.org/10.1145/3236112.3236128
16. Lilleby A, Marstein S (2020) «Chatbot for alle?»-En kvalitativ studie av dyslektikeres opplevelse med chatbot [Chatbots for all? A qualitative study of dyslexic users' experience with chatbos]. Master's thesis, University of Oslo, Norway. https://www.duo.uio.no/bitstream/handle/10852/79751/1/Master_Lilleby_Marstein_2020.pdf
17. Simonsen K (2020) Universell utforming av chatbots [Universal design of chatbots]. Funka, Stockholm. https://www.funka.com/no/forskning-og-innovasjon/arkiv---forskningsoppdrag/universell-utforming-av-chatbots/chatbottar-for-alle/. Accessed 22 December 2020
18. Bigham JP, Aller MB, Brudvik JT, Leung JO, Yazzolino LA, Ladner RE (2008) Inspiring blind high school students to pursue computer science with instant messaging chatbots. SIGCSE Bull 40:449–453. https://doi.org/10.1145/1352322.1352287
19. Boshoff A (2018) Accessibility and chatbots: how to make your chatbot user-friendly for everyone. Rogerwilco. https://www.rogerwilco.co.za/blog/accessibility-and-chatbots-how-make-your-chatbot-user-friendly-everyone. Accessed 17 December 2020
20. Bureau of internet accessibility: five key accessibility considerations for chatbots (2020) https://www.boia.org/blog/five-key-accessibility-considerations-for-chatbots. Accessed 17 December 2020

21. Campbell B (2017) Tips for designing accessibility in voice user interfaces. UX design. https://uxdesign.cc/tips-for-accessibility-in-conversational-interfaces-8e11c58b31f6. Accessed 17 December 2020
22. CANAXESS: accessible chatbot design (2020) https://www.canaxess.com.au/infocard/chatbots/. Accessed 17 December 2020
23. De Filippis ML, Federici S, Mele ML, Borsci S, Bracalenti M, Gaudino G, Cocco A, Amendola M, Simonetti E (2020) Preliminary results of a systematic review: quality assessment of conversational agents (Chatbots) for people with disabilities or special needs. In: Miesenberger K, Manduchi R, Covarrubias Rodriguez M, Peňáz P (eds) Computers helping people with special needs, Springer International Publishing, Cham, pp 250–257. doi:https://doi.org/10.1007/978-3-030-58796-3_30
24. Furini M, Mirri S, Montangero M, Prandi C (2020) Do conversational interfaces kill web accessibility? In: 2020 IEEE 17th annual consumer communications networking conference (CCNC), Las Vegas, Nevada, 1–13 January 2020, pp. 1–6. doi:https://doi.org/10.1109/CCNC46108.2020.9045477
25. Microsoft Research: Responsible bots: 10 guidelines for developers of conversational AI (2018) https://www.microsoft.com/en-us/research/publication/responsible-bots/. Accessed 17 December 2020
26. Orange: Designing an accessible chatbot. https://a11y-guidelines.orange.com/en/web/components-examples/chatbot/ Accessed 17 December 2020
27. Restrepo EGY, Baldassarre M, Boticario JG (2019) Accessibility, biases and ethics in chatbots and intelligent agents for education. In: Proceedings of EDULEARN19 conference, Palma, Mallorca, pp 8824–8833
28. State C (2020) Designing conversational AI for digital healthcare in regards to accessibility. Master's thesis, KTH Royal Institute of Technology, Stockholm, Sweden
29. Torres C, Franklin W, Martins L (2019) Accessibility in chatbots: the state of the art in favor of users with visual impairment. In: Ahram T, Falcão C (eds) Advances in usability, user experience and assistive technology. AHFE 2018. Advances in intelligent systems and computing, vol 794. Springer, Cham. doi:https://doi.org/10.1007/978-3-319-94947-5_6
30. Google Trends: "chatbot." (2020) https://trends.google.com/trends/explore?date=all&q=chatbot. Accessed 22 December 2020
31. Melnyk V, Ashok V, Puzis Y, Borodin Y, Soviak A, Ramakrishnan IV (2015) Look Ma, No ARIA: generic accessible interfaces for web widgets. In: Proceedings of the 12th web for all conference, 18 May 2015, pp 1–4. doi:https://doi.org/10.1145/2745555.2746666
32. W3C Accessibility Guidelines Working Group: Accessibility Guidelines Working Group Teleconference (2019) https://www.w3.org/2019/11/05-ag-minutes.html. Accessed 22 December 2020
33. W3C: Understanding Success Criterion 3.2.6: Findable Help (2020) https://w3c.github.io/wcag/understanding/findable-help.html. Accessed 22 December 2020
34. Stevens AC, Carroll DD, Courtney-Long EA et al (2016) Adults with one or more functional disabilities — United States, 2011–2014. MMWR Morb Mortal Wkly Rep 2016(65):1021–1025. https://doi.org/10.15585/mmwr.mm6538a1
35. Bickmore TW, Picard RW (2005) Establishing and maintaining long-term human-computer relationships. ACM Trans Comput Hum Interact 12:293–327. doi:https://doi.org/10.1145/1067860.1067867
36. Brave S, Nass C, Hutchinson K (2005) Computers that care: investigating the effects of orientation of emotion exhibited by an embodied computer agent. Int J Hum Comput Stud 62:161–178. https://doi.org/10.1016/j.ijhcs.2004.11.002
37. Looije R, Neerincx MA, Cnossen F (2010) Persuasive robotic assistant for health self-management of older adults: design and evaluation of social behaviors. Int J Hum Comput Stud 68:386–397. https://doi.org/10.1016/j.ijhcs.2009.08.007
38. Taylor RM (2017) Situational awareness rating technique (Sart): the development of a tool for aircrew systems design. In: Situational awareness, Routledge, pp 111–128. doi:https://doi.org/10.4324/9781315087924-8

39. Caroux L, Consel C, Merciol M et al (2020) Acceptability of notifications delivered to older adults by technology-based assisted living services. Univ Access Inf Soc 19:675–683. https://doi.org/10.1007/s10209-019-00665-y
40. Talby D, Keren A, Hazzan O, Dubinsky Y (2006) Agile software testing in a large-scale project. IEEE Softw 23(4):30–37. https://doi.org/10.1109/MS.2006.93
41. Usability.gov: User-Centered Design Process Map (2020) https://www.usability.gov/how-to-and-tools/resources/ucd-map.html. Accessed 30 December 2020
42. Dam RF, Siang TY (2020) 5 stages in the design thinking process. Interaction Design Foundation. https://www.interaction-design.org/literature/article/5-stages-in-the-design-thinking-process. Accessed 30 December 2020
43. Finstad K (2010) The usability metric for user experience. Interact Comput 22(5):323–327. https://doi.org/10.1016/j.intcom.2010.04.004

# Event or Emergency Case Detection by Human Running

**Mohamed Artan Abdi and Metin Turan**

**Abstract** Recognizing the difference between walking and running is certainly an essential part of standard sports activity classification. Moreover, that can be commented on a security problem in daily life. If someone is running on the street that may show something goes wrong. The purpose of this study is to recognize the everyday actions of the human, whether walking as regular movement or running in the case of emergency or event. In recent years, significant progress has been made in computer vision and machine learning. CNN, a deep learning algorithm for image processing was used for the model. The dataset, a thousand of images, of the study were collected from different sites of the internet or extracted from videos. Classify frequent human movements, whether a regular walk or running action, were separated by 86.85% success in the research.

**Keywords** Road security system · Human activities · Deep learning · CNN

## 1 Introduction

Every common actions of human being, no matter whether it is walking, running, jumping or joking depends on a reason. On the other hand, every single person of the community uses his/her body in order to complete activity, for example, walk or run requires movements of legs and hands. Human brain can predict what this person does. We can guess with a confidence result of this person is running, or walking. After technology came up to routing our life and also machines made more benefits to our modern society, a question came up to researcher's mind. Can a machine recognize human's action like human brain?

Advanced technology machines have been enabling to understand human actions. Many researchers are coming up with different ideas to build a system that can

M. A. Abdi (✉) · M. Turan
Computer Engineering Department, Istanbul Commerce University, Istanbul, Turkey

M. Turan
e-mail: mturan@ticaret.edu.tr



X.-S. Yang et al. (eds.), *Proceedings of Sixth International Congress on Information and Communication Technology*, Lecture Notes in Networks and Systems 216,
https://doi.org/10.1007/978-981-16-1781-2_81

recognize a human's action and understand their intension. Suppose there is a person who escapes from the police officer, he is running inside the crowded. Security surveillance cameras can capture and recognize his action comparing with the regular actions. Such machine can work more intelligent then human and it can reduce, the number of persons, time to recognize, and upcoming risk like this case. It would be a great beneficial to our environmental safety [1].

There are two major components in actions recognition: action classification and action representation.

- Action classification generates the labels.
- Action representation converts into features.

In this manner, our interest is to study methods of automated labeling of a sequence of motion capture data as a further means of facilitating their reuse in long term productions. At this point, classification of data requires analysis of individual candidates through the extraction of a sequence of attributes from them in order to assign a particular class in a collection [2].

Our ultimate point is to find such system that can capture a human movement, and then can make a prediction of what this action is. It is a complicate challenge to tune such a statistical model about human movement, to face this challenge, we require a dataset, appropriate algorithm, and tools to produce such system.

**Dataset**

In this research, the dataset which we use consists of two types: plural images and singular images. They include the following activities and all were collected from the different web sites on the internet.

- Walking
- Running
- Other actions

singular image was collected from different environments to face the challenge of every single image and also to make sure it can deal with one by one.
plural images are more than two images in one group had same action or same person, one action can be capturing from various viewpoints, example front side, angles, back side or even on top.

**Algorithm**

Convolutional neural network is a powerful model in deep learning, which applies images to biases, various aspects, and be able to recognize one from the other, as an input image or video. Nowadays CNN is become more popular that basis a lot of modern machine learning applications.

**Experimental Tools**

This system is developed using Python libraries, Keras, and TensorFlow. The primary reason for this selection was owing to the ease availed by this equipment, which made it increasingly passible for fast prototyping.

The traditional challenges of computer vision and image processing is to determine whether or not the image comprises of particular object, feature or activity. Recognition can be handled in three categories [3].

The first category called object recognition, otherwise known as object classification, which is one or multiple pre-specified objects that could be recognized normally with their 2D positions within an image or 3D poses identified in a scene [4].

The second category is known as identification, whereby a specific example of an object is recognized. It includes identification of person facial features, fingerprints, handwritten digits, or a specific vehicle [5].

The third category is known as detection it involves scanning for an individual condition with examples detecting likely abnormality in cells or tissues within medical imagery equipment or detecting vehicles through automated road toll systems [6].

## 2 Literature Review

Human movement recognition and action classification problem has been worked by many researchers before and recent important researches on the topic were reviewed separately. In this section, we looked a similarity of these works and the different technics they applied and we compare against one to another evaluating their experimental.

According to research [1], the interest of CNN in image classification is to study methods of automated labeling of a sequence of motion capture data as a further means of facilitating their reuse in long term productions.

The walking/running image categorization technique proposed by Julieta and her colleagues seeks to bring about an improvement in the accuracy of categorizing the image files through the process of combining the information concerning the relationships [2].

In research [3] which shows that the learning of statistical models concerning human movement is a challenging activity due to the increased measurement, nonlinear forces at work as well as the stochastic outline of humanoid walking or running.

Most researches prioritized focusing on natural activities classification [4, 5] and [6] surveyed the recent methods, there studies and datasets committed on regular common actions.

Furthermore, research [7] states that classification of data entails analysis of individual candidates through the extraction of a sequence of attributes from attempting to equal them to a particular class in a collection of the closed session (a researcher) which is also supported by research undertaken by [1].

Chen and his colleagues have demonstrated the effectiveness of the proposed tube convolutional neural network for action detection, they evaluate T-CNN on three trimmed video datasets including UCF-Sports [8].

Research [9] is more different than the former researches. Rui and his colleagues proposed a new approach that can recognize the action while it is streaming. In reality, it is more complicated work to recognize action during the flow. They recognize the action from multiple capturing videos. It is clearer to understand the action and it's easier to detect action. This approach is similar to plural action detecting.

On research [10], Yuqing and Yang proposed an approach to the spatiotemporal localization (detection) and classification of multiple concurrent actions within temporally untrimmed videos through localizing and scoring actions from color images.

As early mentioned, many researchers had studied on human activity recognition, those approaches were designed to recognize, human body types, there is also previous works design to recognize actions of the human, such as research by [9] it can recognize actions before it capturing, on other hand [8] and [11] which may take the decisions as well as we did. However, even though they obtained successful results on both recognizing human movements after it happened or while it was happening in live, as far as we know they didn't make classification of walking and running which is difficult to be categorized because of being similar actions.

Our approach is to detect a person if he/she is walking in regular way or he/she is running in emergency case. Recognizing two different actions and categorizing them from the common human movement is more challenging than recognizing only one action movement.

## 3 Methodology

In recent years, CNN has been growing up faster than other approaches on deep learning, it has obtained successful results on classification of text and images. Convolutional Neural Network (CNN) is a specialized form of the neural networks developed to sequence challenges, where old fashion artificial neural network is working like feed-forward. On other hand, we are dealing with two thousand frames of dataset, CNN can process large capacity data than traditional ANNs [12].

Since the work is to classifying an image from the frame to running or walking categories, CNN is one of the main approaches to develop such challenge. CNN classify such challenge by design, as all propagated information from the system should first pass through input, processing, and output. Such activated functions aims to function on the data so only to permit significant evidences to sustain propagation in the course of training [5].

*Implementation of the CNN model*

Project was implemented in Python with the use of Keras and Tensorflow libraries. The purpose of project is image classification and CNN best feature extraction sequence makes its implementation better. The implementation model with parameters is shown on bellow Fig. 1.

The experiments have shown that a deeper system can perform better as it predicts unclear images [13]. Because of the actions have so similar attitudes, it is wise to use convolutional layers repeatedly in order to extract enough features to separate the actions from images successfully. It is seen on the schema given in Figure.

*Data Collection.*

Every single image collected from the internet or extracted from videos, are preprocessed and divided into classes of similar size. The details of the dataset are given in Table 1. The technique has the benefit of classifying individual files in terms of the likelihood of the data.

At this point, we can take into consideration of an image where an image shows a human character start running as he prepares to run. Now imagine how the researcher would then search for either walk or run cycles such that the system would be able to realize that the specific file would be a satisfactory response in either instance [2].

Even as the practical improvement of the human brain shows to be relative to the system, it makes us confident that at least this is a hint of the model logical. Also,

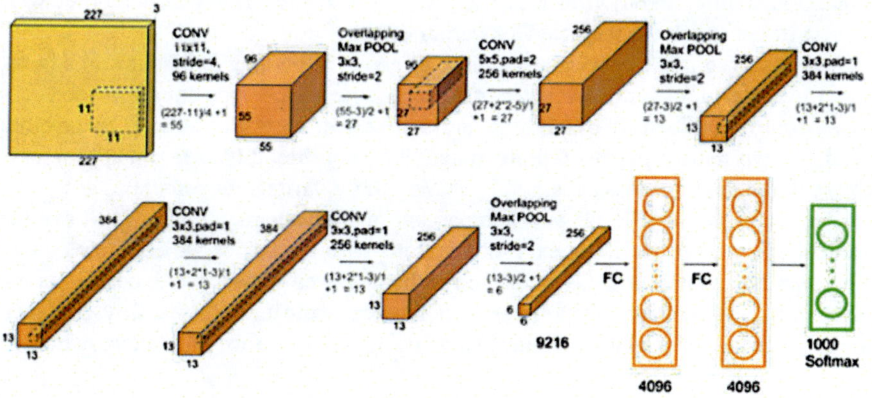


**Fig. 1** CNN parameter structure of implemented model

**Table 1** *Resource* details of the dataset

| Dataset type | From the internet | Extract from video | Total dataset |
|---|---|---|---|
| Walking | 720 | 280 | 1000 |
| Running | 670 | 330 | 1000 |

**Table 2** Proportion details of data for train and test

| Partition | Training (%) | Testing (%) |
|---|---|---|
| Running | 70 | 30 |
| Walking | 70 | 30 |

with bigger datasets, the systems should have a better performance concerning the accuracy classification and data, thereby solving the under-fitting challenge.

The data consists of two types of images: unique and duplicated images. Around 50 percentages of total data are unique images (different characters), they are different scenes. Approximately 50 percentages of total data are duplicated images, where same scene is duplicated in various angles and sides. The proportion of data used in train and test phases are given at Table 2.

## 4 Result

Multiple experimental tests are undertaken with the tool providing coverage of varied network architectures. this is an experimental overload to the most machine learning problem to get a better assessment of the models and architectures over data set.

**Experiments**

Table 3 is a summarization of a careful evaluation of experimental outcomes with respect to the epoch for the best CNN architecture.

It is observed that a particular increase of epoch more than 3000 showed a lousy result. As a result of limited images (data set), more epoch does not affect the accuracy. Moreover, it leads to overfitting, and drops down the test accuracy. On the other hand, it is worth taking note of, these images have considerably different appearance, and still, the network was capable of catching those various movements.

The chart below (Fig. 2) is a comparison of the accuracy, which is to say the vertical axis gained from the experimentations with the use of varied sample sizes along the horizontal axis. Overall, the results achieved with these experiments are taken into consideration as a promise which indicates the model's viability, as shown in the stacked Convolutional Neural Network that could learn with success how to

**Table 3** Experimental results

| Epochs | Accuracy | Test Accuracy |
|---|---|---|
| 1000 | 0.7833 | 0.7300 |
| 1500 | 0.8439 | 0.8300 |
| 2000 | 0.8521 | 0.8479 |
| 2500 | 0.8585 | 0.8553 |
| 3000 | 0.8632 | 0.8610 |

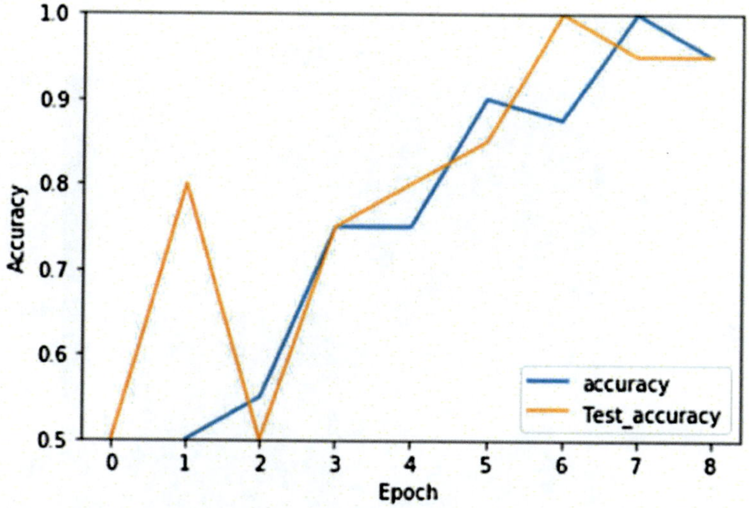


**Fig. 2** Accuracy chart for the outcome

classify the actions of humans from images downloaded from the internet or extracted from video [4].

The Table 4 is a representation of the confusion matrix gained by taking as an instance of the best outcome. When the matrix data evaluated, approximately 87% of the test data labeled correctly to classes walking and running.

The other important metrics for evaluation the model is precision (1), recall (2), and F1 (3). Evaluation is given in the Table 5 for both classes separately.

$$\text{Precision} = \frac{\text{TruePositive}}{\text{TruePositive} + \text{FalsePositive}} \tag{1}$$

$$\text{Recall} = \frac{\text{TruePositive}}{\text{TruePositive} + \text{FalsePositive}} \tag{2}$$

$$\text{F1} = \frac{2 \times \text{Precision} \times \text{Recall}}{\text{Precision} + \text{Recall}} \tag{3}$$

## 5 Conclusion

In this research, the study seeks to bring about an improvement in the accuracy of classifying the image files through the process of combining the actions concerning the movement. These are inclusive of the training and testing comprised within the

**Table 4** Confusion matrix is given for the best result

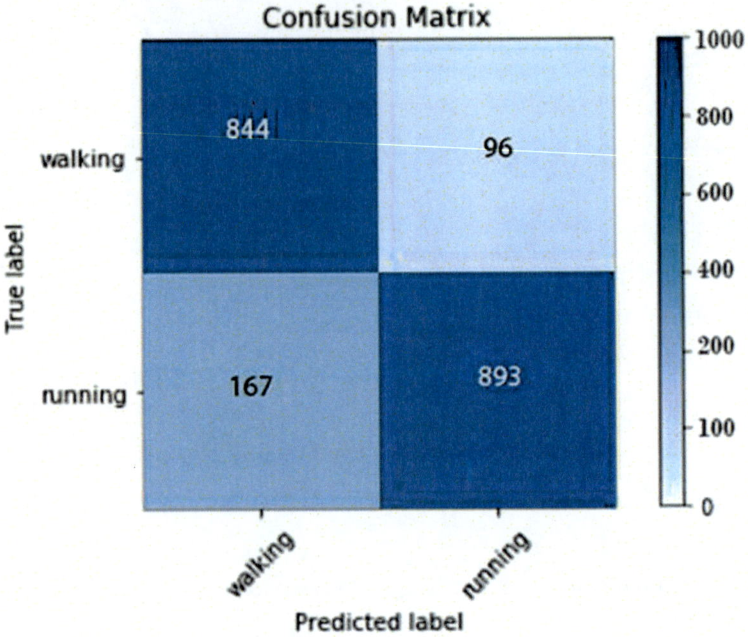


**Table 5** Comparison of literature

| Referenced article number | Accuracy rate (%) | Year | Dataset |
|---|---|---|---|
| 10 | 92.7 | 2017 | Own dataset |
| 11 | 93.8 | 2015 | Samples |
| 12, 14 | 87.9 | 2019 | Own dataset |
| 13 | 97.0 | 2018 | Collection |

images. This study is based on a design of a system reliant on deep learning more particularly a CNN.

Different articles in different ideas, and all have high accuracy given at Table 5. It seems that the accuracy obtained in this research can be considered as successful under so similar activity separation and limited data set.

Our results indicate that the performance is well structured. Although current research uses limited resources, the trained model seems successful to separate movements moderately that substantially differ from each other. At this time, we can hypothesize that this is due to the nature of those particular movements that substantially differ from each other, which makes it modest for the network to differentiate them from each other.

In a future, it is expected that training will be improved by the increasing number of epochs as well as extending the scope and the capacity of the project by adding other actions, such as taking into consideration active postures of the body, the head of running people marked, and processed for emotion analysis that could be of wonderful.

## References

1. Yu K, Yun F (2018) Human action recognition and prediction: a survey. J Latex Cl Files 13(9)
2. Martinez J, Black MJ, Romero J (2017) On human motion prediction using recurrent neural networks. In: Proceedings of the IEEE conference on computer vision and pattern recognition 2017 (pp 2891–2900)
3. Affonso C, Rossi ALD, Vieira FHA, de Leon Ferreira ACP (2017) Deep learning for biological-image classification. Expert Syst Appl 85:114–122
4. Bilal A, Jourabloo A, Ye M, Liu X, Ren L (2017) Do convolutional neural networks learn class hierarchy? IEEE Trans Vis Comput Graph 24(1):152–162
5. Heba E-FX, Nima S, Kathryn K, Min W (2018) Convolution neural networks for person identification and verification using steady state visual evoked potential. In: IEEE international conference on systems, man, and cybernetics. https://doi.org/10.1109/SMC
6. Peng Y, Yin H (2017) Markov random field-based convolutional neural networks for image classification. In: International conference on intelligent data engineering and automated learning (pp 387–396)
7. Lapuschkin S, Binder A, Muller KR, Samek W (2017) Understanding and comparing deep neural networks for age and gender classification. In: Proceedings of the IEEE international conference on computer vision workshops 2017, (pp 1629–1638)
8. Ding C, Li Y, Xia Y, Wei W, Zhang L, Zhang Y (2017) Convolutional neural networks based hyperspectral image classification method with adaptive kernels. Remote Sens 9(6):618
9. Hou R, Chen C, Shah M (2017)Tube convolutional neural network (T-CNN) for action detection in videos. In: IEEE international conference on computer vision (ICCV), Venice, pp 5823–5832. https://doi.org/10.1109/ICCV.2017.620.
10. Chen Y, Xue Y (2015) A deep learning approach to human activity recognition based on single accelerometer, IEEE. 978-1-4799-8697-2/15
11. Lee S-M, Yoon SM, Cho H (2017) Human activity recognition from accelerometer data using convolutional neural network, IEEE. 978-1-5090-3015-6/17
12. Sakib S, Ahmed N, Kabir AJ, Ahmed H An overview of convolutional neural network: its architecture and applications. https://doi.org/10.20944/preprints201811.0546.v1
13. Albawi S, Mohammed TA, Al-Zawi S (2017) Understanding of a convolutional neural network. In: 2017 international conference on engineering and technology (ICET), Antalya, 2017, pp 1–6. https://doi.org/10.1109/ICEngTechnol.2017.8308186.
14. Huang J, Lin S, Wang N, Dai G, Xie Y, Zhou J (2020) TSE-CNN: a two-stage end-to-end CNN for human activity recognition. IEEE J Biomed Health Inform 24(1):292–299. https://doi.org/10.1109/JBHI.2019.2909688
15. El-Fiqi H, Wang M, Salimi N, Kasmarik K, Barlow M, Abbass H (2018) Convolution neural networks for person identification and verification using steady state visual evoked potential. In: 2018 IEEE international conference on systems, man, and cybernetics (SMC), Miyazaki, Japan, pp 1062–1069. https://doi.org/10.1109/SMC.2018.00188

# Towards Iris-Based Authentication for Smart Devices in the Cloud

**Mustafa A. Al Sibahee, Songfeng Lu, Zaid Ameen Abduljabbar, Ke Xu, Enas Wahab Abood, Ahmed Abdulelah Jasim Al-Ashoor, Samet Yilmaz, and Erasmus(Xin) Liu**

**Abstract** Cloud computing is a promising development in the field of information technology. However, it faces many issues, particularly in the area of security. Authentication and integrity is a critical issue in the field of data security and many interesting issues have been raised to realize or ensure any modification with data exchange between a pair of smart devices within a cloud system. Exciting methods in such a field can be powerless against well-known modification and malicious attacks. Therefore, a powerful method is needed to prevent tampering of data for E2E smart devices transmission in the cloud. In this work, we present a robust and one-time biometric message authentication code based on combining biometric feature extracted from the user's iris and double steganography-based discrete wavelet transformation steganography. The result of the combination is that it preserves the user's message integrity, ensures authentication, and prevents known attacks. This approach enjoys several security merits, such as phase key agreement, biometric key management, a

M. A. Al Sibahee (✉) · S. Lu · K. Xu
Shenzhen Institute of Huazhong University of Science and Technology, Shenzhen, China
e-mail: mustafa.a@hust.edu.cn

S. Lu
e-mail: lusongfeng@hust.edu.cn; I201522098@alumni.hust.edu.cn

M. A. Al Sibahee
Iraq University College, Basrah, Iraq

S. Lu · Z. A. Abduljabbar
Hubei Engineering Research Center on Big Data Security, School of Cyber Science and Engineering, Huazhong University of Science and Technology, Wuhan, China
e-mail: alsulamizaid@gmail.com

Z. A. Abduljabbar · E. W. Abood · A. A. J. Al-Ashoor
University of Basrah, Basrah, Iraq

Z. A. Abduljabbar · E. Liu
Neusoft Institute Guangdong, Guangdong, China
e-mail: liuxin@nuit.edu.cn

S. Yilmaz
Department of Computer Science, Hunan University, Changsha, China
e-mail: sametyil@gmail.com

 
X.-S. Yang et al. (eds.), *Proceedings of Sixth International Congress on Information and Communication Technology*, Lecture Notes in Networks and Systems 216,
https://doi.org/10.1007/978-981-16-1781-2_82

one-time biometric key and message anonymity for each specific session, one-time message authentication code for each specific user's session, double steganography, and mutual authentication. Moreover, the efficiency and security of this method have been proved through security analysis and experimental results.

**Keywords** Cloud computing · One-time biometric key · One-time biometric MAC · Smart devices

## 1 Introduction

In the past decades, several data types with huge sizes can be distributed and transmitted via the Internet, due to the qualitative leap in information digitization technologies, such as cloud computing and the number of smart device users has also increased significantly [1]. In addition to sound, image, and video, texts are among the most widely used means to transfer data via smart devices [2]. Unfortunately, smart devices face the issue of limited resources [3]. Cloud computing is the most promising next-generation technology to enable smart device users to easily access large resources available on-demand that such devices lack [1, 3]. However, cloud computing and smart devices face many challenges, and security is the most important one [1, 3]. Therefore, the widespread availability of these devices and their applications depend on the existence of efficient security mechanisms. Due to the large number of messages being sent between smart devices within the cloud computing system, a safe method must be provided to prevent tampering with the contents of messages and verify their authentication. However, several solutions have been proposed that provide authentication and integrity of transmission messages as a pressing issue, and many achievements have been made in this aspect in the last two decades [4–9]. A common way of preventing the manipulation of messages during transmission between two endpoints exist, which is a cryptography of one-way hash functions [4–9].

Some drawbacks related to Message Authentication Code (MAC) research are discussed in detail in the Related Works Section. By studying the existing proposed solutions, it became evident that the main disadvantage of MAC may be, in fact, that it does not meet the level of security required in unsecured communications when used alone for E2E smart devices in the cloud.

The above problems can be overcome by combining more powerful and assurance factors with MAC. Thus, in our work, we proposed a safe, effective, and lightweight method that prevents data of the transmitted message from being tampered with for E2E smart devices in the cloud. The proposed method works by merging three factors, that is, by combining the irises of the smart device sender (*SDS*) and the smart device receiver (*SDR*) to generate a shared biometric image to which a 2D Gabor filter is applied to extract the biometric features [10–12], crypto-hash function SHA-1 [13], and double steganography [14]. These factors are used together to protect the user's message from being modified. The biometric features are used to generate a one-time biometric key for each message delivery session that combines with MAC to generate one-time biometric MAC. Thereafter, the one-time biometric hashed value is hidden in a cover image using double steganography. The authentication scheme can meet the requirements of the E2E secure message transmission [15, 16].

The contributions of this work to the cloud environment and smart devices in general, and to the security of message communication, message authentication and integrity, in particular, are as follows: (1) in the proposed model, all the drawbacks of the previous methods are reviewed and taken into account to establish a secure, lightweight authentication, and integrity system for the messages sent through E2E smart devices in the cloud. Practically, this work involves the powerful biometric features based on irises of both parties, keyed hash function, and double steganography to maintain message integrity and authentication. (2) Smart device users and cloud service provider (CSP) can launch secure authenticated phase keys. (3) Biometric shared information of a pair of smart devices users can provide responsive mutual authentication. (4) It is computationally efficient and can combat many attacks, such as MITM, reply, forgery, parallel session, insider, and reflection attacks.

The structure of this manuscript is as follows. Section 2 describes in-depth the most significant and widespread text authentication solutions. In Sect. 3, we present the mechanism of biometric MAC in terms of initialization, verification, and authentication phases. In Sect. 4, we show how our scheme can combat the well-known attacks. In Sect. 5, the performance of the authentication scheme is proven. Finally, Sect. 6 concludes this work.

## 2 Related Works

Various authors have proposed different MAC schemes to ensure authentication and integrity of the transmitted messages.

In 2006, [4], Jin et al. presented an efficient One-Key Carter-Wegman Message Authentication Code called a One-key Galois Message Authentication Code. This scheme uses a one-time key and a universal hash function, instead of two keys. Our scheme is made more robust by embedding the iris features with a cryptographic one-way hash function along with double steganography to provide simplicity and security.

In 2008, the idea of integrating the function of a smart card with a one-way hash function was presented by Zi-ming et al. [5] who proposed creating an efficient user-to-user authentication scheme in a peer-to-peer environment. Despite using public key infrastructure, the authors first used a one-way hash function and then a smart card to provide strong security and minimal computational cost scheme. The drawback of a smart card is that it is a complex device and that a card a reader would be needed as an extra cost. It also requires an additional middleware application to create a match between a smart card and communication standards. Our work oversteps this drawback by using iris biometric features, where the user's iris data are taken once only and can be used for the following valid user logins. Moreover, the smart card could be stolen or lost, while in our scheme, no sender can guess or steal the iris of the receiver and vice versa, because we use a biometric shared image generated from the intersection between the sender's iris and receiver's iris.

In 2011, the concept of message anonymity was represented by Zhenxing et al. [6] who suggested a hash-based secure interface between two entities over the Internet, which integrated the MAC with the timestamp factor. This concept allows the hashed value to be changed once and every user's message to be used once. The weakness of this scheme is that the authors only briefly discussed the security analysis and the types of attackers that could be withstood were not explained.

In 2013, the concept of a biometric key was proposed by Al-Assam et al. [7] who suggested a scheme that combines steganography with biometric cryptosystems effectively to ensure robust remote mutual authentication between two parties and key exchange that facilitates one-time stego-keys. The aim behind this scheme was to hide the one-time biometric key through transfer over an insecure channel also to prevent replay attacks. The weakness of this approach is that the steganography alone is not sufficient to meet security requirements. Our scheme overcomes this issue by concealing the encrypted start point and the endpoint of the iris features vector and sending them between sender and receiver, instead of sending an explicit key between them, so that the steganography is more secure. Moreover, in our design, because the sender and receiver have shared information about their irises, which is a biometric shared image that offers mutual authentication and establishes a trusting relationship between them. Thus, the user will be able to tell whether the message is coming from an authenticated user.

In 2014, another idea for a one-time key was introduced by Castiglione et al. [8] who proposed a robust one-time authentication protocol, based on two cryptographically strong building blocks, an authenticated key exchange and a keyed hash message authentication code between two endpoints, which enables transparent mutual authentication between two endpoints. Moreover, Key Setup, Key Scheduling, and Key Update operations are accomplished independently by both endpoints. Therefore, this scheme suffers from drawbacks in the form of complexity in which more operations are required (Key Setup, Key Scheduling, and Key Update). In particular, our scheme presents a one-time authentication scheme by using a one-time key and random numbers, in which the MAC is valid only for one user's login and involves good coordination between simplicity and security. Moreover, our scheme is more robust, in that, it integrates an iris biometric key with a keyed-has function that provides robust message anonymity.

In 2018, Abduljabbar [9] introduced the idea of a one-time biometric MAC based on a one-time biometric key extracted from a manual signature. In contrast, we generated the one-time biometric key from the features of the iris, which is more powerful than the manual signature. In addition, both the one-time once biometric key and the biometric MAC have been hidden in a doubled manner based on DWT.

# 3 Proposed Scheme

The initialization, and, verification and authentication phases compose the mechanism of protection E2E message transmission for smart devices in the cloud. In the former phase, the shared biometric image is created, and the biometric shared key is generated and sent to the *SDS* and *SDR*. The verification and authentication phase will be performed when both parties want to exchange messages between them in the cloud while maintaining integrity and authentication.

## 3.1 Initialization Phase

The initialization phase performs the following steps:

- The private and public keys of the cloud service provider (*CSPpu, CSPpr*) are generated through the implementation of the ECC algorithm. These keys secure the process of sending the irises of the *SDS* and *SDR* ($IR_{SDS}$, $IR_{SDR}$) to the *CSP*. The *CSP* shares the public key $CSP_{pu}$ with the *SDS* and *SDR*. Both parties use this key to encrypt their irises ($IR_{SDS}$, $IR_{SDR}$) and send them securely to the *CSP*.
- Once the encrypted irises are received ($IR_{SDS}$, $IR_{SDR}$) by the *CSP*, the latter stores and decrypts both of them using its private key $CSP_{pr}$. Then, the biometric shared image is generated by applying the intersection process (*Bio-Sh* = $IR_{SDS} \bigcap IR_{SDR}$) to generate a biometric shared key *Bio-Shk* = *Fx*(*Bio-Sh*). The feature extraction process *Fx* is performed on the normalized iris data by employing a 2-D Gabor filter [10–12] to generate a biometric feature vector, and then selects a sub-vector that represent *Bio-Shk*.
- The *SDS* and *SDR* implement an ECC algorithm to generate their public and private keys ($SDS_{pu}$, $SDS_{pr}$, $SDR_{pu}$, $SDR_{pr}$). The *SDS* and *SDR* shares their public key with the *CSP*. The (*Bio-Sh, Bio-Shk*) is secured and sent to the *SDS* and *SDR* by encrypting them by ($SDS_{pu}$, $SDR_{pu}$). Upon receipt of encrypted (*Bio-Sh, Bio-Shk*) by both parties, it is decrypted using their private keys ($SDS_{pr}$, $SDR_{pr}$) to obtain the plaintext version. To complete the verification process, the sender uses the biometric shared image to extract biometric features and then generates a one-time anonymous biometric salt-key *Bio-Sk*. This process has an essential role in generating a one-time biometric MAC and will be explained in the following paragraphs.

  **REMARKS**: ECC is used because it is the most compatible with smart devices and the Internet of Things devices [17, 18] in that it is effective with devices have limited energy and memory, and does not require a high computational cost compared to RSA. Moreover, the ECC is applied only once during the initialization phase and its role is finalized.

### 3.2 Verification and Authentication Phase

The following steps describe the verification and authentication phase.

1. *SDS* Side: The following steps are performed to ensure the integrity and authentication of the message to be sent.

   - Given that the *Messg* is the *SDS*'s message and who wants to send it in a way that achieves integrity and authentication when it is received by the *SDR* side.
   - Use *Bio-Sh* and features extraction process (*Fx*) to generate once biometric salt-key $Bio\text{-}Sk = FX(Bio\text{-}Sh) \rightarrow St, Ed$. Hence, *FX* is a process to generate a vector of biometric feature extraction. *St* and *Ed* refer to the starting and ending points that are generated randomly every time for the specific session. Both points must be within the length of the biometric features vector (1024 bytes).
   - A large prime random number *R* is then generated to compute one-time message authentication code for each message delivery between the *SDS* and $SDR(Bio - Messg'{=}h(Messg\|Bio\text{-}Sk\|Bio\text{-}Shk\|R))$, thereby preventing communication attacks if the same message was sent by the *SDS* to *SDR*.
   - Protect the *St* and *Ed* by encrypting them in a light way as $St' = St \bigoplus Bio\text{-}Shk$ and $Ed' = Ed \bigoplus St'$.
   - Embed the $(Messg')$ into the cover image using a double steganography mechanism [15].
   - $St'Ed'$, and *R* can be embedded separately in the cover image as in the previous step.4
   - Send *Messg* and the cover image, which consists of ($Messg'$, $St'$, $Ed'$,and *R*), to *SDR*.

2. *SDR* Side: The following steps are performed to ensure the integrity and authenticity of the message received.

   - $Messg'$ can be extracted from the cover image separately by applying the reverse of the double steganography mechanism [14].
   - $St'$, $Ed'$ and *R* can be extracted from the cover image separately in the same way as in the step above.
   - Compute $St'' = St' \bigoplus Bio - Shk$ and $Ed' = Ed'' \bigoplus St''$.
   - Regenerate biometric salt-key $Bio - Sk' = FX(Bio - sh) \longrightarrow St'', Ed''$.
   - Recompute message authentication code $Messge''{=}\ h(Messg\|Bio - Sk'\ \|Bio - Shk\|R)$ if $Messge'$ matches $Messge''$, the *SDR* verifies the authentication, integrity and origin of the message sent by the *SDS*. Otherwise, it terminates the verification and authentication phase and sends a message to the *SDS* informing him that the sent data have been tampered.

## 4 Security Analysis

We prove that our work is designed well to prevent well-known attacks and enjoys several security merits.

**Theorem 1** *The authentication scheme employs users' iris to provide one-time and robust biometric message authentication code.*

***Proof*** The biometric operator identifies a person based on particular physiological features, such as his/her iris. The iris is one of the most commonly used security measures in biometric topics; it can prevent famous communication attacks. In the initialization stage, the *SDS* and *SDR* send their irises ($IR_{SDS}$ and $IR_{SDR}$) to *CSP* through a secure transmittal using applied ECC. Afterwards, *CSP* saves $IR_{SDS}$ and $IR_{SDR}$, generates a biometric shared image ($Bio - Sh = IR_{SDS} \bigcap IR_{SDR}$) and then sends it to the *SDS* and *SDR*. During the verification phase, a one-time biometric message authentication code $Bio - Messg' = h(Messg \| Bio - Sk \| Bio - Shk \| R)$ should be generated by the *SDS* or *SDR* based on a one-time biometric salt-key ($Bio - Sk$)$\in$ $FX(Bio - Sh) \longrightarrow St, Ed$. Such biometric MAC will be used when both smart devices want to send a message securely while maintaining authentication and integrity in the cloud system. Thus, this method provides a one-time and powerful biometric MAC.

**Theorem 2** *The authentication scheme can combat forgery and parallel session attacks.*

***Proof*** The adversary may attempt to impersonate an authorized smart device user by creating a legal login message by accessing a valid session of message transmission. In the proposed scheme, if an attacker attempts to impersonate, then the attacker can access a valid specific session of message delivery service ($Messg, Messg', St', Ed'$) by taking advantage of the sensitive parameters, *Bio-Sk*, *R*, *St*, and *Ed*. The attacker does not have any information on $Bio - Sh$, $IR_{SDS}$, and $IR_{SDR}$ to compute ($Messg, Messg', St', Ed''$) and impersonate the valid user *SDS/SDR*. Lastly, an attacker will have difficulty forging a valid specific session of message transmission for a pair of smart devices in the cloud. Hence, the proposed method is reliably designed to prevent these attacks.

**Theorem 3** *The authentication scheme uses shared biometric information to prevent replay and MITN attacks.*

***Proof*** As noted, the proposed method adopted the principle of merging the irises of both parties to generate a shared biometric image ($Bio - Sh = IR_{SDS} \cap IR_{SDR}$). This shared biometric image is used to extract a shared biometric key *Bio-Shk* and is also sent as a *Bio-Sh* to both parties securely. Here, each party had a copy of *Bio-Sh* and *Bio-Shk* that contained the biometric information they shared. Therefore, the authentication code for the sent message is a message authentication code of shared biometric information, which achieves responsive mutual authentication of the message transmission for both parties in the cloud. This mutual authentication has an effective and essential role in preventing attacks that might not be prevented without it, such as replay and MITN.

**Theorem 4** *The proposed scheme uses steganography to avoid attracting the attention of an eavesdropper.*

***Proof*** The most important feature of hiding information when compared with the encryption is that it does not attract the attention of attackers during the process of transmitting data between the *SDS* and *SDR*. However, encrypted messages are visible and can attract attraction. The encryption process protects the data while keeping it publicly visible. Meanwhile, the steganography process hides delivered data and does not allow viewing it, thereby avoiding attracting the attention of eavesdroppers and attackers. In the authentication and verification phase, important parameters $(Messg', St', Ed', R)$ are hidden within the cover image using double steganography D-Stego $(Messg', St', Ed', R)$. Therefore, this scheme does not attract the attention of an eavesdropper, and messages remain unknown to possible attackers. Moreover, by using double steganography, it is very difficult for the attacker to extract the one-time parameters $(Messg', St', Ed', R)$ from the cover image. Consequently, the authentication scheme maintains its security merit by diverting the attention of eavesdroppers.

**Theorem 5** *The authentication scheme employs one-time parameters to prevent reflection and insider attacks.*

***Proof*** The basic idea of this attack is to deceive the target to generate the correct answer to its challenge. Specifically, when the valid login request was sent from the authorized *SDS* to the *SDR*, the attacker takes advantage of this process, trying to eavesdrop on the login request $(Messg', St', Ed', R)$ and forwards to the *SDR* with a MAC similar to that of the $SDS$ $(Bio - Messg' = h(Messg \| Bio - Sk \| Bio - Shk \| R))$. In the authentication process, an attacker cannot deceive a pair of smart devices because he/she does not know *Bio-Shk* and *Bio-Sk*. *Bio-Sk* is used to generate a one-time biometric MAC. *Bio-Shk* is used to encrypt and decrypt $(St, Ed)$. Thus, this mechanism can strongly deflect these attacks.

## 5 System Evaluation

Several experiments are conducted in terms of consuming the time for authentication to assess the efficiency of the lightweight authentication scheme. A PC with Intel Core i3 M380 2.5 GHz, 8GB RAM, Windows 10 is adopted and Matlab is employed because of its ability to deal with image processing in terms of the steganography. We adopted the number of users to be 10,000, with an average time of about 200 execution cycles, to evaluate the authentication time. The average authentication

**Table 1** Evaluation parameters

| Symbol | Definition |
|---|---|
| $T_{ECC}$ | Consuming time of ECC process |
| $T_h$ | Consuming time of a hash process |
| $T_{Xor}$ | Consuming time of Xor |
| $T_{Opr}$ | Consuming time of multiplication, addition and subtraction |
| $T_{\|}$ | Consuming time of concatenation function |
| $T_{ST}$ | Consuming time of double steganography |

**Table 2** Performance evaluation

| Phase | CSP | SDS | SDR |
|---|---|---|---|
| Initialization | $T_{ECC} + T_{Opr}$ | $T_{ECC}$ | $T_{ECC}$ |
| Verification and authentication | | $4T_{Opr} + 1T_h + 3T_{\|} + 2T_{Xor} + T_{ST}$ | $2T_{Opr} + 1T_h + 3T_{\|} + 2T_{Xor} + T_{ST}$ |
| Total | $T_{ECC} + T_{Opr}$ | $T_{ECC} + 4T_{Opr} + 1T_h + 3T_{\|} + 2T_{Xor} + 1T_{ST}$ | $T_{ECC} + 2T_{Opr} + 1T_h + 3T_{\|} + 2T_{Xor} + 1T_{ST}$ |

time per user is 0.268 s, which represents a very reasonable and fair consuming time in terms of the provided secure message transmission for a pair of smart devices in the cloud, as shown in Fig. 1. The performance of our work was also measured in terms of accuracy. Figure 2 shows that the accuracy obtained is 100%, taking into account 10,000 users. For purposes of higher visibility, results for 2,000 users are shown in Fig. 1 and for 5,000 users in Fig. 2. Table 1 shows the evaluation parameters. The estimated time of process for initialization and, verification and authentication phases is shown in Table 2.

## 6 Conclusion

Given the drawbacks in the previous methods, we propose a more effective and consistent scheme to ensure the integrity and authentication of message transmission for smart devices in the cloud. This method has the advantage of generating a one-time biometric authentication code and then concealing it using double steganography for further security. It also provides a responsive mutual authentication based on the shared biometric features of both parties, thereby preventing known attacks in a robust manner, such as MITM, reply, forgery, parallel session, insider, and reflection. Such a scheme provides biometric key management. The efficiency of the time consumed in terms of completing the authentication and integrity processes of the message

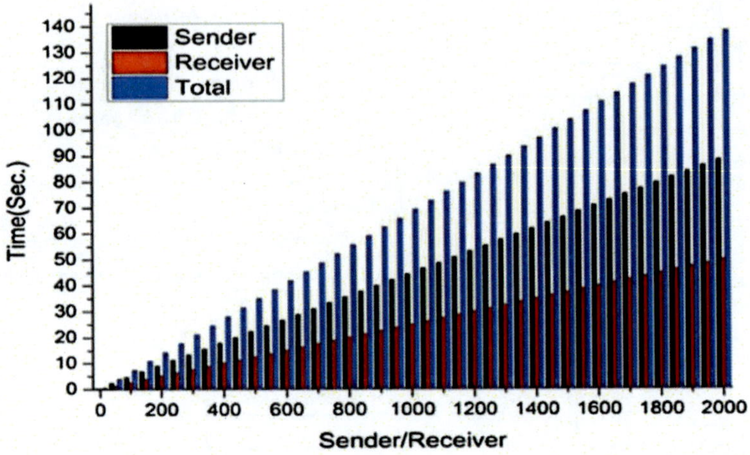


**Fig. 1** Shows the performance of our proposed scheme

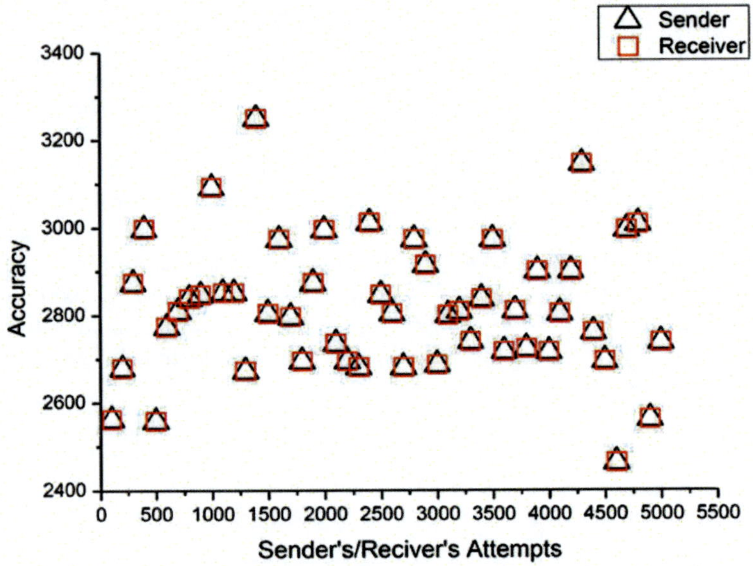


**Fig. 2** Shows the accuracy result of our proposed scheme

sent between the two parties has been proven to be compatible with limited-source devices. Overall, our scheme provides simplicity of use and security.

**Acknowledgements** This work is supported by the Hubei Provincinal Science and Technology Major Project of China under Grant No. 2020AEA011 and the Key Research & Development Plan of Hubei Province of China, under Grant No. 2020BAB100. This work is also supported by the Science and Technology Program of Shenzhen of China under Grant No. JCYJ20180306124612893 and JCYJ20170818160208570.

## References

1. Abduljabbar ZA, Jin H, Ibrahim A, Hussien ZA, Hussain MA, Abbdal SH, Zou D (2016) Privacy-preserving image retrieval in IoT-cloud. In: Proceedings of IEEE Trustcom BigData SEISPA. IEEE Press, Hong Kong, pp 799-806. https://doi.org/10.1109/TrustCom.2016.0141
2. Blaine A, Mahesh B, Daqing H, Stephanie S (2020) Fast free-text authentication via instance-based keystroke dynamics. IEEE Trans Biom Behav Identity Sci 2:377–387. https://doi.org/10.1109/TBIOM.2020.3003988
3. Abduljabbar ZA, Ibrahim A, Hussain MA, Abbdal H, Mustafa ZA, Al Sibahee A, Songfeng L (2019) EEIRI: efficient encrypted image retrieval in IoT-cloud. KSII Trans Internet Inf Syst 13:5692–5716. https://doi.org/10.3837/tiis.2019.11.023
4. Jin X, Dayin W, Dongdai L, Wenling W (2006) An efficient one-key Carter-Wegman message authentication code. In: Proceedings of the international conference on computational intelligence and security. IEEE Press, Guangzhou, China, pp 1331–1334. https://doi.org/10.1109/ICCIAS.2006.295275
5. Zi-ming Z, Yan-fei L, Hui L, Yi-xian Y (2008) An efficient user-to-user authentication scheme in peer-to-peer system. In: Proceedings of the first international conference on intelligent networks and intelligent systems ICINIS. IEEE Press, Wuhan, China, pp 263–266. https://doi.org/10.1109/ICINIS.2008.142
6. Zhenxing L, Lallie HS, Lu L, Yongzhao Z, Kaigui W (2011) A hash-based secure interface on plain connection. In: Proceedings of the sixth international ICST conference on communications and networking in China CHINACOM. IEEE Press, Harbin, China, pp 1236–1239. https://doi.org/10.1109/ChinaCom.2011.6158347
7. Al-Assam H, Rashid R, Jassim S (2013) Combining steganography and biometric cryptosystems for secure mutual authentication and key exchange. In: Proceedings of the 8th international conference for internet technology and secured transactions ICITST. IEEE Press, London, UK, pp 369–374. https://doi.org/10.1109/ICITST.2013.6750224
8. Castiglione A, De S, Castiglione A, Palmieri F (2014) An efficient and transparent one-time authentication protocol with non-interactive key scheduling and update. In: Proceedings of the 28th international conference on advanced information networking and applications AINA. IEEE Press, Victoria, Canada, pp 351–358. https://doi.org/10.1109/AINA.2014.45
9. Abduljabbar ZA, Mohammed RJ (2018) Towards one-time biometric-message authentication code in cloud computing. J Eng Appl Sci JEAS 13:8027–8033. https://doi.org/10.3923/jeasci.2018.8027.8033
10. Nashwan JH (2020) Robust iris recognition framework using computer vision algorithms. In: 4th international conference on smart grid and smart cities ICSGSC. IEEE, Osaka, Japan, pp 18–21. https://doi.org/10.1109/ICSGSC50906.2020.9248564
11. Daugman J (2004) How iris recognition works. IEEE Trans Circuits Syst Video Technol 14:21–30. https://doi.org/10.1109/TCSVT.2003.818350
12. Hariprasath S, Mohan V (2008) Biometric personal identification based on iris recognition using complex wavelet transformations. In: Proceedings of the international conference on

computing, communication and networking ICCCN. Thomas, VI, USA, pp 1–5. https://doi.org/10.1109/ICCCNET.2008.4787736
13. Stallings W (2014) Cryptography and network security: principles and practice. In: Cryptography and network security: principles and practice, 6th edn. New York, NY, USA, pp 1–760
14. Nadiya PV, Imran BM (2013) Image steganography in DWT domain using double-stegging with RSA encryption. In: International conference on signal, image processing and pattern recognition. ACM, New York, NY, USA, pp 283–287. https://doi.org/10.1109/ICSIPR.2013.6497941
15. International Telecommunication Union (2014) M2M service layer: APIs and protocols overview. Focus group technical report of ITU-T focus group on M2M service layer, telecommunication standardization sector of ITU, FG M2M, D3.1 Version 1.0
16. Park J, Kang N (2013) Designing a secure service manager for internet of things. Adv Sci Technol Lett 43:162–165; Multimedia
17. Shen H, Shen J, Khan MK, Lee J-H (2017) Efficient RFID authentication using elliptic curve cryptography for the internet of things. In: Wireless personal communications, vol 96. Springer, pp 5253–5266. https://doi.org/10.1007/s11277-016-3739-1
18. Alizai ZA, Tareen NF, Jadoon I (2018) Improved IoT device authentication scheme using device capability and digital signatures. In: International conference on applied and engineering mathematics (ICAEM). IEEE Xplore, London, U.K., pp 115–119. https://doi.org/10.1109/ICAEM.2018.8536261

# DRLAS: Digital Record Keeping in Land Administration System Relying on Blockchain

Milon Biswas, Tajim Md. Niamat Ullah Akhund, and M. Shamim Kaiser

**Abstract** On every developed or developing world, the Land Administration System (LAS) is a salient infrastructure. Both the digital (traditional database system) and the manual methods are applied in the LAS (paper-based documentations). Both of these systems provide an authority with monopoly power that can increase unethical conduct in land administration. Again, while the Blockchain technology was initially used to maintain a financial ledger, it is possible to expand the use of this technology to incorporate any decentralized computing structures, including the automated record keeping and management framework. Recent studies have found that the new and largely untested applications of Blockchain technology in land administration remain. In this work, the implementation of a blockchain-based system is proposed to build a System for Land Administration. In the Ethereum blockchain platform, the proposed framework was simulated and showed that the framework contributed to developing a stable, secure, effective and efficient system of land administration.

**Keywords** Blockchain · Smart contract · Land administration system · Mutation

M. Biswas (✉)
Department of Computer Science and Engineering, Bangladesh University of Business and Technology, Dhaka, Bangladesh
e-mail: milon@ieee.org

T. Md. Niamat Ullah Akhund
Department of Computer Science and Engineering, Daffodil International University, Dhaka, Bangladesh

M. S. Kaiser
Institute of Information Technology, Jahangirnagar University, Savar Union, Bangladesh
e-mail: mskaiser@juniv.edu



X.-S. Yang et al. (eds.), *Proceedings of Sixth International Congress on Information and Communication Technology*, Lecture Notes in Networks and Systems 216,
https://doi.org/10.1007/978-981-16-1781-2_83

# 1 Introduction

One of the most important divisions of any nation is land administration. The system of land administration (LAS) keeps a record of real (land) properties, primarily of location, (historical) possession, value and usage. Often, the LAS Maintain data on physical, spatial, and topographical property characteristics. Long-term availability, reliability and compliance with proper law should be available in the LAS documents [19]. The main threats to the digital land management system are networks, cyber attacks, and data leakage as well. A blockchain is a time-stamped collection of permanent data records that are maintained by computer clusters that do not belong to any single entity. Blockchain technology is tied to paper published in 2008 by person or persons under the alias of Satoshi Nakamoto [16]. Blockchains are decentralized, distributed, and fault-tolerant databases that can be shared by any network user, but no individual can share them. Controlling.

The purpose of this paper is to propose a blockchain-based platform for the creation of a safe, trusted, and effective system of land administration. A structure is proposed for LAS to achieve this goal, considering the related properties of blockchain technology and main LAS functionalities. Centered on the proposed structure and then a prototypical system is also built Via simulation, assessed its performance.

# 2 Related Works

According to a recent study [14], blockchain can be defined as "A concept map of Blockchain which represents that a Blockchain consists of blocks containing messages, proof of work, and reference of previous block and stored in shared database, which is able to perform transactions over P2P network maintaining irreversible historical records and transparency". Blockchain technology has attracted tremendous interest from wide range of stakeholders, which include finance, healthcare, utilities, real estate, and government agencies [12]. A very few studies show integration of blockchain technology in LAS [17]. A number of studies have been conducted on IT-based Land Management System, where the database systems and web-based technologies were the key concern to make it automated as well as hyper ledger system is also introduced [15]. Choudhury et al. [9] proposed a web-based land management system for Bangladesh which could scan the paper-based land maps. In an another work [21] proposed an automated digital archival system for land registration and keeping records. Talukder et al. [20] Introduced digital land management system which included GPS-based land surveying. A few studies introduced the Blockchain technology in land administration systems [19] and proposed a blockchain-based Land Administration System to increase its transparency, immutability, security, and accountability. Authors of [11] proposed a blockchain-based E-voting system to reduce database manipulation and maintain data integrity. The study by [13] proposed a framework to develop and evaluate a tamper-resistant m-Health system using blockchain technology which enabled trusted and auditable computation using a decentralized network. Peterson [17] proposed an idea of implementing the blockchain in health information sharing. In this recent world, everything is automated like E-voting [6, 18], stock market [7], NID managements [10], Hotel managements [8], Poultry farm [4], virus affected people management [2], remote sensing [3], secure payment [1], supply chain management [5], and so more. Why not automate the management of the land? A new consensus algorithm was implemented in this research to promote interoperability of data. With some pseudo code of mining, the solution was theoretically defined.

## 3 Proposed Framework

In order to create a realistic system,the system is proposed to take three problems into account. Second, the main LAS features, which include land-related updating, Data, verification of land information, application for a mutation, and transfer process of ownership. Second, the LAS stakeholders, i.e., the owner of the land, consumers, the authority (LAS management representative), and the government. Third, the blockchain technology's assets.

The Blocks and the nodes are the two key components of the proposed system. The blocks are the data structures that hold information about any land transaction and changes. In the proposed framework, the blocks include:

1. Previous hash—creates link to previous block;
2. Seller ID—user ID of the seller of the land of this particular transaction;
3. Purchaser ID—user id of the purchaser of the same land;
4. Land ID—the ID of the land in question;
5. Transaction money—price of the land according to market value; and
6. Hash—unique ID for a particular block.

Again, the nodes are the networks' endpoints. As the nodes, the stakeholders involved in the network act. It is possible to classify nodes into two classes. First one: standard nodes that can only make requests for mutation and are unable to engage in the verification process and Second one: nodes that can be authenticated, which are capable of Authenticate a block, which means that it can mine a block that mines them. New transactions are verified by miners and reported on the global ledger. He smart contract is the software where the blockchain behavior is written. After mining one block, it essentially determines the steps needed. It may, under certain circumstances, directly regulate the transfer of digital currencies or assets between parties.

The key tasks that are required for LAS are included in Fig. 1 and gives a graphical representation.

***Updating data related to land***: It may be appropriate to adjust from time to time general information that includes measurement, climate, infrastructure and price

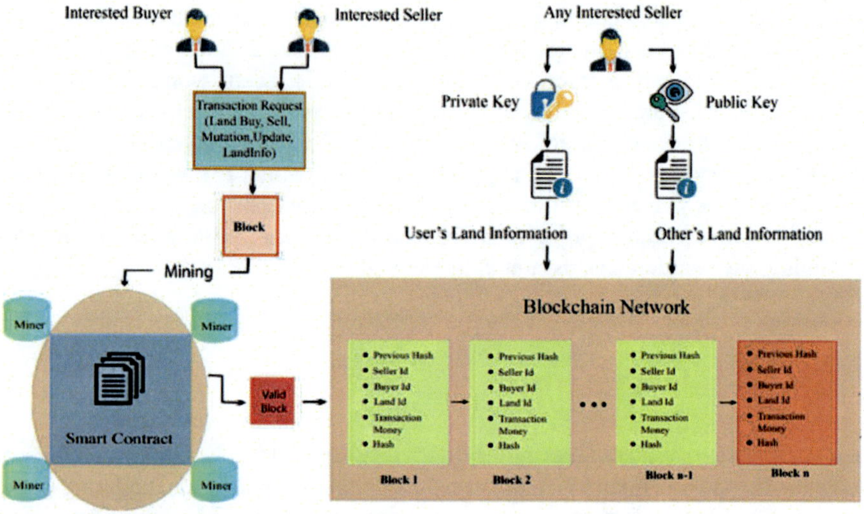


**Fig. 1** Proposed framework for blockchain-based LAS

related information. It will be checked by valid and precise modification to ensure valid and precise change.

***Informational Checking***: Before buying a house, any buyer would like to check the details relating to the land which would be secured by a private key inside the proposed scheme. The private key is open to the owner only.

***Request for Mutation Submission***: The order for mutation relates to the beginning of the transfer process of ownership. Both parties can file a digital application in the proposed system using their corresponding private key. A block will be generated after that, containing all the details about the phase of land transfer and will be broadcast for mining.

***Phase for Ownership Transfer***: In this method, the block generated containing mutation information is checked by consensus algorithm miner nodes and it is checked if the value of the created block is compatible with the current blockchain. Finally, the distributed ledger structure and the central blockchain would be added to it.

## 4 The Proposed Framework Simulation

The section covers from setting up the environment to coding and deploying the smart contract.

## 4.1 Simulation Environment

In order to simulate the proposed framework in local environment, the following steps are followed:

1. *Setting up personal blockchain*: A local personal blockchain Ganache was used to simulate the proposed framework.
2. *Setting up development environment*: Truffle development environment creates the local environment and the file system for blockchain project.
3. *Opening digital wallet*: Digital wallet allows the accounts in the personal blockchain to connect with the browser and finally to the blockchain network.
4. *Creating project and initiating truffle*: The next step is to create the project directory and initiating truffle environment in the directory.
5. *Adding project to ganache*: After the project is created, it has to be added to Ganache in order to visualize the current state of the blockchain.

## 4.2 Deployment and Features of Smart Contract

Solidity programming language, which specializes in the creation of blockchain. After establishing the contract features, such as updating information, verifying information, sending requests for mutation and transfer of ownership, As the member method of the contract. The smart contract's member methods and block structure are as follows:

1. *Block*: The next step is to create the structure of the blocks. The blocks are basically a type data structure. In solidity, it is declared as *struct*.
2. *Posting a free land for sell*: Initially, if any land has a survey mechanism, it can be posted for sale by the survey office or other government entity without a real owner. The selling feature is called Createland and looks like an Algorithm 1. The role obtains the position of land and the price of land and as parameters.
3. *Posting a land for sale by its owner*: If the owner of a land wants to sell the land, it can be posted. To post it, land for sell function is called. This function makes the land block available for potential buyers to create the buying block and put up for mining. The function is shown in Algorithm 2.
4. *Purchase land and mutation*: The purchase land function is the function that actually performs the mutation operation. Once any buyer puts in a mutation request, purchase Land function gets called. The mutation block created by the function is uploaded via metamask for mining in Ethereum. The process of mutation is shown as a pseudo code in Algorithm 3.

**Algorithm 1** Add New Record of Land

```
function CREATELAND(_landLocation, _landPrice)
    Require landLocation.length > 0
    Require _landPrice > 0
    landCount = landCount + 1
    lands[landCount] ← Add New Record of Landendtate emit LandCreated
```

**Algorithm 2** Change Land Status For Sell

```
function LANDFORSELL(landID)
    Require Land ID is valid
    purchase flag ← false
    emit LandforSell
```

**Algorithm 3** Purchase Land

```
function PURCHASELAND(landID)
    Require Land ID is valid
    CurrentLand ← Fetch Land from given Land ID
    Owner ← Get the Land owner
    Require Account Balance land Price
    Require Land is not Purchased
    Require Seller and Buyer are not same
    CurrentLand.Owner ← Change the Owner
    Transfer the fund
    purchase flag ← true
    emitLandPurchased
```

### *4.3 Ethereum Cost and Gas Fee*

To deploy the smart contract in Etheruem, a certain amount of computing power is required. Similarly, calling every function also requires some computing power. In Ethereum blockchain platform, the computing power is measured as gas fee. Also, for issuing a block by triggering a function call, the concerned user must spent some Ether, which is treated as a digital currency. Computing power for five operations has been calculated. They are: deploy migration function, deploy LAS smart contract, create land function, purchase land function, and land for sell function. The histogram in Fig. 2 shows the gas use for each function and also showed that deploying the contract consumes the maximum amount of computing power. The next histogram in Fig. 2 represents ether costs of the same functions. As observed in the Fig. 2, deploying the contract consumes more power, as a result, this is the most costly function.

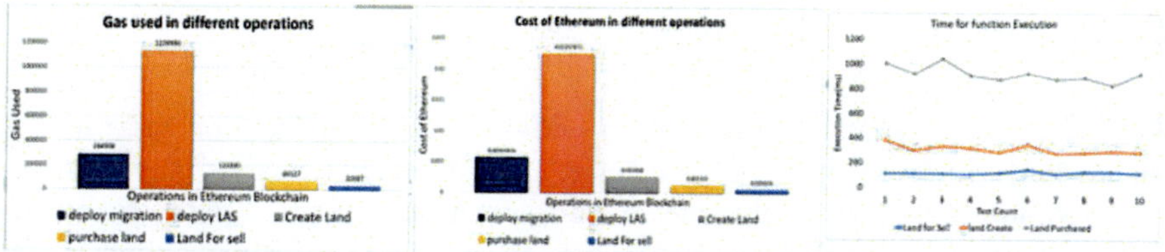


**Fig. 2** Gas used for each functions, ether used for each functions, execution for each functions

## *4.4 Execution Time*

The graph in Fig. 2 shows the execution time of the following functions for several test cases in milliseconds: (1) Land for sell, (2) Land create, and (3) Land purchase

It is evident from the test vs time graph that purchasing land takes the most amount of time and it spans around 900 ms. The second most execution time is taken by creating land which spans around 300–400 ms. Then the lowest execution time is taken by land for sell function.

# 5 Results Discussion

A local web application was developed on the basis of the proposed framework using Ganache to host the blockchain locally. The simulation results showed that the blockchain-based proposed LAS system would bring the following benefits:

1. The proposed architecture provides LAS with extensive security features. The consensus algorithm used in the proposed system ensures that the blockchain does not add an invalid block. In addition, it also guarantees that a consumer may only access his or her details.
2. In an ethereum blockchain platform that utilizes proof of work consensus algorithm, the proposed architecture is simulated. Every change of knowledge is then checked and then updated using the evidence of work in every local copy of the blockchain.
3. Suggested framework stores change logs in the form of chain blocks that often ensure the system's trustworthiness.
4. The computing power needed to mine each block is considerably lower than the smart contract being uploaded. For example, the createLand, purchase land (mutation), land for sale showed less computational power in Fig. 2, for instance.
5. Finally, such practices like posting a (land) sale notice and creating a new one an average execution time, that is, less than 400 milliseconds, was shown by the land block. Again, a comparatively higher execution time is showed for the land purchase (mutation), see Fig. 2, it is around one second (1000 milliseconds), and can still be treated as relatively faster.

## 6 Conclusion

A blockchain-based framework is proposed for land administration system. The functionality includes updating land-related information, checking land information, submitting the mutation request, and transferring process of land ownership. The proposed framework is evaluated through simulation. The simulation results showed that a secure, reliable, effective, and efficient land administration system could be developed based on the proposed framework. However, the performance of the system in real-time transaction is yet to be tested. Our future work will focus on this direction.

## References

1. Ahamed S, Siddika M, Islam S, Anika S, Anjum A, Biswas M (2021) Bps: Blockchain based decentralized secure and versatile light payment system. Asian J Res Comput Sci 12–20
2. Akhund TMNU, Jyoty WB, Siddik MAB, Newaz NT, Al Wahid SA, Sarker MM (2020) Iot based low-cost robotic agent design for disabled and covid-19 virus accepted people. In: 2020 fourth world conference on smart trends in systems, security and sustainability (WorldS4). IEEE, pp 23–26
3. Akhund TMNU, Newaz NT, Hossain MR (2020) Low-cost remote sensing iot based smartphone controlled robot for virus affected people
4. Akhund TMNU, Snigdha SR, Reza MS, Newaz NT, Saifuzzaman M, Rashel MR (2020) Self-powered iot-based design for multi-purpose smart poultry farm. In: International conference on information and communication technology for intelligent systems. Springer, pp 43–51
5. Al-Amin S, Sharkar SR, Kaiser MS, Biswas M (2021) Towards a blockchain-based supply chain management for e-agro business system. In: Proceedings of international conference on trends in computational and cognitive engineering. Springer, pp 329–339
6. Biswas M, Mahi M, Nayeen J, Hossen R, Acharjee UK, Md W (2020) Buvots: a blockchain based unmanipulated voting scheme. Rakib and Acharjee, Uzzal Kumar and Md, Whaiduzzaman (eds) BUVOTS: a blockchain based unmanipulated voting scheme (November 23, 2020)
7. Biswas M, Shome A, Islam MA, Nova AJ, Ahmed S (2021) Predicting stock market price: A logical strategy using deep learning. In: 2021 IEEE 11th IEEE symposium on computer applications & industrial electronics (ISCAIE). IEEE, pp 218–223
8. Biswas M, Whaiduzzaman M (2018) Efficient mobile cloud computing through computation offloading. Int J Adv Technol 10(2)
9. Choudhury E, Ridwan M, Awal MA, Hosain S (2011) A web-based land management system for bangladesh. In: 14th international conference on computer and information technology (ICCIT 2011). IEEE, pp 321–326
10. Datta P, Bhowmik A, Shome A, Biswas M (2020) A secured smart national identity card management design using blockchain. In: 2020 2nd international conference on advanced information and communication technology (ICAICT), pp 291–296. IEEE
11. Hanifatunnisa R, Rahardjo B (2017) Blockchain based e-voting recording system design. In: 2017 11th international conference on telecommunication systems services and applications (TSSA). IEEE, pp 1–6
12. Hossain MP, Khaled M, Saju SA, Roy S, Biswas M (2020) Vehicle registration and information management using blockchain based distributed ledger from Bangladesh perspective. In: 2020 IEEE region 10 symposium (TENSYMP). IEEE
13. Ichikawa D, Kashiyama M, Ueno T (2017) Tamper-resistant mobile health using blockchain technology. JMIR mHealth and uHealth 5(7):e111

14. Islam I, Munim KM, Oishwee SJ, Islam AN, Islam MN (2020) A critical review of concepts, benefits, and pitfalls of blockchain technology using concept map. IEEE Access 8:68333–68341
15. Mukherjee PP, Boshra AA, Ashraf MM, Biswas M (2020) A hyper-ledger fabric framework as a service for improved quality e-voting system. In: 2020 IEEE region 10 symposium (TENSYMP). IEEE, pp 394–397
16. Patil VT, Acharya A, Shyamasundar R (2018) Landcoin: a land management system using litecoin blockchain protocol. In: Proceedings of the symposium on distributed ledger technology
17. Peterson K, Deeduvanu R, Kanjamala P, Boles K (2016) A blockchain-based approach to health information exchange networks. Proceedings of NIST workshop blockchain healthcare 1:1–10
18. Sarker MM, Akhund TMNU The roadmap to the electronic voting system development: a literature review. Int J Adv Eng Manag Sci 2(5):239465
19. Stefanović M, Pržulj D, Ristic S, Stefanović D (2018) Blockchain and land administration: possible applications and limitations. In: International scientific conference on contemporary issues in economics business and management
20. Talukder SK, Sakib MII, Rahman MM (2014) Digital land management system: a new initiative for bangladesh. In: 2014 international conference on electrical engineering and information & communication technology. IEEE, pp 1–6
21. Toaha M, Khan S (2008) Automated digital archive for land registration and records. In: 2008 11th international conference on computer and information technology. IEEE, pp 46–51

# An Ontology for Fire Building Evacuation

Check for updates

**Joaquim Neto, A. Jorge Morais, Ramiro Gonçalves, and António Leça Coelho**

**Abstract** Guiding the building occupants under fire emergency to a safe place is an open research problem. Finding solutions to address the problem requires a perfect knowledge of the fire building evacuation domain. The use of ontologies to model knowledge of a domain allows a common and shared understanding of that domain, between people and heterogeneous systems. This paper presents an ontology that aims to build a knowledge model to better understand the referred domain and to help develop more capable building evacuation solutions and systems. The herein proposed ontology considers the different variables and actors involved in the fire building evacuation process. We followed the *Methontology* methodology for its developing, and we present all the development steps, from the specification to its implementation with the *Protégé* tool.

**Keywords** Fire building evacuation · Ontology · Knowledge model · Ontology development · Fire emergency · Ontologies · Knowledge representation

---

J. Neto (✉) · A. Jorge Morais
Universidade Aberta, Lisboa, Portugal
e-mail: jfn@lnec.pt

A. Jorge Morais
e-mail: jorge.Morais@uab.pt

J. Neto · A. L. Coelho
Laboratório Nacional de Engenharia Civil, Lisboa, Portugal
e-mail: alcoelho@lnec.pt

A. Jorge Morais
LIAAD - INESC TEC, Lisboa, Portugal

R. Gonçalves
UTAD (Universidade de Trás-Os-Montes E Alto Douro), Vila Real, Portugal
e-mail: ramiro@utad.pt


X.-S. Yang et al. (eds.), *Proceedings of Sixth International Congress on Information and Communication Technology*, Lecture Notes in Networks and Systems 216,
https://doi.org/10.1007/978-981-16-1781-2_84

# 1 Introduction

The evacuation of buildings in fire emergencies is a problem for which solutions must be found to help the occupants of these spaces, guiding them on their route until they are safe. The present research work's motivation falls within the scope of our ongoing PhD thesis, which proposes a multi-agent recommender system for real-time guiding the occupants to a safe place [1]. Considering that one of the proposed solution pillars is an ontological model to support interoperability between the different actors involved in the process, a thorough knowledge of buildings' evacuation under a fire emergency is required. The need to apprehend and consolidate knowledge in the referred domain motivated us to create the herein proposed ontology, whose innovative purpose is to build a knowledge model that contributes to a better understanding of that domain and help develop solutions or systems related to the fire building evacuation problem.

Section 2 presents some of the relevant research works that propose ontologies to model knowledge in the emergency domain. Section 3 briefly introduces the ontology concept and describes the main steps in developing an ontology. Section 4 presents the work developed to build the ontology step by step and Sect. 5 describes the ontology evaluation process. Finally, in Sect. 6, we write the conclusions and future work.

# 2 Ontologies for the Building Emergency Domain: Related Work

Many research works have proposed ontologies to represent knowledge in the emergency area. Here we will highlight some of those relevant research works, particularly addressing the fire building emergency field. In their paper, [2] present the ontology SEMA4A, which aims to include information sharing and support the interoperability between systems and between people in an emergency. The ontology provides knowledge according to three domains: accessibility guidelines, emergencies and communication technologies. Later on, based on SEMA4A ontology and focusing on the evacuee's notification about safe places and evacuation routes, [3] develop an extension of the SEMA4A ontology, thus compiling a fourth area related to evacuation. Liu et al. [4] present an ontology that defines fire control concepts and the relationship between these concepts, from the community's perspective, defining how the community members come together to control a fire. Intelligent emergency response applications must interpret and filter relevant information from a wide range of heterogeneous data. To support the development of that kind of applications, [5] present the ontology "Emergency Response Ontology". Casado et al. [6] present the EMERGEL (Emergency ELements) ontology, developed in the scope of the DISASTER project (Data Interoperability Solution in Emergency Reaction Stakeholders). The EMERGEL ontology contains knowledge and concepts related to emergencies. In its article, [7] presents the Emergency Fire ontology, whose scope is to share knowledge of the domain of fire emergency response in buildings. The ontology aims to contribute to the organisations' tactical and strategic plans to respond to fire emergencies in buildings.

## 3 Ontologies: Methodological Approaches

According to Gruber [8], in computing and information science, the term ontology refers to an artefact designed to model knowledge about a domain, be it real or imaginary. According to the same author, in the 1980s, the term was adopted by Artificial Intelligence (AI) researchers, used to refer to a theory of a modelled world or a component of knowledge systems. According to [9], it is possible to see multiple and contradictory definitions for ontology in AI literature. In a widely cited article, [10] defines ontology in the context of AI as an "explicit specification of a conceptualization". Objects, concepts or other entities, and the relationships among them, are used to represent knowledge within a domain.

As referred by [11], ontologies provide a common and shared understanding of a domain to help communicate between heterogeneous systems and people, consisting of a set of terms representing concepts (hierarchically organised) and some specification of its meaning. Ontologies are also promoters and facilitators of interoperability between information systems, intelligent processing by agents, or the sharing and reuse of knowledge between systems.

Although the comparative study of the different methodologies used in the development of ontologies is not within the scope of this paper, it is essential to mention the most representative and commonly used ones [11]: TOVE [12], ENTERPRISE [13] and METHONTOLOGY [14]. Another methodology widely followed is the "Ontology Development 101" [9]. Concerning the ontology development, and independently of the methodology, five stages may be identified in the life cycle developing of an ontology [11, 14].

- The first stage is the ***specification stage***. It is crucial to identify the objective and scope of the ontology. The following answers must be addressed: What is the domain to be covered by the ontology? Why build the ontology? What are the intended uses? What kind of answers should the ontology give?
- At the ***conceptualisation stage***, the ontology is described through a conceptual model to meet the specification defined before. It consists of concepts in the domain and relationships between those concepts (using mind maps, for example).
- In the ***formalisation*stage**, a formal model is constructed from the conceptual model of the previous stage. Concepts are generally defined through axioms that restrict possible interpretations to the meaning of those concepts.
- The fourth stage is the ***implementation*** of the previously formalised ontology in a knowledge representation language.

- The last stage refers to ***maintenance***, during which it is essential to update and correct the implemented ontology.

Also, according to [11, 14], three other activities should also be considered during the whole life cycle:

- ***Acquisition*of knowledge** about the domain, whether using elicitation techniques, listening to specialists or consulting the relevant bibliography.
- Technically ***evaluate*** the ontology, namely using tools already available on the Web.
- ***Documentation***: Registration of what is being done, namely through the annotation and description of the terms represented in the ontology, in order to make it clearer to understand, as well as to facilitate maintenance and future reuse.

## 4 Building an Ontology for Fire Building Evacuation

To develop the ontology presented here, we adopted the METHONTOLOGY development methodology. The development process followed four main steps, already described in the previous section: (i) specification, (ii) conceptualisation, (iii) formalisation and (iv) implementation. During the development process, the activities carried out were also supported by acquiring and updating knowledge of the domain under study. We document and annotate the terms, concepts and relationships represented in the ontology and technically evaluate it.

### *4.1 Specification Stage*

The specification stage is a step of particular importance, as it is time to define the domain, scope and purpose of the ontology. The literature review carried out allowed the identification of different ontological approaches in treating fire emergencies in buildings. We have approaches that focus on the response of the different actors involved in emergencies; approaches that focus on fire control knowledge; and other approaches, whose purpose is to model knowledge about fire safety, including the building evacuation problem. Our focus is the knowledge modelling of the building evacuation process, considering all the variables involved.

Follows from the above referred that the domain of ontology fits the building evacuation under fire emergency. As for the second question, the purpose is to represent knowledge about the evacuation of buildings, considering the different variables involved in the fire building evacuation process. For future use of the ontology, we highlight the ability to strengthening and consolidating knowledge about the domain of evacuation in buildings under fire emergency, as well as a support for the development of evacuation systems and applications, namely concerning the interoperability between the different actors involved in an evacuation process. Regarding the ontology's answers, we consider a set of competency questions [12] that help in the ontology's specification and scope and test the ontology at the evaluation stage. We identify a set of competency questions based on Portuguese legislation and regulations[1] and the document of the Autoridade Nacional de Emergência e Proteção Civil (ANEPC) [15], as well as in the feedback of experts in the field. We present those competency questions and the most relevant aspects of the responses in Table 1. We better define our ontology's scope and objective according to the competency questions table in Table 2.

## 4.2 Conceptualisation Stage

At this stage, we will use the knowledge acquired in the specification phase, summarised in Tables 1 and 2, to describe the ontology through a conceptual model, consisting of a set of concepts or terms and the relationships between them, that we represent through the conceptual map [16], built using the *mindomo*[2] framework. To build the conceptual map, we start with three main terms, *Building*, *Occupant* and *Fire*, that immediately follow what we have already defined as the ontology's domain and scope.

## 4.3 Formalisation and Implementation Stages

Throughout this section, we transform the conceptual model into a formal model, and we implement the formalised ontology in a knowledge representation language. For the implementation, we used *Protégé* [17], a largely used open-source platform that provides a suite of tools to construct and describe ontologies.

Based on the conceptual model [16], we create and describe 52 classes and 57 object properties, that characterise the relationships identified in the conceptualisation stage. In Fig. 1, we exemplify the classes characterisation with the *AutomaticFireDetectionSystem* class, with a *Protégé* screen capture.

Object properties relate instances of two classes [18]. In Fig. 2, we present the example of the property *hasAutomaticFireDetectionSystemComponent*, that defines a relationship between an individual of the class *AutomaticFireDetectionSystem* (domain) and an individual of the class *AutomaticFireDetectionSystemComponent* (range) establishing that an automatic fire detection system has components, represented by instances of the *ControlUnit*, *FireAlarm* and *SensorUnit* subclasses.

---

[1] Decreto-Lei n.° 220/2008 de 12/11—Regime Jurídico da Segurança Contra Incêndios em Edifícios (SCIE).

[2] www.mindomo.com.

**Table 1** Competency questions and answers

| Competency questions | |
|---|---|
| Questions | Answers |
| How to detect the start of the fire? | Depending on the type of building, the law requires installing fire detection systems, that may be triggered automatically by sensors or manually by people in the building |
| How to transmit the alarm to the building occupants? | They are notified through audible, sound or previously recorded messages, or visual alarms: (i) Local alarm (directed to occupants of a part of the building), (ii) General alarm (for the general evacuation of the building); (iii) Restricted alarm (only for security personnel) |
| Who are the occupants? How do they behave in an emergency? | Occupants are people inside the building during evacuation. The evacuee's behaviour depends on age, gender, mobility and psychological characteristics. Also, crucial is the evacuee's knowledge of the building and familiarisation with evacuation drills |
| What is the type of building? Where are the safe zones located? What is the building's capacity? | The answers to these types of questions are in the internal emergency plan. That plan includes the action plan, evacuation plan, safety instructions, emergency plans and the organisation to be adopted in case of emergency |
| How is the building evacuated? What are the evacuation routes, and are they identified? What type of emergency signage? | The evacuation must take place following the building's organisation and management of fire safety, which includes the internal emergency plan, and all prevention procedures and plans, in case of emergency. The emergency signs must respect the building's emergency plan. Occupants follow emergency signs and seek the support of building security staff, firefighters and the police officer |
| What are the types of hazards? | The congestion of the evacuation routes, the toxic gases, the smoke, and the high temperatures are examples of hazards that influence occupants' behaviour and the evacuation of the building. |

Datatype properties relate an individual of a class to a primitive value [18]. In Fig. 3, we show the example of the property *isFamiliarWithBuiding*, which relates an individual of the class evacuee to a primitive type Boolean.

**Table 2** Ontology specification and requirements

| Ontology specification and requirements | |
|---|---|
| Ontology domain: | The domain is the evacuation of buildings under fire emergency |
| Ontology goal: | To develop a knowledge model about buildings' evacuation under fire emergency |
| Ontology contribution: | • Provide a knowledge representation about the domain of evacuation of buildings under fire emergency;<br>• Support the development of fire evacuation applications and systems |
| Answers that the ontology should give: | • How is the start of the fire detected?<br>• How to notify the building's occupants?<br>• What is the impact of occupant behaviour?<br>• What are the types of hazards?<br>• How to identify evacuation routes?<br>• How to find exits and safe locations?<br>• How are occupants guided to a safe place? |

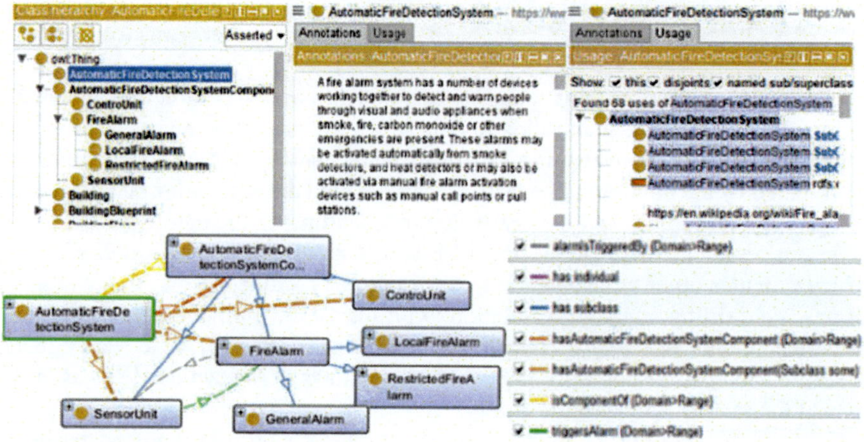


**Fig. 1** Class representation on *Protégé*

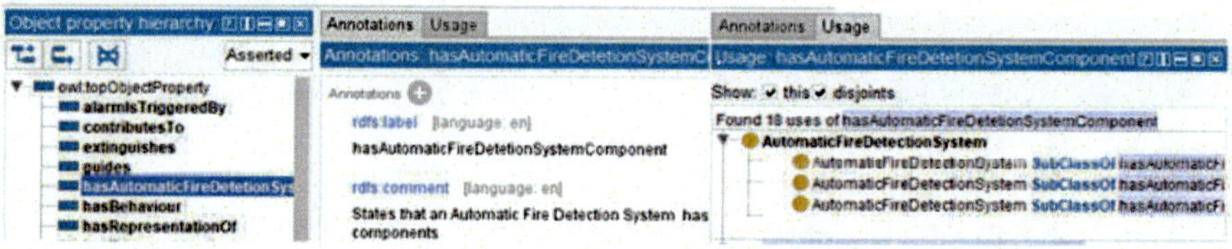


**Fig. 2** The property has automatic fire detection system component

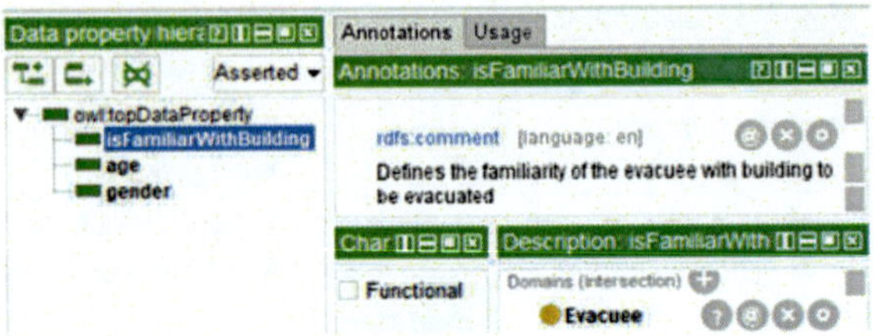


**Fig. 3** Datatype properties in *Protégé*

# 5 Evaluation

According to [19], an ontology evaluation is a technical judgement of its content, against a referential. The evaluation is an iterative process that occurs throughout the ontology's development cycle. The competency questions and the specification requirements, summarised in Tables 1 and 2, embody that referential.

## 5.1 Technical Evaluation

The technical evaluation deals with aspects related to ontology structure and architecture. It considers syntactic validation of classes and properties and aspects related to their documentation, to ensure an adequate understanding of the knowledge model represented by the ontology. The tests carried out with the tools referred above were completed with success and are available for consulting on the ontology URL.[3]

In our research work, we use the Web tool, *Ontology Improvement Tool (V2)*,[4] which provides other tools and services for ontology validation and improvement, namely concerning syntactic validation, with the RDF *Triple-Checker*,[5] ontology consistency, with the *OOps! (OntOlogy Pitfall Scanner)*[6] [20] or to verify if the semantic Web data is correctly published and follows best practices,[7,8,9], as is the case of the service provided by *Vapour*[10].

Concerning ontology documentation, we used *WIDOCO*[11] [21] a wizard that helps to document an ontology by identifying missing metadata, and by integrating other

---

[3] https://www.1000palavras.pt/ontology/fbevac/FireBuildingEvacuation-en.html.

[4] http://perfectsemanticweb.appspot.com/?p=ontologyValidation.

[5] http://graphite.ecs.soton.ac.uk/checker/.

[6] OOPS!—OntOlogy Pitfall Scanner! (linkeddata.es).

[7] Linked Data—Design Issues (w3.org).

[8] Best Practice Recipes for Publishing RDF Vocabularies (w3.org).

[9] Cool URIs for the Semantic Web (w3.org).

[10] http://linkeddata.uriburner.com:8000/vapour.

[11] http://dgarijo.github.io/Widoco/.

**Table 3** Axioms obtained with SPARQL and related graphical representation

| Question | Axioms |
|---|---|
| How to transmit the alarm to the occupants? | • *Sensor* => isIncorporatedOn => *SensorUnit*<br>• *SensorUnit* => *triggersAlarm* => FireAlarm<br>• *FireAlarm* => *warns* => *Occupant* |

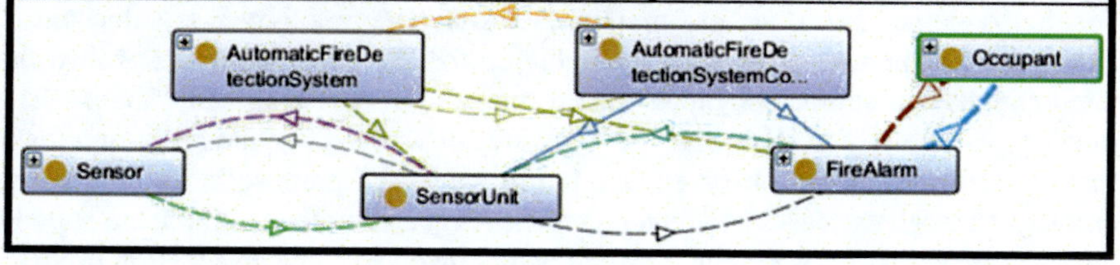


existing tools that allow ontology validation (with *OOPS*!), ontology terms documentation (with *LODE*[12]) and ontology visualisation (with *WebVowl*[13]). The *WIDOCO* tool produces the Web pages' ontology documentation, ready for publishing.

## 5.2 *Evaluating Against a Referential*

Another way of evaluating an ontology is their evaluation against a referential, embodied by the specification requirements and the competency questions, which must be answered by the ontology. It is an evaluation dimension that assesses whether classes, properties and axioms can answer the questions and requirements, that were at the origin of the creation and development of the ontology. To assess whether the ontology proposed here answers the set of competency questions and requirements presented in Tables 1 and 2, we query our ontology using the *SPARQL* language. Table 3 presents an example, in which axioms obtained with a *SPARQL* query are complemented with a graphical visualisation provided by *Protégé's OntoGraf* plug-in.

# 6 Conclusion

By enabling a knowledge representation of a domain, ontologies provide a common and shared understanding for communication between people and heterogeneous systems. The ontology herein proposed aims to contribute to a better understanding of fire building evacuation domain. Another purpose of this ontology is the support for developing more capable building evacuation solutions and systems. This ontology

[12] https://essepuntato.it/lode/.

[13] http://vowl.visualdataweb.org/webvowl.html.

considers the different and relevant variables and actors in building evacuation under fire emergency. The knowledge acquisition that supported the conceptual model considers the Portuguese legislation and regulation analysis and expert feedback. The ontology development followed the METHONTOLOGY methodology and took place over several months, and we present the different stages of the development. For the implementation, we use the *Protégé* tool. The ontology was subjected to a technical evaluation, using tools available on the Web, and validated against the requirements and competency questions formulated in the specification phase. The ontology is publicly available at the URI: https://www.1000palavras.pt/ontology/fbevac/FireBuildingEvacuation-en. As future work, the purpose is deepening the ontology through use cases and reusing it to develop an ontological model to support the interoperability between the different actors involved in the evacuation process, as proposed by [1].

## References

1. Neto J, Morais AJ, Gonçalves R, Leça Coelho A (2019) A multi-agent system for recommending fire evacuation routes in buildings, based on context and IoT, In: Communications in computer and information science, pp 343–347
2. Malizia A, Onorati T, Diaz P, Aedo I, Astorga-Paliza F (2010) SEMA4A: an ontology for emergency notification systems accessibility. Expert Syst Appl 37(4):3380–3391
3. Onorati T, Malizia A, Diaz P, Aedo I (2014) Modeling an ontology on accessible evacuation routes for emergencies. Expert Syst Appl 41(16):7124–7134
4. Liu G, Xu B, Tu Q, Sha Y, Xu Z (2011) Towards building ontology for community-based fire management. In: Proceedings 2011 international conference on transportation, mechanical, and electrical engineering, TMEE 2011, pp 1366–1369
5. Hristoskova A, Ongenae F, De Turck F (2013) Semantic reasoning for intelligent emergency response applications. In: IEEE international conference on industrial informatics (INDIN), pp 547–554
6. Casado R, Rubiera E, Sacristan M, Schütte F, Peters R (2015) Data interoperability software solution for emergency reaction in the Europe Union. Nat Hazards Earth Syst Sci 15(7):1563–1576
7. Bitencourt K, Durão F, Mendonça M (2015) Emergency fire: an ontology for fire emergency situations. In: Proceedings of the 21st Brazilian symposium on multimedia and the web – web media, pp 73–76
8. Gruber T (2009) Ontology. In: Liu L, Özsu MT (eds) Encyclopedia of database systems. Springer US, Boston, pp 1963–1965
9. Noy N, McGuinness D (2001) Ontology development 101: a guide to creating your first ontology. Protege 32:1–25
10. Gruber TR (1995) Toward principles for the design of ontologies used for knowledge sharing. Int. J. Hum. - Comput. Stud. 43(5–6):907–928
11. Pinto HS, Martins JP (Jul. 2004) Ontologies: how can they be built? Knowl Inf Syst 6(4):441–464
12. Grüninger M, Fox MS (1995) Methodology for the design and evaluation of ontologies.
13. Uschold M, King M (1995) Towards a Methodology for Building Ontologies
14. Fernández-López M, Gomez-Perez A, Juristo N (1997) Methontology: from ontological art towards ontological engineering. Eng Work Ontol Eng
15. Pinheiro J (2012) Medidas De Autoproteção De Segurança Contra Incêndio Em Edifícios. Autoridade Nacional de Protecção Civil

16. Neto J (2020) Conceptual map of the fire building evacuation ontology. https://www.mindomo.com/pt/mindmap/38f6214ddb874641b486f3fdc2a06f4a. Accessed 18 Jan 2021
17. Musen MA (2015) The protégé project. A look back and a look forward. *AI Matters, 1*(4):4–12
18. W3C (2004) OWL web ontology language overview. https://www.w3.org/TR/2004/REC-owl-features-20040210/#property. Accessed 08 Dec 2020
19. Gómez-Pérez A, Fernández-López M, Corcho O (2004) Ontological engineering: with examples from the areas of knowledge management, e-commerce and the semantic web. Springer
20. Poveda-Villalón M, Gómez-Pérez A, Suárez-Figueroa MC (2014) OOPS! (OntOlogy Pitfall Scanner!): an on-line tool for ontology evaluation. Int J Semant Web Inf Syst 10(2):7–34 (2014)
21. Garijo D (2017) WIDOCO: awizard for documenting ontologies. In: Lecture notes in computer science (including subseries Lecture Notes in Artificial Intelligence and Lecture Notes in Bioinformatics), vol. 10588. LNCS, pp 94–102

# Classification of Smartphone Users Based on Demographic and Technological Properties

**Rajitha Manellanga and Erunika Dayaratna**

**Abstract** Smartphones have become extremely popular regardless of any demographic factor. Users download different kinds of apps of a wide scope of categories, including finance, social networking, games, etc. The growth of smartphone usage generates valuable information for service providing companies. Service providers can analyze customer behaviors through smartphone usage data. However, most such data are not available to the public due to privacy. This study proposes a classification model to predict the gender and age of a smartphone user, based on usage of mobile applications, phone brand, and the location based on the AdaBoost algorithm, which is an ensemble learning method to build classification models. This model infers the demographics and technological properties from the dataset provided by TalkingData company. The outcomes show a promising accuracy of 97% for gender classification and a 0.89 coefficient of determination for predicting age. The proposed approach is eligible to profile smartphone users without disclosing privacy. Therefore, it is applicable to promote business services for targeted groups and individuals.

**Keywords** Smartphone user classification · User profiling · Gender classification · Mobile advertising · AdaBoost

## 1 Introduction

In the modern age, smartphone usage has grown tremendously, with smartphones becoming more and more sophisticated by the day. As a result, mobile applications are becoming popular among users. Social networking, proceeding online bookings, internet browsing, online payments, and document management are some of the examples of those applications. The competition between those companies is

R. Manellanga (✉) · E. Dayaratna
Department of Statistics and Computer Science, Faculty of Science, University of Peradeniya, Peradeniya, Sri Lanka
e-mail: rajitham@sci.pdn.ac.lk

E. Dayaratna
e-mail: erunika.dayaratna@sci.pdn.ac.lk

 
X.-S. Yang et al. (eds.), *Proceedings of Sixth International Congress on Information and Communication Technology*, Lecture Notes in Networks and Systems 216,
https://doi.org/10.1007/978-981-16-1781-2_85

growing with the increment of communication technologies. Therefore, it is valuable if they can identify their customers' behaviors. This also opens wisdom towards target advertising with minimal cost. Also, customizing their hardware and software technologies according to the user's needs is possible. The growth of mobile applications generates enormous amounts of data. i.e., the recorded events occurring by mobile applications usage interaction. Those data contain vast amounts of technological information such as application category, event, and users' demographic data. Age, gender, and location are common demographic data. Due to privacy issues, not all this information is disclosed to the public. However, anonymous data are open for reference. Those raw data are valuable sources to analyze the social behavior among mobile phone users [1]. Regarding the marketing and sales domain, the analyzed social behavior creates a path to intelligent advertising campaigns based on the distribution of customers. To keep existing customers and to attract new customers, it is essential to implement proper advertising campaigns. Identifying the different user needs within different geographical locations is important to find the target audience of the marketing campaigns. This can be achieved through both studying the patterns of customer distribution and consumer user profiling. Through that, it can be optimized to get better outcomes from geo-marketing and real-time marketing. In this study, we work to identify the demographic and technological facts which are applicable to classify smartphone users.

## 2 Related Works

In different domains, mobile phone data were used to analyze the behavior. Patterns of mobility and seasonal behavior related to subsistence zones in Senegal were analyzed through mobile phone data [2]. In the health sector, the relation between human mobility and the prevalence of a disease based on mobile data was studied [3]. A machine learning-based approach to predict the locations of mobile phone users was proposed by the authors of [4]. An attempt was made for user profiling based on mobile internet adopters in Japan [5]. In that study, cluster analysis and classification techniques were applied. Several studies were conducted to classify smartphone users based on different domains. In the study [6], a method was implemented to classify smartphone users by their gender, smoking habits, software programming experience, and other characteristics. An extensive study of social interactions in Mexico, focusing on gender and age, based on mobile phone usage was done by [7]. Even though some previous studies worked on the domain of classifying smartphone users, in this study, we consider the usage of several mobile applications within a wide geographical area. On the other hand, the smartphone brands are considered in this work as a feature to classify the users. Yalcin [8] proposes a probabilistic model to predict the gender and age of a customer, based on usage behaviors of mobile applications, phone brand, and model. However, we use the AdaBoost algorithm to classify the users of smartphones based on their gender and age. The study [9] represents a review of the AdaBoost algorithm as a classifier.

## 3 Dataset Description

In this section, we represent an overview of the smartphone usage datasets provided by TalkingData Corporation [10]. The data were obtained between the 30th of April 2016 and the 8th of May 2016. The data schema that has been used in this study can be described as follows:

- Event: When a user is involved with the TalkingData SDK, an event is recorded. Each event contains an event id, location (longitude and latitude) and the respective user's device id, and the timestamp.
- App labels: The used apps and their labels that define the subcategories
- Label categories: The categories of the corresponding app labels
- Mobile Phone Users: The users of TalkingData SDK are recorded. Each user has a corresponding device id, gender, age, and age group.
- Mobile Phone Brands: Each device has a corresponding brand

It is expected that the information obtained from smartphone users can identify the connection between demographic facts and smartphone usage.

## 4 Methodology

In this section, our approach for classifying smartphone users based on demographic and technological properties is described. Figure 1 shows the proposed method

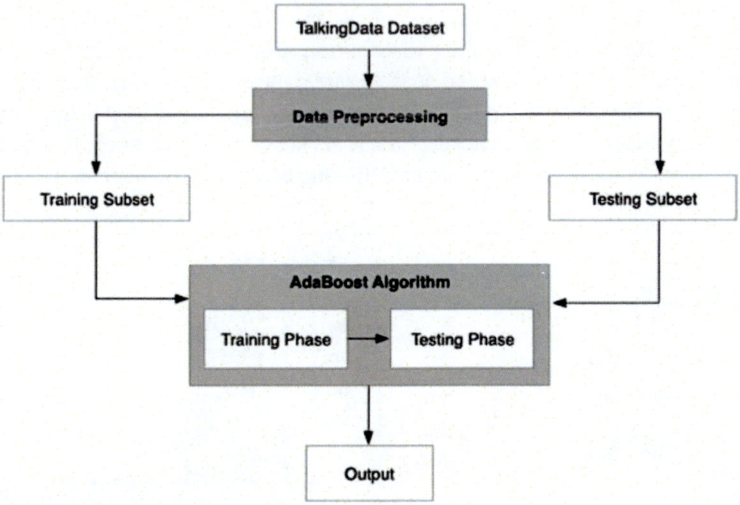


**Fig. 1** The structure of the proposed model

to classify the users. The AdaBoost algorithm is applied to train and evaluate a classification model on the data observed from the aggregated dataset.

## *4.1 Data Preprocessing*

The datasets provided by the TalkingData company contain several demographic properties and data observed through smartphone usage. In this study, we considered the locations, age, and gender of the users as demographic facts. As technological facts, mobile application usage and smartphone brands were considered. Due to the inconsistency and ineligibility to filter the meaningful records, the data about the mobile app labels were discarded. The data obtained from these sources contained missing entries. Since the dataset contains millions of records, those irregular entries were discarded. In this study, the events that occurred only within the geographical boundaries of China were filtered out to localize the model within a selected area. Among the smartphone users, only the users who were involved with the events were considered for further analytics purposes. It is noticeable that some brands of mobile phones have a significantly small number of users. Therefore, the smartphone brands which have at least 100,000 records have been filtered for further purposes. A significantly imbalanced distribution among the age and gender of the users could be observed.

Figure 2 gender and age distribution of the smartphone users. The majority of the users were male and the age distribution between 20 and 40 years contains most of the users. Therefore, this dataset could be identified as a highly imbalanced dataset.

Considering the limitations of the computation, a sample of 200,000 records is selected for the training dataset using the random sampling technique without replacement. Due to the imbalanced distribution of the gender among the users, the training dataset was undersampled. A separate dataset without undersampling was obtained to test the proposed model. To construct the prediction model, demographic factors, age, gender, and the location of the users are considered. As the technological facts, the mobile application and the smartphone brand are considered. The data set

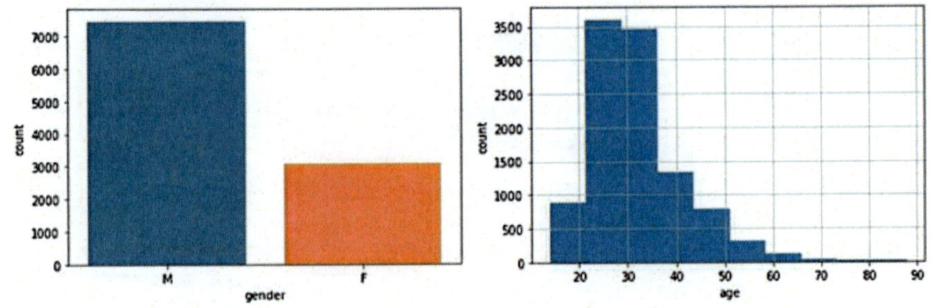


**Fig. 2** Gender and age distribution of smartphone users

**Table 1** Attributes of the aggregated dataset

| Attribute | Description |
|---|---|
| Age | Age of the user |
| Gender | Gender of the user |
| Phone brand | Brand of the user's smartphone |
| Event id | Unique identification number of an event that occurred within a timestamp |
| App id | Unique identification number of the mobile app that currently using |
| Longitude | Longitude of the user's current location |
| Latitude | Latitude of the user's current location |

is reconstructed using distinct records according to the selected demographic and technological variables. Table 1 shows the attributes of the reconstructed dataset.

### 4.2 *Classification Model*

In order to classify the smartphone users, the age and gender of the users were considered. The AdaBoost Classifier was applied for classification purposes. It is an ensemble learning method which is trying to construct a strong classifier from a variety of weak classifiers [11]. As a boosting method, the AdaBoost algorithm showed a higher precision than the other machine learning algorithms. Therefore, the AdaBoost algorithm was selected in this work. We implemented the model in a Python programming environment using the Scikit Learn machine learning library. The smartphone users are classified based on gender by applying the AdaBoost Classifier. Here, the Decision Tree Classifier was applied as the base estimator. Then, the users were classified based on age by using the "AdaBoost Regressor". For that, Decision Tree Regressor was applied as the base estimator. To evaluate the regression method, the coefficient of determination ($R^2$) was calculated using Eq. 1, where RSS is the sum of squares of residuals and TSS is the total sum of squares

$$R^2 = 1 - \frac{RSS}{TSS} \tag{(1)}$$

## 5 Results and Discussions

We observed that the distribution of smartphone users based on gender and age is imbalanced. In the distribution of smartphone brands, Huawei and Xiaomi were the most popular brands among both genders. Figure 3 shows the distribution of smartphone brands among the users. Regardless of gender, both Huawei and Xiaomi

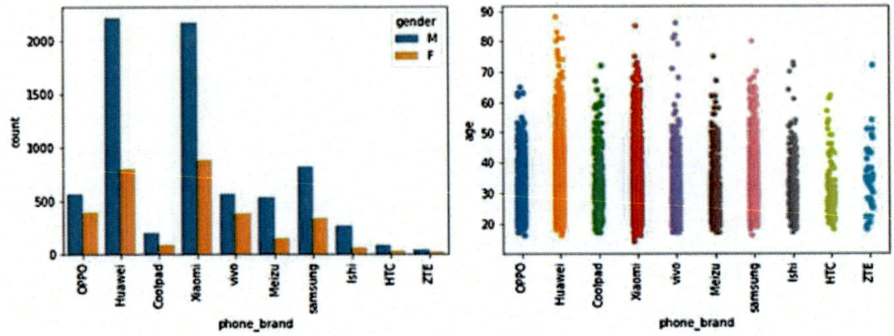

Fig. 3 Distribution of smartphone brands among the users

brands have almost equal levels of popularity among the users. Moreover, every brand was popular among men rather than women. All the brands had an acceptance among the age group of 20 to 40 years. However, the Huawei brand showed a clear popularity among the age group of 50 to 80 years rather than the other brands.

Therefore, it is identical that some age groups affect smartphone brand preference. It is observed that the geographical distribution of the users was centered on a specific area regardless of gender and age. Figure 4 shows the geographical distribution of the users within different hours on selected days.

The squared area shows the centered specific area of the distribution. Therefore, based on the results obtained we can see that users within the squared area have more influence on the outcome of the model.

The proposed AdaBoost Classifier was able to classify the smartphone users by their gender with an accuracy of 97%. We were able to classify the users by their age with an $R^2$ score of 0.89. Since AdaBoost is an iterative algorithm, it reweights the instances of the dataset for focusing the next classifier on incorrectly classified ones. Table 2 shows the accuracies of both models by the metrics. Figure 5 shows the distribution of the actual and predicted values of age. Around 89% of the data fit with the regression model. To exhibit the benefit of AdaBoost Regressor, Decision

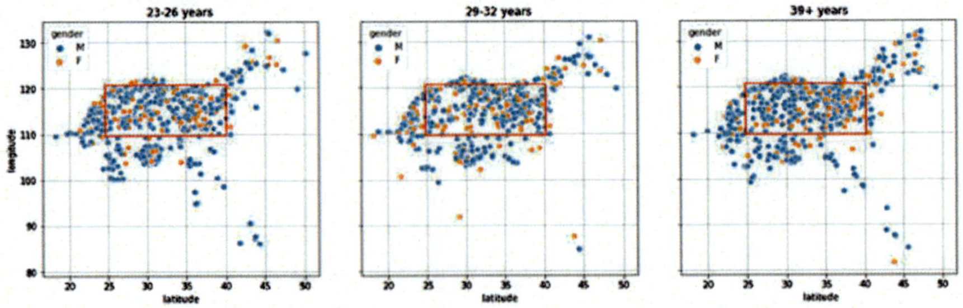

Fig. 4 Geographical distribution of smartphone users

**Table 2** Accuracies of the AdaBoost models

| Model | Evaluation metric | Value |
|---|---|---|
| AdaBoost classifier | Accuracy | 97% |
| AdaBoost regressor | $R^2$ value | 0.89 |

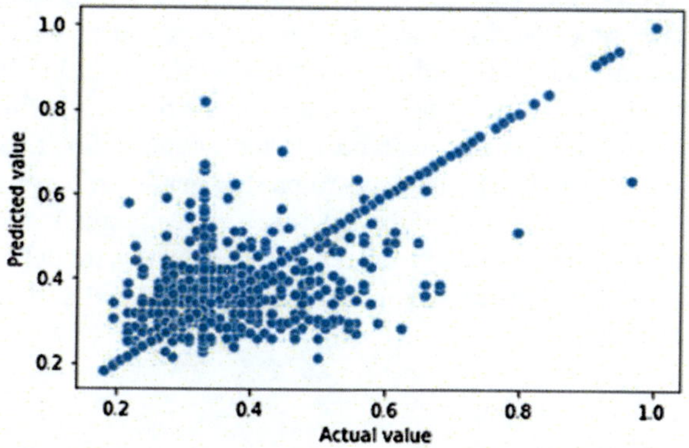


**Fig. 5** Actual and predicted values of age among smartphone users

Tree Regressor was used as a baseline model for predicting the age. An $R^2$ score of 0.83 was obtained from the baseline model. Therefore, it is clear that the AdaBoost algorithm is more accurate here. This algorithm was able to predict the age more precisely for the age group above 50 years rather than the age group between 20 and 40 years. Since the observed test results are satisfactory, the trained AdaBoost model is eligible for classifying smartphone users. Based on the results, we can see that the obtained results are satisfactory in classifying the smartphone users based on demographic and technological properties. The results in this study were obtained without revealing any privacy issues of a user. Since those results are based on demographic properties and mobile phone usage, it is beneficial to build mobile advertising strategies. Especially, focusing on personalized mobile advertising and mobile app usage among different geo-localized age groups.

Since these accuracies are obtained with a smaller number of attributes, the final accuracies could be increased by increasing the number of demographic and technological attributes.

## 6 Conclusion

In this paper, we utilize different demographic and technological properties to classify smartphone users based on their gender and age. The tremendous data provided by TalkingData company was used to propose our approach. Since most of the smartphone users were male and between 20 and 40 years, the dataset is significantly imbalanced. Due to the computing power limit, a sample of records was obtained to train the proposed model. The AdaBoost algorithm was applied to classify the smartphone users by gender and age, respectively. The observed results prove that some age groups can be identified as a fact that affects smartphone brand preference for some groups. It was clear that the geographical distribution of the users was centered within a specific area. It is valid for both genders and different age groups. The classification accuracies obtained by the model were satisfactory to classify the users based on their gender and age. Since we used a real-world imbalanced dataset, our proposed approach is eligible to classify the smartphone users to identify customer behavior. Even though the features including smartphone brands were used to profile the customers in a previous study, this study provides a descriptive analysis and separate ways to predict gender and age rather than making combined groups. The results of this approach were obtained without any disclosure of a user's privacy. Therefore, it is beneficial to apply this model for building mobile advertising strategies. Particularly, promoting personalized and targeted services is possible as outcomes.

## References

1. Felbo B, Sundsøy P, Pentland A, Lehmann S, de Montjoye YA (2017) Modeling the temporal nature of human behavior for demographics prediction. In: Altun Y, Das K, Mielikäinen T, Malerba D, Stefanowski J, Read J, Žitnik M, Ceci M, Džeroski S (eds) Machine learning and knowledge discovery in databases 2017. LNSC, vol 10536, pp 140–152. Springer, Cham. https://doi.org/10.1007/978-3-319-71273-4_12
2. Zufiria PJ, Pastor-Escuredo D, Úbeda-Medina L, Hernandez-Medina MA, Barriales-Valbuena I, Morales AJ, Jacques DC, Nkwambi W, Diop MB, Quinn J, Hidalgo-Sanchís P, Luengo-Oroz M (2018) Identifying seasonal mobility profiles from anonymized and aggregated mobile phone data. Application in food security. PLoS ONE 13(4):e0195714. https://doi.org/10.1371/journal.pone.0195714
3. Wesolowski A, Eagle N, Tatem AJ, Smith DL, Noor AM, Snow RW, Buckee CO (2012) Quantifying the impact of human mobility on malaria. Science 338(6104):267–270. https://doi.org/10.1126/science.1223467
4. Anagnostopoulos T, Anagnostopoulos C, Hadjiefthymiades S, Kyriakakos M, Kalousis A (2009) Predicting the location of mobile users. In: Proceedings of the 2009 international conference on pervasive services -ICPS '09, pp 65–72. Association for Computing Machinery, New York. https://doi.org/10.1145/1568199.1568210
5. Okazaki S (2006) What do we know about mobile Internet adopters? A Clust Anal Inf Manag 43(2):127–141. https://doi.org/10.1016/j.im.2005.05.001
6. Finkelstein A, Biton R, Puzis R, Shabtai A (2017) Classification of smartphone users using internet traffic. abs/1701.00220
7. Sarraute C, Blanc P, Burroni J (2014) A study of age and gender seen through mobile phone usage patterns in Mexico. In: 2014 IEEE/ACM international conference on advances in social networks analysis and mining (ASONAM 2014), pp 836–843. IEEE, Beijing. https://doi.org/10.1109/asonam.2014.6921683
8. Yalcin H (2017) Extracting user profiles from mobile data. In: IEEE international black sea conference on communications and networking (BlackSeaCom), pp 1–5. IEEE, Istanbul. https://doi.org/10.1109/BlackSeaCom.2017.8277701
9. Wang R (2012) AdaBoost for feature selection, classification and its relation with SVM, a review. Phys Procedia 25:800–807. https://doi.org/10.1016/j.phpro.2012.03.160

10. TalkingData Mobile User Demographics (2021). https://www.kaggle.com/c/talkingdata-mobile-user-demographics. Accessed 01 April 2021
11. Schapire RE (2013) Explaining AdaBoost. In: Schölkopf B, Luo Z, Vovk V (eds) Empirical inference, pp 37–52. Springer, Berlin. https://doi.org/10.1007/978-3-642-41136-6_5

# A Low Cost Spatial Calibration Method for Distributed Microphone Arrays

**Erik Verreycken, Walter Daems, and Jan Steckel**

**Abstract** Passive acoustic localization is a non-intrusive method that can be used for animal localization on species that emit vocalizations. Low-cost microphone arrays are an excellent tool in these experiments due to their ability to spatially localize the vocalizations. Accurate spatial calibration of the microphone positions remains an expensive or labor intensive step. Especially given the fact that bio-acoustic experiments can be performed in a wide range of environments such as indoor flight cages with easy accessible walls and a ceiling, as well as outdoor environments, where walls and ceilings are hard to come by. In this document, we present a low-cost tool that can be used to create an acoustic calibration signal. This tool allows for a faster and more robust calibration of the microphone positions while reducing the computational complexity of the calibration phase. This tool enables any researcher to calibrate the sensor array without knowledge of the sensing system.

**Keywords** Passive acoustic localization · Array geometry calibration · Acoustic self-calibration · 6DoF · Transducer array · Acoustic simulation

## 1 Introduction

Passive acoustic localization has many benefits in bio-acoustic experiments in contrast to active localization where a localization device is attached to the animal. The latter often involves capturing the animal which may be very stressful for the animal and requires specialized equipment and trained personnel [1]. As a general rule, the device attached to the animal should not exceed 5% of the body weight of the animal [2, 3]. Devices over this weight limit may change the behavior of the animal. This puts very tight constraints on what type of devices can be used and how powerful the

E. Verreycken · W. Daems · J. Steckel (✉)
FTI-CoSys Lab, University of Antwerp, Antwerp, Belgium
e-mail: jan.steckel@uantwerpen.be

Flanders Make Strategic Research Centre, Lommel, Belgium

 
X.-S. Yang et al. (eds.), *Proceedings of Sixth International Congress on Information and Communication Technology*, Lecture Notes in Networks and Systems 216,
https://doi.org/10.1007/978-981-16-1781-2_86

batteries of the device can be [4]. As an example, the weight of an adult Daubentons bat is approximately 10g [5], limiting the weight of the localization device to 0.5g.

Many animals use vocalizations for communication or navigation. By constructing a distributed microphone array, we can measure the arrival time of these vocalizations at different spatial locations. Given the speed of sound in air, we can translate these arrival times into distances and with enough measurements, at spatially diverse positions, we can find the origin position of the sound [6]. This is a form of multilateration called Time of Arrival (ToA). In fact, as no exact time of emission is known, we can only use the relative time difference between the arriving sound waves, referred to as Time Difference of Arrival (TDoA). Acoustic measurements can be performed using a microphone array constructed by a set of single microphones that connect to a central base station, using a kind of distributed sensor network or a combination thereof [6]. The maximal achievable accuracy of TDoA. Localization is largely determined by two factors [7, 8]: (1) Being able to make synchronized measurements. In previous work we discussed methods for synchronization of distributed sensor networks [9–11]. (2) Accurate knowledge of the 3D position of the microphones. This document focuses on the latter, for which several methods for calibrating the node locations can already be found in literature [12, 13]. **Manual Calibration** After placing the nodes, a researcher measures the location of each node. This is a tedious and error prone process that, as for all methods, has to be repeated every time the experimental setup changes. **GPS** The addition of GPS sensors to each of the nodes has the potential to automate the spatial calibration process. The accuracy of GPS, which for a single measurement lies in the meter range, even when averaging out over a period of an hour still lies in the centimeter range [14]. This is not accurate enough for a TDoA localization algorithm. **External Hardware** Several commercial camera based systems exist that provide sub-millimeter accuracy for localization and allow for spatial calibration of the sensor nodes, e.g., ViCon, Qualisys, etc. These type of systems typically cost in the order of several 10 000€, which is an order of magnitude more than the cost of our TDoA localization system. **Array Self-Calibration** This method involves creating an acoustic calibration signal, which is recorded by the microphone array [15]. The calibration signal can then be used to find the microphone positions in post-processing. This technique requires no extra hardware and adds no complexity to the network when the array is constructed. It is this method that is investigated in this document.

In Sect. 2, we explain how the calibration method works and what practical considerations can be made. Then, the proposed calibration tool is described in Sect. 3. Next, in Sect. 4, we show the simulation results of calibration using the tool. Finally, we propose our conclusions in Sect. 5.

# 2 Calibration

## 2.1 Principle

Calibration of the node positions begins with recording a calibration signal[15–17]. While theoretically this could be done by directly recording the animal sounds[16], often an explicit calibration phase is used. During this phase, an external calibration signal is used so that the node positions are known before the animal experiments start. A typical calibration signal can be recorded by creating a sound at several locations, e.g., by a person walking around with a clicker, i.e., a device that makes a known sound of sufficient bandwidth [18] that can be identified using a cross-correlation algorithm [19]. As a first step in the calibration algorithm, the unknowns are collected in vector $\vec{X}$ where

$$\vec{X} = \begin{bmatrix} \vec{M} \\ \vec{S} \end{bmatrix}$$

with $\vec{M}$ the positions of the microphones and $m$ the amount of microphones

$$M_i \in \mathbb{R}^3, \ i = 1, \ldots, m$$

and $\vec{S}$ the sound positions of $n$ sound emissions

$$S_i \in \mathbb{R}^3, \ i = 1, \ldots, n$$

so that $\vec{X}$ has $3 \cdot (n + m)$ degrees of freedom (DoF). The recorded calibration signal is converted in a set of arrival times, i.e., for each sound the relative arrival time, or difference in arrival time, to microphone 1. This set will be called the measurement vector $\vec{D}_m$ and has size $n \cdot (m - 1)$ as $M_1$ will always have arrival time 0 it is not collected in the measurement vector. From a known vector $\vec{X}$, we can calculate a single measurement vector $\vec{D}_m$, based on the physics of the sound, which means $\vec{D}_m$ is a function of $\vec{X}$

$$f(\vec{X}) = \vec{D}_m$$

A solution for an unknown $\vec{X}$ can be found using a least squares method that minimizes $||f(\vec{X}) - \vec{D}_m||$ [20, 21].

## 2.2 Practical Considerations

Optimal placement strategies for the microphones can be found in [22, 23]. These strategies optimize the amount of information from each microphone so a minimal amount of microphones is needed for adequate localization. While these strategies can be helpful for choosing an array topology, it is often not possible to create the exact topology in the experimental environment. Furthermore, the position of the microphones are not known beforehand which is the premise of this document. For these reasons, field researchers will often prefer to have an overdetermined microphone array which can simply be created by adding more sensors.

The same is true for the calibration signal. The researcher that creates the calibration signal will not be able to create the signal in the most mathematically efficient way. The solution again is to add more sound signals to the calibration signal which results in an oversized calibration signal. As a result of these practical implications, both $\vec{M}$ and $\vec{S}$ are overdetermined, each adding 3 degrees of freedom to $\vec{X}$ which quickly results in a large number of DoF to solve for.

### 2.3 *Previous Work*

In the previous work, we reduced the amount of DoF by grouping several microphones from the larger microphone array into smaller arrays [11, 24]. Each of the small-scale arrays can then be described by 6 DoF, i.e., a $3D$ translation and a $3D$ rotation. In our research, 5 microphones ($5 \times 3Dof(x, y, z) = 15$ unknowns) are grouped together. The optimization problem changes from estimating microphone positions for each microphone to the position and rotation for each group, thus reducing the DoF from 15 to 6 per small-scale array.

## 3 Calibration Tool

In this document, we explore the possibility to further reduce the DoF by creating a tool that can send multiple calibration signals simultaneously from a known constellation of transducers as seen in Fig. 1. This tool is conceived as a handheld device that can be handled by a biological researcher in the same way as the clicker mentioned earlier. Each of the transducers emits a different sound, i.e., a frequency sweep in a certain range so that a cross-correlation algorithm can distinguish between the different transducers [11]. Each sound $S_i$ in the calibration signal can now be described with 6 DoF, 3 rotations, and 3 translations, thus further reducing the total DoF assuming more than 2 transducers are present on the calibration tool. Vector $\vec{X}$ which previously only contained positional data $[x, y, z]$ for the sound positions now also contains rotational data $[\phi, \theta, \psi]$ for the wand. To be able to use $f(\vec{X})$ in its original form, we first need to translate the position and orientation of the group of speakers back to their individual $3D$ positions before $\vec{D}_m$ can be calculated.

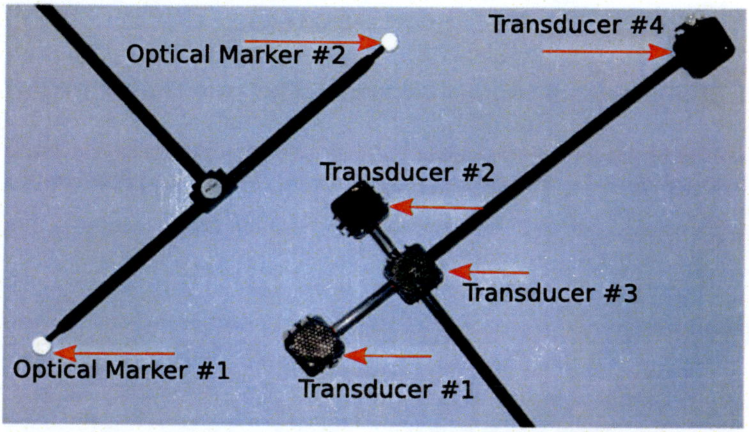


**Fig. 1** On the left is an optical calibration tool such as the one used for the Qualisys system. On the right is a version of our spatial calibration tool with 4 ultrasonic transducers. In both tools, the distances between the markers, optical or acoustic, and their relative positions are known very accurately

## 4 Simulation Results

We have built a simulator to test the calibration tool in different configurations and under different circumstances. We selected 2 realistic use cases for further investigation [25–27]. In the first scenario, the microphones are placed around the localization area so that everything encapsulated by the microphone array is part of the capture volume. For the second scenario, the microphones are placed on a flat surface so they can monitor anything that approaches the surface, e.g., a bat flying towards a wall. We incorporated noise on the detected arrival times of the sound. This noise comes from the limited level of inter-array synchronization and the accuracy to which we can identify a click using the cross-correlation algorithm. We selected a value that is an order of magnitude worse than the worst case we detected using our physical hardware as seen in [11].

$$\sigma_{noise} = 0.1\,\text{ms}$$

For the topology of the transducer array on the calibration tool, we got the best results by creating some variation in all axes. It is also important not to create symmetries in the array, e.g., speakers at equal distances or angles from each other as this can also create ambiguities in the calibration algorithm. This leaves 2 important parameters for the simulator: (1) amount of speakers on the tool, (2) amount of observations, i.e., the amount of clicks. We will do a sweep over several values for these parameters to test their influence on the calibration accuracy. In our simulation, we first create a certain amount of observations. The calibration tool is moved and rotated while making these observations. Each observation contains a number of sound events depending

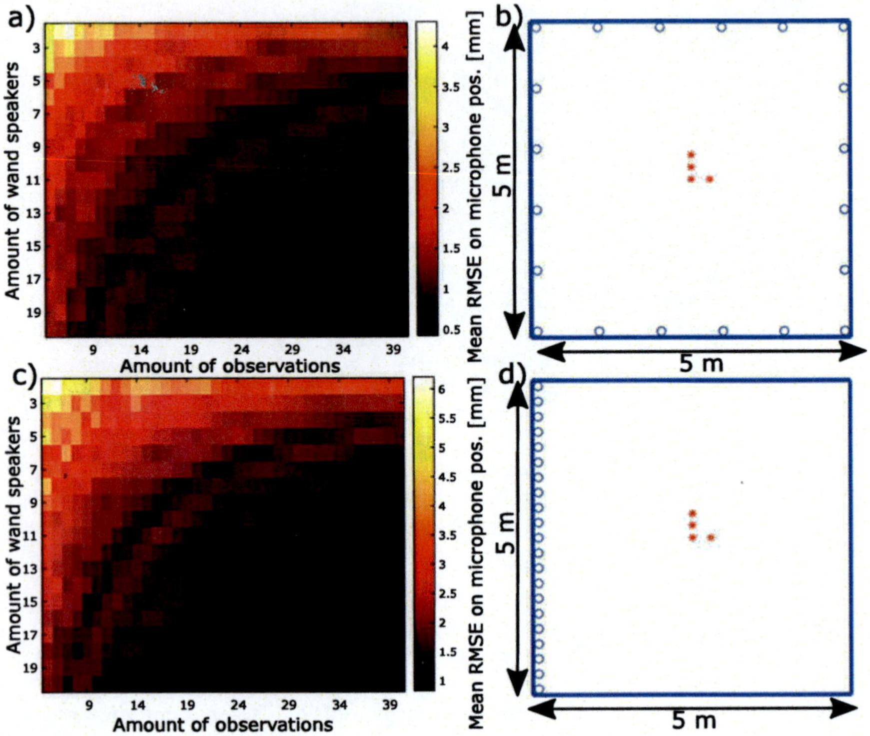


**Fig. 2** Simulation results. **a** Simulation results for the encapsulated area. This figure shows the relationship between the number of wand observations, the number of speakers on the tool, and the mean RMS error of the position of the microphones in mm after calibration of the array. A low amount of observations ($\leq$3) or a single wand speaker result in very larger mean RMS errors. These results have been omitted to show more detail in this figure. **b** The topology of the microphone array for the encapsulated area. A total of 20 microphones ('o') are distributed evenly on the walls of a room. The calibration tool is shown in its first observation position. The speakers are shown as '*' **c** Simulation results for the scenario where the microphones are placed on a single wall or plane. 21 microphones are placed on a wall. **d** The topology of the microphone array for the scenario with all microphones ('o') on a single wall. The calibration tool is shown in its first observation position. The speakers are shown as '*'

on the amount of transducers on the tool. All sound events are then translated to their relative arrival times $\vec{D}_m$. Next the algorithm tries to find a vector $\vec{X}$, which contains the microphone positions and the tool positions and orientations, that minimizes $||f(\vec{X}) - \vec{D}_m||$. The algorithm then returns a guess for the microphone positions and the tool positions. We evaluate the result by calculating the mean RMS error between the guessed microphone positions and the ground truth. The resulting mean RMS error is reported in Fig. 2. The simulation is repeated several times so the reported results in Fig. 2 are a mean over 30 simulations.

### 4.1 Scenario 1 - Encapsulated Capture Area

For this scenario, we placed 5 microphones on each wall of a room as can be seen in Fig. 2b. This scenario corresponds to an experiment in a lab [26]. In Fig. 2a, we show the mean RMS error on the microphone positions in mm. From Fig. 2a, it is clear that both an increase in the amount of observations as an increase in the amount of speakers on the tool have a positive effect on the accuracy to which we can calibrate the positions of the microphones in the array.

Each of the transducers emits a different sound with a large enough bandwidth [18] so that a distinction between the different transducers can be made in post production. Due to the limited bandwidth of both the transducers and the microphones, we cannot keep adding transducers to the calibration tool.

### 4.2 Scenario 2 - Towards a Wall

In this scenario, all microphones are placed in a single plane as can be seen in Fig. 2d. This will give us more information when the animal is moving towards the wall, especially if the vocalizations of the animal have a high directivity [27]. Positional calibration in this scenario is more difficult as there is less spatial diversity in the microphone array. As a result, more clicks are needed in the calibration signal to get positional accuracy that is comparable to the results from the first topology.

## 5 Conclusion

In this document, we have discussed the practical implications of passive acoustic localization for biological experiments. These practical implications can lead to a high number of DoF which results in a large optimization problem for the spatial calibration of the microphone positions. We have suggested a low-cost tool that can reduce the computational complexity of the optimization problem that needs to be solved in the calibration phase while also reducing the required skill and effort of the person calibrating the array. We have simulated a sweep of the 2 most important parameters that influence the accuracy of the spatial calibration. Using the simulation, we show that adding more transducers to the tool indeed has a similar effect to making more measurements in terms of localization accuracy while reducing the computational complexity of the localization algorithm.

## References

1. Barron DG, Brawn JD, Weatherhead PJ (2010) Meta-analysis of transmitter effects on avian behaviour and ecology. Methods Ecol Evol 1(2):180–187
2. O'Mara MT, Wikelski M, Dechmann DK (2014) 50 years of bat tracking: device attachment and future directions. Methods Ecol Evol 5(4):311–319
3. Aldridge HDJN, Brigham RM (1988) Load carrying and maneuverability in an insectivorous bat: a test of the 5% "rule" of radio-telemetry. J Mammal 69(2):379–382
4. Stidsholt L, Johnson M, Beedholm K, Jakobsen L, Kugler K, Brinkløv S, Salles A, Moss CF, Madsen PT (2019) A 2.6-g sound and movement tag for studying the acoustic scene and kinematics of echolocating bats. Methods Ecol Evol 10(1):48–58
5. Dietz M, Kalko EK (2006) Seasonal changes in daily torpor patterns of free-ranging female and male Daubenton's bats (Myotis daubentonii). J Comp Physiol B 176(3):223–231
6. Verreycken E, Daems W, Steckel J (2017) Distributed low-cost microphone array for 3D localization for bio-acoustic applications. In: IEEE sensors 2017 proceedings
7. Scheuing J, Yang B (2008) Disambiguation of TDOA estimation for multiple sources in reverberant environments. IEEE Trans Audio Speech Lang Process 16(8):1479–1489
8. Rabinkin DV, Renomeron RJ, French JC, Flanagan JL (1997) Optimum microphone placement for array sound capture. In: Luk FT (ed) International society for optics and photonics, vol 3162, p 227
9. Verreycken E, Laurijssen D, Daems W, Steckel J (2016) Firefly based distributed synchronization in wireless sensor networks for passive acoustic localization. In: 2016 international conference on indoor positioning and indoor navigation (IPIN). IEEE
10. Laurijssen D, Verreycken E, Geipel I, Daems W, Peremans H, Steckel J (2018) Low-cost synchronization of high-speed audio and video recordings in bio-acoustic experiments. J Exp Biol 221(Pt 4):jeb173724
11. Verreycken E, Daems W, Steckel J (2018) Passive acoustic sound source tracking in 3D using distributed microphone arrays. In: 2018 international conference on indoor positioning and indoor navigation (IPIN). IEEE, pp 1–8
12. Xiao J, Liu Z, Yang Y, Liu D, Han X (2011) Comparison and analysis of indoor wireless positioning techniques. In: 2011 international conference on computer science and service system (CSSS). IEEE, pp 293–296
13. Farid Z, Nordin R, Ismail M (2013) Recent advances in wireless indoor localization techniques and system. J Comput Netw Commun 1–12
14. Lin M, Song X, Qian Q, Li H, Sun L, Zhu S, Jin R (2019) Robust gaussian process regression for real-time high precision GPS signal enhancement
15. Thrun S, Thrun S (2005) Affine structure from sound. IN NIPS 18:1353–1360
16. Gaubitch ND, Kleijn WB, Heusdens R (2013) Auto-localization in ad-hoc microphone arrays. In: 2013 IEEE international conference on acoustics, speech and signal processing. IEEE, pp 106–110
17. Pollefeys M, Nister D (2008) Direct computation of sound and microphone locations from time-difference-of-arrival data. In: 2008 IEEE international conference on acoustics, speech and signal processing. IEEE, pp 2445–2448
18. Velasco J, Taghizadeh MJ, Asaei A, Bourlard H, Martiin-Arguedas CJ, Macias-Guarasa J, Pizarro D (2015) Novel GCC-PHAT model in diffuse sound field for microphone array pairwise distance based calibration. In: 2015 IEEE international conference on acoustics, speech and signal processing (ICASSP). IEEE, pp 2669–2673
19. Knapp C, Carter G (1976) The generalized correlation method for estimation of time delay. IEEE Trans Acoust Speech Signal Process 24(4):320–327
20. Vanwynsberghe C, Challande P, Ollivier F, Marchal J, Marchiano R (2019) Geometric calibration of very large microphone arrays in mismatched free field. J Acoust Soc Am 145(1):215–227
21. Hennecke MH, Fink GA (2011) Towards acoustic self-localization of ad hoc smartphone arrays. In: 2011 joint workshop on hands-free speech communication and microphone arrays. IEEE, pp 127–132

22. Schenck A, Walsh E, Reijniers J, Ooijevaar T, Yudanto R, Hostens E, Daems W, Steckel J (2018) Information theoretic framework for the optimization of UWB localization systems. In: IPIN 2018 - 9th international conference on indoor positioning and indoor navigation. IEEE, pp 1–8
23. Catovic A, Sahinoglu Z (2004) The cramer-rao bounds of hybrid TOA/RSS and TDOA/RSS location estimation schemes. IEEE Commun Lett 8(10):626–628
24. Laurijssen D, Truijen S, Saeys W, Steckel J (2015) Three sources, three receivers, six degrees of freedom: an ultrasonic sensor for pose estimation & motion capture. In: 2015 IEEE SENSORS. IEEE, pp 1–4
25. Motoi K, Sumiya M, Fujioka E, Hiryu S (2017) Three-dimensional sonar beam-width expansion by Japanese house bats (Pipistrellus abramus) during natural foraging. J Acoust Soc Am 141(5):EL439–EL444
26. Surlykke A, Ghose K, Moss CF (2009) Acoustic scanning of natural scenes by echolocation in the big brown bat, Eptesicus fuscus. J Exp Biol 212(Pt 7):1011–20
27. Gaudette JE, Kloepper LN, Warnecke M, Simmons JA (2014) High resolution acoustic measurement system and beam pattern reconstruction method for bat echolocation emissions. J Acoust Soc Am 135(1):513–520

# Type-2 Fuzzy Set Aggregation and Health Status Mining from Condition-Monitoring Applications

Roberto Bodo, Matteo Bertocco, and Alberto Bianchi

**Abstract** The study focuses on health monitoring applications performing Fault Identification (FI) and Maintenance Indication (MI). In industry, third-party algorithms based on artificial intelligence could be supplied in closed-source. For this reason, they may lack of customization, or retraining, that leads to a possible poor performance. The work proposes a system called Fuzzy Set Aggregator (FSA) receiving in input unstable classifier predictions over time and provides a stable MI signal with a reasonable degree of anticipation. Uncertainty quantification and management are embedded in the aggregation process by using the type-2 fuzzy set theory. Four system variants are tested through simulation with data extracted from a real-case study. Two FSA variants were able to anticipate the expected reference values, providing valuable signals to plan maintenance at a reduced cost, and showing a performance improvement when the uncertainty due to expert opinion is greater.

**Keywords** Type-2 fuzzy set · Fault diagnosis · Fault identification · Fault classification · Predictive maintenance · Condition monitoring · Maintenance indication · Maintenance planning

## 1 Introduction

The integration of artificial intelligence techniques inside industrial systems and plants gives the chance to offer new smart services. Among them, the maintenance of cyber-physical systems receives great benefits from Predictive Maintenance (PM,

R. Bodo (✉) · M. Bertocco
Department of Information Engineering, Università degli Studi di Padova, Padova, Italy
e-mail: roberto.bodo@phd.unipd.it

M. Bertocco
e-mail: matteo.bertocco@unipd.it

A. Bianchi
Carel Industries Spa, Padova, Italy
e-mail: alberto.bianchi@carel.com



X.-S. Yang et al. (eds.), *Proceedings of Sixth International Congress on Information and Communication Technology*, Lecture Notes in Networks and Systems 216,
https://doi.org/10.1007/978-981-16-1781-2_87

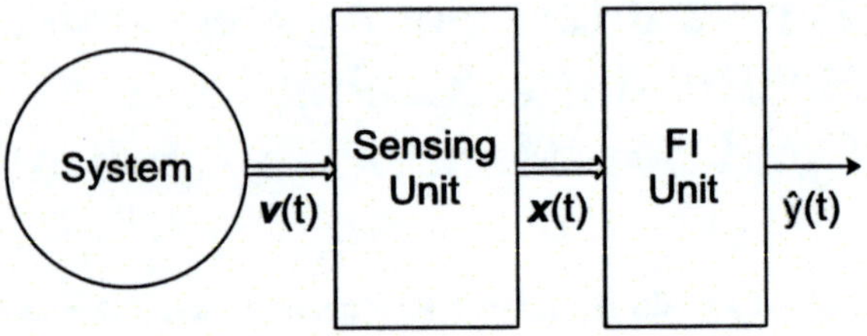


**Fig. 1** Classification scheme for a sensing and fault identifying application

[1]) paradigms, enabling to devise an effective Prognostics and Health Management (PHM, [2]) or Condition-Based Maintenance (CBM, [3]). Several tasks belong to the PM class, from Anomaly Detection (AD) to Remaining Useful Life (RUL) estimation. This study focuses on Fault Identification (FI) and Maintenance Indication (MI) tasks, i.e. to schedule the technician intervention based on the machine health status [4], but can be extended to other tasks as well. In the industrial application context, a third-party monitoring application can embed the PM functionality directly on the edge node, providing a close system with full functionality. This situation, exemplified in Fig. 1, depicts an edge computing node embedding units for sensing, processing, and MI. In such context, the algorithms developed by the third party for FI and MI could suffer from generalization error [5], and under-perform on the real application.

Several methods have been recently applied to solve the FI task: they can be classified into model-based, signal-based , and data-driven. Model-based techniques exploit mathematical models to draw conclusions about the machine health status. However, the strength and weakness of such methods reside in the accuracy of the models built. Signal-based approaches instead compare acquired signals with prior information about the system to detect faults [6]. It may happen that such prior information is stuck in the third-party algorithm, so that less or no service improvement can be done on the field application. The majority of recent methods reside in the data-driven class that ranges from statistical methods to Deep Learning, and Fuzzy Logic. The state of the art of data-driven methods includes those capable of dealing with different types and degrees of uncertainty, called knowledge-based fault diagnosis. Among them, static and dynamic Bayesian Networks (BN) are one of the most promising. However, a correct cause-effect relational model estimation is crucial to provide an accurate and fast estimate. Nonetheless, intermittent faults and sudden changes of health status still represent an open issue [7]. Dempster-Shafter, or evidence theory, enable to manage uncertainty in decision-making coming from multiple sources, but comes with inference computational burden. In particular, hypothesis tests needed for $n$ faults scale up to $2^n$ [8]. Fuzzy Logic methods often provide good application results at the price of a high computational cost. However, these methods provide a powerful model-agnostic and context-free alternative to the above methods, with a high flexibility of hybridation [9]. In some cases, simplifications can be made in order to use closed-form results and speed-up the calculations, below described.

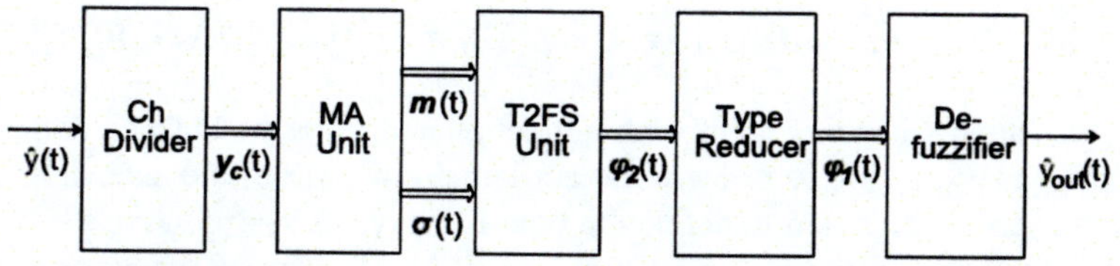


**Fig. 2** Block diagram of the proposed fuzzy system aggregator (FSA)

In order to provide a more stable MI prediction and to mine additional information on the health status from the monitored system, the paper proposes the usage of a Type-2 Fuzzy Set (T2FS, [10]) classification unit suitable for time-varying systems. Such unit, followed by proper type-reduction and defuzzification, enables to devise a fuzzy logic system, here called Fuzzy System Aggregator (FSA) and schematically reported in Fig. 2. The FSA becomes a prediction aggregator over time, having the domain and co-domain equal to the maintenance class set of the third-party MI estimator, that aims:

- to manage the input uncertainty without knowing prior information about the third-party classifier (model agnostic) or the machine being monitored (context-free),
- to free the application expert from designing adaptive fuzzy numbers to describe the dynamic observation of the system and to keep the computational cost of the model as low as possible,
- to anticipate the expected instant of the MI request to a desired degree dependent on the given sample rate and noise level, in order to improve scheduling margin.

## 2 Background

The first publications of fuzzy set theory by Zadeh [11] were focused on the generalization of the classical notion of sets and propositions in order to mathematically cover fuzziness [12]. A Type-1 Fuzzy Set (T1FS) can be defined as:

$$A = \{(x, \mu_1(x)) | x \in X \wedge \mu_1(x) \in (0, 1]\} \quad (1)$$

where an element $x$, taking value in its domain $X$, belongs to the fuzzy set A for a certain degree specified by $\mu_1(x)$. This function takes the name of (primary) membership function and its co-domain is bounded between 0 and 1.

*Type-2 Fuzzy-Sets:* $\mu_1(x)$ takes only one precise value for each input data $x$. This requires the perfect knowledge of the belonging degree of an element $x$ to the set $A$. In order to embed the uncertainty in the knowledge of $\mu_1(x)$, called Footprint-of-Uncertainty (FOU, [13]), the previous definition has been extended by Zadeh by introducing a second membership function $\mu_2(\mu_1)$, according to the following definition:

$$A = \{(x, \mu_1(x), \mu_2(\mu_1)) | x \in X \wedge \mu_1(x) \in (0, 1] \wedge \mu_2(\mu_1) \in (0, 1]\} \quad (2)$$

$\mu_2(\mu_1)$ is called secondary membership function and maps the FOU by defining an additional T1FS, taking as domain the co-domain of $\mu_1(x)$. The extension of the definition can be reiterated to define the type-L fuzzy set, representing the concept "*... of uncertainty of uncertainty of x belonging to A*". Various functions are used in literature [14] to model the shape of $\mu_1(x)$ and $\mu_2(\mu_1)$. Among them, due to the presence of closed-form results, Gaussian functions are widely used, and hence in this work too. Fuzzy sets are the main blocks in Fuzzy Logic Systems (FLS). In such architecture, the input value $x$ is mapped into a type-L fuzzy set and processed with an inference unit based on lexical rules [15]. Then, the inferred result, if needed, is reduced from type-L to type-1 and furtherly reduced to the crisp domain of output values $y \in Y$. Type-reduction deals with the uncertainty modeled in the type-L domain and produces a value in the type-(L-1) domain. Many techniques have been proposed in literature [16], optimizing different tasks and calculus efficiency. Among them, the centroid method is widely used for general purpose tasks. It will be later used as FSA type-reduction unit.

*Fuzzy Set Classification* can be realized by IF-THEN rules. Crisp classification consists in precise output values associated to sharp threshold rules [12]. On the other side, Fuzzy Set Classification (FSC) employs fuzzy sets in the IF part of the rules and fuzzy sets or precise (crisp) values for the THEN part of the rules. When the output takes a single value from a finite set of classes, the winner criteria may be applied. The overall output is thus assigned to the THEN part of the rule having the highest firing value. This process belongs to the family of the defuzzification methods and is named max-membership defuzzifier [17]. The max-membership method can be used directly with T2FS, but comes with the drawback of not embedding the uncertainty modeled in $\mu_2(\mu_1)$ [18]. In order to overcome this limitation, the classification process with T2FS requires, after the inference step, a type-reduction to a T1FS and finally the defuzzification step.

*Applications* Among industry and system maintenance, several applications are found in manufacturing [19], and safety engineering [20]. In system maintenance, FSC is used together with PM and condition monitoring in order to solve the FI task [21], Anomaly Detection (AD) [22], and fault prediction. The maturity of the fuzzy set theory and the performance robustness of FLS often claimed in literature, here only summarized, offers a promising method for the estimation of the machine health status [9].

## 3 Research Method

A MI system is exemplified in Fig. 1. The sensing unit is responsible for collecting input data $\boldsymbol{v}(t)$, representing the physical variables of the system, from the sensors installed. This unit is also responsible for processing data to build the vector of features $\boldsymbol{x}(t)$. The classification unit then use $\boldsymbol{x}(t)$ to make a maintenance class prediction $\hat{y}(t)$, $y \in Y$, among a discrete set of possible maintenance actions $Y$. The MI prediction suffers from the instability caused by the monitored system disturbances $\epsilon_s$ and the classifier model error $\epsilon_{clf}$, according to the model

$$\hat{y}(t) = y(t) + \epsilon_s + \epsilon_{clf} \quad \epsilon_s >> \epsilon_{clf}, \quad \epsilon_s \sim \mathcal{N}(0, \sigma_s^2) \tag{3}$$

being $y(t)$ the true MI class. This study is focused on the system noise, so the classifier uncertainty is supposed to be negligible: this assumption can be a-priori checked by evaluating the algorithm performance declared by the third party. The system noise $\epsilon_s$ is modeled as additive white Gaussian noise reported at the output, expecting the major contribution coming from the physical noise collected by sensors. The system of Fig. 2 uses the classifier predictions $\hat{y}(t)$ to produce a MI signal $\hat{y}_{out}(t)$ and to mine information about the system health status. Such status modeling starts from the number of all possible classes $Y$, e.g. 4, that defines the universe of discourse $H$ [12]. A T2FS is then built for each maintenance class, in order to create a partition of $H$, according to

$$\begin{cases} A, B, C, D \subset H | \cup_{A,B,C,D} \equiv H \\ \boldsymbol{x}(h) = [\mu_2(\mu_A(h)), \mu_2(\mu_B(h)), \mu_2(\mu_C(h)), \mu_2(\mu_D(h))] \\ \mu_2(\mu_{(\cdot)}(h)), \mu_{(\cdot)}(h) \in (0, 1], \forall h \in H \end{cases} \tag{4}$$

Figure 3 provides a graphical intuition of the model. Each health status $h \in H$ can be represented by a vector of secondary membership functions $\boldsymbol{x}$ that model the belonging of $h$ to each fuzzy set $A, B, C, D$ related to an output class of the Y domain. The FOU between $h$ and the output classes is here represented not as a point estimation, but a particle shell centered at $h$ with a Gaussian decreasing magnitude, according to Eq. 3.

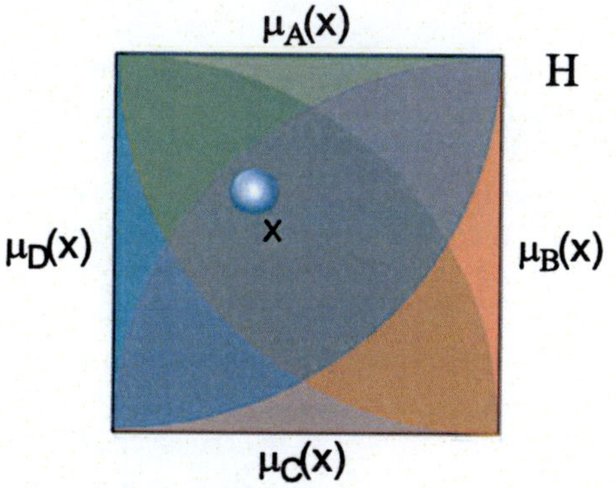


**Fig. 3** Fuzzy system model intuition

### 3.1 FSA Description

The proposed FSA is a sequential composition of various unit, each solving a specific task.

*Channel Divider*: it is responsible for producing a vector of time-series $\boldsymbol{y_c}(t)$, one for each maintenance class label. Each time-series is a dichotomic variable of value 1 when the classifier prediction $\hat{y}(t = \tau) = i$ selects the $i$-th maintenance class for a given instant $\tau$, while it is 0 when the other classes are selected. Each component of $\boldsymbol{y_c}(t)$ is similar to the signal of a 1-bit $\Sigma/\Delta$ modulator [23] before interpolation.

*Moving Average Unit*: it solves the task of estimating the mean $\boldsymbol{m}(t)$ and the standard deviation $\boldsymbol{\sigma}(t)$ for each channel of $\boldsymbol{y_c}(t)$ over a time-rolling window of length $n$. The simplest way of doing such estimates is through the Simple Moving Average (SMA) algorithm [24]. Such method suffers from an intrinsic delay in the algorithm structure, equal to $0.5n$ [25]. It will be later shown, in Fig. 8, that the effect of this lag leads to a poor MI signal. The Exponential Moving Average (EMA) estimator [26], a common adaptation of the SMA estimator with minor lag, assigns exponential weights among the data-points in the time-window. Weights give more importance to the recent samples while they exponentially decrease over the past data-points. The SMA and EMA estimators are later compared as MA unit types for different variants of the FSA.

*Type-2 Fuzzy-Set Unit* It acts as a fuzzifier unit by using the $\boldsymbol{m}(t)$ and $\boldsymbol{\sigma}(t)$ estimates to build a T2FS for each channel; each fuzzy set represents a component of the $\boldsymbol{\varphi_2}(t)$ vector. For each instant $\tau$, the pair $(m_i,\sigma_i)$, where $m_i,\sigma_i$ represent the components of, respectively, $\boldsymbol{m}(t)$ and $\boldsymbol{\sigma}(t)$ belonging to the $i$-th channel, is employed to build the T2FS. $m_i$ is used as primary membership value $\mu_1$, while $\sigma_i$ is used to model the FOU around $\mu_1$, namely $\mu_2(\mu_1(x))$

$$\begin{cases} \mu_1(x) = m_i \\ \mu_2(\mu_1(x)) = e^{-\frac{1}{2}(\frac{x-m_i}{\sigma_i})^2} \end{cases} \tag{5}$$

This choice provides efficient calculations, and keeps interesting properties in closed forms, as quoted in [18]. A graphical exemplification of this process is displayed in Fig. 7, for each channel considered.

*Type-reducer* This unit is responsible for reducing the T2FS to a T1FS, or to embed the FOU previously modeled into a new T1FS whose (primary) membership function $\mu_c(x)$ is calculated starting from the Gaussian curves previously defined. Among several type reducer methods presented in literature, the centroid method has been chosen. It solves, for a discrete domain [17]

$$\mu_c = \frac{\sum_i \mu_2(\mu_1(x_i))\mu_1(x_i)}{\sum_i \mu_2(\mu_1(x_i))} \tag{6}$$

**Table 1** System variant parameters under analysis

| Variant | Parameters |
|---|---|
| A | ma_unit='sma', $n$=24, |
| | hyst_unit='hyst', $th$=0.2 |
| B | ma_unit='ema', $n$=24, |
| | hyst_unit='hyst', $th$=0.2 |
| C | ma_unit='sma', $n$=24, |
| | hyst_unit='hyst_bt', $th$=0.2, $\beta\_th$= 2, $t_w$=360 |
| D | ma_unit='ema', $n$=24, |
| | hyst_unit='hyst_bt', $th$=0.2, $\beta\_th$= 2, $t_w$=360 |

Basically, it calculates the average height of the area below a given curve. It is adaptable for all types of curves defined as FOU and efficient to calculate by using discretization of continuous domains. In the proposed setup, $\mu_c$ is close to $m_i$, theoretically coincident if using ideal Gaussian curves, but it accounts for truncation of $\mu_1(x)$ in the interval [0, 1]. The type reducer returns the value of $\mu_c(t)$ belonging to each channel inside the vector $\boldsymbol{\varphi}_1(t)$.

*Defuzzifier:* this unit uses the centroids values of $\boldsymbol{\varphi}_1(t)$ to make a classification in the same domain of $\hat{y}(t)$ and to provide the MI signal $\hat{y}_{out}(t)$. As reported in Sect. 2, a widely used method to perform classification is the max-membership reduction, from which the $max$ of the $\mu_c$ values is selected. This method, in its simplest formulation, has not been able to directly reduce the system noise contribution $\epsilon_s$ to a reasonable level, so the two following adaptations using hysteresis are made. Algorithm 1, implemented in the A,B variants of Table 1, uses the classical hysteresis concept. A symmetric interval parametrized by the threshold $th$ is defined; if the $max(\boldsymbol{\varphi}_1(t))$ falls inside it, then the output value is not updated, otherwise the opposite. Blue adaptations of Alg. 1, used in the C,D variants of Table 1, extends the concept of threshold hysteresis by adding information about the trend dynamics of $max(\boldsymbol{\varphi}_1(t))$. Function $beta(\cdot, t_w)$ uses a Ordinary Least Square (OLS, [27]) linear regression over a rolling window of length $t_w$ to calculate the linear slope coefficient $\beta$. If the centroid value falls into the hysteresis interval but exhibits a strong growing trend $beta(max(\boldsymbol{\varphi}_1(t)), t_w)$, then the output is updated.

**Algorithm 1** Defuzzification unit: in black the A,B variants while in blue the adaptations added to C,D variants (Table 1).

**Input:** $\varphi_1(t), th, \beta\_th, t_w$
**Output:** $\hat{y}_{out}(t)$

```
hold_y ← 0 % current prediction
hold_μc ← 0 % current centroid
hold_β ← 0
for τ ∈ t do
  if not(1 − th)hold_μc<max(φ1(τ))<(1 + th)hold_μc
  or beta(max(φ1(τ)), tw) >β_th hold_β
   then
    hold_μc ← max(φ2(τ))
    hold_y ← argmax(φ1(τ))
    hold_β ← beta(max(φ1(τ)))
  end if
  ŷout(τ) = hold_y
end for
return ŷout(t)
```

## 4 Experimental Design

In order to check the system performance, a simulation based on a MI system, previously analyzed by the authors in [4, 28], have been carried out, where a multiple-class indication system has been analyzed. The considered application monitors a humidifying system composed by a high pressure pump and an inductive inverter-driven motor for HVAC/R applications.

*Aging-data Generation* The MI system classes are: $0 - baseline$ or healthy system, $1 - abnormal$ or anomalous condition, $2 - mild$ or ordinary maintenance, and $3 - full$ or full system restoration. The $1 - abnormal$ class is already recognized by the machine and proper auto-reset operations are used. Thus, for the ongoing analysis, this class is not involved in the MI task, but still provides a source of mis-classification error, and hence maintained and modeled for experimentation. The class predictions over time $\hat{y}(t)$ are modeled as a time-varying discrete random variable $X(t)$, whose probabilities $p_w(t)$ are given in graphical form in Fig. 4. Such probabilities are obtained from an expert elicitation process [29, 30] and consider a run-to-failure situation, where no maintenance is performed in the monitored system. For each instant $\tau$, the $p_w(\tau)$ values are defined in such a way that $\sum_i p_{w_i}(\tau) = 1$. Accordingly the probability of the $0 - baseline$ is defined as 1 minus the sum of the other class probabilities. The non-monotonic behavior of the $0 - baseline$ of $p_w(t)$ is hence a consequence of the uncertainty intrinsic to the elicitation process. In Fig. 4, it is also shown a time interval where the $p_w(t)$ of the $2 - mild$ class decreases over time; at the same time, the $p_w(t)$ associated to $3 - full$ class increases. This time interval reflects the experts' uncertainty about which correct maintenance action to take, because the system degradation does not require anymore an ordinary main-

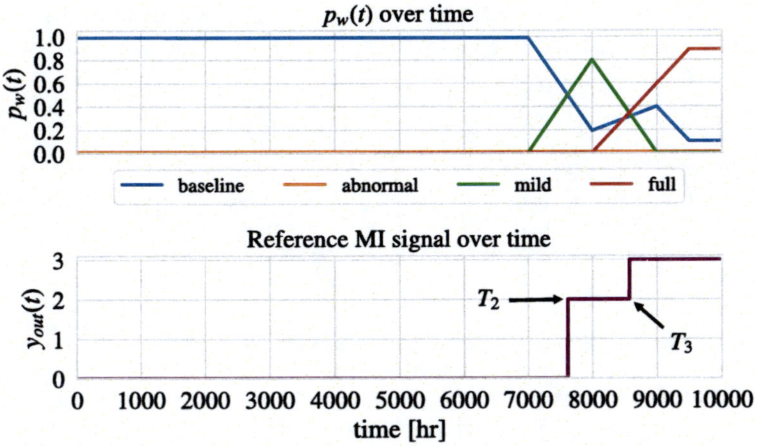


**Fig. 4** Probabilities dynamics $p_w(t)$ used for the data generation and the reference maintenance signal $y_{out}(t)$, T2 and T3 are the first istants reporting, respectively, the *mild* and the *full* classes

tenance action but a forthcoming full restoration. $X(t)$ is then used to draw the time-series $\hat{y}(t)$ corresponding to one system monitored for $10000hr$ sampled at 1 sample per hr. This process is later reiterated to obtain average results.

## 4.1 Design

The research work has led to an experimental design [31] with the following main goal: *"to benchmark the system variants of Table 1 on the first indication of* the, *mild and full classes"*. It is called first indication, on the system run-to-failure life, the first instant $T_i$ when a certain maintenance classes is given by $\hat{y}_{out}$. As reference, the expected first instants $T_2$ and $T_3$, respectively, for the *mild* and the *full* classes are reported in Fig. 4.

*Hypotheses and variables definition* The main goal is thus translated into the two following hypotheses:

$$\begin{cases} H_0 : median(T_i, var) = T_i \\ H_1 : median(T_i, var) < T_i \end{cases}$$
$$i \in \{2 - mild, 3 - full\}, var \in Table\ 1 \quad (7)$$

$$\begin{cases} H_0 : median(T_x, var_{L+1}) = median(T_x, var_L) \\ H_1 : median(T_x, var_{L+1}) < median(T_x, var_L) \end{cases}$$
$$T_x \in \{T_2, T_3\}, var_{L,L+1} \in Table\ 1 \quad (8)$$

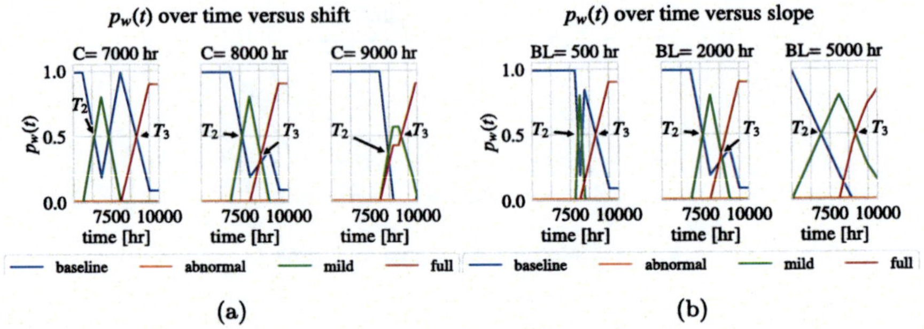


**Fig. 5** Sensitivity analysis: $p_w(t)$, $T_2$, and $T_3$ changes due to: (a) the shift of the *mild* maintenance class, modeled as a triangle function with center in C and base length BL=2000hr; (b) the slope change of the *mild* class, modeled as a triangle function centered in C=8000hr and base length BL

A MI performance is considered good if the FSA variant gives the first indication instants $T_2$ and $T_3$ before the expected reference (Eq. 7). Hypotheses in Eq. (8) check the performance improvement given the different design choices of Table 1 about the MA unit and the hysteresis algorithm used during the defuzzification. In order to check the above hypotheses, the $T2$ and $T3$ median values among system replicates are chosen as response variables.

*Processing* Four system variants are instantiated according to the parameters reported on Table 1. Each system variant is applied to the same time-series and the process is reiterated 100 times in order to simulate a consistent population of monitored systems.

*Evaluation* Each response variable is statistically described and the population indicators: mean, median, and $\sigma$ are extracted. Each response population is then tested for the Gaussian assumption by using the Anderson-Darling test [32] (confidence $C = 95\%; \alpha = 0.05$), that indicates a non-Gaussian distribution for all the population tested. Hence, the 1-sample Sign test [27] has been used to check hypotheses of Eq. 7. The Mann-Withney non-parametric test [33] (confidence $C = 95\%; \alpha = 0.05$) is then employed to check each hypothesis of Eq. 8 disjointedly. Processing has been supported by a Python 3.6.9 platform with Scipy 1.3.1 and Scikit-learn 0.21.3 packages. The statistical hypotheses have been tested using Minitab® Statistical Software v. 19.1.1.

*Sensitivity Analysis* In order to test the system performance under an uncertain expert judgment, the $2 - mild$ distribution of $p_w(t)$ is modified. First, the time instant at which $p_w(t)$ rises up is changed in such a way that the intersection with the curve associated to the $3 - full$ class occurs at different time instants and for different values of the intersection amplitude itself. Then, the rate of change of the curve $p_w(t)$ relative to the $2 - mild$ class has been let varying so that the corresponding effects are investigated. Examples of the corresponding shapes of the curves $p_{w_i}(t)$ are seen in the Fig. 5.

## 5 Experimental Results

Plot of Fig. 6 shows the dynamics of the centroids values of $\varphi_1(t)$. Since plots of Figs. 4 and 6 should be ideally the same, despite a reasonable agreement between reference and estimates, the estimation process that uses one machine time-series is affected by an inner relevant noisy contribution. Moreover, the noisy behavior of the curves does not allow a clear identification of the relevant instants T2 and T3. The second plot in Fig. 6 shows the MI signal, $\hat{y}_{out}(t)$ obtained by the FSA after the defuzzification, variant A in Table 1. Despite a moderate lag is observed due to moving average and hysteresis, the relevant time instants T2 and T3 are clearly identified. Figure. 7 further investigates the T2FS $\varphi_2(\tau = 8670hr)$ observed just after the FSA has been able to decide for the T3 instant. The noisy behavior of $\varphi_1(t)$ components lets the curves to be superimposed each other in an intricate way. At this time boundary, the *abnormal* class has a negligible effect on the later classification decision, while the other classes have a comparable mean $m_i$ and standard deviation $\sigma_i$. In this case, due to the limited number of available samples, a statistical test is unfortunately not able to effectively distinguish among the adjacent classes. The centroid type-reducer is instead able to enhance the distinction between the classes since it exploits all the area under the curves, but it is not able to sufficiently attenuate the incoming noise that need to be anyway processed in the defuzzification unit.

Table 2 collects the statistical description of the $T_2$ and $T_3$ estimates for the FSA variants tested. The Anderson-Darling test p-values ($H_0$ equal to Gaussian distribution) rejects the null hypotheses for the estimates of $T_2$ and $T_3$ highlighted in red, thus explaining the need to use non-parametric tests to compare the median times. The boxplot of Fig. 8 shows the first indication instants T2 and T3 obtained by the FSA variants tested. The dashed lines represent the reference values. The plot and

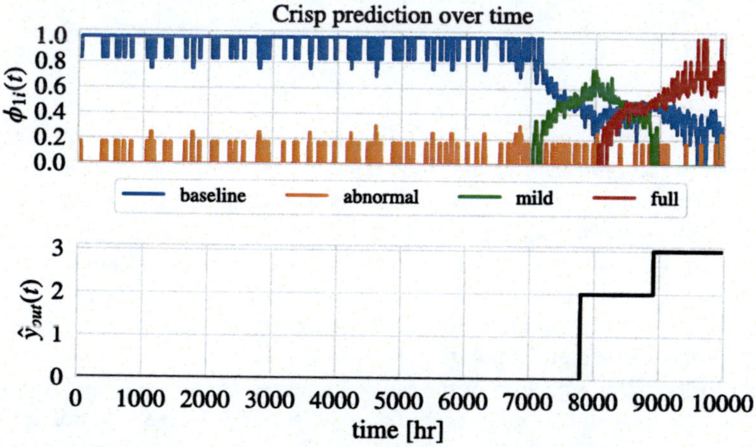


**Fig. 6** Centroid dynamics and the maintenance indication signal, or system output, over time. Variant A of Table 1

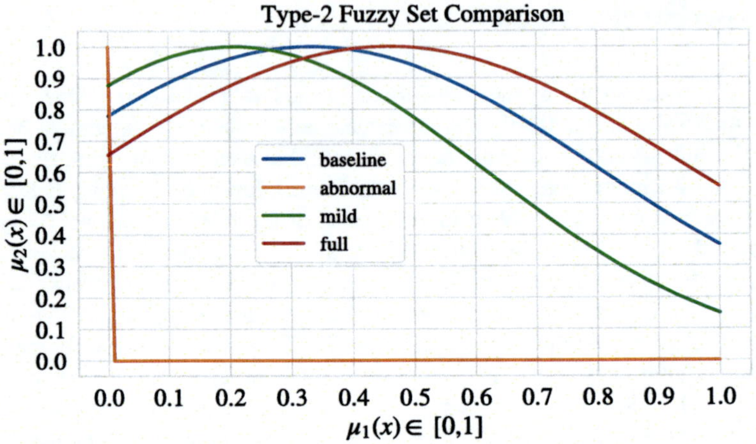


**Fig. 7** Type-2 fuzzy set comparison at $\tau = 8670hr$, right after the expected maintenance indication point (T3) of Fig. 4

**Table 2** Statistical description of first maintenance indication istants of classes *mild* (T2) and *full* (T3) for the system variants of Table 1. In green, the alternative hypotheses is accepted, while in red, the null hypotheses is not rejected

| Metrics Variant | T2 | | | T3 | | |
|---|---|---|---|---|---|---|
| | Median [hr] | $\sigma$ [hr] | AD test p-value | Median [hr] | $\sigma$ [hr] | AD test p-value |
| Reference | 7620 | – | – | 8575 | – | – |
| A | 7785 | 87 | 0,007 | 8664 | 109 | 0,164 |
| B | 7680 | 127 | 0,014 | 8611 | 150 | $< 0,005$ |
| C | 7466 | 73 | 0,728 | 8503 | 185 | $< 0,005$ |
| D | 7425 | 71 | 0,344 | 8410 | 194 | $< 0,005$ |

the table show that the use of the EMA instead of the SMA unit provide a median more close to the reference values, thus reducing the time-lag effect of the SMA algorithm. This come at the expense of additional variance. Additionally, the use of the beta-trend adaptation of the hysteresis defuzzification unit (var C and D, Table 1) enables to compensate the lag effects and anticipate the reference values. It also provides a variance containment benefit, especially for the EMA unit. Variant D (Table 1) contains all the population under the T2 theoretical point while leaving only some outliers for the T3 theoretical point.

Table 3 reports the rejection of the null hypotheses $H_0$ of Eqs. 7 and 8 in green. From the table, it is statistically verified that the A and B variants do not anticipate the reference values of $T_2$ and $T_3$, while variants C and D do, on the average (median) behavior. It can also be stated that the FSA variants of Table 1 are ordered by ascend-

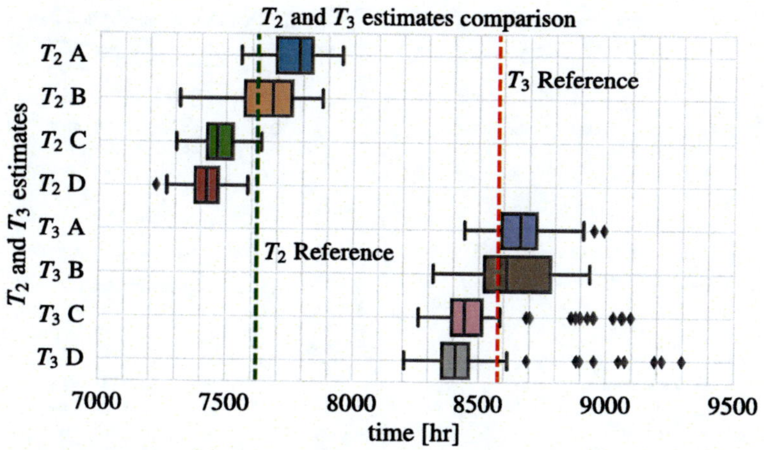


**Fig. 8** $T2$ and $T3$ estimates comparison for the FSA variants of Table 1. Dashed lines report the reference values

**Table 3** Hypotheses (7) and (8) checking. In green the alternative hypotheses accepted while in red the null hypotheses is not rejected

| Metric variant | H1 of Eq. (7) | | H1 of Eq. (8) | |
|---|---|---|---|---|
| | T2 | T3 | T2 | T3 |
| | p-value | p-value | p-value | p-value |
| A | 1 | 1 | – | – |
| B | 0,996 | 0,933 | < 1E-4 | 0,036 |
| C | <1E-4 | <1E-4 | <1E-4 | <1E-4 |
| D | <1E-4 | <1E-4 | <1E-4 | 4E-4 |

ing anticipation performance of both $T_2$ and $T_3$. It is thus preferable to design the FSA by choosing the EMA algorithm for the MA unit and the beta-trend hysteresis for the defuzzification process. Plots in Fig. 9 report the results obtained for the sensitivity analysis. On the x-axis of the a,b plots, the center of the triangle function modeling the *mild* class distribution is reported, as depicted in Fig. 5a. The same is done for c,d graphs, whose x-axes report the base length given to the triangle function, centered at $C = 8000hr$, as illustrated in Fig. 5b. The y-axes reports the anticipation referred to T2 and T3 values of Table 2. Each data-point represents a boxplot, showing the median values and the 95% confidence interval, each obtained by a population of 100 experimental replicates for each FSA variant of Table 1. From the comparison of plots (Fig. 9), all the FSA variants show pretty robust $T_2$ indications until extreme shift values greater than 9000 hr. Plot b show the same behavior with exception of the C variant that exhibits a greater sensitivity and uncertainty. Also variant D shows increased sensitivity and variance, but limited to values greater than $8500hr$. Plots c,d show the anticipation against the slope of $p_w(t)$. It is clearly visible the effect

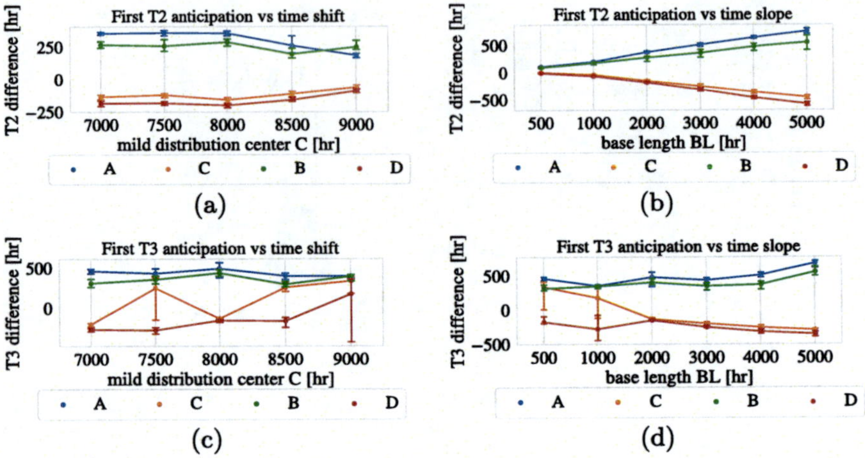


**Fig. 9** Sensitivity analysis: T2 and T3 delay for each system variant of Table 1 with respect to the reference signals

of different designs of the hysteresis mechanism. At triangles with very small base length (sharp slopes), the beta-trend hysteresis is not able to fully capture the trend dynamics and gives comparable results to the classical hysteresis one. As the slope decreases, the dynamics information embedded into $\beta$ coefficients is able to enhance the $T_2$ and $T_3$ anticipation. The classical hysteresis variants A,B instead show an increasing poor performance. The sensitivity analysis also support the hypotheses verified of Table 3, even if a careful design is needed with the C variant when the first indication point present a very unstable decision boundary. It has also been found that the beta-trend hysteresis mechanism is able to counteract the dynamic (linear slope) uncertainty, thus giving an effective design factor that enables effective maintenance indications.

## 6 Conclusions

On the authors' experience, industrial products may suffer from a very limited possibility of retraining and redeploying of the algorithms developed by the third party, e.g. when a design freeze becomes mandatory due to product certification. In other cases, limited connectivity severely impacts updates. Costs associated to modifications of products in other situations do not justify such task, hence a simpler way is searched for such as a retrofit of products. The difficulties associated to a retraining and update phase hence may lead to a reduced performance or prediction instability of employed ML algorithms. This affects critical tasks, like Maintenance Indication, responsible for improving the intervention scheduling of technicians. A post-processing system is thus required in order to reduce the instability propagated from the input classifier. The work proposes a system, called Fuzzy System Aggregator (FSA) that exploits type-2 Fuzzy Set potentiality to embed the estimation uncertainty to provide more stable predictions. Four FSA variants are tested, whose most important design choices regards the moving average unit type and the hysteresis mechanism. Verified hypotheses (Table 3) show that the FSA variant using the EMA unit and the beta-trend hysteresis mechanism provides the most promising performance in terms of maintenance indication anticipation. The beta-trend hysteresis mechanism also provides a growing anticipation against a growing dynamic uncertainty. The FSA tuning enables to design predictive maintenance applications able to anticipate the maintenance indication with a valuable margin and robustness to noise. This lowers the maintenance costs and improves the health management of the system.

**Acknowledgements** The authors wish to thank Carel Industries SpA for the domain-knowledge support associated to the aging of the monitored system.

## References

1. Zonta T, da Costa CA, da Rosa Righi R, de Lima MJ, da Trindade ES, Li GP (2020) Comput Ind Eng 150:106889. https://doi.org/10.1016/j.cie.2020.106889
2. Baur M, Albertelli P, Monno M (2020) Int J Adv Manuf Technol 107(5):2843. https://doi.org/10.1007/s00170-020-05202-3
3. Peng Y, Dong M, Zuo MJ (2010) Int J Adv Manuf Technol 50(1):297. https://doi.org/10.1007/s00170-009-2482-0
4. Bodo R, Bertocco M, Bianchi A (2020) 2020 IEEE international workshop on metrology for industry 4.0 & IoT (MetroInd4.0&IoT) (2020 IEEE MetroInd40&IoT) (Rome, Italy, 2020). https://doi.org/10.1109/MetroInd4.0IoT48571.2020.9138294
5. Mohri M, Rostamizadeh A, Talwalkar A (2018) Foundations of machine learning, 2nd edn. Adaptive computation and machine learning. MIT Press, Cambridge
6. Gao Z, Cecati C, Ding SX (2015) IEEE Trans Ind Electron 62(6):3757. https://doi.org/10.1109/TIE.2015.2417501
7. Cai B, Huang L, Xie M (2017) IEEE Trans Ind Inform 13(5):2227. https://doi.org/10.1109/TII.2017.2695583
8. Verbert K, Babuška R, De Schutter B (2017) Eng Appl Artif Intell 60:136. https://doi.org/10.1016/j.engappai.2017.01.011
9. Mittal K, Jain A, Vaisla KS, Castillo O, Kacprzyk J (2020) Eng Appl Artif Intell 95:103916. https://doi.org/10.1016/j.engappai.2020.103916
10. Mendel JM, John RIB (2002) IEEE Trans Fuzzy Syst 10(2):117. https://doi.org/10.1109/91.995115
11. Zadeh L (1965) Inf Control 8(3):338. https://doi.org/10.1016/S0019-9958(65)90241-X
12. Hudec M (2016)Fuzziness in information systems. Springer International Publishing, p 198. https://doi.org/10.1007/978-3-319-42518-4
13. Mo H, Wang FY, Zhou M, Li R, Xiao Z (2014) Inf Sci 272:96. https://doi.org/10.1016/j.ins.2014.02.092
14. Sadollah A (2018) Sadollah A (ed.) Fuzzy logic based in optimization methods and control systems and its applications. IntechOpen, Rijeka. https://doi.org/10.5772/intechopen.79552
15. Mendel J (2007) Inf Sci 177:84
16. Torshizi AD, Zarandi MHF, Zakeri H (2015) Appl Soft Comput 27:614. https://doi.org/10.1016/j.asoc.2014.04.031

17. Chakraverty S, Sahoo DM, Mahato NR (2019) Defuzzification. Springer, Singapore, pp 117–127. https://doi.org/10.1007/978-981-13-7430-2_7
18. Karnik N, Mendel J, Liang Q (2000) IEEE Trans Fuzzy Syst 7:643. https://doi.org/10.1109/91.811231
19. Azadegan A, Porobic L, Ghazinoory S, Samouei P, Saman Kheirkhah A (2011) Int J Prod Econ 132(2):258. https://doi.org/10.1016/j.ijpe.2011.04.018
20. Kabir S, Papadopoulos Y (2018) Int J Approx Reason 100:29. https://doi.org/10.1016/j.ijar.2018.05.005
21. Cao Q, Samet A, Zanni-Merk C, De Bertrand F, de Beuvron C, Reich, (2019) Procedia Comput Sci 159:630. https://doi.org/10.1016/j.procs.2019.09.218
22. Decker L, Leite D, Giommi L, Bonacorsi D (2020) Real-time anomaly detection in data centers for log-based predictive maintenance using an evolving fuzzy-rule-based approach
23. Arnaldi I (2019) Design of sigma-delta converters in MATLAB®/Simulink®. Springer International Publishing AG. https://doi.org/10.1007/978-3-319-91539-5
24. Hansun S (2013) 2013 conference on new media studies (CoNMedia), pp 1–4. https://doi.org/10.1109/CoNMedia.2013.6708545
25. Wynn HP (1985) J Forecast 4(3):315. https://doi.org/10.1002/for.3980040310
26. Klinker F (2011) Math Semesterber 58(1):97. https://doi.org/10.1007/s00591-010-0080-8
27. Neil S (2007) Encyclopedia of measurement and statistics, vol 1–3. SAGE Publications Inc, Thousand Oaks, California. https://doi.org/10.1007/978-3-540-87481-2_21
28. Bodo R, Bertocco M, Bianchi A (2021) Submitted to: 2021 IEEE international instrumentation and measurement technology conference (I2MTC) (I2MTC 2021). Glasgow, Scotland
29. Bedford T, Quigley J, Walls L (2006) Stat Sci 21(4):428. http://www.jstor.org/stable/27645780
30. Hanratty TP, Hammell R (2019) 2019 IEEE 17th international conference on software engineering research, management and applications (SERA), pp 5–12. https://doi.org/10.1109/SERA.2019.8886799
31. Wohlin C, Runeson P, Hst M, Ohlsson MC, Regnell B, Wessln A (2012) Experimentation in software engineering. Springer Publishing Company, Incorporated. https://doi.org/10.1007/978-3-642-29044-2
32. Anderson TW (2011) Anderson–darling tests of goodness-of-fit. Springer, Berlin, pp 52–54. https://doi.org/10.1007/978-3-642-04898-2_118
33. McKnight PE, Najab J (2010) Mann-whitney U test. American Cancer Society, p 1. https://doi.org/10.1002/9780470479216.corpsy0524

# Ensuring Communication and Information Security of Energy Storage Systems

**Dimitar Arnaudov and Ivan Nedyalkov**

**Abstract** In the present work, the use of the "dark fiber" service to build the communication subnet of a network for charging electric vehicles has been proposed. The advantages and disadvantages of the possible implementation of the proposal for the construction of a communication network through dark fiber are listed.

**Keywords** Charging station · Dark fiber · Power electronics

## 1 Introduction

Electric cars or all types of electrically powered non-rail vehicles are increasingly entering our daily lives. For this reason, it is necessary to build the appropriate infrastructure needed to charge these electric vehicles—charging stations. Charging stations can be combined into a common network of charging stations—similar to a retail chain of gas stations. Charging stations must be able to exchange information both with each other and with a control center. Therefore, it is necessary to build a communication subnetwork for exchanging of data, service information, information related to the payment, and the payment itself for the charging service.

### *1.1 Condition of the Problem*

Most researches in this area are mainly focused on the development of new circuit solutions for converters performing the processes of charging or transferring energy between charging stations and the vehicle, new topologies of the charging network,

D. Arnaudov (✉)
Technical University - Sofia, 8 Kl. Ohridski Blvd., 1000 Sofia, Bulgaria
e-mail: dda@tu-sofia.bg

I. Nedyalkov
66 Ivan Mihailov str., 2700 Blagoevgrad, Bulgaria

 
X.-S. Yang et al. (eds.), *Proceedings of Sixth International Congress on Information and Communication Technology*, Lecture Notes in Networks and Systems 216,
https://doi.org/10.1007/978-981-16-1781-2_88

new elements for energy storage elements, increasing efficiency, and others [1–7]. However, charging networks for electric vehicles, which are part of energy storage systems, have another side—communication and information, which also needs in-depth research. These studies should focus on two main aspects.

In the first one, studies should focus on the communication traffic generated by these devices. These studies include monitoring of the communication part of the network, in particular, the communication devices from which it is built. The results of these studies can be used for the design of the communication network [8].

The second aspect of the studies should be focused on securing of the information—providing secure information exchange of data between devices, access to this data, and others.

The purpose of this paper is to do a review of the main groups of cyber attacks that could be applied to the communication part of the network for charging energy. An extravagant method has been proposed to ensure communication security by using the "dark fiber" service.

## 2 Charging Network for Electric Vehicles

The network can be conditionally divided into two main parts—a power part, which is used to charge/transfer energy to electric vehicles and a communication part, intended for exchange of service information and/or data related to the payment of the consumed energy, used to charge an electric vehicle.

### *2.1 Communication Part of the Network*

This subnet is mainly made up of communication (network) devices that exchange data with each other continuously. This data can be service information, control commands, statistics, information related to the payment for the charging service, and much more. Figure 1 shows a simplified model of the communication subnet. The subnet is made up mainly of the following network devices—routers and switches. The Internet is used to provide connection between the remote devices.

### *2.2 Power Converters*

A modular structure is used to increase the output power of the charging stations. Parallel units are used. Communication between units is protected from external interference by using specific communications between modules that are not available to users. The vulnerable point of attack is the user interface. The transfer of energy from the charging station to the electric car and from the electric car to the

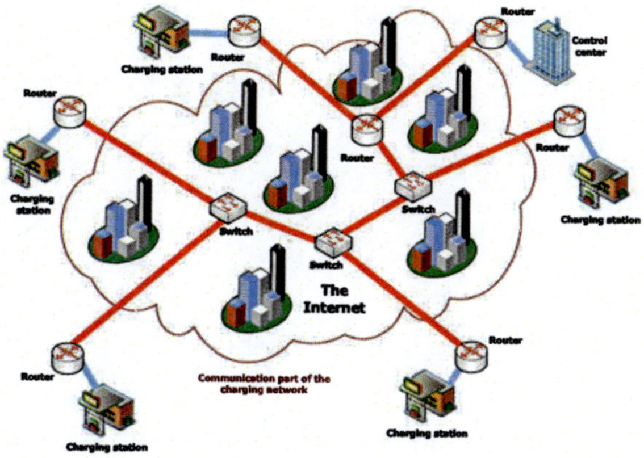


**Fig. 1** The communication part of the charging network

charging station, in cases when the electric car is used as an energy storage system, is controlled by the data that is transmitted according to a standard protocol. The power converters in the charging station, in order to accomplish the energy transfer in one or the other direction must receive a request from the electric vehicle and a permission from the charging infrastructure management center. A potential cyber-attack over the information exchange between the converters of the electric vehicle and the charging station would not significantly affect the system, as this is a lower level in the hierarchy of information exchange. The control of the operating modes of the converters of the charging station is carried out by the secured communication through the dark fiber. Reducing the possibilities for cyber attacks can also be achieved by limiting the possible modes of operation of the converters in the charging station. This is accomplished by the working algorithms of the control systems of the power units. These algorithms does not allow operation in emergency mode, set by the electric car. So the possibilities for cyber attacks over the information exchange transmitted from the electric car to the charging station are reduced.

## 3 Types of Cyber Attacks

Some of the power electronic devices used in the charging network also have communication (network) functions—the ability to connect to various communication networks, mainly IP networks. They can be subjected to the existing cyber attacks, which can be applied to any device that has network functions and is connected to the Internet [9–11]. Basically, attacks can be grouped into two large groups.

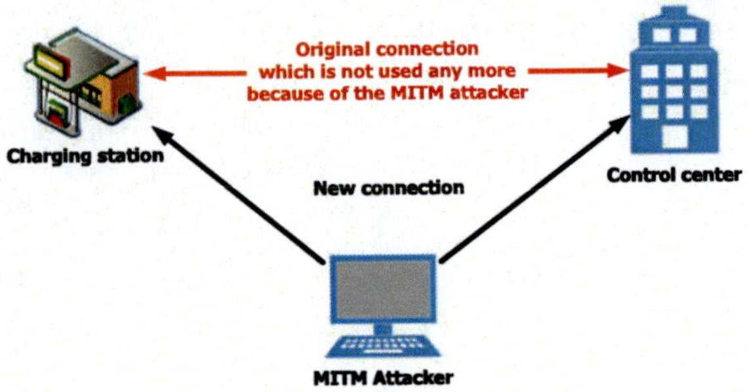


**Fig. 2** MITM session hijack attack

### *3.1 Attacks Used to Collect and/or Replace the Exchanged Data*

The purpose of these attacks is to intercept the information exchanged or to intercept and alter the exchanged information in their favor. A large group of this type of attacks are the "Man in the Middle" (MITM). MITM is a type of cyber attack where the hacker/attacker implements himself in the communication session between two parties (users or servers). The attacking device deceives and pretends to be at the both sides at the same time of the communication session, without the charging station and the control center know of it. This gives access to the information that is exchanged between them. Through this type of attack, the hacker intercepts, receives, and sends altered/false information to both parties, without them knowing it until it is too late.

The MITM attack operates in real time and can be used to track, intercept or change financial transactions related to the payment of the electric vehicle charging service. Figure 2 shows a MITM attack.

### *3.2 Attacks Used to Restrict Access to Devices on the Network*

In this type of attacks, the goal is to prevent remote device to be accessed—to be blocked indefinitely. These attacks do not collect or modify the exchanged information.

## 4 Application of the "Dark Fiber" Service for Building a Safe and Secure Communication Network

There are various technologies for ensuring security in communication networks and in particular in IP networks. Such technologies are VPN, Blockchain, use of Intrusion Detection Systems (IDS)/Intrusion Prevention Systems, and others. In all of them, the transmission of the data is carried out via an IP network, which must be connected to the Internet, because this is the easiest and cheapest way to provide connectivity between the end devices, located at great distances from each other. Thus, the probability of a breach in the system remains high. Therefore, in order to reduce the probability of a breach to zero, it is necessary to build a communication network separate from the Internet, allowing communication over long distances without the need to be built from the scratch. This requirement can be met by the existing optical networks and especially by a service that has been available for years—the dark fiber.

The aim of the paperwork is to present an option for using the dark fiber service to build a safe and secure communication network for data exchange between the charging stations and other elements of a network for charging electrical vehicles.

## 4.1 Description of the Technology

More precisely, the dark fiber is not a technology, but a service that is offered in the optical communications. Dark fiber is a service in which a fiber is leased to users who use it to transmit their own data between two points. Due to the characteristics and specifics of optical cables, these distances between the two points can be hundreds of kilometers. These fibers are found in optical cables that are installed in urban or suburban areas. This service allows the company that owns the fiber—in most cases a telecommunication operator—to sell access to telecommunication services. The dark fibers are continuously rented, then returned to the operator, and rented again to other users. One of the reasons for the existence of the so-called dark fibers is that a large part of the cost of building and installing cable routes is for construction and design. These costs include: planning, mapping, obtaining building permits, digging ditches and laying cable pipes, and finally laying the cables themselves. These activities, according to some economic estimates, usually amount to more than 60% of the cost of building optical networks. Therefore, significantly more optical fibers are planned than necessary. There are many other owners of optical cables such as railway companies or energy distribution companies who also add additional leasing fibers to other operators.

The dark fiber is an optical transport between two points without active equipment between them, using the optical infrastructure of a telecommunication operator, using own or leased active devices at both ends of the line (Fig. 3). This provided fiber or several fibers are not observed/monitored by the telecommunication operator. Any information can be transmitted on this fiber / fibers. These fibers may not be connected to the global network (Internet), so anyone who wants can build their own communication network without investing much in the design and construction of their own cable network.

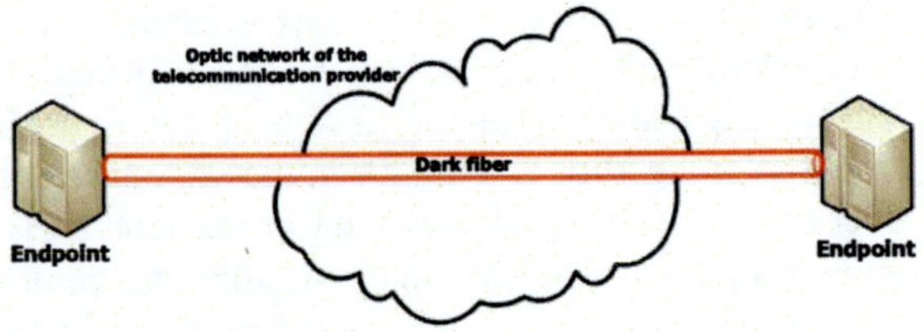


**Fig. 3** The dark fiber concept

## 4.2 Application of the Service into the Power Electronics

Electric cars and other vehicles powered by electricity are entering the everyday life. Therefore, charging networks must be built for them. This charging process must be paid. The payment operation must be secured, such as the data and the service information that is exchanged between the devices in the charging network. In today's world, all communication networks connected to the Internet are subject to constant cyber attacks. These attacks can be harmless, without malicious action. The communication network of the charging network may also be subject to such attacks. Therefore, it must be secured. The question is how to build your own communication network, separated from the Internet, which is both safe and secure, and on the other hand, does not require a huge financial investment for its construction.

To solve this task—safe and secured data exchange and small capital investments, the dark fiber service can be used. From Sect. 4.1, it can be seen that the dark fiber provides an opportunity to build a communication network, which is separated from the Internet. Any kind of information can be exchanged in such communication network. The optical fiber/fibers provider will not monitor the exchanged information and only the fiber tenant will have access to this information.

## 4.3 Proposed Idea

Figure 4 presents the proposed concept of using the dark fiber to build a communication network for the charging network of electric vehicles. The figure shows schematically and conceptually the optical network of a telecommunications operator located throughout the country. The charging stations are located in different places in the country. Nevertheless, they are interconnected in their own optical network built of dark fibers. The connection between the individual dark fibers, in order to realize the independent network, within the optical network of the telecommunication provider, takes place in the optical network of the provider, by redirecting the fibers by using optical switches.

In order to be able to build such a communication network, it is necessary each power electronic device (charging stations), to be able to connect to a communication network. In order to have access to the built communication network by using the dark fiber service, each charging station must be connected to a media converter,

**Fig. 4** Proposed conception

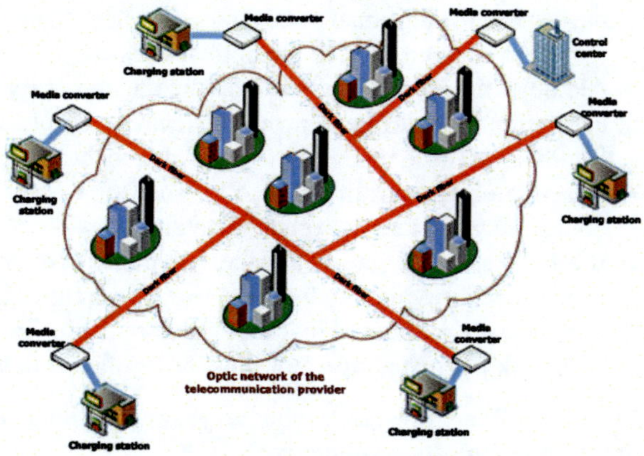


and it to the rented dark fiber. The media converter converts the signal between the various transmission media—from optical fiber to copper cable (converts the signal from light to electrical). This increases the transmission distances far beyond the capabilities of the twisted pair (LAN cable), and on the other hand, makes the network more independent of ambient electromagnetic interference.

The proposed concept for building a safe and secure communication network has the following advantages:

- The biggest advantage of this service is the ability to build a completely separate, independent network that does not have an access to the Internet. A locked, inaccessible network is created through the use of the dark fiber. Only the user of the leased fibers has access to it. Access to this network is only physical. Thus, no other object outside the network can track or record the data and information transmitted through the fiber. Additionally, for even greater security, additional encryption of the transmitted information can be used;
- The delay in the network is small, even it can be rejected under some conditions;
- The transmission speed is almost limitless. Because the tenant of the dark fiber is the sole user, the bandwidth is not shared between the users of the fiber. The bandwidth can be further increased by using active equipment, which achieves even higher transmission speeds;
- The tenant has full control over the dark fiber communication network;
- An own system for payment for the used energy to charge the electric car can be built;
- The exchanged data are not monitored by the owner of the optical network—the telecommunications operator

The proposed concept also has its drawbacks such as:

- In some places where charging stations will be built, the selected optical network provider may not be accessible. From which the dark fibers will be hired. This means that an additional optical path will have to be built in order for the charging station to be connected to the optical network;

- Although the initial investment to hire the necessary fibers will be much less than the option to build your own physical communication network, it will still be significant. These investments must also include the investment for the purchase of active equipment in order to be able to connect the charging stations to the optical network and to exchange data;
- If a failure or problem with access to the network occurs, the problem may not be rectified immediately by the network operator. This is because the leased dark fibers are not monitored by the network operator. To find out if there are problems with dark fibers, the operator must be notified if there are any problems.

In [12, 13] are presented other ways for providing safe and secure data exchange in the communication subnetwork.

## 5 Conclusions

The researches that has been done and is being done in the field of charging of electric vehicles and in part in charging networks are briefly reviewed.

An overview of the structure of the network for charging electric vehicles—communication part and power part has been presented.

The main groups of cyber attacks to which the communication subnet of the network for charging electric vehicles would be subjected are presented.

A thorough review of the dark fiber service has been made. The application of the service in power electronics is presented. A concept for the use of the dark fiber service has been proposed. The dark fiber will be used to build the communication subnetwork of the charging network for electric vehicles. The main advantages and disadvantages of the proposed concept are noted.

**Acknowledgements** This research is carried out within the frames of the project "Optimal design and management of electrical energy storage systems", КП-06-Н37/25/18.12.2019, Bulgarian National Scientific Fund.

## References

1. Huang X, Chen J, Yang H, Cao Y, Guan W, Huang B Economic planning approach for electric vehicle charging stations integrating traffic and power grid constraints. In: IET generation, transmission & distribution, vol 12, p 17
2. Wang S, Bi S, Zhang YA, Huang J (Oct. 2018) Electrical vehicle charging station profit maximization: admission, pricing, and online scheduling. IEEE Trans Sustain Energy 9(4):1722–1731

3. Shuanglong S, Zhe Y, Shuaihua L, Bin L, Yuheng X , Huan X (2018) Design and practice of integrated platform for operation and management of fast charging station. In: 2018 2nd IEEE conference on energy internet and energy system integration (EI2), Beijing, pp 1–5
4. Madzharov ND, Nemkov VS (2017) Technological inductive power transfer systems. J Electr Eng 68(3):235–244
5. Madzharov N, Petkov L (2020) Analyzing contactless transmission of energy and information and communication signals via a common inductive link. In: PCIM Europe digital days 2020; international exhibition and conference for power electronics, intelligent motion, renewable energy and energy management, Germany, pp 1–8
6. Zarkov Z, Bachev I, Lazarov V (2017) Small wind generator system with non-inverting buck-boost converter and battery storage. In: 2017 15th international conference on electrical machines, drives and power systems (ELMA), Sofia, pp 298–303. https://doi.org/10.1109/ELMA.2017.7955452.
7. Vuchev AS, Grigorova TG, Maradzhiev IP (2019) A unified analysis of LLC resonant DC/DC converter with capacitive output filter. In: 2019 X national conference with international participation (ELECTRONICA), Sofia, Bulgaria, pp 1–4. https://doi.org/10.1109/ELECTRONICA.2019.8825604
8. Georgiev GP (2015) Performance evaluation of internettraffic by network measurements. In: Elektrotechnica & Elektronica E+E, vol. 50. No 3–4/2015
9. Fraiji Y, Ben Azzouz L, Trojet W, Saidane LA (2018) Cyber security issues of Internet of electric vehicles. In: 2018 IEEE wireless communications and networking conference (WCNC), Barcelona, pp 1–6
10. Soltan S, Yannakakis M, Zussman G (2019) REACT to cyber attacks on power grids. IEEE Transactions on Network Science and Engineering 6(3):459–473. https://doi.org/10.1109/TNSE.2018.2837894. Accessed , 1 July–Sept 2019
11. Jin M, Lavaei J, Johansson KH (May 2019) Power grid AC-based state estimation: vulnerability analysis against cyber attacks. IEEE Trans Autom Control 64(5):1784–1799. https://doi.org/10.1109/TAC.2018.2852774
12. Jeong S, Dao N, Lee Y, Lee C, Cho S (2018)Blockchain based billing system for electric vehicle and charging station. In: 2018 tenth international conference on ubiquitous and future networks (ICUFN), Prague, pp 308–310
13. Baza M, Nabil M, Ismail M, Mahmoud M, Serpedin E, Ashiqur Rahman M (2019)Blockchain-based charging coordination mechanism for smart grid energy storage units. In: 2019 IEEE international conference on blockchain (Blockchain), Atlanta, GA, USA, pp 504–509

# Design of an On-Body Rectangular Microstrip Patch Antenna for the Diagnosis of Breast Cancer Using S-Band

**Rashedul Islam, Fardeen Mahbub, Sayed Abdul Kadir Al-Nahiun, Shouherdho Banerjee Akash, Raja Rashidul Hasan, and Md. Abdur Rahman**

**Abstract** A common cause of death of women is Breast Cancer and Tumors, and it is difficult to detect both of these at the early stage due to their minimal structure. Microwave Imaging is one of the effective techniques among the current and past methods. In this paper, a compact size microstrip patch antenna has been designed in CST Studio Suite Software for microwave imaging to diagnose breast cancer using the FR-4 (Lossy) substrate material with a resonant frequency of 2.3 GHz (S-band), varying from 1.5 GHz to 3 GHz. A feedline fed the antenna's radiating patch, which is rectangular in size. The breast model with tumors was designed in the CST consisting of skin, fat, glandular, and cancer affected tissues. A Return Loss ($S_{1,1}$) value of −37.28 dB and −45.72 dB were obtained in free space and after applying the antenna in the cancer affected breast phantom, respectively. Other obtained output parameter values such as Directivity (2D and 3D), Radiation Efficiency, Polar Radiation, etc., are also presented in this paper, showing that the antenna model would be a better option for breast cancer diagnosis.

**Keywords** Breast cancer · Breast tumor · Microstrip patch antenna · S-Band · Microwave imaging · Return loss · Directivity · Radiation · CST

## 1 Introduction

According to the World Health Organization (WHO) report in 2018, Breast Cancer is one of the biggest global health challenges. This disease causes 15% of death of women related to Cancer, which is nearly 6,27,000 [1]. Compared with normal

R. Islam (✉) · F. Mahbub · S. Abdul Kadir Al-Nahiun · S. Banerjee Akash · R. Rashidul Hasan · Md. Abdur Rahman
American International University-Bangladesh (AIUB), Dhaka, Bangladesh

R. Rashidul Hasan
e-mail: hemal@aiub.edu

Md. Abdur Rahman
e-mail: arahman@aiub.edu



X.-S. Yang et al. (eds.), *Proceedings of Sixth International Congress on Information and Communication Technology*, Lecture Notes in Networks and Systems 216,
https://doi.org/10.1007/978-981-16-1781-2_89

tissues, breast tissues affected by Cancer can exhibit higher metabolic activities than other cancer tissues [2]. Previously, numerous methods such as Ultrasound, Radiography, Magnetic Resonance Imaging (MRI), X-Ray Mammography, etc., were used to identify tumors or cancer cells in the breast. These previously used methods have some drawbacks; for example, they cannot differentiate between benign and malignant tumors, resulting in excessive breast biopsies. Some other disadvantages of these methods include lengthy working processes, harmful to human health due to ionizing radiation, and its high cost. These concerns lead researchers to establish a different approach for the diagnosis of breast cancer [3].

Among the recently invented techniques for breast cancer diagnosis, Microwave Imaging (MI) using microstrip patch antenna is the most efficient technique. It has numerous benefits such as convenience, potentially low prices, a non-ionizing, and a secured option. The Microwave Imaging technique's functioning process relies on the dielectric constant between the harmful tumor tissues and stable tissues [4]. In the MI system, a microstrip patch antenna is an essential element that irradiates the body under test with microwaves that travel through the body and then detected another antenna working as a receiver. The receiver antenna contains the reflected information from a tumor that has been recorded and analyzed using a suitable signal processing technique for receiving 3D images of the body under test [1]. Several frequency bands can be used to design Microstrip Patch Antenna. Some microstrip patch antennas' output parameters should be appropriately analyzed for better diagnosis, such as Return Loss ($S_{1,1}$), Radiation Pattern, Directivity, and Polarization. This paper aims to design a microstrip patch antenna in free space and human body application for detecting breast cancer through microwave imaging technique.

## 2 Methodology and Modeling of Microstrip Patch Antenna

Figure 1 represents the Geometry of the proposed Rectangular Microstrip Patch Antenna, while Fig. 2 depicts the 3D model of the patch antenna in free space.

At first, the frequency spectrum was selected after which, the substrate material FR-4 (lossy) and the operating frequency of 2.3 GHz were also chosen, which was used simultaneously for measuring the width of the patch (W) using Eq. 1 [5].

$$\mathrm{W} = \frac{c_0}{2f_r\sqrt{\frac{\varepsilon_r+1}{2}}} \tag{1}$$

Here, $\varepsilon_r = 4.3$ and the Effective Dielectric Constant, $\varepsilon_{reff}$ is calculated using Eq. (2) [5].

$$\varepsilon_{\mathrm{reff}} = \frac{\varepsilon_r + 1}{2} + \frac{\varepsilon_r - 1}{2}\left(1 + 12\frac{\mathrm{h}}{\mathrm{w}}\right)^{-0.5} \tag{2}$$

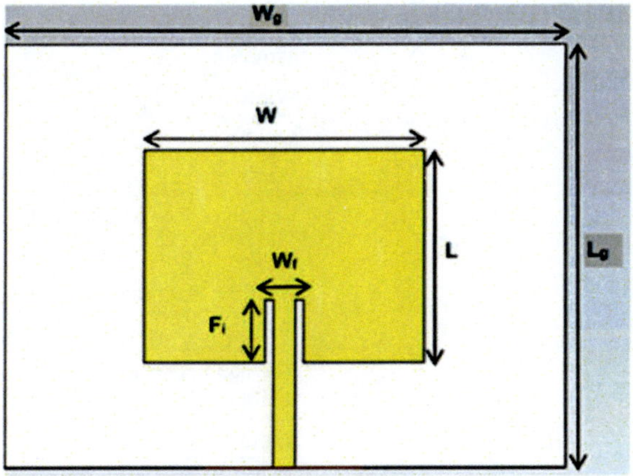


**Fig. 1** The geometry of the proposed microstrip patch antenna (in free space)

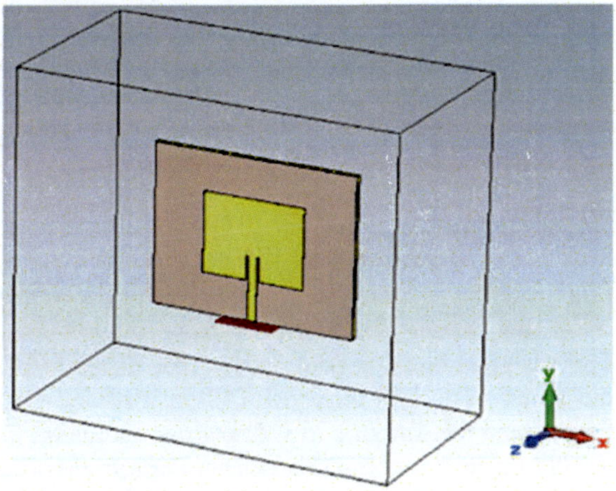


**Fig. 2.** 3D model of microstrip patch antenna in free space

Then, the Effective Length, $L_{eff}$, is calculated using Eq. (3) [5].

$$L_{eff} = \frac{c_0}{2f_r\sqrt{\varepsilon_{reff}}} \tag{3}$$

The fringing factor is subtracted from the effective length to determine the patch's actual length [5].

$$\Delta L = 0.412\frac{\left(\frac{w}{h} + 0.264\right)(\varepsilon_{reff} + 0.3)}{(\varepsilon_{reff} - 0.258)\left(\frac{w}{h} + 0.813\right)} \tag{4}$$

The patch's actual length is determined by using Eq. (5) [5].

Table 1 Parameters of the proposed model

| Parameter descriptions along with their symbols | Dimension (mm) |
|---|---|
| Width of patch, W | 38.5 |
| Length of patch, L | 30.23 |
| Width of the ground plane, $W_g$ | 77 |
| Length of the ground plane, $L_g$ | 60.46 |
| Width of feedline, $W_f$ | 3.2 |
| Feedline insertion, $F_i$ | 8.9295 |
| Height of substrate, $h_s$ | 1.7 |
| The gap between the patch and the feedline, $G_{pf}$ | 1 |
| Ground thickness, $h_t$ | 0.035 |

$$L = L_{eff} - 2\Delta L \tag{5}$$

Following that, the Ground Plane Width's respective values, $W_g = 2*W$, and the Ground Plane Length, $L_g = 2*L$, were calculated [5]. The proposed model of the antenna was connected to a 50-Ω inset feed transmission feedline. Table 1 represents the free space parameter values of the antenna.

## 3 Breast Model with the Microstrip Patch Antenna

The designed breast model carries four different parts: Skin, Fat, Glandular, and the Tumor. The first three parts (Skin, Fat, and Glandular) are known as normal tissues, and the Tumor part is known as malignant tissue. The dielectric properties used to create the normal tissues are Permittivity, Electrical Conductance, Density, Heat Capacity, Thermal Conductance, Diffusivity, Permeability, and the first three properties have been used to create the tumor tissue. The breast model consists of skin with a radius of 30 mm and a thickness of 4 mm. Also, the fat has a radius of 26 mm with a thickness of 5 mm. Moreover, the thickness of the glandular is 21 mm, and the size of the tumor is considered 5 mm. Figure 3 shows the Breast Model, and Table 2 shows the parameter values used to design the human breast model. To study the impact of radiofrequency interaction on the Phantom proximity, different dielectric properties (Electrical Conductance, Relative Permittivity, etc.) have been used. The dielectric properties (Electrical conductance, Relative Permittivity, etc.) of breast tissues (malignant and normal tissues) can be changed significantly according to the frequency spectrum variation. This happens because different tissues have different microwaves' absorption capabilities, and their functions are strongly nonlinear with frequency. The dielectric properties can also be changed for numerous types of breast tissues according to their structure and contents [1]. From the parameters given in

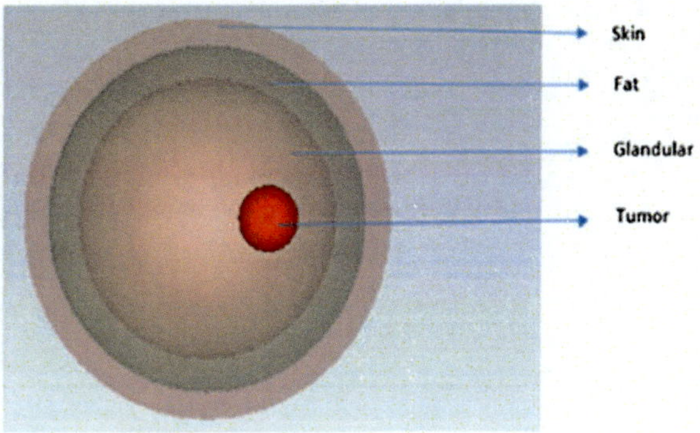


**Fig. 3** Different parts of the designed breast model

**Table 2** Different parameter values of breast tissues

| Tissues properties | Skin | Adipose tissue | Glandular | Tumor |
|---|---|---|---|---|
| Permittivity conductance S/m | 37.7 | 4.9 | 51 | 55.8 |
| Electrical conductance S/m | 2.4 | 0.272 | 3.56 | 4.1 |
| Density Kg/m^3 | 1100 | 920 | 1050 | 1070 |
| Heat capacity Kj/Kg/K | 3.391 | 2.348 | 2.960 | ~ |
| Thermal conductance W/m/K | 37 | 21 | 33 | ~ |
| Diffusivity $m^2/s$ | 9.84e-06 | 9.82e-06 | 1.07e-05 | ~ |
| Permeability | 1 | 1 | 1 | ~ |

Table 2, it can be seen that at an operating frequency of 2.3 GHz, the Permittivity, Electrical Conductance, Density, Heat Capacity, Thermal Conductance of the three normal tissues are different, whereas the Permeability ($\mu$) of those tissues are same.

After creating the cancer affected breast model in CST, the designed free space antenna was applied to the phantom at an operating frequency of 2.3 GHz. Figures 4 and 5 show the Cancer affected Breast model and the location of the tumor, respectively.

## 4 Results and Analysis

Return Loss ($S_{1,1}$) is the measurement of the amount of radiofrequency (RF) energy that can be accepted by an antenna. From the designed model, a Return Loss of −37.276 dB has been obtained in free space. In the phantom model with a tumor, a Return Loss of −45.720 dB has been received. The Bandwidth of 74 MHz in open

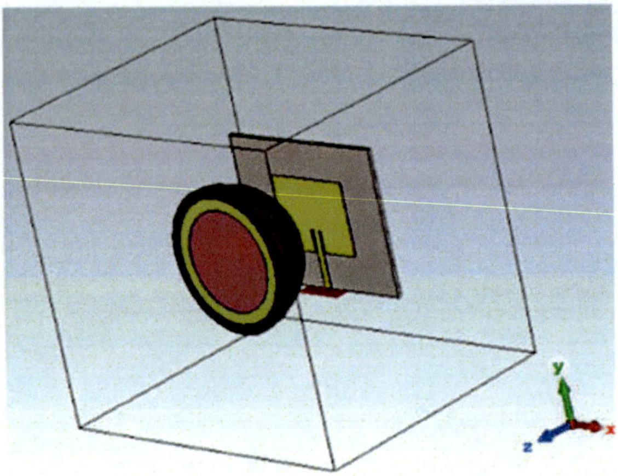


**Fig. 4** Cancer affected breast model with microstrip patch antenna

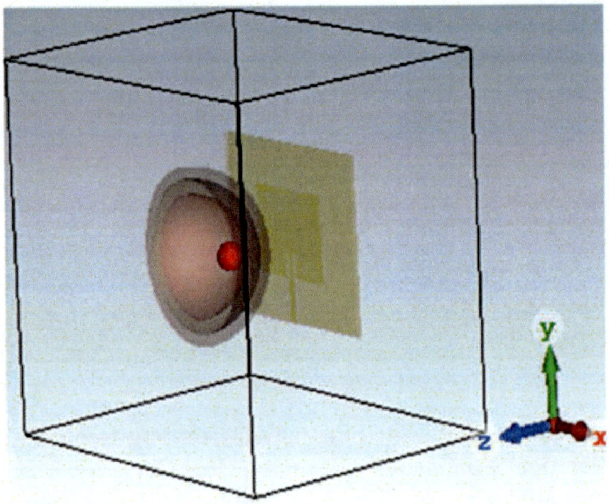


**Fig. 5** Location of tumor after applying the microstrip patch antenna

space and 77.5 MHz in the phantom were obtained. The results are shown in Figs. 6 and 7.

A lower value of the VSWR means that more power is delivered to the antenna. Thus, it can be deduced that the lower the VSWR, the better the antenna is. But VSWR must be greater or equal to 1 and less than 2. A VSWR of 1.0104066 was obtained by applying the antenna in the breast tumor model and a VSWR of 1.0277465 in free space. The results are shown in Figs. 8 and 9.

Figure 10 shows the Farfield Radiation in the Cancer affected Breast Phantom (polar form, Phi = 90°). From there, the main lobe magnitude of 1.14dBi and a side lobe level of −3.4 dB have been achieved. Figure 11 also shows the Surface Current in the Cancer affected Breast Phantom, and a surface current of 30.5 A/m has been achieved in this paper.

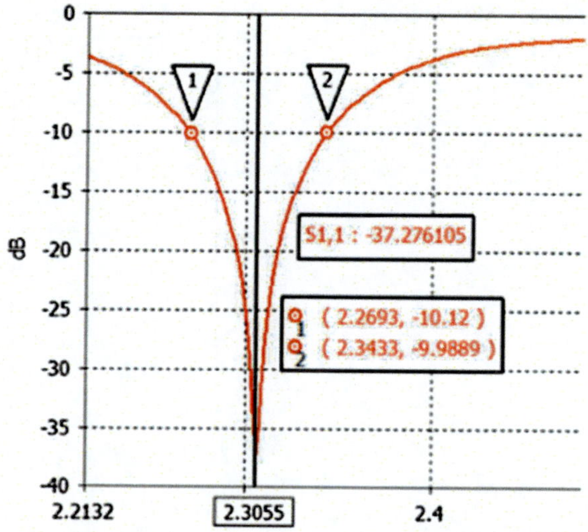


**Fig. 6** Return loss in free space

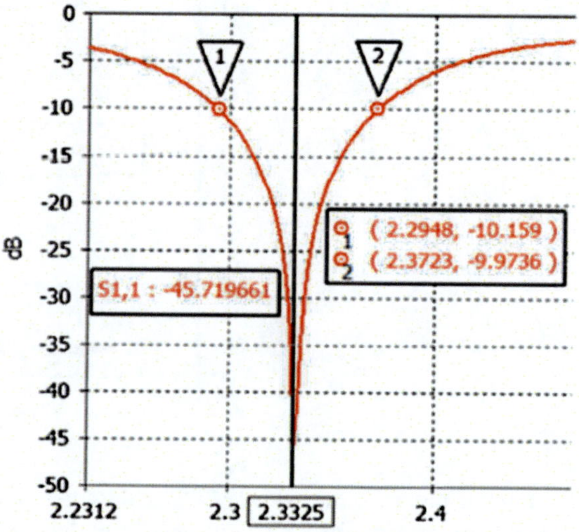


**Fig. 7** Return loss with tumor

As the antenna is for body application, a balanced radiation efficiency needs to be achieved; otherwise, the antenna will not be applicable for breast cancer diagnosis. In this paper, a Radiation Efficiency of −4.0005933 dB and −5.5490247 dB were successfully achieved both in free space and in the Cancer affected body phantom, respectively (Figs. 12 and 13).

The Specific Absorption Rate (SAR) value has also been determined for 1 g of tissue. In Fig. 14, it can be seen that the maximum value of SAR for 1 g of tissue is 0.483 W/kg. Due to the patients' safety purpose, there is a criterion of the Federal

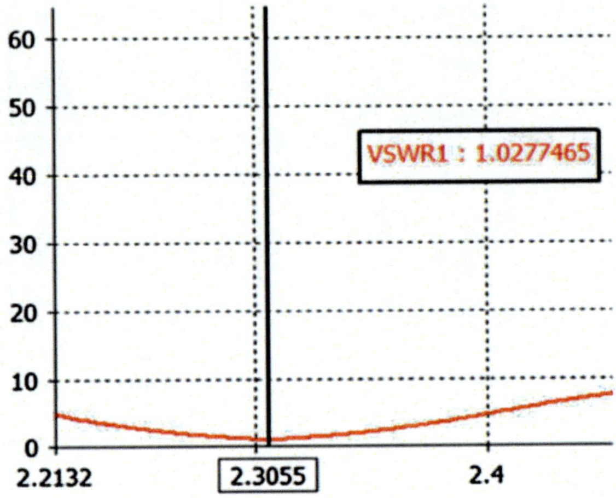


**Fig. 8** VSWR in free space

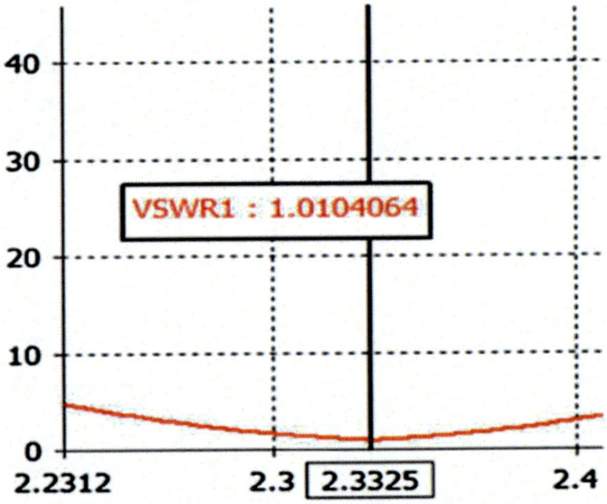


**Fig. 9** VSWR with tumor

Communications Commission (FCC) for the antenna's experimental purpose. As per American standards, the SAR value must be under 1.6 W/kg according to the FCC [8]. Figure 15 shows the 3D Radiation Pattern (Directivity) in the Cancer affected Breast Model in which a directivity value of 6.684 dBi has been successfully achieved.

Moreover, Fig. 16 shows the 2D Radiation Pattern (Directivity) in the Cancer affected Breast Model in which a similar directivity value has been successfully achieved.

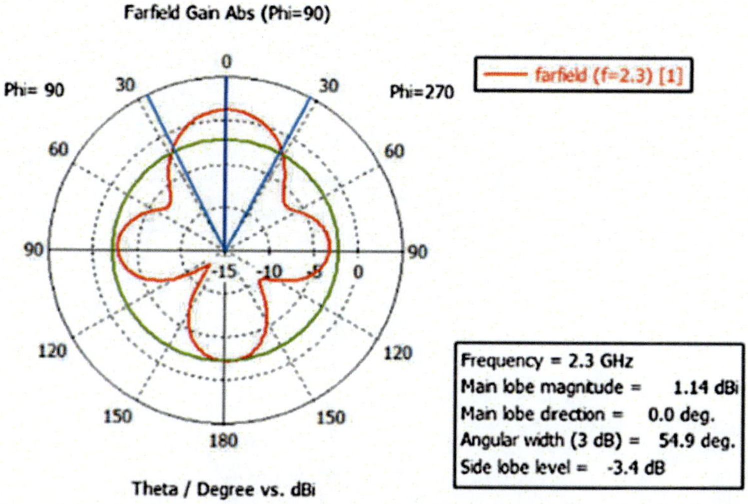


**Fig. 10** Farfield radiation in the cancer affected breast phantom (polar form).

**Fig. 11** Surface current in the cancer affected breast phantom

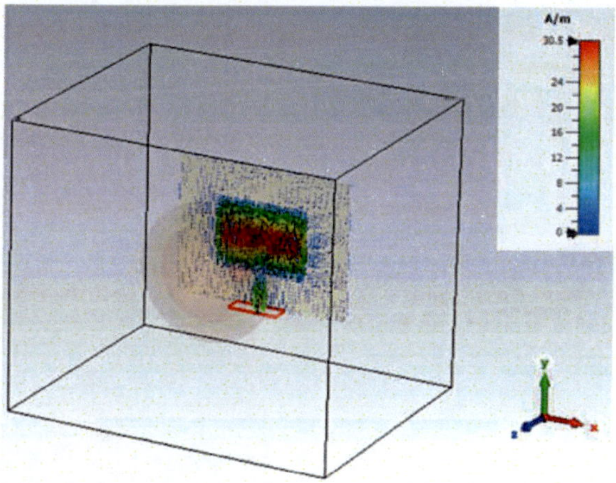


## 5 Comparative Study

Different past research papers have been thoroughly analyzed from where the past antenna models' performance has been compared with the current ones. Therefore, the proposed antenna model is very much effective and compact. The antenna can also detect the low size tumor more precisely than the other models. Table 3 represents the elaborate comparison between the past and the proposed antenna models.

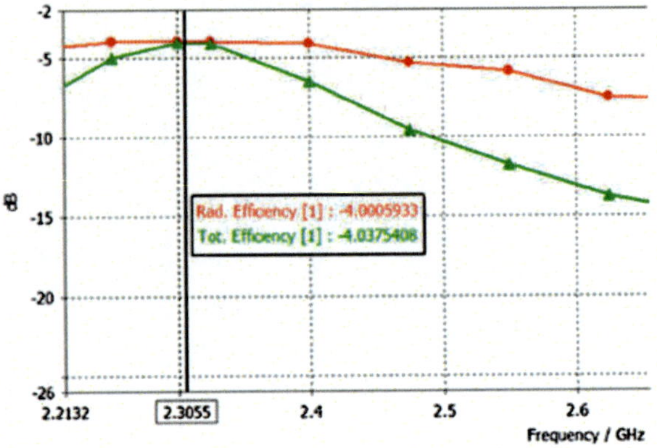


**Fig. 12** Radiation efficiency Pattern in free space

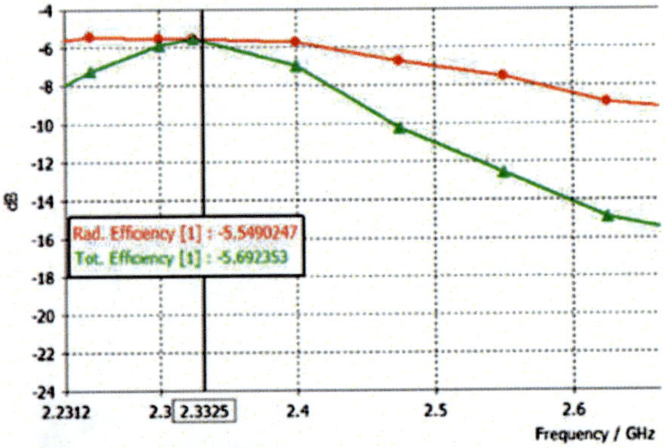


**Fig. 13** Radiation efficiency pattern for the cancer affected breast phantom

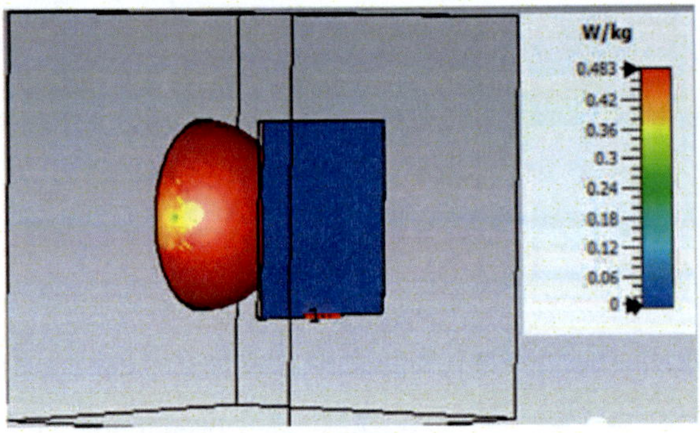


**Fig. 14** SAR calculation (1 g tissue)

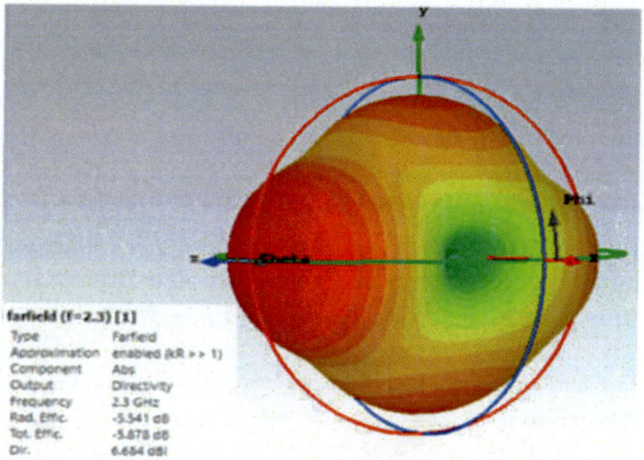


**Fig. 15.** 3D radiation pattern (directivity) in cancer affected breast model

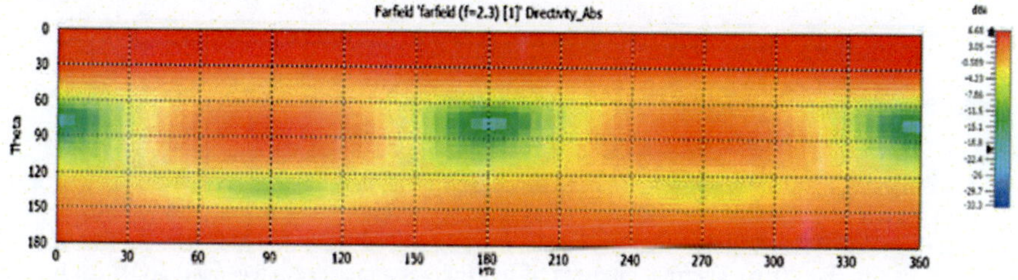


**Fig. 16.** 2D radiation pattern (directivity) in cancer affected breast model

**Table 3** Comparison of the antenna model with previous research paper

| Return Loss ($S_{1,1}$), dB | Directivity, dBi | Antenna dimensions length*width (in mm$^2$) | Tumor size (in mm) | Refs. |
|---|---|---|---|---|
| −28 | 6.17 | 90*40 | 5 | [1] |
| −20 | 6.6 | 25*25 | 5 | [4] |
| −20 | 11 | 22*22 | 5 | [6] |
| −15 | 7 | 76*78 | 10 | [7] |
| −8.5 | 6.5 | 48.46*42.46 | 6 | [8] |
| ~ | 3.5 | 25*16 | 5 | [9] |
| −12 | ~ | 125*51 | 7 | [10] |
| −45.72 | 6.684 | 30.23*38.5 | 5 | Proposed model |

## 6 Conclusion

In this paper, a rectangular-shaped microstrip patch antenna has been designed which has been implemented both in free space and Cancer affected Breast model for the diagnosis of early-stage Breast Cancer. The antenna operates at a frequency of 2.3 GHz. Based on a real human breast's dielectric properties, a breast model was designed to determine the antenna's return loss −37.28 dB in free space and −45.72 dB with the tumor-affected breast model. The Directivity and Radiation Efficiency was found to be 6.95 dBi, −4.0006 dB in free space, and 6.68dBi, −5.55 dB with Cancer affected breast model, respectively. A surface current of 30.5 A/m was obtained with Cancer affected breast model. Lastly, from all the obtained data, it has been understood that the antenna was capable of diagnosing breast cancer from the cancer affected breast through Microwave imaging. However, considering the future demand, these parameters can be upgraded as per the requirements for overcoming the challenges related to the Breast Cancer identification process.

## References

1. Shekhawat S, Sharma V, Jain PK, Sharma BR, Saxena VK, Bhatnagar D (2019) An off-diagonal feed elliptical patch antenna with ring shaped slot in ground plane for microwave imaging of breast. In: 2019 IEEE Indian conference on antennas and propogation (InCAP), pp 1–4. IEEE.
2. Li Y, Fahimi B (2016) Thermal analysis of multiple-antenna-excited breast model for breast cancer detection. In: 2016 38th annual international conference of the IEEE engineering in medicine and biology society (EMBC), pp 1058–1061. IEEE
3. Kahwaji A, Arshad H, Sahran S, Garba AG, Hussain RI (2016) Hexagonal microstrip antenna simulation for breast cancer detection. In: 2016 international conference on industrial informatics and computer systems (CIICS), pp 1–4. IEEE
4. Rokunuzzaman M, Samsuzzaman M, Islam MT (2016) Unidirectional wideband 3-D antenna for human head-imaging application. IEEE Antennas Wirel Propag Lett 16:169–172
5. Balanis CA (2016) Antenna theory: analysis and design. Wiley
6. Alibakhshikenari M, Virdee BS, Shukla P, Parchin NO, Azpilicueta L, See CH, Abd-Alhameed RA, Falcone F, Huynen I, Denidni TA, Limiti E (2020) Metamaterial-inspired antenna array for application in microwave breast imaging systems for tumor detection. IEEE Access 8:174667–174678
7. Mahmud MZ, Islam MT, Misran N, Kibria S, Samsuzzaman M (2018) Microwave imaging for breast tumor detection using uniplanar AMC based CPW-fed microstrip antenna. IEEE Access 6:44763–44775
8. Aziz A, Ahmad D, Shila TA, Rana S, Hasan RR, Rahman MA (2019) On-body circular patch antenna for breast cancer detection. In: 2019 IEEE international electromagnetics and antenna conference (IEMANTENNA), pp 029–034. IEEE.
9. Karli R, Ammor H, Shubair RM, AlHajri MI, Alkurd R, Hakam A (2016) Miniature planar ultra-wide-band microstrip antenna for breast cancer detection. In 2016 16th mediterranean microwave symposium (MMS), pp. 1–4. IEEE
10. Aldhaeebi MA, Almoneef TS, Attia H, Ramahi OM (2019) Near-field microwave loop array sensor for breast tumor detection. IEEE Sens J 19(24):11867–11872

# Revealing Characteristic IoT Behaviors by Performing Simple Energy Measurements via Open Hardware/Software Components

**Dimitrios Loukatos, Nikos Dimitriou, Ioannis Manolopoulos, Kimon Kontovasilis, and Konstantinos G. Arvanitis**

**Abstract** The vast progress in the electronics industry, being experienced in the last decade, made available many devices of amazing characteristics, at very affordable price levels. This dynamic has drastically affected the way that people live and work. In this context, the modern computers tend to be of credit card size and, quite often, can be found in non-conventional places, like "smart" home appliances, autonomous vehicles or in rural area sensor-actuator equipment. Apparently, the underlying diversity in computing and networking activities creates new opportunities but is also posing new challenges, as a huge number of interconnected devices have to operate for long periods, while many of them cannot be powered by fixed lines. For this reason, the energy optimization/minimization problem of the participating equipment is of key importance for providing fluent functionality and autonomy. The first step towards minimizing the energy footprint of these devices is to measure and study their energy requirements in a comprehensive and simple manner. This work is trying to highlight the feasibility of performing similar measurements by exploiting some cost-effective recently appeared hardware and software components. The proposed method is accurate, generic, and flexible enough, and thus can be easily followed by non-expert personnel and assist in making them more sensitive about energy consumption issues.

D. Loukatos (✉) · K. G. Arvanitis
Agricultural University of Athens, Athens, Greece
e-mail: dlouka@aua.gr

K. G. Arvanitis
e-mail: karvan@aua.gr

D. Loukatos · N. Dimitriou · I. Manolopoulos · K. Kontovasilis
National Centre for Scientific Research "Demokritos", Athens, Greece
e-mail: nikodim@iit.demokritos.gr

I. Manolopoulos
e-mail: imanolo@iit.demokritos.gr

K. Kontovasilis
e-mail: kkont@iit.demokritos.gr

 
X.-S. Yang et al. (eds.), *Proceedings of Sixth International Congress on Information and Communication Technology*, Lecture Notes in Networks and Systems 216,
https://doi.org/10.1007/978-981-16-1781-2_90

**Keywords** Energy monitoring · IoT · Performance evaluation · Embedded systems · LoRa · ESP32 · Arduino · Real-time dynamics · Video streaming

## 1 Introduction

Internet of Things applications and networks are expanding rapidly over the last years and they are covering an increasing number of different use cases that rely on sensing, measurement collection, intelligent decision-making, and targeted actuation. This trend is correlated with the phenomenal progress in the electronics industry over the last decade that continuously introduces devices with enhanced capabilities at decreasing costs. These dynamics have drastically affected the way that people live and work. In this context, modern computation units tend to have sizes as small as a credit card and, quite often, can be found in non-conventional places, like "smart" home appliances, autonomous vehicles or in rural area sensor-actuator equipment. Apparently, a huge number of interconnected devices have to operate for long periods, while many of them cannot be powered by fixed lines. For this reason, the energy optimization/minimization problem of the participating equipment is of key importance for providing the required functionality and autonomy and demands prior measuring and studying the corresponding energy requirements, in a comprehensive and simple manner.

In theory, the wattage of an IoT node in action remains comparatively low, e.g., less than 5 W, and given the pricing policy, per kWh, of the energy suppliers, the monetary cost of keeping such a node active is almost negligible. Unfortunately, in practice, things are far from being ideal, as the number of nodes can be quite large and, beyond that, many of them can be powered only by batteries of a limited capacity. According to the relevant calculations, for a battery of moderate capacity, e.g., of 3000 mA–5000 mA, our node would last less than a day. For this reason, the minimization of the nodal energy requirements, in conjunction with an on/off operation pattern that exploits the sleep mode capabilities of the modern microprocessors, is absolutely necessary in order to shift the autonomy barrier from the below-day to the above-month scale. A key factor for achieving this is the meticulous monitoring of the nodes' power consumption. Various works in literature focus on energy and power consumption measurements as well as on related metrics of wireless devices. In [1], the authors presented a traffic generator and analyzer for accurately measuring delays and other packet network loss-related metrics that were employed in a heterogeneous network test-bed; in [2] a similar approach was demonstrated in order to identify vertical handover delay and especially the time during which target air interfaces are activated, which affects the energy consumption of the terminals. In the same direction, the authors in [3] present a model for calculating the energy consumption of wireless network devices, considering components related to networking, processing, and sensing. They provide analytical calculations for assessing the energy costs for periodic reporting applications using Sigfox/LoRa as well as for time slotted channel hopping schemes such as IEEE 802.15.4e and WirelessHART. Additionally, the work [4] shows the simulation and experimental assessment of the energy consumption of Wi-Fi IoT devices for low bandwidth applications, as part of an end-end energy calculation modeling approach that also considers energy costs on the cloud and network infrastructure and provides parametric expressions for determining the IoT device's energy consumption. In [5], the authors present an energy model for LoRa/LoRaWAN technologies considering parameters from the physical layer, as well as MAC packet characteristics and related signaling patterns for different LoRaWAN scenarios and operation modes. In [6], the authors present a platform for enabling low-cost, programmable energy measurements for a range of IoT devices, showing results for IEEE 802.11 and IEEE 802.15.4. Additionally, in [7], the authors present a low-cost, real-time power measurement platform for detecting real-time power for LoWPAN devices. The work [8] focuses on the energy efficiency of the vertical handover operation among heterogeneous networks based on the IEEE 802.21 media independent handover standard and presents an energy measurement methodology that relies on monitoring the rate with which battery is being discharged through the Advanced Configuration and Power Interface (ACPI) standard [9]. The work [10] presented an architectural framework for performing secure decentralized in device energy consumption measurements for IoT end devices that can address device mobility among different measurement aggregators. In [11], the authors proposed an energy calculation model including three causes of energy dissipation for one bit; a fixed and a distance-dependent energy consumption term for transmission and the energy for reception. In [12], a similar energy model is presented for mobile wireless sensor network nodes that considers the operations of transmitting and receiving packets. In [13], the authors presented a model for assessing the power consumption in device-device communications for assessing the power consumption of LTE and Wi-Fi network nodes. Finally, the work [14] focuses on a simple testbed for metering the performance of radio transceivers via different configuration settings by measuring the electric charge instead of the electric current, in order to identify abrupt energy discharges due to the transmitting operations.

The proposed method is more generic than the approach described in [7] and capable of capturing system's real-time dynamics, extending the methods in [8, 14] that are using mainly aggregate quantities. The feasibility of performing efficient energy measurements via simple, widely applicable, and still accurate techniques that utilize low-cost electronic components and pairing software is highlighted. Furthermore, as modern IoT appliances are following very specific activity patterns and use known processor and transceiver modules, their behavior can be very well recorded in a cost-effective manner.

Apart from this introductory section, the rest of this paper is organized as follows. Section 2 contains useful design and implementation details. Section 3 provides characteristic evaluation results. Finally, Sect. 4 is dedicated to concluding remarks.

## 2 Methodology and Implementation Details

For estimating the energy spent during the diverse activity stages of an IoT node, the monitoring of the corresponding current consumption provides an equivalent metric, as in most cases the underlying battery voltage remains almost constant during the various operations. Traditionally, this process would require an ammeter and visual inspection of the results, but this task is drastically simplified grace to the recent advances in the electronic component production.

More specifically, an arduino uno board [15] can be used to measure the voltage drops over a resistor of known value, connected in series with the load of interest. The arduino uno incorporates an analog-to-digital converter (ADC) module of 10-bit resolution. As the six available analog inputs of the arduino are sharing the same ADC module, the capturing of differential measurements would be a quite tricky process. For this reason, it is better to use a separate ADC module providing true differential measurements of greater resolution (i.e., of 12 bits, instead of 10 bits) and adjustable gain options, like the ADS1015 [16] differential amplifier and digitizer. The communication of this module with the hosting arduino board can be done easily through an I2C [17] interface. The presence of the arduino unit allows for additional data processing and fast graphical inspection, via the accompanying software, as explained in Sect. 3.

While the easily replaceable resistor is connected in series with the whole node under testing or with the separate component of it, under testing, a capacitor is also present for improved results. More specifically, a capacitor module, in parallel with this load of interest, after the current measuring point, guarantees that the effect of potential peaks in current consumption will not be overlooked. For better results, the modules used for performing the measurements should have a separate supplying circuit. According to the aforementioned arrangement, "fresh" readings can be acquired every 2 ms while an average of 5 points approximately can be reported through the serial (over USB) port of the hosting arduino module, thus resulting in a 100sps sampling rate.

The methodology being presented has been ported to fit on a WeMos [18] board, instead of an arduino uno. The arduino uno board is a standard for easy prototyping development, but lacks a Wi-Fi interface (IP connectivity), which is a quite mandatory option, especially in conjunction with the widely-used smart phones and tablets. On the other hand, the WeMos board, which is based on the ESP8266 chip, incorporates standard Wi-Fi functionality. Furthermore, it is much faster and with much more memory than the arduino uno board, at the same price-level, and has a GPIO pin arrangement very close to the arduino board, thus maintaining compatibility with several arduino-specific electronic add-ons. The presence of the Wi-Fi interface encourages the cooperation with fast deployable applications for smart phones/tablets through user-friendly environments like the MIT app inventor [19], exactly as the more powerful and popular but energy greedy raspberry pi unit [20] does, following the guidelines in [14]. The experimental setup that contains the discussed arrangements is depicted in Fig. 1, where an ESP32-CAM [21] video streaming module is

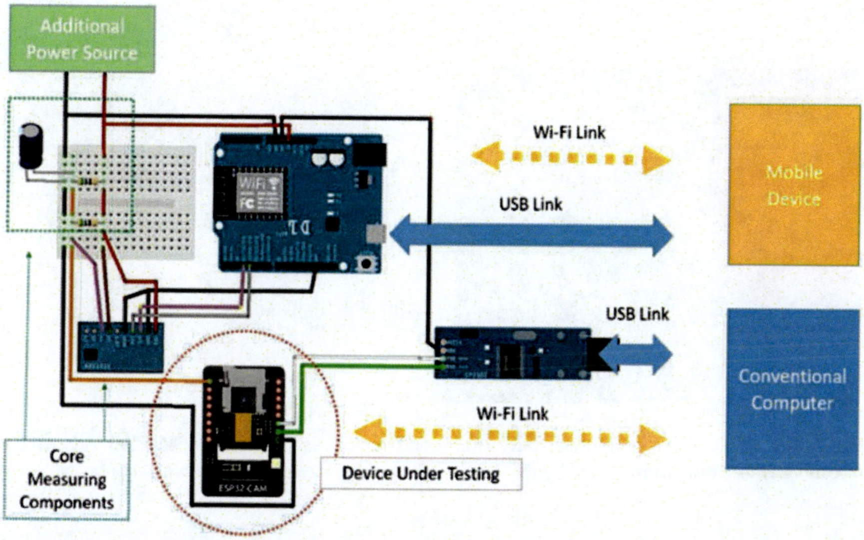


**Fig. 1** The experimental setup for performing current measurements, involving resistors, capacitor, an ADS1015 module and a WeMos board. An ESP32-CAM is the device under testing

used as the IoT device under testing. Further details on how the latter arrangement can be exploited are given in Sect. 3.

## 3 Evaluation Results

A satisfactory use scenario incudes the capturing of the activity alteration from idling to delivering MJPEG video (at the maximum frame rate), of an ESP32-CAM device, as recorded by the ADS1015 module connected on the WeMos board. Figure 2 depicts the user (client) controlling web interface (left) and the corresponding current consumption in mA, both for the ESP32-CAM device. The time trace is easily provided via the serial plotter component of the arduino IDE environment [22], which, apart from textual output (via the serial monitor tool), also provides a handy graphical inspection option.

Figure 3 provides detailed analysis of the ESP32-CAM behavior after restarting the module, with alterations between periods of video streaming activity at 2fps (top right part) and idling and with the Wi-Fi transceiver initially disabled and re-enabled (top left part).

The ESP32-CAM works in a minimal way, and thus the very specific frame preparation and delivery actions can be captured, adjusted at a low rate of 2fps, for better inspection. As the serial plotter window visualizes the 500 most recent readings, the latest 5000 ms of activity are depicted. The right half of the top left part

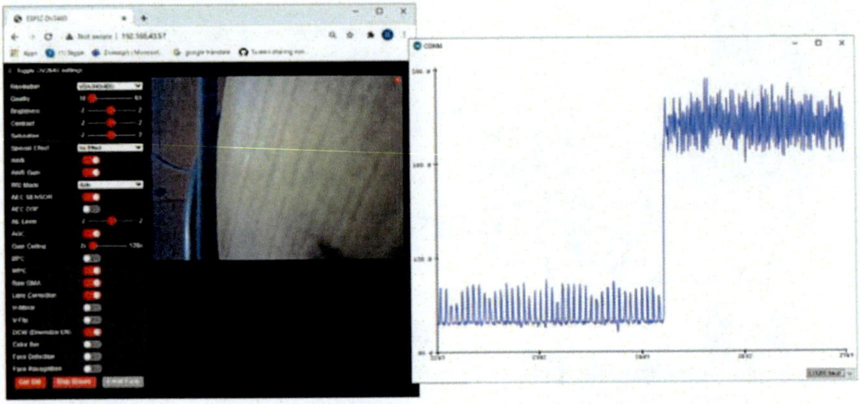

**Fig. 2** The ESP32-CAM module activity alteration from idling to delivering MJPEG video at the maximum frame rate, as recorded by the ADS1015 module connected on the WeMos board

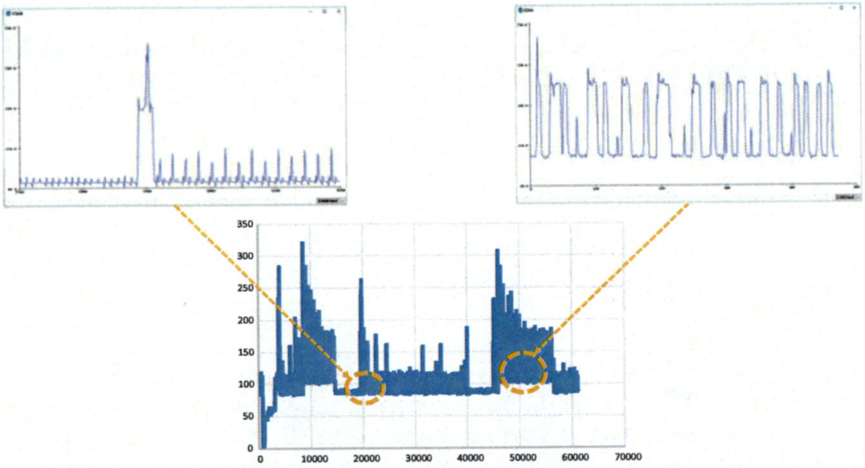


**Fig. 3** Analysis of the ESP32-CAM behavior after restarting the module, with alterations between periods of Wi-Fi radio and video streaming activity and idling

of Fig. 3 contains a regular pattern of peaks every 200 ms, as the DTIM beacons are intercepted by the Wi-Fi interface of the ESP32 module.

Figure 4 depicts the current consumption of a second WeMos module under testing, as a real-time trace, while it is hosting a LoRa Dragino [23] shield and is programmed to transmit at different transmit power (TX) and spreading factor (SF) settings [24]. More specifically, bigger SF numbers result in longer transmission times (top part SF value is 10, while bottom part SF value is 12). Similarly, TX power changes, from 23dBm (left part) to 5dbm (right part), have a drastic impact on the instantaneous total amperage consumption of the module, i.e., from 180 to

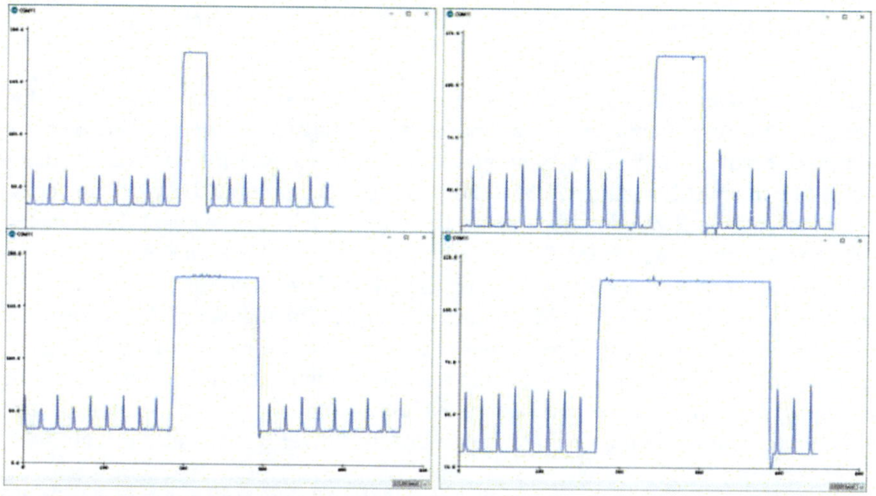

**Fig. 4** The current consumption of a second WeMos module under testing that hosts as LoRa Dragino shield which is programmed to transmit at different radio settings arrangements

114 mA, respectively. The LoRa protocol is a very challenging communication alternative, as it is exhibiting remarkable long-range and low-power capabilities. For this reason, facilitating the study of the idiosyncrasies of this protocol is a key factor for optimizing its performance.

The aforementioned indicative results highlight the feasibility to provide fast and accurate measurements in a comparatively simple manner, in order to capture characteristic activity moments of typical modern IoT devices. The method being presented tried to achieve a trade-off between the exploitation of cheap and easily comprehensive components and the delivery accurate and customizable measurements, revealing real-time dynamics. Resistors from 0.1Ω to 0.3Ω, at 5 W, have been selected for the aforementioned testing scenarios, at 5 V supplying voltage. For accurately capturing comparatively weaker amperage activity (i.e., bellow 1 mA), aggregate methods can be utilized, as in [8, 14], while for currents more than 1A, hall-effect-based sensors devices can also deliver satisfactory non-invasive results. The discussed approach, at the cost of 10–15$, provides a fully-working system while the overall measuring activity is controlled/monitored by a mobile device using a tiny HTTP server hosted on the WeMos device, as authors in [14] did for the raspberry pi unit.

## 4 Conclusions

This work highlighted the feasibility of capturing the short-time dynamics of the energy behavior of characteristic IoT modules, by using some cost-effective recently appeared hardware and software components. More specifically, hardware components like the WeMos unit, an ESP8266 based board, and the ADS1015 converter, combined with software implemented using the user-friendly and well-documented MIT app inventor and Arduino IDE environments, were the backbone of the proposed measurement method. A tiny video server called ESP32-CAM, equipped with Wi-Fi interface and a secondary WeMos board hosting a LoRa dragino shield were the IoT modules under testing. The proposed methods can be easily followed by non-expert personnel and can make them more aware of energy consumption issues. The hiring of well-documented and widely available components makes the approach also suitable for enriching several activities of the university lessons curriculum.

**Acknowledgements** This work was supported by the SYNTELESIS Project, that is funded by the Operational Programme "Competitiveness, Entrepreneurship and Innovation" (NSRF 2014-2020) and co-financed by Greece and the EU (European Regional Development Fund) under Grant MIS 5002521.

## References

1. Loukatos D, Sarakis L, Kontovasilis K, Skianis C, Kormentzas G (2007) Tools and practices for measurement-based network performance evaluation. In: 18th IEEE international symposium on personal, indoor and mobile radio communications, pp 1–5. IEEE, Athens. https://doi.org/10.1109/PIMRC.2007.4394164
2. Dimitriou N, Sarakis L, Loukatos D, Kormentzas G, Skianis C (2011) Vertical handover (VHO) framework for future collaborative wireless networks. Int J Netw Manag 21(6):548–564. https://doi.org/10.1002/nem.786
3. Martinez B, Montón M, Vilajosana I, Prades JD (2015) The power of models: modeling power consumption for IoT devices. IEEE Sens J 15(10):5777–5789. https://doi.org/10.1109/JSEN.2015.2445094
4. Guegan L, Orgerie AC (2019) Estimating the end-to-end energy consumption of low-bandwidth IoT applications for WiFi devices. In: 11th IEEE international conference on cloud computing technology and science CloudCom, pp 287–294, Sydney, Australia
5. Bouguera T, Diouris JF, Chaillout JJ, Jaouadi R, Andrieux G (2018) Energy consumption model for sensor nodes based on LoRa and LoRaWAN. Sensors 18(7):2104. https://doi.org/10.3390/s18072104
6. Dezfoul B, Amirtharaj I, Li CC (2018) EMPIOT: an energy measurement platform for wireless IoT Devices. J Netw Comput Appl 121(1):135–148. https://doi.org/10.1016/j.jnca.2018.07.016
7. Liu Y, Wang Y, Gao W, Zhang W, Qian H, Yang Y (2017) A novel low-cost real-time power measurement platform for LoWPAN IoT devices. Mob Inf Syst. https://doi.org/10.1155/2017/8713873
8. Wassie DA, Loukatos D, Sarakis L, Kontovasilis K, Skianis C (2012) On the energy requirements of vertical handover operations: Measurement-based results for the IEEE 802.21 framework. In: IEEE 17th international workshop on computer aided modeling and design of

communication links and networks (CAMAD), pp 145–149, Barcelona. https://doi.org/10.1109/CAMAD.2012.6335316
9. Duflot L, Levillain O, Morin B (2009) ACPI: design principles and concerns. In: Chen L, Mitchell CJ, Martin A (eds) TRUST 2009, LNCS, vol 5471. Springer, Berlin, pp 14–28
10. Shivaraman N, Saki S, Liu Z, Ramanathan S, Easwaran A, Steinhorst S (2020) Real-time energy monitoring in IoT-enabled mobile devices. In: Design, automation & test in Europe conference & exhibition (DATE), pp 991–994, Grenoble. https://doi.org/10.23919/DATE48585.2020.9116577
11. Neugebauer M, Ploennigs J, Kabitzsch K (2006) Evaluation of energy costs for single hop vs. multi hop with respect to topology parameters. In: IEEE international workshop on factory communication systems, pp175–182, Torino. https://doi.org/10.1109/WFCS.2006.1704148
12. Marta M, Cardei M (2009) Improved sensor network lifetime with multiple mobile sinks. Elsevier Pervasive Mob Comput 5(5):542–555. https://doi.org/10.1016/j.pmcj.2009.01.00
13. Höyhtyä M, Apilo O, Lasanen M (2018) Review of latest advances in 3GPP standardization: D2D communication in 5G systems and its energy consumption models. Future Internet 10(1):3. https://doi.org/10.3390/fi10010003
14. Loukatos D, Manolopoulos I, Arvaniti ES, Arvanitis KG, Sigrimis NA (2018) Experimental testbed for monitoring the energy requirements of LPWAN equipped sensor nodes. IFAC-PapersOnLine 51(17):309–313. https://doi.org/10.1016/j.ifacol.2018.08.196
15. Arduino Uno board description on the official Arduino site (2020). https://store.arduino.cc/arduino-uno-rev3. Accessed 28 Dec 2020
16. The ADS1015 Analog-to-Digital Converter circuit (2020). https://cdn-shop.adafruit.com/datasheets/ads1015.pdf. Accessed 25 Nov 2020
17. I2C Inter-Integrated Circuit description on Wikipedia (2020). https://en.wikipedia.org/wiki/I2C. Accessed 28 Dec 2020
18. The WeMos D1 R2 board (2020). https://wiki.wemos.cc/products:d1:d1. Accessed 25 Nov 2020
19. The MIT App Inventor programming environment (2020). http://appinventor.mit.edu/explore/. Accessed 28 Dec 2020
20. Raspberry Pi 3 Model B board description on the official Raspberry site (2020). https://www.raspberrypi.org/products/raspberry-pi-3-model-b/. Accessed 27 Dec 2020
21. Description of the ESP32-CAM module (2020). http://www.ai-thinker.com/pro_view-24.html. Accessed 28 Dec 2020
22. Arduino IDE software description on the official Arduino site (2020). https://www.arduino.cc/en/software. Accessed 28 Dec 2020
23. The LoRa Dragino shield for Arduino (2020). http://www.dragino.com/products/module/item/102-lora-shield.html. Accessed 28 Dec 2020
24. Semtech, LoRa FAQs (2020). https://docplayer.net/9473940-Lora-faqswww-semtech-com-1-of-4-semtech-semtech-corporation-lora-faq.html. Accessed 28 Dec 2020

# Author Index





X.-S. Yang et al. (eds.), *Proceedings of Sixth International Congress on Information and Communication Technology*, Lecture Notes in Networks and Systems 216,
https://doi.org/10.1007/978-981-16-1781-2